SUPPLEMENTS AND NEW
For the Student

 W9-BKC-427

Audesirk Live! WWW

(http://www.prenhall.com/audesirk)

Launch your exploration of biology on the Web with *Audesirk Live!* Participate in a discussion of cloning, read the latest findings about the new fountain of youth, or just challenge yourself with this fun and informative Web site. The icons positioned throughout the textbook will clue you in to related resources on the Web. This innovative on-line resource center is designed specifically to support and enhance *Biology: Life on Earth, Fifth Edition.*

Biology on the Internet — A Student's Guide, 1997–1998

By Andrew T. Stull (0-13-266537-9)

The perfect tool to help your students take advantage of *Audesirk Live!* on the Web and open up the world of the Internet for exploration. This practical resource gives clear steps for accessing our regularly updated biology resource area as well as an overview of general navigation strategies.

Audesirk Interactive CD-ROM

(free when shrinkwrapped with text)

Explore life on Earth with the *Audesirk Interactive* student CD-ROM and visualize key processes in biology. The tools in *Audesirk Interactive* will help you understand important concepts and test your knowledge. Interactive exercises and animations will allow you to experience how dynamic the study of biology really is.

Study Guide (0-13-081039-8)

By Joseph Chinnici, *Virginia Commonwealth University,* and Susan M. Wadkowski, *Lakeland Community College*

This essential study tool will help students think through the biological concepts and reinforce key concepts presented in the text. It offers a wide range of study exercises and self-tests.

The New York Times Themes of the Times

Coordinated by Kathryn Podwall, *Nassau Community College*

This newspaper-format supplement brings together recent articles on the hottest issues in biology taken directly from the pages of *The New York Times.* This free supplement, available in quantity through your local representative, helps students make connections between the classroom and the world around them.

AIDS Update 1999 (0-13-080352-9)

By Gerald J. Stine, *University of North Florida*

This paperback reader, updated annually, answers many of the questions students have about today's AIDS crisis.

For the Laboratory

Explorations in Basic Biology, Eighth Edition (0-13-079990-4)

By Stanley E. Gunstream, *Pasadena City College*

This best-selling laboratory manual can be used with *Biology: Life on Earth.* It includes 41 self-contained, easy-to-understand experiments that blend traditional experiments with new investigative exercises.

Instructor's Manual to Explorations in Basic Biology, Eighth Edition (0-13-927575-4)

Thinking About Biology: An Introductory Biology Laboratory Manual (0-13-633033-9)

By Mimi Bres and Arnold Weisshaar, both of *Prince George's Community College*

The unique inquiry-based format of this lab manual promotes an active learning style. Extensively class tested by more than 7000 students, it emphasizes problem solving.

Instructor's Manual to Thinking About Biology (0-13-952319-7)

BIOLOGY
Life on Earth

FELIX

Fifth Edition

BIOLOGY

Life on Earth

Teresa Audesirk
Gerald Audesirk

University of Colorado at Denver

Prentice Hall
Upper Saddle River, New Jersey 07458

Library of Congress Cataloging-in-Publication Data

Audesirk, Teresa
    Biology: Life on Earth / Teresa Audesirk, Gerald Audesirk.—
  5th ed.
        p. cm.
      Includes bibliographical references and index.
      ISBN 0-13-792615-4 (hardcover)
      1. Biology. I. Audesirk, Gerald. II. Title.
      QH308.2.A93 1999   98-23688
      570–dc21   CIP

Senior Editor: Teresa Ryu
Executive Editor: Sheri L. Snavely
Senior Development Editor: Karen Karlin
Production Editor: Donna F. Young
Page Layout and Template: Richard Foster; Progressive Information Technologies
Art Director: Heather Scott
Assistant to Art Director: John Christiana
Art Manager: Gus Vibal
Editor in Chief: Paul F. Corey
Editorial Director: Tim Bozik
Marketing Manager: Jennifer Welchans
Marketing Director: John Tweeddale
Marketing Assistant: Cheryl Adam
Assistant Vice President of Production & Manufacturing: David W. Riccardi
Executive Managing Editor: Kathleen Schiaparelli
Director of Creative Services: Paula Maylahn
Associate Creative Director: Amy Rosen
Manufacturing Manager: Trudy Pisciotti
Buyer: Benjamin Smith
Editor in Chief of Development: Ray Mullaney
Associate Editor in Chief of Development: Carol Trueheart
Product Manager: Andrew T. Stull
Media Production Editor: Amy Reed
Associate Editor: Mary Hornby
Special Projects Manager: Barbara Murray
Supplements Production Editors: Dawn Blayer and James Buckley
Editorial Assistant: Lisa Tarabokjia
Copy Editor: James Tully
Proofreader: Marilyn Knowlton
Cover Designer: Bruce Kenselaar
Interior Designers: Donovan & Green; Tom Nery; Judith Matz-Coniglio
Illustrators: Rolando Corujo; Hudson River Studios; Howard S. Friedman; David Mascaro;
            Edmund Alexander; Page Two, Inc.; Roberto Osti
Art Editor: Karen Branson
Project Support: Rhoda Sidney
Photo Research: Yvonne Gerin
Cover Photograph: "Young Vervet Monkey" © Heinrich van den Berg

©1999, 1996 by Prentice-Hall, Inc.
Simon & Schuster/A Viacom Company
Upper Saddle River, New Jersey 07458

All rights reserved. No part of this book may be reproduced, in any form
or by any other means, without permission in writing from the publisher.

Earlier editions ©1993, 1989, 1986 by Macmillan Publishing Company, a division of Macmillan, Inc.

Printed in the United States of America
10 9 8 7 6 5 4 3 2 1

ISBN   0-13-792615-4

Prentice-Hall International (UK) Limited, *London*
Prentice-Hall of Australia Pty. Limited, *Sydney*
Prentice-Hall Canada Inc., *Toronto*
Prentice-Hall Hispanoamericana, S.A., *Mexico*
Prentice-Hall of India Private Limited, *New Delhi*
Prentice-Hall of Japan, Inc., *Tokyo*
Simon & Schuster Asia Pte. Ltd., *Singapore*
Editora Prentice-Hall do Brasil, Ltda., *Rio de Janeiro*

About the Authors

Both Terry and Gerry Audesirk grew up in New Jersey, where they met as undergraduates. After marrying in 1970, they moved to California, where Terry earned her doctorate in marine ecology at the University of Southern California and Gerry earned his doctorate in neurobiology at the California Institute of Technology. As postdoctoral students at the University of Washington's marine laboratories, they worked together on marine mollusks, as a model system for studying the neural bases of behavior.

Terry and Gerry are now professors of biology at the University of Colorado at Denver, where they have taught introductory biology and neurobiology since 1982. In their research lab, funded by the National Institutes of Health, they investigate the mechanisms by which neurons are harmed by low levels of environmental pollutants, such as lead.

Terry and Gerry share a deep appreciation of nature and of the outdoors. They enjoy hiking in the Rockies, running near their home in the foothills west of Denver, and attempting to garden at 7000 feet in the presence of hungry deer and elk. They are long-time members of many conservation organizations. Their daughter, Heather, has added another focus to their lives.

Biology: Life on Earth is the natural outgrowth of the authors' enjoyment of writing, their fascination with all aspects of biology, and their eagerness to share this fascination with their students.

"Our students are a constant source of inspiration as we revise this book. Their questions continuously remind us of the need to link biology to real-life experience. It is a joy to do this; after all, what could be more relevant than learning about life on Earth?"

To our daughter, Heather, and to our parents, Lori, Jack, and Joe

Panel of Biology Educators
Biology: Life on Earth, 5/E

We express sincere gratitude to the contributors who worked closely with the authors to ensure
Biology: Life on Earth's continuing tradition of readability, accuracy, and relevance.

Kerri L. Armstrong
Community College of Philadelphia

Research Area: Teacher preparation; teachers' and preservice teachers' attitudes toward science

"One of the reasons I love to teach undergraduate biology is that it is so wonderful to watch students' enthusiasm for science grow. Students become very involved in the process of science and realize that we all have a part in the scientific process."

George S. Ellmore
Tufts University

Research Area: Plant development

"Presenting biology to nonmajors makes me an ambassador to young professionals headed for influential careers in a global setting. Having that privilege allows me to project our respect for life, for organisms, and for their needs and values into disciplines seldom touched by scientists, including law, finance, and government. Teaching students molds our future in powerful ways."

Bruce Byers
University of Massachusetts, Amherst

Research Area: Behavior of birds

"The best thing about teaching undergraduates is that their thinking is not constrained by prior scientific training. Uninhibited by a major's fear of revealing ignorance, nonmajors ask wild, imaginative questions that force a teacher to see biology from a new perspective and to devise equally imaginative ways of explaining the mysteries of biology. Nonmajors are unpredictable, and that's what makes it fun to teach them."

George W. Gilchrist
University of Washington

Research Area: Evolutionary ecology

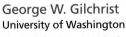

"My career in science began as an English major in a nonmajors biology course, so teaching undergraduates has always been special for me. Nonmajors bring a great diversity of backgrounds to the classroom, which enriches everyone's experience. I especially enjoy seeing my students grow in appreciation of the living world as they learn more about how and why organisms function as they do."

Lewis E. Deaton
University of Southwestern Louisiana

Research Area: Comparative physiology

"I enjoy teaching undergraduates because it gives me the opportunity to try to instill a lifelong interest in biology in my students. They often take the class only because they have to, and by the end of the year I am sure that they have a much better understanding of the complexity of biological systems, of how their bodies function, and of their place in the biosphere. I hope that they have learned enough to make intelligent decisions on pressing scientific issues."

Kate Lajtha
Oregon State University

Research Area: Nutrient cycling in natural and human-disturbed ecosystems

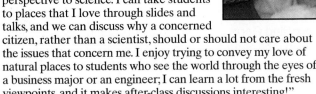

"I thoroughly enjoy teaching undergraduate biology because the students bring a fresh and different perspective to science. I can take students to places that I love through slides and talks, and we can discuss why a concerned citizen, rather than a scientist, should or should not care about the issues that concern me. I enjoy trying to convey my love of natural places to students who see the world through the eyes of a business major or an engineer; I can learn a lot from the fresh viewpoints, and it makes after-class discussions interesting!"

Robin Wright
University of Washington

Research Area: Cell biology and genetics

"Teaching undergraduates is exciting to me because they bring fresh ideas and questions to problems that I frequently take for granted. These new perspectives are often valuable in helping to reveal preconceived notions in my own research approaches. In addition, since most of us AREN'T biologists, by bringing the excitement and wonder of biology to a more general audience, I hope to make a larger impact on how we, as a society, view science and research."

Preface

The AIDS epidemic, global warming, habitat destruction, cloning, genetic engineering, DNA fingerprinting, cancer, environmental estrogens. . . . Biology is in the news. The need for and the demand for an understanding of the basic concepts and issues of biology have never been greater. We believe that this new edition of *Biology: Life on Earth* meets that need more effectively than ever.

In response to increasing student interest and concern, educators have been shifting the focus of introductory biology education. The changes implemented in this edition are based on our own teaching experience and on input from in-depth consultations with biology educators throughout the United States. These instructors combine exceptional teaching skills with expertise in diverse areas of biology. Such collaboration has been at the heart of this revision and of our effort to create a text that is responsive to the needs of today's students. Here's what educators have told us and how we've responded:

An Effective Biology Textbook Actively Engages Students

In response, we converted our major headings within each chapter to numbered questions that encourage students to seek answers as they read. Our full-sentence, conceptual subheadings both suggest answers to the questions posed in the major headings and help students focus on the major points in each section.

Next, we revised each chapter introduction to relate the chapter topic more clearly to familiar experiences and issues and to raise interesting questions that are answered within the chapter. We also added icons that identify topics in the text that students can actively pursue (1) in our chapter Web sites and their relevant links on the Internet (*Audesirk Live!*) or (2) on the Interactive CD-ROM (*Audesirk Interactive*), which contains interactive exercises and animations that engage students in learning and help them visualize biological concepts.

Finally, we added "Group Activities" to each chapter. These exercises, used successfully by our Panel of Biology Educators, encourage students to work out interesting problems in small groups and thereby become active participants in the learning process. The Instructor's Guide to Print and Media includes additional Group Activities and provides suggestions about how instructors can incorporate these exercises into the classroom.

An Effective Biology Texbook Conveys That Biology Is Everywhere

In addition to engaging students' interest by means of new chapter introductions that relate more directly to real-life experiences, we have related key biological concepts in the rest of each chapter both to everyday experiences and to important issues facing society. These applications are interwoven into the text, introduced in end-of-chapter critical-thinking questions ("Applying the Concepts"), and highlighted in updated boxed essays. The essays cover a wide range of current topics in biology, from environmental issues ("Earth Watch"), to clinical discussions ("Health Watch"), to in-depth views of certain processes ("A Closer Look"), to the procedures biologists use in doing their work ("Scientific Inquiry"). For example:

Earth Watch—These environmental essays explore pressing issues such as biodiversity, ozone depletion, and population growth.

Health Watch—These clinical essays investigate topics such as sexually transmitted diseases, the dangers of artificial steroids, and how smoking damages the lungs.

Our chapter Web sites are updated frequently, allowing students to explore topics relevant to each chapter on the World Wide Web. In *Audesirk Live!*, students will discover that biology is everywhere by exploring our "Issues in Biology" and "Bizarre Facts in Biology" sections. *The New York Times Themes of the Times,* our free newspaper-format supplement, provides a collection of recent science articles written by world-renowned science writers for this outstanding newspaper. Instructors can receive our *ABC NEWS/Prentice Hall Video Library for Biology*, containing short news segments from award-winning news programs. These effective teaching enhancements connect biology in the classroom to what's happening in the world today. *College Newslink* provides instructors with e-mail messages highlighting biology news articles and linking the stories to sites on the World Wide Web.

An Effective Biology Textbook Is Flexible and Easy to Use

Our chapter organizing system keeps students aware of the central concepts in each chapter and their relationships to one another. At the start of each chapter, the conceptual questions and their responsive subheadings are brought together, summarizing each chapter's topics "At a Glance." A consistent numbering system identifies the major conceptual headings all the way from "At a Glance," through the body of the chapter in major headings, and into the "Summary of Key Concepts." A "Key Terms" list identifies all the boldface terms that appear in the chapter, along with the page on which each term is introduced; each key term is defined in the glossary at the back of the book.

The chapters in *Biology: Life on Earth* have been written to allow instructors flexibility in the order of use. Chapter 1 sets the stage for the coverage of topics in different orders by reviewing the diversity of life and the principles of evolution. Cross-referencing among chapters allows students to seek additional information on specific topics in other parts of the book.

An Effective Biology Textbook Has Supporting Technology That Enhances Learning and Is Integrated with the Textbook

The explosion of techological developments has allowed publishers to provide textbook users with an amazing number and diversity of supporting materials. We have worked closely with biology educators to ensure that our supplements really work both for students and for instructors.

Student learning is enhanced by the free *Audesirk Interactive* CD-ROM, which accompanies each text and provides interactive exercises and animations. For easy reference and for a tight integration with our multimedia supplements, the "WWW" and CD-ROM icons have been placed next to numbered headings throughout the textbook that have relevant multimedia counterparts. "Net Watch" at the beginning of each chapter also reminds students about their Web site. Students and instructors will find easy access to current information through chapter Web sites, *Themes of the Times*, and the *ABC NEWS* video library. See the front endpapers for a complete listing of supplements.

As an instructor, you can create multimedia lectures with our *Life on Earth Presentation CD-ROM*, which also includes all the line art from the textbook, or the *Prentice Hall Laser Disc for Biology*, created specifically for use with *Biology: Life on Earth*. Choose 250 full-color acetates and color slides supplemented by transparency masters that cover all the line art that is included in the fifth edition. A selection of 2400 sample test questions in a variety of styles, combined with custom testing software, allows you to create and edit exams electronically.

The Fundamental Philosophy of *Biology: Life on Earth*

Although our textbook has evolved in response to the changing needs of our audience, there are fundamental issues in the teaching of biology that don't change. *Biology: Life on Earth* still does the following:

Focuses on Concepts

Our conceptual questions and subheadings cast as sentences, "At a Glance" chapter-opening outlines, and end-of-chapter "Summary of Key Concepts" sections keep students focused on the important themes in each chapter. Figure captions are separated into caption titles, which give the theme of each image, and more-specific information below. Students can lose sight of the underlying concepts in the welter of technical detail, so we provide a general overview of complex subjects in the text itself and offer the details of such topics in "A Closer Look" essays:

A Closer Look—These essays focus on the more challenging details of topics such as chemiosmosis, cellular respiration, and urine formation in the nephron.

Communicates the Scientific Process

Biology is not just a compendium of facts and ideas; rather, it is the outgrowth of a dynamic process of inquiry and human endeavor. In many instances, we describe *how* scientists discovered specific facts. The scientific process is further highlighted in "Scientific Inquiry" essays:

Scientific Inquiry—With these essays, students will learn how fossils are dated and how PET scans are performed. They will follow the development of the science of genetics from Mendel's experiments in a monastery garden to the laboratory of Watson and Crick and finally to the Roslin Center in Scotland, where Ian Wilmut cloned a sheep from adult DNA.

Emphasizes Unifying Themes

As Theodosius Dobzhansky so aptly put it, "Nothing in biology makes sense, except in the light of evolution." Throughout the text, students will find examples of how natural selection has produced organisms that are adapted to specific environments. In addition, many chapters end with an "Evolutionary Connections" section:

Evolutionary Connections—These lively discussions link chapter concepts into the broader perspective of evolution.

Our own concern for the environment also threads its way throughout this text. Whenever appropriate, we have tried to present students with the biological rationale for making sound environmental decisions in their daily lives.

A course in introductory biology may be a student's first real exposure to the fascinating complexity of life. As teachers, we recognize how easily a student can become mired in an overwhelming number of new facts and terms while losing sight of the underlying concepts of biology. We have carefully revised *Biology: Life on Earth* to reduce unnecessary detail and excess terminology and to emphasize the ways in which an understanding of biology can enrich and enlighten day-to-day living. Why study biology? Maybe we're biased, but what can be more fascinating than learning about Life on Earth?

Acknowledgments

Biology: Life on Earth is truly a team effort, and we wish to acknowledge and thank our teammates.

The dauntingly complex task of putting together a book of this magnitude has been handled with unflagging dedication and zeal by the skilled development team at Prentice Hall. Karen Karlin, our Developmental Editor, has made sure the text is clear, consistent, and student-friendly; she has also contributed substantially to the photographic selection and to the consistency and clarity of the art program. Her knowledge of biology and eye for detail have markedly improved the textbook. Donna Young, our Production Editor, patiently dealt with the challenge of turning typed pages into final printed pages; she also assisted in managing and evaluating the art program. Photo Researcher Yvonne Gerin skillfully and doggedly tracked down excellent photos. Copyeditor James Tully tackled the job of copyediting with meticulous care. Proofreader Marilyn Knowlton did an excellent job of reading the final typeset pages to help ensure accuracy.

We also wish to thank Art Director Heather Scott, Associate Creative Director Amy Rosen, and Director of Creative Services Paula Maylahn for guiding the text and cover design with flair and talent; Art Manager Gus Vibal and Art Editors Karen Branson and Rhoda Sidney for coordinating such an immense art program; and Assistant Formatting Manager Richard Foster for his formatting expertise. Mary Hornby, our Associate Editor, has skillfully and patiently overseen the production of transparencies and print supplements. Andy Stull, our Product Manager, did a fantastic job of coordinating the multimedia program, especially our exciting new student CD-ROM, *Audesirk Interactive*, and our student Web site, *Audesirk Live!*; he has worked tirelessly with Teresa Ryu to keep them updated. Thanks also to David Riccardi, Vice President of Production and Manufacturing, and Wendy Allen, Director of Publishing Technology, for patiently providing Macintosh instruction, and to Teresa Ryu for making it possible.

Colleagues at other institutions have helped us enormously. Many, listed in the following pages, have stimulated us to rethink our presentation with careful, thoughtful reviews. Our Panel of Biology Educators, highlighted on page vi, has made a special contribution to the fifth edition. In particular, Bruce Byers provided major assistance with this revision, especially in Units Three and Five. With his extensive work with nonmajors and his expertise in evolutionary biology, Bruce has brought a valuable perspective to this revision. Without his very skilled and conscientious work, this timely revision would have been impossible. He has our sincere respect and thanks.

A wonderful supplements package was prepared by Joseph Chinnici and Susan Wadkowski (Study Guide); Gail Gasparich, Charles Good, Kathleen Murray, and DeLoris Wenzel (Test Item File); and Timothy Metz (Instructor's Guide to Print and Media Resources). We thank Dan Doak, Scott Freeman, Florence Juillerat, Jerri Lindsey, Richard Mortensen, Rhoda Perozzi, Bill Stark, and Gerald Summers for contributing thought-provoking end-of-chapter questions. We are grateful as well to Robert K. Alico, Kerri Armstrong, Amadu D. Ayebo, W. Barclay Butler, Bruce Byers, Lewis Deaton, George Ellmore, George Gilchrist, Martha J. Jack, Kate Lajtha, Thomas R. Lord, Carl S. Luciano, Timothy Metz, Sandra J. Newell, Russell L. Peterson, Ray L. Winstead, and Robin Wright for their valuable suggestions for our "Group Activities."

Marketing is being handled with talent and enthusiasm by Jennifer Welchans, our Senior Marketing Manager. We also would like to thank Paul F. Corey, Editor in Chief of Science, for his insight and support through the past three editions. Finally, but most importantly, our editors: Executive Editor Sheri Snavely and Senior Editor Teresa Ryu have nurtured and supported this project through thick and thin. Teresa has played a special role in smoothing rough spots and boosting our morale with her good humor, optimism, and energy.

Terry and Gerry Audesirk
Golden, Colorado

Panel of Multimedia Advisors and Reviewers

Media Content Contributors:
Sylvester Allred, *Northern Arizona University*
Mike Bell, *Richland College, Dallas*
Bonnie Brenner, *City Colleges of Chicago*
Michael Cunningham, *Ohio State University, Lima*
Julie Froehlig, *University of California, Berkeley*
Gail Gasparich, *Towson University*
Dave Hurley, *University of Washington*
J. Eric Juterbock, *Ohio State University, Lima*
Kerry Kilburn, *Old Dominion University*
Shannon McWeeney, *University of California, Berkeley*
Linda O'Connor, *Richland College, Dallas*

Paul Ramp, *Pellissippi State Technical College*
Jackie Reynolds, *Richland College, Dallas*
Robin Wright, *University of Washington*

Media Consultants:
Arthur L. Buikema, Jr., *Virginia Polytechnic Institute*
Thomas W. Jurik, *Iowa State University*
Michael McKinley, *Glendale Community College*
Frank Schwartz, *Cuyahoga Community College*
Charlene Waggoner, *Bowling Green University*

Media Photo Coordinator:
Blanche C. Haning, *North Carolina State University*

Media Reviewers:
Dawn Adams, *Baylor University*
Kerri Lynn Armstrong, *Community College of Philadelphia*
Sarah Barlow, *Middle Tennessee State University*
Susan Brawley, *University of Maine*
Bonnie Brenner, *City Colleges of Chicago*
Don Buckley, *University of Hartford*
Sarah Bush, *University of Missouri, Columbia*
Guy Cameron, *University of Houston*
Michael Cunningham, *Ohio State University, Lima*
Susan A. Dunford, *University of Cincinnati*
Ronald Edwards, *University of Florida*
Patrick Elvander, *University of California, Santa Cruz*
David Huffman, *Southwest Texas State University*

Florence Juillerat, *Indiana University/Purdue University*
Arnold Karpoff, *University of Louisville*
Terry Keiser, *Ohio Northern University*
Donald Potts, *University of California, Santa Cruz*
Paul Ramp, *Pellissippi State Technical College*
Donald Reinhardt, *Georgia State University*
Jason Rosenblum, *Austin, Texas*
Dan Skean, Jr., *Albion College*
Patrick A. Thorpe, *Grand Valley State University*
Lloyd W. Turtinen, *University of Wisconsin, Eau Claire*
Robin Tyser, *University of Wisconsin, LaCrosse*
Roberta Williams, *University of Nevada, Las Vegas*
Robin Wright, *University of Washington*

Fifth Edition Reviewers

Judith Keller Amand, *Delaware County Community College*
Steve Arch, *Reed College*
Kerri Lynn Armstrong, *Community College of Philadelphia*
Michael C. Bell, *Richland College*
Arlene Billock, *University of Southwestern Louisiana*
Brenda C. Blackwelder, *Central Piedmont Community College*
Robert Burkholter, *Louisiana State University*
W. Barclay Butler, *Indiana University of Pennsylvania*
Bruce Byers, *University of Massachusetts, Amherst*
David H. Davis, *Asheville-Buncombe Technical Community College*
Lewis Deaton, *University of Southwestern Louisiana*
Jean DeSaix, *University of North Carolina, Chapel Hill*
Matthew M. Douglas, *The University of Kansas*
George Ellmore, *Tufts University*
Irja Galvan, *Western Oregon University*
Gail E. Gasparich, *Towson University*
George W. Gilchrist, *University of Washington*

Charles W. Good, *Ohio State University, Lima*
Blanche C. Haning, *North Carolina State University*
Kate Lajtha, *Oregon State University*
Graeme Lindbeck, *University of Central Florida*
Ann S. Lumsden, *Florida State University*
Daniel D. Magoulick, *The University of Central Arkansas*
Paul Mangum, *Midland College*
Kathleen Murray, *University of Maine*
Marcy Osgood, *University of Michigan*
Rhoda E. Perozzi, *Virginia Commonwealth University*
Linda Simpson, *University of North Carolina, Charlotte*
Kathleen M. Steinert, *Bellevue Community College*
Dan Tallman, *Northern State University*
Robin W. Tyser, *University of Wisconsin, LaCrosse*
DeLoris Wenzel, *University of Georgia*
Carolyn Wilczynski, *Binghamton University*
Robin Wright, *University of Washington*

Previous Editions Reviewers

W. Sylvester Allred, *Northern Arizona University*
William Anderson, *Abraham Baldwin Agriculture College*
G. D. Aumann, *University of Houston*
Vernon Avila, *San Diego State University*
J. Wesley Bahorik, *Kutztown University of Pennsylvania*
Bill Barstow, *University of Georgia, Athens*
Colleen Belk, *University of Minnesota, Duluth*
Gerald Bergtrom, *University of Wisconsin*
Raymond Bower, *University of Arkansas*
Marilyn Brady, *Centennial College of Applied Arts & Technology*
Virginia Buckner, *Johnson County Community College*
Arthur L. Buikema, Jr., *Virginia Polytechnic Institute*
William F. Burke, *University of Hawaii*
Kathleen Burt-Utley, *University of New Orleans*
Linda Butler, *University of Texas, Austin*
W. Barkley Butler, *Indiana University of Pennsylvania*
Nora L. Chee, *Chaminade University*
Joseph P. Chinnici, *Virginia Commonwealth University*
Dan Chiras, *University of Colorado, Denver*
Bob Coburn, *Middlesex Community College*
Martin Cohen, *University of Hartford*
Mary U. Connell, *Appalachian State University*
Joyce Corban, *Wright State University*
Ethel Cornforth, *San Jacinto College, South*
David J. Cotter, *Georgia College*

Lee Couch, *Albuquerque Technical Vocational Institute*
Donald C. Cox, *Miami University of Ohio*
Patricia B. Cox, *University of Tennessee*
Peter Crowcroft, *University of Texas, Austin*
Carol Crowder, *North Harris Montgomery College*
Donald E. Culwell, *University of Central Arkansas*
Robert A. Cunningham, *Erie Community College, North*
Jerry Davis, *University of Wisconsin, LaCrosse*
Douglas M. Deardon, *University of Minnesota*
Fred Delcomyn, *University of Illinois, Urbana*
Lorren Denney, *Southwest Missouri State University*
Katherine J. Denniston, *Towson State University*
Charles F. Denny, *University of South Carolina, Sumter*
Jean DeSaix, *University of North Carolina, Chapel Hill*
Ed DeWalt, *Louisiana State University*
Daniel F. Doak, *University of California, Santa Cruz*
Ronald J. Downey, *Ohio University*
Ernest Dubrul, *University of Toledo*
Michael Dufresne, *University of Windsor*
Susan A. Dunford, *University of Cincinnati*
Mary Durant, *North Harris College*
Ronald Edwards, *University of Florida*
Joanne T. Ellzey, *University of Texas, El Paso*
Wayne Elmore, *Marshall University*
Carl Estrella, *Merced College*

Nancy Eyster-Smith, *Bentley College*
Gerald Farr, *Southwest Texas State University*
Rita G. Farrar, *Louisiana State University*
Marianne Feaver, *North Carolina State University*
Linnea Fletcher, *Austin Community College, Northridge*
Charles V. Foltz, *Rhode Island College*
Douglas Fratianne, *Ohio State University*
Scott Freeman, *University of Washington*
Donald P. French, *Oklahoma State University*
Don Fritsch, *Virginia Commonwealth University*
Teresa Lane Fulcher, *Pellissippi State Technical Community College*
Michael Gaines, *University of Kansas*
Farooka Gauhari, *University of Nebraska, Omaha*
David Glenn-Lewin, *Iowa State University*
Elmer Gless, *Montana College of Mineral Sciences*
Margaret Green, *Broward Community College*
Martin E. Hahn, *William Paterson College*
Madeline Hall, *Cleveland State University*
Blanche C. Haning, *North Carolina State University*
Helen B. Hanten, *University of Minnesota*
John P. Harley, *Eastern Kentucky University*
Stephen Hedman, *University of Minnesota*
Jean Helgeson, *Collins County Community College*
Alexander Henderson, *Millersville University*
Alison G. Hoffman, *University of Tennessee, Chattanooga*
Laura Mays Hoopes, *Occidental College*
Michael D. Hudgins, *Alabama State University*
Donald A. Ingold, *East Texas State University*
Jon W. Jacklet, *State University of New York, Albany*
Florence Juillerat, *Indiana University-Purdue University at Indianapolis*
Thomas W. Jurik, *Iowa State University*
Arnold Karpoff, *University of Louisville*
L. Kavaljian, *California State University*
Hendrick J. Ketellapper, *University of California, Davis*
William H. Leonard, *Clemson University*
Jerri K. Lindsey, *Tarrant County Junior College, Northeast*
John Logue, *University of South Carolina, Sumter*
William Lowen, *Suffolk Community College*
Steele R. Lunt, *University of Nebraska, Omaha*
Michael Martin, *University of Michigan*
Margaret May, *Virginia Commonwealth University*
D. J. McWhinnie, *De Paul University*
Gary L. Meeker, *California State University, Sacramento*
Thoyd Melton, *North Carolina State University*
Karen E. Messley, *Rockvalley College*
Glendon R. Miller, *Wichita State University*
Neil Miller, *Memphis State University*
Jack E. Mobley, *University of Central Arkansas*
Richard Mortenson, *Albion College*
Gisele Muller-Parker, *Western Washington University*
Robert Neill, *University of Texas*
Harry Nickla, *Creighton University*
Daniel Nickrent, *Southern Illinois University*
Jane Noble-Harvey, *University of Delaware*

James T. Oris, *Miami University, Ohio*
C. O. Patterson, *Texas A & M University*
Fred Peabody, *University of South Dakota*
Harry Peery, *Tompkins-Cortland Community College*
Rhoda E. Perozzi, *Virginia Commonwealth University*
Bill Pfitsch, *Hamilton College*
Ronald Pfohl, *Miami University, Ohio*
Bernard Possident, *Skidmore College*
James A. Raines, *North Harris College*
Mark Richter, *University of Kansas*
Robert Robbins, *Michigan State University*
Paul Rosenbloom, *Southwest Texas State University*
K. Ross, *University of Delaware*
Mary Lou Rottman, *University of Colorado, Denver*
Albert Ruesink, *Indiana University*
Alan Schoenherr, *Fullerton College*
Edna Seaman, *University of Massachusetts, Boston*
Linda Simpson, *University of North Carolina, Charlotte*
Russel V. Skavaril, *Ohio State University*
John Smarelli, *Loyola University*
Shari Snitovsky, *Skyline College*
Jim Sorenson, *Radford University*
Mary Spratt, *University of Missouri, Kansas City*
Benjamin Stark, *Illinois Institute of Technology*
William Stark, *Saint Louis University*
Barbara Stotler, *Southern Illinois University*
Gerald Summers, *University of Missouri, Columbia*
Marshall Sundberg, *Louisiana State University*
Bill Surver, *Clemson University*
Eldon Sutton, *University of Texas, Austin*
David Thorndill, *Essex Community College*
William Thwaites, *San Diego State University*
Professor Tobiessen, *Union College*
Richard Tolman, *Brigham Young University*
Dennis Trelka, *Washington & Jefferson College*
Sharon Tucker, *University of Delaware*
Gail Turner, *Virginia Commonwealth University*
Glyn Turnipseed, *Arkansas Technical University*
Lloyd W. Turtinen, *University of Wisconsin, Eau Claire*
Robert Tyser, *University of Wisconsin, La Crosse*
Kristin Uthus, *Virginia Commonwealth University*
F. Daniel Vogt, *State University of New York, Plattsburgh*
Nancy Wade, *Old Dominion University*
Jyoti R. Wagle, *Houston Community College, Central*
Michael Weis, *University of Windsor*
Jerry Wermuth, *Purdue University, Calumet*
Jacob Wiebers, *Purdue University*
P. Kelly Williams, *University of Dayton*
Roberta Williams, *University of Nevada, Las Vegas*
Sandra Winicur, *Indiana University, South Bend*
Bill Wischusen, *Louisiana State University*
Chris Wolfe, *North Virginia Community College*
Colleen Wong, *Wilbur Wright College*
Wade Worthen, *Furman University*
Tim Young, *Mercer University*

HOW TO USE THIS TEXT:

NEW!

Chapter Organizing System
A consistent numbering system leads you through each chapter, from **At a Glance** in the beginning, to **conceptual headings** throughout, and finally to the **Summary of Key Concepts** at the end.

At a Glance
These overviews provide a road map to guide you through the major concepts of the chapter.

Patterns of Inheritance **12**

At a Glance

1) How Did Gregor Mendel Lay the Foundations for Modern Genetics?
   Doing It Right: The Secrets of Mendel's Success

2) How Are Single Traits Inherited?
   The Inheritance of Dominant and Recessive Alleles on Homologous Chromosomes Can Explain the Results of Mendel's Crosses
   Mendel's Hypothesis Can Be Used to Predict the Outcome of New Types of Single-Trait Crosses

3) How Are Multiple Traits on Different Chromosomes Inherited?
   Mendel Hypothesized That Genes on Different Chromosomes Are Inherited Independently
   In an Unprepared World, Genius May Go Unrecognized

4) How Are Genes Located on the Same Chromosome Inherited?
   Genes on the Same Chromosome Tend to Be Inherited Together
   Crossing Over May Separate Linked Genes

5) How Is Sex Determined, and How Are Sex-Linked Genes Inherited?
   Sex-Linked Genes Are Found Only on the X or Only on the Y Chromosome

6) What Are Some Variations on the Mendelian Theme?
   In Incomplete Dominance, the Phenotype of Heterozygotes Is Intermediate between the Phenotypes of the Homozygotes
   A Single Gene May Have Multiple Alleles
   Many Traits Are Influenced by Several Genes
   Single Genes Typically Have Multiple Effects on Phenotype
   The Environment Influences the Expression of Genes

7) How Are Human Genetic Disorders Investigated?

8) How Are Human Disorders Caused by Single Genes Inherited?
   Most Human Genetic Disorders Are Caused by Recessive Alleles
   A Few Human Genetic Disorders Are Caused by Dominant Alleles
   Some Human Disorders Are Sex-Linked

9) How Do Errors in Chromosome Number Affect Humans?
   Some Genetic Disorders Are Caused by Abnormal Numbers of Sex Chromosomes
   Some Genetic Disorders Are Caused by Abnormal Numbers of Autosomes

www Net Watch
On-line resources for this chapter are on the World Wide Web at:
http://www.prenhall.com/audesirk
(click on the Table of Contents link and then select Chapter 12).

The tiny Chihuahua weighs about 3 pounds and has large, batlike ears, a short coat, and huge eyes. Next to it stands a massive, 125-pound Great Dane. Despite enormous differences in body size and shape, both of these animals are members of the same species, *Canis familiaris*. The astonishing differences between the two reflect human intervention in breeding over hundreds of generations.

Have you ever speculated as to what would happen if a male Chihuahua somehow managed to mate with a female Great Dane? No? Well, you have probably wondered what determines whether a human baby will be male or female or how two brown-eyed parents can produce a blue-eyed child. Although you may not realize it, you have already come a long way toward understanding the answers to questions such as these. You have learned how the elegantly simple structure of DNA can carry a blueprint for the overwhelming complexity of even simple forms of life. You now know that segments of DNA, called **genes**, carry instructions for synthesizing proteins—and that these proteins, in complex ways, determine how an organism develops, how it looks, and how it functions.

201

The dramatic differences between the Chihuahua and the Great Dane are the result of centuries of selective breeding.

WWW

Net Watch
Meet us on the Web! The *Audesirk Live!* home page offers resources and activities to aid your study and broaden your understanding of biology.

Opening Vignettes
The chapter introductions convey the significance of biological concepts couched in an interesting, relevant story.

Conceptual Headings

These numbered headings, posed as questions to stimulate your curiosity and motivate your quest for knowledge, cover the most fundamental concepts in biology.

These subheadings answer and support the questions previously posed and also provide a context for the discussion that follows.

variability among organisms of the same species.

5 How Is Sex Determined, and How Are Sex-Linked Genes Inherited?

In mammals and many insects, males have the same number of chromosomes as females do, but one "pair," the **sex chromosomes**, are very different in appearance and genetic composition. Females have two identical sex chromosomes, called *X chromosomes*, whereas males have one X chromosome and one *Y chromosome* (Fig. 12-8). Although the Y chromosome normally carries far fewer genes than does the X chromosome, a small part of both sex chromosomes is homologous, so

(22 p........omes), and dogs ... (38 pairs of autosomes).

For organisms in which males are XY and females are XX, the sex chromosome carried by the sperm determines the sex of the offspring (Fig. 12-9). During sperm formation, the sex chromosomes segregate, and each sperm receives either the X or the Y chromosome (plus one member of each pair of autosomes). The sex chromosomes also segregate during egg formation, but because females have two X chromosomes, every egg receives one X chromosome. An offspring is male if an egg is fertilized by a Y-bearing sperm or female if an egg is fertilized by an X-bearing sperm.

Sex-Linked Genes Are Found Only on the X or Only on the Y Chromosome

Genes that are on one sex chromosome but not on the other are said to be **sex-linked**, or *X-linked*. In many animals, the Y chromosome carries relatively few genes other

Summary of Key Concepts

This section, organized by the same numbered headings, helps you begin your review.

Summary of Key Concepts

1) How Did Gregor Mendel Lay the Foundations for Modern Genetics?

Homologous chromosomes carry the same genes located at the same loci, but the genes at a particular locus can exist in alternate forms called alleles. An organism whose homologous chromosomes carry the same allele at a given locus is homozygous for that particular ... if the alleles differ, the organism is heterozygous. ... duced many principles of inheritance ... fore the discovery of DNA, genes, ch... He did this by choosing an appropr...

ject, designing his experiments carefully, and analyzing his data accurately.

2) How Are Single Traits Inherited?

The inheritance of each individual trait is determined by discrete physical units called genes. Each organism possesses a pair of similar, but not necessarily identical, genes that influence each trait. However, only one of each pair of genes is included in any one gamete (the law of segregation). An offspring ... ve alleles, A by the fusion of two gametes therefore receives pairs ... one of each pair inherited from each parent. Differ... ...e of a trait (for example, purple ...ve allele... the same gene, which allele is included in any given gamete is determined by chance (the law of independent assortment). Therefore, we can predict the relative proportions of offspring through the laws of probability.

The physical appearance of an organism (its phenotype) may not always be a reliable indicator of its alleles (the genotype) because of the masking of recessive alleles by dominant alleles. Organisms with two dominant alleles (homozygous dominant) have the same phenotype as do organisms with one dominant and one recessive allele (heterozygous).

3) How Are Multiple Traits on Different Chromosomes Inherited?

If the genes for two traits are located on separate chromosomes, then the F₂ generation from a cross between parents that are heterozygous for two traits will have four different

...romosomes (and hence the alleles) during meiosis.

4) How Are Genes Located on the Same Chromosome Inherited?

Genes located on the same chromosome are said to be linked. In the absence of crossing over, the F₂ offspring will express only the two parental phenotypes. If crossing over does occur, then some of the F₂ offspring will express recombined phenotypes.

5) How Is Sex Determined, and How Are Sex-Linked Genes Inherited?

Sex is determined by sex chromosomes, designated X and Y. The rest of the chromosomes, identical in the two sexes, are called autosomes. In many animals, females have two X chromosomes, whereas males have one X and one Y chromosome. The Y chromosome has many fewer genes. Consequently, males have only one copy of most X chromosome genes, and

ILLUSTRATIONS
Focus on Key Concepts

If you can visualize a concept, you will remember it more easily. Our art and photographs will help you grasp concepts.

Process Art

"Process art" shows clearly how the various stages of a dynamic process in biology unfold. Complex biological processes are broken down step by step to help you visualize and understand how the process occurs.

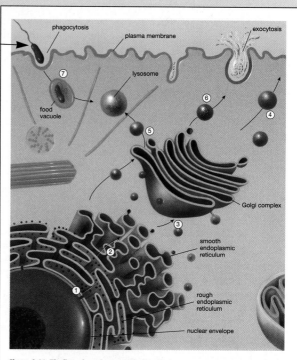

Figure 6-11 The flow of membrane within the cell
Membrane, regardless of its destination, is synthesized by the endoplasmic reticulum.
① Some of this membrane moves inward to form new nuclear envelope, but most of it...

A basilisk lizard runs across a pond, putting the water's surface tension to good use.

Photographs

Captivating photographs provide vivid illustrations of concepts throughout.

Macro-to-Micro Figures

To understand biology, you must be able to visualize structures at several different scales. Macro-to-micro figures help you bridge the gaps among them.

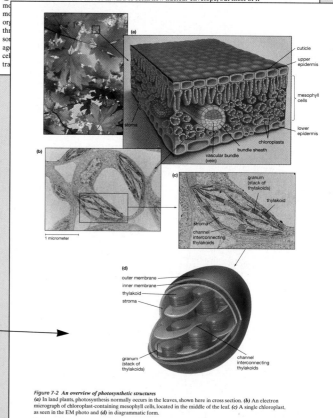

Figure 7-2 An overview of photosynthetic structures
(a) In land plants, photosynthesis normally occurs in the leaves, shown here in cross section. (b) An electron micrograph of chloroplast-containing mesophyll cells, located in the middle of the leaf. (c) A single chloroplast, as seen in the EM photo and (d) in diagrammatic form.

RELEVANCY
Biology Is Everywhere!

You have come into this course already knowing a lot about life. The concepts you are learning will give you a better understanding of your own life and of the world around you.

Earth Watch www
The Ozone Hole—A Puncture in Our Protective Shield

A small fraction of the radiant energy produced by the sun, called *ultraviolet*, or UV, radiation, is so highly energetic that it can damage biological molecules. In small quantities, UV radiation helps human skin produce vitamin D and causes tanning in fair-skinned people. But in larger doses, UV causes sunburn and premature aging of skin, skin cancer, and cataracts, a condition in which the lens of the eye becomes cloudy.

Fortunately, most of the UV radiation is filtered out by ozone in the stratosphere, a layer of atmosphere extending from 10 to 50 kilometers (6 to 30 miles) above Earth. In pure form, *ozone* (O_3) is a bluish, explosive, and highly poisonous gas. In the stratosphere, the normal concentration of ozone is about 0.1 part per million (ppm), compared with 0.02 ppm in the lower atmosphere. This ozone-enriched layer is called the **ozone layer**. Ultraviolet light striking ozone and oxygen causes reactions that break down as well as regenerate ozone. In the process, the UV radiation is converted to heat, and the overall level of ozone remains reasonably constant—or so it did until humans intervened.

In 1985, British atmospheric scientists published a startling discovery: The springtime levels of stratospheric ozone over Antarctica had declined by more than 40% since 1977. A hole had been punctured in Earth's protective shield. In the localized *ozone hole* over Antarctica, ozone now dips to about one-third of its pre-depletion levels (Fig. E41-1). Although depletion of the ozone layer is most severe over Antarctica, the ozone layer

is somewhat reduced over most of the world, including nearly all of the continental United States. Recently, springtime ozone levels over Greenland, Scandinavia, and western Siberia were reduced up to 40%. Evidence of ozone thinning over the Arctic has led researchers to predict that we will soon have an Arctic, as well as an Antarctic, ozone hole.

Satellite data reveal that, since the early 1970s, UV radiation per decade has risen by nearly 7% in the Northern Hemisphere and by almost 10% in the Southern Hemisphere. Skin cancer rates are predicted to rise as a result. Studies using mice show that exposure to UV light in dosages similar to a day at the beach without sunscreen can depress their immune systems, making them more vulnerable to a variety of diseases. Human health effects are only one cause for concern. Photosynthesis by phytoplankton, the producers for marine ecosystems, is reduced under the ozone hole above Antarctica. Some types of trees and farm crops are also harmed by increased UV radiation.

The decline in the thickness of the ozone layer is caused by increasing levels of chlorofluorocarbons (CFCs), compounds that contain chlorine, fluorine, and carbon. Developed in 1928, these gases were widely used as coolants in refrigerators and air conditioners, aerosol spray propellants, agents for producing Styrofoam®, and cleansers for electronic parts. These chemicals are very stable

Figure E41-1 Satellite image of the Antarctic ozone hole
The ozone hole is seen in gray and pink on this image recorded by satellite in October 1997, when ozone levels dipped to less than one-third of normal. Dobson units measure the thickness of the ozone (in hundredths of millimeter in a given layer of atmosphere if it were purified and held at standard temperature and pressure.

Health Watch
Steer Clear of Steroids

Perhaps you know someone who is taking "steroids" in an attempt to boost muscle mass or athletic performance. Although few will admit to taking them, as many as 30% of college and professional athletes may be on steroids, even though steroids are banned by organizations that regulate college, professional, and Olympic sports. The steroids that athletes unwisely consume are actually synthetic male hormones, one of the many substances synthesized from cholesterol. These hormones are normally closely regulated by the body, and the overdoses consumed by athletes can be disastrous. In addition to increasing muscle mass (Fig. E3-1), they may lead to liver disease; enlargement of the prostate gland, which then interferes with urination; atrophy of the testicles; and lowered sperm count. Although the verdict is not yet in, these synthetic male hormones may also contribute to heart attacks and strokes. This is a high price to pay for results that could be achieved through a healthy exercise routine.

Cholesterol itself is a steroid with a bad reputation. Why are so many products now advertising themselves as "cholesterol free" or "low in cholesterol"? Although this steroid is crucial to life, doctors have found that individuals with excessively high levels of cholesterol in their blood are much more likely than other individuals to suffer from heart attacks and strokes. Cholesterol contributes to the formation of obstructions in arteries, called **plaques**. Arterial plaques can reduce blood flow to the heart, and their rough edges stimulate blood clot formation inside the arteries. These clots may eventually block an artery, causing a heart attack (if the blocked artery had carried blood to the heart) or a stroke (if the blocked artery had supplied the brain). Plaque formation is described in more detail in "Health Watch: Matters of the Heart" in Chapter 27. Your body can synthesize

its own cholesterol from other lipids, and some bodies manufacture more than others. People with high cholesterol can often reduce their levels by eating a diet low in cholesterol and saturated fats. Doctors regularly prescribe cholesterol-reducing drugs for individuals with extremely high cholesterol who are unable to reduce it adequately through dietary restrictions.

Figure E3-1 Is this body healthy?
Years of hard work produced this body. But to achieve the same look, many body builders and athletes endanger their health by taking "steroids," or synthetic ma...

Earth Watch

These essays present key environmental issues such as biodiversity, ozone depletion, and population growth.

Health Watch

These essays investigate important health-related issues such as prenatal diagnosis, sexually transmitted diseases, and osteoporosis.

Several options from our supplement package will also make your study of biology more meaningful and will therefore improve your understanding.

WWW
Audesirk Live!
http://www.prenhall.com/audesirk

Launch your exploration of biology on the Web with *Audesirk Live!* Participate in a discussion of cloning, read the latest findings about the new fountain of youth, or just challenge yourself with this fun and informative Web site. The icons positioned throughout the textbook will clue you in to related resources on the Web.

Audesirk Interactive
A Student CD-ROM

Do it! Explore *Life on Earth* with the *Audesirk Interactive* student CD-ROM and visualize the way things really happen in biology. The tools in *Audesirk Interactive* will help you envision and understand important concepts. Interactive exercises and animations will allow you to experience how dynamic the study of biology is.

 College Newslink:
Bring each day's biology news to your desktop!

 The ABC News Video Library

 The New York Times
Themes of the Times

INTEGRATED TECHNOLOGY:

Expand your understanding of biology with *Audesirk Interactive*, a new cutting-edge student CD-ROM, and *Audesirk Live!*, a text-specific Web site, which are truly integrated with this textbook to provide a complete *learning system*.

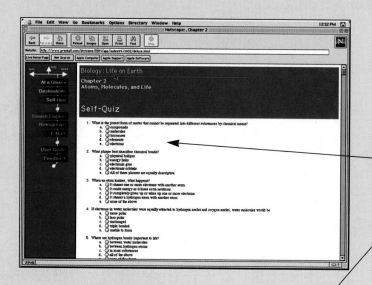

Audesirk On-line:
Audesirk Live!

http://www.prenhall.com/audesirk

For both instructors and students, this resource—organized by chapter and integrated with *Biology: Life on Earth* and *Audesirk Interactive*—provides a powerful tool set for the study of biology. The *Audesirk Live!* Web site includes:

- **Self-Quizzes**, incorporating three challenge levels, unique but complementary to the *Audesirk Interactive* CD-ROM

- **Bizarre Facts in Biology**, encouraging student interest in often-overlooked biological phenomena

- **Issues in Biology**, emphasizing the relevancy of biology

- **Search Terms**, connecting students with powerful Internet search tools

- **Destinations**, compiling the most interesting and relevant Web resources

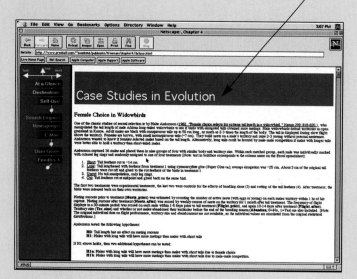

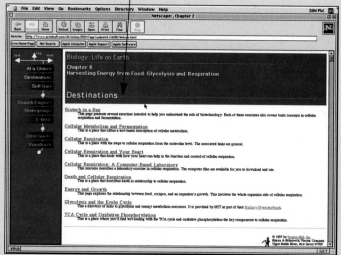

quietly and alone in a monastery garden in the middle of the nineteenth century, lay the foundations for modern genetics?

1 How Did Gregor Mendel Lay the Foundations for Modern Genetics?

www

Before settling down as a monk in the monastery of St. Thomas in Brünn (now Brno, in the Czech Republic), Gregor Mendel (Fig. 12-1) tried his hand at several pursuits, including health care and teaching. To earn his teaching certificate, Mendel attended the University of Vienna for 2 years, where he studied botany and mathematics, among other subjects. This training proved crucial to his later experiments, which were the foundation for the modern science of genetics. At St. Thomas in the mid-1800s, Mendel carried out both his monastic duties and a groundbreaking series of experiments on inheritance in the common edible pea. Although Mendel worked without knowledge of genes or chromosomes, we can more easily follow his experiments after a brief look at some modern genetic concepts.

A gene's specific physical location on a chromosome is called a **locus** (plural, *loci*) (Fig. 12-2). Homologous chromosomes carry the same genes, located at the same loci. Although the nucleotide sequence at a given gene locus is always *similar* on homologous chromosomes, it may not be *identical*. These differences allow different nucleotide sequences at the same gene locus on two homologous chromosomes to produce alternate forms of the gene, called **alleles**. Human A, B, and O blood types, for example, are produced by three alleles of the same gene.

If both homologous chromosomes in an organism have *the same* allele at a given gene locus, the organism is said to be **homozygous** at that gene locus. ("Homozygous" comes from Greek words meaning "same pair.") If two homologous chromosomes have *different* alleles at a given gene locus, the organism is **heterozygous** ("different pair") at that locus and is called a **hybrid**. Recall from Chapter 11 that, during meiosis, homologous chromosomes are separated, so each gamete receives one member of each pair of homologous chromosomes. Therefore, all the gametes produced by an organism that is homozygous at a particular gene locus will contain the same allele. Gametes produced by an organism that is heterozygous at the same gene locus are of two kinds:

WWW The World Wide Web icon shown here indicates that Web destinations are available on *Audesirk Live!* that will help enrich your knowledge and understanding of Mendelian genetics.

Student CD-ROM: *Audesirk Interactive*

Integrated with *Biology: Life on Earth* and independent but complementary to *Audesirk Live!* is the *Audesirk Interactive* student CD-ROM.

Audesirk Interactive includes:

- **Integrating Activity**, offering students a vehicle for exploration of broad but related biological concepts

- **Self-Quizzes**, incorporating three challenge levels, unique but complementary to the *Audesirk Live!* Web site

- **Bizarre Facts in Biology**, encouraging student interest in often-overlooked biological phenomena

- **Glossary**, allowing rapid access to information and incorporating numerous audio pronunciations for commonly mispronounced terms

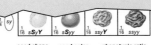

¼ (*sy*)	1/16 *sSyY*	1/16 *sSyy*	1/16 *ssyY*	1/16 *ssyy*

(b)

	seed shape		seed color		phenotypic ratio (9:3:3:1)
	¾ smooth	×	¾ yellow	=	9/16 smooth yellow
	¾ smooth	×	¼ green	=	3/16 smooth green
	¼ wrinkled	×	¾ yellow	=	3/16 wrinkled yellow
	¼ wrinkled	×	¼ green	=	1/16 wrinkled green

Figure 12-6 Predicting genotypes and phenotypes for a cross between gametes that are heterozygous for two traits
Here we are working with both seed color and shape, with yellow (*Y*) dominant to green (*y*), and smooth (*S*) dominant to wrinkled (*s*). *(a)* Punnett square analysis. Assume that both parents are heterozygous for each trait or that a single individual heterozygous for both traits self-fertilizes. There are now 16 boxes in the Punnett square. In addition to predicting all the genotypic combinations, the Punnett square predicts ¾ yellow seeds, ¼ green seeds, ¾ smooth seeds, and ¼ wrinkled seeds, just as we would expect from crosses made of each trait separately. *(b)* Probability theory can be used to predict phenotypes that result from a cross between gametes that are heterozygous for two traits. The fraction of genotypes from each sperm and egg combination is illustrated within each box of the Punnett square. Adding the fractions for the same genotypes will give the genotypic ratios. Converting each genotype to a phenotype and then adding their numbers reveals that ¾ of the offspring will be smooth and ¼ will be wrinkled and that ¾ will be yellow and ¼ will be

...heritance. No doubt to ...tense disappointment, when they searched the scientific literature before publishing their results, they found that Mendel had scooped them more than 30 years earlier. To their credit, they graciously acknowledged the important work of the Augustinian monk, who had died in 1884.

4 How Are Genes Located on the Same Chromosome Inherited?

Gregor Mendel knew nothing about the physical nature of genes or chromosomes. Once scientists had seen chromosomes by microscope and figured out how they are distributed during mitosis and meiosis, it became obvious that chromosomes are the vehicles of inheritance. It also became obvious that there are many more traits (and therefore many more genes) than there are chromosomes. That genes are parts of chromosomes and that each chromosome bears many genes have important implications for inheritance.

Genes on the Same Chromosome Tend to Be Inherited Together

If chromosomes assort independently during meiosis I, then only genes located on *different chromosomes* will assort independently into gametes. Genes on the *same chromosome* are said to be *linked*. **Linkage** is the inheri...

This CD-ROM icon indicates that *Audesirk Interactive* contains an activity and/or animation of gene linkage.

EVOLUTION AND ADAPTATION
See the Connections

Important patterns and themes unify the discipline of biology. Evolution and adaptation to the environment are two related themes you'll revisit throughout the text. You will also be able to connect the concepts with the way the scientific process is carried out in the real world.

Evolutionary Connections

This feature, at the end of many chapters, ties the subject matter of each chapter into the broader perspective of evolution.

Scientific Inquiry

The processes by which scientists answer questions are as important as the questions themselves. These essays emphasize how we know what we know and give you insight into the dynamic process of discovery.

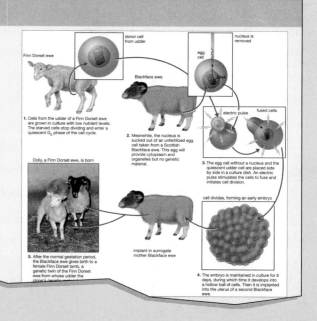

Evolutionary Connections
Fungal Ingenuity—Pigs, Shotguns, and Nooses

Natural selection, operating over millions of years on the diverse forms of fungi, has produced some remarkable adaptations by which fungi disperse their spores and obtain nutrients. A few of these are highlighted here.

The Rare, Sexy Truffle
Although many fungi are prized as food, none are as avidly sought as the truffle (Fig. 20-13). A single specimen resembling a small, shriveled, blackened apple may sell for $100. Truffles are the spore-containing structures of an ascomycete that forms a mycorrhizal association with the roots of oak trees. Although it develops underground, the truffle has evolved a fascinating mechanism to entice animals, especially wild pigs, to dig it up. The truffle releases a chemical that closely resembles the pig's sex attractant. As aroused pigs dig up and devour the truffle, millions of spores are scattered to the winds. Truffle collectors use muzzled pigs to hunt their quarry; a good truffle-pig can smell an underground truffle 50 meters (about 150 feet) away!

The Shotgun Approach to Spore Dispersal
If you get close enough to a pile of horse manure to scrutinize it, you may be fortunate enough to observe the beautiful, delicate reproductive structures of the zygomycete *Pilobolus* (Figure 20-14). Despite their daintiness, these are actually fungal shotguns. The clear bulbs, capped with sticky black spore cases, extend from hy-

Figure 20-14 An explosive zygomycete
The delicate, translucent reproductive structures of the zygomycete *Pilobolus* will literally blow their tops when ripe, dispersing the black caps with their payload of spores.

phae that penetrate the dung. As the bulbs mature, the sugar concentration inside them increases, and water is drawn in by osmosis. Meanwhile, the bulb begins to weaken just below its cap. Suddenly, like an overinflated balloon, the bulb bursts and blows its spore-carrying top up to a meter (3 feet) away. *Pilobolus* bends toward the light, thereby increasing the chances that its spores will land on open pasture when the bulb bursts. There the spores adhere to grass until consumed by a grazing herbivore, perhaps a horse. Later (some distance away) the spores are deposited unharmed in a pile of their favorite food: manure. Growing hyphae penetrate this rich source of nutrients, sending up new projectiles to continue this ingenious cycle.

The Nematode Nemesis
Microscopic *nematodes* (roundworms) abound in rich soil, and fungi have evolved several fascinating forms of nematode-nabbing hyphae that allow them to exploit this rich source of protein. Some fungi produce sticky pods that adhere to passing nematodes, penetrate the worm body with hyphae, and then begin digesting it from within. One species shoots a microscopic harpoonlike spore into passing nematodes; the spore develops into a new fungus inside the worm. The fungal strangler *Arthrobotrys*, a member of the imperfect fungi, produces nooses formed from three hyphal cells. When a nematode wanders into the noose, its contact with the inner parts of the noose stimulates noose cells to swell with water (Fig. 20-15). In a fraction of a second, the [...] the worm. Fungal hyphae th[...]

Figure 20-13 The truffle
Truffles, rare ascomycetes (each about the size of a small apple), are a gastronomic delicacy.

Scientific Inquiry www
Much Ado About Dolly

"And what is the Scientific Community doing...? THEY'RE CLONING SHEEP! Great! Just what we need! Sheep that look MORE ALIKE than they already do!"
    Dave Barry, 1997

Early in 1997, a short article in the journal *Nature* caused headlines throughout the world. Dr. Ian Wilmut and associates of the Roslin Institute in Edinburgh, Scotland, had done what many scientists believed to be impossible: They cloned a mammal, using a nucleus taken from adult tissue. **Cloning** is the process of artificially creating a clone—a new individual that is genetically identical to an existing individual. As you are aware, every body cell has a complete set of genes. Researchers had already discovered that the nucleus of a cell in extremely early stages of embryonic development, before cells have specialized, could be used to replace the nucleus of an egg cell. The embryonic cell nucleus then directs the development of a new individual, with no need for a fusion of sperm and egg. Until Dr. Wilmut's breakthrough, attempts to make clones from nuclei taken from older tissues (from embryos that had more than 16 cells) had failed. Many researchers hypothesized that the inactivation of selected genes that make, say, neurons different from skin cells and liver cells different from fat cells, must be permanent. What Dr. Wilmut and associates demonstrated is that differentiation is reversible under the right conditions (Fig. E11-1). They used cells taken from the udder of a 6-year-old ewe and grown in culture dishes. Depriving these cells of nutrients forced the cells into the nondividing (G$_0$) stage of the cell cycle—a factor that seems crucial to allowing all the genes to be expressed when the nucleus is transplanted into an egg cell.

Cloning is a highly artificial form of asexual reproduction, because the nucleus that directs the development of the offspring was produced mitotically from a diploid cell and not by the fusion of haploid gametes produced by meiosis. Like the hydra budding from its parent, the cloned mammal is literally "a chip off the old block." The advantage of cloning cells from an adult rather than from an embryo is that you know what you are getting. A prize wool-producing sheep could theoretically be cloned to produce a whole herd. The same might be done with a cow that has exceptional

milk production. Cloning an adult allows scientists to use asexual reproduction to take advantage of the natural variability provided by sexual reproduction. By observing the adults, the most advantageous gene combinations produced during meiosis and the union of gametes can be selected and then perpetuated in clones.

The increasing ability of scientists to manipulate individual genes makes cloning even more advantageous. For example, let's say genetic engineering produces a cow that secretes large quantities of a specialized antibiotic in its milk. This antibiotic could be extracted inexpensively to fight hard-to-treat infections. Such a cow would represent an enormous investment in money, time, and ingenuity. But what if the cow were infertile? Or what if none of its offspring carried the gene to produce the antibiotic? Enter adult cloning. Tiny amounts of tissue from the engineered cow would yield millions of cells, each containing the full genetic "blueprint" for an antibiotic-producing cow.

The implications of Dr. Wilmut's research are far-reaching. Scientists now have proof that the long-inactivated genes of specialized adult cells can be made functional again. Perhaps neurons in the brain region destroyed by Parkinson's disease could be made to divide again, replacing those that have perished. Perhaps more genes for red blood cell production could be switched on in an anemia patient.

Dolly and other cloned mammals also promise to provide insights into aging. Do the genes of each species dictate a maximum life span? If so, Dolly may have been born middle-aged, because her genes came from the cell of a 6-year-old ewe. Will her life be shorter as a result? Will she be able to reproduce normally? She is now pregnant. The scientific world is watching.

What really caught the attention of people everywhere is that Dolly raises the possibility of human cloning. Although the technique required 277 attempts before one healthy lamb was born, the technology is relatively simple. There is no theoretical reason why the method could not be applied to humans. For example, adult cloning could produce a child who had exactly the same set of genes as its mother, a beloved grandparent, a famous sports figure, a Nobel Prize winner, or a despot. Think about it. We'll revisit this topic in Chapter 13.

1. Cells from the udder of a Finn Dorset ewe are grown in culture with low nutrient levels. The starved cells stop dividing and enter a quiescent G$_0$ phase of the cell cycle.

2. Meanwhile, the nucleus is sucked out of an unfertilized egg cell taken from a Scottish Blackface ewe. This egg will provide cytoplasm and organelles but no genetic material.

3. The egg cell without a nucleus and the quiescent udder cell are placed side by side in a culture dish. An electric pulse stimulates the cells to fuse and initiates cell division.

4. The embryo is maintained in culture for 6 days, during which time it develops into a hollow ball of cells. Then it is implanted into the uterus of a second Blackface ewe.

5. After the normal gestation period, the Blackface ewe gives birth to a female Finn Dorset lamb, a genetic twin of the Finn Dorset ewe from whose udder the clone's [...]

END-OF-CHAPTER MATERIAL
Check Your Understanding. Expand it. Explore!

Summary of Key Concepts

1) What Is Biotechnology?
Biotechnology is any industrial or commercial use or alteration of organisms, cells, or biological molecules to achieve specific practical goals. Modern biotechnology generates altered genetic material via genetic engineering. Genetic engineering may involve the production of recombinant DNA by combining DNA from different organisms. DNA is transferred between organisms by means of vectors, such as bacterial plasmids. The resulting organisms are termed *transgenic.* Some major goals of genetic engineering are to increase our understanding of gene function, to treat disease, and to improve agriculture.

2) How Does DNA Recombination Occur in Nature?

(3) Amplification of the gene: To produce multiple copies of a recombinant gene, the bacteria containing the gene may be grown in culture, or the polymerase chain reaction (PCR) technique may be used.

4) What Are Some Applications of Biotechnology?
Once sufficient quantities of a gene or DNA fragment are available, the nucleotide sequence can be determined. Researchers can then synthesize and study the coded protein or devise tests for mutated forms of the gene. Amplified DNA might be used in DNA fingerprinting, in which restriction enzymes are used to break the DNA into fragments whose different lengths are characteristic of different individuals (a phenomenon called restriction fragment length polymor... RFLP). DNA fingerprints based on RFLPs are

Summary of Key Concepts

Part of the Chapter Organizing System, the Summary is arranged in numbered sections that correspond to the major concepts that are covered in each chapter.

Thinking Through the Concepts*

Multiple-choice and review questions check your basic understanding of chapter material. Answers to Multiple-Choice Questions appear at the end of each chapter.

genetic recombination p. 238
genotype p. 233
law of independent assortment p. 235
law of segregation p. 232
recessive p. 232

Thinking Through the Concepts

Multiple Choice
1. An organism is described as Rr:red. The Rr is the organism's [A]; red is the organism's [B]; and the organism is [C].
   a. [A] phenotype; [B] genotype; [C] degenerate
   b. [A] karyotype; [B] hybrid; [C] recessive
   c. [A] genotype; [B] phenotype; [C] heterozygous
   d. [A] gamete; [B] linkage; [C] pleiotropic
   e. [A] zygote; [B] epistasis; [C] homozygous

the only possible combination supporting this hypothetical circumstance? (Answers in the order: mother:father:child)
   a. A:B:O
   b. A:O:B
   c. AB:A:O
   d. AB:O:AB
   e. B:O:A

4. A heterozygous red-eyed female Drosophila mated with a white-eyed male would produce
   a. red-eyed females and white-eyed males in the F₁
   b. white-eyed females and red-eyed males in the F₁
   c. half red- and half white-eyed females and all white-eyed males in the F₁
   d. all white-eyed females and half red- and half white-eyed males in the F₁
   e. half red- and half white-eyed females as well as males in

? Review Questions
1. Define the following terms: *gene, allele, dominant,* *true-breeding, homozygous, heterozygous, cross-fer... self-fertilization.*
2. Explain why genes located on one chromosome a... be linked. Why do linked genes sometimes separat... meiosis?
3. Define *polygenic inheritance.* Why could polygenic... tance allow parents to produce offspring that are... different in eye or skin color than either parent?
4. What is sex linkage? In mammals, which sex would b... likely to show recessive sex-linked traits?
5. What is the difference between a phenotype and a...

Applying the Concepts

1. Sometimes the term *gene* is used rather casually. Compare and contrast use of the terms *allele* and *locus* as alternatives to *gene.*
2. Using the information in the chapter, explain why AB individuals are referred to as "universal recipients" in terms of blood transfusions and why people with type O blood are called "universal donors."
3. Mendel's numbers seemed almost too perfect to be real; some believe he may have cheated a bit on his data. Perhaps he continued to collect data until the numbers matched his predicted ratios, then stopped. Recently, there has been much publicity over violations of scientific ethics, including researchers' plagiarizing others' work, using other scientists' methods to develop lucrative patents, or just plain fabricating data. How important an issue is this for society? What are ... ethical scientific behavior? How

5. *Eugenics* is the term applied to the notion that the human condition might be improved by improving the human genome. Do you think there are both good and bad sides to eugenics? What examples can you think of to back up your stand? What would a eugenicist think of the medical advances that have ameliorated the problems of hemophilia?
6. Think about some of the personal, religious, and medical economic issues related to prenatal counseling and diagnosis. Would you avoid having children if you knew that both you and your spouse were heterozygous for a recessive disorder that is fatal at an early age and may involve considerable suffering? What would you do if you or your spouse were pregnant and learned that your offspring, if born, would be homozygous for such a disorder? Is the situation qualitatively different for D...

Applying the Concepts*

Expand your understanding by applying your critical-thinking skills to real-world scenarios.

NEW!
Group Activities

These activities are designed to launch your desire to explore important concepts of biology and their impact on the world around you. They offer a great opportunity for sharing ideas with your peers in a non-intimidating format.

Group Activity

Form a group of four students. Imagine that you've been working in a museum collection, and you discover a dusty old cabinet that hasn't been opened in 150 years. Inside are some mysterious specimens, previously unknown to science. Your job is to classify these organisms. Develop a classification, draw an evolutionary tree, and explain your reasoning. The specimens are pictured below.

For More Information
Crow, J. F. "Genes That Violate Mendel's Rules." *Scientific American,* February 1979. A close look at genetic recombination and its role in evolution.

Hoyt, E. "Wild Relatives." *Living Wilderness,* Summer 1990. Undisturbed ecosystems are a valuable gene bank with great potential for improving crop plants.

Horgan, J. "Eugenics Revisited." *Scientific American,* June 1993. Determining patterns of inheritance, let alone finding suspect genes, for behavioral disorders is notoriously difficult. What's more, it isn't at all clear what one should do with such data, if they become available.

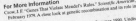

Pines, M. "In the Shadow of Huntington's." *Science 84,* May 1984; Grady, D. "The Ticking of a Time Bomb in the Genes." *Discover,* June 1987; Roberts, L., "Huntington's Gene: So Near, Yet So Far." *Science* 1990, 247:624–627; and Morell, V. "Huntington's Gene Finally Found." *Science* 1993, 260:28–30. These articles focus on the story of the scientific detective work involved in the diagnosis of Huntington's disease by using recombinant DNA techniques, the social dilemma it created, the frustrations of searching for the gene, and, finally, finding it. Particularly poignant because one of the lead investigators may herself be a victim.

Sapienza, C. "Parental Imprinting of Genes." *Scientific American,* October 1990. It is not quite true that all genes are equal, regardless of whether they have been inherited from mother or father. In some cases which parent a gene comes from greatly alters its expression in

For More Information

These suggestions for further readings on chapter-related topics refer you to books and periodicals—the latest as well as the classics.

*Both *Audesirk Interactive* and *Audesirk Live!* include and expand on the "At a Glance," "Summary of Key Concepts," "Thinking Through the Concepts," and "Applying the Concepts" sections. This consistency provides you with an integrated **learning system** that will help you succeed in this course!

Brief Contents

Essays

*See our Web site at http://www.prenhall.com/audesirk for
additional "A Closer Look" essays:

Contents

UNIT TWO
INHERITANCE 141

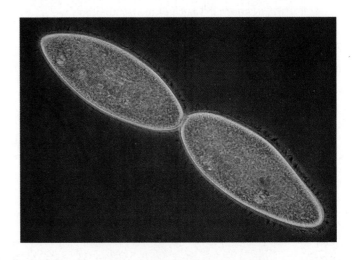

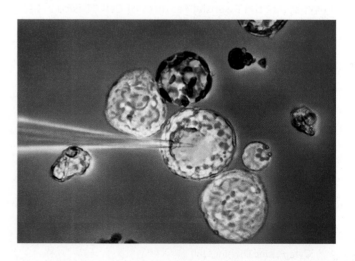

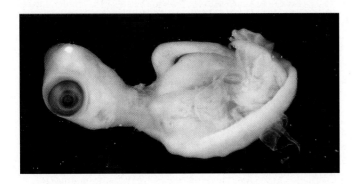

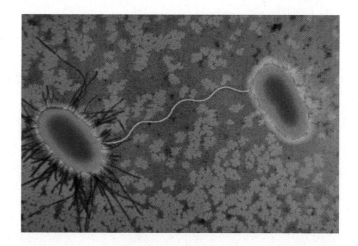

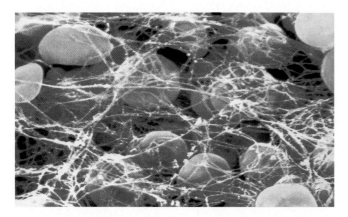

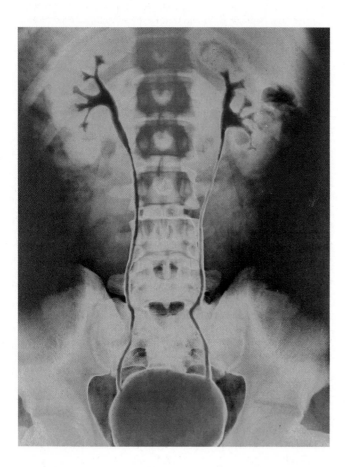

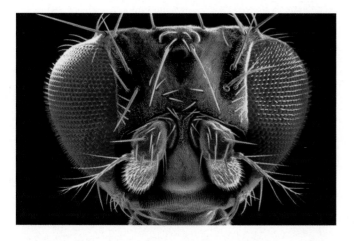

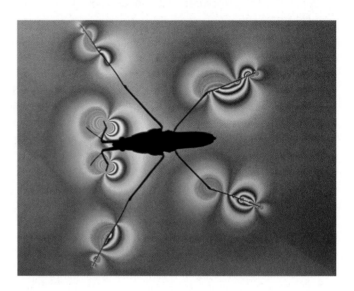

UNIT SIX
ECOLOGY 789

"Viewed from the distance of the moon, the astonishing thing about the earth, catching the breath, is that it is alive. The photographs show the dry, pounded surface of the moon in the foreground, dead as an old bone. Aloft, floating free beneath the moist, gleaming surface of bright blue sky, is the rising earth, the only exuberant thing in this part of the cosmos."

Lewis Thomas in The Lives of a Cell *(1974)*

Life on Earth is confined to a thin film encompassing Earth's surface: the biosphere. Earth, seen from the Moon, is an oasis of life in our solar system.

An Introduction to Life on Earth 1

At a Glance

Net Watch

On-line resources for this chapter are on the World Wide Web at:
http://www.prenhall.com/audesirk
(click on the Table of Contents link and then select Chapter 1).

On your way across campus tomorrow morning, take a moment to look around you. Although you might not notice at first, an astonishing array of creatures thrives even in a place as domesticated as a college campus. Sparrows and squirrels scurry along branches, chirping and chattering. Trees, bushes, grass, and moss blanket the campus with green. Whether you're walking across campus or through the countryside, take a closer look. You may also see honeybees flitting from flower to flower, earthworms—or their winding tracks—at the edges of a small pond, a spider spinning its web, glistening mushrooms in the grass or beneath a bush. If you listen carefully, you may hear creatures you don't see: insects humming, birds calling in the distance, perhaps a frog croaking. But these are only the obvious *organisms*—individual living things. Countless tiny ones, most too small to be seen with the naked eye, swim in the puddles left by last night's rain. Whole communities of microscopic living things thrive in the soil. And living on, in, and around all these organisms, and the humans observing them, are billions of bacteria—simple, single-celled organisms that have survived nearly unchanged for billions of years.

How did such an astounding variety of living things evolve on Earth? In what ways do they interact with one another? How are these bacteria, fungi, plants, and animals alike and how do they differ? What processes must occur in each organism for it to survive in its environment and to reproduce? And perhaps the most basic question of all: How do living things differ from nonliving things? Common questions such as these—asked by curious people for centuries—form the basis of the science of biology. They also relate to the very survival of the human species, because we too are organisms. We evolved in response to the same sorts of survival needs. Our bodies and those of all other living creatures consist of the same basic set of chemicals. The same processes allow us to survive and reproduce. As we journey through this book together, we hope you begin to share with us the excitement and awe that come from understanding the nature of Earth and its inhabitants and from appreciating the beauty of it all.

What Are the Characteristics of Living Things?

To study the nature of life on Earth, we should begin at the beginning: *What is life?* If you look up *life* in a dictionary, you will find definitions such as "the quality that distinguishes a vital and functioning being from a dead body," but you won't find out what that "quality" is.

All of us have an intuitive understanding of what it means to be alive. Nevertheless, defining *life* is difficult, partly because living things are so diverse and nonliving matter is in some cases so lifelike. A more fundamental difficulty in defining life is that living things cannot be described as the sum of their parts. The quality of life emerges as a result of the incredibly complex, ordered interactions among these parts. Because it is based on these **emergent properties**, life is a fundamentally intangible quality that defies any simple definition. We can, however, describe some of the characteristics of living things that, taken together, are not shared by nonliving objects. These characteristics are as follows:

1. Living things have a complex, organized structure that consists largely of *organic* (carbon-based) molecules.
2. Living things acquire and use materials and energy from their environment and convert them into different forms.
3. Living things actively maintain their complex structure and their internal environment, a process called *homeostasis*.
4. Living things grow.
5. Living things respond to stimuli from their environment.
6. Living things reproduce themselves, using a molecular blueprint called *DNA*.
7. Living things, as a whole, have the capacity to evolve.

Let's explore these characteristics in more detail.

Living Things Are Both Complex and Organized

Compared with nonliving matter of similar size, living things are highly complex and organized. A crystal of table salt (Fig. 1-1a), for example, consists of just two chemical elements, sodium and chlorine, arrayed in a precise cubical arrangement: The salt crystal is organized but simple. The oceans (Fig. 1-1b) contain some atoms of all the naturally occurring elements, but these atoms are randomly distributed: The oceans are complex but not organized. In contrast, even the tiny waterflea (Fig. 1-1c) contains dozens of different elements linked together in thousands of specific combinations that are further organized into ever larger and more-complex assemblies to form structures such as eyes, legs, a digestive tract, and even a small brain.

Life on Earth is composed of a hierarchy of structures, with each level of the hierarchy based on the one below it and providing the foundation for the one above it (Fig. 1-2). All of life is built on a chemical foundation based on

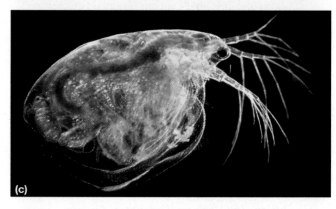

Figure 1-1 *Life is both complex and organized*
(a) Each crystal of table salt, sodium chloride, is a cube, showing great organization but minimal complexity. *(b)* The water and dissolved materials in the ocean represent complexity but very little organization. *(c)* Living things have both complexity and organization. The waterflea, *Daphnia pulex*, is only 1 millimeter (1/1000 meter) long, yet it has legs, a mouth, a digestive tract, reproductive organs, light-sensing eyes, and even a rather impressive brain in relation to its size.

substances called **elements**, each of which is a unique type of *matter*. An **atom** is the smallest particle of an element that retains the properties of that element. For example, a diamond is made of the element carbon. The smallest possible unit of the diamond is an individual carbon

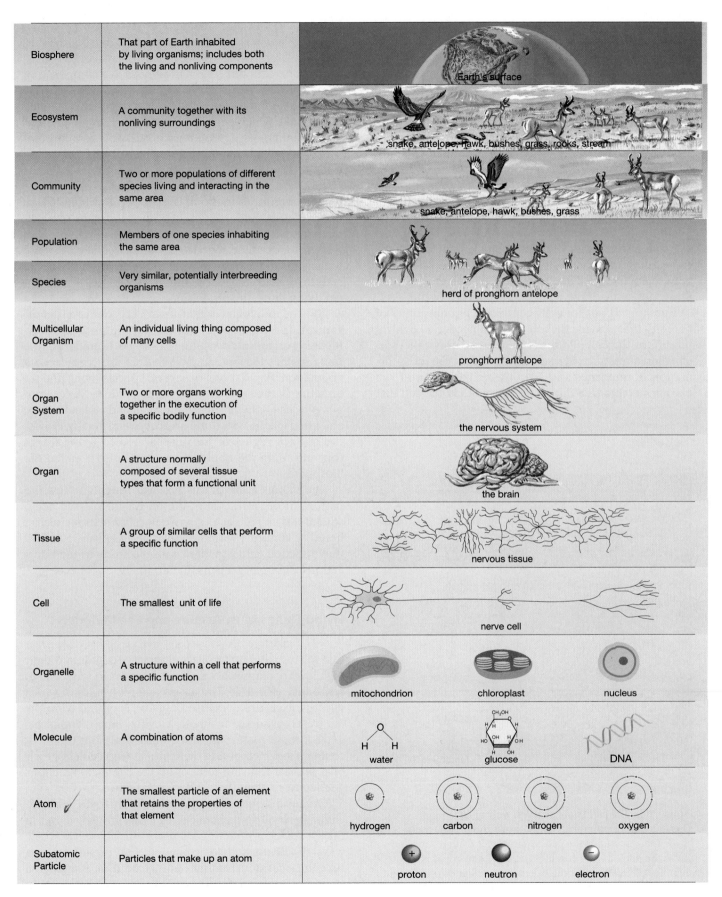

Biosphere	That part of Earth inhabited by living organisms; includes both the living and nonliving components	
Ecosystem	A community together with its nonliving surroundings	
Community	Two or more populations of different species living and interacting in the same area	
Population	Members of one species inhabiting the same area	
Species	Very similar, potentially interbreeding organisms	
Multicellular Organism	An individual living thing composed of many cells	
Organ System	Two or more organs working together in the execution of a specific bodily function	
Organ	A structure normally composed of several tissue types that form a functional unit	
Tissue	A group of similar cells that perform a specific function	
Cell	The smallest unit of life	
Organelle	A structure within a cell that performs a specific function	
Molecule	A combination of atoms	
Atom	The smallest particle of an element that retains the properties of that element	
Subatomic Particle	Particles that make up an atom	

Figure 1-2 *Levels of organization of matter*

All life has a chemical basis, but the quality of life itself emerges on the cellular level. Interactions among the components of each level and the levels below it allow the development of the next-higher level of organization.

atom; any further division would produce isolated **subatomic particles** that would no longer be carbon. Atoms may combine in specific ways to form assemblies called **molecules**; for example, one carbon atom can combine with two oxygen atoms to form a molecule of carbon dioxide. Although many simple molecules form spontaneously, extremely large and complex molecules are manufactured only by living things. The bodies of living things are composed primarily of complex molecules. The molecules of life, which are carbon-based, are called **organic molecules**.

Although the chemical arrangement and interaction of atoms and molecules are the building blocks of life, the quality of life itself emerges on the level of the cell. Just as an atom is the smallest unit of an element, so too the **cell** is the smallest unit of life (Fig. 1-3). The difference between a living cell and a conglomeration of chemicals illustrates some of the emergent properties of life.

Fundamentally, all cells contain (1) **genes**, units of heredity that provide the information needed to control the life of the cell; (2) subcellular structures called **organelles**, miniature chemical factories that use the information in the genes and keep the cell alive; and (3) a **plasma membrane**, a thin sheet surrounding the cell that both encloses a watery medium (the **cytoplasm**) that contains the organelles and separates the cell from the outside world. Some life-forms, mostly microscopic, consist of just one cell, but larger life-forms are composed of many cells whose functions are differentiated. In these multicellular life-forms, cells of similar type are combined into **tissues**, which perform a particular function. Examples of animal tissues are nervous tissue—composed of *neurons* (individual nerve cells), which produce electrical signals—and connective tissue, which commonly surrounds and supports other tissue types. Various tissue types combine to make up a structural unit called an **organ** (for example, the brain). Several organs that collectively perform a single function are called an **organ system**; for example, together the brain, spinal cord, sense organs, and nerves form the nervous system. All the organ systems functioning cooperatively make up an individual living thing, the **organism**.

Beyond individual organisms are broader levels of organization. A group of very similar, potentially interbreeding organisms constitutes a **species**. Members of the same species that live in a given area are considered a **population**. Populations of several species living and interacting in the same area form a **community**. A community plus its nonliving environment, including land, water, and atmosphere, constitute an **ecosystem**. Finally, the entire surface region of Earth inhabited by living things (but including the nonliving components) is called the **biosphere**.

The organization of this text will roughly follow the pattern of organization of life on Earth. We will begin with atoms and molecules, move to cells and principles of heredity, continue with the range of organisms and how they function, and conclude with the study of the interactions among organisms that are the focus of ecology.

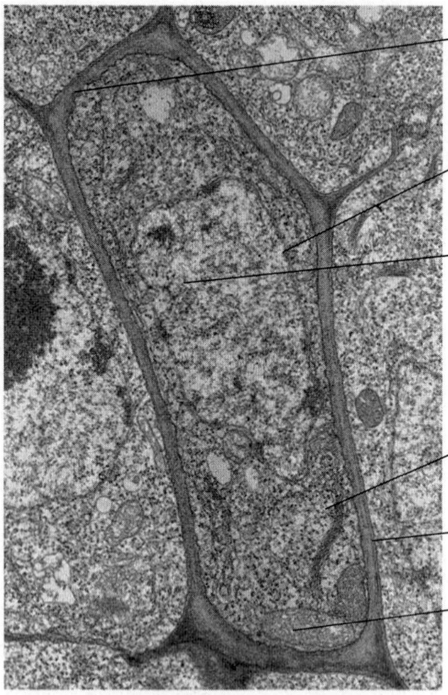

plasma membrane

nuclear envelope (double membrane)

nucleus

cytoplasm

cell wall

mitochondrion (an organelle)

1 micrometer

Figure 1-3 The cell is the smallest unit of life
This electron micrograph clearly shows the plasma membrane that surrounds the cell, separating it from its environment; the nucleus that contains the cell's genes and is surrounded by a membrane called the nuclear envelope; and many other specialized structures, called organelles, that perform particular functions in the life of the cell. It also shows the cell wall, present in many plant cells, which surrounds and supports the cell while allowing free exchange of materials.

Living Things Must Acquire and Use Materials and Energy

Organisms need materials and energy to maintain their high level of complexity and organization, to grow, and to reproduce (Fig. 1-4). The atoms and molecules of which all organisms are composed may be acquired from the air, water, or soil or from other living things. Organisms extract these materials, called **nutrients**, from the environment and incorporate them into the molecules of their own bodies. The sum total of all the chemical reactions needed to sustain life is called **metabolism**.

Organisms obtain **energy**—the ability to do work, including carrying out chemical reactions, growing leaves in the spring, or contracting a muscle—in one of two basic ways. (1) Plants and some single-celled organisms capture the energy of sunlight and store it in energy-rich sugar molecules, a process called **photosynthesis**. (2) In contrast, neither fungi nor animals can photosynthesize, nor can most bacteria; these organisms must consume the energy-rich molecules contained in the bodies of

Figure 1-4 *Living things acquire energy and nutrients from the environment*
The plants in this Kenyan grassland capture energy from the sun and nutrients from the air, water, and soil. In contrast, the grazing Cape buffalo extract both energy and nutrients from the plants. The lion will extract energy and nutrients from its animal prey, indirectly derived from plants. Without a continuous supply of solar energy, nearly all life would cease.

other organisms. In either case, energy taken in is converted into a form that the organism can use or store for future use.

Ultimately, the energy that sustains nearly all life comes from sunlight, captured by photosynthetic organisms and incorporated into energy-rich molecules. Organisms that cannot photosynthesize depend on photosynthetic life-forms for food, either directly or indirectly. Thus, energy flows from the sun through nearly all forms of life. That energy is eventually released again as heat.

Homeostasis Maintains Relatively Constant Internal Conditions

Complex, organized structures are not easy to maintain. Whether we consider the molecules of your body or the books and papers on your desk, organization tends to disintegrate into chaos unless energy is used to sustain it. (We will explore this tendency more fully in Chapter 4.) To stay alive and function effectively, organisms must keep the conditions within their bodies fairly constant, a process called **homeostasis** (derived from Greek words meaning "to stay the same"). One of the many conditions regulated is body temperature. Among warm-blooded animals, for example, vital organs such as the brain and

heart are kept at a warm, constant temperature despite wide fluctuations in environmental temperature.

Maintaining homeostasis is accomplished by a variety of automatic mechanisms. In the case of temperature regulation, these mechanisms include sweating during hot weather, metabolizing more food in cold weather, and behaviors such as basking in the sun or even adjusting the thermostat in the room. Of course, not everything stays the same throughout an organism's life. Major changes, such as growth and reproduction, occur, but these are not failures of homeostasis. Rather, they are specific, genetically programmed parts of the organism's life cycle.

Growth Is a Property of All Organisms

At some time in its life cycle, every living thing becomes larger—that is, it *grows*. This characteristic is obvious for plants, birds, and mammals, all of which start out very small and undergo tremendous growth during their lives. Even single-celled bacteria, however, grow to about double their original size before they divide. In all cases, growth involves the conversion of materials acquired from the environment into the specific molecules of the organism's body.

Living Things Respond to Stimuli

Organisms perceive and respond to stimuli in their internal and external environments. Animals have evolved elaborate sensory organs and muscular systems that allow them to detect and respond to light, sound, chemicals, and many other stimuli from their surroundings. Internal stimuli are perceived by receptors for stretch, temperature, pain, and various chemicals. For example, when you feel hungry, you are perceiving contractions of your empty stomach and low levels of sugars and fats in your blood. You then respond to external stimuli by choosing appropriate objects to eat, such as a piece of pie rather than the plate and fork. Yet, animals, with their elaborate nervous systems and motile bodies, are not the only organisms to perceive and respond to stimuli. The plants on your windowsill grow toward light, and even the bacteria in your intestines manufacture a different set of digestive enzymes depending on whether you drink milk, eat candy, or both.

Living Things Reproduce Themselves

The *continuity of life* occurs because organisms reproduce, giving rise to offspring of the same type (Fig. 1-5). The processes for producing offspring are varied, but the result—the perpetuation of the parents' genetic material—is the same. The *diversity of life* occurs in part because offspring, though arising from the genetic material provided by their parents, are normally somewhat different from their parents, as explained briefly below and in Unit III. The mechanism by which traits are passed

Figure 1-5 Living things reproduce
As they grow, these polar bear cubs will resemble, but not be identical to, their parents. The similarity and variability of offspring are crucial to the evolution of life.

from one generation to the next, through a "genetic blueprint," produces these variable offspring.

DNA Is the Molecule of Heredity

All known forms of life use a molecule called **deoxyribonucleic acid**, or **DNA**, as the repository of hereditary information (Fig. 1-6). Genes are segments of the DNA molecule. Much of Unit II will be devoted to exploring the structure and function of this remarkable molecule. For now, we will simply note that an organism's DNA is its genetic blueprint or molecular instruction manual, a

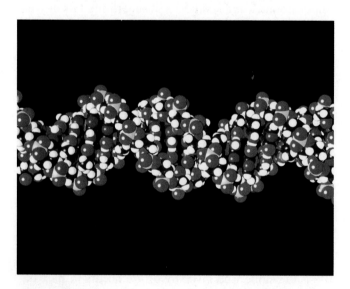

Figure 1-6 DNA
A computer-generated model of DNA, the molecule of heredity.

guide to both the construction and, at least in part, the operation of its body. When an organism reproduces, it passes a copy of its DNA to its offspring. The accuracy of the DNA copying process is astonishingly high: Only about one mistake occurs for every billion bits of information contained in the DNA molecule. But chance accidents to the genetic material also bring about changes in the DNA. The occasional errors and accidental changes, called **mutations**, are crucial. Without mutations, all life-forms might be identical. Indeed, there is reason to believe that, without mutations, there would be no life. Mutations in DNA are the ultimate source of genetic variations. These variations, superimposed on a background of overall genetic fidelity, make possible the final property of life, the capacity to evolve.

Living Things Have the Capacity to Evolve

Although the genetic makeup of a single organism remains essentially the same over its lifetime, the genetic composition of a species as a whole changes over many lifetimes. Over time, mutations and variable offspring provide diversity in the genetic material of a species. In other words, the species *evolves*. The theory of **evolution** states that modern organisms descended, with modification, from preexisting life-forms. The most important force in evolution is **natural selection**, the process by which organisms with *adaptations* (characterstics that help them cope with the rigors of their environment) survive and reproduce more successfully than do others that lack those traits. Adaptive traits arising from genetic mutation are passed on to the next generation.

② How Do Scientists Categorize the Diversity of Life?

Although all living things share the general characteristics discussed earlier, evolution has brought forth an amazing variety of life-forms. In the following brief description of the features used to classify organisms, we introduce you to the diversity of life on Earth. We describe the classification and structures of organisms in detail in Chapters 18 through 22.

Organisms can be grouped into three major categories, called **domains**: (1) Bacteria, (2) Archaea, and (3) Eukarya. This classification is based on fundamental differences among the cell types that comprise these organisms. Members of both the Bacteria and the Archaea normally consist of single, simple cells. Members of the Eukarya have bodies composed of one or more highly complex cells and are subdivided into four **kingdoms**: Protista, Fungi, Plantae, and Animalia (Fig. 1-7). There are exceptions to any simple set of criteria used to characterize the domains and kingdoms, but three characteristics are particularly useful: cell type, the number of cells

Figure 1-7 The domains and kingdoms of life

(a)

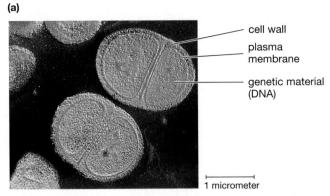

cell wall

plasma membrane

genetic material (DNA)

1 micrometer

(a) The domain Bacteria. A color-enhanced electron micrograph of a dividing bacterium. Bacteria are unicellular and prokaryotic; most are surrounded by a thick cell wall. Some bacteria photosynthesize, but most absorb food from their surroundings.

(b)

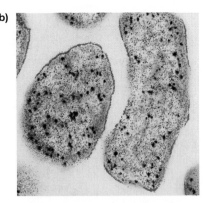

(b) The domain Archaea. A color-enhanced electron micrograph of an archaean. The cell wall appears red, and DNA is scattered inside. Many archaeans can survive extreme conditions. This Antarctic species lives at temperatures as low as −2.5°C.

(c)

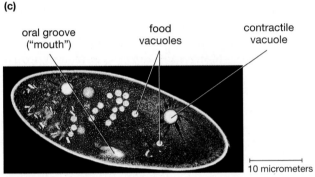

oral groove ("mouth")

food vacuoles

contractile vacuole

10 micrometers

(c) The kingdom Protista (domain Eukarya). This light micrograph of a *Paramecium* illustrates the complexity of these large, normally single, eukaryotic cells. Some protists photosynthesize, but others ingest or absorb their food. Many, including *Paramecium*, are mobile, moving with cilia or flagella.

(d)

(d) The kingdom Fungi (domain Eukarya). An exotic mushroom found in Peru. Most fungi are multicellular. Fungi generally absorb their food, which is usually the dead bodies or wastes of plants and animals. The food is digested by enzymes secreted outside the fungal body. Most fungi cannot move.

(e)

(e) The kingdom Plantae (domain Eukarya). This butterfly weed represents the flowering plants, the dominant members of the kingdom Plantae. Flowering plants owe much of their success to mutually beneficial relationships with animals, such as these pearl crescent butterflies, in which the flower provides food and the insect carries pollen from flower to flower, fertilizing them. Plants are multicellular, nonmotile eukaryotes that acquire nutrients by photosynthesis.

(f)

(f) The kingdom Animalia (domain Eukarya). A wrasse rests on a soft coral. Animals are multicellular; animal bodies consist of a wide assortment of tissues and organs composed of specialized cell types. Most animals can move and respond rapidly to stimuli. The coral is a member of the largest group of animals: the invertebrates, which lack a backbone. This group also includes insects and mollusks. The wrasse is a vertebrate; like humans, it has a backbone.

Table 1-1 Some Characteristics Used in Classification of Organisms				
Domain	Kingdom	Cell Type	Cell Number	Major Mode of Nutrition
Bacteria	(Under discussion)	Prokaryotic	Unicellular	Absorption, photosynthesis
Archaea	(Under discussion)	Prokaryotic	Unicellular	Absorption
Eukarya	Protista	Eukaryotic	Unicellular	Absorption, ingestion, or photosynthesis
	Fungi	Eukaryotic	Multicellular	Absorption
	Plantae	Eukaryotic	Multicellular	Photosynthesis
	Animalia	Eukaryotic	Multicellular	Ingestion

in each organism, and the mode of nutrition—that is, energy acquisition (Table 1-1).

The Domains Bacteria and Archaea Consist of Prokaryotic Cells; the Domain Eukarya Is Composed of Eukaryotic Cells

There are two fundamentally different types of cells: (1) **prokaryotic** and (2) **eukaryotic**. *Karyotic* refers to the **nucleus** of a cell: a membrane-enclosed sac containing the cell's genetic material (see Fig. 1-3). *Eu* means "true" in Greek; eukaryotic cells possess a "true," membrane-enclosed nucleus. Eukaryotic cells are larger than prokaryotic cells and contain a variety of other organelles, many surrounded by membranes. Prokaryotic cells do not have a nucleus; their genetic material resides in their cytoplasm. They are small—only 1 or 2 micrometers long—and lack membrane-bound organelles. *Pro* means "before" in Greek; prokaryotic cells almost certainly evolved before eukaryotic cells (and, as we will see in Chapter 17, eukaryotic cells probably evolved from prokaryotic cells). Bacteria and Archaea consist of prokaryotic cells; the cells of the four kingdoms of Eukarya are eukaryotic.

Bacteria, Archaea, and Members of the Kingdom Protista Are Mostly Unicellular; Members of the Kingdoms Fungi, Plantae, and Animalia Are Primarily Multicellular

Most members of the domains Bacteria and Archaea and members of the kingdom Protista from the domain Eukarya are single-celled, or **unicellular**, although a few live in strands or mats of cells with little communication, cooperation, or organization among them. Most members of the kingdoms Fungi, Plantae, and Animalia are many-celled, or **multicellular**; their lives depend on intimate cooperation among cells.

Members of the Different Kingdoms Have Different Ways of Acquiring Energy

All organisms need energy to live. Photosynthetic organisms capture energy from sunlight and store it in molecules such as sugars and fats. These organisms, including plants, some bacteria, and some protists, are therefore called **autotrophs**, meaning "self-feeders." Organisms that cannot photosynthesize must acquire energy prepackaged in the molecules of the bodies of other organisms; hence, these organisms are called **heterotrophs**, meaning "other-feeders." Many archaeans, bacteria, and protists and all fungi and animals are heterotrophs. Heterotrophs differ in the size of the food they eat. Some, such as bacteria and fungi, absorb individual food molecules; others, including most animals, eat whole chunks of food and break them down to molecules in their digestive tracts (*ingestion*).

 ## What Does the Science of Biology Encompass?

Biology is probably the most diverse of all the sciences. While some biologists are creating new forms of life by manipulating genetic material, others are probing the workings of the brain, tracing the complex interactions within ecosystems, or seeking out new forms of life from the tropical rain forest to the ocean floor. Through the study of biology, you will become familiar with how your body works. You will learn how some diseases are spread and how they are fought by our natural defenses and by medications. You will explore the intricacies of development from single cell to whole human being. You will learn how plants capture the solar energy that ultimately sustains nearly all life. By learning about the relationships among people and other forms of life, you will be better prepared to make informed choices about land use, waste disposal, family size, and many other issues that affect the environment that sustains us.

Biology is a science, and its principles and methods are the same as those of any other science. In fact, a basic tenet of modern biology is that living things obey the same laws of physics and chemistry that govern nonliving matter.

Scientific Principles Underlie All Scientific Inquiry

All scientific inquiry, including biology, is based on a small set of assumptions. Although we can never absolutely prove these assumptions, they have been so thoroughly tested and found valid that we might call them scientific principles. These are the principles of *natural causality*, *uniformity in space and time*, and *common perception*.

Natural Causality Is the Principle That All Events Can Be Traced to Natural Causes

The first principle of science is natural causality. Over the course of human history, two approaches have been taken to the study of life and other natural phenomena. The first assumes that some events happen through the intervention of supernatural forces beyond our understanding. The ancient Greeks believed that the god Zeus hurled thunderbolts from the sky and that the god Poseidon caused earthquakes and storms at sea. In contrast, science adheres to the principle of **natural causality**: All events can be traced to natural causes that are potentially within our ability to comprehend. For example, until relatively recently, epilepsy was commonly thought to be a visitation from the gods. Today we realize that epilepsy is a disease of the brain in which groups of nerve cells are activated uncontrollably.

The principle of natural causality has an important corollary: The evidence we gather about the causes of natural events has not been deliberately distorted to fool us. This corollary may seem obvious, yet not so very long ago some people argued that fossils are not evidence of evolution but were placed on Earth by God as a test of our faith. If we cannot trust the evidence provided by nature, then the entire enterprise of science is futile.

The Natural Laws That Govern Events Apply Everywhere and for All Time

A second fundamental principle of science is that natural laws, laws derived from nature, are uniform in space and time and do not change with distance or time. The laws of gravity, the behavior of light, and the interactions of atoms, for example, are the same today as they were a billion years ago and will hold just as well in Moscow as in New York or even on Mars. Uniformity in space and time is especially vital to biology, because many events of great importance to biology, such as the evolution of today's diversity of living things, happened before humans were around to observe them. Some people believe that all the different types of organisms were individually created at one time in the past by the direct intervention of God, a philosophy called **creationism**. As scientists, we freely admit that we cannot disprove this idea. Creationism, however, is contrary to both natural causality and uniformity in time. The overwhelming success of science in explaining natural events through natural causes has led almost all scientists to reject creationism.

Scientific Inquiry Is Based on the Assumption That People Perceive Natural Events in Similar Ways

A third basic assumption of science is that, as a general rule, all human beings perceive natural events in fundamentally the same way and that these perceptions provide us with reliable information about the natural world. Common perception is, to some extent, a principle peculiar to science. Value systems, such as those involved in the appreciation of art, poetry, and music, do not assume common perception. We may perceive the colors in a painting in a similar way (the scientific aspect of art), but we do not perceive the aesthetic value of the painting identically (the humanistic aspect of art). Values also differ radically among people, commonly owing to their culture or religious beliefs. Because value systems are subjective, not objective, science cannot solve certain types of philosophical or moral problems, such as the morality of abortion.

The Scientific Method Is the Basis for Scientific Inquiry

Given these assumptions, how do biologists study the workings of life? Scientific inquiry is a rigorous method for making observations of specific phenomena and searching for the order underlying those phenomena. Ideally, biology and the other sciences use the **scientific method**, which consists of four interrelated operations: (1) *observation*; (2) *hypothesis*; (3) *experiment*; and (4) *conclusion*. All scientific inquiry begins with an **observation** of a specific phenomenon. The observation, in turn, leads to questions, such as "How did this come about?" Then in a flash of insight, or more typically after long, hard thought, a hypothesis is formulated. A **hypothesis** is a supposition based on previous observations that is offered as an explanation for the observed phenomenon. To be useful, a hypothesis must be testable by further observations, or **experiments**. These experiments produce results that either support or refute the hypothesis, and a **conclusion** is drawn about its validity. A single experiment is never an adequate basis for a conclusion; the results must be repeatable not only by the original researcher but also by others.

Simple experiments test the assertion that a single factor, or **variable**, is the cause of a single observation. To be scientifically valid, the experiment must rule out a variety of other possible variables as the cause of the observation. So scientists design **controls** into their experiments, in which all the variables remain constant. Controls are then compared with the experimental situation, in which only the variable being tested is changed. In the 1600s, Francesco Redi used the scientific method to demonstrate that flies do not arise spontaneously from rotting meat. (See "Scientific Inquiry: Can Maggots Spring from Spoiled Meat?")

The scientific method can be used not only to solve everyday problems but also to generate new knowledge.

Scientific Inquiry
Can Maggots Spring from Spoiled Meat?

The critical experiments of the Italian physician Francesco Redi (1621–1697) beautifully demonstrate the scientific method and also help illustrate the principle of natural causality, on which modern science is based. Redi investigated why maggots appear on spoiled meat. Before Redi, the appearance of maggots was considered to be evidence of **spontaneous generation**, the production of living things from nonliving matter.

Redi *observed* that flies swarm around fresh meat and that maggots appear on meat left out for a few days. He formed a testable *hypothesis*: The flies produce the maggots. In his *experiment*, Redi wanted to test just one variable: the access of flies to the meat. Therefore, he took two clean jars and filled them with similar pieces of meat. He left one jar open (the *control* jar) and covered the other with gauze to keep out flies (the *experimental* jar). He did his best to keep all the other variables the same (for example, the type of jar, the type of meat, and the temperature). After a few days, he observed that maggots swarmed over the meat in the open jar, but no maggots appeared on the meat in the covered jar. Redi *concluded* that his hypothesis was correct and that maggots are produced by flies, not by the meat itself (Fig. E1-1). Only through controlled experiments could the age-old hypothesis of spontaneous generation be laid to rest.

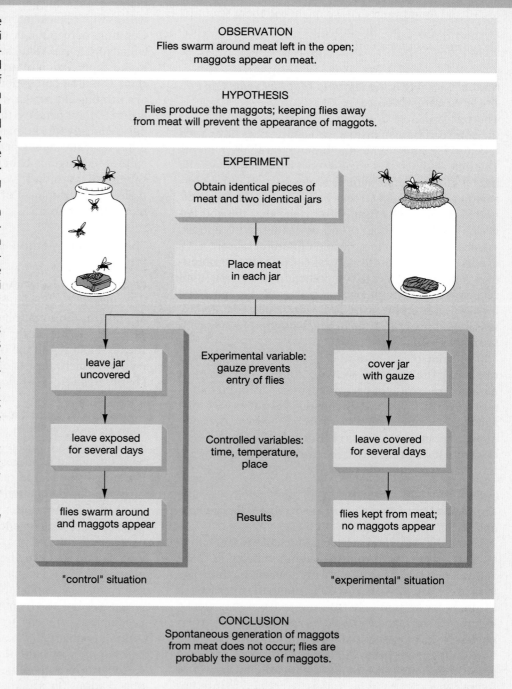

OBSERVATION
Flies swarm around meat left in the open; maggots appear on meat.

HYPOTHESIS
Flies produce the maggots; keeping flies away from meat will prevent the appearance of maggots.

EXPERIMENT
Obtain identical pieces of meat and two identical jars

Place meat in each jar

leave jar uncovered

leave exposed for several days

flies swarm around and maggots appear

"control" situation

Experimental variable: gauze prevents entry of flies

Controlled variables: time, temperature, place

Results

cover jar with gauze

leave covered for several days

flies kept from meat; no maggots appear

"experimental" situation

CONCLUSION
Spontaneous generation of maggots from meat does not occur; flies are probably the source of maggots.

Figure E1-1 *The experiments of Francesco Redi*

Let's consider an everyday situation in which you can apply the scientific method. Late to class, you rush to your car and make the *observation* that it won't start. Immediately, you form a *hypothesis*: The battery is dead. Quick-ly you design an *experiment*: You replace your battery with the battery from your roommate's new car and try to start your car again. The result seems to confirm your hypothesis, because your car starts immediately.

But wait! You haven't provided controls for several variables. Perhaps your battery was fine all along, and you just needed to try to start the car again. Or perhaps the battery cable was loose and simply needed to be tightened. Realizing the need for a good *control*, you replace your old battery, making sure the cables are secured tightly, and attempt to restart the car. If your car repeatedly refuses to start with the old battery but then starts immediately when you put in your roommate's new battery, you have isolated a single *variable*, the battery. And (although you may have missed your class) you can safely draw the *conclusion* that your old battery is dead.

It is important to recognize the limitations of the scientific method. In particular, scientists can seldom be sure that they have controlled *all* the variables other than the one they are trying to study. Therefore, scientific conclusions must always remain tentative and are subject to revision if new observations or experiments demand it.

Science Is a Human Endeavor

Scientists are real people. They are driven by the same ambitions, pride, and fears as other people, and they sometimes make mistakes. As you will read in Chapter 9, ambition played an important role in the discovery by James Watson and Francis Crick of the structure of DNA. Accidents, lucky guesses, controversies with competing scientists, and, of course, the intellectual powers of individual scientists contribute greatly to scientific advances. To illustrate what we might call "real science," let's consider an actual case.

A Good Scientist Is Prepared to Take Advantage of Chance Events

When they study bacteria, microbiologists must use pure *cultures*—that is, plates of bacteria that are free from contamination by other bacteria, molds, and so on. Only by studying a single type at a time can we learn about the properties of that particular bacterium. Consequently, at the first sign of contamination, a culture is normally thrown out, often with mutterings about sloppy technique. On one such occasion, however, in the late 1920s, the Scottish bacteriologist Alexander Fleming turned a ruined culture into one of the greatest medical advances in history.

One of Fleming's bacterial cultures became contaminated with a patch of a mold called *Penicillium*. Before throwing out the culture dish, Fleming noticed that *no bacteria were growing near the mold* (Fig. 1-8). Why not? Fleming hypothesized that perhaps *Penicillium* releases a substance that kills off bacteria growing nearby. To test this hypothesis, Fleming grew some pure *Penicillium* in a liquid medium (a *medium* is a nutritive mixture for grow-

Figure 1-8 Penicillin kills bacteria
Penicillin diffuses outward from a penicillin-soaked disc of paper, creating a pronounced bald spot in the "lawn" of bacteria on this petri dish.

ing microorganisms). He then filtered out the *Penicillium* mold, and then applied the liquid in which the mold had grown to an uncontaminated bacterial culture. Sure enough, the bacteria were killed by the liquid. Further research into these mold extracts resulted in the production of the first *antibiotic*: penicillin, a bacteria-killing substance that has since saved millions of lives. Fleming's experiments are a classic example of the use of scientific methodology. They began with an observation and proceeded to a hypothesis, followed by experimental tests of the hypothesis, which led to a conclusion. But the scientific method alone would have been useless without the lucky combination of accident and a brilliant scientific mind. Had Fleming been a "perfect" microbiologist, he wouldn't have had any contaminated cultures. Had he been less observant, the contamination would have been just another spoiled culture dish. Instead, it was the beginning of antibiotic therapy for bacterial diseases. As the French microbiologist Louis Pasteur said, "Chance favors the prepared mind."

Scientific Theories Have Been Thoroughly Tested

Scientists use the word *theory* in a way that is different from its everyday usage. If Dr. Watson were to ask Sherlock Holmes, "Do you have a theory as to the perpetrator of this foul deed?" in scientific terms, he would be asking Holmes for a hypothesis—an educated guess based on observable evidence, or clues. A **scientific theory** is far more general and more reliable than a hypothesis. Far from being an educated guess, a scientific theory is a general explanation of natural phenomena, developed through extensive and reproducible observations. It is more like a *principle* or a *natural law* in common English usage. For example, the atomic theory (that all matter is composed of atoms) and the theory of gravitation (that objects exert attraction for one another) are fundamental to the science of physics. Likewise, the *cell theory* (that all living things are composed of cells) and the theory of evolution are fundamental to the study of biology. Scientists describe fundamental principles as "theories" because a basic premise of scientific inquiry is that it must be performed with an open mind. If compelling evidence arises, a theory can be modified.

Probably the foremost scientific theory in biology is evolution. Ever since its formulation by two English naturalists, Charles Darwin and Alfred Russel Wallace, in the mid-1800s, the theory of evolution has been supported by fossil finds, geological studies, radioactive dating of rocks, genetics, molecular biology, biochemistry, and breeding experiments. People who refer to evolution as "just a theory" profoundly misunderstand what scientists mean by the word *theory*.

Evolution: The Unifying Concept of Biology

The most important concept in biology is evolution, a unifying theory that explains the origin of diverse forms of life as a result of changes in their genetic makeup. As we noted earlier, the theory of evolution states that modern organisms descended, with modification, from preexisting life-forms. In the words of the biologist Theodosius Dobzhansky, "Nothing in biology makes sense, except in the light of evolution." Why don't snakes have legs? Why are there dinosaur fossils but no living dinosaurs? Why are monkeys so like us, not only in appearance, but even in the structure of their genes and proteins? How are the diverse life-forms you saw on your walk across campus related? The answers to those questions, and thousands more, lie in the processes of evolution (which we shall examine in detail in Chapters 14 through 17). Evolution is so vital to our understanding and appreciation of biology that we must briefly review its important principles before going further.

Three Natural Processes Underlie Evolution

In the mid-1800s, Darwin and Wallace formulated the theory of evolution that is still the basis of our modern understanding. Evolution arises as a consequence of three natural processes: (1) *genetic variation* among members of a population; (2) *inheritance* of those variations by offspring of parents who carry the variation; and (3) *natural selection*, the survival and enhanced reproduction of organisms with favorable variations.

Much of the Variability among Organisms Is Inherited

Look around at your classmates and notice how different they are. Although some of this variation is due to differences in environment and lifestyles, it is mainly influenced by our genes. Most of us could pump iron for the rest of our lives and never develop a body like Arnold Schwarzenegger's. Where does genetic variation come from? The genetic instructions—the genes—of all organisms are segments of molecules of deoxyribonucleic acid, or DNA. Occasionally, the DNA suffers an accident; perhaps radiation strikes the DNA molecule just so, causing a mutation and thereby altering its information content. Mistakes in the copying of DNA during reproduction, although rare, also occur. Such mutations, or changes in a gene, may affect the organism's appearance or ability to function. Many mutations have no effect or are harmless; some make the organism less able to function. But in rare cases mutations may improve an organism's ability to function. As a result of mutations, many of which occurred millions of years ago and have been passed from parent to offspring through countless generations, members of the same species tend to be slightly different from one another.

Natural Selection Tends to Preserve Genes That Help an Organism Survive and Reproduce

On the average, organisms that best meet the challenges of their environment will leave the most offspring. The offspring will inherit the genes that made their parents successful. Natural selection thus preserves genes that help organisms flourish in their environment. For example, a mutated gene providing instructions for larger teeth in beavers will be passed from parent to offspring. These offspring will be able to chew down trees more efficiently, build bigger dams and lodges, and eat more bark than can "ordinary" beavers. Because these big-toothed beavers will obtain more food and better shelter than their smaller-toothed relatives, they will probably raise more offspring. The offspring will inherit their parents' genes for larger teeth. Over time, less-successful, smaller-toothed beavers will become increasingly scarce; after many generations, all beavers will have large teeth.

Structures, physiological processes, or behaviors that aid in survival and reproduction in a particular environment are called **adaptations**. Most of the features that we admire so much in our fellow life-forms, such as the long limbs of deer, the wings of eagles, and the mighty columns of redwood trunks, are adaptations molded by millions of years of mutation and natural selection.

In the long run, however, what helps an organism survive today can become a liability tomorrow. If environments change—for example, as ice ages come and go—then the genetic makeup that best adapts organisms to their environment will also change over time. When new mutations happen to appear that increase the fitness of an organism in the altered environment, these mutations in their turn will spread throughout the population.

Over millennia, the interplay of environment, genetic variation, and natural selection results in evolution: the modification of the genetic makeup of species. In environments that are reasonably constant through time, such as the oceans, some well-adapted forms persist relatively unchanged and are sometimes called "living fossils." For example, sharks (Fig. 1-9) have retained essentially the same body form for tens of millions of years, as their sleek shape, powerful tail, acute sense of smell, and formidable teeth have made them superb predators.

In changing environments, some species do not experience the genetic changes that allow them to adapt. The rate of environmental change outstrips the rate of genetic changes, and those species go *extinct*—that is, no members of those species remain. The dinosaurs (Fig. 1-10) were mighty reptiles that could not withstand changing conditions 65 million years ago. Other species experience chance mutations that adapt them to meet new challenges. The mutation that produced the first stout, fleshy fins in what came to be known as lobefin fishes, for instance, enabled those fish to crawl on the bot-

***Figure 1-10 A fossil of* Triceratops**
This *Triceratops* died in what is now Montana about 70 million years ago. No one is certain what caused the extinction of the dinosaurs, but we do know that the evolution of new adaptations in dinosaurs was unable to keep pace with changes in their habitat.

tom of shallow waters and ultimately to make the "jump" to life on land.

The result of evolution is a tremendous variety of species. Within particular habitats, these species have evolved complex interrelationships with one another and with their nonliving surroundings. The diversity of species and the complex interrelationships that sustain them are encompassed by the term **biodiversity**. In recent decades, the rate of environmental change has been drastically accelerated by a single species, *Homo sapiens* (modern humans). Few species are able to adapt to this rapid change. In habitats most affected by humans, many species are being driven to extinction. This concept is explored further in "Earth Watch: Why Preserve Biodiversity?"

5) How Does Knowledge of Biology Illuminate Everyday Life?

Some people regard science as a "dehumanizing" activity, feeling that too deep an understanding of the world robs us of vision and awe. Nothing could be further from the truth, as we repeatedly discover anew in our own lives. Years ago, we watched a bee foraging at a spike of lupine flowers. Members of the pea family, lupines have a complicated structure, with two petals on the lower half of the flower enclosing the pollen-laden male reproductive structures (*stamens*) and sticky pollen-capturing female reproductive structures (*stigma*). We had recently

Figure 1-9 Ancient adaptations
This tiger shark, photographed in the clear waters of the Bahamas, possesses features that have characterized sharks for tens of millions of years: streamlined shape; a long, powerful tail; an acute sense of smell; and rows of sharp teeth.

Earth Watch
Why Preserve Biodiversity?

"The loss of species is the folly our descendants are least likely to forgive us."
E. O. Wilson, Professor, Harvard University

Ever since the United Nations' 1992 "Earth Summit" in Rio de Janeiro, Brazil, the word *biodiversity* has jumped out at us from magazines and news articles. What is biodiversity, and why should we be concerned with preserving it? Biodiversity refers to the total number of species within an ecosystem and to the resulting complexity of interactions among them; in short, it defines the "richness" of an ecological community.

Over the 3.5-billion-year history of life on Earth, evolution has produced an estimated 8 million to 10 million unique and irreplaceable species. Of these, scientists have named only about 1.4 million, and only a tiny fraction of this number has been studied. Evolution has not, however, merely been churning out millions of independent species. Over thousands of years, organisms in a given area have been molded by forces of natural selection exerted by other living species as well as by the nonliving environment in which they live. The outcome is the community, a highly complex web of interdependent life-forms whose interactions sustain one another. By participating in the natural cycling of water, oxygen, and other nutrients, by producing rich soil and purifying wastes, these communities contribute to the sustenance of human life as well. The concept of biodiversity has emerged as a result of our increasing concern over the loss of countless forms of life and the habitat that sustains them.

The tropics are home to the vast majority of all the species on Earth, perhaps 7 million to 8 million of them, living in complex communities. The rapid destruction of habitats in the tropics, from rain forests to coral reefs, as a result of human activities is producing high rates of extinction of many species (Fig. E1-2). Most of these species have never been named, and others never even discovered. Aside from ethical concerns over eradicating irreplaceable forms of life, as we drive unknown organisms to extinction we lose potential sources of medicine, food, and raw materials for industry.

For example, a wild relative of corn that is not only very disease-resistant but also *perennial* (that is, lasts more than one growing season) was found growing only on a 25-acre plot of land in Mexico that was scheduled to be cut and burned within a week of the discovery. The genes of this plant might one day enhance the disease-resistance of corn or create a perennial corn plant. The rosy periwinkle, a flowering plant found in the tropical forest of the island of Madagascar (off the eastern coast of Africa) produces two substances that are now widely marketed for the treatment of leukemia and Hodgkin's disease, a cancer of the lymphatic organs. Only about 3% of the world's flowering plants have been examined for substances that might fight cancer or other diseases. Closer to home, loggers of the Pacific Northwest frequently cut and burned the Pacific yew tree as a "nuisance species" until the active ingredient that has since gone into making the anticancer drug Taxol® was discovered in its bark. Animals too have proven useful in fighting cancer: In 1997,

Figure E1-2 Biodiversity threatened
Destruction of tropical rain forests by indiscriminate logging threatens Earth's greatest storehouse of biological diversity. Interrelationships such as those that have evolved between this *Heliconia* flower and its hummingbird pollinator sustain these diverse communities and are threatened by human activities.

researchers isolated a potent anticancer compound from a species of coral that dwells in the Indian Ocean.

Many conservationists are also concerned that as species are eliminated, either locally or through total extinction, the communities of which they were a part might change and become less stable and more vulnerable to damage by diseases or adverse environmental conditions. Some experimental evidence supports this viewpoint, but the interactions within communities are so complex that hypotheses are difficult to test. Clearly, some species have a much larger role than do others in preserving the stability of a given ecosystem. Which species are most crucial in each ecosystem? No one knows. Human activities have increased the natural rate of extinction by a factor of at least 100 and possibly by as much as 1000 times the pre-human rate. By reducing biodiversity to support increasing numbers of humans and wasteful standards of living, we have ignorantly embarked on an uncontrolled global experiment, using planet Earth as our laboratory. In their book *Extinction* (1981), Stanford ecologists Paul and Anne Ehrlich compare the loss of biodiversity to the removal of rivets from the wing of an airplane. The rivet-removers continue to assume that there are far more rivets than needed, until one day, upon takeoff, they are proven tragically wrong. As human activities drive species to extinction while we have little knowledge of the role each plays in the complex web of life, we run the risk removing "one rivet too many."

(a)

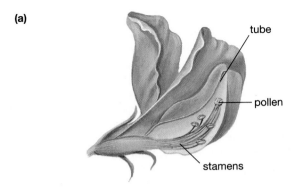

(b)

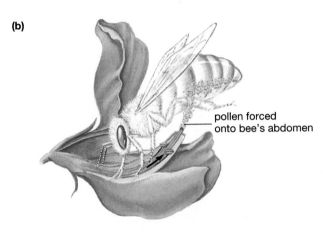

Figure 1-11 *Complex adaptations help ensure pollination*
Lupines, like many other members of the pea family, have complex flowers. *(a)* The reproductive structures are enclosed within the lower petals. In young lupine flowers, the lower petals form a tube within which the stamens fit snugly. The stamens shed pollen within the tube. *(b)* When the weight of a foraging bee pushes on the lower petals, the stamens are thrust forward, and pollen is forced out the tube's end onto the bee's abdomen. Some pollen adheres to the abdomen and may come off on the sticky, pollen-receiving stigma of the next flower that the bee visits.

learned that in young lupine flowers (Fig. 1-11a), the weight of a bee pushing on these petals compresses the stamens, pushing pollen onto the bee's abdomen (Fig. 1-11b). In older flowers, the stigma protrudes through the lower petals; when a pollen-dusted bee visits, it usually leaves behind a few grains of pollen.

Did our new-found insights into the functioning of lupine flowers detract from our appreciation of them? Far from it. Rather, we now looked on lupines with new delight, understanding something of the interplay of form and function, bee and flower, that shaped the evolution of the lupine. A few months later we ventured atop Hurricane Ridge in Olympic National Park in Washington State, where the alpine meadows burst with color in August (Fig. 1-12). As we crouched beside a wild lupine, an elderly man stopped to ask what we were looking at so intently. He listened with interest as we ex-

plained the structure to him; he then went off to another patch of lupines to watch the bees foraging. He too felt the increased sense of wonder that comes with understanding.

We try to convey to you that dual sense of understanding and wonder throughout this text. We also emphasize that biology is not a completed work but an exploration that we have really just begun. Lewis Thomas, a physician and natural philosopher, described our situation in an elegant way: "The only solid piece of scientific truth about which I feel totally confident is that we are profoundly ignorant about nature. Indeed, I regard this as the major discovery of the past hundred years of biology . . . but we are making a beginning."

We cannot urge you strongly enough, even if you are not contemplating a career in biology, to join in the journey of biological discoveries throughout your life. Don't think of biology as just another course to take, just another set of facts to memorize. Biology can be much more than that. It can be a pathway to a new understanding of yourself and of the life on Earth around you.

Figure 1-12 *Wild lupines and subalpine fir trees*
Thousands of people visit Hurricane Ridge in Washington State's Olympic National Park each summer to gaze in awe at Mt. Olympus, but few bother to investigate the wonders at their feet.

Summary of Key Concepts

1) What Are the Characteristics of Living Things?

The quality of life emerges as the result of interactions of emergent properties. Organisms possess the following characteristics: Their structure is complex and organized; they maintain homeostasis; they grow; they acquire energy and materials from the environment; they respond to stimuli; they reproduce; and they have the capacity to evolve.

2) How Do Scientists Categorize the Diversity of Life?

Organisms can be grouped into three major categories, called domains: Archaea, Bacteria, and Eukarya. Within the Eukarya are four kingdoms: Protista, Fungi, Plantae, and Animalia. Some features used to classify organisms are the type of cell(s) the organism possesses, the number of cells in each organism, and the mode of acquiring energy:

Cell Type: The genetic material of eukaryotic cells is enclosed within a membrane-bound nucleus. Prokaryotic cells do not have a nucleus.

Cell Number: Organisms may consist of a single cell (unicellular) or of many cells bound together and working cooperatively (multicellular).

Energy Acquisition: Most autotrophic organisms obtain energy by capturing and storing the energy of sunlight in energy-rich molecules by means of photosynthesis. Heterotrophic organisms normally obtain energy by eating energy-rich molecules (food) synthesized in the bodies of other organisms. The food may be eaten in large chunks and broken down (ingestion) or may be absorbed molecule by molecule from the environment (absorption). The features of the domains and kingdoms are summarized in Table 1-1.

3) What Does the Science of Biology Encompass?

Biology is based on the scientific principles of natural causality, uniformity in space and time, and common perception. These principles are assumptions that cannot be directly proved but that are validated by experience. Knowledge in biology is acquired through the application of the scientific method. First, an observation is made. Then a hypothesis is formulated that suggests a natural cause for the observation. The hypothesis is used to predict the outcome of further observations or experiments. A conclusion is then drawn about the validity of the hypothesis. A scientific theory is a general explanation of natural phenomena, developed through extensive and reproducible experiments and observations.

4) Evolution: The Unifying Concept of Biology

Evolution is the theory that modern organisms descended, with modification, from preexisting life-forms. Evolution occurs as a consequence of (1) genetic variation among members of a population, caused by mutation; (2) inheritance of those variations by offspring; and (3) natural selection of the variations that best adapt an organism to its environment.

5) How Does Knowledge of Biology Illuminate Everyday Life?

The more you know about living things, the more fascinating they become!

Key Terms

adaptation *p. 12*
atom *p. 2*
autotroph *p. 8*
biodiversity *p. 13*
biosphere *p. 4*
cell *p. 4*
community *p. 4*
conclusion *p. 9*
control *p. 9*
creationism *p. 9*
cytoplasm *p. 4*
deoxyribonucleic acid
  (DNA) *p. 6*
domain *p. 6*
ecosystem *p. 4*

element *p. 2*
emergent property *p. 2*
energy *p. 4*
eukaryotic *p. 8*
evolution *p. 6*
experiment *p. 9*
gene *p. 4*
heterotroph *p. 8*
homeostasis *p. 5*
hypothesis *p. 9*
kingdom *p. 6*
metabolism *p. 4*
molecule *p. 4*
multicellular *p. 8*
mutation *p. 6*

natural causality *p. 9*
natural selection *p. 6*
nucleus *p. 8*
nutrient *p. 4*
observation *p. 9*
organ *p. 4*
organelle *p. 4*
organic molecule *p. 4*
organism *p. 4*
organ system *p. 4*
photosynthesis *p. 4*
plasma membrane *p. 4*
population *p. 4*
prokaryotic *p. 8*
scientific method *p. 9*

scientific theory *p. 12*
species *p. 4*
spontaneous generation
  p. 10
subatomic particle *p. 4*
tissue *p. 4*
unicellular *p. 8*
variable *p. 9*

Thinking Through the Concepts

Multiple Choice

1. *Which of the following is paired* incorrectly?
   a. organ—a structure formed of cells of similar types
   b. cell—the smallest unit of life
   c. genes—units of heredity
   d. cytoplasm—watery substance within cells that contains organelles
   e. plasma membrane—surrounds each cell

2. *A scientist examines an organism and finds that it is eukaryotic, heterotrophic, and multicellular and that it absorbs nutrients. She concludes that the organism is a member of the kingdom*
   a. Bacteria
   b. Protista
   c. Plantae
   d. Fungi
   e. Animalia

3. *Which statement is correct?*
   a. Eukaryotic cells are simpler than prokaryotic cells.
   b. *Heterotroph* means "self-feeder."
   c. Mutations are accidental changes in genes.
   d. A scientific theory is similar to an educated guess.
   e. Genes are proteins that produce DNA.

4. *Choose the answer that best describes the scientific method.*
   a. observation, hypothesis, experiment, absolute proof
   b. guess, hypothesis, experiment, conclusion
   c. observation, hypothesis, experiment, conclusion
   d. hypothesis, experiment, observation, conclusion
   e. experiment, observation, hypothesis, conclusion

5. *Which of the following are characteristics of living things?*
   a. They reproduce.
   b. They respond to stimuli.
   c. They are complex and organized.
   d. They acquire energy.
   e. all of the above

6. *The three natural processes that form the basis for evolution are*
   a. adaptation, natural selection, and inheritance
   b. predation, genetic variation, and natural selection
   c. mutation, genetic variation, and adaptation
   d. fossils, natural selection, and adaptation
   e. genetic variation, inheritance, and natural selection

? Review Questions

1. What are the differences between a salt crystal and a tree? Which is living? How do you know? How would you test your "knowledge"? What controls would you use?

2. What is the difference between a scientific theory and a hypothesis? Explain how each is used by scientists.

3. Define and explain the terms *natural selection*, *evolution*, *mutation*, *creationism*, and *population*.

4. Starting with the cell, list the hierarchy of organization of life, briefly explaining each level.

5. Define *homeostasis*. Why must organisms continuously acquire energy and materials from the external environment to maintain homeostasis?

6. Describe the scientific method. In what ways do you use the scientific method in everyday life?

7. What is evolution? Briefly describe how evolution occurs.

Applying the Concepts

1. In the heavily populated state of California, natural occurrences, including earthquakes, heavy rains, and grass and forest wildfires, have reduced the quality of life for many Californians. Look at it another way for a moment. What ecosystems are found in this state? What impacts have humans had on these natural ecosystems? Are changes in these areas the reason for fires and floods and mudslides? What needs to be done to alter the balance of "humans and nature"?

2. Design an experiment to test the effects of a new dog food, "Super Dog," on the thickness and water-shedding properties of the coats of golden retrievers. Include all the parts of a scientific experiment. Design objective methods to assess coat thickness and water-shedding ability.

3. Science is based on principles, including uniformity in space and time and common perception. Assume that humans one day encounter intelligent beings from a planet in another galaxy who evolved under very different conditions. Discuss the two principles mentioned above and how they would affect (1) the nature of scientific observations on the different planets and (2) communications about these observations.

For More Information

Attenborough, D. *Life on Earth*. Boston: Little, Brown, 1979. Gorgeously illustrated and beautifully written introduction to the diversity of life on Earth; the inspiration for our title.

Dawkins, R. *The Blind Watchmaker*. New York: W. W. Norton & Co., 1986. An engagingly written description of the process of evolution, which Dawkins compares to a blind watchmaker.

Ehrlich, P. R. *The Machinery of Nature*. New York: Simon & Schuster, 1986. Using layperson's terms, a foremost ecologist and author explains the science of ecology and the biological rationale for environmental concern.

Leopold, A. *A Sand County Almanac*. New York: Oxford University Press, 1949 (reprinted in 1989). A classic by a natural philosopher; provides an eloquent foundation for the conservation ethic.

Swain, R. B. *Earthly Pleasures*. New York: Charles Scribner's Sons, 1981. Insightful essays stress the interrelatedness and diversity of life.

Thomas, L. *The Medusa and the Snail*. New York: Bantam Books, 1980. The late Lewis Thomas, physician, researcher, and philosopher, shares his awe of the living world in a series of delightful essays.

Wilson, E. O. *The Diversity of Life*. New York: W. W. Norton & Co., 1992. A celebration of the diversity of life, how it evolved, and how humans are impacting it. Wilson's writings have won two Pulitzer Prizes.

Answers to Multiple-Choice Questions
1. a   2. d   3. c   4. c   5. e   6. e

Unit One
The Life of a Cell

Single cells can be complex, independent organisms, such as these two ciliates, of the kingdom Protista. A large (about 300-μm-long) Euplotes prepares to eat a much smaller Paramecium. Both are covered with cilia, short, beating structures used to move and ingest prey.

"If there is magic on this planet, it is contained in water."

Loren Eiseley in **The Immense Journey (1957)**

A basilisk lizard runs across a pond, putting the water's surface tension to good use.

Atoms, Molecules, and Life 2

At a Glance

Net Watch

On-line resources for this chapter are on the World Wide Web at: http://www.prenhall.com/audesirk

(click on the Table of Contents link and then select Chapter 2).

Water! We delight in the crashing waves of the ocean, whose vastness and power bring a sense of perspective to our lives and whose presence moderates the near-shore climate. Children squeal with joy as they run through the spray of a water sprinkler on a hot summer day. The lake whose summer surface reflects a forest sparkles with ice in the winter, while fish swim safely below. Trees absorb water through their roots and, in some cases hundreds of feet away, evaporate it from their leaves. And if you've ever experienced the slap and sting of a belly flop into a swimming pool, you've experienced the joys of surface tension. The basilisk lizard exploits this same property of water. Taking 20 steps per second on specially adapted feet, these lizards, when frightened, actually run across the surface of ponds or lakes!

Life almost certainly originated in water, and the need for water by all forms of life has helped shape their diverse body structures. Your own body is about 60% water by weight; all the diverse molecules that make up your body and allow it to function must operate in a watery environment. To understand water and the biological molecules with which it interacts, we must learn about the building blocks of all forms of matter—atoms. Are all atoms the same? What are they made of? How do they interact with one another to form molecules? Is it possible that a common principle underlies the movement of water to the top of a tree, the sting of a belly flop, and the formation of shimmering droplets of fat on the surface of chicken noodle soup?

This chapter introduces some of the fundamental properties of matter, the structure of atoms, and how atoms interact with one another to form molecules. Later chapters in Unit I will describe the molecules that form the bodies of all living organisms; the structure and function of cells, which are remarkably similar in every organism from bacteria to elephants; and the chemical reactions that allow the acquisition and use of energy by individual cells and entire organisms.

1) What Are Atoms?

Atoms, the Basic Structural Units of Matter, Are Composed of Still Smaller Particles

If you took a diamond (a form of carbon) and cut it into pieces, each piece would still be carbon. If you could continue to make finer and finer divisions, you would eventually produce a pile of carbon atoms. **Atoms** are the fundamental structural units of matter. Atoms themselves, however, are composed of a central **atomic nucleus** (often called just "nucleus," but don't confuse it with the nucleus of a cell!) and subatomic particles called **electrons** outside the atomic nucleus (Fig. 2-1). The nucleus contains two types of subatomic particles of equal weight: positively charged **protons** and uncharged **neutrons**. Electrons are lighter, negatively charged particles that orbit the nucleus. An atom by itself has an equal number of electrons and protons and is therefore electrically neutral.

There are 92 types of atoms that occur naturally. Each type of atom forms the structural unit of a different element. An **element** is a substance that can neither be broken down nor converted to other substances by ordinary chemical means. The number of protons in the nucleus, called the **atomic number**, is a characteristic of each element. For example, every hydrogen atom has one proton in its nucleus, every carbon atom has six protons, and every oxygen atom has eight. Each element has unique chemical properties. Some elements, such as oxygen and hydrogen, are gases at room temperature; others, such as lead, are extremely dense solids. Most elements are quite rare, and only a relative handful are essential to life on

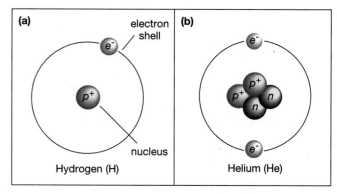

Figure 2-1 Atomic models
Structural representations of the two smallest atoms, **(a)** hydrogen and **(b)** helium. In these models, the electrons are imagined as miniature planets, circling in precisely defined orbits around a nucleus that contains protons and neutrons.

Earth. Table 2-1 lists the most common elements in the universe, on Earth, and in the human body.

Atoms of the same element may have different numbers of neutrons; these atoms are called **isotopes** of each other. The nuclei of some **radioactive** isotopes spontaneously break apart, forming different types of atoms and releasing energy in the process. Radioactive isotopes are extremely useful as "labels" in studying biological processes (see "Scientific Inquiry: The Use of Isotopes in Biology and Medicine").

Electrons Orbit the Nucleus at Fixed Distances, Forming Electron Shells That Correspond to Different Energy Levels

As you may know from experimenting with a magnet, like poles repel each other and opposite poles attract

		Table 2-1 Common Elements Important in Living Organisms			
Element	Symbol	Atomic Number[a]	Percent in Universe[b]	Percent in Earth[b]	Percent in Human Body[b]
Hydrogen	H	1	91	0.14	9.5
Helium	He	2	9	Trace	Trace
Carbon	C	6	0.02	0.03	18.5
Nitrogen	N	7	0.04	Trace	3.3
Oxygen	O	8	0.06	47	65
Sodium	Na	11	Trace	2.8	0.2
Magnesium	Mg	12	Trace	2.1	0.1
Phosphorus	P	15	Trace	0.07	1
Sulfur	S	16	Trace	0.03	0.3
Chlorine	Cl	17	Trace	0.01	0.2
Potassium	K	19	Trace	2.6	0.4
Calcium	Ca	20	Trace	3.6	1.5
Iron	Fe	26	Trace	5	Trace

[a]Atomic number = number of protons in the atomic nucleus.

[b]Approximate percentage of atoms of this element, by weight, in the universe, in Earth's crust, and in the human body.

Scientific Inquiry
The Use of Isotopes in Biology and Medicine

As you read this text, you will encounter many statements that may cause you to wonder, How do they know *that*? How do biologists know that DNA is the genetic material of cells (Chapter 9)? How do paleontologists measure the ages of fossils (Chapter 17)? How do botanists know that sugars made in plant leaves during photosynthesis are transported to other parts of the plant in the sieve tubes of phloem (Chapter 23)? These discoveries, and many more, have been possible only through the use of isotopes.

Although all atoms of a particular element have the same number of protons, the number of neutrons may vary. Neutrons don't affect the chemical reactivity of an atom very much, but they do make their presence felt in other ways. First, neutrons add to the atom's mass, which can be detected by sophisticated instruments such as mass spectrometers. Second, nuclei with "too many" neutrons break apart spontaneously, or *decay*, often emitting radioactive particles in the process. Those particles can also be detected—for example, with Geiger counters. The process in which a radioactive isotope spontaneously breaks apart is called *radioactive decay*.

A particularly fascinating and medically important application of radioactive isotopes is *positron emission tomography*, more commonly known as *PET scans* (Fig. E2-1a). In a common application of PET scans, a subject is given the sugar glucose that has been labeled with (that is, attached to) a harmless radioactive isotope of fluorine, ^{18}F. When the nucleus of ^{18}F decays, it emits two bursts of energy that travel in opposite directions along the same line. Energy detectors arranged in a ring around the subject record the nearly simultaneous arrival of the two energy bursts (Fig. E2-1b). A powerful computer then calculates the location within the subject at which the decay must have occurred and generates a map of the frequency of ^{18}F decays.

Because the ^{18}F is attached to glucose molecules, this map reflects the glucose concentrations within the subject's brain. The brain uses prodigious amounts of this sugar for energy; the more active a brain cell is, the more glucose it uses. How can this information be used in biological research?

Let's suppose that a neuroscientist is trying to locate the areas of the brain that are involved in memory. The researcher might give ^{18}F-labeled glucose to a few volunteer subjects and then ask them to memorize a word list, which is read aloud. Because brain regions that are active during this process would need more energy and would take up more ^{18}F-glucose molecules than would inactive regions, the active regions would have more ^{18}F decays. The PET scans taken during the memorization would pinpoint brain regions active in storing memories of words.

Physicians also use PET scans in the diagnosis of brain disorders. For example, brain regions in which epileptic seizures originate generally have excessively high glucose utilization and show up in PET scans as "hot spots." Many brain tumors also light up in PET scans (Fig. E2-1c). Abnormal metabolism of certain brain regions may also appear in patients with some mental disorders, such as schizophrenia.

Biology and medicine have profited immensely from close interactions with the other sciences, especially chemistry and physics. The development of PET scans required close cooperation with chemists (developing and synthesizing the radioactive probes), physicists (understanding interactions between electrons and short-lived, positively charged particles called *positrons* as well as the geometry of the resulting energy emissions), and engineers (designing and building the electronic apparatus). Continued teamwork among scientists promises further advances in both the fundamental understanding of biological processes and applications in medicine and agriculture.

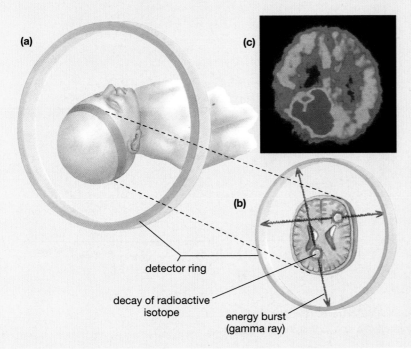

(a)

(c)

(b)

detector ring

decay of radioactive isotope

energy burst (gamma ray)

Figure E2-1 *How positron emission tomography works*
(a) A subject is given a substance containing a harmless radioactive isotope. The subject's head is inserted within a ring of detectors. Computerized analysis provides an image of the "slice" of the head that lies within the plane of the detector ring. *(b)* Inside the "head slice," radioactive decay releases energetic particles that activate the detectors. Computer analysis locates the site of the radioactive decay. *(c)* The number of decays at each location is color-coded and projected onto a television screen: Usually red and yellow mean many decays, and green and blue mean few decays. The resulting image provides a visual representation of the activity and structure of the brain. A brain tumor consists of actively dividing, metabolically energetic cells. It normally has a rich blood supply and consumes both oxygen and glucose voraciously. Therefore, a brain tumor "lights up" in a PET scan. In this scan, a malignant brain tumor shows clearly as a red area.

each other. In a similar way, electrons repel one another, owing to their negative electrical charge, and they are drawn to the positively charged protons of the nucleus. Because of their mutual repulsion, however, only limited numbers of electrons can occupy the space closest to the nucleus. Large atoms accommodate many electrons by having electrons orbit at increasing distances from their nucleus. Because the electrons orbit through a three-dimensional space, their orbits are called **electron shells** (Figs. 2-1 and 2-2).

The electron shell closest to the atomic nucleus is the smallest; it can hold only two electrons. The second shell can hold up to eight electrons. The electrons in an atom normally fill the shells closest to the nucleus. Thus, a carbon atom, with six electrons, has two electrons in the first shell, closest to the nucleus, and four electrons in its second shell (see Fig. 2-2). Electrons in shells closest to the nucleus have the least energy, and those farthest away have the most. For this reason, electron shells are often called **energy levels**.

Nuclei and electron shells play complementary roles in atoms. Nonradioactive nuclei provide stability, and the electron shells allow interactions with other atoms. Nuclei resist disturbance by outside forces. Ordinary sources of energy, such as heat, electricity, and light, hardly affect them at all. Because its nucleus is stable, a carbon atom remains carbon whether it is part of a diamond, carbon dioxide, or sugar. Electron shells, however, are dynamic; as you will soon see, atoms interact with one another by gaining, losing, or sharing electrons.

2 How Do Atoms Interact to Form Molecules?

Atoms Will Interact with Other Atoms Only When There Are Vacancies in Their Outermost Electron Shells

A **molecule** consists of two or more atoms (of the same element or different elements) held together by interactions among their outermost electron shells. A substance whose molecules are formed of different types of atoms is called a **compound**. Atoms interact with one another according to two basic principles:

1. An atom will not react with other atoms when its outermost electron shell is completely full or empty. Such an atom is described as being *stable*.
2. An atom will react with other atoms when its outermost electron shell is only partially filled. Atoms of this sort are described as *reactive*.

To demonstrate these principles, consider the two smallest atoms, hydrogen and helium (see Fig. 2-1). Hydrogen has one proton in its nucleus and one electron in its single (and therefore outermost) electron shell, which can hold up to two electrons. Helium has two protons in its nucleus, and two electrons fill its single electron shell. Therefore, we predict that hydrogen atoms, with a half-empty shell, should be reactive and that helium atoms, with a full shell, should be stable. Our predictions are correct. Hydrogen is highly flammable; that is, it reacts readily with oxygen in an explosive way. For example, the space shuttle and many other rockets use liquid hydrogen

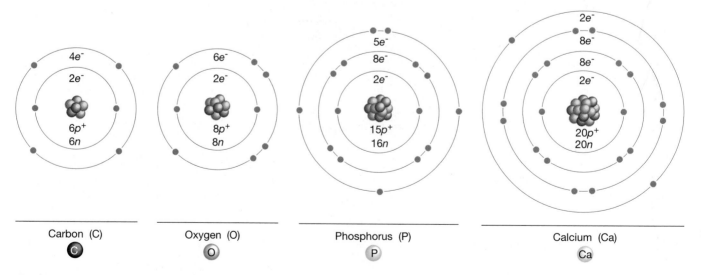

Carbon (C)	Oxygen (O)	Phosphorus (P)	Calcium (Ca)

Figure 2-2 Electron shells in atoms
Most biologically important atoms have at least two shells of electrons. The first shell, closest to the nucleus, can hold two electrons; the next shell can hold a maximum of eight electrons. More-distant shells can hold even more electrons.

as fuel to power lift-off. Normally, helium is almost completely unreactive.

An atom with an outermost electron shell that is partially full can gain stability by losing electrons (emptying the shell completely), by gaining electrons (filling the shell), or by sharing electrons with another atom (with both atoms behaving as though they had full outer shells). Losing, gaining, and sharing electrons result in forces called **chemical bonds**, which hold atoms together in molecules. Chemical bonds are not *things*, such as glue, but are *attractive forces* that hold atoms together. Each element has bonding properties that result from the configuration of electrons in its outer shell (Table 2-2). *Chemical reactions*, the making and breaking of chemical bonds, are essential for the maintenance of life and for the working of modern society. Whether they occur in a plant cell as it captures solar energy, your brain as it forms new memories, or your car's engine as it guzzles gas, chemical reactions consist of making new chemical bonds and/or breaking existing ones.

Charged Atoms Called Ions Interact to Form Ionic Bonds

Both atoms that have an almost empty outermost electron shell and atoms that have an almost full outermost shell can become stable by losing electrons (emptying their outermost shells) or by gaining electrons (filling their outermost shells). The formation of table salt (sodium chloride) demonstrates this principle. Sodium (Na) has only one electron in its outermost electron shell, and chlorine (Cl) has seven electrons in its outer shell—one electron short of being full (Fig. 2-3a). Sodium, therefore, can become stable by losing the electron from its outer shell, leaving that shell empty; chlorine can fill its outer shell by gaining an electron. Atoms that have lost or gained electrons, altering the balance between protons and electrons, are *charged*. These charged atoms are called **ions**. To form sodium chloride, sodium loses an electron and thereby becomes a positively charged sodium ion (Na^+); and chlorine picks up an electron and becomes a negatively charged chloride ion (Cl^-) (Fig. 2-3b,c).

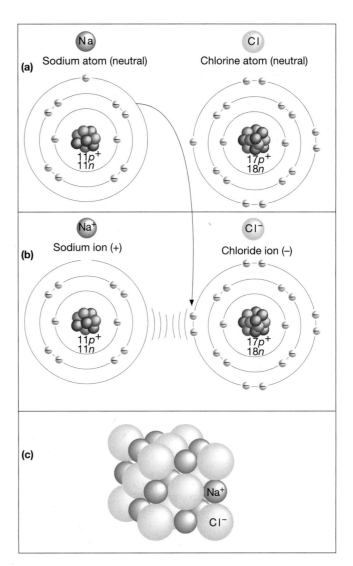

Figure 2-3 The formation of ions and ionic bonds
(*a*) Sodium has only one electron in its outer electron shell; chlorine has seven. (*b*) Sodium can become stable by losing an electron, and chlorine can become stable by gaining an electron. Sodium becomes a positively charged ion, and chlorine a negatively charged ion. (*c*) Because oppositely charged particles attract one another, the resulting sodium ions (Na^+) and chloride ions (Cl^-) nestle closely together in a crystal of salt, NaCl.

Table 2-2	
Type of Bond	**Bond Forms:**
Weak Bonds: allow interactions between individual atoms or molecules	
Hydrogen bonds	Between a hydrogen atom involved in a polar covalent bond and another atom involved in a polar covalent bond
Ionic bonds	Between positive and negative ions
Hydrophobic interactions	Because interactions between water molecules exclude hydrophobic molecules
Strong Bonds: hold atoms together within molecules	
Covalent bonds	By the sharing of electron pairs; equal sharing produces nonpolar covalent bonds; unequal sharing produces polar covalent bonds

Opposite charges attract; therefore, sodium ions and chloride ions tend to stay near one another. They form crystals that contain repeating orderly arrangements of the two ions (Fig. 2-3c). The electrical attraction between oppositely charged ions that holds them together in crystals is called an **ionic bond**. Ionic bonds are weak and easily broken, as occurs when salt is dissolved in water (Table 2-2).

Uncharged Atoms Can Become Stable by Sharing Electrons, Forming Covalent Bonds

An atom with a partially full outermost electron shell can also become stable by sharing electrons with another atom, forming a **covalent bond**. Consider the hydrogen atom, which has one electron in a shell built for two. A hy-

drogen atom can become reasonably stable if it shares its single electron with another hydrogen atom, forming a molecule of hydrogen gas, H_2 (Fig. 2-4a). Because the two hydrogen atoms are identical, neither nucleus can exert more attraction and capture the other's electron. So both electrons spend part of the time in the shell of both hydrogen atoms in a **single covalent bond**; each hydrogen behaves almost as if it had two electrons in its shell. Two oxygen atoms also share electrons equally, producing a molecule of oxygen gas, O_2, with a **double covalent bond** (Fig. 2-4b). In such a bond, each atom contributes two electrons. If atoms share three pairs of electrons, a **triple covalent bond** is formed. Nitrogen gas, N_2, forms such a bond.

All covalent bonds are strong compared with ionic bonds, but some are stronger than others, depending on

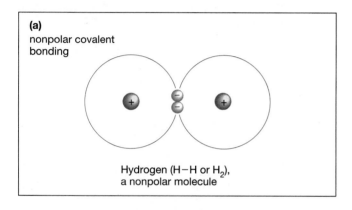

(a)
nonpolar covalent bonding

Hydrogen (H—H or H_2),
a nonpolar molecule

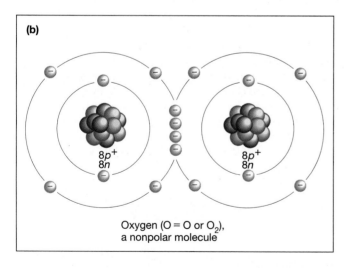

(b)

Oxygen (O＝O or O_2),
a nonpolar molecule

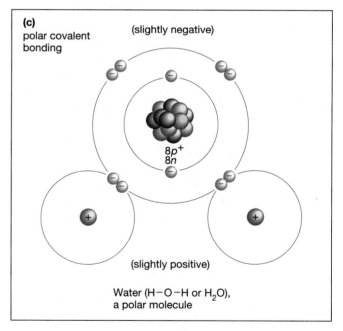

(c)
polar covalent bonding

(slightly negative)

$8p^+$
$8n$

(slightly positive)

Water (H—O—H or H_2O),
a polar molecule

Figure 2-4 Covalent bonds
Electrons are shared between atoms to form covalent bonds. **(a)** In hydrogen gas, one electron from each hydrogen atom is shared, forming a single covalent bond. The resulting molecule of hydrogen gas is represented as H–H or H_2. **(b)** In oxygen gas, two oxygen atoms share four electrons, forming a double bond (O=O or O_2). **(c)** Oxygen lacks two electrons to fill its outer shell, so oxygen can make two single bonds, one with each of two hydrogen atoms to form water (H–O–H or H_2O). Oxygen exerts a greater pull on the electrons than does hydrogen, so the oxygen end of the molecule has a slight negative charge and the hydrogen end has a slight positive charge. This is an example of polar covalent bonding. The water molecule with its slightly charged ends is called a polar molecule.

the atoms involved (Table 2-2). Some covalent bonds, such as those in water (H_2O; Fig. 2-4c) and carbon dioxide (CO_2), are extremely stable—that is, it takes a lot of energy to break the bonds. Other bonds, such as those in oxygen gas (Fig. 2-4b) or gasoline, are less stable, coming apart more easily. When a chemical reaction occurs in which less-stable bonds are broken and more-stable bonds are formed (such as burning gasoline with oxygen to form carbon dioxide and water), energy is released, as we shall describe in Chapter 4.

Most Biological Molecules Utilize Covalent Bonding

Covalent bonds are crucial to life because the atoms in most biological molecules are joined by covalent bonds. The molecules in proteins, sugars, bone, and cellulose are formed of atoms held together by covalent bonds. Hydrogen, carbon, oxygen, nitrogen, phosphorus, and sulfur are the most common atoms found in biological molecules. Except for hydrogen, each of these atoms needs at least two electrons to fill its outermost electron shell and can share electrons with two or more other atoms. Hydrogen can form a covalent bond with one other atom; oxygen and sulfur with two other atoms; nitrogen with three; and phosphorus and carbon with up to four. This diversity of bonding arrangements permits biological molecules to be constructed in an almost infinite variety and complexity. Double and triple bonds increase the variety of shapes and functions of biological molecules. Table 2-3 summarizes bonding patterns in biological molecules.

Polar Covalent Bonds Form When Atoms Share Electrons Unequally

In hydrogen gas, the two nuclei are identical, and the shared electrons spend equal time near each nucleus. Therefore, not only is the molecule as a whole electrically neutral, but each end, or *pole*, of the molecule is also electrically neutral. Such an electrically symmetrical bond is called a **nonpolar covalent bond**. But electron sharing in covalent bonds is not always equal. In many molecules, one nucleus may initially have a larger positive charge, and therefore attract the electrons more strongly, than does the other nucleus. This situation produces a **polar covalent bond**. Although the molecule as a whole is electrically neutral, it has charged parts: The atom that attracts the electrons more strongly then picks up a slightly negative charge (the negative pole of the molecule), and the other atom has a slightly positive charge (the positive pole). In water, for example, oxygen attracts electrons more strongly than does hydrogen, so the oxygen end of a water molecule is negative and each hydrogen is positive (Fig. 2-4c). Water with its charged ends is a polar molecule.

Because of the polar nature of their covalent bonds, nearby water molecules attract one another. The partially negatively charged oxygens of some water molecules attract the partially positively charged hydrogens of other water molecules. This electrical attraction is called a **hydrogen bond** (Fig. 2-5; Table 2-2). As we shall shortly, hydrogen bonds between molecules give water several unusual properties that are essential to life on Earth.

Many biological molecules form hydrogen bonds. Both nitrogen and oxygen atoms attract electrons more strongly than do hydrogen atoms. Therefore, the nitrogen or oxygen pole of a nitrogen–hydrogen or oxygen–hydrogen bond is slightly negative, and the hydrogen pole is slightly positive. The resulting polar parts of the molecules can form hydrogen bonds with water, with other biological molecules, or with distant, polar parts of the same molecule. Although individual hydrogen bonds are quite weak, many of them working together are quite strong. As we shall see in Chapter 3, hydrogen bonds play crucial roles in shaping the three-dimensional structures of proteins and DNA.

Table 2-3 Bonding Patterns of Atoms Commonly Found in Biological Molecules

Atom	Capacity of Outer Electron Shell	Electrons in Outer Shell	Number of Covalent Bonds Normally Formed	Common Bonding Patterns
Hydrogen	2	1	1	—H
Carbon	8	4	4	—C— —C= =C= —C≡
Nitrogen	8	5	3	—N— —N= N≡
Oxygen	8	6	2	—O— O=
Phosphorus	8	5	5	—P=
Sulfur	8	6	2	—S—

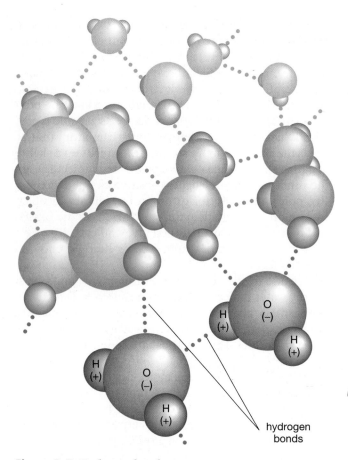

Figure 2-5 Hydrogen bonds
The partial charges on different parts of water molecules produce weak attractive forces called hydrogen bonds (dotted lines) between the hydrogens of one water molecule and the oxygens of other molecules.

3 Why Is Water So Important to Life?

www

Water is extraordinarily abundant on Earth, has unusual properties, and is so essential to life that it merits special consideration. Life is very likely to have arisen in the waters of the primeval Earth. The first life-form was probably a small sac enclosing water with an array of dissolved enzymes and simple genetic material. Living organisms still contain about 60% to 90% water, and all life depends intimately on the properties of water. Why is water so crucial to life?

Water Interacts with Many Other Molecules

Water enters into many of the chemical reactions that occur in living cells. The oxygen that green plants release into the air is derived from water during photosynthesis. In manufacturing a protein, fat, nucleic acid, or sugar, your body produces water in the process; conversely, when you digest proteins, fats, and sugars in the foods

you eat, water is used in the reactions. Why is water so important in biological chemical reactions?

Water is an extremely good **solvent**—that is, it is capable of dissolving a wide range of substances, including protein, salts, and sugars. Water or other solvents containing dissolved substances are called *solutions*. Recall that a crystal of table salt is held together by the electrical attraction between positively charged sodium ions and negatively charged chloride ions (see Fig. 2-3c). Because water is a polar molecule, it has positive and negative poles. If a salt crystal is dropped into water, the positively charged hydrogen ends of water molecules will be attracted to and will surround the negatively charged chloride ions, and the negatively charged oxygen poles of water molecules will surround the positively charged sodium ions. As water molecules enclose the sodium and chloride ions and shield them from interacting with each other, the ions separate from the crystal and drift away in the water—the salt dissolves (Fig. 2-6).

Water also dissolves molecules held together by polar covalent bonds. Its positive and negative poles are attracted to oppositely charged regions of dissolving molecules. Ions and polar molecules are termed **hydrophilic** (Greek for "water-loving") because of their electrical attraction for water molecules. Many biological molecules, including sugars and amino acids, are hydrophilic and dissolve readily in water (Fig. 2-7). Water also dissolves gases such as oxygen and carbon dioxide. The fish swimming below the ice on a frozen lake rely on oxygen that dissolved before the ice formed, and they release CO_2

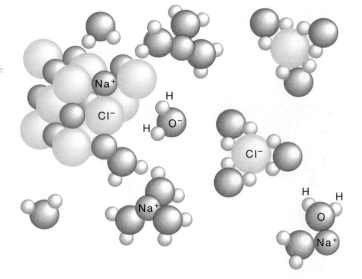

Figure 2-6 Water as a solvent
The polarity of water molecules allows water to dissolve polar and charged substances. When a salt crystal is dropped into water, the water surrounds the sodium and chloride ions with oppositely charged poles of the water molecules. Thus insulated from the attractiveness of other molecules of salt, the ions float away, and the whole crystal gradually dissolves.

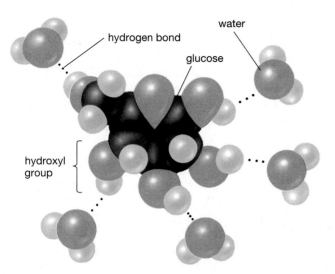

Figure 2-7 Water dissolves many biological molecules
Many biological molecules dissolve in water because they have polar parts—for example, OH⁻ (hydroxyl) groups—that can form hydrogen bonds with water molecules. As shown, hydrogen bonds can form between the hydroxyl groups on a glucose molecule (a simple sugar) and surrounding water molecules.

into solution in the water. By dissolving such a wide variety of molecules, the watery substance inside a cell provides a suitable environment for the countless chemical reactions essential to life on Earth.

Molecules that are uncharged and nonpolar, such as fats and oils, usually do not dissolve in water and hence are called **hydrophobic** ("water-fearing"). Nevertheless, water has an important effect on such molecules. Oils, for example, form globules when spilled into water. Oil molecules in water disrupt the hydrogen bonding among adjacent water molecules. When oil molecules encounter one another in water, their nonpolar surfaces nestle closely together, surrounded by water molecules that form hydrogen bonds with one another but not with the oil. To separate again, the oil molecules would have to break apart the hydrogen bonds that link surrounding water molecules. Thus, the oil molecules remain together, forming the glistening spheres on your soup surface. The tendency for hydrophobic molecules to clump together in watery solutions is often termed a **hydrophobic interaction** (Table 2-2). As we shall discuss in Chapter 5, the membranes of living cells owe much of their structure to the organizing effect of water on hydrophobic molecules.

Water Can Form H⁺ and OH⁻ Ions

Although water is generally regarded as a stable compound, individual water molecules constantly gain, lose, and swap hydrogen atoms. As a result, at any given time about two of every billion water molecules are ionized—

that is, broken apart into hydrogen ions (H⁺) and hydroxide ions (OH⁻):

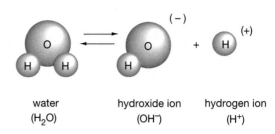

water
(H_2O)

hydroxide ion
(OH^-)

hydrogen ion
(H^+)

Pure water contains equal concentrations of hydrogen ions and hydroxide ions. In many solutions, though, the concentrations of H⁺ and OH⁻ are not the same. If the concentration of H⁺ exceeds the concentration of OH⁻, the solution is **acidic**. An **acid** is a substance that gives off hydrogen ions. When hydrochloric acid (HCl), for example, is added to pure water, almost all of the HCl molecules separate into H⁺ and Cl⁻. Therefore, the concentration of H⁺ greatly exceeds the concentration of OH⁻, and the resulting solution is acidic. (Notice in Figure 2-8 that many of the acidic substances, such as lemon juice and vinegar, have a sour taste. The sour-taste receptors on your tongue are specialized to respond to the excess of H⁺ in such substances.)

If the concentration of OH⁻ is greater, the solution is **basic**. A **base** is a substance that combines with hydrogen ions, reducing their number. If, for instance, sodium hydroxide (NaOH) is added to water, the NaOH molecules separate into Na⁺ and OH⁻. The OH⁻ combine with H⁺, reducing their number. The solution is then basic.

The degree of acidity is expressed on the **pH scale** (Fig. 2-8), in which neutrality (equal numbers of H⁺ and OH⁻) is assigned the number 7. Acids have a pH below 7; bases have a pH above 7. Pure water, with equal concentrations of H⁺ and OH⁻, has a pH of 7. Each unit on the pH scale represents a tenfold change in the concentration of H⁺. Thus, a cola drink (pH = 3) has a concentration of H⁺ 10,000 times that of water (pH = 7).

A Buffer Helps Maintain a Solution at a Relatively Constant pH

In most mammals, including humans, both the cell interior (cytoplasm) and the fluids that bathe the cells are nearly neutral (pH about 7.3 to 7.4). Small increases or decreases in pH may cause drastic changes in both the structure and function of biological molecules, leading to the death of cells or entire organisms. Nevertheless, living cells seethe with chemical reactions that take up or give off H⁺. How, then, does the pH remain constant overall? The answer lies in the many buffers found in living organisms. A **buffer** is a compound that tends to maintain a solution at a constant pH by accepting or releasing H⁺ in response to small changes in H⁺ concentration. If the H⁺ concentration rises, buffers combine with them;

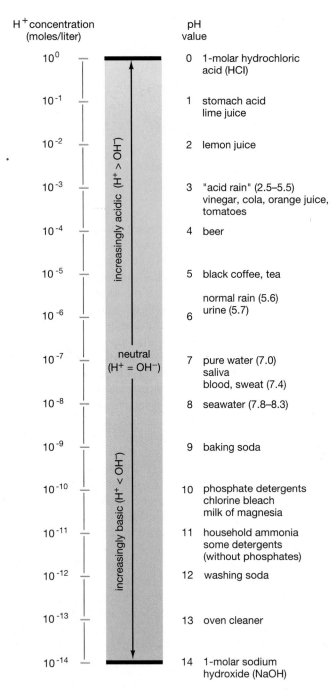

Figure 2-8 The pH scale
The pH scale expresses the concentration of hydrogen ions in a solution on a scale of 0 (very acidic) to 14 (very basic). Each unit of change in pH on the pH scale represents a tenfold change in the concentration of hydrogen ions. Lemon juice, for example, is about 10 times more acidic than orange juice, and the most severe acid rains in the northeastern United States are almost 1000 times more acidic than normal rainfall. Except for the insides of your stomach, nearly all the fluids in your body are finely adjusted to a pH of 7.4.

if the H^+ concentration falls, buffers release H^+. The result is that the concentration of H^+ is restored to its original level. Common buffers in living organisms include bicarbonate (HCO_3^-) and phosphate ($H_2PO_4^-$ and HPO_4^{2-}), both of which can accept or release H^+, depending on the circumstances. If the blood becomes too acidic, for example, bicarbonate accepts H^+ to form carbonic acid:

$$HCO_3^- \quad + \quad H^+ \quad \rightarrow \quad H_2CO_3$$
(bicarbonate) (hydrogen ion) (carbonic acid)

If the blood becomes too basic, carbonic acid liberates hydrogen ions, which combine with the excess hydroxide ions, forming water:

$$H_2CO_3 \quad + \quad OH^- \quad \rightarrow \quad HCO_3^- \quad + \quad H_2O$$
(carbonic acid) (hydroxide ion) (bicarbonate) (water)

In either case, the end result is that the blood pH remains near its normal value.

Water Moderates the Effects of Temperature Changes

Organisms can survive only within a limited temperature range. As we shall see in Chapter 4, high temperatures may damage enzymes that guide the chemical reactions essential to life. Low temperatures are also dangerous, because enzyme action slows as the temperature drops. Subfreezing temperatures within the body are normally lethal, because cells are ruptured by spearlike ice crystals.

Water has three properties that moderate the effects of temperature changes. These properties help keep the bodies of organisms within tolerable temperature limits and also cause large lakes and the oceans to have a moderating effect on the climate of nearby land. First, water has a high **specific heat**, the amount of energy needed to raise the temperature of 1 gram of a substance by 1°C. Temperature reflects the speed of molecules: The higher the temperature, the greater the average speed. Generally speaking, if heat energy enters a system, the molecules of that system move more rapidly, and the temperature of the system rises. Individual water molecules, however, are weakly linked to one another by hydrogen bonds (see Fig. 2-5). When heat enters a watery system such as a lake or a living cell, much of the heat energy goes into breaking hydrogen bonds rather than speeding up individual molecules. Thus, 1 **calorie** of energy will heat 1 gram of water 1°C, whereas it takes only 0.6 calorie per gram to heat alcohol 1°C, 0.2 calorie for table salt, and 0.02 calorie for common rocks such as granite or marble. As a result, because the human body is mostly water, a sunbather can absorb a lot of heat energy without sending his or her body temperature soaring.

Second, water moderates the effects of high temperatures through its great **heat of vaporization**, the amount of heat required to convert liquid water to water vapor. Water has one of the highest heats of vaporization known, 539 calories per gram. This too is due to the hydrogen bonds that interconnect individual water molecules. For a water molecule to evaporate, it must move quickly enough to break all the hydrogen bonds that hold it to the other water molecules in the solution. Only the fastest-moving water molecules, carrying the most energy, can break their hydrogen bonds and escape into the air as water vapor. The remaining liquid is cooler for the loss of these high-energy molecules. As children romp through a sprinkler on a hot summer day, water coats their bodies. Heat energy is transferred from their skin to the water and from the water to the vapor as the water evaporates. Evaporating just 1 gram of water cools 539 grams of a person's body 1°C, so water is a very effective coolant. This also means that evaporating perspiration produces a great loss of heat without much loss of water.

Third, water moderates the effects of low temperatures, because it has a high **heat of fusion**, the energy that must be removed from molecules of liquid water before they form the precise crystal arrangement of ice. Therefore, water, both in a living organism or in a lake, freezes more slowly than do many other liquids at a given temperature.

Water Forms an Unusual Solid: Ice

Water, of course, will become a solid after prolonged exposure to temperatures below its freezing point. But even solid water is unusual. Most liquids become denser when they solidify, and the solid sinks. Ice is less dense than liquid water. When a pond or lake starts to freeze in winter, the ice stays on top, forming an insulating layer that delays the freezing of the rest of the water. This insulation allows fish and other lake residents to survive below. If ice were to sink, ponds and lakes in much of North America would freeze solid during the winter, killing fish and underwater plants and making drinking water far less available to animals.

Water Molecules Tend to Stick Together

Because hydrogen bonds interconnect individual water molecules, liquid water has high **cohesion**—that is, a tendency to stick together. Cohesion among water molecules at the surface of a lake or pond produces **surface tension**, the tendency for the water surface to resist being broken. In general, objects that are denser than water sink. Because the water molecules at the surface of a pond stick to one another, however, the surface film acts almost as a solid, supporting relatively dense objects such as fallen leaves, water striders (Fig. 2-9a), and the basilisk lizard (see the chapter-opening photo).

(a)

(b)

Figure 2-9 Cohesion among water molecules
(a) Cohesion among water molecules allows water striders to skate across the surface of still waters. *(b)* In giant redwoods, cohesion holds water molecules together in continuous strands from the roots to the topmost leaves even 100 meters above the ground.

A more important role of cohesion occurs in the life of land plants (Fig. 2-9b). A plant absorbs water through its roots. How does the water reach the above-ground parts, especially if the plant is a hundred-meter-tall redwood? As we shall see in Chapter 26, water molecules are pulled up by the leaves. Water fills tiny tubes that connect the leaves, stem, and roots. Water molecules that evaporate from the leaves pull water up the tubes, much like a chain being pulled up from the top. The system works because the hydrogen bonds inter-connecting water molecules are stronger than the weight of the water in the tubes, even a hundred meters' worth; thus, the water "chain" doesn't break. Without the cohesion of water, there would be no land plants as we know them, and terrestrial life would undoubtedly have evolved quite differently. You may have realized by now that the "common bond" producing the sting of a belly flop, the globules of fat on soup, and the movement of water up a tree is, in fact, the hydrogen bond between water molecules.

Summary of Key Concepts

1) What Are Atoms?

An element is a substance that can neither be broken down nor converted to different substances by ordinary chemical means. The smallest possible particle of an element is the atom, which is itself composed of a central nucleus, containing protons and neutrons, and electrons outside the nucleus. All atoms of a given element have the same number of protons, which is different from the number of protons in the atoms of every other element.

Electrons orbit the nucleus in electron shells, at specific distances from the nucleus. Each shell can contain a fixed maximum number of electrons. The chemical reactivity of an atom depends on the number of electrons in its outermost electron shell: An atom is most stable, and therefore least reactive, when its outermost shell is either completely full or empty.

2) How Do Atoms Interact to Form Molecules?

Atoms may combine to form molecules. The forces holding atoms together in molecules are called chemical bonds. Atoms that have lost or gained electrons are negatively or positively charged particles called ions. Ionic bonds are electrical attractions between charged ions, holding them together in crystals.

When two atoms share electrons, covalent bonds form. In a nonpolar covalent bond, the two atoms share electrons equally. In a polar covalent bond, one atom may attract the electron more strongly than the other atom does; in this case, the strongly attracting atom bears a slightly negative charge, and the weakly attracting atom bears a slightly positive charge. Some polar covalent bonds give rise to hydrogen bonding, the attraction between charged regions of individual polar molecules or distant parts of a large polar molecule.

3) Why Is Water So Important to Life?

Properties of the water molecule that are important to living organisms include its ability to interact with many other molecules and to dissolve many polar and charged substances; to force nonpolar substances, such as fat, to assume certain types of physical organization; to participate in chemical reactions; to maintain a fairly stable temperature in the face of wide temperature fluctuations in the environment; and to cohere to itself by using hydrogen bonds between water molecules.

Key Terms

acid *p. 29*	compound *p. 24*	hydrophobic *p. 29*	pH scale *p. 29*
acidic *p. 29*	covalent bond *p. 26*	hydrophobic interaction	polar covalent bond *p. 27*
atom *p. 22*	double covalent bond *p. 26*	*p. 29*	proton *p. 22*
atomic nucleus *p. 22*	electron *p. 22*	ion *p. 25*	radioactive *p. 22*
atomic number *p. 22*	electron shell *p. 24*	ionic bond *p. 26*	single covalent bond *p. 26*
base *p. 29*	element *p. 22*	isotope *p. 22*	solvent *p. 28*
basic *p. 29*	energy level *p. 24*	molecule *p. 24*	specific heat *p. 30*
buffer *p. 29*	heat of fusion *p. 31*	neutron *p. 22*	surface tension *p. 31*
calorie *p. 30*	heat of vaporization *p. 31*	nonpolar covalent bond	triple covalent bond *p. 26*
chemical bond *p. 25*	hydrogen bond *p. 27*	*p. 27*	
cohesion *p. 31*	hydrophilic *p. 28*		

Thinking Through the Concepts

Multiple Choice

1. *What is the purest form of matter that cannot be separated into different substances by chemical means?*
   a. compounds   b. molecules   c. atoms
   d. elements   e. electrons

2. *Which phrase best describes chemical bonds?*
   a. physical bridges   b. energy links
   c. electronic glue   d. atomic forces
   e. all of these phrases are equally descriptive

3. *When an atom ionizes, what happens?*
a. It shares one or more electrons with another atom.
b. It emits energy as it loses extra neutrons.
c. It gives up or takes up one or more electrons.
d. It shares a hydrogen atom with another atom.
e. none of the above

4. *If electrons in water molecules were equally attracted to hydrogen nuclei and oxygen nuclei, water molecules would be*
a. more polar   b. less polar
c. unchanged   d. triple bonded
e. unable to form

5. *A covalent bond forms*
a. when two ions are attracted to one another
b. between adjacent water molecules, producing surface tension
c. when one atom gives up its electron to another atom
d. when two atoms share electrons
e. between water molecules and fat globules

6. *What is the defining characteristic of an acid?*
a. It donates hydrogen ions.
b. It accepts hydrogen ions.
c. It will donate or accept hydrogen ions, depending on the pH.
d. It has an excess of hydroxide ions.
e. It has a pH greater than 7.

? **Review Questions**

1. What are the six most abundant elements that occur in living organisms?

2. Distinguish among atoms and molecules; elements and compounds; and protons, neutrons, and electrons.

3. Compare and contrast covalent bonds and ionic bonds.

4. Why can water absorb a great amount of heat with little increase in its temperature?

5. Describe how water dissolves a salt. How does this phenomenon compare with the effect of water on a hydrophobic substance such as corn oil?

6. Define *acid*, *base*, and *buffer*. How do buffers reduce changes in pH when hydrogen ions or hydroxide ions are added to a solution? Why is this phenomenon important in organisms?

Applying the Concepts

1. Many "over-the-counter" substances are sold to bring relief from "acid stomach" or "heartburn." What is the chemical basis for these compounds? Why do they work?

2. Fats and oils do not dissolve in water; polar and ionic molecules dissolve easily in water. Detergents and soaps help clean by dispersing fats and oils in water so that they can be rinsed away. From your knowledge of the structure of water and the hydrophobic nature of fats, what general chemical structures (for example, polar or nonpolar parts) must a soap or detergent have, and why?

3. What would the effects be for aquatic life if the density of ice were greater than that of liquid water? What would be the impact on terrestrial organisms?

4. How does sweating help you regulate your body temperature? Why do you feel hotter and more uncomfortable on a hot, humid day than on a hot, dry day?

Group Activity

Take the number of your birth month as your atomic number. Determine the configuration of electrons in your valence shell, and see how you might interact with the other "atoms" in your group to form ions or new molecules.

For More Information

Atkins, P. W. *Molecules*. New York: Scientific American Library, 1987. A layperson's introduction to atoms and molecules, with superb illustrations.

Glasheen, J. W., and McMahon, T. A. "Running on Water." *Scientific American*, September 1997. Answers the question, "How does the basilisk lizard run on water?"

Morrison, P., and Morrison, P. *Powers of Ten*. New York: W. H. Freeman, 1982. A fascinating journey from the universe to the nucleus of an atom.

Storey, K. B., and Storey, J. M. "Frozen and Alive." *Scientific American*, December 1990. By triggering ice formation here, suppressing it there, and loading up their cells with antifreeze molecules, some animals, including certain lizards and frogs, can survive with 60% of their body water frozen solid.

Answers to Multiple-Choice Questions
1. d   2. b   3. c   4. b   5. d   6. a

A large meal such as this breakfast contains representatives of most types of biological molecules.

Biological Molecules 3

At a Glance

Think back to breakfast this morning—or did you oversleep and rush off to class without it? In either case, imagine a big breakfast, say, pancakes with maple syrup, blueberries and butter, a glass of orange juice, sausage and bacon, eggs, and coffee with milk and sugar. As you learned in Chapter 2, the biological molecules contained in this food are held together with chemical bonds. Salt on the eggs dissociates into ions in the watery environment of your mouth, and these ions are detected by taste receptors on your tongue. Sugar from the berries and syrup is also dissolved in your mouth and carried to taste receptors. The orange juice, although sweet, is also a bit sour, an indication that it is acidic.

Most of the nutrients in your breakfast were once molecules within other organisms. Think about it; your pancakes were once seeds of wheat. The blueberries, as well as the oranges in your juice, are the fruit of plants. Your eggs were the reproductive structures of a chicken, and the milk in your coffee was synthesized in a cow's glands that normally provide food for the cow's own offspring. Your bacon and sausage were contractile proteins that helped a pig walk, along with some fat the pig was using to store energy. How are we able to thrive by consuming other organisms?

All of the diverse forms of life, from bacteria to mushrooms to redwood trees to sea urchins to humans, evolved from a distant common ancestor. As a result, the bodies of all organisms use the same basic types of molecules. For this reason, humans can obtain nutrients from other organisms, and our bodies, in turn, can become nutrients for other organisms after we die.

What sorts of basic biological molecules make up living things? How are these molecules formed? What roles do these biological molecules play in our bodies and in the bodies of plants, insects, and fungi? In this chapter we examine the basic types of organic molecules—the carbohydrates, lipids, proteins, and nucleic acids—that form the basis of life on Earth.

Net Watch
On-line resources for this chapter are on the World Wide Web at:
http://www.prenhall.com/audesirk
(click on the Table of Contents link and then select Chapter 3).

1) Why Is Carbon So Important in Biological Molecules?

www

You have probably seen "organic" fruits and vegetables in your grocery store. To a chemist, this phrase is redundant; all produce is organic, because it is formed from biological molecules. In chemistry, the term **organic** is used to describe molecules that have a carbon skeleton and also contain some hydrogen atoms. The term is derived from the ability of living *organisms* to synthesize and use these molecules. **Inorganic** molecules include carbon dioxide and all molecules without carbon, such as water.

Although the common structure and function of the types of organic molecules among organisms afford unity, the tremendous range of organic molecules accounts for the diversity of living organisms and for the diversity of structures within single organisms and even within individual cells. This vast array of organic molecules, in turn, is possible because the carbon atom is so versatile. A carbon atom has four electrons in its outermost shell, with room for eight. Therefore, carbon atoms are able to form many bonds. They become stable by sharing four electrons with other atoms, forming up to four single covalent bonds or fewer double or triple covalent bonds. Molecules with many carbon atoms can assume complex shapes, including chains, branches, and rings.

Organic molecules are much more than just complicated skeletons of carbon atoms, however. Attached to the carbon backbone are groups of atoms, called **functional groups**, that determine the characteristics and chemical reactivity of the molecules. These functional groups are far less stable than the carbon backbone and are more likely to participate in chemical reactions. The common functional groups found in organic molecules are shown in Table 3-1.

The similarity among organic molecules from all forms of life is a consequence of two main features: (1) the use of the same set of functional groups in virtually all organic molecules in all types of organisms and (2) the use of the "modular approach" to synthesizing large organic molecules.

Table 3-1 Important Functional Groups in Biological Molecules

Group	Structure	Properties	Types of Molecules
Hydrogen (–H)	—H	Polar or nonpolar, depending on which atom hydrogen is bonded to; involved in condensation and hydrolysis	Almost all organic molecules
Hydroxyl (–OH)	—O—H	Polar; involved in condensation and hydrolysis	Carbohydrates, nucleic acids, alcohols, some acids, and steroids
Carboxyl (–COOH)	—C(=O)—O—H	Acidic; negatively charged when H$^+$ dissociates; involved in peptide bonds	Amino acids, fatty acids
Amino (–NH$_2$)	—N(H)(H)	Basic; may bond an additional H$^+$, becoming positively charged; involved in peptide bonds	Amino acids, nucleic acids
Phosphate (–H$_2$PO$_4$)	O—P(=O)(O—H)(O—H)	Acidic; up to two negative charges when H$^+$ dissociates; links nucleotides in nucleic acids; energy-carrier group in ATP	Nucleic acids, phospholipids
Methyl (–CH$_3$)	—C(H)(H)(H)	Nonpolar; tends to make molecules hydrophobic	Many organic molecules; especially common in lipids

2 How Are Organic Molecules Synthesized?

Many organic molecules are large and complex. In principle, there are two ways to manufacture a large, complex molecule: by combining atom after atom following an extremely detailed blueprint, or by preassembling smaller molecules and hooking them together. Just as trains are made by coupling engines to various train cars, so too does life take the modular approach. Small organic molecules (for example, *sugars*) are used as **subunits** with which to synthesize longer molecules (for example, *starches*), like cars in a train. The individual subunits are often called **monomers** (from Greek words meaning "one part"); long chains of monomers are called **polymers** ("many parts"). Most organic molecules in our bodies are in the form of either monomers or polymers.

Biological Molecules Are Joined Together or Broken Apart by Adding or Removing Water

You already learned some of the reasons why water is so important to life; here you will learn that water plays a central role in reactions that break down biological molecules (such as those in your breakfast) to liberate subunits that your body can use. In addition, when complex biological molecules are synthesized in the body, water is often produced as a by-product.

The subunits that make up large biological molecules are almost always linked together by a chemical reaction called a **dehydration synthesis** (literally, "to form by removing water"). In a dehydration synthesis, a hydrogen atom (−H) is removed from one subunit and a hydroxyl group (−OH) is removed from a second subunit, creating openings in the outer electron shells of the two subunits. These openings are filled by sharing electrons between the subunits, creating a covalent bond that links them. The free hydrogen and hydroxyl ions then combine to form a molecule of water (H_2O):

dehydration synthesis *called condensation reaction*

The reverse reaction, **hydrolysis** ("to break apart with water"), can split the molecule into individual subunits:

hydrolysis

Considering how complicated living things are, it might surprise you that nearly all biological molecules fall into one of only four general categories: carbohydrates, lipids, proteins, or nucleic acids (Table 3-2).

3 What Are Carbohydrates?

Carbohydrates are molecules composed of carbon, hydrogen, and oxygen in the approximate ratio of 1:2:1. All carbohydrates are either small, water-soluble **sugars** (*glucose* or *fructose*, for example) or chains, such as starch or *cellulose*, that are made by stringing sugar subunits together. If a carbohydrate consists of just one sugar molecule, it is called a **monosaccharide** (Greek for "single sugar"). When two or more monosaccharides are linked, they form a **disaccharide** ("two sugars") or a **polysaccharide** ("many sugars").

Carbohydrates such as sugars and starches are important energy sources for most organisms. Your pancakes consist mainly of carbohydrates stored in the seeds of wheat or other grains. The sugar that sweetens your syrup, blueberries, and orange juice is also stored by plants as an energy source. Other carbohydrates, such as cellulose and similar molecules, provide structural support for individual cells or even for the entire bodies of organisms as diverse as plants, fungi, bacteria, and insects.

Most single sugars (monosaccharides) have a backbone of three to seven carbon atoms. Most of the carbon atoms have both a hydrogen (−H) and a hydroxyl group (−OH) attached to them, so carbohydrates generally have the approximate chemical formula $(CH_2O)_n$. (Here n is the number of carbons in the backbone.) This formula explains the origin of the name *carbohydrate*, which literally means "carbon plus water." When dissolved in water, such as in the cytoplasm of a cell, the carbon backbone of a sugar usually "circles up" into a ring. Here are a linear representation and two ways the circular structure of a carbohydrate may be drawn, all for the sugar glucose:

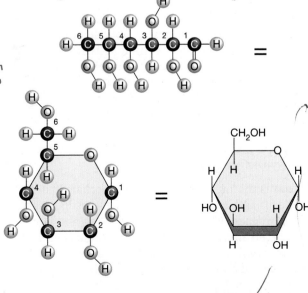

Table 3-2 The Principal Biological Molecules

Class of Molecule	Principal Subtypes	Example	Function
Carbohydrate: Normally contains carbon, oxygen, and hydrogen, in the approximate formula $(CH_2O)_n$	*Monosaccharide*: Simple sugar	Glucose	Important energy source for cells; subunit of which most polysaccharides are made
	Disaccharide: Two monosaccharides bonded together	Sucrose	Principal sugar transported throughout bodies of land plants
	Polysaccharide: Many monosaccharides (normally glucose) bonded together	Starch Glycogen Cellulose	Energy storage in plants Energy storage in animals Structural material in plants
Lipid: Contains high proportion of carbon and hydrogen; usually nonpolar and insoluble in water	*Triglyceride*: Three fatty acids bonded to glycerol	Oil, fat	Energy storage in animals, some plants
	Wax: Variable numbers of fatty acids bonded to long-chain alcohol	Waxes in plant cuticle	Waterproof covering on leaves and stems of land plants
	Phospholipid: Polar phosphate group and two fatty acids bonded to glycerol	Phosphatidylcholine	Common component of membranes in cells
	Steroid: Four fused rings of carbon atoms with functional groups attached	Cholesterol	Common component of membranes of eukaryotic cells; precursor for other steroids such as testosterone, bile salts
Protein: Chains of amino acids; contain carbon, hydrogen, oxygen, nitrogen, and sulfur		Keratin	Helical protein, principal component of hair
		Silk	Protein produced by silk moths and spiders
		Hemoglobin	Globular protein composed of four subunit peptides; transport of oxygen in vertebrate blood
Nucleic acid: Made of nucleotide subunits; may consist of a single nucleotide or long chain of nucleotides	*Long-chain nucleic acids*	Deoxyribonucleic acid (DNA)	Genetic material of all living cells
		Ribonucleic acid (RNA)	Genetic material of some viruses; in living cells, essential in transfer of genetic information from DNA to protein
	Single nucleotides	Adenosine triphosphate (ATP)	Principal short-term energy-carrier molecule in cells
		Cyclic adenosine monophosphate (cyclic AMP)	Intracellular messenger

It is in the ring form that sugars link together to make disaccharides (see Fig. 3-1) and polysaccharides.

Most small carbohydrates are soluble in water. As in water molecules, the O–H bond in a hydroxyl group is polar, because oxygen attracts electrons more strongly than hydrogen does. Hydrogen bonds between water molecules and the polar hydroxyl groups of the carbohydrate keep the carbohydrate in solution (see Fig. 2-7).

There Are a Variety of Monosaccharides with Slightly Different Structures

Glucose is the most common monosaccharide in living organisms and is the subunit of which most polysaccharides are made. Glucose has six carbons, and hence its chemical formula is $C_6H_{12}O_6$. Many organisms synthesize other monosaccharides that have the same chemical formula as glucose but have slightly different structures.

These include *fructose* (the "corn sugar" found in corn syrup, also the molecule that makes your orange juice taste sweet) and *galactose* (part of lactose, or "milk sugar"), shown below:

fructose   galactose

Some other common monosaccharides, such as *ribose* and *deoxyribose*, have five carbons:

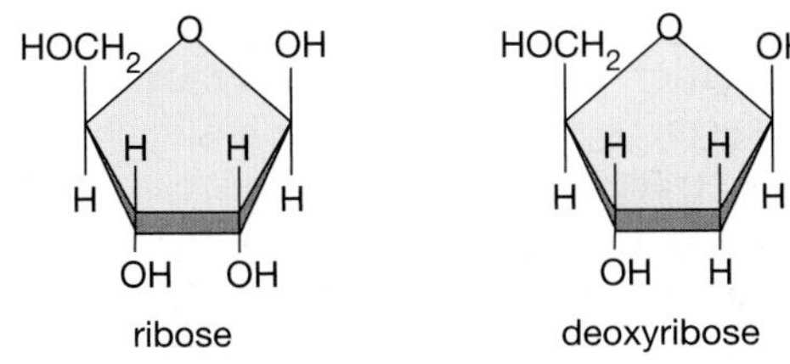

Ribose and deoxyribose are parts of the genetic molecules *ribonucleic acid* (RNA) and *deoxyribonucleic acid* (DNA), respectively.

Disaccharides Consist of Two Single Sugars Linked by Dehydration Synthesis

Monosaccharides, especially glucose and its relatives, have a short life span in a cell. Most are either broken down to free their chemical energy for use in various cellular activities or are linked by dehydration synthesis to form disaccharides or polysaccharides (Fig. 3-1). Disaccharides are often used for short-term energy storage, especially in plants. Common disaccharides include **sucrose** (glucose plus fructose), which you stirred into your coffee; **lactose** (milk sugar: glucose plus galactose), found in the milk you poured in your coffee; and **maltose** (glucose plus glucose, which will form in your digestive tract as you break down the starch in your pancakes). When energy is required, the disaccharides are broken apart into their monosaccharide subunits by hydrolysis.

Polysaccharides Are Chains of Single Sugars

For long-term energy storage, monosaccharides, normally glucose, are joined together into polysaccharides, forming **starch** (in plants; Fig. 3-2) or **glycogen** (in animals).

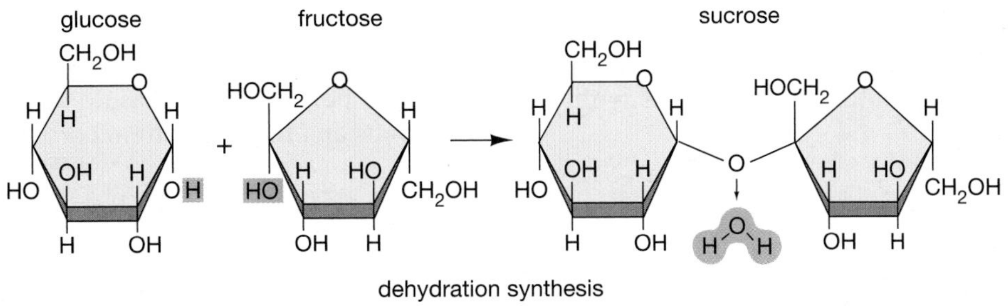

Figure 3-1 Synthesis and breakdown of a disaccharide
(a) The disaccharide sucrose is synthesized by a dehydration synthesis reaction in which a hydrogen (–H) is removed from glucose and a hydroxyl group (–OH) is removed from fructose. A water molecule is formed in the process, leaving the two monosaccharide rings joined by single bonds to the remaining oxygen atom. **(b)** Hydrolysis of sucrose is just the reverse of its synthesis, as water is split and added to the monosaccharides.

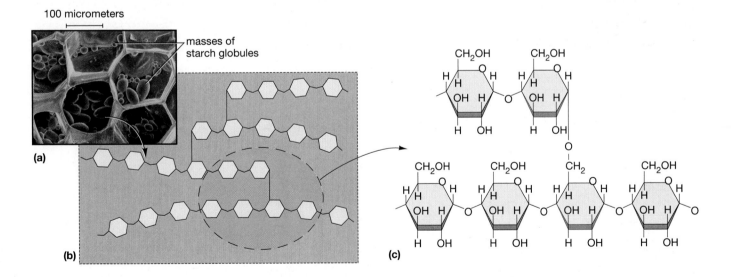

Figure 3-2 Starch is an energy-storage polysaccharide made of glucose subunits
(a) Starch globules inside individual potato cells. Most plants synthesize starch, which forms water-insoluble globules consisting of many starch molecules, which make up the bulk of this potato. **(b)** A small portion of a single starch molecule. Starch commonly occurs as branched chains of up to half a million glucose subunits. **(c)** The precise structure of the blue highlighted portion of the starch molecule in (b). Note the linkage between the individual glucose subunits for comparison with cellulose (Fig. 3-3).

Starch is commonly formed in roots and seeds; in the case of your pancakes, from the seeds of wheat. Starch may occur as coiled, unbranched chains of up to 1000 glucose subunits or, more commonly, as huge branched chains of up to half a million glucose monomers. Try chewing a pancake (without syrup) for a long time to allow enzymes in your saliva to break it down into its component sugars. Does it taste sweeter the longer you chew? Glycogen, stored as an energy source in the liver and muscles of animals such as humans, is generally much smaller than starch and has branches every 10 to 12 glucose subunits. Having many small branches probably makes it easier to split off the glucose subunits for quick energy release.

Many organisms also use polysaccharides as structural materials. One of the most important structural polysaccharides is **cellulose**, which makes up most of the cell walls of plants and about half the bulk of a tree trunk (Fig. 3-3). When you picture the vast fields and forests that blanket much of our planet, you will not be surprised to learn that there is probably more cellulose on Earth than all other organic molecules put together. Ecologists estimate that about a trillion tons of cellulose is synthesized each year!

Like starch, cellulose consists of glucose subunits bonded together. However, whereas most animals can easily digest starch, only a few microbes, such as those in the digestive tracts of cows or termites, can digest cellulose. Why is this the case, given that both starch and cellulose consist of glucose? The orientation of the bonds between subunits is different in the two polysaccharides. In cellulose, every other glucose is "upside down" (compare Fig. 3-2c with Fig. 3-3). This bond orientation prevents the digestive enzymes of animals from attacking the bonds between glucose subunits. Enzymes synthesized by certain microbes, though, can break these bonds. As a result, cellulose is food for these microbes. But for most animals, cellulose is *roughage* or *fiber*, material that passes undigested through the digestive tract.

The hard outer coverings (exoskeletons) of insects, crabs, and spiders are made of **chitin**, a polysaccharide in which the glucose subunits have been chemically modified by the addition of a nitrogen-containing functional group (Fig. 3-4). Interestingly, chitin also stiffens the cell walls of many fungi. Bacterial cell walls contain other types of modified polysaccharides, as do the lubricating fluids in our joints and the transparent corneas of our eyes.

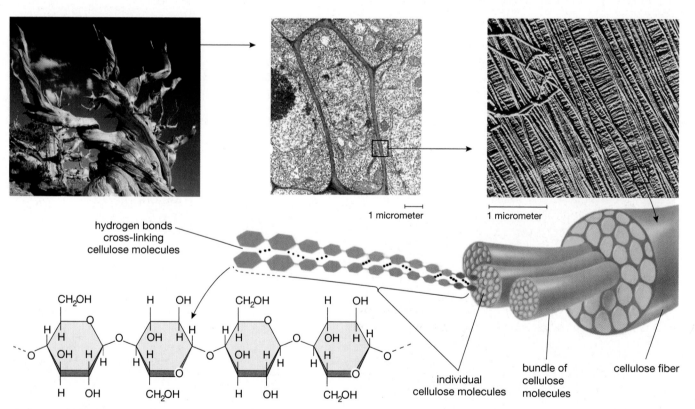

1 micrometer   1 micrometer

hydrogen bonds cross-linking cellulose molecules

CH₂OH   H   OH   CH₂OH   H   OH

individual cellulose molecules

bundle of cellulose molecules

cellulose fiber

Figure 3-3 Cellulose structure and function
Cellulose, like starch, is composed of glucose subunits, but the orientation of the bond between subunits in cellulose is different (compare with Fig. 3-2c) such that every other glucose molecule is "upside down." Unlike starch, cellulose has great structural strength, due partly to the difference in bonding and partly to the arrangement of parallel molecules of cellulose into long, cross-linked fibers. Plant cells often lay down cellulose fibers in layers that run at angles to each other, resulting in resistance to tearing in both directions. The final product can be incredibly tough, as this 3000-year-old bristlecone pine in California's White Mountains testifies.

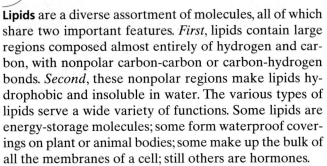

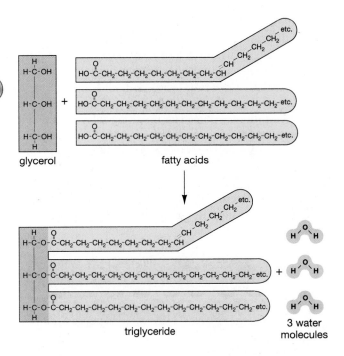

Figure 3-4 Chitin: A unique polysaccharide
Chitin has the same "alternating upside down" bonding of glucose molecules as cellulose has, but in chitin the glucose subunits are modified by the replacement of one of the hydroxyl groups with a nitrogen-containing functional group (yellow). Tough, slightly flexible chitin supports the otherwise soft bodies of arthropods (insects, spiders, and their relatives) and fungi.

Many other molecules, including *mucus*, some chemical messengers called *hormones*, and many molecules in the plasma membrane, consist in part of carbohydrate. Perhaps the most important of these molecules are the nucleic acids (discussed later in this chapter), which carry hereditary information.

4) What Are Lipids?

Lipids are a diverse assortment of molecules, all of which share two important features. *First*, lipids contain large regions composed almost entirely of hydrogen and carbon, with nonpolar carbon-carbon or carbon-hydrogen bonds. *Second*, these nonpolar regions make lipids hydrophobic and insoluble in water. The various types of lipids serve a wide variety of functions. Some lipids are energy-storage molecules; some form waterproof coverings on plant or animal bodies; some make up the bulk of all the membranes of a cell; still others are hormones.

Lipids are classified into three major groups: (1) oils, fats, and waxes, which are similar in structure and contain only carbon, hydrogen, and oxygen; (2) phospholipids, structurally similar to oils but also containing phosphorus and nitrogen; and (3) the fused-ring family of steroids.

Oils, Fats, and Waxes Are Lipids Containing Only Carbon, Hydrogen, and Oxygen

Oils, fats, and waxes are related in three ways: (1) They contain only carbon, hydrogen, and oxygen; (2) they contain one or more **fatty acid** subunits, which are long chains of carbon and hydrogen with a *carboxyl group* (–COOH) at one end; and (3) they normally do not have ring structures. **Fats** and **oils** are formed by dehydration synthesis from three fatty acid subunits and one

molecule of **glycerol**, a short, three-carbon molecule with one hydroxyl group (–OH) per carbon:

This structure of three fatty acids joined to one glycerol molecule gives fats and oils their chemical name, **triglyceride**. Notice that a double bond between two carbons in the fatty acid subunit forms a kink in the chain.

Fats and oils have a high concentration of chemical energy, about 9.3 Calories per gram, compared with 4.1 for sugars and proteins. (A *Calorie* with a capital *C* equals 1000 calories; the Calorie is used in measuring the energy content of foods.) Fats and oils are therefore used for long-term energy storage, especially in animals. For

Figure 3-5 Lipids
(a) A European brown bear ready to hibernate. Fat is an efficient way to store energy. If this bear stored the same amount of energy in carbohydrates instead of fat, she probably would be unable to walk! (b) Wax is a highly saturated lipid that remains very firm at normal outdoor temperatures. Its rigidity allows it to be used to form the strong but thin-walled hexagons of this honeycomb.

(a)

(b)

example, bears that feast during summer and fall put on fat to tide them over during their winter hibernation (Fig. 3-5a). Because fats store the same energy with less weight than do carbohydrates, fat is an efficient way for animals to store energy.

The difference between a fat (such as the butter on your pancakes), which is a solid at room temperature, and an oil (such as you used to fry your eggs) lies in their fatty acids. Fats have fatty acids with all single bonds in their carbon chains. Hydrogens occupy all the other bond positions on the carbons. The resulting fatty acid is said to be **saturated**, because it is "saturated" with hydrogens—it has as many hydrogens as possible. Lacking double bonds between carbons, the carbon chain of the fatty acid is straight. The saturated fatty acids of fats (such as the beef fat molecule illustrated below) can nestle closely together, forming solid lumps at room temperature:

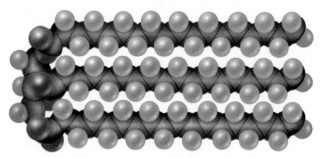

Beef fat (saturated)

If there are double bonds between some of the carbons, and consequently fewer hydrogens, the fatty acid is said to be **unsaturated**. Oils have mostly unsaturated fatty acids. The double bonds in the unsaturated fatty acids of oils produce kinks in the fatty acid chains, as illustrated by the linseed oil molecule next:

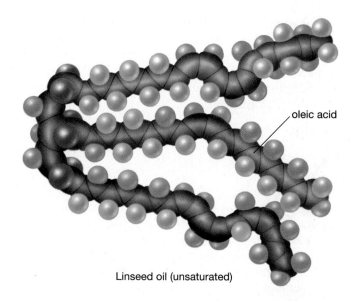

oleic acid

Linseed oil (unsaturated)

The kinks keep oil molecules apart; as a result, oil is liquid at room temperature. An oil can be converted to a fat by breaking the double bonds between carbons, replacing them with single bonds, and adding hydrogens to the remaining bond positions. This is the "hydrogenated oil" listed in the ingredients on a box of margarine.

Waxes are chemically similar to fats. They are highly saturated, making them solid at normal outdoor temperatures. Waxes form a waterproof coating over the leaves and stems of land plants. Animals synthesize waxes as waterproofing for mammalian fur and insect exoskeletons and, in a few cases, to build elaborate structures such as beehives (Fig. 3-5b).

Phospholipids Have Water-Soluble "Heads" and Water-Insoluble "Tails"

The plasma membrane that separates the inside of a cell from the outside world contains several types of **phos-**

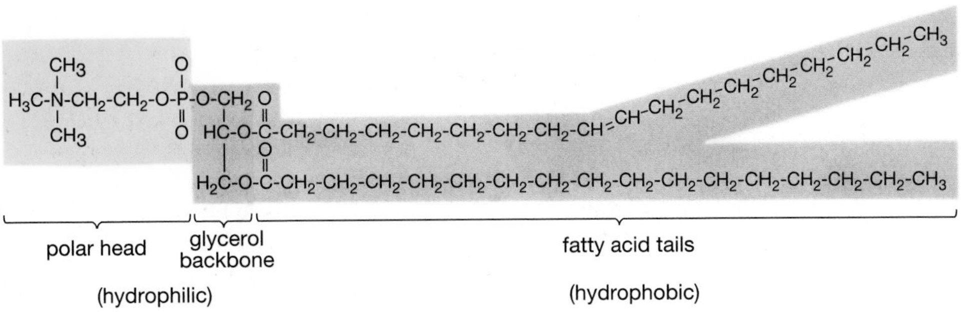

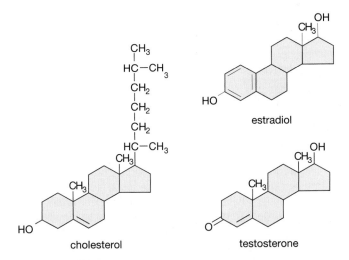

polar head | glycerol backbone | fatty acid tails
(hydrophilic) | | (hydrophobic)

Figure 3-6 Phospholipids
Phospholipids are similar to fats or oils, except that only two fatty acid tails are attached to the glycerol backbone. The third position on the glycerol is occupied by a polar head composed of a phosphate group ($-PO_4^-$) to which is attached a second, typically nitrogen-containing, functional group. The phosphate group is negatively charged, and the nitrogen-containing group is positively charged.

pholipids. A phospholipid is similar to an oil, except that one of the three fatty acids is replaced by a phosphate group with a short, polar functional group, typically a nitrogen-containing group, attached to the end (Fig. 3-6). Unlike the two fatty acid "tails," which are insoluble in water, the phosphate-nitrogen "head" is polar, or charged, and is water soluble. Thus, a phospholipid has two dissimilar ends: a hydrophilic head attached to hydrophobic tails. As you will see in Chapter 5, this dual nature of phospholipids is crucial to the structure and function of the plasma membrane.

Steroids Consist of Four Carbon Rings Fused Together

Steroids are structurally different from all the other lipids. All steroids are composed of four rings of carbon fused together, with various functional groups protruding from them (Fig. 3-7). One type of steroid is *cholesterol*; your egg yolk supplies more than half your recommended daily allotment. Cholesterol is a vital component of the membranes of animal cells and is also used by cells to synthesize other steroids. Other steroids the body synthesizes from cholesterol include male and female sex hormones, hormones that regulate salt levels, bile that assists in fat digestion, and insect hormones that stimulate the shedding of exoskeletons. Why then, have "cholesterol" and "steroids" gotten so much bad publicity? Find out in "Health Watch: Steer Clear of Steroids."

Figure 3-7 Steroids
Steroids are synthesized from cholesterol. All steroids have almost the same molecular structure (colored rings; note that the carbon atoms at the corners of the rings and the hydrogen atoms attached to the rings have been omitted from these drawings). Great differences in steroid function can be a result of even minor differences in functional groups attached to the rings. Notice the similarity in structure between the male sex hormone testosterone and the female sex hormone estradiol (a type of estrogen).

5) What Are Proteins? *www*

Proteins are molecules composed of one or more chains of *amino acids*. Proteins perform many functions (Table 3-3). Protein **enzymes** guide almost all the chemical reactions that occur inside cells, as you will learn in Chapter 4. Because each enzyme assists only one or a few specific reactions, most cells contain hundreds of differ-

Table 3-3 The Many Functions of Proteins

Function	Example
Structure	Collagen in skin; keratin in hair, nails, horns
Movement	Actin and myosin in muscle
Defense	Antibodies in bloodstream
Storage	Zeatin in corn seeds
Signals	Growth hormone in bloodstream
Catalysis	Enzymes: catalyze nearly every chemical reaction in our cells; DNA polymerase (makes DNA); pepsin (digests protein); amylase (digests carbohydrates); ATP synthase (makes ATP)

Health Watch
Steer Clear of Steroids

Perhaps you know someone who is taking "steroids" in an attempt to boost muscle mass or athletic performance. Although few will admit to taking them, as many as 30% of college and professional athletes may be on steroids, even though steroids are banned by organizations that regulate college, professional, and Olympic sports. The steroids that athletes unwisely consume are actually synthetic male hormones, one of the many substances synthesized from cholesterol. These hormones are normally closely regulated by the body, and the overdoses consumed by athletes can be disastrous. In addition to increasing muscle mass (Fig. E3-1), they may lead to liver disease; enlargement of the prostate gland, which then interferes with urination; atrophy of the testicles; and lowered sperm count. Although the verdict is not yet in, these synthetic male hormones may also contribute to heart attacks and strokes. This is a high price to pay for results that could be achieved through a healthy exercise routine.

Cholesterol itself is a steroid with a bad reputation. Why are so many products now advertising themselves as "cholesterol free" or "low in cholesterol"? Although this steroid is crucial to life, doctors have found that individuals with excessively high levels of cholesterol in their blood are much more likely than other individuals to suffer from heart attacks and strokes. Cholesterol contributes to the formation of obstructions in arteries, called **plaques**. Arterial plaques can reduce blood flow to the heart, and their rough edges stimulate blood clot formation inside the arteries. These clots may eventually block an artery, causing a heart attack (if the blocked artery had carried blood to the heart) or a stroke (if the blocked artery had supplied the brain). Plaque formation is described in more detail in "Health Watch: Matters of the Heart" in Chapter 27. Your body can synthesize

its own cholesterol from other lipids, and some bodies manufacture more than others. People with high cholesterol can often reduce their levels by eating a diet low in cholesterol and saturated fats. Doctors regularly prescribe cholesterol-reducing drugs for individuals with extremely high cholesterol who are unable to reduce it adequately through dietary restrictions.

Figure E3-1 Is this body healthy?
Years of hard work produced this body. But to achieve the same look, many body builders and athletes endanger their health by taking "steroids," or synthetic male hormones.

---

ent enzymes. Other proteins are used for structural purposes, such as *elastin*, which gives skin its elasticity; *keratin*, the principal protein of hair, horns, and claws; and the silk of spider webs and silk moth cocoons (Fig. 3-8). Still other types of proteins are used for energy and material storage (*albumin* in your egg white, *casein* in the milk in your coffee), transport (*hemoglobin* to carry oxygen in the blood), and cell movement (contractile proteins in muscle—the meat in your bacon and sausage). Some hormones (insulin, growth hormone), *antibodies* (which help fight disease and infection), and many poisons (rattlesnake venom) are also proteins.

Proteins Are Formed from Chains of Amino Acids

Proteins are polymers of **amino acids**. All amino acids have the same fundamental structure, consisting of a central carbon bonded to four different functional groups: a

nitrogen-containing *amino group* ($-NH_2$); a carboxyl, or carboxylic acid, group ($-COOH$); a hydrogen; and a variable group (represented by the letter R):

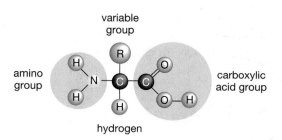

The R group differs among amino acids and gives each its distinctive properties (Fig. 3-9). Twenty amino acids are commonly found in the proteins of organisms.

(a)

(b)

(c)

Figure 3-8 Structural proteins
Common structural proteins include those of (*a*) hair,
(*b*) horn, and (*c*) spider web silk.

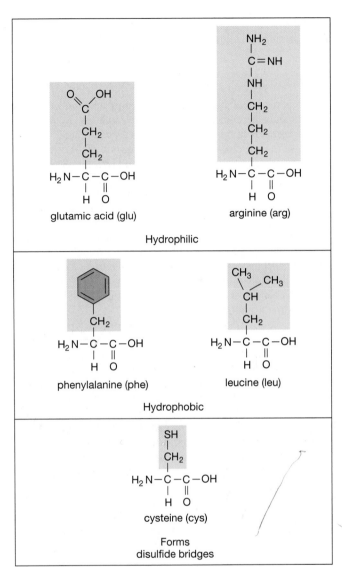

glutamic acid (glu)

arginine (arg)

Hydrophilic

phenylalanine (phe)

leucine (leu)

Hydrophobic

cysteine (cys)

Forms
disulfide bridges

Figure 3-9 Amino acid diversity
The diversity of amino acid structures is a consequence of
differences in the variable R group (colored blue). Some
amino acids are classified according to the variable R group
as hydrophilic, others as hydrophobic. Cysteine stands in a
class by itself. Two cysteines in distant parts of a protein
molecule can form a covalent bond between their sulfur
atoms, making a disulfide bridge that brings the cysteines
very close together and bends the protein chain.

Some amino acids are hydrophilic; their R groups are
polar and soluble in water. Others are hydrophobic, with
nonpolar R groups that are insoluble in water. Another
type of amino acid, cysteine, has sulfur in its R group and
can form bonds with other cysteines, linking protein
chains together. These bonds between the R groups of
cysteine are called **disulfide bridges** (see "A Closer Look:
Protein Structure—A Hairy Subject").

A Closer Look
Protein Structure—A Hairy Subject

A single strand of human hair, thin and not even alive, is nonetheless a highly organized, complex structure. Hair is composed mostly of a single, helical protein called keratin. If we look closely at the structure of hair, we can learn a great deal about biological molecules, chemical bonds, and why human hair behaves as it does.

A single hair consists of a hierarchy of structures (Fig. E3-2). The outermost layer is a set of overlapping shingle-like scales that protect the hair and keep it from drying out. Inside the hair lie closely packed, cylindrical dead cells, each filled with long strands called *microfibrils*. Each microfibril is a bundle of *protofibrils*, and each protofibril contains helical keratin molecules twisted together. As a hair grows, living cells in the hair follicle embedded in the skin whip out new keratin at the rate of 10 turns of the protein helix every second.

Pull the ends of a hair, and you will notice that it is rather strong. Hair gets its strength from three types of chemical bonds. First, the individual molecules of keratin are held in their helical shape by many hydrogen bonds. Before a hair will break, all the hydrogen bonds of all the keratin molecules in one cross-sectional plane of the strand must break to allow the helix to be stretched to its maximal extent. Second, each molecule is cross-linked to neighboring keratin molecules by disulfide bridges between cysteines (particular amino acids). Some of these bridges must break as the hair stretches. Finally, at least one peptide bond in each keratin molecule must break before the strand as a whole breaks.

Hair is also fairly stiff. The stiffness arises from hydrogen bonds within the individual helices of keratin and from disulfide bridges that hold neighboring keratin molecules together. When hair gets wet, however, the hydrogen bonds between turns of the helices are replaced by hydrogen bonds between the amino acids and the water molecules surrounding them, so the helices collapse. Wet hair is therefore very limp. If wet hair is rolled onto curlers and allowed to dry, the hydrogen bonds re-form in slightly different places, holding the hair in a curve. The slightest moisture, even humid air, allows these hydrogen bonds to rearrange

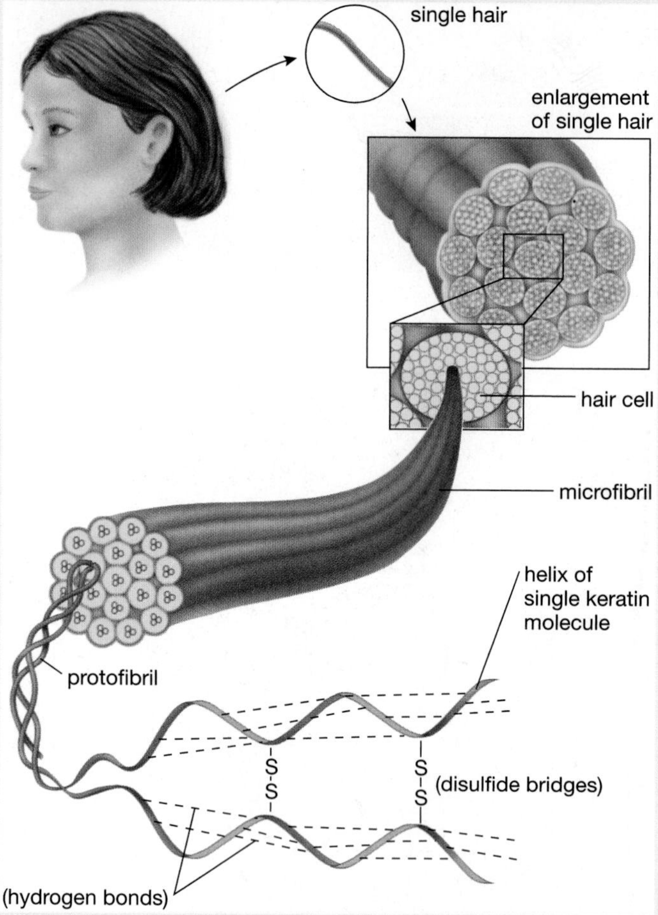

Figure E3-2 *The organization of hair*
At the microscopic level, a single hair is organized into bundles of fibers within further bundles of fibers. Hydrogen bonds and disulfide bridges between cysteines impart strength and elasticity to individual hairs.

Amino acids differ in their chemical and physical properties—size, water solubility, electrical charge—because of their different R groups. Therefore, the exact sequence of amino acids dictates the function of each protein: whether it is water soluble or not, whether it is an enzyme or a hormone or a structural protein. Scrambled sequences of amino acids are useless. In some cases, just one wrong amino acid can cause a protein to function incorrectly.

Amino Acids Are Joined to Form Chains by Dehydration Synthesis

Like lipids and polysaccharides, proteins are formed by dehydration synthesis. The nitrogen of the amino group ($-NH_2$) of one amino acid is joined to the carbon of the carboxyl group ($-COOH$) of another amino acid by a single covalent bond (Fig. 3-10). This bond is called a **peptide bond**, and the resulting chain of two amino acids is called a **peptide**. More amino acids are added, one by one, until the protein is complete. Amino acid chains in living cells vary in length from three to thousands of amino acids. Often, the word *protein* or *polypeptide* is reserved for long chains—say, 50 or more amino acids in length—and *peptide* is used for shorter chains.

A Protein Can Have Up to Four Levels of Structure

The phrase "amino acid chains" may evoke images of proteins as floppy, featureless structures, but they are not. Proteins are, instead, highly organized molecules that

into their natural configuration, and normally straight hair straightens out.

Pull gently, and you will discover still another property of hair. It stretches and then springs back into shape when you release the tension. When hair stretches, many of the hydrogen bonds within each keratin helix are broken, allowing the helix to be extended. Most of the covalent disulfide bonds between different levels of the helices, in contrast, are distorted by stretching but do not break. When tension is released, these disulfide bridges contract, returning the hair to its normal length.

Finally, each hair has a characteristic shape: It may be straight, wavy, or curly. The curliness of hair is genetically specified and is determined biochemically by the arrangement of disulfide bridges (see Fig. E3-2). Curly hair has disulfide bridges cross-linking the various keratin molecules at *different levels*, whereas straight hair has bridges mostly at the *same level*. When straight hair is given a "permanent," two lotions are applied. The first lotion breaks disulfide bonds between neighboring helices. The hair is then rolled tightly onto curlers, and a second solution, which re-forms the bridges, is applied. The new disulfide bridges connect helices at different levels, holding the strands of hair in a curl. These new bridges are more or less permanent, and

genetically straight hair can be transformed into biochemically curly hair. As new hair grows in, it will have the genetically determined arrangement of bridges and will not be curly:

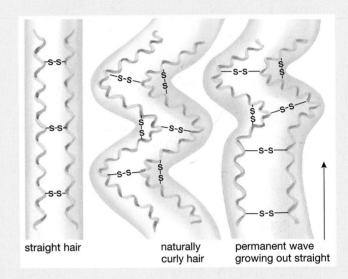

straight hair   naturally curly hair   permanent wave growing out straight

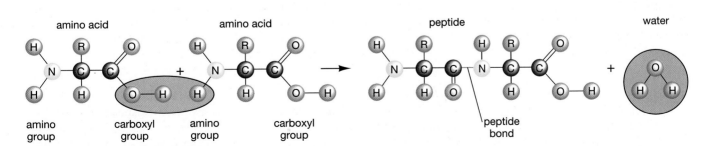

Figure 3-10 Protein synthesis
In protein synthesis, a dehydration synthesis joins the carbon of the carboxyl acid group of one amino acid to the nitrogen of the amino group of a second amino acid. The resulting covalent bond is called a peptide bond.

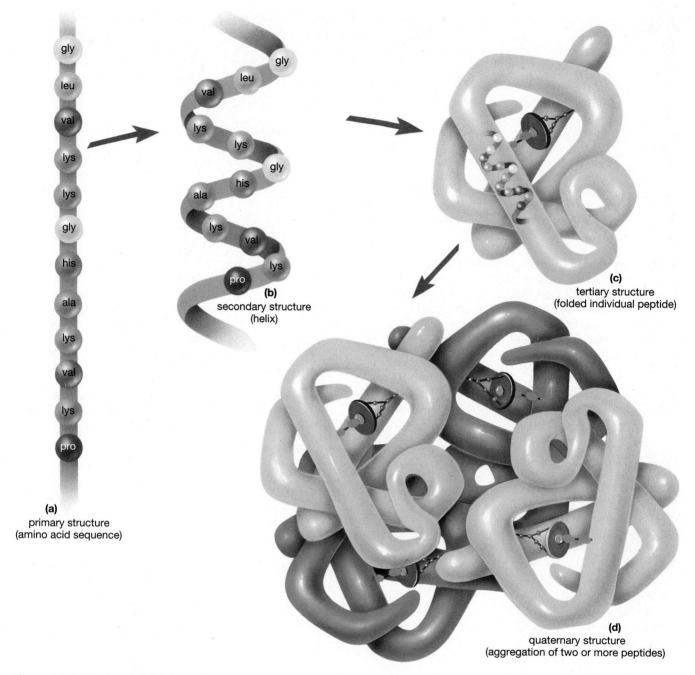

(a)
primary structure
(amino acid sequence)

(b)
secondary structure
(helix)

(c)
tertiary structure
(folded individual peptide)

(d)
quaternary structure
(aggregation of two or more peptides)

Figure 3-11 The four levels of protein structure
Levels of protein structure are represented here by hemoglobin, the oxygen-carrying protein in red blood cells.
All levels of protein structure are determined by the amino acid sequence of the protein, interactions among
the R groups of the amino acids (primarily hydrogen bonds and disulfide bridges between cysteines), and
interactions between the R groups and their surroundings (generally water or lipids).

come in a variety of shapes. Biologists recognize four levels of organization in protein structure. A single molecule of hemoglobin, the oxygen-carrying protein in red blood cells, exhibits all four structural levels (Fig. 3-11). The **primary structure** is the sequence of amino acids that make up the protein (Fig. 3-11a). This sequence is coded by the genes. Different types of proteins have different sequences of amino acids.

Hydrogen bonds cause many protein chains to form **secondary structures**. Many proteins, such as the hair protein keratin, and subunits of the hemoglobin molecule (Fig. 3-11b) have a coiled, springlike secondary structure called a **helix**. Hydrogen bonds between the relatively negative oxygen of the –C=O and the relatively positive hydrogen of the –N–H groups of amino acids hold the turns of the coils together.

In addition to their secondary structures, most proteins assume complex, three-dimensional **tertiary structures** (Fig. 3-11c). Disulfide bridges formed between cysteine amino acids may bring otherwise distant parts of a single peptide close together. In keratin, helices are held together in various ways by disulfide bonds, depending on whether the hair is straight or curly (see the in-text figure in the essay on p. 47).

Probably the most important influence on the tertiary structure of a protein is its cellular environment—specifically, whether the protein is dissolved in the water of the cytoplasm, in the lipids of the membranes, or half in one and half in the other. Hydrophilic amino acids can form hydrogen bonds with nearby water molecules, whereas hydrophobic amino acids cannot. Therefore, a protein dissolved in water folds into an irregular glob, with its hydrophilic amino acids facing the outside watery environment and its hydrophobic ones clustered in the center of the molecule.

Peptides may sometimes join into aggregations to form the fourth level of protein organization, called **quaternary structure**. Hemoglobin consists of two pairs of very similar peptides, held together by hydrogen bonds (Fig. 3-11d). Each peptide holds an iron-containing organic molecule called a *heme* (the red disks in Fig. 3-11d) that can bind one molecule of oxygen.

The Functions of Proteins Are Linked to Their Three-Dimensional Structures

Within a protein, the exact type, position, and number of amino acids bearing specific R groups determine both the structure of the protein and its biological function. In any given protein, however, some amino acids are more important than others. In hemoglobin, for example, certain amino acids bearing specific R groups must be present in precisely the right places to hold the iron-containing heme group that binds oxygen. Some of the other amino acids are interchangeable if they are functionally equivalent. For instance, the amino acids on the outside of a hemoglobin molecule serve mostly to keep it dissolved in the cytoplasm of a red blood cell. Therefore, as long as they are hydrophilic, it doesn't matter too much exactly which amino acids are where. As we will see in Chapter 12, however, replacing a hydrophilic amino acid with a hydrophobic amino acid can have catastrophic effects on the solubility of the hemoglobin molecule. In fact, such a substitution is the molecular cause of a painful and sometimes life-threatening disorder called sickle-cell anemia.

6 What Are Nucleic Acids?

Nucleic acids are long chains of similar but not identical subunits called **nucleotides**. All nucleotides have a three-part structure: (1) a five-carbon sugar (*ribose* or

deoxyribose), (2) a phosphate group, and (3) a nitrogen-containing base that differs among nucleotides:

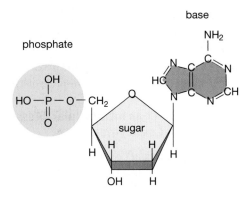

Deoxyribose nucleotide

There are two types of nucleotides, the ribose nucleotides (containing the sugar ribose) and the deoxyribose nucleotides (containing the sugar deoxyribose). Deoxyribose nucleotides bond to four types of nitrogen-containing bases—adenine, guanine, cytosine, and thymine. Similarly, ribose nucleotides bond to four types of bases—adenine, guanine, cytosine, and uracil instead of thymine.

Nucleotides may be strung together in long chains as nucleic acids, with the phosphate group of one nucleotide covalently bonded to the sugar of another:

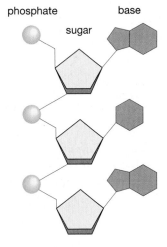

Nucleotide chain

There are two types of nucleic acids: (1) *deoxyribonucleic acid* and (2) *ribonucleic acid*.

DNA and RNA, the Molecules of Heredity, Are Nucleic Acids

Deoxyribose nucleotides form chains millions of units long called **deoxyribonucleic acid**, or **DNA**. DNA is found in the chromosomes of all living things. Its sequence of nucleotides, like the dots and dashes of a biological Morse code, spells out the genetic information needed to construct the proteins of each organism. Chains of ribose nucleotides, called **ribonucleic acid**, or **RNA**, are copied from the central repository of DNA in the nucleus of each cell. RNA carries DNA's genetic code into the cell's cytoplasm and directs the synthesis of proteins. (We shall cover DNA and RNA in more detail in Chapters 9 and 10.)

Other Nucleotides Act as Intracellular Messengers, Energy Carriers, or Coenzymes

Not all nucleotides are part of nucleic acids. Some exist singly in the cell or occur as parts of other molecules. **Cyclic nucleotides**, such as *cyclic adenosine monophosphate (cyclic AMP*; Fig. 3-12a), are intracellular messengers that carry information from the plasma membrane to other molecules in the cell. Cyclic AMP is synthesized

when certain hormones come in contact with the plasma membrane. Cyclic AMP then stimulates essential reactions in the cytoplasm or nucleus.

Some nucleotides have extra phosphate groups. These diphosphate and triphosphate nucleotides, such as **adenosine triphosphate** (*ATP*; Fig. 3-12b), are unstable molecules that carry energy from place to place within a cell. They capture energy where it is produced (during photosynthesis, for example) and give it up to drive energy-demanding reactions elsewhere (say, to synthesize a protein).

Finally, certain nucleotides assist enzymes in their role of promoting and guiding chemical reactions. These nucleotides are called **coenzymes**. Most coenzymes consist of a nucleotide combined with a vitamin (Fig. 3-12c). You will learn more about energy-carrier nucleotides, enzymes, and coenzymes in Chapter 4, where we discuss energy production and use in the cell.

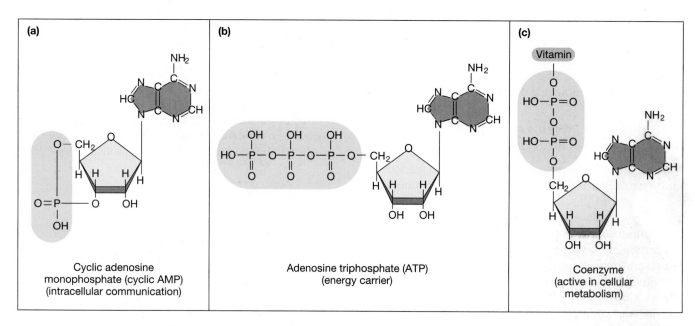

Cyclic adenosine monophosphate (cyclic AMP) (intracellular communication)

Adenosine triphosphate (ATP) (energy carrier)

Coenzyme (active in cellular metabolism)

Figure 3-12 A sampling of the diversity of nucleotides
Individual nucleotides, constructed of a sugar, a phosphate group, and a nitrogen-containing base, are often modified by the addition of different functional groups and serve a variety of cellular functions.

Summary of Key Concepts

1) Why Is Carbon So Important in Biological Molecules?
Organic molecules are so diverse because the carbon atom is able to form many types of bonds. This ability, in turn, allows organic molecules (molecules with a backbone of carbon and hydrogen atoms) to form many complex shapes, including chains, branches, and rings.

2) How Are Organic Molecules Synthesized?
Most large biological molecules are polymers synthesized by linking together many smaller subunits, or monomers. Chains of subunits are connected by covalent bonds through dehydration synthesis; the chains may be broken apart by hydrolysis reactions. The most-important organic molecules fall into one of four classes: carbohydrates, lipids, proteins, and nucleic acids. Their major characteristics are summarized in Table 3-2.

3) What Are Carbohydrates?
Carbohydrates include sugars, starches, chitin, and cellulose. Sugars (monosaccharides and disaccharides) are used for temporary storage of energy and for the construction of other molecules. Starches and glycogen are polysaccharides that serve for longer-term energy storage in plants and animals, respectively. Cellulose and related polysaccharides form the cell walls of bacteria, fungi, plants, and some microorganisms.

4) What Are Lipids?
Lipids are nonpolar, water-insoluble molecules of diverse chemical structure that include oils, fats, waxes, phospholipids, and steroids. Lipids are used for energy storage (oils and fats), as waterproofing for the outside of many plants and animals (waxes), as the principal component of cellular membranes (phospholipids), and as hormones (steroids).

5) What Are Proteins?
Proteins are chains of amino acids. Both the structure and the function of a protein are determined by the sequence of amino acids in the chain. Proteins can be enzymes (which guide chemical reactions), structural molecules (hair, horn), hormones (insulin), or transport molecules (hemoglobin).

6) What Are Nucleic Acids?
Nucleic acid molecules are chains of nucleotides. Each nucleotide is composed of a phosphate group, a sugar group, and a nitrogen-containing base. The two types of nucleic acids are deoxyribonucleic acid (DNA) and ribonucleic acid (RNA). Other nucleotides include intracellular messengers (cyclic AMP), energy-carrier molecules (ATP), and coenzymes.

Key Terms

adenosine triphosphate *p. 50*	**enzyme** *p. 43*	**monosaccharide** *p. 37*	**ribonucleic acid (RNA)** *p. 50*
amino acid *p. 44*	**fat** *p. 41*	**nucleic acid** *p. 49*	**saturated** *p. 42*
carbohydrate *p. 37*	**fatty acid** *p. 41*	**nucleotide** *p. 49*	**secondary structure** *p. 48*
cellulose *p. 40*	**functional group** *p. 36*	**oil** *p. 41*	**starch** *p. 39*
chitin *p. 40*	**glucose** *p. 38*	**organic** *p. 36*	**steroid** *p. 43*
coenzyme *p. 50*	**glycerol** *p. 41*	**peptide** *p. 46*	**subunit** *p. 37*
cyclic nucleotide *p. 50*	**glycogen** *p. 39*	**peptide bond** *p. 46*	**sucrose** *p. 39*
dehydration synthesis *p. 37*	**helix** *p. 48*	**phospholipid** *p. 42*	**sugar** *p. 37*
deoxyribonucleic acid (DNA) *p. 49*	**hydrolysis** *p. 37*	**plaque** *p. 44*	**tertiary structure** *p. 49*
	inorganic *p. 36*	**polymer** *p. 37*	**triglyceride** *p. 41*
disaccharide *p. 37*	**lactose** *p. 39*	**polysaccharide** *p. 37*	**unsaturated** *p. 42*
disulfide bridge *p. 45*	**lipid** *p. 41*	**primary structure** *p. 48*	**wax** *p. 42*
	maltose *p. 39*	**protein** *p. 43*	
	monomer *p. 37*	**quaternary structure** *p. 49*	

Thinking Through the Concepts

Multiple Choice

1. *Which of the following is* not *a function of polysaccharides in organisms?*
a. energy storage
b. storage of hereditary information
c. formation of cell walls
d. structural support
e. formation of exoskeletons

2. *Characteristics of carbon that contribute to its ability to form an immense diversity of organic molecules include its*
a. tendency to form covalent bonds
b. ability to bond with up to four other atoms
c. capacity to form single and double bonds
d. ability to bond together to form extensive, branched or unbranched carbon skeletons
e. all of the above

3. *Foods that are high in fiber are most likely to be derived from*
a. plants
b. dairy products
c. meat
d. fish
e. all of the above

4. *Proteins differ from one another because*
a. the peptide bonds linking amino acids differ from protein to protein
b. the sequence of amino acids in the polypeptide chain differs from protein to protein
c. each protein molecule contains its own unique sequence of sugar molecules
d. the number of nucleotides in each protein varies from molecule to molecule
e. the number of nitrogen atoms in each amino acid differs from the number in all others

5. *Which, if any, of the following choices does not properly pair an organic compound with one of its building blocks (subunits)?*
a. polysaccharide–monosaccharide
b. fat–fatty acid
c. nucleic acid–glycerol
d. protein–amino acid
e. all are paired correctly

6. *Which of the following statements about lipids is false?*
a. A wax is a lipid.
b. Unsaturated fats are liquid at room temperature.
c. The body doesn't need any cholesterol.
d. Both male and female sex hormones are steroids.
e. Beef fat is highly saturated.

? Review Questions

1. Which elements are common components of biological molecules?

2. List the four principal types of biological molecules, and give an example of each.

3. What roles do nucleotides play in living organisms?

4. One way to convert corn oil to margarine (solid at room temperature) is to add hydrogen atoms, decreasing the number of double bonds in the molecules of oil. What is this process called? Why does it work?

5. Describe and compare dehydration synthesis and hydrolysis. Give an example of a substance formed by each chemical reaction, and describe the specific reaction in each instance.

6. Distinguish among the following: monosaccharide, disaccharide, and polysaccharide. Give two examples of each and their functions.

7. Describe the synthesis of a protein from amino acids. Then describe primary, secondary, tertiary, and quaternary structures of a protein.

8. Most structurally supportive materials in plants and animals are polymers of special sorts. Where would we find cellulose? Chitin? In what way(s) are these two polymers similar? Different?

9. Which kinds of bonds or bridges between keratin molecules are altered when hair is (a) wet and allowed to dry on curlers; (b) given a permanent wave?

Applying the Concepts

1. A preview question for Chapter 5: In Chapter 2 you learned that hydrophobic molecules tend to cluster when immersed in water. In this chapter, you discovered that a phospholipid has a hydrophilic head and hydrophobic tails. What do you think would be the configuration of phospholipids that are immersed in water?

2. Many birds must store large amounts of energy to power flight during migration. Which type of organic molecule would be the most advantageous for energy storage? Why?

3. Remember the nuclear accident at Chernobyl in 1986? A scientist suspects that the food in a nearby ecosystem may

have been contaminated with radioactive nitrogen over a period of months. Which substances in plants and animals could be examined for radioactivity to test his hypothesis?

4. Fat contains twice as many calories per unit weight as carbohydrate does, so fat is an efficient way for animals, who must move about, to store energy. Compare the way fat and carbohydrates interact with water, and explain why this interaction also gives fat an advantage for weight-efficient energy storage.

Group Activity

Work in groups of three or four to play "molecular 20 questions." Each of you should select a molecule important for life (water, oxygen, carbohydrate, sugar, protein, amino acid, nucleic acid, DNA, RNA, lipid, phospholipid, cholesterol, and so

on) without telling the others in the group. Then, the others in the group can ask "yes-or-no" type questions to try to identify each molecule.

For More Information

Atkins, P. W. *Molecules*. New York: W. H. Freeman, 1987. A fascinating, nontechnical introduction to the beauties of chemistry.

Doolittle, R. F. "Proteins." *Scientific American*, October 1985. Proteins are the molecular tools that, directly or indirectly, carry out almost all the functions of a cell.

Dushesne, L. C., and Larson, D. W. "Cellulose and the Evolution of Plant Life." *BioScience*, April 1989. Chemical and natural history aspects of the most plentiful organic molecule on Earth.

Goodsell, D. S. *The Machinery of Life*. New York: Springer, 1993. Goodsell depicts the molecules of a cell in all their three-dimensional, interactive glory. A great way to gain a feel for the beauty and intricacy of the organic molecules of life.

Hill, J. W. *Chemistry for Changing Times*. 8th ed. Englewood Cliffs, NJ: Prentice Hall, 1998. A chemistry textbook for nonscience majors that is both clearly readable and thoroughly enjoyable.

Radetsky, P. "Kim's Coils." *Discover*, June 1995. Biochemist Peter Kim is uncovering the relationship between structure and function in the elaborate coiling of proteins.

Sharon, N., and Lis, H. "Carbohydrates in Cell Recognition." *Scientific American*, January 1993. Sugar-protein complexes on the surfaces of cells regulate cell identification and interaction between cells. See also Sharon's earlier article, "Carbohydrates," *Scientific American*, November 1980.

Answers to Multiple-Choice Questions
1. b   2. e   3. a   4. b   5. c   6. c

"Disorder spreads through the universe, and life alone battles against it."

G. Evelyn Hutchinson

The bodies of these runners are efficiently converting energy stored in chemicals to the energy of movement and heat. Their pounding footsteps noticeably shake the Verrazano Bridge during the New York Marathon.

Energy Flow in the Life of a Cell 4

At a Glance

Net Watch

On-line resources for this chapter are on the World Wide Web at:
http://www.prenhall.com/audesirk
(click on the Table of Contents link and then select Chapter 4).

Lungs heaving, legs pumping, the runners push themselves to the limits of their endurance—not only running 26 miles, but doing it fast. When over 20,000 people run the New York Marathon, collectively they convert more than 50 million Calories of energy into moving their bodies, shaking the Verrazano Bridge, and heating the air around them. At the end of the race, they douse their overheated bodies with water and refuel on high-energy snacks—after all, together they've "burned up" the equivalent of 30,000 pounds of chocolate cake! Finally, they board cars, buses, and airplanes to return to homes throughout the world.

What exactly is energy? Do the same principles govern energy use in the engines of cars and airplanes as in our bodies and those of other organisms? Why do our bodies generate heat, and why do we give off more heat when we exercise than when we watch TV? Many runners don't eat much immediately before a race, so how do the runners store energy and make it available to their muscles when they need it? We often talk about "burning" Calories. In fact, whether you set a spoonful of sugar on fire or eat it and allow your body to "burn" it, the reactions have much in common. In both cases, oxygen is combined with the sugar to produce carbon dioxide, water, and heat. A major difference is that our bodies aren't roasted when we metabolize food, and some of the energy is harnessed in other molecules that can be used to power muscle movement or a huge variety of processes within our cells. How do we control the breakdown of high-energy molecules to produce useful energy?

In this chapter we discuss the physical laws that govern energy flow in the universe, how energy flow in turn governs chemical reactions, and how the chemical reactions within living cells are controlled by the cell's own molecules. All organisms have the ability to direct chemical reactions within their bodies along certain pathways. The chemical reactions, which either require or release energy, serve many functions: to synthesize the molecules that make up the organism's body, to reproduce, to move, even to think. We will return to our discussion of energy in Chapters 7 and 8, which focus on photosynthesis, the chief "port of entry" for energy into the biosphere, and on glycolysis and cellular respiration, the most important sequences of chemical reactions that release energy and that power all the reactions in the life of a cell.

1 What Is Energy?

WWW

Energy can be defined simply as the *capacity to do work*, including synthesizing molecules, moving things around, and generating heat and light. There are two forms of energy: *kinetic energy* and *potential energy*. Both of these, in turn, exist in many different forms. **Kinetic energy**, or *energy of movement*, includes light (movement of photons), heat (movement of molecules), electricity (movement of electrically charged particles), and movement of large objects. **Potential energy**, or *stored energy*, includes chemical energy stored in the bonds that hold atoms together in molecules, electrical energy stored in a battery, and positional energy stored in a diver poised to spring (Fig. 4-1). Under the right conditions, kinetic energy can be transformed into potential energy, and vice versa. For ex-

ample, the diver converted kinetic energy of movement into potential energy of position when he climbed up to the platform; when he eventually jumped off, the potential energy was converted back into kinetic energy.

To understand energy flow, we need to know two things: (1) the quantity of available energy and (2) the usefulness of the energy. These are the subjects of the laws of thermodynamics, which we will now explore.

The Laws of Thermodynamics Describe the Basic Properties of Energy

The **laws of thermodynamics** define the basic properties and behavior of energy. The **first law of thermodynamics** states that, assuming there is no influx of energy, the total amount of energy remains constant. Energy can neither be created nor destroyed, but it can be changed in form (for example, from chemical energy to heat energy). The first law is therefore often called the *law of conservation of energy*. When you drive a car, you convert the potential chemical energy of gasoline into the kinetic energy of movement and heat (Fig. 4-2). The total amount of energy remains the same, although it is changed in form. Likewise, a runner is converting the potential chemical energy from food into equal amounts of the kinetic energy of movement and heat.

The **second law of thermodynamics** states that when energy is converted from one form to another, the

Figure 4-1 From potential to kinetic energy
The body of a diver perched atop the platform has potential energy, because the heights of the platform and the pool are different. As he dives, the potential energy is converted to the kinetic energy of motion of the diver's body. Finally, some of this kinetic energy is transferred to the water, which itself is set in motion.

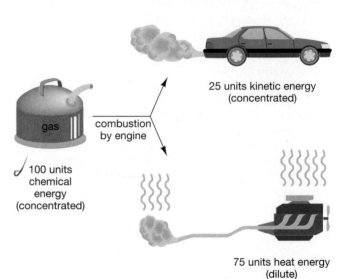

Figure 4-2 Thermodynamics of a car
According to the first law of thermodynamics, the quantity of energy originally present in the gasoline equals the energy in the resulting products: the molecules in the exhaust, the moving car, and heat. The form of the energy changes from chemical energy into energy of movement and heat. According to the second law of thermodynamics, some of the energy in the products is in a less-concentrated, less-usable form than the chemical energy of the gasoline. In autos, about 75% of the chemical energy of the gas is given off as waste heat to the environment, contributing to random movement of molecules in the air.

amount of useful energy decreases. There is a tendency for all systems to achieve the lowest possible energy. Put another way, the second law states that all spontaneous changes result in a more uniform distribution of energy, reducing the energy differences that are essential for doing work; in other words, energy is spontaneously converted from more-useful into less-useful forms.

In the case of a car engine, the kinetic energy of the moving vehicle is much less than the chemical energy the gasoline had contained (see Fig. 4-2). Where is the "missing" energy? The burning gas not only moved the car but also heated up the engine, the exhaust system, and the air around the car. The friction of tires on the pavement heated the road up a little, too. So, as the first law dictates, there isn't any missing energy. However, the energy released as heat is in a less-usable form; it merely increased the random movement of molecules in the engine, the road, and the air. So, too, the energy of heat that runners liberate to the air when food "burns" in their cells cannot be harnessed to allow them to run farther or faster. Thus, the second law tells us that no process—including processes carried on in the body—is 100% efficient.

The second law of thermodynamics also tells us something about the organization of matter. Regions of concentrated energy tend to be regions of great orderliness. The eight carbon atoms in a single molecule of gasoline have a much more orderly arrangement than do the carbon atoms of the eight separate, randomly moving molecules of carbon dioxide that are formed when the gasoline burns. Therefore, we can also phrase the second law in terms of the organization of matter: Unless energy is added to the system, processes that proceed spontaneously result in an increase in randomness and disorder. This tendency toward loss of orderliness and high-level energy and an increase in randomness, disorder, and low-level energy is called **entropy**. The tendency toward entropy can be seen in your dorm or apartment. Frequent inputs of your energy are required to keep dirt and debris confined to the trash bin, newspapers to folded stacks, books to their shelves, and clothes to drawers and closets. Without your energetic cleaning efforts, many of these items end up in their lowest-energy state, scattered on the floor! When G. Evelyn Hutchinson said "Disorder spreads through the universe, and life alone battles against it," he was making an eloquent reference to entropy and the second law of thermodynamics.

Living Things Use the Energy of Sunlight to Create the Low-Entropy Conditions Characteristic of Life

If you think about the second law of thermodynamics, you may wonder how life can exist at all. If chemical reactions, including those inside living cells, cause the amount of low-level, unusable energy to increase, and if matter tends toward increasing randomness and disorder, how can organisms accumulate the concentrated energy and precisely ordered molecules that characterize living things? The answer is that nuclear reactions in the

sun produce concentrated energy (sunlight) along with vast increases in entropy. Living things on Earth use a continuous input of concentrated solar energy to synthesize complex molecules and maintain orderly structures—to "battle against disorder." The highly organized low-entropy systems that characterize life do not violate the second law; they are achieved at the expense of an enormous loss of useful energy from sunlight. The entropy of the solar system as a whole constantly increases.

2 How Does Energy Flow in Chemical Reactions?

A chemical reaction converts one set of substances, called the **reactants**, into another set, the **products**. All chemical reactions are either *exergonic* or *endergonic*. A reaction is **exergonic** (Greek for "energy out") if the reactants have more energy than the products. Consequently, the reaction releases energy:

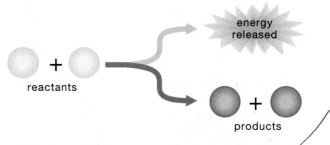

Conversely, a reaction is **endergonic** (Greek for "energy in") if the products have more energy than the reactants. According to the second law of thermodynamics, endergonic reactions require an influx of energy from some outside source:

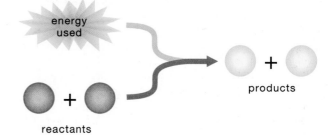

Let's look at two processes that illustrate these types of reactions: burning sugar and photosynthesis.

Exergonic Reactions Release Energy

In an exergonic reaction, the reactants contain more energy than the products. The sugar that the runners' bodies use as fuel contains more energy than do the carbon dioxide and water that are produced when that sugar breaks down. The extra energy is liberated as muscular movement and heat. Sugar can also be burned, as any cook can tell you. When sugar is burned by a flame, similar to the way it is burned in the body, it reacts with oxygen (O_2) to produce carbon dioxide (CO_2) and water (H_2O) and energy, as described by this equation:

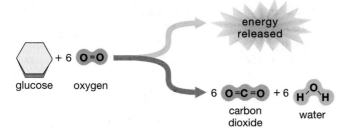

This reaction illustrates two important concepts, diagrammed in Figure 4-3a. First, molecules of sugar and oxygen contain much more energy than do molecules of carbon dioxide and water, so the reaction releases energy. Energy release allows exergonic reactions to occur without a net input of energy. Once ignited, sugar will continue to burn spontaneously. It may be helpful to think of exergonic reactions as running "downhill," from high energy to low energy.

Although the burning of sugar releases energy, a spoonful of sugar doesn't burst into flames by itself. This observation leads to the second important concept: All chemical reactions require an initial input of energy, called the **activation energy**, to get started. Chemical reactions require activation energy because atoms and molecules are surrounded by a shell of negatively charged electrons. For two molecules to react with each other, their electron shells must be forced together, despite their mutual electrical repulsion. That forcing requires energy. The usual source of activation energy is the kinetic energy of movement. Only molecules moving with sufficient speed collide hard enough to force their electron shells to mingle and react. Because molecules move faster as the temperature increases, most reactions occur more readily at high temperatures. The initial heat provided by setting sugar on fire allows these reactions to begin. The combination of sugar with oxygen then releases its own heat, sustaining the reaction.

Endergonic Reactions Require an Input of Energy

In contrast to what happens when sugar is burned, many reactions in living systems result in products that contain *more energy* than do the reactants. Sugar, produced in photosynthetic organisms, contains far more energy than do the carbon dioxide and water from which it is formed. The protein in a muscle cell contains more energy than do the individual amino acids that were joined together to synthesize it. In other words, synthesizing complex biological molecules requires an input of energy (Fig. 4-3b). As we shall see in Chapter 7, photosynthesis in green plants takes low-energy water and carbon dioxide and produces high-energy sugar and oxygen from them:

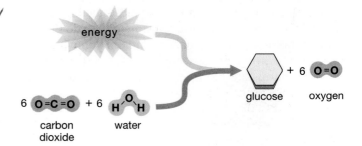

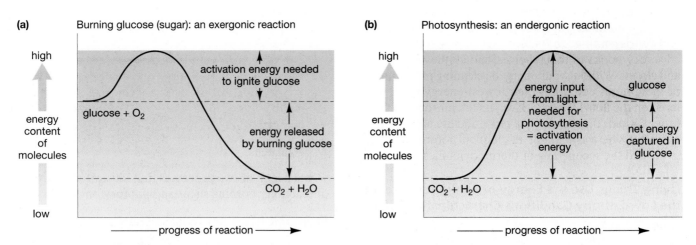

Figure 4-3 Energy relations in exergonic and endergonic reactions
(a) An exergonic ("downhill") reaction, such as the burning of sugar, proceeds from high-energy reactants (here, glucose) to low-energy products (CO_2 and H_2O). The energy difference between the chemical bonds of the reactants and products is released as heat. To start the reaction, however, an initial input of energy—the activation energy—is required. **(b)** An endergonic ("uphill") reaction, such as photosynthesis, proceeds from low-energy reactants (CO_2 and H_2O) to high-energy products (glucose) and therefore requires a net input of energy, in this case from sunlight.

Photosynthesis requires energy, which plants (and algae and some types of bacteria) obtain from sunlight. We might call endergonic reactions "uphill" reactions, going from low energy to high energy. But where do we get the energy to synthesize muscle protein and other complex biological molecules?

Coupled Reactions Link Exergonic with Endergonic Reactions

Endergonic reactions require energy from other sources; they obtain this energy from exergonic, or energy-releasing, reactions. In a **coupled reaction** (Fig. 4-4), an

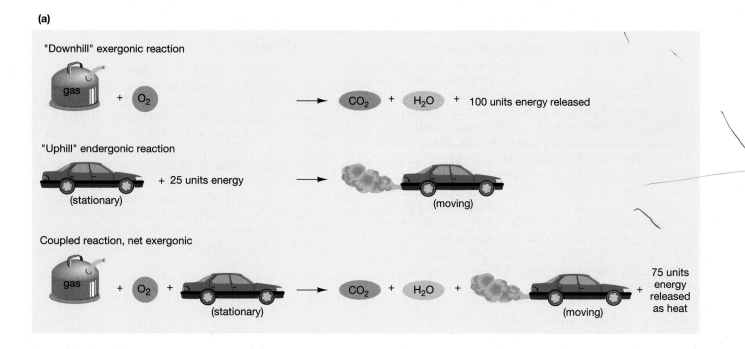

(a)

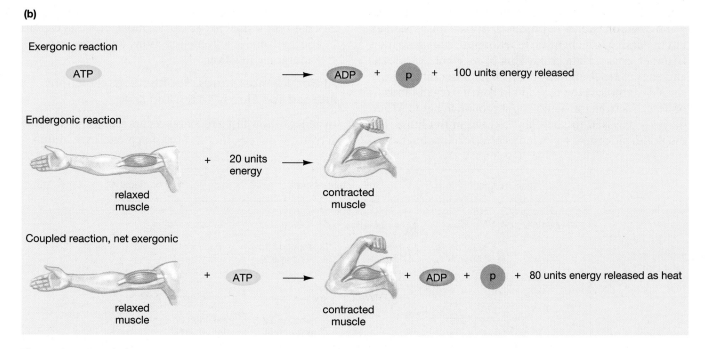

(b)

Figure 4-4 *Coupled reactions*

(a) The burning of gasoline is an exergonic reaction that can be coupled to the endergonic reaction of moving a car. According to the second law of thermodynamics, the quantity of concentrated, useful energy decreases in the process. Therefore, the exergonic reaction must release more energy than the endergonic reaction requires. The excess energy is given off to the environment as heat. *(b)* Muscle movement is an endergonic reaction coupled to the exergonic reaction of ATP breakdown. The energy released by ATP as it is broken down into ADP plus phosphate (P) exceeds the energy put into muscle contraction, so the overall reaction is exergonic.

exergonic reaction provides the energy needed to drive an endergonic reaction. When you drive a car, the exergonic reaction of burning gasoline provides the energy for the endergonic reaction of starting a stationary car into motion; much energy is lost as heat (Fig. 4-4a). Photosynthesis is another coupled reaction. In photosynthesis, the exergonic reaction occurs in the sun, and the endergonic reaction occurs in the plant. Most of the energy liberated by the sun is lost as heat, and some of the energy liberated by the exergonic reaction is also lost as heat, so the second law of thermodynamics still applies.

Living organisms constantly use the energy given off by exergonic reactions (such as the chemical breakdown of sugar, some of whose energy is trapped in ATP) to drive essential endergonic reactions (such as brain activity, muscular contraction or other types of movement, or the synthesis of complex molecules), as shown in Figure 4-4b. The exergonic and endergonic halves of coupled reactions typically occur in different places, so there must be some way to transfer the energy from the exergonic reaction to the endergonic reaction. In coupled reactions occurring within cells, energy is normally transferred from place to place by *energy-carrier* molecules, of which the most common is *adenosine triphosphate*, or *ATP*. We will examine the synthesis and use of ATP later in this chapter.

Chemical Reactions Are Reversible

In the preceding discussions, we have treated chemical reactions as if they always proceed in one direction: "Forward" from reactants to products. In fact, this is usually not the case. Most chemical reactions are *reversible*; that is, under appropriate conditions of reactants, products, and energy, reactions may proceed in either direction. A simple example is the reaction in which oxygen binds to or detaches from the hemoglobin protein in blood. When blood is exposed to abundant oxygen in the lungs, hemoglobin picks up oxygen molecules, forming oxyhemo-

globin. When blood enters a muscle where oxygen is being used rapidly, the low concentration of oxygen causes the reaction to reverse. Oxygen detaches from the hemoglobin molecule and enters the oxygen-starved muscle cell. Our lives are completely dependent on the reversibility of this reaction!

3 How Do Cells Control Their Metabolic Reactions?

Cells are miniature, incredibly complex chemical factories. The **metabolism** of a cell is the sum of all its chemical reactions. Many of these reactions are linked in sequences called **metabolic pathways** (Fig. 4-5). Photosynthesis is one such pathway. *Glycolysis*, the series of reactions that begin the digestion of glucose (see Chapter 8), is another. Different metabolic pathways may utilize the same molecules; as a result, all the reactions and all the molecules of a cell are interconnected in a single, enormously complicated metabolic pathway.

The chemical reactions in a cell are governed by the same laws of thermodynamics that control other reactions. How, then, do orderly metabolic pathways arise? The biochemistry of cells is finely tuned in three ways:

1. Cells regulate chemical reactions through the use of proteins called *enzymes*.
2. Cells couple reactions together, driving energy-requiring endergonic reactions with the energy released by exergonic reactions.
3. Cells synthesize energy-carrier molecules that capture energy from exergonic reactions and transport it to endergonic reactions.

At Body Temperatures, Spontaneous Reactions Proceed Too Slowly to Sustain Life

In general, how fast a reaction occurs is determined by its activation energy, that is, how much energy is required

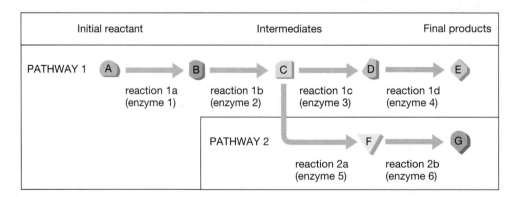

Figure 4-5 Simplified view of metabolic pathways
The original reactant molecule, A, undergoes a series of reactions, each catalyzed by a specific enzyme. The product of each reaction serves as the reactant for the next reaction in the pathway. Metabolic pathways are commonly interconnected such that the product of a step in one pathway may serve as a reactant for the next reaction in that pathway or for a reaction in another pathway.

to start the reaction (see Fig. 4-3). Reactions with low activation energies can proceed swiftly at body temperature, whereas reactions with high activation energies, such as the reaction of coal with oxygen, are practically nonexistent at similar temperatures. Most reactions can be accelerated by raising the temperature, thereby increasing the speed of molecules.

The reaction of sugar with oxygen to yield carbon dioxide and water is exergonic, but it has an enormous activation energy. At the temperatures found in living organisms, sugar and many other energetic molecules would almost never break down spontaneously and give up their energy. Because of the enzymes produced by living cells, however, sugar is an important energy source for life on Earth. Let's see how enzymes and other *catalysts* influence chemical reactions.

Catalysts Reduce Activation Energy

Catalysts are molecules that speed up the rate of a reaction without themselves being used up or permanently altered. Catalysts speed up reactions by reducing the activation energy (Fig. 4-6). As an example of catalytic action, let's consider the catalytic converters in many automobile engines. When gasoline is burned completely, the final products are carbon dioxide and water:

$$2\ C_8H_{18} + 25\ O_2 \rightarrow 16\ CO_2 + 18\ H_2O + \text{energy}$$
(octane)

However, flaws in the combustion process generate other substances, including poisonous carbon monoxide (CO).

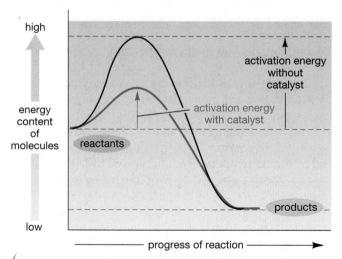

Figure 4-6 Activation energy controls the rate of chemical reactions
High activation energy (black curve) means that reactant molecules must collide very forcefully in order to react. Only very fast-moving molecules will collide hard enough to react, so reactions with high activation energies proceed slowly at low temperatures, where most molecules move relatively slowly. Catalysts lower the activation energy of a reaction (red curve), so a much higher proportion of molecules move fast enough to react when they collide. Therefore, the reaction proceeds much more rapidly.

Carbon monoxide reacts spontaneously but slowly with oxygen in the air to form carbon dioxide:

$$2\ CO + O_2 \rightarrow 2\ CO_2 + \text{energy}$$

In large cities, so many vehicles emit so much CO that the spontaneous reaction of CO with O_2 can't keep up, and unhealthy levels of carbon monoxide accumulate. Enter the catalytic converter. Platinum catalysts in the converter hasten the conversion of CO to CO_2, thereby reducing air pollution.

Note three important principles about all catalysts:

1. Catalysts speed up reactions.
2. Catalysts can speed up only those reactions that would occur spontaneously anyway, but at a much slower rate.
3. Catalysts are not consumed in the reactions they promote. No matter how many reactions they accelerate, the catalysts themselves are not permanently changed.

4) What Are Enzymes?

Enzymes are biological catalysts, normally proteins, synthesized by living organisms. Enzymes possess the characteristics of catalysts that we just described. But enzymes have two additional attributes that set them apart from inorganic catalysts:

1. Enzymes are normally very specific, catalyzing at most only a few types of chemical reactions. In most cases, an enzyme catalyzes a single reaction that involves one or two specific molecules but leaves even quite similar molecules untouched. (You may recall from Chapter 3, for example, that animals have enzymes that break apart starch molecules but leave cellulose intact, despite the fact that both starch and cellulose are composed of glucose subunits.)
2. Enzyme activity is regulated—that is, enhanced or suppressed—in many cases by the very molecules whose reactions they catalyze.

The Structure of Enzymes Allows Them to Bind Specific Molecules and Catalyze Specific Reactions

Why are enzymes specific, and how are they regulated? Enzyme function is intimately related to enzyme structure. Enzymes are proteins with complex three-dimensional shapes. Each enzyme has a dimple or groove, called the **active site**, into which reactant molecules, called **substrates**, can enter.

The active site of each enzyme has a distinctive shape and distribution of electrical charge. The active site's shape and charge distribution are complementary to those of its substrate, like the fit between a lock and its key. The enzyme allows only certain molecules to enter. For example, complete digestion of all the proteins in the human diet requires several enzymes, because each enzyme breaks apart only a specific sequence of amino acids.

Some molecules may be able to enter the active site of an enzyme but do not have chemical bonds on which the enzyme can act, so no reaction occurs. Many poisons, including some insecticides, enter the active site of enzymes essential to brain function and never leave. The active site remains plugged up, so no substrate molecule can enter; the reactions that such enzymes normally catalyze cannot occur. Parts of the brain turn off or become wildly hyperactive, in some cases with fatal consequences.

How does an enzyme catalyze a reaction? First, both the shape and the charge of the active site force substrates to enter the enzyme in specific orientations (Fig. 4-7, step 1). Second, when substrates enter the active site, both substrate and active site change shape (step 2). Certain amino acids that form the active site may temporarily bond with atoms of the substrates; or electrical interactions between the active site and substrates may distort the chemical bonds within the substrates. The combination of substrate selectivity, substrate orientation, temporary chemical bonds, and bond distortion promotes the specific chemical reaction catalyzed by a particular enzyme. When the final reaction between the substrates is finished, the product(s) no longer fit properly into the active site and are expelled (step 3). The temporary changes in shape, charge, and bonding patterns within the enzyme revert to their original configuration, and the enzyme is ready to accept another set of substrates (back to step 1).

How do enzymes speed up the rate of chemical reactions? The interactions between substrate and enzyme are like a series of mini-reactions, each with a very low activation energy. Just as carving a staircase into a cliff allows the cliff to be climbed, this series of low activation energy steps allows the overall reaction to surmount its otherwise high activation energy "cliff," so the reaction can proceed at body temperature.

The Activity of Enzymes Is Influenced by Their Environment

Enzymes have very complex, three-dimensional structures that are sensitive to environmental conditions. Each enzyme has evolved to function optimally at a particular pH, temperature, and salt concentration. Some also require the presence of other molecules called *coenzymes*, typically derived from water-soluble vitamins, in order to function.

Most enzymes function optimally at a pH between 6 and 8, the level found in most body fluids and maintained within cells. An exception is the protein-digesting enzyme *pepsin*. Pepsin is converted from an inactive to an active form by the highly acidic (pH 2) conditions within the stomach. At this pH, the excess of hydrogen ions causes hydrogen to attach to certain locations on the enzyme, altering its configuration and exposing the active site. In proteins that function best at neutral pH (7), such interactions would distort the enzyme's structure and destroy its normal function.

Before the advent of refrigeration, foods such as meat were commonly preserved by using concentrated salt solutions. These solutions kill most bacteria, partly by interfering with enzyme function. Salts dissociate into ions, which form bonds with enzymes that interfere with the enzymes' normal three-dimensional structure, destroying enzyme activity. Dill pickles are very well-preserved in a vinegar-salt solution, which combines both salty and acidic conditions. Organisms that live in highly salty environments, as you might predict, have enzymes whose configuration *depends* on the presence of salt ions.

Temperature also affects the rate of enzyme-catalyzed reactions. Because molecules move more rapidly at higher temperatures, their random movements are then more likely to bring them into contact with the active site of an appropriate enzyme. Thus, these reactions are accelerated by higher temperatures and slowed by lower temperatures. Stories abound of people who have fallen into frozen lakes for extended periods and survived. In one true story, a boy who fell through the ice was rescued and survived unharmed after 20 minutes under water. Although normally the brain would die after about 4 minutes without oxygen, the lowering of his body temperature slowed all his metabolic reactions and drastically reduced his need for oxygen. When temperatures rise too high, the bonds that regulate enzyme shape may be broken apart by the excessive molecular motion. Think of the protein in egg white and how completely it is altered in color and texture by cooking. Far lower temperatures than those required to fry an egg can still be too hot for normal enzyme function.

Some enzymes require helper molecules called **coenzymes** to function. These organic molecules bind to

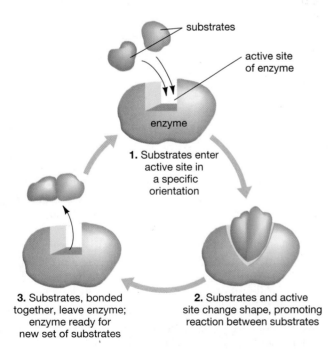

1. Substrates enter active site in a specific orientation

2. Substrates and active site change shape, promoting reaction between substrates

3. Substrates, bonded together, leave enzyme; enzyme ready for new set of substrates

Figure 4-7 The cycle of enzyme–substrate interactions

the enzyme and interact with the substrate molecule. Coenzymes help weaken the bonds of the substrate, allowing it to react with an enzyme. Many water-soluble vitamins (such as the B vitamins) are essential to humans because they are used by the body to synthesize coenzymes.

Cells Regulate the Amount and Activity of Their Enzymes, Thus Precisely Regulating Their Metabolic Reactions

To be useful, the metabolic reactions within cells must be carefully controlled—they must occur at the proper rate and with the proper timing. Cells have evolved the following three ways of regulating enzyme activity:

1. A cell may regulate the synthesis of enzymes to meet its changing needs. Reactions can occur only if the necessary enzymes are available.
2. A cell may synthesize an enzyme in an inactive form and activate it only when needed. An example is the protein-digesting enzyme pepsin described earlier. Pepsin doesn't digest the proteins of the cells that produce it, because it becomes active only in the acidic environment of your stomach.
3. A cell can inhibit an enzyme when adequate amounts of the enzyme's product are available. In **feedback inhibition**, the activity of an enzyme is inhibited by its product or by a subsequent product produced further along in a metabolic pathway (Fig. 4-8). For example, assume that an enzyme catalyzes the conversion of one amino acid to another. The concentration is regulated automatically by feedback inhibition if the end-product amino acid inhibits the enzyme once it reaches sufficiently high concentrations.

The ability of an enzyme to catalyze reactions is controlled by many factors, including the concentration of active enzyme, the concentration of inhibitor molecules, and the concentration of substrates. The interactions among these molecules usually maintain suitable concentrations of both substrates and products within a cell.

Enzymes are precisely regulated and promote only very specific reactions. Thus, the breakdown or synthesis of a molecule within a cell normally occurs in many discrete steps, each catalyzed by a different enzyme (see Fig. 4-5). The reason we are not roasted by the burning of sugar in our bodies is that, unlike the situation in which sugar is set on fire, our cells break down sugar in many small steps. Small, usable quantities of energy are liberated at several of these steps. As you will learn in Chapter 8, the breakdown of sugar by enzymes actually takes about 16 steps. Along the way, energy is inevitably lost as heat, but much is also captured in energy-carrier molecules, as described next.

5 How Is Cellular Energy Carried between Coupled Reactions?

As we saw earlier, cells couple reactions so that the energy released by exergonic reactions is used to drive endergonic reactions. In the case of a runner, the breakdown of a sugar (glucose) releases energy; this energy release is coupled to energy-consuming reactions that cause muscle contraction. But glucose cannot be used directly for muscle contraction. Instead, the energy from glucose must be transferred to an **energy-carrier molecule** for transfer to the muscle protein that uses energy to contract. Energy carriers work something like rechargeable batteries, picking up an energy charge at an exergonic reaction, moving to another location within the cell, and releasing the energy to drive an endergonic reaction. Because energy-carrier molecules are unstable, they are used only for temporary energy transfer, not for long-term energy storage. Muscles store energy in glycogen, which consists of chains of glucose molecules. When the

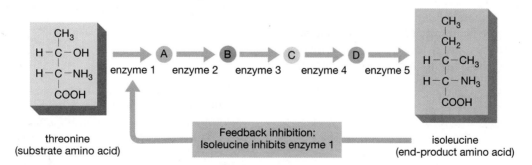

Figure 4-8 Enzyme regulation by feedback inhibition
In this example, the first enzyme in the metabolic pathway that converts threonine (an amino acid substrate) to isoleucine (an amino acid product) is inhibited by high concentrations of isoleucine. If a cell lacks isoleucine, the first enzyme is not inhibited, and the pathway proceeds rapidly. As isoleucine concentrations build up, the isoleucine binds to the first enzyme and gradually shuts down the pathway. When concentrations of isoleucine drop and fewer molecules are around to inhibit the enzyme, the pathway resumes production.

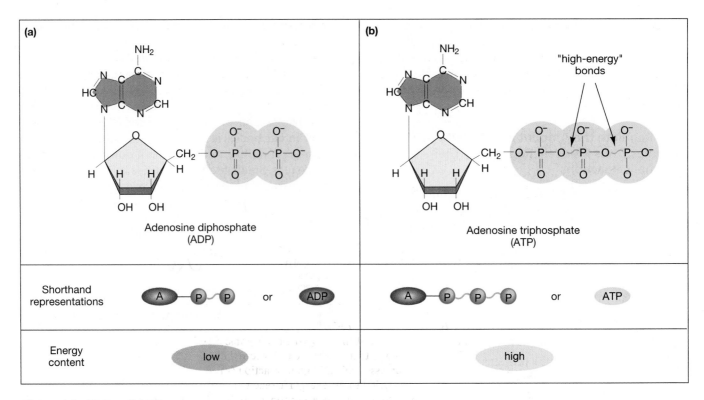

Figure 4-9 ADP and ATP
A phosphate group is added to **(a)** ADP (adenosine diphosphate) to make **(b)** ATP (adenosine triphosphate). In most cases, only the last phosphate group and its high-energy bond are used to carry energy and transfer it to endergonic reactions within a cell. In much of the remainder of the text, we will employ the shorthand representations of ATP and ADP.

marathon begins, however, the glycogen in the body is broken down by enzymes first to glucose, then to carbon dioxide and water, and its energy is captured and transferred to muscle protein molecules by energy-carrier molecules such as ATP.

ATP Is the Principal Energy Carrier in Cells

The most common energy-carrier molecule in cells is **adenosine triphosphate**, or **ATP**. As you learned in Chapter 3, ATP is a nucleotide composed of the nitrogen-containing base adenine, the sugar ribose, and three phosphate groups (Fig. 4-9). Energy released in cells through glucose breakdown is used to drive the synthesis of ATP from **adenosine diphosphate (ADP)** and inorganic phosphate:

ATP synthesis: Energy is stored in ATP

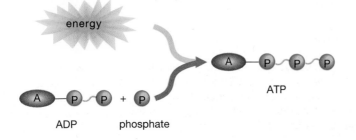

ATP carries this energy to sites in the cell that perform energy-requiring reactions, such as the synthesis of proteins or muscle contraction (see Fig. 4-4b). The ATP is then broken down to form ADP and inorganic phosphate:

ATP breakdown: Energy of ATP is released

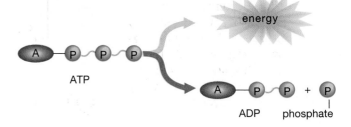

During these energy transfers, heat is given off at each stage, and there is a loss of usable energy (Fig. 4-10).

ATP is admirably suited to carry energy within cells. The bonds joining the last two phosphate groups of ATP to the rest of the molecule (sometimes called *high-energy bonds*) require a large amount of energy to form, so considerable energy can be trapped from exergonic reactions by synthesizing ATP molecules. ATP is also unstable; it readily releases its energy in the presence of appropriate enzymes. Under most circumstances, only the bond joining the last phosphate group (the one joining

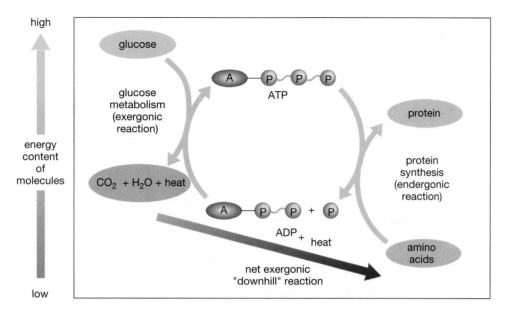

Figure 4-10 *Coupled reactions within living cells*
Exergonic reactions (such as glucose metabolism) drive the endergonic reaction of ATP synthesis from ADP. The ATP molecule moves to a part of the cell where the breakdown of ATP liberates some of this energy to drive an essential endergonic reaction (such as protein synthesis). The ADP and phosphate are recycled to the exergonic reactions; they will be converted back to ATP. The overall reaction is "downhill": More energy is produced by the exergonic reaction than is needed to drive the endergonic reaction. The extra energy is released as heat.

phosphate to ADP to form ATP) carries energy from exergonic to endergonic reactions.

The life span of an ATP molecule is very short, because this energy carrier is continuously formed, broken apart to ADP, and resynthesized. If the molecules of ATP that you use just sitting at your desk all day could be captured (instead of recycled), they would weigh 40 kg—nearly 90 pounds! A marathon runner may use a pound of ATP every minute. (The ADP must be quickly converted back to ATP, or it would be a very brief run.) As you can see, ATP is *not* a long-term energy-storage molecule. More-stable molecules, such as sucrose, glycogen, starch, or fat, store energy in your body for hours, days, or months.

Electron Carriers Also Transport Energy within Cells

Energy can also be transported within a cell by other carrier molecules. In some exergonic reactions, including both glucose metabolism and the light-capturing stage of photosynthesis, some energy is transferred to electrons. These energetic electrons (in some cases, along with hydrogen atoms) are captured by **electron** carriers (Fig. 4-11). Common electron carriers include *nicotinamide adenine dinucleotide* (NAD$^+$) and its relative *flavin adenine dinucleotide* (FAD). Loaded electron carriers then donate the electrons, along with their energy, to other molecules. You will learn more about electron carriers and their role in cellular metabolism in Chapters 7 and 8.

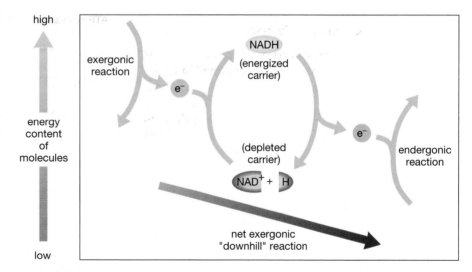

Figure 4-11 *Electron carriers*
Electron-carrier molecules such as nicotinamide adenine dinucleotide (NAD$^+$) pick up electrons generated by exergonic reactions and hold them in high-energy outer electron shells. Hydrogen atoms are often picked up simultaneously. The electron is then deposited, energy and all, with another molecule to drive an endergonic reaction, typically the synthesis of ATP.

Summary of Key Concepts

1) What Is Energy?

Energy is the capacity to do work. Kinetic energy is the energy of movement (light, heat, electricity, movement of large particles). Potential energy is stored energy (chemical energy, positional energy). The flow of energy among atoms and molecules obeys the laws of thermodynamics. The first law of thermodynamics states that, assuming there is no influx of energy, the total amount of energy remains constant, although it may change in form. The second law of thermodynamics states that any use of energy causes a decrease in the quantity of concentrated, useful energy and an increase in the randomness and disorder of matter. Entropy is a measure of disorder within a system.

2) How Does Energy Flow in Chemical Reactions?

Chemical reactions fall into two categories. In exergonic reactions, the product molecules have less energy than do the reactant molecules, so the reaction releases energy. In endergonic reactions, the products have more energy than do the reactants, so the reaction requires an input of energy. Exergonic reactions can occur spontaneously, but all reactions, including exergonic ones, require an initial input of energy (the activation energy) to overcome electrical repulsions between reactant molecules. Exergonic and endergonic reactions may be coupled such that the energy liberated by an exergonic reaction drives the endergonic reaction. Organisms couple exergonic reactions such as light-energy capture or sugar metabolism with endergonic reactions such as the synthesis of organic molecules.

3) How Do Cells Control Their Metabolic Reactions?

Cellular reactions are linked in interconnected sequences called metabolic pathways. The biochemistry of cells is regulated in three ways: first, through the use of protein catalysts called enzymes; second, by coupling endergonic with exergonic reactions; and third, through the use of energy-carrier molecules that transfer energy within cells.

High activation energies slow many reactions, even exergonic ones, to an imperceptible rate under normal environmental conditions. Catalysts lower the activation energy and thereby speed up chemical reactions. Catalysts are not permanently altered during the reaction.

4) What Are Enzymes?

Organisms synthesize protein catalysts called enzymes that promote one or a few specific reactions. The reactants temporarily bind to the active site of the enzyme, making it easier to form the new chemical bonds of the products. Enzyme action is regulated on the cellular level in three ways: (1) by altering the rate of enzyme synthesis, (2) by activating previously inactive enzymes, and (3) by inhibiting enzyme activity through feedback inhibition.

5) How Is Cellular Energy Carried between Coupled Reactions?

Energy released by chemical reactions within a cell is captured and transported about the cell by energy-carrier molecules such as ATP and electron carriers. These molecules are the major means by which cells couple exergonic and endergonic reactions that occur at different places in the cell.

Key Terms

activation energy *p. 58*	coupled reaction *p. 59*	exergonic *p. 57*	metabolism *p. 60*
active site *p. 61*	electron carrier *p. 65*	feedback inhibition *p. 63*	potential energy *p. 56*
adenosine diphosphate (ADP) *p. 64*	endergonic *p. 57*	first law of thermodynamics *p. 56*	product *p. 57*
	energy *p. 56*		reactant *p. 57*
adenosine triphosphate (ATP) *p. 64*	energy-carrier molecule *p. 63*	kinetic energy *p. 56*	second law of thermodynamics *p. 56*
catalyst *p. 61*	entropy *p. 57*	laws of thermodynamics *p. 56*	substrate *p. 61*
coenzyme *p. 62*	enzyme *p. 61*	metabolic pathway *p. 60*	

Thinking Through the Concepts

Multiple Choice

1. *According to the first law of thermodynamics, the total amount of energy in the universe*
   a. is always increasing
   b. is always decreasing
   c. varies up and down
   d. is constant
   e. cannot be determined

2. *What is predicted by the second law of thermodynamics?*
   a. Energy is always decreasing.
   b. Disorder cannot be created or destroyed.
   c. Systems always tend toward greater states of disorder.
   d. All potential energy exists as chemical energy.
   e. all of the above

3. *Which statement about exergonic reactions is true?*
   a. The products have more energy than do the reactants.
   b. The reactants have more energy than do the products.
   c. They will not proceed spontaneously.
   d. Energy input reverses entropy.
   e. none of the above

4. *ATP is important in cells because*
   a. it transfers energy from exergonic reactions to endergonic reactions.
   b. it is assembled into long chains that make up cell membranes.
   c. it acts as an enzyme.
   d. it accelerates diffusion.
   e. all of the above

5. *How does an enzyme increase the speed of a reaction?*
a. by changing an endergonic to an exergonic reaction
b. by adding activation energy
c. by lowering activation energy requirements
d. by decreasing the concentration of reactants
e. by increasing the concentration of products

6. *Which of the following statements about enzymes is (are) true?*
a. They interact with specific reactants (substrates).
b. Their three-dimensional shapes are closely related to their activities.
c. They change the shape of the reactants.
d. They have active sites.
e. all of the above

Applying the Concepts

1. A preview question for ecology: When a brown bear eats a salmon, does the bear acquire all the energy contained in the body of the fish? Why or why not? What implications do you think this answer would have for the relative abundance (by weight) of predators and their prey? Does the second law of thermodynamics help explain the title of the book *Why Big Fierce Animals Are Rare*?

2. Many people in sub-Saharan Africa have experienced the effects of malnutrition and starvation, but the very young are most severely affected. Some individuals suffer permanent disability even if food is provided. How could a lack of food intake interfere with functions of individual cells and tissues? Which tissues are likely to suffer the most irreversible damage?

Group Activity

Shake hands with the student next to you, noting the relative temperature of his or her hand. Now, rub your own hands together and clasp and unclasp them for 30 seconds. Shake hands

1. Explain why organisms do not violate the second law of thermodynamics. What is the ultimate energy source for most forms of life on Earth?

2. Define *metabolism*, and explain how reactions can be coupled to one another.

3. What is activation energy? How do catalysts affect activation energy? How does this change the rate of reactions?

4. Describe some exergonic and endergonic reactions that occur in plants and animals very regularly.

5. Describe the structure and function of enzymes, and compare them to inorganic catalysts. How is enzyme activity regulated?

3. As you learned in Chapter 3, the subunits of virtually all organic molecules are joined by condensation reactions and can be broken apart by hydrolysis reactions. Why, then, does your digestive tract have separate enzymes to digest proteins, fats, and carbohydrates—and in fact several of each type?

4. Suppose someone tried to disprove the existence of evolution with the following argument: "According to evolutionary theory, organisms have increased in complexity through time. However, such evolution of increased biological complexity contradicts the second law of thermodynamics. Therefore, evolution is impossible." If you were to support the theory of evolution, what would be your response to this argument?

again; you should detect a warmer handshake this time. Discuss how the additional heat was generated.

For More Information
Baker, J. J. W., and Allen, G. E. *Matter, Energy, and Life.* 4th ed. Reading, MA: Addison-Wesley, 1981. An excellent introduction to chemical-energy principles for those interested in biology.

Fenn, J. *Engines, Energy, and Entropy.* New York: W. H. Freeman, 1982. Elegantly simple introduction to the laws of thermodynamics and their relationship to everyday life.

Koshland, D. E., Jr. "Protein Shape and Biological Control." *Scientific American*, October 1973. Enzyme function is intimately related to structure. Koshland discusses enzyme specificity and regulation in terms of its protein structure.

Sackheim, G. *Introduction to Chemistry for Biology Students.* Redwood City, CA: Benjamin/Cummings Publishing, 1991. This book is designed specifically for biology students who don't have much chemistry background.

Answers to Multiple-Choice Questions
1. d   2. c   3. b   4. a   5. c   6. e

> "To stay alive, you have to be able to hold out against equilibrium, maintain imbalance, bank against entropy, and you can only transact this business with membranes in our kind of world."

Lewis Thomas in The Lives of a Cell *(1974)*

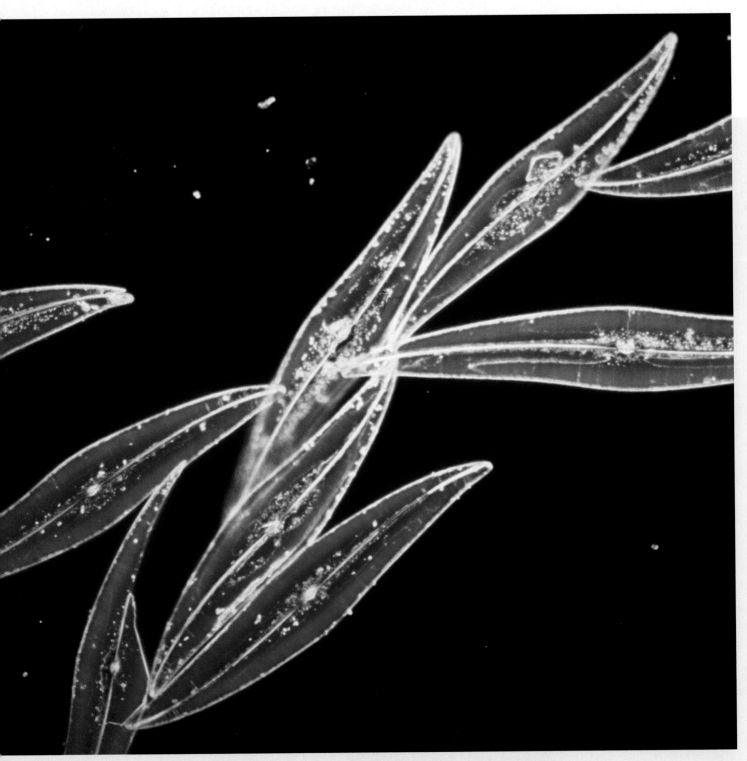

Diatoms, seen under the light microscope. Different species of diatoms are found in both fresh and salt waters. A glassy outer shell and a selectively permeable plasma membrane help cells maintain relatively constant internal conditions.

Cell Membrane Structure and Function 5

Net Watch
On-line resources for this chapter are on the World Wide Web at:
http://www.prenhall.com/audesirk
(click on the Table of Contents link and then select Chapter 5).

Consider the challenge presented to a unicellular protist drifting about in a tide pool, stranded by the receding tide (Fig. 5-1). The protist inhabits an environment whose conditions change quickly and often. As the tide recedes, the sun warms the pool. As the day wears on, some of the water evaporates, boosting the concentration of dissolved salts. A sudden rain or the returning tide dilutes the water again. When night falls, the water temperature drops sharply. Just surviving requires a protective barrier between the inside and outside of the cell.

Merely withstanding environmental changes is not enough, however. The cell must absorb sunlight for photosynthesis or engulf other organisms to obtain the nutrients it needs to make new cellular components. To

Figure 5-1 A challenging environment
A tide pool contains numerous life-forms, including the photosynthetic single-celled diatom in the chapter-opening photograph. All cells face formidable challenges in coping with their external environment. The membrane that both isolates the cell from its environment and regulates interactions with that environment is the focus of this chapter.

maintain a consistent internal environment in which its components function best, the cell must shuttle salt, water, and other molecules in and out. It must excrete waste products. It must detect chemical clues to the location of food or to healthy and unhealthy environmental conditions. Finally, if it reproduces sexually, it must communicate with at least one other cell of its own species sometime during its life.

A membrane as thick as only two lipid molecules stands between the delicate inner workings of the cell and the extracellular environment. This membrane selectively permits the absorption and release of chemicals, sometimes using cellular energy. It allows molecules outside the cell to initiate chemical reactions inside the cell. And it may deform to engulf food or enable the cell to move. All these processes rely on a variety of proteins that are embedded in the lipid membrane or attached to its surfaces.

Some single-celled organisms, such as the amoeba, survive with no barrier between life and the environment except a cellular membrane. Cells of plants, bacteria, and many protists, such as diatoms, have various types of walls and other protective coverings outside the membrane that lend mechanical support. Cells of multicellular animals live surrounded by other cells and bathed in a bloodlike fluid. But all these cells possess very similar cellular membranes and carry on many of the same processes across those membranes.

In this chapter, we explore the membrane structures that make defense, exchange of materials with the environment, and communication between cells possible.

1) How Is the Structure of a Membrane Related to Its Function?

www

The Plasma Membrane Protects and Isolates the Cell While Allowing Extensive Communication with Its Surroundings

In Chapter 1, we defined the *cell* as the smallest unit of life. Each cell is surrounded by a thin **plasma membrane**, which can be considered a gatekeeper, allowing only specific substances in or out and passing chemical messages from the external environment to the cell's interior. As gatekeeper, the plasma membrane must perform three general functions:

1. Isolate the cell's contents from the external environment.
2. Regulate the exchange of essential substances between the cell's contents and the external environment.
3. Communicate with other cells.

These are formidable tasks for a structure so thin that 10,000 plasma membranes stacked atop one another would scarcely equal the thickness of this page. The key to membrane function lies in membrane structure. Membranes are not simply homogeneous sheets: They are complex, heterogeneous structures with different parts

performing very distinct functions, and they change dynamically in response to their surroundings.

Whereas the plasma membrane surrounds the cell, most cells have internal membranes as well. These internal membranes form compartments in which specialized biochemical activities can occur. All the membranes of a cell have a similar basic structure: proteins floating in a double layer of lipids. Lipids are responsible for the isolating function of membranes, whereas proteins regulate the exchange of substances and communication with the environment. Although this chapter focuses on the plasma membrane, the information presented here applies also to other cellular membranes.

Membranes Are "Fluid Mosaics" in Which Proteins Move within Layers of Lipids

The **fluid mosaic model** of cellular membranes was developed by cell biologists S. J. Singer and G. L. Nicolson in 1972. According to this model, a membrane, when viewed from above, looks something like a lumpy, constantly shifting mosaic of tiles (Fig. 5-2). A double layer of phospholipids forms a viscous, fluid "grout" for the mosaic; an assortment of proteins are the "tiles," often sliding within the phospholipid layers. Thus, the overall structure slowly changes over time, though the components remain relatively constant. As strange as this model may seem, it captures something of the dynamic quality of real membranes. With this model in mind, let's look more closely at the structure of membranes.

The Phospholipid Bilayer Is the Fluid Portion of the Membrane

As you learned in Chapter 3, a phospholipid consists of two very different parts: (1) a polar, hydrophilic head (attracted to water) and (2) a pair of nonpolar, hydrophobic tails (repelled by water):

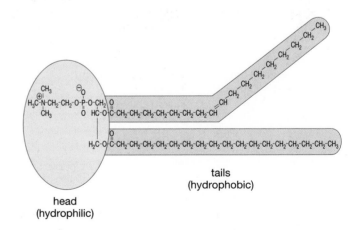

tails
(hydrophobic)

head
(hydrophilic)

All cells are surrounded by water, whether it is the tide pool that strands our protist or the *extracellular fluid* that bathes animal cells. The **cytoplasm** consists of all of a cell's internal contents (including all organelles but the nucleus, in eukaryotes); it is mostly water. Plasma membranes

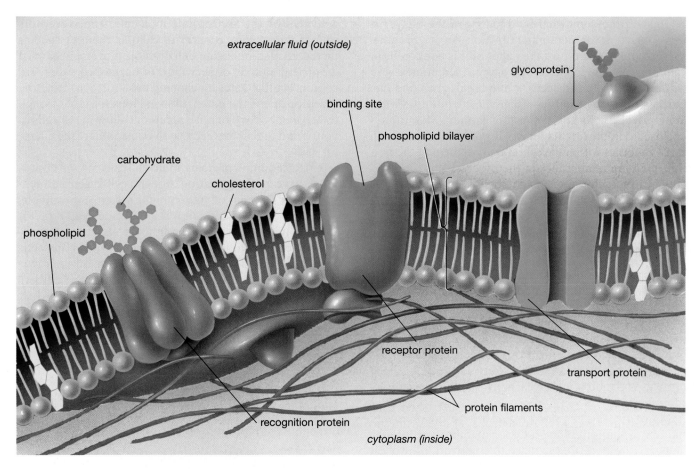

Figure 5-2 *The plasma membrane is a fluid mosaic*
The plasma membrane is a bilayer of phospholipids in which various proteins are embedded. Many proteins have carbohydrates attached to them, forming glycoproteins. The wide variety of membrane proteins fall mostly into three categories: transport proteins, receptor proteins, and recognition proteins.

separate a watery cytoplasm from a watery external environment or surround watery compartments within the cell. Under these conditions, phospholipids spontaneously arrange themselves in a double layer called a **phospholipid bilayer**, in which the hydrophilic heads form the outer borders and the hydrophobic tails "hide" inside:

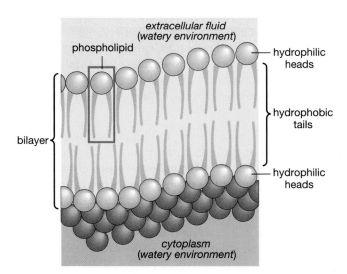

Hydrogen bonds can form between water and the phospholipid heads, so the heads face the cytoplasm or the extracellular fluid. Hydrophobic interactions (see Chapter 2) cause the phospholipid tails to hide inside the bilayer. Because individual phospholipid molecules are not bonded to one another, this double layer is quite fluid; individual phospholipids move about easily within each layer.

Most biological molecules, such as salts, amino acids, and sugars, are polar and water soluble. In fact, most substances that contact a cell are water soluble. They therefore cannot pass easily through the nonpolar, hydrophobic fatty acid tails of the phospholipid bilayer. The bilayer acts as a barrier to the entrance of these molecules. It is largely responsible for the first of the three membrane functions listed earlier—isolating the cell's contents from the external environment.

The phospholipid bilayer of membranes also contains cholesterol. Some cellular membranes have just a few cholesterol molecules; others have as many cholesterol molecules as they do phospholipids. Cholesterol affects membrane structure and function in several ways: It makes the bilayer stronger, more flexible but less fluid, and less permeable to water-soluble substances such as ions or monosaccharides.

The flexible, somewhat fluid nature of the bilayer is very important for membrane function. As you breathe, move your eyes, and turn the pages of this book, cells in your body change shape. If their plasma membranes were stiff instead of flexible, cells would break open and die. Further, as you will learn in Chapter 6, membranes within eukaryotic cells are in constant motion. Membrane-enclosed compartments ferry substances into the cell, carry materials within the cell, and expel them to the outside, merging membranes in the process. This flow and merger of membranes is made possible by the fluid nature of the lipid bilayer.

A Mosaic of Proteins Is Embedded in the Membrane

Various proteins are embedded within or attached to the surface of a membrane's phospholipid bilayer. Collectively, these proteins regulate the movement of substances through the membrane and communicate with the environment. Many of the proteins in plasma membranes have carbohydrate groups attached to them, especially to the parts that stick outside the cell. These proteins and their attached carbohydrates are called **glycoproteins**.

Many proteins can move about within the relatively fluid phospholipid bilayer. Other membrane proteins, however, are anchored in place to a network of protein filaments within the cytoplasm. The attachments between plasma membrane proteins and the underlying protein filaments produce the characteristic shapes of animal cells, from the dimpled discs of red blood cells to the elaborate branching of nerve cells.

There are three major categories of membrane proteins, each of which serves a different function: *transport proteins, receptor proteins,* and *recognition proteins* (see Fig. 5-2).

(1) **Transport proteins** regulate the movement of water-soluble molecules through the plasma membrane. Some, called **channel proteins**, form pores or channels that allow small water-soluble molecules to pass through the membrane (see Fig. 5-2). Every plasma membrane bears a large assortment of protein channels, each lined with specific amino acids that allow certain molecules, normally ions such as potassium (K^+), sodium (Na^+), and calcium (Ca^{2+}), to pass through. Other transport proteins, called **carrier proteins**, have binding sites, much like the active sites of enzymes, that can grab onto specific molecules on one side of the membrane. The transport protein then changes shape, in some cases through the use of cellular energy, and moves the molecule across the membrane. We will discuss the mechanisms whereby transport proteins move molecules across the membrane in the next section.

(2) **Receptor proteins** trigger cellular responses when specific molecules in the extracellular fluid, such as hormones or nutrients, bind to them. Most cells bear dozens of types of receptors on their plasma membranes. When activated by the appropriate molecule, some receptors set off elaborate sequences of cellular changes, such as increased metabolic rate, cell division, movement toward a nutrient source, or secretion of hormones. Other receptors act like gates on channel proteins; activating the receptor opens the gates, allowing ions to flow through the channels. For example, receptors allow nerve cells in your brain to communicate with one another (see Chapter 33).

(3) **Recognition proteins**, many of which are glycoproteins, serve as identification tags and cell-surface attachment sites. The cells of your immune system, for example, recognize a bacterium as a foreign invader and target it for destruction. These same immune cells ignore the trillions of cells in your own body because your body cells have different identification glycoproteins on their surfaces. During development, the growth of nerve fibers from your spinal cord down to the muscles in your feet is guided by attachments between recognition proteins on the nerve cell and the other cells it traverses on its way to the muscle.

As you can see from these brief descriptions, membrane proteins are largely responsible for moving substances across the membrane and for communicating with other cells.

2 How Are Substances Transported across Membranes?

Molecules in Fluids Move in Response to Gradients

Because the plasma membrane separates the fluid in the cell's cytoplasm from its fluid extracellular environment, let's begin our study of membrane transport with a brief look at the characteristics of fluids. We must start with a few definitions:

1. A **fluid** is a liquid or a gas—that is, any substance that can move or change shape in response to external forces without breaking apart.
2. The **concentration** of molecules in a fluid is the number of molecules in a given unit of volume.
3. A **gradient** is a physical difference between two regions of space such that molecules tend to move from one region to the other. Cells frequently encounter gradients of concentration, pressure, and electrical charge.

The individual molecules in a fluid move continuously, bouncing off one another in random directions. These random movements enable molecules (or ions) to move, on average, from regions of high concentration to regions of low concentration. For example, when sugar dissolves in coffee or perfume molecules move from the open bottle into the air, these substances are moving in response to the **concentration gradient**, a difference in concentration between one region and another. By analogy with

Table 5-1 Transport across Membranes

Passive transport	Movement of substances across a membrane, going down a gradient of concentration, pressure, or electrical charge. Does not require the cell to expend energy.
Simple diffusion	Diffusion of water, dissolved gases, or lipid-soluble molecules through the phospholipid bilayer of a membrane.
Facilitated diffusion	Diffusion of (normally water-soluble) molecules through a channel or carrier protein.
Osmosis	Diffusion of water across a differentially permeable membrane—that is, a membrane that is more permeable to water than to dissolved molecules.
Energy-requiring transport	Movement of substances across a membrane, usually against a concentration gradient, using cellular energy.
Active transport	Movement of individual small molecules or ions through membrane-spanning proteins, using cellular energy, normally ATP.
Endocytosis	Movement of large particles, including large molecules or entire microorganisms, into a cell by engulfing extracellular material, as the plasma membrane forms membrane-bound sacs that enter the cytoplasm.
Exocytosis	Movement of materials out of a cell by enclosing the material in a membranous sac that moves to the cell surface, fuses with the plasma membrane, and opens to the outside, allowing its contents to diffuse away.

gravity, we will refer to such movements as going "down" the gradient.

Movement across Membranes Occurs by Both Passive and Active Transport

The cytoplasm of a cell is very different from the extracellular fluid. Thus, significant gradients exist across the plasma membrane. In its role as gatekeeper of the cell, the plasma membrane provides for two types of movement: (1) *passive transport* and (2) *active transport* (Table 5-1).

During **passive transport**, substances move into or out of cells down concentration gradients. For example, if the tide pool has a higher concentration of oxygen than the body of the protist does, oxygen will tend to move into the cell from the surrounding water. Carbon dioxide, higher in the body of the protist, will tend to diffuse out into the tide pool. This movement by itself requires no expenditure of energy. The gradients provide the potential energy that drives the movement and controls the direction of movement, into or out of the cell. The lipids and protein pores of the plasma membrane regulate which molecules can cross, but they do not influence the direction of movement.

During **active transport**, the cell uses energy to move substances *against* a concentration gradient. In this case, transport proteins do control the direction of movement. A helpful analogy for understanding the difference between passive and active transports is to consider what happens when you ride a bike. If you don't pedal, you can go only downhill, as in passive transport. However, if you put enough energy into pedaling, you can go uphill as well, as in active transport.

In Passive Transport, Substances Move down Their Concentration Gradients

Most materials that move passively across plasma membranes do so in response to concentration gradients. **Diffusion** is the net movement of molecules in a fluid down

a concentration gradient, from regions of high concentration to regions of low concentration. Diffusion can occur from one part of a fluid to another or across a membrane separating two fluid compartments. We will first examine the simplest case, the diffusion of molecules within a fluid, with no membrane.

To see how concentration gradients cause molecules to diffuse from one place to another, let's consider what happens when we place a drop of dye in a glass of water. (You might want to try this yourself, using a drop of food coloring.) With time, the drop will seem to spread out and become paler, until eventually, even without stirring, the entire glass of water will be uniformly faintly colored. Why?

A drop of dye consists of many individual dye molecules moving in all directions (Fig. 5-3). Some molecules of dye, simply owing to random motion, spread out into the water. Thus, the net movement of dye molecules is from the region of high dye concentration (the drop) to the region of low dye concentration (the water). The same thing happens with water molecules. Random motion causes some water molecules to enter the dye droplet, and the net movement of water is from the high water concentration outside the drop into the low water concentration inside the drop.

To the observer, dye molecules that move beyond the original border of the drop make the drop appear larger; water molecules that invade the drop dilute the dye, making the drop paler. At first, there is a very steep concentration gradient, and the dye diffuses rapidly. As the concentration differences lessen, the dye diffuses more and more slowly. In other words, the greater the concentration gradient, the faster the rate of diffusion.

However, as long as the concentration of dye within the expanding drop is greater than the concentration of dye in the rest of the glass, the net movement of dye will be from drop to water, until the dye becomes uniformly dispersed in the water. Then, with no concentration gradient of either dye or water, diffusion stops. Individual

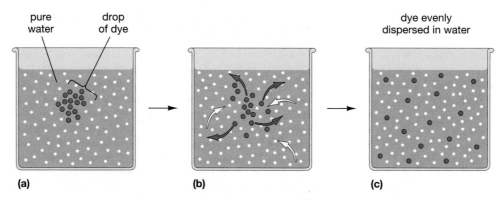

Figure 5-3 Diffusion of a dye in water
(a) A drop of pure dye (red dots) is placed in a glass of pure water (white dots in blue background). *(b)* Although individual molecules move at random, concentration gradients cause dye molecules to diffuse into the water and water molecules into the drop of dye. *(c)* Eventually, both dye and water are evenly dispersed. Individual molecules still move, but there is no longer any concentration gradient and therefore no further diffusion.

molecules still move about, but no changes occur in concentration of either water or dye.

As you can appreciate from this simple experiment, diffusion cannot move molecules rapidly over long distances. Although the drop of dye immediately begins to diffuse into the water, uniform dispersion may take many minutes. As you will learn in Chapter 6, the slow rate of diffusion over long distances is one of the reasons cells are small.

Summing Up
The Principles of Diffusion

1. Diffusion is the net movement of molecules down a gradient from high to low concentration.
2. The greater the concentration gradient, the faster the rate of diffusion.
3. If no other processes intervene, diffusion will continue until the concentration gradient is eliminated.
4. Diffusion cannot move molecules rapidly over long distances.

Molecules Diffuse across Membranes down Their Concentration Gradients

Many molecules cross plasma membranes by diffusion, driven by differences between their concentration in the cytoplasm and in the external environment. Depending on their properties, molecules cross the plasma membrane at different locations and at different rates. Therefore, plasma membranes are said to be **differentially permeable**—that is, they allow some molecules to pass through, or *permeate*, but prevent other molecules from passing. A barrier that prevents the passage of all molecules is said to be *impermeable*.

Some Molecules Move across Membranes by Simple Diffusion

Water, dissolved gases such as oxygen and carbon dioxide, and lipid-soluble molecules such as ethyl alcohol and vitamin A easily diffuse across the phospholipid bilayer. This process is called **simple diffusion** (Fig. 5-4a). Generally, the rate of simple diffusion is a function of the concentration gradient across the membrane, the size of the molecule, and how easily it dissolves in lipids (its lipid *solubility*): Large concentration gradients, small molecule size, and high lipid solubility all increase the rate of simple diffusion.

Other Molecules Cross the Membrane by Facilitated Diffusion, with the Help of Membrane Transport Proteins

Most water-soluble molecules, such as ions (K^+, Na^+, Ca^{2+}), amino acids, and monosaccharides, cannot move through the phospholipid bilayer on their own. These molecules can diffuse across only with the aid of one of two types of transport proteins: channel proteins and carrier proteins. This process is called **facilitated diffusion**.

Channel proteins form permanent pores, or channels, in the lipid bilayer through which certain ions can cross the membrane (Fig. 5-4b). Most channel proteins have a specific interior diameter and distribution of electrical charges that allow only particular ions to pass through. Nerve cells, for example, have separate channels for sodium ions, potassium ions, and calcium ions.

Carrier proteins bind specific molecules, such as particular amino acids, sugars, or small proteins, from the cytoplasm or extracellular fluid. Binding triggers a change in the shape of the carrier that allows the molecules to pass through the protein and across the plasma membrane. Facilitated diffusion occurs through carrier pro-

(a) simple diffusion

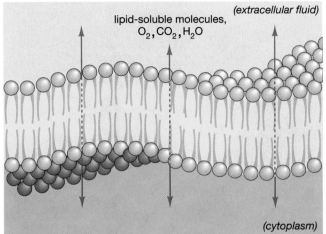

lipid-soluble molecules, O_2, CO_2, H_2O
(extracellular fluid)

(cytoplasm)

(b) facilitated diffusion through a channel

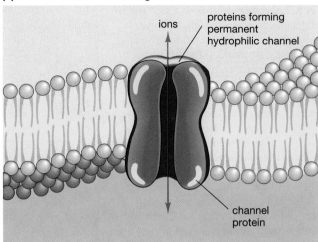

ions

proteins forming permanent hydrophilic channel

channel protein

(c) facilitated diffusion through a carrier

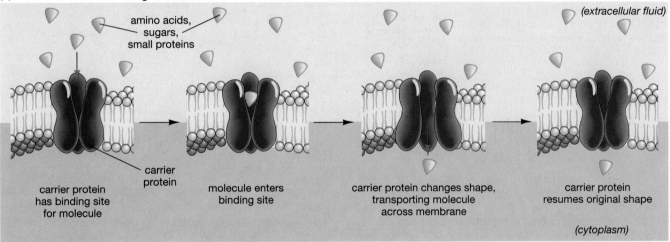

amino acids, sugars, small proteins

(extracellular fluid)

carrier protein

carrier protein has binding site for molecule

molecule enters binding site

carrier protein changes shape, transporting molecule across membrane

carrier protein resumes original shape

(cytoplasm)

Figure 5-4 Diffusion through the plasma membrane
(a) Simple diffusion through the phospholipid bilayer: Gases such as oxygen and carbon dioxide and lipid-soluble molecules can diffuse directly through the phospholipids. *(b)* Facilitated diffusion through a channel: Some molecules cannot pass through the bilayer on their own. Protein channels (pores) allow some water-soluble molecules, principally ions, to enter or exit the cell. *(c)* Facilitated diffusion through a carrier: Carrier proteins may bind a specific molecule and, as a result, change their own shape, passing the molecule through the middle of the protein to the other side of the membrane.

teins that do not use cellular energy. These carrier proteins move molecules only down their concentration gradients (Fig. 5-4c).

Because they must rely on transport proteins, molecules that cross the membrane by facilitated diffusion usually do so more slowly than do those that cross by simple diffusion through the lipid bilayer.

Osmosis Is the Diffusion of Water across Membranes

Water also diffuses from regions of high water concentration to regions of low water concentration. However, the diffusion of water across differentially permeable membranes has such dramatic and important effects on cells that we refer to it by a special name: **osmosis.**

What do we mean when we describe a solution as having a "high water concentration" or a "low water concentration"? The answer is simple: Pure water has the highest water concentration. Any substance added to pure water displaces some of the water molecules; the resulting solution will have a lower water content than pure water. Dissolved substances may form weak bonds with some of the water molecules, making these water molecules unavailable to diffuse across the membrane (Fig. 5-5a). The higher the concentration of dissolved substances, the lower the concentration of water. A very simple differentially permeable membrane might have pores just large enough for water to pass through but small enough to be impermeable to sugar molecules.

Figure 5-5 Osmosis
(a) Membrane pores allow "free" water molecules to pass through, but sugar molecules are too large. "Bound" water molecules, attracted to the sugars by hydrogen bonds, are thus also kept from passing through the pore. *(b)* A membrane is differentially permeable to free water molecules (white dots) but not to larger molecules such as sugar (yellow hexagons) or water molecules held to the sugars by hydrogen bonds. If a bag made of such a membrane is filled with a sugar solution and suspended in pure water, free water molecules will diffuse down their concentration gradient from the high concentration of water outside the bag to the lower concentration of water inside the bag. The bag swells up as water enters. If the bag is weak enough, the increasing water pressure will cause it to burst.

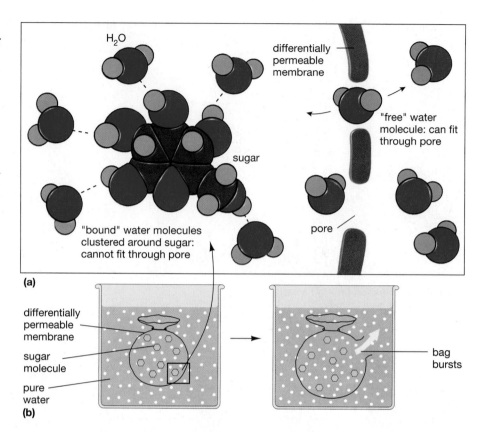

Consider a bag made of a differentially permeable membrane. What would happen if we were to place a sugar solution in the bag, then immerse the sealed bag in pure water? The principles of osmosis tell us that the bag will swell. If it is weak enough, it will burst (Fig. 5-5b).

Summing Up
The Principles of Osmosis

1. Osmosis is the diffusion of water across a differentially permeable membrane.
2. Water moves across a membrane from a high concentration of free water molecules to a low concentration of free water molecules (or from high pressure to low pressure).
3. Dissolved substances will reduce the concentration of free water molecules in a solution.

Osmosis across the Plasma Membrane Plays an Important Role in the Lives of Cells

Most plasma membranes are highly permeable to water. Because all cells contain dissolved salts, proteins, sugars, and so on, the flow of water across the plasma membrane depends on the concentration of water in the liquid that bathes the cells. The extracellular fluids of animals are usually **isotonic** ("having the same strength") to the insides of the body cells; that is, the concentration of water

inside is the same as that outside the cells, so there is no net tendency for water to enter or leave the cells (Fig. 5-6a). Note that the *types of dissolved particles* are seldom the same inside and outside the cells, but the *total concentration of all dissolved particles* is equal, with the result that the water concentration inside is equal to that outside the cells.

If red blood cells, for example, are taken out of the body and immersed in salt solutions of varying concentrations, the effects of the differential permeability of the plasma membrane to water and dissolved particles become dramatically apparent. If the solution has a higher salt concentration than the cytoplasm (that is, if the solution has a lower water concentration), water will leave the cells by osmosis. The cells will shrivel up until the concentrations of water inside and outside become equal (Fig. 5-6b). Solutions that have a higher concentration of dissolved particles than does a cell's cytoplasm, and thus cause water to leave the cell by osmosis, are termed **hypertonic** ("having greater strength").

Conversely, if the solution has little or no salt, water will enter the cells, causing them to swell (Fig. 5-6c). If red blood cells are placed in pure water, they will burst. Solutions that have a lower concentration of dissolved particles than a cell's cytoplasm does, and thus cause water to enter the cell by osmosis, are called **hypotonic** ("having lesser strength"). This process is why your fingers wrinkle up after a long bath. It may seem like your fingers are shrinking, but they're not. Instead, water is

(a) Isotonic solution

equal movement of water
into and out of cells

(b) Hypertonic solution

net water movement
out of cells

(c) Hypotonic solution

net water movement
into cells

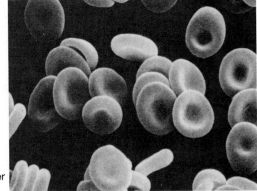

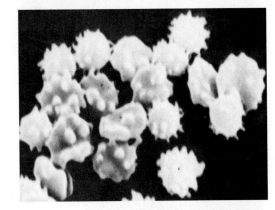

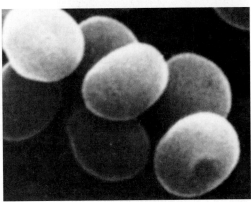

⊢————⊣
10 micrometers

Figure 5-6 The effects of osmosis
Red blood cells are normally suspended in the fluid environment of the blood and cannot regulate water flow across their plasma membranes. **(a)** If red blood cells are immersed in an isotonic salt solution, which has the same concentration of dissolved substances as the blood cells do, there is no net movement of water across the plasma membrane. The red blood cells keep their characteristic dimpled disk shape. **(b)** A hypertonic solution, with too much salt, causes water to leave the cells, shriveling them up. **(c)** A hypotonic solution, with less salt than is in the cells, causes water to enter, and the cells swell.

diffusing into the outer skin cells of your fingers, swelling them more rapidly than the cells underneath and causing the wrinkling.

The swelling caused by osmosis can have considerable effects on cells. As we shall see in Chapter 6, protists such as *Paramecium* that live in fresh water have special structures called *contractile vacuoles* to eliminate the water that continuously leaks in. In contrast, water entry into *central vacuoles* of plant cells helps support the plant. Osmosis across plasma membranes is crucial to the functioning of many biological systems, including water uptake by plant roots (Chapter 23), absorption of dietary water from your intestines (Chapter 29), and reabsorption of water and minerals in your kidneys (Chapter 30).

Active Transport Uses Energy to Move Substances against Their Concentration Gradients

All cells need to move some materials "uphill" across their plasma membranes, against concentration gradients. For example, every cell requires some nutrients that are less concentrated in the environment than in the cell's cytoplasm. Therefore, diffusion would cause the cell to lose, not gain, these nutrients. Other substances, such as sodium and calcium ions in your brain cells, must be maintained at much lower concentrations inside the cells than in the extracellular fluid. When these ions diffuse into the cells, they must be pumped out again against their concentration gradients.

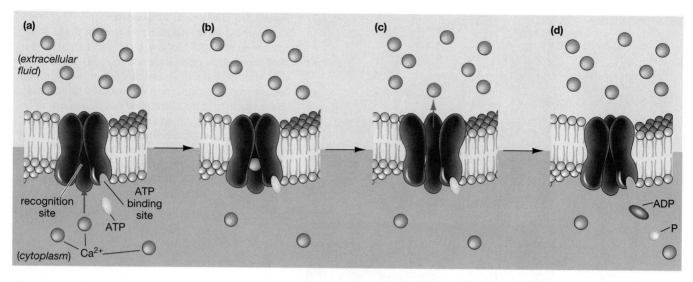

(a) (extracellular fluid) recognition site ATP binding site ATP Ca²⁺ (cytoplasm)

(b)

(c)

(d) ADP P

Figure 5-7 Active transport
Active transport uses cellular energy to move molecules across the plasma membrane, often against a concentration gradient. **(a)** A transport protein (blue) has an ATP binding site and a recognition site for the molecules to be transported, calcium ions (Ca²⁺) in this case. **(b)** The transport protein binds ATP and Ca²⁺. **(c)** Energy from ATP changes the shape of the transport protein and moves the ion across the membrane. **(d)** The carrier releases the ion and the remnants of the ATP (ADP and P) and resumes its original configuration.

In active transport, membrane proteins use cellular energy to move individual molecules across the plasma membrane, usually against their concentration gradient (Fig. 5-7). Active-transport proteins span the membrane and have two active sites. One active site (which may be either on the face of the plasma membrane in contact with the cytoplasm or on the face in contact with the extracellular fluid, depending on the transport protein) recognizes a particular molecule, say a calcium ion, and binds it. The second site (always on the inside of the membrane) binds an energy-carrier molecule, normally adenosine triphosphate (ATP). The ATP donates energy to the protein, causing it to change shape and move the calcium ion across the membrane. Active-transport proteins are often called *pumps*, in analogy to water pumps, because they use energy to move molecules "uphill" against a concentration gradient. We will see that plasma membrane pumps are vital in mineral uptake by plants (Chapter 23), mineral absorption in your intestines (Chapter 29), and maintaining concentration gradients essential to nerve cell functioning (Chapter 33).

Cells Engulf Particles or Fluids by Endocytosis

Many cells acquire or expel particles or substances that are too large to diffuse across a membrane regardless of concentration gradients. Cells have evolved several processes that use cellular energy to move materials into or out of the cell. Cells can acquire fluids or particles, especially large proteins or entire microorganisms such as bacteria, by a process called **endocytosis** (Greek for "into the cell"). During endocytosis, the plasma membrane engulfs the fluid droplet or particle and pinches off a membranous sac called a **vesicle**, with the fluid or particle inside, into the cytoplasm (Fig. 5-8). We can distinguish three types of endocytosis on the basis of the size of the particle acquired and the method of acquisition: (1) *pinocytosis*, (2) *receptor-mediated endocytosis*, and (3) *phagocytosis*.

Pinocytosis Moves Liquids into the Cell

In **pinocytosis** ("cell drinking"), a very small patch of plasma membrane dimples inward as it surrounds extracellular fluid and buds off into the cytoplasm as a tiny vesicle (Fig. 5-8a). Pinocytosis moves a droplet of extracellular fluid, contained within the dimpling patch of membrane, into the cell. Therefore, the cell acquires materials in the same concentration as in the extracellular fluid.

Receptor-Mediated Endocytosis Moves Specific Molecules into the Cell

Cells can take up certain molecules (cholesterol, for example) most efficiently by a process known as **receptor-mediated endocytosis** (Fig. 5-8b). Most plasma membranes bear many receptor proteins on their outside surfaces, each with a binding site for a particular nutrient molecule. In most cases, these receptors move through the phospholipid bilayer and accumulate in depressions of the plasma membrane called *coated pits* (Fig. 5-9). If

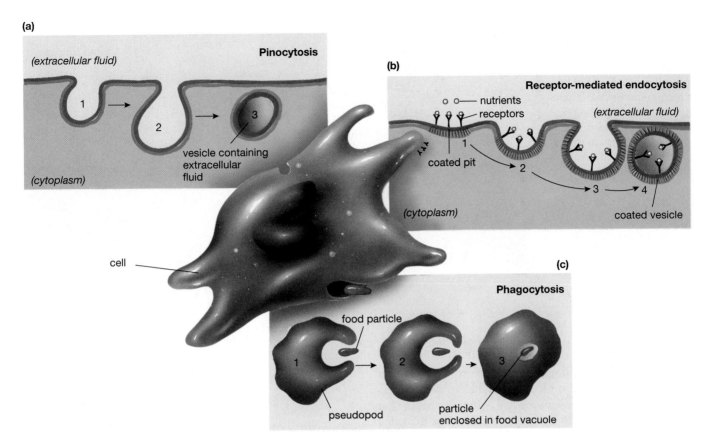

Figure 5-8 Three types of endocytosis
(a) Pinocytosis: A dimple in the plasma membrane deepens and eventually pinches off as a fluid-filled vesicle, which contains a random sampling of the extracellular fluid. **(b)** Receptor-mediated endocytosis: Receptor proteins selectively bind molecules (for example, nutrients) in the extracellular fluid. The receptors migrate along the phospholipid bilayer of the plasma membrane to dimpling sites (coated pit). The membrane dimples inward, carrying the receptor-captured molecule complexes with it. The end of the coated pit buds off a coated vesicle into the cell's cytoplasm. The vesicle contains both extracellular fluid and a high concentration of the molecules that bind to the receptors. **(c)** Phagocytosis: Extensions of the plasma membrane, called pseudopodia, encircle an extracellular particle (for example, food). The ends of the pseudopodia fuse, forming a large vesicle (a food vacuole) containing the engulfed particle.

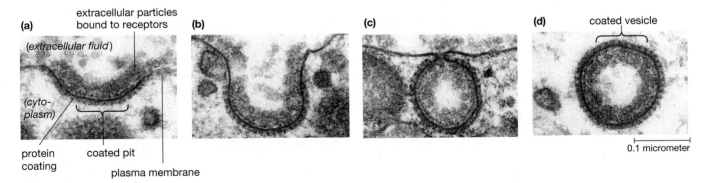

Figure 5-9 Receptor-mediated endocytosis
These electron micrographs illustrate the sequence of events in receptor-mediated endocytosis. **(a)** This type of endocytosis begins with a shallow depression in the plasma membrane, coated on the inside with a protein (dark, fuzzy substance in the micrographs) and bearing receptor proteins on the outside (not visible). **(b,c)** The pit deepens and **(d)** eventually pinches off as a coated vesicle. The protein coating is eventually recycled back to the plasma membrane.

Figure 5-10 Exocytosis
Exocytosis is functionally the reverse of endocytosis. The material to be ejected from the cell is encapsulated into a membrane-bound vesicle that moves to the plasma membrane and fuses with it. As the vesicle opens to the outside, its contents leave by diffusion.

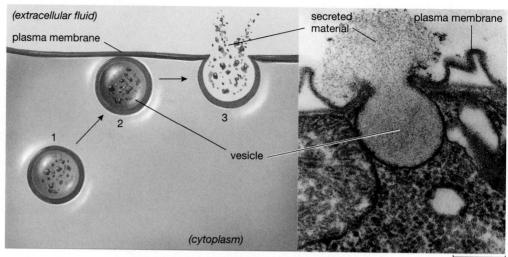

(extracellular fluid)
plasma membrane

secreted material

plasma membrane

vesicle

(cytoplasm)

0.2 micrometer

the right molecule contacts a receptor protein in one of these coated pits, that molecule attaches to the binding site. The coated pit deepens into a U-shaped pocket that eventually pinches off into the cytoplasm as a *coated vesicle*. Both the receptor–nutrient complex and a bit of extracellular fluid move into the cell in the coated vesicle.

Phagocytosis Moves Large Particles into the Cell

Phagocytosis (which means "cell eating") is used to pick up large particles, including whole microorganisms (Fig. 5-8c). When the freshwater protist *Amoeba*, for example, senses a tasty *Paramecium*, *Amoeba* extends parts of its surface membrane. These membrane extensions are called **pseudopodia** (Latin for "false foot"; singular, **pseudopod**). The pseudopodia surround the luckless *Paramecium*; their ends fuse, and the prey is carried into the interior of the *Amoeba* inside a vesicle called a *food vacuole* for digestion. Like *Amoeba*, white blood cells also use phagocytosis and intracellular digestion to engulf and destroy bacteria that have invaded your body.

Exocytosis Moves Material Out of the Cell

Cells often use the reverse of endocytosis, a process called **exocytosis** (Greek for "out of the cell"), to dispose of unwanted materials, such as the waste products of digestion, or to secrete materials, such as hormones, into the extracellular fluid (Fig. 5-10). During exocytosis, a membrane-enclosed vesicle carrying material to be expelled moves to the cell surface, where the vesicle's membrane fuses with the cell's plasma membrane. The vesicle then opens to the extracellular fluid, and its contents diffuse out.

3 How Are Cell Surfaces Specialized?

A Variety of Junctions Allow Connection and Communication among Cells

In multicellular organisms, plasma membranes hold together clusters of cells and provide avenues through which cells communicate with their neighbors. Depending on the organism and the cell type, four types of connection may occur between cells: (1) *desmosomes*, (2) *tight junctions*, (3) *gap junctions*, and (4) *plasmodesmata*.

Desmosomes Attach Cells Together

Animals, as you know, tend to be flexible, mobile organisms. Many of an animal's tissues are stretched, compressed, and bent as the animal moves. If the skin, intestines, stomach, urinary bladder, and other organs are not to tear apart under the stresses of movement, their cells must adhere firmly to one another. Such animal tissues have junctions called **desmosomes,** which hold adjacent cells together (Fig. 5-11a). In a desmosome, the membranes of adjacent cells are glued together by proteins and carbohydrates. Protein filaments attached to the insides of the desmosomes extend into the interior of each cell, further strengthening the attachment.

Tight Junctions Make the Cell Leakproof

The animal body contains many tubes or sacs that must hold their contents without leaking; a leaky urinary bladder would spell disaster for the rest of the body. Spaces between the cells that line such tubes or sacs are sealed with strands of protein to form **tight junctions** (Fig. 5-11b). The membranes of adjacent cells nearly fuse along a series of ridges, effectively forming leakproof gaskets between cells. Continuous tight junctions sealing each cell to its neighbors prevent molecules from escaping between cells.

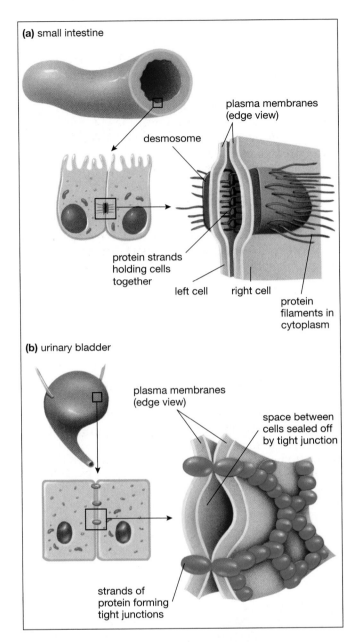

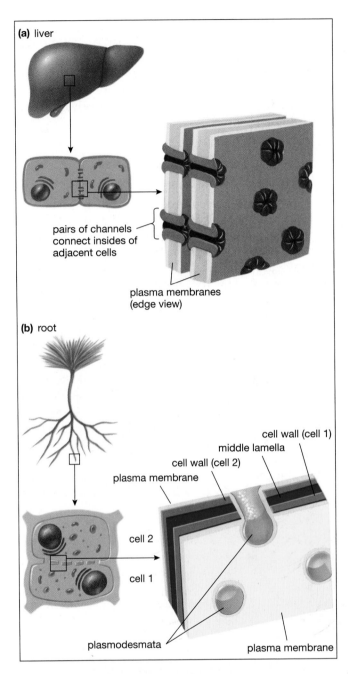

Figure 5-11 Cell attachment structures
(a) Cells lining the small intestine are firmly attached to one another by desmosomes. Protein filaments bound to the inside surface of each desmosome extend into the cytoplasm and attach to other filaments inside the cell, strengthening the connection between cells. *(b)* Leakage between cells of the urinary bladder is prevented by close-fitting tight junctions.

Figure 5-12 Cell communication structures
(a) Gap junctions, such as those between cells of the liver, contain cell-to-cell channels that interconnect the cytoplasm of adjacent cells. *(b)* Plant cells are interconnected by plasmodesmata, which pass through openings in the walls of adjacent plant cells.

Gap Junctions and Plasmodesmata Allow Communication between Cells

Multicellular organisms must coordinate the actions of their component cells. In animals, many cells, including heart muscle cells, most gland cells, some brain cells, and every cell of very young embryos, communicate through protein channels that directly connect the insides of adjacent cells (Fig. 5-12a). These cell-to-cell channels are clustered in specialized regions called **gap junctions**. Hormones, nutrients, ions, and even electrical signals can pass through the channels at gap junctions.

Virtually all of the living cells of plants are connected to one another by **plasmodesmata** (singular: **plasmodesma**). Plasmodesmata are cytoplasmic strands, surrounded by plasma membrane, that pass through openings in the walls of adjacent plant cells (Fig. 5-12b). Plasmodesmata form a continuous bridge between the cytoplasm of one cell and the cytoplasm of its neighbor. Many plant cells have thousands of plasmodesmata. As a result, water, nutrients, and hormones pass quite freely from one cell to another.

Some Cells Are Supported by Cell Walls

The outer surfaces of the cells of bacteria, plants, fungi, and some protists are covered with nonliving, typically stiff coatings called **cell walls**. The protist in our tide pool may have an outer covering of cellulose, protein, or glassy silica that protects the delicate plasma membrane. Plant cell walls are composed of cellulose and other *polysaccharides*, whereas fungal cell walls are made of the modified polysaccharide *chitin*. (We described both polysaccharides and chitin in Chapter 3.) Bacterial cell walls have a chitinlike framework to which short chains of amino acids and other molecules are attached.

Cell walls are produced by the cells they surround. Plant cells secrete cellulose through their plasma membranes, forming the **primary cell wall**. Many plant cells later secrete more cellulose and other polysaccharides beneath the primary wall to form a thick **secondary cell wall**, pushing the primary cell wall away from the plasma membrane. In some plant cells, the secondary wall becomes thicker than the cell inside it. The primary cell walls of adjacent cells are joined by the **middle lamella**, a

layer made primarily of *pectin*. Pectin is the polysaccharide that makes jelly solidify (Fig. 5-13).

Cell walls support and protect otherwise fragile cells. For example, cell walls allow plants and mushrooms to resist the forces of gravity and wind and to stand erect on land. Tree trunks, composed almost entirely of cellulose and other materials laid down over the years and capable of supporting impressive loads, are the ultimate proof of the strength of cell walls.

Although strong, cell walls are normally porous, permitting easy passage of small molecules such as minerals, water, oxygen, carbon dioxide, amino acids, and sugars. (Otherwise, the cell within would soon die.) However, the structure that really governs the interactions that occur between a cell and its external environment is the plasma membrane.

Evolutionary Connections
Caribou Legs and Membrane Diversity

The membranes of all cells are similar in structure, reflecting the common evolutionary heritage of all life on Earth. Membrane function varies tremendously from organism to organism, however, and even from cell to cell within a single organism. This diversity arises largely from the different proteins and phospholipids in the mem-

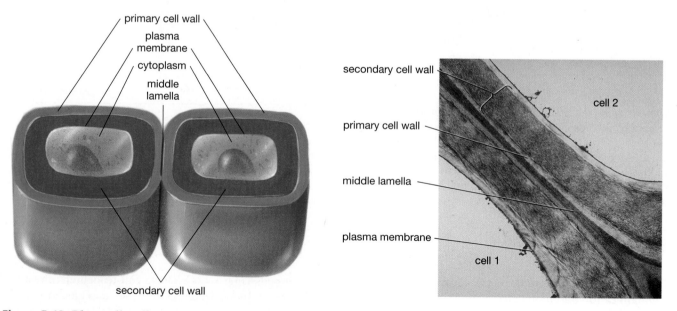

Figure 5-13 Plant cell walls
Each plant cell secretes cellulose and other carbohydrates to form its primary cell wall, just outside the plasma membrane. Many cells may then secrete additional cellulose and other polysaccharides beneath the primary wall, forming a secondary cell wall, which pushes the primary cell wall away from the plasma membrane. The middle lamella separates adjacent plant cells.

brane, which have evolved under differing environmental pressures.

Our discussion of membranes emphasized the unique functions of membrane proteins. Consequently, you may think that the phospholipids are just a waterproof place for the proteins to reside. This isn't quite true, as we can see by examining the plasma membrane phospholipids in cells of the legs of caribou, animals that live in very cold regions of North America (Fig. 5-14). During the long arctic winters of these regions, temperatures plummet far below freezing. For caribou to keep their legs and feet really warm would waste precious energy. Through evolution by natural selection, these conditions have favored the development of specialized arrangements of arteries and veins in caribou legs that allow the temperature of their lower legs to drop almost to freezing (0°C). The upper legs and main trunk of the body, in contrast, remain at about 37°C. Further, the phospholipids in the membranes of cells in the upper legs of caribou are very different from those near the hooves.

Remember, the membrane of a cell needs to be somewhat fluid to allow the membrane proteins to move to sites where they are needed. The fluidity of a membrane is a function of the fatty acid tails of its phospholipids: Unsaturated fatty acids remain more fluid at lower temperatures than do saturated fatty acids (see Chapter 3). Caribou have a range of phospholipids in the plasma membranes of the cells in their legs. The membranes of cells near the chilly hoof have lots of unsaturated fatty acids, whereas the membranes of cells near the warmer trunk have more saturated ones. This arrangement gives the plasma membranes throughout the leg the proper fluidity despite great differences in temperature.

As important as the phospholipids are, the membrane proteins probably play the major roles in determining cell function and in governing the interactions between a cell and its neighbors. The extreme complexity of the protein molecule makes it more susceptible than other types of molecules to mutations that alter its amino acid composition, its shape, and hence its function. Over billions of years, an incredible diversity of proteins has evolved. Every nerve cell in your body, for instance, has membrane proteins essential for producing electrical signals and conducting them along the nerves to various parts of your body. Other membrane proteins receive chemical messages from neighboring nerve cells or from hormones and other chemicals in the blood. Each cell in the brain has a specific set of membrane proteins, allowing it to respond to some stimuli while ignoring others. In fact, your ability to read this page depends on the proteins that reside in the membranes of your brain cells.

As we progress through this book, we shall return many times to the concepts of membrane structure presented in this chapter. Understanding the diversity of membrane lipids and proteins is the key to understanding not just the isolated cell, but also entire organs, which function as they do largely because of the properties of the membranes of their component cells.

Figure 5-14 *Caribou browse on the frozen Alaskan tundra*
The lipid composition of the membranes in the cells of a caribou's legs varies with distance from the animal's trunk. Unsaturated phospholipids predominate in the lower leg; more saturated phospholipids are found in the upper leg.

Summary of Key Concepts

1) How Is the Structure of a Membrane Related to Its Function?

The plasma membrane has three major functions: (1) isolate the cytoplasm from the external environment, (2) regulate the flow of materials into and out of the cell, and (3) communicate with other cells. The membrane, according to the fluid mosaic model, consists of a bilayer of phospholipids in which a variety of proteins are embedded. There are three major categories of membrane proteins: (1) transport proteins, which regulate the movement of most water-soluble substances through the membrane; (2) receptor proteins, which bind molecules in the external environment, triggering changes in the metabolism of the cell; and (3) recognition proteins, which serve as identification tags and attachment sites.

2) How Are Substances Transported across Membranes?

Diffusion is the movement of particles from regions of higher concentration to regions of lower concentration. In simple diffusion, water, dissolved gases, and lipid-soluble molecules diffuse through the phospholipid bilayer. In facilitated diffusion, water-soluble molecules cross the membrane through protein channels or with the assistance of protein carriers. In both cases, molecules move down their concentration gradients and cellular energy is not required.

Osmosis is the diffusion of water across a differentially permeable membrane and down its concentration gradient.

Dissolved substances decrease the concentration of free water molecules. Osmosis does not require cellular energy.

In active transport, protein carriers in the membrane use cellular energy (ATP) to drive the movement of molecules across the plasma membrane, usually against concentration gradients. Large molecules (for example, proteins), particles of food, microorganisms, and extracellular fluid may be acquired by endocytosis, either by pinocytosis, receptor-mediated endocytosis, or phagocytosis. The secretion of substances such as hormones and the excretion of wastes from a cell are accomplished by exocytosis.

3) How Are Cell Surfaces Specialized?

Cells may be connected to one another by a variety of junctions. Desmosomes attach cells firmly to one another, preventing the tearing of a tissue during movement or stress. Tight junctions seal off the spaces between adjacent cells, leakproofing organs such as the urinary bladder. Gap junctions in animals and plasmodesmata in plants are the locations at which the cytoplasm of two adjacent cells are interconnected.

The outer surfaces of some protist cells and of each bacterial, plant, and fungal cell are surrounded by a rigid cell wall outside the plasma membrane. The cell wall, produced by the cell that it surrounds, protects and supports that cell.

Key Terms

active transport *p. 73*
carrier protein *p. 72*
cell wall *p. 82*
channel protein *p. 72*
concentration *p. 72*
concentration gradient *p. 72*
cytoplasm *p. 70*
desmosome *p. 80*
differentially permeable *p. 74*

diffusion *p. 73*
endocytosis *p. 78*
exocytosis *p. 80*
facilitated diffusion *p. 74*
fluid *p. 72*
fluid mosaic model *p. 70*
gap junction *p. 81*
glycoprotein *p. 72*
gradient *p. 72*
hypertonic *p. 76*
hypotonic *p. 76*

isotonic *p. 76*
middle lamella *p. 82*
osmosis *p. 75*
passive transport *p. 73*
phagocytosis *p. 80*
phospholipid bilayer *p. 71*
pinocytosis *p. 78*
plasma membrane *p. 70*
plasmodesma *p. 81*
primary cell wall *p. 82*
pseudopod *p. 80*

receptor-mediated endocytosis *p. 78*
receptor protein *p. 72*
recognition protein *p. 72*
secondary cell wall *p. 82*
simple diffusion *p. 74*
tight junction *p. 80*
transport protein *p. 72*
vesicle *p. 78*

Thinking Through the Concepts

Multiple Choice

1. *Active transport through the plasma membrane occurs through the action of*
   a. diffusion   b. membrane proteins
   c. DNA   d. water
   e. osmosis

2. *The following is a characteristic of a plasma membrane:*
   a. It separates the cell contents from its environment.
   b. It is permeable to certain substances.
   c. It is a lipid bilayer with embedded proteins.
   d. It contains pumps for moving molecules against their concentration gradient.
   e. all of the above

3. *If an animal cell is placed into a solution whose concentration of dissolved substances is higher than that inside the cell,*
   a. the cell will swell
   b. the cell will shrivel
   c. the cell will remain the same size
   d. the solution is described as hypertonic
   e. both (b) and (d) are correct

4. *Small, nonpolar hydrophobic molecules such as fatty acids*
   a. pass readily through a membrane's lipid bilayer
   b. diffuse very slowly through the lipid bilayer
   c. require special channels to enter a cell
   d. are actively transported across cell membranes
   e. must enter the cell via endocytosis

5. *Which of the following would be least likely to diffuse through a lipid bilayer?*
   a. water
   b. oxygen
   c. carbon dioxide
   d. sodium ions
   e. the small, nonpolar molecule butane

6. *Which of the following processes causes substances to move across membranes without the expenditure of cellular energy?*
   a. endocytosis   b. exocytosis
   c. active transport   d. diffusion
   e. pinocytosis

? Review Questions

1. Describe and diagram the structure of a plasma membrane. What are the two principal types of molecules in plasma membranes? What are the four principal functions of plasma membranes?

2. What are the three types of proteins commonly found in plasma membranes, and what is the function of each?

3. Define *diffusion*, and compare that process to osmosis. How do these two processes help plant leaves remain firm?

4. Define *hypotonic*, *hypertonic*, and *isotonic*. What would be the fate of an animal cell immersed in each of the three types of solution?

5. Describe the following types of transport processes: simple diffusion, facilitated diffusion, active transport, pinocytosis, receptor-mediated endocytosis, phagocytosis, and exocytosis.

6. Name four types of cell-to-cell junctions and the function of each. Which junctions allow communication between the interiors of adjacent cells?

Applying the Concepts

1. Different cells have somewhat different plasma membranes. The plasma membrane of a *Paramecium*, for example, is only about 1% as permeable to water as the plasma membrane of a human red blood cell. Referring back to our discussion of the effects of osmosis on red blood cells and the role of contractile vacuoles in *Paramecium,* what do you think is the function of the low water permeability of *Paramecium*? What molecular differences do you think might account for this low water permeability?

2. A preview question for Chapter 31: The integrity of the plasma membrane is essential for cellular survival. Could the immune system utilize this fact to destroy foreign cells that have invaded the body? How might cells of the immune system disrupt membranes of foreign cells? (Two hints: Virtually all cells can secrete proteins, and some proteins form pores in membranes.)

3. A preview question for Chapter 24: Plant roots take up minerals (inorganic ions such as potassium) that are dissolved in the water of the soil. The concentration of such ions is usually much lower in the soil water than in the cytoplasm of root cells. Design the plasma membrane of a hypothetical mineral-absorbing cell, with special reference to mineral-permeable channel proteins and mineral-transporting active transport proteins. Justify your choice of channels and active-transport proteins.

4. Red blood cells will swell up and burst when placed in a hypotonic solution such as pure water. Why don't we swell up and burst when we swim in water that is hypotonic to our cells and body fluids?

Group Activity

Work in groups of three or four to discuss the following situation that cells may encounter:

A person becomes very dehydrated, so the concentration of water in her blood decreases. Which direction will water move across the plasma membranes of her blood cells? What will happen to the volume of the cells as a consequence? Why is proper hydration important for multicellular organisms?

For More Information

Bretscher, M. S. "The Molecules of the Cell Membrane." *Scientific American*, October 1985. Beautifully illustrated, this article explores the structure and function of cell membranes, with special attention to cell junctions.

Dautry-Varsat, A., and Lodish, H. "How Receptors Bring Proteins and Particles into Cells." *Scientific American*, May 1984. Receptor-mediated endocytosis is an important pathway both for cell nutrition and for cell-to-cell signaling.

McNeil, P. L. "Cell Wounding and Healing." *American Scientist*, May–June 1991. The fluidity of plasma membrane phospholipids makes cells able to withstand minor damage.

Rothman, J. E., and Orci, L. "Budding Vesicles in Living Cells." *Scientific American*, March 1996. Membranes within cells form small containers called vesicles, which transport materials inside the cell. Researchers are discovering the mechanisms by which these containers are formed.

Sharon, N., and Lis, H. "Carbohydrates in Cell Recognition." *Scientific American*, January 1993. Carbohydrates, normally attached to proteins as part of glycoproteins, identify cells, serve as parts of receptors in hormone binding, and regulate the attachment and movement of cells.

Answers to Multiple-Choice Questions
1. b   2. e   3. e   4. a   5. d   6. d

"It is all too easy now to underestimate cells. We have known about them for such large fractions of our lives that, for the most part, we cease being aware of how remarkable they really are."

Bruce Alberts, Dennis Bray, Julian Lewis, Martin Raff, Keith Roberts, and James Watson *from the Prologue to* **Molecular Biology of the Cell** *(1983)*

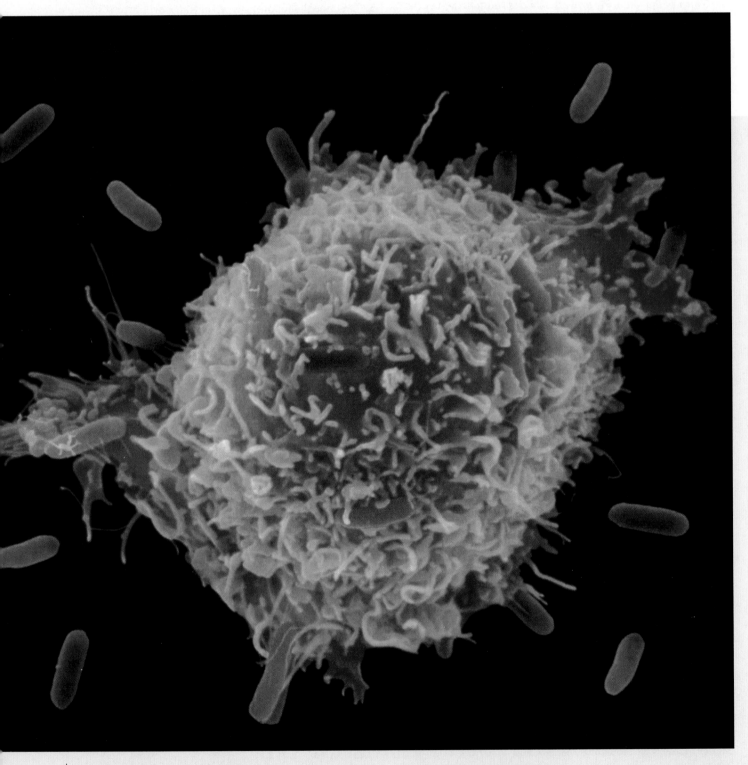

When you become ill, a battle rages as specialized white cells in your blood attack any bacterial invaders. These two cell types represent the two fundamentally different types of cells: The white blood cell (in red) is eukaryotic, and the bacterial cells (E. coli, in green) it is attacking are prokaryotic.

Cell Structure and Function 6

At a Glance

Net Watch

On-line resources for this chapter are on the World Wide Web at: http://www.prenhall.com/audesirk

(click on the Table of Contents link and then select Chapter 6).

Have you ever awakened with an awful headache and a sore throat that just won't go away? Later, you become feverish and feel as if a porcupine were dancing on your tonsils. At the clinic, a nurse scrapes the back of your throat with a swab and tells you to wait. Five minutes later, she's back with the news: You have strep throat. In biological terms, a certain variety of the bacterium *Streptococcus* has taken up residence in your throat. The good news is that antibiotics will kill it, and in a few days you'll be up and around, feeling almost normal.

What are bacteria, and how can antibiotics kill them without harming the cells of your body? A fascinating fact about life on Earth is that, despite the incredible diversity of organisms, there are only two basic types of cells: prokaryotic cells—bacteria and members of the domain Archaea (see Chapter 1)—and eukaryotic cells, which make up your body and the bodies of all the rest of life-forms. In this chapter, we explore the stuctures of these two basic cell types, from the relative simplicity of bacterial cells, which resemble the first life-forms to appear about 3½ billion years ago, to the amazing complexity of the eukaryotic cell.

1) What Are the Basic Features of Cells?

All Life on Earth Is Composed of One or More Cells

In the late 1850s, the Austrian pathologist Rudolf Virchow wrote, "Every animal appears as a sum of vital units, each of which bears in itself the complete characteristics of life." Furthermore, Virchow predicted, "All cells come from cells." The three principles of modern cell theory evolved directly from Virchow's statements:

1. Every living organism is made up of one or more cells.
2. The smallest living organisms are single cells, and cells are the functional units of multicellular organisms.
3. All cells arise from preexisting cells.

All living things, from the *Streptococcus* bacteria that can infect your throat to your own body, are composed of cells. Whereas each bacterium consists of a single, relatively simple cell, your body consists of trillions of complex cells, each specialized to perform a specific function. To survive, all cells must obtain energy and nutrients from their environment. They must synthesize a variety of proteins and other molecules necessary for their growth and repair, and they must get rid of wastes. Many cells need to interact with other cells. To ensure the continuity of life, cells must also reproduce. These activities are accomplished by specialized parts of each cell, described in the following sections.

The Plasma Membrane Encloses the Cell and Mediates Interactions between the Cell and Its Environment

As we saw in Chapter 5, the **plasma membrane** consists of a double layer of phospholipids in which a wide variety of proteins are embedded. The plasma membrane performs three major functions: (1) It isolates the cell's contents from the external environment; (2) it regulates the flow of materials into and out of the cell, for example, acquiring nutrients and expelling wastes; and (3) it allows interaction with other cells.

DNA Determines Cell Structure and Function and Allows the Cell to Reproduce

Each cell has its own hereditary blueprint that stores the instructions for making all the other parts of the cell and for producing new cells. The genetic material in all cells is **deoxyribonucleic acid (DNA)**. In *eukaryotic* cells (plants, animals, fungi, and protists), the DNA is contained within a separate, membrane-bound structure called the **nucleus**. In *prokaryotic* cells (bacteria and archaeans), the DNA, although localized to a particular place within the cell called the **nucleoid**, is not separated by membranes from the rest of the cell's interior.

All Cells Contain Cytoplasm, in Which Metabolic Reactions Occur

The **cytoplasm** consists of all the material inside the plasma membrane and outside the DNA-containing region. The cytoplasm includes water, salts, and an assortment of organic molecules, including many enzymes that catalyze reactions. The cytoplasm of eukaryotic cells also contains a variety of discrete membrane-enclosed structures called **organelles**, each of which performs a distinct cellular function, and specialized proteins that provide an intracellular support system.

Cell Function Limits Cell Size

Most cells are small, ranging from about 1 to 100 micrometers (millionths of a meter) in diameter (Fig. 6-1). Because cells are so small, their discovery awaited the invention of the microscope. Ever since the first cells were seen in the late 1600s, scientists have devised in-

creasingly sophisticated ways of observing them, as described in "Scientific Inquiry: The Search for the Cell."

Why are most cells small, and why do large organisms consist of many cells rather than one large cell? The answer lies in the need for cells to exchange nutrients and wastes with their external environment through the plasma membrane. As you learned in Chapter 5, many nutrients and wastes move into, through, and out of cells by *diffusion*, the movement of molecules from places of high concentration of those molecules to places of low concentration. As a roughly spherical cell becomes larger, its innermost regions become farther away from the membrane. For a giant cell 20 centimeters in diameter (8.5 inches), oxygen molecules would take more than *200 days* to diffuse to the center of the cell! Another difficulty is that as a cell enlarges, its volume increases more rapidly than its surface area does. For example, a cell that doubles in radius becomes eight times greater in volume but only four times greater in surface area (Fig. 6-2, p. 90). So a larger cell has a greater need for the exchange of nutrients and wastes with the environment but a relatively smaller expanse of plasma membrane through which to make these exchanges. In a very large, roughly spherical cell, the surface area of the membrane would be too small to keep up with the cell's metabolic needs. Some cells, such as your neurons and muscle cells, can get very large because they have an elongated shape that increases their membrane surface area.

There Are Two Basic Types of Cells: Prokaryotic and Eukaryotic

There are two fundamentally different types of cells. The first type, represented by the bacteria and archaeans, is called **prokaryotic**, Greek for "before the nucleus" (Fig. 6-3a, p. 90). The second type of cell, which probably evolved from the prokaryotic cell and makes up the bodies of protists, plants, fungi, and animals, is called **eukaryotic**, for "true nucleus" (Fig. 6-3b). As their names imply, one striking difference between prokaryotic cells and eukaryotic cells is that the genetic material of eukaryotic cells is contained within a membrane-enclosed nucleus, whereas the genetic material of prokaryotic cells is not enclosed within a membrane. In the following sections, we discuss the features of prokaryotic and eukaryotic cells.

2) What Are the Features of Prokaryotic Cells?

Most prokaryotic cells are very small (less than 5 micrometers long), with a relatively simple internal structure (Fig. 6-3a). Most are surrounded by a relatively stiff cell wall. The cell wall confers shape and protects the bacterial cell. Antibiotics such as penicillin (or amoxicillin, which might be prescribed for your strep throat) fight bacterial infections by inhibiting cell wall synthesis, causing the bacteria to rupture. Some bacteria can move,

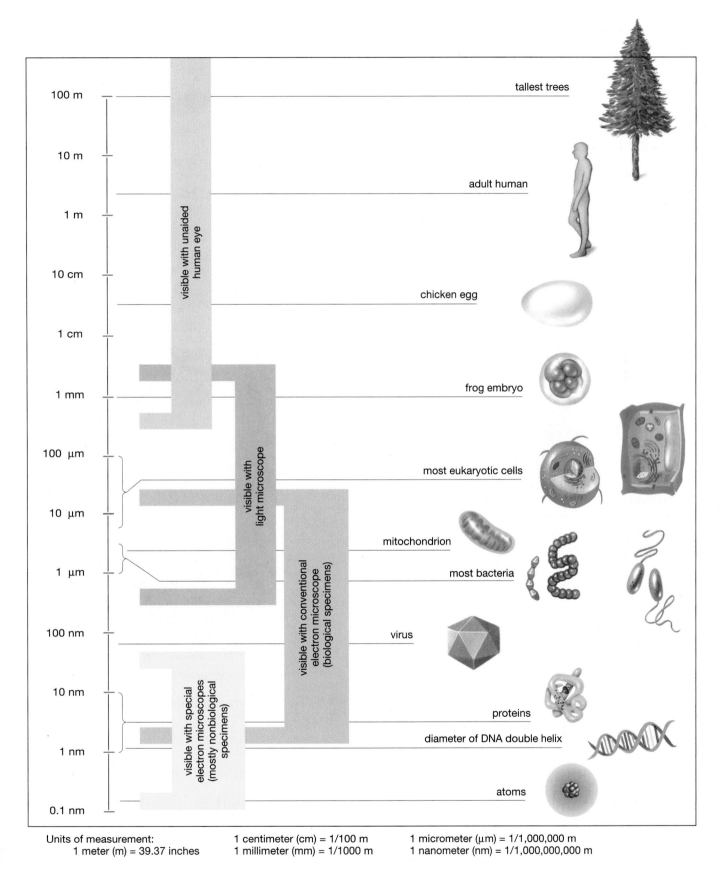

Units of measurement:
1 meter (m) = 39.37 inches

1 centimeter (cm) = 1/100 m
1 millimeter (mm) = 1/1000 m

1 micrometer (μm) = 1/1,000,000 m
1 nanometer (nm) = 1/1,000,000,000 m

Figure 6-1 *Relative sizes*
Dimensions commonly encountered in biology range from about 100 meters (the height of the tallest red-woods) through a few micrometers (the diameter of most cells) to a few nanometers (the diameter of many large molecules). Note that, in the metric system (used almost exclusively in science), separate names are given to dimensions that differ by factors of 10, 100, and 1000.

Figure 6-2 Geometrical considerations limit the size of relatively spherical cells

As a cell enlarges, the distance from the center of the cell to the outside world increases. Further, the volume increases much more rapidly than the surface area. As these spherical cells illustrate, doubling the radius halves the ratio of surface area to volume. Thus, each unit volume of cytoplasm in a cell has only half the membrane area available to exchange nutrients and wastes with the external environment. However, if the volume of the largest sphere is divided among cells the size of the smallest sphere, the overall ratio of surface area to volume of the resulting "multicellular organism" remains large.

distance to center (r)	1.0	2.0	3.0	1.0
surface area $(4\pi r^2)$	12.6	50.3	113.1	339.4
volume $(4/3\, \pi r^3)$	4.2	33.5	113.1	113.1
area/volume	3.0	1.5	1.0	3.0

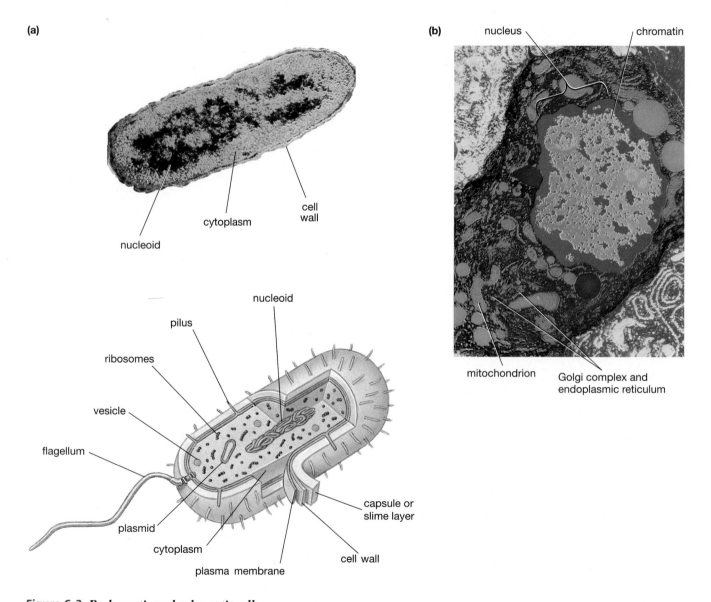

(a)

nucleoid

cytoplasm

cell wall

pilus

nucleoid

ribosomes

vesicle

flagellum

plasmid

cytoplasm

plasma membrane

capsule or slime layer

cell wall

(b)

nucleus

chromatin

mitochondrion

Golgi complex and endoplasmic reticulum

Figure 6-3 Prokaryotic and eukaryotic cells

(a) The simple internal structure of a prokaryotic cell, shown in a color-enhanced electron micrograph (the bacterium *Escherichia coli*) and diagram. (b) The complex internal structure of a eukaryotic cell.

propelled by flagella different in structure from those of eukaryotic cells. **Pili** (singular, **pilus**), surface projections made of protein, are used to attach some types of bacteria to surfaces or to exchange genetic material. **Capsules** or **slime layers** are polysaccharide or protein coatings that some disease-causing bacteria secrete outside their cell wall. These coatings help the bacteria attach to their hosts and may allow them to evade attack by immune cells. In its normal capsulated form, *Streptococcus pneumoniae*, a close relative of the bacteria responsible for your sore throat, will readily kill mice; a mutant lacking the capsule, however, is harmless. We discuss bacterial shapes and cell walls in more detail in Chapter 19.

The cytoplasm of most prokaryotic cells is relatively homogeneous in appearance (although some photosynthetic bacteria have elaborate internal membranes). Prokaryotic cells have a single, circular strand of DNA. It is usually coiled, attached to the plasma membrane, and concentrated in the nucleoid region of the cell. It is not, however, physically separated from the rest of the cytoplasm by a membrane.

Not only do prokaryotic cells lack nuclei but they also lack the membrane-enclosed organelles that eukaryotic cells possess. Despite their simplicity, prokaryotic organisms carry on all the necessary reactions to sustain and perpetuate life. Bacterial cytoplasm contains **ribosomes**, structures composed of proteins and **ribonucleic acid (RNA)**, on which proteins are synthesized. Photosynthetic bacteria possess internal membranes in which are embedded the light-capturing proteins and enzymes that catalyze the synthesis of high-energy molecules. In prokaryotic cells, reactions that harvest energy from the breakdown of sugars are catalyzed by enzymes that may be localized along the inner plasma membrane or that may be free in the cytoplasm. Table 6-1 compares prokaryotic cells with the eukaryotic cells of plants and animals. The diversity and specialized structures of bacteria are covered in more detail in Chapter 19.

Table 6-1 Cell Structures, Their Functions, and Their Distribution in Living Cells

Structure	Function	Prokaryotes	Plants	Animals
Cell surface				
Cell wall	Protects, supports cell	Present	Present	Absent
Plasma membrane	Isolates cell contents from environment; regulates movement of materials into and out of cell; communicates with other cells	Present	Present	Present
Organization of genetic material				
Genetic material	Encodes information needed to construct cell and control cellular activity	DNA	DNA	DNA
Chromosomes	Contain and control use of DNA	Single, circular, no proteins	Many, linear, with proteins	Many, linear, with proteins
Nucleus	Membrane-bound container for chromosomes	Absent	Present	Present
Nuclear envelope	Encloses nucleus; regulates movement of materials into and out of nucleus	Absent	Present	Present
Nucleolus	Synthesizes ribosomes	Absent	Present	Present
Cytoplasmic structures				
Mitochondria	Produce energy by aerobic metabolism	Absent	Present	Present
Chloroplasts	Perform photosynthesis	Absent	Present	Absent
Ribosomes	Provide site of protein synthesis	Present	Present	Present
Endoplasmic reticulum	Synthesizes membrane components and lipids	Absent	Present	Present
Golgi complex	Modifies and packages proteins and lipids; synthesizes carbohydrates	Absent	Present	Present
Lysosomes	Contain intracellular digestive enzymes	Absent	Present	Present
Plastids	Store food, pigments	Absent	Present	Absent
Central vacuole	Contains water and wastes; provides turgor pressure to support cell	Absent	Present	Absent
Other vesicles and vacuoles	Contain food obtained through phagocytosis; contain secretory products	Absent	Present (some)	Present
Cytoskeleton	Gives shape and support to cell; positions and moves cell parts	Absent	Present	Present
Centrioles	Synthesize microtubules of cilia and flagella; may produce spindle in animal cells	Absent	Absent (in most)	Present
Cilia and flagella	Move cell through fluid or move fluid past cell surface	Present[a]	Absent (in most)	Present

[a]Many prokaryotes have structures called flagella, but these are not made of microtubules and move in a fundamentally different way than eukaryotic cilia or flagella do.

Scientific Inquiry
The Search for the Cell

Human understanding of the cellular nature of life came slowly. In 1665, English scientist and inventor Robert Hooke reported observations with a primitive microscope. He aimed his instrument at an "exceeding thin...piece of Cork" and saw "a great many little Boxes" (Fig. E6-1). Hooke called the boxes "cells," because he thought they resembled the tiny rooms, or cells, occupied by monks. Cork comes from the dry outer bark of the Mediterranean oak. Hooke wrote that in the living oak and other plants, "These cells [are] fill'd with juices."

In the 1670s, the Dutch microscopist Anton van Leeuwenhoek was constructing his own simple microscopes and observing a previously unknown world. His descriptions of myriad "animalcules" (his term for protists) in rain, pond water, and well water caused quite an uproar, because water was consumed without treatment in those days. He made careful observations of an enormous range of microscopic specimens, including red blood cells, sperm, and the eggs of small insects such as weevils, aphids, and fleas. His discoveries struck a blow to the then-common belief in spontaneous generation; at that time, fleas were believed to emerge spontaneously from sand or dust, and the weevil from grain!

More than a century passed before biologists began to understand the role of cells in life on Earth. Microscopists first noted that many plants consist entirely of cells. The thick wall surrounding all plant cells, first observed by Hooke, made their observations easier. Animal cells, however, escaped notice until the 1830s, when German zoologist Theodor Schwann saw that cartilage contains cells that "exactly resemble [the cells of] plants." In 1839, after studying cells for years, Schwann was confident enough to publish his theory, calling cells the elementary particles of both plants and animals. By the mid-1800s, German botanist Matthias Schleiden had further refined science's view of cells when he wrote; "It is . . . easy to perceive that the vital process of the individual cells must form the first, absolutely indispensable fundamental basis" of life.

Ever since the pioneering efforts of Robert Hooke and Anton van Leeuwenhoek, biologists, physicists, and engineers have collaborated in the development of a variety of advanced microscopes to view the cell and its components:

Light microscopes use lenses, normally made of glass, to focus and magnify light rays that either pass through or bounce off a specimen. Light microscopes provide a wide range of images, depending on how the specimen is illuminated and whether it has been stained (Fig. E6-2a). The *resolving power* of light microscopes—that is, the smallest structure that can be seen—is about 1 micrometer (a millionth of a meter).

Electron microscopes use beams of electrons instead of light. The electrons are focused by magnetic fields rather than by lenses. Some types of electron microscopes can resolve structures as small as a few nanometers (billionths of a meter). *Transmission electron microscopes* (TEMs) pass electrons through a thin specimen and can reveal minute subcellular structures, including organelles and plasma membranes (Fig. E6-2b). *Scanning electron microscopes* (SEMs) bounce electrons off

Figure E6-1 Early observations of cells
Robert Hooke's drawings of the cells of cork, as he viewed them with an early light microscope similar to the one shown here. The cells of cork are not living, and only the cell walls remain to outline the cell.

specimens that have been coated with metals and provide three-dimensional images. SEMs can be used to view structures ranging in size from entire insects down to cells and even organelles (Fig. E6-2c,d).

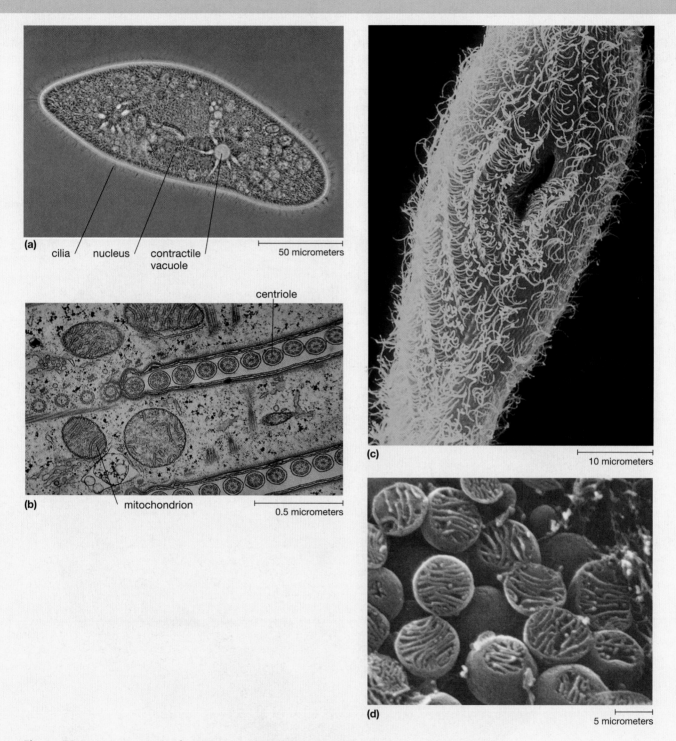

(a) cilia / nucleus / contractile vacuole

50 micrometers

centriole

(b) mitochondrion

0.5 micrometers

(c)

10 micrometers

(d)

5 micrometers

Figure E6-2 A comparison of microscope images
(*a*) A living *Paramecium* viewed through a light microscope. (*b*) A false-color TEM photo showing the centrioles at the bases of the cilia that cover *Paramecium*'s single-celled body; mitochondria are also visible. (*c*) A false-color SEM photo of a *Paramecium*. (*d*) An SEM photo at much higher magnification, showing mitochondria, many of which are sliced open.

3) What Are the Features of Eukaryotic Cells?

Eukaryotic Cells Contain Organelles

Eukaryotic cells differ from prokaryotic cells in many ways. For one thing, they are normally larger than prokaryotic cells—typically more than 10 micrometers in diameter. The fluid portion of the cytoplasm of prokaryotic and eukaryotic cells contains similar types of chemical substances; in each case it is a thick soup of proteins, lipids, carbohydrates, salts, sugars, amino acids, and nucleotides. However, the cytoplasm of eukaryotic cells also houses a variety of membrane-enclosed organelles that lend structural and functional organization to the cell. Giving shape and organization to the cytoplasm of eukaryotic cells is a network of protein fibers, the **cytoskeleton**. Many of the organelles of the cytoplasm are attached to the cytoskeleton.

Eukaryotic cells, however, are not all alike. Figures 6-4 and 6-5 illustrate the structures that are found in an-

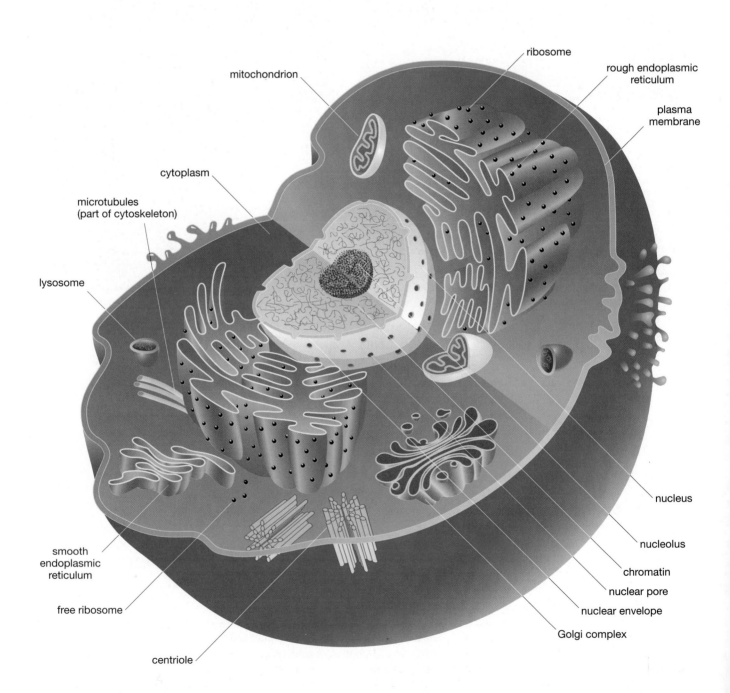

Figure 6-4 A generalized animal cell

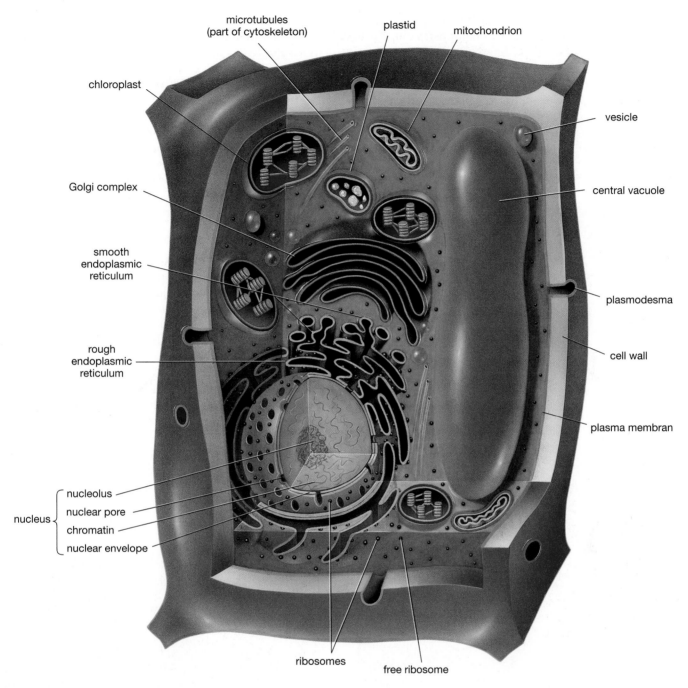

Figure 6-5 A generalized plant cell

imal and plant cells, respectively, although few individual cells possess all the features shown in either drawing. Each type of cell has a few unique organelles not found in the other. You may want to refer to these illustrations as we describe the structures of the cell in more detail. The major components of eukaryotic cells are listed in Table 6-1 and explained in more detail in the following sections.

The Nucleus Is the Control Center of the Eukaryotic Cell

Deoxyribonucleic acid (DNA) is the genetic material of all living cells. A cell's DNA is the repository of information needed to construct the cell and direct the countless chemical reactions necessary for life and reproduction. The hereditary information in DNA is used selectively by the cell, depending on its stage of development and its environmental conditions. In eukaryotic cells, the DNA is housed within the nucleus.

The nucleus consists of three readily distinguishable components (Fig. 6-6). The **nuclear envelope** separates the nuclear material from the cytoplasm. Inside the nuclear envelope, the nucleus contains a granular-looking material called **chromatin** and a darker region called the **nucleolus**.

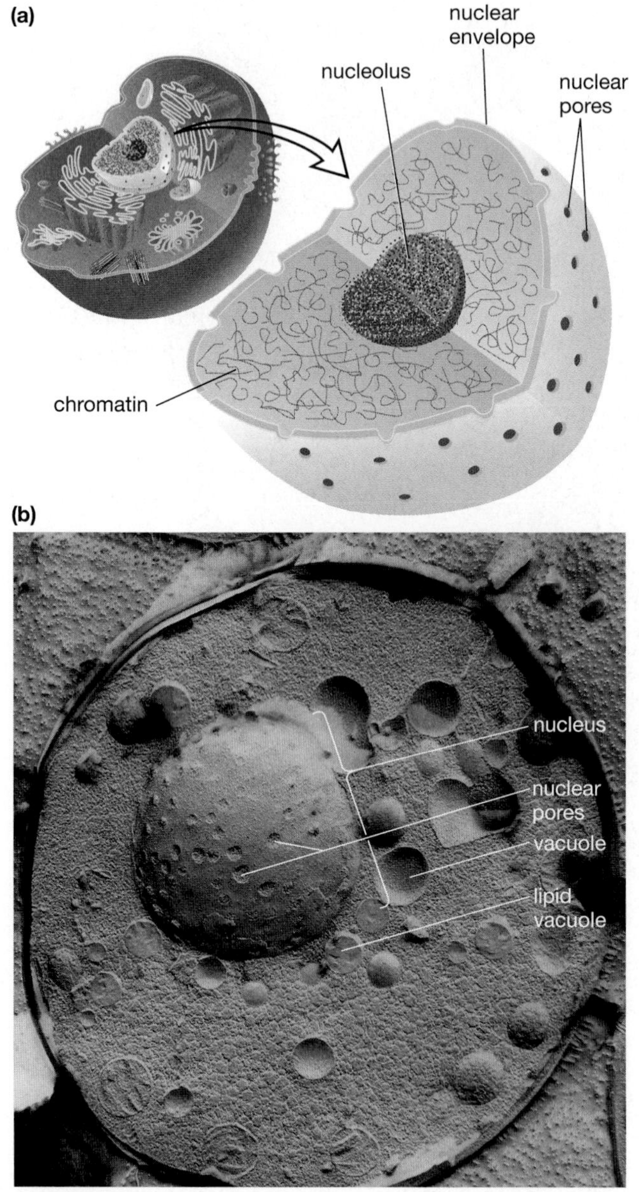

The Nuclear Envelope Allows Selective Exchange of Materials

The nucleus is isolated from the rest of the cell by a nuclear envelope that consists of two membranes riddled with pores. Water, ions, and small molecules such as ATP can pass freely through the pores, but the passage of large molecules, particularly proteins and RNA, is regulated. Consequently, the pores help control the flow of information that is carried in large molecules between the nucleus and cytoplasm.

Chromatin Consists of DNA and Its Associated Proteins

Because the nucleus is highly colored by common stains used in light microscopy, early microscopists named the nuclear material *chromatin*, meaning "colored substance." Biologists have since learned that chromatin consists of DNA associated with proteins. Eukaryotic DNA and its associated proteins form long strands called **chromosomes** ("colored bodies"). When cells divide, each chromosome coils upon itself, becoming thicker and shorter. The resulting "condensed" chromosomes are easily visible even with light microscopes (Fig. 6-7).

Chemical reactions within the cell that are responsible for growth and repair, nutrient and energy acquisition and use, and reproduction are governed by the informa-

(a)

nuclear envelope

nucleolus

nuclear pores

chromatin

(b)

nucleus

nuclear pores

vacuole

lipid vacuole

Figure 6-6 The nucleus
(a) The nucleus is bounded by a double outer membrane. Inside are chromatin (chromosomes in an uncondensed state) and a nucleolus, which contains DNA coding for ribosomal RNA, ribosomes in various stages of synthesis, and associated proteins. **(b)** An electron micrograph of a yeast cell that has been frozen and broken open to reveal its internal structures. The large nucleus, with nuclear pores penetrating its nuclear membrane, is clearly visible.

Figure 6-7 Chromosomes
Chromosomes, seen here in a light micrograph of a dividing cell in an onion root tip, are the same material (DNA and proteins) as the chromatin seen in nondividing cells but in a more compact state.

tion encoded in DNA. Because the DNA stays in the nucleus whereas most of the chemical reactions that it controls occur in the cytoplasm, information molecules must be exchanged between the nucleus and the cytoplasm. Genetic information is copied from DNA into molecules of RNA, which move through the pores of the nuclear envelope into the cytoplasm. This information is then used to direct the synthesis of cellular proteins. These proteins include enzymes, which catalyze and regulate chemical reactions; membrane proteins, which govern interactions between the cell and its environment; and a variety of structural proteins. Some of these proteins pass from the cytoplasm into the nucleus and regulate the transfer of information from DNA to RNA, depending on what is happening in the cytoplasm and in the extracellular environment. We take a closer look at these processes in Chapter 10.

The Nucleolus Is the Site of Ribosome Assembly

Most eukaryotic nuclei have one or more darkly staining regions called *nucleoli* ("little nuclei"; one nucleolus is shown in Fig. 6-6). The nucleolus consists of *ribosomal RNA*, proteins, ribosomes in various stages of synthesis, and DNA (bearing genes that specify the blueprint for ribosomal RNA).

Nucleoli are the sites of ribosome synthesis. A ribosome is a small particle composed of RNA and several proteins that serves as a workbench for the synthesis of proteins. Like many workbenches, ribosomes are nonspecific—the same set of tools can be used to construct many different objects. Any ribosome can be used to synthesize any of the thousands of proteins made by a cell. In electron micrographs, ribosomes appear as dark granules, either distributed in the cytoplasm (Fig. 6-8) or clustered along the membranes of the nuclear envelope and the endoplasmic reticulum (see Fig. 6-9).

Eukaryotic Cells Contain a Complex System of Internal Membranes

All eukaryotic cells have an elaborate system of membranes. This membrane system is composed of the *plasma membrane* and several organelles within the cytoplasm, including the endoplasmic reticulum, nuclear envelope, Golgi complex, and a variety of membrane-enclosed sacs such as lysosomes. The various parts of the membrane system of the cells can exchange membrane material with one another (see Figs. 6-4 and 6-5).

The Plasma Membrane Both Isolates the Cell and Allows Selective Interactions between the Cell and Its Environment

The plasma membrane forms the outer boundary of the living part of a cell. It is a marvelously complex structure that must perform the seemingly contradictory functions of separating the cytoplasm of the cell from the outside

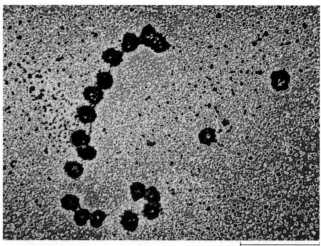

0.05 micrometers

Figure 6-8 Ribosomes
Ribosomes may be found free in the cytoplasm either singly or strung along messenger RNA molecules as they participate in protein synthesis. Ribosomes also stud the rough endoplasmic reticulum, giving it a rough appearance and allowing the synthesis of proteins within the ER.

environment while providing for the transport of selected substances into or out of the cell. In plants, fungi, and some protists, a *cell wall* is secreted through the plasma membrane and forms an outer, protective coating. Both plasma membrane and eukaryotic cell wall structure and function were discussed in Chapter 5.

The Endoplasmic Reticulum Forms Membrane-Enclosed Channels within the Cytoplasm

The **endoplasmic reticulum (ER)** is a series of interconnected membrane-enclosed tubes and channels in the cytoplasm (Fig. 6-9); the ER membrane is continuous with the nuclear membrane. Eukaryotic cells have two forms of ER: rough and smooth. Numerous ribosomes stud the outside of the **rough endoplasmic reticulum**; **smooth endoplasmic reticulum** lacks ribosomes.

The different structures of smooth and rough ER reflect different functions. Enzymes embedded in the membranes of the smooth ER are the major site of lipid synthesis, including the phospholipids and cholesterol used in membrane formation. Smooth ER in liver cells contains enzymes that detoxify harmful drugs and metabolic by-products. In some cells the smooth ER synthesizes other types of lipids as well, such as the steroid hormones testosterone and estrogen, which are produced in the reproductive organs of mammals.

The ribosomes on the outside of rough ER synthesize proteins, including membrane proteins. Therefore, the ER can synthesize itself, both lipid and protein

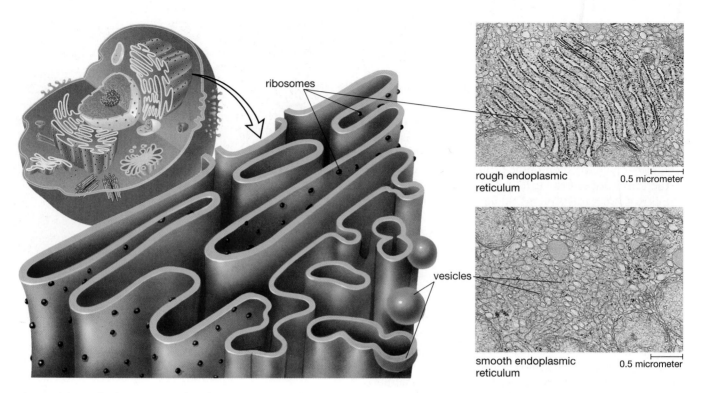

ribosomes

rough endoplasmic
reticulum   0.5 micrometer

vesicles

smooth endoplasmic
reticulum   0.5 micrometer

Figure 6-9 Endoplasmic reticulum
There are two types of endoplasmic reticulum: rough ER, coated with ribosomes, and smooth ER, without ribosomes. Although in electron micrographs the ER looks like a series of tubes and sacs, it is actually a maze of folded sheets and interlocking channels. In many cells the rough and smooth ER are thought to be continuous, as depicted in the drawing. Ribosomes (black) stud the cytoplasmic face of the rough ER membrane.

components. Although most of the membrane synthesized in the ER forms new or replacement ER membrane, some of it moves inward to replace nuclear membrane or outward to form the Golgi complex, lysosomes, and the plasma membrane.

Ribosomes on rough ER also manufacture the proteins that some secretory cells export into their surroundings, including digestive enzymes and protein hormones (for example, insulin secreted by cells of the pancreas). These proteins are synthesized by the ribosomes on the outside of the ER and transported into the channels within the ER.

The proteins synthesized for secretion or use within the cell then move through the ER and accumulate in pockets at the ends of the ER. These pockets then bud off, forming membrane-bound sacs called **vesicles** that carry their protein cargo to the Golgi complex.

The Golgi Complex Sorts, Chemically Alters, and Packages Important Molecules

The **Golgi complex** is a specialized set of membranous sacs derived from the endoplasmic reticulum. In fact, the Golgi complex looks very much like a stack of smooth ER that has been squashed in the middle, mak-

ing the ends bulge out (Fig. 6-10). Vesicles from the smooth ER fuse with one side of the Golgi complex, adding their membrane to the Golgi complex and emptying their contents into the Golgi sacs. Other vesicles bud off the Golgi complex on the opposite side of the stack, carrying away proteins, lipids, and other complex molecules. The Golgi complex performs the following three major functions:

1. It separates proteins and lipids received from the ER according to their destinations; for example, the Golgi separates digestive enzymes that are bound for lysosomes from hormones that the cell will secrete.
2. It modifies some molecules—for instance, it adds sugars to proteins to make glycoproteins.
3. It packages these materials into vesicles that are then transported to other parts of the cell or to the plasma membrane for export.

Lysosomes Serve as the Cell's Digestive System

Some of the proteins manufactured in the ER and sent to the Golgi complex are intracellular digestive enzymes that can break down proteins, fats, and carbohydrates into their component subunits. In the Golgi, these en-

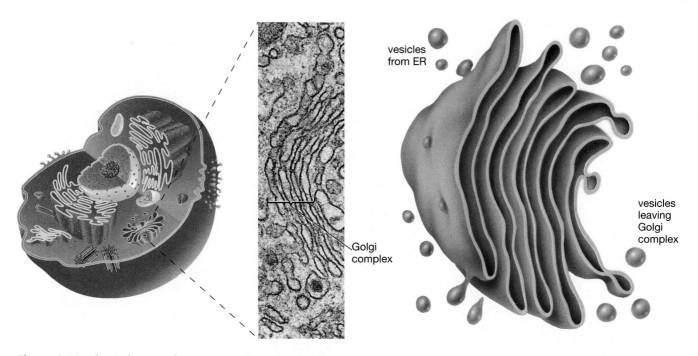

Figure 6-10 *The Golgi complex*
The Golgi complex is a stack of flat membranous sacs derived from the endoplasmic reticulum. Vesicles constantly bud off from and fuse with the Golgi and ER, transporting material from the ER to the Golgi and back again, and from the Golgi to plasma membrane, lysosomes, and vesicles.

zymes are packaged in membranous vesicles called **lysosomes** (Fig. 6-11). The major function of lysosomes is to digest food particles, which range from individual proteins to complete microorganisms.

As you have seen in Chapter 5, many cells "eat" by *phagocytosis*—that is, by engulfing extracellular particles with extensions of the plasma membrane. The food particles are then moved into the cytoplasm, enclosed within membranous sacs, now called **food vacuoles.** Lysosomes recognize these food vacuoles and fuse with them. The contents of the two vesicles mix, and the lysosomal enzymes digest the food into amino acids, monosaccharides, fatty acids, and other small molecules. These simple molecules then diffuse out of the lysosome and into the cytoplasm to nourish the cell. Cell biologists continue to search for the key to how lysosomes recognize these food vacuoles.

Lysosomes also digest defective or malfunctioning organelles, such as *mitochondria* or *chloroplasts.* After identifying these organelles, the cell encloses them in vesicles made of membrane from the ER. These vesicles fuse with lysosomes, and digestive enzymes within the lysosome enable the cell to recycle valuable materials from the defunct organelles. How the cell identifies organelles that have outlived their usefulness is a topic of continuing research.

Membrane Synthesized in the Endoplasmic Reticulum Flows through the Membrane System of the Cell in an Orderly Way

The nuclear envelope, rough and smooth ER, Golgi complex, lysosomes, food vacuoles, and plasma membrane all form an integrated membrane system. Membrane is synthesized in the ER and flows back and forth among these structures in an orderly way. As an example, let's look at the movement of materials destined for inclusion in the plasma membrane (Fig. 6-11). The ER synthesizes the phospholipids and proteins that make up the plasma membrane and buds off a vesicle whose membrane includes these plasma membrane components. The vesicle fuses with the Golgi complex. Plasma membrane material continues on through the Golgi, where it may be modified—for example, by adding sugars to make glycoproteins or *glycolipids* (lipids to which a sugar is attached). Eventually the plasma membrane material becomes an "outward-bound" vesicle that buds off the far side of the Golgi and moves to the cell surface. The vesicle fuses with the plasma membrane, replenishing and enlarging the membrane.

The Golgi complex processes and packages all membrane-enclosed materials produced by the cell. Many of the vesicles pinched off from the Golgi contain secretory products (for example, hormones) that are released

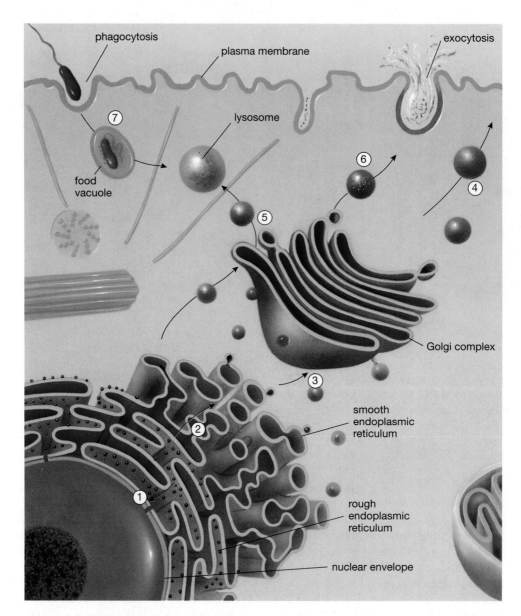

Figure 6-11 *The flow of membrane within the cell*
Membrane, regardless of its destination, is synthesized by the endoplasmic reticulum.
① Some of this membrane moves inward to form new nuclear envelope, but most of it
moves outward to form ② smooth ER and ③ Golgi membrane. From the Golgi, membrane
moves to form ④ new plasma membrane and ⑤ the membranes that surround other cell
organelles such as lysosomes. Many proteins are synthesized in the rough ER and move
through the smooth ER to the Golgi. There the proteins are sorted according to function;
some remain as parts of the Golgi membrane, some are returned to the ER, some are packaged in vesicles bound for the plasma membrane, where they will be ⑥ secreted from the
cell, and some are packaged in lysosomes. ⑦ Lysosomes may fuse with food vacuoles for intracellular digestion of food particles.

outside the cell. Lysosomes contain digestive enzymes
and often fuse with food vacuoles to carry out intracellular digestion. How these diverse materials are
recognized, purified, modified properly, separated out,
and individually packaged remains a challenge to cell
biologists.

Vacuoles Serve Many Functions, Including Water Regulation, Support, and Storage

Most cells contain one or more **vacuoles**, which are fluid-
filled sacs surrounded by a single membrane. Some, such
as the food vacuoles that form during phagocytosis, are

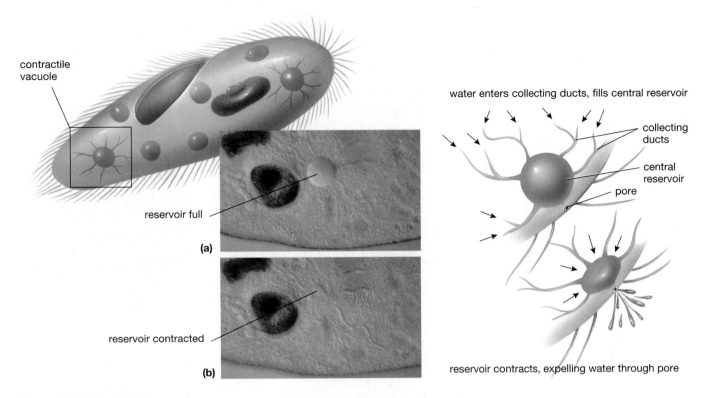

Figure 6-12 Contractile vacuoles
Many freshwater protists contain contractile vacuoles. **(a)** Water constantly enters the cell by osmosis. In the cell, water is taken up by collecting ducts and drains into the central reservoir of the vacuole. **(b)** When full, the reservoir contracts, expelling the water through a pore in the plasma membrane.

temporary features of cells. However, many cells contain permanent vacuoles that have important roles in maintaining the integrity of the cell, especially by regulating the cell's water content.

Freshwater Microorganisms Have Contractile Vacuoles

Freshwater protists such as *Paramecium* consist of a single, eukaryotic cell. Many of these organisms possess complex **contractile vacuoles** composed of collecting ducts, a central reservoir, and a tube leading to a pore in the plasma membrane (Fig. 6-12). Because fresh water is hypotonic to the cytoplasm of these organisms, water constantly enters the cell by osmosis. The increasing volume of incoming water might soon burst the fragile creature if it did not have a mechanism to excrete the water. Cellular energy is used to pump salts into collecting ducts. Water follows by osmosis and drains into the central reservoir. When the reservoir is full, it contracts, squirting the water out through a pore in the plasma membrane.

Plant Cells Have Central Vacuoles

Three-quarters or more of the volume of many plant cells is occupied by a large **central vacuole** (Fig. 6-13a; see also Fig. 6-4). The central vacuole has several functions. Filled mostly with water, the central vacuole is involved in the cell's water balance. It also provides a dump site for hazardous wastes, which plant cells often cannot excrete. Some plant cells store extremely poisonous substances, such as sulfuric acid, in their vacuoles, which deter animals from munching on the otherwise tasty leaves. Vacuoles may also store sugars and amino acids not immediately needed by the cell. Blue or purple pigments stored in central vacuoles are responsible for the colors of many flowers.

These dissolved substances make the vacuole contents hypertonic to the cell cytoplasm, which in turn is usually hypertonic to the extracellular fluid that bathes the cells. Water therefore enters the vacuole by osmosis, which tends to make it swell. The pressure of the water within the vacuole, called **turgor pressure**, pushes the fluid portion of the cytoplasm up against the cell wall with considerable force (Fig. 6-13b). Cell walls are usually somewhat flexible, so both the overall shape and the rigidity of the cell depend on turgor pressure within the cell. Turgor pressure thus provides support for the non-wood parts of plants. If you forget to water your houseplants, the central vacuoles and cytoplasm lose water and the cells shrink away from their cell walls. Just as a balloon goes limp when its air leaks out, so too the plant droops as its cells lose turgor pressure.

Figure 6-13 The central vacuole and turgor pressure in plant cells
(a) (top) When water is plentiful, it fills the central vacuole and pushes the cytoplasm against the cell wall, helping maintain the cell's shape. (bottom) The pressure of water inside each cell supports the leaves of this impatiens plant. **(b)** (top) When water is scarce, the central vacuole shrinks, the cytoplasm pulls away from the cell wall, and the cell may become soft and shrunken. (bottom) Deprived of the support of water, the plant wilts.

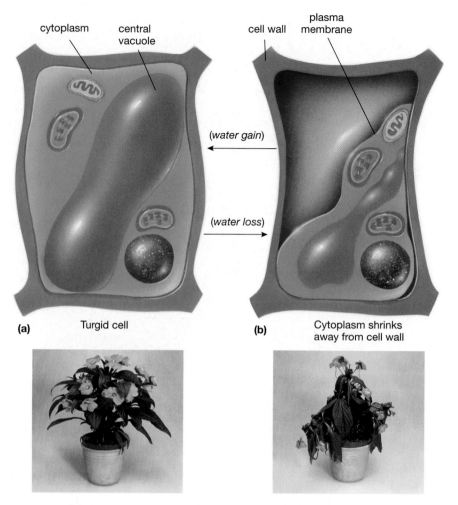

Mitochondria Extract Energy from Food Molecules, and Chloroplasts Capture Solar Energy

Every cell has prodigious energy needs: to manufacture materials, to pick things up from the environment and throw other things out, to move, and to reproduce. The structures that provide the energy are the **chloroplasts** and **mitochondria**, which biologists believe evolved from prokaryotic bacteria that took up residence long ago within fortunate eukaryotic cells (see Chapter 17).

Mitochondria and chloroplasts are similar to each other in many ways. Both are about 1 to 5 micrometers in diameter and are surrounded by a double membrane. Both have enzyme assemblies that synthesize ATP, although the assemblies are used in a very different manner. Finally, both have many characteristics, including their own DNA, that seem to be remnants of their probable evolution from free-living organisms. However, they also have many differences, corresponding to their vastly different roles in cells: Mitochondria convert the energy of sugar into ATP for use by the cell; chloroplasts capture the energy of sunlight during photosynthesis and store it in sugar.

Mitochondria Use Energy Stored in Food Molecules to Produce ATP

Mitochondria extract energy from food molecules and store it in the high-energy bonds of ATP. All eukaryotic cells, including plants, have mitochondria, which are often called the "powerhouses of the cell." As you shall see in Chapter 8, different amounts of energy can be released from a food molecule, depending on how it is metabolized. The breakdown of food molecules is initiated without the use of oxygen by enzymes in the fluid portion of the cytoplasm. This **anaerobic** (without oxygen) metabolism does not convert very much food energy into ATP energy. Mitochondria are the only places in a eukaryotic cell where oxygen can be used in the breakdown of high-energy molecules. These **aerobic** (with oxygen) reactions are much more effective in generating energy than are the anaerobic reactions; 18 or 19 times more ATP is generated by aerobic metabolism in the mitochondria than by anaerobic metabolism. Not surprisingly, mitochondria are found in large numbers in metabolically active cells, such as muscle, and are less abundant in cells that are less metabolically active, such as those of bone and cartilage.

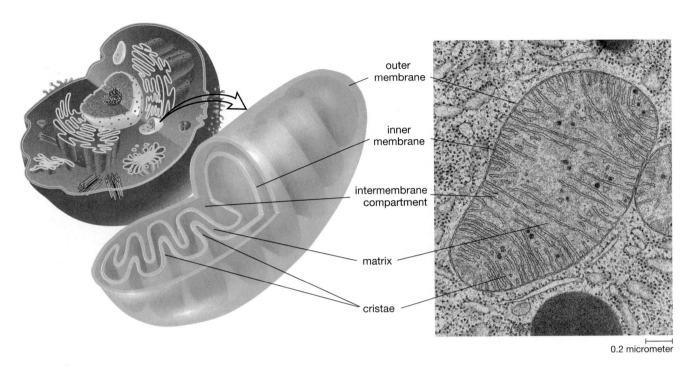

outer
membrane

inner
membrane

intermembrane
compartment

matrix

cristae

0.2 micrometer

Figure 6-14 A mitochondrion
Mitochondria consist of a pair of membranes enclosing two fluid compartments, the intermembrane compartment between the outer and inner membranes and the matrix within the inner membrane. The outer membrane is smooth, but the inner membrane forms deep folds called cristae. Mitochondria are the site of aerobic metabolism.

Mitochondria are round, oval, or tubular sacs made of a pair of membranes (Fig. 6-14). Although the outer mitochondrial membrane is smooth, the inner membrane loops back and forth to form deep folds called **cristae** (singular, **crista**, meaning "crest"). As a result, the mitochondrial membranes enclose two fluid-filled spaces: the **intermembrane compartment** between the inner and outer membranes and the **matrix**, or inner compartment, within the inner membrane. Some of the reactions of food metabolism occur in the fluid matrix contained within the inner membrane; the rest are conducted by a series of enzymes attached to the membranes of the cristae.

Chloroplasts Are the Site of Photosynthesis
Chloroplasts (Fig. 6-15) are specialized organelles, surrounded by a double membrane, that are the sites of photosynthesis in the eukaryotic cells of plants and of photosynthetic protists. The inner membrane of the chloroplast encloses a semifluid material called the **stroma**. Embedded within the stroma are interconnected stacks of hollow membranous sacs. The individual sacs are called **thylakoids**, and a stack of sacs is a **granum** (plural, **grana**). During chloroplast development, thylakoids probably bud off from the inner membrane into the stroma; in mature chloroplasts, thylakoids are not connected to the inner membrane.

The thylakoid membranes contain the green pigment **chlorophyll** (which gives plants their green color) as well as other pigment molecules. During photosynthesis, chlorophyll captures the energy of sunlight and transfers it to other molecules in the thylakoid membranes. These molecules in turn transfer the energy to ATP and other energy-carrier molecules. The energy carriers diffuse into the stroma, where their energy is used to drive the synthesis of sugar from carbon dioxide and water. Photosynthesis is described in more detail in Chapter 7.

Plants Use Plastids for Storage

Chloroplasts are just one type of **plastid.** Plastids are organelles, found only in plants and photosynthetic protists, that are surrounded by a double membrane and serve a variety of functions. Plants and photosynthetic protists use other types of plastids as storage containers for various types of molecules, including pigments that give ripe fruits their yellow, orange, or red colors. In plants that continue growing from one year to the next, plastids store photosynthetic products from the summer for use during the following winter and spring. Most plants convert the sugars made during photosynthesis into starch, which is stored in plastids (Fig. 6-16). Potatoes, for example, are masses of cells stuffed with starch-filled plastids.

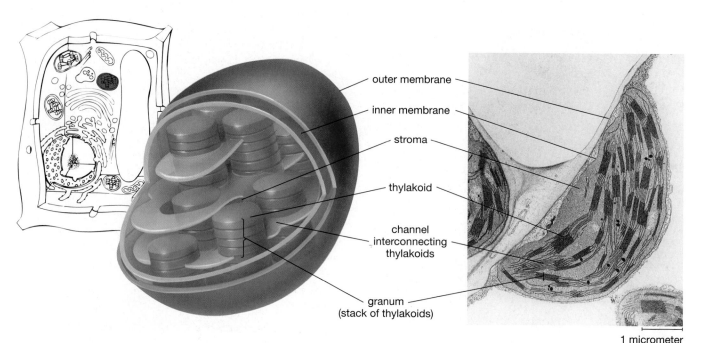

1 micrometer

Figure 6-15 *A chloroplast*
Chloroplasts are surrounded by a double membrane, although the inner membrane is not usually visible in electron micrographs. Enclosed by the inner membrane is the semifluid stroma, in which are embedded stacks of sacs collectively referred to as grana. The individual sacs of the grana are called thylakoids. Chlorophyll is embedded in the membranes of the thylakoids.

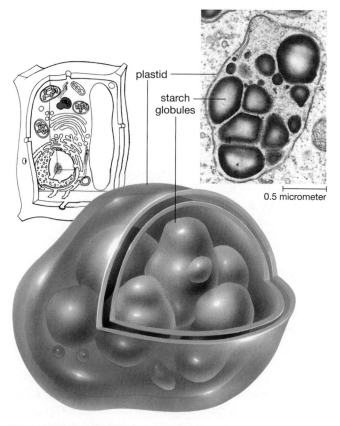

0.5 micrometer

Figure 6-16 *A plastid*
Plastids, found in the cells of plants and plantlike protists, are organelles surrounded by a double outer membrane. Chloroplasts are the most familiar plastids; other types store various materials, such as the starch filling these plastids in potato cells.

The Cytoskeleton Provides Shape, Support, and Movement

Organelles do not drift about the cytoplasm haphazardly; most are attached to a network of protein fibers called the cytoskeleton (Fig. 6-17). Even individual enzymes, which are often parts of complex metabolic pathways, may be fastened in sequence to the cytoskeleton, so that molecules can be passed from one enzyme to the next. Several types of protein fibers, including thin **microfilaments**, medium-sized **intermediate filaments**, and thick **microtubules**, make up the cytoskeleton (Table 6-2).

The cytoskeleton performs the following important functions:

1. *Cell shape.* In cells without cell walls, the cytoskeleton, especially networks of intermediate filaments, determines the shape of the cell.
2. *Cell movement.* The assembly, disassembly, and sliding of microfilaments and microtubules cause cell movement. Cell movement includes both the "crawling" of white blood cells, the contraction of muscle cells, and the migration and shape changes that occur during the development of multicellular organisms.
3. *Organelle movement.* Microtubules and microfilaments move organelles from place to place within a cell. For example, microfilaments attach to vesicles formed during *endocytosis* (see Chapter 5), when large

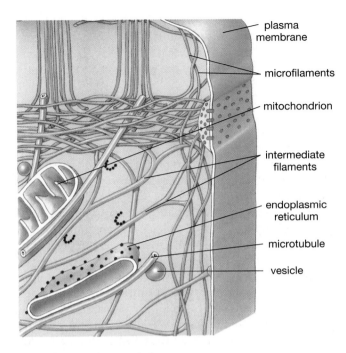

plasma membrane

microfilaments

mitochondrion

intermediate filaments

endoplasmic reticulum

microtubule

vesicle

Figure 6-17 *The cytoskeleton*
Eukaryotic cells are given shape and organization by the cytoskeleton, which consists of three types of proteins: microtubules, intermediate filaments, and microfilaments.

particles are engulfed by the plasma membrane and pull the vesicles into the cell. Vesicles budded off the ER and Golgi complex are probably guided by the cytoskeleton as well.

4. *Cell division.* Microtubules and microfilaments are essential to cell division in eukaryotic cells. First, when eukaryotic nuclei divide, microtubules move the chromosomes into the daughter nuclei. Second, in animal cells, division of the cytoplasm of a single parent cell into two new daughter cells results from the contraction of a ring of microfilaments that pinch the "waist" of the parent cell around the middle. Cell division will be covered in detail in Chapter 11.

Cilia and Flagella Move the Cell or Move Fluid Past the Cell

Both **cilia** (Latin for "eyelash") and **flagella** ("whip") are slender extensions of the plasma membrane. Each cilium and flagellum contains a ring of nine fused pairs of microtubules, with an unfused pair of microtubules in the center of the ring (forming what is called a *"9 + 2" arrangement*; Fig. 6-18). This pattern of microtubules is

Table 6-2 Components of the Cytoskeleton

	Structure	Protein Structure	Function
Microfilaments	Twisted double strands, each consisting of a string of protein subunits; about 7 nm in diameter and up to several centimeters long (in muscle cells)	Actin actin subunits	Muscle contraction; changes in cell shape, including cytoplasmic division in animal cells; cytoplasmic movement; movement of pseudopodia
Intermediate Filaments	Consist of eight subunits composed of ropelike protein strands; 8–12 nm in diameter and 10–100 µm in length	Protein varies with tissue type subunit	Maintenance of cell shape; attachments of microfilaments in muscle cells; support of nerve cell extensions; attach cells together (desmosomes)
Microtubules	Tubes consisting of spiraling two-part protein subunits; about 25 nm in diameter and can be 50 µm in length	Tubulin tubulin subunit	Movement of chromosomes during cell division; movement of organelles within cytoplasm; movement of cilia and flagella

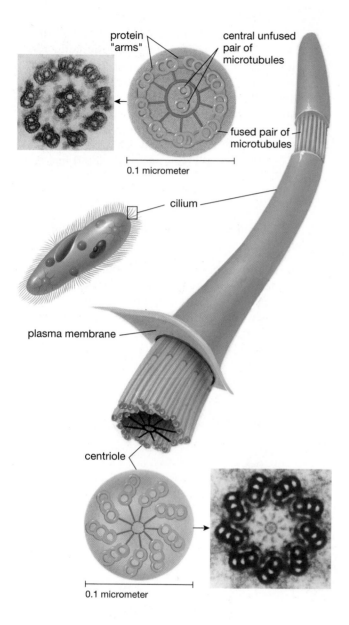

protein "arms"

central unfused pair of microtubules

fused pair of microtubules

0.1 micrometer

cilium

plasma membrane

centriole

0.1 micrometer

Figure 6-18 Cilia and flagella
Both cilia and flagella contain microtubules arranged in an outer ring of nine fused pairs of microtubules surrounding a central unfused pair (a 9 + 2 arrangement). The nine outer pairs have "arms" made of protein that interact with adjacent pairs to provide the force for bending. Cilia and flagella arise from centrioles located just beneath the plasma membrane. Centrioles have nine fused triplets of microtubules, from which the fused pairs arise, and an indistinct central hub. It is not known how the centriole organizes the somewhat different structure of the cilium or flagellum.

produced by a centriole located just beneath the plasma membrane. A **centriole** is a short, barrel-shaped ring consisting of nine microtubule triplets. Two of the members of each triplet give rise to the pairs of microtubules in the cilium or flagellum.

Tiny "arms" of protein attach neighboring pairs of microtubules in cilia and flagella. When these arms flex, they slide one pair of microtubules relative to the next pair. The centriole anchors the "bottom" of the structure, so the cilia and flagella move during microtubule sliding. The energy of ATP powers the movement of the protein arms during microtubule sliding. Cilia and flagella often move almost continuously and, consequently, require enormous supplies of ATP, which are generated by mitochondria that are normally found in abundance near the centrioles.

The main differences between cilia and flagella lie in their length, number, and the direction of the force they generate. In general, cilia are short (about 10 to 25 micrometers long) and numerous. They provide force in a direction parallel to the plasma membrane, like the oars in a canoe. This is accomplished through a fairly stiff "rowing" motion during the forceful *power stroke*, with most of the bending occurring at the base of the cilium, and a flexible *return stroke* that brings the cilium back to its original position (Fig. 6-19a). Flagella are long (50 to 75 micrometers), normally few in number, and provide force perpendicular to the plasma membrane, like the engine on a motorboat. Flagella undulate with a continuous bending wavelike motion, without distinct power-and-return strokes (Fig. 6-19b).

Some unicellular organisms, such as *Paramecium* and *Euglena*, use cilia or flagella to move about. Most animal sperm also rely on flagella for movement. In small multicellular animals, cilia are occasionally used for locomotion, moving the animal through water. Many small aquatic invertebrates, for example, swim by the coordinated beating of rows of cilia, like the oars on a Roman galley. More commonly, however, the organism stays put, and its cilia move fluids and suspended particles past a surface. Ciliated cells line such diverse structures as the gills of oysters (moving food- and oxygen-rich water), the oviducts of female mammals (moving the eggs along from the ovary to the uterus), and the respiratory tracts of most land vertebrates (clearing mucus that carries debris and microorganisms from the windpipe and lungs; see Fig 6-19a).

Prokaryotic cells may also bear slender protrusions that undulate or spin, thereby enabling the cell to move about. However, the "flagella" of prokaryotic cells do not contain microtubules and have no evolutionary relationship to eukaryotic flagella or cilia.

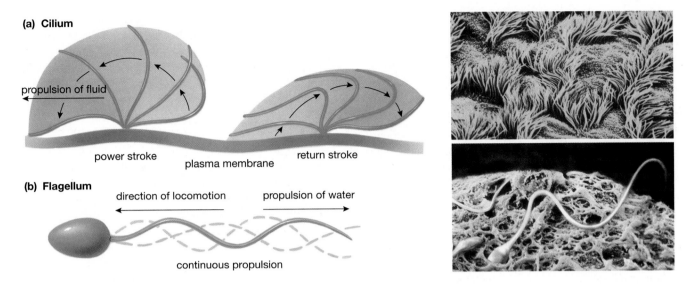

(a) Cilium

propulsion of fluid

power stroke   plasma membrane   return stroke

(b) Flagellum

direction of locomotion   propulsion of water

continuous propulsion

Figure 6-19 How cilia and flagella move
(a) (left) Cilia usually "row" along, providing a force of movement parallel to the plasma membrane, just as oars provide movement parallel to the sides of a rowboat. Their movement resembles the arms of a swimmer doing the breast stroke. (right) SEM photo of cilia lining the trachea (which conducts air to the lungs); they sweep out mucus and trapped particles. *(b)* (left) Flagella often move in a wavelike motion, with a continuous bending that starts at the base and moves up to the tip. This motion provides a force of movement perpendicular to the plasma membrane. In this way a flagellum attached to a sperm can move the sperm straight ahead. (right) A human sperm cell on the surface of a human egg cell.

Summary of Key Concepts

1 What Are the Basic Features of Cells?
The principles of the cell theory are:

1. Every living organism is made up of one or more cells.
2. The smallest living organisms are single cells, and cells are the functional units of multicellular organisms.
3. All cells arise from preexisting cells.

Cells are limited in size by two constraints. First, if a cell were too large, the rate of diffusion of essential materials from the outer surface of the cell to the center of the cell would be too slow to sustain life. Second, as a roughly spherical cell enlarges, its volume increases more rapidly than its surface area. Therefore, the surface area of a very large cell would be too small to service the metabolic needs of the cell's cytoplasm.

All cells are either prokaryotic or eukaryotic. Prokaryotic cells, represented by the bacteria, are small and relatively simple in structure. Eukaryotic cells make up protists, plants, fungi, and animals.

2 What Are the Features of Prokaryotic Cells?
Prokaryotic cells are generally very small with a relatively simple internal structure. Most are surrounded by a relatively stiff cell wall. The cytoplasm of prokaryotic cells lacks membrane-enclosed organelles (although some photosynthetic bacteria have elaborate internal membranes). A single, circular strand of DNA is found in the nucleoid. Table 6-1 compares prokaryotic cells to the eukaryotic cells of plants and animals.

3 What Are the Features of Eukaryotic Cells?
Genetic material (DNA) is contained within the nucleus, which is bounded by the double membrane of the nuclear envelope. Pores in the nuclear envelope regulate the movement of molecules between nucleus and cytoplasm. The genetic material of eukaryotic cells is organized into linear strands called chromosomes, which consist of DNA and proteins. The nucleolus consists of the genes that code for ribosome synthesis, together with ribosomal RNA and ribosomal proteins. Ribosomes are particles of ribosomal RNA and protein that are the sites of protein synthesis.

The membrane system of a cell consists of the plasma membrane, endoplasmic reticulum (ER), Golgi complex, and vesicles derived from these membranes. Endoplasmic reticulum with ribosomes, called rough ER, manufactures many cellular proteins. Endoplasmic reticulum without ribosomes, called smooth ER, manufactures lipids. The ER is the site of all membrane synthesis within the cell. The Golgi complex is a series of membranous sacs derived from the ER. The Golgi complex processes and modifies materials synthesized in the rough and smooth ER. Some substances in the Golgi are packaged into vesicles for transport elsewhere in the cell. Lysosomes are vesicles that contain digestive enzymes, which digest food particles and defective organelles.

All eukaryotic cells contain mitochondria, organelles that use oxygen to complete the metabolism of food molecules, capturing much of their energy as ATP. Cells of plants and some protists contain plastids, including chloroplasts, which capture the energy of sunlight during photosynthesis, enabling the cells to manufacture organic molecules, particularly sugars, from simple inorganic molecules. Both mitochondria and chloroplasts probably originated from bacteria. Storage plastids store pigments or starch.

Many cells contain sacs, called vacuoles, that are bounded by a single membrane and that store food or wastes, excrete water, or support the cell. Some protists have contractile vacuoles, which collect and expel water. Plants use central vacuoles to support the cell as well as to store wastes and toxic materials.

The cytoskeleton organizes and gives shape to eukaryotic cells and moves and anchors organelles. The cytoskeleton is composed of microfilaments, intermediate filaments, and microtubules (see Table 6-2). Cilia and flagella are whiplike extensions of the plasma membrane that contain microtubules in a characteristic pattern. These structures move fluids past the cell or move the cell through its fluid environment.

Study Note

Figures 6-4 and 6-5 illustrate the overall structure of animal and plant cells, respectively. Table 6-1 lists the principal organelles, their functions, and their occurrence in prokaryotic cells, animal cells, and plant cells.

Key Terms

aerobic   p. 102
anaerobic   p. 102
capsule   p. 91
central vacuole   p. 101
centriole   p. 106
chlorophyll   p. 103
chloroplast   p. 102
chromatin   p. 96
chromosome   p. 96
cilium   p. 105
contractile vacuole
   p. 101
crista   p. 103
cytoplasm   p. 88

cytoskeleton   p. 94
deoxyribonucleic acid
   (DNA)   p. 88
endoplasmic reticulum (ER)
   p. 97
eukaryotic   p. 88
flagellum   p. 105
food vacuole   p. 99
Golgi complex   p. 98
granum   p. 103
intermediate filament
   p. 104
intermembrane
   compartment   p. 103

lysosome   p. 99
matrix   p. 103
microfilament   p. 104
microtubule   p. 104
mitochondrion   p. 102
nuclear envelope   p. 96
nucleoid   p. 88
nucleolus   p. 96
nucleus   p. 88
organelle   p. 88
pilus   p. 91
plasma membrane   p. 88
plastid   p. 103
prokaryotic   p. 88

ribonucleic acid (RNA)
   p. 91
ribosome   p. 91
rough endoplasmic reticulum
   p. 97
slime layer   p. 91
smooth endoplasmic reticulum
   p. 97
stroma   p. 103
thylakoid   p. 103
turgor pressure   p. 101
vacuole   p. 100
vesicle   p. 98

Thinking Through the Concepts

Multiple Choice

1. *The outermost boundary of an animal cell is the*
   a. plasma membrane   b. nucleus
   c. cytoplasm   d. cytoskeleton
   e. cell wall

2. *Which organelle contains a eukaryotic cell's genetic material?*
   a. Golgi complex   b. ribosomes
   c. nucleus   d. mitochondria
   e. chloroplast

3. *Most of the cell's ATP is synthesized in the*
   a. Golgi complex   b. ribosomes
   c. nucleus   d. mitochondria
   e. chloroplast

4. *Which organelle sorts, chemically modifies, and packages newly synthesized proteins?*
   a. Golgi complex   b. ribosomes
   c. nucleus   d. mitochondria
   e. chloroplast

5. *Membrane-enclosed digestive organelles that contain enzymes are called*
   a. lysosomes
   b. smooth endoplasmic reticulum
   c. cilia
   d. Golgi complex
   e. mitochondria

6. *A series of membrane-enclosed channels studded with ribosomes are called*
   a. lysosomes
   b. Golgi complex
   c. rough endoplasmic reticulum
   d. mitochondria
   e. smooth endoplasmic reticulum

? Review Questions

1. Diagram "typical" prokaryotic and eukaryotic cells, and describe their important similarities and differences.

2. Which organelles are common to both plant and animal cells, and which are unique to each?

3. Define *stroma* and *matrix.*

4. Describe the nucleus, including the nuclear envelope, chromatin, chromosomes, DNA, and the nucleolus.

5. What are the functions of mitochondria and chloroplasts? In each case, explain how the internal structure of these organelles supports its function.

6. What is the function of ribosomes? Where in the cell are they typically found?

7. Describe the structure and function of the endoplasmic reticulum and Golgi complex.

8. How are lysosomes formed? What is their function?

9. Diagram the structure of cilia and flagella.

Applying the Concepts

1. If muscle biopsies (samples of tissue) were taken from the legs of a world-class marathon runner and a typical couch potato, which would you expect to have a higher density of mitochondria? Why? What about a muscle biopsy from the biceps of a weight lifter?

2. One of the functions of the cytoskeleton in animal cells is to give shape to the cell. Plant cells have a fairly rigid cell wall surrounding the plasma membrane. Does this mean that a cytoskeleton is superfluous for a plant cell? Defend your answer in terms of other functions of the cytoskeleton.

3. Most cells are very small. What physical and metabolic constraints limit cell size? What problems would an enormous cell encounter? What adaptations might help a very large cell survive?

Group Activity

Work in groups of two to estimate the number of cells in your little finger. Assume that your finger is a cylinder, for which volume ≈ 3(radius)^2(height). Assume also that the average size of one of your body cells is 1000 cubic micrometers.

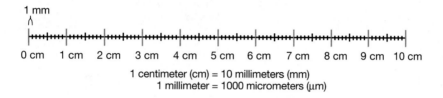

1 centimeter (cm) = 10 millimeters (mm)
1 millimeter = 1000 micrometers (μm)

For More Information

de Duve, C. "The Birth of Complex Cells." *Scientific American*, April 1996. Describes the mechanisms by which the first eukaryotic cells were produced from prokaryotic ancestors.

Ford, B. J. "The Earliest Views." *Scientific American*, April 1998. The author used the original microscopes of Anton van Leeuwenhoek to see the microscopic world as Leeuwenhoek saw it. Photographic images taken through these early and very primitive instruments reveal remarkable detail.

Glover, D. M., Gonzalez, C., and Raff, J. W. "The Centrosome." *Scientific American*, June 1993. The centrosome is the origin of much of a cell's cytoskeleton and therefore regulates cell shape, movement, and cell division.

Goodsell, D. S. *The Machinery of Life*. New York: Springer, 1993. Wonderful, drawn-to-scale images of the organelles and molecules of the cell.

Ingber, D. E. "The Architecture of Life." *Scientific American*, January 1998. Counteracting forces stabilize the design of organic structures, from carbon compounds to the cytoskeleton-reinforced architecture of the cell.

Kiester, E., Jr. "A Bug in the System." *Discover*, February 1991. Mitochondria may be the descendants of bacteria that live within our cells. Because mitochondria are essential for human life, defects in mitochondrial genes can cause some devastating diseases.

Murray, M. "Life on the Move." *Discover*, March 1991. Many cells, including *Amoeba* and white blood cells, crawl about by means of tiny protein "motors" to manipulate their cytoskeleton.

Rothman, J. "The Compartmental Organization of the Golgi Apparatus." *Scientific American*, September 1985. Although they look homogeneous in electron micrographs, the sacs of the Golgi actually form three biochemically distinct processing units.

Rothman, J. E., and Orci, L. "Budding Vesicles in Living Cells." *Scientific American*, March 1996. Researchers are uncovering the mechanisms by which cells form vesicles.

Symmons, M., Prescott, A., and Warn, R. "The Shifting Scaffolds of the Cell." *New Scientist*, February 18, 1989. Good overview showing the dynamic nature of the cytoskeleton.

Answers to Multiple-Choice Questions

1.a   2.c   3.d   4.a   5.a   6.c

" . . .the cells of life
Bound themselves into clans, a multitude of cells
To make one being—as the molecules before
Had made of many one cell. Meanwhile they had invented
Chlorophyll and ate sunlight, cradled in peace
On the warm waves."

Robinson Jeffers in The Beginning and the End *(1963)*

A sunburst through the leaves of a fern depicts the harvesting of light energy by photosynthesis, the ultimate source of energy for nearly all life on Earth.

Capturing Solar Energy: Photosynthesis

7

At a Glance

Net Watch

On-line resources for this chapter are on the World Wide Web at:
http://www.prenhall.com/audesirk
(click on the Table of Contents link and then select Chapter 7).

Late one winter in the 1950s, our family decided to make maple syrup. We drove taps into several maple trees on our property in the wilds of New Jersey and hung small buckets from them. Each morning I rushed around to every tree and was thrilled by the accumulation of watery, faintly sweet liquid. We learned several things from that experience, such as don't try to make syrup in your kitchen. Sap must be boiled down to one thirtieth of its original volume; the walls and ceiling of our small, poorly insulated house literally dripped with water condensed from the boiling pots. But the syrup was special, and with a child's unquestioning acceptance of things that naturally are, I never wondered why organisms as different from one another as I and a maple tree could both be energized by sugar.

To shed light on this puzzle, we must revisit Earth at least 2 billion years ago when some cells, through chance mistakes in their genetic machinery, acquired the ability to harness the energy in sunlight. These cells combined simple inorganic molecules—carbon dioxide and water—into more complex organic molecules such as glucose. In the process of *photosynthesis*, these cells captured a small fraction of the sunlight's energy, storing it as chemical energy in these complex organic molecules. Exploiting this new source of energy without competition, early photosynthetic cells filled the seas, releasing oxygen as a by-product. A new element in the atmosphere, free oxygen was harmful to many organisms. But the endless variation produced by random genetic mistakes (*mutations*) eventually gave rise to some cells that could survive in oxygen and, later, cells that made use of oxygen to break down glucose in a new, more efficient process: *cellular respiration*. Today, most forms of life on Earth—including you—depend on the sugar produced by photosynthetic organisms as an energy source, and release the energy from that sugar through cellular respiration, using the photosynthetic by-product, oxygen (Fig. 7-1). Sunlight powers life on Earth and is captured only by photosynthesis, which we explore in this chapter. In Chapter 8, we examine the processes by which all living things break down the sugary energy-storage

Figure 7-1 The relationships among photosynthesis and cellular respiration *(a)* Chloroplasts in green plants use the energy of sunlight to synthesize high-energy carbon compounds such as glucose from low-energy molecules of water and carbon dioxide. Plants themselves, and other organisms that eat plants or one another, extract energy from these organic molecules by cellular respiration, yielding water and carbon dioxide once again. This energy in turn drives all the reactions of life. *(b)* The bubbles released by the leaves of this aquatic plant (*Elodea*) are filled with oxygen, the by-product of photosynthesis.

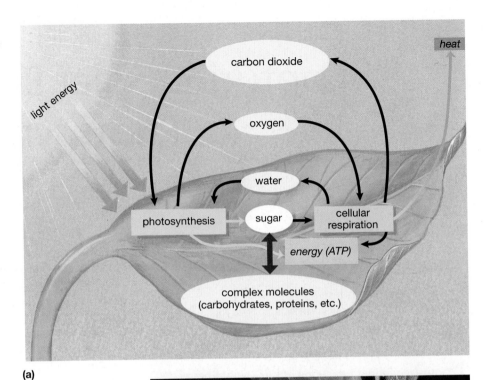

(a)

molecules produced by photosynthesis and reclaim the energy for powering other metabolic reactions.

Next time you see sunlight reflected off a green leaf, think of the equally stunning processes occurring below the leaf's surface as it absorbs the sun's energy. How does photosynthesis combine simple molecules of carbon dioxide and water to form sugar? Why are many leaves broad and flat, and why are they green? Why is the lush bluegrass of your lawn in springtime increasingly invaded by crabgrass as the summer progresses? Look for answers as you read this chapter.

(b)

1) What Is Photosynthesis? www

Starting with the simple molecules of carbon dioxide (CO_2) and water, **photosynthesis** converts the energy of sunlight into chemical energy stored in the bonds of glucose ($C_6H_{12}O_6$) and oxygen. The overall chemical reaction for photosynthesis is:

$$6\ CO_2 + 6\ H_2O + \text{light energy} \rightarrow C_6H_{12}O_6 + 6\ O_2$$

Photosynthesis occurs in plants, algae, and certain types of bacteria, all of which are *autotrophs* (literally, "self-feeders"). In this chapter, we will restrict our discussion of photosynthesis to plants, particularly land plants. Photosynthesis in plants takes place within the chloroplasts, most of which are contained in leaf cells. Let's begin, then, with a brief look at the structures of leaves and chloroplasts. Chapter 23 will examine leaf structure and function more thoroughly.

Leaves and Chloroplasts Are Adaptations for Photosynthesis

The leaves of most land plants are only a few cells thick (Fig. 7-2a). Both the upper and lower surfaces of a leaf consist of a layer of transparent cells, the *epidermis*. The outer surface of both epidermal layers is covered by the *cuticle*, a waxy, waterproof covering that reduces the evaporation of water from the leaf. A leaf obtains CO_2 for photosynthesis from the air; adjustable pores in the epidermis, called **stomata** (singular, **stoma**; Greek for mouth"), open and close at appropriate times to admit CO_2. Inside the leaf are a few layers of cells collectively called **mesophyll** (which means "middle of the leaf"). The mesophyll cells contain the vast majority of a leaf's chloroplasts (Fig. 7-2b), and consequently photosynthesis occurs principally in these cells. *Vascular bundles*, or veins, supply water and minerals to the mesophyll cells and carry the sugars they produce to other parts of the plant.

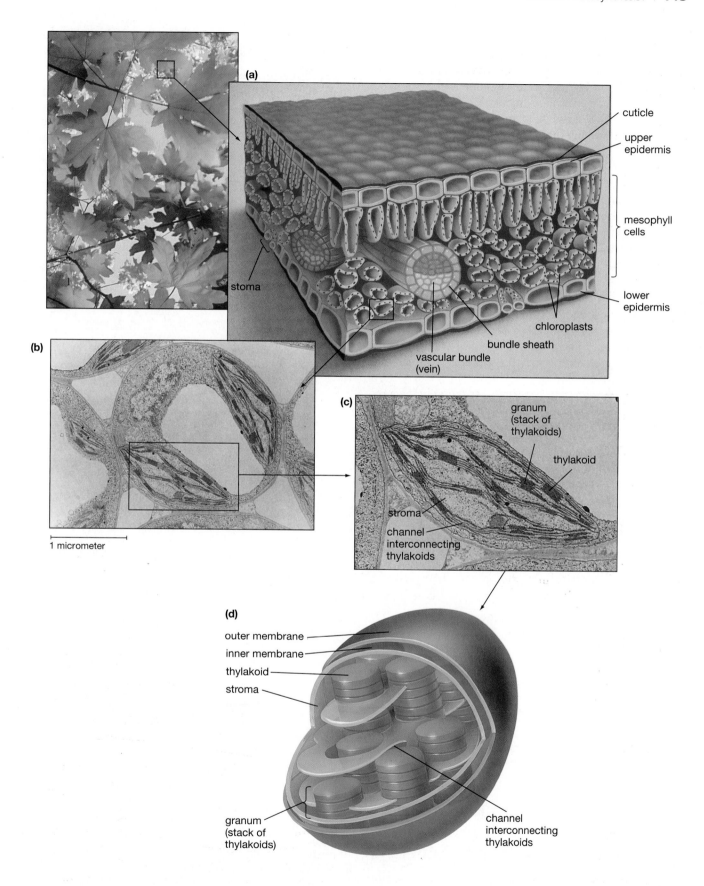

Figure 7-2 An overview of photosynthetic structures
(a) In land plants, photosynthesis normally occurs in the leaves, shown here in cross section. *(b)* An electron micrograph of chloroplast-containing mesophyll cells, located in the middle of the leaf. *(c)* A single chloroplast, as seen in the EM photo and *(d)* in diagrammatic form.

As we discussed in Chapter 6, chloroplasts are organelles that consist of a double outer membrane enclosing a semifluid medium, the **stroma** (Fig. 7-2c,d). Embedded in the stroma are disk-shaped, interconnected membranous sacs called **thylakoids**. In most chloroplasts, the thylakoids are piled atop one another in stacks called **grana** (singular, **granum**).

Photosynthesis Consists of Light-Dependent and Light-Independent Reactions

The simple overall chemical reaction of photosynthesis actually involves dozens of enzymes that catalyze dozens of individual reactions. The process of photosynthesis can be summarized in terms of two stages of reactions coupled by energy-carrier molecules: (1) *light-dependent reactions* and (2) *light-independent reactions*. Each reaction occurs in a different site in the chloroplast.

1. In the **light-dependent reactions**, chlorophyll and other molecules in the membranes of the thylakoids capture sunlight energy and convert some of it into the chemical energy of energy-carrier molecules (ATP and NADPH).
2. In the **light-independent reactions**, enzymes in the stroma use the chemical energy of the carrier molecules to drive the synthesis of glucose or other organic molecules.

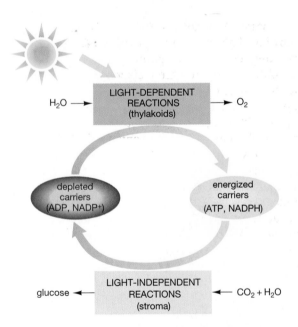

During Photosynthesis, Light Is First Captured by Pigments in Chloroplasts

The sun emits energy in a broad spectrum of electromagnetic radiation. The *electromagnetic spectrum* ranges from short-wavelength gamma rays, through ultraviolet, visible, and infrared light, to very long wavelength radio waves (Fig. 7-3a). Light and the other types of radiation are composed of individual packets of energy called **photons**. The energy of a photon corresponds to its wave-

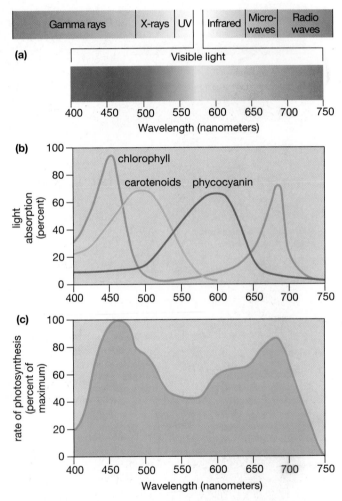

Figure 7-3 Light, chloroplast pigments, and photosynthesis
(a) Visible light, a small part of the electromagnetic spectrum, consists of wavelengths that correspond to the colors of the rainbow. *(b)* The first step in photosynthesis is the absorption of light by pigment molecules in the thylakoid membranes of chloroplasts. Different pigments selectively absorb certain colors; the height of the curves represents the ability of each pigment to absorb light of each color. Chlorophyll (green curve) strongly absorbs violet, blue, and red light and therefore looks green. The other pigments in chloroplasts absorb other colors of light. *(c)* Photosynthesis is driven to some extent by all colors of light, owing to the absorption of light by several thylakoid pigments.

2) Light-Dependent Reactions: How Is Light Energy Converted to Chemical Energy?

www

The light-dependent reactions convert the energy of sunlight into the chemical energy of two different carrier molecules: the familiar energy carrier ATP and the electron carrier *NADPH* (*nicotinamide adenine dinucleotide phosphate*).

length: Short-wavelength photons are very energetic, whereas longer-wavelength photons have lower energies. Visible light consists of wavelengths with energies that are strong enough to alter the shape of certain pigment molecules but weak enough not to damage crucial molecules such as DNA.

One of three processes might occur when light strikes an object such as a leaf: The light may be (1) *absorbed*, (2) *reflected* (bounced back again), or (3) *transmitted* (passed through). Light that is absorbed can heat up the object or drive biological processes, such as photosynthesis. Light that is reflected or transmitted gives an object its color.

Chloroplasts contain several types of molecules that absorb different wavelengths of light. **Chlorophyll**, the key light-capturing molecule in thylakoid membranes, strongly absorbs violet, blue, and red light but reflects green, and therefore it appears green (Fig. 7-3b). The predominance of chlorophyll in most leaves gives them their green color.

Thylakoids also contain other molecules, called **accessory pigments**, that capture light energy and transfer it to chlorophyll. **Carotenoids** absorb blue and green light and thus appear yellow, orange, or red; **phycocyanins** absorb green and yellow and thus appear blue or purple. Because all wavelengths of light are absorbed to some degree by either chlorophyll, carotenoids, or phycocyanins, all wavelengths can drive photosynthesis to some extent in plants with those pigments (Fig. 7-3c).

The Light-Dependent Reactions Occur in Clusters of Molecules Called Photosystems

In the thylakoid membranes, proteins including chlorophyll, accessory pigment molecules, and electron-carrier molecules form highly organized assemblies termed **photosystems**. Each thylakoid contains thousands of copies of two kinds of photosystems, named *photosystem I* and *photosystem II* (Fig. 7-4a). Each consists of two major parts: (1) a *light-harvesting complex* and (2) an *electron transport system*.

Each **light-harvesting complex** contains about 300 chlorophyll and accessory pigment molecules. These molecules absorb light and pass the energy to a specific chlorophyll molecule called the **reaction center**. By analogy with television reception, the light-absorbing pigments are called *antenna molecules*, because they gather energy and transfer it to the energy-processing reaction center. The reaction-center chlorophyll is located next to the **electron transport system**, a series of electron carrier molecules embedded in the thylakoid membrane.

When the reaction-center chlorophyll receives energy from the antenna molecules, one of its electrons absorbs the energy, leaves the chlorophyll, and jumps over to the electron transport system. This energetic electron moves from one carrier to the next. At some of the transfers, the electron releases energy. The energy drives reactions that result in the synthesis of ATP from ADP or NADPH

from $NADP^+$. (*NADP* is the electron carrier NAD, described in Chapter 4, plus a phosphate group.) With this overall scheme in mind, let's look more closely at the actual sequence of events in the light-dependent reactions (Fig. 7-4b).

Photosystem II Generates ATP

For historical reasons, the photosystems are numbered "backward": The usual process of light-energy capture is most easily understood by starting with photosystem II and following the events caused by the capture of two photons of light. The light-dependent reactions begin when photons of light are absorbed by the light-harvesting complex of photosystem II (step ① in Fig. 7-4b). Each photon's energy passes from molecule to molecule until it reaches the reaction center, where it boosts an electron out of the chlorophyll molecule (step ②). The first electron carrier of the adjacent electron transport system instantly accepts these energized electrons (step ③). The electrons move from one carrier molecule to the next, releasing energy as they go. Some of the energy is used to produce a hydrogen ion gradient within the thylakoid. This gradient drives the synthesis of ATP by a process known as **chemiosmosis** (step ④). "A Closer Look: Chemiosmosis—ATP Synthesis in Chloroplasts" (p. 117) describes chemiosmosis in more detail.

Photosystem I Generates NADPH

Meanwhile, light rays have also been striking the light-harvesting complex of photosystem I (step ⑤ in Fig. 7-4b). Each photon there ejects an electron from its reaction-center chlorophyll (step ⑥). These electrons jump to the electron transport system of photosystem I (step ⑦). Photosystem I's reaction-center chlorophyll immediately obtains replacements for its lost electrons from the last electron carrier in photosystem II's electron transport system. Photosystem I's high-energy electrons move through its electron transport system to the electron carrier $NADP^+$. Each $NADP^+$ molecule picks up two energetic electrons and one hydrogen ion, forming NADPH (step ⑧). Both $NADP^+$ and NADPH are water-soluble molecules dissolved in the chloroplast stroma.

Splitting Water Maintains the Flow of Electrons through the Photosystems

Overall, electrons flow from the reaction center of photosystem II, through the photosystem II electron transport system, to the reaction center of photosystem I, through the photosystem I electron transport system, and on to NADPH. To sustain this one-way flow of electrons, photosystem II's reaction center must be continuously supplied with new electrons to replace the ones it gives up. These replacement electrons come from water (step ⑨ in Fig. 7-4b). In a series of reactions that scientists are still deciphering, photosystem II's reaction-center chlorophyll attracts electrons from water molecules within the

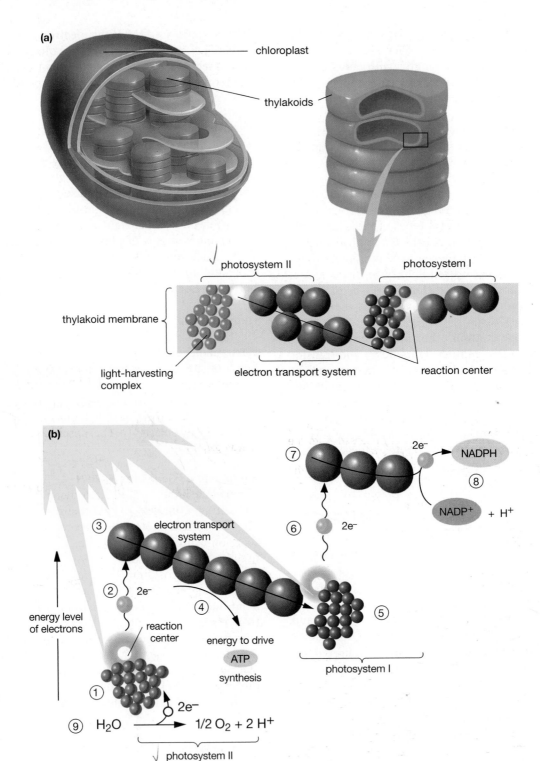

Figure 7-4 Thylakoid structure and the light-dependent reactions of photosynthesis
(a) The thylakoid membranes contain many copies of photosystems I and II. Each photosystem consists of a light-harvesting complex of pigment molecules and an adjacent electron transport system. *(b)* A summary of the light-dependent reactions. ① Light is absorbed by the light-harvesting complex of photosystem II, and the energy is passed to the reaction-center chlorophyll molecule. ② This energy ejects electrons out of the reaction center. ③ The electrons pass to the adjacent electron transport system. ④ The transport system passes the energetic electrons along, and some of their energy is used to pump hydrogen ions into the thylakoid interior. The hydrogen ion gradient thus generated can drive ATP synthesis. ⑤ Light strikes photosystem I, ⑥ causing it to emit electrons. ⑦ The electrons are captured by the photosystem I electron transport system. The electrons lost from the reaction center of photosystem I are replaced by those from the transport system of photosystem II. ⑧ The energetic electrons from photosystem I are captured in molecules of NADPH. ⑨ The electrons lost from the reaction center of photosystem II are replaced by electrons obtained from splitting water. This reaction also releases oxygen.

A Closer Look
Chemiosmosis—ATP Synthesis in Chloroplasts

In the electron transport system of photosystem II, energetic electrons move from carrier to carrier. Until about 1960, it was thought that ATP synthesis was directly coupled to these electron transfers: Where the exergonic steps (reactions in which there is a net release of energy) were large enough, the energy given up by the electrons drove ATP synthesis. However, it turns out that ATP synthesis in chloroplasts is *not* a simple coupled reaction. The electron transfers do not directly drive ATP synthesis; rather, the energy released during the transfers is used to generate a concentration gradient of hydrogen ions across the thylakoid membrane. In a separate reaction, the energy stored in this gradient then powers ATP synthesis. Let's follow these reactions.

In the first step of the light-dependent reactions of photosynthesis, a photon strikes the light-harvesting complex of photosystem II; the energy is absorbed and passed to the reaction-center chlorophyll. There, an electron absorbs the energy and is ejected out of the chlorophyll molecule. Within a billionth of a second, the electron is captured by the first electron carrier of the adjacent electron transport system.

As the electron moves from one carrier molecule to the next, it loses energy at each transfer. The energy released from the exergonic reaction of electron movement is used to transport hydrogen ions actively across the thylakoid membrane from the stroma into the thylakoid interior:

This transport raises the concentration of hydrogen ions (and therefore also the positive charge) inside the thylakoid, creating a gradient of both hydrogen ions and positive charge across the thylakoid membrane. The thylakoid membrane does not allow hydrogen ions to leak out, except at specific protein channels that are coupled to ATP-synthesizing enzymes. When hydrogen ions flow through these channels, down their gradients of charge and concentration, the energy released drives the synthesis of ATP:

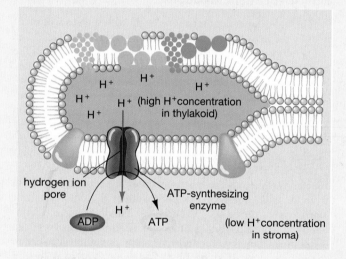

How does a gradient of hydrogen ions generate energy to drive ATP synthesis? Compare the hydrogen ion gradient across the thylakoid membrane to water stored behind a dam at a hydroelectric plant. When gates within the dam are opened, the rapid flow of water can be used to turn turbines and generate electricity. The thylakoid operates in an analogous way. Hydrogen ions in the thylakoid interior can move down their gradients into the stroma only through the ATP-synthesizing enzyme channels. This flow of hydrogen ions can do work—namely, drive ATP synthesis.

This mechanism of ATP synthesis was first proposed in 1961 by the British biochemist Peter Mitchell, who called it *chemiosmosis*. Chemiosmosis has been shown to be the mechanism of ATP generation in chloroplasts, mitochondria (as we shall see in Chapter 8), and bacteria. For his brilliant hypothesis, Mitchell was awarded the Nobel Prize for Chemistry in 1978.

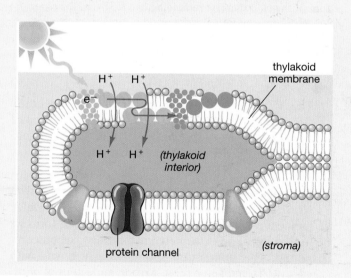

thylakoid compartment, causing the water molecules to split apart:

$$H_2O \rightarrow \frac{1}{2} O_2 + 2 H^+ + 2 e^-$$

For every two photons captured by photosystem II, two electrons are boosted out of the reaction-center chlorophyll and are replaced by the two electrons obtained by splitting one water molecule. As water molecules are split, their oxygen atoms combine to form molecules of oxygen gas, O_2. The oxygen may be used directly by the plant in its own respiration or given off to the atmosphere.

Summing Up
Light-Dependent Reactions

① The light-dependent reactions begin with the absorption of light by the light-harvesting complex of photosystem II. ② The light energy energizes electrons from the reaction center of the complex, causing them to be ejected. ③ The electrons are transferred to photosystem II's electron transport system. As the electrons pass through the transport system, they release energy. ④ Some of the energy is used to create a hydrogen ion gradient that drives ATP synthesis. ⑤ Meanwhile, light is absorbed by the light-harvesting complex of photosystem I. ⑥ The light energy ejects electrons from the reaction center that ⑦ are picked up by photosystem I's electron transport system. Electrons lost from the reaction center are replaced by those from the transport system of photosystem II. ⑧ Some of this energy is captured as NADPH. ⑨ Finally, the "electron-deprived" chlorophyll of photosystem II attracts electrons from water molecules. A water molecule splits, donating electrons to the photosystem II chlorophyll and generating oxygen as a by-product.

 ### Light-Independent Reactions: How Is Chemical Energy *www* Stored in Glucose Molecules?

The ATP and NADPH synthesized during the light-dependent reactions are dissolved in the stroma. There, they provide the energy to power the synthesis of glucose from carbon dioxide and water. The reactions that eventually produce glucose are called the light-independent reactions because they can occur independently of light as long as ATP and NADPH are available.

The C₃ Cycle Captures Carbon Dioxide

Carbon dioxide capture occurs in a set of reactions known as the **Calvin-Benson cycle** (after its discoverers), or the **C₃ (three-carbon) cycle**, because some of the important molecules in the cycle have three carbon atoms in them (Fig. 7-5). The C₃ cycle requires ① CO_2 (normally from the air); ② a CO_2-capturing sugar, *ribulose bisphosphate* (RuBP); ③ enzymes to catalyze all the reactions; and ④ energy in the form of ATP and NADPH (normally from the light-dependent reactions).

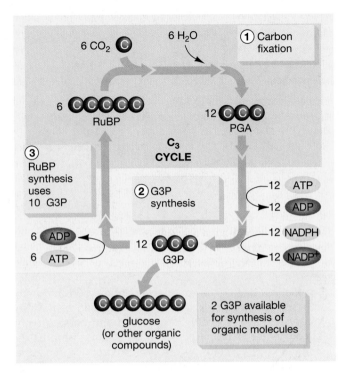

Figure 7-5 The C₃ cycle of carbon fixation
① Six molecules of RuBP react with six molecules of CO_2 and six molecules of H_2O to form 12 molecules of PGA. This reaction is carbon fixation, the capture of carbon from CO_2 into organic molecules. ② The energy of 12 ATPs and the electrons and hydrogens of 12 NADPHs are used to convert the 12 PGA molecules to 12 G3Ps. ③ Energy from six ATPs is used to rearrange 10 G3Ps into six RuBPs, completing one turn of the C₃ cycle. The remaining two G3P molecules are further processed into glucose or other organic molecules such as glycerol, fatty acids, or the carbon skeleton of amino acids, depending on the needs of the plant.

The C₃ cycle is best understood if we mentally divide it into ① carbon fixation, ② the synthesis of *glyceraldehyde-3-phosphate* (G3P), and ③ the regeneration of RuBP. Keep track of the carbons as you follow Figure 7-5.

1. *Carbon fixation.* The C₃ cycle begins (and ends) with RuBP, a five-carbon sugar. Enzymes combine six RuBP molecules with CO_2 from the atmosphere to form six molecules of an extremely unstable six-carbon compound. This compound spontaneously reacts with water to form twelve three-carbon molecules of *phosphoglyceric acid* (PGA), whose three carbons give the C₃ cycle its name. The capturing of CO_2 is called **carbon fixation**, because it "fixes" gaseous CO_2 into a relatively stable organic molecule, PGA (step ① in Fig. 7-5).
2. *Synthesis of G3P.* In a series of enzyme-catalyzed reactions, energy donated by ATP and NADPH (generated during the light-dependent reactions) is used to convert PGA to G3P (step ②).
3. *Regeneration of RuBP.* Through complex reactions requiring ATP energy, 10 molecules of G3P (10 × 3 car-

bons) regenerate the six molecules of RuBP (6 × 5 carbons; step ③) used at the start of carbon fixation.

Carbon Fixed during the C$_3$ Cycle Is Used to Synthesize Glucose

If the C$_3$ cycle starts with RuBP, adds carbon from CO$_2$, and ends one "turn" with RuBP again, there is carbon left over from the captured CO$_2$. Using the simplest "carbon accounting" numbers shown in Figure 7-5, if you start and end one turn of the cycle with six molecules of RuBP, two molecules of G3P are left over. These two G3P molecules (three carbons each) are combined to form one molecule of glucose (six carbons). Later, glucose may be broken down during cellular respiration or linked together in chains to form starch (a storage molecule) or cellulose (a major component of cell walls), or modified further into amino acids, lipids, and other cellular constituents.

Summing Up
Light-Independent Reactions

For the synthesis of one molecule of glucose through one turn of the C$_3$ cycle: Six molecules of CO$_2$ are captured by six molecules of RuBP. A series of reactions driven by energy from ATP and NADPH produces 12 molecules of G3P. Two molecules join to become one molecule of glucose. Further ATP energy is used to regenerate the six RuBP molecules from 10 G3P molecules.

What Is the Relationship between Light-Dependent and Light-Independent Reactions?

www

As Figure 7-6 illustrates, the light-dependent and light-independent reactions are closely coordinated. The light-dependent reactions in the thylakoids use light energy to "charge up" the energy-carrier molecules ADP and NADP$^+$ to form ATP and NADPH. These energized carriers move to the stroma, where their energy is used to drive glucose synthesis by means of the light-independent reactions. The depleted carriers, ADP and NADP$^+$, then return to the light-dependent reactions for recharging to ATP and NADPH.

Water, CO$_2$, and the C$_4$ Pathway

www

Photosynthesis requires light and carbon dioxide. Therefore, you might think that an ideal leaf should have a large surface area to intercept lots of sunlight and be very porous to allow lots of

CO$_2$ to enter the leaf from the air. For land plants, however, being porous to CO$_2$ also allows water to evaporate easily from the leaf. Loss of water from leaves is a prime cause of stress to land plants and may be fatal.

Many plants have evolved leaves that are a compromise between obtaining adequate CO$_2$ supplies and reducing water loss: large, waterproof leaves with adjustable pores, the stomata, that readily admit carbon dioxide (see Fig. 7-2; Chapter 23 supplies more details of leaf structure). When water supplies are adequate, the stomata open, letting in CO$_2$. If the plant is in danger of drying out, the stomata close. Closing the stomata reduces evaporation but has two disadvantages: It reduces CO$_2$ intake and restricts the release of O$_2$, which is produced during photosynthesis.

When Stomata Are Closed to Conserve Water, Wasteful Photorespiration Occurs

What happens to carbon fixation when the stomata close, CO$_2$ levels drop, and O$_2$ levels rise? As it happens, the enzyme that catalyzes the reaction of RuBP with CO$_2$ is not very selective: It can cause *either* CO$_2$ *or* O$_2$ to combine with RuBP (Fig. 7-7a). When O$_2$, rather than CO$_2$, is combined with RuBP, the process is called **photorespiration**, rather than *carbon fixation*. During photorespiration (as in cellular respiration), O$_2$ is used up and CO$_2$ is generated. But unlike cellular respiration, photorespiration does not produce any useful cellular energy, and it prevents the C$_3$ pathway from synthesizing glucose.

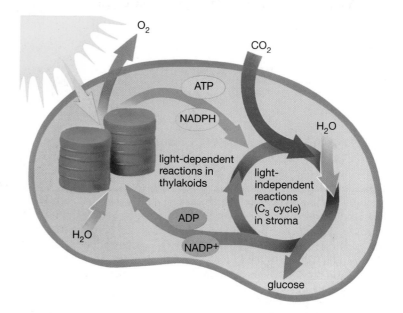

Figure 7-6 A summary diagram of photosynthesis
The light-dependent reactions in the thylakoids convert the energy of sunlight into the chemical energy of ATP and NADPH. Part of the sunlight energy is also used to split H$_2$O, forming O$_2$. In the stroma, the light-independent reactions (C$_3$ cycle) use the energy of ATP and NADPH to convert CO$_2$ and H$_2$O to glucose. The depleted carriers, ADP and NADP$^+$, return to the thylakoids to be recharged by the light-dependent reactions.

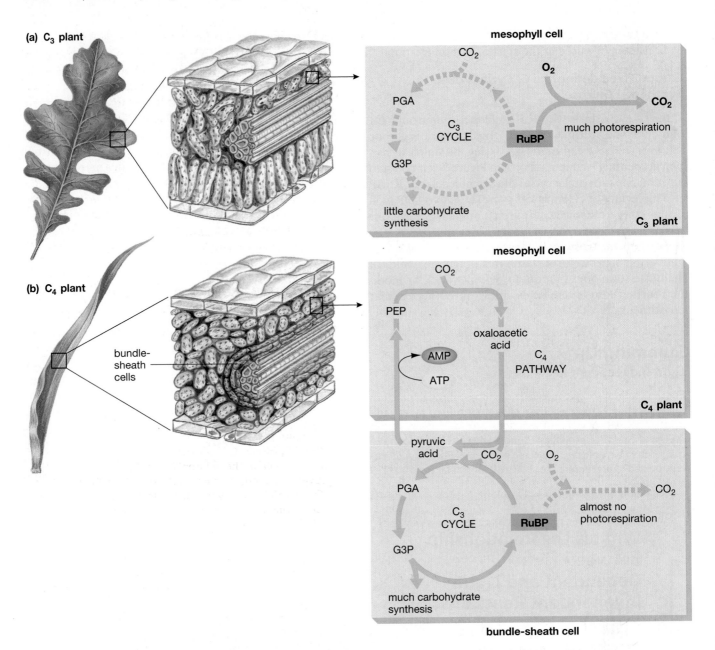

Figure 7-7 Comparison of C_3 and C_4 plants

(a) In C_3 plants, only the mesophyll cells carry out photosynthesis. All carbon fixation occurs by the C_3 pathway. With low CO_2 and high O_2 levels, photorespiration dominates in C_3 plants, because the enzyme that should catalyze the RuBP + CO_2 reaction catalyzes the RuBP + O_2 reaction instead. *(b)* In C_4 plants, both the mesophyll cells and bundle-sheath cells contain chloroplasts and participate in photosynthesis. The initial carbon-fixation step in the mesophyll cells is a reaction between phosphoenolpyruvic acid (PEP) and CO_2, with which O_2 does not compete. A four-carbon molecule of oxaloacetate is produced and releases CO_2 in the bundle-sheath cells, thus maintaining a high CO_2 concentration in their chloroplasts. Higher CO_2 levels allow efficient carbon fixation in the C_3 pathway of the bundle-sheath cells with little photorespiration. Notice that the regeneration of PEP requires energy: Two phosphates are removed from ATP to produce AMP (adenosine monophosphate).

Some photorespiration occurs all the time, even under the best of conditions. Recent research suggests that this process, in moderation, actually helps protect C_3 plants (plants that utilize the C_3 pathway) from the damaging effects of too much light. But during hot, dry weather the stomata seldom open; CO_2 from the air can't get in, and the O_2 generated by photosynthesis can't get out. With CO_2 levels low and O_2 levels very high, photorespiration dominates (Fig. 7-7a). Plants, especially seedlings, may die during hot, dry weather because they are unable to capture enough energy in glucose to meet their metabolic needs.

C₄ Plants Reduce Photorespiration by Means of a Two-Stage Carbon-Fixation Process

Some plants, described as *C₄ plants*, have evolved a way to reduce photorespiration and boost photosynthesis during dry weather. In the leaves of "regular" (C_3) plants, such as a maple, almost all the chloroplasts reside in the mesophyll cells. In plants such as corn and crabgrass, which are adapted to hot, dry conditions, both mesophyll cells and **bundle-sheath cells** (which surround the vascular bundles) contain chloroplasts (Fig. 7-7b). These plants use a two-stage carbon-fixation pathway, the **C₄ pathway**.

In C_4 plants, the mesophyll cells contain a three-carbon molecule called *phosphoenolpyruvate* (PEP) instead of RuBP. The CO_2 reacts with PEP to form a four-carbon molecule of *oxaloacetate* (for which C_4 plants are named). This reaction is highly specific for CO_2 and is not hindered by high O_2 concentrations. Oxaloacetate is used as a shuttle to transport carbon from mesophyll cells to bundle-sheath cells. There it breaks down, releasing CO_2 again. The high CO_2 concentration created in the bundle-sheath cells now allows the regular C_3 cycle to proceed with less competition from oxygen. The remnant of the shuttle molecule returns to the mesophyll cells, where ATP energy is used to regenerate the PEP molecule.

C₃ and C₄ Plants Are Each Adapted to Different Environmental Conditions

Plants using the C_4 process to fix carbon are locked into this pathway, which uses up more energy to produce glucose than the C_3 pathway. C_4 plants have an advantage when light energy is abundant but water is not. However, if water is plentiful (so the stomata of C_3 plants can stay open and let in lots of CO_2) or if light levels are low, the C_3 carbon fixation pathway is more efficient.

Consequently, C_4 plants thrive in deserts and in midsummer in temperate climates, when light energy is plentiful but water is scarce. The C_3 plants have the advantage, however, in cool, wet, cloudy climates, because the C_3 pathway is more energy-efficient. This is why your lawn of lush Kentucky bluegrass (a C_3 plant) may be taken over by spiky crabgrass (a C_4 plant) during a long, hot, dry summer.

Summary of Key Concepts

1) What Is Photosynthesis?
Photosynthesis uses the energy of sunlight to convert the inorganic molecules of carbon dioxide and water into high-energy organic molecules such as glucose. In plants, photosynthesis takes place in the chloroplasts, in two major steps: the light-dependent and the light-independent reactions. Photosynthesis is summarized in Figure 7-6.

2) Light-Dependent Reactions: How Is Light Energy Converted to Chemical Energy?
In the light-dependent reactions, light excites electrons in chlorophyll molecules and transfers the energetic electrons to electron transport systems. The energy of these electrons drives three processes:

1. *Photosystem II generates ATP.* Some of the energy from the electrons is used to pump hydrogen ions into the thylakoids. The hydrogen ion concentration is therefore higher inside the thylakoids than in the stroma outside. Hydrogen ions move down this concentration gradient through ATP-synthesizing enzymes in the thylakoid membranes, providing the energy to drive ATP synthesis.
2. *Photosystem I generates NADPH.* Some of the energy, in the form of energetic electrons, is added to electron-carrier molecules of $NADP^+$ to make the highly energetic carrier NADPH.
3. *Splitting water maintains the flow of electrons through the photosystems.* Some of the energy is used to split water, generating electrons, hydrogen ions, and oxygen.

3) Light-Independent Reactions: How Is Chemical Energy Stored in Glucose Molecules?
In the stroma of the chloroplasts, both ATP and NADPH provide the energy that drives the synthesis of glucose from CO_2 and H_2O. The light-independent reactions occur in a cycle of chemical reactions called the Calvin-Benson, or C_3, cycle. The C_3 cycle has three major parts:

1. *Carbon fixation.* Carbon dioxide and water combine with ribulose bisphosphate (RuBP) to form phosphoglyceric acid (PGA).
2. *Synthesis of G3P.* PGA is converted to glyceraldehyde-3-phosphate (G3P), using energy from ATP and NADPH. G3P may be used to synthesize organic molecules such as glucose.
3. *Regeneration of RuBP.* Ten molecules of G3P are used to regenerate six molecules of RuBP, again using ATP energy.

4) What Is the Relationship between Light-Dependent and Light-Independent Reactions?
The light-dependent reactions produce the energy carrier ATP and the electron carrier NADPH. Energy from these carriers is used in the synthesis of organic molecules during the light-independent reactions. The depleted carriers, ADP and $NADP^+$, return to the light-dependent reactions for recharging.

5) Water, CO$_2$, and the C$_4$ Pathway

The enzyme that catalyzes the reaction between RuBP and CO$_2$ may also catalyze a reaction, called photorespiration, between RuBP and O$_2$. Photorespiration prevents carbon fixation and does not generate ATP. If CO$_2$ concentrations drop too low or if O$_2$ concentrations rise too high, photorespiration may exceed carbon fixation. Some plants have evolved an additional step for carbon fixation that minimizes photorespiration. In the mesophyll cells of these C$_4$ plants, CO$_2$ combines with phosphoenolpyruvic acid (PEP) to form the four-carbon molecule oxaloacetate. Oxaloacetate is transported into adjacent bundle-sheath cells, where it releases CO$_2$, thereby maintaining a high CO$_2$ concentration in those cells. This CO$_2$ is then fixed in the C$_3$ cycle.

Key Terms

accessory pigments   *p. 115*
bundle-sheath cell   *p. 121*
C$_3$ cycle   *p. 118*
C$_4$ pathway   *p. 121*
Calvin-Benson cycle   *p. 118*
carbon fixation   *p. 118*
carotenoid   *p. 115*

chemiosmosis   *p.115*
chlorophyll   *p. 115*
electron transport system   *p. 115*
granum   *p. 114*
light-dependent reactions   *p. 114*

light-harvesting complex   *p. 115*
light-independent reactions   *p. 114*
mesophyll   *p. 112*
photon   *p. 114*
photorespiration   *p. 119*

photosynthesis   *p. 112*
photosystem   *p. 115*
phycocyanin   *p. 115*
reaction center   *p. 115*
stoma   *p. 112*
stroma   *p. 114*
thylakoid   *p. 114*

Thinking Through the Concepts

Multiple Choice

1. *Photosynthesis is measured in the leaf of a green plant exposed to different wavelengths of light. Photosynthesis is*
   a. highest in green light
   b. highest in red light
   c. highest in blue light
   d. highest in red and blue light
   e. the same at all wavelengths

2. *Where do the light-dependent reactions of photosynthesis occur?*
   a. in the stomata
   b. in the chloroplast stroma
   c. within the thylakoid membranes of the chloroplast
   d. in the leaf cell cytoplasm
   e. in leaf cell mitochondria

3. *The oxygen produced during photosynthesis comes from*
   a. the breakdown of CO$_2$
   b. the breakdown of H$_2$O
   c. the breakdown of both CO$_2$ and H$_2$O
   d. the breakdown of oxaloacetate
   e. photorespiration

4. *The role of accessory pigments is to*
   a. provide an additional photosystem to generate more ATP
   b. allow photosynthesis to occur in the dark
   c. prevent photorespiration
   d. donate electrons to chlorophyll reaction centers
   e. capture additional light energy and transfer it to the chlorophyll reaction centers

5. *The generation of ATP by electron transport in photosynthesis and cellular respiration depends on*
   a. a proton gradient across a membrane
   b. proton pumps driven by electron transport chains
   c. an ATP-synthesizing enzyme complex
   d. a, b, and c are all required for ATP generation
   e. none of the above are required for ATP generation

6. *Where do the light-independent, carbon-fixing reactions occur?*
   a. in the guard cell cytoplasm
   b. in the chloroplast stroma
   c. within the thylakoid membranes
   d. at night in the thylakoids
   e. in mitochondria

? Review Questions

1. Write the overall equation for photosynthesis. Does the overall equation differ between C$_3$ and C$_4$ plants?

2. Draw a diagram of a chloroplast, and label it. Explain specifically how chloroplast structure is related to its function.

3. Briefly describe the light-dependent and light-independent reactions. In what part of the chloroplast does each occur?

4. What is the difference between carbon fixation in C$_3$ and in C$_4$ plants? Under what conditions does each mechanism of carbon fixation work most effectively?

5. Describe the process of chemiosmosis in chloroplasts, tracing the flow of energy from sunlight to ATP.

Applying the Concepts

1. Many lawns and golf courses are planted with bluegrass, a C_3 plant. In the spring, the bluegrass grows luxuriously. In the summer, crabgrass, a weed and a C_4 plant, often appears and spreads rapidly. Explain this sequence of events, given the normal weather conditions of spring and summer and the characteristics of C_3 versus C_4 plants.

2. Suppose an experiment is performed in which plant I is supplied with normal carbon dioxide but with water that contains radioactive oxygen atoms. Plant II is supplied with normal water but with carbon dioxide that contains radioactive oxygen atoms. Each plant is allowed to perform photosynthesis, and the oxygen gas and sugars produced are tested for radioactivity. Which plant would you expect to produce radioactive sugars, and which plant would you expect to produce radioactive oxygen gas? Why?

3. You continuously monitor the photosynthetic oxygen production from the leaf of a plant illuminated by white light. Explain what will happen (and why) if you place (a) red, (b) blue, and (c) green filters between the light source and the leaf.

4. A plant is placed in a CO_2-free atmosphere in bright light. Will the light-dependent reactions continue to generate ATP and NADPH indefinitely? Explain how you reached your conclusion.

5. You are called before the Ways and Means Committee of the House of Representatives to explain why the U.S. Department of Agriculture should continue to fund photosynthesis research. How would you justify an expensive project to produce by genetic engineering the enzyme that catalyzes the reaction of RuBP with CO_2 to prevent RuBP from reacting with oxygen as well as CO_2. What are the potential applied benefits of this research?

Group Activity

Work in groups of two. Each of you should write down each component of the last meal you ate. Then, for each food item your partner should explain how that food contains energy derived orignally from the sun. Do the same with your partner's list.

For More Information

Bazzazz, F. A., and Fajer, E. D. "Plant Life in a CO_2-Rich World." *Scientific American*, January 1992. Burning fossil fuels is increasing CO_2 levels in the atmosphere (see Chapter 40). This increase could tip the balance between C_3 and C_4 plants.

Govindjee, and Coleman, W. J. "How Plants Make Oxygen." *Scientific American*, February 1990. The generation of oxygen during photosynthesis is just beginning to be understood.

Grodzinski, B. "Plant Nutrition and Growth Regulation by CO_2 Enrichment." *BioScience*, 1992. How higher CO_2 levels influence plant metabolism.

Hall D. O., and Rao, K. K. *Photosynthesis*. 5th ed. New Series in Biology. New York: Cambridge University Press, 1994. An excellent short book recommended to any student interested in finding out more about photosynthesis.

Hinkle, P. C., and McCarthy, R. E. "How Cells Make ATP." *Scientific American*, March 1978. A good explanation of chemiosmosis, which is a difficult concept for many students.

Monastersky, R. "Children of the C_4 World." *Science News*, January 3, 1998. What role did a shift in global vegetation toward C_4 photosynthesis play in the evolution of humans?

Mooney, H. A., Drake, B. G., Luxmoore, R. J., Oechel, W. C., and Pitelka, L. F. "Predicting Ecosystems' Responses to Elevated CO_2 Concentrations." *BioScience*, 1994. What effects will CO_2 enrichment of the atmosphere due to human activities have on ecosystems?

Zimmer, C. "The Processing Plant." *Discover*, September 1995. Describes organisms inhabiting the watery digestive chamber of the pitcher plant, which is both photosynthetic and carnivorous.

Answer to Multiple-Choice Questions

1. d   2. c   3. b   4. e   5. d   6. b

With wings beating 60 times per second, the ruby-throated hummingbird has a metabolic rate 50 times that of a human. The muscles of its wings are packed with mitochondria, which supply the ATP needed to meet the bird's energy demands.

Harvesting Energy: Glycolysis and Cellular Respiration

8

At a Glance

Net Watch

www

On-line resources for this chapter are on the World Wide Web at: http://www.prenhall.com/audesirk

(click on the Table of Contents link and then select Chapter 8).

Nearly all forms of life on Earth—from a maple tree, to a child eating pancakes drenched in maple syrup, to a polar bear feeding on a seal, to a hummingbird sipping nectar from a flower—rely directly or indirectly on light energy captured during photosynthesis. Photosynthesis converts the energy of sunlight into the chemical energy of organic molecules such as the sugar in nectar. The hummingbird must eat its weight in nectar daily to provide its enormous energy demands. The cells of the hummingbird must efficiently extract energy from the glucose in the nectar. When the hummingbird's cells break down glucose molecules in the presence of oxygen, the cells produce the molecules the plant started with—carbon dioxide and water—while making chemical energy available for rapid muscle movements, cellular repair, growth, and reproduction.

However, the hummingbird's cells cannot directly use the chemical energy derived from the breakdown of glucose. Its wing muscles require energy stored in the molecule adenosine triphosphate (ATP). Its brain uses ATP in conducting nerve signals, and its ovaries use ATP in making eggs. The plant, for its part, uses ATP to make the flower petals, pigments, and fragrance that attract the bird, and the leaves and chlorophyll molecules that trap the sun's energy. How is ATP produced from the energy stored in sugar? Why do many organisms, including the hummingbird and plant, die rapidly if deprived of oxygen? We know that some organisms live in environments that lack oxygen; can energy be extracted from glucose without oxygen? What role do the mitochondria play in energy extraction? In this chapter, we explore the metabolic processes of glycolysis and respiration by which organisms convert the energy of organic molecules into the usable energy of ATP.

1 How Is Glucose Metabolized?

www

Most cells can metabolize a variety of organic molecules to produce ATP. We focus on the metabolism of glucose for three reasons. First, virtually all cells metabolize glucose for energy at least part of the time. Some, such as nerve cells in your brain, rely predominantly on glucose as a source of energy. Second, glucose metabolism is less complex than the metabolism of most other organic molecules. Finally, when using other organic molecules as energy sources, cells normally convert the molecules to glucose or other compounds that enter the pathways of glucose metabolism (see "Health Watch: Why Can You Get Fat by Eating Sugar?").

As you learned in Chapter 7, during photosynthesis, photosynthetic organisms harvest and store the energy of sunlight in glucose. During glucose breakdown, that energy is released and converted to ATP. The chemical equations for glucose formation by photosynthesis and for the complete metabolism of glucose back to CO_2 and H_2O (the original reactants in photosynthesis) are almost perfectly symmetrical:

Photosynthesis:

$$6 CO_2 + 6 H_2O + \text{sunlight energy} \rightarrow C_6H_{12}O_6 + 6 O_2$$

Complete Glucose Metabolism:

$$C_6H_{12}O_6 + 6 O_2 \rightarrow 6 CO_2 + 6 H_2O$$
$$+ \text{ chemical and heat energies}$$

This symmetry might lead you to believe that a cell can convert all of the chemical energy contained in a glucose molecule to high-energy bonds of ATP. Unfortunately, according to the second law of thermodynamics, "you can't break even"; in other words, the conversion of energy into different forms always results in the decrease of the amount of concentrated, useful energy. In fact, most of the energy listed on the right-hand side of the glucose metabolism equation is heat energy, not the chemical energy of ATP. Nevertheless, a cell can extract a great deal of energy, in the form of ATP, from glucose if the glucose molecule is completely broken down to CO_2 and H_2O.

Figure 8-1 summarizes the major steps of glucose metabolism in eukaryotic cells. The first stage, *glycolysis*, does not require oxygen and proceeds in exactly the same way under both aerobic (with oxygen) and anaerobic

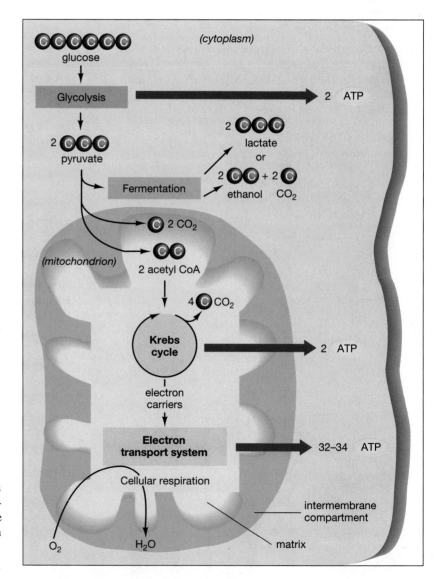

Figure 8-1 A summary of glucose metabolism
Refer to this diagram as we progress through the reactions of glycolysis (in the fluid portion of the cytoplasm) and cellular respiration (in the mitochondria). The breakdown of glucose occurs in stages, with various amounts of energy harvested as ATP along the way. The vast majority of the ATP is produced in the mitochondria, justifying their nickname, "powerhouse of the cell."

Health Watch
Why Can You Get Fat by Eating Sugar?

As you know, humans do not live by glucose alone. Nor does the typical diet contain exactly the required amounts of each nutrient. Accordingly, the cells of the human body seethe with biochemical reactions, synthesizing one amino acid from another, making fats from carbohydrates, and channeling surplus organic molecules of all types into energy storage or release. Let's look at two examples of these metabolic transformations: the production of ATP from fats and proteins and the synthesis of fats from sugars.

How Are Fats and Proteins Metabolized? Even the leanest people have some fat in their bodies. During fasting or starvation, the body mobilizes these fat reserves for ATP synthesis, because even the bare maintenance of life requires a continuous supply of ATP, and seeking out new food sources demands even more energy. Fat metabolism flows directly into the pathways of glucose metabolism.

Chapter 3 illustrated the structure of a fat: three fatty acids connected to a glycerol backbone. In fat metabolism, the bonds between the fatty acids and glycerol are hydrolyzed (broken into subunits by the addition of water). The glycerol part of a fat, after activation by ATP, feeds directly into the middle of the glycolysis pathway (Fig. E8-1). The fatty acids are transported into the mitochondria, where enzymes in the inner membrane and matrix chop them up into acetyl groups. These groups attach to coenzyme A to form acetyl CoA, which enters the Krebs cycle.

In individuals with severe starvation or those feeding almost exclusively on protein, amino acids can be used to produce energy. First, the amino acids are converted to pyruvate, acetyl CoA, or the compounds of the Krebs cycle. These molecules then proceed through the remaining stages of cellular respiration, yielding amounts of ATP that vary with their point of entry into the pathway.

How Is Fat Synthesized from Sugar? The body, in addition to having developed ways of coping with fasting or starvation, has also evolved strategies for coping with situations in which food intake exceeds current energy needs. The sugars and starches in corn flakes or candy bars can be converted into fats for energy storage. Complex sugars, such as starches and sucrose, are first hydrolyzed into their monosaccharide subunits (see Chapter 3). The monosaccharides are broken down to pyruvate and converted to acetyl CoA. If the cell needs ATP, the acetyl CoA will enter the Krebs cycle. If the cell has plenty of ATP, acetyl CoA will be used to make fatty acids by a series of reactions that are essentially the reverse of fatty-acid breakdown. In humans, the liver synthesizes fatty acids, but fat storage is relegated to fat cells, with their all-too-familiar distribution in the body, particularly around the waist and hips.

Energy use, fat storage, and eating are usually precisely balanced. Where the balance point lies, however, varies from person to person. Some people seem able to eat nearly continuously without ever storing much fat; other people crave high-calorie foods even when they have a lot of fat already stored. Before the

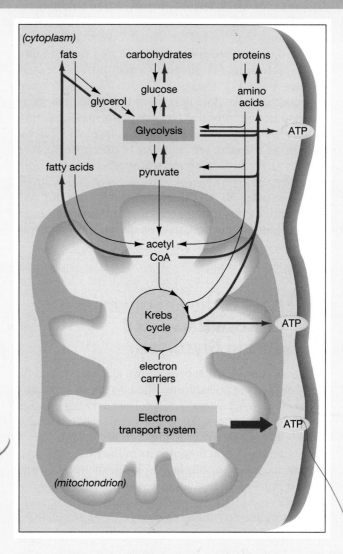

Figure E8-1 How various nutrients yield energy
Fats, carbohydrates such as starches, and proteins can all be broken down metabolically into molecules that enter glycolysis or the Krebs cycle, where they are used to generate ATP.

fashionably thin become too smug about their svelte condition, they should realize that, from an evolutionary point of view, overeating during times of easy food availability is highly adaptive behavior. If hard times come, the plump may easily (if hungrily) survive while the trim succumb to starvation. Only recently (over evolutionary time) have people in societies such as ours had continuous access to inexpensive, high-calorie food. Under these conditions, the drive to eat and the adaptation of storing excess food as fat will lead to obesity, a growing health problem in the United States.

(without oxygen) conditions. Glycolysis splits apart a single glucose molecule (a six-carbon sugar) into two three-carbon molecules of **pyruvate**. This splitting releases a small fraction of the chemical energy stored in the glucose, some of which is used to generate two ATP molecules. The presence of oxygen becomes an issue only in the processes that follow glycolysis. Under anaerobic conditions, the pyruvate is usually converted by fermentation into lactate or ethanol. Fermentation does not produce more ATP energy. Both glycolysis and fermentation occur in the fluid portion of the cytoplasm.

The pyruvate produced by glycolysis may also enter the mitochondria. There, if oxygen is available, cellular respiration uses oxygen to break pyruvate down completely to carbon dioxide and water, generating an additional 34 or 36 ATP molecules (the amount differs among cells). The extra ATP produced by cellular respiration is so important to most organisms that anything that interferes with its production, such as the poison cyanide or lack of oxygen, quickly results in death.

2) How Is the Energy in Glucose Harvested during Glycolysis?

www

The initial reactions that break down glucose without the use of oxygen are collectively called **glycolysis** (in Greek, "to break apart a sweet"). Glycolysis is believed to be one of the most ancient of all biochemical pathways, because it is used by every living creature on the planet. This sequence of reactions occurs in the fluid portion of the cytoplasm and results in a molecule of glucose being cleaved into two molecules of pyruvate. Glycolysis produces relatively little energy, only two molecules of ATP and two molecules of the electron carrier NADH. But without it, life would rapidly be extinguished. Reduced to

its essentials, glycolysis consists of two major parts (each with several steps): (1) glucose activation and (2) energy harvest (Fig. 8-2).

Glycolysis Breaks Down Glucose to Pyruvate, Releasing Chemical Energy

Before glucose is broken down, it must be activated. During glucose activation, a molecule of glucose undergoes two enzyme-catalyzed reactions, each of which uses ATP energy (Fig. 8-2a). These reactions convert a relatively stable glucose molecule into a highly unstable molecule of *fructose bisphosphate*. Fructose is a sugar molecule similar to glucose; *bisphosphate* refers to the two phosphate groups acquired from the ATP molecules. Forming fructose bisphosphate costs the cell two ATP molecules. But this initial consumption of energy is necessary to produce much greater energy returns in the long run.

In the energy harvest steps, fructose bisphosphate splits apart into two three-carbon molecules of glyceraldehyde 3-phosphate (G3P; Fig. 8-2b). (In Chapter 7 we encountered G3P in the C_3 cycle of photosynthesis.) Each G3P molecule then goes through a series of reactions that convert it to pyruvate. During these reactions, two ATP are generated for each G3P, for a total of four ATPs. Because two ATPs were used to activate the glucose molecule in the first place, *there is a net gain of only two ATPs per glucose molecule*. At another step along the way from G3P to pyruvate, two high-energy electrons and a hydrogen ion are added to the "empty" electron carrier NAD^+ to make the "energized" carrier NADH (this is a slightly different electron carrier from the $NADP^+$ used in photosynthesis). Two G3P molecules are produced per glucose molecule, so two NADH carrier molecules are formed when those G3P molecules are converted to pyruvate. For a complete description of glycolysis, see "A Closer Look: Glycolysis."

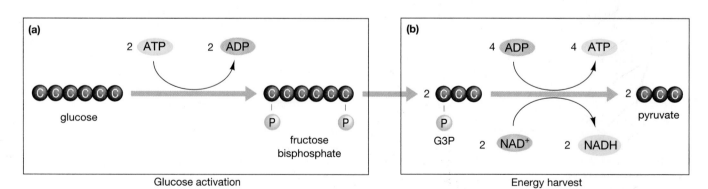

Figure 8-2 The essentials of glycolysis
(a) Glucose activation: The energy of two ATP molecules is used to convert glucose to the highly reactive fructose bisphosphate. Fructose bisphosphate splits into two smaller, but still reactive, molecules of glyceraldehyde 3-phosphate (G3P). **(b)** Energy harvest: The two G3P molecules undergo a series of reactions that generate four ATP and two NADH molecules. Therefore, glycolysis results in a net harvest of two ATP and two NADH molecules per glucose molecule.

A Closer Look
Glycolysis

Glycolysis is a series of enzyme-catalyzed reactions that break down a single molecule of glucose into two molecules of pyruvate. To help you follow the reactions, we show only the "carbon skeletons" of glucose and the molecules produced during glycolysis. Each blue arrow represents a reaction catalyzed by at least one enzyme.

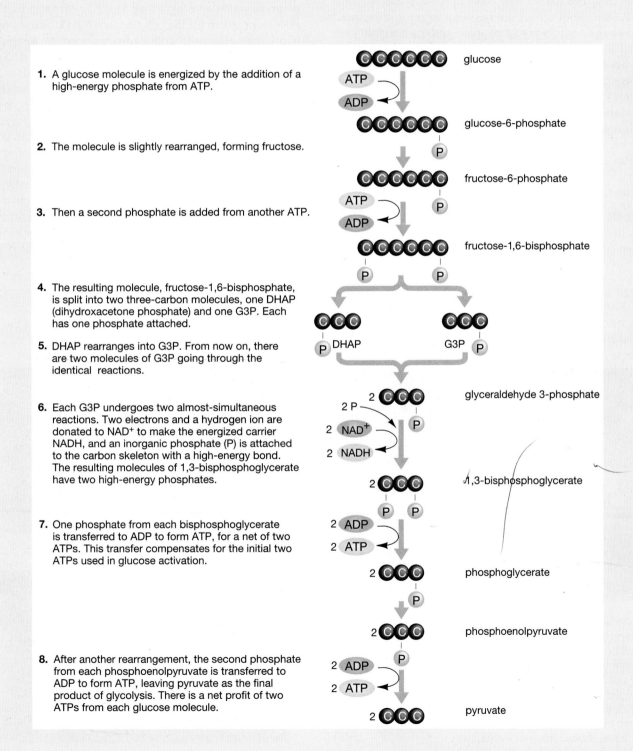

1. A glucose molecule is energized by the addition of a high-energy phosphate from ATP.

2. The molecule is slightly rearranged, forming fructose.

3. Then a second phosphate is added from another ATP.

4. The resulting molecule, fructose-1,6-bisphosphate, is split into two three-carbon molecules, one DHAP (dihydroxyacetone phosphate) and one G3P. Each has one phosphate attached.

5. DHAP rearranges into G3P. From now on, there are two molecules of G3P going through the identical reactions.

6. Each G3P undergoes two almost-simultaneous reactions. Two electrons and a hydrogen ion are donated to NAD+ to make the energized carrier NADH, and an inorganic phosphate (P) is attached to the carbon skeleton with a high-energy bond. The resulting molecules of 1,3-bisphosphoglycerate have two high-energy phosphates.

7. One phosphate from each bisphosphoglycerate is transferred to ADP to form ATP, for a net of two ATPs. This transfer compensates for the initial two ATPs used in glucose activation.

8. After another rearrangement, the second phosphate from each phosphoenolpyruvate is transferred to ADP to form ATP, leaving pyruvate as the final product of glycolysis. There is a net profit of two ATPs from each glucose molecule.

glucose

glucose-6-phosphate

fructose-6-phosphate

fructose-1,6-bisphosphate

DHAP

G3P

glyceraldehyde 3-phosphate

1,3-bisphosphoglycerate

phosphoglycerate

phosphoenolpyruvate

pyruvate

Summing Up
Glycolysis

Each molecule of glucose is broken down to two molecules of pyruvate. During these reactions, two ATP molecules and two NADH electron carriers are formed.

Carrier molecules such as NAD^+ capture energy by accepting high-energy electrons. Carriers can transport these electrons to sites where their energy is used to form ATP. One major difference between anaerobic (without oxygen) and aerobic (with oxygen) glucose breakdown is the use that is made of these high-energy electrons. In the absence of oxygen, pyruvate acts as the electron acceptor from NADH, producing molecules (ethanol or lactate) that the cell cannot use; this process is called **fermentation**. During cellular respiration, which occurs in the presence of oxygen, oxygen becomes the electron acceptor, allowing the pyruvate to be fully broken down and its energy harvested as ATP.

Some Cells Ferment Pyruvate to Form Lactate

The electrons carried in NADH are highly energetic, but their energy can be used to synthesize ATP only when oxygen is available, via cellular respiration (de-scribed later in this chapter). Many organisms (particularly microorganisms) thrive in the stomach and intestine of animals, deep in soil, in sediments underlying lakes and oceans, or in bogs and marshes where oxygen is rare or absent. Even some of our own body cells must cope without oxygen for brief periods. In anaerobic conditions (the conditions in which life, and probably glycolysis, evolved), NADH production is *not* used as a method of energy capture; it is actually a way of getting rid of hydrogen ions and electrons produced during the breakdown of glucose to pyruvate. But this disposal method poses a problem for the cell, because NAD^+ is used up as it accepts electrons and hydrogen ions in becoming NADH. Without a way to regenerate NAD^+ and to dispose of the electrons and hydrogen ions, glycolysis would have to stop once the supply of NAD^+ was exhausted.

Fermentation solves this problem by enabling pyruvate to act as the final acceptor of electrons and hydrogen ions from NADH. Thus, NAD^+ is regenerated for use in further glycolysis. There are two main types of fermentation; one type converts pyruvate to lactate, and the other converts pyruvate to carbon dioxide and ethanol.

Fermentation to lactate occurs in your muscles when you exercise vigorously, such when you race to class after you've overslept, or in the muscles of a runner sprinting through the finish line (Fig. 8-3a). (You may hear this

(a)

(b)

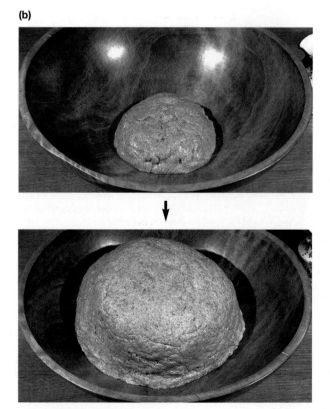

Figure 8-3 Fermentation
(a) Sprinters at the end of a race. A runner's respiratory and circulatory systems cannot supply oxygen to her leg muscles fast enough to keep up with the demand for energy, so glycolysis and lactate fermentation must provide the ATP. Panting after the race brings in the oxygen needed to remove the lactate through cellular respiration. ***(b)*** Bread rises as CO_2 is liberated by fermenting yeast, which converts glucose to ethanol via the alcoholic fermentation pathway.

compound called "lactic acid"; lactate is the ionized form of lactic acid that is in solution in the cytoplasm.) Even though working muscles need lots of ATP and cellular respiration generates much more ATP than does glycolysis, cellular respiration is limited by the organism's ability to provide oxygen (by breathing, for example). While you exercise vigorously, you may not be able to get enough air into your lungs and enough oxygen into your blood to supply your muscles with sufficient oxygen to allow cellular respiration to meet all their energy needs.

When deprived of adequate oxygen, your muscles do not immediately stop working. Instead, glycolysis continues for a while, providing its meager two ATP molecules per glucose and generating both pyruvate and NADH. Then, to regenerate NAD^+, muscle cells ferment pyruvate molecules to lactate, using electrons and hydrogen ions from NADH:

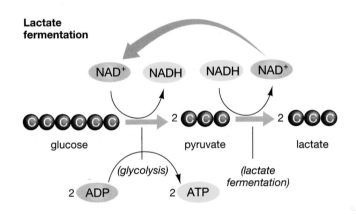

Lactate fermentation

In high concentrations, however, lactate is toxic to your cells. Soon, it causes intense discomfort and fatigue, forcing you to stop or at least slow down. As you rest, breathing rapidly after your sprint, oxygen once more becomes available and the lactate is converted back to pyruvate.

Interestingly enough, the conversion from lactate to pyruvate occurs not in the muscle cells, which lack the necessary enzymes, but in the liver. This pyruvate is then broken down by cellular respiration to carbon dioxide and water.

Various microorganisms, including the bacteria that produce yogurt, sour cream, and cheese, also use lactate fermentation. As you may know, acids taste sour, and lactate (lactic acid) contributes to the distinctive tastes of these foods. Some microorganisms ferment even when oxygen is present; others are poisoned by oxygen.

Other Cells Ferment Pyruvate to Alcohol

Many microorganisms use another process to regenerate NAD^+ under anaerobic conditions: *alcoholic fermentation*. These reactions produce ethanol and CO_2 (rather than lactate) from pyruvate, using hydrogen ions and electrons from NADH:

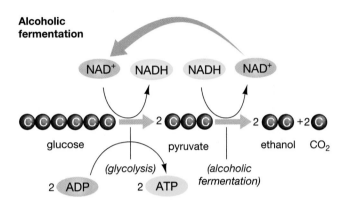

Alcoholic fermentation

Sparkling wines, such as champagne, are bottled while the yeasts are still alive and fermenting, trapping both the alcohol and the CO_2. When the cork is removed, the pressurized CO_2 is released, sometimes explosively. Baker's yeast in bread dough produces CO_2, making the bread rise; the alcohol generated by the yeast evaporates while the bread is baking (Fig. 8-3b).

3) How Does Cellular Respiration Generate Still More Energy from Glucose?

Cellular respiration is a series of reactions, occurring under aerobic conditions, in which large amounts of ATP are produced. During cellular respiration, the pyruvate produced by glycolysis is broken down to carbon dioxide and water. The final reactions of cellular respiration require oxygen because oxygen acts as the final acceptor of electrons.

In eukaryotic cells, cellular respiration occurs in the mitochondria. Recall from Chapter 6 that a mitochondrion has two membranes that produce two compartments: an inner compartment that is enclosed by the inner membrane and that contains the fluid **matrix**, and an **intermembrane compartment** between the two membranes (Fig. 8-4). Most of the ATP produced during cellular respiration is generated by enzyme-catalyzed reactions in the matrix, by electron transport proteins in the inner membrane, and by the movement of hydrogen ions through ATP-synthesizing proteins in the inner membrane.

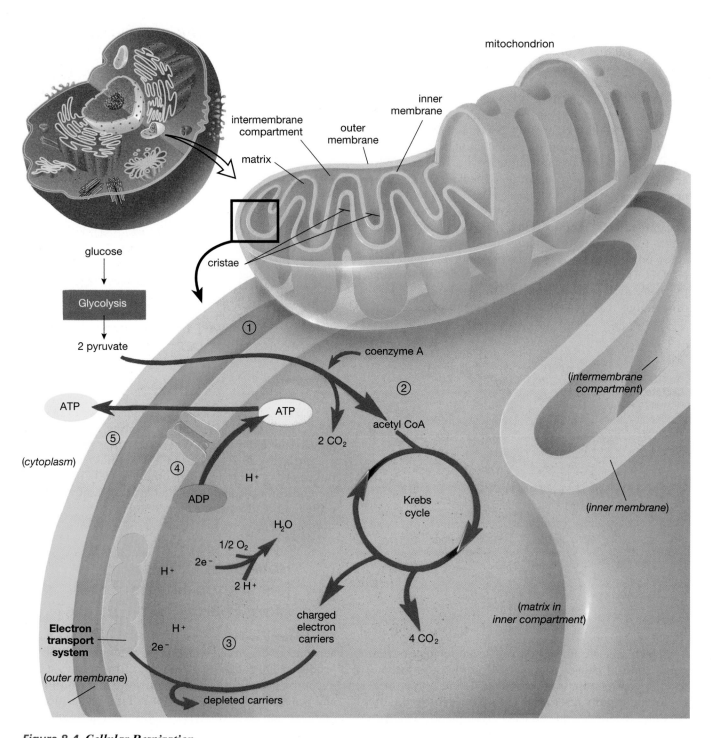

Figure 8-4 Cellular Respiration
Cellular respiration takes place in the mitochondria, whose structure, like that of a chloroplast, reflects the compartmentalized reactions that occur there. The inner membrane separates the inner compartment, containing the soluble enzymes of the matrix, from the intermembrane compartment (between the inner and outer membranes). The diagram summarizes the essential steps of glucose metabolism, from glycolysis in the cytoplasm to the transport of ATP out of the mitochondrion and back into the cytoplasm.

Figure 8-4 summarizes the main events of cellular respiration:

1. The two molecules of pyruvate produced by glycolysis are transported across both mitochondrial membranes and into the matrix.
2. Each pyruvate is split into CO_2 and a two-carbon acetyl group, which enters the *Krebs cycle* (discussed in the next section). The Krebs cycle releases the remaining carbons as CO_2, produces one ATP, and donates energetic electrons to several electron-carrier molecules.
3. The electron carriers donate their energetic electrons to the electron transport system of the inner membrane. There the energy of the electrons is used to transport H^+ from the matrix to the intermembrane compartment. At the end of the system, the electrons combine with O_2 and H^+ to form H_2O.
4. In chemiosmosis, the hydrogen ion gradient created by the electron transport system discharges through ATP-synthesizing enzymes in the inner membrane, and the energy is used to produce large amounts of ATP.
5. The ATP is transported back out of the mitochondrion into the fluid of the cytoplasm, where the ATP is available to drive needed reactions in the rest of the cell.

We have already discussed glycolysis; now let's look a little more closely at the processes of cellular respiration in the mitochondria. These two processes are summarized in Table 8-1.

Pyruvate Is Transported to the Mitochondrial Matrix, Where It Is Broken Down via the Krebs Cycle

Recall that pyruvate is the end product of glycolysis and that it is synthesized in the fluid portion of the cytoplasm. The pyruvate diffuses down its concentration gradient into the mitochondria, through pores in the mitochondrial membranes, until it reaches the mitochondrial matrix, where it is used in cellular respiration.

In the matrix, pyruvate reacts with a molecule called *coenzyme A* (Fig. 8-5, step 1). Each pyruvate is split into CO_2 and a two-carbon molecule called an *acetyl group*, which immediately attaches to coenzyme A (CoA), forming an acetyl–coenzyme A complex (*acetyl CoA* for short). During this reaction, two energetic electrons and a hydrogen ion are transferred to NAD^+, forming NADH.

The two acetyl CoA molecules then enter a cyclic pathway known as the **Krebs cycle**, named after its discoverer, Hans Krebs (Fig. 8-5, step 2). Each acetyl CoA briefly combines with a molecule of oxaloacetate. The two-carbon acetyl group is donated to the four-carbon *oxaloacetate* (which, as you may remember from Chapter 7, also functions in the carbon fixation stage of the C_4 pathway) to form the six-carbon *citrate*. Coenzyme A is released once again. (Coenzyme A, like an enzyme, is not permanently altered during these reactions and is reused many times.) Mitochondrial enzymes then lead each citrate through a number of rearrangements that regenerate the oxaloacetate, give off two CO_2 molecules, and capture most of the energy of the acetyl group as one

TABLE 8-1 Summary of Glycolysis and Cellular Respiration

Process	Location	Reactions	Electron Carriers Formed	ATP Yield (per glucose molecule)
Glycolysis	Fluid cytoplasm	Glucose broken down to the two pyruvates	2 NADH	2 ATP
Acetyl CoA formation	Matrix of mitochondrion	Each pyruvate combined with coenzyme A to form acetyl CoA and CO_2	2 NADH	
Krebs cycle	Matrix of mitochondrion	Acetyl group of acetyl CoA metabolized to two CO_2	6 NADH, 2 $FADH_2$	2 ATP
Electron transport	Inner membrane, intermembrane compartment	Energy of electrons from NADH, two $FADH_2$ used to pump H^+ into intermembrane compartment, H^+ gradient used to synthesize ATP: three ATP per NADH, two ATP per $FADN_2$		32–34 ATP*

*Glycolysis produces two NADH molecules in the fluid portion of the cytoplasm of the cell. Unlike the NADH and $FADH_2$ molecules generated in the matrix of the mitochondrion, the electrons from these two NADH molecules must be transported into the matrix before they can enter the electron transport system of the inner membrane. In most eukaryotic cells, the energy of one ATP molecule is used to transport the electrons from one NADH molecule into the matrix. Thus, the two "glycolytic NADH" molecules net only two ATPs, not the usual three, during electron transport. The heart and liver cells of mammals, however, use a different transport system, one that does not consume ATP. In these cells, then, the "glycolytic NADH" molecules net three ATPs apiece, just as the "mitochondrial NADH" molecules do.

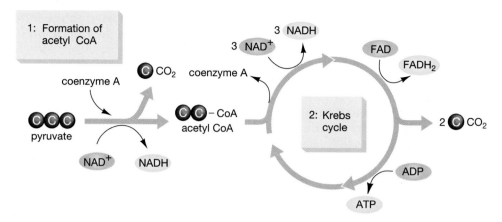

Figure 8-5 *The essential reactions in the mitochondrial matrix*
Pyruvate reacts with coenzyme A to form CO_2 and acetyl CoA. During this reaction, an energetic electron is added to NAD^+ to form NADH. When acetyl CoA enters the Krebs cycle, the acetyl group combines with oxaloacetate to form citrate, and coenzyme A is released. One turn through the reactions of the cycle produces 3 molecules of NADH, 1 of $FADH_2$, 2 of CO_2, and 1 of ATP for each acetyl CoA. Because each glucose molecule yields 2 pyruvates, the total energy harvest per glucose molecule in the matrix is 2 ATP, 8 NADH (2 from the synthesis of acetyl CoA, 6 from the Krebs cycle), and 2 $FADH_2$.

ATP and four electron carriers—one $FADH_2$ (flavin adenine dinucleotide) and three NADH.

The essay "A Closer Look: The Mitochondrial Matrix Reactions" shows the complete set of reactions that occur in the mitochondrial matrix, from acetyl CoA formation through the Krebs cycle.

Summing Up
The Mitochondrial Matrix Reactions

The synthesis of acetyl CoA produces one CO_2 and one NADH per pyruvate. The Krebs cycle produces two CO_2, one ATP, three NADH, and one $FADH_2$ per acetyl CoA. Therefore, at the conclusion of the matrix reactions, the two pyruvates that are produced from a single glucose molecule have been completely broken down by the addition of oxygen to form six CO_2 molecules. In the process, two ATPs, eight NADH, and two $FADH_2$ electron carriers have been produced.

Energetic Electrons Produced by the Krebs Cycle Are Carried to Electron Transport Systems in the Inner Mitochondrial Membrane

At this point the cell has gained only four ATP molecules from the original glucose molecule: two during glycolysis and two during the Krebs cycle. The cell has, however, captured many energetic electrons in carrier molecules: 2 NADH during glycolysis plus 8 more NADH and 2 $FADH_2$ from the matrix reactions, for a total of 10 NADH and 2 $FADH_2$. The carriers deposit their electrons in **electron transport systems** located in

the inner mitochondrial membrane (Fig. 8-6). These electron transport systems are similar in function to those embedded in the thylakoid membrane of chloroplasts. The energetic electrons move from molecule to molecule along the transport systems. Energy released by the

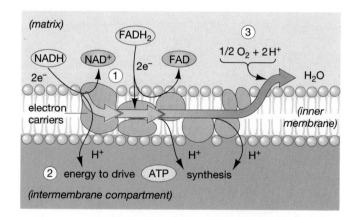

Figure 8-6 *The electron transport system of mitochondria*
① The electron carrier molecules NADH and $FADH_2$ deposit their energetic electrons with the carriers of the transport system located in the inner membrane. ② The electrons move from carrier to carrier within the transport system. Some of their energy is used to pump hydrogen ions across the inner membrane from the matrix into the intermembrane compartment. This movement out of the inner compartment creates a hydrogen ion gradient that can be used to drive ATP synthesis (for details, see "A Closer Look: Chemiosmosis in Mitochondria"). ③ At the end of the electron transport system, the energy-depleted electrons combine with oxygen and hydrogen ions in the matrix to form water.

A Closer Look
The Mitochondrial Matrix Reactions

Mitochondrial matrix reactions occur in two stages: (1) the formation of acetyl coenzyme A and (2) the Krebs cycle. Recall that glycolysis produces two pyruvates from each glucose molecule, so each set of matrix reactions occurs twice during the metabolism of a single glucose molecule.

First Stage: Formation of Acetyl Coenzyme A   Pyruvate is split to CO_2 and an acetyl group. The acetyl group attaches to coenzyme A to form acetyl CoA. Simultaneously, NAD^+ receives two electrons and a hydrogen ion to make NADH. The acetyl CoA enters the second stage of the matrix reactions.

Second Stage: The Krebs Cycle

① Acetyl CoA donates its acetyl group to oxaloacetate to make citrate.
② Citrate is rearranged to form isocitrate.
③ Isocitrate loses a carbon to CO_2, forming α-ketoglutarate; NADH is formed from NAD^+.
④ Alpha-ketoglutarate loses a carbon to CO_2, forming succinate; NADH is formed from NAD^+, and additional energy is stored in ATP. By this point, two molecules of CO_2 have been given off. (These two CO_2 molecules, along with the one that was released during the formation of acetyl CoA, account for the three carbons of the original pyruvate.)
⑤ Succinate is converted to fumarate, and the electron carrier FAD is charged up to $FADH_2$.
⑥ Fumarate is converted to malate.
⑦ Malate is converted to oxaloacetate, and NADH is formed from NAD^+.

The Krebs cycle produces three molecules of CO_2 and NADH, one $FADH_2$, and one ATP per acetyl CoA. NADH and $FADH_2$ will donate their electrons to the electron transport system of the inner membrane, where the energy of the electrons will be used to synthesize ATP by chemiosmosis.

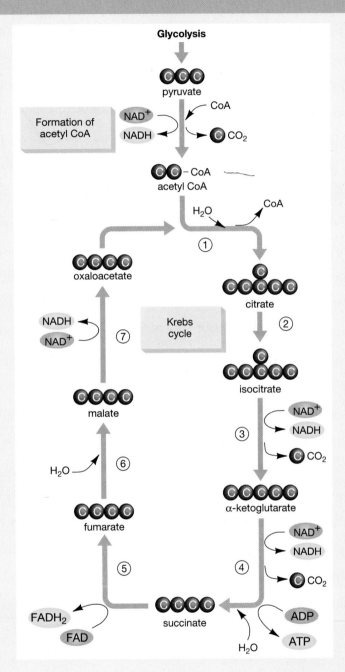

electrons during these transfers is used to pump hydrogen ions from the matrix, across the inner membrane, and into the intermembrane compartment during *chemiosmosis* (discussed in the next section).

Finally, at the end of the electron transport system, oxygen and hydrogen ions accept the energetically depleted electrons: Two electrons, one oxygen atom, and two hydrogen ions combine to form water. This step clears out the transport system, leaving it ready to run more electrons through. Without oxygen, the electrons would "pile up" in the transport system; the hydrogen ions would not be pumped across the inner membrane. The hydrogen ion gradient would soon dissipate, and ATP synthesis would stop.

Chemiosmosis Captures Energy Stored in a Hydrogen Ion Gradient and Produces ATP

Hydrogen ion pumping across the inner membrane generates a large H^+ concentration gradient—that is, a high concentration of hydrogen ions in the intermembrane compartment and a low concentration in the matrix. The inner membrane is impermeable to hydrogen ions except at protein channels that are part of ATP-synthesizing enzymes. In the process of **chemiosmosis**, hydrogen ions move down their concentration gradient from the inter-

membrane compartment to the matrix by means of these ATP-synthesizing enzymes. The flow of hydrogen ions provides the energy to synthesize 32 to 34 molecules of ATP from ADP (adenosine diphosphate) and phosphate in the matrix. The box "A Closer Look: Chemiosmosis in Mitochondria" examines chemiosmosis in more detail.

The ATP that was synthesized from ADP and phosphate in the matrix during chemiosmosis is transported across the inner membrane from the matrix to the intermembrane compartment. It then diffuses out of the mitochondrion to the surrounding cytoplasm through the

A Closer Look
Chemiosmosis in Mitochondria

ATP synthesis in mitochondria is similar to the process of chemiosmosis that we described in chloroplasts in Chapter 7. The inner membrane of a mitochondrion has an electron transport system that functions similarly to the one in the thylakoids. Further, the intermembrane compartment between the outer and inner membranes of a mitochondrion is analogous to the interior of a thylakoid.

Anatomically, the arrangement in mitochondria looks like this:

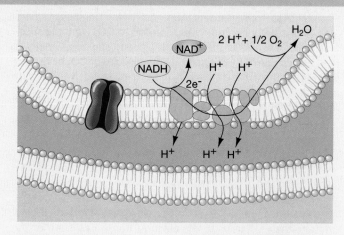

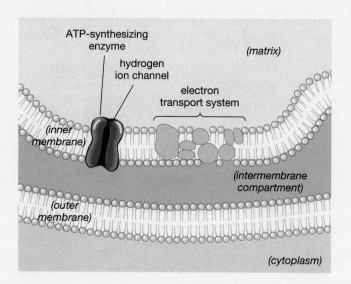

The electron carriers formed during glycolysis and the Krebs cycle—NADH and $FADH_2$—deposit their electrons with the electron transport system of the inner membrane. (For clarity, $FADH_2$ is not shown in the illustration.) As they pass through the electron transport system, the electrons provide the energy to pump hydrogen ions (H^+) across the inner membrane, from the matrix to the intermembrane compartment:

This pumping process increases the H^+ concentration in the intermembrane compartment and decreases the H^+ concentration in the matrix; therefore, a H^+ gradient is produced across the inner membrane. Like the thylakoid membrane of a chloroplast, the inner membrane of a mitochondrion is permeable to H^+ only at channels that are coupled with ATP-synthesizing enzymes. The movement of hydrogen ions down their concentration gradient through these channels drives ATP synthesis:

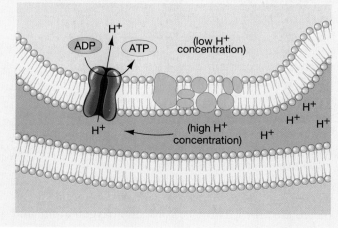

outer membrane, which is very permeable to ATP. These ATP molecules provide most of the energy needed by the cell. ADP simultaneously diffuses from the fluid of the cytoplasm across the outer membrane and is transported across the inner membrane to the matrix, replenishing the supply of ADP.

Summing Up
Electron Transport and Chemiosmosis

Electrons from the electron carriers NADH and $FADH_2$ enter the electron transport system of the inner mitochondrial membrane. Here their energy is used to generate a hydrogen ion gradient. Movement of hydrogen ions down their gradient through the pores of ATP-synthesizing enzymes drives the synthesis of 32 to 34 molecules of ATP. At the end of the electron transport system, two electrons combine with one oxygen atom and two hydrogen ions to form water.

Glycolysis and Cellular Respiration Influence the Way Entire Organisms Function

Many students believe that the details of glycolysis and cellular respiration are hard to learn and don't really help them understand the living world around them. Have you ever read a murder mystery and wondered how cyanide could kill a person almost instantly? Cyanide reacts with one of the proteins in the electron transport system, immediately blocking the movement of electrons through the system and bringing cellular respiration to a screeching halt. Under normal conditions, metabolic processes within individual cells have enormous impacts on the functioning of entire organisms. To take just two examples, let's consider the migration of ruby-throated hummingbirds and Olympic track events.

Hummingbird Migration Requires Efficient Energy Storage and Use

As the days shorten in August, many birds in North America prepare to migrate to Central and South America, thereby escaping the cold weather and food shortages of the northern winter. For some, migration is a very dangerous undertaking. Ruby-throated hummingbirds, for example, fly across the Gulf of Mexico from the southeastern United States to Mexico and Central America. On this journey of more than 1000 kilometers (621 miles) of open sea, there is nowhere to stop and no food.

This trek presents real difficulties for a hummingbird. With its short wings, a hummer couldn't fly if it weighed too much. Yet flying 1000 kilometers requires a lot of energy that must be stored in the bird's body. Hummers solve this dilemma with a twofold strategy of storing the highest-energy molecules possible—fat—and extracting the maximum usable energy during flight.

A ruby-throated hummingbird weighs 2 to 3 grams (0.11 to 0.16 ounces) before it puts on weight for migration. It adds as much as 2 grams of fat in late summer, nearly doubling its weight. As you recall from Chapter 3, fats contain more than twice as much energy per unit weight as do proteins or carbohydrates. If a hummer had to store 4 or 5 grams (0.21 to 0.26 ounces) of glycogen or protein, it would be too heavy to lift off.

Even so, the hummer must still squeeze every ATP molecule possible out of each fat molecule. The hummer that just makes it to Guatemala on 2 grams of fat by using cellular respiration would collapse before reaching the Gulf Coast if it used lactate fermentation instead. The cells of a hummingbird's flight muscles are packed with mitochondria, so each cell is capable of producing large quantities of ATP.

Furthermore, the hummingbird's respiratory system is exquisitely designed to extract oxygen out of the air even while exhaling, enabling the lungs to provide a constant supply of oxygen to the cells. Therefore, even during strenuous flight, cellular respiration never falters for lack of oxygen.

Human Runners Face Similar Challenges

Humans, like hummingbirds, must regulate energy reserves and energy use. Why is the average speed of the 5000-meter run in the Olympics slower than that of the 100-meter dash? During the dash, or during the sprint across the finish line of a marathon, runners' leg muscles use more ATP than cellular respiration can supply, because their bodies cannot deliver enough oxygen to keep up with the demand. Glycolysis and lactate fermentation can keep the muscles supplied with ATP for a short time, but soon the toxic effects of lactate buildup cause fatigue and cramps. Although runners can do a 100-meter dash anaerobically, distance runners must pace themselves, using cellular respiration to power their muscles for most of the race and saving the anaerobic sprint for the finish.

Marathon runners face somewhat the same dilemma that migrating hummingbirds do. A marathon may require 3000 kilocalories of stored energy with cellular respiration supplying nearly all the ATP. Marathoners train by running 50 or 100 miles a week, not so much to build up their leg muscles as to build up the capacity of their respiratory and circulatory systems to deliver enough oxygen to their muscles. An efficient transport of oxygen to the cells is necessary to support the cellular respiration that such vigorous exercise demands.

As you can see, sustaining life depends on efficiently obtaining, storing, and using energy. By gaining an understanding of the principles of cellular respiration, you can more fully appreciate the energy-related adaptations of living organisms.

Summary of Key Concepts

1 How Is Glucose Metabolized?

Cells produce usable energy by breaking down glucose to lower-energy compounds and capturing some of the released energy as ATP. In glycolysis, glucose is metabolized in the fluid portion of the cytoplasm to two molecules of pyruvate, generating two ATP molecules. In the absence of oxygen, pyruvate is converted by fermentation to lactate or ethanol and CO_2. If oxygen is available, the pyruvates are metabolized to CO_2 and H_2O through cellular respiration in the mitochondria, generating much more ATP than does fermentation.

Figure 8-7 and Table 8-1 summarize the locations, major mechanisms, and overall energy harvest for the complete metabolism of glucose from glycolysis through cellular respiration.

2 How Is the Energy in Glucose Harvested during Glycolysis?

During glycolysis, a molecule of glucose is activated by the addition of phosphates from two ATP molecules, to form fructose bisphosphate. In a series of reactions, the fructose bisphosphate is broken down into two molecules of pyruvate. These reactions produce four ATP molecules and two NADH electron carriers. Because two ATPs were used in the activation steps, the net yield from glycolysis is two ATPs and two NADHs. Glycolysis, in addition to providing a small yield of ATP, uses up NAD^+ to produce NADH. Once the cell's supply of NAD^+ is consumed, glycolysis must stop. NADH may be regenerated by fermentation, with no additional ATP gain, or by cellular respiration, which also produces additional ATP.

If oxygen is absent, the pyruvates undergo either lactate fermentation, producing lactate, or alcoholic fermentation, producing ethanol and CO_2. In both cases, no new ATP is formed. However, both types of fermentation regenerate NAD^+ from NADH, replenishing the cell's supply of NAD^+.

3 How Does Cellular Respiration Generate Still More Energy from Glucose?

If oxygen is available, cellular respiration can occur. The pyruvates are transported into the matrix of the mitochondria. In the matrix, each pyruvate reacts with coenzyme A to form acetyl CoA plus CO_2. One NADH is also formed at this step. The two-carbon acetyl group of acetyl CoA enters the Krebs cycle, which releases the remaining two carbons as CO_2. One ATP, three NADHs, and one $FADH_2$ are also formed for each acetyl group that goes through the cycle. At this point, each glucose molecule has produced 4 ATPs (two from glycolysis and one from each acetyl CoA via the Krebs cycle), 10 NADHs (two from glycolysis, one from each pyruvate during the formation of acetyl CoA, and 3 from each acetyl CoA during the Krebs cycle), and 2 $FADH_2$s (one from each acetyl CoA during the Krebs cycle).

The NADHs and $FADH_2$s deposit their energetic electrons in the electron transport system embedded in the inner mitochondrial membrane. The energy of the electrons is used to pump hydrogen ions across the inner membrane from the matrix to the intermembrane compartment. At the end of the electron transport system, the

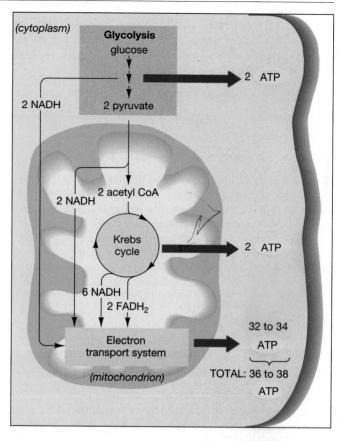

Figure 8-7 The energy harvest from the complete metabolism of one glucose molecule
Glycolysis and the Krebs cycle each produce 2 ATP molecules. The reactions within the mitochondrial matrix produce 8 NADH molecules and 2 $FADH_2$ molecules. By donating its electrons to the electron transport system, each NADH molecule yields 3 ATP molecules, for a total of 24 ATPs. Each $FADH_2$ molecule yields 2 ATP molecules, for a total of 4 ATPs. The electrons from the 2 NADH molecules produced in the cytoplasm during glycolysis must be transported into the mitochondrion to reach the electron transport system. In heart and liver cells, this transport is "free"; in most cells, transport costs 1 ATP per NADH. The 2 "glycolytic NADH" molecules therefore yield either 4 or 6 ATP molecules, depending on the cell type. Consequently, the energy harvest from electron transport is 32 to 34 ATPs. Including 2 ATPs from glycolysis and 2 ATPs from the Krebs cycle, the total energy yield from glucose metabolism is 36 to 38 ATPs.

depleted electrons combine with hydrogen ions and oxygen to form water. This is the oxygen-requiring step of cellular respiration. During chemiosmosis, the hydrogen ion gradient created by the electron transport system is used to produce ATP, as the hydrogen ions diffuse back across the inner membrane through channels in ATP-synthesizing enzymes. Electron transport and chemiosmosis yield 32 to 34 additional ATPs, for a net yield of 36 to 38 ATPs per glucose molecule.

Key Terms

cellular respiration   *p. 131*
chemiosmosis   *p. 136*
electron transport system
   p. 134

fermentation   *p. 130*
glycolysis   *p. 128*
intermembrane
   compartment   *p. 131*

Krebs cycle   *p. 133*
matrix   *p. 131*
pyruvate   *p. 128*

Thinking Through the Concepts

Multiple Choice

1. *Where does glycolysis occur?*
   a. cytoplasm
   b. matrix of mitochondria
   c. inner membrane of mitochondria
   d. outer membrane of mitochondria
   e. stroma of chloroplast

2. *Where does respiratory electron transport occur?*
   a. cytoplasm
   b. matrix of mitochondria
   c. inner membrane of mitochondria
   d. outer membrane of mitochondria
   e. stroma of chloroplast

3. *What is the product of the fermentation of sugar by yeast in bread dough that is essential for the rising of the dough?*
   a. lactate   b. ATP
   c. ethanol   d. CO_2
   e. O_2

4. *The majority of ATP produced in aerobic respiration comes from*
   a. glycolysis
   b. the Krebs cycle
   c. chemiosmosis
   d. fermentation
   e. photosynthesis

5. *The process that converts glucose into two molecules of pyruvate is*
   a. glycolysis
   b. fermentation
   c. the Krebs cycle
   d. respiratory electron transport
   e. the Calvin-Benson cycle

6. *The process that causes lactate buildup in muscles during strenuous exercise is*
   a. glycolysis
   b. fermentation
   c. the Krebs cycle
   d. respiratory electron transport
   e. the Calvin-Benson cycle

? Review Questions

1. Starting with glucose ($C_6H_{12}O_6$), write the overall reactions for (a) aerobic respiration and (b) fermentation in yeast.

2. Draw a labeled diagram of a mitochondrion, and explain how its structure is related to its function.

3. What role do the following play in respiratory metabolism: (a) glycolysis, (b) mitochondrial matrix, (c) inner membrane of mitochondria, (d) fermentation, and (e) NAD^+?

4. Outline the major steps in (a) aerobic and (b) anaerobic respiration, indicating the sites of ATP production. What is the overall energy harvest (in terms of ATP molecules generated per glucose molecule) for each?

5. Describe the Krebs cycle. In what form is most of the energy harvested?

6. Describe the mitochondrial electron transport system and the process of chemiosmosis.

7. Why is oxygen necessary for cellular respiration to occur?

Applying the Concepts

1. Some years ago a freight train overturned, spilling a load of grain. Because the grain was spoiled, it was buried in the embankment. Although there is no shortage of other food locally, the local bear population has created a nuisance by continually uncovering the grain. What do you think has happened to the grain to make them do this, and how is it related to human cultural evolution?

2. In detective novels, "the odor of bitter almonds" is the telltale clue to murder by cyanide poisoning. Cyanide works by attacking the enzyme that transfers electrons from respiratory electron transport to O_2. Why is it not possible for the victim to survive by using anaerobic respiration? Why is cyanide poisoning almost immediately fatal?

3. More than a century ago, the French biochemist Louis Pasteur described a phenomenon, now called "the Pasteur effect," in the wine-making process. He observed that in a sealed container of grape juice containing yeast, the yeast will consume the sugar very slowly as long as oxygen remains in the container. As soon as the oxygen is gone, however, the rate of sugar consumption by the yeast increases greatly and the alcohol content in the container rises. Discuss the Pasteur effect, on the basis of what you know about aerobic and anaerobic cellular respiration.

4. Some species of bacteria that live at the surface of sediment on the bottom of lakes are facultative anaerobes; that is, they are capable of either aerobic or anaerobic respiration. How will their metabolism change during the summer when the deep water becomes anoxic (deoxygenated)? If the bacteria continue to grow at the same rate, will glycolysis increase, decrease, or remain the same after the lake becomes anoxic? Explain why.

5. The dumping of large amounts of raw sewage into rivers or lakes typically leads to massive fish kills, although sewage itself is not toxic to fish. Similar fish kills also occur in shallow lakes that become covered in ice during the winter. What kills the fish? How might you reduce fish mortality after raw sewage is accidentally released into a small pond containing large bass?

6. Although respiration occurs in all living cells, different cells respire at different rates. Explain why. How could you predict the relative respiratory rates of different tissues in a fish by microscopic examination of cells?

Group Activity

Bo and Luke are competitive body builders who train at the same Florida gym but are bitter enemies. At a workout a week before an upcoming national competition, Bo begins to feel very hot, sweats profusely, then collapses. When the paramedics arrive, they note that Bo is feverish, has a rapid heart beat and a rapid breathing rate, and appears blue around the mouth. On the way to the hospital, Bo goes into a coma; he dies a day later.

During the subsequent investigation, crystals of some substance are discovered on the top of Bo's water bottle. In addition, the police discover that Luke has an illegal tropical fish business in which he uses cyanide to stun fish that swim near the reefs around the Florida Keys. When they float to the top, he scoops them up in a net. Those fish that recover are sent to aquarium stores throughout the United States.

The police suspect that, to increase his chances of winning the competition, Luke has poisoned Bo with cyanide. Luke maintains that he is innocent and thinks that Bo has accidentally poisoned himself with dinitrophenol, an illegal and highly toxic drug that some body builders take to lose weight. Break up into groups of three or four to determine whether Bo has died of cyanide poisoning or of dinitrophenol poisoning.

HINT 1: Both cyanide and dinitrophenol interfere with the function of mitochondria, but cyanide targets the electron transport chain whereas dinitrophenol targets the functioning of ATP synthase.

HINT 2: Which drug will cause oxygen consumption to increase? Increased oxygen consumption by your cells will result in rapid breathing and a bluish skin color as oxygen is removed from the blood faster than it can be replaced. Decreased oxygen consumption will result in shortness of breath and bright red veins and arteries.

For More Information

Calder, W. A. "Red-Hot Hummers." *Nature Conservancy,* March/April 1998. Beautiful photos and lively writing describe the trials of the tiny rufous hummingbird as it fuels up for its long migration South.

Hinkle, P. C., and McCarty, R. E. "How Cells Make ATP." *Scientific American*, March 1978. A summary of cellular respiration, with an emphasis on electron transport in mitochondria.

McCarty, R. E. "H⁺-ATPases in Oxidative and Photosynthetic Phosphorylation." *BioScience*, January 1985. A description of the structure and function of the ATP-synthesizing enzymes in mitochondria and chloroplasts.

Nelson, M., Burgess, T. L., Alling, A., Alverez-Romo, N., Dempster, W. F., Walford, R. L., and Allen, J. P. "Using a Closed Ecological System to Study Earth's Biosphere." *BioScience*, 1993. An artificial ecosystem allows scientists to learn more about how natural ecosystems function.

Answers to Multiple-Choice Questions
1. a   2. c   3. d   4. c   5. a   6. b

Unit Two
Inheritance

Inheritance provides for both similarity and difference. Both of these animals are tigers because they inherited a vast majority of identical genes from common ancestors; they differ in color as a result of differences in a single gene.

"A structure this pretty just had to exist."

James Watson in **The Double Helix** *(1968)*

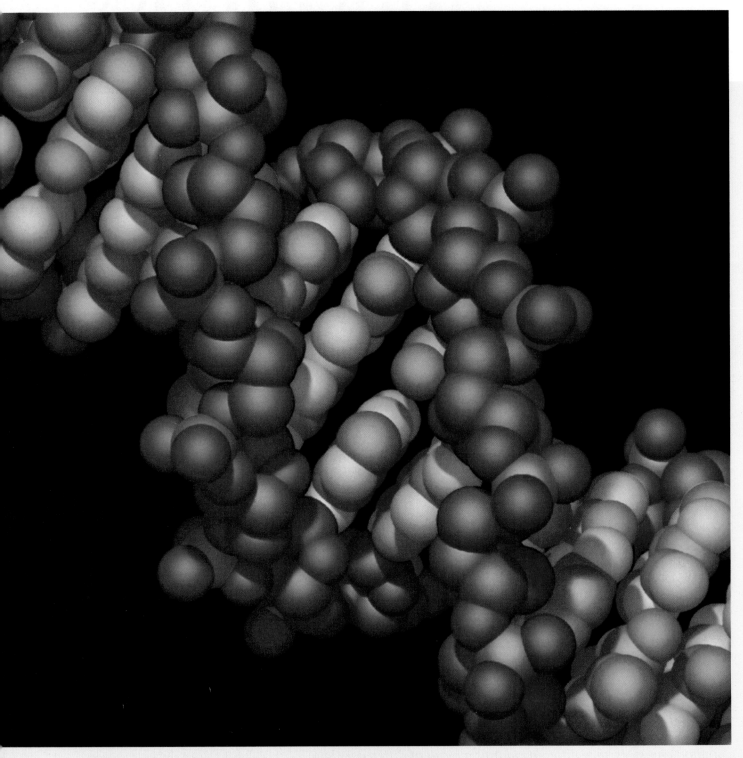

A computer-generated model of the structure of DNA. Like a twisted ladder formed of repeating subunits, this elegantly simple molecule permits nearly endless variation, allowing it to provide the blueprints for all forms of life on Earth.

DNA: The Molecule of Heredity

9

At a Glance

Have you ever watched a hawk soar effortlessly—and wished you had wings? As you juggle too many books on your way to class, or try to scratch the middle of your back, have you ever wished for extra arms—perhaps some strong, flexible tentacles like those of an octopus? If not, you've probably looked at your friends and classmates and admired some of their characteristics, such as hair color, eye color, or a certain size or shape of nose.

What makes you what you are—a person rather than an octopus, a hawk, or even bread mold? Why do you have arms and not wings, tentacles, or even tendrils like the vines clinging to an old building? What gives you the traits that make you unique among people—your skin, hair, and eye colors, your height, and the shape of your nose? Is it possible that a single type of molecule is at the root of why you are a human being and not a plant?

As you know, every organism is composed of cells. Individually and collectively, these cells confer the traits that identify each organism as a particular species and a unique individual. The inherited molecular "instructions" that direct the life of each cell and confer on each cell its specialized characteristics are contained in a molecule whose elegantly simple structure can exist in almost endless variations. Just 50 years ago, no one knew that *deoxyribonucleic acid*, or *DNA*, found in structures called *chromosomes*, is the molecule of heredity—a molecule that carries the blueprints for all forms of life. What is the structure of DNA? How does this structure allow DNA to copy itself exactly prior to cell division, passing precise copies of itself to daughter cells and ultimately to future generations?

In this chapter, as we begin to provide answers to these questions, you will also learn a bit about how scientists work and how they unraveled pieces of the complex fabric of inheritance, starting with the structure of DNA.

1) What Is the Composition of Chromosomes?

Eukaryotic Chromosomes Consist of DNA and Protein

Chromosomes, long strands of DNA that, in eukaryotes, are complexed with protein, become clearly visible

Net Watch

On-line resources for this chapter are on the World Wide Web at:

http://www.prenhall.com/audesirk

(click on the Table of Contents link and then select Chapter 9).

in the nucleus during cell division. A **gene** is a functional segment of DNA located at a particular place on a chromosome. The composition of chromosomes was first studied around 1870, when the biochemist Friedrich Miescher extracted from cell nuclei a previously unknown chemical substance, an acidic material with an unusually high phosphorus content. Because of its location in the nucleus and its acidic properties (which derive from its phosphate groups), this material came to be known as *nucleic acid*. In the early 1900s, the discovery that chromosomes are the carriers of hereditary information focused attention on their chemical composition. Increasingly sophisticated biochemical techniques showed that nucleic acid consists of four very similar subunits called **nucleotides**. The specific type of nucleic acid in all chromosomes of eukaryotic cells is called **deoxyribonucleic acid** or **DNA**, because each nucleotide contains the sugar *deoxyribose* (see Chapter 3). Each nucleotide of DNA consists of three parts: (1) a phosphate group, (2) deoxyribose, and (3) a nitrogen-containing **base** that has a single-ringed or double-ringed structure:

Nucleotide

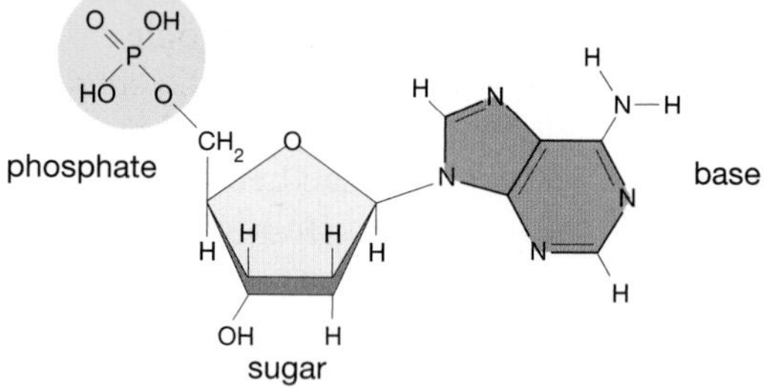

The four types of nucleotides have the same phosphate and sugar but different bases. The bases in DNA occur in two classes: purines and pyrimidines. The **pyrimidine** bases *thymine* (abbreviated T) and *cytosine* (C) have single rings:

Nucleotides with pyrimidine bases:

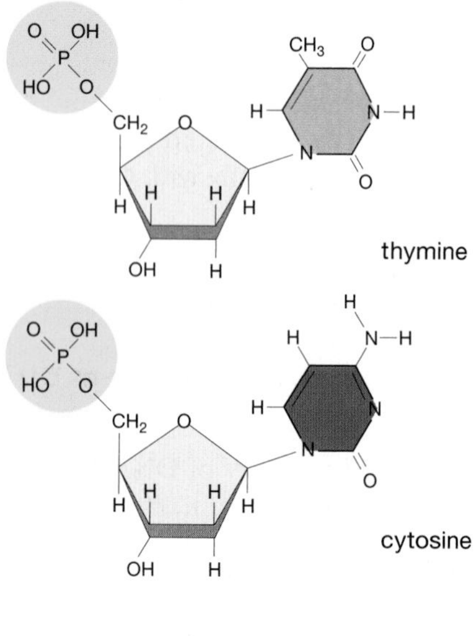

thymine

cytosine

The **purine** bases *adenine* (A) and *guanine* (G) have double rings:

Nucleotides with purine bases:

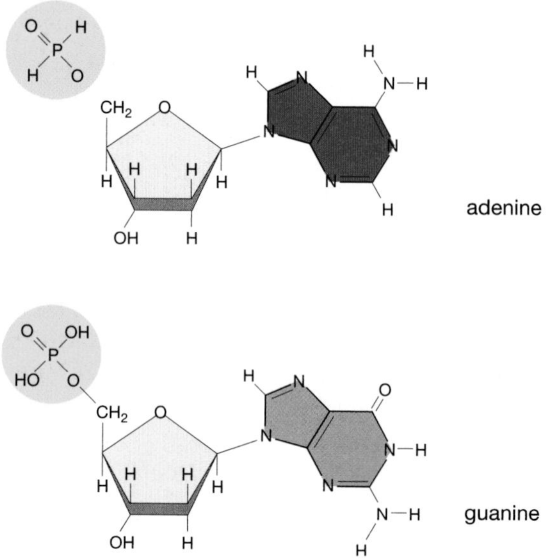

adenine

guanine

Biochemists determined that a eukaryotic chromosome is composed only of protein and DNA. From the mid-1920s through the early 1950s, various experiments by several investigators who studied bacteria and viruses led to the important conclusion that the DNA of chromosomes, and not the protein of chromosomes, is the molecule of heredity.

The Sequence of Bases in DNA Can Encode Vast Amounts of Information

Most of the substance of living cells, aside from water, is protein or is synthesized through the action of enzymes, which are proteins. But there are about 50,000 different types of protein in your body. Most proteins contain more than 50 amino acids, and some have as many as 2000! How can just four types of bases in DNA encode all of the information needed to produce tens of thousands of proteins, each with various combinations of 20 amino acids?

The four types of bases can be arranged in any linear order along a strand of DNA. Each sequence of bases represents a unique set of genetic instructions, like a biological Morse code (which, after all, can produce the entire alphabet, and therefore the English language, with only two symbols). A stretch of DNA just 10 nucleotides long can have more than a million possible sequences of the four bases. Because an "average" chromosome has millions (in bacteria) to billions (in plants or animals) of nucleotides, DNA molecules can encode a staggering amount of information.

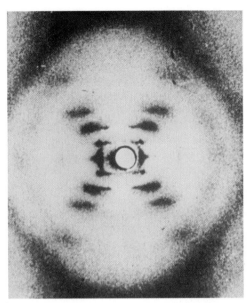

(a)   (b)

Figure 9-1 The X-ray diffraction pattern of DNA, taken by Rosalind Franklin
(a) The X formed of dark spots is characteristic of helical molecules such as DNA. Measurements of various aspects of the pattern indicate the dimensions of the DNA helix; for example, the distance between the dark spots corresponds to the distance between turns of the helix.
(b) Rosalind Franklin, before her untimely death at the age of 37 in 1958, published about 40 scientific papers.

2) What Is the Structure of DNA?

The demonstration that DNA carries genetic information stimulated a flurry of research on the chemical nature of that molecule. In the early 1950s, Alfred Mirsky studied the amount of DNA in the cells of various tissues of several organisms. He found that, although the quantity of DNA varies among species, it is constant in each nucleus of any one species, no matter what tissue the nucleus comes from. *Gametes* (sex cells, such as sperm and eggs) are a notable exception; they have half as much DNA as do the other cells of the body. When we discuss meiosis and gamete formation in Chapter 11, you will see why this is to be expected with DNA as the hereditary material.

What seemed equally significant, if only someone could figure out what it meant, were the results of experiments by Erwin Chargaff. In the 1940s, Chargaff analyzed the relative amounts of the four nucleotides of DNA in a variety of species and found a curious consistency. Although the amounts of each of the four bases vary considerably from species to species, *the DNA of any given species contains equal amounts of adenine and thymine and equal amounts of cytosine and guanine.*

Several major questions still remained unanswered: How does DNA encode genetic information? How is it duplicated before mitosis, so that each daughter cell receives exactly the same genetic information? Biologists in the early 1950s agreed that the secrets of DNA function, and therefore of heredity itself, could be found only by understanding the structure of the molecule.

Determining the structure of any biological molecule is no simple task. Even the most powerful electron microscopes cannot reveal the structure of molecules in atomic detail. To study the structure of DNA, Maurice Wilkins and Rosalind Franklin turned to X-ray diffraction. They bombarded crystals of purified DNA with X-rays and photographed the resulting diffraction patterns (Fig. 9-1). As you can see, the diffraction pattern of DNA does not provide a picture of the molecule's structure. However, Wilkins and Franklin inferred from the pattern that the DNA molecule is *helical* (that is, twisted like a corkscrew), has a uniform diameter of 2 nanometers (2 billionths of a meter), and consists of repeating subunits.

The DNA of Chromosomes Is a Double Helix

The chemical and X-ray diffraction data were not nearly enough information with which to work out the complete structure of DNA; some good guesses were also needed. (See the essay "Scientific Inquiry: The Discovery of the Double Helix.") Combining a knowledge of how complex organic molecules bond with one another with an intuition that "important biological objects come in pairs," James Watson and Francis Crick proposed that the DNA molecule consists of two strands, each composed of a series of nucleotides. In each single strand of

Scientific Inquiry
The Discovery of the Double Helix

In the early 1950s, many biologists realized that the key to understanding inheritance lay in the structure of DNA. They also knew that whoever deduced the correct structure of DNA would receive recognition from fellow biologists, fame in the popular press, and very possibly the Nobel Prize. Less obvious were the best methods to employ and who would be the person to do it.

The betting favorite in the race to discover the structure of DNA had to be Linus Pauling of Caltech. Pauling probably knew more about the chemistry of large organic molecules than did any person alive, and he had realized that accurate models could aid in deducing molecular structure. Like Rosalind Franklin and Maurice Wilkins, Pauling was an expert in X-ray diffraction techniques. Finally, he was almost frighteningly brilliant. In 1950, he demonstrated these traits by showing that many proteins were coiled into single-stranded helices (see Chapter 3). Pauling, however, had two main handicaps. First, for years he had concentrated on protein research, and therefore he had little data about DNA. Second, he was active in the peace movement. During the early 1950s, some government officials, including Senator Joseph McCarthy, considered such activity to be potentially subversive and possibly dangerous to national security. This latter handicap may have proved decisive, as we shall see.

The second most likely competitors were Wilkins and Franklin, the English scientists who had set out to determine the structure of DNA by the most direct procedure, namely the careful study of the X-ray diffraction patterns of DNA. They were the only scientists who had very good data about the general shape of the DNA molecule. Unfortunately for them, their methodical approach was also slow.

The door was open for the eventual discoverers of the double helix, James Watson and Francis Crick, two young scientists (American and English, respectively) with neither Pauling's tremendous understanding of chemical bonds nor Franklin and Wilkins's expertise in X-ray analysis. They did have three crucial advantages: (1) the knowledge that models could be enormously helpful in studying molecular structure, a lesson learned from Pauling's work on proteins; (2) access to the X-ray data; and (3) a driving ambition to be first.

Watson and Crick did no experiments in the ordinary sense of the word; rather, they spent their time thinking about DNA, trying to construct a molecular model that made sense and fit the data. Because they were based in England and because Wilkins was very open about his and Franklin's data, Watson and Crick were familiar with all the X-ray information relating to DNA. This information was just what Pauling lacked. Because of Pauling's presumed subversive tendencies, the U.S. State Department refused to issue him a passport to leave the United States, so he could neither attend meetings at which Wilkins presented the X-ray data nor visit England to talk with Franklin and Wilkins directly.

Watson and Crick knew that Pauling was working on DNA structure and were terrified that he would beat them to it. In his book *The Double Helix*, Watson recounts his belief that, if Pauling could have seen the X-ray pictures, "in a week at most, Linus would have the structure."

You might be thinking by now, "But wait just a minute! That's not fair. If the goal of science is to advance knowledge, then everyone should have access to all the data. If Pauling was the best, he should have discovered the double helix first." Perhaps so. But science is an activity of scientists, who, after all, are people too. Although virtually all scientists want to see the advancement and benefit of humanity, each individual also wants to be the one responsible for that advancement and to receive the credit and the glory. The ambition to be first helps inspire the intense concentration, the sleepless nights, and the long days in the laboratory that ultimately produce results.

Linus Pauling remained in the dark about the correct X-ray pictures of DNA and was beaten to the correct structure (Fig. E9-1). When Watson and Crick discovered the base-pairing rule that was the key to DNA structure, Watson wrote a letter about it to Max Delbruck, a friend and advisor at Caltech. He asked Delbruck not to reveal the contents of the letter to Pauling until their structure was formally published. Delbruck, perhaps more of a model scientist, firmly believed that scientific discoveries belong in the public domain and promptly told Pauling all about it. With the class of a great scientist and a great person, Pauling graciously congratulated Watson and Crick on their brilliant solution to the DNA structure. The race was over.

Figure E9-1 *The discovery of DNA*
James Watson and Francis Crick with a model of the structure of DNA.

DNA, the phosphate group of one nucleotide bonds to the sugar of another nucleotide. This bonding pattern forms a long strand consisting of a "backbone" of alternating sugars and phosphates, with the bases protruding from the backbone. Notice that the DNA strand has a "free" sugar on one end and a "free" phosphate on the opposite end.

Single strand of DNA:

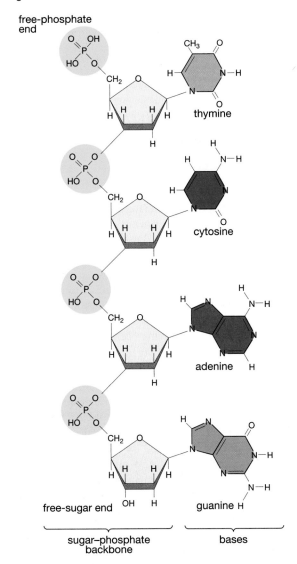

Watson and Crick proposed that the two strands of nucleotides are twisted about each other into a **double helix**, much like a ladder twisted lengthwise into a corkscrew shape (Fig. 9-2). The sugar–phosphate backbones of the two DNA strands are on the outside of the double helix, like the uprights of the ladder. Notice that the sugar–phosphate uprights run in opposite directions, so that one upright has its free sugar on one end of the DNA molecule while the other has its free sugar on the opposite end. The bases are packed into the middle, paired up to form the rungs of the ladder.

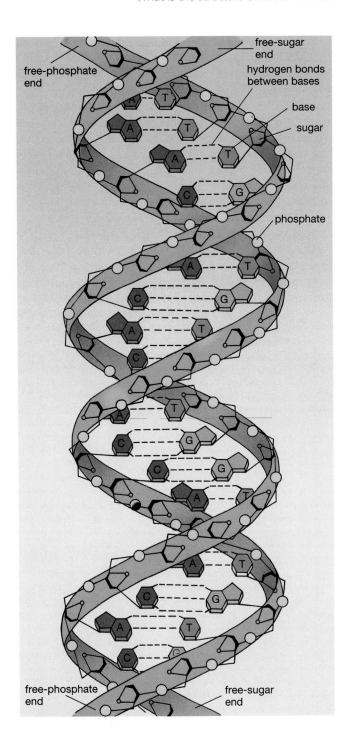

Figure 9-2 The Watson-Crick model of DNA
In this model, two strands of DNA are wound about each other in a double helix, like a ladder twisted about its lengthwise axis. Specific patterns of hydrogen bonding allow complementary bases to pair together in the center of the helix. Three hydrogen bonds hold guanine to cytosine; two hydrogen bonds hold adenine to thymine. In this and Figure 9-3, the "arrowheads" (with the oxygen at the point of the arrow) represent the sugars, and the yellow circles represent the phosphate groups. The two DNA strands run in opposite directions: Note that the free sugar is at the bottom of one strand and that the free phosphate is at the bottom of the other strand.

Watson and Crick suggested that each rung of the double helix ladder is composed of a pair of nucleotides held together by hydrogen bonds. Adenine–thymine and guanine–cytosine pairs make sense of Erwin Chargaff's data—that the amount of adenine in DNA equals the amount of thymine and that the amount of guanine equals the amount of cytosine. In nucleic acids, bases that pair together, held by hydrogen bonds, are called **complementary base pairs**. In DNA, adenine is complementary to thymine and guanine is complementary to cytosine. This is called the "base-pairing rule." The hydrogen bonds between complementary base pairs are indicated by the dashed lines in Figures 9-2 and 9-3.

The structure of DNA was solved. Although further data would be needed to confirm its details, "a structure this pretty just had to exist," as Watson later put it. Watson and Crick published their double helix model of

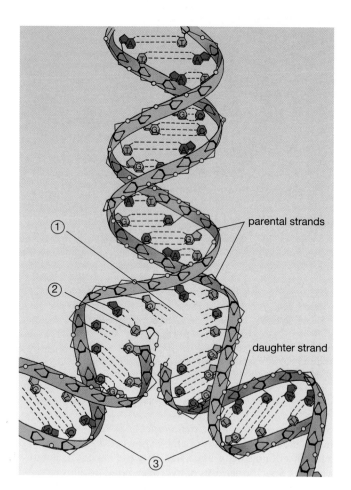

Figure 9-3 A summary of DNA replication
The sugar–phosphate backbones of the parental DNA strands are light tan and the backbones of the daughter DNA strands are purple. ① The parental DNA strands unwind, and the hydrogen bonds between complementary bases are broken. ② Complementary nucleotides are added to the growing daughter strand. ③ Each parental strand and its new daughter strand form a new double helix.

DNA in 1953, and that model revolutionized not only genetics but also much of biology in just a few years. In 1962, Watson, Crick, and Maurice Wilkins shared the Nobel Prize for Physiology and Medicine in recognition of their brilliant work in deciphering the structure of DNA. Rosalind Franklin, whose work was equally deserving of a Nobel award, died in 1958. The Nobel Prize is not awarded posthumously, so her contributions have not received the recognition they deserve.

What remains is to find out how the structure *works*: how identical copies are formed during DNA *replication* and how the sequence of nucleotides in DNA directs the synthesis of cellular components. In the following sections, we will explore how the Watson-Crick model of DNA helps provide answers to these questions.

3) How Does DNA Replication Ensure Genetic Constancy?

Prior to cell division, chromosomes are duplicated so each daughter cell can inherit an exact copy of all the genetic information of the parent cell. Each chromosome consists of a long strand of DNA, together with proteins that help organize and fold up the DNA. When a chromosome is duplicated, its double-helical strand of DNA is copied in a process called **replication** to produce two identical DNA double helices.

In classically understated scientific style, Watson and Crick included the following comment in their paper describing the double helix: "It has not escaped our notice that the specific [base] pairing we have postulated immediately suggests a possible copying mechanism for the genetic material." In other words, the base-pairing rule offers a simple hypothesis for the replication of the DNA of chromosomes prior to mitosis and meiosis: A chromosome could be replicated by separating the two DNA strands and synthesizing new strands from nucleotides with bases complementary to the parental strands. Each new chromosome would then consist of one parental strand and one complementary daughter strand.

DNA Replication Occurs in Three Basic Steps

The DNA replication in all cells occurs in three fundamental steps, each catalyzed by an enzyme (study the corresponding numbered steps in Figure 9-3 as you read this):

① The two original, or *parental*, DNA strands of the double helix unwind and separate.
② Each parental strand is used as a template for the formation of a new daughter strand of DNA. The daughter strand is formed by connecting nucleotides in an order determined by the nucleotide sequence of the parental strand according to the base-pairing

rule: Adenine pairs with thymine, and cytosine pairs with guanine.

(3) Finally, one parental DNA strand and its newly synthesized daughter strand wind together into one double helix, while the other parental strand and its daughter strand wind together into a second double helix. In forming a new double helix, the process of DNA replication conserves one parental DNA strand and produces one newly synthesized strand. Hence the process is called **semiconservative replication**. Each of these stages will be described in more detail in the following sections.

DNA Replication Begins by Separating and Unwinding Segments of the Parental DNA Strands

The first step of DNA synthesis is to separate the two parental DNA strands and thereby unwind the double helix (Fig. 9-4a). To do so, an enzyme named *DNA helicase* ("an enzyme that breaks apart the helix") works its way between the two strands, breaking the hydrogen bonds that hold them together. DNA helicase then "walks" along one strand, nudging the other strand out of the way as it goes. The result is that the two DNA strands separate, thus exposing their bases.

Complementary Daughter DNA Strands Are Synthesized

The second step of DNA synthesis requires the enzyme **DNA polymerase**. As the name "polymerase" suggests, this enzyme is responsible for joining nucleotide subunits to form the new strand of DNA, which is a polymer of nucleotides. Two DNA polymerase molecules now bind to the unwound strands, one to each strand (Fig. 9-4b). During DNA replication, DNA polymerase performs a dual function. First, it recognizes bases exposed in a parental strand and matches them up with free nucleotides that have complementary bases: Adenine forms hydrogen bonds with an exposed thymine from the parental DNA strand, and guanine with an exposed cytosine. Second, DNA polymerase bonds together the sugars and phosphates of the complementary nucleotides to form the backbone of the daughter strand (Fig. 9-4c).

DNA polymerase molecules treat the parental DNA strands as one-way streets: DNA polymerase can travel only in one direction on a DNA strand, from the "free sugar" end to the "free phosphate" end. Thus, the two DNA polymerase molecules, one on each parental strand, move in opposite directions. This process necessitates a third step of DNA synthesis.

As the DNA helicase molecule continues to separate the parental DNA strands, one polymerase molecule simply follows behind it, synthesizing a long, continuous complementary daughter strand as it goes (Fig. 9-4c,d). The

polymerase that settled on the second strand also moves in the "free sugar" to "free phosphate" direction; thus, that enzyme travels *away from the DNA helicase* (Fig. 9-4c). Therefore, as the helicase continues to separate the parental strands, this polymerase cannot reach the newly separated parts of the second strand (Fig. 9-4d). Soon, however, a new DNA polymerase molecule attaches to the second strand close behind the helicase. Like the first polymerase, it moves away from the helicase. What happens when this second polymerase reaches the place where the first polymerase started? The two short DNA strands synthesized by the two DNA polymerases must be joined to make a continuous DNA strand. An enzyme called *DNA ligase* bonds the two strands together (Fig. 9-4e). This process is repeated perhaps 10 million or so times for a human chromosome until the second strand has been replicated.

Proofreading Produces Almost Error-Free Replication of DNA

What processes help ensure that DNA is copied accurately? In preparation for cell division, the DNA of each chromosome unwinds and replicates, producing identical paired structures called *chromatids* (these identical twins are often referred to as sister chromatids). Each chromatid consists of a double helix of DNA, composed of one of the parental DNA strands plus one new strand that is an exact copy of the other parental strand. When sister chromatids separate during cell division and are parceled out to the daughter cells, each daughter cell receives an exact copy of each parental chromosome. Thus, if there are no mistakes in the entire process, the integrity of the genetic information will be maintained from cell division to cell division and from parent to offspring.

Hydrogen bonding between complementary base pairs makes DNA replication highly accurate. Nevertheless, DNA replication is not perfect. Partly because replication is so fast (50 to 500 nucleotides per second) and partly because chemical flip-flops in the bases occur spontaneously, DNA polymerase occasionally matches bases incorrectly, making perhaps one mistake in every 10,000 base pairs. In mammalian cells, however, the completed DNA strands contain only about one mistake in every billion base pairs. This phenomenal accuracy is ensured by several DNA repair enzymes, including some forms of DNA polymerase, that "proofread" each daughter strand during and after its synthesis.

But Mistakes Do Happen

Despite this amazing accuracy, neither we nor any other life-forms are free of DNA damage. In addition to mistakes made during DNA replication, each cell in your body loses about 10,000 bases each day due to spontaneous chemical breakdown at body temperature. A variety of environmental conditions can also damage DNA.

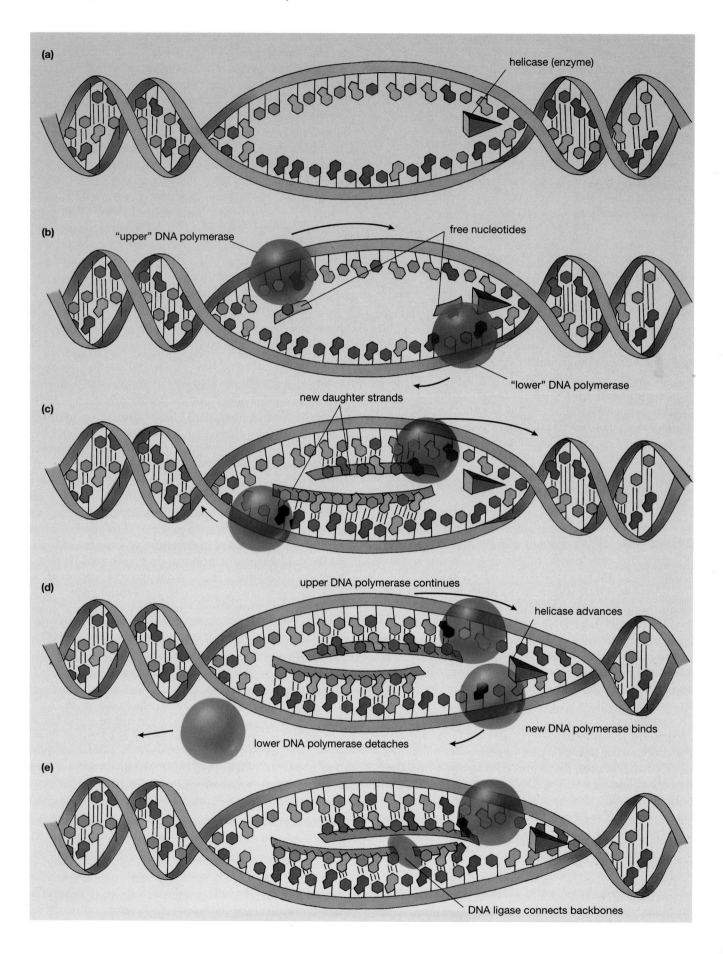

(a)

helicase (enzyme)

(b)

"upper" DNA polymerase

free nucleotides

"lower" DNA polymerase

(c)

new daughter strands

(d)

upper DNA polymerase continues

helicase advances

new DNA polymerase binds

lower DNA polymerase detaches

(e)

DNA ligase connects backbones

Figure 9-4 The details of DNA replication
(a) The enzyme DNA helicase separates the two DNA strands, unwinding a small portion of the double helix.
(b) Two DNA polymerase molecules attach to the separated parental strands. The one on the upper strand follows the helicase; the polymerase on the lower strand must move away from the helicase. *(c)* The two DNA polymerase molecules match up free nucleotides with the parental DNA strands, using the base-pairing rules, and synthesize daughter DNA strands. *(d)* The helicase advances, unwinding more parental DNA, followed by the DNA polymerase on the upper strand. The DNA polymerase on the lower strand, however, reaches the end of the unwound DNA and detaches from the DNA, having synthesized only a short daughter DNA segment. Meanwhile, a new DNA polymerase molecule attaches just behind the helicase and proceeds away from the helicase toward the first daughter DNA segment. *(e)* When the new DNA polymerase has synthesized a daughter DNA segment that reaches the end of the first daughter DNA segment, that enzyme also detaches. Another enzyme, DNA ligase, attaches the two short daughter DNA segments.

For example, whenever you go out in the sunshine, DNA in some of your skin cells is damaged by high-energy rays of ultraviolet light. Several DNA repair enzymes are continuously on call to repair this damage, but inevitably, some errors are overlooked. A cell with DNA damage may function normally (for example, if the damage occurs in a noncritical stretch of DNA), it may survive but not function as efficiently as before, or it may die. On rare occasions, described in more detail in Chapter 31, errors in DNA replication may cause a cell to divide uncontrollably, resulting in cancer. Studies of people with genetic diseases that cause premature aging suggest that some aspects of aging may be caused by a deterioration in the accuracy of DNA replication, a topic that is explored in Chapter 10 in "Health Watch: Sex, Aging, and Mutations."

Summary of Key Concepts

1) What Is the Composition of Chromosomes?
Eukaryotic chromosomes are composed of deoxyribonucleic acid (DNA) and protein. DNA is the molecule that carries genetic information. It is composed of subunits called nucleotides, linked together into long strands. Each nucleotide consists of a phosphate group, the five-carbon sugar deoxyribose, and a nitrogen-containing base. Four types of bases occur in DNA: adenine, guanine, thymine, and cytosine.

2) What Is the Structure of DNA?
The DNA of chromosomes is composed of two strands, wound about one another in a double helix. The sugars and phosphates that link one nucleotide to the next form a backbone on each side of the double helix. The bases from each strand pair up in the middle of the helix, held together by hydrogen bonds. Only specific pairs of bases, called complementary base pairs, can bond together in the helix: adenine with thymine, and guanine with cytosine.

3) How Does DNA Replication Ensure Genetic Constancy?
When a chromosome replicates prior to cell division, the two DNA strands of each double helix unwind. The enzyme DNA polymerase moves along each strand, linking up free nucleotides into new DNA strands. The sequence of nucleotides in each newly formed strand is complementary to the sequence on a parental strand. During this semiconservative replication process, two double helices are synthesized, each consisting of one parental DNA strand and one newly synthesized, complementary strand that is an exact copy of the other parental strand. The two daughter DNA molecules are therefore duplicates of the parental DNA molecule.

Key Terms

base *p. 144*
chromosome *p. 143*
complementary base pair
  p. 148
deoxyribonucleic acid
  (DNA) *p. 144*
DNA polymerase *p. 149*
double helix *p. 147*
gene *p. 144*
nucleotide *p. 144*
purine *p. 144*
pyrimidine *p. 144*
replication *p. 148*
semiconservative replication
  p. 149

Thinking Through the Concepts

Multiple Choice

1. *How many different possible base sequences are there in a nucleotide chain three nucleotides in length?*
a. 1   b. 3   c. 9   d. 64   e. more than 64

2. *Erwin Chargaff, in analyzing DNA, found that the amount of*
a. cytosine equals that of guanine
b. cytosine equals that of thymine
c. cytosine equals that of adenine
d. each nucleotide is unrelated to all others
e. each nucleotide is equal to all others

3. *Semiconservative replication refers to the fact that*
a. each new DNA molecule contains two new single DNA strands
b. DNA polymerase uses free nucleotides to synthesize new DNA molecules
c. certain bases pair with specific bases
d. each parental DNA strand is joined with a new strand containing complementary base pairs
e. mistakes are made during DNA replication

4. *DNA helicase*
a. cleaves hydrogen bonds that join the two strands of DNA
b. converts two single strands of DNA into a double helix
c. is a unique form of DNA
d. adds nucleotides to newly forming DNA molecules
e. proofreads the newly formed DNA strand

5. *DNA polymerase*
a. can advance in either direction along a single strand of DNA
b. cleaves hydrogen bonds that join the two strands of DNA
c. creates a polymer that consists of many molecules of DNA
d. adds appropriate nucleotides to a newly forming DNA strand
e. is the protein found in conjunction with DNA in eukaryotic chromosomes

6. *Which of the following are incorrectly matched?*
a. purines ↔ adenine and thymine
b. bases ↔ adenine, thymine, cytosine, and guanine
c. nucleotide ↔ phosphate and sugar and base
d. eukaryotic chromosome ↔ DNA and protein
e. enzymes involved in DNA replication ↔ DNA polymerase and DNA helicase

? Review Questions

1. Draw the general structure of a nucleotide. Which parts are identical in all nucleotides, and which can vary?

2. Name the four types of nitrogen-containing bases found in DNA.

3. Which bases are complementary to one another? How are they held together in the double helix of DNA?

4. Describe the structure of DNA in a chromosome. Where are the bases, sugars, and phosphates in the structure?

5. Describe the process of DNA replication.

Applying the Concepts

1. As you learned in "The Discovery of the Double Helix," scientists in different laboratories often compete with one another to make new discoveries. Do you think this competition helps promote scientific discoveries? Sometimes, researchers in different laboratories collaborate with one another. What advantages does collaboration offer over competition? What factors might provide barriers to collaboration and lead to competition?

2. During the Inquisition, the Church forced Galileo to disavow some of his scientific statements. Today, scientific advances are being made at an astounding rate, and nowhere is this more evident than in our understanding of the biology of heredity. Using DNA as a starting point, do you believe there are limits to the knowledge people *should* acquire? Defend your answer.

Group Activity

Make a class model of DNA: Form two lines of students, with the students in each line facing a different direction. Each of you in the line should put your left hand on the shoulder of the person in front of you and extend your right hand out toward the other line, palm facing forward. Each of you is a nucleotide. The hand on the shoulder represents the sugar–phosphate link. The hands extended, almost touching, represent the base pairs. It is easy to see the antiparallel nature of DNA in this model.

As an additional exercise, have one student play the role of DNA polymerase by matching "complementary bases"—that is, by pairing up students in opposite lines on the basis of eye color or some other basis. For example, pair brown eyes with green eyes, and pair blue eyes with gray or black eyes.

For More Information

Crick, F. *What Mad Pursuit: A Personal View of Scientific Discovery.* New York: Basic Books, 1998. Another view of the race to determine the structure of DNA, by Francis Crick himself.

Gibbs, W. W. "Peeking and Poking at DNA." *Scientific American* (Explorations), March 31, 1997. An update of new techniques for studying the DNA molecule, such as atomic force microscopy.

Judson, H. F. *The Eighth Day of Creation.* Cold Spring Harbor, NY: Cold Spring Harbor Laboratory Press, 1993. A very readable historical perspective on the development of genetics.

Leutwyler, K. "Turning Back the Strands of Time." *Scientific American* (Explorations), February 2, 1998. A brief discussion of telomeres, the repeating DNA strings that tie up the ends of chromosomes.

Mirsky, A. E. "The Discovery of DNA." *Scientific American*, June 1968. The early history of DNA research.

Olby, R. *The Path to the Double Helix.* Seattle: University of Washington Press, 1975. A historical perspective on the development of genetics in the twentieth century.

Radman, M., and Wagner, R. "The High Fidelity of DNA Duplication." *Scientific American*, August 1988. Faithful duplication of chromosomes requires both reasonably accurate initial replication of DNA sequences and final proofreading.

Rennie, J. "DNA's New Twists." *Scientific American*, March 1993. A reprise of new information on DNA structure and function.

Watson, J. D. *The Double Helix.* New York, Atheneum, 1968. If you still believe the Hollywood images that scientists are either maniacs or cold-blooded, logical machines, be sure to read this book. Although hardly models for the behavior of future scientists, Watson and Crick are certainly human enough!

Answers to Multiple-Choice Questions
1. d   2. a   3. d   4. a   5. d   6. a

Is the calico cat confused—why isn't the fur of the calico uniform black or orange? The various color patches arose from cells early in development that expressed either a gene for orange fur or a gene for black fur.

Gene Expression and Regulation 10

At a Glance

1) How Are Genes and Proteins Related?

Red Bread Mold Provided Insight into the Role of Genes

Most Genes Encode the Information for the Synthesis of a Protein

DNA Provides Instructions for Protein Synthesis via RNA Intermediaries

Genetic Information Flows in a Cell from DNA to RNA in Order to Direct Protein Synthesis

In the Genetic Code, Base Sequences Stand for Amino Acids

2) What Is the Role of RNA in Protein Synthesis?

Transcription Produces mRNA Molecules That Are Complementary Copies of One Strand of DNA

Messenger RNA Conveys the Code for Protein Synthesis from the Nucleus to the Cytoplasm

Ribosomal RNA Forms an Important Part of the Protein-Synthesizing Machinery of a Ribosome

Transfer RNA Molecules Decode the Sequence of Bases in mRNA into the Amino Acid Sequence of a Protein

Translation: mRNA, tRNA, and Ribosomes Cooperate to Synthesize Proteins

Messenger and Transfer RNAs Decode the Sequence of Bases in DNA to the Sequence of Amino Acids in Protein

3) How Do Mutations in DNA Affect the Function of Genes?

Mutations Result from Nucleotide Substitutions, Insertions, or Deletions

Mutations Differ in Their Effects on Protein Structure and Function

Mutations Provide the Raw Material for Evolution

4) How Are Genes Regulated?

Eukaryotic Cells May Regulate the Transcription of Individual Genes, Regions of Chromosomes, or Entire Chromosomes

Net Watch

On-line resources for this chapter are on the World Wide Web at:
http://www.prenhall.com/audesirk
(click on the Table of Contents link and then select Chapter 10).

The beautiful color pattern of the calico cat is a striking example of gene expression and gene regulation. Calicos have patches of orange, black, and white fur. *Genes*, which are functional segments of DNA molecules, are biological blueprints that specify the structure of complex molecules, primarily proteins. Most of the proteins coded by genes are enzymes, which catalyze reactions that produce substances used throughout the body. The orange and black fur colors of the calico cat arise from the presence of pigments, substances that reflect light in different ways and produce color in fur (or scales or skin or feathers, for that matter). A given pigment is synthesized in a series of reactions, each catalyzed by a specific enzyme. But look at the cat in the chapter-opening photo. Not all of her skin cells are producing the same pigment, and some are producing no pigment, creating the areas of white fur. All her cells developed from a single fertilized egg and contain the same DNA, so why aren't they all producing the same enzymes and making the same pigment molecules? Clearly, different genes are being expressed in different patches of skin.

How does a gene direct the synthesis of an enzyme or other protein? Because chromosomes are confined to the nucleus in eukaryotic cells, how do proteins reach the cytoplasm? And finally, how can different cells of the body have different structures and functions? Why, for example, do hair follicle cells synthesize hair but not hormones? Why are some cells of a calico cat specialized not only to synthesize hair, but to produce different colors of hair?

In this chapter we explore the amazing process by which the *genetic code*—the sequence of nucleotides in DNA—is ultimately translated into the sequence of amino acids in protein. We will also introduce some of the mechanisms by which this process is regulated to give rise to the diverse cell types in your body, to make you a male or female, and to give the calico cat her "coat of many colors."

1) How Are Genes and Proteins Related?

A **gene** is a segment of DNA whose sequence of nucleotides specifies the sequences of amino acids in a particular protein. Each chromosome, which contains one long DNA molecule, includes many thousands of genes. Although genes store the information that is used to produce a protein, they don't really *do* anything. Rather, they are much like a blueprint that tells you how to build a structure but doesn't saw any boards or pound any nails. The information in DNA specifies the properties of each cell and, ultimately, of the entire organism. Proteins, particularly enzymes, guide the synthesis of all the parts of a cell and promote its other metabolic reactions. So it seems logical that, somehow, the information in DNA must be used to synthesize proteins. In the following sections, we describe how this assumption was verified and how the process proceeds.

Red Bread Mold Provided Insight into the Role of Genes

The common red bread mold, *Neurospora crassa*, proved to be an ideal organism for studying the relationship between genes and enzymes. Although we commonly see this mold on stale bread, *Neurospora* is an extremely independent organism that can synthesize almost all the organic compounds it needs. It can grow on a simple nutrient solution (minimal medium) that contains a few minerals, a single vitamin, and an energy source such as sucrose.

Besides being easy to grow, *Neurospora* is genetically ideal as an experimental subject. For most of its life, *Neurospora* has just one copy of each chromosome and therefore just one copy of each gene. Most plants and animals, in contrast, have two copies of each chromosome and thus two copies of each gene. Consequently, the effects of a defective gene may be masked by a normal gene on the second chromosome. In *Neurospora* the effects of a defective gene cannot be masked, because there is no other copy of that gene.

In the early 1940s, geneticists George Beadle and Edward Tatum bombarded *Neurospora* with X-rays. The high energy of X-rays causes *mutations*, which are changes in the base sequence of DNA. Eventually the X-rays produced hundreds of different mutations that influenced the nutritional requirements of the mold. Each mutant mold was no longer able to grow on minimal medium unless a specific nutrient—for example, one of the B vitamins or a certain amino acid—was added. Beadle and Tatum concluded that each of these mutations inactivated a specific enzyme that normally allowed the mold to synthesize a nutrient. Their experiments supported the hypothesis that each gene codes for a single enzyme. A few years later, to determine the series of chemical reactions by which normal molds synthesize the amino acid arginine, biochemists used mutant *Neurospora* that couldn't grow on minimal medium unless arginine was added. They found that the mold uses the following biochemical pathway:

$$\text{ornithine} \xrightarrow{\textit{enzyme A}} \text{citrulline} \xrightarrow{\textit{enzyme B}} \text{arginine}$$

By starting with minimal medium and then adding one precursor molecule in the pathway of arginine synthesis at a time, the researchers found that the mutant lacked a single enzyme that catalyzed one specific step in arginine synthesis (Fig. 10-1). This finding further supported the hypothesis that each gene encodes the information needed for the synthesis of a specific enzyme.

Most Genes Encode the Information for the Synthesis of a Protein

Geneticists have since learned that not all genes encode the information needed by a cell to produce enzyme proteins. Some genes carry information for the synthesis of structural proteins, such as collagen in skin or keratin in hair, of hormones, such as insulin, or of receptors (proteins to which molecules such as hormones bind, initiating a response in a cell). For a few genes, the final product isn't protein but a nucleic acid called *ribonucleic acid* (RNA), described below. Nevertheless, the ultimate cellular products encoded by most genes are proteins or parts of proteins. As a generalization, each gene encodes the information for a single protein; this generalization is called the **one-gene, one-protein hypothesis**.

Different genes have different base sequences, and different proteins have different amino acid sequences. Therefore, the sequence of bases in DNA must encode the sequence of amino acids in a protein. How does a cell convert the information stored in its DNA base sequences into proteins?

DNA Provides Instructions for Protein Synthesis via RNA Intermediaries

The DNA of a eukaryotic cell is located in the cell's nucleus, but protein synthesis occurs on ribosomes in the cytoplasm. Therefore, DNA cannot directly guide protein synthesis. There must be an intermediary, a molecule that carries the information from DNA in the nucleus to the ribosomes in the cytoplasm. This molecule is **ribonucleic acid**, or **RNA**.

RNA is similar to DNA but differs structurally in three respects: (1) RNA is normally single-stranded; (2) RNA has the sugar ribose instead of deoxyribose in its backbone; and (3) the base uracil in RNA replaces thymine in DNA (Table 10-1).

There are three types of RNA in a cell: (1) *messenger RNA* (mRNA), (2) *ribosomal RNA* (rRNA), and (3) *transfer RNA* (tRNA) (Fig. 10-2). Each is involved in

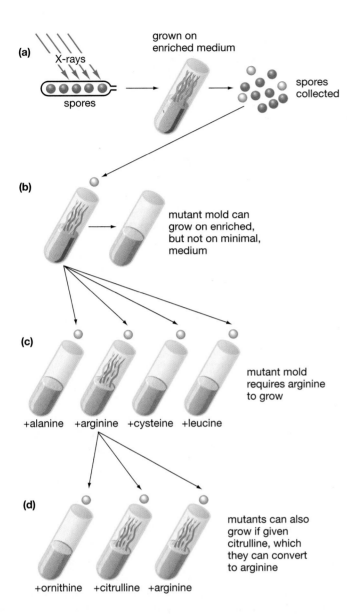

Table 10-1 A Comparison of DNA and RNA

	DNA	RNA
Number of Strands	2	1
Type of Sugar	deoxyribose	ribose
Base Pairs	adenine (A)–thymine (T) cytosine (C)–guanine (G)	adenine (A)–uracil (U) cytosine (C)–guanine (G)
Function	Composed of genes whose base sequence creates a code for protein synthesis	*messenger RNA:* carries the code from genes to ribosomes *ribosomal RNA:* combines with protein to form ribosomes, on which protein synthesis occurs *transfer RNA:* carries amino acids to ribosome

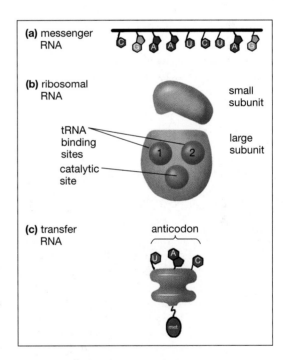

Figure 10-1 Beadle and Tatum's experiments with the red bread mold **Neurospora**

These experiments showed that the mutation of a single gene can produce a single defective enzyme. *(a)* Mold spores (each of which can sprout into a new mold) are bombarded with X-rays to induce mutations. Molds are grown from these spores on an enriched medium containing all the amino acids; then their spores are collected. *(b)* The spores are grown individually in enriched medium, then a growing portion is placed on minimal medium—lacking amino acids. If the mold cannot grow on minimal medium, a mutation has occurred. *(c)* Pieces of mutated mold are placed in tubes containing minimal medium plus one amino acid each. This mutant can grow only if arginine is added. Therefore, it cannot synthesize arginine. *(d)* Molds synthesize arginine via the pathway ornithine —*enzyme A* → citrulline —*enzyme B* → arginine. Mutants are tested in media to which ornithine and citrulline—precursors in the synthesis pathway of arginine—have been added. Because the mold can grow if supplied with citrulline or arginine but not with ornithine, the mutation must have ruined *enzyme A*, which normally catalyzes the conversion of ornithine to citrulline.

Figure 10-2 The three types of RNA

(a) Messenger RNA consists of a single strand of nucleotides whose bases are complementary to those of the template DNA from which the mRNA molecule was transcribed. *(b)* Ribosomal RNA, in conjunction with proteins, forms ribosomes. Each ribosome consists of a large and a small subunit, which join during protein synthesis. The small subunit binds the mRNA; the large subunit binds tRNA and catalyzes the formation of bonds between amino acids to form a protein. *(c)* Transfer RNA has an anticodon of three nucleotides that is complementary to a codon in mRNA. A binding site at the base of the tRNA molecule binds a particular amino acid that is specified by the anticodon and by the corresponding codon of the mRNA. Different tRNAs have different anticodons and bind to different amino acids.

translating the genetic information in DNA into the amino acid sequence of proteins. All three types of RNA are coded by DNA. We will examine their functions in more detail shortly.

Genetic Information Flows in a Cell from DNA to RNA in Order to Direct Protein Synthesis

Information from DNA is used to direct the synthesis of proteins in a two-step process (Fig. 10-3):

1. In **transcription**, the information contained in the DNA of a specific gene is copied into **messenger RNA (mRNA)**. The sequence of bases in messenger RNA carries information from the nucleus to the ribosomes. This information specifies the sequence of amino acids in the protein to be manufactured.

2. In **translation**, **transfer RNA (tRNA)** and **ribosomal RNA (rRNA)** convert the information of the base sequence in messenger RNA into a specific amino acid sequence and thereby help synthesize the protein.

To understand the molecular mechanisms of this sequence of events from DNA to protein, geneticists first had to break the language barrier: How does the language of base sequences in DNA and messenger RNA translate into the language of amino acid sequences in

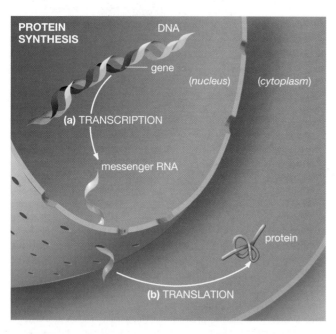

Figure 10-3 *From DNA to RNA to protein synthesis*
During transcription, a gene (located on one DNA strand) that encodes a protein provides the information for the synthesis of mRNA. During translation, the information in mRNA is used to specify the sequence of amino acids in the protein.

proteins? This translation relies on a "dictionary" called the *genetic code*.

In the Genetic Code, Base Sequences Stand for Amino Acids

We have used the word "code" several times to refer to the information that is stored in DNA and ultimately translated into the amino acid sequence of proteins. This **genetic code** is conceptually similar to Morse code: a set of symbols (bases in nucleic acids, dots and dashes in Morse code) that can be translated into another set of symbols (amino acids in proteins, letters of the alphabet). What combinations of bases stand for which amino acids?

The Genetic Code Uses Three Bases to Specify Each Amino Acid

There are four types of bases in DNA and four types of bases in RNA (see Table 10-1). However, there are 20 different amino acids in proteins. Therefore, one base cannot code for just one amino acid because there are simply not enough types of bases. The genetic code must rely on a short sequence of bases to encode each amino acid, just as Morse code relies on a short sequence of dots and dashes to encode each letter of the alphabet. If a sequence of two bases codes for an amino acid, then there would be 16 (4 × 4) possible combinations of bases. This isn't enough to account for all 20 amino acids. A three-base sequence, however, gives 64 (4 × 4 × 4) possible combinations of bases, which is more than enough. Under the assumption that nature operates as economically as possible, biologists hypothesized that the genetic code must be a triplet code: Three bases specify a single amino acid. In 1961, Francis Crick and three co-workers demonstrated that this hypothesis is correct (see "Scientific Inquiry: Cracking the Genetic Code").

For any language to be understood, the users must know what the words mean, where words start and stop, and where sentences begin and end. Crick's experiments demonstrated that all the "words" of the genetic code are three bases long and that a set of three bases signifies one amino acid. Shortly after this discovery, researchers began to decipher the genetic code. They ground up bacteria and isolated the components needed to synthesize proteins. To this mixture, they added artificial mRNA, which allowed them to control what "words" were to be transcribed. Researchers could then see which amino acids were incorporated into the resulting proteins. For example, an RNA strand composed entirely of uracil (UUUUUUUU . . .) directed the mixture to synthesize a protein composed solely of the amino acid phenylalanine. Therefore, the triplet specifying phenylalanine must be UUU. Because the genetic code was deciphered by

Scientific Inquiry
Cracking the Genetic Code

The hypothesis that three bases in DNA code for one amino acid in a protein is an attractive and logical possibility. But how could you prove that nature actually uses a triplet code?

Francis Crick and his co-workers exposed *bacteriophages* (or simply *phages*), viruses that multiply inside bacteria, to a chemical called acridine. Acridine causes one, two, or three nucleotides to be inserted into the viral DNA molecule at random places near the beginning of a particular gene. The researchers found that inserting one or two nucleotides into the DNA caused the synthesis of defective enzymes that prevented the viruses from multiplying inside their host bacteria. Inserting three nucleotides, however, sometimes produced phages that synthesized normal or nearly normal enzymes.

Crick concluded that during RNA synthesis, a gene's DNA is "read" in a linear order, starting at the beginning. Each set of three bases makes up a "word"; that is, three bases in DNA encode a single amino acid. The code must specify the beginning and end of a gene, but within a gene there are no spaces or punctuation between words. How did he arrive at these conclusions?

To understand Crick's reasoning, let's suppose that we have a gene with the repetitive sequence of bases (TAG) as shown in Fig. E10-1a. If DNA always has three-letter words, then this base sequence spells out the word "TAG" over and over again. You don't need spaces to separate the words if they all have three and only three letters.

Now suppose that the acridine inserts another nucleotide with the base cytosine near the beginning of the gene (Fig. E10-1b). Most of the gene now reads GTA . . ., which is the code for a different amino acid and produces a nonfunctional enzyme. Inserting another nucleotide nearby again changes the amino acid specified, as most of the gene now reads AGT . . ., and results in another nonfunctional enzyme (Fig. E10-1c). A third insertion, however, results in most of the gene reading TAG . . . again (Fig. E10-1d). In this case, most of the enzyme would be synthesized correctly and might work well enough to allow the viruses to reproduce.

Crick concluded that only a triplet code could account for the fact that three insertions, and not one or two, restore near-normal enzyme function. Further, only a code without punctuation or spaces between words would be confused by any of the insertions. Finally, a code without punctuation between words could work only if the start points and stop points of protein synthesis are clearly marked. All three conclusions have been found to be correct, as we describe in the text.

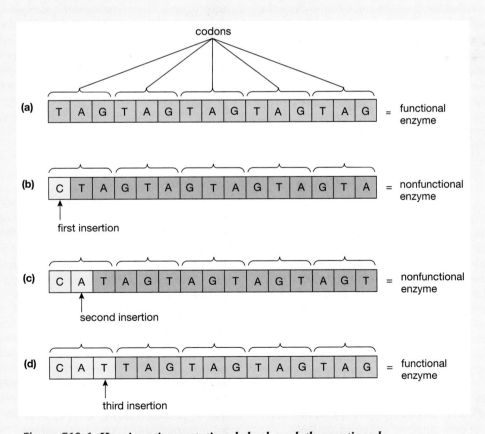

Figure E10-1 *How insertion mutations helped crack the genetic code*

Table 10-2 The Genetic Code (Codons of mRNA)

Second Base

First Base		U		C		A		G		Third Base
U		UUU	Phenylalanine	UCU	Serine	UAU	Tyrosine	UGU	Cysteine	U
		UUC	Phenylalanine	UUC	Serine	UAC	Tyrosine	UGC	Cysteine	C
		UUA	Leucine	UCA	Serine	UAA	Stop	UGA	Stop	A
		UUG	Leucine	UCG	Serine	UAG	Stop	UGG	Tryptophan	G
C		CUU	Leucine	CCU	Proline	CAU	Histidine	CGU	Arginine	U
		CUC	Leucine	CCC	Proline	CAC	Histidine	CGC	Arginine	C
		CUA	Leucine	CCA	Proline	CAA	Glutamine	CGA	Arginine	A
		CUG	Leucine	CCG	Proline	CAG	Glutamine	CGG	Arginine	G
A		AUU	Isoleucine	ACU	Threonine	AAU	Asparagine	AGU	Serine	U
		AUC	Isoleucine	ACC	Threonine	AAC	Asparagine	AGC	Serine	C
		AUA	Isoleucine	ACA	Threonine	AAA	Lysine	AGA	Arginine	A
		AUG	Start (Methionine)	ACG	Threonine	AAG	Lysine	AGG	Arginine	G
G		GUU	Valine	GCU	Alanine	GAU	Aspartic Acid	GGU	Glycine	U
		GUC	Valine	GCC	Alanine	GAC	Aspartic Acid	GGC	Glycine	C
		GUA	Valine	GCA	Alanine	GAA	Glutamic Acid	GGA	Glycine	A
		GUG	Valine	GCG	Alanine	GAG	Glutamic Acid	GGG	Glycine	G

using these artificial RNAs, the code is usually written in terms of the base triplets in mRNA (rather than in DNA) that code for each amino acid (Table 10-2). These mRNA triplets are called **codons**.

What about punctuation? Given that one mRNA molecule may contain thousands of bases, how does the cell recognize where the code for a given protein starts and stops? Research showed that the codon AUG is a **start codon**—that is, it codes for an amino acid that signals the beginning of a protein. Three codons—UAG, UAA, and UGA—are **stop codons**, signaling the ribosome to release the mRNA and the new protein.

There are 60 codons, including the one start codon and three stop codons, but only 20 amino acids for which to code. All 60 codons are used in the genetic code. Thus, a single amino acid may be specified by several codons. For example, six different codons all code for arginine (see Table 10-2). However, each codon specifies one, and only one, amino acid.

What Is the Role of RNA in Protein Synthesis?

www

Earlier we noted that genetic information flows from DNA to protein synthesis in a two-step sequence: (1) transcription followed by (2) translation. In the following sections, we describe how the three types of RNA are involved in these two processes.

Transcription Produces mRNA Molecules That Are Complementary Copies of One Strand of DNA

The transcription of DNA into mRNA is restricted in two major ways. First, in any cell, transcription normally copies the DNA of only selected genes into mRNA. For example, hair follicle cells transcribe the DNA that en-

codes the protein keratin, which forms hair; insulin-secreting cells of the pancreas transcribe the genes that encode insulin; and so on. Second, transcription normally copies only one of the two strands of DNA into mRNA. In most cases, the useful information of a given gene resides on only one strand of the DNA double helix. Why? Remember, the two strands of DNA are *complementary*, not *identical*. The gene in question lies on only one strand. The sequence of bases on that strand codes for a sequence of amino acids that forms a functional protein, but the complementary strand will have a different base sequence, one that is very unlikely to code for a useful protein. The DNA strand that contains the gene and is transcribed into mRNA is called the **template strand**: It is the template from which the complementary RNA strand is made. A chromosome, which is one long DNA molecule, contains many genes. In the two-stranded DNA molecule, one strand may be the template strand for some genes, and the other strand may be the template strand for other genes.

With these constraints in mind, we can view transcription as a three-step process consisting of (1) *initiation*, (2) *elongation* of the RNA molecule, and (3) *termination* (Fig. 10-4). These three steps correspond to the three major parts of most genes in both eukaryotes and prokaryotes: (1) a *promoter* region at the beginning of the gene; (2) the "body" of the gene, consisting of the DNA bases that code for amino acids in the protein to be synthesized; and (3) a termination signal at the end of the gene.

Initiation: RNA Synthesis Begins at the Promoter of a Gene

The synthesis of all three types of RNA is carried out by an enzyme called **RNA polymerase**. This enzyme uses genes on DNA as templates to specify the sequence of

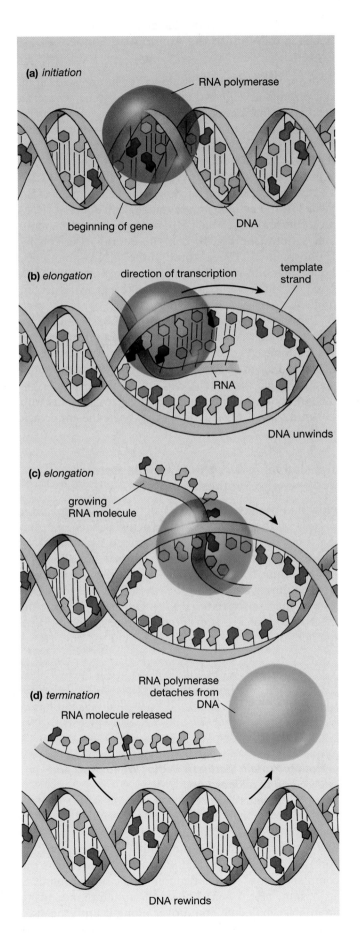

(a) *initiation*

RNA polymerase

beginning of gene

DNA

(b) *elongation*

direction of transcription

template strand

RNA

DNA unwinds

(c) *elongation*

growing RNA molecule

(d) *termination*

RNA polymerase detaches from DNA

RNA molecule released

DNA rewinds

RNA nucleotides. RNA polymerase must locate the beginning of the gene for transcription to begin. The **promoter** region of a gene is a short sequence of DNA bases that marks the beginning of the gene. RNA polymerase binds to the DNA at the promoter site, initiating transcription (Fig. 10-4a). The "decision" to transcribe a particular gene is based on conditions inside and outside the cell that signal a need for a particular enzyme or other protein.

Elongation and Termination: RNA Synthesis Proceeds from the Promoter to the End of the Gene

Once the RNA polymerase has bound to the promoter site, the enzyme changes shape, forcing the DNA double helix to unwind at the beginning of the gene. RNA polymerase then travels along the template strand, much as DNA polymerase does. Using free RNA nucleotides present in the nucleus, RNA polymerase synthesizes a single strand of RNA that is complementary to the template strand of DNA (Fig. 10-4b). The same base-pairing rule applies to RNA as to DNA, except that in RNA, uracil, rather than thymine, pairs with adenine (see Table 10-1).

After about 10 nucleotides have been added to the growing RNA chain, the beginning of the RNA molecule separates from the DNA and pairing between the bases on either DNA strand is re-established (Fig. 10-4c). As the RNA continues to elongate, it forms a long "tail" that drifts away from the DNA (Fig. 10-5).

RNA polymerase continues along the template strand until it reaches the termination signal, a sequence of DNA bases that triggers two events (see Fig. 10-4d). First, the RNA molecule separates from both the DNA and the RNA polymerase. Second, the RNA polymerase detaches from the DNA template strand.

Messenger RNA Conveys the Code for Protein Synthesis from the Nucleus to the Cytoplasm

Although all three forms of RNA are produced by transcription from DNA, only mRNA carries the code for the amino acid sequence of a protein (see Fig. 10-2a). In eukaryotic cells, mRNA molecules are synthesized in the nucleus and enter the cytoplasm through the pores in the nuclear envelope. In the cytoplasm, mRNA binds to ribosomes, where the codons of mRNA are translated into

Figure 10-4 A diagram of RNA transcription
(a) Initiation: The enzyme RNA polymerase binds to the promoter region of DNA near the beginning of a gene. *(b)* and *(c)* Elongation: The DNA double helix unwinds. RNA polymerase travels along the DNA template strand, catalyzing the formation of a strand of mRNA from complementary free RNA nucleotides. The process continues until *(d)* Termination: The RNA polymerase reaches the end of the gene, where the enzyme detaches from the DNA. The DNA molecule rewinds, and the RNA molecule is released.

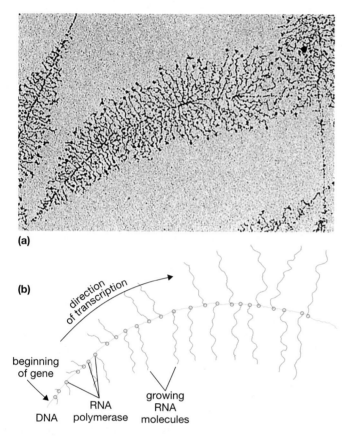

(a)

(b)

direction of transcription

beginning of gene

DNA RNA polymerase growing RNA molecules

Figure 10-5 RNA transcription in action
(a) This electron micrograph shows the progress of RNA transcription in the egg of an African clawed toad. *(b)* In each treelike structure, the central "trunk" is DNA and the "branches" are RNA molecules. A series of RNA polymerase molecules are traveling down the DNA, synthesizing RNA as they go. The beginning of the gene is on the left. Therefore, the short RNA molecules on the left have just begun to be synthesized; the long RNA molecules on the right are almost finished.

the language of amino acids in proteins. The gene remains safely stored in the nucleus, like a valuable document in a library, while mRNA carries the information to the cytoplasm to be used in protein synthesis.

Ribosomal RNA Forms an Important Part of the Protein-Synthesizing Machinery of a Ribosome

Ribosomes are composites of rRNA and many different proteins. A single eukaryotic cell has tens of thousands of ribosomes. Each ribosome is composed of two subunits— one small and one large. Unless they are actively synthesizing proteins, the two subunits remain separate (see Fig. 10-2b). The small subunit recognizes and binds mRNA and part of tRNA. The large ribosomal subunit contains an enzymatic region that catalyzes the addition of amino acids to the growing protein chain, and it bears two other sites that bind to tRNA. Ribosomal RNA probably plays the major part in recognizing mRNA and in catalyzing the formation of peptide bonds between the amino acids that form the growing protein.

Transfer RNA Molecules Decode the Sequence of Bases in mRNA into the Amino Acid Sequence of a Protein

Transfer RNA molecules bind to free amino acids in the cytoplasm and deliver them to the ribosome, where, according to instructions from mRNA, they are incorporated into protein chains. There are many types of tRNA molecules, at least one type for each amino acid. Transfer RNAs are like "code books"; they are the only molecules in the cell that can decipher the codons in mRNA and can translate them into an amino acid sequence. Enzymes in the cytoplasm recognize each specific tRNA molecule and attach the correct amino acid to the molecule's stem (see Fig. 10-2c). The energy of adenosine triphosphate (ATP) is stored in the tRNA–amino acid bond. That energy will be used later to forge a peptide bond when the amino acid is added to a growing protein molecule.

Transfer RNA bears three exposed bases, called the **anticodon**, that decipher the code of the mRNA codon. Anticodon bases of each tRNA pair in a complementary manner to the mRNA codon bases that specify the amino acid to which that tRNA can attach. For example, the mRNA codon AUG is complementary to, and pairs with, the anticodon UAC of a tRNA that bears the amino acid methionine.

Translation: mRNA, tRNA, and Ribosomes Cooperate to Synthesize Proteins

Now that we have introduced all the actors involved in protein synthesis, let's look at the actual events. First, during transcription, mRNA is transcribed from the DNA template of the genes in the nucleus (see Fig. 10-4). The mRNA then travels to a ribosome in the cytoplasm. Second, during translation, the ribosome binds to mRNA and to the appropriate tRNAs. On the ribosome, the mRNA codons are translated into the amino acid sequence of a protein, with the help of the tRNA anticodons.

We described transcription earlier in this chapter; here we will examine translation (Fig. 10-6). Like transcription, translation has three steps: (1) *initiation* of protein synthesis, (2) *elongation* of the protein chain, and (3) *termination*.

Initiation: Protein Synthesis Begins When tRNA and mRNA Bind to a Ribosome

As you'll recall, the genetic code signals where protein synthesis is to begin by means of a specific start codon (AUG on mRNA; see Table 10-2). In eukaryotic cells, the first step in translation is the binding of several protein "initiation factors" and an *initiator tRNA* that bears the complementary "start anticodon" to the small subunit of a ribosome (Fig. 10-6a). The small subunit then binds to an mRNA molecule and moves along it until the start codon is encountered. At this point, the tRNA base pairs

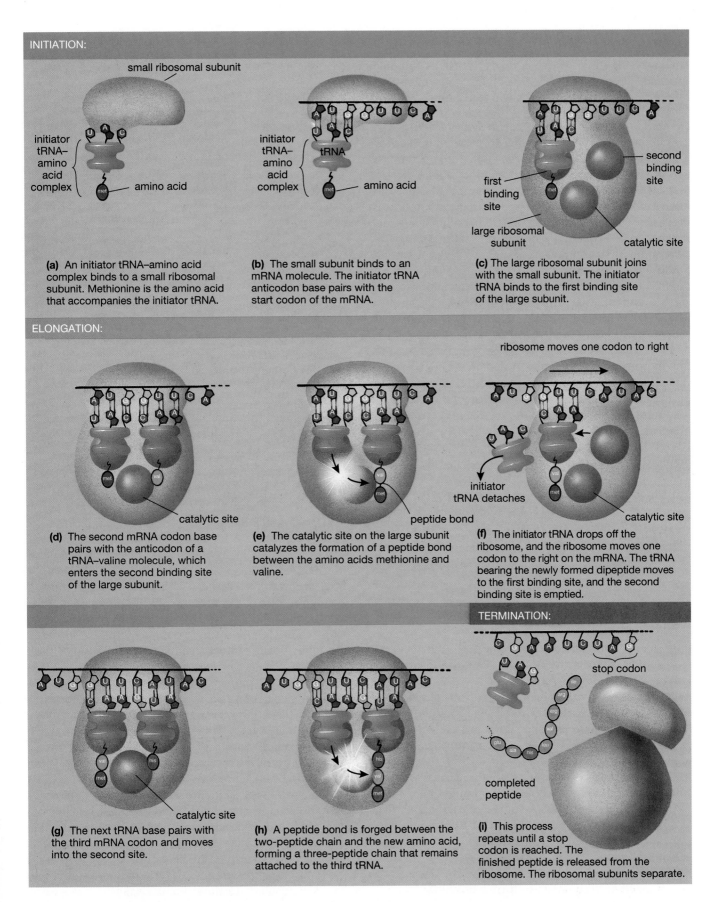

INITIATION:

(a) An initiator tRNA–amino acid complex binds to a small ribosomal subunit. Methionine is the amino acid that accompanies the initiator tRNA.

(b) The small subunit binds to an mRNA molecule. The initiator tRNA anticodon base pairs with the start codon of the mRNA.

(c) The large ribosomal subunit joins with the small subunit. The initiator tRNA binds to the first binding site of the large subunit.

ELONGATION:

ribosome moves one codon to right

(d) The second mRNA codon base pairs with the anticodon of a tRNA–valine molecule, which enters the second binding site of the large subunit.

(e) The catalytic site on the large subunit catalyzes the formation of a peptide bond between the amino acids methionine and valine.

(f) The initiator tRNA drops off the ribosome, and the ribosome moves one codon to the right on the mRNA. The tRNA bearing the newly formed dipeptide moves to the first binding site, and the second binding site is emptied.

TERMINATION:

(g) The next tRNA base pairs with the third mRNA codon and moves into the second site.

(h) A peptide bond is forged between the two-peptide chain and the new amino acid, forming a three-peptide chain that remains attached to the third tRNA.

(i) This process repeats until a stop codon is reached. The finished peptide is released from the ribosome. The ribosomal subunits separate.

Figure 10-6 *Translation: protein synthesis*
Protein synthesis is the translation of the sequence of bases in mRNA to the sequence of amino acids in the encoded protein.

with the mRNA start codon (Fig. 10-6b). The large ribosomal subunit then attaches to the small subunit. Simultaneously, the initiator tRNA binds to the first binding site on the large subunit (Fig. 10-6c). The ribosome is now fully assembled and ready to begin translation.

Elongation and Termination: Protein Synthesis Proceeds One Amino Acid at a Time Until a Stop Codon Is Reached

The completed ribosome is large enough to encompass two mRNA codons. The anticodon of a second tRNA, carrying another amino acid, recognizes the second mRNA codon and moves into the second binding site on the large subunit (Fig. 10-6d). The two amino acids carried by the two tRNAs are now next to one another. The large subunit has a catalytic site that breaks the bond holding the first amino acid to its tRNA and forms a peptide bond between the two adjacent amino acids. At the end of this step, the first tRNA is "empty," whereas the second tRNA bears a two-amino-acid chain (Fig. 10-6e).

The empty tRNA then drops off the ribosome, and the ribosome shifts to the next codon on the mRNA molecule (Fig. 10-6f). The tRNA holding the elongating chain of amino acids shifts too, from the second to the first binding site of the ribosome. A new tRNA, carrying another amino acid, binds to the empty second site (Fig. 10-6g). The catalytic site on the large subunit now links the third amino acid onto the growing chain (Fig. 10-6h). The "empty" tRNA leaves the ribosome, the ribosome shifts to another codon, and the process repeats.

Near the end of the mRNA, a stop codon is reached. At this point, special enzymes cut the finished protein chain off the last tRNA, releasing the chain from the ribosome (Fig. 10-6i).

Messenger and Transfer RNAs Decode the Sequence of Bases in DNA to the Sequence of Amino Acids in Protein

We can now understand how a cell decodes the genetic information stored in its DNA to synthesize a protein. Follow these steps in Figure 10-7:

(a) The DNA contains many genes, each consisting of dozens to thousands of bases. Each gene codes for a protein (the genes for tRNA and rRNA are exceptions). A sequence of three bases of DNA specifies one amino acid of the protein.

(b) A codon of mRNA consists of three bases that are complementary to the three bases of the DNA.

(c) An anticodon of tRNA, in turn, is complementary to a specific codon of mRNA. The tRNA is attached to a specific amino acid by enzymes that can "read" the anticodon.

(d) Amino acids carried to the mRNA by tRNAs are linked together in sequence to form a protein.

This "decoding chain," from bases in DNA to mRNA codon to tRNA anticodon to amino acid, results in the amino acid sequence in the newly formed protein.

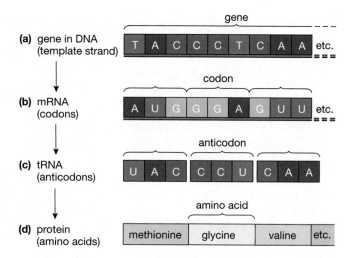

(a) gene in DNA (template strand)

T A C C C T C A A etc.

(b) mRNA (codons)

A U G G G A G U U etc.

(c) tRNA (anticodons)

U A C C C U C A A

(d) protein (amino acids)

methionine | glycine | valine | etc.

Figure 10-7 From DNA to protein
(a) The template strand of DNA contains a gene for a particular protein. **(b)** Triplets of bases in DNA are transcribed into codons of mRNA. **(c)** Each mRNA codon is recognized by the anticodon of a tRNA molecule that carries a specific amino acid. **(d)** On a ribosome, the amino acids are linked together, forming the protein.

3 How Do Mutations in DNA Affect the Function of Genes?

Up to now, we have emphasized the precision and fidelity of DNA replication and its transcription and translation into proteins: DNA molecules replicate just prior to cell division; mRNA molecules are complementary copies of the DNA of a gene; and the information in mRNA is used to produce a protein with the desired amino acid sequence. Of course, nothing alive is perfect. Mistakes can be made in any of these processes. A single faulty copy of mRNA or a single defective protein molecule normally doesn't affect a cell very much, because there are many correct molecules in the cell at the same time to carry out the proper cellular functions. However, a single faulty copy of a *gene* is much more serious, because a cell may have only one or two copies of the gene and, as a result, may synthesize defective proteins. We explore the far-reaching results of a single defective gene for the male sex hormone receptor in the essay "Health Watch: Sex, Aging, and Mutations" (p. 166).

Changes in the sequence of bases in DNA are called **mutations**. How can a base sequence change? One way for a mutation to occur is through a mistake in base pairing during replication. A few base-pairing mistakes occur spontaneously; despite the best efforts of proofreading enzymes, the replication of several billion bases results in a few mistakes. Certain chemicals (such as aflatoxins, synthesized by some molds that live on grain and peanuts) and some types of radiation (such as X-rays and ultraviolet rays in sunlight) increase the frequency of base-pairing errors during replication or even induce changes in DNA composition between replications.

Random changes in DNA composition are unlikely to code for improvements in the functioning of the gene products, much as randomly replacing words in the midst of a script of *Hamlet* will be unlikely to improve on Shakespeare's work. Some mutations, however, have no effect or (in very rare instances) are even beneficial, as you will learn later in this chapter.

Mutations Result from Nucleotide Substitutions, Insertions, or Deletions

In a **point mutation**, a pair of bases becomes incorrectly matched. Normally, repair enzymes recognize the mismatch, cut out the incorrect nucleotide, and replace it with a nucleotide that bears a complementary base. Occasionally, however, the enzymes replace the *correct* nucleotide instead of the incorrect one. The resulting base pair is complementary, but it is the wrong pair of nucleotides.

An **insertion mutation** occurs when one or more new nucleotide pairs are inserted into a gene. A **deletion mutation** occurs when one or more nucleotide pairs are removed from a gene.

Mutations Differ in Their Effects on Protein Structure and Function

If a mutation occurs in cells whose offspring become gametes (sperm or eggs), that mutation may be passed on to future generations. But how does a change in base sequence affect the organism that inherits the mutated DNA? As the box "Cracking the Genetic Code" (p. 159) points out, deletions and insertions can have particularly catastrophic effects on a gene, because all the codons that follow the deletion or insertion will be misread. The protein synthesized from such misread directions is almost certain to be nonfunctional.

Four categories of effects may result from point mutations (Table 10-3). As a concrete example, let's consider possible mutations of the DNA sequence CTC, which codes for glutamic acid.

1. *The protein is unchanged:* Recall that one amino acid may be encoded by several different codons. For example, if a mutation changes CTC to CTT, the new triplet still codes for glutamic acid. Therefore, the protein synthesized from the mutated gene remains the same.

2. *The new protein is equivalent:* Many proteins have large "background" regions whose exact amino acid sequence is relatively unimportant. For example, in hemoglobin, the amino acids on the outside of the protein must be hydrophilic to keep the protein dissolved in the cytoplasm of red blood cells. Exactly *which* hydrophilic amino acids are on the outside doesn't much matter. A mutation from CTC to CTA, replacing glutamic acid (hydrophilic) with aspartic acid (also hydrophilic), probably wouldn't affect the solubility of hemoglobin. Mutations that do not detectably change the function of the encoded protein are called **neutral mutations**.

3. *Protein function is changed by an altered amino acid sequence:* A mutation from CTC to CAC replaces glutamic acid (hydrophilic) with valine (hydrophobic). This substitution, which is the genetic defect in sickle-cell anemia (see Chapter 12), causes hemoglobin molecules to stick to each other, clumping up and distorting the shape of the red blood cells. These changes can cause serious illness.

4. *Protein function is destroyed by a stop codon:* A mutation from CTC to ATC will produce a stop codon. An inappropriate stop codon will cut short the translation of mRNA before the protein is finished. A mutation that creates a misplaced stop codon is almost certainly catastrophic to protein functioning. If the protein is essential to life, as hemoglobin is, the mutation will be lethal.

Table 10-3 Examples of Functional Outcomes of Single Substitutions in the Glutamic Acid Codon of DNA

	DNA	mRNA	Amino Acid	Properties	Functional Effect
Original sequence	CTC	GAG	Glutamic acid	Hydrophilic, acidic	—
Mutation 1	CT**T**	GA**A**	Glutamic acid	Hydrophilic, acidic	Neutral
Mutation 2	CT**A**	GA**U**	Aspartic acid	Hydrophilic, acidic	Neutral
Mutation 3	C**A**C	G**U**G	Valine	Hydrophobic, neutral	Lose water solubility; possibly catastrophic
Mutation 4	**A**TC	**U**AG	Stop codon	Ends translation	Synthesize only part of protein; catastrophic

Health Watch
Sex, Aging, and Mutations

Imagine yourself a girl in her mid-teens. Your breasts have filled out, but you have not begun to menstruate. Accompanied by a concerned parent, you visit your family doctor, who takes a small blood sample to do a chromosome test. When the results come in, you and your parents are called back to the office, where a genetic counselor gives you amazing news: Your sex chromosomes are XY, a combination that would normally give rise to a male. The reason you have not begun to menstruate is that you lack ovaries and a uterus but instead have testes that have remained inside your abdominal cavity. You have levels of male hormones (collectively called androgens, such as testosterone) circulating in your blood that are normal for a male. In fact, these male hormones, produced by the testes, have been present since early in your development. The problem is that your cells have never responded to them—a rare condition called **androgen insensitivity**. This condition was a problem for Maria Jose Martinez Patino, an outstanding Spanish athlete who reached the Olympics, only to be barred from the hurdles competition because her chromosomes indicated that she was male. After three years of struggle, the fact that she had developed as a female was finally recognized, and she was allowed to compete against others of her gender.

Many of the features of a male, including the formation of a penis, the descent of the testes into sacs outside the body cavity, and sexual characteristics that develop at puberty, such as a beard and increased muscle mass, all occur because various body cells are responding to male sex hormones produced by the testes. Many body cells in normal males have androgen–receptor proteins in their cytoplasm. These proteins bind with male hormones such as testosterone and escort them into the nucleus, where the hormone–receptor complex binds to DNA and influences the transcription of genes into mRNA. The mRNA molecules will serve as templates for the synthesis of specific proteins that contribute to maleness. In different groups of cells, the androgen receptor–testosterone complex influences gene transcription in different ways, giving rise to a wide range of male characteristics. Androgen receptors are proteins coded by specific genes. A mutation in one of those genes in the fertilized egg cell that produced an individual will cause the protein to have an improper amino acid sequence, which in turn may prevent the protein from binding androgens. Thus, a change in the nucleotide sequence of a single gene, causing a single type of defective protein to be produced, can cause a person who is genetically male (XY) to look like and feel like a woman (Fig. E10-2).

A second type of mutation is providing clues to the mystery of why people age. Why will your hair whiten, skin wrinkle, joints ache, and eyes cloud as you become elderly? A small number of individuals carry a defective gene that causes **Werner syndrome**, which causes premature aging (Fig. E10-3).

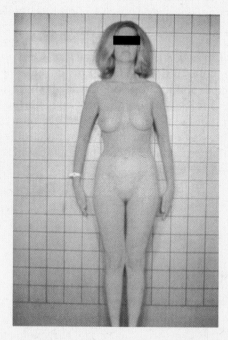

Figure E10-2 Androgen insensitivity leads to female features This individual has the male sex chromosome. As a result, she has testes in her abdominal cavity and lacks ovaries and a uterus. Female appearance is the result of a mutation in a single gene for the receptor protein for androgens, so body cells do not respond to male hormones.

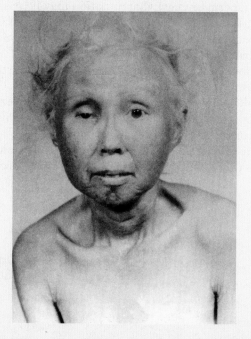

Figure E10-3 A 48-year-old woman with Werner syndrome This condition, most common among people of Japanese ancestry, is the result of a mutation that interferes with proper DNA replication, increasing the incidence of mutations throughout the body.

People with this disorder die of age-related conditions, typically by age 50. Recent research has localized the mutations in most people with Werner syndrome to a gene that codes for proteins that are parts of enzymes responsible for DNA replication. As you have seen, the accurate replication of DNA is crucial to the production of normally functioning cells. If a mutation interferes with the ability of enzymes to promote accurate DNA replication and to proofread and repair errors in DNA, then mutations will accumulate progressively in cells throughout the body.

The fact that an overall increase in mutations caused by defective replication enzymes produces symptoms of old age provides support for one hypothesis about the way many of the symptoms of normal aging occur. During a typical long (say, 80-year) life span, there is a gradual accumulation of mutations caused by mistakes in DNA replication and by environmentally induced DNA damage. Eventually, these mutations interfere with nearly every aspect of body functioning and contribute to death from "old age."

Disorders such as androgen insensitivity and Werner syndrome provide profound insights into the impact of mutations, the function of specific genes and their protein products, the ways hormones regulate gene transcription, and even the mystery of aging.

Mutations Provide the Raw Material for Evolution

If mutations in gametes are not lethal, they may be passed on to future generations. The rate of mutation in gametes varies not only among organisms, but also among genes within an organism. In humans, mutation rates in genes studied thus far range from 1 in every 100,000 gametes to 1 in 1,000,000 gametes (for reference, a man releases about 300–400 million sperm per ejaculation). Although most mutations are potentially harmful or neutral, mutations are essential for evolution, because ultimately all genetic variation originates as these random changes in DNA base sequence. Natural selection tests new base sequences in the crucible of competition for survival and reproduction. Occasionally, a mutation proves beneficial in the organism's interactions with its environment. The mutant base sequence may spread throughout the population and become common as organisms that possess it outcompete rivals that bear the original, unmutated base sequence. This process is described in detail in Unit III.

4) How Are Genes Regulated?

www

Knowing how proteins are synthesized and the likely effects of gene mutations on protein structure and function does not provide a complete understanding of how an organism's genes are used to produce its structures and behaviors. Most of the cells of your body have the same DNA—but they don't use all the DNA all the time. Each cell in your body has the same full set of genes. No one is sure how many, but recent estimates range from 60,000 to more than 100,000 genes. Individual cells *express* (transcribe and translate) only a small fraction of their genes—namely, those genes that are appropriate to the function of that particular cell type. Thus, the transcription of much of a cell's DNA is restricted. Muscle cells, for example, synthesize the contractile proteins actin and myosin but not insulin or hair proteins. Gene expression also changes over time, depending on the body's needs from moment to moment. For example, during pregnancy, milk-producing cells in a woman's breasts multiply tremendously. Immediately after the baby's birth, those cells begin producing copious amounts of milk protein. These events are due to changes in gene expression.

An organism's environment can also help determine which genes are transcribed. For example, in temperate climates, an increase in day length stimulates an increase in the size of the sex organs (testes or ovaries) in birds. The sex organs in turn produce sex hormones that influence a variety of physiological processes as well as behavior. The proliferation of cells in the sex organs, the production of hormones by these cells, and the effects of those hormones on other cells throughout the body all result, directly or indirectly, from changes in gene expression.

The use of genetic information by a cell is a multistep process, beginning with the transcription of DNA and commonly ending with an enzyme catalyzing a needed reaction. Regulation can occur at any of these steps, four of which are illustrated in Figure 10-8, as follows:

1. The rate of transcription of individual genes may be regulated.

The rate of transcription of specific genes depends on the type of cell and on the metabolic activity of both the

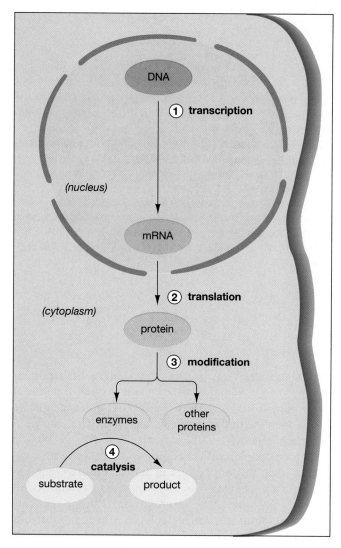

Figure 10-8 An overview of "information flow" in a cell
This simplified diagram shows the major steps from DNA to protein to chemical reactions catalyzed by enzymes. Regulation may occur at any step.

cell and the entire organism. Some genes are never transcribed in certain cell types; for example, the gene for insulin is not transcribed in muscle cells. Transcription of other genes is turned on or off, according to the demand for them.

2. Messenger RNAs may be translated at different rates.

Messenger RNAs vary in stability and in the rate at which they are translated into protein. Some mRNAs are extremely stable, which offers the possibility for repeated translation, whereas others rapidly degrade. Further, depending on metabolic requirements, a cell may block the translation of certain mRNAs.

3. Proteins may require modification before they can carry out their functions in a cell.

Many proteins must be modified before they become active. For instance, the protein-digesting enzymes produced by cells of your stomach wall and pancreas are initially synthesized in an inactive form (to keep the enzymes from digesting the very cells that produce them). After these inactive forms are secreted into the digestive tract, portions of the enzymes are then snipped out to unveil the active site.

4. The rate of enzyme activity may be regulated.

Enzyme activity is commonly controlled by inhibition, as discussed in Chapter 4.

All of these methods of regulating gene activity are important and are probably used to some extent by virtually all eukaryotic cells. We will restrict our discussion, however, to the transcription of DNA in eukaryotic cells. DNA transcription is complicated by the fact that eukaryotic genes consist of coding DNA segments interrupted by sequences of nucleotides that do not code for protein, as described in "A Closer Look: Eukaryotic Genes Are Found in Bits and Pieces."

Eukaryotic Cells May Regulate the Transcription of Individual Genes, Regions of Chromosomes, or Entire Chromosomes

All cells can regulate the rate of transcription of genes. Transcriptional regulation can operate on three levels: (1) the individual gene, (2) regions of chromosomes, or (3) entire chromosomes.

The Transcription of Individual Genes Is Altered by Regulatory Proteins

Some of the best-known examples of transcriptional regulation at the level of the individual gene are the cellular actions of steroid hormones, such as sex hormones. In female birds, estrogen stimulates the production of albumin (egg white) protein. During the breeding season, estrogen secreted by the ovaries enters the cells that produce albumin and binds to estrogen receptor proteins in the cytoplasm. The estrogen–receptor complex then enters the nucleus, where it binds to DNA, probably near the albumin gene. The binding of estrogen makes it easier for RNA polymerase to begin transcribing the albumin gene, so large amounts of albumin are produced. Similar activation of genes by steroid hormones occurs in other animals, including humans. Evidence for the importance of hormones in gene regulation and for the importance of gene regulation in development is provided by genetic defects in which receptors for sex hormones are nonfunctional (see "Health Watch: Sex, Aging, and Mutations" on p. 166).

A Closer Look
Eukaryotic Genes Are Found in Bits and Pieces

In the 1970s, molecular geneticists discovered that most eukaryotic *structural genes* (genes whose products are structural parts of the cell) have much more DNA than is needed to encode the amino acids of proteins. These genes consist of two or more base sequences that encode for a protein, interrupted by other base sequences that are not translated into a protein. The researchers called the coding segments **exons**, because they are **ex**pressed in protein, and the noncoding segments **introns**, because they **int**ervene between the exons (Fig. E10-4a). Most genes examined since introns were first discovered have been found to contain these noncoding segments. In fact, one gene coding for a particular type of connective tissue in chickens has about 50 introns!

When a eukaryotic gene is transcribed, a very long molecule of RNA is synthesized, starting before the first exon and ending after the last exon (Fig. E10-4b). The resulting RNA generally contains nucleotide sequences that code for introns as well as exons. To convert this RNA molecule into true mRNA that contains only the exons needed to code for the protein, enzymes in the nucleus precisely cut the molecule apart, splice together the coding sections, and discard the rest.

Why are eukaryotic genes split up like this? Gene fragmentation appears to serve at least two functions. The first function is to allow a cell to produce multiple proteins from a single gene by splicing exons together in different ways. Rats, for example, have a gene that is transcribed in both the thyroid and the brain. In the thyroid, one splicing arrangement results in the synthesis of a hormone called calcitonin, which helps regulate calcium concentrations in the blood. In the brain, another splicing arrangement results in the synthesis of a peptide that is probably used as a molecular messenger for communication among brain cells.

The second function is more speculative but is supported by some good experimental evidence: Fragmented genes may provide a quick and efficient way for eukaryotes to evolve new proteins with new functions. Chromosomes sometimes break apart, and their parts may reattach to different chromosomes. If the breaks occur within the noncoding introns of genes, exons may be moved intact from one chromosome to another. Most such errors would be harmful. But some of these shuffled exons might code for a protein subunit that has a specific function (binding ATP, for example). In rare instances, adding this subunit to an existing gene may cause the gene to code for a new protein with useful functions. The accidental exchange of exons among genes produces new eukaryotic genes that will, on occasion, enhance survival and reproduction of the organism that carries them.

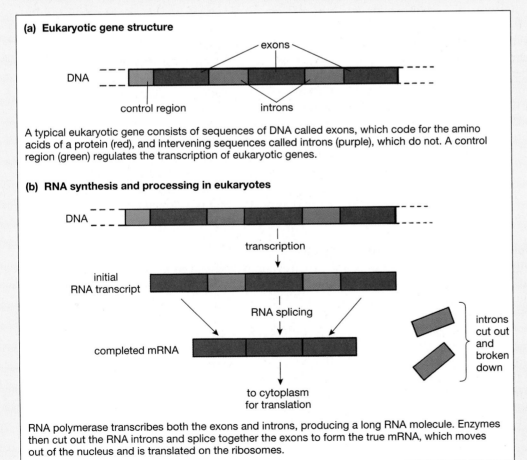

(a) Eukaryotic gene structure

exons

DNA

control region   introns

A typical eukaryotic gene consists of sequences of DNA called exons, which code for the amino acids of a protein (red), and intervening sequences called introns (purple), which do not. A control region (green) regulates the transcription of eukaryotic genes.

(b) RNA synthesis and processing in eukaryotes

DNA

transcription

initial RNA transcript

RNA splicing

introns cut out and broken down

completed mRNA

to cytoplasm for translation

RNA polymerase transcribes both the exons and introns, producing a long RNA molecule. Enzymes then cut out the RNA introns and splice together the exons to form the true mRNA, which moves out of the nucleus and is translated on the ribosomes.

Figure E10-4 Eukaryotic genes contain introns and exons

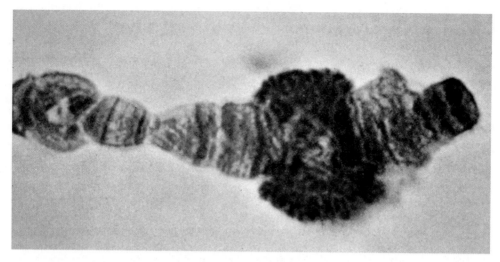

Figure 10-9 *A giant chromosome of a midge (a small fly)*
Most of the chromosome is in a tight, condensed state, but a few regions have puffed out in a much looser configuration. The orange-red stain shows the RNA, which is transcribed mostly from the loose, puffed regions of DNA. The puffing and condensation of the chromosomes change in a pattern that reflects different genes turning on and off during development.

Regions of Chromosomes Are Condensed and Not Normally Transcribed

Certain parts of eukaryotic chromosomes are in a highly condensed, compact state in which the DNA seems to be inaccessible to RNA polymerase. Some of these regions are structural parts of chromosomes that don't contain genes. Other tightly condensed regions are functional genes that are not currently being transcribed. When the product of a gene is needed, the portion of the chromosome containing that gene becomes "decondensed"—loosened so that the nucleotide sequence is accessible to RNA polymerase and transcription can occur (Fig. 10-9).

Entire Chromosomes May Be Inactivated, Thereby Preventing Transcription

In some cases an entire chromosome may be condensed and inaccessible to RNA polymerase. An example occurs in the sex chromosomes of females. Sex chromosomes are either XY (male) or XX (female). Although female mammals have two X chromosomes, only one X chromosome is available for transcription in any given cell. The other X chromosome is condensed into a tight mass. In a light microscope, the inactivated, condensed X chromosome shows up in the nucleus as a dark spot called a **Barr body**, named after its discoverer, Murray

Barr (Fig. 10-10a). Apparently, both X chromosomes are in the "loose" state in fertilized eggs. After a few cell divisions, one or the other condenses and forms a Barr body. (The cells in the ovaries that will give rise to eggs are an exception: Both X chromosomes are in the uncondensed state.) Which X chromosome is inactivated in any given cell is random, but all its daughter cells will then have the same condensed chromosome. As a result, female mammals (including human females) are mosaics: Patches of cells with one X chromosome active are interspersed with patches in which the other X chromosome is active. In calico cats, each X chromosome bears a gene that codes for an enzyme that results in a different fur pigment. Patches of orange and black fur represent areas of skin that developed from cells in the early embryo in which different X chromosomes were inactivated (Fig. 10-10b). Calico coloring is therefore unique to female cats.

To prevent males from competing against women unfairly at the Olympics, officials verify that athletes who compete in women's events are truly female by performing a sex test. Barr bodies are sought, normally in cells obtained by gently scraping the mouth inside the cheek. This test caused major problems for a female hurdler, Maria Jose Martinez Patino, when no Barr bodies were found in her cells. Learn why in "Health Watch: Sex, Aging, and Mutations" (p. 166).

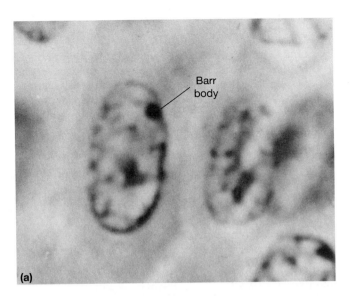

Figure 10-10 *Inactivation of the X chromosome regulates gene expression*
(a) In female mammals, one X chromosome is inactivated, forming a condensed mass called a Barr body. These nuclei, from a normal XX woman, each have one Barr body. Nuclei from men (XY) lack Barr bodies. *(b)* These cats are from the same litter. The male (left) has a single X chromosome carrying the gene for orange fur. The calico female (right) carries genes for both orange and black fur, one on each of her two X chromosomes. Inactivation of different X chromosomes produces the black and orange patches. The white color is due to an entirely different gene, expressed in other cells, that prevents pigment formation.

Summary of Key Concepts

1) How Are Genes and Proteins Related?

Genes are segments of DNA on chromosomes. The ultimate cellular product encoded by a gene is in most cases a protein. Thus, with a few exceptions (such as transfer RNA and ribosomal RNA), the specific base sequence of a gene encodes the amino acid sequence of a protein or part of a protein.

Ribonucleic acid (RNA) is a single-stranded nucleic acid that helps transcribe and translate genetic information in DNA to amino acid sequences. The three types of RNA are: (1) messenger RNA (mRNA), (2) transfer RNA (tRNA), (3) and ribosomal RNA (rRNA). Information flows from DNA to proteins in a two-step process: (1) In transcription, the information contained in a gene's DNA is copied into mRNA; (2) in translation, the sequence of bases in mRNA provides the information needed by tRNA and rRNA to synthesize a protein with the amino acid sequence specified by the base sequence of the gene's DNA.

The sequence of bases in mRNA carries the genetic code for the amino acid sequence in a protein. Sequences of three bases in mRNA, called codons, specify the amino acids of the protein. Start codons and stop codons signal the beginning and end of protein synthesis, respectively.

2) What Is the Role of RNA in Protein Synthesis?

Synthesizing proteins from the information in DNA requires RNA molecules as intermediaries. Messenger RNA is transcribed from one DNA strand (the template strand) by the enzyme RNA polymerase. Recognizing the promoter region of DNA as the beginning of a gene, RNA polymerase uses free ribose nucleotides to synthesize an mRNA strand that is complementary to the gene's DNA. Thus, the sequence of bases in mRNA carries the information needed to determine the amino acid sequence of a protein.

Together, rRNA and proteins form ribosomes, consisting of large and small subunits. Each tRNA binds a specific amino acid and transports it to a ribosome. A set of three bases in tRNA, the anticodon, is complementary to the mRNA codon that specifies the amino acid to which that tRNA is bound.

Protein synthesis occurs in the following sequence:

1. Messenger RNA is transcribed from a gene. The mRNA leaves the nucleus and travels to a ribosome, where it binds to the small ribosomal subunit.
2. Transfer RNAs carry their amino acids to the mRNA. The tRNA anticodons pair with the mRNA codons, the large and small ribosomal subunits join, and the tRNAs bind to the large ribosomal subunit.
3. The large ribosomal subunit catalyzes the formation of a peptide bond between the amino acids carried by the tRNA molecules.
4. As each new amino acid is attached, a tRNA detaches and is replaced by another tRNA that carries the next amino acid in the growing protein.
5. This process continues until a stop codon is reached, whereupon the mRNA and the newly formed protein leave the ribosome.

3) How Do Mutations in DNA Affect the Functions of Genes?

A mutation is a change in the nucleotide sequence of a gene. Mutations are caused by mistakes in base pairing during replication, by chemical agents, and by environmental factors such as radiation. Common types of mutations include point mutations, insertions, and deletions. Most mutations are neutral or harmful, but in rare cases a mutation will promote better adaptation to the environment and thus will be favored by natural selection.

4) How Are Genes Regulated?

Which genes are transcribed in a cell at any given time is regulated by the function of the cell, the developmental stage of the organism, and the environment. The mechanisms of gene regulation are numerous, complex, and only partially understood. Some examples follow: Messenger RNA molecules vary in stability, affecting how many protein molecules can be synthesized from each mRNA molecule. Even after they are synthesized, some proteins must be modified by enzymes before they can function; the availability of these enzymes is a factor that cells can regulate. Some proteins, such as sex hormones, alter the transcription rate of individual genes. Entire chromosomes or parts of chromosomes may be condensed and inaccessible to RNA polymerase, whereas other portions are expanded, allowing transcription to occur.

Key Terms

androgen insensitivity *p. 166*
anticodon *p. 162*
Barr body *p. 170*
codon *p. 160*
deletion mutation *p. 165*
exon *p. 169*
gene *p. 156*
genetic code *p. 158*

insertion mutation *p. 165*
intron *p. 169*
messenger RNA (mRNA) *p. 158*
mutation *p. 164*
neutral mutation *p. 165*
one-gene, one-protein hypothesis *p. 156*

point mutation *p. 165*
promoter *p. 161*
ribonucleic acid (RNA) *p. 156*
ribosomal RNA (rRNA) *p. 158*
ribosome *p. 162*
RNA polymerase *p. 160*

start codon *p. 160*
stop codon *p. 160*
template strand *p. 160*
transcription *p. 158*
transfer RNA (tRNA) *p. 158*
translation *p. 158*
Werner syndrome *p. 166*

Thinking Through the Concepts

Multiple Choice

1. *A gene*
   a. is synonymous with a chromosome
   b. is composed of mRNA
   c. is a specific segment of nucleotides on DNA
   d. contains only those nucleotides required to synthesize a protein
   e. specifies the sequence of nutrients required by the body

2. *Which of the following is a single-stranded molecule assembled by using DNA as a template that contains the information for assembly of a specific protein?*
   a. transfer RNA
   b. messenger RNA
   c. exon DNA
   d. intron DNA
   e. ribosomal RNA

3. Anticodon *is the term applied to*
   a. the list of amino acids that corresponds to the genetic code
   b. the concept that multiple codons sometimes code for a single amino acid
   c. the part of the tRNA that interacts with the codon
   d. the several three-nucleotide stretches that code for "stop"
   e. the control mechanism of prokaryotic genes

4. *Eukaryotic DNA*
   a. takes part directly in protein synthesis by leaving the nucleus and being translated on the ribosome
   b. takes part indirectly in protein synthesis; the DNA itself stays in the nucleus
   c. has nothing to do with protein synthesis; it is involved only in cell division
   d. is involved in protein synthesis that takes place in the nucleus
   e. codes for mRNA but not for tRNA or rRNA

5. *The making of a protein from a template of messenger RNA*
   a. is catalyzed by DNA polymerase
   b. is catalyzed by RNA polymerase
   c. is called translation
   d. is called transcription
   e. occurs in the nucleus

6. *A Barr body is*
   a. a condensed X chromosome
   b. an organelle involved in protein synthesis
   c. another term for chromosomes as they are seen during cell division
   d. visible in cells of both males and females
   e. present only in males, because their female chromosomes are inactivated

? Review Questions

1. How does RNA differ from DNA?

2. What are the three types of RNA? What is the function of each?

3. Define the following terms: *genetic code; codon; anticodon.* What is the relationship among the bases in DNA, the codons of mRNA, and the anticodons of tRNA?

4. How is mRNA formed from a eukaryotic gene?

5. Diagram and describe protein synthesis.

6. Describe some mechanisms of gene regulation.

7. Define *mutation*, and give one example of how a mutation might occur. Would you expect most mutations to be beneficial or harmful? Explain your answer.

Applying the Concepts

1. Suppose mammalian cells are grown and divide many times in a culture medium that contains thymine made radioactive with a form of hydrogen (^{3}H). The cells are then removed from the radioactive medium and allowed to replicate several times in a normal culture medium. Daughter chromosomes are tested each generation to determine whether they contain radioactive thymine. While in the radioactive medium, every strand of DNA of each chromosome had radioactive thymine. Assuming that each chromosome contains a single long, double-stranded molecule of DNA, predict the radioactive status of daughter chromosomes after one, two, and three rounds of cell division in the normal medium. Explain how your predictions are consistent with the Watson-Crick explanation of semiconservative DNA replication.

2. Think of the translation of proteins on ribosomes as analogous to the production of a variety of widgets in a factory. Each type of widget is composed of the same materials, but those materials are assembled in a different way. Now you extend the analogy. How are the widget components delivered? What are the instructions for making widgets? Extend the analogy as far as you can—be imaginative!

3. If a single base-pair deletion occurs near the beginning of the coding sequence of a gene, why might transcription to mRNA be prematurely halted by a stop codon? Make a simple diagram showing how this could occur in a short nucleotide sequence.

4. As you have learned in this chapter, many factors influence gene expression, including hormones. The use of anabolic steroids and growth hormones among athletes has created controversy in recent years. Hormones certainly affect gene expression, but, in the broadest sense, so do vitamins and foods. What do you think are appropriate guidelines for the use of hormones? Should athletes take steroids or growth hormones? Should children at risk of being unusually short be given growth hormones? Should parents be allowed to request growth hormones for their children of normal height in the hope of producing a future basketball player?

Group Activity

Form groups of three or four students. Discuss and answer the following question: Will a change in a single DNA nucleotide (a nucleotide substitution mutation) always result in a changed protein? Why or why not? Next, design a random 15-nucleotide DNA sequence, starting with the sequence TAC. Transcribe and translate this sequence to determine the amino acid sequence for which it codes. Use the genetic code, Table 10-2. Now "mutate" your DNA sequence by changing the first nucleotide; repeat transcription and translation of this new sequence. Note whether the resulting amino acid sequence is different. Repeat the mutation process by mutating every other nucleotide (the third, fifth, seventh, and so on) of your original sequence one at a time, and repeating transcription and translation. Each time, your original DNA nucleotide sequence should differ from the original DNA sequence by only one nucleotide. Note any differences in the resulting protein each time. What conclusions can you derive? Do your results support your answer to the original question?

For More Information

Beardsley, T. "Smart Genes." *Scientific American*, August 1991. How genes are regulated is just as important as which genes are on the chromosomes. New findings are beginning to outline the regulation of gene expression during development.

Crick, F. H. C. "The Genetic Code." *Scientific American*, October 1962. The determination of the triplet nature of the genetic code.

Crick, F. H. C. "The Genetic Code: III." *Scientific American*, October 1966. The genetic code is completely solved.

Darnell, J. E. "RNA." *Scientific American*, October 1985. A description of the types of RNA, their synthesis, and their processing.

Felsenfeld, G. "DNA." *Scientific American*, October 1985. The structure of DNA and how its use is regulated in a cell.

Grunstein, M. "Histones as Regulators of Genes." *Scientific American*, October 1992. Histones are proteins associated with DNA in eukaryotic chromosomes. Once thought to be merely a scaffold for DNA, they are actually important in gene regulation.

Nirenberg, M. W. "The Genetic Code: II." *Scientific American*, March 1963. Nirenberg describes some of the experiments in which he deciphered much of the genetic code.

Tjian, R. "Molecular Machines That Control Genes." *Scientific American*, February 1995. Complexes of proteins regulate which genes are transcribed in a cell and therefore help determine the cell's structure and function.

Travis, J. "Biology's Periodic Table." *Science News*, March 22, 1997. A brief synopsis of the Human Genome Project.

Answers to Multiple-Choice Questions
1. c   2. b   3. c   4. b   5. c   6. a

"All cells come from cells."

Rudolf Virchow, German physician and early proponent of the cell theory (1858)

Resembling an undersea flower, this sea anemone (Metridium senile) is surrounded by its offspring at its base. The small anemones, formed by budding that uses mitotic cell division, are genetically identical to their parent.

The Continuity of Life: Cellular Reproduction

11

At a Glance

 Net Watch
On-line resources for this chapter are on the World Wide Web at:
http://www.prenhall.com/audesirk
(click on the Table of Contents link and then select Chapter 11).

If you have an opportunity to explore the fascinating life beneath the ocean, keep your eyes open for a sea anemone. Look closely, and you may see a cluster of small anemones surrounding it. If you probe gently, you may find some still attached to the parent. You are seeing an intriguing form of reproduction known as *budding*, in which tiny miniatures grow directly from the adult's body. No sex is involved here; the cells of the offspring are identical to those of the single parent.

"All cells come from cells." This simple statement captures the crucial importance of cellular reproduction for the continuity of life on Earth. When a cell reproduces itself, it gives rise to two daughter cells that—if all goes well—have exactly the same genetic information as the parent cell, carried in *chromosomes* that are exact copies of the parent cell's chromosomes. Each daughter cell also inherits about half of the parent cell's cytoplasm, including a full complement of organelles. How is cell reproduction accomplished? As a shoot springs up in your garden, as a child grows and develops, as a sea anemone grows miniature versions of itself from the sides of its body, how are copies of perhaps 100,000 genes accurately and evenly parceled out to each of millions or trillions of daughter cells? This chapter will introduce you to the precise and beautifully choreographed movements of the chromosomes as genetic material is duplicated and then shared evenly between two daughter cells during *mitotic cell division.*

An organism's **life cycle** is the sequence of events that occur from one generation to the next. For most multicellular organisms, the life cycle includes sexual reproduction. Even most of those capable of reproducing without sex, such as the sea anemone, can engage in sexual reproduction as well. How do organisms such as plants, fungi, anemones, hummingbirds, and humans combine the genes from two parents to produce genetically unique offspring? As you learn that *meiotic cell division*—the process of producing sex cells with half the original number of chromosomes—has twice as many steps as mitotic cell division does, you might ask, "Why go to all this trouble? What are the advantages of sexual reproduction?"

Finally, you might remember Dolly. Her blank, cud-chewing visage was shown in the news around the world in 1997. A sheep had been cloned from the nucleus of an udder cell from a 6-year-old ewe—Dolly's identical twin could have been her mother. The media went wild. "*Udderly* amazing—we will see *ewe* again!" reporters quipped. Why did the birth of this sheep shake up the scientific community, inspire impassioned philosophical debates, and elicit acts of Congress? Let's find out.

1) What Are the Essential Features of Cell Division?

www

As we discussed in previous chapters, the hereditary information of all cells is carried in *deoxyribonucleic acid* (*DNA*), consisting of two long chains of smaller subunits called *nucleotides.* Segments of DNA a few hundred to many thousand nucleotides long are the units of inheritance—the genes—that carry the genetic information to produce specific traits. The sequence of nucleotides in a gene encodes information for the synthesis of the RNA and protein molecules that are needed to build a cell and carry out its metabolic activities.

The DNA of all cells is packaged into one or more structures called **chromosomes** (see Chapter 5). Under the light microscope, nuclei of nondividing cells appear as large, darkly staining masses without much internal structure. Early microscopists named this material in the nucleus **chromatin**, from the Greek word for color, because chromatin is stained by some commonly used dyes. During cell division in eukaryotic cells, threadlike structures form where the nucleus used to be. The microscopists called these threads "chromosomes," meaning "colored bodies." With the advent of the electron microscope and advances in molecular biology, we now know that chromatin and chromosomes are actually the same thing (DNA and its associated proteins) but in different stages of condensation.

For any cell to survive, it must have a complete set of genetic instructions. Therefore, when a cell divides, it cannot simply split its set of genes in half and give each daughter cell half a set. Rather, the cell must first replicate (duplicate) its DNA, much like making a photocopy of an instruction manual. It does this by duplicating its chromosomes. Each daughter cell then receives a complete set of chromosomes, in effect a "DNA manual" containing all the genes. When a cell divides, it must transmit to its offspring two essential requirements for life: (1) hereditary information that directs life processes, and (2) materials in the cytoplasm that the "new" cells need to survive and to utilize their hereditary information (Fig. 11-1).

The activities of a cell from one cell division to the next constitute the **cell cycle**. Newly formed cells usually acquire nutrients from their environment, synthesize more of their own materials, and grow larger. After a variable amount of time, depending on the organism, the type of cell, and the nutrients available, the cell divides. This general description applies to both eukaryotic and prokaryotic cells. However, prokaryotic cells are much simpler than eukaryotic cells, and their cell cycles differ in many respects. Thus, we will discuss prokaryotic and eukaryotic cell cycles separately.

2) What Are the Events of the Prokaryotic Cell Cycle?

www

The DNA of a Prokaryotic Cell Is Contained in a Single, Circular Chromosome

The DNA of a prokaryotic cell is contained in a single, circular chromosome (Fig. 11-2). Prokaryotic cells lack nuclei, so the chromosome is found within the cytoplasm. One point on the chromosome is attached to the plasma membrane of the cell. A prokaryotic chromosome normally has relatively little protein associated with the DNA.

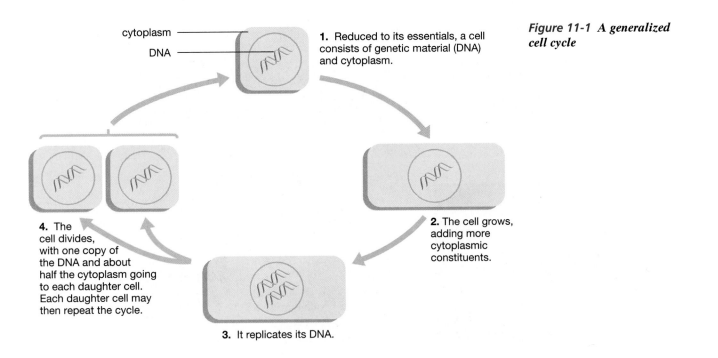

cytoplasm

DNA

1. Reduced to its essentials, a cell consists of genetic material (DNA) and cytoplasm.

Figure 11-1 A generalized cell cycle

2. The cell grows, adding more cytoplasmic constituents.

3. It replicates its DNA.

4. The cell divides, with one copy of the DNA and about half the cytoplasm going to each daughter cell. Each daughter cell may then repeat the cycle.

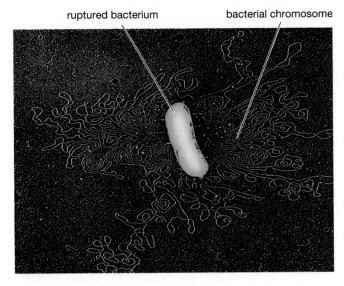

ruptured bacterium   bacterial chromosome

Figure 11-2 *The bacterial chromosome*
This bacterial cell is about 3 micrometers long. The cell was ruptured to free its chromosome. If you look carefully, you will see that there are no free ends to the chromosome: It is a single thin strand. If stretched out, the chromosome would be a circular molecule of DNA 1 or 2 millimeters in circumference.

Cell Division in Prokaryotic Cells Is Called Binary Fission

With only a single chromosome and no separate nucleus, prokaryotic cells (such as bacteria) utilize a simple method of cell division called **binary fission** (literally, "splitting in two"). Under favorable conditions, the cell

cycle of many prokaryotes proceeds rapidly. The common intestinal bacterium *Escherichia coli*, for example, can complete its cell cycle in 30 minutes or less. During most of this time, the cell absorbs nutrients from the environment, grows, and replicates its DNA. As soon as the DNA is replicated, the bacterium divides; in fact, some bacterial cells begin dividing even before DNA replication is quite finished.

The cell cycle of prokaryotic organisms is described next and illustrated in Figure 11-3:

1. Prior to DNA replication, one point on the chromosome attaches to the plasma membrane.
2. Binary fission begins when the chromosome replicates. The resulting pair of identical chromosomes attach to the plasma membrane at nearby, but distinct, points.
3. The cell elongates, and new plasma membrane is added between the two attachment points, pushing them apart.
4. As the two chromosomes move toward opposite ends of the cell, the plasma membrane around the middle of the cell grows inward.
5. Finally, two new daughter cells are formed, completing binary fission.

Each daughter cell thus receives one of the replicated chromosomes and about half the cytoplasm. With a complete set of genes and enough materials to work with, the cell cycle continues as the cell enters a temporary nondividing state. If environmental conditions are favorable, they will acquire nutrients and increase in size prior to undergoing another binary fission.

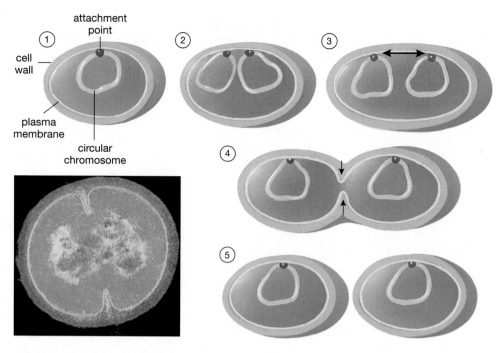

Figure 11-3 Prokaryotic cells divide by binary fission
① The circular DNA attaches to the plasma membrane at one point. ② The chromosome is replicated. The two copies are attached to the membrane at nearby points. ③ The cell elongates; new plasma membrane is added between the attachment points. ④ The plasma membrane grows inward at the middle of the cell. ⑤ The parent cell has divided into two daughter cells.

③ What Is the Structure of the Eukaryotic Chromosome?

The Eukaryotic Chromosome Consists of a Strand of DNA Complexed with Proteins

Each eukaryotic chromosome contains one single long double-helix molecule of DNA, complexed with proteins called *histones:*

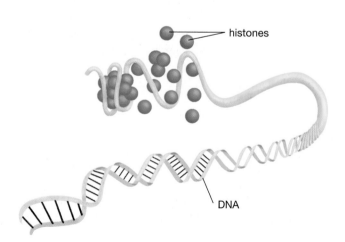

During most of its life—except during the actual process of cell division—a cell must be able to use the information coded in its genes (segments of DNA within the chromosome) to produce the molecules that it needs to perform its metabolic activities and sustain life. The cell can use and replicate its DNA only when the chromosomes are extended. In this extended state, individual chromosomes are too thin to be visible, even under powerful microscopes. The total length of all the chromosomes in a human cell is about 200 centimeters (about 6 feet 6 inches).

During cell division, however, the chromosomes must be sorted out and moved into two daughter nuclei. Just as thread is easier to organize when it is wound onto spools rather than tangled, sorting and transporting chromosomes is easier if they are condensed and shortened (Fig. 11-4).

Most chromosomes consist of two arms that extend out from a specialized region of DNA called the **centromere** (meaning "middle body"):

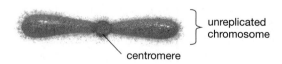

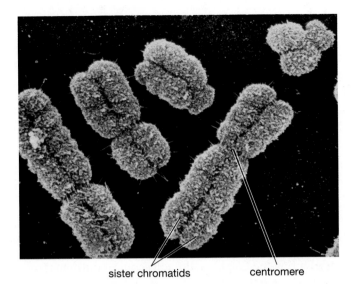

sister chromatids   centromere

Figure 11-4 Human chromosomes during mitosis
A scanning electron micrograph of duplicated human chromosomes during mitosis. The DNA and associated proteins have coiled up into the thick, short sister chromatids attached at the centromere. Each strand of "texture" is a loop of chromosome. During cell division, the condensed chromosomes are about 5 to 20 micrometers long. At other times, the same chromosomes uncoil into thin strands about 10,000 to 40,000 micrometers long.

Before a cell divides, it duplicates its chromosomes. The two copies remain attached at their centromeres. As long as they are attached to one another, the copies are called sister **chromatids**. Each chromatid is a single DNA molecule identical to the DNA of the original chromosome before replication. We will call the two sister chromatids attached at their centromeres a "duplicated chromosome":

replicated chromosome

sister chromatids

Figure 11-4 shows several duplicated chromosomes, each consisting of two sister chromatids.

During cell division, the two sister chromatids separate, and each chromatid becomes an independent daughter chromosome:

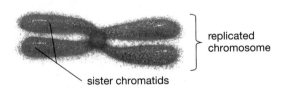

independent daughter chromosomes, each in unreplicated state

Eukaryotic Chromosomes Usually Occur in Pairs with Similar Genetic Information

The chromosomes of each species have characteristic shapes, sizes, and staining patterns (Fig. 11-5). The nonreproductive cells of many organisms contain *pairs of chromosomes*; both members of each pair are the same length and stain in the same pattern. The chromosomes of each pair also have similar, although normally not identical, genetic information. Therefore, the members of a pair are called **homologues**, meaning "to say the same thing" in Greek. Cells with pairs of homologous chromosomes are called **diploid**. Mitosis preserves the diploid number of chromosomes as they are packaged into the newly formed daughter cell nuclei.

The 46 chromosomes of the nonreproductive cells of human beings occur as 23 pairs of homologues. In fact, the chromosomes of many organisms come in pairs. Why? As you learned in Chapter 10, mutations in DNA typically result in the production of defective proteins. Cells with two copies of each gene are more likely to survive a mutation in one of them, because the remaining gene can still produce a normally functioning protein.

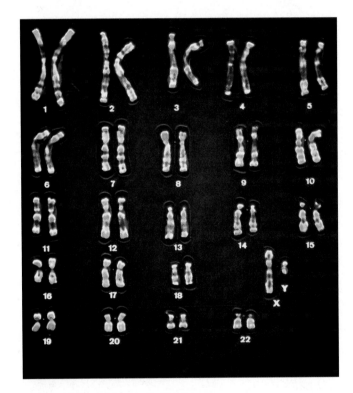

Figure 11-5 A human male karyotype
A *karyotype* is produced by staining and photographing the chromosomes of a dividing cell. Pictures of the individual chromosomes are cut out and arranged in descending order of size. The chromosomes occur in pairs (homologues) that are similar in both size and staining patterns and have similar genetic material. If these chromosomes were from a female, there would be two X chromosomes. Notice that the Y chromosome is much smaller than the X chromosome.

Not all cells have pairs of homologous chromosomes. At some point in the life cycle of all sexually reproducing organisms, a special type of cell division called *meiosis* produces cells that have only one copy of each type of chromosome; these cells are called **haploid**. In animals, these haploid cells are the sex cells, or **gametes** (sperm and eggs).

In biological shorthand, the number of types of chromosomes is termed the *haploid number* and is designated *n*. For humans, $n = 23$—that is, we have 23 types of chromosomes; for other organisms, *n* ranges from 1 to several hundred. Diploid cells, with two homologues of each type, contain *2n* chromosomes. Each human body cell thus has 46 chromosomes altogether. Diploid cells of fruit flies have eight chromosomes, whereas those of ferns contain hundreds.

4) What Are the Events of the Eukaryotic Cell Cycle?

The eukaryotic cell cycle consists of two major phases: interphase and mitotic cell division (Fig. 11-6). During **interphase**, the cell acquires nutrients from its environment, grows, and replicates its chromosomes. During mitotic cell division, one copy of each chromosome, and

approximately half the cytoplasm (with its mitochondria, ribosomes, and other organelles), is parceled out into each of the two daughter cells.

Growth, Replication of Chromosomes, and Most Cell Functions Occur during Interphase

Most of the life of a normal eukaryotic cell is spent in interphase, between cell divisions. (Notable exceptions occur when cancer develops, and cells divide out of control, a topic explored in detail in "Health Watch—Cancer: Mitotic Cell Division Runs Wild." DNA synthesis occurs only during a discrete part of interphase (see Fig. 11-6).

The period immediately after cell division and before duplication of the chromosomes is the G_1 *phase* (short for "first gap," referring to the first gap in DNA synthesis). During the G_1 phase, the cell acquires nutrients from its environment, carries out its specialized functions (for example, hormone synthesis and secretion), and grows.

Toward the end of the G_1 phase, cells take one of two paths. They may enter a nondividing state called the G_0 *phase*, which varies in length depending on the cell type and the developmental state of the organism. (Most neurons and muscle cells remain in this state for the adult life of the organism.) Alternatively, they continue into the *S phase* ("S" stands for "synthesis") in preparation for cell division. DNA replication, duplicating each chromosome, occurs during the S phase.

The period after DNA synthesis but before the next cell division is the G_2 *phase* (the "second gap"). The cell is already committed to cell division, and most of the G_2 phase is spent in synthesizing molecules other than DNA that are required for cell division.

Mitotic Cell Division Consists of Nuclear Division and Cytoplasmic Division

Mitotic cell division has two major parts: mitosis and cytokinesis. In eukaryotic cells, the process of nuclear division that preserves the diploid number of chromosomes is called **mitosis**. (*Mitosis* comes from the Greek word for thread; the chromosomes appear to be threadlike structures at this time.) In most cases, when a cell divides, its cytoplasm is divided about equally between the two daughter cells, a process called **cytokinesis** (from the Greek word for "cell movement"). Cytokinesis normally provides both daughter cells with all the organelles, nutrients, enzymes, and other molecules they need to maintain life, grow, and reproduce. The two daughter cells produced by mitotic cell division are essentially identical to each other as well as genetically identical to the parent cell.

Most cells undergo mitosis and cytokinesis together. But these two processes are potentially independent events. Some cells, including some fungi and the skeletal muscle cells of vertebrates, undergo mitosis *without* cytokinesis. This process produces single cells with many nuclei.

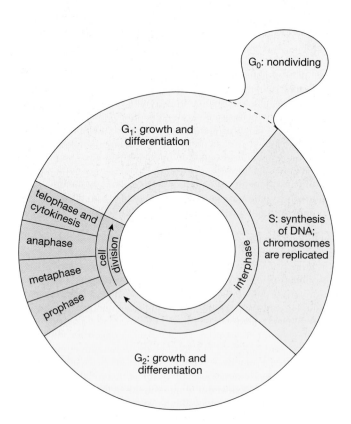

Figure 11-6 The eukaryotic cell cycle
The eukaryotic cell cycle consists of two major phases, interphase and cell division. Each is divided into subphases.

Health Watch
Cancer: Mitotic Cell Division Runs Wild

How does a cell know when to divide and when not to? What keeps adult nerve cells and muscle cells from dividing while skin cells divide continuously? The answers lie in the same amazing stretches of DNA that determine all the specialized structures and functions of the different cell types—the genes. The cell cycle itself is under genetic control. When activated, some genes suppress cell division whereas other genes promote it. *Cancer* occurs when the precise, tissue-specific regulation of the cell cycle is disrupted, commonly through mutation of these cell cycle-regulating genes. For example, if a gene that normally represses cell division suffers a mutation that renders its protein product ineffective, the cell will divide out of control. In the fastest-growing cancers, a mass the size of a very small pea will contain a billion cells and produce about a million new cells each hour. Because mitosis preserves the genetic makeup of each parent cell, all the progeny of a cancerous cell will be cancerous as well. If a cancerous cell is carried by the circulatory system to another part of the body, it may start a new cancerous growth there.

Although different cancers require from two to twenty mutations in different genes to develop, recent research has found that a certain gene, called the *p53* gene, may be mutated in as many as 50% of all cancers. The *p53* gene codes for a protein that suppresses cell division in the G_1 phase, before DNA replication occurs. The *p53* gene is activated when DNA is damaged and must be repaired before it is replicated. Cells in which the *p53* protein is defective do not have the luxury of time to fix errors in DNA. If these cells experience a mutation that damages other genes that regulate the cell cycle and prevent uncontrolled cell division, the mutation is less likely to be repaired, resulting in unregulated cell division: cancer.

Damage to *p53* and other genes may be the result of environmental assaults on DNA, such as are caused by too much exposure to sunlight, certain viruses, and certain chemicals that contaminate food, water, and air. For example, when cells grown in the laboratory are exposed to benzo(a)pyrene, one of the many *carcinogens* (cancer-causing agents) in cigarette smoke, a mutation typically occurs in the *p53* gene. This same type of mutation is found in a common form of lung cancer seen in heavy smokers. Carcinogens such as those in cigarette smoke exert a "double whammy" effect—increasing the numbers of errors in DNA replication and simultaneously damaging the *p53* gene that would normally allow many of these errors to be repaired. We shall discuss cancer in more detail in Chapter 31.

A multicellular organism produced by sexual reproduction grows from a single *fertilized egg* (the result of the fusion of sperm and egg) by repeated mitotic cell divisions. Therefore, with a few exceptions, such as sex cells and the cells of the immune system (as we shall see in Chapter 31), every cell of a multicellular organism is genetically identical. As incredible as it seems, a brain cell and a liver cell have the same genes. These cells have undergone **differentiation**, the process by which cells assume their specialized structures and functions because they have used different genes as they develop.

 ## 5) What Are the Phases of Mitosis?

www

For convenience, we divide mitosis into four phases: (1) *prophase*, (2) *metaphase*, (3) *anaphase*, and (4) *telophase*. As with most biological processes, however, these phases are not really discrete events. Rather, they form a continuum, each phase merging into the next.

During Prophase, the Chromosomes Condense and the Spindle Fibers Form and Attach to the Chromosomes

The first phase of mitosis is called **prophase** (meaning "the stage before" in Greek). Prophase follows inter-phase, as three major events occur: (1) the chromosomes condense, (2) the spindle fibers form, and (3) the chromosomes are captured by the spindle (Figs. 11-7a,b and 11-8a–c, p. 183).

Recall that the chromosomes are duplicated during interphase. Therefore, when mitosis begins, each chromosome *already* consists of two sister chromatids attached to one another at the centromere. During prophase, the chromosomes coil up and condense (Figs. 11-7b and 11-8b), and the nucleolus disappears.

Near the end of prophase, after the chromosomes are condensed, the **spindle fibers** begin to form. The spindle fibers are specialized microtubules that guide the movements of chromosomes during mitosis. Although the spindle fibers play a similar role during mitosis in all eukaryotic cells, in animal cells (but not in plant cells) the spindle fibers are guided by a pair of microtubule-containing **centrioles**. During interphase, the two centrioles are duplicated, and during prophase one pair migrates to each pole. There each pair serves as a central point from which the spindle fibers radiate upward toward the equator of the cell.

When the spindle fibers are fully formed, the nuclear envelope disintegrates, releasing the duplicated chromosomes. Each sister chromatid has a structure called a **kinetochore**, which is an attachment site for spindle fibers, located at the centromere (Fig. 11-8c). The kinetochore of each sister chromatid captures spindle fibers that lead

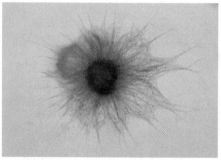

(a) Interphase in a cell of the endosperm (a food-storage organ in the seed) before mitosis begins: The chromosomes are in the thin, extended state and appear as a mass in the center of the cell. The microtubules extend outward from the nucleus to all parts of the cell.

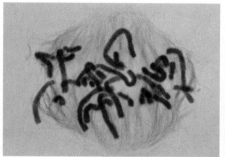

(b) Late prophase: The chromosomes have condensed and attached to microtubules of the spindle fibers.

(c) Metaphase: The spindle fibers have pulled the chromosomes to the equator of the cell.

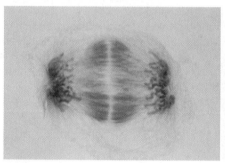

(d) Anaphase: The spindle fibers are moving one set of chromosomes to each pole of the cell.

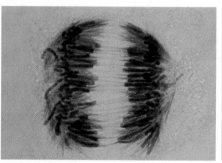

(e) Telophase: The chromosomes have been gathered into two clusters, one at the site of each future nucleus.

(f) Resumption of interphase: The chromosomes are relaxing again into their extended state. The spindle fibers are disappearing, and the microtubules of the two daughter cells are rearranging into the interphase pattern.

Figure 11-7 The cell cycle in a plant cell
Interphase and mitosis in the African blood lily. The chromosomes are stained bluish-purple, and the spindle microtubules are stained pink to red. Compare these micrographs with the drawings of mitosis in an animal cell shown in Figure 11-8. (Each micrograph is of a different single cell that has been fixed and stained at a particular stage of mitosis. Only chromosomes and microtubules have been stained.)

to opposite poles of the cell (Fig. 11-8d). When the sister chromatids separate later in mitosis, the spindle fibers will pull the newly independent chromosomes to opposite poles.

Other spindle fibers do not attach to chromosomes; rather, they retain free ends that overlap along the cell's equator. As we will see, these "free" spindle fibers will push the two spindle poles apart.

During Metaphase, the Chromosomes Become Aligned along the Equator of the Cell

During **metaphase** (the "middle stage"), the spindle fibers running to opposite poles engage in a tug of war, each pulling toward its own pole. It appears that long spindle fibers pull harder than do short ones, with the result that each chromosome ends up midway between the poles.

Metaphase ends when all the chromosomes have lined up along the equator of the cell, with one kinetochore facing each pole (Figs. 11-7c and 11-8d).

During Anaphase, Sister Chromatids Separate and Are Pulled to Opposite Poles of the Cell

At the beginning of **anaphase** (Figs. 11-7d and 11-8e), the sister chromatids are pulled apart and separate into two independent daughter chromosomes. The kinetochores now perform some fancy molecular moves: Protein "motors" in the kinetochores pull the kinetochores (and their attached chromosomes) poleward along the spindle fibers. Simultaneously, the spindle fibers disassemble inside the kinetochores, shortening as the kinetochores move toward the poles. As the kinetochores tow their chromosomes toward the poles, the free spindle fibers

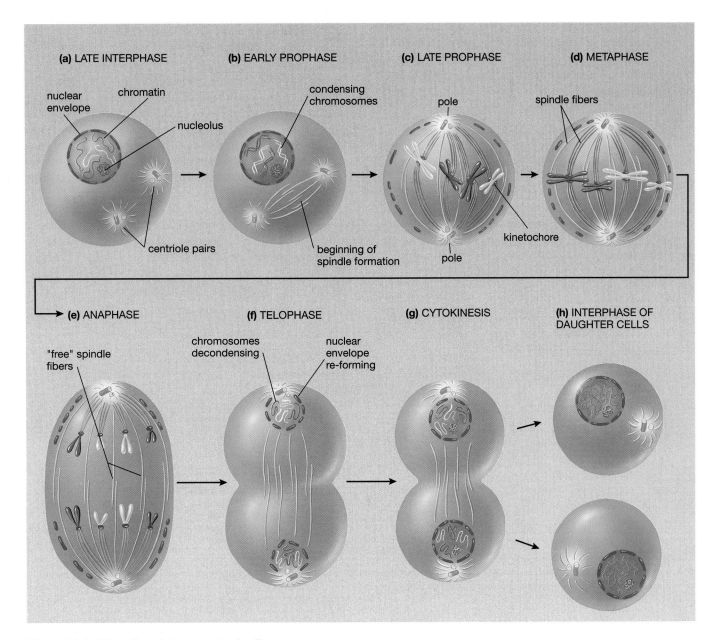

Figure 11-8 The cell cycle in an animal cell
(a) **Late interphase:** The chromosomes have been replicated but remain elongated and re-laxed within the nucleus. The centrioles have also been replicated. *(b)* **Early prophase:** The chromosomes condense and shorten. The centrioles begin to move apart, and the spindle fibers begin to form. *(c)* **Late prophase:** The nucleolus disappears; the nuclear envelope breaks down, and the spindle fibers invade the nuclear region and attach to each chromo-some at its kinetochore (red spot). *(d)* **Metaphase:** The spindle fibers move the chromo-somes to the cell's equator. *(e)* **Anaphase:** Chromatids separate at the centromere, becoming independent chromosomes. The spindle fibers pull one chromosome toward each pole of the cell. The free spindle fibers slide past one another, pushing the poles farther apart. *(f)* **Telophase:** One complete set of chromosomes has reached each pole. The chromosomes relax into their extended state, the spindle fibers begin to disappear, and the nuclear en-velopes begin to re-form. *(g)* **Cytokinesis:** At the end of telophase, the cytoplasm is divided along the equator of the parent cell, with each daughter cell receiving one nucleus and about half the original cytoplasm. *(h)* **Interphase of daughter cells:** The daughter cells enter inter-phase. The spindle fibers disappear, the nuclear envelope re-forms, the chromosomes finish extending, and the nucleolus reappears.

grow longer and push the poles of the cell apart, forcing the cell into an oval shape (Fig. 11-8e). Because the sister chromatids are identical copies of the original chromosomes, both clusters of chromosomes that form on opposite poles of the cell contain one copy of every chromosome.

During Telophase, Nuclear Envelopes Form around Both Groups of Chromosomes

When the chromosomes reach the poles, **telophase** (the "end stage") has begun (Figs. 11-7e and 11-8f). The spindle fibers disintegrate, and nuclear envelopes form around both groups of chromosomes. The chromosomes revert to their extended state, and the nucleoli reappear. In most cells, cytokinesis occurs during telophase, enclosing each daughter nucleus into a separate cell.

6 What Are the Events of Cytokinesis?

 www

In animal cells (normally during telophase), microfilaments attached to the plasma membrane form a ring around the equator of the cell. During cytokinesis, the ring contracts and constricts the cell's equator, much like pulling the drawstring around the waist of a pair of sweatpants. Eventually the "waist" contracts down to nothing, dividing the cytoplasm into two new daughter cells (Figs. 11-8g and 11-9).

Cytokinesis in plant cells is quite different, perhaps because the stiff cell wall makes it impossible to divide one cell into two by pinching at the waist. Instead, the Golgi complex buds off carbohydrate-filled vesicles that line up along the cell's equator between the two nuclei (Fig. 11-10). The vesicles fuse, producing a structure called

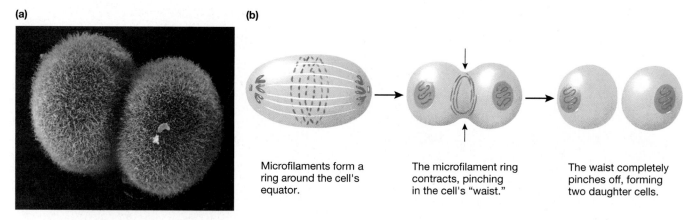

Microfilaments form a ring around the cell's equator.

The microfilament ring contracts, pinching in the cell's "waist."

The waist completely pinches off, forming two daughter cells.

Figure 11-9 Cytokinesis in an animal cell
(*a*) A ring of microfilaments just beneath the plasma membrane contracts around the equator of the cell, pinching it in two. (*b*) The mechanism of cytokinesis in animal cells.

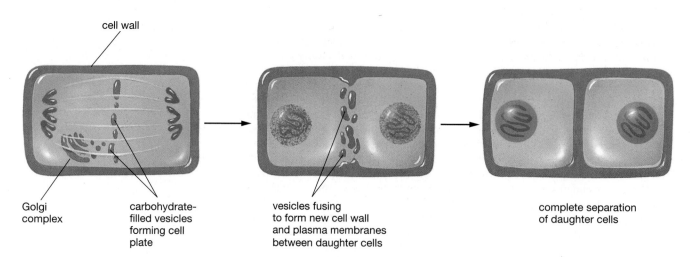

cell wall

Golgi complex

carbohydrate-filled vesicles forming cell plate

vesicles fusing to form new cell wall and plasma membranes between daughter cells

complete separation of daughter cells

Figure 11-10 Cytokinesis in a plant cell
Carbohydrate-filled vesicles produced by the Golgi complex congregate at the equator of the cell, forming the cell plate. The vesicles will fuse to form the two plasma membranes separating the daughter cells; their carbohydrate contents form part of the cell wall.

the **cell plate**, shaped like a flattened sac, surrounded by plasma membrane and filled with sticky carbohydrates. When enough vesicles have fused, the edges of the cell plate merge with the original plasma membrane around the circumference of the cell. The carbohydrate formerly contained in the vesicles remains between the plasma membranes as part of the cell wall.

Cytokinesis is followed by interphase, thus completing the cell cycle (Figs. 11-7f and 11-8h).

7) What Are the Functions of Mitotic Cell Division?

Mitotic Cell Division Results in Growth, Maintenance, and Repair of Body Tissues

Mitotic cell division plays several roles in the lives of multicellular eukaryotic organisms. First, in conjunction with the differential expression of genes in different cells, it allows a fertilized egg eventually to become an adult consisting of perhaps trillions of individual cells.

Mitotic cell division also allows an organism to maintain its tissues, many of which require frequent replacement. For example, your skin cells live for only about

2 weeks. As dead skin flakes off imperceptibly but constantly, skin cells are continuously replaced by cell division. Your red blood cells become worn out after about 4 months. Through cell division, specialized cells in bone marrow give rise to new red blood cells, which enter the bloodstream at the rate of 3 million per second! Cells of your stomach lining, exposed to acid and digestive enzymes, survive only about 3 days before they must be replaced through the division of underlying cells.

Mitotic Cell Division Forms the Basis of Asexual Reproduction

Mitotic cell division provides the basis of **asexual reproduction**, in which offspring are formed from a single parent without the uniting of male and female gametes. This mode of reproduction is normal for many unicellular organisms, such as *Tetrahymena*, and yeasts (Fig. 11-11a,b). Many multicellular organisms can also reproduce asexually. Small replicas of the parent grow by means of cell division. Like its relative the sea anemone, a *Hydra* can reproduce by growing a miniature replica of itself as a bud (Fig. 11-11c). Eventually the bud separates from its parent, going off to live independently. Because mitosis

(a)

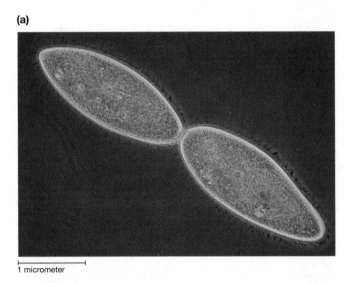

1 micrometer

(b)

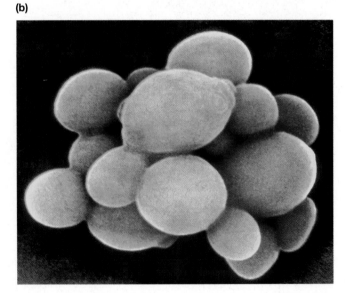

(c)

Figure 11-11 Modes of asexual reproduction
(a) In unicellular microorganisms, such as the protist *Tetrahymena*, mitotic cell division produces two new, independent organisms. *(b)* The yeast, an unusual unicellular fungus, reproduces by mitotic cell division. *(c)* *Hydra*, a freshwater relative of jellyfish, grows a miniature replica of itself (a bud) protruding from its side. When fully developed, the bud breaks off and assumes independent life.

produces genetically identical cells, these offspring are genetically identical to their parents; they are called **clones**.

Many plants reproduce both asexually and sexually. The beautiful aspen groves of Colorado, Utah, and New Mexico (Fig. 11-12) develop asexually from shoots growing up from the root system of a single parent tree. The entire grove, although seeming to be a population of separate trees to the admiring visitor, may actually be considered to be a single individual, with its multiple trunks interconnected by a common root system. Recently, biologists at the University of Colorado reported that one of the single largest organisms yet discovered on Earth is a huge aspen grove in Utah, covering 106 acres and including about 47,000 trunks with a total mass of up to 6 million kilograms (13 million pounds). Although individual trunks age and die, the grove lives on; some groves are thought to be hundreds of thousands of years old. If you plant an aspen tree in your backyard, who knows what you may be starting!

As you will discover later in this chapter and as you explore the diversity of life in later units, plants, many fungi, and some unicellular algae produce sex cells by mitosis. Mitosis also gave rise to the nucleus that produced Dolly. As you will learn in "Scientific Inquiry: Much Ado About Dolly" (pp. 194–195), researchers removed the nucleus of a cell from the udder of a sheep and used it to produce a whole new lamb. This type of asexual reproduction in mammals can occur only in the laboratory!

8 What Are the Advantages of Sexual Reproduction?

If asexual reproduction can produce one of the largest organisms on Earth, it must work pretty well! Why, then, have all known forms of life, even the simplest, evolved ways of exchanging genetic material? This evolution is evidence of the tremendous advantage that DNA exchange among individuals confers on a species. The reshuffling of genes among individuals to create genetically unique offspring is the principle underlying **sexual reproduction**.

Mutations in DNA Are the Ultimate Source of Genetic Variability

As we saw in Chapter 10, the fidelity of DNA replication and proofreading minimizes the number of errors, but changes in DNA base sequences *do* occur, producing mutations. Although most mutations are neutral or harmful, they are also the raw material for evolution. Bacteria are different from bison, and you are different from your predecessors, because of differences in DNA sequence that originally arose as mutations.

Mutations are perpetuated by DNA replication. If they are not lethal, mutations are passed to offspring and become a part of the genetic makeup of each species. Such mutations form **alleles**, alternate forms of a given gene that confer variability on individuals, such as brown or blue eyes.

Reshuffling Genes May Combine Different Alleles in Beneficial Ways

Consider a simple hypothetical example of two individuals of the same species. Each possesses a particularly beneficial allele of a different gene. One allele causes individual *A* to remain motionless when a predator is near, making *A* less likely to be noticed; the second allele provides individual *B* with camouflage color. Clearly it would be advantageous for both al-

Figure 11-12 *The trees in aspen groves are genetically identical*
Each tree grows up from the roots of a single ancestral tree. This photo shows three separate groves near Aspen, Colorado. The genetic identity within a grove, and the genetic difference between groves, is shown by their leaves in fall. One entire grove is still green (bottom), the second has turned bright gold (middle), and the third has already lost its leaves (top).

leles to be combined in a single individual. How might this be accomplished? Not through asexual reproduction, which produces offspring that are genetically identical to the single parent. These useful alleles cannot be combined in a single offspring unless two parent organisms combine their DNA. Enter sexual reproduction.

Meiotic Cell Division Produces Haploid Cells That Can Merge to Combine Genetic Material from Two Parents in a Diploid Offspring

Most eukaryotic cells are either *haploid*, containing a single copy of each type of chromosome, or *diploid*, containing pairs of homologous chromosomes. Homologous chromosomes have the same sequence of genes, but a given gene may exist in a slightly different form (allele) on each homologue. For example, in a hypothetical organism, the gene for eye color (located in the same position on each of two homologous chromosomes) might have two alleles—one coding for red eyes and one for white eyes, one allele on each of the two chromosomes.

The first eukaryotic cells to evolve, about 1 billion to 1.5 billion years ago, were probably haploid. Relatively early on, two events evolved in single-celled eukaryotic organisms that allowed the organisms to shuffle and recombine genetic information. First, two haploid (parental) cells fused, resulting in a temporary, large diploid cell. In this diploid cell were two copies of each chromosome, one copy from each of the parental cells. Second, this diploid cell soon underwent a type of cell division, called *meiotic cell division* (discussed in the next section). The result was haploid daughter cells, each containing one copy of each type of chromosome—one of each of the pairs of homologues:

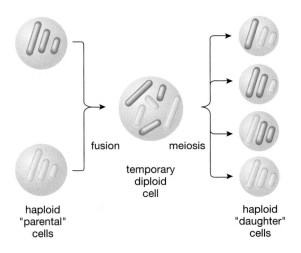

haploid "parental" cells   fusion   meiosis   temporary diploid cell   haploid "daughter" cells

Notice that a single daughter cell formed by meiosis could contain some chromosomes from each parental cell. The daughter cell could therefore combine two or more useful alleles that would allow it to survive and reproduce more successfully in its environment.

Although we will never know for sure, diploidy probably evolved when a "temporarily" diploid cell delayed

meiosis for a while, remaining diploid for a significant portion of its life cycle. Eventually, some of these diploid cells underwent mitotic divisions as well, producing diploid daughter cells. From there, it was a relatively short evolutionary leap to multicellular, diploid, sexually reproducing organisms in which meiosis produces haploid gametes (sperm or eggs) that cannot live independently. Sperm fuses with egg to yield a diploid cell—the fertilized egg, or **zygote**. Through repeated mitotic cell divisions, the zygote gives rise to a new, multicellular organism.

9 What Are the Events of Meiosis?

Meiosis Separates Homologous Chromosomes in a Diploid Nucleus, Producing Haploid Daughter Nuclei

The key to sexual reproduction in eukaryotic cells is **meiosis**, the production of haploid nuclei with unpaired chromosomes from diploid parent nuclei. The word *meiosis* comes from a Greek word meaning "to diminish"; meiosis reduces the number of chromosomes by half. In **meiotic cell division** (meiosis followed by cytokinesis), each daughter cell receives one member from each pair of homologous chromosomes. For example, each diploid cell in your body contains 23 *pairs* of chromosomes; meiotic cell division produces sex cells (sperm or eggs) with one copy of each pair, for a total of 23 chromosomes. Because meiosis evolved from mitosis, many of the structures and events in meiosis are similar or identical to those of mitosis.

Let's begin with an overview of the primary events of meiosis (Fig. 11-13). Meiosis involves two nuclear divisions: *meiosis I* and *meiosis II*. The overall process proceeds as follows:

1. In preparation for meiosis, all the chromosomes are duplicated during interphase, producing sister chromatids joined at the centromere.
2. During meiosis I, the duplicated homologous chromosomes are separated into two daughter cells. Each daughter nucleus receives one homologue of each pair of chromosomes. The sister chromatids do *not* separate during meiosis I. Thus, at the completion of meiosis I, each chromosome remains duplicated—still consisting of two sister chromatids. Because each daughter nucleus contains one homologue of each pair of chromosomes, the daughter nuclei are considered to be haploid. The chromosomes *do not replicate again* between meiosis I and meiosis II.
3. In meiosis II, the sister chromatids of each daughter cell are separated. Cytokinesis may occur, producing four haploid cells, each with one set of unduplicated chromosomes.

The number of chromosomes is reduced from diploid to haploid during meiosis I. Meiosis II is essentially the

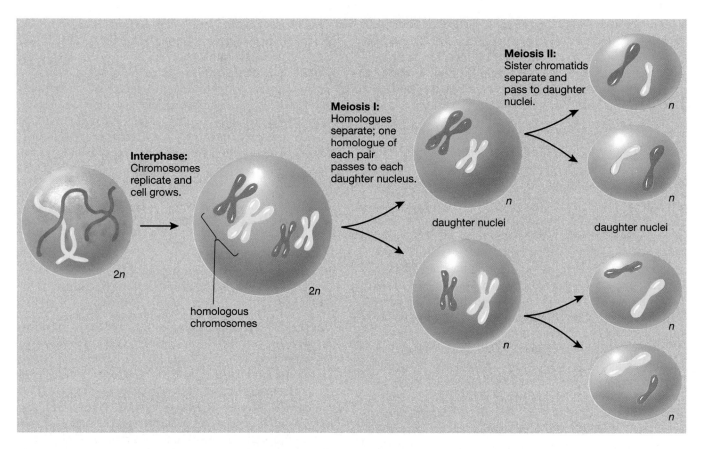

Meiosis II: Sister chromatids separate and pass to daughter nuclei.

Meiosis I: Homologues separate; one homologue of each pair passes to each daughter nucleus.

Interphase: Chromosomes replicate and cell grows.

2n

homologous chromosomes

2n

daughter nuclei

n

n

daughter nuclei

n

n

n

n

n

n

Figure 11-13 *An overview of meiotic cell division, starting with interphase*
Meiosis consists of two nuclear divisions: meiosis I and meiosis II. Together, the two divisions separate homologous chromosomes and produce haploid nuclei that each contain one member of each pair of homologues. The number of different types of chromosomes in a cell is designated by the letter *n*. Haploid cells, with only one copy of each type of chromosome, are indicated by *n*. Diploid cells, with two copies of each type, are indicated by 2*n*.

same as mitosis in a haploid cell; the number of chromosomes remains the same. Because each nuclear division is normally accompanied by cytokinesis, a cell that undergoes both meiosis I and meiosis II produces a total of four haploid cells. Homologous chromosomes in the parental diploid cell generally have different alleles of the same genes, so haploid daughter cells produced by meiosis are genetically similar, but not identical, to one another.

The phases of meiosis have the same names as the roughly equivalent phases in mitosis, followed by a I or II to distinguish the two stages of meiosis. As you shall see, important differences exist between mitosis and meiosis I (Fig. 11-14). In the descriptions that follow, we assume that cytokinesis accompanies the nuclear divisions.

Meiosis I Separates Homologous Chromosomes into Two Daughter Nuclei

In multicellular organisms, meiosis occurs only in a relatively few, specialized cells in the reproductive organs,

such as the testes and ovaries in animals. Meiosis begins in these cells with the replication of the chromosomes. (Because one function of meiosis is to *reduce* the number of chromosomes from diploid to haploid, you may believe that to begin by duplicating the chromosomes is an unnecessary step. However, the cellular mechanisms of meiosis evolved from those of mitosis, and the replication of chromosomes is probably leftover evolutionary baggage. As we will see many times throughout this text, evolution is not an engineer; it's more like a tinkerer.) As in mitosis, the sister chromatids of each chromosome are attached to one another at the centromere. From this point on, meiosis I differs tremendously from mitosis.

During Prophase I, Homologous Chromosomes Pair Up and Exchange DNA

During *prophase I*, homologous chromosomes line up side by side, exchange segments of DNA, and condense (Fig. 11-14a). In a way that is not yet understood, homologous chromosomes find one another early in

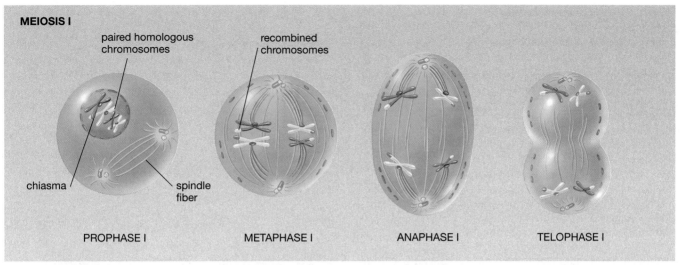

MEIOSIS I

paired homologous chromosomes

recombined chromosomes

chiasma

spindle fiber

PROPHASE I   METAPHASE I   ANAPHASE I   TELOPHASE I

(a) Prophase I. Chromosomes thicken and condense. Homologous chromosomes come together in pairs, and chiasmata occur as chromatids of homologues exchange parts. The nuclear envelope disintegrates, and spindle fibers form.

(b) Metaphase I. Paired homologous chromosomes line up along the equator of the cell. One homologue of each pair faces each pole of the cell. Both chromatids become attached to spindle fibers at their kinetochores (in red).

(c) Anaphase I. Homologues separate, one member of each pair going to each pole of the cell. Sister chromatids do not separate.

(d) Telophase I. Spindle fibers disappear. Two clusters of chromosomes have formed, each containing one member of each pair of homologues. The daughter nuclei are therefore haploid. Cytokinesis commonly occurs at this stage. There is little or no interphase between meiosis I and meiosis II.

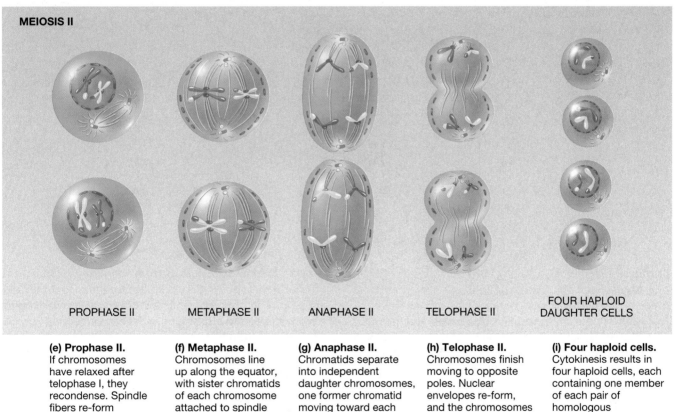

MEIOSIS II

PROPHASE II   METAPHASE II   ANAPHASE II   TELOPHASE II   FOUR HAPLOID DAUGHTER CELLS

(e) Prophase II. If chromosomes have relaxed after telophase I, they recondense. Spindle fibers re-form and attach to the sister chromatids.

(f) Metaphase II. Chromosomes line up along the equator, with sister chromatids of each chromosome attached to spindle fibers that lead to opposite poles.

(g) Anaphase II. Chromatids separate into independent daughter chromosomes, one former chromatid moving toward each pole.

(h) Telophase II. Chromosomes finish moving to opposite poles. Nuclear envelopes re-form, and the chromosomes become extended again.

(i) Four haploid cells. Cytokinesis results in four haploid cells, each containing one member of each pair of homologous chromosomes.

Figure 11-14 The details of meiotic cell division
In meiotic cell division (meiosis and cytokinesis), the homologous chromosomes of a diploid cell are separated, producing four haploid daughter cells. Each daughter cell contains one member of each pair of parental homologous chromosomes. In these diagrams, two pairs of homologous chromosomes are shown, large and small. The yellow chromosomes are from one parent (for example, the father), and the violet chromosomes are from the other parent.

Figure 11-15 The mechanism of crossing over
① Homologous chromosomes pair up side by side. ② One end of each chromosome binds to the nuclear envelope. Protein strands "zip" homologous chromosomes together. ③ Homologous chromosomes are fully joined by protein strands. Recombination enzymes bind to the chromosomes. ④ Recombination enzymes snip chromatids apart and reattach the chromatids. Chiasmata are formed when one end of a chromatid of a paternal chromosome (yellow) is attached to the other end of a chromatid of a maternal chromosome (violet). ⑤ The protein strands and recombination enzymes leave as the chromosomes condense. The chiasmata remain as locations where homologous chromosomes are twisted around each other, helping to hold homologues together.

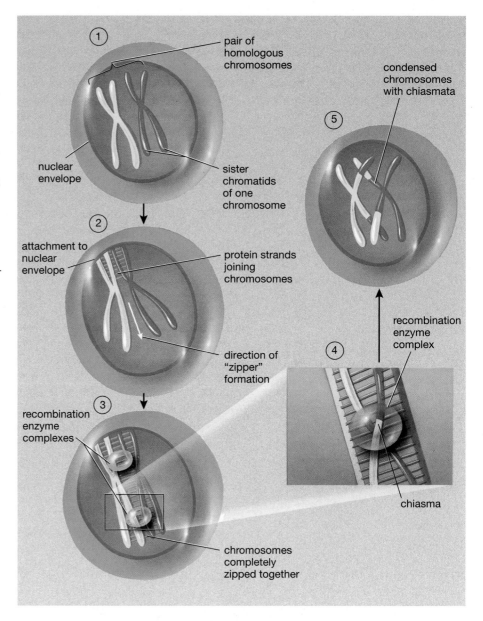

prophase I of meiosis (Step ① of Fig. 11-15). We'll call one homologue the "maternal chromosome" and the other the "paternal chromosome," because one was originally inherited from the organism's mother and the other from the organism's father. The ends of the maternal and paternal chromosomes attach to the nuclear envelope right next to each other (Step ②). Protein strands join the homologous chromosomes so that they match up exactly along their entire length, much like closing a zipper. Then, large enzyme complexes assemble at random places along the paired chromosomes (Step ③). The enzymes snip apart the chromosomes and graft the broken ends together again. However, these recombination enzymes almost always graft a maternal "stump" onto a paternal "stem" and vice versa (Step ④). Thus, the maternal and paternal chromosomes now intertwine, forming crosses, or **chiasmata** (singular, **chiasma**). Eventually the

enzyme complexes detach from the chromosomes, and the protein zippers that held homologues together disappear. The chromosomes become visible with a light microscope, and chiasmata are seen connecting the homologues (Step ⑤).

Chiasmata serve two functions. First, the maternal and paternal chromosomes have swapped some segments of their DNA, an event called **crossing over**. If (as is likely) the chromosomes had slightly different versions of the same genes, then neither chromosome is quite the same as it was before the chiasmata formed. The result of crossing over, then, is genetic **recombination**: the formation of new combinations of the different alleles of each gene on a chromosome. Second, chiasmata hold the homologous chromosomes together during their attachment to the spindle fibers. As we will see, this continued pairing is essential.

As in mitosis, the spindle fibers begin to assemble outside the nucleus during prophase I. Near the end of prophase I, the nuclear envelope breaks down and the spindle fibers capture the chromosomes by attaching to their kinetochores.

During Metaphase I, Paired Homologous Chromosomes Move to the Equator of the Cell

During *metaphase I*, the spindle fibers move the chromosomes to the equator of the cell (Fig. 11-14b). Unlike in mitosis, in which *individual chromosomes* line up along the equator, during metaphase I of meiosis *homologous pairs of chromosomes* line up along the equator. Each of the two homologous chromosomes that make up a chromosome pair is attached to spindle fibers that lead to opposite poles of the cell.

Because the attachment of the chromosomes to the spindle fibers is the key to understanding the movement of chromosomes during meiosis I, let's compare the spindle fiber attachments of mitosis to those of meiosis I (Fig. 11-16). In mitosis, each sister chromatid of each duplicated chromosome attaches to spindle fibers that pull toward opposite poles (Fig. 11-16a). Thus, when the centromeres split during anaphase, the former sister chromatids are pulled apart to opposite poles. In meiosis I, both chromatids of each homologue attach to spindle fibers that pull toward opposite poles, so the spindle fibers will separate the paired homologous chromosomes but will not pull apart the sister chromatids (Fig. 11-16b).

Which member of a pair of chromosomes faces which pole of the cell is random. The maternal chromosome may face "north" for some pairs and "south" for other pairs. This randomness (also called *independent assortment*), together with genetic recombination caused by crossing over, is responsible for the genetic diversity of haploid cells produced by meiosis.

During Anaphase I, Homologous Chromosomes Separate

In *anaphase I*, the paired homologous chromosomes separate from one another and move to opposite poles of the cell (see Fig. 11-14c), pulled by the spindle fibers. One chromosome of each homologous pair (still consisting of two sister chromatids) moves to both poles of the dividing cell. At the end of anaphase I, there is a cluster of chromosomes at both poles. Each cluster contains one member of each pair of homologous chromosomes.

During Telophase I, Two Haploid Clusters of Chromosomes Are Formed

In *telophase I*, the spindle fibers disappear. Cytokinesis commonly occurs at this phase, although the nuclear envelope normally does not reappear. Completion of meiosis I is usually followed immediately by meiosis II, with little or no intervening interphase and no replication of the chromosomes. Typically, the chromosomes remain in their condensed, shortened state.

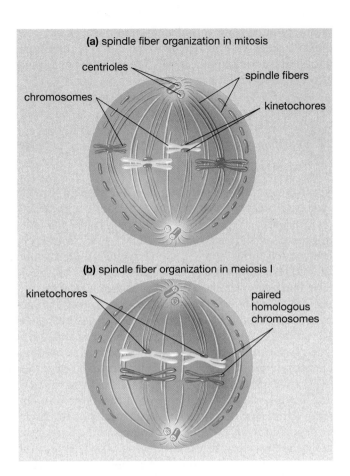

Figure 11-16 A comparison of the spindles formed during mitosis and meiosis I
(a) In mitosis, homologous chromosomes are not paired. The kinetochores of sister chromatids are attached to kinetochore microtubules that lead to opposite poles. When the sister chromatids separate during anaphase, the newly independent daughter chromosomes move to opposite poles of the cell. *(b)* In meiosis I, homologous chromosomes are paired. Both kinetochores of the sister chromatids of a single chromosome are attached to kinetochore microtubules that lead to the same pole. During anaphase I, sister chromatids of each chromosome remain together, moving to the same pole, but homologous chromosomes separate and move to opposite poles.

Meiosis II Separates Sister Chromatids

In meiosis II, the sister chromatids of each chromosome separate. Meiosis II is easy to follow, because it is virtually identical to mitosis, except that it occurs in a haploid nucleus. During *prophase II*, the spindle fibers re-form (see Fig. 11-14e). The chromosomes attach to spindle fibers as they did in mitosis: *Each sister chromatid of each chromosome attaches to spindle fibers that extend to opposite poles of the cell.* During *metaphase II*, spindle fibers pull the chromosomes to the cell's equator. During *anaphase II*, the centromeres holding sister chromatids together split, and the spindle fibers pull each chromatid, now an independent chromosome, to opposite poles. *Telophase II* and cytokinesis conclude meiosis II as nuclear envelopes re-form, the chromosomes relax into their extended state, and the cytoplasm divides. If both daughter cells produced in meiosis I undergo meiosis II, a total of four haploid cells will be formed from a single parent cell by meiotic cell division.

10 What Are the Roles of Mitosis and Meiosis in Eukaryotic Life Cycles?

www

You have probably already figured out where in the life cycle of humans—and all other animals—mitosis and meiosis occur (Fig. 11-17). As you explore the diversity of life on Earth in later units, you will find a variety of life cycles among single-celled organisms, plants, and fungi. Despite this variability, the life cycles of almost all eukaryotic organisms have a common overall pattern. First, two haploid cells fuse, bringing together genes from two parent organisms and endowing the resulting diploid cell with new gene combinations. Second, at some point in the life cycle, meiosis occurs, re-creating haploid cells. Third, at some point, mitosis of either haploid or diploid cells, or both, results in the growth of multicellular bodies and/or asexual reproduction. The seemingly vast differences among life cycles of, say, fungi, ferns, and humans are due to variations in the part of the life cycle in which mitosis and meiosis occur and in the relative proportions of the life cycle spent in the diploid and haploid states, as shown in Figure 11-18 and described next.

Some eukaryotes, such as many fungi and some unicellular algae, have haploid life cycles, in which the dominant organisms consist of haploid cells, with single copies of each type of chromosome (Fig. 11-18a). Under certain environmental conditions, specialized "sexual" haploid cells are produced. Two of these sexual haploid cells fuse, forming a diploid cell. This cell immediately undergoes meiosis, producing haploid cells that undergo mitotic cell division to form a new adult. This life cycle is probably fairly similar to the life cycle of eukaryotic organisms in which meiosis and sexual reproduction first evolved.

In contrast, virtually the entire life cycle of most animals is spent in the diploid state (Fig. 11-18b). Haploid gametes (sperm in males and eggs in females) are formed through meiosis. These gametes fuse to form a diploid fertilized egg, the zygote. Growth and development of the zygote into the adult organism result from mitosis in diploid cells.

The life cycle of most plants, described as **alternation of generations**, includes both multicellular diploid and multicellular haploid body forms. In the typical pattern (Fig. 11-18c), a multicellular diploid body gives rise to haploid cells called **spores**, through meiosis. The spores then undergo mitosis to create a multicellular haploid stage (the "haploid generation"). Certain haploid cells differentiate into gametes, two of which fuse to form a diploid zygote. The zygote grows by mitosis into a diploid multicellular body (the "diploid generation").

In "primitive" plants, such as ferns, both the haploid and diploid stages are free-living, independent plants. The flowering plants, however, have reduced the haploid stage to a minimum, represented only by the pollen grain and a small cluster of cells in the ovary of the flower.

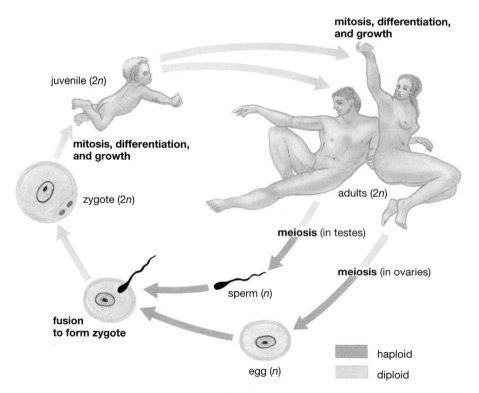

Figure 11-17 The human life cycle
Through meiosis, the two sexes produce different types of gametes—sperm in the testes of males and eggs in the ovaries of females—that fuse to form a diploid zygote. The haploid state lasts only a few hours to a few days; the diploid state may survive for a century. Mitosis and specialization of cells produce an embryo, child, and ultimately a sexually mature adult.

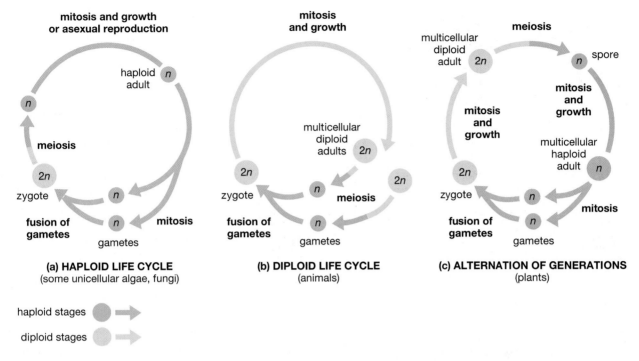

**mitosis and growth
or asexual reproduction**

**mitosis
and growth**

meiosis

(a) HAPLOID LIFE CYCLE
(some unicellular algae, fungi)

(b) DIPLOID LIFE CYCLE
(animals)

(c) ALTERNATION OF GENERATIONS
(plants)

haploid stages

diploid stages

Figure 11-18 Schematic diagrams of the three major types of eukaryotic life cycles
The lengths of the arrows correspond roughly to the proportion of the life cycle spent in each stage. *(a)* Haploid life cycle (some unicellular algae and fungi): Most of the life cycle is spent with haploid cells. Mitosis of haploid cells results in either asexual reproduction of unicellular organisms or growth of multicellular organisms. At some point, specialized reproductive haploid cells fuse, and the resulting diploid cell almost immediately undergoes meiosis. *(b)* Diploid life cycle (animals): Most of the life cycle is spent with diploid cells. When haploid cells are produced by meiosis, they fuse to form a diploid zygote. Mitosis of the zygote produces multicellular bodies composed of diploid cells. *(c)* Alternation of generations (plants): Both diploid and haploid stages consist of multicellular organisms. Cells of the diploid stage undergo meiosis to form spores, which then undergo mitosis to produce a multicellular haploid body. Later, specialization of haploid cells produces gametes, which fuse to form a zygote. Through mitosis, the zygote develops into a multicellular diploid form.

11) What Are the Roles of Meiosis and Sexual Reproduction in Producing Genetic Variability?

WWW

As we noted earlier, genetic variability among organisms is essential for survival and reproduction in a changing environment, and therefore for evolution. Over millions of years, the mutation of DNA is the ultimate source of genetic variability within populations of organisms. Over a single generation, however, the shuffling of homologous chromosomes, crossing over, and the fusion of gametes from separate parents during sexual reproduction provide most of the genetic variability within a species.

Shuffling of Homologues Creates Novel Combinations of Chromosomes

Maternal and paternal homologues are randomly shuffled to the daughter cells of meiosis I. Remember, at metaphase I the paired homologues line up at the cell's equator. For each pair of homologues, the maternal chromosome faces one pole and the paternal chromosome faces the opposite pole, but which homologue faces which pole is random. Let's consider an organism with three pairs of homologous chromosomes—large, medium, and small in size. Further, let's color-code the maternal homologue of each pair yellow and the paternal homologues violet. The chromosomes at metaphase I can align in four configurations:

Anaphase I can therefore produce eight possible sets of chromosomes ($2^3 = 8$):

Scientific Inquiry *www*
Much Ado About Dolly

"And what is the Scientific Community doing…? THEY'RE CLONING SHEEP. Great! Just what we need! Sheep that look MORE ALIKE than they already do!"

Dave Barry, 1997

Early in 1997, a short article in the journal *Nature* caused headlines throughout the world. Dr. Ian Wilmut and associates of the Roslin Institute in Edinburgh, Scotland, had done what many scientists believed to be impossible: They cloned a mammal, using a nucleus taken from adult tissue. **Cloning** is the process of artificially creating a *clone*—a new individual that is genetically identical to an existing individual. As you are aware, every body cell has a complete set of genes. Researchers had already discovered that the nucleus of a cell in extremely early stages of embryonic development, before cells have specialized, could be used to replace the nucleus of an egg cell. The embryonic cell nucleus then directs the development of a new individual, with no need for a fusion of sperm and egg. Until Dr. Wilmut's breakthrough, attempts to make clones from nuclei taken from older tissues (from embryos that had more than 16 cells) had failed. Many researchers hypothesized that the inactivation of selected genes that make, say, neurons different from skin cells and liver cells different from fat cells, must be permanent. What Dr. Wilmut and associates demonstrated is that differentiation is reversible under the right conditions (Fig. E11-1). They used cells taken from the udder of a 6-year-old ewe and grown in culture dishes. Depriving these cells of nutrients forced the cells into the nondividing (G_0) stage of the cell cycle—a factor that seems crucial to allowing all the genes to be expressed when the nucleus was transplanted into an egg cell.

Cloning is a highly artificial form of asexual reproduction, because the nucleus that directs the development of the offspring was produced mitotically from a diploid cell and not by the fusion of haploid gametes produced by meiosis. Like the hydra budding from its parent, the cloned mammal is literally "a chip off the old block." The advantage of cloning cells from an adult rather than from an embryo is that you know what you are getting. A prize wool-producing sheep could theoretically be cloned to produce a whole herd. The same might be done with a cow that has exceptional milk production. Cloning an adult allows scientists to use asexual reproduction to take advantage of the natural variability provided by sexual reproduction. By observing the adults, the most advantageous gene combinations produced during meiosis and the union of gametes can be selected and then perpetuated in clones.

The increasing ability of scientists to manipulate individual genes makes cloning even more advantageous. For example, let's say genetic engineering produces a cow that secretes large quantities of a specialized antibiotic in its milk. This antibiotic could be extracted inexpensively to fight hard-to-treat infections. Such a cow would represent an enormous investment in money, time, and ingenuity. But what if the cow were infertile? Or what if none of its offspring carried the gene to produce the antibiotic? Enter adult cloning. Tiny amounts of tissue from the engineered cow would yield millions of cells, each containing the full genetic "blueprint" for an antibiotic-producing cow.

The implications of Dr. Wilmut's research are far-reaching. Scientists now have proof that the long-inactivated genes of specialized adult cells can be made functional again. Perhaps neurons in the brain region destroyed by Parkinson's disease could be made to divide again, replacing those that have perished. Perhaps more genes for red blood cell production could be switched on in an anemia patient.

Dolly and other cloned mammals also promise to provide insights into aging. Do the genes of each species dictate a maximum life span? If so, Dolly may have been born middle-aged, because her genes came from the cell of a 6-year-old ewe. Will her life be shorter as a result? Will she be able to reproduce normally? She is now pregnant. The scientific world is watching.

What really caught the attention of people everywhere is that Dolly raises the possibility of human cloning. Although the technique required 277 attempts before one healthy lamb was born, the technology is relatively simple. There is no theoretical reason why the method could not be applied to humans. For example, adult cloning could produce a child who had exactly the same set of genes as its mother, a beloved grandparent, a famous sports figure, a Nobel Prize winner, or a despot. Think about it. We'll revisit this topic in Chapter 13.

Each of these chromosome clusters will then undergo meiosis II to produce two gametes. Therefore, our hypothetical parent organism with three pairs of homologous chromosomes could produce gametes with eight different chromosome sets.

Crossing Over Creates Chromosomes with Novel Combinations of Genes

Crossing over produces chromosomes that differ from those of either parent. Consequently, offspring may receive new combinations of genes that may never before have occurred on single chromosomes. Each time our hypothetical parent produces gametes, its homologous chromosomes cross over in new and different places. Therefore, it can produce many more than eight types of gametes. *Because of crossing over, the parent probably never produces any two gametes that are completely identical!*

Fusion of Gametes Adds Further Genetic Variability to the Offspring

Finally, two gametes, possibly genetically unique, fuse to form a diploid offspring. For our hypothetical organisms

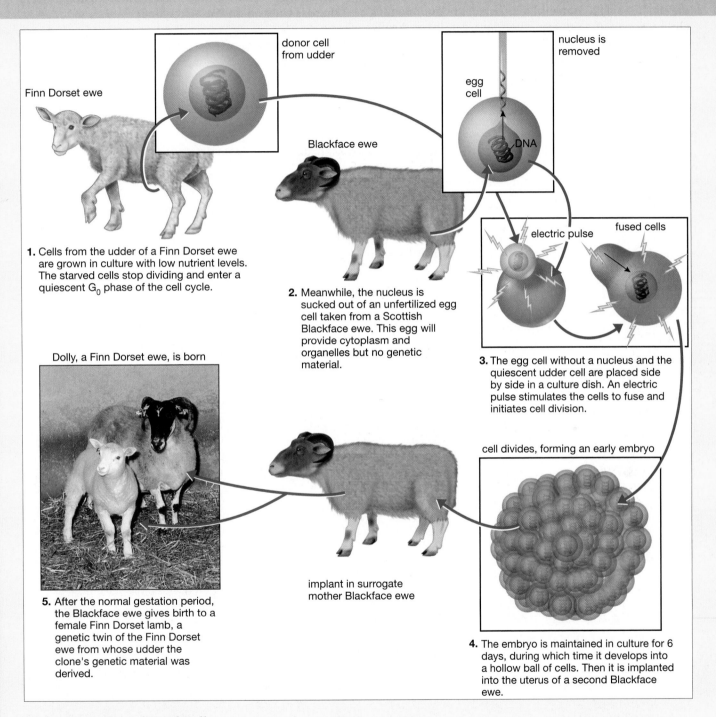

1. Cells from the udder of a Finn Dorset ewe are grown in culture with low nutrient levels. The starved cells stop dividing and enter a quiescent G_0 phase of the cell cycle.

2. Meanwhile, the nucleus is sucked out of an unfertilized egg cell taken from a Scottish Blackface ewe. This egg will provide cytoplasm and organelles but no genetic material.

3. The egg cell without a nucleus and the quiescent udder cell are placed side by side in a culture dish. An electric pulse stimulates the cells to fuse and initiates cell division.

4. The embryo is maintained in culture for 6 days, during which time it develops into a hollow ball of cells. Then it is implanted into the uterus of a second Blackface ewe.

5. After the normal gestation period, the Blackface ewe gives birth to a female Finn Dorset lamb, a genetic twin of the Finn Dorset ewe from whose udder the clone's genetic material was derived.

Figure E11-1 *The making of Dolly*

with three types of chromosomes, the union of gametes could result in $(2^3) \times (2^3) = 64$ different combinations of these unique gametes.

Now let's look at the potential for genetic variability in human beings. Humans have 23 pairs of homologous chromosomes. One person, therefore, can theoretically produce 2^{23} (about 8 million) different gametes, based solely on independent assortment (random separation of the homologues). The fusion of gametes from just two people could produce 8 million $\times$ 8 million = 64 trillion genetically different children! When you add to this the almost endless variability produced by crossing over, is it any wonder that, with the exception of identical twins, there is truly no one just like you?

The two forms of cell division are summarized in Table 11-1 below.

Table 11-1 A Comparison of Mitotic and Meiotic Cell Divisions in Animal Cells

Feature	Mitotic Cell Division	Meiotic Cell Division
Cells in which it occurs	Body cells	Gamete-producing cells
Final chromosome number	Diploid—$2n$ (46 for humans); has two copies of each chromosome (homologous pairs)	Haploid—$1n$ (23 for humans); has one member of each homologous pair
Number of daughter cells	Two, identical to the parent cell and to each other	Four, containing recombined chromosomes due to crossing over
Number of cell divisions per DNA replication	One	Two
Function in animals	Development, growth, repair and maintenance of tissues, asexual reproduction	Gamete production for sexual reproduction

MITOSIS

interphase   no stages comparable to meiosis I   prophase   metaphase   anaphase   telophase   2 diploid cells

MEIOSIS

MEIOSIS I   MEIOSIS II

interphase   prophase I   metaphase I   anaphase I   telophase I   prophase II   metaphase II   anaphase II   telophase II   4 haploid cells

In these diagrams, comparable phases are aligned. In both mitosis and meiosis, chromosomes are replicated during interphase. Meiosis I, with the pairing of homologous chromosomes, formation of chiasmata, exchange of chromosome parts, and separation of homologues to form haploid daughter nuclei, has no counterpart in mitosis. Meiosis II, however, is similar to mitosis in a haploid cell.

Summary of Key Concepts

1) What Are the Essential Features of Cell Division?

Every cell must begin life with (1) a complete set of genetic information (DNA) and (2) the materials necessary to survive and use that genetic information. Cellular reproduction involves replication of the DNA, with one copy packaged into each daughter cell, and cytoplasmic division, with about half the parental cytoplasm passed to each daughter cell.

2) What Are the Events of the Prokaryotic Cell Cycle?

The DNA of a prokaryotic cell is contained in a single, circular chromosome within the cytoplasm, and attached to the plasma membrane. Prokaryotic cells divide by binary fission. First, the chromosome replicates, and the new chromosome attaches to the plasma membrane nearby. The two chromosomes move toward opposite ends of the cell as new plasma membrane is added between their attachment points. The plasma membrane around the middle of the cell grows inward, meeting in the middle and forming two new daughter cells. A temporary, nondividing state during which growth may occur completes the cycle.

3) What Is the Structure of the Eukaryotic Chromosome?

The chromosomes of eukaryotic cells consist of both DNA and proteins. During interphase, the chromosomes are in an extended form, accessible for use in reading their genetic instructions. During cell division, the chromosomes condense into short, thick structures. At some point in all eukaryotic life cycles, cells contain pairs of chromosomes called homologues. Homologues have virtually identical appearance and similar genetic information. Cells with pairs of homologous chromosomes are diploid. Cells with only a single member of each pair are haploid.

4) What Are the Events of the Eukaryotic Cell Cycle?

The eukaryotic cell cycle consists of two major phases: (1) interphase and (2) cell division. During interphase, the cell takes in nutrients, grows, and duplicates its chromosomes. Cells spend variable amounts of time in this nondividing state, in some cases remaining in interphase for the adult life of the organism. Mitotic cell division consists of two processes: (1) mitosis (nuclear division) and (2) cytokinesis (cytoplasmic division). Mitosis parcels out one copy of each chromosome into two separate nuclei, and cytokinesis subsequently encloses each nucleus in a separate cell.

5) What Are the Phases of Mitosis?

The chromosomes are duplicated during interphase, prior to mitosis. The two identical copies, called chromatids, are attached to one another at the centromere. During mitosis, the chromatids separate, one going to each daughter nucleus. The two daughter nuclei thus receive identical genetic information. Mitosis consists of four phases (see Fig. 11-8):

(1) **Prophase:** The chromosomes condense, and the spindle fibers form. The nuclear membrane breaks down. Spindle fibers attach to the chromosomes at kinetochores located at the centromeres. Free spindle fibers extend from the cell's poles past the cell's equator.

(2) **Metaphase:** The chromosomes are pulled by the spindle fibers to the equator.

(3) **Anaphase:** The two chromatids of each chromosome separate. The spindle fibers move the newly independent chromosomes to opposite poles of the cell.

(4) **Telophase:** The chromosomes relax into their extended state, and nuclear envelopes re-form around each new daughter nucleus.

6) What Are the Events of Cytokinesis?

Cytokinesis normally coincides with the end of telophase and divides the cytoplasm into approximately equal halves, each containing a nucleus. In animal cells, the plasma membrane is pinched in along the equator by a ring of microfilaments. In plant cells, new plasma membrane forms along the equator by the fusion of vesicles produced by the Golgi complex.

7) What Are the Functions of Mitotic Cell Division?

Mitotic cell division plays several roles in the eukaryotic life cycle: In conjunction with the differential expression of genes, it allows the fertilized egg to grow and develop into a mature adult. Mitotic cell division replaces many tissues that suffer wear and tear, such as skin and blood. It also contributes to reproduction. Some organisms reproduce asexually—that is, without the fusion of gametes. Asexual reproduction normally occurs by the growth of a new organism from the body of a single parent organism, by repeated cell divisions, and differentiation of daughter cells. Therefore, the offspring are genetically identical to the parent. In plants, in many fungi, and in some unicellular algae, mitosis produces sex cells.

8) What Are the Advantages of Sexual Reproduction?

For evolution to occur, there must be genetic differences among organisms. These genetic differences originate as mutations. Mutations preserved within a species produce alternate forms of genes called alleles. Alleles in different individuals of a species may be combined in offspring through sexual reproduction.

9) What Are the Events of Meiosis?

Meiosis separates homologous chromosomes and produces haploid cells with only one homologue from each pair. These haploid cells or their descendants fuse to form diploid cells that receive one homologue of each pair from each parent, reestablishing pairs of homologous chromosomes. During interphase before meiosis, chromosomes are replicated. During meiosis, a diploid cell undergoes two specialized cell divisions—meiosis I and meiosis II—to produce four haploid daughter cells.

Meiosis I: During prophase I, homologous chromosomes, each consisting of two chromatids, pair up and exchange parts by crossing over. During metaphase I, paired homologues move together to the cell's equator, one member of each pair facing opposite ends of the cell. Homologous chromosomes separate during anaphase I, and two nuclei form during telophase I. Each daughter nucleus receives only one member of each pair of homologues and is therefore haploid. However, sister chromatids do *not* separate during meiosis I, so each chromosome is still duplicated.

Meiosis II: Meiosis II occurs in both daughter nuclei and resembles mitosis in a haploid cell. The chromosomes move to the cell's equator during metaphase II. The two chromatids of each chromosome separate and move to opposite ends of the cell during anaphase II. This second division produces four haploid nuclei. Cytokinesis normally occurs during or shortly after telophase II, producing four haploid cells.

Meiosis separates pairs of homologous chromosomes in diploid cells to form unpaired chromosomes in haploid cells. Homologous chromosomes contain similar, but generally slightly different, genetic information. The separation of homologues during meiosis I is random: Any one member of each pair of homologues ("maternal" or "paternal") may end up in either daughter nucleus. This independent assortment, plus the recombination of genes during crossing over, produces haploid cells that differ in genetic composition.

Table 11-1 compares the processes of mitosis and meiosis in animal cells. In the accompanying illustration comparable stages are aligned. Note that no stages in mitosis are comparable to meiosis I. Meiosis II is essentially identical to mitosis in a haploid cell.

10 What Are the Roles of Mitosis and Meiosis in Eukaryotic Life Cycles?

Most eukaryotic life cycles are similar in three respects: (1) Sexual reproduction combines haploid gametes to form a diploid cell; (2) at some point in the life cycle, diploid cells undergo meiosis to produce haploid cells once again; and (3) mitosis of either the haploid or diploid cells, or both, results in the growth of multicellular bodies and/or sexual reproduction. Figure 11-18 summarizes the three major types of life cycles.

11 What Are the Roles of Meiosis and Sexual Reproduction in Producing Genetic Variability?

From one generation to the next, the shuffling of homologous chromosomes, crossing over, and the fusion of gametes from separate parents during sexual reproduction provide most of the genetic variability within a species. The random shuffling of homologous maternal and paternal chromosomes creates new chromosome combinations. Crossing over creates chromosomes with gene combinations that may never before have occurred on single chromosomes. Because of crossing over, a parent probably never produces any two gametes that are completely identical. Finally, the fusion of two genetically unique gametes adds further genetic variability to the offspring.

Key Terms

allele *p. 186*	**chiasma** *p. 190*	**gamete** *p. 180*	**mitotic cell division** *p. 180*
alternation of generations *p. 192*	**chromatid** *p. 179*	**haploid** *p. 180*	**prophase** *p. 181*
	chromatin *p. 176*	**homologue** *p. 179*	**recombination** *p. 190*
anaphase *p. 182*	**chromosome** *p. 176*	**interphase** *p. 180*	**sexual reproduction** *p. 186*
asexual reproduction *p. 185*	**clone** *p. 186*	**kinetochore** *p. 181*	**spindle fiber** *p. 181*
binary fission *p. 177*	**cloning** *p. 194*	**life cycle** *p. 176*	**spore** *p. 192*
cell cycle *p. 176*	**crossing over** *p. 190*	**meiosis** *p. 187*	**telophase** *p. 184*
cell plate *p. 185*	**cytokinesis** *p. 180*	**meiotic cell division** *p. 187*	**zygote** *p. 187*
centriole *p. 181*	**differentiation** *p. 181*	**metaphase** *p. 182*	
centromere *p. 178*	**diploid** *p. 179*	**mitosis** *p. 180*	

Thinking Through the Concepts

Multiple Choice

1. *At which stage of mitosis are chromosomes arranged along a plane at the midline of the cell?*
a. anaphase   b. telophase   c. metaphase
d. prophase   e. interphase

2. *A diploid cell contains in its nucleus*
a. an even number of chromosomes
b. an odd number of chromosomes
c. one copy of each homologue
d. either an even or an odd number of chromosomes
e. two sister chromatids of each chromosome during G_1

3. *Synthesis of new DNA occurs during*
a. prophase   b. interphase   c. mitosis
d. cytokinesis   e. formation of the cell plate

4. *Which statement is most correct?*
a. All mutations are harmful.
b. Both mitosis and meiosis add to genetic diversity.
c. Crossing over helps each gamete get a different set of genes in meiosis.
d. Mitosis always makes diploid daughter cells; meiosis always produces gametes.
e. The only haploid cells are gametes.

5. *When do homologous chromosomes pair up?*
a. only in mitosis   b. only in meiosis I
c. only in meiosis II   d. in both mitosis and meiosis
e. in neither mitosis nor meiosis

6. *Curiously, there is no crossing over of any chromosome in the male fruit fly* Drosophila, *which has four pairs of chromosomes. How many different combinations of maternal vs. paternal chromosomes are possible in a male fruit fly's sperm?*
a. 2   b. 4   c. 8
d. 16   e. many more than the above

? Review Questions

1. Diagram and describe the eukaryotic cell cycle. Name the various phases, and briefly describe the events that occur during each. What is the role of the cell cycle in a human?

2. Define *mitosis* and *cytokinesis*. Do these two processes always occur together?

3. Diagram the stages of mitosis. How does mitosis ensure that each daughter nucleus receives a full set of chromosomes?

4. Define the following terms: *homologous chromosome, centromere, kinetochore, chromatid, diploid, haploid.*

5. Describe and compare the process of cytokinesis in animal cells and in plant cells.

6. Diagram the events of meiosis. At which stage do haploid nuclei first appear?

7. Describe homologue pairing and crossing over. At which stage of meiosis do they occur? Name two functions of chiasmata.

8. In what ways are mitosis and meiosis similar? In what ways are they different?

9. Describe how meiosis provides for genetic variability. If an animal had a haploid number of 2 (no sex chromosomes), how many genetically different types of gametes could it produce? (Assume no crossing over.) If it had a haploid number of 5?

Applying the Concepts

1. Both nerve cells in the adult human central nervous system and heart muscle cells remain in the G_0 portion of interphase. In contrast, cells lining the inside of the small intestine divide frequently. Discuss this difference in terms of why damage to the nervous system and heart muscle cells (such as caused by a stroke or heart attack) is so dangerous. What do you think might happen to tissues such as the intestinal lining if some disorder or drug blocked mitoses in all cells of the body?

2. Do you think that each type of cell in your body steadily progresses through one cell division after another, or do you think that the rate of division is likely to vary from time to time? Explain and justify your answer in terms of some common events in human life: childhood growth, healing a wound in your skin, menstruation, replacing donated blood, and cancer.

3. Cancer cells divide out of control. Chemotherapy and radiation therapy typically cause the loss of hair and of the gastrointestinal lining, producing severe nausea. What can you infer about the mechanisms of these treatments? What would you look for in an improved cancer therapy?

4. Suppose you were a researcher using tissue from a deceased person who had donated his body to science. You discovered a chromosomal abnormality that might affect the health of surviving offspring. Should you be allowed to communicate this finding to relatives of the deceased?

5. Some animal species can reproduce either asexually or sexually, depending on the state of the environment. Asexual reproduction tends to occur in stable, favorable environments; sexual reproduction is more common in unstable and/or unfavorable circumstances. Discuss the advantages or disadvantages this behavior might have on survival of the species in an evolutionary sense or on survival of individuals.

6. Would you predict that a clone produced as Dolly was would be a carbon copy of the animal from which the nucleus was obtained? What factors might cause the two sheep to differ? What variables might cause a human clone to differ from the individual who donated the genetic material?

Group Activity

Your instructor will choose several traits (such as a widow's peak, the ability to curl your tongue, and gender) that you can easily evaluate and will show you the appropriate symbols for those traits and which traits are dominant or recessive. If you have the dominant phenotype for a given trait and can't determine whether you are heterozygous dominant or homozygous dominant, assume that you are heterozygous dominant for that trait. On one side of a strip of paper, label one allele (for example, T); on the other side, label the other allele (say, t, if you are heterozygous). For simplicity's sake, assume that each allele is on a separate chromosome, so label each trait on separate paper strips; you'll have as many strips of paper (chromosomes) as traits chosen. Put your initials on each paper strip so that you'll be able to identify it as yours. Label each strip as chromosome 1, 2, 3, and so on according to the traits assigned. (In other words, each student should show the same trait on the same numbered chromosome.)

Drop your "chromosome 1" strip onto the floor so that the allele that faces up is determined at random. Do the same thing for each trait. The alleles that face up are the ones you'll pass on to your offspring—one per trait. Record those alleles, indicating the trait for which each applies. Then "mate" with—that is, share chromosomes with—another student. Record your partner's alleles for each trait. You'll get pairs of alleles that result from random mating. Then determine the genotype and phenotype of your offspring. If you have time, try to get the results of a few different random matings, and compare them.

For More Information

Anderson, W. F. "Gene Therapy." *Scientific American*, September 1995. Describes the exciting new prospects for correcting genetic disorders by using viruses to deliver copies of healthy genes to affected cells.

Grant, M. C. "The Trembling Giant." *Discover*, October 1993. Aspen groves are really single individuals: huge, slowly spreading from the roots of the original parent tree, and potentially almost immortal.

Kahn, P. "Gene Hunters Close in on Elusive Prey." *Science*, March 8, 1996. Using high technology and sampling large families, researchers are homing in on the roots of diseases that involve complex interactions between heredity and the environment.

Kinoshita, J. "Swap Meet." *Discover*, April 1991. A discussion of a hypothesis about the evolution of sex.

Murray, A. W., and Kirschner, M. W. "What Controls the Cell Cycle." *Scientific American*, March 1991. The synthesis, processing, and degradation of a group of proteins seems to control the progression of a cell through the stages of the cell cycle.

Nash, M. "The Age of Cloning." *Time*, March 10, 1997. Well-written and well-illustrated description of the technique and implications of using an adult cell to clone a sheep.

Stahl, F. "Genetic Recombination." *Scientific American*, February 1987. A look at the mechanisms of crossing over.

Travis, J. "A Fantastical Experiment." *Science News*, April 5, 1997. A clear description of the cloning of Dolly the sheep and some of its implications.

The dramatic differences between the Chihuahua and the Great Dane are the result of centuries of selective breeding.

Patterns of Inheritance

12

 Net Watch

On-line resources for this chapter are on the World Wide Web at: http://www.prenhall.com/audesirk (click on the Table of Contents link and then select Chapter 12).

The tiny Chihuahua weighs about 3 pounds and has large, batlike ears, a short coat, and huge eyes. Next to it stands a massive, 125-pound Great Dane. Despite enormous differences in body size and shape, both of these animals are members of the same species, *Canis familiaris*. The astonishing differences between the two reflect human intervention in breeding over hundreds of generations.

Have you ever speculated as to what would happen if a male Chihuahua somehow managed to mate with a female Great Dane? No? Well, you have probably wondered what determines whether a human baby will be male or female or how two brown-eyed parents can produce a blue-eyed child. Although you may not realize it, you have already come a long way toward understanding the answers to questions such as these. You have learned how the elegantly simple structure of DNA can carry a blueprint for the overwhelming complexity of even simple forms of life. You now know that segments of DNA, called **genes**, carry instructions for synthesizing proteins—and that these proteins, in complex ways, determine how an organism develops, how it looks, and how it functions.

You have a basic understanding of how cells copy the vital instructions carried in DNA and pass these copies on to daughter cells. Finally, you have learned how the special type of cell division called meiosis creates sex cells with only half of the total genetic material. This process allows the sperm and egg of two individuals to fuse and restore the full complement of genetic material, producing offspring that are different from either parent—and from each other.

But how are these differences transmitted? What rules govern **inheritance**, the genetic transmission of characteristics from parent to offspring? How is sex determined, and why are color blindness and hemophilia seen almost exclusively in males? Why does human skin come in such a wide range of shades? What is Down syndrome, and how does it occur? In this chapter we explore the ways in which traits are passed from parent to offspring, how different forms of the same gene arise, and how they interact when brought together by the fusion of sperm and egg.

The Chihuahua and the Great Dane illustrate that, for centuries, people have used the basic principle that spontaneously arising variations can be passed from parent to offspring. Yet the fundamental rules of inheritance remained obscure until the beginning of the twentieth century. Interestingly, the basic principles of inheritance had been described in the 1860s by the Augustinian monk Gregor Mendel, but they were ignored or misunderstood for nearly three decades. Mendel's rediscovered studies then became the basis for understanding how genes are inherited, long before anyone knew what a gene was. How did a monk, working quietly and alone in a monastery garden in the middle of the nineteenth century, lay the foundations for modern genetics?

1) How Did Gregor Mendel Lay the Foundations for Modern Genetics?

Before settling down as a monk in the monastery of St. Thomas in Brünn (now Brno, in the Czech Republic), Gregor Mendel (Fig. 12-1) tried his hand at several pursuits, including health care and teaching. To earn his teaching certificate, Mendel attended the University of Vienna for 2 years, where he studied botany and mathematics, among other subjects. This training proved crucial to his later experiments, which were the foundation for the modern science of genetics. At St. Thomas in the mid-1800s, Mendel carried out both his monastic duties and a groundbreaking series of experiments on inheritance in the common edible pea. Although Mendel worked without knowledge of genes or chromosomes, we can more easily follow his experiments after a brief look at some modern genetic concepts.

Figure 12-1 Gregor Mendel
A portrait painted about the time of Mendel's pioneering genetics experiments.

A gene's specific physical location on a chromosome is called a **locus** (plural, *loci*) (Fig. 12-2). Homologous chromosomes carry the same genes, located at the same loci. Although the nucleotide sequence at a given gene locus is always *similar* on homologous chromosomes, it may not be *identical*. These differences allow different nucleotide sequences at the same gene locus on two homologous chromosomes to produce alternate forms of the gene, called **alleles**. Human A, B, and O blood types, for example, are produced by three alleles of the same gene.

If both homologous chromosomes in an organism have *the same* allele at a given gene locus, the organism is said to be **homozygous** at that gene locus. ("Homozygous" comes from Greek words meaning "same pair.") If two homologous chromosomes have *different* alleles at a given gene locus, the organism is **heterozygous** ("different pair") at that locus and is called a **hybrid**. Recall from Chapter 11 that, during meiosis, homologous chromosomes are separated, so each gamete receives one member of each pair of homologous chromosomes. Therefore, all the gametes produced by an organism that is homozygous at a particular gene locus will contain the same allele. Gametes produced by an organism that is heterozygous at the same gene locus are of two kinds:

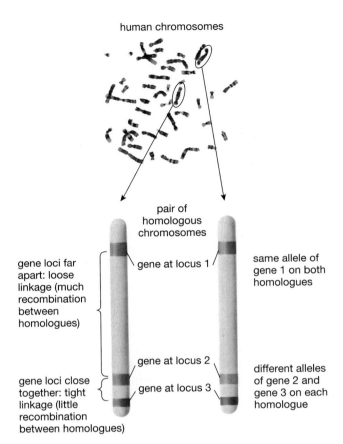

human chromosomes

pair of homologous chromosomes

gene loci far apart: loose linkage (much recombination between homologues)

gene at locus 1

same allele of gene 1 on both homologues

gene at locus 2

gene at locus 3

gene loci close together: tight linkage (little recombination between homologues)

different alleles of gene 2 and gene 3 on each homologue

Figure 12-2 The relationships among genes, alleles, and chromosomes
A given gene is a segment of DNA at the same location (locus) on a pair of homologous chromosomes. Differences in nucleotide sequences at the same gene locus produce different alleles of the gene (depicted by different colors).

Half of the gametes contain one allele, and half contain the other.

Interestingly, both the patterns of inheritance and many essential facts about genes, alleles, and the distribution of alleles in gametes and zygotes during sexual reproduction were deduced by Gregor Mendel long before DNA, chromosomes, or meiosis had been discovered. Because his experiments are a succinct, elegant example of science in action, let's follow Mendel's paths of discovery.

Doing It Right: The Secrets of Mendel's Success

There are three key steps to any successful experiment in biology: (1) choosing the right organism to work with, (2) designing and performing the experiment correctly, and (3) analyzing the data properly. Mendel was the first geneticist to complete all three steps.

Mendel's choice of the edible pea as an experimental subject was critical to the success of his experiments. In plants, a male gamete, which for simplicity we'll call the sperm, is contained in each pollen grain. The structure of the pea flower normally prevents another flower's pollen

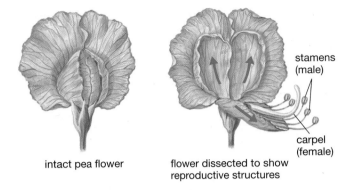

stamens (male)

carpel (female)

intact pea flower

flower dissected to show reproductive structures

Figure 12-3 The flower of the edible pea
In the intact flower (left), the lower petals form a container enclosing the reproductive structures—the stamens (male) and carpel (female). Pollen normally cannot enter the flower from outside, so peas normally self-fertilize.

from entering (Fig. 12-3). Instead, each pea flower normally supplies its own pollen, so the egg cells in each flower are fertilized by sperm from the pollen of the same flower. This process is called **self-fertilization**. Even in Mendel's time, commercial seed dealers sold many types of peas that were **true-breeding**. In true-breeding plants, all the offspring produced through self-fertilization are homozygous for a given trait and are essentially identical to the parent plant.

Although peas normally self-fertilize, plant breeders can also mate two plants by hand, a process called **cross-fertilization**. Breeders pull apart the petals and remove the stamens, preventing self-fertilization (Fig. 12-3). By dusting the carpels with pollen they have selected, breeders can control cross-fertilization. In this way, two true-breeding plants can be mated to see what types of offspring they produce.

In contrast to earlier scientists, Mendel chose to study *traits*—heritable characteristics—that had unmistakably different forms, such as white flowers versus purple flowers, and he worked with one trait at a time. These factors allowed Mendel to see through to the underlying principles of inheritance. Equally important was the fact that Mendel counted the numbers of offspring with each type of trait and analyzed the numbers. The use of statistics as a tool to verify the validity of results has since become an extremely important practice in biology.

2 How Are Single Traits Inherited?

www

Mendel started as simply as possible. He raised varieties of pea plants that were true-breeding for different forms of a single trait and cross-fertilized them. Mendel saved the resulting hybrid seeds and grew them the following year to determine their characteristics.

In one of these experiments, Mendel cross-fertilized a white-flowered pea with a purple-flowered one. This was the *parental generation*, denoted by the letter P. When he grew the resulting seeds, he found that all the first-generation offspring (the "first filial," or F_1, generation) produced purple flowers:

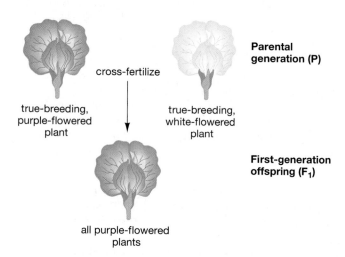

What had happened to the white color? The flowers of the hybrids were just as purple as their parent. The white color seemed to have disappeared in the F_1 offspring.

Mendel then allowed the F_1 flowers to self-fertilize, collected the seeds, and planted them the next spring. In the second generation (F_2), about three-fourths of the plants had purple flowers and one-fourth had white flowers:

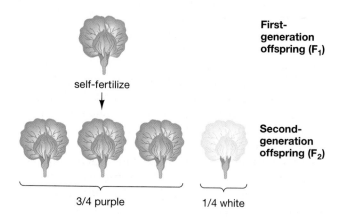

The exact numbers were 705 purple and 224 white, or a ratio of about 3 purple to 1 white. This result showed that the gene that produced white flowers had not disappeared but had only been "hidden."

Mendel allowed the F_2 plants to self-fertilize and produce yet a third (F_3) generation. He found that all the white-flowered F_2 plants produced white-flowered offspring; that is, they bred true. For as many generations as he had time and patience to raise, white-flowered parents always gave rise to white-flowered offspring. The purple-flowered F_2 plants were of two types: About one-third of these were true-breeding for purple; the remaining two-

thirds were hybrids that produced both purple- and white-flowered offspring, again in the ratio of 3 to 1. Therefore, the F_2 generation included ¼ true-breeding purple plants, ½ hybrid purple, and ¼ true-breeding white.

The Inheritance of Dominant and Recessive Alleles on Homologous Chromosomes Can Explain the Results of Mendel's Crosses

Mendel's results, supplemented by our knowledge of genes and homologous chromosomes, allow us to develop a five-part hypothesis:

1. Each trait is determined by pairs of discrete physical units, which we now call genes. Each individual has two genes for a given trait (such as flower color), one on each of the homologous chromosomes. The allele for the gene that controls flower color is different for white-flowered peas, for example, than for true-breeding purple-flowered peas.

2. Pairs of genes on homologous chromosomes separate from each other during gamete formation, so each gamete receives only one allele of an organism's pair of genes. This conclusion is known as Mendel's **law of segregation**: Each gamete receives only one of each parent's pair of genes for each trait. When a sperm fertilizes an egg, the resulting offspring receives one allele from the father and one from the mother.

3. Which member of a pair of genes becomes included in a gamete is determined by chance. This randomness occurs because the separation of pairs of homologous chromosomes during meiosis is random.

4. When two alternate forms of a gene are present, one (the **dominant** allele) may mask the expression of the other (the **recessive** allele). The dominant allele does not, however, alter the physical presence of the recessive allele, which is passed unchanged to the individual's gametes. In Mendel's experiments with flower color, the allele for purple flowers is dominant, and the allele for white flowers is recessive.

5. True-breeding organisms have two of the same alleles for a given trait (these organisms are homozygous); hybrids have two different alleles for that trait (heterozygous). A homozygous organism, with only one type of allele, can produce only one type of gamete:

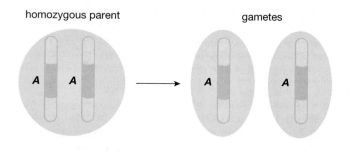

A heterozygous organism, with two different alleles, produces equal numbers of gametes with each of the two alleles:

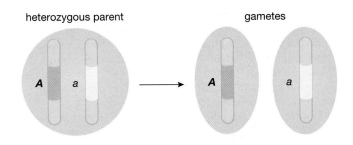

heterozygous parent   gametes

Let's see how Mendel's hypothesis explains the results of his experiments with flower color. Using letters to represent the different alleles, we will assign the uppercase letter *P* to the allele for purple and the lowercase letter *p* to the allele for white (recessive). (By convention, the dominant allele is represented by a capital letter.) A true-breeding (homozygous) purple-flowered plant has two alleles for purple (*PP*), whereas a white-flowered plant has two alleles for white (*pp*). All the sperm and eggs produced by a *PP* plant carry the *P* allele; all the sperm and eggs of a *pp* plant carry the *p* allele:

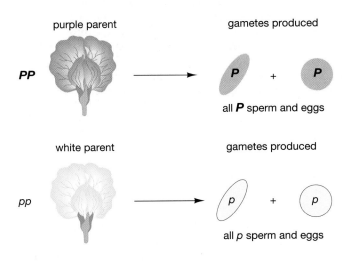

purple parent   gametes produced

PP   all **P** sperm and eggs

white parent   gametes produced

pp   all *p* sperm and eggs

The F₁ hybrid offspring are produced when *P* sperm fertilize *p* eggs or when *p* sperm fertilize *P* eggs. In either case, the F₁ offspring are *Pp*. Because *P* is dominant to *p*, all the offspring are purple:

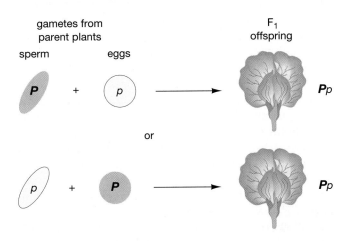

gametes from parent plants

sperm   eggs   F₁ offspring

or

Each gamete produced by a heterozygous *Pp* plant has an equal chance of receiving either the *P* allele or the *p* allele (that is, the plant produces equal numbers of *P* and *p* sperm and equal numbers of *P* and *p* eggs). When a *Pp* plant self-fertilizes, each type of sperm has an equal chance of fertilizing each type of egg:

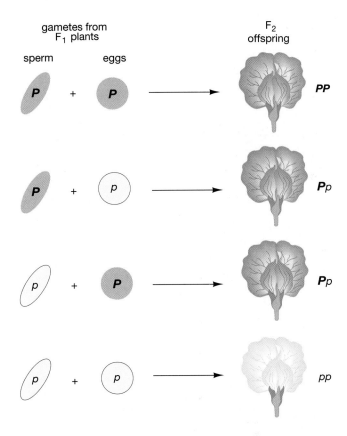

gametes from F₁ plants   F₂ offspring

sperm   eggs

Therefore, three types of offspring can be produced: *PP*, *Pp*, and *pp*. The three types occur in the approximate proportions of ¼ *PP*, ½ *Pp*, and ¼ *pp*.

The actual combination of alleles carried by an organism (for example, *PP* or *Pp*) is its **genotype**. The organism's traits, including its outward appearance, behavior, digestive enzymes, blood type, or any other observable feature, make up its **phenotype**. As we have seen, plants with both genotypes *PP* and *Pp* bear purple flowers. Thus, even though they have different genotypes, they have the same phenotype. Therefore, the F₂ generation consists of three genotypes (¼ *PP*, ½ *Pp*, and ¼ *pp*) but only two phenotypes (¾ purple and ¼ white).

The **Punnett square method**, named after a famous geneticist of the early 1900s, R. C. Punnett, is an intuitive way to predict the genotypes and phenotypes of offspring. Figure 12-4 shows how to use a Punnett square to determine the proportion of offspring that arise from the self-fertilization of a flower that is heterozygous for color (or from any two gametes that are heterozygous for a single trait). This figure also provides the fractions that allow you to calculate the same outcomes based on probability theory. As you use these "genetic bookkeeping"

(a)

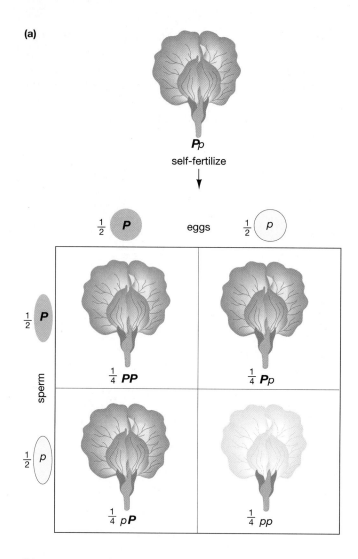

Figure 12-4 *The Punnett square method*
The Punnett square method allows you to predict both genotypes for specific crosses; here we use it for a cross between gametes that are heterozygous for a single trait.
(a) To generate a Punnett square:
(1) Assign letters to the different alleles; use uppercase for dominant and lowercase for recessive.
(2) Determine all the types of genetically different gametes that can be produced by the male and female parents.
(3) Draw the Punnett square, with each row and column labeled with one of the possible genotypes of sperm and eggs, respectively. (We have included the fractions of these genotypes with each label.)
(4) Fill in the genotype of the offspring in each box by combining the genotype of sperm in its row with the genotype of the egg in its column. (We have placed the fractions in each box.)
(5) Count the number of offspring with each genotype. (Note that Pp is the same as pP.)
(6) Convert the number of offspring of each genotype to a fraction of the total number of offspring. In this example, out of four fertilizations, only one is predicted to produce the pp genotype, so ¼ of the total number of offspring produced by this cross is predicted to be white. To determine phenotypic fractions, add the fractions of genotypes that would produce a given phenotype. For example, purple flowers are produced by ¼ PP + ¼ Pp + ¼ pP, for a total of ¾ of the offspring.
(b) Probability theory. Probabilities can be calculated by determining the fractions of eggs and sperm of each genotype and then multiplying these numbers together to arrive at fractions of offspring genotypes. The appropriate genotypic ratios can then be added to predict phenotypic ratios.

(b)

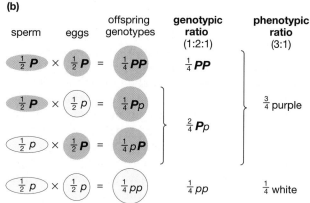

techniques, keep in mind that, in a real experiment, the offspring will occur only in *approximately* the predicted proportions. This inexactness reflects the fact that not every sperm fertilizes an egg, nor is every egg fertilized.

Mendel's Hypothesis Can Be Used to Predict the Outcome of New Types of Single-Trait Crosses

You have probably recognized that Mendel used the scientific method, observing results and formulating a hypothesis based on them. The scientific method has a third

step: to use the hypothesis to predict the results of other experiments and to see if those experiments support or refute the hypothesis. For example, if the hybrid F_1 flowers have one allele for purple and one for white (Pp), then Mendel could predict the outcome of cross-fertilizing these Pp plants with homozygous recessive white plants (pp). Can you? Mendel predicted that there would be equal numbers of Pp (purple) and pp (white) offspring, and this is indeed what he found.

This type of experiment also has practical uses. Cross-fertilization of an individual with an unknown genotype but a dominant phenotype (in this case, a purple flower) with a homozygous recessive individual (a white flower) is called a **test cross**, because it can be used to test whether the unknown genotype is homozygous or heterozygous (in this case, for purple color). When crossed with a homozygous recessive (pp), a homozygous dominant (PP) produces all phenotypically dominant offspring, whereas a heterozygous dominant (Pp) yields offspring with both dominant and recessive phenotypes in a 1:1 ratio:

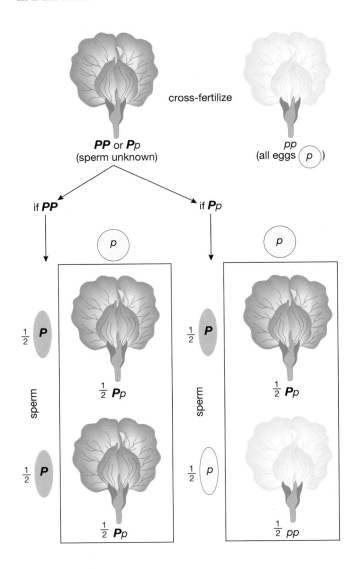

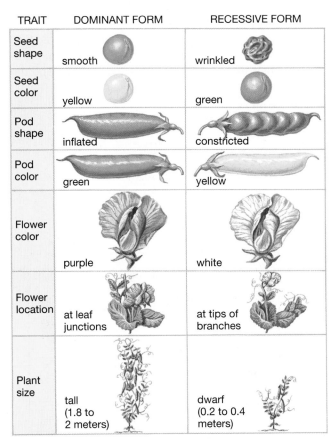

Figure 12-5 Traits of pea plants that Mendel studied

3 How Are Multiple Traits on Different Chromosomes Inherited?

www

Mendel Hypothesized That Genes on Different Chromosomes Are Inherited Independently

Having determined the mode of inheritance of single traits, Mendel then turned to the more complex question of multiple traits by using a variety of traits in peas (Fig. 12-5). He began by cross-breeding plants that differed in two traits—for example, seed color (yellow or green) and seed shape (smooth or wrinkled). If he crossed a plant that was homozygous for smooth yellow seeds with one that was homozygous for wrinkled green seeds, all the F₁ offspring bore smooth yellow seeds. This was no surprise, because from separate crosses of each of these traits, Mendel already knew that smooth (*S*) is dominant to wrinkled (*s*) and that yellow (*Y*) is dominant to green (*y*) (see Fig. 12-4).

All the F₁ offspring, therefore, are genotypically *SsYy*. Allowing these F₁ plants to self-fertilize, Mendel found that the F₂ generation consisted of 315 smooth yellow seeds, 101 wrinkled yellow seeds, 108 smooth green seeds, and 32 wrinkled green seeds, in a ratio of about 9:3:3:1. The F₂ generations produced from other crosses of gametes that were heterozygous for two traits had similar phenotypic ratios.

Mendel realized that these results could be explained if the genes for seed color and seed shape are inherited independently of each other and do not influence each other during gamete formation. Thus, a 3:1 ratio of offspring would be expected for each trait. The laws of probability state that the independent combination of two 3:1 ratios yields a 9:3:3:1 ratio, and we can see from Mendel's results that this is just what happened. There were 423 smooth seeds (of either color) to 133 wrinkled ones (about 3:1) and 416 yellow seeds (of either shape) to 140 green ones (about 3:1). Figure 12-6 shows how a Punnett square can be used to determine the outcome of a cross between gametes that are heterozygous for two traits and how two independent 3:1 ratios combine to form an overall 9:3:3:1 ratio.

The independent inheritance of two or more distinct traits is called the **law of independent assortment**. It states that the alleles for one trait may be distributed to the gametes independently of the alleles for other traits. Recall the events of meiosis in Chapter 11; how might independent assortment of alleles for two different genes occur? Independent assortment will occur if the gene loci are on different chromosomes. When paired homologous chromosomes line up during metaphase I, which homologue faces which pole of the cell is random, and the

(a)

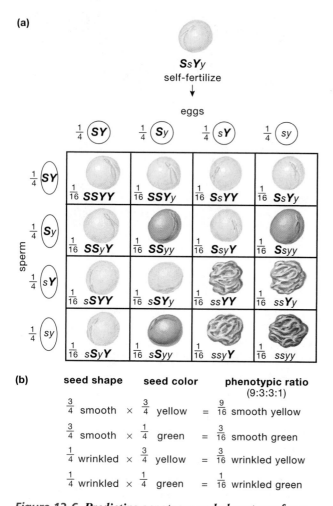

(b)

seed shape		seed color		phenotypic ratio (9:3:3:1)
$\frac{3}{4}$ smooth	×	$\frac{3}{4}$ yellow	=	$\frac{9}{16}$ smooth yellow
$\frac{3}{4}$ smooth	×	$\frac{1}{4}$ green	=	$\frac{3}{16}$ smooth green
$\frac{1}{4}$ wrinkled	×	$\frac{3}{4}$ yellow	=	$\frac{3}{16}$ wrinkled yellow
$\frac{1}{4}$ wrinkled	×	$\frac{1}{4}$ green	=	$\frac{1}{16}$ wrinkled green

Figure 12-6 Predicting genotypes and phenotypes for a cross between gametes that are heterozygous for two traits Here we are working with both seed color and shape, with yellow (Y) dominant to green (y), and smooth (S) dominant to wrinkled (s). *(a)* Punnett square analysis. Assume that both parents are heterozygous for each trait or that a single individual heterozygous for both traits self-fertilizes. There are now 16 boxes in the Punnett square. In addition to predicting all the genotypic combinations, the Punnett square predicts ¾ yellow seeds, ¼ green seeds, ¾ smooth seeds, and ¼ wrinkled seeds, just as we would expect from crosses made of each trait separately. *(b)* Probability theory can be used to predict phenotypes that result from a cross between gametes that are heterozygous for two traits. The fraction of genotypes from each sperm and egg combination is illustrated within each box of the Punnett square. Adding the fractions for the same genotypes will give the genotypic ratios. Converting each genotype to a phenotype and then adding their numbers reveals that ¾ of the offspring will be smooth and ¼ will be wrinkled and that ¾ will be yellow and ¼ will be green. Multiplying these independent probabilities produces predictions for the phenotype of offspring. These ratios are identical to those generated by the Punnett square.

orientation of one homologous pair does not influence other pairs. Therefore, when the homologues separate during anaphase I, the alleles of genes on different chromosomes are distributed (or "assorted") independently (Fig. 12-7). Amazingly, each of the several traits Mendel had chosen to study happened to be controlled by a single gene located on a different chromosome.

In an Unprepared World, Genius May Go Unrecognized

In 1865, Gregor Mendel presented his theories of inheritance to the Brünn Society for the Study of Natural Science, and they were published the following year. His paper did *not* mark the beginning of genetics. In fact, it didn't make any impression at all on the biology of his time. Mendel's experiments, which eventually spawned one of the most important scientific theories in all of biology, simply vanished from the scene. Apparently, very few biologists read his paper, and those who did failed to recognize its significance or discounted it because it contradicted prevailing ideas of inheritance.

It was not until 1900 that three biologists—Carl Correns, Hugo de Vries, and Erich Tschermak—working independently and knowing nothing of Mendel's work, rediscovered the principles of inheritance. No doubt to their intense disappointment, when they searched the scientific literature before publishing their results, they found that Mendel had scooped them more than 30 years earlier. To their credit, they graciously acknowledged the important work of the Augustinian monk, who had died in 1884.

4 How Are Genes Located on the Same Chromosome Inherited?

Gregor Mendel knew nothing about the physical nature of genes or chromosomes. Once scientists had seen chromosomes by microscope and figured out how they are distributed during mitosis and meiosis, it became obvious that chromosomes are the vehicles of inheritance. It also became obvious that there are many more traits (and therefore many more genes) than there are chromosomes. That genes are parts of chromosomes and that each chromosome bears many genes have important implications for inheritance.

Genes on the Same Chromosome Tend to Be Inherited Together

If chromosomes assort independently during meiosis I, then only genes located on *different chromosomes* will assort independently into gametes. Genes on the *same chromosome* are said to be *linked*. **Linkage** is the inheritance of certain genes as a group because they are on the same chromosome. Linked genes tend to be inherited together and do not assort independently. One of the first

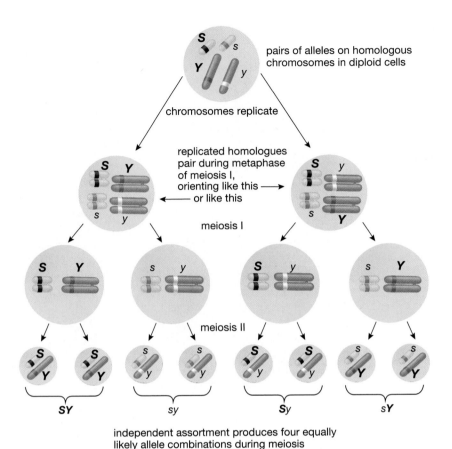

independent assortment produces four equally likely allele combinations during meiosis

Figure 12-7 Independent assortment of alleles
Chromosome movements during meiosis produce independent assortment of alleles of two different genes. Because meiosis occurs in many reproductive cells in a plant, each combination is equally likely to occur. Therefore, an F_1 plant would produce gametes in the predicted proportions ¼ *SY*, ¼ *sy*, ¼ *sY*, and ¼ *Sy*.

pairs of linked genes to be discovered was the gene for flower color (red vs. purple) and the gene for pollen grain shape (long vs. round) in the sweet pea, a type different from the edible pea. Both of these genes are carried on the same chromosome, so they normally assort together during meiosis and are inherited together.

Crossing Over May Separate Linked Genes

There is one thing wrong with this tidy scheme: Genes on the same chromosome do not always stay together. In the sweet pea cross just described, for example, the F_2 generation will commonly include a few plants in which the genes for flower color and pollen shape have been inherited as if they were not linked. How can this be?

As you learned in Chapter 11, during prophase I of meiosis, the nonsister chromatids of homologous chromosomes intertwine at sites called chiasmata. At chiasmata, segments of homologous chromosomes are exchanged with each other, a process called **crossing over**. This is a common event; there is usually at least one exchange between homologues during any meiosis. The exchange of corresponding segments of DNA during crossing over forms new gene combinations on both homologous chromosomes. Then, when homologous chromosomes separate at anaphase I, the chromosomes that each haploid daughter cell receives will be different from those of the parent cell.

Crossing over during meiosis explains the appearance of new combinations of genes that were previously

linked. In the case of the sweet pea, let's assume that a pea had this pair of homologous chromosomes:

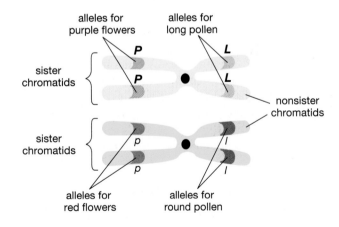

Assume that in a few reproductive cells, crossing over occurs at a point between the genes for flower color and pollen shape. Thus, nonsister chromatids of homologous chromosomes exchange alleles for flower color:

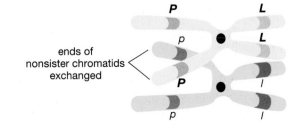

At anaphase I, the separated homologous chromosomes will have this gene composition:

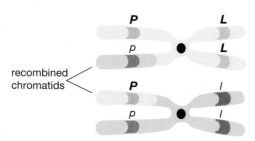

recombined chromatids

Four types of chromosomes are then distributed to the haploid daughter cells during meiosis II:

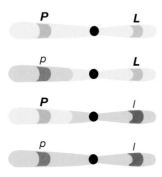

Therefore, some gametes will be produced with each of four chromosome configurations: *PL* and *pl* (the original parental types), and *Pl* and *pL* (recombined chromosomes).

This **genetic recombination** is the generation of new combinations of alleles by the exchange of DNA between homologous chromosomes during crossing over. In most organisms, offspring receive one homologous chromosome from each of two genetically different parents. Recall that a human couple can theoretically generate 64 trillion different combinations of chromosomes when sperm and egg combine. This phenomenal number does not even take genetic recombination into account. Together, genetic recombination and fertilization provide enormous genetic variability among organisms of the same species.

5) How Is Sex Determined, and How Are Sex-Linked Genes Inherited?

www

In mammals and many insects, males have the same number of chromosomes as females do, but one "pair," the **sex chromosomes**, are very different in appearance and genetic composition. Females have two identical sex chromosomes, called *X chromosomes*, whereas males have one X chromosome and one *Y chromosome* (Fig. 12-8). Although the Y chromosome normally carries far fewer genes than does the X chromosome, a small part of both sex chromosomes is homologous, so X and Y chromosomes pair up during prophase of meiosis I and separate during anaphase I. All other chromo-

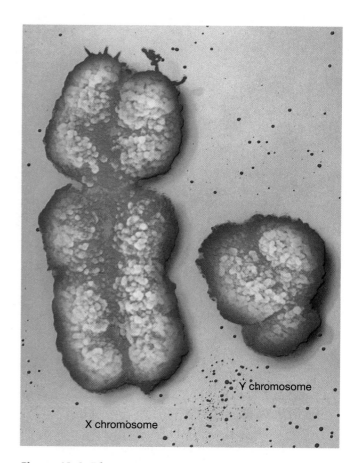

Figure 12-8 Photomicrograph of human sex chromosomes Notice the small size of the Y chromosome, which carries relatively few genes.

somes, which occur in pairs of identical appearance in both males and females, are called **autosomes**. Numbers of chromosomes vary tremendously among species, but there is always only one pair of sex chromosomes. For example, the fruit fly *Drosophila* has four pairs of chromosomes (three pairs of autosomes), humans have 23 pairs (22 pairs of autosomes), and dogs have 39 pairs (38 pairs of autosomes).

For organisms in which males are XY and females are XX, the sex chromosome carried by the sperm determines the sex of the offspring (Fig. 12-9). During sperm formation, the sex chromosomes segregate, and each sperm receives either the X or the Y chromosome (plus one member of each pair of autosomes). The sex chromosomes also segregate during egg formation, but because females have two X chromosomes, every egg receives one X chromosome. An offspring is male if an egg is fertilized by a Y-bearing sperm or female if an egg is fertilized by an X-bearing sperm.

Sex-Linked Genes Are Found Only on the X or Only on the Y Chromosome

Genes that are on one sex chromosome but not on the other are said to be **sex-linked**, or *X-linked*. In many animals, the Y chromosome carries relatively few genes other

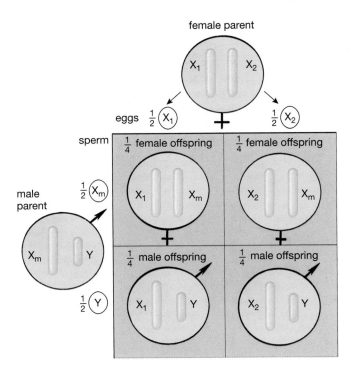

female parent

eggs

sperm

male parent

Figure 12-9 Sex determination
Male offspring receive the Y chromosome from the father; female offspring receive the father's X chromosome (labeled X_m). The mother passes one of her X chromosomes (X_1 or X_2) to both male and female offspring.

than those that determine maleness, whereas the X chromosome bears many genes that have nothing to do with specifically female traits. The human X chromosome, for example, contains genes for color vision, blood clotting, and certain structural proteins in muscles that have no counterpart on the Y chromosome. Therefore, because they have two X chromosomes, females can be either homozygous or heterozygous for genes on the X chromosome. Normal dominant versus recessive relationships among alleles will be expressed. Males, in contrast, must fully express all alleles they have on their single X chromosome, whether those alleles are dominant or recessive. For this reason, in humans, most cases of color blindness, hemophilia, and certain types of muscular dystrophy occur in males. We will return to this concept later in the chapter.

How does sex linkage affect inheritance? Let's look at the first example of sex linkage to be discovered, the inheritance of eye color in the fruit fly *Drosophila*. Because these flies are small, have a rapid reproductive rate, are easy to grow in the laboratory, and have few chromosomes, *Drosophila* have been favored subjects for genetics studies for more than a century. Normally, *Drosophila* have red eyes. In the early 1900s, researchers in the laboratory of Thomas Hunt Morgan at Columbia University discovered a male fly with white eyes. This white-eyed male was mated to a virgin, true-breeding, red-eyed female. All the resulting offspring were red-eyed flies, suggesting that white eye color (r) is recessive to red (R). The F$_2$ generation, however, was a surprise:

There were nearly equal numbers of red-eyed males and white-eyed males but no females with white eyes! A test cross of the F$_1$ red-eyed females and the original white-eyed male yielded roughly equal numbers of red-eyed and white-eyed males and females.

From these data, could you figure out how eye color is inherited? Morgan made the brilliant hypothesis that *the gene for eye color must be located on the X chromosome and that the Y chromosome has no corresponding gene* (Fig. 12-10a). In the F$_1$ generation, both male and female offspring received an X chromosome, with its R allele for red eyes, from their mother. The F$_1$ males received a Y chromosome from their father with no allele for eye color, and so the males had a -R genotype. (Here "-" indicates no gene for eye color on the Y chromosome.) The F$_1$ females received the father's X chromosome with its r allele, so the females had an Rr genotype. Thus, all male and female F$_1$ offspring had red eyes (Fig. 12-10b).

Crossing two F$_1$ flies resulted in an F$_2$ generation with the chromosome distribution shown in Figure 12-10b. All the F$_2$ females received one X chromosome from their F$_1$ male parent, with its R allele, and therefore had red eyes. All the F$_2$ males inherited their single X chromosome from their F$_1$ female parent, heterozygous for eye color (Rr). So the F$_2$ males had a 50:50 chance of receiving an X chromosome with the R allele or one with the r allele. *With no corresponding gene on the Y chromosome, the F$_2$ males displayed the phenotype determined by the allele on the X chromosome.* Therefore, half the F$_2$ males had red eyes, and half had white eyes.

Before you read on, consider this question: How much crossing over would you expect between X and Y chromosomes? If you concluded "very little" or "none," you're right. Most of the X and Y chromosomes are not homologous, so the normal mechanisms of crossing over do not apply.

6 What Are Some Variations on the Mendelian Theme?

In our discussion of patterns of inheritance thus far, we have made some major simplifying assumptions: that each trait is completely controlled by a single gene, that there are only two possible alleles of each gene, and that one allele is completely dominant to the other, recessive, allele. Most traits, however, are influenced in more varied and subtle ways than this.

In Incomplete Dominance, the Phenotype of Heterozygotes Is Intermediate between the Phenotypes of the Homozygotes

In his pea experiments, Mendel encountered a particularly simple situation—heterozygotes and homozygous dominants had the same phenotype—but this is commonly not the case. In snapdragons, for example, crossing homozygous red-flowered plants (RR) with homozygous white-flowered ones ($R'R'$) does not produce red-

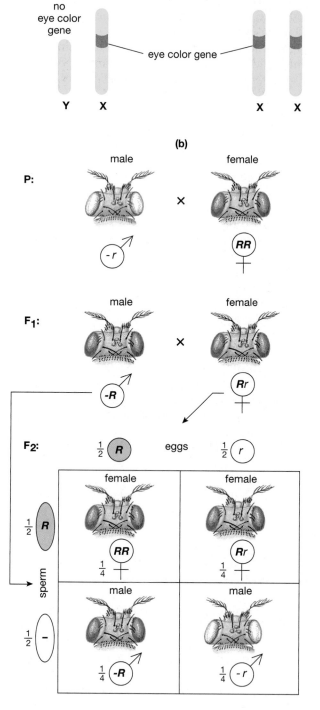

flowered F_1 hybrids. Instead, the F_1 flowers are pink. When the heterozygous phenotype is intermediate between the two homozygous phenotypes, the pattern of inheritance is called **incomplete dominance**. This apparent blending of the phenotype is not, however, the result of any change in alleles. In homozygous flowers of the F_2 generation, the red and white colors are as strong as ever (Fig. 12-11). The F_2 offspring include about ¼ red (RR), ½ pink (RR'), and ¼ white ($R'R'$) flowers.

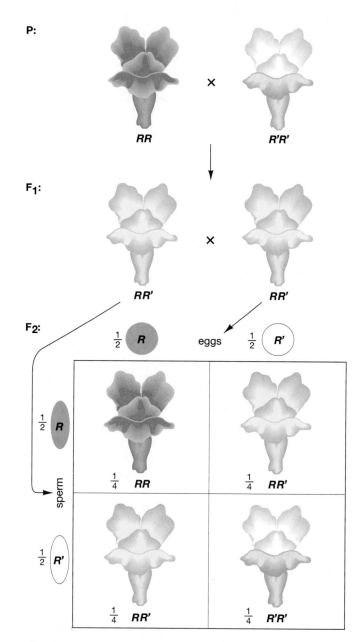

Figure 12-10 Sex-linked inheritance of eye color in fruit flies
(a) The gene for eye color is carried on the X chromosome; there is no corresponding gene on the Y chromosome.
(b) Normal red eyes (R) is dominant to the mutant allele for white eyes (r). When a white-eyed male is mated to a homozygous red-eyed female, all members of the F_1 generation have red eyes. All females are heterozygous, receiving the r allele from their father and the R allele from their mother. Male offspring receive only the R allele from their mother (the "-" indicates the lack of any allele for eye color on the Y chromosome). The genotypes and phenotypes of the F_2 generation are calculated by using a Punnett square.

Figure 12-11 Incomplete dominance
The inheritance of flower color in snapdragons is an example of incomplete dominance. (In such cases, we will use capital letters for both alleles, here R and R'.) Hybrids (RR') have pink flowers, whereas the homozygotes are red (RR) or white ($R'R'$). Because heterozygotes can be distinguished from homozygous dominants, the distribution of phenotypes in the F_2 generation (¼ red: ½ pink: ¼ white) is the same as the distribution of genotypes (¼ RR: ½ RR': ¼ $R'R'$).

What causes incomplete dominance? It occurs when *both* copies of a functional allele are necessary in order to produce enough protein to give rise to the dominant phenotype. In snapdragons, for example, the *R* allele codes for an enzyme that catalyzes the formation of red pigment; the *R′* allele codes for a defective enzyme. When only one copy of the *R* allele is present, less red pigment is produced and the resulting flower appears pink. Plants with the *RR* genotype produce lots of red pigment and appear red, but those homozygous for the nonfunctional enzyme (*R′R′*) are white.

A Single Gene May Have Multiple Alleles

Alleles arise through mutation, and the same gene in different individuals may have different mutations, each producing a new allele. If we could sample all the individuals of a species, we would typically find **multiple alleles**—as many as dozens of alleles for every gene. The gene for eye color in fruit flies, for example, has many alleles, each recessive to normal red eyes and producing various shades of white, yellow, or pink when homozygous. Remember, however, that an *individual* diploid organism can have at most only two different alleles for a given gene.

The blood types of humans are another example of multiple alleles. The blood types A, B, AB, and O arise as a result of three different alleles (for simplicity, we will designate them *A, B,* and *o*) of a single gene located on chromosome 9. This gene codes for an enzyme responsible for adding sugar molecules to the ends of glycoproteins that protrude from the surfaces of red blood cells. Alleles *A* and *B* code for enzymes that add different sugars to the glycoproteins (we will call the

resulting molecules glycoproteins *A* and *B*, respectively). Allele *o* codes for a nonfunctional enzyme that doesn't add any sugar molecule. An individual may have one of six genotypes: *AA, BB, AB, Ao, Bo,* or *oo* (Table 12-1). Alleles *A* and *B* are dominant to *o*. Therefore, individuals with genotypes *AA* or *Ao* have type A glycoproteins and have type A blood. Those with genotypes *BB* or *Bo* synthesize type B glycoproteins and have type B blood. Homozygous recessive *oo* individuals lack either glycoprotein and have type O blood. Alleles *A* and *B* are said to be *codominant* to one another. In **codominance**, *both phenotypes are expressed in heterozygotes*. In individuals with type AB blood, both enzymes are present, and both A and B glycoproteins are present on red blood cells.

People with any of the blood groups make antibodies to the type of glycoprotein(s) that they lack. These antibodies are proteins in blood plasma that bind foreign glycoproteins by recognizing different end-sugar molecules. The antibodies cause red blood cells that bear the foreign glycoproteins to clump together and also to rupture. The resulting clumps and fragments can clog small blood vessels and damage vital organs such as the brain, heart, lungs, or kidneys. This means that blood type must be determined and matched carefully before a blood transfusion is made. Type O blood, lacking any end sugars, is not attacked by antibodies in A, B, or AB blood, so it can be transfused safely to all other blood types. (The antibodies present in transfused blood become too diluted to cause problems.) People with type O blood are called "universal donors." But O blood carries antibodies to both A and B glycoproteins, so type O individuals can receive transfusions of only type O blood. Can you pre-

Table 12-1 Human Blood Group Characteristics

Genotype	Blood Type and Frequency in U.S.	Red Blood Cells	Plasma Antibodies	Can Receive From:	Can Donate To:
AA *Ao*	A 40%	A glycoprotein	anti-B	A O	A AB
BB *Bo*	B 10%	B glycoprotein	anti-A	B O	B AB
AB (universal recipient)	AB 4%	Both A and B glycoproteins	none	A B AB O	AB
oo (universal donor)	O 46%	Neither A nor B glycoprotein	Both anti-A and anti-B	O	A B AB O

dict the blood type of people called "universal recipients"? Table 12-1 summarizes blood types and transfusion characteristics.

Many Traits Are Influenced by Several Genes

If you look around your class, you are likely to see people of varied heights, skin colors, and body builds, to consider just a few obvious traits. Traits such as these are not governed by single genes but are influenced by interactions among two or more genes—as well as by interactions with the environment. Many traits, such as human skin color, may have several phenotypes or even seemingly continuous variation that cannot be split up into convenient, easily defined categories. This is an example of **polygenic inheritance**, a form of inheritance in which the interaction of two or more functionally similar genes contribute to a single phenotype.

Eye color is another good example of polygenic inheritance. The color of the human iris varies from very pale blue through green to almost black, depending on the amount of pigment, yellowish-brown melanin, in the outer layer of the iris. If there is little or no pigment here, the iris appears blue, owing to the same type of light scattering that makes the sky appear blue. People with more melanin in the outer iris may have eyes that appear green (blue plus yellowish-brown), brown, or almost black. At least two (and probably more) genes direct the synthesis of melanin in the front of the iris, with each gene having two alleles that show incomplete dominance. The simplest scheme of two genes can create five shades of eye colors from one set of parents (Fig. 12-12).

As you might imagine, the more genes that contribute to a single trait, the greater the number of phenotypes and the finer the distinctions among them. When more than three pairs of genes contribute to a trait, differences between phenotypes are small, and it is extremely difficult to classify the phenotypes reliably. Skin color is believed to be controlled by three or four genes.

Single Genes Typically Have Multiple Effects on Phenotype

We have just seen that a single phenotype may result from the interaction of several genes. The reverse is also true: Single genes commonly have multiple phenotypic effects, a phenomenon called **pleiotropy**. A good example is the SRY gene, discovered in 1990 on the Y chromosome. The

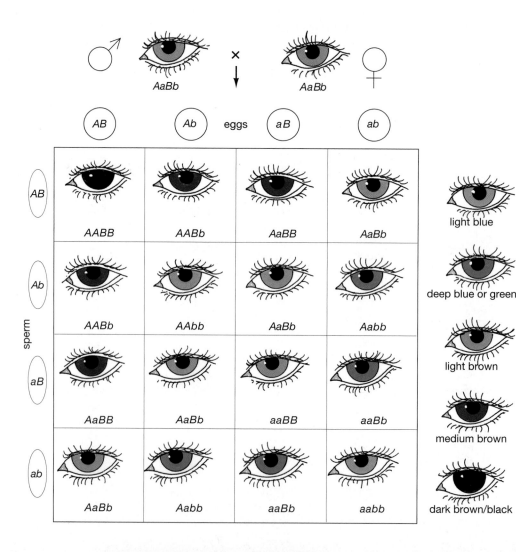

Figure 12-12 Human eye color
At least two separate genes, each with two incompletely dominant alleles, govern human eye color. A man and a woman, each heterozygous for both genes, could have children with five different eye colors, ranging from light blue (no dominant alleles) through light brown (two dominants) to almost black (all four alleles dominant).

SRY gene (short for "sex-determining region of the Y chromosome") codes for a protein that activates other genes; those genes in turn code for proteins that switch on male development in an embryo. Under the influence of the genes activated by the SRY protein, sex organs develop into testes. The testes, in turn, secrete sex hormones that stimulate the development of both internal and external male reproductive structures, such as the epididymides, seminal vesicles, prostate gland, penis, and scrotum.

Pleiotropy is a widespread phenomenon. As you have learned, many of the proteins coded by genes are enzymes. A single enzyme might participate in the production of several important substances, each of which may have multiple effects on the phenotype. For example, one enzyme converts the amino acid tyrosine into DOPA, a molecule that other enzymes act on to produce (1) dopamine (a neural signaling molecule important in the central nervous system control of movement); (2) the hormones norepinephrine and epinephrine (also called adrenalin), both of which participate in your responses to stress or danger; and (3) the pigment melanin, which gives skin, hair, and eyes their color.

The Environment Influences the Expression of Genes

An organism is not just the sum of its genes. In addition to the genotype, the environment in which an organism lives profoundly affects its phenotype. A striking example of environmental effects on gene action occurs in the Himalayan rabbit, which, like the Siamese cat, has pale body fur but black ears, nose, tail, and feet (Fig. 12-13). The Himalayan rabbit actually has the genotype for black fur all over its body. The enzyme that produces the black pigment, however, is temperature sensitive; above about

34°C (93°F), the enzyme is inactive. At typical ambient temperatures, extremities such as the ears and feet are cooler than the rest of the body, so black pigment can be produced there. The main body surface is warmer than 34°C, so the fur there is pale.

Most environmental influences are more complicated and subtle. This is particularly true of human characteristics. The polygenic trait of skin color is modified by the environmental effects of sun exposure, which stimulates melanin production. Height, another polygenic trait, can be reduced by poor nutrition, an environmental factor.

Intelligence, too, has both genetic and environmental components. Dozens of studies have compared IQ levels in people of varying degrees of relatedness. Even when separated at birth and raised in different environments, identical twins make similar scores on IQ tests, although not as similar as identical twins who have been raised together. Brothers and sisters who are not twins differ more than do twins but are still fairly similar. Thus, the more genetically related two people are, the more similar their IQ scores. These findings indicate that intelligence (or whatever abilities IQ tests measure) has a genetic component. However, unrelated people who have been raised together as children (for example, adoptees) show more similarity on IQ tests than do unrelated people reared apart. Therefore, it is fair to say that *both heredity and environment play major roles in the development of intelligence* and almost certainly other personality traits as well.

The interactions between complex genetic systems and varied environmental conditions can create a continuum of phenotypes that defies analysis into genetic and environmental components. The human generation time is long, and the number of offspring per couple is small. Add to these factors the countless subtle ways in which people respond to their environments, and you can see why it may be impossible to determine precisely the genetic basis of complex human traits such as intelligence or musical or athletic ability.

7) How Are Human Genetic Disorders Investigated?

www

Because experimental crosses with humans are out of the question, human geneticists search medical, historical, and family records to study past crosses. Records extending across several generations can be arranged in the form of family **pedigrees**, diagrams that show the genetic relationships among a set of individuals (Fig. 12-14). Careful analysis of pedigrees shows that certain traits, such as an unattached earlobe, are inherited as simple dominants, whereas others, such as albinism, are inherited as recessives, and still others, such as color blindness, are sex-linked.

Since the mid-1960s, great strides have been made in understanding gene function on the molecular level. For

Figure 12-13 Environmental influence on phenotype
The expression of the gene for black fur in the Himalayan rabbit is a simple case of interaction between genotype and environment in producing a particular phenotype. Cool areas (nose, ears, and feet) allow expression of the gene for black fur.

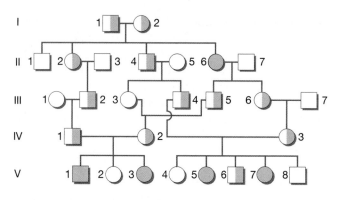

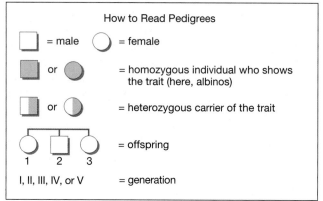

Figure 12-14 A family pedigree
This pedigree is for a recessive trait, such as albinism. Both of the original parents are carriers. Because the allele for albinism is rare, pairing between carriers is an unlikely event. However, the chance that each of two *related* people will carry a rare recessive allele (inherited from a common ancestor) is much higher than normal. As a result, pairings between cousins or even closer relations are the cause of a disproportionate number of recessive diseases. In this family, pairings between cousins were common—between III 3 and III 5, III 4 and IV 3, and IV 1 and IV 2.

instance, geneticists now know the genes responsible for several dozen inherited diseases, such as sickle-cell anemia, prostate cancer, and cystic fibrosis. Research in molecular genetics promises to increase our ability to predict genetic diseases and perhaps even to cure them, a topic we explore further in Chapter 13.

 8 **How Are Human Disorders Caused by Single Genes Inherited?**

Many common human traits, such as freckles, long eyelashes, cleft chin, and the ability to roll one's tongue, are inherited in a simple Mendelian fashion; that is, each appears to be controlled by a single gene with a dominant and a recessive allele. Here, we will concentrate on a few examples of medically important genetic disorders and the ways in which they are transmitted from one generation to the next.

Most Human Genetic Disorders Are Caused by Recessive Alleles

The human body depends on the integrated actions of thousands of enzymes and other proteins. A mutation in the gene coding for one of these enzymes almost always impairs or destroys enzyme function. However, the presence of one normal allele may generate enough functional enzyme or other protein that heterozygotes with a defective allele are phenotypically indistinguishable from homozygous normals. Therefore, most normal alleles are inherited as dominant traits and mutant alleles as recessive traits. In other words, both alleles must be defective for the problem to show up in the offspring.

Heterozygous **carriers** are phenotypically normal but can pass on their defective recessive allele to their offspring. Although many people are carriers of a serious genetic defect, an unrelated man and woman who marry are unlikely to possess the *same* defective allele and to produce a homozygous child with the disorder. Related couples, however (especially first cousins or closer), have inherited some of their genes from recent common ancestors. Therefore, they are much more likely to carry the same defective allele and, if they bear children, to pass on the defect (see Fig. 12-14).

Albinism Results from a Defect in Melanin Production
If both homologous chromosomes carry a mutation that interferes with melanin production, albinism results (Fig. 12-15). *Albinism* in humans and other mammals is manifested as white skin and hair and pink eyes (because blood vessels in the retina are visible in the absence of masking melanin pigment).

Sickle-Cell Anemia Is Caused by a Defective Allele for Hemoglobin Synthesis
The multiple disabilities caused by **sickle-cell anemia**, a recessive disease in which defective hemoglobin is produced, have been traced to a mutation in a single gene. The protein hemoglobin, which gives red blood cells their color, transports oxygen in the blood. In sickle-cell anemia, the substitution of one nucleotide results in a single incorrect amino acid at a crucial position in hemoglobin, altering the properties of the hemoglobin molecule. Under conditions of low oxygen (such as in muscles during exercise), masses of hemoglobin molecules in each red blood cell clump together. The clumps force the red blood cell out of its normal disk shape (Fig. 12-16a) into a longer, sickle shape (Fig. 12-16b). The sickled cells are more fragile than normal red blood cells, making them likely to break; they also tend to aggregate, clogging capillaries. Tissues "downstream" do not receive enough oxygen or have their wastes removed. This can cause pain, especially in joints. Paralyzing strokes can occur if blood vessels in the brain are blocked. Other symptoms include anemia, because so many red blood cells are destroyed, and reduced immunity to disease. Although heterozygotes have about half normal and half abnormal hemoglobin, they usually have

(a)   (b)   (c)

Figure 12-15 Albinism
Albinism is controlled by a single, recessive allele. Melanin is found throughout the animal kingdom, and albinos of many species have been observed: *(a)* human, *(b)* rattlesnake, and *(c)* wallaby. The female wallaby, having mated with a normally pigmented male, carries a normal offspring in her pouch.

few sickled cells and are not disabled by the disease—in fact, many world-class athletes are heterozygotes for the sickle-cell allele.

About 7% of the African-American population is heterozygous for sickle-cell anemia, reflecting a genetic legacy of African origins. In some regions of Africa, 15% to 20% of the population carries the allele. The prevalence of the sickle-cell allele in Africa is explained by the fact

that heterozygotes have some resistance to the parasite that causes malaria. We will explore this benefit further in Chapter 13.

Heterozygous carriers of the sickle-cell allele can be detected by a blood test, but until fairly recently there was no way to learn whether a fetus is homozygous or heterozygous. Recombinant DNA techniques have been devised that can distinguish chromosomes with the nor-

(a)   (b)

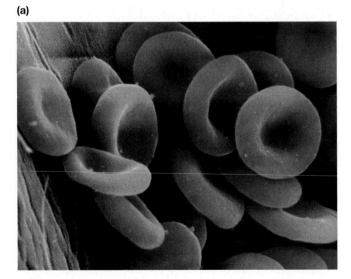
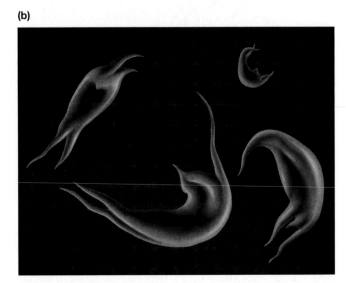

Figure 12-16 Sickle-cell anemia
(a) Normal red blood cells are disc-shaped with indented centers. *(b)* Sickled red blood cells in a person with sickle-cell anemia occur when blood oxygen is low. In this shape they are fragile and tend to clump together, clogging capillaries.

mal hemoglobin allele from those with the sickle-cell allele. Analysis of fetal cells now allows medical geneticists to diagnose sickle-cell anemia in fetuses (see "Health Watch: Prenatal Genetic Screening" in Chapter 13).

A Few Human Genetic Disorders Are Caused by Dominant Alleles

Many normal physical traits, including cleft chin and freckles, are inherited as dominants. However, few people have serious genetic diseases caused by dominant alleles, because everyone bearing a dominant defective allele will develop the disease. Before the advent of modern medicine, if the disease was serious, these people died without reproducing and did not pass on their defective allele to future generations.

An exception to this rule is **Huntington's disease**. An incurable disease, Huntington's causes a slow, progressive deterioration of parts of the brain, resulting in the loss of motor coordination, flailing movements, personality disturbances, and eventual death. Huntington's disease is particularly insidious because symptoms typically do not appear until 30 to 50 years of age. Therefore, many people pass the allele to their children before they suffer the first symptoms. In 1983 both molecular genetic technology and painstaking pedigree analysis were used to localize the Huntington's gene to a small part of chromosome 4. (See "Scientific Inquiry: In Search of the Huntington's Gene," p. 220) Geneticists finally isolated the Huntington's gene in 1993 and a few years later identified the gene's product, a protein dubbed "huntingtin."

The next steps—determining the function of the normal protein and how the improperly constructed protein damages the brain—are now under way.

Some Human Disorders Are Sex-Linked

As we described earlier, the X chromosome bears many genes that have no counterpart on the Y chromosome. Because males have only one X chromosome, all the alleles on the X chromosome that have no Y counterpart are expressed, a phenomenon called *sex-linked inheritance*.

A son receives his X chromosome from his mother and can pass it only to his daughters. Thus, sex-linked disorders caused by a recessive allele have a unique pattern of inheritance. Such disorders appear far more frequently in males and typically skip generations; an affected male passes the trait on to a phenotypically normal, carrier daughter, who in turn bears affected sons. The most familiar genetic defects due to recessive alleles of X-chromosome genes are *red-green color blindness* (Fig. 12-17) and **hemophilia** (Fig. 12-18). Hemophilia is caused by a recessive allele on the X chromosome that results in a deficiency of one of the substances important in blood clotting. People with hemophilia bleed excessively from a wound or from mild damage to internal structures and are subject to excessive bruising. Hemophiliacs often have anemia due to blood loss. But even before modern treatment with clotting factor derived from donor blood, some hemophiliac males survived to pass on their defective allele to their daughters, who carried the allele and could pass it to their sons.

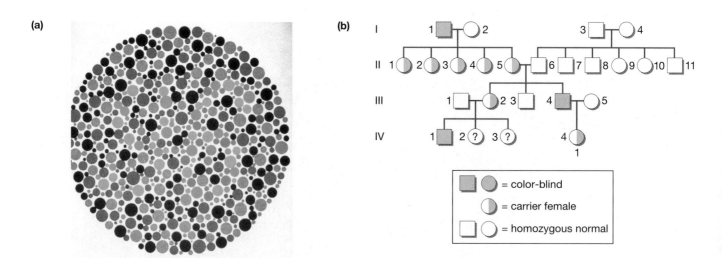

Figure 12-17 Color blindness, a sex-linked recessive trait
(a) This figure, called an Ishihara chart after its inventor, distinguishes color-vision defects. People with red-deficient vision see a 6, and those with green-deficient vision see a 9. People with normal color vision see 96. *(b)* Pedigree of one of the authors (G. Audesirk: III 4), showing sex-linked inheritance of red-green color blindness. Both the author and his maternal grandfather (I 1) are color deficient; his mother and her four sisters carry the trait but have normal color vision. This pattern of more-common occurrence in males and of transmission from affected male to carrier female to affected male is typical of sex-linked recessive traits.

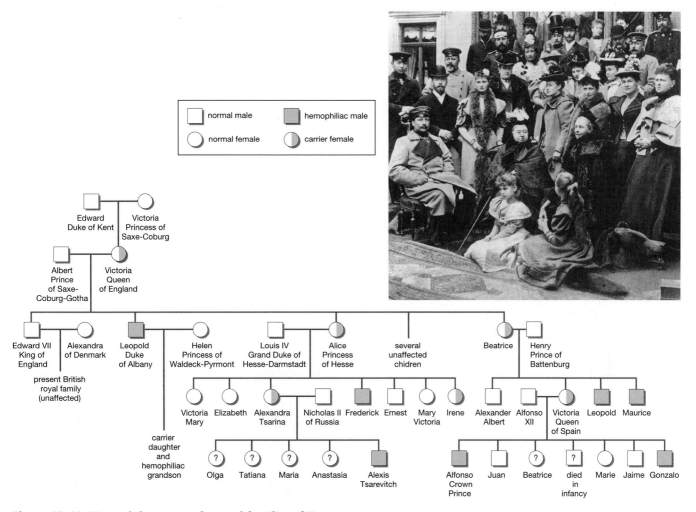

Figure 12-18 *Hemophilia among the royal families of Europe*
The most famous genetic pedigree involves the transmission of sex-linked hemophilia from Queen Victoria of England (seated center front, with cane, 1885) to her offspring and eventually to virtually every royal house in Europe. Because Victoria's ancestors were free of hemophilia, the hemophilia allele must have arisen as a mutation either in Victoria herself or in one of her parents. Extensive intermarriage among royalty spread Victoria's hemophilia allele throughout Europe. Her most famous hemophiliac descendant was great-grandson Alexis, Tsarevitch (crown prince) of Russia. The Tsarina Alexandra (Victoria's granddaughter) believed that the monk Rasputin, and no one else, could control Alexis' bleeding. Rasputin may actually have used hypnosis to cause Alexis to cut off circulation to bleeding areas by muscular contraction. The influence that Rasputin had over the imperial family may have contributed to the downfall of the Tsar during the Russian Revolution. In any event, hemophilia was not the cause of Alexis' death; he was killed with the rest of this family by the Bolsheviks (Communists) in 1918.

 ## How Do Errors in Chromosome Number Affect Humans? *www*

In Chapter 11 we examined the intricate mechanisms of meiosis, which ensure that each sperm and egg receive only one homologous chromosome of each pair. Not surprisingly, this elaborate dance of the chromosomes occasionally misses a step, resulting in gametes that have too many or too few chromosomes. Such errors in meiosis, called **nondisjunction**, can affect the distribution of both sex chromosomes and autosomes. Most embryos that arise from the fusion of gametes with abnormal chromosome numbers spontaneously abort, accounting for 20% to 50% of all miscarriages, but some survive.

Some Genetic Disorders Are Caused by Abnormal Numbers of Sex Chromosomes

In men, nondisjunction of sex chromosomes produces sperm that are O (lacking any sex chromosome), XX,

Scientific Inquiry
In Search of the Huntington's Gene

As a young boy in 1860, George Huntington encountered a mother and daughter, both emaciated and uncontrollably twitching, grimacing, twisting, and bowing. "I stared in wonderment, almost in fear," he later wrote. Huntington went on to study medicine at Columbia University and spent the rest of his life investigating this disease, now called Huntington disease (HD) in his honor. Dr. Huntington showed that HD occurs in about half the children of an affected parent, evidence that it is inherited as a dominant allele of a gene.

More than 100 years later, in the late 1960s, Leonore Wexler was diagnosed with HD. In response, her husband, Milton Wexler, a clinical psychologist, founded the Hereditary Disease Foundation in 1968, hoping to find a cure or an effective treatment for the disease that eventually killed his wife. Nancy Wexler, a young woman of 22, was profoundly affected by her mother's suffering and especially by her mother's progressive loss of mental function. Trained as a clinical psychologist like her father, Dr. Wexler decided to devote her life to finding a cure for HD and now serves as president of the Hereditary Disease Foundation. She taught herself genetics by attending lectures, workshops, and seminars on the subject, by holding many informal discussions with geneticists, and by reading extensively. Dr. Wexler learned that in several villages near Lake Maracaibo in Venezuela, an unusually high percentage of the villagers died of HD. Visiting these villages, she found that about 1 in 5 people there were stricken with the disease (Fig. E12-1). All of them descended from a Portuguese sailor who married a local woman seven generations earlier, in the 1800s.

By 1981, Dr. Wexler began assembling a pedigree showing the relationships of this huge, genetically interconnected Venezuelan population, where families with 13 or more children are common. More than 10,000 individuals have been added to this extensive pedigree. Wexler also collected blood samples from the villagers with HD and shipped the samples to the laboratory of Dr. James Gusella at Massachusetts General Hospital in Boston. In 1993, Gusella's lab, using molecular genetic technology, identified a region of DNA near the tip of human chromosome 4 as the location of the gene for HD. Comparing the genetic information in the Venezuelan blood samples with the relationships depicted in the Venezuelan pedigree and others,

Gusella's group devised a screening test that very accurately determines whether a person has inherited the gene for HD from an affected parent or is free of the "genetic time bomb," as Dr. Wexler describes the Huntington gene.

Ironically, although Nancy Wexler was instrumental in the development of the screening test, she has decided not to be tested to determine whether she has inherited her mother's Huntington gene. As Dr. Wexler explains, each individual must make a deeply personal decision as to whether the pain of knowing the gene is present outweighs the joy of knowing the gene is absent, or whether living with the anxiety of not knowing is preferable. The availability of this test is of special importance to an individual who may have inherited the Huntington's gene but wants to have children. Genetic screening coupled with the voluntary decision not to risk passing the gene to the next generation could rapidly eliminate this tragic condition.

Figure E12-1 Nancy Wexler tracks the gene for Huntington's disease
Nancy Wexler, right, with one of the many inhabitants of the Lake Maracaibo region of Venezuela who have Huntington's disease.

YY, or XY instead of the normal X or Y. In women, nondisjunction produces O or XX eggs instead of the normal X. When normal gametes fuse with these defective sperm or eggs, the zygotes have abnormal numbers of sex chromosomes (Table 12-2). The most common abnormalities are XO, XXX, XXY, and XYY. (Genes on the X chromosome are essential to survival, and all embryos with no X chromosome spontaneously abort very early in development.)

Turner Syndrome (XO)

About one in every 5000 phenotypically female babies has only one X chromosome. At puberty, an XO female fails

to menstruate or develop normal secondary sexual characteristics, such as breasts. These **Turner syndrome** women are sterile, generally short in stature, and commonly have folds of skin around their necks. Under microscopic examination, their nuclei are shown to lack Barr bodies (see Fig. 10-10). Mentally, they tend to be normal but weak in mathematics and spatial perception. The differences between XO and XX women suggest that the hypothesis of X chromosome inactivation in females (Chapter 10), which states that in a normal female one X chromosome is condensed and "inactivated," is oversimplified. Some genes on the inactivated X chromosome must be functional in XX females, preventing the Turner syndrome traits.

Table 12-2 Effects of Nondisjunction of the Sex Chromosomes during Meiosis

Nondisjunction in Father

Sex Chromosomes of Defective Sperm	Sex Chromosomes of Normal Egg	Sex Chromosomes of Offspring	Phenotype
O	X	XO	Female—Turner syndrome
XX	X	XXX	Female—Trisomy X
YY	X	XYY	Male—"XYY male"
XY	X	XXY	Male—Klinefelter syndrome

Nondisjunction in Mother

Sex Chromosomes of Normal Sperm	Sex Chromosomes of Defective Egg	Sex Chromosomes of Offspring	Phenotype
X	O	XO	Female—Turner syndrome
Y	O	YO	Dies as embryo
X	XX	XXX	Female—Trisomy X
Y	XX	XXY	Male—Klinefelter syndrome

Trisomy X (XXX)

About one in every 1000 women have three copies (trisomy) of X chromosomes. Most such women have no detectable defects, except for a higher incidence of below-normal intelligence. Unlike women with Turner syndrome, **trisomy X** women are fertile and, interestingly enough, almost always bear normal XX and XY children. Some unknown mechanism must operate during meiosis to prevent the extra X chromosome from being included in the egg.

Klinefelter Syndrome (XXY)

About one male in every 1000 is born with two X chromosomes and one Y chromosome. At puberty, these men show mixed secondary sexual characteristics, including partial breast development, broadening of the hips, and small testes. Men with **Klinefelter syndrome** are sterile but generally not impotent. As is common in people with abnormal chromosome numbers, XXY males have an increased incidence of mental deficiency; in fact, about 1% of all people institutionalized for mental retardation are XXY males.

XYY Males

Another common type of sex chromosome abnormality is XYY, occurring in about one male in every 1000. You might expect that having an extra Y chromosome, which presumably has few genes, would not make very much difference, and this seems to be true in most cases. However, XYY males may have below-average intelligence and above-average height (about two-thirds of XYY males are over 6 feet tall, compared with the average male height of 5 feet 9 inches). There is some debate about whether XYY males are genetically predisposed to violence, because several studies have shown that a higher-than-expected percentage of men in prison are XYY. In several countries men accused of murder have attempted to use their XYY makeup as a defense, like the insanity plea. They were not acquitted; only a tiny percentage of XYY males ever commit any sort of crime, so an extra Y chromosome certainly doesn't force anyone into a life of violence.

Some Genetic Disorders Are Caused by Abnormal Numbers of Autosomes

Nondisjunction of the autosomes may also occur, producing eggs or sperm with a missing autosome or two copies of an autosome. Fusion with a normal gamete (bearing one copy of each autosome) leads to an embryo with either one or three copies of the affected autosome. With only one copy of any of the autosomes, the embryo aborts so early in development that the woman never knows she was pregnant. In most cases the presence of three copies of an autosome (trisomy) also causes spontaneous abortion, but typically later in the pregnancy. In some cases, however, trisomic babies are born, especially those with three copies of chromosomes 13, 18, or 21. Of these, trisomy 21 is the most common.

Trisomy 21, or Down Syndrome

In about one of every 900 births, the child has an extra copy of the 21st chromosome. Children with **trisomy 21**, also called **Down syndrome**, have several distinctive physical characteristics, including weak muscle tone, a small mouth held partially open because it cannot accommodate the tongue, and distinctively shaped eyelids (Fig. 12-19). Much more serious defects include low resistance to infectious diseases, heart malformations, and mental

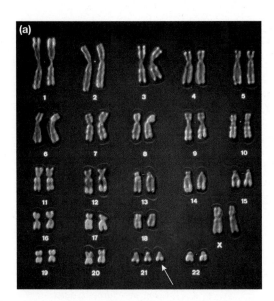

Figure 12-19 Trisomy 21, or Down syndrome
(a) Karyotype of a Down syndrome child reveals three copies of chromosome 21. (b) These girls have the relaxed mouth and distinctively shaped eyes typical of Down syndrome.

retardation so severe that only about one in 25 ever learns to read and only one in 50 learns to write.

The frequency of nondisjunction increases with the age of the parents, especially that of the mother (Fig. 12-20). Nondisjunction in sperm accounts for about 25% of the cases of Down syndrome, and there is a small increase in such defective sperm with increasing age of the father. Since the 1970s, it has become more common for couples to delay having children, increasing the probability of trisomy 21. Fetal trisomy can be diagnosed by examining the chromosomes of fetal cells.

Figure 12-20 Down syndrome frequency increases with maternal age
The increase in frequency of Down syndrome after maternal age 35 is quite dramatic.

Summary of Key Concepts

1) How Did Gregor Mendel Lay the Foundations for Modern Genetics?

Homologous chromosomes carry the same genes located at the same loci, but the genes at a particular locus can exist in alternate forms called alleles. An organism whose homologous chromosomes carry the same allele at a given locus is homozygous for that particular gene. If the alleles differ, the organism is heterozygous. Gregor Mendel deduced many principles of inheritance in the mid-1800s, before the discovery of DNA, genes, chromosomes, or meiosis. He did this by choosing an appropriate experimental sub-

ject, designing his experiments carefully, and analyzing his data accurately.

2) How Are Single Traits Inherited?

The inheritance of each individual trait is determined by discrete physical units called genes. Each organism possesses a pair of similar, but not necessarily identical, genes that influence each trait. However, only one of each pair of genes is included in any one gamete (the law of segregation). An offspring formed by the fusion of two gametes therefore receives pairs of genes, one of each pair inherited from each parent. Different alleles cause different forms of a trait (for example, purple

Earth Watch
Crops, Livestock, and Wild Genes

The science of genetics, which began in a monastery more than a century ago, is critically important to humankind today. Advances in the diagnosis, prevention, and cure of inherited diseases, advances in forensic science, and the creation of bioengineered drugs and crops are changing our society. Research into crop and livestock improvement represents some of the greatest successes, as well as challenges, for geneticists. Virtually all the corn grown in the United States today, for example, is hybrid corn, the result of crossing inferior parent strains. Faced with increasing populations and diminishing farmland, agricultural geneticists are constantly trying to develop strains of plants or animals that produce more food per unit input of energy, labor, money, and space.

Scientists take three principal approaches to improving crops and livestock. First, breeders can search for individual plants or animals with the desired characteristics. The characteristics selected for preservation through selective breeding could arise from mutations or from chance recombinations of preexisting alleles. This approach relies on "good" alleles appearing spontaneously and on our ability to recognize them. Second, molecular geneticists can try to create superior genes in the laboratory or transplant them from one species to another, as we shall describe in Chapter 13. The third approach is to look for desirable traits in wild populations of the same or closely related species. Breeders might then cross-breed or otherwise incorporate the desired genetic information into crop plants or livestock. This last approach shows great promise and, in fact, is how many crops and livestock breeds were developed in the first place, thousands of years ago. Wheat, for example, was a cross between varieties of wild grass, aided by irregularities in meiosis that resulted in a plant with multiple sets of chromosomes, producing large, edible kernels.

Rare species and local varieties of widespread species have been important in crop development in the past and hold promise for the future. In the 1960s, for example, plant geneticists cross-bred commercial strawberries with a wild variety growing in Cottonwood Canyon, Utah. The result was strawberries that produce fruit year-round. More recently, Florida breeders have developed a heat-tolerant blueberry bush by cross-breeding domestic blueberries with the rabbit-eye blueberry of the South. Geneticists at Oregon State University are developing a wildflower called meadowfoam into a source of high-quality commercial oil for the pharmaceutical and electronics industries. One of the most useful meadowfoam species is found in a 6-square-mile area near Medford, Oregon, and nowhere else in the world.

Today, our reserves of wild genes are diminishing at a frightening rate, largely due to the pressures of human population and development. Most of the estimated 10 million species of organisms on Earth have never even been scientifically named or described, much less studied. Sadly, we are in danger of losing these species without ever having gotten to know them. With the destruction of rain forests and other previously undisturbed tracts of land all over the world (perhaps even in our own backyards), countless species and genetic varieties are disappearing rapidly. For example, the anticancer drug Taxol® was discovered in the Pacific yew tree, which grows in restricted areas of the Pacific Northwest. In 1987 a researcher brought back samples of a rain-forest tree from Borneo. Scientists at the National Cancer Institute tested an extract from the tree, a member of the genus *Calophyllum*, against the AIDS virus and found the extract to be extremely effective in stopping viral replication in cultured cells. When the researcher returned to Borneo to collect more, the tree he had sampled had been cut down, and the other local *Calophyllum* trees were of a different species with only weak anti-viral activity. Fortunately, the same species of tree was protected in the Singapore Botanic Garden. Researchers were able to isolate and then synthesize the compound, which began clinical trials in human AIDS patients in 1997.

Preserving the genes of the resident plants and animals is an important, but little recognized, reason to preserve wilderness. This preservation is also an important goal of efforts to save endangered species, for once a species becomes extinct, its genes are lost forever. Genes, with all their various alleles, have evolved over hundreds of millions of years and represent one of our most valuable and irreplaceable natural resources.

---

flowers versus white flowers). Dominant alleles mask the expression of recessive alleles. In individuals with two different alleles of the same gene, which allele is included in any given gamete is determined by chance (the law of independent assortment). Therefore, we can predict the relative proportions of offspring through the laws of probability.

The physical appearance of an organism (its phenotype) may not always be a reliable indicator of its alleles (the genotype) because of the masking of recessive alleles by dominant alleles. Organisms with two dominant alleles (homozygous dominant) have the same phenotype as do organisms with one dominant and one recessive allele (heterozygous).

3) How Are Multiple Traits on Different Chromosomes Inherited?

If the genes for two traits are located on separate chromosomes, then the F_2 generation from a cross between parents that are heterozygous for two traits will have four different phenotypes, resulting from the independent assortment of the chromosomes (and hence the alleles) during meiosis.

4) How Are Genes Located on the Same Chromosome Inherited?

Genes located on the same chromosome are said to be linked. In the absence of crossing over, the F_2 offspring will express only the two parental phenotypes. If crossing over does occur, then some of the F_2 offspring will express recombined phenotypes.

5) How Is Sex Determined, and How Are Sex-Linked Genes Inherited?

Sex is determined by sex chromosomes, designated X and Y. The rest of the chromosomes, identical in the two sexes, are called autosomes. In many animals, females have two X chromosomes, whereas males have one X and one Y chromosome. The Y chromosome has many fewer genes. Consequently, males have only one copy of most X chromosome genes, and recessive traits on the X chromosome are more likely to be phenotypically expressed in males.

6) What Are Some Variations on the Mendelian Theme?
Not all inheritance follows the simple dominant–recessive pattern:

1. In incomplete dominance, heterozygotes have a phenotype intermediate between the two homozygous phenotypes. This would occur if both copies of the allele are required to produce sufficient enzyme or other protein to express the trait fully.

2. Multiple alleles for the same gene may exist within a population, as in the case of human A, B, and O blood types. Blood types A and B result from codominant alleles, both of which are phenotypically detectable in heterozygotes. This occurs when both types of proteins, each coded by a different allele, contribute to the phenotype.

3. Many traits are determined by several genes with similar actions, a phenomenon called polygenic inheritance. Traits that appear to exist in a continuum of finely graded forms are typically determined polygenically. Human eye color and skin color are good examples.

4. Many genes have multiple phenotypic effects (pleiotropy). The same enzymes, for example, may participate in the synthesis of several different end products. Alternatively, the same end product (such as melanin) may influence several aspects of the phenotype.

5. The environment influences the phenotypic expression of most, if not all, traits.

7) How Are Human Genetic Disorders Investigated?
The genetics of humans is similar to the genetics of other animals, except that experimental crosses are not feasible. Analysis of family pedigrees and, more recently, molecular genetic techniques must be used to determine the mode of inheritance of human traits.

8) How Are Human Disorders Caused by Single Genes Inherited?
Many genetic disorders are inherited as recessives; therefore, only homozygous recessive persons show symptoms of the disease. Heterozygotes are called carriers, because they carry the allele but do not express it. Recessive genetic disorders include albinism and sickle-cell anemia.

Many normal traits, and a few diseases, such as Huntington's disease, are inherited as simple dominants. Both heterozygous and homozygous dominant individuals show the trait.

The human Y chromosome bears few genes other than those that determine maleness; therefore, men phenotypically display whichever allele they carry on their single X chromosome, a phenomenon called sex-linked inheritance. Sex-linked conditions include red-green color blindness and hemophilia.

9) How Do Errors in Chromosome Number Affect Humans?
Errors in meiosis can result in gametes with abnormal numbers of sex chromosomes or autosomes. Many people with abnormal numbers of sex chromosomes have physical deficiencies and may be of below-average intelligence. Abnormal numbers of autosomes typically lead to spontaneous abortion early in pregnancy. In rare instances, the fetus may survive to birth, but severe mental and physical deficiencies always occur, as is the case with Down syndrome, or trisomy 21. The likelihood of abnormal numbers of chromosomes increases with increasing age of the mother, and, to a lesser extent, the father.

Key Terms

allele   p. 202
autosome   p. 210
carrier   p. 216
codominance   p. 213
cross-fertilization   p. 203
crossing over   p. 209
dominant   p. 204
Down syndrome   p. 221
gene   p. 201
genetic recombination   p. 210
genotype   p. 205

hemophilia   p. 218
heterozygous   p. 202
homozygous   p. 202
Huntington's disease   p. 218
hybrid   p. 202
incomplete dominance   p. 212
inheritance   p. 202
Klinefelter syndrome   p. 221
law of independent
   assortment   p. 207
law of segregation   p. 204

linkage   p. 208
locus   p. 202
multiple alleles   p. 213
nondisjunction   p. 219
pedigree   p. 215
phenotype   p. 205
pleiotropy   p. 214
polygenic inheritance   p. 214
Punnett square method
   p. 205
recessive   p. 204

self-fertilization   p. 203
sex chromosome   p. 210
sex-linked   p. 210
sickle-cell anemia   p. 216
test cross   p. 206
trisomy 21   p. 221
trisomy X   p. 221
true-breeding   p. 203
Turner syndrome   p. 220

Thinking Through the Concepts

Multiple Choice

1. *An organism is described as* Rr:red. *The* Rr *is the organism's [A]; red is the organism's [B]; and the organism is [C].*
   a. [A] phenotype; [B] genotype; [C] degenerate
   b. [A] karyotype; [B] hybrid; [C] recessive
   c. [A] genotype; [B] phenotype; [C] heterozygous
   d. [A] gamete; [B] linkage; [C] pleiotropic
   e. [A] zygote; [B] epistasis; [C] homozygous

2. *The 9:3:3:1 ratio is a ratio of*
   a. phenotypes in a test cross
   b. phenotypes in a cross of individuals that differ in one trait
   c. phenotypes in a cross of individuals that differ in two traits
   d. genotypes in a cross of individuals that differ in one trait
   e. genotypes in a cross of individuals that differ in two traits

3. *A lawyer tells a male client that blood type cannot be used in a paternity suit against the client because the child could, in fact, be his according to blood type. Which of the following is*

the only possible combination supporting this hypothetical circumstance? (Answers in the order: mother:father:child)
a. A:B:O
b. A:O:B
c. AB:A:O
d. AB:O:AB
e. B:O:A

4. *A heterozygous red-eyed female* Drosophila *mated with a white-eyed male would produce*
a. red-eyed females and white-eyed males in the F$_1$
b. white-eyed females and red-eyed males in the F$_1$
c. half red- and half white-eyed females and all white-eyed males in the F$_1$
d. all white-eyed females and half red- and half white-eyed males in the F$_1$
e. half red- and half white-eyed females as well as males in the F$_1$

5. *Which is NOT true of sickle-cell anemia?*
a. It is most common in African Americans.
b. It involves a one–amino acid change in hemoglobin.
c. It involves red blood cells.
d. It is lethal in heterozygotes because it is dominant.
e. It confers some resistance to malaria.

6. *Sex-linked disorders such as color blindness and hemophilia are*
a. caused by genes on the X chromosome
b. caused by genes on the autosome
c. caused by genes on the Y chromosome
d. the same as sex-influenced traits such as baldness; they are simply more likely to be expressed in men
e. expressed only when two chromosomes are homozygous recessive

? Review Questions

1. Define the following terms: *gene, allele, dominant, recessive, true-breeding, homozygous, heterozygous, cross-fertilization, self-fertilization.*

2. Explain why genes located on one chromosome are said to be linked. Why do linked genes sometimes separate during meiosis?

3. Define *polygenic inheritance*. Why could polygenic inheritance allow parents to produce offspring that are notably different in eye or skin color than either parent?

4. What is sex linkage? In mammals, which sex would be most likely to show recessive sex-linked traits?

5. What is the difference between a phenotype and a genotype? Does knowledge of an organism's phenotype always allow you to determine the genotype? What type of experiment would you perform to determine the genotype of a phenotypically dominant individual?

6. If one (heterozygous) parent of a couple has Huntington's disease, calculate the fraction of that couple's children that would be expected to develop the disease. What if both parents were heterozygous?

7. Why are most genetic diseases inherited as recessives rather than dominants?

8. Define *nondisjunction,* and describe the common syndromes caused by nondisjunction of sex chromosomes and autosomes.

Applying the Concepts

1. Sometimes the term *gene* is used rather casually. Compare and contrast use of the terms *allele* and *locus* as alternatives to *gene.*

2. Using the information in the chapter, explain why AB individuals are referred to as "universal recipients" in terms of blood transfusions and why people with type O blood are called "universal donors."

3. Mendel's numbers seemed almost too perfect to be real; some believe he may have cheated a bit on his data. Perhaps he continued to collect data until the numbers matched his predicted ratios, then stopped. Recently, there has been much publicity over violations of scientific ethics, including researchers' plagiarizing others' work, using other scientists' methods to develop lucrative patents, or just plain fabricating data. How important an issue is this for society? What are the boundary lines of ethical scientific behavior? How should the scientific community or society "police" scientists? What punishments would be appropriate for violations of scientific ethics?

4. Although American society has been described as a "melting pot," people often engage in "assortative mating," in which they marry others of similar height, socioeconomic status, race, and IQ. Discuss the consequences to society of assortative mating among humans. Would society be better off if people mated more randomly? Discuss why or why not.

5. *Eugenics* is the term applied to the notion that the human condition might be improved by improving the human genome. Do you think there are both good and bad sides to eugenics? What examples can you think of to back up your stand? What would a eugenicist think of the medical advances that have ameliorated the problems of hemophilia?

6. Think about some of the personal, religious, and medical economic issues related to prenatal counseling and diagnosis. Would you avoid having children if you knew that both you and your spouse were heterozygous for a recessive disorder that is fatal at an early age and may involve considerable suffering? What would you do if you or your spouse were pregnant and learned that your offspring, if born, would be homozygous for such a disorder? Is the situation qualitatively different for Down syndrome? (Down syndrome children have a life expectancy of 20 to 30 years with mental retardation but have a generally pleasant disposition.) If you were heterozygous for Huntington's disease, would you want to avail yourself of the medical diagnostic tests now available to find out? If the answer is "no" and you were a man, what would you think about your wife's potential decision to undergo diagnostic tests?

Genetics Problems

(Note: An extensive group of genetics problems, with answers, can be found in the Study Guide.)

1. In certain cattle, hair color can be red (homozygous *RR*), white (homozygous *R'R'*), or roan (a mixture of red and white hairs, heterozygous *RR'*).
   a. When a red bull is mated to a white cow, what genotypes and phenotypes of offspring could be obtained?
   b. If one of the offspring in (a) were mated to a white cow, what genotypes and phenotypes of offspring could be produced? In what proportion?

2. The palomino horse is golden in color. Unfortunately for horse fanciers, palominos do not breed true. In a series of matings between palominos, the following offspring were obtained:

   65 palominos, 32 cream-colored,
   34 chestnut (reddish brown)

   What is the probable mode of inheritance of palomino coloration?

3. In the edible pea, tall (*T*) is dominant to short (*t*), and green pods (*G*) are dominant to yellow pods (*g*). List the types of gametes and offspring that would be produced in the following crosses:
   a. *TtGg × TtGg*   b. *TtGg × TTGG*   c. *TtGg × Ttgg*

4. In tomatoes, round fruit (*R*) is dominant to long fruit (*r*), and smooth skin (*S*) is dominant to fuzzy skin (*s*). A true-breeding round, smooth tomato (*RRSS*) was cross-bred with a true-breeding long, fuzzy tomato (*rrss*). All the F_1 offspring were round and smooth (*RrSs*). When these F_1 plants were bred, the following F_2 generation was obtained:

   Round, smooth: 43   Long, fuzzy: 13

   Are the genes for skin texture and fruit shape likely to be on the same chromosome or on different chromosomes? Explain your answer.

5. In the tomatoes of problem 4, an F_1 offspring (*RrSs*) was mated with a homozygous recessive (*rrss*). The following offspring were obtained:

   Round, smooth: 583   Round, fuzzy: 21
   Long, fuzzy: 602   Long, smooth: 16

   What is the most likely explanation for this distribution of phenotypes?

6. In humans, hair color is controlled by two interacting genes. The same pigment, melanin, is present in both brown-haired and blond-haired people, but brown hair has much more of it. Brown hair (*B*) is dominant to blond (*b*). Whether any melanin can be synthesized depends on another gene. The dominant form (*M*) allows melanin synthesis; the recessive form (*m*) prevents melanin synthesis. Homozygous recessives (*mm*) are albino. What will be the expected proportions of phenotypes in the children of the following parents?
   a. *BBMM × BbMm*   b. *BbMm × BbMm*
   c. *BbMm × bbmm*

7. In humans, one of the genes determining color vision is located on the X chromosome. The dominant form (*C*) produces normal color vision; red-green color blindness (*c*) is recessive. If a man with normal color vision marries a color-blind woman, what is the probability of their having a color-blind son? A color-blind daughter?

8. In the couple described in problem 7, the woman gives birth to a color-blind daughter. The husband sues for a divorce on the grounds of adultery. Will his case stand up in court? Explain your answer.

Answers to Genetics Problems

1. a. A red bull (*RR*) is mated to a white cow (*R'R'*). The bull will produce all *R* sperm; the cow will produce all *R'* eggs. All the offspring will be *RR'* and will have roan hair (codominance).
   b. A roan bull (*RR'*) is mated to a white cow (*R'R'*). The bull produces half *R* and half *R'* sperm; the cow produces *R'* eggs. Using the Punnett square method:

   eggs

	R'
R	R R'
R'	R' R'

   (sperm)

   Using probabilities:

sperm	egg	offspring
½ *R*	*R'*	½ *RR'*
½ *R'*	*R'*	½ *R'R'*

   The predicted offspring will be ½ *RR'* (roan) and ½ *R'R'* (white).

2. The offspring occur in three types, classifiable as dark (chestnut), light (cream), and intermediate (palomino). This distribution suggests incomplete dominance, with the alleles for chestnut (*C*) combining with the allele for cream (*C'*) to produce palomino heterozygotes (*CC'*). We can test this hypothesis by examining the offspring numbers. There are approximately ¼ chestnut (*CC*), ½ palomino (*CC'*), and ¼ cream (*C'C'*). If palominos are heterozygotes, we would expect the cross *CC' × CC'* to yield ¼ *CC*, ½ *CC'*, and ¼ *C'C'*. Our hypothesis is supported.

3. a. *TtGg × TtGg*. This is a "standard" cross for differences in two traits. Both parents produce *TG*, *Tg*, *tG*, and *tg* gametes. The expected proportions of offspring are 9/16 tall green, 3/16 tall yellow, 3/16 short green, 1/16 short yellow.
   b. *TtGg × TTGG*. In this cross, the heterozygous parent produces *TG*, *Tg*, *tG*, and *tg* gametes. However, the homozygous dominant parent can produce only *TG* gametes. Therefore, all offspring will receive at least one *T* allele for tallness and one *G* allele for green pods, and thus all the offspring will be tall with green pods.
   c. *TtGg × Ttgg*. The second parent will produce two types of gametes, *Tg* and *tg*. Using a Punnett square:

   eggs

	Tg	*tg*
TG	*TTGg*	*TtGg*
Tg	*TTgg*	*Ttgg*
tG	*TtGg*	*ttGg*
tg	*Ttgg*	*ttgg*

   (sperm)

The expected proportions of offspring are ⅜ tall green, ⅜ tall yellow, ⅛ short green, ⅛ short yellow.

4. If the genes are on separate chromosomes—that is, assort independently—then this would be a typical two-trait cross with expected offspring of all four types (about 9/16 round smooth, 3/16 round fuzzy, 3/16 long smooth, and 1/16 long fuzzy). However, only the parental combinations show up in the F_2 offspring, indicating that the genes are on the same chromosome.

5. The genes are on the same chromosome and are quite close together. On rare occasions, crossing over occurs between the two genes, producing recombination of the alleles.

6. a. *BBMM* (brown) × *BbMm* (brown). The first parent can produce only *BM* gametes, so all offspring will receive at least one dominant allele for each gene. Therefore, all offspring will have brown hair.

   b. *BbMm* (brown) × *BbMm* (brown). Both parents can produce four types of gametes: *BM, Bm, bM,* and *bm*. Filling in the Punnett square:

eggs

	BM	Bm	bM	bm
BM	BBMM	BBMm	Bb MM	Bb Mm
Bm	BBMm	BBmm	Bb Mm	Bb mm
bM	Bb MM	Bb Mm	bb MM	bb Mm
bm	Bb MM	Bb mm	bb Mm	bb mm

(sperm)

All *mm* offspring are albino, so we get the expected proportions 9/16 brown-haired, 3/16 blond-haired, 4/16 albino.

c. *BbMm* (brown) × *bbmm* (albino):

eggs

bm

	bm
BM	Bb Mm
Bm	Bb mm
bM	bb Mm
bm	bb mm

(sperm)

The expected proportions of offspring are ¼ brown-haired, ¼ blond-haired, ½ albino.

7. A man with normal color vision is *CY* (remember, the Y chromosome does not have the gene for color vision). His color-blind wife is *cc*. Their expected offspring will be:

eggs

c

	c
C	Cc
Y	cY

(sperm)

We therefore expect that all the daughters will have normal color vision and all the sons will be color-blind.

8. The husband should win his case. All his daughters must receive one X chromosome, with the *C* allele, from him, and therefore should have normal color vision. If his wife gives birth to a color-blind daughter, her husband cannot be the father (unless there was a new mutation for color blindness in his sperm line, which is very unlikely).

Group Activity

Genetic counseling is a field that has been growing with our increased knowledge of the human genome. Individuals and prospective parents can now obtain help so as to make informed decisions. Counselors provide information and discuss the consequences of various decision scenarios but do not tell their clients what they should do. Your instructor will provide your group with a case study for you to discuss and to determine how you, as a genetic counselor, might proceed.

For More Information

Crow, J. F. "Genes That Violate Mendel's Rules." *Scientific American,* February 1979. A close look at genetic recombination and its role in evolution.

Hoyt, E. "Wild Relatives." *Living Wilderness,* Summer 1990. Undisturbed ecosystems are a valuable gene bank with great potential for improving crop plants.

Horgan, J. "Eugenics Revisited." *Scientific American,* June 1993. Determining patterns of inheritance, let alone finding suspect genes, for behavioral disorders is notoriously difficult. What's more, it isn't at all clear what one should do with such data, if they become available.

Kahn, P. "Gene Hunters Close in on Elusive Prey." *Science,* March 8, 1996. Using pedigrees and biotechnology, researchers are uncovering the causes of diseases that involve complex interactions between genes and the environment.

McGue, M. "Nature–Nurture and Intelligence." *Nature* 1989; 340:507–508. Nature works via nurture in producing the outward manifestations of intelligence.

O'Brien, S. J. "A Role for Molecular Genetics in Biological Conservation." *Proceedings of the National Academy of Sciences,* June 1994, Vol. 91. Habitat destruction has reduced many wild populations to tiny remnants of their former size. Genetic analysis is used to determine how much loss of genetic diversity has occurred due to inbreeding.

Pines, M. "In the Shadow of Huntington's." *Science 84,* May 1984; Grady, D. "The Ticking of a Time Bomb in the Genes." *Discover,* June 1987; Roberts, L., "Huntington's Gene: So Near, Yet So Far." *Science* 1990, 247:624–627; and Morell, V. "Huntington's Gene Finally Found." *Science* 1993, 260:28–30. These articles focus on the story of the scientific detective work involved in the diagnosis of Huntington's disease by using recombinant DNA techniques, the social dilemma it created, the frustrations of searching for the gene, and, finally, finding it. Particularly poignant because one of the lead investigators may herself be a victim.

Sapienza, C. "Parental Imprinting of Genes." *Scientific American,* October 1990. It is not quite true that all genes are equal, regardless of whether they have been inherited from mother or father. In some cases, which parent a gene comes from greatly alters its expression in the offspring.

Stern, C., and Sherwood, E. R. *The Origin of Genetics: A Mendel Source Book.* San Francisco: W. H. Freeman, 1966. There is no substitute for the real thing, in this case a translation of Mendel's original paper to the Brünn Society.

Wivel, N. A., and Walters, L. "Germ-line Gene Modification and Disease Prevention: Some Medical and Ethical Perspectives." *Science* 1993, 262: 533–538. If we can fix a defective allele in an egg or zygote, should we?

Answers to Multiple-Choice Questions

1. c   2. c   3. a   4. e   5. d   6. a

"Woe to that child which when kissed on the forehead tastes salty. He is bewitched and soon must die."

British saying dated to the seventeenth century

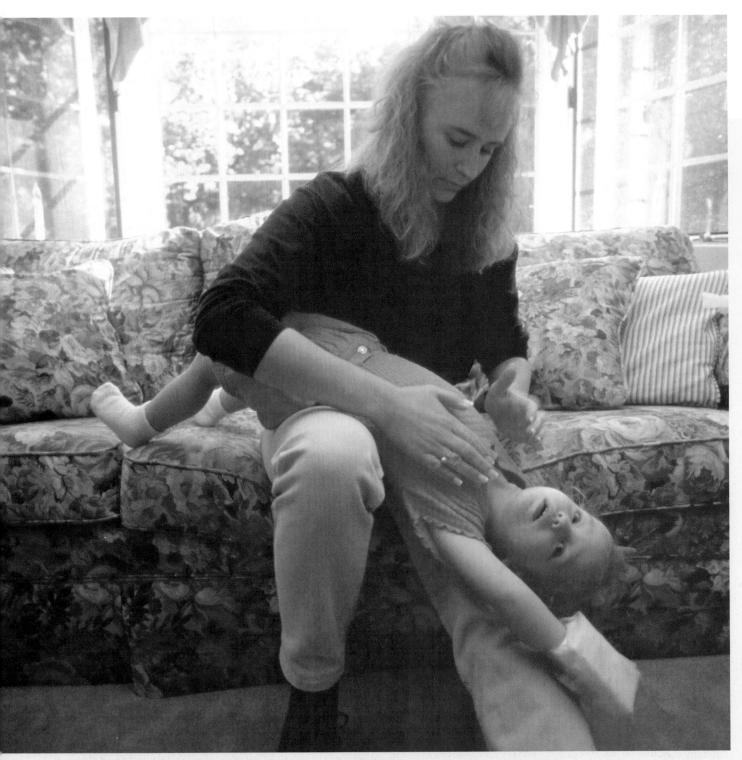

A child is treated for cystic fibrosis. Gentle pounding on the chest and back while the child is held upside-down helps dislodge mucus from the lungs. A device on the child's wrist injects antibiotics into a vein. These treatments help combat the numerous lung infections to which cystic fibrosis patients are vulnerable. Treatments based on biotechnology offer hope for such children.

Biotechnology 13

At a Glance

Net Watch

On-line resources for this chapter are on the World Wide Web at:
http://www.prenhall.com/audesirk
(click on the Table of Contents link and then select Chapter 13).

Can a kiss at birth foretell an early death, or does our opening quote from the 1600s describe a meaningless superstition, a relic of an age of medical ignorance? Or might it refer to cystic fibrosis? **Cystic fibrosis** (CF) is most prevalent among people who came from northwest Europe, where the old saying originated. One of every 46 Hispanics, 1 in 63 African Americans, and 1 in 150 Asian Americans carry the cystic fibrosis gene. But among Caucasians of European descent, 1 out of every 25 adults carries the gene, and 1 child out of every 2000 is born with CF. A single copy of this recessive gene produces no symptoms. But when two carriers marry, their children have a one in four chance of inheriting both defective alleles and living with—and probably dying of—CF.

What is CF, and what is its link to an ancient superstition? Cystic fibrosis begins when a person inherits two copies of a defective gene that codes for a protein that transports chloride ions across cell membranes. The defective functioning of this protein sets off a chain of events that leads to a buildup of salt in the lungs and the production of thick, sticky mucus that clogs the airways and restricts air exchange. The salty fluid that bathes the lungs inhibits the natural defensive secretion produced by the lungs to fight infection; the thick mucus holds viruses and provides an environment where bacteria flourish. Therefore, people with CF endure a continuing series of respiratory infections, including pneumonia. In other parts of the body, abnormally thick mucus coating the digestive tract inhibits nutrient absorption and may block the duct that carries digestive enzymes from the pancreas to the intestine, so affected individuals are malnourished as well. Prior to 1940, the average child with CF died before age two. Whence the saying? People with CF produce sweat with an abnormally high salt content, making their skin taste unusually salty. The "sweat test" has long been an important diagnostic tool for CF—a test that apparently dates back to the 1600s!

What is the link between CF and biotechnology? This common, fatal genetic disorder is under attack on several fronts. Using newly developed tools of biotechnology, scientists have located the CF gene on chromosome 7 and discovered its nucleotide sequence. They have produced a mouse with the same defective gene, allowing them to study causes and cures for the disorder in a laboratory animal. Researchers have devised a test that can detect the CF gene in carriers with no symptoms. In a controversial move, a panel of experts has recommended offering the test to *everyone* who contemplates having children. The estimated 25,000 Americans with CF, who now live to an average age of 30, are hoping for new treatments based on biotechnology. Some are inhaling a genetically engineered virus that carries the corrected human gene in a study to determine if the gene will be incorporated into cells that line the subject's respiratory tracts, alleviating the symptoms.

As you read this chapter, you will be introduced to the amazing promise of biotechnology and alerted to a number of concerns as well. In a very short time, scientists have learned a great deal about how genes control basic life processes and about how defective genes cause disease. The common genetic heritage of all life on Earth means that genes from one species can be decoded and used by another species. Biotechnology allows researchers to alter genes and transfer them between cells and even between species. Some observers have compared deciphering and manipulating the genetic code to opening Pandora's box—in our curiosity, are we releasing unforeseen dangers or conferring immense benefits—or both?

1) What Is Biotechnology?   *www*

In its broadest sense, **biotechnology** is defined as any industrial or commercial use or alteration of organisms, cells, or biological molecules to achieve specific practical goals. By this definition, biotechnology is nothing new—it is as old as the use of yeast to make bread rise or to ferment grape juice into wine, a process that originated 10,000 years ago. It is as old as selective breeding of plants and animals in agriculture. Squash fragments preserved in a dry cave in Mexico were recently dated as 8000–10,000 years old. Their seeds are larger and their rinds thicker and more colorful than those of wild varieties, providing evidence of selective breeding—a very early form of genetic manipulation by humans. Prehistoric art and animal remains suggest that dogs, sheep, goats, and camels were domesticated 10,000–12,000 years ago.

Modern biotechnology commonly utilizes **genetic engineering**, the modification of genetic material to achieve specific goals. Genetically engineered cells may have genes deleted, added, or replaced. The major goals of genetic engineering are threefold:

1. to understand more about the processes of inheritance and gene expression;
2. to provide better understanding and treatment of various diseases, particularly genetic disorders; and

3. to generate economic benefits, including improved plants and animals for agriculture and efficient production of valuable biological molecules.

An important tool in genetic engineering is **recombinant DNA**, that is, DNA that has been altered by the incorporation of genes from a different organism, typically from a different species. Recombinant DNA can be transferred into animals or plants by using carriers such as bacteria or viruses. Any carrier that introduces a foreign gene into cells is called a **vector**. A **transgenic** animal or plant expresses DNA derived from another species. Since their introduction in the 1970s, the techniques, applications, and promises of recombinant DNA technology have grown explosively.

2) How Does DNA Recombination Occur in Nature?   *www*

Many recombinant DNA technologies make use of naturally occurring DNA recombination processes. To understand what goes on in modern biotechnology laboratories, let's start with a survey of some naturally occurring methods of DNA recombination.

DNA Recombination Occurs Naturally through Processes Such as Sexual Reproduction, Bacterial Transformation, and Viral Infection

A natural form of DNA recombination within a species begins every time two gametes fuse during sexual reproduction. The actual recombination then occurs through crossing over during meiosis I, when genes from a maternal chromosome and a paternal chromosome are exchanged, producing a chromosome with a new combination of alleles derived from two individuals. Recombination, by whatever mechanism, changes the genetic makeup of organisms. As is the case with mutations, natural recombination is random and undirected. Naturally occurring DNA recombinations are tested, just as mutations are, by natural selection. Most recombinations probably prove to be neutral, and some are harmful. A few, however, prove helpful in a particular environment; organisms with these new DNA combinations thrive and pass on the new combinations to their offspring.

Many people tend to consider recombination within a species during sexual reproduction as "natural" and therefore good but think of recombinations performed in the laboratory between different species as "unnatural" and therefore intrinsically bad. However, interspecies recombinations also occur in nature, as described next.

Transformation May Combine DNA from Different Bacterial Species

Bacteria employ several methods of recombination that allow gene transfer between unrelated species. A process called **transformation** allows bacteria to pick up free DNA from the environment. The free DNA may be part of the chromosome of another bacterium, including DNA from a bacterium of another species. Transformation may also occur when bacteria pick up tiny circular DNA molecules called **plasmids** (Fig. 13-1). Plasmids, which range in size from about 1000 to 100,000 nucleotides, are self-replicating lengths of DNA normally found in the cytoplasm of many types of bacteria and some yeasts. A single bacterium—a *host cell*—may contain dozens or even hundreds of copies of a plasmid. Although the bacterium's "own" chromosome contains all the genes the cell normally needs for survival, the genes carried by plasmids may also be useful. For example, some plasmids contain genes that code for enzymes that digest certain antibiotics, such as penicillin. In environments where exposure to antibiotics is high, such as hospitals, these plasmids quickly spread, conferring a major advantage to their bacterial hosts and making antibiotic-resistant infections a serious problem (see Chapter 19). A bacterium may acquire plasmids from its own strain or from other types of bacteria. These plasmids are either liberated into the environment when a bacterium dies or are exchanged among living bacteria.

Viruses May Transfer DNA between Bacteria and between Eukaryotic Species

Viruses, which are little more than genetic material encased in a protein coat, transfer their genetic material to cells. There viral genes replicate and direct the synthesis of viral proteins. New virus particles are assembled inside the cell, then released to repeat the cycle (Fig. 13-2). Viruses may transfer genes among bacteria and among eukaryotic organisms, such as plants, as well. **Bacteriophages**, specialized viruses that infect bacteria, occasionally acquire pieces of bacterial DNA. The bacteriophages, or *phages*, then release this DNA into other bacteria that they infect. In some cases, the transferred bacterial DNA becomes incorporated into the host bacterial chromosome, adding new genetic material.

Recent research suggests that gene transfer by viruses among eukaryotic organisms occurs in a generally similar manner. Certain types of viruses can insert their DNA into a chromosome of their eukaryotic host and exist there quietly for days, months, or even years. When the viral DNA finally leaves the chromosome, it may take a bit of the eukaryotic DNA along with it. The viral DNA then directs the synthesis of new viruses that will also include some host DNA. If one of these offspring infects a

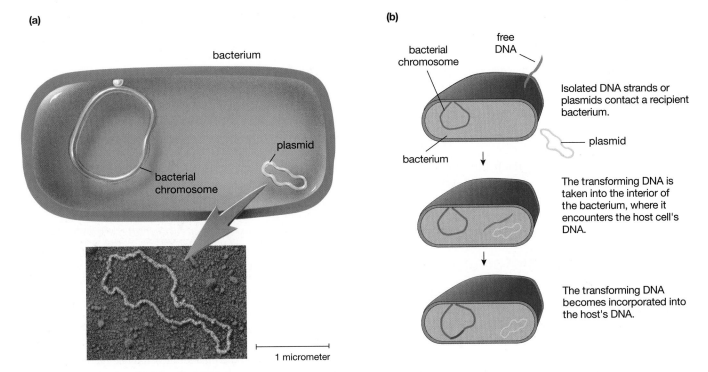

(a)

bacterium

plasmid

bacterial chromosome

1 micrometer

(b)

free DNA

bacterial chromosome

bacterium

plasmid

Isolated DNA strands or plasmids contact a recipient bacterium.

The transforming DNA is taken into the interior of the bacterium, where it encounters the host cell's DNA.

The transforming DNA becomes incorporated into the host's DNA.

Figure 13-1 Recombination in bacteria
(a) In addition to their single, circular chromosome, bacteria commonly possess rings of DNA called plasmids, which may carry useful genes. Plasmids can replicate within the bacterium and can be liberated to the environment and picked up by different types of bacteria.
(b) Bacterial transformation occurs when bacteria pick up DNA from their surroundings. This DNA includes plasmids and chromosomes liberated by the death of other bacteria.

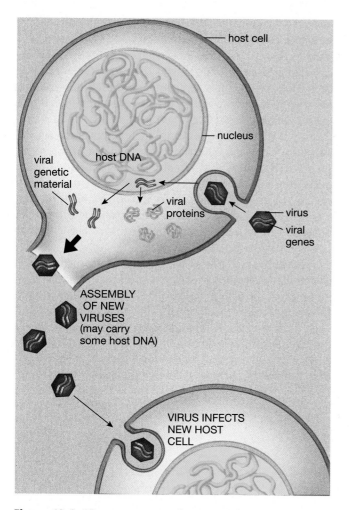

Figure 13-2 Viruses may transfer genes
A virus infecting a eukaryotic cell uses its own genetic material and the host's cellular machinery to produce copies of the virus, which break free and infect new host cells. Segments of the host DNA may "accidentally" be incorporated into the viral genome and transferred to hosts of other species.

different species and inserts itself into a chromosome, the new host may acquire some genes that originally belonged to an unrelated species.

3) What Are Some of the Methods of Biotechnology?

Within the past few years, the technologies of recombinant DNA have mushroomed. We will follow a typical sequence of procedures that might be used to solve a particular problem or to produce a specific product.

A DNA Library Is Produced by Inserting DNA from a Selected Organism into Bacterial Plasmids

The first task in recombinant DNA technology is to produce a **DNA library**—a readily accessible, easily duplicable assemblage of all the DNA of a particular organism.

The entire set of genes carried by a member of any given species is called a **genome**. Why build a DNA library of a species' genome? A DNA library organizes the DNA in a way that researchers can use it. *Restriction enzymes*, plasmids, and bacteria are the most commonly used tools in assembling a DNA library.

Restriction Enzymes Cut DNA at Specific Nucleotide Sequences

Many bacteria produce **restriction enzymes**, which sever DNA at particular nucleotide sequences. In nature, restriction enzymes defend bacteria against viral infections by cutting apart the viral DNA. (The bacteria protect their own DNA, probably by attaching methyl ($-CH_3$) groups to some of the DNA nucleotides.) Researchers have isolated restriction enzymes and use them to break DNA into shorter strands at specific sites.

Most restriction enzymes recognize and sever *palindromic* sections of DNA, in which the nucleotide order is the same in one direction on one strand as in the reverse direction on the other strand. (A palindrome is a word that reads the same forward and backward, such as "madam.") For the enzyme illustrated in Figure 13-3, this cutting results in two pieces of DNA, one a single-stranded piece reading TTAA and one a single-stranded piece reading AATT. These single-stranded cut pieces of the DNA fragment are called "sticky ends," because they will stick to (form hydrogen bonds with) other single-stranded cut pieces of DNA with the complementary series of bases. If the appropriate DNA repair enzyme (called

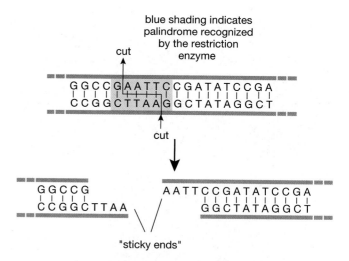

Figure 13-3 Restriction enzymes sever DNA at specific sites
Many of the most useful restriction enzymes sever DNA at palindromes, where the nucleotide sequence reads the same on one DNA strand as it does in the opposite direction on the other strand. Two short, single-stranded "sticky ends" (here, AATT on one strand and TTAA on the other) are created. Because these sticky ends are complementary, they can base-pair, holding together any strands with the same sticky-end sequences.

DNA ligase) is added, DNA from different sources cut by the same restriction enzyme can be joined as if the DNA had occurred naturally. *Segments of DNA from fundamentally different types of organisms, such as bacteria and humans, can be joined if they have complementary sticky ends.*

Many different restriction enzymes have been isolated from various species of bacteria. Each cuts DNA apart at different but specific palindromic nucleotide sequences. The variety of restriction enzymes has enabled molecular geneticists to identify and isolate specific segments of DNA from many organisms, including humans.

Restriction Enzymes Are Used to Insert DNA into Plasmids to Build a DNA Library

Suppose that human DNA is isolated from white blood cells and is cut apart into many small fragments with a restriction enzyme (Fig. 13-4). The same restriction enzyme is then used to sever the DNA of bacterial plasmids. Now both human and plasmid DNA have complementary sticky ends that, when mixed, form hydrogen bonds. When DNA ligase is added, it bonds the sugar–phosphate backbones together, inserting segments of human DNA into plasmids.

The new rings of plasmid–human DNA (recombinant DNA) are mixed with bacteria, which take up the recombinant DNA. Millions or billions of plasmids collectively could incorporate DNA from the entire human genome. Usually, 100 to 1000 times more bacteria than plasmids are used, so that no individual bacterium ends up with more than one recombinant DNA molecule. The resulting population of bacteria containing recombined plasmid–human DNA constitutes a human DNA library.

A DNA Probe May Be Used to Locate the Gene within the DNA Library

Researchers find genes of interest by using **DNA probes**, sequences of nucleotides that are complementary to those genes. DNA probes are generated in specific ways. For example, if you know the amino acid sequence of the protein encoded by a gene, you can work backward through the genetic code to determine the corresponding DNA nucleotide sequences. Researchers synthesize a complementary copy of the nucleotide sequence for all or part of the gene, incorporating a marker such as radioactivity. This synthetic sequence can then be used to identify bacteria from the DNA library that contain a plasmid with the gene in question (Fig. 13-5).

Messenger RNA can also be used as a DNA probe. Immature red blood cells, for example, synthesize lots of hemoglobin, so mRNA for hemoglobin can easily be extracted from them. This mRNA is complementary to the DNA of the hemoglobin gene and can be labeled and used as a DNA probe to locate the plasmid-containing bacteria (see Fig. 13-5). The amino acid sequences for

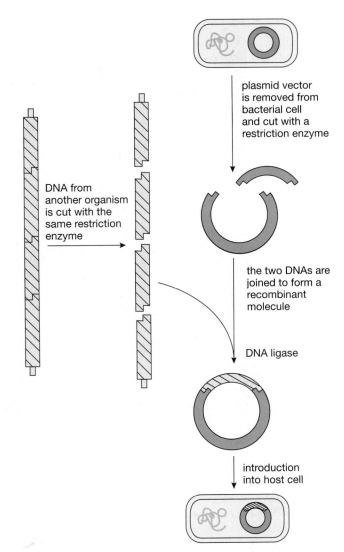

Figure 13-4 The use of restriction enzymes in building a human DNA library
Both human chromosomes and bacterial plasmids are cut with the same restriction enzyme. Thus, both types of DNA have sticky ends. When the severed DNA molecules are mixed, plasmid–human combinations are held together by complementary base pairing of sticky ends, and DNA ligase bonds together the backbones of the plasmid–human DNA molecules. The new plasmid–human rings are mixed with bacteria, some of which take up a recombinant ring.

many human hormones are known; these sequences can be used to create DNA probes for the genes for human hormones.

Once mRNA for the desired gene is obtained, it can also be used to generate the corresponding complementary strand of DNA (*cDNA*) by using a viral enzyme called *reverse transcriptase*. Geneticists can then use the cDNA as a DNA probe to identify the bacterium that contains the appropriate plasmid. The cDNA can instead serve as a template for generating double-stranded DNA, which can then be copied by techniques described next.

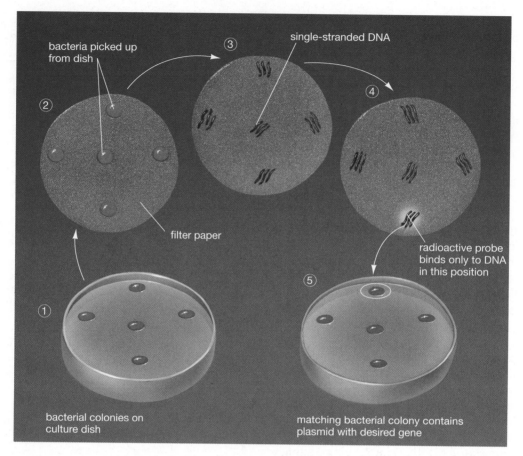

Figure 13-5 Searching a human DNA library for a specific gene
① Bacteria from the DNA library are sparsely distributed over a culture dish that contains growth medium. Each bacterium multiplies into a visible bacterial population (called a *colony*) containing a single type of plasmid–human DNA combination. ② A sheet of special filter paper is pressed onto the culture dish. It picks up a few bacteria from each colony, preserving the colony positions. The original culture dish is saved. ③ The filter paper is placed in a basic solution, breaking open the bacteria, freeing the plasmids, and separating the double-stranded DNA of the plasmids into single strands. ④ The paper is bathed in solution of neutral pH that contains a DNA probe (yellow), a sequence of nucleotides complementary to the desired gene. The probe hydrogen-bonds only to plasmid DNA that is complementary to the nucleotide sequence of the probe and that therefore contains the human gene. ⑤ The locations of radioactivity on the paper are matched to the locations of bacterial colonies on the original culture dish. Colonies in the same position consist of bacteria that contain plasmids with the desired gene. Samples of these bacteria are now cultured in new dishes, replicating the desired gene.

Once Located, the Gene Must Be Amplified

One major problem with all the recombinant DNA technologies described in this chapter is that they require *lots* of DNA: millions or billions of copies. Geneticists must *amplify*, or copy, DNA to make studying it feasible. After a specific gene or other interesting fragment of DNA has been isolated, two fundamentally different techniques are used to duplicate it for study: (1) growth in bacteria or other living organisms, and (2) the *polymerase chain reaction*, or *PCR*, in which DNA is replicated in test tubes by using the enzyme DNA polymerase.

The Bacteria Containing the Gene of Interest May Be Cultured to Create Multiple Copies

Once the bacterial colony containing the plasmid with the desired gene has been identified, the transformed bacteria are grown in culture medium, producing multiple copies of the human DNA segment. Many biotechnology firms use huge vats to produce many kilograms of bacteria, all containing a specific gene that may code for a human protein such as insulin (Fig. 13-6).

PCR Is an Important Technique for Amplifying DNA Fragments

Another method for making copies of specific stretches of DNA, from single cells as well as from DNA libraries, is called the **polymerase chain reaction (PCR)** technique. Developed in 1986 by Kary B. Mullis of the Cetus Corporation, PCR allows billions of copies of selected genes to be made, faster and more cheaply than they can be grown in bacteria. PCR is so elegant and so crucial to continued advances in molecular biology that it earned Mullis a share in the Nobel Prize for Chemistry in 1993.

Figure 13-6 A gene-copying "factory"
A culturing vat in which a commercial biotechnology company raises huge numbers of bacteria, each containing a plasmid that incorporates a valuable human gene.

PCR is based on the use of the enzyme DNA polymerase. Recall from Chapter 10 that during DNA replication, the double-stranded helix is unwound by enzymes and that each strand is used as a template to produce its complementary strand. DNA polymerase is responsible for linking up appropriate nucleotides to each single-stranded molecule of DNA. The process is initiated by adding a short synthetic sequence of DNA called a *primer*, which bonds with the original DNA molecule just upstream of the stretch of DNA to be copied. This primer "tells" the DNA polymerase where to start copying.

PCR is accomplished by the following basic steps, which are repeated for as many cycles as are necessary to generate sufficient copies of the DNA segment:

1. DNA is separated into single strands by heating to about 90°C.
2. Specific primers, one for each complementary strand, start the reaction at the appropriate segment of DNA.
3. Heat-resistant DNA polymerase (isolated from bacteria that thrive in hot springs or hot ocean vents) adds nucleotides, constructing the new DNA segment.

With appropriate mixtures of primers, free nucleotides, and DNA polymerase, a PCR machine runs heating and cooling cycles over and over again. Each cycle takes only a few minutes, so PCR can produce billions of copies of a gene or DNA segment in a single afternoon, starting, if necessary, from a single molecule of DNA (Fig. 13-7).

4) What Are Some Applications of Biotechnology?

www

Once sufficient copies of a gene or other DNA segments have been collected, they can be put to a variety of uses. The use of genes in biotechnology is having an increasing impact on our society. Some applications of gene technology are described next.

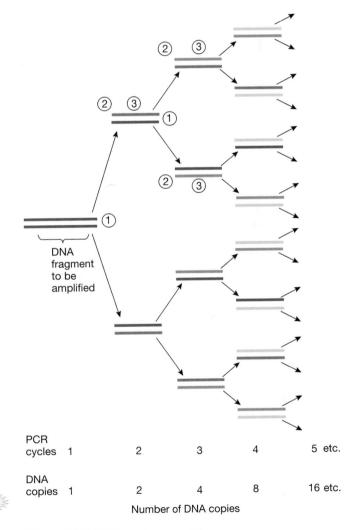

PCR cycles	1	2	3	4	5 etc.
DNA copies	1	2	4	8	16 etc.

Number of DNA copies

Figure 13-7 PCR copies a specific gene
① DNA is separated into single strands by heating. ② Primers start the reaction at the appropriate segment of DNA. ③ Heat-resistant DNA polymerase adds complementary nucleotides, constructing a new DNA segment.

Amplified Genes Provide Enough DNA for Gene Sequencing

Once a gene has been amplified, its exact nucleotide sequence can be determined by using new automated techniques that link labeled sequences of unknown DNA with known sequences arranged in a specific pattern on a microchip. Why is knowing the nucleotide sequence of a gene useful? For one thing, the nucleotide sequence determines the amino acid sequence of the encoded protein. Once the protein is known, its function in the cell can be studied and defective versions of the protein caused by mutation can be investigated. The gene for cystic fibrosis was sequenced in late 1989; that sequence was the starting point for pinpointing the underlying genetic defect.

Further, knowing a gene's nucleotide sequence helps allow geneticists to design rapid, definitive tests for diseases caused by a defective version of the gene. Such tests are the basis for prenatal diagnosis of sickle-cell anemia (see "Health Watch: Prenatal Genetic Screening," p. 248).

In one of the most remarkable instances of cooperation and collaboration in modern biology, scientists worldwide are undertaking the ultimate sequencing task—to sequence the entire human genome, described later this chapter.

DNA Fingerprinting Facilitates Genetic Detection on Many Fronts

DNA fingerprinting has rapidly emerged as a tool for genetic inquiry in many different contexts. Like recombinant DNA technology, DNA fingerprinting uses restriction enzymes. There are a large number of known restriction enzymes, and each cuts DNA at a different sequence of nucleotides. If a pair of homologous chromosomes is exposed to a given restriction enzyme, the DNA is cut up into pieces called **restriction fragments** (Fig. 13-8a). These fragments are separated by size through a technique known as **gel electrophoresis**. In gel electrophoresis, DNA fragments are placed on restricted tracks on a thin sheet of gelatinous material. The application of an electric field to the gel forces the DNA fragments to migrate along their tracks. The rate at which the fragments migrate through the gel depends on their length (Fig. 13-8b). After a specific period of time, the movement is stopped and the DNA is marked with a radioactive substance. The radioactive pattern on the gel is used to expose photographic film, creating dark bands where the DNA fragments are located. The band locations correlate with the lengths of the restriction fragments. Differences in restriction fragment lengths are called **restriction fragment length polymorphisms**, or **RFLP**s (*polymorphism* means "many forms" in Greek). Because restriction enzymes cut DNA only at particular nucleotide sequences, RFLPs reflect differences in nucleotide sequences on the homologous chromosomes.

Just as each person has a unique set of fingerprints, so is each person's DNA cut into a unique set of restriction fragments. Although the DNA sequences of all humans are fundamentally similar, each of us is genetically unique. Each of our 100,000 or so genes can have multiple alleles, and the alleles are combined in unique ways in each person (or in any member of a sexually reproducing species). The slight differences in nucleotide sequence that distinguish each of us create different cutting sites for restriction enzymes, producing restriction fragments of different lengths for stretches of DNA that contain the same genes. The pattern that appears when these fragments are sep-

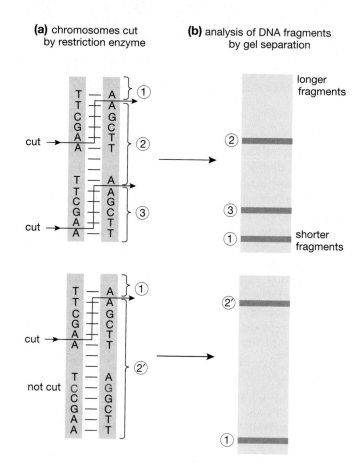

(a) chromosomes cut by restriction enzyme

(b) analysis of DNA fragments by gel separation

Figure 13-8 Restriction fragment length polymorphisms (a) One of two homologous chromosomes (here, the lower one) has a mutated allele (in red) that causes a difference in a single nucleotide. A restriction enzyme cuts DNA at AAGCTT/TTCGAA palindromes. The upper chromosome has two of these sequences and is cut into three fragments—two short (1 and 3) and one medium long (2). Because of the mutation, the lower chromosome has only one sequence of this type and thus is cut into only two fragments—one short (1) and one very long (2'). *(b)* The DNA fragments are separated by electrophoresis as they are pushed along by an electric field. Small pieces (1 and 3) slip through the gel more easily than do large pieces (2 and 2') and migrate farther. Differences in fragment lengths caused by different cut points in alleles of the same genes are called *restriction fragment length polymorphisms* (RFLPs).

arated by length on an electrified gel creates a unique "DNA fingerprint" that can be used to identify an individual (see Fig. E13-1 in "Scientific Inquiry: DNA Fingerprinting: A Tool for Medical Detectives," p. 238).

RFLPs Can Be Used to Locate Mutated Genes that Cause Genetic Disorders

To use RFLPs to detect a gene that causes a human genetic disorder, researchers collect samples of DNA from as many affected and nonaffected family members as possible. Because related people share more nucleotide sequences, differences caused by a mutated gene are easier to detect. Then researchers look for a uniquely sized DNA fragment whose presence correlates with the disease, indicating that a unique restriction enzyme "cut site" must be within, or very near, the defective gene causing the disorder. In the late 1980s, RFLP analysis was used to identify the approximate locations of the genes for cystic fibrosis and for Huntington's disease, the latter based in part on hundreds of blood samples from the extended family of Huntington's victims in a remote Venezuelan village. (See "Scientific Inquiry: In Search of the Huntington's Gene" in Chapter 12.)

Once a gene responsible for a genetic disorder has been localized on a chromosome and amplified by PCR, RFLP analysis can also be used for the prenatal diagnosis of individuals who carry that gene. In "Health Watch: Prenatal Genetic Screening" (p. 248), we describe how RFLPs are used to diagnose diseases such as sickle-cell anemia. DNA fingerprinting is also used increasingly in forensic medicine, as explained in the essay "Scientific Inquiry: DNA Fingerprinting: A Tool for Medical Detectives" (p. 238).

Biotechnology Is Revolutionizing Agriculture

Genetic engineering technology is rapidly replacing traditional breeding techniques to produce improved strains of crops. The number of field tests of genetically engineered organisms is growing exponentially, and experts predict that all major farm crops in the United States are likely to express genetically engineered traits within the next few decades. Some of the applications of genetic engineering to crop plants and farm animals are shown in Table 13-1.

Recombinant plants are produced in several ways. One commonly used technique is via a plasmid (called the *Ti plasmid*) found in a certain soil bacterium that infects plants. The plasmid, which carries genes that cause plant tumors, inserts itself into the plant DNA. Geneticists can use restriction enzymes to cleave the tumor-directing section of plasmid and substitute other genes. When the plasmid is returned to its host bacterium and allowed to infect a plant cell in a culture dish, the recombined genetic material from the plasmid becomes incorporated into the plant DNA. Under proper laboratory conditions, single plant cells grow and divide to form a mass of cells that researchers can coax into forming many individual

Table 13-1 Genetically Engineered Traits in Agricultural Products (Accomplished or Under Development)

Trait	Advantage	Sample Product
Pest-resistance	Less damage by insects, viruses, bacteria, etc.	Corn-borer-resistant corn
Herbicide-resistance	Herbicides will kill only weeds, not crops	Cotton
Delayed ripening	Can be shipped with less damage	Tomato
Miniature size	Improved eating quality	Watermelon
Improved sweetness	Better tasting	Sweet peas
Cold-resistance	Withstands freezing and thawing	Strawberries
High starch	Absorbs less oil when fried	Potato
Polyester gene added	Better fiber properties	Cotton
Growth hormone added	Faster growth	Salmon
Human gene for alpha-1-antitrypsin	Can be extracted for human use	Sheep
Hepatitis B virus protein added	May provide immunity to hepatitis	Bananas

plants by cloning. (Recall from Chapter 11 that **cloning** is the production of a *clone*, a new organism that is genetically identical to an existing organism. *Cloning* also refers to the production of identical copies of a gene.) All the cloned plants will carry the recombinant DNA. The tobacco plant in Figure 13-9a (p. 239) was engineered with the protein that makes fireflies light up: namely, the enzyme luciferase. The plant's eerie glow clearly demonstrates that the gene has been incorporated successfully into the plant genome and that the firefly protein is being synthesized by the plant cells. Geneticists have transferred genes to plants from North Atlantic fish that confer cold tolerance and from bacteria that direct synthesis of a natural insecticide.

Another method of getting new genes into plants is much more direct: blasting the foreign DNA into the plant cells directly by using a "gene gun" (Fig. 13-9b). Because genes are shot directly into seedlings, this approach is useful for plants that do not grow well in culture. It can introduce several genes at once, although the success rate is low. Scientists at Agracetus, Inc., recently used the gene gun to inject cotton plants with a gene that causes the plants to produce cotton fibers whose normally hollow

Scientific Inquiry www
DNA Fingerprinting: A Tool for Medical Detectives

The use of restriction fragment length polymorphisms (RFLPs) in DNA fingerprinting has caused a revolution in many fields of biology. Nowhere has the impact of this technique been felt more than in forensics, the collection of information to be used as evidence in court proceedings. First used in a court case in 1986 in which blood samples from males in an entire British village were tested to solve the rape and murder of two young women, DNA testing has since entered the mainstream of forensic science. Because the PCR (polymerase chain reaction) technique allows amplification of minute quantities of DNA, enough DNA to produce a DNA fingerprint can easily be obtained from a semen stain, cells at the base of a hair shaft, skin fragments found under a victim's fingernails after a struggle, or a speck of dried blood (Fig. E13-1). Forensic scientists have recently discovered that they can swab off objects regularly handled by a suspect, such as a telephone handset, a briefcase handle, or the inside of vinyl gloves, and use DNA fingerprinting on the collected "fingerprints" themselves!

Some fascinating cases have used DNA from other species to solve murders. When a Phoenix, Arizona, woman was found strangled near a palo verde tree in 1992, the prime suspect denied having been at the crime scene. However, palo verde seeds were found in the bed of the suspect's pickup truck. Homicide investigators turned to Dr. Timothy Helentjaris, then at the University of Arizona. In his research on the evolution of crop plants, he had found considerable differences in the DNA fingerprints of individual plants.

Helentjaris was given the seeds from the suspect's truck along with seeds collected from a dozen palo verde trees in the area, only one of which was near the body. Without knowing which was which, Helentjaris extracted DNA from the seeds and used PCR to amplify it. DNA fingerprinting showed that the pattern from the seeds found in the truck exactly matched the pattern from only one of the dozen trees—the one nearest the body. In an important additional test, Helentjaris found that this pattern was also different from that of seeds collected from 18 trees at random sites around Phoenix. This information was crucial in demolishing the suspect's alibi, and he was found guilty of first-degree murder.

When a Canadian woman was murdered in 1994, suspicion fell on her estranged, common-law husband, who was living with his parents and a white cat named Snowball. When investigators found that a discarded man's jacket stained with the victim's blood also had a few white cat hairs clinging to it, they called on Marilyn Menotti-Raymond and co-workers at the National Cancer Institute, who had compiled an extensive map of domestic cat DNA. The researchers extracted DNA from the root of one of the hairs on the jacket, amplified it using PCR, and compared it to hair and blood samples from Snowball. The resulting perfect match was instrumental in convicting the suspect of murder in 1996.

DNA fingerprinting, combined with PCR, promises to revolutionize other areas of biology and medicine as well. For example, a group of British and Russian scientists used PCR to identify the remains of the Russian royal family, Czar Nicholas Romanov II, his wife, Alexandra, and three of their five children. (The pedigree of this family is shown in Figure 12-18). The bodies of the family, executed by the Bolsheviks (Communists) in 1918, were discovered in a unmarked grave in Russia in 1979 and positively identified in 1992.

In 1997, Oxford researchers using DNA fingerprinting announced the longest human lineage ever traced. They extracted DNA from the bones of a 9000-year-old skeleton found in 1903 in a cave near the town of Cheddar, England. Wondering if descendants of the "Cheddar Man" might still live in the area, they analyzed DNA from local families; amazingly, they found a match in a Cheddar schoolteacher.

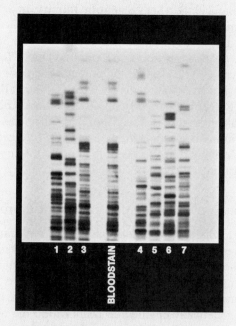

Figure E13-1 DNA fingerprinting in forensics
The differences in restriction fragment lengths provide a DNA fingerprint as individual and unique as a normal fingerprint. Here, DNA fingerprinting clearly shows that the blood of only one suspect matches that found at the crime scene. Can you identify the guilty suspect? [Cellmark Diagnostics, Germantown, MD]

inner core contains a small amount of the plastic polyester, producing a "natural" synthetic blend. Research is under way to increase the polyester content and perhaps incorporate other materials, such as keratin (the protein that makes up hair), into cotton fibers. Fabric made from this engineered cotton might be stronger, more heat-retaining, and faster-drying than normal cotton.

Bioengineering Can Make Plants Resistant to Pests
By far the largest agricultural application of genetic engineering lies in using genes that give crops an edge in the battle against insects, diseases, weeds, and other pests. The most common approach is to incorporate genes that allow crop plants to resist the effects of herbicides. Thus, herbicides can be applied after the crop plants have sprouted

(a)

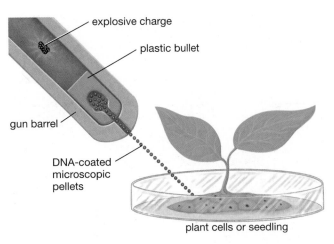

(b)

Figure 13-9 *Inserting genes into plant cells*
(***a***) Genetic engineering techniques have been used to insert a firefly gene that codes for the enzyme luciferase into this tobacco plant. The enzyme breaks down the chemical luciferin, releasing light in the process. This plant has taken up water containing luciferin. (***b***) A "gene gun" literally blasts microscopic DNA-coated pellets directly into the cells of plants.

to kill competing weeds. Strains of cotton have already been genetically engineered to resist four types of herbicide. Transgenic crop plants, including lettuce, corn, tomatoes, and wheat, are made resistant to the popular herbicide Roundup®.

To create resistance to pests, the most common strategy is to incorporate genes from a bacterium (*Bacillus thuringiensis,* or Bt) that code for the synthesis of a chemical toxic to insects. The Bt toxin gene has been introduced into more than 50 crop plants, including corn and soya beans. Genes that confer resistance to plant viruses are also being engineered into plants such as squash (Fig. 13-10a). In 1997, researchers announced that they had isolated the first gene conferring resistance to nematodes, microscopic roundworms that destroy $100 billion worth of crops worldwide each year. The gene, found in wild beets, has been transferred into sugar beets. Farmers can also plant potatoes that have been engineered with a Bt gene to resist the Colorado potato beetle, a major insect pest (Fig. 13-10b).

Genetic Engineering May Improve Other Qualities of Food Plants and Animals

Nothing can compare with the taste of vine-ripened tomatoes, but with modern harvesting and shipping tech-

niques, vine-ripened tomatoes would resemble tomato paste by the time they reached your local produce department. The traditional "solution" has been to pick and ship tomatoes when they are firm and green, then artificially ripen them with ethylene gas—producing soft, reddish tomatoes with all the flavor of wet cardboard. Enter genetic engineers, who have manipulated tomato genes to inhibit production of the enzyme that causes softening. Other advances on the horizon include the development of vegetables with higher protein or oil content and crops that will be more drought-resistant and salt-tolerant.

Progress in the genetic engineering of livestock has been much slower and less successful. In an attempt to produce faster-growing, leaner hogs, researchers transferred human and cow growth-hormone genes into pigs. On a high-protein diet, the pigs indeed grew faster, but they developed arthritis, ulcers, and sterility and died prematurely. Growth-hormone genes from rainbow trout have been introduced into carp, causing them to grow 40% faster than normal. Is this good? We discuss some of the environmental implications of biotechnology in "Earth Watch: Do Recombinants Hold Environmental Promise or Peril?" (p. 242).

(a)

(b)

Figure 13-10 *Transgenic crops*
(a) This squash has been genetically engineered to resist two common plant viruses. *(b)* NewLeaf® potatoes (middle row), which have been engineered to produce a protein that resists the Colorado potato beetle, show a clear advantage over their nonengineered relatives planted on either side.

Will Biotechnology Create a Real Jurassic Park?

In the early 1990s, scientists from each of three laboratories made stunning announcements that they had isolated DNA from insects preserved in amber (fossilized tree sap) 30 million and 125 million years ago and from a dinosaur fossil. Could the DNA in dinosaur blood be found in the stomachs of blood-sucking insects almost perfectly preserved in amber (Fig. 13-11)? Can DNA really be extracted from fossil bones? If so, could dinosaur DNA be inserted into lizard eggs and used to produce a living dinosaur, as in the movie *Jurassic Park*? Recent research has dashed hopes of finding even small, intact pieces of DNA from dinosaurs, or from amber-preserved insects. The most exacting techniques have failed to replicate the insect findings, and the reported "dinosaur DNA" turned out to be a small piece of human DNA that had accidentally contaminated the PCR process used to amplify it.

Figure 13-11 *A source of prehistoric DNA?*
Insects (such as this biting midge) some 100 million years old have been found almost perfectly preserved in amber, which dehydrates them. Recent research has dashed hopes of finding intact DNA from these insects, from their stomach contents, or from the bones of fossilized dinosaurs.

5) What Are Some Medical Uses of Biotechnology?   *WWW*

Biotechnology applied to the human genome promises to have a profound and growing impact on human health. One increasingly available application is the use of this technology to screen for genetic defects. Potential parents can learn if they are carriers of a genetic disorder; an embryo can be diagnosed early in a pregnancy (see "Health Watch: Prenatal Genetic Screening," p. 248). In the following sections we discuss some other important medical applications of biotechnology for today and tomorrow.

Knockout Mice Provide Models of Human Genetic Diseases

When the normal gene whose mutated state causes a specific human disease has been located and sequenced, genetic similarities among species commonly allow the same gene to be identified in laboratory animals, such as mice. Genetic-engineering techniques can then be used to substitute a disrupted version of the gene into the mouse chromosome. First performed in 1989, the production of such a *knockout mouse* takes about a year and costs about $1 million. It combines recombinant DNA techniques, cell culturing, and manipulation of early-stage embryos, followed by selective breeding. The resulting mice lack a specific gene. By studying these mice and their defects, researchers can clarify the role of the normal gene. Scientists can also study the effects of different treatments on the disorder before the treatments are applied to humans. One of the more than 100 knockout mice produced to date provides a model for cystic fibrosis.

Genetic Engineering Allows Production of Therapeutic Proteins

Using giant vats of bacteria that carry and express human genes (see Fig. 13-6), biotechnology firms generate large quantities of human proteins. The first human protein made by recombinant DNA technology was insulin, grown in a nondisease-causing form of *Escherichia coli*, a common intestinal bacteria. Prior to 1982, when recombinant human insulin was first licensed for use, diabetics used insulin extracted from the pancreas of cattle slaughtered for beef. Although cow insulin is very similar to human insulin, slight differences caused an allergic reaction in about 5% of diabetics—a problem avoided with the recombinant insulin. Human growth hormone, also grown in *E. coli*, became available in 1985. Other human proteins produced by recombinant DNA technology are listed in Table 13-2.

Table 13-2 Genetically Engineered Pharmaceutical Products Available or in Clinical Testing

Gene Product	Condition Being Treated	Produced by
Alpha-1-antitrypsin	Emphysema	Bacteria, mammalian cells
Epidermal growth factor	Burns, skin transplants	Bacteria
Erythropoietin	Anemia	Bacteria
Factor VIII	Hemophilia	Bacteria
Gamma interferon	Cancer	Yeast
Granulocyte colony-stimulating factor	Cancer	Mammalian cells
Hepatitis B vaccine	Hepatitis	Yeast, banana
Human growth hormone	Dwarfism	Bacteria
Insulin	Diabetes	Bacteria
Interleukin-2	Cancer	Bacteria
Superoxide dismutase	Transplants	Bacteria
Tissue plasminogen activator	Heart attacks	Mammalian cells

If vats of *E. coli* can serve as "biofactories," why can't groves of banana trees, fields of corn, or herds of grazing mammals, such as sheep, cows, or goats? One fascinating application of genetic engineering is to create "edible" vaccines. To accomplish this, researchers engineer a protein from an infectious virus into an edible plant. The protein is expected to cause an immune response that allows the body to fight off a later infection by a real virus. To date, a protein from the hepatitis B virus has been successfully engineered into a banana, and volunteers will soon be testing this tasty vaccine. Agracetus, a genetic engineering corporation, recently announced the creation of "plantibodies"—human antibody proteins grown in plants. Researchers have transplanted into corn the human gene for antibodies that bind tumor cells. These antibodies will carry radioisotopes to cancer cells, selectively killing them. Although human antibodies can be grown in animal cell cultures, the process is expensive and the yield is small. In contrast, Agracetus scientists believe they could produce enough of this anticancer antibody for the entire U.S. market from just 30 acres of corn. Clinical trials of these plantibodies will begin in 1998.

Cows, goats, pigs, rabbits, rats, and sheep have all been engineered to produce various therapeutic human proteins. For example, people who inherit a potentially fatal form of emphysema have a defective gene for the production of the protein alpha-1-antitrypsin (AAT), which they must take as a drug. The gene for the production of human AAT has now been introduced into sheep, who then secrete AAT in their milk—as much as 35 grams per liter (just over 1 oz per quart of milk).

To create transgenic "pharm" animals (animals used in generating pharmaceutical products), human genes of interest are cloned and introduced into the DNA of livestock. The human gene is inserted at a specific site that regulates the production of milk proteins and is activated only in the mammary tissue. Thus, the human protein is secreted in milk but not expressed elsewhere in the animal's body. The recombinant DNA is injected into the nucleus of a fertilized egg of the appropriate animal (Fig. 13-12); the egg is then implanted and develops in the uterus of a surrogate mother of the same species. The resulting offspring are tested for the gene and then bred with each other to produce animals homozygous for the human gene.

Human Gene Therapy Is Just Beginning

More than 4000 genetic diseases in humans are caused by damage to a single gene; many more, including cancer, arthritis, and asthma, involve impaired functioning of one or more genes. Thus, the possibilities for human gene therapy are enormous. There are two possible approaches to human gene therapy: (1) replacing defective genes in somatic (body) cells or (2) altering the genetic composition of the fertilized egg, thus permanently repairing the genetic defect in the individual. Ever since federally

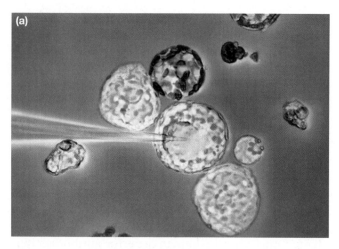

Figure 13-12 Transgenic animals
(a) A glass tube with a fine point is used to inject cloned DNA into the nucleus of a mammalian fertilized egg. The egg will be transferred to the uterus of a surrogate mother. *(b)* A transgenic cow has been genetically engineered to secrete lactoferrin in its milk. Lactoferrin is a protein normally found in human breast milk.

Earth Watch
Do Recombinants Hold Environmental Promise or Peril?

Scarcely a century ago, the towering American chestnut tree dominated many eastern hardwood forests (Fig. E13-2a). Today the splendor of the chestnuts is barely a memory; few notice the struggling sprouts emerging from the roots of the former giants. The American chestnut has been all but driven to extinction by the ravages of the chestnut blight fungus, inadvertently imported from Asia around 1900 on Asian chestnuts. However, there is a possibility that the American chestnut will reappear, thanks to the techniques of biotechnology. Scientists have engineered a weakened chestnut blight fungus that carries a fungal virus. They hope that after they infect the trees with the recombinant fungus, the debilitating virus will spread from it to

the lethal fungus, allowing the chestnuts to recover (Fig. E13-2b). Because Asian chestnuts resist the blight, scientists are also trying to identify and transfer the gene conferring blight-resistance from the Asian chestnut to the American variety.

In 1975, Kurt Benirschke of the San Diego Zoo initiated a program to freeze tissue samples from a variety of endangered species. Although he never envisioned this application, the cloning of Dolly the sheep (see "Scientific Inquiry: Much Ado About Dolly" in Chapter 11) raises the possibility of increasing the populations of endangered animals that may be nearly extinct in the wild and do not reproduce well in zoos. Surrogate mothers of related species might bear the young of endan-

Figure E13-2 Biotechnology may help the American chestnut
(a) A healthy American chestnut, which escaped the blight because it had been transplanted far west of the chestnut's normal range before the invasion of the blight fungus. *(b)* Chestnut blight fungus has produced a lesion on this chestnut tree. The lesioned area was treated with infected fungus, and the tree is walling off the infection with a growth of new tissue.

approved human gene therapy trials first began in the United States in 1990, all trials have been directed to body cells. Many disappointing results have been off-set in part by some dramatic success stories. The field is young, and the challenges are staggering.

A major challenge is to deliver functional genes into the proper cells so that the genes will be expressed, pro-ducing the functional protein when and where it is needed. Two approaches are used. One approach is to introduce gene-carrying vectors directly into appropriate sites in the body; this approach is being tested in cystic fibrosis patients. The second and more common approach is to remove cells in which the defective gene is needed, cul-ture them in test tubes with vectors that deliver the normal gene into the cells, then return the cells to an appropriate site in the body. This therapy has been used to combat SCID, described below.

Several vectors for human genes are under study, each with advantages and problems (Table 13-3). One promis-ing artificial vector is the liposome, a hollow sphere of lipid encasing the therapeutic gene. Because lipids are compatible with cell membranes, they can carry DNA into cells where it may enter the nucleus and be ex-pressed. Some cystic fibrosis patients are inhaling an aerosol spray that contains a normal copy of their de-fective gene, encased in liposomes. Medical researchers hope that the normal gene will make its way into cells that line the lungs, alleviating the deadly symptoms of this disease.

A retroviral vector has successfully been used in treat-ing the genetic disorder called *SCID* (*severe combined immunodeficiency*). In this disorder, a defective gene pre-vents the synthesis of an enzyme (adenosine deaminase; ADA) crucial to immune-system function. Until the 1980s, when physicians began administering ADA iso-lated from cattle, children with SCID died in infancy, un-able to fight off simple infections. In 1990, researchers performed the first test of recombinant gene therapy on an SCID patient, 4-year-old Ashanti DeSilva. Her white blood cells were removed, genetically altered with a

gered—or even recently extinct—species from which living tis-sue had been saved and frozen. The hurdles are enormous, but so are the possibilities.

What perils do engineered species pose? One concern is that with crop plants now resistant to herbicides, farmers will use far more of these chemicals, jeopardizing human health and the environment. Further, plant genes are occasionally carried be-tween species by plant viruses or bacteria, so genes for herbicide resistance might be transferred from crops to weeds. Even if the gene is not transferred, increased herbicide use will select for weeds with their own mutations for herbicide-resistance. In ei-ther of these scenarios, herbicides might ultimately become much less effective. Similarly, widespread cultivation of plants containing the Bt toxin would favor the evolution of insects that could tolerate or inactivate the toxin, rendering it useless.

Successful experiments with domestic crops have led to re-search into the genetic engineering of species that provide wood, fiber, animal forage, ornamental landscaping, or food (such as faster-growing salmon, Fig. E13-3). These types of genetically modified species are much more likely than crops (such as cot-ton or corn) to invade natural ecosystems, with unforeseen and unpredictable consequences. Will they spread and displace na-tive species, disrupting the ecological balance?

Genetic engineers correctly assert that humans have prac-ticed crude genetic engineering for millennia, breeding plants and animals with desired properties. Modern biotechnology is merely a faster, more precise version of standard agricultural practice. What's more, various forms of genetic recombination occur all the time in nature. The features of biotechnology that provide both vast promise and potential threat are the specifici-ty with which biotechnology can (or will soon be able to) direct genetic changes; the speed with which genetic changes can be made; and the ability to transfer genetic material between species in ways that could never occur in the wild.

Figure E13-3 Transgenic fish could displace natural varieties
Growth hormone has been genetically engineered into several types of fish, such as the Atlantic salmon (bottom; compare it with the two small siblings, top). If they escape into natural eco-systems, they might displace slower-growing wild varieties and have a greater impact on other species.

Table 13-3 Possible Vectors for Human Genes

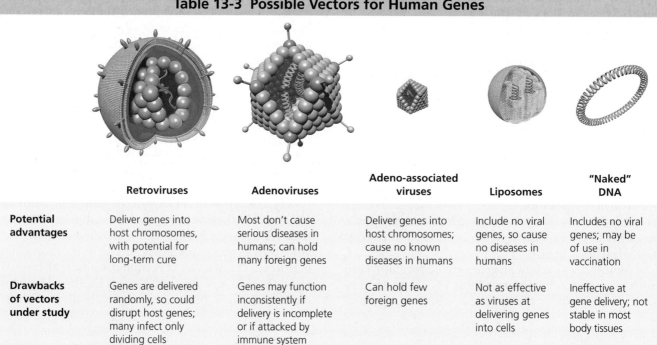

	Retroviruses	Adenoviruses	Adeno-associated viruses	Liposomes	"Naked" DNA
Potential advantages	Deliver genes into host chromosomes, with potential for long-term cure	Most don't cause serious diseases in humans; can hold many foreign genes	Deliver genes into host chromosomes; cause no known diseases in humans	Include no viral genes, so cause no diseases in humans	Includes no viral genes; may be of use in vaccination
Drawbacks of vectors under study	Genes are delivered randomly, so could disrupt host genes; many infect only dividing cells	Genes may function inconsistently if delivery is incomplete or if attacked by immune system	Can hold few foreign genes	Not as effective as viruses at delivering genes into cells	Ineffective at gene delivery; not stable in most body tissues

retrovirus vector for the functional gene, then replaced. With periodic treatments to replace the altered immune cells that die, Ashanti can live the life of a normal, active girl. In 1993, researchers attempted a more permanent cure: They added the functional ADA gene to stem cells that produce white blood cells. Stem cells, which normally reside in bone marrow, are also found in umbilical-cord blood. When Crystal and Leonard Gobea, who had already lost a child to SCID, learned from amniocentesis that their second fetus was also affected, they agreed to participate in a bold new experiment. Stem cells were isolated from newborn Andrew Gobea's blood, genetically modified, and injected back into the baby 4 days later (Fig. 13-13a,b). During the next several years, while Andrew received supplemental ADA injections, white cells with the functional genes gradually replaced the defective ones. His physicians hope that the ADA injections will soon be unnecessary.

Many other human genetic diseases are currently being treated in clinical trials that employ gene therapy. These diseases include at least a dozen types of cancer, hemophilia, rheumatoid arthritis, emphysema, and AIDS.

The Human Genome Project Aims to Sequence the Entire Human Genome by the Year 2005

The Human Genome Project was conceived in 1986 and first launched in 1990 by the National Institutes of Health (NIH). The project has become an international effort to determine the exact location of the sequence of nucleotides in each of the roughly 100,000 human genes. Researchers hope to achieve a 99.9% accuracy rate and to complete the project by the year 2005. If this book were to contain the entire nucleotide sequence, it would be about 350,000 pages long! When first conceived, the time, effort, and cost (an estimated $3 billion) of sequencing the human genome seemed daunting. Recent developments in RFLP analysis, PCR, and increasingly rapid automated DNA sequencing technologies now make the goals of the project attainable. At the time of this writing, researchers predict that well over 200 million bases will be sequenced by May 1998, considerably ahead of earlier projections.

The project has already provided insights into genetic disease. Researchers have found that some diseases, such as Huntington's, arise as the result of repeating triplets of nucleotides that code for an amino acid. In the Huntington's gene, for example, the CAG triplet (coding for glutamine) is repeated as many as 121 times, in contrast to the normal range of 10 to 37 repeats. Another finding is that genes may contain mutations at many different sites, all producing disease. More than 600 different mutations have been found in the enormous gene that is defective in cystic fibrosis; some of these give rise to milder forms of the disease than do others.

The Human Genome Project is probably science's most ambitious response to the command of the ancient Oracle at Delphi: "Know thyself." The answers to a host of questions about humankind reside in the nucleotide sequences of our genes. Further, humanity is heir to about 4000 genetic disorders. Knowing the nucleotide sequence of the entire human genome will eventually allow rapid screening and prenatal diagnosis of a variety of genetic disorders and better therapies for many diseases.

(a)

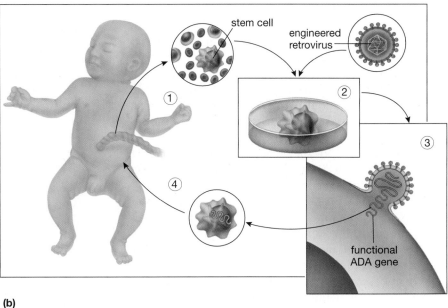

(b)

Figure 13-13 Hope through gene therapy
(a) Andrew Gobea (shown with his mother) received gene therapy on stem cells as a newborn. Eventually, he may produce adequate numbers of normal immune cells on his own. *(b)* Genetic engineering of stem cells may permanently replace the defective ADA gene. ① Stem cells are harvested from umbilical cord blood. ② A retrovirus engineered to contain normal human ADA genes is mixed with the stem cells in culture. ③ The retrovirus transmits its DNA, including the functional gene, into the stem cells. ④ The engineered stem cells are injected into the same newborn to take up residence in bone marrow and produce normal immune cells.

6 What Are Some Ethical Implications of Human Biotechnology?

www

As a direct or indirect result of the Human Genome Project, the ability to screen for human diseases is increasing at an exponential rate. One company has developed a technology that can simultaneously analyze DNA from 500 patients for 106 different mutations in seven genes. Physicians will soon have access to dozens of easily administered tests that will reveal whether a patient carries a specific gene. These new powers demand a new set of decisions, both ethical and economic, by individuals, physicians, and society.

Genetic Tests for Cystic Fibrosis and Inherited Forms of Breast Cancer Illustrate Potential Problems

Cystic fibrosis is the most common fatal genetic disease in the United States. To reduce the 850 or so CF babies born in the United States each year (most to families with no known history of CF), an advisory panel of the National Institutes of Health has recommended that testing for CF be offered to *all* couples who consider having a baby. This is the first genetic test to be recommended to the general public.

If this policy is adopted, many couples who have never even heard of CF will learn that they both carry the CF gene and that their chances are 1 in 4 of producing a child with CF. Will they get the extensive medical information and counseling they need to deal with the results? They will need to be educated about the symptoms, treatments, and monetary and emotional costs of the disorder, about the "genetic lottery" they are about to enter, and about their options. Should they remain childless or adopt a child? Should they conceive and then use prenatal genetic diagnosis to determine whether their child will develop CF? If tests are positive, then what? Is abortion acceptable to the couple? If not, they will give birth to a child who will struggle for breath and be hospitalized repeatedly for respiratory infections despite daily doses of antibiotics. The child will need to lie down and be pounded thoroughly on the chest and back twice a day to dislodge the thick buildup of mucus (see the chapter-opening photo), and will require a special diet fortified with digestive enzymes. Despite these treatments, their child will have a 50% chance of dying before he or she reaches age 30. The costs will be staggering. Should society bear these costs for a family who knowingly takes this risk? Are insurance companies justified in denying coverage for CF treatment to the children of parents who both carry the CF gene?

Further dilemmas are posed by a genetic test recently developed for a gene linked to a specific type of hereditary breast and ovarian cancer. The presence of this gene does not condemn a woman to cancer, but it does mean that her chances of developing cancer are very high. Some women find living with this "genetic time bomb," and the risk of passing it to their children, very distressing. Many have chosen not to have the test. Others, finding that they carry the gene, have made the painful decision to have both breasts and/or both ovaries surgically removed. Genetic counselors are concerned that women who find they have not inherited the disease might become lax about regular screening procedures; these women still have the same risks as others for noninherited types of breast cancer. Many women who choose to be tested for the gene do so anonymously or under a false name to prevent the information from reaching their insurance companies.

The increasing availability and use of tests for genetic diseases raise major concerns about genetic discrimination by insurance companies and employers. Many people worry that, under pressure to hold down rising costs, insurance companies will require information about genetic diseases and will use this information as a basis for denying or restricting the availability of life insurance or health insurance. Many people with a family history of inherited disease have already been denied insurance and in some cases employment as a result. As more genetic information becomes available and treatments become more costly, this trend is sure to intensify.

Sickle-Cell Anemia and Tay-Sachs Disease Illustrate the Hazards and Benefits of Genetic Screening Programs

An example of the hazards of widespread genetic screening without adequate education occurred in the 1970s when a major screening program for **sickle-cell anemia** was launched, and even made mandatory in some areas. Although the intentions were good, the results were misused through ignorance and prejudice. Healthy carriers of the sickle-cell trait (most of whom are black) were led to believe they were sick. Some black carriers were denied insurance, others were blocked from admission to the U.S Air Force Academy or from jobs as flight attendants—on the inaccurate assumption that they were more likely to be affected by the lack of oxygen at high altitudes. Suggestions that carriers (who make up almost 7% of the African-American population) should avoid having children led to fears that the testing was intended as a means of reducing the numbers of this racial minority.

Testing for **Tay-Sachs disease**, which is prevalent among Jews of Eastern European descent, is a contrasting success story. Tay-Sachs is a fatal degenerative disease in which the victim gradually loses mental and physical abilities, becomes blind and deaf, and dies in early childhood (Fig. 13-14). A recessive disease, it is caused by a deficiency of enzymes that regulate lipid breakdown in the brain. It is such a devastating illness that few people will knowingly give birth to such a child. Voluntary genetic testing for Tay-Sachs within the Jewish community, coupled with prenatal testing and careful genetic counseling, has dramatically reduced the incidence of this fatal disorder.

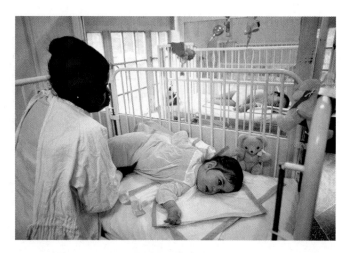

Figure 13-14 Tay-Sachs disease
Apparently bright, normal children with Tay-Sachs disease begin to deteriorate at about 6 months of age. As the nervous system is progressively destroyed, the children gradually lose all ability to function, and most die in early childhood. There is no treatment.

The Potential to Clone Humans Raises Further Ethical Issues

The cloning of Dolly the sheep from the nucleus of an adult cell introduced into the cytoplasm of an egg (see "Scientific Inquiry: Much Ado About Dolly" in Chapter 11) led to a flurry of debate and proposed legislation regarding human cloning. There are no known technical barriers to the eventual development of human cloning. Are there ethical barriers that should restrict this technology to nonhuman species? One argument is that there is no valid reason to clone humans. Perhaps a small number of infertile couples would want to perpetuate their genes in this way, but there are already so many reproductive options (including sperm or egg donors) that this is not a pressing need. For some people, the possibility that cloning would be used by unscrupulous individuals in positions of power to perpetuate themselves or to create copies of certain people to fill specific functions raises the specter of a "Brave New World" type of society. Some argue that biological uniqueness is a fundamental human right and intrinsic to human dignity.

Countering these arguments are those who point out that identical twins are more similar to one another than clones would ever be, because twins share not only the same genes, but also cytoplasmic factors from the same original egg cell, the same uterine environment, and the same home environment and upbringing (Fig. 13-15). An individual trying to produce a self-copy might be sorely disappointed at the differences these environmental variables (not to mention growing up in a different generation) might produce. An important spin-off of human cloning techniques would be the ability to "reawaken" dormant genes in a differentiated human cell. Researchers might then regenerate damaged tissues, such as neurons, that are not normally replaceable in adults. A genetic defect in an embryo might be permanently corrected by starting with *in vitro* fertilization and culture of embryonic cells (already performed routinely), followed by gene therapy and cloning techniques (Fig. 13-16).

As governments debate whether to enact laws forbidding research on human cloning, one fact is painfully clear: Advances in our understanding and ability to manipulate the genes of humans and other species threaten to outpace the ability of society to understand and utilize the resulting technologies effectively. The next decade will see considerable upheaval as society struggles to come to grips with this new knowledge and the power that accompanies it.

Figure 13-15 *Identical twins*
Identical twins are more alike than a person would be to his or her cloned offspring, because the cells of identical twins received cytoplasm from the same original egg cell and developed in the same uterus, and because twins share the same parents, home, and social environment.

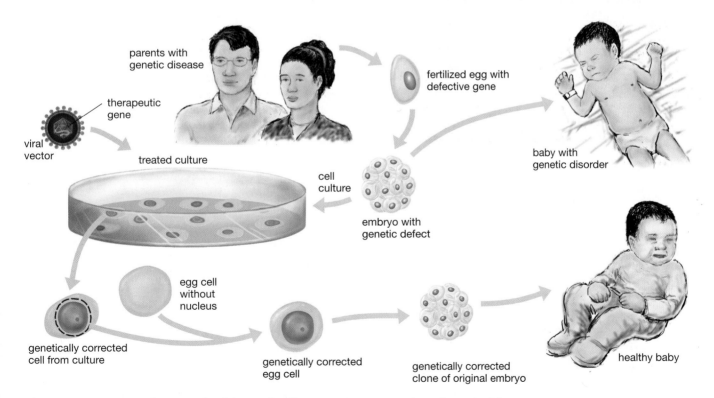

Figure 13-16 *Human cloning technology might allow permanent correction of genetic defects*
Researchers might derive human embryos from eggs fertilized in culture dishes, using sperm and eggs from the natural parents, one or both of whom have a genetic disorder. When an embryo containing a defective gene grows into a small cluster of cells, a single cell could be removed from the embryo and the defective gene replaced by means of an appropriate vector. Then the repaired nucleus could be implanted into another egg (taken from the mother) whose nucleus had been removed. The repaired, diploid egg cell could then be implanted in the mother's uterus for normal development.

Health Watch *WWW*
Prenatal Genetic Screening

Prenatal diagnosis of a variety of genetic disorders, including cystic fibrosis, sickle-cell anemia, Tay-Sachs disease, and Down syndrome, requires samples of fetal cells or chemicals produced by the fetus. Presently, there are two main techniques used to obtain these samples—*amniocentesis* and *chorionic villus sampling*—and a technique for analyzing fetal cells from maternal blood is under development. Once samples are collected, several tests, including some that use recombinant DNA techniques, can be performed that allow prenatal diagnosis of many genetic disorders.

Amniocentesis The human fetus, like all animal embryos, develops in a watery environment. As you'll see in Chapter 36, a waterproof membrane called the *amnion* surrounds the fetus and holds the fluid. As the fetus develops, it sheds some of its own cells into the fluid, which is called *amniotic fluid*. When a fetus is 16 weeks or older, amniotic fluid can be collected safely by a procedure called **amniocentesis**. A physician determines the position of the fetus by ultrasound scanning (Fig. E13-4a), inserts a sterilized needle through the abdominal wall,

(a)

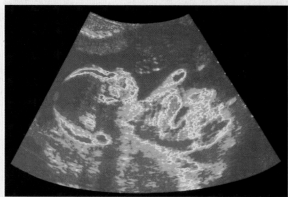

the uterus, and amnion, and withdraws 10 to 20 milliliters of fluid (Fig. E13-4b). Biochemical analysis may be performed on the fluid immediately, but there are very few cells in the fluid sample. For many analyses, such as karyotyping for Down syndrome, the cells must first be grown in culture, where they multiply. After a week or two, there are normally enough cells for karyotyping or other analyses.

Chorionic Villus Sampling As you'll see in Chapter 36, the *chorion* is a membrane that is produced by the fetus and becomes part of the placenta. The chorion produces many small projections, called *villi*; the loss of a few villi seems to cause no harm. In **chorionic villus sampling (CVS)**, a physician inserts a small tube into the uterus through the mother's vagina and suctions off a few fetal villi for analysis (Fig. E13-4b). CVS has two major advantages over amniocentesis. First, it can be done much earlier in pregnancy—as early as the eighth week. This factor is especially important if the woman is contemplating a therapeutic abortion in case the fetus has a major defect. Second, the sample contains a much higher concentration of fetal cells than amniocentesis can obtain, so analyses can be performed immediately rather than a week or two later. However, chorionic cells tend to be more likely to have abnormal numbers of chromosomes (even when the fetus is normal), which complicates karyotyping.

Fetal Cells from Maternal Blood Fetal cells become detectable in maternal blood between the sixth and twelfth weeks of pregnancy. The goal of a new procedure is to harvest and analyze fetal cells collected from a 20-milliliter sample (just over a tablespoon) of maternal blood. If successful, this procedure might replace both amniocentesis and CVS as a means of obtaining fetal cells. By eliminating the need to penetrate the uterus, blood sampling greatly reduces the risk of injury and infection to both mother and fetus. But only about 1 in every 100,000 cells in ma-

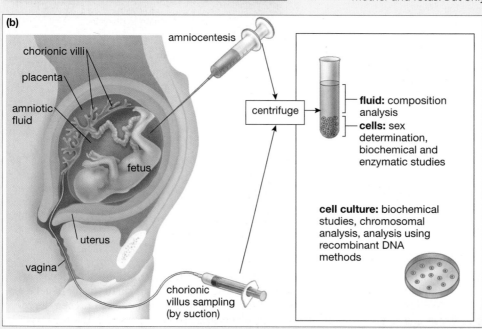

Figure E13-4 Prenatal cell sampling techniques **(a)** Ultrasound imaging reveals the position of the fetus (21 weeks). **(b)** Two methods of obtaining fetal cell samples—amniocentesis and chorionic villus sampling—and some of the tests performed on the fetal cells.

ternal blood comes from the fetus, so isolating fetal cells efficiently is a major challenge. A large study is currently in progress to determine the feasibility of this technique.

Analyzing the Samples   Several types of analyses can be performed on the fetal cells or on the amniotic fluid (see Fig. E13-4b). Biochemical analysis is used to determine the concentration of chemicals in the amniotic fluid. For example, Tay-Sachs disease and many other metabolic disorders can be detected by the low concentration of the enzymes that normally catalyze specific metabolic pathways or by the abnormal accumulation of precursors or by-products. Analysis of the chromosomes of the fetal cells can show if all the chromosomes are present, if there are too many or too few of some, and if any chromosomes show structural abnormalities.

Recombinant DNA techniques can be used to analyze the DNA of fetal cells to detect many defective alleles, such as those for cystic fibrosis, Tay-Sachs, or sickle-cell anemia. Prior to the development of PCR, fetal cells typically had to be grown in culture

for as long as 2 weeks before cells had multiplied sufficiently. Now, the second step in prenatal diagnosis is to extract the DNA from a few cells and to use PCR to amplify the region containing the gene in question. After a few hours, enough DNA is available for techniques such as RFLP analysis, which is used to detect the allele that causes sickle-cell anemia. Technicians cut the PCR-amplified DNA with restriction enzymes. Because the single-nucleotide substitution that causes sickle-cell anemia happens to destroy a cutting site for a particular restriction enzyme, there will be a recognizable difference in the length of certain restriction fragments. This difference allows the hemoglobin genotype of the fetus to be easily determined. If the infant is homozygous for the sickle-cell allele, some therapeutic measures can be taken. In particular, regular doses of penicillin greatly reduce bacterial infections that otherwise kill about 15% of homozygous children. Further, knowing that a child has the disorder ensures correct diagnosis and rapid treatment, should "sickling crises" occur.

Summary of Key Concepts

1) What Is Biotechnology?

Biotechnology is any industrial or commercial use or alteration of organisms, cells, or biological molecules to achieve specific practical goals. Modern biotechnology generates altered genetic material via genetic engineering. Genetic engineering may involve the production of recombinant DNA by combining DNA from different organisms. DNA is transferred between organisms by means of vectors, such as bacterial plasmids. The resulting organisms are termed *transgenic*. Some major goals of genetic engineering are to increase our understanding of gene function, to treat disease, and to improve agriculture.

2) How Does DNA Recombination Occur in Nature?

DNA recombination occurs naturally through processes such as sexual reproduction (during crossing over), bacterial transformation, in which bacteria acquire DNA from plasmids or other bacteria, and viral infection, in which viruses incorporate fragments of DNA from their hosts and transfer the fragments to members of the same or other species.

3) What Are Some of the Methods of Biotechnology?

The identification and amplification of a specific gene may involve (1) constructing a DNA library: The entire genome is broken into fragments by restriction enzymes, which are also used to insert these fragments into bacterial plasmids. The plasmids are incorporated into bacteria, where they are maintained and copied. (2) Localizing the gene of interest: The nucleotide sequence of a gene may be identified if the amino acid sequence of the protein for which the gene codes is known, or if mRNA from the protein can be collected.

(3) Amplification of the gene: To produce multiple copies of a recombinant gene, the bacteria containing the gene may be grown in culture, or the polymerase chain reaction (PCR) technique may be used.

4) What Are Some Applications of Biotechnology?

Once sufficient quantities of a gene or DNA fragment are available, the nucleotide sequence can be determined. Researchers can then synthesize and study the coded protein or devise tests for mutated forms of the gene. Amplified DNA might be used in DNA fingerprinting, in which restriction enzymes are used to break the DNA into fragments whose different lengths are characteristic of different individuals (a phenomenon called restriction fragment length polymorphism, or RFLP). DNA fingerprints based on RFLPs are used to locate mutated genes within large families and then to diagnose the presence of these genes in other individuals. DNA fingerprints can establish relatedness of individuals and is used in forensics to analyze biological material at a crime scene and link it to the suspect.

Biotechnology is revolutionizing agriculture by allowing scientists to engineer plants that resist pests and herbicides and that have superior qualities for shipping and eating, and to produce better fibers.

5) What Are Some Medical Uses of Biotechnology?

Knockout mice, in which a specific gene has been rendered nonfunctional, provide laboratory animal models of human genetic diseases that can be studied to determine disease mechanisms and to test various treatments. Genetic engineering allows production of therapeutic proteins, such as human insulin, by culturing huge quantities of bacteria that contain the recombinant gene and extracting the protein. Farm animals can also be genetically engineered to secrete proteins of interest in milk.

Human gene therapy is in its infancy. Vectors such as viruses and liposomes have been used to deliver functional genes to specific cells that lack them. The vector may be delivered directly into the body, or cells may be removed from the body, mixed with the vector, which carries the genes into the cells, and the cells then returned to the body. This technique has been used successfully in treating children with a deficiency in adenosine deaminase, an enzyme essential in immune-system function.

An ambitious collaborative project is under way to sequence the entire human genome by the year 2005. The Human Genome Project has already provided insights into the nature of mutations and facilitated the development of screening tests for many defective genes. Knowledge of normal and mutated gene sequences will also promote the development of new and better therapies against disease.

6 What Are Some Ethical Implications of Human Biotechnology?

Physicians will soon have access to dozens of tests for genetic diseases. To administer these tests appropriately, the physician or genetic counselor must decide who should receive the tests, then educate the patients about the disorder and assist them in making a series of difficult decisions about their own health or that of potential offspring. Questions include whether to conceive or adopt a child, whether to have prenatal testing, and whether to abort a fetus with a fatal or permanently debilitating disorder. Should insurance companies pay the lifelong costs to treat children with genetic diseases born to parents who know they carry the defective gene? How can society protect individuals from various types of discrimination on the basis of their genetic makeup? Should research on human cloning techniques be permitted? Both the rapid acquisition of knowledge about genetics and the power to manipulate the genome of humans and other organisms threaten to outpace our ability to deal with the consequences.

Key Terms

amniocentesis *p. 248*	DNA library *p. 232*	recombinant DNA *p. 230*	transformation *p. 231*
bacteriophage *p. 231*	DNA probe *p. 233*	restriction enzyme *p. 232*	transgenic *p. 230*
biotechnology *p. 230*	gel electrophoresis *p. 236*	restriction fragment *p. 236*	vector *p. 230*
chorionic villus sampling (CVS) *p. 248*	genetic engineering *p. 230*	restriction fragment length polymorphism (RFLP) *p. 236*	
cloning *p. 237*	genome *p. 232*		
cystic fibrosis *p. 229*	plasmid *p. 231*	sickle-cell anemia *p. 246*	
DNA fingerprinting *p. 236*	polymerase chain reaction (PCR) *p. 234*	Tay-Sachs disease *p. 246*	

Thinking Through the Concepts

Multiple Choice

1. *Restriction enzymes are:*
   a. isolated from bacterial cells
   b. used to produce DNA fingerprints
   c. used to create recombinant plasmids
   d. used to create a DNA library
   e. all of the above

2. *Restriction fragment length polymorphisms*
   a. can be used to detect differences in DNA among individuals
   b. have been used to identify the Huntington's gene
   c. can be used to analyze fetal cells to detect disorders prenatally
   d. are produced through the use of restriction enzymes
   e. all of the above

3. *The polymerase chain reaction*
   a. is a method of synthesizing human protein from human DNA
   b. takes place naturally in bacteria
   c. can produce billions of copies of a DNA fragment in several hours
   d. uses restriction enzymes
   e. is relatively slow and expensive compared with other types of DNA amplification

4. *A gene gun*
   a. blasts DNA into cells, using microscopic pellets
   b. is used to break DNA apart for analysis
   c. is used in forensic medicine
   d. is used to introduce genes into bacteria
   e. creates RFLPs

5. *Fetal cells can be isolated from maternal blood*
   a. only if there is something wrong with the placenta
   b. in the first week of pregnancy
   c. but this process is still under development
   d. in very large numbers, making this the best way to collect fetal DNA
   e. by running the blood through a filter

6. *A child with a single allele for sickle-cell anemia*
   a. is likely to die before age 30 from this disease
   b. will have difficulty getting enough oxygen at high altitudes
   c. will have difficulty participating in sports
   d. will need immediate treatment at birth
   e. will be able to lead a perfectly normal life

? Review Questions

1. Describe three natural forms of genetic recombination, and discuss the similarities and differences between recombinant DNA technology and these natural forms of genetic recombination.
2. What is a plasmid? How are plasmids involved in bacterial transformation?
3. What is a restriction enzyme? How can restriction enzymes be used to splice a piece of human DNA within a plasmid?
4. What is a DNA library? Briefly describe the steps involved in creating a DNA library of a mouse genome.
5. What is a restriction fragment length polymorphism (RFLP)? Describe the use of RFLPs in locating a gene on a human chromosome, finding a murder suspect, and determining relatedness among people.

6. Describe the polymerase chain reaction. Why is it so useful in the prenatal diagnosis of genetic defects?
7. Describe several uses of genetic engineering in agriculture.
8. Describe several uses of genetic engineering in human medicine.
9. Describe amniocentesis and chorionic villus sampling, including the advantages and disadvantages of each. What are their medical uses?

Applying the Concepts

1. Discuss the ethical issues that surround the release of bioengineered organisms (plants, animals, or bacteria) into the environment. What could go wrong? What precautions might prevent the problems you listed from occurring? What benefits do you think would justify the risks?
2. Do you think that using recombinant DNA technologies to change the genetic composition of a human egg cell is ever justified? If so, what restrictions should be placed on such a use? What about human cloning?
3. Are there any conditions under which insurance companies or employers should be allowed access to the results of tests for genetic defects? Discuss this issue from the standpoint of the company or employer, the individual who was tested, and society as a whole.

4. In what ways was the program to test for sickle-cell anemia in the 1970s flawed? What criteria should be met (education, privacy guarantees, etc.) before any genetic screening test is administered to the general public?
5. If you were contemplating having a child, would you want both yourself and your spouse tested for the cystic fibrosis gene? If both of you were carriers, how would you deal with this decision?

Group Activity

DNA evidence is used to identify individuals unambiguously in a variety of situations, including criminal investigations and paternity suits. The uniqueness of each person's DNA results in differences in patterns of restriction enzyme digestion. The gel on the right shows DNA evidence from a paternity suit. A child's DNA fragment pattern (DNA profile) will be a mixture of the father's and mother's patterns. In groups of three, answer the questions concerning the DNA evidence in this paternity suit.

(a) Would you expect the child's profile to contain all of the fragments from either parent? Why or why not?
(b) Which male(s) cannot be the father of this child?
(c) Which male(s) might be the father?
(d) Does this evidence prove that one of these men is the child's father? Explain.
(e) Does this evidence prove that one or more of these men could not be the father? Explain.

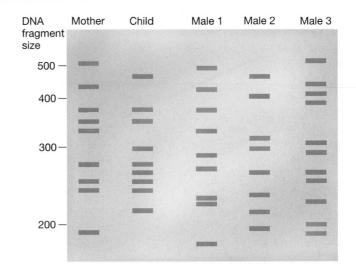

For More Information

Beardsley, T. "Vital Data." *Scientific American*, March 1996. Discusses the Human Genome Project and the problems and prospects of the tests for genetic diseases that are arising as a spin-off of this project.

Cohen, J. "Can Cloning Help Save Beleaguered Species?" *Science*, May 30, 1997. Discusses the prospects of gene cloning to help restore populations of endangered species.

Elmer-Dewitt, P. "Cloning: Where Do We Draw the Line?" *Time*, Nov. 8, 1993. A discussion of the controversy surrounding the potential to clone a human.

Fackelmann, K. A. "DNA Dilemmas." *Science News*, December 17, 1994. Uses case histories to explore the ethical controversies surrounding human applications of genetic engineering.

Gibbs, W. W. "Plantibodies." *Scientific American*, November 1997. Clinical trials using human antibodies engineered into corn will begin in 1998.

Karapelou, J. "Gene Therapy: Special Delivery." *Discover*, January 1996. Describes several methods on trial for delivering therapeutic genes into human cells.

Karapelou, J. "The Fix Sticks." *Discover*, January 1995. Describes an apparently successful use of gene therapy against a deadly disease that causes cholesterol levels to skyrocket.

Miller, R. V. "Bacterial Gene Swapping in Nature." *Scientific American*, January 1998. How likely is it that genes introduced into bioengineered organisms might be inadvertently transferred to wild organisms?

Mirsky, S., and Rennie, J. "What Cloning Means for Gene Therapy." *Scientific American*, June 1997. Could human cloning technology help permanently repair genetic defects?

Roberts, L. "To Test or Not to Test." *Science*, 247:17–19, 1990. Should the entire childbearing population be screened for carriers of cystic fibrosis?

Ronald, P. C. "Making Rice Disease-Resistant." *Scientific American*, November 1997. Rice, a staple crop for millions of people, has finally been engineered for disease resistance.

Scientific American, June 1997. A special issue devoted to the prospects of human gene therapy.

Snow, A. A., and Palma, P. M. "Commercialization of Transgenic Plants: Potential Ecological Risks." *BioScience*, February 1997. Do genetically engineered plants pose threats to native species?

Sutherland, G. R., and Richards, R. I. "Dynamic Mutations." *American Scientist*, March–April 1994. Several genetic disorders, including Huntington's disease, are characterized by large numbers of repeated nucleotide triplets.

Travis, J. "Cystic Fibrosis Controversy." *Science News*, May 10, 1997. Might gene therapy in the womb help CF patients?

Answers to Multiple-Choice Questions

1. e 2. e 3. c 4. a 5. c 6. e

Unit Three
Evolution

Beautifully preserved large and small fossil ammonites, which are mollusks that went extinct at about the same time as the dinosaurs, roughly 65 million years ago.

"When on board H.M.S. Beagle, as naturalist, I was much struck with . . . the distribution of the inhabitants of South America, and . . . the geological relations of the present to the past inhabitants of that continent. These facts seemed to me to throw some light on the origin of species—that mystery of mysteries, as it has been called by one of our greatest philosophers."

Charles Darwin in On the Origin of Species by Means of Natural Selection (1859)

The elephants that roam the savannas of Africa today are far removed from the Siberian prairie where this mummified mammoth (inset) perished more than 10,000 years ago. Nonetheless, the living elephant and the extinct mammoth are linked by descent from a common ancestor.

Principles of Evolution 14

At a Glance

Net Watch

On-line resources for this chapter are on the World Wide Web at: http://www.prenhall.com/audesirk (click on the Table of Contents link and then select Chapter 14).

The living world presents a thoughtful observer with an apparent paradox: Life is simultaneously varied and uniform. It comes in a seemingly infinite array of shapes, sizes, and ways of perpetuating itself, but all living things have some fundamental properties in common. A bacterium on the ocean floor and an elephant on the African savanna could hardly be more different, yet both are constructed of cells and share the same biochemical machinery for specifying and manufacturing the substances required for growth and development. How can this paradox be resolved? How can we explain both the diversity and the unity of life?

The biologist's answer to these questions was neatly summarized by Theodosius Dobzhansky when he wrote, "Nothing in biology makes sense except in the light of evolution." **Evolution** is the process by which each type of organism is descended from ancestors that were similar but not identical to it. This process requires that all life share a common ancestry. After all, if each organism has an unbroken chain of ancestors leading back in time to the first of its kind, then each of those ancient ancestors must in turn have its own chain of ancestors. And these chains link to one another. For example, some distant ancestor of the elephants we know today was also the ancestor of the now-extinct mammoths that once roamed over much of the world. If we follow many such chains of ancestry far enough back in time, they must ultimately link up in a single, huge tree of life, linking all organisms.

The insight that all life-forms must have evolved by the slow accumulation of changes on a single tree of life itself evolved slowly over the history of science. Nonetheless, this pivotal idea assumed its place among the bedrock principles of biology only after Charles Darwin's presentation, in his 1859 book *On the Origin of Species*, of a massive and meticulously documented body of evidence that organisms are indeed the result of a history of descent with modification from common ancestors. Darwin (1809–1882), however, delivered not only a persuasive case for life's historical unity, but he also developed an explanation for its astonishing variety and for the exquisite match between organisms and their modes of survival. What Darwin, and independently his contemporary Alfred Russel Wallace (1823–1913), proposed was nothing less than a mechanism for evolutionary change. Darwin carefully and elegantly described this mechanism, which he called *natural selection*. This revolutionary idea was an immediate and controversial sensation that changed biology permanently and profoundly.

1 How Did Evolutionary Thought Evolve?

Evidence Supporting Evolution Emerged Even before Darwin's Time

Pre-Darwinian science, heavily influenced by theology, held that all organisms were created simultaneously by God and that each distinct life-form remained fixed, immutable, and unchanging from the moment of its creation. This explanation of how life's diversity came to be had its origins among the ancient Greek philosophers, especially Plato and Aristotle. Plato (427–347 B.C.) proposed that each object on Earth was merely a temporary reflection of its divinely inspired "ideal form." Plato's student Aristotle (384–322 B.C.) categorized all organisms into a linear hierarchy that he called the "ladder of Nature."

These two ideas formed the intellectual basis of the view that the form of each type of organism is permanently fixed. This view reigned unchallenged for nearly 2000 years. By the eighteenth century, however, several lines of newly emerging evidence began to erode the dominance of this static view of Creation.

Exploration of New Lands Revealed a Staggering Diversity of Life

As early European naturalists explored the newly discovered lands of Africa, Asia, and America, they found that the number of **species**, or different types of organisms, was much greater than anyone had suspected. Some of these exotic species closely resembled one another yet also differed in some characteristics. This unpredicted expansion of information led some naturalists to consider that perhaps species could change after all and that some of the similar species might have developed from a common ancestor.

Fossils in Rocks Resembled Parts of Living Organisms

As new lands were explored, excavations for roads, mines, and canals revealed that many rocks occur in layers (Fig. 14-1). In some cases, a few strangely shaped rocks or fragments were found embedded within one of these layers. These **fossils** (from the Latin, meaning "dug up") resembled parts of living organisms. At first, fossils were thought to be ordinary rocks that wind, water, or people had worked into lifelike forms. As more and more fossils

Figure 14-1 The Grand Canyon of the Colorado River
Layer upon layer of sedimentary rock forms the walls of the Grand Canyon, exposed in cliffs and mesas. The canyon strata (rock layers) cover more than a billion years of evolutionary history.

were discovered, however, it became obvious that they were in fact the remains of plants or animals that had died long ago and been changed into or in some way preserved in rock (Fig. 14-2). The rapidly accumulating fossil discoveries also revealed that fossils come in many forms. The classic image of a fossil is of bones or other hard parts (such as shells) that have been transformed into rock by eons of geological processing. But fossils also include casts, molds, and other impressions that organisms left in ancient sediments before decomposing. Some of the most interesting and informative fossils are the trails, burrows, tracks, or droppings that organisms left behind. In fact, any tangible trace of an organism that is preserved in rock or sediments is a fossil.

These windows into the past are fascinating in and of themselves, but the distribution of fossils in rock can also be revealing. After studying fossils carefully, the British surveyor William Smith (1769–1839) realized that certain fossils were always found in the same layers of rock. Further, the organization of fossils and rock layers was consistent: Fossil type A could always be found in a rock layer resting atop an older layer containing fossil type B, which in turn rested atop a still older layer containing fossil type C, and so on.

Fossil remains also showed a remarkable progression of form. Most fossils found in the lowest (and therefore oldest) rock layers were very different from modern forms; the resemblance to modern forms gradually increased upward toward younger rocks, as if there were indeed a ladder of Nature stretching back in time. Many of these fossils were the remains of plant or animal species that had gone *extinct*—that is, no members of the

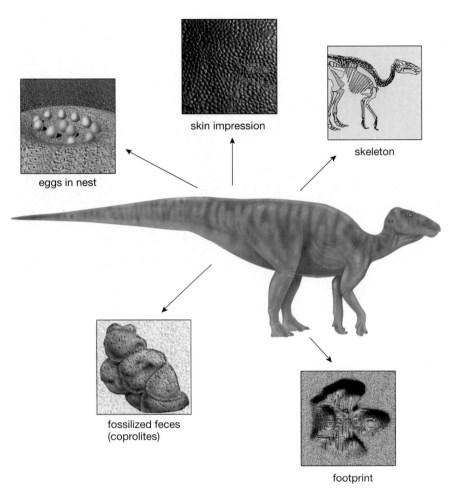

Figure 14-2 Types of fossils
We're used to thinking of fossils as bones or shells of long-dead animals, but many other kinds of evidence may be preserved in rock to form fossils. Impressions of skin or of leaves, footprints, feces, eggs, wood, and animal burrows are just some of the features that can be preserved. Paleontologists use all these types of fossils to discover clues about the appearance and lifestyle of organisms that lived long ago.

skin impression

skeleton

eggs in nest

fossilized feces
(coprolites)

footprint

species still lived on Earth (Fig. 14-3). Putting these facts together, scientists came to the inescapable conclusion that different types of organisms had lived at various times in the past.

But what did this newfound richness of organisms, both living and extinct, mean? Was each organism produced by a separate act of Creation? If so, why? And why bother to create so many types, letting thousands of types go extinct? The French naturalist Georges Louis LeClerc (1707–1788), known by the title Comte de Buffon, suggested that perhaps the original Creation provided a relatively small number of founding species and that some of the modern species had been "conceived by Nature and produced by Time"—that is, they had evolved through natural processes. Most people were not convinced. First, Buffon could not provide a mechanism whereby Nature could "conceive" new species. Second, no one thought that there was time enough for their "production."

Geology Provided Evidence That Earth Is Exceedingly Old

In the early 1700s, few scientists suspected that Earth could be more than a few thousand years old. Counting

generations in the Old Testament, for example, yields a maximum age of 4000 to 6000 years. From the descriptions of plants and animals from ancient writers such as Aristotle, it was clear that wolves, deer, lions, and other European organisms had not changed in over 2000 years. How, then, could whole new species have arisen if Earth was created only a couple of thousand years before Aristotle's time?

To account for a multitude of species, both extinct and modern, while preserving the notion of Creation, Georges Cuvier (1769–1832) proposed the theory of **catastrophism**. Cuvier, a French paleontologist (a scientist who studies fossils), hypothesized that a vast supply of species was created initially. Successive catastrophes (such as the Great Flood described in the Bible) produced the layers of rock and destroyed many species, fossilizing some of their remains in the process. The reduced flora and fauna of the modern world, he theorized, are the species that survived the catastrophes. However, if modern *species* have survived from an original Creation, then many *individuals* of those species should have died in the ancient catastrophes. Surely some of them would have been fossilized, and even the lowest and oldest rock layers should contain at least a few fossils of present-day

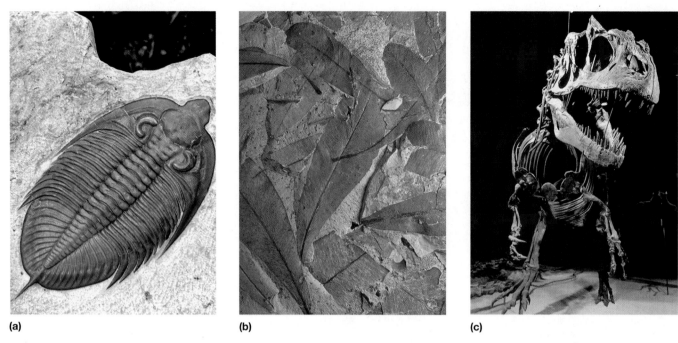

(a)   (b)   (c)

Figure 14-3 Fossils of extinct organisms
Fossils provide strong support for the idea that today's organisms were not created all at once but arose over time by the process of evolution. If all species were created simultaneously, we would not expect a fossil record in which **(a)** trilobites appear earlier than **(b)** seed ferns (*Glossopteris*), which in turn appear before **(c)** dinosaurs, such as *Allosaurus*. Trilobites had gone extinct by about 230 million years ago, seed ferns by about 150 million years ago, and dinosaurs by 65 million years ago.

species. Unfortunately for Cuvier's hypothesis, they do not. The French geologist Louis Agassiz (1807–1873) proposed that there was a new creation after each catastrophe and that modern species result from the most recent creation. The fossil record forced Agassiz to postulate at least 50 separate catastrophes and creations!

Alternatively, perhaps Earth *is* old enough to allow for the production of new species. Geologists James Hutton (1726–1797) and Charles Lyell (1797–1875) contemplated the forces of wind, water, earthquakes, and volcanism. They concluded that there was no need to invoke catastrophes to explain the findings of geology. Don't rivers in flood lay down layers of sediment? Don't lava flows produce layers of basalt? Why, then, should we assume that layers of rock are evidence of anything but ordinary natural processes, occurring repeatedly over long periods of time? This concept, called **uniformitarianism**, satisfies a scientific axiom often called *Occam's Razor*: The simplest explanation that fits the facts is probably correct. The implications of uniformitarianism were profound. If slow, natural processes alone are enough to produce layers of rock thousands of feet thick, then Earth must be old indeed, many millions of years old. Hutton and Lyell, in fact, concluded that Earth was eternal: "No Vestige of a Beginning, no Prospect of an End," in Hutton's words. (Modern geologists estimate that Earth is about 4.5 billion years old; see Chapter 17, "Scientific Inquiry: How Do We Know How Old a Fossil Is?") Thus,

Hutton and Lyell provided the time for evolution to occur. But there was still no convincing mechanism.

Early Biologists Proposed Mechanisms for Evolution

One of the first to propose a mechanism for evolution was the French biologist Jean Baptiste Lamarck (1744–1829). Lamarck was impressed by the progression of forms in the fossil record. Older fossils tend to be simpler, whereas younger fossils tend to be more complex and more like existing organisms. In 1801, Lamarck hypothesized that organisms evolved through the **inheritance of acquired characteristics**: Living organisms modify their bodies through the use or disuse of parts, and these modifications can be inherited by their offspring. (As it turns out, the first part of his hypothesis is correct to some extent, but the second part is not.) Why would organisms modify their bodies? Lamarck proposed that all organisms possess an innate drive for perfection, an urge to climb the ladder of Nature. In his best-known example, Lamarck hypothesized that ancestral giraffes stretched their necks to feed on leaves that grow high up in trees, and as a result their necks became slightly longer. Their offspring would have inherited these longer necks and stretched even farther to reach still higher leaves. Eventually, this process might have produced modern giraffes with very long necks indeed.

Today, Lamarck's theory seems silly: The fact that a prospective father pumps iron doesn't mean that his children will look like Arnold Schwarzenegger. Remember, though, that in Lamarck's day no one had the foggiest idea how inheritance worked. Gregor Mendel wouldn't even be born for another 20 years, and his principles of inheritance weren't incorporated into mainstream biology until the early twentieth century.

Although Lamarck's theory fell by the wayside, by the mid-1800s some biologists were beginning to realize that the fossil record and the similarities between fossil forms and modern species could be best explained if present-day species had evolved from preexisting ones. The question remained: *But how?* In 1858, Charles Darwin and Alfred Russel Wallace independently provided convincing evidence that the driving force behind evolutionary change was natural selection.

Both Darwin and Wallace Proposed That Evolution Occurs by Natural Selection

Although their social and educational backgrounds were very different, Darwin and Wallace were quite similar in some respects. Both had traveled extensively in the tropics and had studied the staggering variety of plants and animals living there. Both found that some species differed only in a few fairly subtle, but ecologically impor-

tant, features (Fig. 14-4). Both were familiar with the fossil record, which showed a trend of increasing complexity through time. Finally, both were aware of the studies of Hutton and Lyell, who had proposed that Earth is extremely ancient. These facts suggested to both Darwin and Wallace that species change over time; that is, species evolve. Both sought a mechanism that might direct evolutionary change over many generations.

In 1858, Darwin and Wallace independently described a mechanism for evolution in remarkably similar papers that were presented to the Linnaean Society in London. Like Gregor Mendel's manuscript on the principles of genetics, their papers had little impact. The secretary of the society, in fact, wrote in his annual report that nothing very interesting happened that year. Fortunately, the next year Darwin published his monumental *On the Origin of Species by Means of Natural Selection*, forcing everyone to take note of the new theory. (See "Scientific Inquiry: Charles Darwin—Nature Was His Laboratory.")

Evolutionary Theory Arises from Scientific Observations and Conclusions Based on Them

How did Darwin and Wallace arrive at the conclusion that life's huge variety of excellent designs arose by a

(a) large ground finch

(b) small ground finch

(c) warbler finch

(d) vegetarian tree finch

Figure 14-4 Darwin's finches, residents of the Galapagos Islands
The outcome of selection is sometimes easier to see in the relatively simple, isolated ecosystems of islands. For example, the Galapagos Islands are home to a group of closely related species of finches, each of which specializes in eating a different type of food. Natural selection has favored those individuals best suited to exploit each food source efficiently. The result is a wide range of beak sizes and shapes among otherwise similar birds. Some beaks are suited to **(a)** cracking large seeds, **(b)** some for small seeds, **(c)** some for insects, **(d)** some for eating leaves, and so on.

Scientific Inquiry www
Charles Darwin—Nature Was His Laboratory

According to the prevailing modern stereotype, the typical scientist is a solitary figure, clad in a white lab coat and hunched over a microscope, surrounded by an array of test tubes, computers, and complex instruments. But one of the greatest scientists in history, Charles Darwin, did his most important work outdoors, equipped with nothing more than a walking stick and a brilliant mind. His chief mode of scientific inquiry consisted of sharp observation of nature, followed by careful, logical inference.

Charles Darwin's Voyage on the *Beagle* Sowed the Seeds for His Theory of Evolution

Like many modern students, Charles Darwin excelled only in subjects that intrigued him. Although his father was a physician, Darwin was uninterested in medicine and unable to stand the sight of surgery. He eventually obtained a degree in theology from Cambridge University, although theology too was of minor interest to him. What he really liked to do was to tramp over the hills, observing plants and animals, collecting new specimens, scrutinizing their structures, and categorizing them. As he himself later put it, "I was a born naturalist." Fortunately for Darwin (and for the development of biology), some of his professors at Cambridge, most notably the botanist John Henslow, had similar interests. So constant was their companionship in field studies that Darwin was sometimes called "the man who walks with Henslow."

In 1831, when Darwin was only 22 years old (Fig. E14-1), the British government sent Her Majesty's Ship *Beagle* on a 5-year surveying expedition that went first along the coastline of South America and then around the world. As was common on such expeditions, the *Beagle* would carry along a naturalist to observe and collect geological and biological specimens encountered along the route. Thanks to Henslow's recommendation to the captain, Robert FitzRoy, Darwin was offered the position of naturalist aboard the *Beagle*. When they first met, FitzRoy almost rejected Darwin for the post because of the shape of Darwin's nose! Apparently, FitzRoy felt that a person's personality could be predicted by the shape of the facial features, and Darwin's nose failed to measure up, as it were. Later, Darwin expressed the opinion that FitzRoy was "afterwards well satisfied that my nose had spoken falsely."

The *Beagle* sailed to South America, making many stops along the coast. There Darwin observed the plants and animals of the tropics and was stunned by the diversity of species compared with that of Europe. Although he boarded the *Beagle* convinced of the permanence of species, his experiences soon led him to doubt this. He discovered a snake with rudimentary hind limbs, calling it "the passage by which Nature joins the lizards to the snakes." Another snake vibrates its tail like a rat-

Figure E14-1 A painting of Charles Darwin as a young man

tlesnake but has no rattles and therefore makes no noise. Penguins use their wings to paddle through the water rather than fly through the air. If a Creator had individually created each animal in its present form, to suit its present environment, what could be the purpose behind these makeshift arrangements?

Perhaps the most significant stopover of the voyage was the month spent on the Galapagos Islands, off the northwestern coast of South America. There Darwin found huge tortoises (Fig. E14-2a); in fact, *galapagos* means "tortoise" in Spanish. Different islands were home to distinctively different types of tortoises. On islands without tortoises, prickly pear cacti grew with their juicy (though spiny) pads and fruits spread out over the ground. On islands where tortoises lived, the prickly pears grew substantial trunks, bearing the fleshy pads and fruits high above the reach of the voracious and tough-mouthed tortoises (Fig. E14-2b). Darwin also found several varieties of mockingbirds and finches; as with the tortoises, different islands supported slightly different forms. Could the differences in these organisms have arisen after they became isolated from one another on separate islands? The diversity of tortoises and birds "haunted" him for years afterward.

(a)

(b)

Figure E14-2 *Tortoises act as agents of selection*
(a) Galapagos tortoises feed on prickly pear cactuses. *(b)* On islands with tortoises, a young cactus quickly grows a tall trunk, which lifts the succulent pads beyond the reach of the tortoises.

In 1836, Darwin returned to England after 5 years on the *Beagle* and became established as one of the foremost naturalists of his time. But constantly gnawing on his mind was the problem of the origin of species. Part of the solution came to him from an unlikely source: the writings of an English economist and clergyman, Thomas Malthus. In his *Essay on Population*, Malthus wrote, "It may safely be pronounced, therefore, that [human] population, when unchecked, goes on doubling itself every 25 years, or increases in a geometrical ratio." Darwin realized that a similar principle holds true for plant and animal populations. In fact, most organisms can reproduce much more rapidly than can humans (consider rabbits, dandelions, and houseflies) and consequently could produce overwhelming populations in short order. Nonetheless, the world is *not* chest-deep in rabbits, dandelions, or flies: Natural populations do *not* grow "unchecked" but tend to remain approximately constant in size. Clearly, vast numbers of individuals must die in each generation, and most must not reproduce.

From his experience as a naturalist, Darwin realized that the individual members of a species typically differ from one another in form and function. Further, *which individuals die in each generation is not arbitrary but depends to some extent on the structures and abilities of the organisms*. This observation was the source of the theory of evolution by natural selection. As Darwin's colleague Alfred Wallace put it, "those which, year by year, survived this terrible destruction must be, on the whole, those which have some little superiority enabling them to escape each special form of death to which the great majority succumbed." Here you see the origin of the expression "survival of the fittest." That "little superiority" that confers greater fitness might be better resistance to cold, more-efficient digestion, or any of hundreds of other advantages, some very subtle. Everything now fell into place. Darwin wrote, "It at once struck me that under these circumstances favorable variations would tend to be preserved, and unfavorable ones to be destroyed." If the favorable variations were inheritable, then the entire species would eventually consist of individuals possessing the favorable trait. With the continual appearance of new variations (due, as we now know, to mutations), which in turn are subject to further selection, "the result . . . would be the formation of new species. Here, then, I had at last got a theory by which to work."

When Darwin finally published *On the Origin of Species* in 1859, his evidence had become truly overwhelming. Although its full implications would not be realized for decades, Darwin's theory of evolution by natural selection has become a unifying concept for virtually all of biology. Events that change the world sometimes hinge on small details, even the shape of a nose!

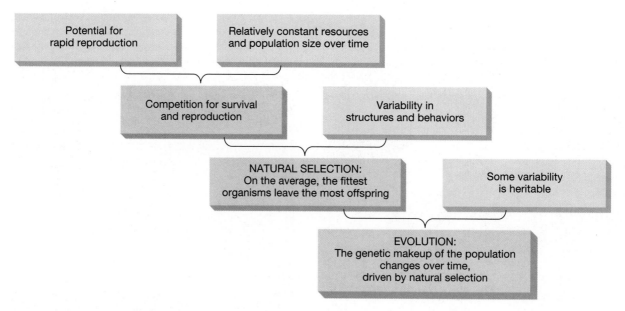

Figure 14-5 *A flowchart of evolutionary reasoning*
This chart is based on the hypotheses of Darwin and Wallace but incorporates ideas from modern genetics.
Observations are in tan; conclusions based on these observations are in blue.

process of descent with modification? The chain of logic leading to this powerful conclusion turns out to be surprisingly simple and straightforward. We summarize their theory in modern terms (Fig. 14-5):

Observation 1: Natural populations of all organisms have the potential to increase rapidly, because organisms can produce far more offspring than are required merely to replace the parents. (A **population** consists of all the individuals of one species in a particular area.)

Observation 2: Nevertheless, the sizes of most natural populations and the resources available to maintain them (such as food and appropriate habitat) remain relatively constant over time.

   Conclusion 1: Therefore, there is competition for survival and reproduction. In each generation, many individuals must die young, fail to reproduce, produce few offspring, or produce less-fit offspring that fail to survive and reproduce in their turn.

Observation 3: Individual members of a population differ from one another in their ability to obtain resources, withstand environmental extremes, escape predators, and so on.

   Conclusion 2: The most well-adapted (the "fittest") individuals in one generation tend to be the ones that leave the most offspring. This is **natural selection**, the process by which the environment selects for those individuals whose traits best adapt them to that particular environment.

Observation 4: At least some of the variation among individuals in traits that affect survival or reproduction is due to genetic differences that may be passed on from parent to offspring.

Conclusion 3: Over many generations, differential, or unequal, reproduction among individuals with different genetic makeup changes the overall genetic composition of the population. This process is evolution by natural selection.

As you know, the principles of genetics had not yet been discovered when Darwin published *On the Origin of Species*. Our observation 4, therefore, was an untested assumption for Darwin and a grave weakness in his theory. Although he could not explain how inheritance operated, Darwin's theory made an important prediction that we now know is correct. According to Darwin, the variations that appear in natural populations arise purely by chance. Unlike Lamarck, he proposed no internal drives for perfection or any other mechanisms that would ensure that variations would be favorable. Molecular genetics has shown that Darwin was correct: Variations arise as a result of chance mutations in DNA (see Chapters 9 and 10).

How might natural selection among chance variations change the makeup of a species? In *On the Origin of Species*, Darwin proposed the following example. "Let us take the case of a wolf, which preys on various animals, securing [them] by . . . fleetness. . . . The swiftest and slimmest wolves would have the best chance of surviving, and so be preserved or selected. . . . Now if any slight innate change of habit or structure benefited an individual wolf, it would have the best chance of surviving and of leaving offspring. Some of its young would probably inherit the same habits or structure, and by the repetition of this process, a new variety might be formed." The same argument would apply to the wolf's prey, in which the fastest or most alert would be the most likely to avoid predation and would pass on these traits to its offspring. Notice that natural selection is acting on individuals within a population. Over generations, the population would change as a greater percentage of its individuals acquired favorable adaptive traits. An individual cannot evolve, but a population can.

Although it is easiest to understand how natural selection would cause *changes within a species*, under the right circumstances, the same principles might produce *entirely new species*. In Chapter 16, we shall discuss the circumstances that give rise to new species. In the rest of this chapter, we briefly review some of the evidence for evolution.

2) How Do We Know That Evolution Has Occurred? *www*

Virtually all biologists consider evolution to be a fact. Although debates still rage over the *mechanisms* of evolutionary change, exceedingly few biologists dispute that evolution occurs. Why? Because an overwhelming body of evidence permits no other conclusion. The key lines of evidence include such sources as the fossil record, comparative anatomy (the study of anatomical structures for comparative purposes among species), embryology, and biochemistry and genetics.

The Fossil Record Provides Evidence of Evolutionary Change over Time

Because fossils are the remains of members of species ancestral to modern species, we would expect to find progressive series of fossils leading from an ancient, primitive organism, through several intermediate stages, and culminating in the modern forms. (Just how fine the gradations between stages should be is currently a subject of debate among evolutionary biologists; see Chapter 16.) Probably the best-known of such series are the fossil horses (Fig. 14-6), but giraffes, elephants, and several mollusks all show a roughly gradual evolution of body form over time, suggesting that species evolved from, and replaced, previous species. Certain sequences of fossil snails have such slight gradations in form between successive fossils that paleontologists cannot easily decide where one species leaves off and the next one begins.

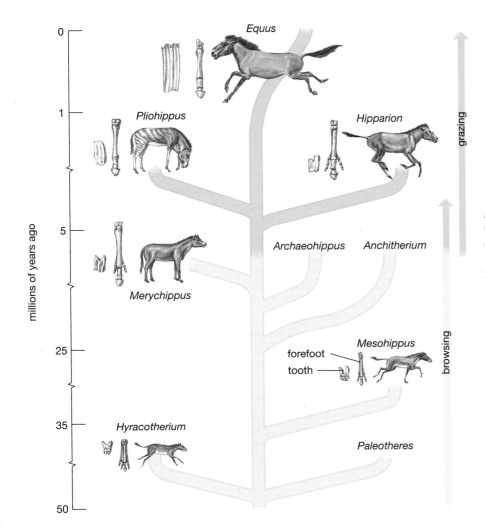

Figure 14-6 The evolution of the horse
Over the past 50 million years, horses evolved from small woodland browsers to large plains-dwelling grazers. Three major changes include size, leg anatomy, and tooth anatomy. No one is certain why horses became larger, but it may have provided an advantage against predators. In any case, a large body running over hard plains favored the evolution of large, hard hooves, attached by springlike, shock-absorbing joints to legs with stout bones. Finally, the teeth became larger, with more enamel, reflecting a change in diet from the relatively soft leaves and buds of bushes to the abrasive blades of grasses, which contain silicon (glass). If a modern horse had teeth like those of *Hyracotherium*, its teeth would be ground away while it was still very young, leaving it to starve.

Figure 14-7 Analogous structures
Similar environmental pressures
acting on unrelated animals may result in the convergent evolution of
outwardly similar structures. The
wings of **(a)** insects and **(b)** birds
and the sleek, streamlined shapes of
(c) seals and **(d)** penguins are examples of such analogous structures.

(a)

(b)

(c)

(d)

Comparative Anatomy Provides Structural Evidence of Evolution

Appearance has long been used as an indicator of the relatedness of organisms. Structure, inexorably tied to function, also provides evidence of descent with modification.
The elephant and the mammoth, for instance, clearly
have similar anatomies and share a common ancestor.

Unrelated Species in Similar Environments Have Evolved Similar Forms

Evolution by natural selection also predicts that, given
similar environmental demands, unrelated species might
independently evolve superficially similar structures, a
process called **convergent evolution**. Such outwardly similar body parts in unrelated organisms, termed **analogous
structures**, may be very different in internal anatomy, because the parts are not derived from common ancestral
structures. The wings of flies and of birds are analogous
structures that have arisen by convergent evolution; the
fat-insulated, streamlined shapes of seals (mammals) and
of penguins (birds) are another example (Fig. 14-7).

Homologous and Vestigial Structures Provide Evidence of Relatedness of Organisms Adapted to Different Environments

Modern organisms are adapted to a wide variety of habitats and lifestyles. The forelimbs of birds and mammals,
for example, are variously used for flying, swimming, running over several types of terrain, and grasping objects
such as branches and tools. Despite this enormous diversity of function, the internal anatomy of all bird and
mammal forelimbs is remarkably similar (Fig. 14-8). It is
inconceivable that nearly the same bone arrangements
could be ideal for such different functions, as we would
expect if each animal had been created separately. Such
similarity is exactly what we would expect, however, if
bird and mammal forelimbs were derived from a common ancestor. Through natural selection, each has been
modified to perform a particular function. Such internally similar structures are called **homologous structures**,
meaning that they have the same evolutionary origin despite possible differences in function. Studies of comparative anatomy have long been used to determine the

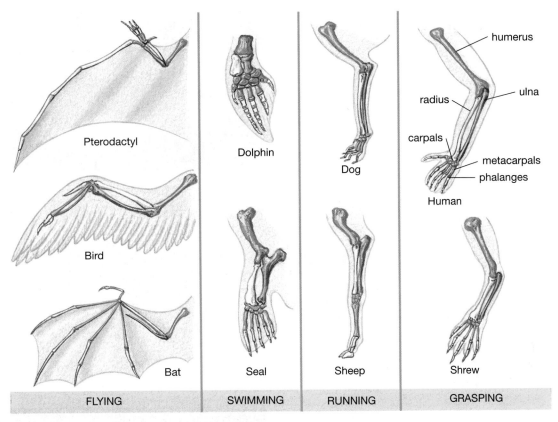

Figure 14-8 Homologous structures
Despite wide differences in function, the forelimbs of all of these animals contain the same set of bones, inherited through evolution from a common ancestral vertebrate. The bones have been tinted different colors to highlight the correspondences among the various species.

relationships among organisms, on the grounds that the more similar the internal structures of two species, the more closely related the species must be; that is, the more recently they must have diverged from a common ancestor.

Evolution by natural selection also helps explain the curious circumstance of **vestigial structures**. These are structures that serve no apparent purpose, including such things as molar teeth in vampire bats (which live on a diet of blood and therefore don't chew their food) and pelvic bones in whales and certain snakes (Fig. 14-9). Both of these vestigial structures are clearly homologous to structures that are found in—and used by—other vertebrates (animals with a backbone). Their continued existence in animals that have no use for them is best explained as a sort of "evolutionary baggage." For example, the ancestral mammals from which whales evolved had four legs and a well-developed set of pelvic bones. Whales do not have hind legs, yet they have small pelvic and leg bones embedded in their sides. During whale evolution, there was a selective advantage to the loss of the hind legs, the better to streamline the body

for movement through water. The result is the modern whale with small, useless, and unused pelvic bones.

Embryological Stages of Animals Can Provide Evidence of Common Ancestry

In the early 1800s, the German embryologist Karl von Baer noted that all vertebrate embryos look quite similar to one another early in their development (Fig. 14-10). In their early embryonic stages, fish, turtles, chickens, mice, and humans all develop tails and gill slits. Only fish go on to develop gills, and only fish, turtles, and mice retain substantial tails. Why do such diverse vertebrates have similar developmental stages? The only plausible explanation is that ancestral vertebrates possessed genes that direct the development of gills and tails. All of their descendants still retain those genes. In fish, these genes are active throughout development, resulting in gill-bearing and tail-bearing adults. In humans and chickens, these genes are active only during early developmental stages, and the structures are lost or inconspicuous in adults.

(a) salamander

(b) baleen whale

(c) boa constrictor

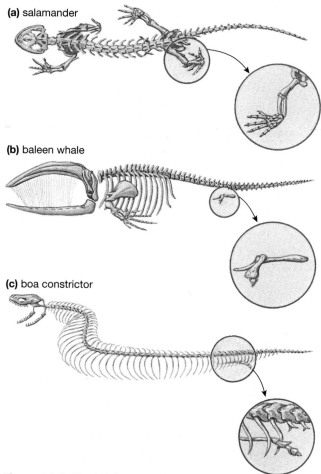

Figure 14-9 *Vestigial structures*

Many organisms have structures that serve no apparent function. Such structures were most likely inherited, in modified form, from ancestors that did use them. For example, the similarities among the bones of the functional hindlimb of **(a)** the salamander (for comparison) and the functionless vestigial bones that appear at similar locations in **(b)** the legless baleen whale and **(c)** boa constrictor are good evidence that all three animals inherited the bones from a common ancestor.

Modern Biochemical and Genetic Analyses Reveal Relatedness among Diverse Organisms

Biochemistry and molecular biology provide striking evidence of the evolutionary relatedness of all living organisms. At the most fundamental biochemical levels, all living cells are very similar. For example, all cells have DNA as the carrier of genetic information; all use RNA, ribosomes, and approximately the same genetic code to translate that genetic information into proteins; all use roughly the same set of 20 amino acids to build proteins; and all use ATP as an intracellular energy carrier. Species share similarities in chromosome structure, in sequences of amino acids in proteins, and in DNA composition; each of these parameters is now used to investigate relatedness among organisms. We shall return to the methods of measuring evolutionary relatedness in Chapter 18.

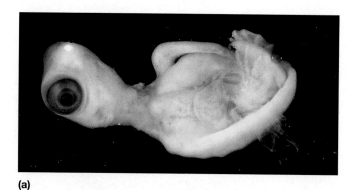

(a)

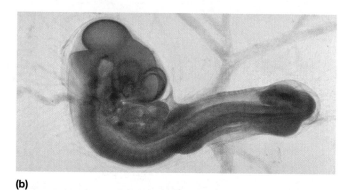

(b)

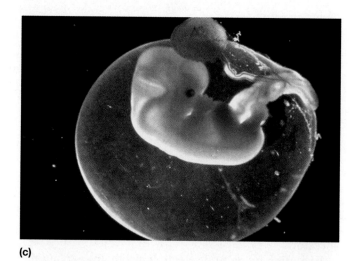

(c)

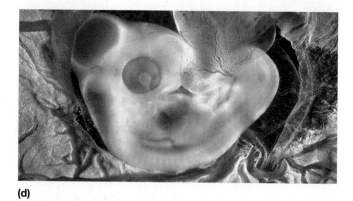

(d)

Figure 14-10 *Embryological stages reveal evolutionary relationships*

Early embryonic stages of a **(a)** turtle, **(b)** mouse, **(c)** human, and **(d)** chicken, showing strikingly similar anatomical features.

3 What Is the Evidence That Populations Evolve by Natural Selection?

We have seen that evidence of evolution comes from several types of sources. But what evidence is there that evolution occurs by the mechanism of natural selection?

Artificial Selection Demonstrates That Organisms May Be Modified by Controlled Breeding

One line of evidence supporting evolution by natural selection that particularly impressed Charles Darwin was **artificial selection**: the breeding of domestic plants and animals to produce specific desirable features. The various breeds of dogs provide a striking example of artificial selection (Fig. 14-11). Dogs descended from wolves, and even today the two will readily cross-breed. With rare exceptions, however, few modern dogs resemble wolves. Some breeds are so different from one another that they would be considered separate species if they were found in the wild; for instance, see the Chihuahua and the Great Dane in the Chapter 12 opening photo. Interbreeding would hardly be possible without a lot of human assistance. If humans could breed such radically different dogs in a few hundred to at most a few thousand years, Darwin reasoned, it seemed quite plausible that natural selection could produce the spectrum of living organisms in hundreds of millions of years.

Evolution by Natural Selection Occurs Today

The logic of natural selection gives us no reason to believe that evolutionary change is limited to the past. After all, heritable variation and competition for access to resources are certainly not limited to the past. So, if Darwin and Wallace were correct that those conditions lead inevitably to evolution by natural selection, then scientific observers and experimenters ought to be able to detect evolutionary change as it occurs. And they have. One example of present-day evolution has been caused by the increasing industrial pollution in the nineteenth and twentieth centuries.

Tree trunks in British woodlands usually support a lush growth of mottled gray lichens (an association of an alga and a fungus that benefits both species; see Chapter 20). Before the Industrial Revolution of the nineteenth century, most peppered moths, *Biston betularia*, were white with scattered specks of black pigment. This coloration matched the color and pattern of the lichens growing on the trees. Because of that coloring, predatory birds could not easily see the moths, which sat quietly on the lichens during the day (Fig. 14-12a). Occasionally, mutant black individuals appeared. These black moths, extremely conspicuous against the pale lichens, were easily spotted by birds and didn't live long.

During the Industrial Revolution, Britain's growing industries began to burn huge quantities of coal for fuel. With no pollution-control technology, soot from the smokestacks soon blanketed the countryside around the

(a)

(b)

Figure 14-11 Dog diversity illustrates artificial selection
A comparison of **(a)** the ancestral dog (the gray wolf, *Canis lupus*) and **(b)** various breeds of dog. Artificial selection by humans has caused a great divergence in size and shape of dogs in only a few thousand years. Huge differences in size would make it extremely difficult for the largest dogs to interbreed with the smallest.

Figure 14-12 *Protective colors differ in different environments* The two color varieties of the peppered moth, resting on *(a)* a lichen-covered tree trunk and *(b)* a soot-blackened trunk. The pale variety is well camouflaged on the lichen but is conspicuous on the sooty trunk; the reverse is true of the black form.

(a)

(b)

mills and factories. The lichens on the trees became covered with pollutants and died off, leaving the trunks sooty black. The pale-colored peppered moths were no longer camouflaged while they sat on the blackened trunks (Fig. 14-12b). Increasingly, pale moths fell prey to birds. Soot-covered trees, however, provided excellent camouflage for black moths. The black moths' color became, in Wallace's words, the "little superiority enabling them to escape" predation by birds—that is, a trait that helped protect the black moths. Black moths, once rare, survived and reproduced, passing on their genes for dark pigmentation to succeeding generations. As the years passed, ever-increasing numbers of black moths appeared, until by the end of the nineteenth century, about 98% of the moths around the industrial city of Manchester were black. Fortunately, pollution-control laws have now dramatically reduced emissions of soot and other pollutants, and lichens grow once again on the trees near British cities. As predicted by the principles of evolution by natural selection, the pale moths are making a comeback in these areas, and the black variety is becoming increasingly rare.

Natural selection has also been documented more recently. In Florida, homeowners were dismayed to realize that roaches were ignoring a formerly effective poison bait called Combat.® Researchers discovered that the bait had acted as an agent of natural selection. Roaches that liked it were consistently killed; those that survived had inherited a rare mutation that caused them to dislike glucose, a type of sugar found in the corn syrup used as bait in Combat. By the time researchers identified

the problem in the early 1990s, the formerly rare mutation had become widespread in Florida's urban roach population.

In a meadow near Carson City, Nevada, biologists have closely monitored a rare subspecies of checkerspot butterfly. In 1983, about 80% of the eggs of the butterfly were laid on a native meadow plant, *Collinsia*. But over the next decade, the meadow was invaded by a weed, *Plantago*. By 1993, over 70% of the checkerspots' eggs were being laid on *Plantago*. When eggs laid on *Plantago* were hatched in the laboratory and the caterpillars reared exclusively on *Collinsia*, the mature butterflies still sought out *Plantago* to lay their eggs, just as their mothers had. The preference for *Plantago* was clearly inherited and, as it offered a selective advantage (as the weeds began to overtake the native plant), had rapidly increased within the population during a single decade.

Note three important points about these examples:

1. The black coloration in British moths, distaste for glucose in Florida cockroaches, and preference for *Plantago* in Nevada butterflies were not *produced* by the pollution, poisoned corn syrup, or the invading weed. The mutations that produced each of these beneficial traits arise spontaneously now and then. *The variations on which natural selection works are produced by chance mutations.*

2. *Selection does not necessarily produce well-adapted species.* Some other species of moths in Britain lacked black mutants; these species went extinct in industrial areas. If the right raw material isn't there (black

moth mutants, in this case), then natural selection may drive a species to extinction.

3. Evolution is commonly regarded as a mechanism for producing the ladder of Nature, with ever-greater degrees of perfection appearing over time. It is not. *Evolution by natural selection selects for organisms that are best adapted to a particular environment.* Under natural conditions, a distaste for glucose would put a cockroach at a disadvantage, causing it to avoid many rich food sources. In the presence of poisoned corn syrup, however, it becomes advantageous. Natural selection does not select for the "best" in any absolute sense, but only in the context of a particular environment, which varies from place to place and which may change over time.

Other examples of modern-day evolution include antibiotic resistance in bacteria (see Chapter 19) and the appearance of the virus that causes AIDS (HIV) (Chapter 31). Evolution is easiest to observe in rapidly reproducing, short-lived organisms such as insects or bacteria, but it occurs constantly in all the species on Earth.

4 A Postscript by Charles Darwin

"It is interesting to contemplate an entangled bank, clothed with many plants of many kinds, with birds singing on the bushes, with various insects flitting about, and with worms crawling through the damp earth, and to reflect that these elaborately constructed forms . . . have all been produced by laws acting around us. These laws, taken in the highest sense, being Growth with Reproduction; Inheritance [and] Variability . . .; a Ratio of Increase so high as to lead to a Struggle for Life, and as a consequence to Natural Selection, entailing Divergence of Character and Extinction of less-improved forms. . . . There is grandeur in this view of life, with its several powers, having been originally breathed into a few forms or into one; and that, whilst this planet has gone cycling on according to the fixed law of gravity, from so simple a beginning endless forms most beautiful and most wonderful have been, and are being, evolved."

These are the concluding sentences of Darwin's *On the Origin of Species.*

Summary of Key Concepts

1 How Did Evolutionary Thought Evolve?

Historically, the most common explanation for the origin of species has been the divine Creation of each species in its present form, and species were believed not to have changed significantly since their creation. Evidence provided by the diversity of living things, by fossils, and by geology challenged this view, although no convincing mechanism for the evolution of present-day species from previous ones was proposed. Since the middle of the nineteenth century, scientists have concluded that species originate by the operation of natural laws, as a result of changes in the genetic makeup of the populations of organisms. This process is called evolution.

Charles Darwin and Alfred Russel Wallace independently proposed the theory of evolution by natural selection. Their theory can be concisely expressed as three conclusions based on four observations. These are summarized in modern biological terms in Figure 14-5.

2 How Do We Know That Evolution Has Occurred?

Many lines of evidence indicate that evolution has occurred, including the following:

1. Fossils of ancient species tend to be simpler in form than modern species. Sequences of fossils have been discovered that show a graded series of changes in form. Both of these facts would be expected if modern species evolved from older species.
2. Species thought to be related through evolution from a common ancestor show many similar anatomical struc-

tures. Examples include the limbs of amphibians, reptiles, birds, and mammals.
3. Stages in embryological development are quite similar among very different types of vertebrates.
4. Similarities in chromosome structure, sequences of amino acids in proteins, and DNA composition all support the notion of descent of related species through evolution from common ancestors.

3 What Is the Evidence That Populations Evolve by Natural Selection?

Similarly, many lines of evidence indicate that natural selection is the chief mechanism driving changes in the characteristics of species over time, including the following:

1. Rapid, heritable changes have been produced in domestic animals and plants by selectively breeding organisms with desired features (artificial selection). The immense variations in species produced in a few thousand years of artificial selection by humans makes it seem likely that much larger changes could be wrought by hundreds of millions of years of natural selection.
2. Evolution can be observed today. Both natural and human activities drastically change the environment over short periods of time. Significant changes in the characteristics of species have been observed in response to these environmental changes. A well-studied example is the evolution of black coloration among moths in response to the darkening of their environment by industrial pollutants.

Key Terms

analogous structures *p. 264*
artificial selection *p. 267*
catastrophism *p. 257*
convergent evolution *p. 264*

evolution *p. 255*
fossil *p. 256*
homologous structures *p. 264*

inheritance of acquired characteristics *p. 258*
natural selection *p. 262*
population *p. 262*

species *p. 256*
uniformitarianism *p. 258*
vestigial structure *p. 265*

Thinking Through the Concepts

Multiple Choice

1. *Your arm is homologous with*
   a. a seal flipper   b. an octopus tentacle
   c. a bird wing   d. a sea star arm
   e. both a and c

2. *All organisms share the same genetic code. This commonality is evidence that*
   a. evolution is occurring now
   b. convergent evolution has occurred
   c. evolution occurs gradually
   d. all organisms are descended from a common ancestor
   e. life began a long time ago

3. *Which of the following are fossils?*
   a. pollen grains buried in the bottom of a peat bog
   b. the petrified cast of a clam's burrow
   c. the impression a clam shell made in mud, preserved in mudstone
   d. an insect leg sealed in plant resin
   e. all of the above

4. *In Africa, there is a species of bird called the yellow-throated longclaw. It looks almost exactly like the meadowlark found in North America, but they are not closely related. This is an example of*
   a. uniformitarianism
   b. artificial selection
   c. gradualism
   d. vestigial structures
   e. convergent evolution

5. *Which of the following are examples of vestigial structures?*
   a. your tailbone
   b. nipples on male mammals
   c. sixth fingers found in some humans
   d. your kneecap
   e. none of the above

6. *Which of the following would stop evolution by natural selection from occurring?*
   a. if humans became extinct because of a disease epidemic
   b. if a thermonuclear war killed most living organisms and changed the environment drastically
   c. if ozone depletion led to increased ultraviolet radiation, which caused many new mutations
   d. if genetic recombination, sexual reproduction, and mutation stopped, so all offspring of all organisms were exact copies of their parents
   e. all of the above

? Review Questions

1. Selection acts on individuals, but only populations evolve. Explain why this is true.

2. Distinguish between catastrophism and uniformitarianism. How did these hypotheses contribute to the development of evolutionary theory?

3. Describe Lamarck's theory of inheritance of acquired characteristics. Why is it invalid?

4. What is natural selection? Describe how natural selection might have caused differential reproduction among the ancestors of a fast-swimming predatory fish, such as the barracuda.

5. Describe how evolution occurs through the interactions among the reproductive potential of a species, the normally constant size of natural populations, variation among individuals of a species, natural selection, and inheritance.

6. What is convergent evolution? Give an example.

7. How do biochemistry and molecular genetics contribute to the evidence that evolution occurred?

Applying the Concepts

1. Does evolution through natural selection produce "better" organisms in an absolute sense? Are we climbing the "ladder of Nature"? Defend your answer.

2. Both the study of fossils and the idea of special creation have had an impact on evolutionary thought. Discuss why one is considered scientific endeavor and the other not scientific.

3. In evolutionary terms, "success" can be defined in many different ways. What are the most successful organisms you can think of in terms of (a) persistence over time, (b) sheer numbers of individuals alive now, (c) numbers of species (for a lineage), and (d) geographical range?

4. In what sense are humans currently acting as "agents" of selection on other species? Name some organisms that are *favored* by the environmental changes humans cause.

5. Darwin and Wallace's discovery of natural selection is one of the great revolutions in scientific thought. Some scientific revolutions spill over and affect the development of philosophy and religion. Is this true of evolution? Does (or should) the idea of evolution by natural selection affect the way humans view their place in the world?

6. In your mind, what scientific question currently represents the "mystery of mysteries"?

Group Activity

Form a group of four students. Your group has been asked to analyze some data from a long-term study in the Galapagos Islands. Another group of researchers spent 8 years measuring beaks, observing feeding behavior, and measuring rainfall in a population of a particular species of Darwin's finch. They found that this species eats only small seeds and that its beak enables the bird to crack and eat such seeds quickly and efficiently. The beak-size data look like this:

Year	1	2	3	4	5	6	7	8
Rainfall	normal	normal	normal	very low	very low	normal	normal	normal
Average beak size (in mm)	10.1	10.3	10.2	12.1	12.5	10.1	10.0	10.1

As a group, develop a hypothesis that explains what caused beak size to increase in dry years. Your hypothesis should specify how the fates of individuals determine changes in the population.

For More Information

Altman, S. A. "The Monkey and the Fig." *American Scientist*, May–June 1989. An amusing, informative discussion of many evolutionary themes, cast as a Socratic dialogue.

Bishop, J. A., and Cook, L. M. "Moths, Melanism and Clean Air." *Scientific American*, January 1975. This follow-up to H. B. D. Kettlewell's 1959 work shows that cleaning the air reverses industrial melanism in moths—that is, becoming pigmented as a consequence of industrial activities—providing graphic evidence that evolution only adapts organisms to existing environments.

Darwin, C. *On the Origin of Species by Means of Natural Selection.* Garden City, NY: Doubleday, 1960 (originally published in 1859). An impressive array of evidence amassed to convince a skeptical world.

Dawkins, R. "God's Utility Function." *Scientific American*, November 1995. One evolutionary biologist's view of the nature of evolutionary change and how it relates to the meaning of life. Dawkins is an articulate advocate of the "selfish gene" view of evolution, in which genes use organisms as tools in a contest for superiority in self-replication.

Dennet, D. *Darwin's Dangerous Idea.* New York: Simon & Schuster, 1995. A philosopher's view of Darwinian ideas and their application to the world outside biology. A thought-provoking book that seems to have inspired admiration and condemnation in roughly equal proportions.

Eiseley, L. C. "Charles Darwin." *Scientific American*, February 1956. An essay on the life of Darwin, by one of his foremost American biographers. Even if you need no introduction to Darwin, read this anyway as an introduction to Eiseley, author of many marvelous essays.

Gould, S. J. *Ever Since Darwin*, 1977; *The Panda's Thumb*, 1980; and *The Flamingo's Smile*, 1985. New York: W. W. Norton. A series of witty, imaginative, and informative essays, mostly from *Natural History* magazine. Many deal with various aspects of evolution.

Grant, P. R. "Natural Selection and Darwin's Finches." *Scientific American*, October 1991. A drought in the Galapagos Islands provides dramatic evidence of natural selection as an agent of evolutionary change.

Kettlewell, H. B. D. "Darwin's Missing Evidence." *Scientific American*, March 1959. Industrial melanism as an example of modern-day evolution.

Weiner, J. "Evolution Made Visible." *Science*, January 6, 1995. A clear summary of modern evidence for evolution in action.

Answers to Multiple-Choice Questions
1. e   2. d   3. e   4. e   5. b   6. d

*"What but the wolf's tooth whittled so fine
The fleet limbs of the antelope?
What but fear winged the birds, and hunger
Jewelled with such eyes the great goshawk's head?"*

Robinson Jeffers in **The Bloody Sire** *(1941)*

This intricate spider web is strong, lightweight, hard to detect, and ruthlessly effective at capturing passing insects. It's a marvel of efficient design, but its "designer" is the simple, mindless process of natural selection.

How Organisms Evolve

Net Watch

On-line resources for this chapter are on the World Wide Web at:

http://www.prenhall.com/audesirk

(click on the Table of Contents link and then select Chapter 15).

Natural selection is a mindless, mechanical process. All that happens is that some individuals reproduce more successfully than others do, and some of the inherited characteristics of these successful individuals are therefore more prevalent in the next generation. Can such a simple process really result in structures as complex as a spider's web, an eagle's eye, or a redwood? Can natural selection really "design" a hunting machine as efficient as a leopard or a thinking machine as fantastic as the human brain? Charles Darwin's answer was yes, if we allow enough time to pass. And ever since Darwin's day, biologists have been theorizing, hypothesizing, observing, and experimenting in an effort to put Darwin's idea to the test and to deepen our understanding of exactly how the processes of evolution have led to the complex, interdependent, well-designed organisms of today's world.

In their quest to understand the process of evolution, present-day biologists have one great advantage over those of Darwin's era. Darwin had no knowledge of the mechanics of heredity. Only recently has knowledge of chromosomes, genes, and DNA been acquired. Even after the biochemical basis of heredity was established, evolutionary biologists were slow to integrate the new information. Today, however, the fruits of genetics and population biology have been brought to bear on evolutionary investigations and have fostered myriad new insights into the nature of evolution and evolutionary mechanisms.

1) How Are Populations, Genes, and Evolution Related?

The changes that we see in an individual organism as it grows and develops are not evolutionary changes. Instead, evolutionary changes are those that occur from generation to generation, the changes that cause descendants to be different from their ancestors.

Furthermore, we can't really detect evolutionary changes across generations by looking at single individuals. For example, if you observed a mature man who stood five feet tall, could you conclude that humans were evolving to become shorter? Obviously not. If you wanted to know about evolutionary change in human height, your first step would be to measure many humans across many generations to see if the average height was changing over time. Clearly, evolution is a property not of individuals but of **populations**, which include all the individuals of a species living in a given area.

The recognition that evolution is a population-level phenomenon was one of Darwin's key insights. But populations are composed of individuals, and it is the actions and fates of individuals that determine which characteristics will be passed to descendant populations. In this fashion, inheritance provides the link between the lives of individual organisms and the evolution of populations. We will therefore begin our discussion of the processes of evolution by reviewing the principles of genetics as they apply to individuals and then extend those principles to the genetics of populations. You may want to refer to Unit II to refresh your memory on specific points.

Genes, Influenced by the Environment, Determine the Traits of Each Individual

Each cell of every organism contains a repository of genetic information encoded in the DNA of its chromosomes. Recall that a *gene* is a segment of DNA located at a particular place on a chromosome. Its sequence of nucleotides encodes the sequence of amino acids of a protein, normally an enzyme that catalyzes one particular reaction in the cell. At a given gene's location, there can be slightly different nucleotide sequences, called *alleles*, as we saw in Chapter 12. Different alleles generate different forms of the same enzyme. In this way, various alleles of the gene for eye color in humans, for example, produce eyes that are brown, or blue, or green, and so on. In any population of organisms, there are usually two or more alleles of each gene. An individual of a diploid or polyploid species whose alleles of a particular gene are all of the same type is *homozygous*, and an individual with alleles of different types for that gene is *heterozygous*. The specific alleles borne on an organism's chromosomes (its *genotype*), interacting with the environment, determine its physical and behavioral traits (its *phenotype*).

Let's illustrate these principles with an example that should be familiar to you from Unit II. A pea flower is colored purple because a chemical reaction in its petals converts a colorless molecule to a purple pigment. When we say that a pea plant has the allele for purple flowers, we mean that a particular stretch of DNA on one of its chromosomes contains a sequence of nucleotides that codes for the enzyme that catalyzes this reaction. A pea with the allele for white flowers has a different sequence of nucleotides at the corresponding position on one of its chromosomes. The enzyme for which that different sequence codes cannot produce purple pigment. If a pea is homozygous for the white allele, its flowers produce no pigment and thus are white.

The Gene Pool Is the Sum of All the Genes in a Population

We can often deepen our understanding of a concept or phenomenon by looking at it from more than one perspective. In studying evolution, looking at the process from the point of view of a gene has proven to be an enormously effective tool. In particular, evolutionary biologists have made excellent use of the tools of a branch of genetics called **population genetics**, which deals with the frequency, distribution, and inheritance of alleles in populations. To take advantage of this potent aid to understanding the mechanisms of evolution, you will need to learn to use a few of the basic concepts of population genetics.

In population genetics, the **gene pool** is defined as the sum of all the genes in a population. In other words, the gene pool consists of all the alleles of all the genes in all the individuals of that population. The gene pool of a population of pea plants consists of alleles for genes that determine flower color, seed shape, and so on. Each particular gene can also be considered to have its own gene pool, which consists of all the alleles of that specific gene in a population. For example, in a population of 100 pea plants, the gene pool of the gene for flower color would consist of 200 alleles (peas are diploid, so there are two color alleles per plant, multiplied by 100 plants). All those alleles can produce only two phenotypes, either purple or white. If we could analyze the genetic composition of every plant in the population, we might find that some have alleles for white flowers, some have alleles for purple flowers, and some have both alleles. If we added up the color alleles of all the plants in the population, we could determine the relative proportions of the different alleles, a number called the **allele frequency**. Let's say that the gene pool for flower color consists of 140 alleles for the purple phenotype and 60 alleles for the white phenotype. The allele frequencies would then be purple, 0.7 (70%), and white, 0.3 (30%).

Evolution Is the Change of Gene Frequencies within a Population

A casual observer might choose to define evolution on the basis of changes in the outward appearance or behaviors of the members of a population. The population geneticist, however, looks at a population and sees a gene pool that just happens to be divided into the packages that we call individual organisms. So any phenotypic changes that we observe in the individuals that make up the population can also be viewed as the outward expression of underlying changes to the gene pool. A population geneticist therefore defines evolution as changes in gene frequencies that occur in a gene pool over time. *Evolution is nothing more or less than a change in the genetic makeup of populations over generations.*

The Equilibrium Population Is a Hypothetical Population in Which Evolution Does Not Occur

It will be easier to understand the forces that cause populations to evolve if we first consider the characteristics of a population that would *not* evolve. In 1908, the English mathematician Godfrey H. Hardy and the German physician Wilhelm Weinberg independently developed a simple mathematical model now known as the **Hardy-Weinberg principle.** This model showed that, under certain conditions, allele frequencies and genotype frequencies in a population will remain constant no matter how many ~~words, evolution will not occur~~ ~~on geneticists call this ideal-~~ ~~tion an~~ **equilibrium popula-** ~~enetic equilibrium~~ as long as

~~n.~~
~~w~~ *between populations*; that ~~nigration of alleles into the~~ ~~igration) or out of the pop-~~ ~~on).~~
~~ery large.~~
~~m,~~ with no tendency for cer-
~~th specific other genotypes.~~
~~selection;~~ that is, all geno-
~~ptive and reproduce equal-

e frequencies within a pop-
indefinitely. If one or more
~~ted,~~ then allele frequencies
~~ccur.~~
~~/ if any natural populations~~
~~n,~~ what is the importance of
~~rg principle?~~ The Hardy-Weinberg conditions are useful starting points for studying the mechanisms of evolution. In the following sections, we will examine each condition, show why it is often violated by natural populations, and illustrate the consequences of its violation. In this way, you can better understand both the inevitability of evolution and the forces that drive evolutionary change.

2) What Causes Evolution? *www*

Population genetics theory predicts that the Hardy-Weinberg equilibrium can be disrupted by deviations from any of its five main underlying conditions. We might therefore predict five major causes of evolutionary

change: (1) mutation, (2) gene flow, (3) small population size, (4) nonrandom mating, and (5) natural selection.

Mutations Are the Ultimate Source of Genetic Variability

A population will remain in genetic equilibrium only if no *mutations* (changes in DNA sequence) occur, but mutations are inevitable. Although cells have efficient mechanisms that protect the integrity of their genes, including enzymes that constantly scan the DNA and repair flaws caused by radiation, chemical damage, or mistakes in copying, some changes in nucleotide sequence nevertheless slip past the checking and repair systems. When one of these changes occurs in a cell that produces gametes, the mutation may be passed to an offspring and enter the gene pool of a population.

How significant is mutation in altering the gene pool of a population? Mutations are rare, occurring once in 100,000 to 1,000,000 genes per individual in each generation. Therefore, mutation by itself is not a major force in evolution. However, *mutations are the source of new alleles,* new heritable variations on which other evolutionary processes can work. As such, they are the foundation of evolutionary change. *Without mutations there would be no evolution and no diversity among life-forms.*

Mutations are not goal-directed. A mutation does not arise as a result of, or in anticipation of, environmental necessities (Fig. 15-1). A mutation simply happens and may in turn produce a change in the structure or function of the organism. Whether that change is helpful or harmful or neutral, now or in the future, depends on environmental conditions over which the organism has little or no control. The mutation provides *potential;* other forces, such as migration and especially natural selection, that act on that potential may favor the spread of a mutation through the population or eliminate it.

Gene Flow between Populations Changes Allele Frequencies

When individuals move from one population to another and interbreed at the new location, alleles are transferred from one gene pool to another. This movement of alleles, or gene flow, between populations alters the distribution of alleles among populations. Baboons, for example, live in social groupings called troops. Within each troop, all the females mate with a handful of dominant males. Juvenile males usually leave the troop. If they are lucky, they join and perhaps even become dominant in another troop. Thus, the male offspring of one troop carry genes to the gene pools of other troops.

Movement of breeding organisms between populations has two significant effects:

1. *Gene flow spreads advantageous alleles throughout the species.* Suppose that a new allele arises in one population and that this new allele benefits the organisms

Freedom Visa
4313 1234 5
EXPIRE 00/00
431
C BARD COLE CARDH
SIN

Request the Freedor
Make the most of y

No Annual Fee
Introductory 5.9%
Annual Percentage Rate
on cash advance checks and ba

***Figure 15-1 Mutations occur
spontaneously***
Experiment supporting the hypothesis
that mutations occur spontaneously and
not in response to specific environmental pressures. Many colonies of bacteria,
each the offspring of a single individual
and thus having the same genetic makeup, are grown on a solid nutrient medium in a dish. These bacteria have never
been exposed to antibiotics. ***(a)*** A piece
of velvet the exact size of the dish is
lightly pressed into the bacterial
colonies and then ***(b)*** touched to the
surface of nutrient medium that contains the antibiotic streptomycin in each
of three dishes. Many bacteria from
each original colony stick to the velvet
and then come off the velvet into the
new dishes, and the exact positions of
the "parent" colonies are duplicated.
(c) After the dishes are incubated,
(d) only a few daughter colonies grow
in the new dishes, but the colonies grow
in the exact same positions in all three
experimental dishes. The identical patterns show that mutations for resistance
to streptomycin must have already been
present in the original dish, before the
appearance of any environmental pressure in the form of streptomycin.

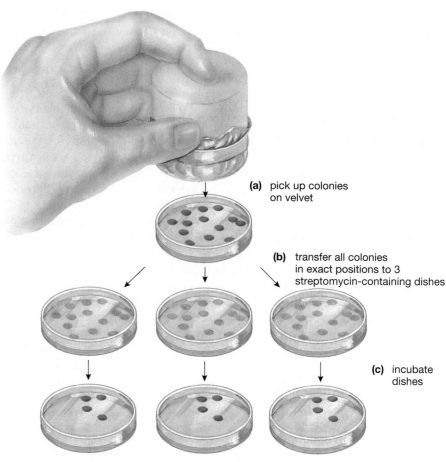

(a) pick up colonies
on velvet

(b) transfer all colonies
in exact positions to 3
streptomycin-containing dishes

(c) incubate
dishes

(d) only streptomycin-resistant colonies
grow on same few locations in each dish

that possess it. Migration can carry this new allele to
other populations of the species.
2. *Gene flow helps maintain all the organisms over a large
area as one species.* If migrants constantly carry genes
back and forth among populations, then the populations can never develop large differences in allele frequencies. Isolation of a population, with no gene flow
to or from other populations of the same species, is a
key factor in the origin of new species, as we will discuss in Chapter 16.

Small Populations Are Subject to Random Changes in Allele Frequencies

To remain in genetic equilibrium, a population must be
so large that chance events have no impact on its overall
genetic makeup. Disaster may befall even the fittest organism. The maple seed that falls into a pond never
sprouts; the deer and elk blasted away by the eruption
of Mount St. Helens left no descendants. If a population
is sufficiently large, chance events are unlikely to alter
the overall gene frequency, because such events would
be expected to interfere equally with the reproduction

of organisms of all genotypes. In a small population, however, certain alleles may be carried by only a few organisms. Chance events could reduce or even eliminate such
alleles from the population, altering its genetic makeup.

Genetic Drift Is an Example of Random Genetic Change in Small Populations

Chance events are much more likely to change allele frequencies in a small population than in a large population, by a process called **genetic drift**. Consider, for
example, two hypothetical populations of haploid ladybugs in which the outer shell is either spotted or solid-
colored, under the control of alternate alleles of a single
gene. In each population, half the ladybugs are spotted
and half are solid-colored (that is, the frequencies of both
alleles are 0.5, or 50%), but one population has only four
bugs and the other has 1000. Let us assume for simplicity that each individual that survives to maturity produces
two offspring identical to itself. For the population sizes
to remain constant, exactly half the individuals must reproduce in each generation. Let us further assume that
whether an individual ladybug reproduces is determined
entirely by chance.

(a)

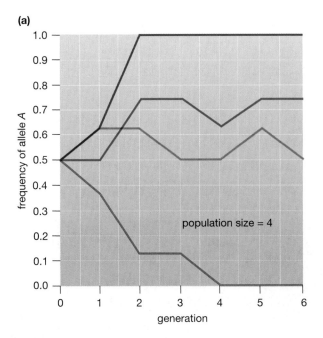

(b)

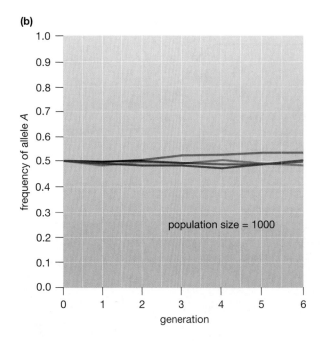

Figure 15-2 *Genetic drift*
Computer-generated graphs illustrating the effect of population size on genetic drift. In both graphs, the initial population was composed of half *A* and half *a* alleles, and six generations were simulated, with individuals chosen at random to contribute alleles to the next generation. Four simulations were run for each population size, producing the four lines on each graph. *(a)* With a population size of 4, one allele sometimes became "extinct" owing to chance. For example, in the top simulation run, the *a* allele went extinct by the second generation (therefore, the frequency of the *A* allele became 1.0). *(b)* With a population size of 1000, allele frequencies remained relatively constant.

In the large population, 500 bugs will be parents to the next generation. Although their survival is random, the odds against all 500 of the reproducing ladybugs being spotted are enormous. In fact, it would be extremely unlikely for even 300 parents to be spotted. In this large population, then, we would not expect a major change in allele frequencies to occur from generation to generation (Fig. 15-2). The effects of chance are minimized by the large population size. But in the small population, only two individuals will reproduce. There is a 25% chance that both parents will be spotted (this is the same likelihood as flipping two coins and having both come up heads). If this happens, then the next generation will consist entirely of spotted ladybugs. Within a single generation, it is possible for the allele for a solid-colored shell to disappear from the population.

Figure 15-2a illustrates two important points about genetic drift:

1. *Genetic drift tends to reduce genetic variability within a small population.* In extreme cases, all members of a population may become genetically identical (Fig. 15-2a, top line).
2. *Genetic drift tends to increase genetic variability between populations.* Purely as a result of chance, separate populations may evolve extremely different allele frequencies (Fig. 15-2a, top versus bottom lines).

Two special cases of genetic drift, called the *population bottleneck* and the *founder effect*, further illustrate the enormous consequences that small population size may have on the allele frequencies of a species.

A Population Bottleneck Is an Example of Genetic Drift
In a **population bottleneck**, a population undergoes a drastic reduction in size—as a result of a natural catastrophe or overhunting, for example. Then only a few individuals are available to contribute genes to the entire future population. As our ladybug example showed, population bottlenecks can cause both *changes in allele frequencies* and *reductions in genetic variability* (Fig. 15-3a). Even if the population then rebounds, these genetic effects of the bottleneck may remain for hundreds or thousands of generations.

Loss of genetic variability has been documented in the northern elephant seal and the cheetah (Fig. 15-3b,c). The elephant seal was hunted almost to extinction in the 1800s; by the 1890s only about 20 had survived. Because elephant seals breed harem-style, with a single male mating with a stable group of females, one male may have fathered all the offspring at this extreme bottleneck point. Since then, elephant seals have increased in number to about 30,000 individuals, but biochemical analysis shows that all northern elephant seals are genetically almost identical. Other species of seals, whose popula-

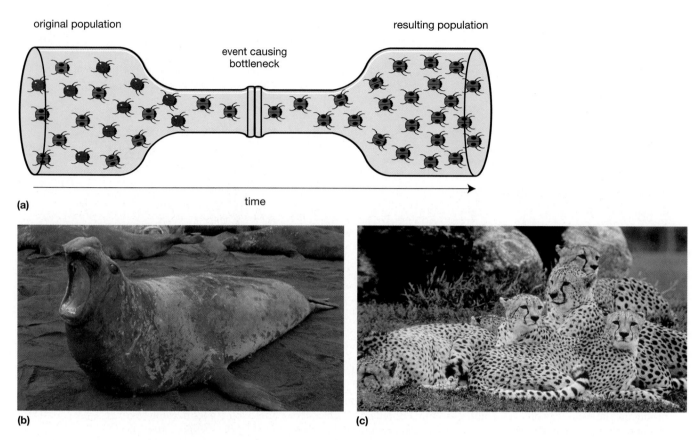

original population

resulting population

event causing
bottleneck

(a)

time

(b)

(c)

Figure 15-3 Genetic bottlenecks reduce variability
(a) If a population is reduced to a very small number of individuals, the gene pool is reduced and a population bottleneck occurs. The recovered population shows reduced genetic and phenotypic variability, because all are offspring of the few organisms that survived the bottleneck. Both *(b)* the northern elephant seal and *(c)* the cheetah passed through a population bottleneck in the recent past, resulting in an almost total loss of genetic diversity. As a result, the ability of these populations to adapt to changing environments is very limited.

tions have historically always remained large, exhibit much more genetic variability. The rescue of the northern elephant seal from extinction is rightly regarded as a triumph of conservation; however, with very little genetic variation, the elephant seal has much less potential to evolve in response to environmental changes. No matter how many elephant seals there are, the species must be considered to be threatened with extinction. Cheetahs are also genetically uniform, although the reason for the bottleneck is unknown. Consequently, cheetahs too could be gravely threatened by small changes in their environment.

A special case of a population bottleneck is the **founder effect**, which occurs when isolated colonies are founded by a small number of organisms. A flock of birds, for instance, may become lost during migration or may be blown off course by a storm. (This is thought to have happened in the case of Darwin's finches in the Galapagos Islands.) Among humans, small groups may migrate for religious or political reasons (Fig. 15-4). Such a small group may have allele frequencies that are very differ-

ent from the frequencies of the parent population because of chance inclusion of disproportionate numbers of certain alleles in the founders. If the isolation of the founders is maintained for a long period of time, a sizable new population may arise that differs greatly from the original population.

How much does genetic drift contribute to evolution? No one really knows. Only rarely are natural populations extremely small or completely cut off from gene flow with other populations. Populations occasionally do become very small, however, and it may be precisely these small populations that contribute most to major evolutionary changes. As we will see in Chapter 16, biologists believe that new species typically arise in such small populations.

Mating within a Population Is Almost Never Random

Organisms seldom mate strictly randomly. For example, most animals have limited mobility and are most likely to

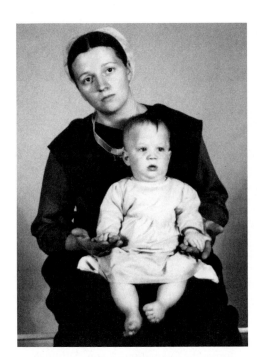

Figure 15-4 A human example of the founder effect
An Amish woman with her child, who suffers from a set of genetic defects known as the Ellis–van Creveld syndrome (short arms and legs, extra fingers, and in some cases heart defects). Fleeing from religious persecution, about 200 members of the Amish religion migrated from Switzerland to Pennsylvania between 1720 and 1770. Since that time, virtually all the Pennsylvania Amish moved to Lancaster County and have remained reproductively isolated from non-Amish Americans. The population increased to about 8000 by 1964. In that year, geneticist Victor McKusick surveyed the Lancaster County Amish and discovered that they had an allele frequency for Ellis–van Creveld of about 0.07, compared with a frequency of less than 0.001 in the general population. Why? One couple who immigrated in 1744 carried the allele. Inbreeding among the Amish passed the allele along to their descendants: a clear example of a founder effect. In addition, by chance the Ellis–van Creveld carriers had more children than the Amish average, further increasing the allele frequency by genetic drift. The combination of an initially high frequency in the immigrants (1 or 2 out of 200) plus genetic drift has resulted in more cases of Ellis–van Creveld syndrome from Lancaster County than from the rest of the world.

Earth Watch
Endangered Species: From Gene Pools to Gene Puddles

Ever since the Endangered Species Act was passed in 1973, the United States has had an official policy of protecting rare species. In fact, the real goal of the act is not protection but recovery; as one U.S. Fish and Wildlife Service official said, "The goal is to get species *off* the list." Wildlife biologists try to determine how large a population a species needs to have before it is no longer in danger of extinction from unpredictable events, such as a couple of years of drought or an epidemic of parasites. If a species reaches this critical population size, it is no longer legally "endangered" with extinction.

Does a "large enough" population (which usually is still very small by historical standards) really ensure a species' survival? From our discussion of genetic drift and population bottlenecks, you probably realize that the answer is no. If the population of a species has been reduced to the point at which it is declared an endangered species, then it probably has lost much of its genetic diversity. As ecologist Thomas Foose aptly put it, loss of habitat and the consequent reduction in population size mean that "gene pools are being converted into gene puddles." Even if the species recovers in numbers, the variety of its original gene pool has been lost. When the forces of natural selection change at some future time, the species may not have the necessary genetic variability to produce individuals adapted to the new environment, and it may then go extinct.

What can be done? The best solution, of course, is to leave enough habitat of diverse types so that species never become endangered in the first place. The human population, however, has grown so large and appropriated so much of Earth's resources that this solution is not possible in many places (Fig. E15-1). For many species, the only solution is to preserve enough habitat so that the remaining population is large enough to retain all or most of the total genetic diversity of the species. If

we are to be the caretakers of the planet and not merely its ultimate consumers, then protection of other life-forms and their genetic heritage will be our continuing responsibility.

Figure E15-1 Endangered by habitat destruction
This orangutan and its young, who live in the tropical rain forests of Borneo, are among the innumerable species whose continued survival is threatened by habitat destruction.

Figure 15-5 *Nonrandom, assortative mating among snow geese*
Some snow geese have white plumage, and others have dark blue plumage. These geese are most likely to mate with another bird of the same color. The preference is apparently established early in life, as individuals almost invariably choose mates that match the plumage of their own parents and siblings.

Figure 15-6 *Male competition promotes nonrandom mating*
Sparring contests between males result in extremely nonrandom mating among many animals, including deer, elk, seals, and many monkeys. Here, two male bighorn sheep square off against each other during the fall mating season. Although the horns are potentially deadly weapons, they are used in ritualized ways that minimize the danger of injury to either contestant.

mate with nearby members of their species. Further, they may choose to mate with certain individuals of their species rather than with others. The snow goose is a case in point. Individuals of this species come in two "color phases"; some snow geese are white, and others are blue (Fig. 15-5). Although both white and blue geese are members of the same species, mate choice is not random with respect to color phase. The birds exhibit a strong tendency to mate with a partner of the same color. This kind of preference for mates that are similar to one's self is known as *assortative mating.*

Another common form of nonrandom mating in animals is harem breeding. In some species, such as elephant seals, baboons, and bighorn sheep, only a few males fertilize all the females. Following some sort of contest, which may involve showing off with loud sounds or flashy colors, making threatening gestures, or actual combat, only certain males succeed in gathering a harem and mating (Fig. 15-6).

In many animal species, mating is not random because one sex, normally the female, controls mate selection and is quite picky about who qualifies for the privilege of breeding. Males display their virtues, such as the bright plumage of a peacock (Fig. 15-7). or the rich territory of a songbird. A female evaluates the males and chooses her mate. We shall explore this phenomenon in more detail later in this chapter.

Any form of nonrandom mating that is based on phenotype will alter allele frequencies in the population where it occurs.

All Genotypes Are Not Equally Adaptive

Genetic equilibrium requires that all genotypes must be equally adaptive—that is, no one genotype has any advantage over the others. It is probably true that some alleles are adaptively neutral, so organisms possessing any of several alleles will be equally likely to survive and reproduce. However, this is clearly not true of all alleles in all environments. Any time an allele confers, in Alfred Russel Wallace's words, "some little superiority," natural selection will favor the enhanced reproduction of the individuals who possess it. This phenomenon is perhaps best illustrated with an example.

Suppose that a flower-eating cow happens upon a field of purple pea flowers and, being enamored of purple flowers, eats all of them before they produce seeds. Because the allele for purple flowers (P) is dominant to the allele for white flowers (p), all the purple alleles in the entire population are in the purple-flowered pea plants (PP or Pp). If none of these plants reproduce but the white-flowered plants do reproduce, then the next generation will consist entirely of white-flowered pea plants (pp). The allele frequency for purple will drop to 0; the allele frequency for white will rise to 1.0 (100%). As a result of the cow's selective eating habits, *evolution will have occurred in that field.* The gene pool of the pea population will have changed, and natural selection, in the form of foraging by the cow, will have caused the change.

This simple example illustrates four important points about evolution:

Figure 15-7 The male peacock's showy tail has evolved through sexual selection
If females are picky when deciding which male to mate with, they might tend to favor males with slightly longer or more colorful tails. Their sons would then inherit tails that are larger and more colorful than average. If females in the next generation also preferred long, colorful tails, the cycle would continue. Given enough cycles of female selectivity, the scene pictured above would become the species' norm.

1. *Natural selection does not cause genetic changes in individuals.* The alleles for purple or white flower color arose spontaneously, long before the cow ever found the field. The cow did not cause white alleles to appear; its dietary preference merely favored the differential survival of white alleles compared with purple alleles.
2. *Natural selection acts on individuals, but evolution occurs in populations.* Individual pea plants either reproduced or did not, but it was the population as a whole that evolved as its gene frequencies changed.
3. *Evolution is a change in the allele frequencies of a population, owing to differential reproduction among organisms bearing different alleles.* In evolutionary terminology, the **fitness** of an organism is measured by its reproductive success: In our example, the white-flowered pea plants had greater fitness than the purple-flowered plants did, because the white-flowered plants produced more viable offspring.

4. *Evolutionary changes are not "good" or "progressive" in any absolute sense.* The white alleles were favored only because of the dietary preferences of this particular cow; in another environment, with other predators, the white allele may well be selected against.

Natural selection is not the *only* evolutionary force. As we have seen, mutation provides initial variability in heritable traits. The chance effects of genetic drift may change allele frequencies, even spawning new species. Further, evolutionary biologists are just now beginning to appreciate the power of random catastrophe in shaping the history of life on Earth—mass extinctions that may exterminate flourishing and floundering species alike. Nevertheless, it is natural selection that shapes the evolution of populations as they adapt to their changing environment. For this reason, we will examine the mechanisms of natural selection in some detail.

3) How Does Natural Selection Work?

To most people, the words *natural selection* are synonymous with the phrase *survival of the fittest*. Natural selection evokes images of wolves chasing caribou, of lions snarling angrily in competition over a zebra carcass. Natural selection, however, is not really about *survival* alone but also about *reproduction*. It is certainly true that an organism must survive at least long enough to reproduce. In some cases, it is also true that a longer-lived organism has more chances to reproduce. But no organism lives forever, and the only way that its genes continue into the future is through successful reproduction. When an organism that fails to reproduce dies, its genes die with it. The organism that reproduces lives on, in a sense, through the genes that it has passed on to its offspring. Therefore, although evolutionary biologists often discuss survival, partly because survival is usually easier to measure than reproduction, natural selection is really an issue of **differential reproduction**: Individuals bearing certain alleles leave more offspring (who inherit those alleles) than do other individuals with different alleles.

Natural Selection Acts on the Phenotype, Which Reflects the Underlying Genotype

The agents of natural selection cannot directly detect an organism's genotype. Rather, selection acts on phenotypes, the actual structures and behaviors displayed by the organisms in a population. Genotype and phenotype, however, are related in the following way. If you were to measure the phenotypes of a specific trait in all the individuals in a population, you would find a range of values. This range of phenotypes would arise from

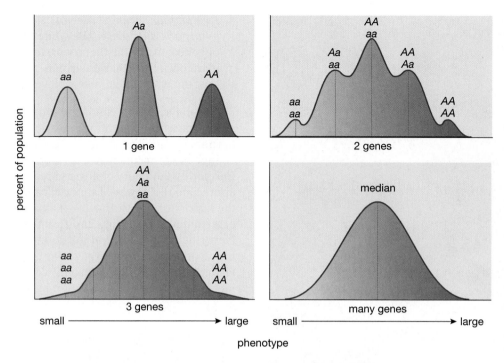

Figure 15-8 Both genes and environment contribute to the phenotype
This series of graphs illustrates the distribution of phenotypes that would be expected if one, two, three, or many genes, each with two incompletely dominant alleles (see Chapter 12), contributed to a particular body characteristic (for example, size). The vertical lines represent the precise size expected on the basis of genotype alone. In each case, environmental conditions (for example, amount of available food) create some variation in size, represented by the colored curves. As the number of genes contributing to the characteristic becomes large, the distribution of phenotypes approximates a smooth curve called a normal distribution (lower right-hand graph). The most common value for the phenotype is the middle value, called the median.

differences both in the genotypes of the organisms and in the environments in which they live (Fig. 15-8). However, environmental differences influencing phenotypes tend to average out in a large population. Thus, on the whole, genotype reflects phenotype: Most large plants will have genes that promote large size, whereas most small plants will have genes that promote small size. In our discussion of selection, therefore, we will ignore environmental causes of variability.

Natural Selection Can Influence Populations in Three Major Ways

Biologists recognize three major categories of natural selection based on its effect on the population over time (Fig. 15-9):

1. **Directional selection** favors individuals who possess values for a trait at one end of the distribution range of a particular trait and selects against both average individuals and individuals at the opposite extreme of the distribution. For example, directional selection may favor small size and select against both average and large individuals in a population.

2. **Stabilizing selection** favors individuals who possess an average value for a trait (for example, intermediate body size) and selects against individuals with extreme values.

3. **Disruptive selection** favors individuals who possess relatively extreme values for a trait at the expense of individuals with average values. Disruptive selection favors organisms at both ends of the distribution of the trait (for example, both large and small body sizes).

Directional Selection Shifts Character Traits in a Specific Direction

If environmental conditions change in a consistent way—for example, if the climate becomes colder—then a species may evolve in a consistent direction in response—for example, with thicker fur. The evolution of long necks in giraffes was almost certainly due to directional selection (Fig. 15-9a): Ancestral giraffes with longer necks obtained more food and therefore reproduced more prolifically than their shorter-necked contemporaries did. Antibiotic resistance in bacteria is another example of directional selection (see Chapter 18).

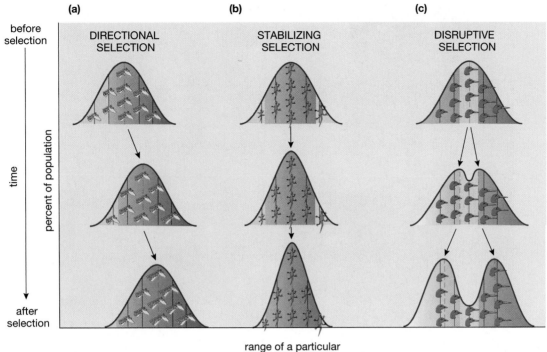

Figure 15-9 *Three ways natural selection affects a population over time*
A graphical illustration of three ways natural selection, acting on a normal distribution of phenotypes (in these examples, size), can affect a population over time. In all graphs, the beige areas represent individuals that are selected against—that is, do not reproduce as successfully as do the individuals in the purple range. *(a)* In directional selection, phenotypes that are either larger or smaller than average (larger is illustrated here) are favored. The average phenotype shifts position over the generations. *(b)* In stabilizing selection, the organisms most likely to reproduce are those with phenotypes close to the average for the population. The variability of phenotypes may decline, but the average value remains the same. *(c)* In disruptive selection, phenotypes that are both larger and smaller than average are favored. The population splits into two phenotypic groups.

Stabilizing Selection Acts against Individuals Who Deviate Too Far from the Average

Directional selection can't go on forever. Once a species is well adapted to a particular environment, and if the environment doesn't change, then most variations that appear through new mutations or recombination of old alleles during sexual reproduction will be harmful. Therefore, the species will typically undergo stabilizing selection, which favors the survival and reproduction of "average" individuals (Fig. 15-9b). Stabilizing selection commonly occurs when a single trait is under opposing environmental pressures from two different sources. Biologist M. K. Hecht, for example, studied lizards of the genus *Aristelliger*. He found that small lizards have a hard time defending territories but that large lizards are more likely to be preyed on by owls. Therefore, *Aristelliger* lizards are under stabilizing selection that favors an "average" body size.

It is widely assumed, although difficult to prove, that many traits are under stabilizing selection. We have already mentioned several. Although the lengths of legs and necks of giraffes probably originated under directional selection for feeding on leaves high up in trees,

they are almost certainly now under stabilizing selection, as a compromise between the advantage of reaching higher leaves for food and the disadvantage of vulnerability while drinking water (Fig. 15-10). Similarly, female mate choice probably drove the evolution of elaborate sexual displays in many birds, but now increased vulnerability to predation may exert stabilizing selection: If a peacock's tail became so long that he couldn't fly, he would be unlikely to live long enough to woo a female with that flashy tail.

Opposing environmental pressures may give rise to **balanced polymorphism**, in which two or more alleles of a gene are maintained in a population because each is favored by a separate environmental force. This seems to have occurred with the hemoglobin alleles in native Africans (see Chapter 12). The hemoglobin molecules of people who are homozygous for sickle-cell anemia (having two alleles for defective hemoglobin) clump up into long chains, distorting and weakening their red blood cells. This distortion causes severe anemia and potentially death. Before the advent of modern medicine, people homozygous for sickle-cell anemia were strongly selected against. Heterozygotes, who have one allele for defective

(a)   **(b)**

Figure 15-10 A compromise between opposing environmental pressures
(a) The long neck and legs of a giraffe are a definite advantage in feeding on acacia leaves high up in trees. **(b)** But a giraffe has to get into an extremely awkward and vulnerable position to drink. Feeding and drinking thus place opposing environmental pressures on the length of neck and legs.

hemoglobin and one allele for normal hemoglobin, suffer only mild anemia, though they may be adversely affected during strenuous exercise. Under these circumstances, you might think that natural selection would eliminate the sickle-cell allele.

Far from being eliminated, however, the sickle-cell allele is carried by nearly half the people in some areas of Africa. This distribution seems to result from the counterbalancing effects of anemia and malaria, a disease that formerly caused high death rates in equatorial Africa. The parasites that cause malaria multiply rapidly within the red blood cells of homozygous normal individuals. Before effective medical treatments were discovered, homozygous normals commonly died of malaria. Heterozygotes, however, enjoy some protection against malaria. Malaria parasites inside a heterozygote's red blood cells use up oxygen, causing the sickle-cell hemoglobin to clump and the cells to become sickle shaped. Infected sickled cells are destroyed by the spleen before the parasites can complete their development. Heterozygotes, therefore, have mild anemia but do not succumb to malaria.

During the evolution of African populations, heterozygotes survived better than either type of homozygote did and reproduced the most. As a result, both the normal hemoglobin allele and the sickle-cell allele have been preserved (Fig. 15-11).

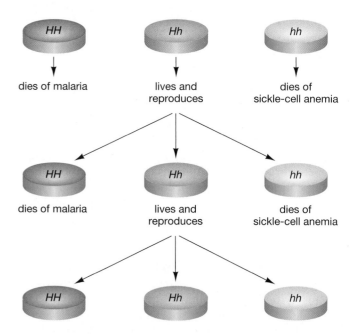

Figure 15-11 Stabilizing selection can produce balanced polymorphism
Two or more alleles, each producing a different phenotype, may be maintained in a population by opposing environmental pressures. The alleles for normal (*H*) and sickle-cell (*h*) hemoglobin are maintained by selection against both homozygotes. Heterozygotes (*Hh*) reproduce the most, thereby keeping both alleles in the population.

Disruptive Selection Adapts Individuals within a Population to Different Habitats

Disruptive selection (Fig. 15-9c) may occur when a population occupies an area that provides different types of resources that can be used by the species. In this situation, different characteristics best adapt individuals to use each type of resource. For example, an island, such as one of the Galapagos, may have several species of plants, some producing large, hard seeds and others small, soft seeds. Large seeds provide the most food per seed, but they can be cracked and eaten only by birds with large bills. Although large birds can easily eat small seeds, they would probably spend too much energy lugging their large bodies about looking for tiny seeds. If a single species of bird colonized such an island, what would happen? We would expect that larger-bodied, larger-beaked birds would specialize on large seeds and that small-bodied, small-beaked birds would specialize on small seeds. Medium-sized birds might not be able to crack open the large seeds and might not get enough energy from small seeds and so would be selected against. Disruptive selection would favor the survival and reproduction of both large and small, but not medium-sized, birds. Disruptive selection has not been extensively studied, although it has been supported by studies of both butterflies and birds.

A Variety of Processes Can Cause Natural Selection

As we've just seen, natural selection consists of nothing more than the plain fact that some phenotypes reproduce more successfully than others do. What makes this simple process such a powerful agent of change is that only the "best" phenotypes pass traits to subsequent generations. But what makes a phenotype "best"? Successful phenotypes are those that have the best *adaptations* to their particular environment. **Adaptations** are characteristics that help an individual survive and reproduce in an environment that includes not only physical factors but also the other organisms with which the individual interacts. These other organisms include predators, prey, and same-species competitors.

The nonliving (*abiotic*) component of the environment includes such factors as climate, availability of water, and minerals in the soil. The abiotic environment provides the "bottom line" requirements that an organism must have to survive and reproduce. However, many of the adaptations that we see in modern organisms have arisen because of interactions with other organisms—the living (*biotic*) component of the environment. As Darwin wrote, "The structure of every organic being is related . . . to that of all other organic beings, with which it comes into competition for food or residence, or from which it has to escape, or on which it preys." A simple example illustrates this concept.

A buffalo grass plant sprouts in a small patch of soil in the eastern Wyoming plains. Its roots must be able to take up enough water and minerals for growth and reproduction, and to that extent it must be adapted to its abiotic environment. Even in the dry prairies of Wyoming, this requirement is relatively trivial provided that the plant is alone and protected in its square meter of soil. In reality, many plants—including other grasses, sagebrush bushes, and annual wildflowers—also sprout in that same patch of soil. If our buffalo grass is to survive, it must compete for resources with the other plants. Its long, deep roots and efficient mineral uptake processes have evolved not so much because the plains are dry, but because it must share the dry prairies with other plants. Further, cattle (and, in the past, bison) graze the prairies. Buffalo grass is extremely tough, with silica (glass) compounds reinforcing the blades, an adaptation that discourages grazing. Over millennia, tougher plants were harder to eat and so survived better and reproduced more than did less-tough plants—another adaptation to the biotic environment.

Competition for Scarce Resources Favors the Best-Adapted Individuals

One of the major agents of natural selection in the biotic environment is **competition** with other members of the same species. As Darwin wrote in *On the Origin of Species*, "The struggle almost invariably will be most severe between the individuals of the same species, for they frequent the same districts, require the same food, and are exposed to the same dangers." In other words, no competing organism has such similar requirements for survival as does another member of the same species. For example, both lazuli buntings and western bluebirds are brightly colored in blue, red, and white, and both nest and rear their young in the foothills of the Rocky Mountains in the summer. But they do not compete very much with each other, because they eat different foods: Bluebirds mostly catch insects, whereas buntings specialize in seeds. Each mosquito picked off by a bluebird makes little difference to a bunting, but makes it harder for other bluebirds to find enough to eat.

Different species may also compete for the same resources, although generally to a lesser extent than do individuals within a species. As we will discuss more fully in Chapter 39, whether a particular plot of prairie is covered with grass, sagebrush, or trees is at least partly determined by competition among those plants for scarce soil moisture.

During Predation, Both Predator and Prey Act as Agents of Selection

When two species interact extensively, each exerts strong selection pressures on the other. When one evolves a new feature or modifies an old one, the other typically evolves new adaptations in response. As the Red Queen told Alice in Lewis Carroll's *Through the Looking Glass*, "Here, you see, it takes all the running you can do to keep in the same place." This constant, mutual feedback between two species is called **coevolution**. Perhaps the most familiar form of coevolution is found in *predator–prey* relationships.

Although we commonly think of predation as one animal preying upon another animal, **predation** includes any situation in which one organism eats another. In some instances, coevolution between *predators* (those who do the eating) and *prey* (those who are eaten) is a sort of "biological arms race," with each side evolving new adaptations in response to "escalations" by the other. Darwin used the example of wolves and deer: Wolf predation selects against slow or careless deer, thus leaving faster, more-alert deer to reproduce and continue the species. In their turn, alert, swift deer select against slow, clumsy wolves, because such predators cannot acquire enough food.

Symbiosis Produces Adaptations for Living in Intimate Association with Another Species

The term **symbiosis** describes a relationship in which individuals of different species live in direct contact with one another for prolonged periods of intimate interaction. Symbiosis incorporates several kinds of ecological relationships, including *parasitism*, in which one species lives and feeds on a larger species; *commensalism*, in which one species benefits and the other remains unharmed; and *mutualism*, in which both species benefit. The different types of symbiosis are described in Chapter 39. From an evolutionary perspective, of all the types of biotic interaction, symbiosis leads to the most-intricate coevolutionary adaptations. Whereas a given predator usually preys on several species and may interact with a particular species only occasionally, partners in symbiosis typically live together virtually their entire lives (Fig. 15-12). At least one of the partners, and normally both, must continually adjust to any evolutionary changes developed by the other.

Sexual Selection Favors Traits That Help an Organism Mate

In many species of animals, males compete for access to matings with females. This competition can take the form of contests for dominance among males, or it may involve behaviors or physical features that females find attractive. In the latter case, males may compete for the attention of females through song, elaborate displays, the defense of large territories, or even by building elaborate structures such as those of the bowerbird (as we shall see in Chapter 37). Choosing a male with a good territory is obviously advantageous, because good territories provide adequate food and shelter to raise young. However, females often also prefer elaborate "fashions" in their mates, such as bright colors and long feathers or fins that may make the male more vulnerable to predation. Why? A popular hypothesis is that structures and colors that do not serve any clear adaptive purpose actually provide the females with an outward sign of the males' fitness. Only vigorous, energetic males can survive when bur-

Figure 15-12 An example of symbiosis
Several species of clownfish live in a symbiotic relationship with anemones, each species of fish favoring a particular species of anemone. The fish nestle within the stinging tentacles of the anemone, thus protected from being eaten by other fish. The clownfish evolved specialized skin secretions and behaviors that protect it from the anemone's stingers. The fish may accidentally drop food onto the anemone once in a while, but the benefits are probably fairly one-sided.

dened with conspicuous coloration or large tails. Similarly, males that are sick or under parasitic attack may be dull and frumpy compared with healthy males. Whatever the exact selective mechanisms, it is thought that many of the elaborate structures and behaviors found only in males have evolved through selection caused by female mate choice: Only the flashy males transmitted their genes to the next generation.

Both the conspicuous structures and bizarre courtship behaviors of some male animals and the apparent willingness of females to base their mate choices on these features frequently seem to be at odds with efficient survival and reproduction. As we've stated above, exaggerated ornamentation may make males more attractive to females, but it also makes the males more vulnerable to predators. Darwin was intrigued by this apparent contradiction. He coined the term **sexual selection** to describe the special kind of natural selection that acts on traits that help an animal mate.

The trade-off between sexual selection and other forms of natural selection has recently been demonstrated through observation and experiment in guppies (small freshwater fish). In streams where predation is a threat, male guppies are inconspicuous, blending with the sandy streambed. But in safer waters, male guppies show more-conspicuous colors, apparently as a result of female preference for these markings. University of California biologist John Endler recently transplanted camouflage-colored male guppies from dangerous waters into safe waters. Within a year (about 20 guppy generations), their protected descendants had evolved conspicuous colors.

Kin Selection Favors Altruistic Behaviors

Evolution is often portrayed as being a bloody and vicious fight for survival. This portrayal, however, is not the complete picture. Although it is true that competitive and predatory interactions influence the evolution of most species, cooperation and even self-sacrifice may be favored by selection. **Altruism** refers to any behavior that endangers an individual organism or reduces its reproductive success but benefits other members of its species. Altruistic behaviors are common in the animal kingdom. A mother killdeer flutters just out of reach of a predator, feigning an injured wing and luring the predator away from her nestlings (Fig. 15-13); female worker bees forego reproduction and devote their lives to raising the offspring of the hive queen; and young male baboons scout around the edges of the troop, even though doing so increases their danger from leopards.

You might think that altruism runs counter to natural selection: If altruism is encoded in an organism's genes, those genes are placed at risk every time the altruist performs one of its brave behaviors. But natural selection can indeed select for altruistic genes, if the altruistic individual helps relatives who possess the same alleles. This special case of natural selection is called **kin selection**; it is explored in "Evolutionary Connections: Knowing Your Relatives: Kin Selection and Altruism."

4 What Causes Extinction?

We now know that natural selection is responsible for the development of such wonders as the thinking machine we call the human brain and the spider's ability to weave its intricate web. But natural selection may also lead to **extinction**, the death of all the members of a species. Trilobites, dinosaurs, saber-tooth cats—all are extinct, known to us only from fossils. Paleontologists estimate that *at least* 99.9% of all the species that ever existed are now extinct. Why? The actual cause of extinction is probably always environmental change, either in the living or the nonliving parts of the environment. Two characteristics seem to predispose a species to extinction when the environment changes: localized distribution and overspecialization. Three major environmental changes that drive a species to extinction are competition among species, the introduction of new predators or parasites, and habitat destruction.

Localized Distribution and Overspecialization Make Species Vulnerable in Changing Environments

Species vary widely in their range of distribution and, hence, in their susceptibility to extinction. Some species, such as herring gulls, white-tailed deer, and humans, inhabit entire continents or even the whole Earth; others, such as the Devil's Hole pupfish (Fig. 15-14), have extremely limited ranges. Obviously, if a species occurs in only a very small area, any disturbance of that area could

Figure 15-14 *Very localized distribution can endanger a species*
The Devil's Hole pupfish is found in only one spring-fed water hole in the Nevada desert. During the last glacial period, the southwestern deserts received a great deal of rainfall, forming numerous lakes and rivers. As the rainfall decreased, pupfish populations were isolated in shrinking small springs and streams. Isolated small populations and differing environmental conditions caused the ancestral pupfish species to split up into several very restricted modern species, all of which are on the brink of extinction.

Figure 15-13 *Altruism between mother and offspring*
A female killdeer lures a predator away from its nest by faking injury. The mother places herself in some small danger (she can always fly away if the predator comes too close) but saves her offspring from much greater danger.

Figure 15-15 *Extreme specialization places species at risk*
The Everglades kite feeds exclusively on the apple snail, found in swamps of the southeastern United States. Such behavioral specialization renders the kite extremely vulnerable to any environmental change that may exterminate its single species of prey.

easily result in extinction. If Devil's Hole dries up due to climatic change or well-drilling nearby, its pupfish will immediately vanish. Conversely, wide-ranging species normally do not succumb to local environmental catastrophes.

Another factor that may make a species vulnerable to extinction is overspecialization. Each species develops a set of genetic adaptations in response to pressures from its particular environment. These adaptations may limit the organism to a very specialized set of environmental conditions. The Everglades kite, for example, is a bird of prey that feeds only on the apple snail, a freshwater snail (Fig. 15-15). As the swamps of the American Southeast are drained for farms and developments, the snail population shrinks. If the snail becomes extinct, the kite will surely go extinct along with it.

In the fossil record, such behavioral specialization is hard to recognize. Structural specializations, however, may be just as restrictive. A case in point is giantism. For poorly understood reasons, many animals evolved to be huge in size, including certain amphibians, dinosaurs, and giant mammals, such as mammoths and ground sloths. To support their bulk, these animals must have consumed enormous amounts of food. If environmental conditions deteriorated, those giants may have been unable to find enough food and thus died out. Smaller animals that ate the same food but needed less of it survived.

Interactions with Other Organisms May Drive a Species to Extinction

As described earlier, interactions such as competition, predation, and parasitism serve as forces of natural selection. In some cases, these same forces can lead to extinction rather than to adaptation.

Competition for limited resources occurs in all environments. If a species' competitors evolve superior adaptations and the species doesn't evolve fast enough to keep up, it may become extinct. A particularly striking example of extinction through competition occurred in South America. For millions of years North America and South America had been isolated from one another, and each had developed a distinctive array of animal life. Then the Isthmus of Panama, a land bridge, formed about 3 million years ago, connecting the two continents, and massive migrations took place. In general, North American animals displaced their South American counterparts, and many South American species went extinct.

When formerly isolated populations encounter one another, it isn't only competitors that migrate between the areas—predators and parasites do too. With the exception of humans, who have exterminated hundreds of species, and of predators introduced into an area by humans, natural predators probably cause few extinctions. Parasites, however, can be devastating. In North America, Dutch elm disease and chestnut blight are well-known instances of introduced parasites that almost completely destroyed widespread native species. We cannot tell much about prehistoric parasite invasions, but the extinction of South American animals 3 million years ago might have been at least partly due to diseases carried south by resistant North American migrants.

Habitat Change and Destruction Are the Leading Causes of Extinction

Habitat change, both contemporary and prehistoric, is the single greatest cause of extinctions. Presently, habitat destruction due to human activities is proceeding at a frightening pace. Perhaps the most rapid extinction in the history of life will occur over the next 50 years, as tropical forests are cut for timber and to clear land for cattle and crops. As many as half the species presently on Earth may be lost because of tropical deforestation. (See "Earth Watch: From Gene Pools to Gene Puddles," p. 279.)

Prehistoric habitat alteration usually occurred over a longer time span but nevertheless had serious consequences. Climate changes, in particular, have caused many extinctions. At various times in the geologic past, moist, warm climates gave way to drier, colder climates with more variable temperatures. Many plants and animals failed to adapt to the harsh new conditions and went extinct. One cause of climate change is *plate tectonics*. Earth's surface is divided into portions called *plates*, which include the continents and the sea floor (Fig. 15-16). The solid plates slowly move above a viscous but fluid layer. As the plates wander, their positions may change in latitude. Further drift eventually resulted in the modern positions of the continents. Continental drift continues today: The Atlantic Ocean, for example, widens by a few centimeters each year.

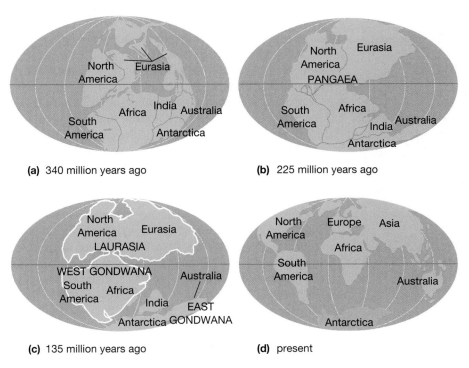

(a) 340 million years ago

(b) 225 million years ago

(c) 135 million years ago

(d) present

Figure 15-16 Plate tectonics is tied to climate change
Although slow, the drifting of continents—passengers on plates moving on Earth's surface as a result of plate tectonics—can cause tremendous environmental changes. The solid surfaces of the continents slide about over a viscous, but fluid, layer. *(a)* About 340 million years ago, much of what is now North America was positioned at the equator. *(b)* All the plates eventually fused together into one gigantic landmass, which geologists call Pangaea. *(c)* Gradually Pangaea broke up into Laurasia and Gondwanaland, which itself eventually broke up into West and East Gondwana. *(d)* Further plate motion eventually resulted in the modern positions of the continents. Plate tectonics continues today; the Atlantic Ocean, for example, widens by a few centimeters each year.

More than 350 million years ago, much of North America was located at or near the equator, an area characterized by consistently warm and wet tropical weather. But plate tectonics carried the continent up into temperate and arctic regions. As a result, the tropical weather was replaced by cooler temperatures, less rainfall, and seasonal changes.

An extreme, and very sudden, type of habitat destruction is caused by catastrophic geological events, such as massive volcanic eruptions. Several prehistoric eruptions that would make the 1980 Mount St. Helens explosion look like a firecracker by comparison wiped out every living thing for dozens of miles around and probably caused global climatic changes as well.

Mass extinctions are disappearances of many varied species in a relatively short period of time and over a wide area. The fossil record reveals episodes of extensive worldwide extinctions, especially among marine life (Fig. 15-17). Enormous meteorites, several kilometers in diameter, may have hit Earth at some of these times. If a huge meteorite struck land, it would kick up enormous amounts of dust. The dust might be thick enough, and spread widely enough, to block out most of the sun's rays. Fires started by the impact might be widespread, adding

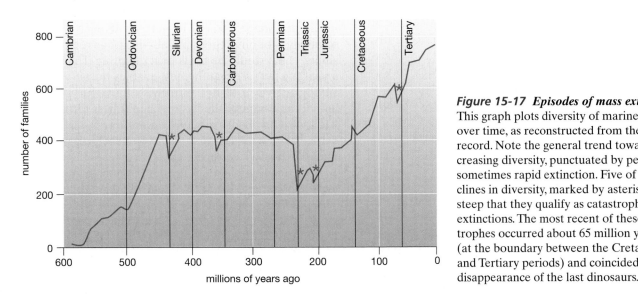

Figure 15-17 Episodes of mass extinction
This graph plots diversity of marine animals over time, as reconstructed from the fossil record. Note the general trend toward increasing diversity, punctuated by periods of sometimes rapid extinction. Five of these declines in diversity, marked by asterisks, are so steep that they qualify as catastrophic mass extinctions. The most recent of these catastrophes occurred about 65 million years ago (at the boundary between the Cretaceous and Tertiary periods) and coincided with the disappearance of the last dinosaurs.

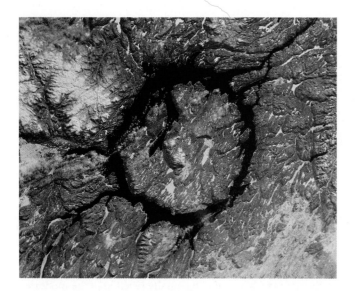

Figure 15-18 *Meteorites may have triggered some mass extinctions*
The Manicouagan crater in Quebec is about 45 miles in diameter. Giant meteorite-impact craters such as this one typically contain a central dome of rock that splashes up after the meteorite has buried itself in the ground, leaving a ring-shaped depression between the outer crater wall and the inner dome. In this satellite photo, water backed up behind a dam fills the crater ring. The Manicouagan meteorite struck a little over 200 million years ago; geologist Paul Olsen suggests that its impact triggered the mass extinction near the Triassic–Jurassic boundary (see Fig. 15-17).

soot to the atmosphere. Many plants would die, because they couldn't photosynthesize. Many animals, all of which ultimately depend on plants for food, would also die. Smaller amounts of dust might still block out enough sunlight to cause global cooling, perhaps even triggering an ice age. Mass extinctions would result.

Could such immense meteorite strikes have caused mass extinctions? No one knows for sure, but considerable evidence points to meteorites as the causes of at least some major extinctions (Fig. 15-18). Recently, two groups of researchers have suggested that the Chicxulub crater near the Yucatan Peninsula of Mexico was the impact site of the meteorite that might have killed the dinosaurs and many other species about 65 million years ago.

Evolutionary Connections
Knowing Your Relatives: Kin Selection and Altruism

Altruism is any behavior that is potentially harmful to the survival and future reproduction of an individual but enhances the reproductive potential of other individuals of the same species. Altruism includes worker bees' rear-ing the offspring of queen bees. Note that altruism does not imply conscious, voluntary decisions to engage in selfless behavior but rather describes a set of behaviors that is expressed in response to particular social and environmental stimuli.

From an evolutionary viewpoint, how can we explain altruism? Surely, if a mutation arose that caused altruistic behavior and the bearers of that mutation died or failed to reproduce because of their self-sacrificing behaviors, their "altruistic alleles" would disappear from the population. Maybe, or maybe not. To understand the evolution of altruism, we will need to introduce a new concept: inclusive fitness. The **inclusive fitness** of an individual is determined by the individual's success at contributing its own genes to the next generation. This contribution has two components: (1) a direct contribution from producing offspring *and* (2) an indirect contribution gained by helping relatives produce more offspring than would otherwise have been possible.

To see how altruism might increase the inclusive fitness of an individual, let's consider the Florida scrub jay. Year-old jays rarely mate and reproduce. Instead, these yearlings remain at their parents' nest and help feed and protect the parents' next brood.

Altruistic yearlings forgo reproduction for at least one year; some die from predation or accidents and never reproduce at all. How, then, can this behavior be adaptive? It all has to do with the reproductive options available to young Florida scrub jays. In their particular ecological setting, suitable breeding habitat is in very short supply. Long-term studies have shown that a young jay that leaves its home territory is fairly unlikely to be successful at finding a territory and raising a brood. But if a young bird stays on home territory, it may eventually inherit the territory or carve out a smaller area for itself. For this reason, selection has favored the behavior of "staying home." Still, this doesn't explain why the stay-at-homes expend valuable energy to help raise their younger siblings. Answering *that* question requires us to recognize that, on average, a diploid, sexually reproducing animal shares 50% of its genes with each sibling. A scrub jay—or any other animal—is just as related to its siblings as it would be to its own offspring. Therefore, if a yearling jay's help at the nest enables the survival of a sibling who would not otherwise have survived, the helper has accomplished as much in fitness terms as if it had raised an offspring of its own. As long as the average extra (indirect) fitness that results from helping is greater than the average (direct) fitness that would have resulted from an attempt at independent breeding, then the altruistic helping behavior is the most adaptive behavior. This phenomenon, whereby the actions of an individual increase the survival or reproductive success of its relatives, is called *kin selection.*

As this example suggests, *kin selection can favor the evolution of altruism if the altruistic behavior benefits relatives that bear the same altruistic allele.* In most cases, an animal will not know if another carries the altruism allele. But the animal must at least be able to distinguish relatives from strangers: Relatives stand a good chance of possessing the altruism allele, but you never can tell with strangers. A yearling jay that helped out at the nest of unrelated adult jays would probably waste its time and effort.

Identification of relatives isn't too hard to imagine in the case of jays and their parents. Many biologists, however, have objected to kin selection as a plausible explanation of other instances of altruistic behaviors, arguing that animals cannot evaluate degrees of relatedness. Two findings seem to address this objection. First, many social groups, including wolf packs and baboon troops, are actually family groups. Therefore, an animal would not have to identify relatives for its altruistic behaviors to benefit them the most. Second, many animals, including birds, monkeys, tadpoles, bees, and even tunicate larvae, can indeed identify relatives (Fig. 15-19). Given the choice between relatives and strangers, these animals preferentially associate with their relatives, even if they were separated at birth and have never seen those relatives before. If animals selectively form related groups, then altruistic behaviors will most likely benefit relatives. Although it is not the only mechanism of natural selection, kin selection has been a powerful environmental force in the evolution of altruism in many species, probably including humans.

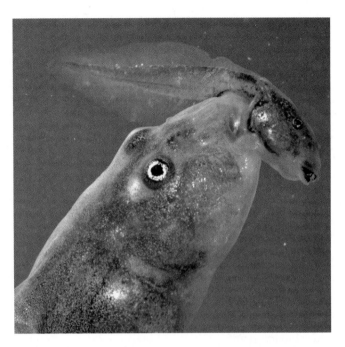

Figure 15-19 *Cannibalistic animals don't eat close relatives*
Spadefoot toad tadpoles, found in transient water holes of the Arizona desert, are cannibalistic. Many of their prey, however, are released unharmed after being tasted briefly. Researchers have discovered that the tadpoles can indeed distinguish, and spit out, their own brothers and sisters, preferring to eat unrelated members of their own species.

Summary of Key Concepts

1) How Are Populations, Genes, and Evolution Related?
The gene pool of a population is the total of all the alleles of all the genes carried by the members of that population. The sources of genetic variability within a population are mutation, which produces new genes and alleles, and recombination during sexual reproduction. In its broadest sense, evolution is a change in the frequencies of alleles in a population's gene pool due to enhanced reproduction by individuals who bear certain alleles.

Allele frequencies in a population will remain constant over generations only if the following conditions are met: (1) There is no mutation; (2) there is no gene flow—that is, no net migration of alleles into or out of the population; (3) the population is very large; (4) all mating is random; (5) all genotypes reproduce equally well (that is, no natural selection). These conditions are rarely, if ever, met in nature. Understanding why they are not met leads to an understanding of the mechanisms of evolution.

2) What Causes Evolution?

1. Mutations are random, undirected changes in DNA composition. Although most mutations are neutral or harmful to the organism, some prove advantageous in certain environments. Mutations are rare and do not change allele frequencies very much, but they provide the raw material for evolution.

2. In gene flow between populations, if the alleles that a migrant carries are different from those in the population

from which the migrant comes or to which it migrates, allele frequencies will change.

3. In any population, chance events kill or prevent reproduction by some of the individuals. If the population is small, chance events may eliminate a disproportionate number of individuals who bear a particular allele, thereby greatly changing the allele frequency in the population. This change is termed genetic drift.

4. Many organisms do not mate randomly. If only certain members of a population can mate, then the next generation of organisms in the population will all be offspring of this select group, whose allele frequencies may differ from those of the population as a whole. Population bottleneck and founder effect, two types of genetic drift, illustrate the consequences of small population size on allele frequency.

5. The survival and reproduction of organisms are influenced by their phenotype. Because phenotype depends at least partly on genotype, natural selection tends to favor the reproduction of certain alleles at the expense of others.

3) How Does Natural Selection Work?
Natural selection is an issue of differential, or unequal, reproduction. Natural selection can affect a population over time in three ways:

1. *Directional selection.* Individuals with characteristics that are different from average in one direction (for example, smaller) are favored both over average individuals and over individuals that differ from average in the opposite direction.

2. *Stabilizing selection.* Individuals with the average "value" for a characteristic are favored over individuals with extreme values.

3. *Disruptive selection.* Individuals with extreme values for a characteristic are favored over individuals with average values.

Natural selection occurs as a result of the interactions of organisms with both the biotic (living) and abiotic (nonliving) parts of their environments. Within a species, sexual selection and altruism are two types of natural selection.

When two or more species interact extensively so as to exert mutual environmental pressures on each other for long periods of time, both of them evolve in response. Such coevolution can occur as a result of any type of relationship between organisms, including competition, predation, and symbiosis.

4) What Causes Extinction?

Two factors that increase the likelihood of extinction, the death of all the members of a species, are localized distribution and overspecialization. Factors that actually cause extinctions include competition among species, the introduction of new predators or parasites, and habitat destruction.

Key Terms

adaptation   *p. 285*
allele frequency   *p. 274*
altruism   *p. 287*
balanced polymorphism
   p. 283
coevolution   *p. 285*
competition   *p. 285*
differential reproduction
   p. 281

directional selection   *p. 282*
disruptive selection   *p. 282*
equilibrium population
   p. 275
extinction   *p. 287*
fitness   *p. 281*
founder effect   *p. 278*
gene flow   *p. 275*
gene pool   *p. 274*

genetic drift   *p. 276*
genetic equilibrium   *p. 275*
Hardy-Weinberg principle
   p. 275
inclusive fitness   *p. 290*
kin selection   *p. 287*
mass extinction   *p. 289*
natural selection   *p. 273*
population   *p. 274*

population bottleneck   *p. 277*
population genetics   *p. 274*
predation   *p. 286*
sexual selection   *p. 286*
stabilizing selection   *p. 282*
symbiosis   *p. 286*

Thinking Through the Concepts

Multiple Choice

1. *Genetic drift is a _____ process.*
   a. random   b. directed
   c. selection-driven   d. coevolutionary
   e. uniformitarian

2. *Most of the 700 species of fruit flies found in the Hawaiian archipelago are each restricted to a single island. One hypothesis to explain this pattern is that each species diverged after a small number of flies had colonized a new island. This mechanism is called*
   a. sexual selection
   b. genetic equilibrium
   c. disruptive selection
   d. the founder effect
   e. assortative mating

3. *You are studying leaf size in a natural population of plants. The second season is particularly dry, and the following year the average leaf size in the population is smaller than the year before. But the amount of overall variation is the same, and the population size hasn't changed. Also, you've done experiments that show that small leaves are better adapted to dry conditions than are large leaves. Which of the following has occurred?*
   a. genetic drift   b. directional selection
   c. stabilizing selection   d. disruptive selection
   e. the founder effect

4. *You have bacteria thriving in your gastrointestinal tract. This is an example of*
   a. inclusive fitness   b. balanced polymorphism
   c. symbiosis   d. kin selection
   e. altruism

5. *Lamarckian evolution, the inheritance of acquired characteristics, could occur*
   a. if each gene had only one allele
   b. if individuals had different phenotypes
   c. if the genotype was altered by the same environmental changes that altered the phenotype
   d. if the phenotype was altered by the environment
   e. under none of these conditions

6. *Of the following possibilities, the best way to estimate an organism's evolutionary fitness is to measure the*
   a. size of its offspring
   b. number of eggs it produces
   c. number of eggs it produces over its lifetime
   d. number of offspring it produces over its lifetime
   e. number of offspring it produces over its lifetime that survive to breed

? Review Questions

1. What is a gene pool? How would you determine the allele frequencies in a gene pool?

2. Define *equilibrium population.* Outline the conditions that must be met for a population to stay in genetic equilibrium.

3. How does population size affect the likelihood of changes in allele frequencies by chance alone? Can significant changes in allele frequencies (that is, evolution) occur as a result of genetic drift?

4. If you measured the allele frequencies of a gene and found large differences from the proportions predicted by the Hardy-Weinberg principle, would that prove that natural selection is occurring in the population you are studying? Review the conditions that lead to an equilibrium population, and explain your answer.

5. People like to say that "you can't prove a negative." Study the experiment in Figure 15-1 again, and comment on what it demonstrates.

6. Describe the three ways in which natural selection can affect a population over time. Which way(s) is (are) most likely to occur in stable environments, and which way(s) in rapidly changing environments?

7. What is sexual selection? How is sexual selection similar to and different from other forms of natural selection?

8. Briefly describe competition, predation, symbiosis, and altruism, and give an example of each.

9. Define *kin selection* and *inclusive fitness*. Can these concepts help explain the evolution of altruism?

Applying the Concepts

1. In North America, the average height of adult humans has been increasing steadily for decades. Is directional selection occurring? What data would justify your answer?

2. Malaria is rare in North America. In populations of African Americans, what would you predict is happening to the frequency of the hemoglobin allele that leads to sickling in red blood cells? How would you go about determining if your prediction is true?

3. By the 1940s the whooping crane population had been reduced to fewer than 50 individuals. Thanks to conservation measures, their numbers are now increasing. But what special evolutionary problems do whooping cranes have now that they have passed through a population bottleneck?

4. In many countries, conservationists are trying to design national park systems so that "islands" of natural area (the big parks) are connected by thin "corridors" of undisturbed habitat. The idea is that this arrangement will allow animals and plants to migrate between refuges. Why would such migration be important?

5. Extinctions have occurred throughout the history of life on Earth. Why should we care if humans are causing a mass extinction event now?

6. A preview question for Chapter 16: A species is all the populations of organisms that potentially interbreed with one another but that are reproductively isolated from (cannot interbreed with) other populations. Using the five assumptions of the Hardy-Weinberg principle as a starting point, what factors do you think would be important in the splitting of a single ancestral species into two modern species?

Group Activity

Form a group of four students. Your group has been asked to study two recently discovered fish species, both found in waters off a small island. The two species are very similar in general structure, but there are two key differences between them. In species A, males are brightly patterned with blue, red, and purple scales, whereas the females are drab, and males are much larger than females. In species B, males and females are the same size, and both are a dull brown color that blends in with the sandy bottom. You've decided to do an in-depth study of these two species, but to get your research grant, you must first develop some predictions about what you'll find. What differences in behavior do you predict (list as many as you can think of)? How might the evolutionary histories of the two species differ? Discuss these questions, and submit a brief report of your answers.

For More Information

Allison, A. C. "Sickle Cells and Evolution." *Scientific American*, August 1956. The story of the interaction between sickle-cell anemia and malaria in Africa.

Alvarez, W., and Asaro, F. "An Extraterrestrial Impact," and Courtillot, V. E. "A Volcanic Eruption." *Scientific American*, October 1990. Leading geologists debate the question, What caused the mass extinction of the dinosaurs and so many other species at the end of the Cretaceous period?

Dawkins, R. *Climbing Mount Improbable*. New York: Norton, 1996. An eloquent book-length tribute to the power of natural selection to design intricate adaptations. The chapter on the evolution of the eye is an instant classic.

Fellman, B. "To Eat or Not to Eat." *National Wildlife*, February–March 1995. How animals with altruistic behaviors identify their relatives.

Gould, S. J. "The Evolution of Life on the Earth." *Scientific American*, October 1994. The importance of chance and catastrophe in shaping modern life.

May, R. M. "The Evolution of Ecological Systems." *Scientific American*, September 1979. Coevolution accounts for much of the structure of natural communities of plants and animals.

O'Brien, S. J., Wildt, D. E., and Bush, M. "The Cheetah in Peril." *Scientific American*, May 1986. According to molecular and immunological techniques, a population bottleneck has reduced the genetic variability of the world's cheetahs almost to zero.

Ryan, M. J. "Signals, Species, and Sexual Selection." *American Scientist*, January–February 1990. Ryan explores a variety of experiments on sexual selection, including the genetic basis of male characteristics and female choice.

Stebbins, G. L., and Ayala, F. "The Evolution of Darwinism." *Scientific American*, July 1985. A synthesis of molecular and classical evolutionary methodologies.

Ward, P. D. *The End of Evolution: On Mass Extinctions and the Preservation of Biodiversity*. New York: Bantam Books, 1994. An engaging first-person account of a paleontologist's investigation of the causes of mass extinction.

Answers to Multiple-Choice Questions
1. a   2. d   3. b   4. c   5. c   6. e

". . . endless forms most beautiful and most wonderful have been, and are being, evolved."

Charles Darwin in On the Origin of Species by Means of Natural Selection *(1859)*

A portion of the stunning diversity of land snail species, all members of the genus Achatinella, *in the Hawaiian Islands. These and hundreds of other land snail species emerged from a single ancestral snail population by the process known as speciation.*

The Origin of Species

16

At a Glance

Why are there so many kinds of organisms? Although Charles Darwin brilliantly explained how evolution builds complex, amazingly well-designed organisms, his ideas did not fully explain life's diversity. In particular, the mechanism of natural selection cannot by itself explain the observation that living things appear to be divided into discrete groups, with each group distinctly different from all other groups. When we look at big cats, we don't see a continuous array of different tiger phenotypes that gradually grades into a lion phenotype. We see lions and tigers as separate, distinct entities with no overlap. Each distinct group is known as a *species*.

It wasn't until some time after Darwin's death that biologists began to ponder seriously the process that led to the existence of so many kinds of organisms. Evolutionary theorists eventually began to unravel puzzles such as the one posed by Hawaiian land snails. Hundreds of species of land snails, most of which have colorful shells with distinctive ornamental markings, inhabit the Hawaiian Islands. Each species is distinguished by its unique shell pattern, which differs from that of all other species. The fossil record tells us that the different shell patterns have remained distinct for many generations. What could possibly have led to the proliferation of so many similar but separate species in one place? How exactly did all of these branches of the snail family tree arise from ancestral Hawaiian snails? The tree of life forms new branches (new species) only when a population of organisms somehow splits into two or more separate groups that do not later merge back into a single group. To understand how such a separation can occur, we must supplement the theory of evolution by natural selection with a few new ideas and hypotheses.

1) What Is a Species?

Before we can study the origin of species, we must first decide what a species is. Throughout most of human history, "species" was a poorly defined concept. When using the word *species*, most Europeans meant one of the

Net Watch

On-line resources for this chapter are on the World Wide Web at:
http://www.prenhall.com/audesirk
(click on the Table of Contents link and then select Chapter 16).

(a)

(b)

Figure 16-1 Interbreeding blurs the distinction between species
(a) The myrtle warbler and **(b)** Audubon's warbler were formerly thought to be two separate species but are now considered to be merely local varieties of one widespread species.

originally created "kinds" referred to in the Bible. But because, in their view, no one was present at the biblical Creation to record the criteria of the Creator, they had to distinguish among species by visible differences in structure; in fact, *species* is Latin for "appearance." Clearly, warblers are different from eagles and ducks. But how do biologists distinguish species of warblers? Today, biologists define **species** as "groups of actually or potentially interbreeding natural populations, which are reproductively isolated from other such groups." This definition, known as the *biological-species concept*, has at least one major limitation. It does not help us discern species boundaries among asexually reproducing organisms, which can't interbreed. Despite this limitation, most biologists accept the biological-species concept and its emphasis on reproductive community as the main criterion for identifying species. Biologists have found that differences in appearance do not always mean that two populations belong to different species. For example, field guides published in the 1970s listed the myrtle warbler and Audubon's warbler (Fig. 16-1) as distinct species; these birds differ in geographic range and in the color of their throat feathers. More recently, scientists decided that these birds are merely local varieties of the same species. The reason: Where their ranges overlap, these warblers interbreed, and the offspring are just as vigorous and fertile as the parents.

2 How Do New Species Form?

Darwin never proposed a complete mechanism of **speciation**, the process by which new species, such as the hundreds of species of Hawaiian land snails, form. If Darwin didn't, then who did? One scientist who played a large role in describing the process of speciation was Ernst Mayr of Harvard University, an ornithologist (expert on birds) and a pivotal figure in the history of evolutionary biology. Mayr developed the species definition given above, and he was among the first to recognize that speciation depends on two factors: the (1) isolation and (2) genetic divergence of two populations.

1. *Isolation of populations.* If two populations are to become sufficiently distinct that interbreeding is difficult or impossible, then there must be relatively little gene flow (migration) between them. If there is a great deal of gene flow, then genetic changes in one population will soon become widespread in the other as well.

2. *Genetic divergence.* It is not sufficient for two populations simply to be isolated. They will become separate species only if, during the period of isolation, they evolve sufficiently large genetic differences so they can no longer interbreed or produce vigorous, fertile offspring if they are reunited. If isolated populations are small, chance events may generate significant genetic differences by genetic drift (see Chapter 15). In both small and large populations, different environmental pressures in separate environments may favor the evolution of large genetic differences.

Speciation has seldom been observed in the wild (with the exception of "instant" speciation in plants by *polyploidy*, described later in this chapter). Nevertheless, evolutionary biologists have synthesized theories, observations, and experiments to devise hypothetical mechanisms for the origin of new species. These mechanisms fall into two broad categories: (1) **allopatric speciation**, in which two populations are geographically separated from one another, and (2) **sympatric speciation**, in which two populations share the same geographical area (Fig. 16-2).

Figure 16-2 *Models of allopatric and sympatric speciations*
(Left column) Allopatric speciation. *(a)* A single species (white mice) occupies a relatively homogeneous habitat. *(b)* An impass-able geographical barrier (here, a river changing course) splits the habitat into two parts, separating the species into two isolated populations. *(c)* Genetic drift or different environmental pressures cause the two populations to diverge genetically (tan vs. white mice). *(d)* The barrier is removed (the river changes course again), and the members of the two populations can share the same habitat. If the genetic differences between the two populations have become large enough that interbreeding cannot occur (that is, they are reproductively isolated from one another), then the two populations constitute separate species (brown vs. white mice). *(Right column)* Sympatric speciation. *(a)* A single species (white mice) occupies a homogeneous habitat. *(b)* Climate change or other factors form two distinctly different habitats that are still physically part of the same general region; that is, there are no barriers to movement between habitats. *(c)* Different environmental pressures in the two habitats lead to genetic diver-gence of organisms living in each (tan vs. white mice). *(d)* Sufficient genetic divergence causes reproductive isolation; former occupants of the two different habitats are now separate species (brown vs. white mice).

At first glance, you might think that sympatric speciation violates our first principle of speciation, isolation of populations, because the speciating populations live in the same locale. However, it is *isolation from gene flow* that is crucial to speciation. Although such isolation may be most commonly imposed by a physical barrier such as a river, two populations living in the same area can also experience restricted gene flow if they occupy different habitats within the area (for example, marshes as opposed to forests) or have differences in chromosome numbers such that they cannot form fertile hybrids. Therefore, the principle still holds: *Isolation from gene flow is the key to both allopatric speciation and sympatric speciation.*

Allopatric Speciation Can Occur in Populations That Are Physically Separated

In Greek, the word *allopatric* means "having a different fatherland." Allopatric speciation occurs when two populations become *geographically isolated* from one another; that is, they are physically separated either by distance or by an impassible barrier, such as a river or ocean (Fig. 16-3). If two populations become physically separated, little or no migration (and therefore little or no gene flow; see Chapter 15) can occur between them. Physical separation could occur if, for example, some members of a population drifted, swam, or flew to a remote island or if part of a population remained in a patch of suitable habitat that became isolated by a change in climate or by geological processes such as volcanism or continental drift. If, after such a separation, the pressures of natural selection differ in the two locations, or if the populations are small enough for genetic drift to

occur, then the two populations may accumulate large genetic differences and become separate species. Founder events, in which a few members of a species become isolated from the main body of the species, may also be important in initiating genetic differences between populations. There is some debate as to whether genetic drift or natural selection normally plays the major role in allopatric speciation. Probably the majority opinion is that natural selection, resulting from different environmental pressures in the two geographical locales, provides the major impetus to speciation. In either case, evolutionary biologists hypothesize that geographical isolation is involved in most cases of speciation, especially in animals.

Sympatric Speciation Can Occur in Populations That Live in the Same Area

Sympatric means "having the same fatherland." As the name implies, sympatric speciation refers to speciation that occurs within a single geographical area, as was the case with Hawaiian land snails. Sympatric speciation, like allopatric speciation, requires limited gene flow. There are two likely mechanisms whereby gene flow can be reduced between members of a single population in a given area: (1) ecological isolation and (2) chromosomal aberrations.

Ecological Isolation Restricts Different Populations to Different Habitats within the Same Area

If the same geographical area contains two distinct types of habitats (for example, distinct food sources, nesting places, and so on), different members of a single species may begin to specialize in one habitat or the other. If con-

(a)

(b)

Figure 16-3 Geographical isolation
To determine if these two squirrels are members of different species, we must know if they're "actually or potentially interbreeding." Unfortunately, it's hard to tell, because *(a)* the Kaibab squirrel lives only on the north rim of the Grand Canyon and *(b)* the Abert squirrel lives exclusively on the south rim. The two populations are geographically separated but still quite similar. Have they diverged enough after their separation to be considered separate species? On the basis of our current knowledge, it's impossible to say.

ditions are right, natural selection for habitat specialization may cause the formerly single species to split into two species. Such a split seems to be occurring right before biologists' eyes, so to speak, in the case of the fruit fly *Rhagoletis pomonella* (Fig. 16-4).

Rhagoletis is a parasite of the American hawthorn tree. This fly lays its eggs in the hawthorn's fruit; when the maggots hatch, they eat the fruit. About 150 years ago, entomologists (scientists who study insects) noticed that *Rhagoletis* began to infest apple trees, which were introduced into North America from Europe. It appears today that *Rhagoletis* is splitting into two species, one that breeds on apples and one that sticks to hawthorns. Substantial genetic differences exist between the flies that live on apples and the flies that live on hawthorns. At least some of these genetic differences, such as the timing of emergence of the adult flies, are important for survival on a particular host plant. Apple trees and hawthorns are typically quite close together, and flies, after all, can fly. So why don't apple-flies and hawthorn-flies interbreed and cancel out any incipient genetic differences? There appear to be at least two reasons. First, female flies usually lay their eggs in the same type of fruit in which they themselves developed. Males also tend to rest on the same type of fruit in which they developed. Therefore, apple-liking males are likely to encounter and mate with apple-liking females. Second, apples mature 2 or 3 weeks later than hawthorn fruits do, and the two types of flies emerge with a timing appropriate for their chosen host fruits. Thus, the two varieties of flies have very little chance of meeting. Although there is still some interbreeding between the two types of flies, it seems they are well on their way to speciation. Will they make it?

Figure 16-4 Sympatric isolation
Individuals of the fruit fly species *Rhagoletis pomonella* lay their eggs in either apples or hawthorn fruits. Although the two sympatric subgroups interbreed freely in the laboratory, in the wild they exhibit genetic differences that suggest gene flow between them is restricted. Will this reduced gene flow eventually slow to a trickle and then stop altogether, leaving two, reproductively isolated species? Only time will tell.

Entomologist Guy Bush suggests, "Check back with me in a few thousand years."

Changes in Chromosome Number Can Cause Immediate Reproductive Isolation of a Population

In some instances, new species can arise nearly instantaneously through changes in chromosome number. A common speciation mechanism in plants is **polyploidy** (Fig. 16-5), the acquisition of multiple copies of each chromosome. As you know from Chapter 11, most plants and animals have paired chromosomes and are described as diploid. Occasionally, especially in plants, a fertilized egg duplicates its chromosomes but doesn't divide into two daughter cells. The resulting cell thus becomes *tetraploid*, with four copies of each chromosome. If all of the subsequent cell divisions are normal, this tetraploid zygote will develop into a plant that consists of tetraploid cells. Most tetraploid plants are vigorous and healthy, and many can successfully complete meiosis to form viable gametes. The gametes, however, are diploid (whereas meiosis normally produces haploid gametes from diploid cells). These diploid gametes can fuse with other diploid gametes to produce new tetraploid offspring, so it's no problem for tetraploids to interbreed with other tetraploids of that species or to self-fertilize (as many plants do).

If, however, a tetraploid interbreeds with a diploid individual from the "parental" species, the outcome is not so successful. For example, if a diploid sperm from a tetraploid plant fertilizes a haploid egg cell of the parental species, the resulting offspring will be *triploid*, with three copies of each chromosome. Many triploid individuals experience problems during growth and development. Even if the triploid offspring develops normally, it will be sterile, because when a triploid cell attempts to undergo meiosis, the odd number of chromosomes makes chromosome-pairing impossible. Meiosis fails, and viable gametes are not formed. Thus, the offspring of diploid–tetraploid matings are inevitably sterile, so tetraploid plants and their diploid parents form distinct reproductive communities that cannot interbreed successfully. A new species can form in a single generation.

Why is speciation by polyploidy common in plants but not in animals? Many plants can either self-fertilize or reproduce asexually, or both. If a tetraploid plant self-fertilizes, then its offspring will also be tetraploid. Asexual offspring, of course, are genetically identical to the parent and are also tetraploid. In either case, the new tetraploid plant may perpetuate itself and form a new species. Most animals, however, cannot self-fertilize or reproduce asexually. Therefore, if an animal produced a tetraploid offspring, the offspring would have to mate with a member of the diploid parental species and would produce all triploid offspring. As we pointed out, the triploid offspring would almost certainly be sterile. Speciation by polyploidy is extremely common in plants; in

Figure 16-5 Speciation by polyploidy
Chromosomal mutations, especially in plants, can result in polyploid individuals with extra copies of each chromosome. This example shows that a tetraploid mutant can successfully self-fertilize (or can interbreed with other tetraploid individuals) to yield a new generation of tetraploids but that matings between tetraploids and normal diploid individuals will yield only sterile offspring. Tetraploid mutants are thus reproductively isolated from their diploid ancestors and may constitute a new species.

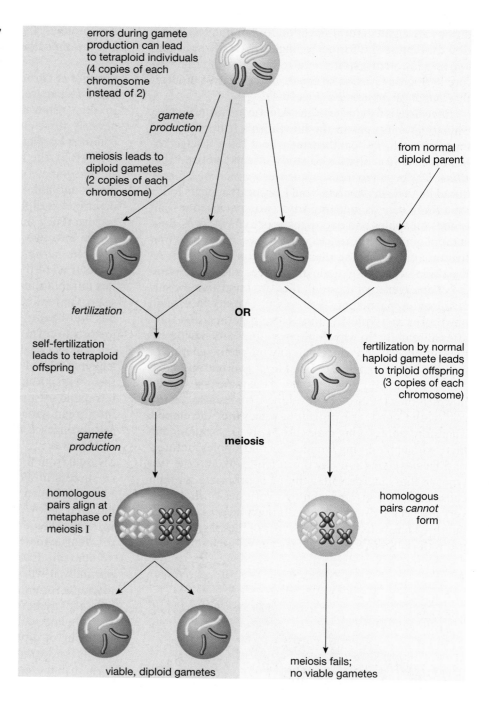

fact, nearly half of all species of flowering plants are polyploid, many of them tetraploid.

Change over Time within a Species Can Cause Apparent "Speciation" in the Fossil Record

The mechanisms of speciation and reproductive isolation that we have described lead to forking branches in the *evolutionary tree* of life (Fig. 16-6), as one species splits into two species. This kind of branching is a key source of evolutionary change, but changes *within* a species over time are also important. As generations pass and evolutionary innovations accumulate, the members of a species

may come to be very different from their distant ancestors, even if no speciation event takes place—that is, even if a new species doesn't form.

When a biologist encounters two populations of living organisms, he or she can, in principle, devise a test to see if the two populations are reproductively isolated and, therefore, separate species. For a paleontologist (a scientist who studies fossils), however, things are not so simple. Fossils cannot breed, so it is difficult to determine whether they were reproductively isolated from other fossils. Furthermore, different fossil organisms may be found in different rock layers that were laid down thou-

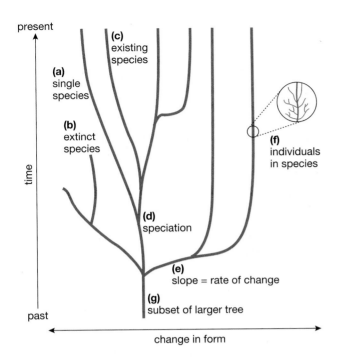

Figure 16-6 *Interpreting an evolutionary tree*
Evolutionary history is often represented by an evolutionary tree, a graph in which the vertical axis plots time and the horizontal axis stands for change in form (the greater the horizontal distance between points, the more phenotypic differences there are). When reading evolutionary trees, keep in mind that *(a)* each line on the tree represents a species; *(b)* lines that fail to reach the top of the graph represent extinct species; *(c)* lines that extend to the top represent existing species; *(d)* forks in the tree denote speciation events; *(e)* the slope of each line indicates the speed of phenotypic change over time (steep lines indicate periods in which species changed slowly, whereas lines that are nearly horizontal indicate periods of rapid change; *(f)* each species consists of many individuals, so an extreme close-up of a line would reveal that the line is made up of many smaller lines representing the lives of individuals; and *(g)* any evolutionary tree that can be drawn for a group is a portion of a much larger tree of life that connects all living things.

sands or even millions of years apart, and it is typically impossible to know if any branching, speciation events occurred between an older organism and a younger one. For these reasons, paleontologists must typically assign extinct organisms to species without reference to the biological-species concept. When comparing differences between extinct organisms, it's often impossible to tell if the differences arose due to branching evolution or to change within a single line of descent. Given the futility of applying the biological-species concept to fossils, many paleontologists choose to use a different system in which it is considered acceptable to assign different species names to anatomically different fossils, even if the two

fossils may simply represent different time points on a single evolutionary branch.

Circumstances Dictate Whether Anatomical Change and Speciation Are Linked

Speciation events may be accompanied by significant anatomical change (Fig. 16-7a), but speciation can occur without dramatic anatomical change, and anatomical change can occur without speciation (Fig.16-7b). For example, a small population that is isolated from gene exchange with other populations may be subjected to new environmental pressures that alter its average phenotype,

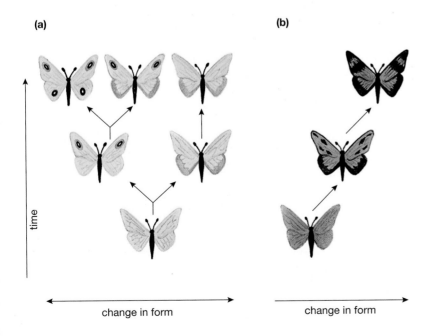

Figure 16-7 *Phenotypic change can occur with or without speciation*
(a) In the aftermath of speciation, which occurs when a species splits into two or more daughter species that exist simultaneously and are reproductively isolated from one another, one or more of the daughter species may undergo phenotypic change. *(b)* Phenotypic change can also occur without speciation. As a species evolves over time, sufficient genotypic changes may occur that later populations are phenotypically distinct from ancestral populations.

but an isolated population might inhabit an environment with environmental pressures no different from those of the parental population. Similarly, if an isolated population is small, then genetic drift may cause considerable anatomical change, but in another isolated population the random genetic changes that result from drift may not cause large changes in phenotype. Thus, speciation events can—but need not—be linked to anatomical change. In particular, if anatomical change in an isolated population causes or is accompanied by changes that prevent interbreeding with the parent population, then major anatomical change will be linked with speciation.

Anatomical change can, of course, take place in the absence of speciation. For example, large populations subjected to widespread directional selection, in which one extreme phenotype is favored, may experience anatomical change over time. If, however, gene flow among populations is adequate, speciation will not occur. In this case, anatomical change will not be linked to speciation.

Both Natural Selection and Random Events Drive Evolution

Is natural selection the paramount force operating during evolution? Some evolutionary biologists emphasize the importance of nonselective, random events during evolution. They have two major objections to the concept that most features evolved as adaptations through natural selection. First, it is extremely difficult to prove that any particular feature—say, the size of a bird's beak—was shaped by natural selection. If a biologist observes a large-billed bird that is cracking a large, tough nut, it is tempting to assume that the large bill evolved because of the advantage it confers in nutcracking. But the bird may have acquired the large bill as an adaptation for courtship, say, or even through genetic drift, and eats nuts because it "accidentally" has a suitable bill. On one of the Galapagos Islands, however, zoologist Peter Grant of Princeton University demonstrated quite conclusively that natural selection strongly influences beak size in a species of finch. When a prolonged wet spell altered the vegetation of the island, greatly increasing the availability of small seeds, the average bill size of breeding birds became longer and narrower over just 4 years.

The second argument against natural selection as the primary force in evolution is that random forces do indeed operate in nature. In Chapter 15, we discussed the likelihood that gigantic meteorites strike Earth from time to time, causing extinctions of many species. Species that were superbly adapted to warm, sunny climates might go extinct within a few years in the cold and gloom after the impact. The survival or extinction of species might have little to do with adaptations to their habitats, because the habitats could disappear. Smaller-scale catastrophes, such as floods, sea-level changes, or massive forest fires, could similarly cause the extinction of well-adapted but localized species.

The consensus is that both natural selection and random forces shape the evolutionary history of life on Earth. No biologist doubts that natural selection is the driving force behind the evolution of the magnificent structures and behaviors that we witness in the living world around us. And no one doubts that a 10-kilometer-wide meteorite would drive many otherwise well-adapted species to extinction. Which processes dominate, over what time scales and over what fraction of species on Earth, is the real debate.

3 How Is Reproductive Isolation between Species Maintained?

The process of speciation is primarily a story about the evolution of mechanisms that prevent interbreeding. Genetic divergence during the period of isolation is a necessary condition for the origin of a new species, but it is not sufficient unless part of that genetic divergence happens to cause the development of something that ensures **reproductive isolation** (Table 16-1). The structural and/or behavioral modifications that prevent interbreeding are called **isolating mechanisms**. Once such mechanisms have arisen, they have a clear adaptive value. Separate species tend to be genetically distinct in ways that adapt them to different environments. Any individual that mates with a member of another species will probably produce unfit

---

Table 16-1 Mechanisms of Reproductive Isolation

Premating Isolating Mechanisms: any structure, physiological function, or behavior that prevents organisms of two populations from mating.

1. **Geographical isolation:** the separation of two populations by a physical barrier.
2. **Ecological isolation:** lack of interbreeding between populations that occupy distinct habitats within the same general area.
3. **Temporal isolation:** the inability of populations to interbreed if they have significantly different breeding seasons.
4. **Behavioral isolation:** lack of interbreeding between populations of animals that differ substantially in courtship and mating rituals.
5. **Mechanical incompatibility:** the inability of male and female organisms to exchange gametes, normally because the reproductive structures are incompatible.

Postmating Isolating Mechanisms: any structure, physiological function, or developmental abnormality that prevents organisms of two populations, once mating has occurred, from producing vigorous, fertile offspring.

1. **Gametic incompatibility:** the inability of sperm from one population to fertilize eggs of another population.
2. **Hybrid inviability:** the failure of a hybrid offspring of two different populations to survive to maturity.
3. **Hybrid infertility:** reduced fertility (or complete sterility) in hybrid offspring of two different populations.

or sterile offspring, thereby "wasting" its genes and contributing nothing to future generations. Thus, there is strong environmental pressure to avoid mating between species. Incompatibilities that prevent mating between species are called **premating isolating mechanisms**.

When premating isolation fails or has not yet evolved, members of different species do mate. Then another isolating mechanism comes into play. If the resulting hybrid offspring die during development, then naturally the two species are still reproductively isolated from one another. In some cases, however, viable hybrid offspring are produced. Even so, if the hybrids are less fit than their parents or are themselves infertile, the two species may still remain separate, with little or no gene flow between them. Incompatibilities that prevent the formation of vigorous, fertile hybrids between species are called **postmating isolating mechanisms**.

Premating Isolating Mechanisms Prevent Mating between Species

Mechanisms that prevent mating between species include geographical isolation, ecological isolation, temporal isolation, behavioral isolation, and mechanical incompatibility.

Geographical Isolation Prevents Members of Different Species from Meeting One Another

Members of different species cannot mate if they never get near one another. As we have already seen, **geographical isolation** typically provides the conditions for speciation in the first place. However, we cannot determine if geographically separated populations necessarily constitute distinct species. Should the barrier separating the two populations disappear (an intervening river changes course, for example), the reunited populations might interbreed freely and not be separate species after all. If they cannot interbreed, then other mechanisms, such as different courtship rituals, must have developed during their isolation. Geographical isolation, therefore, is usually considered to be a mechanism that *allows new species to form* rather than a mechanism that *maintains reproductive isolation between species*.

Ecological Isolation Confines Members of Different Species to Different Habitats

Two populations that have different resource requirements may use different local habitats within the same general area and thus exhibit **ecological isolation**. White-crowned and white-throated sparrows, for example, have extensively overlapping ranges. The white-throated sparrow, however, frequents dense thickets, whereas the white-crowned sparrow inhabits fields and meadows, seldom penetrating far into dense growth. The two species may coexist within a few hundred yards of one another and yet seldom meet during the breeding season. A more dramatic example is provided by the more than 750 species of fig wasp (Fig. 16-8). Each species of fig wasp breeds in (and pollinates) the fruits of a particular species

Figure 16-8 Ecological isolation
This tiny female fig wasp must find her way to a tree of the single fig species to which her reproductive cycle is bound. She must then enter the developing fruit through a pore so small that her passage through it can rip off her wings. Once inside, she will never depart. After laying eggs (and depositing pollen that she carried from the fig where she hatched), she will die.

of fig, and each fig species hosts one and only one species of pollinating wasp. Although ecological isolation may slow down interbreeding, it seems unlikely that it could prevent gene flow entirely. Other mechanisms normally also contribute to interspecific isolation.

Temporal Isolation Occurs between Species That Breed at Different Times

Even if two species occupy similar habitats, they cannot mate if they have different breeding seasons, a phenomenon called **temporal** (time-related) **isolation**. Bishop pines and Monterey pines coexist near Monterey on the California coast (Fig. 16-9). Viable hybrids have been pro-

Figure 16-9 Temporal isolation
Bishop pines, such as these, and Monterey pines coexist in nature. In the laboratory they produce fertile hybrids. In the wild, however, they do not interbreed, because they release pollen at different times of the year.

duced between these two species in the laboratory. In the wild, however, they release their pollen at different times: The Monterey pine releases pollen in early spring, the bishop pine in summer. Therefore, the two species never crossbreed under natural conditions. Hawthorn-liking and apple-liking *Rhagoletis* fruit flies are also partially isolated from one another because they emerge from their host fruits and breed at slightly different times of the year.

Different Courtship Rituals Create Behavioral Isolation

Among animals, the elaborate courtship colors and behaviors that so enthrall human observers have evolved not only as recognition and evaluation signals between male and female; they may also aid in distinguishing among species. These different signals and behaviors create **behavioral isolation**. The striking colors and calls of male songbirds, for example, may attract females of their own species, but females of other species treat them with the utmost indifference. Among frogs, males are often impressively indiscriminate, jumping on every female in sight, regardless of the species, when the spirit moves them. Females, however, approach only male frogs that croak the "ribbet" appropriate to their species. If they do find themselves in an unwanted embrace, they utter the "release call," which causes the male to let go. As a result, few hybrids are produced.

Mechanical Incompatibility Occurs When Physical Barriers between Species Prevent Fertilization

In rare instances, ecological, temporal, and behavioral isolating mechanisms fail, and male and female of different species attempt to mate. Among animal species with internal fertilization (in which the sperm is deposited inside the female's reproductive tract), the male's and female's sexual organs simply may not fit together. Among plants, differences in flower size or structure may prevent pollen transfer between species. For example, different species may attract different pollinators (see Chapter 24 for a description of the interesting deceptions used by flowers to lure specific types of pollinators). Isolating mechanisms of this type are called **mechanical incompatibility**.

Postmating Isolating Mechanisms Prevent Production of Vigorous, Fertile Offspring

In some cases premating isolation fails, and mating occurs between members of different species. However, gene flow between the two species will still be restricted if the mating fails to produce vigorous, fertile hybrid offspring. Postmating isolating mechanisms include gametic incompatibility, hybrid inviability, and hybrid infertility.

Gametic Incompatibility Occurs When Sperm from One Species Are Unable to Fertilize Eggs of Another

Even if a male inseminates a female, his sperm may not fertilize her eggs, a situation called **gametic incompatibility**. For example, the fluids of the female reproductive tract may weaken or kill sperm of other species. Among plants, chemical incompatibility may prevent the germination of pollen from one species that lands on the stigma (pollen-catching structure) of the flower of another species.

Hybrid Inviability Occurs If Hybrid Offspring Survive Poorly

If fertilization does occur, the resulting hybrid may be weak or even unable to survive, a situation called **hybrid inviability**. The genetic programs directing development of the two species may be so different that hybrids abort early in development. Even if the hybrid survives, it may display behaviors that are mixtures of the two parental types. In attempting to do some things the way species *A* does them and other things the way species *B* does them, the hybrid may be hopelessly uncoordinated. Hybrids between certain species of lovebirds, for example, have great difficulty learning to carry nest materials during flight and probably could not reproduce in the wild.

Hybrid Infertility Occurs If Hybrid Offspring Are Unable to Produce Normal Sperm or Eggs

Most animal hybrids, such as the mule (a cross between a horse and a donkey) or the liger (a zoo-based cross between a male lion and a female tiger), are sterile. **Hybrid infertility** prevents hybrids from passing on their genetic material to offspring. A common reason is the failure of chromosomes to pair properly during meiosis, so eggs and sperm never develop. As we saw earlier, among plants that have speciated by polyploidy, any offspring produced by mating between the diploid "parent" species and the tetraploid "daughter" species will be triploid and sterile.

 ## At What Rate Does Speciation Occur?

Over the course of evolutionary history, species continually form, exist for a time, and go extinct. Geographical isolation, genetic drift, mutations, the invasion of new habitats, and coevolution all ensure that speciation never stops. The rate of speciation varies considerably over evolutionary time, however, and bursts of speciation can be seen both in the fossil record and in the distribution of modern organisms.

In Animals, Instantaneous Speciation Is Unlikely

Almost all evolutionary biologists agree that mutations that cause huge anatomical jumps within a single generation are highly unlikely to contribute to evolutionary change. A large, random mutation is much more likely to disrupt an organism's complex, finely tuned machinery than to improve it. Imagine that you opened the hood of an automobile and replaced one random part of the engine with another part randomly chosen from the shelves of an auto-parts store. How likely is this procedure to improve the performance of the engine? Similarly, a large, random mutation in an organism will probably decrease its possessor's chances of survival and reproduction and is unlikely to be passed to subsequent generations. Instead, large evolutionary changes tend to accrue from long series of smaller changes that accumulate over many generations.

Recent discoveries in developmental biology suggest that certain kinds of mutations may have the potential to circumvent the usual selective purge of sudden large changes. It turns out that single regulatory genes, or small numbers of genes, control major developmental pathways, such as how cells become specialized to form different structures. So in theory, mutations in just a few regulatory genes might result in significant changes in body plan. Most such changes would be detrimental, but some theorists now believe that one particular class of changes may have been quite significant in evolutionary history. In particular, mutations that caused the sudden *duplication* of body segments may have been important in the evolution of different body plans. Nonetheless, large, instantaneous changes will be harmful in the vast majority of circumstances, so genetic divergence sufficient for speciation probably cannot occur in a single generation (except for the special case of speciation by polyploidy in plants).

During Adaptive Radiation, One Species Gives Rise to Many

In some cases, a species gives rise to many new species in a relatively short time. This process, called **adaptive radiation**, occurs when populations of a single species invade a variety of new habitats and evolve in response to the differing environmental pressures in those habitats. Adaptive radiation has occurred many times and in many groups of organisms. Adaptive radiation normally results from one of two causes. First, a species may encounter a wide variety of unoccupied habitats—for example, when the ancestors of Darwin's finches colonized the Galapagos Islands, or when marsupial mammals first invaded Australia, or when an ancestral cichlid fish species arrived at Lake Victoria (Fig. 16-10). With no competitors except other members of their own species, all the available habitats and food sources were rapidly utilized by new species that evolved from the original invaders. The ultimate in unoccupied habitats results from mass extinctions, such as those that might follow a meteorite impact (see Chapter 15). With most potential competitors exterminated by the effects of the impact, the survivors speciate prolifically, filling an array of empty habitats.

Adaptive radiation can also occur if a species develops superior adaptations that enable it to displace less well-adapted species from a variety of habitats. Such adaptive radiations have apparently happened several times during evolutionary history. For example, about 2.5 million years ago, the isthmus of Panama rose above sea level

and formed a land bridge between North America and South America. After the previously separated continents were connected, the mammal species that had evolved in isolation on each continent were able to mix. Many species did indeed expand their ranges, as North American mammals moved southward and South American species drifted northward. As they moved, each species encountered resident species that occupied the same kinds of habitats and exploited the same kinds of resources. The ultimate result of the ensuing competition

Figure 16-10 Adaptive radiation
More than 300 species of cichlid fishes inhabit Lake Victoria in East Africa. These species are found nowhere else, and all of them descended from a single ancestral population within a million years. This dramatic adaptive radiation has led to a collection of closely related species with an array of adaptations for exploiting the many different food sources in the lake.

was that the North American species diversified and underwent an adaptive radiation that displaced the vast majority of the South American species. Clearly, evolution had bestowed on the North American species some (as yet unknown) set of adaptations that enabled their descendants to exploit resources more efficiently and effectively than their South American counterparts could.

The Fossil Record Is Ambiguous with Respect to the Tempo of Evolution

Although there is little disagreement among biologists about the basic mechanisms of speciation, controversy abounds over the tempo at which speciation has occurred throughout evolutionary history. At issue is interpretation of the fossil record. In many groups of organisms, if we arrange all known fossils in chronological order, we won't see a record of continuous, gradual change from one form to the next. Instead, we see what seems to be a series of jumps from one stage in the evolution of the group to the next stage. Ever since Darwin's day, the conventional interpretation of these jumps has been that the fossil record is imperfect. When an organism dies, its

chances of being fossilized are tiny; it has to die under just the right conditions to be preserved. And the chances that a particular fossilized organism will be found by a human are likewise very small. So, it's been proposed that the "missing" evolutionary stages existed but are simply not part of the skimpy record that fossils provide.

In a 1972 article, paleontologists Niles Eldredge, of the American Museum of Natural History, and Stephen Jay Gould, of Harvard University, challenged this conventional interpretation of the gaps in the fossil record. Perhaps, suggested Eldredge and Gould, the gaps reflected not an imperfect record but a real phenomenon of evolutionary history. Perhaps major evolutionary shifts in phenotype occurred not at a steady pace but in comparatively brief bursts. And perhaps these bursts were separated by long periods of *stasis*, stable intervals in which few evolutionary changes occurred. Eldredge and Gould called this interpretation of the fossil record the **punctuated equilibrium** theory (Fig. 16-11). In the years after their initial publication of the theory, the two scientists refined it by proposing that speciation was responsible for the bursts of phenotypic change (the "punctuated" part of the theory) between static intervals (the "equilibrium"

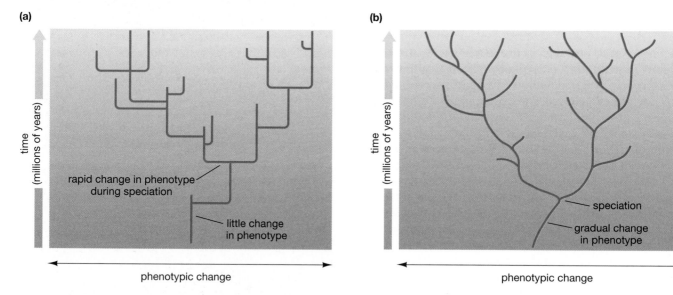

Figure 16-11 Two interpretations of gaps in the fossil record
(a) A chronological sequence of fossils in many lineages reveals sudden jumps between different forms rather than a series of smooth transitions. One way to explain these gaps is the punctuated equilibrium theory, which suggests that almost all phenotypic change occurs rapidly (at least, in a geological time scale) at speciation events; the remainder of each lineage exhibits little change in phenotype. So even if we recovered a fossil from every species, we'd see phenotypic gaps because change occurs so swiftly that fossils don't capture transitional forms. **(b)** Another possibility is that phenotypic change occurs at various times in each species' history and at different rates at different times. In this view, the gaps in the fossil record result from the long odds against finding fossils from multiple points along each lineage.

part). In this view, the generations immediately following a speciation event tend to go through periods of phenotypic change. The new forms that appear suddenly in the fossil record are actually new species, and the forms that disappear are species that were displaced by the new ones.

The concept of punctuated equilibria is easily misinterpreted, so it's worth taking a moment to reflect on what is not stated by the theory:

1. Most importantly, the punctuated equilibrium theory does not imply instantaneous, one-shot speciation in which an offspring is so different from its parents that it constitutes a new species. The "sudden" bursts of evolutionary change that Eldredge and Gould believe are found in the fossil record were sudden only in the context of geologic time. For example, a change that took 50,000 years, which is many thousands of generations for any kind of organism, will appear to be instantaneous in the fossil record, because most of our methods of determining the age of fossils cannot resolve differences as small as 50,000 years.

2. The punctuated equilibrium theory does not take issue with any of the accepted mechanisms of speciation. The theory does not address *how* species arise; rather, it is concerned with the distribution of speciation events over evolutionary time and with the rapidity of phenotypic change immediately after those events. In particular, the theory hypothesizes that the "evolutionary status" of a species is by default a pattern of little outward change from generation to generation and that this stasis persists for long stretches of a species' evolutionary history. A corollary of this hypothesis is that major evolutionary innovations originate mainly at comparatively infrequent speciation events.

Eldredge and Gould's initial publication of their idea unleashed a tidal wave of scientific debate that continues to this day. Opponents of the punctuated equilibrium interpretation of the fossil record are often labeled "gradualists," but this label serves only to confuse the issue by suggesting that those who favor the punctuated equilibrium theory are not gradualists themselves. In fact, evolutionary biologists of all points of view agree that significant evolutionary change requires many generations. In addition, no credible evolutionary biologist believes that evolution always proceeds at the same rate. The pace of evolution depends on the circumstances and on the species involved, so few biologists doubt that a pattern of punctuated equilibrium could, in theory, occur in at least some circumstances. The more vexing question at the heart of the debate is whether punctuated equilibrium is the normal pattern of evolution or a rare phenomenon. The debate about punctuated equilibria essentially reduces to a discussion of whether the gap-filled pattern in the fossil record reflects a basic feature of how evolution works or is simply an artifact of a spotty fossil record.

Evolutionary Connections
Scientists Don't Doubt Evolution

In the popular press, conflicts among evolutionary biologists are sometimes seen as conflicts about evolution itself. We occasionally read statements to the effect that new theories are overthrowing Darwin's and casting doubt on the reality of evolution. Nothing could be further from the truth. Proponents of punctuated equilibrium, gradualism, and everyone in between, unanimously agree that evolution occurred in the past and is still occurring today. The only argument is over the relative importance of the various mechanisms of evolutionary change in the history of life on Earth, their pace, and which forces were most important in shaping the evolution of particular species. Meanwhile, wolves still tend to catch the slowest caribou, small populations still undergo genetic drift, habitats still change or disappear, and perhaps somewhere in our galaxy another meteorite is swinging our way. Evolution continues, still generating, in Darwin's words, "endless forms most beautiful."

Summary of Key Concepts

1) What Is a Species?

According to the biological-species concept, a species is defined as all the populations of organisms that are potentially capable of interbreeding under natural conditions and that are reproductively isolated from other populations.

2) How Do New Species Form?

Speciation, the development of new species, requires that two populations be isolated from gene flow between them and develop significant genetic divergence. Allopatric speciation occurs by geographical isolation and subsequent divergence of the separated populations through genetic drift or natural selection. Sympatric speciation occurs by ecological isolation and subsequent divergence or by rapid chromosomal changes, such as polyploidy. Fossil species must normally be defined on the basis of anatomy alone, as information on reproductive status cannot be retrieved from them.

3) How Is Reproductive Isolation between Species Maintained?

Reproductive isolation between species may be maintained by one or more of several mechanisms, collectively known as premating isolating mechanisms and postmating isolating mechanisms. Premating isolating mechanisms include geographical isolation, ecological isolation, temporal isolation, behavioral isolation, and mechanical incompatibility. Postmating isolating mechanisms include gametic incompatibility, hybrid inviability, and hybrid infertility.

4) At What Rate Does Speciation Occur?

Instantaneous speciation via single, large mutation events is probably rare, especially in animals. Rates of speciation vary greatly over geological time. Invasion of new, unoccupied habitats or the development of fundamentally superior adaptations may trigger extremely rapid speciation, a process called adaptive radiation. The punctuated equilibrium theory, based on interpretation of the fossil record, states that speciation events and other major evolutionary changes in a given lineage occur only infrequently in evolutionary history and are separated by long periods of evolutionary stability.

Key Terms

adaptive radiation *p. 305*	geographical isolation *p. 303*	polyploidy *p. 299*	reproductive isolation *p. 302*
allopatric speciation *p. 296*	hybrid infertility *p. 304*	postmating isolating	speciation *p. 296*
behavioral isolation *p. 304*	hybrid inviability *p. 304*	mechanism *p. 303*	species *p. 296*
ecological isolation *p. 303*	isolating mechanism *p. 302*	premating isolating	sympatric speciation *p. 296*
gametic incompatibility	mechanical incompatibility	mechanism *p. 303*	temporal isolation *p. 303*
p. 304	*p. 304*	punctuated equilibrium *p. 306*	

Thinking Through the Concepts

Multiple Choice

1. *Many closely related species of marine invertebrates exist on either side of the isthmus of Panama. They probably resulted from*
a. premating isolation
b. isolation by distance
c. postmating isolation
d. gametic incompatibility
e. allopatric speciation

2. *Many hybrids are sterile because their chromosomes don't pair up correctly during meiosis. Why aren't polyploid plants sterile?*
a. They backcross to the parental generation.
b. Most are triploid.
c. They cross-pollinate.
d. They self-fertilize, using their diploid gametes.
e. Their eggs develop directly, without fertilization.

3. *In many species of fireflies, males flash to attract females. Each species has a different flashing pattern. This is probably an example of*
a. ecological isolation   b. temporal isolation
c. geographical isolation   d. premating isolation
e. postmating isolation

4. *Under the biological-species concept, the main criterion for identifying a species is*
a. anatomical distinctiveness
b. behavioral distinctiveness
c. geographic isolation
d. reproductive isolation
e. gametic incompatibility

5. *In terms of changes in gene frequencies, founder events result in*
a. gradual accumulation of many small changes
b. large, rapid changes
c. polyploidy
d. hybridization
e. mechanical incompatibility

6. *After the demise of the dinosaurs, mammals evolved rapidly into many new forms because of*
a. the founder effect
b. a genetic bottleneck
c. adaptive radiation
d. geological time
e. genetic drift

? Review Questions

1. Define the following terms: *species*, *speciation*, *allopatric speciation*, and *sympatric speciation*. Explain how allopatric and sympatric speciations might work, and give a hypothetical example of each.

2. Many of the oak tree species in central and eastern North America hybridize (interbreed). Are they "true species"?

3. Review the material on the possibility of sympatric speciation in *Rhagoletis* varieties that breed on apples or hawthorns. What types of genotypic, phenotypic, or behavioral data would convince you that the two forms have become separate species?

4. A drug called *colchicine* affects the mitotic spindle fibers and prevents cell division after the chromosomes have doubled at the start of meiosis. Describe how you would use colchicine to produce a new polyploid species of your favorite garden flower.

5. List and describe the different types of premating and postmating isolating mechanisms.

6. Define *punctuated equilibrium*. Look up the evolutionary history of a lineage (for example, dinosaurs or humans) and describe how the punctuated equilibrium theory might account for the evolution of that lineage.

Applying the Concepts

1. The biological-species concept has no meaning with regard to asexual organisms, and it's difficult to apply to extinct organisms that we know only as fossils. Can you devise a meaningful, useful species definition that would apply in all situations?

2. Seedless varieties of fruits and vegetables, created by breeders, are triploid. Explain why they are seedless.

3. Why do you suppose there are so many *endemic* species—that is, species found nowhere else—on islands? And why have the overwhelming majority of recent extinctions occurred on islands?

4. A contrarian biologist you've met claims that the fact that humans are pushing other species into small, isolated populations is good for biodiversity because these are the conditions that lead to new speciation events. Comment.

5. Southern Wisconsin is home to several populations of gray squirrels (*Sciurus carolinensis*) with black fur. Design a study to determine if they are actually separate species.

6. It is difficult to gather data on speciation events in the past or to perform interesting experiments about the process of speciation. Does this difficulty make the study of speciation "unscientific"? Should we abandon the study of speciation?

Group Activity

Form a group of four students. Imagine that you are the first scientists to explore a large island. In the course of your investigation, you discover that two large populations of flying birds coexist on the island. The two populations seem similar in general appearance, but you can detect small but consistent differences in plumage. Do the populations comprise one species or two? What key evidence would you need if the two populations are allopatric? What about if portions of the two populations inhabit a zone of sympatry? Be sure that your answers include your criteria for assessing what constitutes a species.

For More Information

Cowie, R. H. "Variation in Species, Diversity and Shell Shape in Hawaiian Land Snails." *Evolution* (49), 1996. A discussion of the ecological relationships involved in the sympatric speciation of Hawaiian land snails.

Eldredge, N. *Fossils: The Evolution and Extinction of Species*. New York: H. N. Abrams, 1991. A nicely illustrated exploration of a paleontologist's approach to examining and interpreting the past, including speciation events.

Gould. S. J. "Darwinian Fundamentalism," and "Evolution: The Pleasures of Pluralism." *The New York Review of Books*, June 12, 1997, and June 26, 1997, respectively. One of the originators of the punctuated equilibrium theory mounts a spirited defense against his critics. Subsequent exchanges of letters with Daniel Dennett and Stephen Pinker in the August 14, 1997, and October 9, 1997, issues are also good fun.

Gould, S. J., and Eldredge, N. "Punctuated Equilibrium Comes of Age." *Nature*, November 18, 1993. An overview of the history and evolution of the punctuated equilibrium theory of evolution by two of its chief proponents.

Grant, P. R. "Natural Selection and Darwin's Finches." *Scientific American*, October 1991. A temporary climate change altering the food supply created selection pressures on finches that altered their average beak shape within several generations.

Maynard Smith, J. *Evolution Now: A Century after Darwin*. San Francisco: W. H. Freeman, 1982. The section titled "Evolution—Sudden or Gradual?" contains essays by Russell Lande and John Maynard Smith himself on the "gradualist" side and by Stephen Jay Gould on the "punctuationist" side.

Quammen, D. *The Song of the Dodo*. New York: Scribner, 1996. Beautifully written exposition of the biology of islands. Read this book to understand why islands are known as "natural laboratories of speciation."

Wilson, E. O. *The Diversity of Life*. New York: W. W. Norton, 1992. Elegant description of how species arise, how they disappear, and why we should preserve them.

1. e   2. d   3. d   4. d   5. b   6. c

Answers to Multiple-Choice Questions

"[T]hence life was born,
Its nitrogen from ammonia, carbon from methane,
Water from the cloud and salts from the young seas . . . the cells of life
Bound themselves together into clans, a multitude of cells
To make one being—as the molecules before
Had made of many one cell.

Robinson Jeffers in The Beginning and the End *(1963)*

A group of paleontologists works to extricate a fossilized dinosaur skeleton—an Oviraptor, *a reptile the size of an ostrich. These scientists will use this frozen moment from the past to help reconstruct part of the history of life.*

The History of Life on Earth

17

At a Glance

Some biologists are historians. But instead of studying the short history of human social endeavors, they study the 3.5 billion-year history of life on Earth. Actually, they don't study it so much as try to reconstruct it. Their task is challenging because it's not, of course, possible to observe directly evolutionary changes that occurred in the past. Instead, evolutionary historians must infer the events of the past from whatever evidence is available in the present. Some of the best evidence comes in the form of fossils, and paleontologists (scientists who study fossils) have made a huge contribution to our understanding of life's history. Nonetheless, the historical snapshots frozen in fossils supply only a fragmentary record of evolutionary history, so paleontologists need a lot of help with their task. The complementary efforts of geologists, chemists, molecular biologists, and comparative anatomists have been especially important.

Perhaps the most fundamental question asked by evolutionary historians is also one of the grandest and most profound questions that any human can ask: How did life begin? The transition from nonlife to life is shrouded in mystery, and the nature of the transition will never be known with any certainty. Still, laboratory experiments have yielded results that form the basis for some scientifically informed speculation about life's origin.

Once living things arose, life on Earth unfolded in a slow, spectacular panorama of events. Single-celled organisms have dominated the planet for most of its history. Such organisms were the vehicles for the evolution of the majority of life's most significant innovations, including nucleic acids for encoding information and the processes of photosynthesis and oxygen-based respiration. Multicellular organisms, despite their relatively late arrival on the scene, changed the face of the planet by invading the land and the skies. A long series of evolutionary dynasties has since transpired, with each new dominant group rising, ruling the land or the seas for a period of time and, inevitably, falling into decline and extinction. We humans are only the most recent manifestation of this perpetual dynastic succession. Which evolutionary historian will reconstruct our history? What is our fate?

Net Watch
On-line resources for this chapter are on the World Wide Web at:
http://www.prenhall.com/audesirk
(click on the Table of Contents link and then select Chapter 17).

1) How Did Life Begin?

How and when did life first appear on Earth? Just a few centuries ago, this question would have been considered trivial. Although no one knew how life *first* arose, people thought that new living things appeared all the time, through **spontaneous generation** from both nonliving matter and other, unrelated forms of life. In 1609, a French botanist wrote, "There is a tree . . . frequently observed in Scotland. From this tree leaves are falling; upon one side they strike the water and slowly turn into fishes, upon the other they strike the land and turn into birds." Medieval writings abound with similar observations and delightful recipes for creating life—even human beings. Microorganisms were thought to arise spontaneously from broth, maggots from meat, and mice from mixtures of sweaty shirts and wheat.

In 1668 the Italian physician Francesco Redi disproved the maggots-from-meat hypothesis simply by keeping flies (whose eggs hatch into maggots) away from uncontaminated meat (see Chapter 1). Then in the mid-1800s, Louis Pasteur in France and John Tyndall in England disproved the broth-to-microorganism idea (Fig. 17-1). Although their work effectively demolished the notion of spontaneous generation, it did not address the question of how life on Earth originated in the first place.

For almost half a century, the subject lay dormant. Eventually, biologists returned to the question of the origin of life and began to seek answers. In the 1920s and 1930s, Alexander Oparin in Russia and John B. S. Haldane in England noted that the oxygen-rich atmosphere that we know would not have permitted the spontaneous formation of the complex organic molecules necessary for life. Oxygen reacts readily with other molecules, disrupting chemical bonds and thus tending to keep molecules simple. Oparin and Haldane speculated that the atmosphere of the young Earth was very low in oxygen and rich in hydrogen in the form of hydrogen gas (H_2), methane (CH_4), and ammonia (NH_3). Given these and other conditions, which we will discuss below, Oparin and Haldane proposed that life could have arisen from nonliving matter through ordinary chemical reactions. This process is called chemical evolution, or **prebiotic evolution**: that is, evolution before life existed.

Prebiotic Evolution Was Controlled by the Early Atmosphere and Climate

The primordial Earth differed greatly from the planet we now enjoy. When Earth first formed about 4.5 billion years ago, it was quite hot. A multitude of meteors smashed into the forming planet, and the kinetic energy of these extraterrestrial rocks was converted into heat on impact. Still more heat was released as radioactive atoms decayed. The rock composing Earth melted, and heavier elements such as iron and nickel sank to the center of the planet, where they remain molten today. Gradually Earth cooled, and elements combined to form compounds of many sorts. Virtually all of the oxygen combined with hydrogen to form water, with carbon to form carbon dioxide, or with heavier elements to form minerals. After millions of years, Earth cooled enough to allow water to exist as a liquid; it must have rained for thousands of years as water vapor condensed out of the cooling atmosphere. As the water struck the surface, it dissolved many minerals, forming a weakly salty ocean. Lightning from storms, heat from volcanoes, and intense

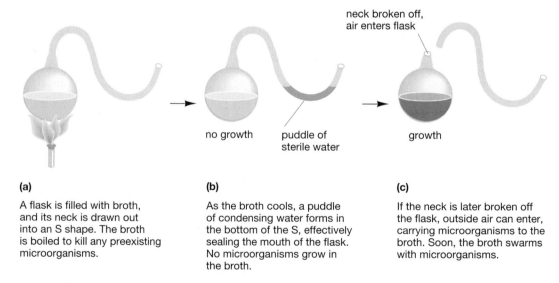

(a)
A flask is filled with broth, and its neck is drawn out into an S shape. The broth is boiled to kill any preexisting microorganisms.

(b)
As the broth cools, a puddle of condensing water forms in the bottom of the S, effectively sealing the mouth of the flask. No microorganisms grow in the broth.

(c)
If the neck is later broken off the flask, outside air can enter, carrying microorganisms to the broth. Soon, the broth swarms with microorganisms.

Figure 17-1 Spontaneous generation refuted
Louis Pasteur's experiment disproving the spontaneous generation of microorganisms in broth.

ultraviolet light from the sun all poured energy into the young seas.

Judging from the chemical composition of the rocks that formed at this time, geochemists have deduced that the primitive atmosphere probably contained carbon dioxide, methane, ammonia, hydrogen, nitrogen, hydrochloric acid, hydrogen sulfide, and water vapor. But there was virtually no free oxygen in the early atmosphere, because oxygen atoms were bound up in water, carbon dioxide, and minerals. The absence of free oxygen is assumed in all hypotheses and experiments that deal with prebiotic evolution.

Organic Molecules Can Be Synthesized Spontaneously under Prebiotic Conditions

Inspired by the ideas of Oparin and Haldane, Stanley Miller, a graduate student, and his advisor, Harold Urey of the University of Chicago, set out in 1953 to simulate prebiotic evolution in the laboratory. They tried to reproduce the atmosphere of early Earth by mixing water vapor, ammonia, hydrogen, and methane in a flask. An electrical discharge mimicked the intense energy of early Earth's lightning storms. In this experimental microcosm, Miller and Urey found that simple organic molecules appeared after just a few days (Fig. 17-2). Other, similar experiments by Miller and others have produced amino acids, short proteins, nucleotides, adenosine triphosphate (ATP), and other molecules characteristic of living things. Interestingly, the exact composition of the "atmosphere" used in these experiments is unimportant, provided that hydrogen, carbon, and nitrogen are available and that free oxygen is excluded. Similarly, a variety of energy sources, including ultraviolet light, electrical discharge, and heat, are all about equally effective. Even though geochemists may never know exactly what the primordial atmosphere was like, organic molecules certainly were synthesized on the ancient Earth.

Prebiotic Conditions Would Allow Organic Molecules to Accumulate

Prebiotic synthesis would not have been very efficient or very fast. Nonetheless, in a few hundred million years, large quantities of organic molecules could accumulate, especially because they didn't break down nearly as fast back then. On Earth today, most organic molecules have a short life; either they are digested by living organisms or they react with atmospheric oxygen. Primeval Earth lacked both life and free oxygen, so these sources of degradation were absent. However, the primordial atmosphere also lacked an ozone layer, a region high in the atmosphere that is enriched with ozone (O_3) molecules, which absorb some of the sun's high-energy ultraviolet (UV) light before it reaches Earth. During the early history of Earth, before the ozone layer formed, UV bombardment, which can break apart organic molecules, must have been fierce. Some places, however, such as those beneath rock ledges or at the bottoms of even fairly shallow

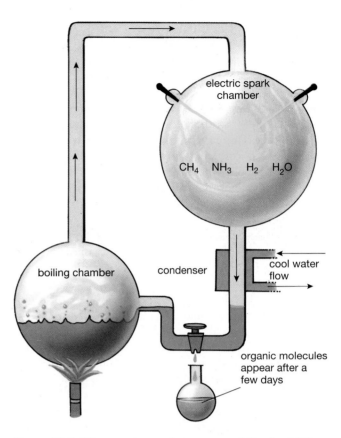

Figure 17-2 The experimental apparatus of Stanley Miller and Harold Urey
Life's very earliest stages left no fossils, so evolutionary historians have pursued a strategy of re-creating in the laboratory the conditions that may have prevailed on early Earth. In this experimental apparatus, a lightning storm in the ancient atmosphere is simulated by the discharge of an electric spark in a mixture of methane, ammonia, hydrogen, and water vapor. Simple organic molecules, potential building blocks of life, then form.

seas, would have been protected from UV radiation. In these locations, organic molecules may have accumulated to relatively high levels.

Even in areas protected from the sun, however, it's unlikely that molecules dissolved in a huge ocean could reach the concentrations necessary to form spontaneously the more-complex molecules that arose in the next stage of prebiotic evolution. The chemical reactions in which simple molecules combined to form larger molecules such as RNA and proteins required that the reacting molecules be packed closely together. Scientists have proposed several mechanisms by which the requisite high concentrations might have been achieved on early Earth. One possibility is that shallow pools at the ocean's edge were filled with water by waves crashing onto the shore. Afterward, some of the water in the pool might have evaporated, concentrating the dissolved substances. Given enough cycles of refilling and evaporation, the molecules in these pools could have become a concentrated "primordial soup" in which spontaneous chemical reactions could generate complex organic molecules. These molecules could then have become the building blocks of the first living organisms.

Was RNA the First Self-Reproducing Molecule?

One of the key features of living things is their capacity to reproduce. This capacity depends on the special ability of the DNA molecule to replicate itself. In modern cells, DNA encodes the information the cells need to synthesize the proteins that carry out most cellular functions. Self-replication of DNA ensures that this information is transferred to subsequent generations of cells. But how did this system of self-copying arise? It is far from obvious how any of the components of a primordial soup could have become endowed with the ability to make copies of themselves.

In the 1980s, Thomas Cech of the University of Colorado and Sidney Altman of Yale University offered an intriguing solution to this question. They discovered that certain small RNA molecules, dubbed **ribozymes**, act as enzymes that catalyze cellular reactions, including the synthesis of more RNA molecules. It seems inevitable that, during hundreds of millions of years of prebiotic chemical synthesis, RNA nucleotides may have occasionally bonded together to form short RNA chains. Let us suppose that, purely by chance, one of these RNA chains was a ribozyme that could catalyze the synthesis of copies of itself from free ribonucleotides in the surrounding waters. This first ribozyme probably wasn't very good at its job and made lots of mistakes. These mistakes were the first mutations. Like modern mutations, most undoubtedly ruined the catalytic abilities of the "daughter molecules," but a few may have been improvements. Molecular evolution could begin, as ribozymes with increased speed and accuracy of replication reproduced faster, making more and more copies of themselves.

Even after small self-replicating molecules arose, the transition to the modern "DNA → RNA → protein" mechanism must have involved a complex series of intermediate steps. One particularly difficult problem is that, although small nucleotide chains can catalyze their own replication, longer nucleotide chains cannot replicate without the assistance of specific protein enzymes that are encoded by the long nucleotide chains themselves. So how could the earliest RNA molecules get long enough to encode the necessary proteins if those proteins couldn't exist until the RNA molecules were long enough? Fortunately, the laws of physics and chemistry allow a solution to this knotty problem, as well as to the similarly tricky problem of how RNA gradually receded into its present role as an intermediary between DNA and protein enzymes. The details are fascinating but too intricate to describe here. Readers interested in exploring this stage of chemical evolution in all its glory should read Manfred Eigen's book *Steps Towards Life* (see "For More Information" on page 335).

Protocells May Have Consisted of Ribozymes within Microspheres

Self-replicating molecules alone do not constitute life; some kind of enclosing membrane is also required. The precursors of the earliest biological membranes may have been simple structures that formed spontaneously from purely physical, mechanical processes. For example, chemists have shown that if water containing proteins and lipids is agitated to simulate waves beating against ancient shores, hollow structures called **microspheres** form (Fig. 17-3). These hollow balls resemble living cells in several respects. They have a well-defined outer boundary that separates internal contents from the external solution. If the composition of the microsphere is right, a "membrane" forms that is remarkably similar in appearance to a real cell membrane. Under certain conditions microspheres can absorb material from the solution ("feed"), grow, and even divide.

If a microsphere happened to surround the right ribozymes, something a bit like a living cell would have been formed. We could call it a **protocell**, structurally similar to a cell but not a living thing. The ribozymes and their protein products would have been protected from free-roaming ribozymes in the primordial soup. Nucleotides and amino acids might have diffused across the membrane and have been used to synthesize new RNA and protein molecules. After sufficient growth, the microsphere may have divided, with a few copies of both ribozymes and proteins becoming incorporated into each daughter microsphere. If so, the path to the evolution of the first cells would be nearly at its end. Was there a particular moment when a nonliving protocell gave rise to something alive? Probably not. Like most evolutionary

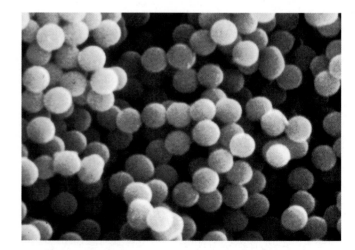

Figure 17-3 Do microspheres resemble the earliest cells? Cell-like microspheres can be formed by agitating proteins and lipids in a liquid medium. Such microspheres can take in material from the surrounding solution, grow, and even "reproduce," as these are doing.

transitions, the change from protocell to living cell was a continuous process, with no sharp boundary between one state and the next.

But Did All This Happen?

The above scenario, although plausible and consistent with many research findings, is by no means certain. One of the most striking aspects of origin-of-life research is the great diversity of assumptions, experiments, and contradictory hypotheses (see the suggested reading by J. Horgan, "In the Beginning . . ." for a taste of the controversies). Researchers disagree as to whether life arose in quiet pools, in the sea, in moist films on the surfaces of clay or iron pyrite (fool's gold), in furiously hot deep-sea vents. A few even argue that life may have arrived on Earth from space. Can we draw any conclusions from the research done so far? No one really knows the answer to that question, but we can offer a few observations.

First, the experiments of Miller and others show that amino acids, nucleotides, and other organic molecules would have formed in abundance on the primordial Earth. Second, no one imagines that prebiotic evolution would have formed any molecules as large or sophisticated as, say, hemoglobin. The first ribozymes, if they existed, may have been only a dozen nucleotides long. Third, ribozymes or other primitive catalysts need not have been very efficient to have accelerated the synthesis of important molecules. The stone tools of early humans weren't as handy as a Swiss army knife, either, but they were better than fingernails. Therefore, to a biologist, the origin of life means the origin of a few small molecules with terribly inefficient catalytic activities, perhaps surrounded by a film vaguely resembling a primitive membrane. Finally, several hundred million years probably elapsed between the appearance of organic molecules and the first primitive cell-like structures, and spontaneous synthesis must have proceeded over huge areas of Earth. Given enough time and space, even extremely rare events may in fact happen quite often.

Most biologists accept that the origin of life is probably an inevitable consequence of the working of natural laws. We should emphasize, however, that *this proposition is not proved and never will be*. Biologists investigating the origin of life have neither millions of years nor trillions of liters of reaction solutions with which to work!

2 What Were the Earliest Organisms Like? *www*

Life began during the *Precambrian* era, an interval designated by geologists and paleontologists, who have devised a hierarchical naming system of eras, periods, and epochs to delineate the vast expanse of geologic time (Table 17-1). The oldest fossil organisms found so far are in Precambrian rocks that are about 3.5 billion years old, based on radiometric dating techniques (see "Scientific Inquiry: How Do We Know How Old a Fossil Is?" on p. 317). Chemical traces in older rocks have led some paleontologists to believe that life is even older, perhaps as old as 3.9 billion years. What were the earliest cells like? The first cells were prokaryotic; that is, their genetic material was not sequestered from the rest of the cell within a membrane-limited nucleus. These cells probably obtained nutrients and energy by absorbing organic molecules from their environment. There was no free oxygen in the atmosphere, so the cells must have metabolized the organic molecules anaerobically. You will recall from Chapter 8 that anaerobic metabolism yields only small amounts of energy.

As you probably have already recognized, the earliest cells were primitive anaerobic bacteria. As these ancestral bacteria multiplied, however, they must have eventually used up the organic molecules produced by prebiotic synthesis. Simpler molecules, such as carbon dioxide and water, were still very abundant, as was energy, in the "dilute" form of sunlight. What was lacking, then, was not *materials* or *energy itself*, but *energetic molecules*—that is, molecules in which energy is stored in chemical bonds. Eventually some cells evolved the ability to use the energy of sunlight to drive the synthesis of their own complex, high-energy molecules from simpler molecules: In other words, photosynthesis appeared.

Several kinds of photosynthetic bacteria evolved, but the ones that proved most important in the evolution of life were the cyanobacteria, sometimes called blue-green algae. Their photosynthetic reactions converted water and carbon dioxide to organic compounds, releasing oxygen as a by-product. (You will undoubtedly breathe in some oxygen molecules today that were expelled about 2 billion years ago by a cyanobacterium!) At first, the oxygen reacted with iron atoms in Earth's *crust*, or surface layer, forming huge deposits of iron oxide, which you know as rust. After all the accessible iron turned to rust, the concentration of free oxygen in the atmosphere increased. Chemical analysis of rocks suggests that significant amounts of free oxygen appeared in the atmosphere about 2.2 billion years ago.

Oxygen is potentially very dangerous to living things, because it reacts with organic molecules, destroying these molecules and releasing their stored energy. The accumulation of oxygen in the atmosphere provided the environmental pressure for the next great advance in the Age of Microbes: the ability to use oxygen in metabolism, channeling its destructive power through aerobic respiration to generate useful energy for the cell. Because the amount of energy available to a cell is vastly increased when oxygen is used to metabolize food molecules, aerobic cells had a significant selective advantage.

Table 17-1 The History of Life on Earth

Era	Period	Epoch	Years Ago* (millions)	Major Events
Precambrian			4600	Origin of solar system and Earth.
			4000–3900	Appearance of first rocks on Earth.
			3900–3500	First living cells (prokaryotes).
			3500	Origin of photosynthesis (in cyanobacteria).
			2200	Accumulation of free oxygen in atmosphere.
			2000–1700	First eukaryotes (single-celled algae).
			By 1000	First multicellular organisms.
			About 1000	First animals (soft-bodied marine invertebrates).
Paleozoic	Cambrian		544–505	Primitive marine algae flourish; origin of most marine invertebrate types.
	Ordovician		505–440	Invertebrates, especially arthropods and mollusks, dominant in sea; first fish, fungi.
	Silurian		440–410	Many fish, trilobites, mollusks in sea; first vascular plants; invasion of land by plants; invasion of land by arthropods.
	Devonian		410–360	Fishes and trilobites flourish in sea; first amphibians and insects.
	Carboniferous		360–286	Swamp forests of tree ferns and club mosses; dominance of amphibians; numerous insects; first reptiles.
	Permian		286–245	First conifers; massive marine extinctions, including last of trilobites; flourishing of reptiles and decline of amphibians; continents aggregated into one land mass, Pangaea.
Mesozoic	Triassic		245–208	First mammals and dinosaurs; forests of gymnosperms and tree ferns; breakup of Pangaea begins.
	Jurassic		208–146	Dominance of dinosaurs and conifers; first birds; continents partially separated.
	Cretaceous		146–65	Flowering plants appear and become dominant; mass extinctions of marine life and some terrestrial life, including last dinosaurs; modern continents well separated.
Cenozoic	Tertiary	Paleocene	65–54	Widespread flourishing of birds, mammals, insects, and flowering plants; continents are shifted into modern positions; mild climate at beginning of period, with extensive mountain building and cooling toward end.
		Eocene	54–38	
		Oligocene	38–23	
		Miocene	23–5	
		Pliocene	5–1.8	
	Quaternary	Pleistocene	1.8–0.01	Evolution of genus *Homo*; repeated glaciations in Northern Hemisphere; extinction of many giant mammals.
		Recent	0.01–present	

* From University of California Museum of Paleontology, December 1997.

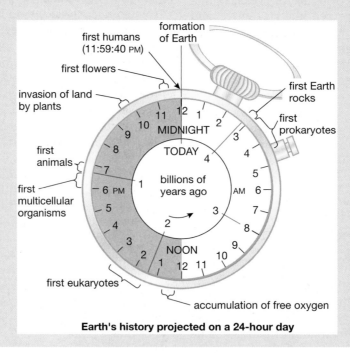

Earth's history projected on a 24-hour day

Scientific Inquiry
How Do We Know How Old a Fossil Is?

Early geologists could date rock layers and their accompanying fossils only in a *relative* way: Fossils found in deeper layers of rock were generally older than those found in shallower layers. With the discovery of radioactivity, it became possible to determine *absolute* dates, within certain limits of uncertainty. The nuclei of radioactive elements spontaneously break down, or *decay*, into other elements. For example, carbon-14 (usually written ^{14}C) emits an electron to become nitrogen-14 (^{14}N). Each radioactive element decays at a rate that is independent of temperature, pressure, or the chemical compound of which the element is a part. The time it takes for half the radioactive nuclei to decay at this characteristic rate is called the *half-life*. The half-life of ^{14}C, for example, is 5730 years.

How are radioactive elements used in determining the age of rocks? If we know the rate of decay and measure the proportion of decayed nuclei to undecayed nuclei, we can estimate how much time has passed since these radioactive elements were trapped in rock. This process is called *radiometric dating*. A particularly straightforward dating technique uses the decay of potassium-40 (^{40}K), which has a half-life of about 1.25 billion years, into argon-40 (^{40}Ar). Potassium is a very reactive element and is a common constituent of volcanic rocks such as granite and basalt. Argon, however, is an unreactive gas. Let us suppose that a volcano erupts with a massive lava flow, covering the countryside. All the ^{40}Ar, being a gas, will bubble out of the molten lava, so when the lava cools and solidifies into rock, it will start out with no ^{40}Ar. Any ^{40}K present in the hardened lava will decay to ^{40}Ar, half the ^{40}K decaying every 1.25 billion years. The ^{40}Ar gas will be trapped in the rock. A geologist could take a sample of the rock and determine the proportion of ^{40}K to ^{40}Ar (Fig. E17-1). If the analysis finds equal amounts of the two elements, the geologist will conclude that the lava hardened 1.25 billion years ago. With appropriate care, such age estimates are quite reliable. If a fossil is found beneath a lava flow dated at, say, 500 million years, then we know that the fossil is at least that old.

Some radioactive elements, as they decay, can even give an estimate of the age of the solar system. Analysis on uranium, which decays to lead, has shown that the oldest meteorites and moon rocks collected by astronauts are about 4.6 billion years old.

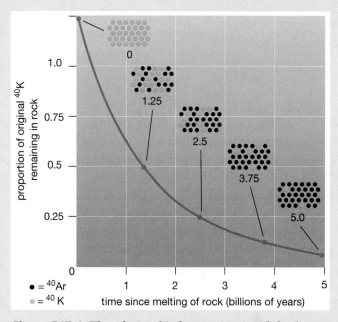

Figure E17-1 The relationship between time and the decay of radioactive ^{40}K to ^{40}Ar

Eukaryotes Developed Membrane-Enclosed Organelles and a Nucleus

Hordes of bacteria would offer a rich food supply to any organism that could eat them. There are no fossil records of the first predatory cells, but paleobiologists speculate that once a suitable prey population (such as these bacteria) appeared, predation would have evolved quickly. These predators would have been specialized prokaryotic cells that lacked cell walls and consequently were able to engulf whole bacteria as prey. According to the most widely accepted hypothesis, these predators were otherwise quite primitive, being capable of neither photosynthesis nor aerobic metabolism. Although they could capture large food particles, namely bacteria, they metabolized them inefficiently. By about 1.7 billion years ago, however, one predator probably gave rise to the first eukaryotic cell.

Eukaryotic cells differ from prokaryotic cells in many ways, but perhaps most fundamental is the presence in eukaryotes of a membrane-bound nucleus that contains the cell's genetic material. Other key eukaryote structures are the organelles used for energy metabolism: mitochondria and (in plants only) chloroplasts. How did these organelles evolve?

Mitochondria and Chloroplasts May Have Arisen from Engulfed Bacteria

The **endosymbiont hypothesis**, championed most forcefully by Lynn Margulis of the University of Massachusetts, proposes that primitive cells acquired the precursors of mitochondria and chloroplasts by engulfing certain types of bacteria. These cells and the bacteria trapped inside them (*endo* means "within") gradually entered into a *symbiotic* relationship, a close association between different types of organisms over an extended time. Let us suppose that an anaerobic predatory cell captured an aerobic bacterium for food, as it often did, but for some reason failed to digest this particular prey (Fig. 17-4a). The aerobic bacterium remained alive and well. In fact,

(a) *The probable origin of mitochondria and chloroplasts*

An anaerobic, prokaryotic cell engulfs an aerobic bacterium. The bacterium is enclosed in a sac of the predatory cell's membrane. The resulting "proto-organelle" has a double membrane, one from the predator and one from the prey.

(b) *Alternative possible origins of the membrane-enclosed nucleus*

1. The nucleus may have originated as an infolding of the plasma membrane that came to surround the DNA.

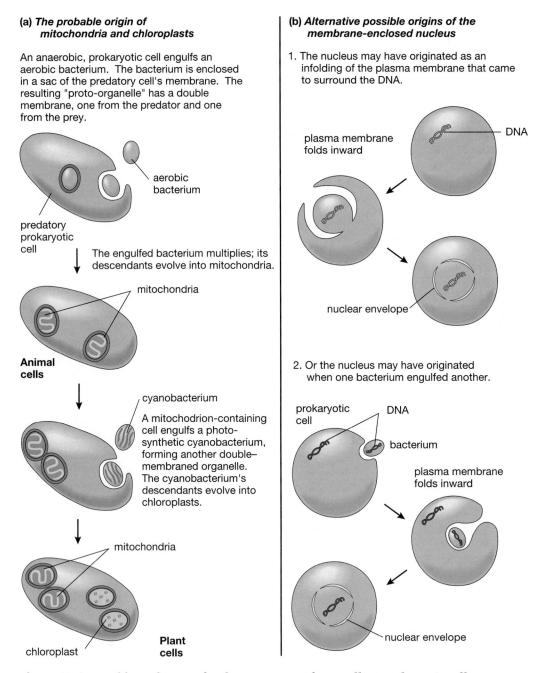

The engulfed bacterium multiplies; its descendants evolve into mitochondria.

A mitochodrion-containing cell engulfs a photosynthetic cyanobacterium, forming another double–membraned organelle. The cyanobacterium's descendants evolve into chloroplasts.

2. Or the nucleus may have originated when one bacterium engulfed another.

Figure 17-4 *Possible mechanisms for the appearance of organelles in eukaryotic cells*

it was better off than ever, because the cytoplasm of its predator/host was chock-full of half-digested food molecules, the remnants of anaerobic metabolism. The aerobe absorbed these molecules and used oxygen to complete their metabolism, gaining enormous amounts of energy as it did so. So abundant were its food resources, and so bountiful its energy production, that the aerobe must have leaked energy, probably as ATP or similar molecules, back into its host's cytoplasm. The anaerobic predatory cell with its symbiotic bacteria could now metabolize food aerobically, gaining a great advantage over its anaerobic compatriots. Soon its progeny filled the seas. Eventually, the endosymbiotic bacterium lost its ability to

live independently of its host, and the mitochondrion had been born.

One of these successful new cellular partnerships must have managed a second feat: It captured a photosynthetic cyanobacterium and similarly failed to digest its prey (see Fig. 17-4a). The cyanobacterium flourished in its new host and gradually evolved into the first chloroplast. Other eukaryotic organelles may have also originated through endosymbiosis. Many paleobiologists believe that cilia, flagella, centrioles, and microtubules may all have evolved from a symbiosis between a spirilla-like bacterium (a form of bacteria with an elongated corkscrew shape) and a primitive eukaryotic cell.

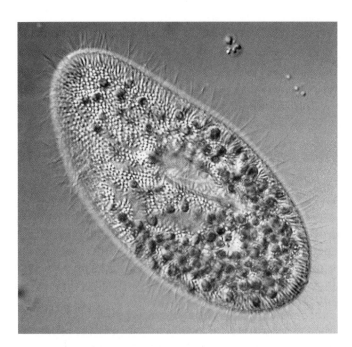

Figure 17-5 A modern intracellular symbiosis
Is an endosymbiotic origin of cellular organelles plausible? The plausibility of any scenario for evolutionary history is enhanced if we can point to living examples of the proposed intermediate stages. The ancestors of the chloroplasts in today's plant cells may have resembled *Chlorella*, the green, unicellular algae living symbiotically within the cytoplasm of the *Paramecium* pictured here.

Several lines of evidence support the endosymbiont hypothesis. For example, mitochondria, chloroplasts, and centrioles each contain their own minute supply of DNA, which some researchers interpret as a remnant of the DNA originally contained within the symbiotic bacterium. Another kind of support comes from *living intermediates*, organisms alive today that are similar to a hypothetical ancestral condition and thus help show that a proposed evolutionary pathway is plausible. For example, the amoeba *Pelomyxa palustris* lacks mitochondria but hosts a permanent population of aerobic bacteria that carry out much the same role. Similarly, a variety of corals, some clams, a few snails, and at least one species of *Paramecium* harbor a permanent collection of algae in their cells (Fig. 17-5). These examples of modern cells that host bacterial endosymbionts suggest that we have no reason to doubt that similar symbiotic associations could have occurred almost 2 billion years ago and led to the first eukaryotic cells.

The Origin of the Nucleus Is More Obscure

The evolution of the nucleus is more obscure. One possibility is that the plasma membrane folded inward, surrounding the DNA. This would create the nuclear membrane (Fig. 17-4b, top). Further infoldings could have produced the endoplasmic reticulum, which is continuous with the nuclear membrane. An alternative hypothesis is that, like many other eukaryotic organelles, the nucleus arose as a result of endosymbiosis. In this case, the engulfed bacterium took control of its host (Fig. 17-4b, bottom). However the nucleus originated, having the DNA sequestered within the nucleus seems to have conferred great advantages in regulating the use of the genetic material. Today, organisms composed of eukaryotic cells are by far the most visible forms of life on Earth.

3 How Did Multicellularity Arise?

Once predation had evolved, increased size became an advantage. A larger cell could more easily engulf a smaller cell, while in turn being more difficult for other predatory cells to ingest. Most larger organisms can also move faster than small ones, making successful predation and escape more likely. But enormous single cells have problems. Oxygen and nutrients going into the cell and waste products going out must diffuse through the plasma membrane. As we pointed out in Chapter 6, the larger a cell becomes, the less surface membrane is available per unit volume of cytoplasm. There are only two ways that an organism larger than a millimeter or so in diameter can survive. First, it can have a low metabolic rate so that it doesn't need much oxygen or produce much carbon dioxide. This seems to work for certain very large unicellular algae. Alternatively, an organism may be multicellular—that is, it may consist of many small cells packaged into a large, unified body.

The first multicellular organisms almost certainly evolved in the sea, but the fossil record reveals little about the origin of multicellularity. Unicellular eukaryotic fossils first appear in rocks that are about 1.7 billion years old, but the first fossil evidence of multicellular organisms is in rocks that are 700 million years younger. Some of these billion-year-old rocks have yielded fossil traces of animal tracks and burrows and of multicellular organisms that may have been the precursors of today's algae. But because the earliest multicellular organisms almost certainly had no skeleton or other hard parts, they left few fossils. Consequently, we may never learn very much about them.

Multicellular Plants Developed Specialized Structures That Facilitated Their Invasion of Diverse Habitats

During the Precambrian era, unicellular eukaryotic cells containing chloroplasts gave rise to the first multicellular plants. Multicellularity would have provided at least two advantages for plants. First, large, many-celled plants would have been difficult for unicellular predators to swallow. Second, specialization of cells would have conferred the potential for staying in one place in the brightly lit waters of the shoreline, as rootlike structures burrowed in sand or clutched onto rocks, while leaflike structures floated above in the sunlight. The green, brown, and red algae lining our shores today—some, such as the brown kelp, over 60 meters (200 feet) long—are the descendants of these early multicellular algae.

Multicellular Animals Developed Specializations That Allowed Them to Capture Prey, Feed, and Escape More Efficiently

A wide variety of Precambrian invertebrate animals (animals lacking backbones) appear in rocks laid down between 650 million and 544 million years ago. Many of these ancient animals are quite different in appearance from any animals that appear later in the fossil record, and some paleontologists believe that they represent a kind of failed evolutionary experiment in body plans that failed to survive the test of time. These Precambrian animals appear rather suddenly in the fossil record and disappear just as suddenly, so it's difficult to determine either their ancestry or the fate of their descendants. A new and different set of fossil animals appears just as suddenly in strata from the Cambrian period, marking the beginning of the Paleozoic era, about 544 million years ago. Unlike the Precambrian animals, the Cambrian fauna includes some animals whose forms are familiar and that are clearly the ancestors of today's animals. Like the Precambrian animals, the first fossils of Cambrian animals reveal an adaptive radiation that had already yielded a diverse array of complex body plans. The evolutionary history that produced such an impressive range of different animal forms remains obscure.

For animals, one of the advantages of multicellularity is the potential for eating larger prey. The first animal capable of consuming large prey probably resembled a jellyfish. A single opening served both as a mouth, to take in food, and as an anus, to expel indigestible remains (Fig. 17-6a). A half-digested prey filling its digestive tract keeps such an animal from feeding again until it is finished with

its first meal. Soon a more-efficient means of feeding evolved: A separate mouth and anus, found today in almost all animals, was developed (Fig. 17-6b). With this design, the animal can feed more or less continuously, as earthworms and sea cucumbers do today.

Coevolution of predator and prey led to increased sophistication in many kinds of animals. By the Silurian period (440 million to 410 million years ago), mud-skimming, armored trilobites were preyed on by (now-extinct) ammonites and the chambered nautilus, which still survives almost unchanged in deep Pacific waters (Fig. 17-7). A major evolutionary trend of this era was toward greater mobility. Predators often need to travel over wide areas in search of suitable prey, and speedy escape is also an advantage for prey. Locomotion is normally accomplished by the contraction of muscles that move body parts through their attachments to some sort of skeleton. Most invertebrates at this time possessed either an internal hydrostatic skeleton, much like a water-filled tube (worms), or an external skeleton covering the body (arthropods such as trilobites). Greater sensory capabilities and more-sophisticated nervous systems evolved along with the ability to move more efficiently. Senses for detecting touch, chemicals, and light became highly developed. The senses were typically concentrated in the head end of the animal, along with a nervous system capable of handling the sensory information and directing appropriate behaviors.

About 500 million years ago, one group of animals developed a new form of support for the body: an internal skeleton. For a hundred million years these were inconspicuous members of the ocean community, but by 400 million years ago some of these animals had evolved into fishes. By and large, the fishes proved to be faster than the invertebrates, with more-acute senses and larger brains. Eventually, they became the dominant predators of the open seas.

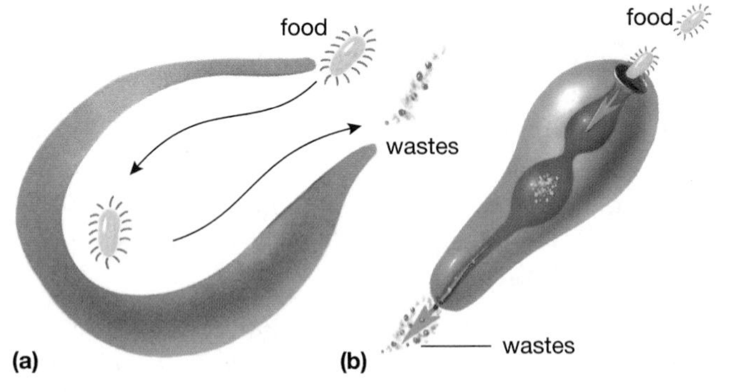

4 How Did Life Invade the Land?

One of the more thrilling subplots in the long tale of life's history is the story of life's invasion of land after more than 3 billion years of a strictly watery existence. In moving to solid ground, organisms had many obstacles to overcome. Life in the sea provides buoyant support against gravity, but on land an organism must bear its weight against the crushing force of gravity. The sea provides ready access to life-sustaining water, but a terrestrial organism must find adequate water. Sea-dwelling plants and animals can reproduce by means of mobile sperm and/or eggs that swim to each other through the water, but land-dwellers must ensure that their gametes be protected from drying out.

Despite the obstacles to life on land, the vast empty spaces of the Paleozoic landmass represented a tremen-

Figure 17-6 Advances in digestive tracts
(a) The earliest animals probably possessed a digestive system with only one opening. Feeding would have had to be periodic, because ingestion of food and expulsion of wastes would interfere with one another. ***(b)*** A major advance in feeding is to have separate openings for ingesting food and for ejecting wastes, allowing more or less continuous feeding.

food

wastes

food

wastes

(a)   (b)

Figure 17-7 Diversity of ocean life during the Silurian period
(a) Characteristic life of the oceans during the Silurian period, 440 million to 410 million years ago. Among the most common fossils from that time are *(b)* the trilobite and its predators *(c)* the ammonites and the nautiloids. Although *(d)* illustrates a living *Nautilus*, the Silurian nautiloids were very similar in structure, showing that a successful body plan may exist virtually unchanged for hundreds of millions of years.

dous evolutionary opportunity. The potential rewards of terrestrial life were especially great for plants. Water strongly absorbs light, so even in the clearest water, photosynthesis is limited to the upper few hundred meters, and usually much less. Out of the water, the sun is dazzlingly bright, permitting rapid photosynthesis. Furthermore, terrestrial soils are rich storehouses of nutrients, whereas seawater tends to be low in certain nutrients, particularly nitrogen and phosphorus. Finally, the Paleozoic sea swarmed with plant-eating animals, but the land was devoid of animal life. The plants that first colonized the land would have ample sunlight, untouched nutrient sources, and no predators.

Some Plants Developed Specialized Structures That Adapted Them to Life on Dry Land

In moist soils at water's edge, a few small green algae began to grow, taking advantage of the sunlight and nutrients. They didn't have large bodies to support against

the force of gravity, and, living right in the film of water on the soil, they could easily obtain water. About 400 million years ago, some of these algae gave rise to the first multicellular land plants. Initially simple, low-growing forms, land plants rapidly developed solutions to two of the main difficulties of plant life on land: (1) obtaining and conserving water and (2) staying upright despite gravity and winds. Waterproof coatings on the aboveground parts reduced water loss by evaporation, and rootlike structures delved into the soil, mining water and minerals. Specialized cells formed tubes called *vascular tissues* to conduct water from roots to leaves. Extra-thick walls surrounding certain cells enabled stems to stand erect.

Primitive Land Plants Retained Swimming Sperm and Required Water to Reproduce

Reproduction out of water presented greater challenges. We will examine plant reproduction in more detail in Chapters 21 and 24. For now, the important point is that, like animals, plants produce sperm and eggs. Primitive marine plants have swimming sperm and in some cases swimming eggs as well. The first land plants retained swimming sperm. Consequently, they were restricted to swamps and marshes, where the sperm and eggs could be released into the water, or to areas with abundant rainfall, where the ground would occasionally be covered with water.

Conifers Encased Sperm in Pollen Grains, Allowing Them to Flourish in Dry Habitats

The reproductive strategy described above sufficed for millions of years. During the Carboniferous period, 360 million to 286 million years ago, the climate was warm and moist, and great stretches of the land were covered with forests of giant tree ferns and club mosses that produced swimming sperm (Fig. 17-8). The coal we mine today is derived from the fossilized remains of those forests. Meanwhile, some plants inhabiting drier regions had begun to evolve reproductive strategies that no longer depended on films of water. These early cone-bearing plants, called **conifers**, arose in the Permian period (286 million to 245 million years ago). Their eggs were retained in the parent plant, and the sperm were encased in drought-resistant pollen grains that the wind blew from plant to plant. Landing on a female cone near the egg, the pollen released sperm cells directly into living tissue, eliminating the need for a surface film of water. About 250 million years ago, mountains rose, swamps drained, and the moist climate dried up. Tree ferns and giant club mosses, with their swimming sperm, largely went extinct; the conifers, which did not depend on water for reproduction, flourished and spread.

Flowering Plants Enticed Animals to Carry Pollen

About 130 million years ago, during the Cretaceous period, the flowering plants appeared, having evolved from a group of conifer-like plants. The initial advantage of the flowering plants seems to have been pollination by insects. The conifers are wind-pollinated; because the vast majority of pollen grains fail to reach their target, conifers must produce an enormous amount of pollen. Flower pollination by insects wastes far less pollen. Flowering plants also evolved other advantages, including more-rapid reproduction and, in some cases, much more rapid growth. Today, flowering plants dominate the land, except in cold northern regions, where conifers still prevail.

Some Animals Evolved Specialized Structures That Adapted Them to Life on Dry Land

Soon after land plants evolved, providing potential food sources for animals, arthropods (probably early relatives of scorpions) emerged from the sea. Why arthropods? The answer seems to be that they were **preadapted** for land life—that is, they already possessed structures, evolved under totally different environmental pressures, that, purely by chance, were suited to life on land. Foremost among these preadaptations was the external skeleton, or **exoskeleton**, a hard covering surrounding the body, such as the shell of a lobster or crab. Exoskeletons are both waterproof and strong enough to support a small animal against the force of gravity.

Arthropods also solved with relative ease another difficulty experienced by land animals: breathing. To allow gas exchange, respiratory surfaces must be kept moist, and doing so is difficult in the dry air. Some arthropods, such as land crabs and spiders, evolved what amounts to an internal gill, kept moist within a waterproof sac. Insects developed *tracheae*, small branching tubes directly penetrating the body, with adjustable openings in the exoskeleton that lead to the outside air.

Figure 17-8 The swamp forest of the Carboniferous period
The treelike plants are tree ferns and giant club mosses, both now mostly extinct.

For millions of years, arthropods had the land and its plants to themselves, and for tens of millions of years more, they were the dominant animals. Dragonflies with a wingspan of 70 centimeters (28 inches) flew among the Carboniferous tree ferns, while millipedes 2 meters (6.5 feet) long munched their way across the swampy forest floor. Eventually, however, the arthropods' splendid isolation came to an end.

Amphibians Evolved from Lobefin Fishes

About 400 million years ago, a group of Silurian fishes called the *lobefins* appeared, probably in fresh water. Lobefins had two important preadaptations to land-life: (1) stout, fleshy fins with which they crawled about on the bottoms of shallow, quiet waters and (2) an outpouching of the digestive tract that could be filled with air, like a primitive lung (Fig. 17-9). The coelacanth seen in Figure 22-32d is a lobefin that was believed to be long extinct before its discovery in 1939. In one group of lobefins, the lung evolved into a swim bladder, with which the fish could regulate their buoyancy and remain suspended in the water without active exertion. Many of these lobefins migrated back to the sea, where they evolved into the modern bony fishes. Another group of lobefins colonized very shallow ponds and streams, which shrank during droughts and whose water often became oxygen-poor. By taking air into their lungs, these lobefins could obtain oxygen anyway. Some of their descendants began to use their fins to crawl from pond to pond in search of prey or water, as some modern fish can do today (Fig. 17-10).

As the arthropods had discovered previously, the land is a rich source of food. Feeding on land and moving from pool to pool favored the evolution of fish that could stay out of water for longer periods and that could move about more effectively on land. With improvements in lungs and legs, the amphibians evolved from lobefins, first appearing in the fossil record about 350 million years ago. If an amphibian could have thought about such things, it would have thought that the Carboniferous swamp forests were heaven itself: no predators to speak of, abundant prey, and a warm, moist climate. As with the insects and millipedes, some amphibians evolved gigantic size, including salamanders over 3 meters (10 feet) long.

Despite their success, the early amphibians were still not fully adapted to life on land. Their lungs were simple sacs without very much surface area, so they had to obtain some of their oxygen through their skin. Therefore, their skin had to be kept moist, which restricted them to swampy habitats where they wouldn't dry out. Further, amphibian sperm and eggs could not survive in dry surroundings and had to be deposited in watery environments. So, although amphibians could move about on land, they could not stray too far from the water's edge. As with the tree ferns and club mosses, when the climate turned dry at the beginning of the Permian period about 286 million years ago, amphibians were in trouble.

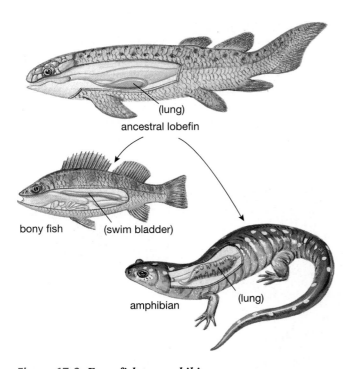

Figure 17-9 *From fish to amphibians*
A group of primitive fish, called the lobefins, gave rise both to the bony fish and the amphibians. Lobefins had a pair of lungs, which arose as outpocketings of the digestive tract. Lobefins probably used their lungs for breathing air when the pools in which they lived became stagnant, foul, and low in oxygen. In one group of lobefins, one lung evolved into a swim bladder, later used by bony fishes to regulate buoyancy. The other lung disappeared. Another group of lobefins became further adapted for breathing air, evolving more complex, efficient lungs and sturdier fins for walking on land. These were the ancestors of the amphibians.

Figure 17-10 *A fish that walks on land*
Some modern fish resemble their lobefin ancestors in their ability to crawl about on land. The walking catfish was imported into Florida several years ago, escaped into local waters, and has spread across much of southern Florida by walking from pond to pond.

Reptiles, Which Evolved from Amphibians, Developed Several Adaptations to Dry Land

Just as the conifers had been evolving on the fringes of the swamp forests, so too was a group of amphibians evolving, developing adaptations to drier conditions. These amphibians evolved into the reptiles, which achieved four great advances over their amphibian relatives. (1) They evolved *internal fertilization*. The reptilian female provides the watery environment for the sperm inside her reproductive tract. Thus, sperm transfer could occur on land, without the reptile having to venture back to the dangerous swamps full of fish and amphibian predators. (2) Reptiles evolved shelled, waterproof eggs that enclosed a supply of water for the developing embryo, again providing freedom from the swamps. (3) The ancestral reptiles evolved scaly, waterproof skin that helped prevent the loss of body water to the dry air. (4) Reptiles evolved improved lungs that were able to provide the entire oxygen supply for an active animal. As the climate dried during the Permian period, reptiles became the dominant land vertebrates, relegating amphibians to swampy backwaters, where most remain today.

A few tens of millions of years later, the climate returned to more moist and equable conditions, providing for lush plant growth. Once again, gigantic size was selected for, as certain families of reptiles evolved into the dinosaurs. The variety of dinosaur forms was enormous—from carnivores (Fig. 17-11) to herbivores; from those that dominated the land, to others that took to the air, to still others that returned to the sea. Dinosaurs were among the most successful animals ever, if we consider persistence as a measure of success. They flourished for more than a hundred million years, until about 65 million years ago, when the last dinosaurs went extinct. No one is certain why they died out, but a climate change, perhaps initiated by a gigantic meteorite impact, seems to have been the final blow (see Chapter 14).

Even during the age of dinosaurs, many reptiles remained quite small. One major difficulty faced by small reptiles is the maintenance of a high body temperature. Being active on land seems to require a rather warm body to maximize the efficiency of the nervous system and muscles. But a warm body loses heat to the environment unless the air is also warm. Heat loss is a bigger problem for smaller animals, which have a larger surface area per unit of weight than do larger animals. Many species of small reptiles have retained slow metabolisms and have coped with the heat-loss problem by developing lifestyles in which they remain active only when the air is sufficiently warm. Two groups of small reptiles, however, independently followed a different evolutionary strategy about 150 million years ago: They developed insulation. One group evolved feathers, and another evolved hair.

Reptiles Gave Rise to Both Birds and Mammals

In the ancestral birds, insulating feathers retained body heat. Consequently, these animals could be active in cool habitats and during the night, when their scaly relatives became sluggish. Later, some ancestral birds evolved longer, stronger feathers on their forelimbs, perhaps allowing them to glide from trees or to assist in jumping after insect prey. From this point, the evolution of flight became possible.

The hair of ancestral mammals also provided insulation. Unlike the birds, which retained the reptilian

Figure 17-11 A reconstruction of a Cretaceous forest
By the Cretaceous era, flowering plants dominated terrestrial vegetation. Dinosaurs, such as the predatory pack of 6-foot-long *Velociraptors* shown here, were the pre-eminent land animals. Although small by dinosaur standards, *Velociraptor* was a formidable predator with great running speed, sharp teeth, and deadly, sickle-like claws on its hind feet.

habit of laying eggs, mammals evolved live birth and the ability to feed their young with secretions of the mammary (milk-producing) glands. Because these structures do not fossilize, we may never know when the uterus, mammary glands, and hair first appeared, or what their intermediate forms looked like. Recently, however, a team of paleontologists found bits of fossil hair preserved in *coprolites*, which are fossilized animal feces. These coprolites, found in the Gobi Desert of China, were deposited by an anonymous predator 55 million years ago, so mammals have presumably had hair at least that long.

The earliest mammals were small creatures, probably living in trees and being active mostly at night. When the dinosaurs went extinct, the mammals adaptively radiated out into the vast array of modern forms. Some stayed small and nocturnal, eating mostly seeds and insects; larger size and different behaviors evolved in others. These latter mammals colonized the habitats left empty by the extinction of the dinosaurs. One group remained in the trees, giving rise to the *primates*.

5) How Did Humans Evolve?

Humans are intensely interested in their own origin and evolution, especially in trying to determine what conditions led to the evolution of the gigantic human brain.

This obsession has triggered sweeping speculations despite an often skimpy fossil record. Therefore, although the outline of human evolution that we will present is a synthesis of current thought on the subject, it is by no means as well understood as, say, the genetic code. Paleontologists disagree about the interpretation of the fossil evidence, and many ideas may have to be revised as new fossils are found.

Primate Evolution Has Been Linked to Grasping Hands, Binocular Vision, and a Large Brain

Fossils of **primates** (lemurs, monkeys, apes, and humans) are relatively rare compared with those of many other animals. The primate fossil record is nonetheless sufficient to show that the most-likely primate ancestors were insect-eating tree shrews, whose fossils are found in rocks about 80 million years old. Nimble, probably nocturnal animals, tree shrews were smaller than all but the tiniest modern primates.

Over the next 50 million years, the descendants of the tree shrews evolved forms that are similar to modern tarsiers, lemurs, and monkeys (Fig. 17-12). Early primates, which probably fed on fruits and leaves, evolved several adaptations for life in the trees. Many modern primates retain the tree-dwelling lifestyle of their ancestors.

(a)

(b)

(c)

Figure 17-12 Representative primates
(a) Tarsier, *(b)* lemur, and *(c)* lion-tail macaque monkey. Note that all have relatively flat faces, with forward-looking eyes providing binocular vision. All also have color vision and grasping hands. These features served as preadaptations for tool and weapon use by early humans.

Some species are nocturnal, but many others are active during the day.

Grasping Hands in Early Primates Allowed Both Powerful and Precise Manipulations

The tree shrews already possessed handlike paws for holding on to small branches, and the primates further refined these appendages. Most primates are much larger than tree shrews, so only relatively large tree limbs could bear their weight. Long, grasping fingers that could wrap around and hold larger limbs made life in the trees a little safer. The primate line that led to humans evolved hands that could perform both delicate maneuvers, using a *precision grip* (manipulating small objects, writing, sewing), and powerful actions, using a *power grip* (swinging a club, thrusting with a spear).

Binocular Vision Provides Accurate Depth Perception

One of the earliest primate adaptations seems to have been large, forward-facing eyes (see Fig. 17-12). Jumping from branch to branch is risky business unless an animal can accurately judge where the next branch is located. Accurate depth perception was made possible by binocular vision provided by forward-facing eyes with overlapping fields of view. Another key adaptation was color vision. We cannot, of course, tell if a fossil animal had color vision, but modern primates have excellent color vision, and it seems reasonable to assume that earlier primates did too. Many primates feed on fruit, and color vision helps in detecting ripe fruit among a bounty of green leaves.

A Large Brain Facilitated Hand-Eye Coordination and Complex Social Interactions

Primates have brains that are larger, relative to their body size, than almost all other animals. No one really knows for certain what environmental forces favored the evolution of large brains. It seems reasonable, however, that controlling and coordinating binocular, color vision, rapid locomotion through trees, and dexterous movements of the hands would be facilitated by increased brain power. Most primates also have fairly complex social systems, which probably would require relatively high intelligence. If sociality promoted increased survival and reproduction, then there would have been environmental pressures for the evolution of larger brains.

Hominids Evolved from Dryopithecine Primates

Between 20 million and 30 million years ago, in the moist tropical forests of Africa, a group of primates called the dryopithecines diverged from the monkey line. The dryopithecines appear to have been ancestral to the **hominids** (humans and their fossil relatives) and *pongids* (the great apes). About 18 million years ago, global climatic cooling began to shrink the vast expanses of forest,

splitting up the woodlands into isolated islands dotted on a sea of grasslands. Diversification of habitat and isolation of small populations led to the diversification of dryopithecines. One of these primates gave rise to the later hominids and pongids.

The Earliest Hominids Could Stand and Walk Upright

The hominid line is believed to have diverged from the ape lineage sometime between 5 million and 8 million years ago, but direct evidence of that transition is absent from the fossil record. The first trace of our hominid ancestors appears in 4.4-million-year-old rocks in Ethiopia, where excavations have yielded fragments of *Ardipithecus ramidus*. *Ardipithecus* was clearly a hominid, sharing several anatomical features with better-known, later members of the group. But this oldest known member of our family also exhibits other features (such as thin tooth enamel and thick arm bones) that are more characteristic of apes, so *Ardipithecus* may represent a point on our family tree that is fairly near the split from the apes.

Ardipithecus is known from only a few bones and teeth. A more extensive record of early hominid evolution does not really begin until about 4 million years ago. That datum marks the beginning of the fossil record of the genus *Australopithecus* ("southern ape," after their original discovery site in southern Africa; Fig. 17-13). The brains of *australopithecines* (as the various species of *Australopithecus* are collectively known) were fairly large but still much smaller than those of modern humans. Although the earliest australopithecines had legs that were shorter, for their height, than those of modern humans, their knee joints allowed them to straighten their legs fully, permitting efficient bipedal (upright, two-legged) locomotion. Footprints almost 4 million years old, discovered in Tanzania by anthropologist Mary Leakey, show that even the earliest australopithecines could, and at least sometimes did, walk upright.

Upright posture is an extremely important feature of hominids, because it freed their hands from use in walking. Later hominids thus were able to carry weapons, manipulate tools, and eventually achieve the cultural revolutions produced by modern *Homo sapiens*. What environmental pressures led to the evolution of bipedalism? The short answer is that no one knows, but British physiologist Peter Wheeler has an intriguing hypothesis: Our ancestors developed the ability to walk upright because this posture exposes a minimum body surface to the blazing savanna sun. Wheeler constructed a small-scale model of an ancient australopithecine and photographed it in bipedal and quadrupedal (four-legged) postures from a series of overhead locations that imitated the pathway of the sun. He found that a four-legged posture exposed the model to 60% more solar radiation than did the upright posture. Further, says Wheeler, standing upright exposes more of the body to cooler

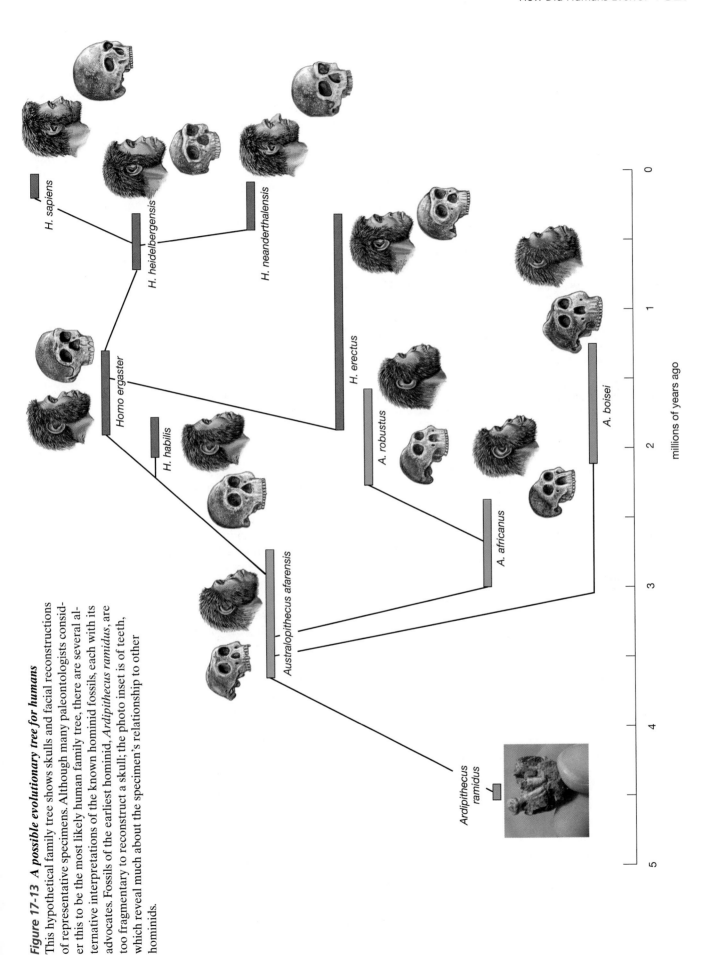

Figure 17-13 A possible evolutionary tree for humans
This hypothetical family tree shows skulls and facial reconstructions of representative specimens. Although many paleontologists consider this to be the most likely human family tree, there are several alternative interpretations of the known hominid fossils, each with its advocates. Fossils of the earliest hominid, *Ardipithecus ramidus*, are too fragmentary to reconstruct a skull; the photo inset is of teeth, which reveal much about the specimen's relationship to other hominids.

H. sapiens

H. heidelbergensis

H. neanderthalensis

Homo ergaster

H. habilis

H. erectus

A. robustus

A. africanus

A. boisei

Australopithecus afarensis

Ardipithecus ramidus

millions of years ago

0

1

2

3

4

5

breezes a few feet above the ground. Together, these factors would increase the time the animal could spend foraging in the sun, allowing it to increase its access to food.

The oldest australopithecine species, represented by fossilized teeth, skull fragments, and arm bones, was unearthed near an ancient lake bed in Kenya from sediments that were dated, by means of radioactive isotopes, as between 3.9 million and 4.1 million years old (see "Scientific Inquiry: How Do We Know How Old a Fossil Is?"). It was named *Australopithecus anamensis* by its discoverers (*anam* means "lake" in the local Ethiopian language). The second most ancient australopithecine, called *Australopithecus afarensis*, was discovered in the Afar region of Ethiopia in 1974, a partial skeleton of a 3.5-foot-tall female nicknamed Lucy. Fossil remains of this species as much as 3.9 million years old have been unearthed. Later, the *A. afarensis* line apparently gave rise to at least two distinct forms: the small, omnivorous *A. africanus* (which in size and eating habits was similar to *A. afarensis*), and the large, herbivorous *A. robustus* and *A. boisei*. All of the australopithecine species had apparently gone extinct by 1.2 million years ago, but one of them (probably either *A. afarensis* or *A. africanus*) first gave rise to a new branch of the hominid family tree, the genus *Homo*.

The Genus Homo *Diverged from the Australopithecines 2.5 Million Years Ago*

Hominids that are sufficiently similar to modern humans to be placed in the genus *Homo* first appear in the fossil record in Africa, beginning about 2.5 million years ago. The earliest African *Homo* fossils are assigned to two species, *H. rudolfensis* and *H. habilis* (see Fig. 17-13), whose body and brain were larger than those of the australopithecines but which retained the apelike long arms and short legs of their australopithecine forbears. In contrast, the skeletal anatomy of *H. ergaster*, a species whose fossils first appear 2 million years ago, has limb proportions more like those of modern humans and is believed by many paleoanthropologists (scientists who study human origins) to be on the evolutionary branch that led ultimately to our own species, *H. sapiens*. In this view, *H. ergaster* was the common ancestor of two distinct branches of hominids. The first branch led to *H. erectus*, which was the first hominid species to leave Africa. The brain of the average *H. erectus* was as large as the smallest modern adult human brains (about 1000 cubic centimeters). The face, unlike that of modern humans (see Fig. 17-13), featured large brow ridges above the eyes, a slightly protruding face, and no chin.

The second branch from *H. ergaster* gave rise to *H. heidelbergensis*. After migrating to Europe, *H. heidelbergensis* gave rise to the Neanderthals, *H. neanderthalensis*. Meanwhile, back in Africa, another branch split off from the *H. heidelbergensis* lineage. This branch ultimately became *H. sapiens*, modern humans.

The Evolution of Homo *Was Accompanied by Advances in Tool Technology*

Hominid evolution is closely tied to the development of tools, a hallmark of hominid behavior. The oldest tools discovered so far were found in 2.5-million-year-old East African rocks, concurrent with the early emergence of the genus *Homo*. Early *Homo*, whose cheek teeth were much smaller than those of its australopithecine ancestors, might have first used stone tools to break and crush tough foods that were hard to chew. Hominids constructed their earliest tools by striking one rock with another to chip off fragments and leave a sharp edge behind. Over the next several hundred thousand years, tool-making techniques in Africa gradually became more advanced; by 1.7 million years ago, tools were more sophisticated, with flakes chipped symmetrically from both sides of a rock to form "biface" tools ranging from hand axes, used for cutting and chopping, to points, probably used on spears (Fig. 17-14a,b). *Homo ergaster* and other bearers of these weapons presumably ate meat, probably both hunting and scavenging for the remains of prey killed by other predators. Biface tools were carried to Europe at least 600,000 years ago by migrating populations of *H. heidelbergensis*, and the Neanderthal descendants of these emigrants took stone tool construction to new heights of skill and delicacy (Fig. 17-14c).

Neanderthals Had Large Brains and Ritualistic Behaviors

Neanderthals first appear in the European fossil record about 150,000 years ago; by about 70,000 years ago, they had spread throughout Europe and western Asia. By 30,000 years ago, however, Neanderthals were extinct. Contrary to the popular image of a hulking, stoop-shouldered "cave man," Neanderthals were quite similar to modern humans in many ways. Although more heavily muscled, Neanderthals walked fully erect, were dexterous enough to manufacture finely crafted stone tools, and had brains, on the average, slightly larger than those of modern humans. Although many of the European fossils show heavy brow ridges and a broad, flat skull, others, particularly from areas around the eastern shores of the Mediterranean Sea, were remarkably like ourselves. Neanderthal remains also show evidence of modern behaviors, particularly ritualistic burial ceremonies (Fig. 17-15). Neanderthal skeletons have been discovered in clearly marked burial sites that were surrounded with stones and typically included offerings of flowers, bear skulls, and food. Altars found at other Neanderthal sites were probably used in "religious" rites associated with a bear cult.

Until recently most researchers believed that Neanderthals were an archaic variety of *H. sapiens*. It is becoming increasingly clear, however, that *H. neanderthalensis* was a separate species. A recent and dramatic piece of evidence in support of this hypothesis was uncovered in 1997, when researchers were able to

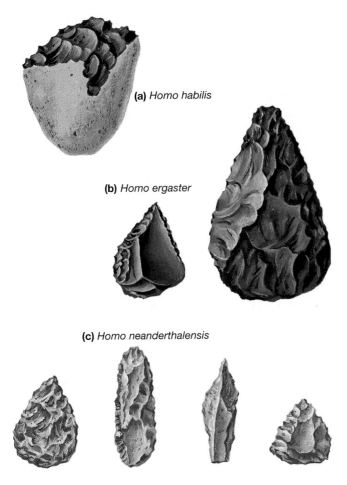

(a) *Homo habilis*

(b) *Homo ergaster*

(c) *Homo neanderthalensis*

Figure 17-14 Representative hominid tools
(a) Homo habilis produced only fairly crude chopping tools called hand axes, usually unchipped on one end to hold in the hand. *(b) Homo ergaster* manufactured much finer tools. The tools were typically sharp all the way around the stone, so at least some of these blades were probably tied to spears rather than held in the hand. *(c)* Neanderthal tools were works of art, with extremely sharp edges made by flaking off tiny bits of stone. In comparing these weapons, note the progressive increase in the number of flakes taken off the blades and the corresponding decrease in flake size. Smaller, more numerous flakes produce a sharper blade and suggest more insight into tool making (perhaps passed down culturally from experienced tool makers to apprentices), more patience, finer control of hand movements, or perhaps all three.

isolate and analyze DNA from a 30,000-year-old Neanderthal bone. The researchers determined the nucleotide sequence of a Neanderthal gene and compared it to the same gene from almost 1000 different modern humans from various parts of the world. The Neanderthal gene sequence was much different from that of the modern humans and indicated that the evolutionary branch leading to *H. neanderthalensis* diverged from the ancestral human line hundreds of thousands of years prior to the emergence of modern *H. sapiens*.

Figure 17-15 Neanderthal burial
Museum reconstruction of a Neanderthal burial site. Some Neanderthals were buried with elaborate ceremonies, with bear skulls placed on the grave of the deceased.

Modern Humans Emerged Only 150,000 Years Ago

Finally, about 150,000 years ago, anatomically modern humans appear in the African fossil record. The location of these fossils suggests that *Homo sapiens* originated in Africa, but most of our knowledge about our own early history comes from European and Middle Eastern fossil humans collectively known as *Cro-Magnons* (after the district in France in which their remains were first discovered). Cro-Magnons, which appear in the fossil record beginning about 90,000 years ago, had domed heads, smooth brows, and prominent chins (just like us!). Their tools were precision instruments not very different from the stone tools used until recently in many parts of the world. Behaviorally, Cro-Magnons seems to have been similar to, but more sophisticated than, Neanderthals. Perhaps the most remarkable accomplishment of Cro-Magnons is the magnificent art left in caves such as Altamira in Spain and Lascaux in France (Fig. 17-16). No one knows exactly why these paintings were made, but they attest to minds fully as human as our own.

Cro-Magnons coexisted with Neanderthals in Europe and the Middle East for perhaps up to 50,000 years before the Neanderthals disappeared. Some researchers believe that Cro-Magnons interbred extensively with Neanderthals, so Neanderthals were essentially absorbed into the human genetic mainstream. Other scientists dis-

Figure 17-16 *The sophistication of Cro-Magnon people*
Cave paintings by Cro-Magnon people remarkably preserved by the relatively constant underground conditions of a cave in Lascaux, France.

agree, citing mounting evidence such as the fossil DNA described earlier, and suggest that later-arriving Cro-Magnons simply overran and displaced the less well adapted Neanderthals. Neither hypothesis, however, does a good job of explaining how the two kinds of hominids managed to occupy the same geographical areas for such a long time. The persistence in one area of two similar but distinct groups for tens of thousands of years seems inconsistent with both interbreeding *and* direct competition. Perhaps *H. neanderthalensis* and *H. sapiens* were indeed different biological species, unable to interbreed successfully but able to coexist in the same habitat.

Africa is where the human family tree has its roots, but hominids found their way out of Africa on numerous occasions. For example, *H. erectus* reached tropical Asia almost 2 million years ago, apparently thrived there, and eventually spread across Asia. Similarly, *H. heidelbergensis* made it to Europe at least 780,000 years ago. It is increasingly clear that the genus *Homo* made repeated long-distance emigrations, beginning just as soon as sufficiently capable limb anatomy evolved. What is less

clear is how all this wandering is related to the origin of modern *H. sapiens*. According to one hypothesis (the basis of the scenario outlined earlier), *H. sapiens* emerged in Africa and dispersed less than 150,000 years ago, spreading into the Near East, Europe, and Asia and supplanting all other hominids (Fig. 17-17a). But some paleoanthropologists believe that populations of *H. sapiens* evolved in many regions simultaneously from the already widespread populations of *H. erectus*. According to this hypothesis, continued migrations and interbreeding among *H. erectus* populations maintained them as a single species as they gradually evolved into *H. sapiens* (Fig. 17-17b). A steadily increasing number of comparative molecular studies of modern humans support the "single African origin" model of the genesis of our species, but both points of view are consistent with the fossil record, and the question remains unsettled.

The Evolution of Human Behavior Is Highly Speculative

Perhaps the most hotly debated subject in human evolution is the development of human behavior. Except in rare instances, such as the Neanderthal burials and the Cro-Magnon cave paintings, there is no direct evidence of the behavior of prehistoric hominids. Biologists and anthropologists have offered a few hypotheses in accordance with the fossil record and the behavior of modern humans and other animals.

The fossil record shows that the development of truly huge brains occurred within the last couple of million years. What were the selective advantages of large brains? This subject continues to stir controversy among paleoanthropologists. The enlarging brain may have been selected both because it provided for improved hand-eye coordination and because it facilitated complex social interactions.

As the early hominids such as *Australopithecus* descended from the trees into the savanna, they began to walk upright. Bipedal locomotion allowed them to carry things in their hands as they walked. Hominid fossils reveal shoulder joints capable of powerful throwing motions and an *opposable thumb* (a thumb that can be placed directly against each of the other fingers) that would enhance the ability to manipulate objects. These early hominids probably could see well and judge depth accurately with binocular vision. The brain expansion that occurred in the australopithecines may have been at least partly related to integration of visual input and control of hand and arm movements.

Homo erectus, H. ergaster, and perhaps *H. habilis* and *H. rudolfensis* before them, were social. So, of course, are many monkeys and apes. The later hominids, however, seem to have engaged in a new type of social activity: cooperative scavenging and hunting. *Homo erectus*, in particular, seems to have hunted extremely large game, which calls for cooperation in all phases of the hunt. Some paleoanthropologists believe that selection favored

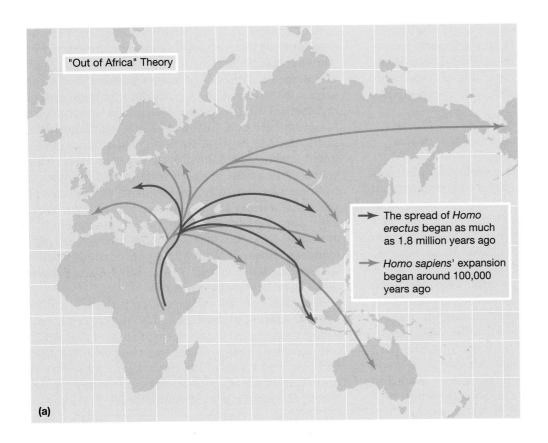

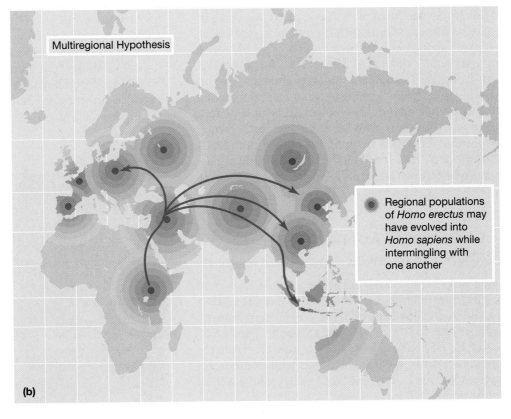

Figure 17-17 *Competing hypotheses for the evolution of* **Homo sapiens**
(a) The "out of Africa" advocates hypothesize that *H. sapiens* evolved in Africa, then migrated throughout the Near East, Europe, and Asia. *(b)* The "multiregional" hypothesis states that populations of *H. sapiens* evolved in many regions simultaneously from the already widespread populations of *H. erectus*.

larger and more-powerful brains as an adaptation for success in cooperative hunting. (Lest you become too impressed with such speculation, remember that lions and wolves also hunt cooperatively yet are not noticeably more intelligent than their relatives, such as leopards and coyotes, who do not.)

The Cultural Evolution of Humans Now Far Outpaces Biological Evolution

In recent millennia, the biological evolution of humans has been far outpaced by our **cultural evolution**: learned behaviors passed down from previous generations. Careful estimates of the human population size since our species originated reveals three major surges of population growth. Each surge was initiated by a cultural–technological revolution: the development of new forms of technology that increased the resources available to sustain human life. The first revolution was the development of tools, beginning with early hominids and extending until 10,000 years ago, when the total human population numbered 5 million. Beginning then and extending over the next 8000 years, humans grew crops and domesticated animals, and the population gradually increased to about 500 million in an agricultural revolution. We are now more than 300 years into the scientific–industrial revolution, and our population is approaching 6 billion.

Human cultural evolution and its accompanying increases in human population have had profound effects on the continuing biological evolution of other life forms. Our agile hands and minds have transformed much of Earth's terrestrial and aquatic habitats. Humans have become the single overwhelming agent of natural selection. In the words of evolutionary biologist Stephen Jay Gould of Harvard University, "We have become, by the power of a glorious evolutionary accident called intelligence, the stewards of life's continuity on Earth. We did not ask for this role, but we cannot abjure it. We may not be suited for it, but here we are."

Evolutionary Connections
Is There Life on Mars?

www

In August 1996, supermarket checkout lines everywhere were lined with tabloids sporting pictures of little green men and doe-eyed aliens that peered out from beneath headlines like "NASA FINDS MARTIANS" and "EXTRATERRESTRIALS FOUND ON MARS!" Even the more conservative media were abuzz with excited pronouncements about life on Mars. What, exactly, was all the fuss about?

The excitement and controversy about life on Mars centers, strangely enough, on an object that was found right here on Earth: a softball-sized meteorite that goes by the unassuming name of ALH84001 and was found in 1984 by a meteorite specialist in Antarctica. More than a decade later, other researchers determined that the

rock originated on Mars (a determination they reached by comparing the chemical composition of the specimen with that of other meteorites believed to be from Mars). Occasionally, chunks of rock are torn from the Martian surface in the aftermath of collisions with large meteors, and some of this flying debris eventually lands on Earth. Scientists have thus far found 12 such Martian rocks on Earth, but ALH84001 is by far the oldest piece of Mars found. Radiometric dating puts its age at about 4.5 billion years. This ancient heritage is important, because astronomers believe that the Mars of 4.5 billion years ago was much warmer and wetter than today's cold, barren, dry planet. Perhaps ALH84001 broke away from Mars at a time when conditions were more conducive to life. A team of scientists at NASA, led by David McKay, believe so. They say that the rock of ALH84001 bears evidence that it once held living organisms.

The McKay team has presented four pieces of evidence in support of their claim that Mars may have supported single-celled organisms in the distant past. (1) The rock contains globules of carbonate that have a structure and chemical composition that suggest they were deposited from an aqueous (water-containing) solution. Water, of course, is a prerequisite for life as we know it. (2) Many of the carbonate globules are coated with deposits of the minerals iron sulfide and magnetite. Similar deposits of these minerals are made by certain species of bacteria on Earth. (3) ALH84001 contains high levels of chemical substances know as *polycyclic aromatic hydrocarbons* (or PAHs), especially in the vicinity of the carbonate globules. On Earth, PAHs are by-products of biological decomposition. (4) Electron microscopy has revealed the presence of tiny, worm-shaped structures in the meteorite (Fig. 17-18). These structures resemble fos-

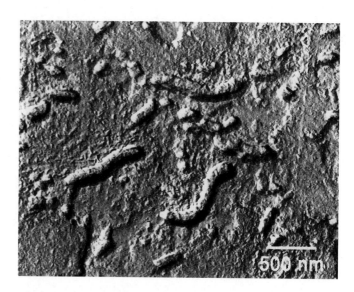

Figure 17-18 Martian bacteria?
This electron micrograph shows the structures found in the Martian rock ALH84001. The tubular forms resemble bacterial fossils from our own planet.

sil bacteria that have been found on Earth (though the Mars versions are far smaller than Earth fossils).

Are the McKay team's findings clear evidence of Martian life? Not really. Skeptics have challenged each of the team's findings, providing convincing evidence that similar features could arise by nonbiological processes or that the meteorite could have been contaminated with terrestrial substances during the 13,000 years that it lay on the Antarctic ice. But the evidence is at the very least suggestive, and biologists, geologists, astronomers, UFO buffs, and other curious humans can't help but be intrigued by these signs that we may not be alone in the universe. Stay tuned.

Summary of Key Concepts

1) How Did Life Begin?

Before life arose, lightning, ultraviolet light, and heat formed organic molecules from water and the components of primordial Earth's atmosphere. These molecules probably included nucleic acids, amino acids, short proteins, and lipids. By chance, some molecules of RNA may have had enzymatic properties, catalyzing the assembly of copies of themselves from nucleotides in Earth's waters. These may have been the forerunners of life. Protein-lipid microspheres enclosing these RNA molecules may have formed the first cell-like organisms, protocells.

2) What Were the Earliest Organisms Like?

The first fossil cells are found in rocks about 3.5 billion years old. These cells were prokaryotes that fed by absorbing organic molecules that had been synthesized in the environment. Because there was no free oxygen in the atmosphere at that time, energy metabolism must have been anaerobic. As the cells multiplied, they depleted the organic molecules that had been formed by prebiotic synthesis. Some cells developed the ability to synthesize their own food molecules by using simple inorganic molecules and the energy of sunlight. These earliest photosynthetic cells were probably ancestors of today's cyanobacteria.

Photosynthesis releases oxygen as a by-product, and by about 2.2 billion years ago significant amounts of free oxygen had accumulated in the atmosphere. Aerobic metabolism, which generates more cellular energy than does anaerobic metabolism, probably arose about this time.

Eukaryotic cells evolved by about 1.7 billion years ago. The first eukaryotic cells probably arose as symbiotic associations between predatory prokaryotic cells and bacteria. Mitochondria may have evolved from aerobic bacteria engulfed by predatory cells. Similarly, chloroplasts may have evolved from photosynthetic cyanobacteria.

3) How Did Multicellularity Arise?

Multicellular organisms evolved from eukaryotic cells, first appearing about 1 billion years ago. Multicellularity offers several advantages, including increased speed of locomotion and greater size. The first multicellular organisms arose in the sea. In plants, increased size due to multicellularity offered some protection from predation. Specialization of cells allowed plants to anchor themselves in the nutrient-rich, well-lit waters of the shore. For animals, multicellularity allowed more-efficient predation and more-effective escape from predators. These in turn provided environmental pressures for faster locomotion, improved senses, and greater intelligence.

4) How Did Life Invade the Land?

The first land organisms were probably algae. The first multicellular land plants appeared about 400 million years ago. Although the land required special adaptations for support of the body, reproduction, and the acquisition, distribution, and retention of water, the land also offered abundant sunlight and protection from aquatic herbivores. Soon after land plants evolved, arthropods invaded the land. Absence of predators and abundant land plants for food probably facilitated the invasion of the land by animals.

The earliest land vertebrates evolved from lobefin fishes, which had leglike fins and a primitive lung. A group of lobefins evolved into the amphibians about 350 million years ago. Reptiles evolved from amphibians, with several further adaptations for life on land: internal fertilization, waterproof eggs that could be laid on land, waterproof skin, and better lungs. About 150 million years ago, birds and mammals evolved independently from separate groups of reptiles. Major advances included a high, constant body temperature and insulation over the body surface.

5) How Did Humans Evolve?

One group of mammals evolved into the tree-dwelling primates. Primates show several preadaptations for human evolution: forward-facing eyes for binocular vision, color vision, grasping hands, and relatively large brains. Between 20 million and 30 million years ago, some primates descended from the trees; these were the ancestors of apes and humans. The australopithecines arose in Africa about 4 million years ago. These hominids walked erect, had larger brains than did their forebears, and fashioned primitive tools. One group of australopithecines evolved into modern humans.

Key Terms

Thinking Through the Concepts

Multiple Choice

1. *Our best estimate for the age of Earth, about 4.5 billion years, is supported by*
   a. the assumption of uniformitarianism
   b. the Big Bang theory
   c. gradualism hypothesis
   d. punctuated equilibrium theory
   e. radioactive dating of the oldest rocks found

2. *There was no free oxygen in the early atmosphere because most of it was tied up in*
   a. water   b. ammonia
   c. methane   d. rock
   e. radioactive isotopes

3. *RNA became a candidate for the first information-carrying molecule when Tom Cech and Sidney Altman discovered that some RNA molecules can act as enzymes that*
   a. degrade proteins
   b. turn light into chemical energy
   c. split water and release oxygen gas
   d. synthesize copies of themselves
   e. synthesize amino acids

4. *Which three of the following observations support Lynn Margulis's endosymbiont hypothesis for the origin of chloroplasts and mitochondria from ingested bacteria?*
   a. Aerobic respiration takes place in mitochondria.
   b. Mitochondria have their own DNA.
   c. Photosynthesis takes place in chloroplasts.
   d. Chloroplasts have their own DNA.
   e. Bacterial plasma membranes are strikingly similar to the inner membrane of mitochondria.

5. *The exoskeleton of early, marine-dwelling arthropods can be considered a preadaptation for life on land because that shell*
   a. can support an animal's weight against the pull of gravity
   b. allows a wide diversity of body types
   c. resists drying
   d. absorbs light
   e. both a and c

6. *The evolution of the shelled, waterproof egg was an important event in vertebrate evolution because it*
   a. led to the Cambrian explosion
   b. was the first example of parents caring for their young
   c. allowed the colonization of freshwater environments
   d. freed organisms from having to lay their eggs in water
   e. allowed internal fertilization of eggs

? Review Questions

1. What is the evidence that life might have originated from nonliving matter on the primordial Earth? What kind of evidence would you like to see before you would accept this hypothesis?

2. If they were so much more efficient at producing energy, why didn't the first cells with aerobic metabolism extinguish cells with only anaerobic metabolism?

3. Explain the endosymbiont hypothesis for the origin of chloroplasts and mitochondria.

4. Name two advantages of multicellularity in both plants and animals.

5. What advantages and disadvantages would terrestrial existence have had for the first plants to invade the land? For the first land animals?

6. Review the material on the evolution of flowering plants. How did flowering plants entice animals to carry pollen? (That is, what was in it for the animals?) Why do conifers still prevail in cold, northern regions?

7. Outline the general trends in the evolution of vertebrates, from fish to amphibians to reptiles to birds and mammals. Explain how these adaptations increased the fitness of the various groups for life on land.

8. Outline the evolution of humans from early primates. Include in your discussion such features as binocular vision, grasping hands, bipedal locomotion, social living, tool making, and brain expansion.

Applying the Concepts

1. What is cultural evolution? Is cultural evolution more or less rapid than biological evolution? Why?

2. Do you think that studying our ancestors can shed light on the behavior of modern humans? Why or why not?

3. A biologist would probably answer the age-old question "What is life?" by saying "the ability to self-replicate." Do you agree with this definition? If so, why? If not, how would you define life in biological terms?

4. Traditional definitions of humans have emphasized "the uniqueness of humans" because we possess language and use tools. But most animals can communicate with other individuals in sophisticated ways, and many vertebrates use tools to accomplish tasks. Pretend that you are a biologist from Mars, and write a taxonomic description of the species *Homo sapiens*.

5. The "out of Africa" and "multiregional" hypotheses of *Homo sapiens* evolution make contrasting predictions about the extent and nature of genetic divergence among human races. One predicts that races are old and highly diverged genetically; the other predicts that races are young and little diverged genetically. What data would help you determine which hypothesis is closer to the truth?

6. In biological terms, what do you think was the most significant event in the history of life? Explain your answer.

Group Activity

Like a grand mosaic, our picture of the evolutionary history of life has been pieced together from a multitude of small clues pulled from the fossil record by paleontologists. To understand a bit about the process of adding a piece to the mosaic, form a group of three students, and examine the accompanying drawing of a skull that paleontologists unearthed in South Dakota. Use the observable features of the skull to develop a hypothesis that describes what this creature's life was like. To which vertebrate group does it belong? What can you predict about this species' feeding habits, posture, sensory abilities, and social behavior? For each prediction that you make, cite the specific anatomical evidence that supports your claim. Finally, list five additional pieces of fossil evidence that, if found, would help confirm your hypothesis.

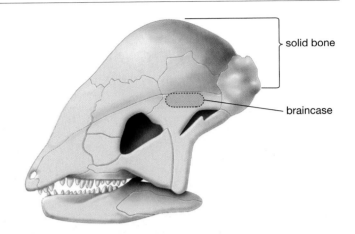

solid bone

braincase

For More Information

Calvin, W. H. "The Emergence of Intelligence." *Scientific American*, October 1994. The evolution of several hallmarks of human intelligence may have been linked to the ability to coordinate rapid movements.

Diamond, J. "How to Speak Neanderthal." *Discover*, January 1990. Could the Neanderthals speak? How could we tell? Jared Diamond lucidly explains the arguments and methodology behind a scientific dispute.

Eigen, M. *Steps Towards Life*. New York: Oxford University Press, 1992. A Nobel laureate lays out a scenario for the origin of life and describes supporting experimental evidence. Challenging material, concisely and precisely explained.

Fenchel, T., and Finlay, B. J. "The Evolution of Life without Oxygen." *American Scientist*, January/February 1994. Clues to the origin of the first eukaryotic cells are provided by symbiotic relationships of organisms in oxygen-free environments.

Gibbs, W., and Powell, C. "Bugs in the Data?" *Scientific American*, October 1996. A report on the issues raised by skeptics who challenge the claim that life once existed on Mars.

Hay, R. L., and Leakey, M. D. "The Fossil Footprints of Laetoli." *Scientific American*, February 1982. The actual footprints of a hominid family were discovered by Hay and Leakey in volcanic ash 3.5 million years old.

Horgan, J. "In the Beginning . . . ." *Scientific American*, February 1991. An exploration of the controversies surrounding research into the origin of life.

Monastersky, R. "The Rise of Life on Earth." *National Geographic*, March 1998. A beautifully illustrated and engaging description of current ideas and evidence of how life arose.

Morell, V. "Announcing the Birth of a Heresy." *Discover*, March 1987. Paleontologists Robert Bakker and Jack Horner speculate that dinosaurs were not the plodding beasts of monster movies but warm-blooded, advanced animals that may even have cared for their young.

Orgel, L. E. "The Origin of Life on Earth." *Scientific American*, October 1994. The evolution of self-reproducing molecules, of which RNA may have been the first, made life possible.

Shreeve, J. "Find of the Century?" *Discover*, January 1997. A brief account of the evidence presented by NASA scientists in support of the proposition that life once existed on Mars.

Tattersall, I. "Out of Africa . . . Again and Again?" *Scientific American*, April 1997. An up-to-date review of the fossil evidence of human evolutionary history.

Waters, T. "Almost Human." *Discover*, May 1990. Artist John Gurche reconstructs fossil hominids by adding clay "muscle" and "skin" to replicas of hominid skulls.

Answers to Multiple-Choice Questions
1. e 2. a 3. d 4. b, d, e 5. e 6. d

"Systems of classification are not hat racks, objectively presented to us by nature. They are dynamic theories developed by us to express particular views about the history of organisms. Evolution has provided a set of unique species ordered by differing degrees of genealogical relationship. Taxonomy, the search for this natural order, is the fundamental science of history."

Stephen Jay Gould in Natural History (1987)

A scaly, massive, earthbound crocodile seems as different as can be from the feathered, delicate, flying egret that shares its habitat. Research by systematists, however, has shown that crocodiles are birds' closest living relatives.

Systematics: Seeking Order Amidst Diversity

18

At a Glance

Why are biologists so obsessed with naming and classifying species? To understand the diligence and zeal of *systematists* (the scientists who do the naming and classifying), imagine that each of the roughly 1.4 million known species on Earth is a book in the library of life. In this imaginary library, none of the books have titles. So when you approach the librarian to ask for a book that you'd like to read, you say, "It's a green-and-red hardcover, about 9 inches tall and pretty thick." Needless to say, the chances of the librarian being able to locate the book you want are quite small. Now imagine a library of life in which each book (species) has its own title. You can stride confidently up to the librarian and say, "I'd like to see a copy of *Casmerodius albus*." You can see why systematists give each species a unique, two-part Latin name that will unambiguously identify it. But as the librarian turns to survey thousands of shelves of randomly arranged books and begins to read their titles one by one, you realize that names alone are not enough. No, to find a book, you need a system of classification.

The careful classifying done by systematists really pays off for locating a specimen in a museum collection or for, say, making a list of the species most similar to the honeybee. It would indeed be impossible to study or even discuss the diversity of life if there were no system of biological classification. But classification also has a second, equally important purpose. Classification schemes describe the evolutionary relationships among species. Systematists strive to categorize organisms according to their evolutionary history, to build classifications that accurately reflect the structure of the tree of life. So, when we say, for example, that humans and chimpanzees are closely related, we mean that the two species share a recent common ancestor from which both species evolved. And when we say that birds are more closely related to crocodiles and alligators than to any other living species, we mean that the common ancestor of birds and crocodilians lived more recently than the common ancestor of birds and any other kind of organism alive today. To classify a group of organisms is to state a hypothesis about their evolutionary history. Systematists are not mere

Net Watch
On-line resources for this chapter are on the World Wide Web at:
http://www.prenhall.com/audesirk
(click on the Table of Contents link and then select Chapter 18).

biological bookkeepers; they are on the front lines of evolutionary biology. This chapter describes their work, which forms the basis of our discussion of the diversity of life in Chapters 19 through 22.

1 How Are Organisms Named and Classified?

Systematics is the science of reconstructing **phylogeny**, or evolutionary history. A key part of systematics is **taxonomy** (from the Greek word *taxis*, meaning "arrangement"), the science of naming organisms and placing them into categories on the basis of their evolutionary relationships. There are seven major categories: (1) **kingdom**, (2) **division** or **phylum** (divisions are used for plants; phyla for animals and most other organisms), (3) **class**, (4) **order**, (5) **family**, (6) **genus**, and (7) **species**. (As we shall soon see, an eighth major category has recently been proposed.) These taxonomic categories form a nested hierarchy in which each level includes all of the other levels below it. Thus, each kingdom includes many phyla, each phylum includes many classes, each class includes many orders, and so on. As we move down the hierarchy, smaller and smaller groups are included. Each category is increasingly narrow and specifies a group whose common ancestor is increasingly recent. Some examples of classifications of specific organisms are given in Table 18-1.

The **scientific name** of an organism is formed from the two smallest taxonomic categories, the genus and the species. For example, the scientific name for humans, *Homo sapiens*, places us in the genus *Homo* ("man") and the species *sapiens* ("thinking") within the genus *Homo*. Note that the species name is never used alone; it must always be paired with its genus name. The genus is a category that includes very closely related species that do not normally interbreed. The species level includes populations of organisms that can potentially interbreed under natural conditions. Thus, the genus *Sialia* (bluebirds) includes the eastern bluebird (*Sialia sialis*), the western bluebird (*Sialia mexicana*), and the mountain bluebird (*Sialia currucoides*)—very similar birds that do not normally interbreed (Fig. 18-1).

Scientific names are always <u>underlined</u> or *italicized*. The first letter of the genus name is always capitalized, and the first letter of the species name is always lower-cased. These names are recognized by biologists worldwide, overcoming language barriers and allowing precise communication. It is an organism's scientific name that allows us to check out a single "book" in the library of life.

Taxonomy Originated as a Hierarchy of Categories

Aristotle (384–322 B.C.) was among the first to attempt to formulate a logical, standardized language for naming living things. Using characteristics such as structural complexity, behavior, and degree of development at birth, he classified about 500 organisms into 11 categories. Aristotle placed organisms into a hierarchy of categories each more inclusive than the one before it, a concept that is still used today.

Building on this foundation more than 2000 years later, the Swedish naturalist Carl von Linné (1707–1778)—who called himself Carolus Linnaeus, a latinized version—laid the groundwork for the modern classification system. He placed each organism into a series of hierarchically arranged categories on the basis of its resemblance to other life-forms, and he also introduced the scientific name composed of genus and species. Nearly 100 years later, Charles Darwin (1809–1882) published *On the Origin of Species*, which demonstrated that all organisms are connected by a common genealogy. Taxonomists then began to recognize that taxonomic categories ought to reflect the pattern of evolutionary relatedness among organisms. The more categories two organisms share, the closer their evolutionary relationship.

Table 18-1 Classification of Selected Organisms, Reflecting Their Degree of Relatedness*

	Human	Chimpanzee	Wolf	Fruit Fly	Sequoia Tree	Sunflower
Domain	**Eukarya**	**Eukarya**	**Eukarya**	**Eukarya**	**Eukarya**	**Eukarya**
Kingdom	**Animalia**	**Animalia**	**Animalia**	**Animalia**	**Plantae**	**Plantae**
Phylum	**Chordata**	**Chordata**	**Chordata**	Arthropoda	Coniferophyta	Anthophyta
Class	**Mammalia**	**Mammalia**	**Mammalia**	Insecta	Coniferosida	Dicotyledoneae
Order	**Primates**	**Primates**	Carnivora	Diptera	Coniferales	Asterales
Family	Hominidae	Pongidae	Canidae	Drosophilidae	Taxodiaceae	Asteraceae
Genus	*Homo*	*Pan*	*Canis*	*Drosophila*	*Sequoiadendron*	*Helianthus*
Species	*sapiens*	*troglodytes*	*lupus*	*melanogaster*	*giganteum*	*annuus*

*Boldface categories are those that are shared by more than one of the organisms classified. Genus and species names are always italicized or underlined. Domains will be introduced later in the chapter.

Figure 18-1 Three species of bluebird
The similarities among these species of bluebirds (genus *Sialia*) are obvious, but the species remain distinct because they do not interbreed.

Sialia sialis
(eastern bluebird)

Sialia mexicana
(western bluebird)

Sialia currucoides
(mountain bluebird)

Modern Systematists Use Numerous Criteria for Classification

Systematists seek to reconstruct the tree of life, but they must do so without much direct knowledge of evolutionary history. Because they can't see into the past, they must infer the past as best they can, on the basis of similarities among living organisms. Not just any similarity will do, however. Some observed similarities stem from convergent evolution in organisms that are not closely related; such similarities are not useful to systematists. Instead, systematists value the similarities that arise when two kinds of organisms share a feature because both inherited it from a common ancestor. Therefore, one of a systematist's main tasks is to distinguish informative similarities caused by common ancestry from the less useful similarities that result from convergent evolution. In the search for informative similarities, systematists look at many different kinds of characteristics.

Historically, the most important and useful of distinguishing characteristics has been anatomy. In addition to obvious similarities in external body structure (see Fig. 18-1), taxonomists look carefully at details such as skeletons and tooth structure. For example, homologous structures, such as the finger bones of dolphins, bats, seals, and humans (see Fig. 14-8), provide evidence of an evolutionarily distant common ancestor. To distinguish between closely related species, taxonomists may use microscopes to discern finer details—the number and shape of the "teeth" on the tonguelike *radula* of a snail,

the shape and position of the bristles on a marine worm, or the fine structure of pollen grains of a flowering plant (Fig. 18-2).

Developmental stages also provide clues to common ancestry. For example, although the sea squirt, a type of tunicate, spends its adult life permanently attached to rocks on the ocean floor, its free-swimming larva has a nerve cord, tail, and gills, placing it in the same phylum as the vertebrates: phylum Chordata.

The developmental and anatomical characteristics shared by related organisms are expressions of underlying genetic similarities, so it stands to reason that evolutionary relationships among species must also be reflected in genetic similarities. Unfortunately, direct genetic comparisons were not possible for most of the history of biology. In the 1990s, however, advances in the techniques of molecular genetics have triggered a revolution in studies of evolutionary relationships. For the first time, the nucleotide sequence of DNA (that is, genotype), rather than phenotypical features such as appearance, behavior, or even proteins, can be used to investigate relatedness among organisms. Some of the key methods and findings of genetic analysis are explored in the essay "Scientific Inquiry: Molecular Genetics Reveals Evolutionary Relationships" (pp. 342–343).

Despite its power, this new direct access to DNA has not yet completely eliminated other ways of making genetic comparisons. For example, relatedness among organisms can also be evaluated by examining the structure of their chromosomes. Among the findings derived from

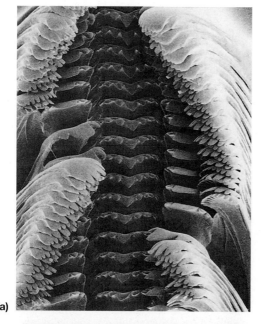

(a)

(b)

(c)

***Figure 18-2 Microscopic structures help
taxonomists classify organisms***
(a) The "teeth" on a snail's tonguelike radula
(a structure used in feeding) or ***(b)*** the bristles
on a marine worm are potential taxonomic criteria. ***(c)*** The shape and surface features of pollen
grains help identify plants, particularly from fossil deposits and sediments.

this technique is that the chromosomes of chimpanzees
and humans are extremely similar, showing that these
two species are very closely related (Fig. 18-3).

 ## What Are the Kingdoms of Life?

Before 1970, taxonomists classified all forms of life into
two kingdoms: Animalia and Plantae. Bacteria, fungi, and
photosynthetic protists were considered plants, and the
protozoa were classified as animals. As scientists learned

more about fungi and microorganisms, however, it became apparent that the two-kingdom system oversimplified the true nature of evolutionary history. To help
rectify this problem, Robert H. Whittaker in 1969 proposed a five-kingdom classification scheme that was
eventually adopted by most systematists.

Whittaker's five-kingdom system divides unicellular
microorganisms into two kingdoms, based primarily on
whether they show prokaryotic or eukaryotic cellular organization. The kingdom Monera consists of generally
single-celled prokaryotic organisms, whereas the kingdom Protista consists of generally single-celled eukary-

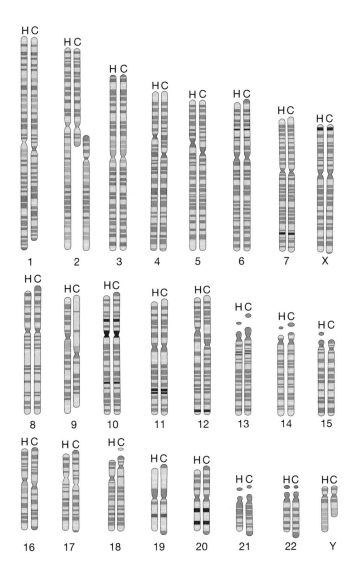

Figure 18-3 Human and chimp chromosomes are similar
Chromosomes from different species can be compared by
means of banding patterns that are revealed by staining.
The comparison illustrated here, between human chromosomes (left member of each pair; H) and chimpanzee
chromosomes (C), reveals that the two species are genetically very similar. In fact, it has been estimated that 99% of
the two genomes is identical. The numbering system shown
is that used for human chromosomes; note that human
chromosome 2 corresponds to a combination of two chimp
chromosomes.

trast, members of the kingdom Animalia ingest their food
and then digest it, either within an internal cavity or within individual cells.

Because it more accurately reflected our understanding of evolutionary history, the five-kingdom system was
an improvement over the old two-kingdom system. As
our understanding has continued to grow, however, it has
proved necessary to change our view of life's most fundamental categories. The pioneering work of microbiologist Carl Woese has shown that systematists have
overlooked a fundamental event in the early history of
life, one that demands a new and more-accurate classification of life.

Since the 1970s, Woese and other biologists interested
in the phylogeny of microorganisms have studied the biochemistry of prokaryotic organisms. These researchers,
focusing on nucleotide sequences of the RNA that is
found in ribosomes, determined that what had been considered the kingdom Monera actually consists of two very
different kinds of organisms. Woese has dubbed these
two groups the Bacteria and the Archaea (Fig. 18-4). Despite superficial similarities in their appearance under

(a)

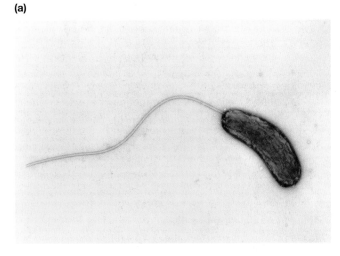

(b)

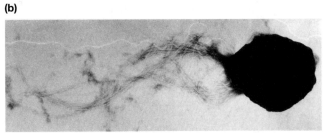

Figure 18-4 Two domains of prokaryotic organisms
Although superficially similar in appearance, **(a)** *Vibrio
cholerae* and **(b)** *Methanococcus jannaschi* are less closely related than a mushroom and an elephant. Each prokaryote belongs to a fundamentally different domain of life. *Vibrio* is in
the domain Bacteria, and *Methanococcus* in the Archaea.

otic organisms. The remaining three kingdoms (Plantae,
Fungi, and Animalia) include only eukaryotic organisms,
most of which are multicellular. These three kingdoms
of multicellular eukaryotes can be distinguished on the
basis of their methods of acquiring nutrients. Members of
the kingdom Plantae photosynthesize, and members of
the kingdom Fungi secrete enzymes outside their bodies
and then absorb the externally digested nutrients. In con-

Scientific Inquiry
Molecular Genetics Reveals Evolutionary Relationships

Evolution is the result of the accumulation of inherited changes in populations. Because DNA is the molecule of heredity, evolutionary changes must be reflected in changes in DNA. Chapter 14 explored a few of the genetic mechanisms that contribute to evolution. In this essay, we will consider two methods of studying DNA and how those methods are applied to determining the evolutionary relatedness of organisms.

Cross-Species DNA Hybridization The DNA in the chromosomes of all living organisms is double stranded, with the two strands held together by hydrogen bonds between complementary base pairs. The two strands of DNA can be separated by heating; when the temperature becomes high enough, the hydrogen bonds break and the two strands separate. The amount of separation can be measured by observing the absorption of ultraviolet (UV) light. Single DNA strands absorb UV light more effectively than double strands do, so the higher the proportion of single strands, the greater the UV absorption.

Suppose now that you had two single strands of DNA in which only half of the bases were complementary. The complementary bases would still hold the two strands together, but with only half the normal number of hydrogen bonds. If you heated this "semidouble helix," you would find that, with fewer hydrogen bonds holding the strands together, the strands would separate at a lower temperature.

The DNA of two closely related species has more-similar nucleotide sequences than does the DNA of two distantly related species. In the technique called **DNA–DNA hybridization**, DNA from each of two species is heated to separate, or melt, the individual nucleotide strands. The strands are then allowed to form "hybrid" double helices by complementary base pairing (Fig. E18-1). The hybrid double helices are then reheated, and the temperature at which the two strands melt is measured. Separation temperatures for strands from different species will necessarily be less than those for the same species. The principle is this: The more-similar the nucleotide sequences, the greater the number of hydrogen bonds between them, and the higher the temperature required for the hybrid (two-species) helices to separate. When the degree of nucleotide matching and the melting temperature for hybrid DNA are very similar to those of the DNA for each single species, the two species are very similar genetically and are very closely related. Using this type of information, we can generate a tree of evolutionary relationships on the basis of the separation temperature of the hybrid helices (Fig. E18-2).

DNA Sequencing: Direct Comparison of Nucleotide Sequences "Melting" DNA is a somewhat indirect way of measuring similarities in nucleotide sequences. A more precise comparison can be made by **DNA sequencing**—determining the sequence of nucleotides in segments of DNA from different species. Fewer differences in nucleotide sequence indicate more closely related organisms. Direct sequencing, however, may yield insufficient information unless a number of different genes are

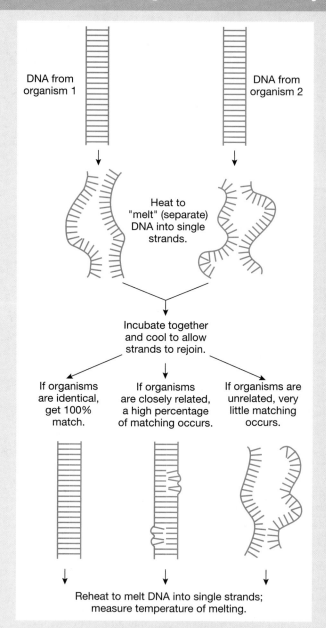

Figure E18-1 DNA–DNA hybridization
As DNA is heated, the two strands gradually separate. DNA strands from different organisms are then combined and cooled, allowing hydrogen bonds to re-form complementarily between the strands. When DNA strands from two species have been "hybridized" in this way, the degree of matching between them indicates how closely related are the species. The hybrid strands are then reheated. The temperatures at which they melt further indicate evolutionary relatedness: Similar melting temperatures are evidence of recent common ancestry.

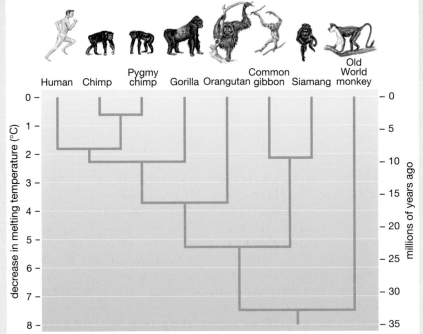

Figure E18-2 Relatedness can be determined by DNA melting temperatures
In this study by Charles Sibley and Jon Ahlquist, double helices of DNA were made by combining single strands from various pairs of primate species. The melting temperatures were then determined and compared with melting temperatures of DNA from a single species (the difference is graphed here as the decrease in melting temperature). The greater the difference in melting temperature of hybrid versus "pure" DNA, the earlier the two species diverged from a common ancestor.

sequenced. Because the nucleotide sequence of one gene may change more rapidly over time than that of another gene, comparisons of a *single* gene may not provide an accurate picture of evolutionary change across the genome. So if only a single, short stretch of DNA is examined, direct sequencing may be even less informative than DNA–DNA hybridization, which detects changes over millions of nucleotides. In the past, sequencing was so costly and time-consuming that it was difficult to gather a data set large enough to be useful for constructing a phylogeny. Today, however, the *polymerase chain reaction* (PCR, see Chapter 13) allows systematists to accumulate large samples of

DNA from organisms easily, and automated machinery makes sequence determination a comparatively simple task. As a result, sequence data are piling up with unprecedented rapidity, and systematists have access to sequences from an ever-increasing number of species. For several species, including some bacteria, an archaean, and a yeast, the entire genome has been sequenced. The Human Genome Project is scheduled to complete its work by the year 2005, and then our own DNA sequences will be a matter of public record. The revolution in molecular biology has fostered a great leap forward in our understanding of evolutionary history.

Figure 18-5 The tree of life
The three domains of life represent the earliest branches in evolutionary history.

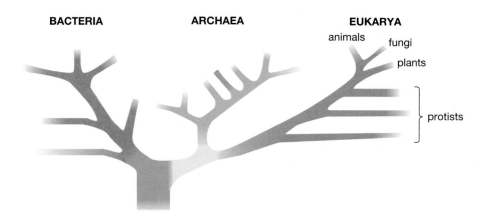

the microscope, the Bacteria (sometimes called *eubacteria*) and the Archaea (also known as *archaebacteria*) are radically different. The members of these two groups are no more closely related to one another than either one is to any eukaryote (Fig 18-5). The tree of life split into three parts very early in the history of life, long before the eukaryotes gave rise to plants, animals, and fungi. So Woese has proposed, and many systematists agree, that we ought to classify life into three broad categories called **domains**: the (1) *Bacteria*, (2) *Archaea*, and (3) *Eukarya*.

If we accept the three-domain classification system, we're left with some taxonomic dilemmas with regard to the status of our five kingdoms. It does not make sense to lump all prokaryotes into the single kingdom Monera, because the Bacteria and the Archaea are only very distantly related. If life has three main branches, we can't justify lumping two of them (Archaea and Bacteria) into one kingdom while splitting the third branch (Eukarya) into four kingdoms. A related problem is that if the domain Eukarya can contain four kingdoms, then the other two domains may also contain more than one kingdom. In fact, there were a number of very early splits both within the Archaea and within the Bacteria that long predated, for example, the split between plants and animals. If the plant–animal split is ancient enough to warrant

kingdom status for each group, then each of the even older splits within the Bacteria and Archaea should also merit kingdom status. Thus, we would need a lot more than five kingdoms to construct a phylogenetically accurate classification of life. Some systematists have even suggested that plants, animals, and fungi don't deserve kingdom status. After all, they represent only tiny twigs way down at the end of the eukaryote branch of the tree of life.

Overall, we're now at a rather awkward moment in the evolution of high-level classification. The systematic data strongly support the three-domain system, but the thorny issues of kingdom-level classification are yet to be definitively worked out. At the same time, it's unsatisfying to downplay the seemingly fundamental differences among protists, plants, animals, and fungi. In this text, our imperfect solution will be to reserve judgment on kingdom-level classification within the two prokaryotic domains and to retain the four kingdoms within the Eukarya. Table 18-2 compares the characteristics within those four kingdoms, and Figure 18-6 shows their evolutionary relationships. This arrangement is useful, but it is important for you to understand that it is derived from a five-kingdom system that may be replaced some time in the future by a more phylogenetically accurate system.

Table 18-2 Some Characteristics of the Eukaryotic Kingdoms

Kingdom	Cell Type	Cell Number	Major Mode of Nutrition	Motility (Movement)	Cell Wall	Reproduction
Protista	Eukaryotic	Unicellular	Absorb, ingest or photosynthesize	Both motile and nonmotile	Present in algal forms: varies	Both sexual and asexual
Animalia	Eukaryotic	Multicellular	Ingest	All motile at some stage	Absent	Both sexual and asexual
Fungi	Eukaryotic	Most multicellular	Absorb	Generally nonmotile	Present: chitin	Both sexual and asexual
Plantae	Eukaryotic	Multicellular	Photosynthesize	Generally nonmotile	Present: cellulose	Both sexual and asexual

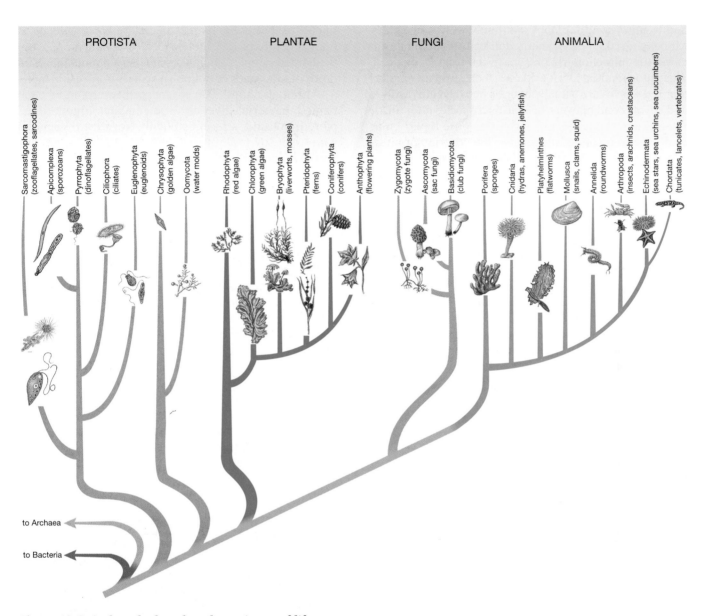

Figure 18-6 A closer look at the eukaryotic tree of life
The four kingdoms within the domain Eukarya and some of the major phyla or divisions within them are illustrated.

3 Why Do Taxonomies Change?

As the emergence of the three-domain system shows, the hypotheses of evolutionary relationships on which taxonomy is based are subject to revision as new data emerge. Even the five kingdoms, representing deep, old branchings of the tree of life and the most fundamental categories of taxonomy, will have to be modified in light of recent advances in systematics. At the other end of the taxonomic spectrum, species designations are revised quite frequently. For example, in 1973 ornithologists officially declared the Baltimore oriole and the Bullock's

oriole to be a single species, which they named the northern oriole. The reason: Where their ranges overlap, the two "species" of birds were found to interbreed. Then, in 1997, ornithologists officially declared that the Baltimore oriole and the Bullock's oriole were indeed separate species. The reason: Closer study of the hybrid zone revealed that there was little gene flow between the two types of oriole.

Another example is the red wolf from the southern United States, which is currently listed as an endangered species. Researchers recently analyzed the DNA from mitochondria taken from red wolves and found it to be identical to mitochondrial DNA from coyotes in some cases and to mitochondrial DNA from gray wolves in others. Mitochondrial DNA is inherited directly from the mother via the mitochondria present in the original egg cell. This mitochondrial DNA evidence, then, strongly suggests that red wolves are actually hybrids between gray wolves and coyotes and may not be a distinct species after all.

Asexually reproducing organisms pose a particular challenge to taxonomists, because the criterion of interbreeding cannot be used to distinguish among species. For instance, some taxonomists recognize 200 species of the British blackberry (a plant that can produce seeds *parthenogenetically*—that is, without fertilization); others recognize only 20 species. Most unicellular organisms reproduce asexually most of the time, making their classification difficult. For example, molecular biologists recently sequenced the DNA of two strains of the bacterium *Legionella pneumophila*, which is responsible for Legionnaire's disease, and found only a 50% correspondence between the strains. This degree of genetic dissimilarity, which occurs within a single species of bacterium, is as great as the genetic difference between mammals and fishes! Because the taxonomy of bacteria is so controversial, we present only general descriptions of bacteria in Chapter 19.

Taxonomic categories are continuously debated and revised as taxonomists learn more and more about evolutionary relationships, particularly with the application of techniques derived from molecular biology. Although the precise evolutionary relationships of many organisms continue to elude us, taxonomy is enormously helpful in ordering our thoughts and investigations into the diversity of life on Earth.

4 Exploring Biodiversity: How Many Species Exist? *www*

Scientists do not know even within an order of magnitude how many species share our world. Each year, between 7000 and 10,000 new species are named, most of them insects, many from the tropical rain forests. The total number of named species is currently about 1.4 million. However, many scientists believe that 7 million to 10 million species may exist, and estimates range as high as 30 million. This total range of species diversity, and the complex interrelationships among the species, is known as **biodiversity**. Of all the species that have been identified thus far, about 5% are prokaryotes and protists. An additional 22% are plants and fungi, and the rest are animals. This distribution has little to do with the actual abundance of these organisms and a lot to do with the size of the organisms, how easy they are to classify, how accessible they are, and the number of scientists studying them. Historically, systematists have chiefly focused on large or conspicuous organisms in temperate regions, but biodiversity is greatest among small, inconspicuous organisms in the tropics. In addition to the overlooked species on land and in shallow waters, an entire new "con-

tinent" of species lies largely unexplored on the deep-sea floor at depths of 1000 meters (3250 feet) and more. From the limited samples available, scientists estimate that hundreds of thousands of unknown species may reside there.

Although about 5000 species of bacteria have been named, bacterial diversity also remains largely unexplored. Consider a recent study by Norwegian scientists who used *DNA–DNA hybridization* (a technique discussed in the Scientific Inquiry essay) to identify the different bacteria in a small sample of forest soil. To distinguish among species, the researchers arbitrarily defined bacterial DNA as coming from separate species if it differed by at least 30% from that of any other bacterial DNA. Using this criterion, they reported more than 4000 types of bacteria in their soil sample and an equal number of new forms in a sample of shallow marine sediment!

Our ignorance of the full extent of life's diversity adds a new dimension to the tragedy of the destruction of the tropical rain forests, discussed in Chapter 41. Although these forests cover only about 6% of Earth's land area, they are believed to be home to two-thirds of the world's existing species, most of which have never been studied or named. Because these forests are being destroyed so rapidly, Earth is losing many species that we will never even know existed! For example, in 1990 a new species of primate, the black-faced lion tamarin, was discovered in a small patch of dense rain forest on an island just off the east coast of Brazil (Fig. 18-7). Only a dozen individuals

Figure 18-7 The black-faced lion tamarin
This rare species of monkey was discovered in 1990 in the vanishing Brazilian rain forest.

of this squirrel-sized monkey have ever been spotted, and captive breeding may be its only hope for survival. At current rates of deforestation, most of the tropical rain forests, with their undescribed wealth of life, will be gone within the next century.

Evolutionary Connections
The Case of the Homeless Kelp

Biologists use classification schemes to place the huge number of organisms on Earth into natural groupings. Ideally, these groupings represent evolutionary relationships, which until recently have been inferred by taxonomists solely on the basis of the presence of shared distinguishing features. But the kingdom Protista, defined as comprising all single-celled eukaryotic organisms, is not a natural grouping, and scientists disagree about which organisms it should include. Plants, animals, and fungi all have close protistan relatives, and the separation of single-celled organisms from multicellular organisms is sometimes problematic. It is especially so for the algae, which have both single-celled and multicellular representatives within most smaller taxonomic groupings. Can closely related organisms be placed into separate kingdoms, Protista and Plantae, simply on the basis of multicellularity? Some textbooks place all the algae—photosynthetic organisms with simple reproduction—into the kingdom Protista; and some textbooks, including this one, split the algae between the Protista and the Plantae depending on whether they are single-celled or multicellular. Some taxonomists split the multicellular algae between two kingdoms, placing the multicellular brown and red algae with the protists and the multicellular green algae into the plant kingdom. These different attempts to classify closely related organisms demonstrate how difficult it is to develop standard criteria for grouping organisms, even at the kingdom level.

One approach to this problem, enthusiastically endorsed by Dr. Lynn Margulis, a biologist at the University of Massachusetts, is the creation of the kingdom Protoctista. This taxonomic category includes single-celled organisms and their close descendants (for example, the multicellular algae but *not* the animals, fungi, or plants). Margulis describes the kingdom Protoctista as "the entire motley and unruly group of non-plant, non-animal, non-fungal organisms representative of lineages of the earliest descendants of the eukaryotes."

It is conceptually difficult to group one of the largest multicellular organisms in the world, the brown algae called giant kelps, with simple microscopic single-celled organisms. Kelps, some of which are up to 60 meters (almost 200 feet) long, possess a tissuelike level of organization that is relatively complex and can transport materials over long distances, as can the tissues of more-advanced plants. The cells in kelps and some other algae are specialized and show division of labor. However, kelps reproduce like other algae and differently from plants. Dr. Thomas Cavalier-Smith of the University of British Columbia has proposed that brown algae merit their own kingdom (kingdom *Chromista*) on the basis of ultrastructural features and molecular comparisons of all algae. Thus, even among the algae alone, clear differences abound that some scientists believe are sufficient to justify the status of a separate kingdom.

As we learn more about the relationships among organisms and refine the criteria used to classify them, classification schemes will keep changing. As the superficially simple question "In which kingdom should we place the algae?" illustrates, the taxonomic categories in textbooks are tentative and subject to revision as we continue to discover more about life on Earth.

Summary of Key Concepts

1) How Are Organisms Named and Classified?
Taxonomy is the science by which organisms are classified and placed into hierarchical categories that reflect their evolutionary relationships. The seven major categories, in order of decreasing inclusiveness, are (1) kingdom, (2) division or phylum, (3) class, (4) order, (5) family, (6) genus, and (7) species. The scientific name of an organism is composed of its genus name and species name. A hierarchical concept was first used by Aristotle, but Linnaeus in the mid-1700s laid the foundation for modern taxonomy. In the 1860s, evolutionary theory proposed by Charles Darwin provided an explanation for the observed similarities and differences among organisms, and modern taxonomists attempt to classify organisms according to their evolutionary relationships. Today, taxonomists use features such as anatomy, developmental stages, and biochemical similarities to categorize organisms. Molecular biological techniques are used to determine the sequences of nucleotides in DNA and RNA and of amino acids in proteins. Biochemical similarities among organisms are a measure of evolutionary relatedness.

2) What Are the Kingdoms of Life?
A five-kingdom classification system consisting of the kingdoms Monera, Protista, Animalia, Fungi, and Plantae is widely used, but a new and increasingly accepted three-domain system will require revisions at the kingdom level of classification. The three domains, representing the three main branches of life, are Bacteria, Archaea, and Eukarya. (See Table 18-2 for characteristics of the four eukaryotic kingdoms.)

3) Why Do Taxonomies Change?

Taxonomic categories are controversial and subject to revision, particularly in the case of asexually reproducing species. However, taxonomy is essential for precise communication and contributes to our understanding of the origins and diversity of species.

4) Exploring Biodiversity: How Many Species Exist?

Although only about 1.4 million species have been named, estimates of the total number of species range up to 30 million. New species are being identified at the rate of 7000 to 10,000 annually, mostly in tropical rain forests.

Key Terms

biodiversity *p. 346*	DNA sequencing *p. 342*	phylogeny *p. 338*	taxonomy *p. 338*
class *p. 338*	family *p. 338*	phylum *p. 338*	
division *p. 338*	genus *p. 338*	scientific name *p. 338*	
domain *p. 344*	kingdom *p. 338*	species *p. 338*	
DNA–DNA hybridization *p. 342*	order *p. 338*	systematics *p. 338*	

Thinking Through the Concepts

Multiple Choice

1. *It is possible to imagine the various levels of taxonomic classification as a kind of "family tree" for an organism. If the kingdom is analogous to the trunk of the tree, which taxonomic category would be analogous to the large limbs coming off that trunk?*
   a. class   b. family
   c. order   d. phylum or division
   e. subfamily

2. *The more taxonomic categories two organisms share, the more closely related those organisms are in an evolutionary sense. Which scientist's work led to this insight?*
   a. Aristotle   b. Darwin
   c. Linnaeus   d. Whittaker
   e. Woese

3. *Which of the following criteria could not be used to determine how closely related two types of organisms are?*
   a. similarities in the presence and relative abundance of specific molecules
   b. DNA sequence
   c. the presence of homologous structures
   d. developmental stages
   e. occurrence of both organisms in the same habitat

4. *Which one of the following habitats appears to have the greatest number of species?*
   a. the sea floor   b. deserts
   c. tropical rain forests   d. grasslands
   e. mountaintops

5. *Which of the following organisms are considered plants by some taxonomists but protists by others?*
   a. algae   b. fungi
   c. mosses   d. archaea
   e. protozoa

6. *An organism is described to you as having many nuclei-containing cells, each surrounded by a cell wall of chitin and absorbing its food. In which kingdom would you place it?*
   a. Plantae   b. Protista
   c. Animalia   d. Fungi
   e. Monera

? Review Questions

1. What contributions did Aristotle, Linnaeus, and Darwin each make to modern taxonomy?

2. What features would you study to determine whether a dolphin is more closely related to a fish or to a bear?

3. What techniques might you use to determine whether the extinct cave bear is more closely related to a grizzly bear or to a black bear?

4. Only a small fraction of the total number of species on Earth has been scientifically described. Why?

5. In England, "daddy long-legs" refers to a long-legged fly, but the same name refers to a spider-like animal in the United States. How do scientists attempt to avoid such confusion?

Applying the Concepts

1. There are many areas of disagreement among taxonomists, for example, whether algae belong in the kingdom Protista or the kingdom Plantae, and whether archaebacteria should be in the kingdom Monera or in an entirely new kingdom. What difference does it make whether biologists consider algae as plants or protists, or archaebacteria as monerans or something else? As Shakespeare put it, "What's in a name?"

2. The pressures created by human population growth and economic expansion place storehouses of biological diversity such as the tropics in peril. The seriousness of the situation is clear when we consider that probably only 1 out of every 20 tropical species is known to science at present. What arguments can you make for preserving biological diversity in poor and developing countries? Does such preservation require that these countries sacrifice economic development? Suggest some solutions to the conflict between the growing demand for resources and the importance of conserving biodiversity.

3. During major floods only the topmost branches of submerged trees may be visible above the water. If you were asked to sketch the branches below the surface of the water, solely on the basis of the positions of the exposed tips, you would be attempting a reconstruction somewhat similar to the "family tree" by which taxonomists link various organisms according to their common ancestors (analogous to branching points). What sources of error do both exercises share? What advantages do modern taxonomists have?

4. The Florida panther, found only in the Florida Everglades, is currently classified as an endangered species, protecting it from human activities that could lead to its extinction. It has long been considered a subspecies of cougar (mountain lion), but recent mitochondrial DNA studies have shown that the Florida panther may actually be a hybrid between American and South American cougars. Should the Florida panther be protected by the Endangered Species Act?

Group Activity

Form a group of four students. Imagine that you've been working in a museum collection, and you discover a dusty old cabinet that hasn't been opened in 150 years. Inside are some mysterious specimens, previously unknown to science. Your job is to classify these organisms. Develop a classification, draw an evolutionary tree, and explain your reasoning. The specimens are pictured below.

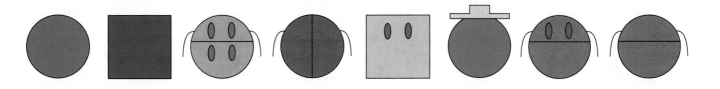

For More Information

Avise, J. C. "Nature's Family Archives." *Natural History*, March 1989. Shows how evolutionary relationships can be determined by analyzing differences in DNA contained in mitochondria.

Gould, S. J. "What Is a Species?" *Discover*, December 1992. Discusses the difficulties of distinguishing separate species.

Lowenstein, J. M. "Molecular Approaches to the Identification of Species." *American Scientist*, 73: 541–547, 1985. A basic introduction to immunological methods of taxonomy.

Mann C., and Plummer, M. *Noah's Choice: The Future of Endangered Species*. New York: Knopf, 1995. A thought-provoking look at the hard choices we must make with regard to protecting biodiversity. Which species will we choose to preserve? What price are we willing to pay?

Margulis, L., and Schwartz, K. *Five Kingdoms*. 2nd ed. New York: W. H. Freeman, 1988. An illustrated paperback guide to the diversity of life.

Margulis, L., and Sagan, D. *What Is Life?* London: Weidenfeld & Nicolson, 1995. A lavishly illustrated survey of life's diversity. Also includes an account of life's history and a mediation on the question posed by the title.

May, R. M. "How Many Species Inhabit the Earth?" *Scientific American*, October 1992. Although no one knows the answer to this question, an accurate count is crucial in an effort to manage our biological resources.

Moffett, M. W. *The High Frontier: Exploring the Tropical Rainforest Canopy*. Cambridge, MA: Harvard University Press, 1994. The tremendous diversity of life in the rainforest treetops is only now becoming known to us. This book documents the unexpected and spectacular diversity of animals in the upper reaches of this endangered habitat.

Wilson, E. O. *The Diversity of Life*. Cambridge, MA: Harvard University Press, 1992. An outline of the processes that created the diversity of life, and a discussion of the threats to that diversity and the steps required to preserve it.

Answers to Multiple-Choice Questions

1. d   2. b   3. e   4. c   5. a   6. d

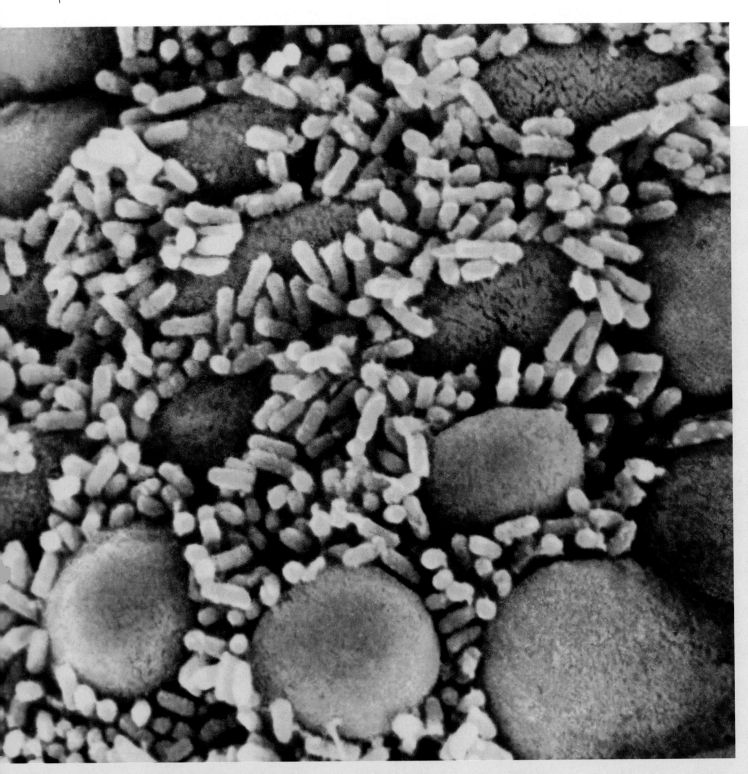

An up-close and personal view of Escherichia coli *(yellow), one of the microbial inhabitants of folds in the human intestine (pink). At least 400 species of bacteria live in intimate association with humans.*

The Hidden World of Microbes 19

At a Glance

Net Watch
On-line resources for this chapter are on the World Wide Web at:
http://www.prenhall.com/audesirk
(click on the Table of Contents link and then select Chapter 19).

It's easy to overlook *microbes*—that is, microorganisms. They are, after all, invisible to the naked eye. Despite their inconspicuous stature, however, single-celled organisms play a crucial role in life on our planet. In fact, for most of the history of life on Earth, all life was microbial. Only after 3 billion years of microbial rule did larger organisms arrive on the scene. And even now the lives of plants and animals (including humans) are utterly dependent on microbes. For example, if not for the decomposition and recycling of wastes and dead organisms by microbes, Earth's finite supply of nutrient molecules would remain locked away and unavailable to living things.

Microbes are uncountably numerous and occur in virtually every imaginable habitat, including the bodies of larger creatures, such as humans. We live in intimate association with microbes; our skin is covered with a veritable garden of bacteria, and our intestines are home to hundreds of millions of microbial inhabitants. We don't think much about these vast cities of bacteria that live in and on us, but our health and survival are intimately tied to theirs. Ordinarily, this intimacy is mutually beneficial, but it also has a darker side: Some of humanity's most deadly diseases stem from microbes. This chapter introduces the hidden world of microbes. For convenience, we discuss viruses and viruslike particles here, although they are not considered to be organisms.

1) What Are Viruses, Viroids, and Prions?

Viruses possess no membranes of their own, no ribosomes on which to make proteins, no cytoplasm, and no source of energy. They cannot move or grow, and they can reproduce only inside a **host** cell—the cell a virus or other infectious agent infects. The simplicity of viruses makes it impossible to call them cells and, indeed, seems to place them outside the realm of living things.

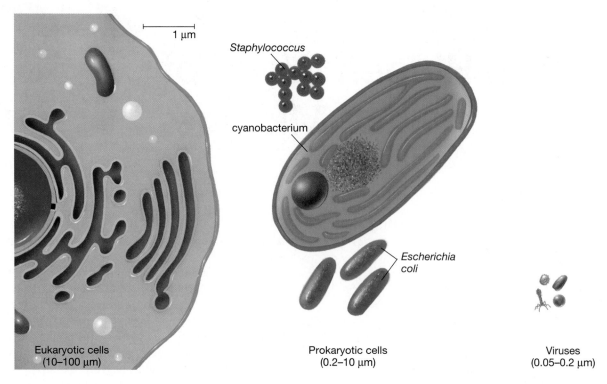

Figure 19-1 *The sizes of microorganisms*
The relative sizes of eukaryotic cells, prokaryotic cells, and viruses (1 μm = ¹⁄₁₀₀₀ millimeter).

A Virus Consists of a Molecule of DNA or RNA Surrounded by a Protein Coat

A virus particle is so small (0.05–0.2 micrometer [μm: 1/1000 of a millimeter] in diameter) that seeing it requires the enormous magnification of an electron microscope (Fig. 19-1). Viruses consist of two major parts: (1) a molecule of hereditary material, either DNA or RNA; and (2) a coat of protein surrounding the molecule. The protein coat may itself be surrounded by an envelope formed from the plasma membrane of the host cell (Fig. 19-2a).

Viruses are unable to grow or reproduce on their own, even if placed in a rich broth of nutrients at optimal temperature. They lack the complex cellular organization that these activities require. A virus's protein coat, however, is specialized to enable the virus to penetrate the cells of a specific host. After a virus enters a host cell, the viral genetic material takes command. The hijacked host cell is forced to "read" the viral genes and to use the instructions encoded there to produce the components of new viruses. The pieces are rapidly assembled, and an army of new viruses bursts forth to invade and conquer neighboring cells (Fig. 19-2b; see "A Closer Look: Viruses—How the Nonliving Replicate," pp. 354–355).

Viral Infections Cause Diseases That Are Difficult to Treat

Each type of virus is specialized to attack a specific host cell (Fig. 19-3). As far as we know, no organism is immune to all viruses. Even bacteria fall victim to viral invaders; viruses that infect bacteria are called **bacteriophages** (Fig. 19-4).

Within a particular organism, viruses specialize in attacking particular cell types. Viruses responsible for the common cold, for example, attack the membranes of the respiratory tract; those causing measles infect the skin, and the rabies virus attacks nerve cells. One type of herpes virus specializes in the mucous membranes of the mouth and lips, causing cold sores; a second type, transmitted through sexual contact, produces similar sores on or near the genitals. Herpes viruses take up permanent residence in the body, erupting periodically (typically during times of stress) as infectious sores. The devastating disease AIDS (acquired immune deficiency syndrome), which cripples the body's immune system, is caused by a virus that attacks a specific type of white blood cell that controls the body's immune response (as we shall see in Chapter 31). Viruses have also been linked to some types of cancer, such as T-cell leukemia, a cancer of the white blood cells. The recent identification of the papilloma virus, which causes genital warts, in 90% of cervical cancers sampled suggests that this virus may cause these cancers. Because viruses are intracellular infectious agents that require the cellular machinery of their host, the illnesses they cause are difficult to treat: Antiviral agents may destroy host cells as well as viruses. The antibiotics so effective against bacterial infections are useless against viruses, although some promising antiviral drugs are being developed.

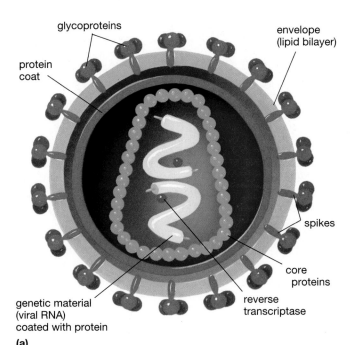

glycoproteins

protein coat

envelope (lipid bilayer)

spikes

core proteins

genetic material (viral RNA) coated with protein

reverse transcriptase

(a)

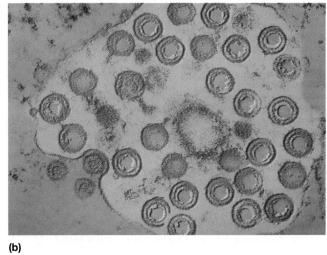

(b)

Figure 19-2 Viral structure and replication
(a) Cross section of the virus that causes AIDS. Inside is genetic material surrounded by a protein coat and molecules of reverse transcriptase, an enzyme that catalyzes the transcription of DNA from the viral RNA template after the virus enters the host cell. Some viruses, including those that cause herpes, rabies, or AIDS, have an outer envelope that may be formed from the host cell's plasma membrane. Spikes made of glycoprotein (protein and carbohydrate) may project from the envelope; some viruses use these spikes to attach to their host cell. **(b)** In this electron micrograph, herpes viruses are seen packed into an infected cell.

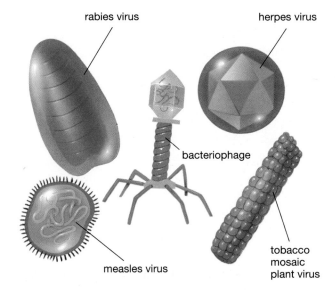

rabies virus

herpes virus

bacteriophage

measles virus

tobacco mosaic plant virus

Figure 19-3 Viruses come in a variety of shapes
Viral shape is determined by the nature of the virus' protein coat. Viruses such as the rabies and herpes viruses are surrounded by an extra envelope derived from membranes of the host cell.

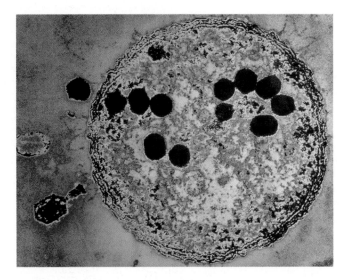

Figure 19-4 Some viruses infect bacteria
In this electron micrograph, bacteriophages are seen attacking a bacterium. They have injected their genetic material inside, leaving their protein coats clinging to the bacterial cell wall. The black objects inside the bacterium are newly forming viruses.

A Closer Look
Viruses—How the Nonliving Replicate

Viruses multiply, or "replicate," by using their own genetic material, which consists of single-stranded or double-stranded RNA or DNA, depending on the virus. This material serves as a template (a "blueprint") for the viral proteins and genetic material required to make new viruses. Viral enzymes may participate in replication as well, but the overall process depends on the biochemical machinery that the host cell uses to make its own proteins.

Viral replication follows a general sequence:

1. **Penetration.** Viruses may be engulfed by their host cell (endocytosis). Some viruses have surface proteins that bind to receptors on the host cell's plasma membrane and stimulate endocytosis. Other viruses are coated with an envelope that can fuse with the host's membrane. The viral genetic material is then released into the cytoplasm.

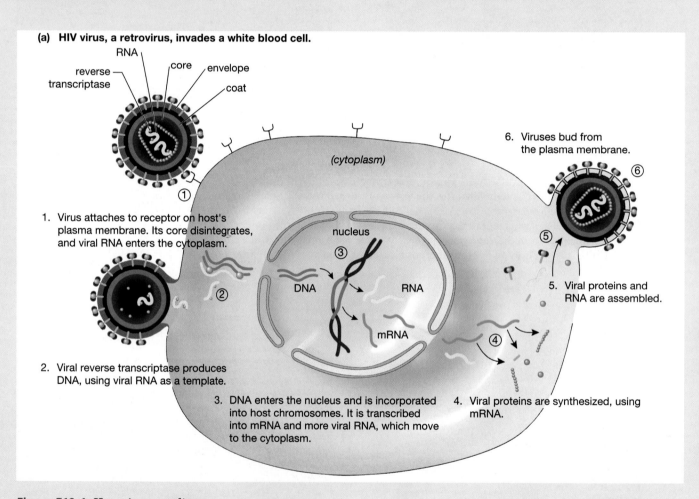

(a) **HIV virus, a retrovirus, invades a white blood cell.**

1. Virus attaches to receptor on host's plasma membrane. Its core disintegrates, and viral RNA enters the cytoplasm.

2. Viral reverse transcriptase produces DNA, using viral RNA as a template.

3. DNA enters the nucleus and is incorporated into host chromosomes. It is transcribed into mRNA and more viral RNA, which move to the cytoplasm.

4. Viral proteins are synthesized, using mRNA.

5. Viral proteins and RNA are assembled.

6. Viruses bud from the plasma membrane.

Figure E19-1 How viruses replicate

Some Infectious Agents Are Even Simpler Than Viruses

In 1971, the plant pathologist T. O. Diener discovered that some plant diseases are caused by particles only one-tenth the size of normal plant viruses. Called **viroids**, these particles are merely short strands of RNA, lacking even a protein coat. Viroids apparently enter the nucleus of the infected cell, where they direct the synthesis of new viroids. About a dozen crop diseases, including cucumber pale fruit disease, avocado sunblotch, and potato spindle tuber disease, have been attributed to viroids.

Prions are even more puzzling than viroids. In the 1950s, physicians studying the Fore, a primitive tribe in New Guinea, were puzzled to observe numerous cases of a fatal degenerative disease of the nervous system, which the Fore called kuru. The symptoms of **kuru**—a

2. **Replication.** The viral genetic material is copied many times.
3. **Transcription.** Viral genetic material is used as a blueprint to make messenger RNA (mRNA).
4. **Protein synthesis.** In the host cytoplasm, viral mRNA is used to synthesize viral proteins.
5. **Viral assembly.** The viral genetic material and enzymes are surrounded by their protein coat.
6. **Release.** Viruses emerge from the host cell by "budding" from the cell membrane or by bursting the cell.

Here we illustrate two types of viral life cycle. In Figure E19-1a, the *human immunodeficiency virus (HIV)*, which causes AIDS, is a *retrovirus*. Retroviruses use single-stranded RNA as a template to make double-stranded DNA by using a viral enzyme called *reverse transcriptase*. Many other retroviruses exist, and several cause cancers or tumors. In Figure E19-1b, the *herpes virus* contains double-stranded DNA that is transcribed into mRNA. The *influenza virus* (not illustrated) has a replication cycle somewhat like that of the herpes virus, but it uses single-stranded RNA as a template for mRNA and derives its envelope from the host cell's plasma membrane.

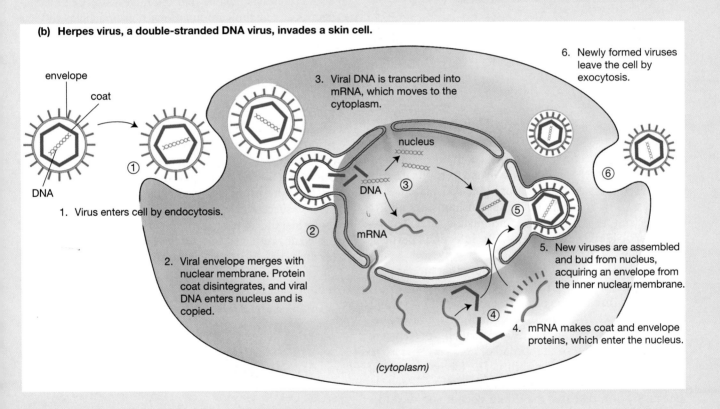

(b) Herpes virus, a double-stranded DNA virus, invades a skin cell.

envelope

coat

DNA

1. Virus enters cell by endocytosis.

2. Viral envelope merges with nuclear membrane. Protein coat disintegrates, and viral DNA enters nucleus and is copied.

3. Viral DNA is transcribed into mRNA, which moves to the cytoplasm.

nucleus

DNA

mRNA

4. mRNA makes coat and envelope proteins, which enter the nucleus.

5. New viruses are assembled and bud from nucleus, acquiring an envelope from the inner nuclear membrane.

6. Newly formed viruses leave the cell by exocytosis.

(cytoplasm)

loss of coordination, dementia, and ultimately death—were similar to those of the rare but more widespread *Creutzfeldt-Jakob disease* in humans and of *scrapie*, a disease of domestic livestock. Each of these diseases typically results in brain tissue that is spongy—riddled with holes. The researchers in New Guinea eventually determined that kuru was transmitted by ritual cannibalism; members of the Fore tribe honored their dead by consuming their brains. This practice has since stopped, and kuru has virtually disappeared. Clearly, kuru was caused by an infectious agent transmitted by infected brain tissue—but what was that agent?

In 1982, the Nobel prize-winning neurologist Stanley Prusiner published evidence that scrapie (and, by extension, kuru, Creutzfeldt-Jakob disease, and a number of other, similar afflictions) is caused by an infectious agent that consists of nothing but protein. This idea seemed preposterous at the time, because most scientists

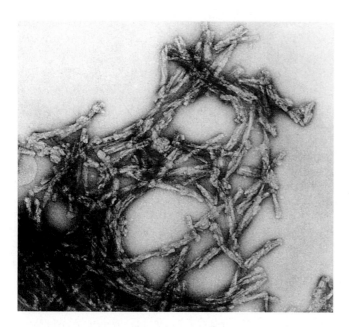

Figure 19-5 Prions: puzzling proteins
A section from the brain of a hamster infected with scrapie (a degenerative disease of the nervous system) contains fibrous clusters of prion proteins.

believed that infectious agents must contain genetic material such as DNA or RNA in order to replicate. But Prusiner and his colleagues were able to isolate the infectious agent from scrapie-infected hamsters and to demonstrate that it contained no nucleic acids. The researchers called these protein-only, infectious particles prions (Fig. 19-5).

Of current concern is the possibility that humans can be infected with *bovine spongiform encephalopathy* ("mad cow disease"), a prion disease that affects cattle, by eating beef from infected animals. Laboratory research indicates that prions could potentially infect across species boundaries, so the spread of mad cow disease to humans is a very real possibility. The likelihood that it has already happened is difficult to determine, however, because prion diseases may take decades to develop. Nonetheless, in 1996 British researchers reported 15 cases of a previously unknown variant of Creutzfeldt-Jakob disease that has since been shown to be caused by the same agent that causes mad cow disease.

How can a protein replicate itself and be infectious? Not all researchers are convinced that it's even feasible, but recent research findings have sketched the outline of a possible mechanism for replication. It turns out that prions consist of a single protein produced by normal nerve cells. Some copies of this normal protein molecule, for reasons still poorly understood, become folded into the wrong shape and are thus transformed into infectious prions. Once present, prions can apparently induce other, normal copies of the protein molecule to become trans-

formed into prions. Eventually, the concentration of prions in nerve tissue may get high enough to cause cell damage and degeneration. Why would a slight alteration to a normally benign protein turn it into a dangerous cell-killer? No one knows.

Another peculiarity of the prion diseases is that they can be inherited as well as transmitted by infection. Recent research has shown that certain small mutations in the gene that codes for the "normal" prion protein increase the likelihood that the protein will fold into its abnormal form. If one of these mutations is genetically passed on to offspring, a tendency to develop a prion disease may also be inherited.

No One Is Certain How These Infectious Particles Originated

The origin of viruses, viroids, and prions is obscure. Some scientists believe that the huge variety of mechanisms for self-replication among these particles reflects their status as evolutionary remnants of the very early history of life, before the main line of evolution settled on the more familiar large, double-stranded DNA molecules. Virus genomes in particular are amazingly diverse, encompassing forms that use DNA, RNA, or a combination of the two, in molecules than may be single-stranded or double-stranded, linear or circular. Another possibility is that viruses, viroids, and prions may be the degenerate descendants of parasitic cells. *Parasites* are organisms that live in or on host organisms, harming their hosts in the process. These ancient parasites my have been so successful at exploiting their hosts that they eventually lost the ability to synthesize the full complement of molecules required for survival and became dependent on the host's biochemical machinery. Whatever the origin of these infectious particles, their success poses a continuing challenge to living things.

2 Which Organisms Make Up the Prokaryotic Domains— Bacteria and Archaea?

Two of life's three domains (see Chapter 18) consist entirely of prokaryotes, single-celled microbes that lack organelles such as the nucleus, chloroplasts, and mitochondria. (See Chapter 6 for a comparison of prokaryotic and eukaryotic cells.) In the past, all prokaryotes were categorized as bacteria. On the basis of an improved understanding of evolutionary history, however, we now recognize two fundamentally different prokaryotic domains, **Bacteria** and **Archaea**. Both bacteria and archaea are normally very small, ranging from about 0.2 to 10 micrometers in diameter, compared with eukaryotic cells, whose diameters range from about 10 to 100 micrometers. About 250,000 average-sized bacteria or archaea could congregate on the period at the end of this sentence.

Prokaryotes Are Difficult to Classify

Bacteria and archaea are superficially similar in appearance under the microscope, but, as we shall see later, they differ strikingly in numerous structural and biochemical features, such as the structure and composition of their cell walls and plasma membranes. Within each domain, however, classification poses particular challenges to taxonomists. Because their reproduction is usually asexual, prokaryotic species cannot be defined on the basis of their ability to interbreed. In addition, the fossil record of prokaryotes is quite sparse. Consequently, taxonomists classify bacteria and archaea according to a variety of criteria: shape, means of locomotion, pigments, staining properties, nutrient requirements, sequence of DNA and RNA molecules, and the appearance of *colonies* (groups of bacteria that descend from a single cell). As mentioned in Chapter 18, bacterial diversity is almost unexplored. Although about 5000 prokaryotic species have been described, there may be 1000 times that many yet undescribed.

Bacteria Possess a Remarkable Variety of Shapes and Structures

A Cell Wall Gives Bacteria Protection and Their Characteristic Shapes

Nearly all bacteria are encased in a porous but rigid cell wall that protects them from osmotic rupture in watery environments and gives different types of bacteria their characteristic shapes. The most-common bacterial shapes are rodlike **bacilli** (singular, **bacillus**), spheres called **cocci** (singular, **coccus**), and the corkscrew-shaped **spirilla** (singular, **spirillum**) (Fig. 19-6). The cell wall contains a material called **peptidoglycan**, which is unique to bacteria. Peptidoglycan is composed of chains of sugars crosslinked by peptides (short chains of amino acids). The **Gram stain**, a staining technique, distinguishes two types of cell-wall construction in bacteria, enabling us to classify them as either *gram-positive* or *gram-negative*. The cell walls of gram-negative bacteria include an additional outer membrane resembling a plasma membrane in structure. This outer membrane is sometimes toxic to mammals, providing one mechanism by which some of these bacteria cause disease. The antibiotic penicillin works best on gram-positive bacteria.

Outside the Cell Wall, Capsules, Slime Layers, and Pili Help Bacteria Survive in Specialized Environments

Surrounding the cell walls of some bacteria are sticky **capsules** or **slime layers**, composed of polysaccharide or protein. Capsules help certain disease-causing bacteria escape detection by their host animal's immune system. Slime layers allow the bacteria that cause tooth decay to adhere in masses to the smooth surface of a tooth. This slime forms the basis of dental plaque (Fig. 19-7). Some bacteria cover themselves with a fuzz of hairlike projections called **pili** (singular, **pilus**). Pili are made of protein and generally serve to attach the bacterium to other cells. The pili of some infectious bacteria, such as those causing the venereal disease gonorrhea, attach to the cell membranes of their host, facilitating infection. Some bacteria produce special sex pili, described later.

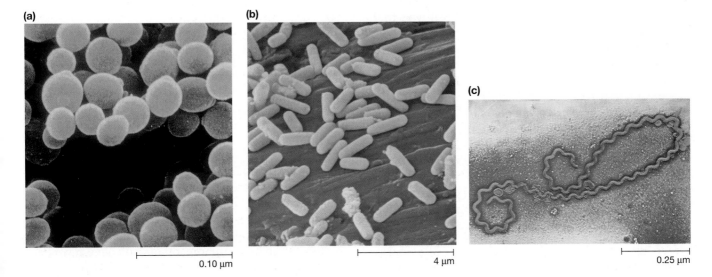

(a)   **(b)**

0.10 μm   4 μm   **(c)**   0.25 μm

Figure 19-6 Three common bacterial shapes
(a) Spherical bacteria, called cocci, of the genus *Micrococcus*; *(b)* rod-shaped bacteria, called bacilli, shown here growing on the point of a pin; and *(c)* corkscrew-shaped bacteria, called spirilla; of the species *Leptospirosis interrogans*.

Figure 19-7 The cause of tooth decay
The human body provides a safe environment for a diverse assortment of bacteria, including these bacteria that inhabit the mouth. These bacteria possess a slime layer that allows them to cling to tooth enamel, where they can cause tooth decay unless they are removed by their chief antagonist, a toothbrush (seen here as green bristles).

Some Bacteria Can Move by Using Simple Flagella

Some bacteria are equipped with **flagella** (singular, **flagellum**). These are simpler in structure than the flagella seen in some eukaryotic cells, as discussed in Chapter 6. Bacterial flagella, which may cover the cell or form a tuft at one end (Fig. 19-8a), can rotate rapidly, propelling the bacterium through its liquid environment. Recent research has revealed a unique wheel-like structure embedded in the bacterial membrane and cell wall that allows the flagellum to rotate (Fig. 19-8b). Flagella allow bacteria to disperse into new habitats, to migrate toward nutrients, and to leave unfavorable environments. Flagellated bacteria may orient toward various stimuli, a behavior called a **taxis**. Some bacteria are **chemotactic**, moving toward chemicals given off by food or away from toxic chemicals. Some are **phototactic**, moving toward or away from light, depending on the habitat they require. Other flagellated bacteria are **magnetotactic**. These forms detect Earth's magnetic field by means of tiny magnets formed from iron crystals within their cytoplasm. They use this unique sensory system to direct their beating flagella to move them downward into aquatic sediments.

Protective Endospores Allow Some Bacteria to Withstand Adverse Conditions

When environmental conditions become inhospitable, many rod-shaped bacteria form protective resting structures called **endospores**. The endospore (literally, "inside

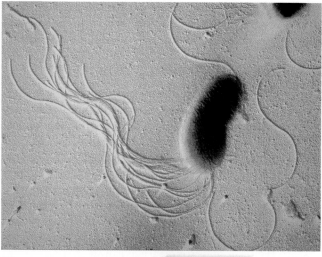

(a)

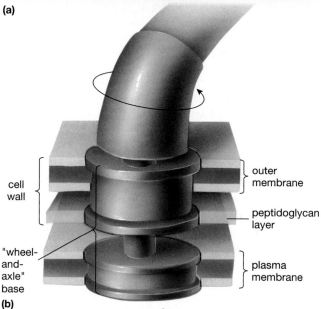

cell wall

outer membrane

peptidoglycan layer

"wheel-and-axle" base

plasma membrane

(b)

Figure 19-8 The bacterial flagellum
(a) A flagellated bacterium of the genus *Pseudomonas* uses its flagella to move toward favorable environments. **(b)** A unique "wheel-and-axle" arrangement anchors the bacterial flagellum within the cell wall and plasma membrane, allowing the flagellum to rotate rapidly.

spore") forms inside the bacterium (Fig. 19-9). It contains genetic material and a few enzymes encased within a thick protective coat. Metabolic activity ceases. Endospores are resistant structures that can survive extremely unfavorable conditions. Some can withstand boiling for an hour or more; others, still alive, have been found in the intestines of mummies 2000 years old. Endospores are important agents of bacterial dispersal, because they can be carried for long distances in air or water and then produce new bacteria rapidly when they encounter favorable conditions.

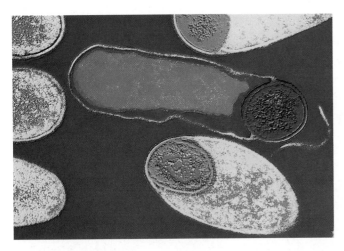

Figure 19-9 Spores protect some bacteria
Resistant endospores, here colored red, have formed inside bacteria of the genus *Clostridium*, which causes the potentially fatal food poisoning called botulism.

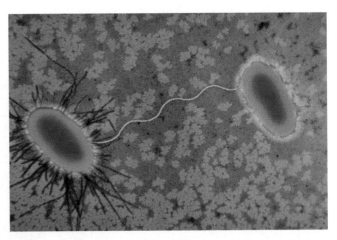

Figure 19-11 Conjugation: bacterial "mating"
Bacterial conjugation—the transfer of genetic material—occurs through a special large, hollow sex pilus, shown here connecting a pair of *Escherichia coli*. One bacterium (at top right) acts as a donor, transferring DNA to the recipient. In this photo, the donor bacterium is bristling with nonsex pili that probably help it attach to surfaces.

Bacteria Reproduce by Cell Division Called Binary Fission

Most bacteria reproduce asexually by a simple form of cell division called **binary fission** (see Chapter 11), which produces genetically identical copies of the original cell (Fig. 19-10). Under ideal conditions, a bacterium can divide about once every 20 minutes, potentially giving rise to sextillions (1×10^{21}) of offspring in a single day! This rapid reproduction allows bacteria to exploit temporary habitats such as a mud puddle or warm pudding. Recall that many mutations, the source of genetic variability,

occur as a result of mistakes in DNA replication during cell division (see Chapter 10). Thus, the rapid reproductive rate of bacteria provides ample opportunity for new forms to arise and also allows mutations that enhance survival to spread quickly. (See "Health Watch: Antibiotics—Miracle Drugs Rendered Useless?" on p. 363.)

Some bacteria transfer genetic material from a donor bacterium to a recipient bacterium during a process called **bacterial conjugation**. Some conjugating bacteria use specialized hollow *sex pili* to transfer genetic material (Fig. 19-11). The genetic material that is transferred during bacterial conjugation is located outside the single, circular bacterial chromosome, in a structure called a **plasmid**. A plasmid is a small, circular DNA molecule that may carry genes for antibiotic resistance or even alleles of genes also found on the main bacterial chromosome. Researchers in molecular genetics have made extensive use of bacterial plasmids, as described in Chapter 13. Conjugation produces new genetic combinations that may allow the resulting bacteria to survive under a greater variety of conditions.

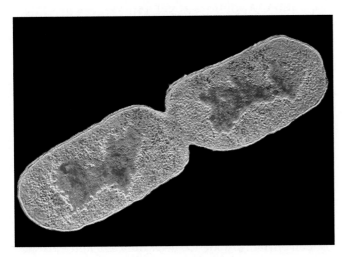

Figure 19-10 Reproduction in prokaryotes
Prokaryotic cells reproduce by a simple form of cell division called binary fission. In this color-enhanced electron micrograph, *Escherichia coli*, a normal component of the human intestine, is dividing. Red areas are genetic material.

Bacteria Are Specialized for Specific Habitats

Bacteria occupy virtually every habitat, including those where extreme conditions prevent occupation by other forms of life. For example, some bacteria thrive in very hot environments such as the near-boiling hot springs of Yellowstone National Park (Fig. 19-12) or the even hotter deep-ocean vents, where superheated water is spewed through cracks in Earth's crust at temperatures of up to 110°C (230°F). It's also pretty warm 2.8 kilometers (1.7 miles) below Earth's surface, which is where scientists

Figure 19-12 Some bacteria thrive in extreme conditions
Hot springs harbor bacteria that are both heat-tolerant and mineral-tolerant. Some cyanobacteria can tolerate temperatures up to 85°C (186°F). Several species of cyanobacteria paint these hot springs in Yellowstone National Park with vivid colors, and each is confined to a specific area determined by temperature range. The bacterial pigments aid in photosynthesis. For scale, note the footpath at upper left.

recently discovered a new bacterial species. Bacteria are also found in very cold environments, such as Antarctic sea ice that remains frozen nearly year-round. Even extreme chemical conditions fail to impede bacterial invasion. Thriving colonies of prokaryotes live in the Dead Sea, where a salt concentration seven times that of the oceans precludes all other life, and in waters that are as acidic as vinegar or as alkaline as household ammonia. Of course, rich bacterial communities also reside in a full range of more-moderate habitats, including in and on the healthy human body. An animal need not remain healthy to harbor bacteria, though. Recently, a colony of bacteria was found dormant within the intestinal contents of a mammoth that had lain in a peat bog for 11,000 years.

No single species of bacteria, however, is as versatile as these examples may suggest. In fact, bacteria are specialists. Those that inhabit deep-sea vents, for example, stop growing at temperatures below 90°C (194°F) and could thrive nowhere else. Bacteria that live on the human body are also specialized; different species colonize the skin, the mouth, the respiratory tract, the large intestine, and the urogenital tract.

Bacteria Perform Many Functions That Are Important to Other Forms of Life

Bacteria are able to colonize such diverse habitats partly because they are able to use a wide variety of nutrient sources. Some bacteria, such as the **cyanobacteria** (also known as blue-green bacteria, because *cyan* is Greek for "dark blue"; Fig. 19-13a), engage in plantlike photosynthesis. Depending on the amount and type of pigments they contain, cyanobacteria can also be purple, red, or yellow. Like green plants, cyanobacteria possess chlorophyll, produce oxygen as a by-product of photosynthesis, and can exist only where light and oxygen are available. Some cyanobacteria form chains of cells that

are unique among prokaryotes, because they have a rudimentary division of labor. In these filamentous colonies (Fig. 19-13b), a few cells capture atmospheric oxygen while the rest photosynthesize. When volcanic eruptions form new islands, cyanobacteria are among the first colonizers of the cooled lava; their activities prepare the way for more-complex but less self-sufficient organisms.

Other bacteria are **chemosynthetic**, deriving energy through reactions that combine oxygen with inorganic molecules such as sulfur, ammonia, or nitrite. In the process, they release *sulfates* or *nitrates*, crucial plant nutrients, into the soil. Many bacteria, called **anaerobes**, do not depend on oxygen to extract energy. Some, such as the bacterium that causes *tetanus*, are poisoned by oxygen. Others are opportunists, engaging in fermentation when oxygen is lacking and switching to aerobic respiration (a more efficient process) when oxygen becomes available. Anaerobes such as the sulfur bacteria obtain energy by a unique type of bacterial photosynthesis. They use hydrogen sulfide (H_2S) instead of water (H_2O) in photosynthesis, releasing sulfur instead of oxygen.

Certain bacteria have the unusual ability to break down cellulose, the principal component of plant cell walls. Some of these bacteria have entered into a **symbiotic** (literally, "living together") relationship with a group of mammals called *ruminants* (including cows, sheep, and deer). These bacteria live in the digestive tracts of ruminants and help liberate nutrients from plant fodder that the animals are unable to break down themselves. Symbiotic bacteria also inhabit your intestines. These bacteria feed on undigested food and synthesize nutrients such as vitamin K and vitamin B_{12}, which the human body absorbs. Another form of bacterial symbiosis of enormous ecological and economic importance is the growth of **nitrogen-fixing bacteria** in specialized *nodules*, small, rounded lumps on the roots of certain plants (**legumes**,

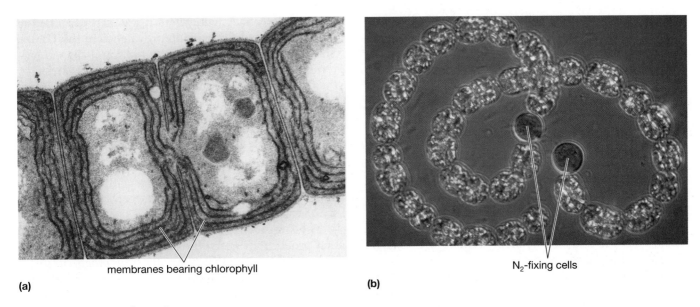

membranes bearing chlorophyll

(a)

N_2-fixing cells

(b)

Figure 19-13 Cyanobacteria
(a) Electron micrograph of a section through a cyanobacterial filament (genus *Oscillatoria*). Chlorophyll is located on the membranes visible within the cells. **(b)** Simple division of labor, rare among prokaryotic cells, is seen in this filamentous cyanobacterium (genus *Nostoc*). The larger cells are specialized for nitrogen fixation; the rest of the cells photosynthesize.

which include alfalfa, soybeans, lupines, and clover; Fig. 19-14). These bacteria capture nitrogen gas (N_2, which the plant cannot use directly) from air trapped in the soil and combine it with hydrogen to produce ammonium (NH_4^+), an important nutrient for plants. Bacteria are also important in the production of human foods, including cheese, yogurt, and sauerkraut. The aging of meat tenderizes it by means of controlled bacterial digestion.

Most bacteria are heterotrophic, obtaining energy by breaking down complex organic (carbon-containing)

(a)

(b)

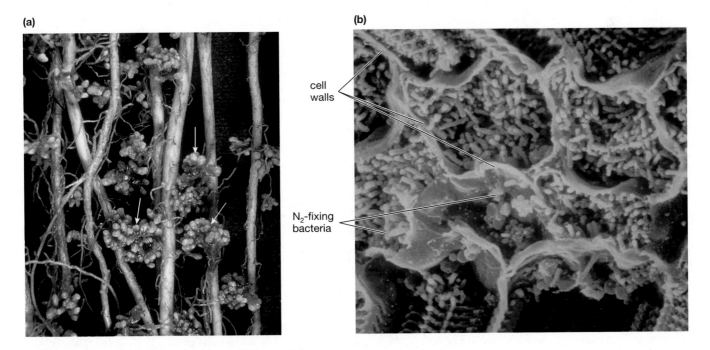

cell walls

N_2-fixing bacteria

Figure 19-14 Nitrogen-fixing bacteria in root nodules
(a) Special chambers called nodules on the roots of a legume (alfalfa) provide a protected and constant environment for nitrogen-fixing bacteria. **(b)** This scanning electron micrograph shows the nitrogen-fixing bacteria inside cells within the nodules.

molecules. The range of compounds attacked by bacteria is staggering. Nearly anything that human beings can synthesize, including detergents and the poisonous solvent benzene, some bacteria can destroy. The term *biodegradable* (meaning "broken down by living things") refers largely to the work of bacteria. Even oil is biodegradable. Soon after the tanker *Valdez* dumped 11 million gallons of crude oil into Prince William Sound, Alaska, researchers from Exxon sprayed oil-soaked beaches with a fertilizer that encouraged the growth of natural populations of oil-eating bacteria. Within 15 days the oil deposits were noticeably reduced compared with unsprayed areas. Bacteria are much less successful, however, at degrading most kinds of plastic. The search for useful, economical plastics that are also biodegradable is therefore a major focus of industrial research.

The appetite of some bacteria for nearly any organic compound is the key to their important role as decomposers in ecosystems. While feeding themselves, bacteria break down the waste products and dead bodies of plants and animals, freeing nutrients for reuse. The recycling of nutrients provides the basis for continued life on Earth, as we shall see in Chapter 40.

Some Bacteria Pose a Threat to Human Health

Despite the benefits some bacteria provide, the feeding habits of certain bacteria threaten our health and well-being, and bacterial infections are on the rise. (See "Health Watch: Antibiotics—Miracle Drugs Rendered Useless?") These **pathogenic** ("disease-producing") bacteria synthesize toxic substances that cause disease symptoms. Some bacteria, such as *Clostridium tetani* and *C. botulinum*, which cause tetanus and *botulism* (a lethal form of food poisoning), respectively, produce toxins that attack the nervous system. These bacteria are anaerobes that survive as spores until introduced into a favorable environment. A deep puncture wound, through which tetanus bacteria enter the body, also protects the bacteria from contact with oxygen. As they multiply, the bacteria release their paralyzing poison into the bloodstream. A sealed container of canned food that has been improperly sterilized provides a haven for botulism bacteria. These anaerobes produce a toxin so potent that a single gram could kill 15 million people.

The *plague*, or "Black Death," which killed 100 million people during the mid-fourteenth century, is caused by highly infectious bacteria, *Yersinia pestis*, which is spread by fleas that feed on infected rats and then move to human hosts. In 1994, an outbreak of plague occurred in India for the first time in 30 years. Tuberculosis, a bacterial disease once almost vanquished in developed countries, is again on the rise in the United States and elsewhere. Two bacterial diseases, *gonorrhea* and *syphilis*, transmitted through direct sexual contact, have reached epidemic proportions in modern society (as we shall see in Chapter 35). *Lyme disease*, named after the town of

Old Lyme, Connecticut, where it was first described in 1975, is rapidly increasing in prevalence in the United States. This disease is caused by the spiral-shaped bacterium *Borrelia burgdorferi*. The bacterium is carried by the deer tick and transmitted to humans who are bitten by the tick. At first, the symptoms resemble flu, with chills, fever, and body aches. If untreated, weeks or months later the victim may experience rashes, bouts of arthritis, and in some cases abnormalities of the heart and nervous system. Both physicians and the general public are becoming more familiar with the disease, so more victims are receiving treatment before serious symptoms develop.

The streptococcus bacterium comes in many forms that produce a variety of diseases. One type of streptococcus causes strep throat. Another, *Streptococcus pneumoniae*, causes pneumonia by stimulating an allergic reaction that clogs the lungs with fluid. Yet another form seems to have developed unusual virulence. A small percentage of people who become infected with strep bacteria experience severe symptoms, described luridly in the headlines of a 1994 British tabloid newspaper that read, "Killer Bug Ate My Face." In Britain, 11 people died from these "flesh-eating" bacteria, and a few cases were also reported in the United States. When these streptococci enter broken skin, they spew out toxins that either destroy flesh directly or stimulate an overwhelming and misdirected attack by the immune system against the body's own cells. A limb can be destroyed in hours, and in some cases only amputation can halt the rapid tissue destruction. In other cases, these rare strep infections sweep through the body, causing death within a matter of days; Jim Henson, creator of the Muppets, died as a result of such an infection.

Although some bacteria assault the human body, most are harmless and many are beneficial. For example, the normal bacterial community in the female vagina creates an environment that is hostile to infections by parasites such as yeasts. Bacteria harmlessly inhabiting our intestines are an important source of vitamin K. As the late physician, researcher, and author Lewis Thomas so aptly put it, "Pathogenicity is, in a sense, a highly skilled trade, and only a tiny minority of all the numberless tons of microbes on the Earth has ever been involved in it; most bacteria are busy with their own business, browsing and recycling the rest of life."

Archaea Are Fundamentally Different from Bacteria

Until recently, the archaea were thought to be part of the same evolutionary lineage as bacteria and were therefore inaccurately called "archaebacteria." Since the 1970s, however, it has become increasingly clear that the archaea diverged from the bacteria very early in the history of life and that the two groups are only very distantly related. Among the distinctive features of the archaea are their

Health Watch
Antibiotics—Miracle Drugs Rendered Useless?

In the early 1950s, several antibacterial drugs, including penicillin, streptomycin, and tetracycline, became available to treat bacterial infections. Penicillin was added to toothpaste, mouthwash, and chewing gum, and it was used to treat mild infections of all types. It soon became apparent that certain bacteria were becoming difficult to kill with these drugs—in other words, the pathogens had developed **antibiotic resistance**. By continuously exposing bacteria to antibiotics, humans have unwittingly introduced a strong agent of natural selection into the microbial world. Bacteria reproduce and mutate rapidly. Those bacteria whose mutations provide resistance to the effects of the drug survive, flourish, and soon dominate bacterial populations. When resistant bacteria cause disease, antibiotics are useless. To make matters worse, the genes that confer antibiotic resistance are typically carried on plasmids, which can be transferred between bacterial cells by conjugation and other processes. This bacterial ability to "share" genetic material, even across species boundaries, dramatically shortens the time necessary for drug resistance to spread within and between bacterial populations.

Antibiotic-resistant pathogens now arise with predictable, if alarming, regularity. For example, a number of foodborne pathogens have developed antibiotic resistance. A resistant strain of *Salmonella* causes a virulent form of food poisoning, and a new resistant strain of *Escherichia coli* recently sickened hundreds of Americans who consumed contaminated apple juice or hamburger meat. One likely contributor to the evolution of strains with antibiotic resistance is the widespread use of antibiotics as growth-promoting additives to animal feeds. As a result, the microbial communities inhabiting farm animals are continuously exposed to antibiotics and inevitably come to be dominated by resistant strains.

Other agents of infectious disease are also becoming increasingly resistant to antibiotics. A penicillin-resistant strain of gonorrhea has developed as a result of the regular use of penicillin as a preventive measure by prostitutes in Southeast Asia. The resistant gonorrhea bacteria are now spreading through the United States. Tetracycline and other antibiotics are also losing their effectiveness against gonorrhea, as well as against meningitis and urinary and respiratory tract infections. An especially ominous new development is the emergence of strains of tuberculosis bacterium that are resistant to most antibiotics. Once considered vanquished in developed countries, tuberculosis (TB) is now spreading, especially among the homeless and persons with AIDS. Twenty-two thousand cases of TB were reported in the United States in 1996, and most of these cases involved infection by strains with multidrug resistance. Vigorous treatment of AIDS patients with antibiotics is believed to have hastened the emergence of these deadly new strains, which typically kill 50% of the people they infect.

Most public health researchers believe that the problem of resistant pathogens is exacerbated by excessive and sometimes unnecessary medical use of antibiotics. Antibiotics are commonly prescribed inappropriately, for example, in the treatment of viral infections (such as colds) that may have symptoms similar to those of bacterial diseases, or in the treatment of minor bacterial infections from which a patient would recover without antibiotics. This overuse of antibiotics means that susceptible bacteria are under constant attack and that resistant strains have little competition.

At least one strain of bacteria is resistant to every existing antibiotic drug. Even vancomycin, the potent "drug of last resort" against the most deadly bacterial infections, has met its match in a recently identified vancomycin-resistant strain of *Enterococcus*. Researchers believe it is only a matter of time until vancomcyin resistance spreads to other, even more dangerous species of bacteria. Researchers are working to develop new antibiotics, but drug development and testing take many years, while resistant bacterial strains spread rapidly. Unfortunately, most of the new drugs are more toxic to humans and far more expensive than those they replace. Clearly we must restrict the use of antibiotics to situations in which they are urgently required rather than relying on a steady flow of new ones. Awareness and restraint may yet enable our children to benefit from some of the same "miracle drugs" that protected our parents.

---

cellular membrane lipids, which differ considerably from those of both eukaryotic and bacterial cells. The composition of archaeal cell walls and the sequence of subunits in their ribosomal RNA are also distinctive.

The domain Archaea includes species that are **methanogens**, organisms that convert carbon dioxide to methane (sometimes called "swamp gas"). Methanogens are found in such diverse habitats as swamps, sewage-treatment plants, hot springs, deep-sea vent communities (as we shall see in Chapter 41), and the stomachs of cows. Other archaea include **halophiles**, organisms that thrive in concentrated salt solutions such as the Dead Sea, and **thermoacidophiles**, which, as their name implies, live in hot, acidic environments such as hot sulfur springs. Archaea were first discovered in these environments that are hostile to most other forms of life; for many years researchers believed these species were confined to such locations. Recently, however, microbiologists have found thriving populations of archaea in ocean water samples from all over the world. Why were they never detected before? Archaea are almost impossible to grow in the laboratory, so when researchers attempted to identify marine prokaryotes from cultures grown from seawater samples, nothing but bacteria showed up. Only when researchers used molecular genetics techniques to look for archaeal RNA sequences in seawater did they find evidence of the presence and abundance of archaea.

Recent analysis of archaeal RNA nucleotide sequences has revealed that the archaea are more closely related to eukaryotes than are the bacteria. It thus appears that life split into archaeal and bacterial lineages very early in its history and that eukaryotes are derived from a later branch of the archaeal line.

3) Which Organisms Make Up the Kingdom Protista?

The third domain, the Eukarya, includes one group of microbes, the kingdom **Protista**, as well as multicellular organisms (the kingdoms Fungi, Plantae, and Animalia, which we shall discuss in Chapters 20 through 22, respectively). To appreciate the single-celled eukaryotes of the kingdom Protista, we have to overcome our size bias. Protists are largely invisible to us as we go about our daily lives. If we could somehow inhabit their microscopic scale, however, we might be more impressed with their spectacular and beautiful forms, their varied and active lifestyles, their astonishingly diverse modes of reproduction, and the variety of structural and physiological innovations that are possible within the limits of a single cell. Eukaryotic cells contain many membrane-bound organelles that prokaryotic cells lack. Organelles such as mitochondria and chloroplasts are similar in size to typical bacteria and indeed probably evolved from bacteria (see Chapter 17). Within the kingdom Protista are some of the most complex eukaryotic cells in existence, with organelles taking on functions served by organs in multicellular organisms.

Since Anton van Leeuwenhoek first observed protists through his simple homemade microscope in 1674, at least 50,000 species have been described. Although some protists form colonies, most consist of a single eukaryotic cell. Most protists can reproduce asexually by mitotic cell division, but many are also capable of a form of sexual reproduction called *conjugation* (Fig. 19-15). All three major modes of nutrition are represented in this kingdom: The unicellular algae trap solar energy through photosynthesis; predatory protists ingest their food; and parasitic forms, some flagellates, and the versatile euglenoids can absorb nutrients from their surroundings.

Protists Are a Diverse Group Including Funguslike, Plantlike, and Animal-like Forms

The kingdom Protista is an extremely diverse group, including funguslike, plantlike, and animal-like forms, which have sometimes been classified as fungi, plants, or animals. The presence of both *phyla* (a taxonomic category of animals) and *divisions* (a taxonomic category of

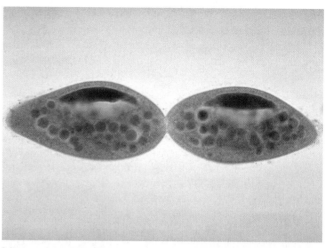

(a)

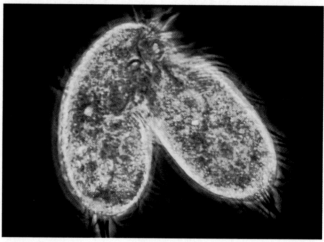

(b)

Figure 19-15 Two modes of protistan reproduction
(a) Paramecium, a ciliate, reproduces asexually by cell division that results in two daughters identical to the original parent. *(b)* Mating in a ciliate, *Euplotes*. Genetic material is exchanged across a cytoplasmic bridge. After the exchange occurs, new individuals formed by cell division will have gene combinations different from those of either parent cell.

plants) within this kingdom reflects the difficulty of classifying protists. To the confusion of taxonomists, many protists (such as the euglenoids; see Fig. 19-22) fit equally well into animal-like or plantlike categories. In fact, many protist groups are not closely related to any of the multicellular kingdoms but have been evolving independently for a long time. If we used the genetic differences among fungi, plants, and animals as a guide, we'd recognize at least a dozen kingdoms of protists. The difficulty of untangling the evolutionary history of protists means that protistan taxonomy remains subject to revision and controversy. Here we discuss members of the kingdom Protista in three categories: the funguslike slime molds and water molds, the plantlike unicellular algae, and the animal-like protozoa (Table 19-1).

Table 19-1 The Major Groups of Protists

General Category	Division/Phylum	Locomotion	Nutrition	Representative Features	Representative Genus
Plantlike protists: unicellular algae	Dinoflagellates (Division Pyrrophyta)	Swim with two flagella	Autotrophic; photosynthetic	Many bioluminescent; often have cellulose wall; most marine	*Gonyaulax* (causes red tide)
	Diatoms (Division Chrysophyta)	Glide along surfaces	Autotrophic; photosynthetic	Have silica shells; most marine	*Navicula* (glides toward light)
	Euglenoids (Division Euglenophyta)	Swim with one flagellum	Autotrophic; photosynthetic	Have an eyespot; all freshwater	*Euglena* (common (pond-dweller)
Funguslike protists: water molds and slime molds	Water molds	Swim with flagella (gametes)	Heterotrophic	Filamentous bodies	*Plasmopara* (causes downy mildew)
	Acellular (plasmodial) slime molds (Division Myxomycota)	Sluglike mass oozes over surfaces	Heterotrophic	Form multinucleate plasmodium	*Physarum* (forms a large bright orange mass)
	Cellular slime molds (Division Acrasiomycota)	Amoeboid cells extend pseudopodia; sluglike mass crawls over surfaces	Heterotrophic	Form pseudoplasmodium with individual amoeboid cells	*Dictyostelium* (often used in laboratory studies)
Animal-like protists: protozoa	Zooflagellates (Phylum Sarcomastigophora)	Swim with flagella	Heterotrophic	Inhabit soil or water or may be parasitic	*Trypanosoma* (causes African sleeping sickness)
	Sarcodines (Phylum Sarcomastigophora)	Extend pseudopodia	Heterotrophic	Both naked and shelled forms exist	*Amoeba* (common pond-dweller)
	Sporozoans (Phylum Apicomplexa)	Nonmotile	Heterotrophic; all parasitic	Form infectious spores	*Plasmodium* (causes malaria)
	Ciliates (Phylum Ciliophora)	Swim with cilia	Heterotrophic	Most complex single cells	*Paramecium* (fast-moving pond-dweller)

The Water Molds and Slime Molds Are Funguslike Protists

Some protists are characterized by physical similarity to the filaments or fruiting bodies of fungi and have a funguslike mode of nutrition. Like fungi, they absorb nutrients from the soil, water, or tissues of other organisms, and typically help decompose dead organisms. The funguslike protists form three groups: (1) the water molds, (2) the acellular slime molds, and (3) the cellular slime molds.

The Water Molds Have Had Important Impacts On Humans

The **water molds**, or *oomycetes*, form a small division (Oomycota) of filamentous protists that includes both inoffensive species that live in water and damp soil and some species of profound economic importance. For example, a water mold causes the disease known as downy mildew of grapes (Fig. 19-16). Its inadvertent introduction into France from the United States in the late 1870s nearly destroyed the French wine industry. Another oomycete has destroyed millions of avocado trees in California; still another is responsible for *late blight*, a devastating disease of potatoes. When accidentally introduced into Ireland about 1845, this protist destroyed nearly the entire potato crop, causing the devastating potato famine during which a million people in Ireland starved and many more emigrated to the United States.

Figure 19-16 A parasitic water mold
Downy mildew, a plant disease caused by the water mold *Plasmopara*, nearly destroyed the French wine industry in the 1870s.

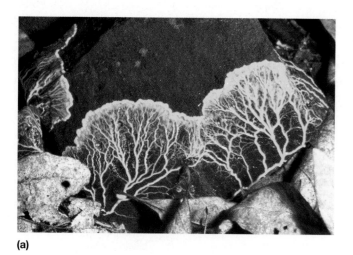

(a)

(b)

Figure 19-17 The acellular slime mold **Physarum**
(a) Physarum oozes over a stone on the damp forest floor. *(b)* When food becomes scarce, the mass differentiates into black fruiting bodies in which spores are formed.

Sexual reproduction in water molds involves the fertilization of a large egg cell, hence the name Oomycota, which means "egg fungi." (These protists were once thought to be fungi.) Water molds also reproduce asexually by means of actively swimming flagellated cells called **zoospores**. The swimming spores require very moist conditions.

The Acellular Slime Molds Form a Multinucleate Mass of Cytoplasm Called a Plasmodium

The life cycle of the *slime mold* consists of two phases: a mobile feeding stage and a stationary reproductive stage called a **fruiting body**. The **acellular**, or **plasmodial**, **slime molds** of the division Myxomycota consist of a mass of cytoplasm that may spread thinly over an area of several square meters. Although the mass contains thousands of diploid nuclei, the nuclei are not confined in discrete cells surrounded by plasma membranes, as in most multicellular organisms. This structure, called a **plasmodium**, explains why these protists are described as "acellular" (without cells). The plasmodium oozes through decaying leaves and rotting logs, engulfing food such as bacteria and particles of organic material. The mass may be bright yellow or orange—a large plasmodium can be rather startling (Fig. 19-17a). Dry conditions or starvation stimulate the plasmodium to form a fruiting body, on which haploid spores are produced (Fig. 19-17b). The spores are dispersed and germinate under favorable conditions, eventually giving rise to a new plasmodium.

The Cellular Slime Molds Live as Independent Cells but Aggregate into a Pseudoplasmodium When Food Is Scarce

The **cellular slime molds** of the division Acrasiomycota live in soil as independent haploid cells that move and

feed by producing extensions called **pseudopods** (literally, "false feet"). These extensions surround and engulf food such as bacteria. In the best-studied genus, *Dictyostelium*, individual cells release a chemical signal when food becomes scarce. This signal attracts nearby cells into a dense aggregation that forms a sluglike mass called a **pseudoplasmodium** ("false plasmodium") because it actually consists of individual cells (Fig. 19-18). The pseudoplasmodium then behaves like a multicellular organism. After crawling toward a source of light, the cells in the aggregation take on specific roles, forming a fruiting body. Haploid spores formed within the fruiting body are dispersed by wind and germinate directly into new amoeboid individuals.

The Unicellular Algae, or Phytoplankton, Are Plantlike Protists

Often called **phytoplankton** (literally, "floating plants"), these photosynthetic protists are widely distributed in oceans and lakes. Although they are microscopic, phytoplankton are immensely important. Marine phytoplankton account for nearly 70% of all the photosynthetic activity on Earth, absorbing carbon dioxide, recharging the atmosphere with oxygen, and supporting the complex web of aquatic life. There are three major divisions of plantlike protists: (1) the division Pyrrophyta, also called dinoflagellates; (2) the division Chrysophyta, the diatoms; and (3) the division Euglenophyta, the euglenoids.

Dinoflagellates Swim by Means of Two Whiplike Flagella

Dinoflagellates are named for the motion created by their two whiplike flagella (*dino* is Greek for "whirlpool").

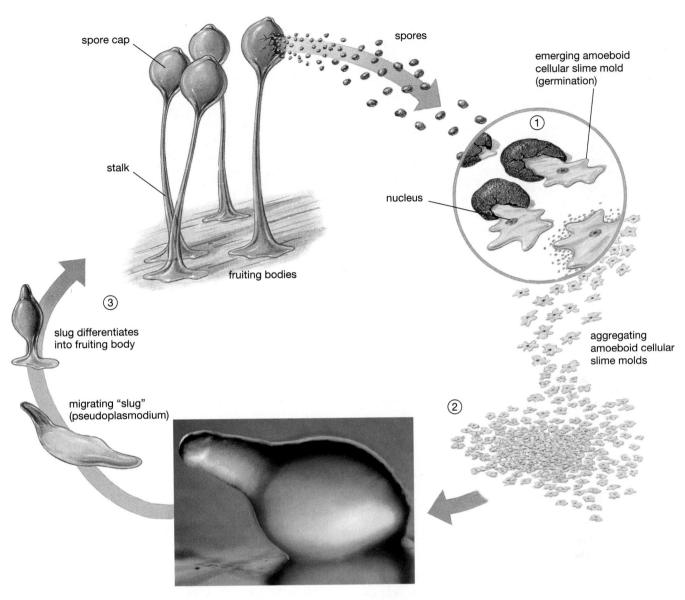

Figure 19-18 *The life cycle of a cellular slime mold*
① Single amoebalike cells emerge from spores, crawl, and feed. ② When food becomes scarce, they aggregate into a sluglike mass called a pseudoplasmodium (inset). ③ The pseudoplasmodium migrates toward the light and then forms a fruiting body, in which haploid spores are produced.

One flagellum encircles the cell, and the second projects behind it. Some dinoflagellates are covered only by a cell membrane; others have cellulose walls that resemble armor plates (Fig. 19-19). Although some types live in fresh water, dinoflagellates are especially abundant in the ocean, where they are an important food source for larger organisms. Many dinoflagellates are bioluminescent, producing a brilliant blue-green light when disturbed. Specialized dinoflagellates known as *zooxanthellae* live within the tissues of corals, some clams and even other protists, where the algae provide photosynthetic nutrients and remove carbon dioxide. Reef-building coral are found only in shallow, well-lit waters in which their zooxanthellae can survive.

The green chlorophyll in dinoflagellates is typically masked by red pigments that help trap light energy. When the water is warm and rich in nutrients, a population explosion occurs. Dinoflagellates can become so numerous that the water is dyed red by the color of their bodies, causing a "red tide" (Fig. 19-20). During red tides, fish die by the thousands, suffocated by clogged gills or by oxygen depletion as a result of the decay of the bodies of billions of dinoflagellates. But oysters, mussels, and clams have a feast, filtering millions from the water for food. In the process, however, they concentrate a nerve poison the dinoflagellates produce in their bodies. Humans who eat these mollusks may be stricken with potentially lethal paralytic shellfish poisoning.

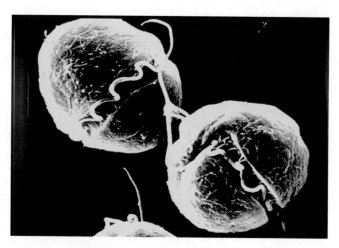

Figure 19-19 Dinoflagellates
Two dinoflagellates covered with protective cellulose armor.
Two flagella lie within the grooves that encircle the body.

Figure 19-21 Some representative diatoms
This photomicrograph illustrates the intricate, microscopic
beauty and variety of the glassy walls of diatoms.

Figure 19-20 A red tide
The explosive reproductive rate of certain dinoflagellates
under the right environmental conditions can produce dinoflagellate concentrations so great that their microscopic
bodies dye the seawater red or brown, as in this bay in
Mexico.

Diatoms Encase Themselves within Glassy Walls

The **diatoms**, photosynthesizers found in both fresh and
salt water, are so important to marine food webs that they
have been called the "pastures of the sea." They produce
protective shells of *silica* (glass), some of exceptional
beauty (Fig 19-21). These shells consist of top and bottom
halves that fit together like a pillbox or petri dish. Accumulations of the glassy walls of diatoms over millions of
years have produced fossil deposits of "diatomaceous
earth" that may be hundreds of meters thick. This slightly abrasive substance is widely used in products such as
toothpaste and metal polish. Diatoms store reserve food
as oil; their buoyancy in water helps their bodies float near
the surface, where light for photosynthesis is abundant.
Prehistoric accumulations of diatoms and their stored oil
might have contributed to today's petroleum reserves.

Euglenoids Lack a Rigid Covering and Swim by Means of Flagella

The **euglenoids** are named after the group's best-known
representative, *Euglena* (Fig. 19-22), a complex single cell
that moves about by whipping its flagellum through
water. Its simple light-sensing organelles consist of a photoreceptor, called an *eyespot*, at the base of the flagellum
and an adjacent patch of pigment. The pigment shades
the photoreceptor only when light strikes from certain directions, enabling *Euglena* to determine the direction of
the light source. Using information from the photoreceptor, the flagellum propels the protist toward light levels
appropriate for photosynthesis. Most euglenoids live in
fresh water. Unlike diatoms or dinoflagellates, they lack
a rigid outer covering, so some can move by wriggling as
well as by whipping the flagellum. If *Euglena* is maintained in darkness, it loses its chloroplasts but can still absorb nutrients from its surroundings. In this state it closely
resembles the animal-like zooflagellates described later.

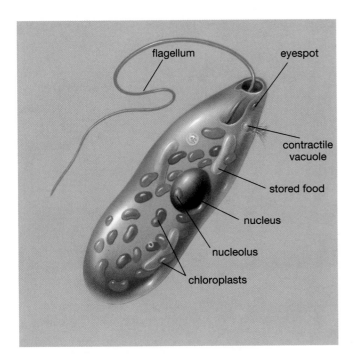

Figure 19-22 Euglena, *a representative euglenoid*
Euglena's elaborate, single cell is packed with green chloroplasts, which will disappear if the protist is kept in darkness.

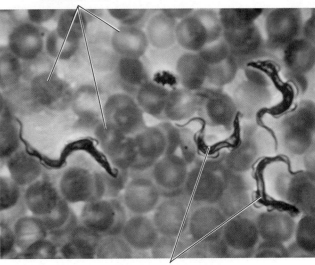

red blood cells

Trypanosoma parasites

Figure 19-23 A *disease-causing zooflagellate*
This photomicrograph shows human blood that is heavily infested with the parasitic zooflagellate *Trypanosoma*, which causes African sleeping sickness.

The Protozoa Are Animal-like Protists

The **protozoa** (literally, "first animals") are described as animal-like because they can move and they obtain their food from other organisms. They are placed into three major phyla: (1) the phylum Sarcomastigophora, which includes the zooflagellates and sarcodines (amoebae); (2) the phylum Apicomplexa, the sporozoans; and (3) the phylum Ciliophora, the ciliates. All are unicellular, eukaryotic, and heterotrophic; they differ in methods of locomotion.

Zooflagellates Possess a Single Versatile Flagellum

All **zooflagellates** possess at least one flagellum. This versatile organelle may propel the organism, sense the environment, or ensnare food. The zooflagellates are a diverse group believed to be ancestral to the other protists. Many are free living, inhabiting soil and water; others are symbiotic, living inside other organisms in a relationship that may be either mutually beneficial or parasitic. One symbiotic form can digest cellulose and lives in the gut of termites, where it helps them extract energy from wood. A zooflagellate of the genus *Trypanosoma* is responsible for African sleeping sickness, a potentially fatal disease (Fig. 19-23). Like many parasites, this organism has a complex life cycle, part of which is spent in the tsetse fly. While feeding on the blood of a mammal, the fly transmits the trypanosome to the mammal. The parasite then develops in the new host (which may be a human) and enters the bloodstream. It may then be ingested by another tsetse fly that bites the host, thus beginning a new cycle of infection.

Another parasitic zooflagellate, *Giardia*, is an increasing problem in the United States, particularly to hikers who drink from apparently pure mountain streams. *Cysts* (tough resting structures) of this flagellate are released in the feces of infected humans, dogs, or other animals (a single gram of feces may contain 300 million cysts) and enter freshwater streams and even community reservoirs. Cysts develop into the adult form (Fig. 19-24) in the small intestine of their mammalian host. In humans, infections can cause severe diarrhea, dehydration, nausea, vomiting, and cramps. Fortunately, these

Figure 19-24 Giardia: *the curse of campers*
A zooflagellate (genus *Giardia*) that may infect drinking water, causing gastrointestinal disorders, shown here in the human small intestine.

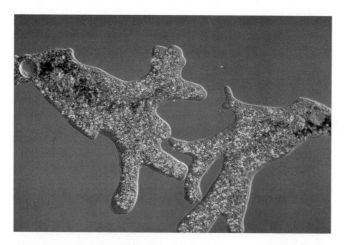

Figure 19-25 *The amoeba*
An amoeba uses cytoplasmic projections called pseudopodia to move about and to capture prey.

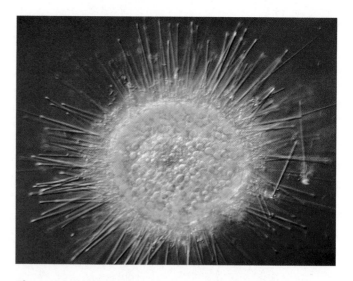

Figure 19-26 *Heliozoans*
Heliozoans are beautiful freshwater sarcodines; the needle-like pseudopodia are clearly visible in this specimen (genus *Acanthocystis*).

infections can be cured with drugs, and deaths from *Giardia* infections are uncommon.

Sarcodines, Including the Amoebae, Move by Means of Pseudopods

Sarcodines possess flexible plasma membranes that they can extend in any direction to form pseudopodia, which are used for locomotion and for engulfing food (Fig. 19-25). One parasitic form causes amoebic dysentery, a particular problem in warm climates. Multiplying in the intestinal wall, this parasite causes severe diarrhea and may pierce the intestine, occasionally causing fatal infections.

Sarcodines called **amoebae** are common in freshwater lakes and ponds. Amoebae lack many of the specialized organelles found in flagellates and ciliates, but they have a complex internal structure and sophisticated ability to sense and capture prey. **Heliozoans** ("sun animals"), a striking form of freshwater sarcodine, may be found floating in ponds or attached by stalks to an underwater plant or rock (Fig. 19-26). They have stiff, needle-like pseudopodia, each of which is supported internally by a bundle of microtubules. Some heliozoans cover themselves with intricate and delicate shells of silica. The **foraminiferans** and **radiolarians** are primarily marine sarcodines that also produce beautiful and elaborate shells. The shells of foraminiferans are constructed mostly of calcium carbonate (chalk; Fig. 19-27); those of radiolarians are of silica (Fig. 19-27b). These elaborate shells are

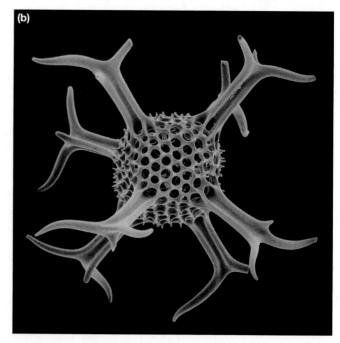

Figure 19-27 *Foraminiferans and radiolarians*
(a) The chalky shells of foraminiferans show numerous interior chambers. *(b)* The delicate, glassy shells of radiolarians. Pseudopodia, which sense the environment and capture food, extend out through the openings in the shells.

pierced by myriad openings through which pseudopods extend. The chalky shells of foraminiferans, accumulating over millions of years, have resulted in immense deposits of limestone such as those that form the famous White Cliffs of Dover, England.

Sporozoans Are Parasitic and Have No Means of Locomotion

All **sporozoans** are parasitic, living inside the bodies and sometimes inside the individual cells of their hosts. They are named after their ability to form infectious spores, resistant structures transmitted from one host to anoth-er through food, water, or the bite of an infected insect. As adults, sporozoans have no means of locomotion. Many have complex life cycles, a common feature of parasites. A well-known example is the malarial parasite *Plasmodium* (Fig. 19-28). Parts of its life cycle are spent in the stomach, and later the salivary glands, of the female *Anopheles* mosquito. When the mosquito bites a human, it passes the *Plasmodium* to the unfortunate victim. The sporozoan develops in the liver, then enters the blood, where it reproduces rapidly in red blood cells. The release of large quantities of spores through the rupture of the blood cells causes the recurrent fever of malaria.

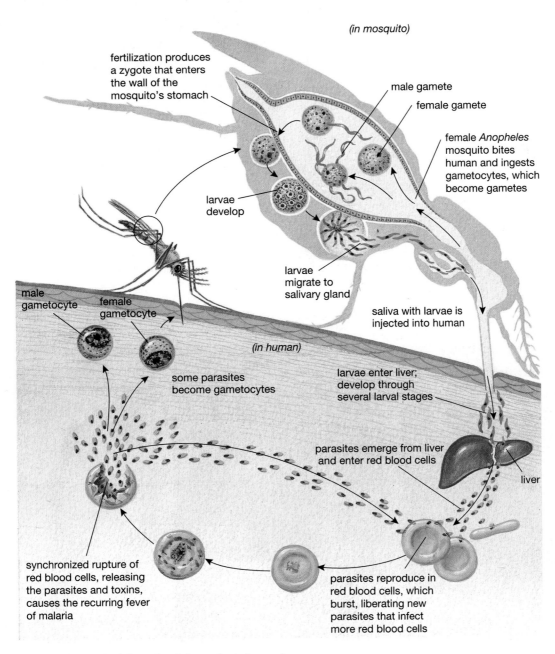

Figure 19-28 *The life cycle of the malarial parasite*
In the complex life cycle of the malarial parasite *Plasmodium*, humans and mosquitoes serve as alternate hosts.

Uninfected mosquitoes may acquire the parasite by feeding on the blood of a malaria victim, spreading the parasite when it bites another person, thus continuing the infectious cycle.

Although the drug chloroquine kills the malarial parasite, drug-resistant populations of *Plasmodium* are, unfortunately, spreading rapidly throughout Africa, where the disease is prevalent. Programs to eradicate mosquitoes have failed because the mosquitoes rapidly evolve resistance to pesticides. Researchers hope to use genetic engineering to breed strains of mosquito that kill, rather than transmit, the sporozoan parasite.

Ciliates, Named for Their Locomotory Cilia, Are the Most Complex of the Protozoa

Ciliates, which inhabit fresh or salt water, represent the peak of unicellular complexity. They possess many specialized organelles, including the **cilia** (singular, **cilium**), the short hairlike outgrowths after which they are named. The cilia may cover the cell, or they may be localized. In the well-known freshwater genus *Paramecium* (Fig. 19-29), rows of cilia cover the entire body surface. Their coordinated beating propels the cell through the water at a protistan speed record of a millimeter per second. Although only a single cell, *Paramecium* responds to its environment as if it had a well-developed nervous system. Confronted with some noxious chemical or with a physical barrier, the cell immediately backs up by reversing the beating of its cilia and then proceeds in a new direction.

Ciliates are accomplished predators (Fig. 19-30). Some, including *Paramecium* and *Didinium*, immobilize their

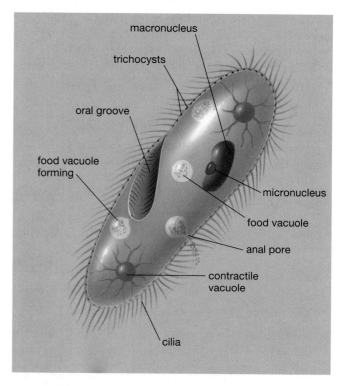

Figure 19-29 The complexity of ciliates
The ciliate *Paramecium* illustrates some important ciliate organelles. The oral groove acts as a mouth, and food vacuoles, miniature digestive systems, form at its apex; waste is expelled by exocytosis through an anal pore. The contractile vacuoles regulate water balance. Trichocysts help this predator stun its prey, whereas cilia propel it rapidly through the water.

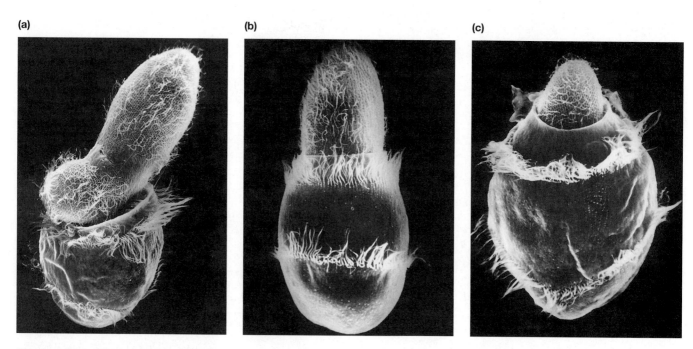

Figure 19-30 Microscopic predator
In this series of scanning electron micrographs, *Paramecium*, with cilia all over its body, is *(a)* stung and *(b,c)* gradually engulfed by another predatory ciliate, *Didinium*, whose cilia are confined to two bands that encircle the egg-shaped body. This microscopic drama could occur on a pinpoint with room to spare.

prey with explosive darts called **trichocysts** embedded in the outer covering of the cell. Prey is escorted to a mouth-like opening, the *oral groove*. It is digested in a food vacuole, which forms a temporary "stomach," and is excreted by exocytosis. Excess water is accumulated in a contractile vacuole, which periodically contracts, emptying the fluid to the outside through a hole called the *anal pore*.

Evolutionary Connections
Our Unicellular Ancestors

Today's microbes include the types of cellular organization that gave rise to the complex multicellular organisms with which we are most familiar. In external appearance, modern prokaryotes have changed little from their ancestors, whose fossilized remains date back 3.5 billion years. Life might still consist solely of prokaryotic single cells if the protists, with their radical eukaryotic design, had not appeared on the scene nearly 2 billion years ago. As you learned in the discussion of the endosymbiont theory in Chapter 17, eukaryotic cells may have originated when one prokaryote, perhaps a bacterium capable of aerobic respiration, took up residence inside a partner, forming the first "mitochondrion." A separate but equally crucial merger may have occurred when a photosynthetic bacterium (probably resembling a cyanobacterium) took up residence within a nonphotosynthetic partner and became the first "chloroplast." The foundations of multicellularity were laid with the eukaryotic cell, whose intricacy allowed specialization of entire cells for specific functions within a multicellular aggregation. Thus, primitive protists, some absorbing nutrients from the environment, some photosynthesizing, and others consuming their food in chunks, almost certainly followed divergent evolutionary paths that led to the three multicellular kingdoms—the fungi, plants, and animals—which are the subjects of the following three chapters.

Summary of Key Concepts

Microorganisms are classified in three domains: Archaea, Bacteria, and Eukarya. Archaea and Bacteria consist of tiny prokaryotic cells that lack organelles such as nuclei, mitochondria, and chloroplasts. Eukarya contains organisms with eukaryotic cells that possess the full range of organelles and resemble the cells of multicellular organisms. Within Eukarya, a diverse assemblage of single-celled species forms the kingdom Protista. Without the photosynthetic, nitrogen-trapping, and decomposing abilities of the prokaryotes and the photosynthetic activities of protists, life as we know it would grind to a halt.

1) What Are Viruses, Viroids, and Prions?

Viruses are parasites consisting of a protein coat that surrounds genetic material. They are noncellular and unable to move, grow, or reproduce outside a living cell. They invade cells of a specific host and use the host cell's energy, enzymes, and ribosomes to produce more virus particles, which are liberated when the cell ruptures. Many viruses are pathogenic to humans, including those causing colds and flu, herpes, AIDS, and certain forms of cancer.

Viroids are short strands of RNA that can invade a host cell's nucleus and direct the synthesis of new viroids. To date, viroids are known to cause only certain diseases of plants.

Prions have been implicated in diseases of the nervous system, such as kuru, Creutzfeld-Jacob disease, and scrapie. Prions are unique in that they lack genetic material. They are composed solely of mutated prion protein, which may act as an enzyme, catalyzing the formation of more prions from normal prion protein.

2) Which Organisms Make Up the Prokaryotic Domains—Bacteria and Archaea?

Members of the domains Bacteria and Archaea—the bacteria and archaea—are unicellular and prokaryotic. Among the bacteria, a cell wall of peptidoglycan determines their characteristic shape: coccus (round), bacillus (rodlike), or spiral. Many bacteria form spores that disperse widely and withstand inhospitable environmental conditions. Bacteria obtain energy in a variety of ways. Some, including the cyanobacteria, rely on photosynthesis. Others are chemosynthetic, breaking down inorganic molecules to obtain energy. Heterotrophic forms are capable of consuming a wide variety of organic compounds. Many are anaerobic, able to obtain energy from fermentation when oxygen is not available.

Some bacteria are pathogenic, causing disorders such as pneumonia, tetanus, botulism, and the venereal diseases gonorrhea and syphilis. Most bacteria, however, are harmless to humans and play important roles in natural ecosystems. Bacteria have colonized nearly every habitat on Earth. Some live in the digestive tracts of larger organisms such as cows and sheep, where these bacteria break down cellulose. Nitrogen-fixing bacteria enrich the soil and aid in plant growth; many others live off the dead bodies and wastes of other organisms, liberating nutrients for reuse.

The archaea are a unique and diverse group that flourish under extreme conditions, including hot, acidic, very salty, and anaerobic environments. They differ in several ways from bacteria, including cell wall composition, ribosomal RNA sequence, and membrane lipid structure.

3) Which Organisms Make Up the Kingdom Protista?

The kingdom Protista consists of organisms composed of single, highly complex eukaryotic cells. They are classified as funguslike, plantlike, and animal-like.

The funguslike acellular (plasmodial) slime molds form a multinucleate plasmodium that crawls in amoeboid fashion, ingesting decaying organic matter. Drought or starvation stimulates the formation of a fruiting body on which spores are formed. Cellular slime molds exist as independent amoeboid cells. Under adverse conditions, these cells aggregate in response to a chemical signal and form a pseudoplasmodium that differentiates into a spore-forming fruiting body.

The plantlike unicellular algae are important photosynthetic organisms in marine and freshwater ecosystems. They include dinoflagellates, diatoms, and the exclusively freshwater euglenoids.

Protozoa are nonphotosynthetic protists that absorb or ingest their food. They are widely distributed in soil and water; some are parasitic. They include the zooflagellates, the amoeboid sarcodines, the parasitic sporozoans, and the predatory ciliates.

Key Terms

acellular slime mold   *p. 366*
amoeba   *p. 370*
anaerobe   *p. 360*
antibiotic resistance   *p. 363*
Archaea   *p. 356*
bacillus   *p. 357*
Bacteria   *p. 356*
bacterial conjugation   *p. 359*
bacteriophage   *p. 352*
binary fission   *p. 359*
capsule   *p. 357*
cellular slime mold   *p. 366*
chemosynthetic   *p. 360*
chemotactic   *p. 358*
ciliate   *p. 372*
cilium   *p. 372*

coccus   *p. 357*
cyanobacterium   *p. 360*
diatom   *p. 368*
dinoflagellate   *p. 366*
endospore   *p. 358*
euglenoid   *p. 368*
flagellum   *p. 358*
foraminiferan   *p. 370*
fruiting body   *p. 366*
Gram stain   *p. 357*
halophile   *p. 363*
heliozoan   *p. 370*
host   *p. 351*
kuru   *p. 354*
legume   *p. 360*
magnetotactic   *p. 358*

methanogen   *p. 363*
nitrogen-fixing bacterium
   p. 360
pathogenic   *p. 362*
peptidoglycan   *p. 357*
phototactic   *p. 358*
phytoplankton   *p. 366*
pilus   *p. 357*
plasmid   *p. 359*
plasmodial slime mold
   p. 366
plasmodium   *p. 366*
prion   *p. 354*
Protista   *p. 364*
protozoan   *p. 369*
pseudoplasmodium   *p. 366*

pseudopod   *p. 366*
radiolarian   *p. 370*
sarcodine   *p. 370*
slime layer   *p. 357*
spirillum   *p. 357*
sporozoan   *p. 371*
symbiotic   *p. 360*
taxis   *p. 358*
thermoacidophile   *p. 363*
trichocyst   *p. 373*
viroid   *p. 354*
virus   *p. 351*
water mold   *p. 365*
zooflagellate   *p. 369*
zoospore   *p. 366*

Thinking Through the Concepts

Multiple Choice

1. *Which of the following is true?*
   a. Viruses cannot reproduce outside a host cell.
   b. Prions are infectious proteins.
   c. Viroids lack a protein coat.
   d. Some viruses cause cancer.
   e. all of the above

2. *Which structure is used to transfer genetic material between bacteria?*
   a. flagellum   b. pilus   c. peptidoglycan
   d. spore   e. capsule

3. *Most pathogenic bacteria cause disease by*
   a. directly destroying individual cells of the host.
   b. fixing nitrogen and depriving the host of this nutrient.
   c. producing toxins that disrupt normal functions.
   d. depleting the energy supply of the host.
   e. depriving the host of oxygen.

4. *Cyanobacteria*
   a. have chlorophyll   b. are chemosynthetic
   c. can live without oxygen   d. are archaea
   e. all of the above

5. *Which organisms are sometimes called the "pastures of the sea"?*
   a. dinoflagellates   b. foraminiferans
   c. diatoms   d. radiolarians
   e. amoebae

6. *Which of the following pairs of organism and disease is incorrect?*
   a. sporozoan: malaria   b. prion: kuru
   c. bacterium: syphilis   d. archaean: gonorrhea
   e. virus: AIDS

? Review Questions

1. Describe the structure of a typical virus. How do viruses replicate?

2. List the major differences between prokaryotes and protists.

3. Describe some of the ways in which bacteria obtain energy and nutrients.

4. What are nitrogen-fixing bacteria, and what role do they play in ecosystems?

5. What is an endospore? What is its function?

6. Why do bacteria readily become resistant to antibiotics, and what steps can be taken to prevent this situation?

7. Describe some examples of bacterial symbiosis.

8. What is the importance of dinoflagellates in marine ecosystems? What happens when they reproduce rapidly?

9. What is the major ecological role played by unicellular algae?

10. What protozoan group consists entirely of parasitic forms?

11. Describe the life cycle and mode of transmission of the malarial parasite.

12. Which protozoan group's shells form the white limestone cliffs of Dover, England?

13. How do sarcodines feed? What disease is caused by a parasitic sarcodine?

14. In which protozoan group are the most complex single cells found? How do these cells feed?

15. Name three divisions of plantlike protists, three of animallike protists, and two of funguslike protists, and give an example of each.

Applying the Concepts

1. In some developing countries, antibiotics can be purchased without a prescription. Why do you think this is done? What biological consequences would you predict?

2. Before the discovery of prions, many (perhaps most) biologists would have agreed with the statement, "It is a fact that no infectious organism or particle can exist that lacks nucleic acid (such as DNA or RNA)." What lessons do prions have to teach us about nature, science, and scientific inquiry? You may wish to review Chapter 1 to help answer this question.

3. Recent research shows that ocean water off southern California has warmed by 2–3°F (1–1.5°C) over the past four decades, possibly due to the greenhouse effect. This warming has led indirectly to a depletion of nutrients in the water and thus a decline in photosynthetic protists such as diatoms. What effects is this warming likely to have for the life in the oceans?

4. Argue for and against the statement, "Viruses are alive."

5. The internal structure of many protists is much more complex than that of cells of multicellular organisms. Does this mean that the protist is engaged in more complex activities than the multicellular organism? If not, why should the protistan cell be much more complicated?

6. Why would the lives of multicellular animals be impossible if prokaryotic and protistan organisms did not exist?

Group Activity

Form a group of four students. Imagine that you're working in a laboratory at a remote field station. Someone hands you a vial of microscopic organisms and asks you to determine if the organisms in it are bacteria, algae, or protozoans. Unfortunately, your only microscope is broken. Fortunately, the lab is well stocked with everything you need to do chemical tests for the presence of particular molecules and to break cells down into their components. Discuss your research goals, and prepare a list of at least four tests that you'll perform on the sample to allow you to determine its nature. Then prepare a table of the test results that you would expect for bacteria, algae, and protozoans.

For More Information

Diamond, J. "The Return of Cholera." *Discover*, March 1992. The cholera bacterium is causing epidemics in some developing countries.

Dixon, B. "Overdosing on Wonder Drugs." *Science 86*, May 1986. The dangers of overuse of antibiotics are explored.

Greene, W. C. "AIDS and the Immune System." *Scientific American*, September 1993. Studies of the life cycle of the AIDS virus suggest possible avenues for treatment.

Knoll, A. "Life's Expanding Realm." *Natural History*, June 1994. The role of cyanobacteria in modifying Earth's surface and their impact on the evolution of life on Earth.

Madigan, M., and Marrs, B. "Extremophiles." *Scientific American*, April 1997. Prokaryotes that prosper under extreme conditions and potential industrial uses of the enzymes that they use to do so.

McEvedy, C. "The Bubonic Plague." *Scientific American*, February 1988. The story of the bacterium that killed nearly one-third of the world's population in the fourteenth century.

Prusiner, S. "The Prion Diseases." *Scientific American*, January 1995. A description of prions and the research that led to their *Discovery*, from the point of view of the most influential scientist in the field.

Simons, K., Garoff, H., and Helenius, A. "How an Animal Virus Gets Into and Out of Its Host Cell." *Scientific American*, February 1982. (Offprint No. 1511).

Woese, C. "Archaebacteria." *Scientific American*, June 1981. An introduction to what is now known as the domain Archaea, written by a scientist who's at the forefront of efforts to understand these distinctive prokaryotes.

Zimmer, C. "Triumph of the Archaea." *Discover*, February 1995. An account of how the archaea, once thought to be a small and minor group, has been shown to be astonishingly diverse and widespread.

Answers to Multiple-Choice Questions
1. e 2. b 3. c 4. a 5. c 6. d

A ripe puffball touched by rain, a breeze, or a scurrying animal releases clouds of spores that are dispersed far and wide on air currents. Here, a rounded earthstar puffball emits spores when hit by a raindrop.

The Fungi 20

What do mushroom pizza, penicillin, wine, bread, mildew, and athlete's foot have in common? All are made possible by the activities of fungi. Many members of the kingdom Fungi have been harnessed by humans for medical miracles or culinary enjoyment. Other fungi have turned the tables—they feast on our crops, building materials, and bodily tissues.

Whether friend, foe, or completely indifferent to humans, all fungi share properties that, taken together, separate them from denizens of the other kingdoms. For example, fungi eat by secreting enzymes that digest food outside their bodies. This mode of securing nutrition has served the fungi well. They are diverse and widespread in terrestrial habitats, in large part because fungi can digest such a wide variety of different foods. Pretty much every biological material that has ever evolved can be consumed by one or another fungal species, so nutritional support for fungi is likely to be at hand nearly anywhere.

The ubiquity of fungi is also enhanced by their mode of reproduction. Unlike plants and animals, fungi form no embryos. Instead, fungi propagate by means of tiny, lightweight reproductive packages known as *spores*, which are extraordinarily mobile, although they lack a means for self-propulsion. Spores are distributed far and wide as hitchhikers on the outside of animal bodies, as passengers inside the digestive systems of animals that have eaten them, or as airborne drifters, cast aloft by chance or shot into the atmosphere by elaborate reproductive structures.

It is the pervasiveness of spores that accounts for the inevitable growth of fungi on every uneaten sandwich and container of leftovers. That pervasiveness has also enabled the fungi to assume their role as Earth's undertakers, consuming the dead of all kingdoms and returning their component substances to the ecosystems from which they came.

Net Watch

On-line resources for this chapter are on the World Wide Web at:
http://www.prenhall.com/audesirk
(click on the Table of Contents link and then select Chapter 20).

1) What Are the Main Adaptations of Fungi?

Most Fungi Have Filamentous Bodies

When you think of a fungus, you probably picture a mushroom. Mushrooms, however, are just temporary reproductive structures that extend from the main bodies of certain kinds of fungus. The body of almost all fungi is a **mycelium** (Fig. 20-1a), which is an interwoven mass of threadlike filaments one-cell thick called **hyphae** (singular, **hypha**; Fig. 20-1b). Depending on the species, hyphae either consist of single elongated cells with numerous nuclei or are subdivided by partitions called **septa** (singular, **septum**) into many cells, each containing from one to many nuclei. Pores in the septa allow cytoplasm to stream between cells, distributing nutrients. Like plant cells, fungal cells are surrounded by cell walls. Unlike plant cells, however, fungal cell walls are strengthened by chitin, the same substance found in the exoskeletons of arthropods (as we shall see in Chapter 22).

Fungi are not able to move about. They compensate for this lack of mobility with filaments that can grow rapidly in any direction within a suitable environment. The fungal mycelium quickly infuses itself into aging bread or cheese, beneath the bark of decaying logs, or into the soil. Periodically, the hyphae grow together and differentiate into reproductive structures that project above the surface beneath which the mycelium grows. These structures, including mushrooms, puffballs, and the powdery molds on unrefrigerated food, which represent only a fraction of the complete fungal body, are typically the only part of the fungus that we can easily see.

Fungi Obtain Their Nutrients from Other Organisms

Like animals, fungi are heterotrophic, surviving by breaking down nutrients stored in the bodies or wastes of other organisms. Some fungi are **saprobes**, which digest the bodies of dead organisms. Others are parasitic, feeding on living organisms and causing disease. Others, including *lichens* and *mycorrhizae*, live in mutually beneficial symbiotic relationships with other organisms. There are even a few predatory fungi, which attack tiny worms in soil.

Fungal nutrient absorption resembles that of bacteria, which also have cell walls that prevent them from ingesting food. Like bacteria, fungi must therefore secrete enzymes outside their bodies that digest complex molecules into smaller subunits that can then be absorbed. The fungal body, composed of long filaments only one cell thick, presents an enormous surface area through which to secrete enzymes and absorb nutrients.

Rapidly growing filaments penetrate deeply into a source of nutrients, such as a ripe apple, digesting the flesh and hastening its decomposition. Some of the resulting small nutrient molecules diffuse through the fungal cell walls and into the fungal cells. Other nutrients remain outside, providing nourishment for other organisms.

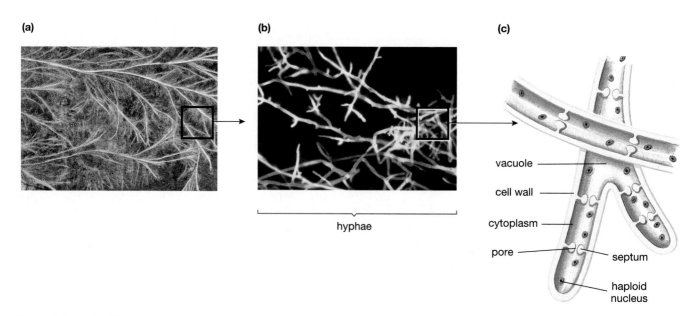

(a) **(b)** **(c)**

vacuole

cell wall

cytoplasm

pore

septum

haploid nucleus

hyphae

Figure 20-1 The filamentous body of a fungus
(a) A fungal mycelium spreads over decaying vegetation. The mycelium is composed of *(b)* a tangle of microscopic hyphae, only one cell thick, *(c)* portrayed in cross-section to show their internal organization.

Most Fungi Can Reproduce Both Sexually and Asexually

Fungal reproduction is highly variable, taking different forms across species and even within species. We can, however, make some generalizations about fungal reproduction. It is typically complex, involving both asexual and sexual processes. During simple asexual reproduction, a mycelium breaks into pieces, each of which grows into a new individual. Many species of fungi reproduce both asexually and sexually through different types of **spores**, which are small, resistant structures that can disperse and then produce new fungi. Spores are formed on or within special structures that project above the mycelium. These special structures allow the spores to be dispersed by wind or water (see the chapter-opening photo).

Fungal body cells typically contain haploid nuclei that, unlike the diploid nuclei of animal cells, contain only a single copy of each chromosome. Mitotic divisions of haploid fungal cells can give rise to haploid asexual spores. If the asexual spores are deposited in favorable habitats, the spores begin mitotic divisions. These divisions produce a new haploid mycelium that is a genetic replica of the parent cells.

Sexual spore formation begins when two haploid nuclei of compatible mating types fuse, resulting in a diploid zygote. This zygote then undergoes meiosis to form haploid sexual spores. These spores are dispersed, germinate, and divide mitotically to form a new haploid mycelium. The reproductive capacity of fungi is prodigious: A single giant puffball may contain 5 trillion sexual spores (see Fig. 20-6a).

Although the health of ecosystems depends on the relentless nature of fungal feeding, penetrating fungal filaments can have adverse consequences as well. Parasitic fungi, for example, cause disease. In humans, such fungi cause a range of diseases such as ringworm and athlete's foot, which infect the skin; valley fever and histoplasmosis, which infect the lungs; and common vaginal yeast infections. Fungi also cause the majority of plant diseases. The fungi that cause chestnut blight and Dutch elm disease have drastically reduced American chestnut and elm tree populations. Fungal parasites also result in billions of dollars in crop losses annually from diseases such as corn smut (see Fig. 20-6d).

The fungal impact on agriculture is not entirely negative, however. Fungal parasites that attack insects and other arthropod pests (Fig. 20-2) can be an important ally in pest control. Farmers who wish to reduce their dependence on toxic and expensive chemical pesticides are increasingly turning to biological methods of pest control, including the application of "fungal pesticides." Fungal pathogens are currently used to control a variety of pests, including termites, rice weevils, tent caterpillars, aphids, and citrus mites.

Fungi, of course, also make an important contribution to human nutrition. This contribution goes far beyond the obvious use of wild and cultivated mushrooms. Other fungi, such as the rare and prized truffle, are also consumed directly. Of greater importance, however, are the less visible manifestations of fungal activities. In particular, fungi are responsible for making bread rise, for converting grape juice to wine, for the distinctive flavor of

2 How Do Fungi Affect Humans? *www*

Edible mushrooms are the most obvious fungal contribution to human welfare, but fungi have many other, less visible but more important impacts as well. Many of these impacts are positive, and some fungal benefits extend far beyond mere gastronomic concerns. For example, as decomposers, fungi make an incalculable contribution to ecosystems. The extracellular digestive activities of many fungi liberate nutrients such as carbon, nitrogen, and phosphorus compounds and minerals that can be used by plants. If fungi and bacteria were suddenly to disappear, the consequences would be disastrous. Nutrients would remain locked in the bodies of dead plants and animals, the recycling of nutrients would grind to a halt, soil fertility would rapidly decline, and waste and organic debris would accumulate. In short, ecosystems would collapse.

Figure 20-2 A helpful fungal parasite
Although some parasitic fungi attack crops, others are used by farmers to control insect pests. Here, a *Cordyceps* species has killed a grasshopper, which can reduce rangeland productivity.

many cheeses, and for the bubbles (and alcohol) in beer. Our diets would certainly be a lot duller without the help we get from fungal partners.

 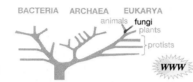

3) How Are Fungi Classified?

Although nearly 100,000 species of modern fungi have been described, biologists have only begun to comprehend the diversity of these organisms—at least 1000 additional species are described each year. Like plants, fungi are grouped into divisions, which are comparable to animal phyla. The major divisions of fungi are the Zygomycota (zygote fungi), Ascomycota (sac fungi), Basidiomycota (club fungi), and Deuteromycota (imperfect fungi) (Table 20-1).

The Zygote Fungi (Division Zygomycota) Can Reproduce by Forming Diploid Zygospores

The *zygomycetes*, also called the **zygote fungi**, include about 600 species. Familiar—and annoying—zygomycetes are those of the genus *Rhizopus*, which cause soft fruit rot and black bread mold. The life cycle of the black bread mold, showing both sexual and asexual reproduction, is depicted in Figure 20-3a. The haploid hyphae of zygomycetes appear identical but are actually two different mating types. The two types "mate sexually," fusing their nuclei to produce diploid **zygospores** (Fig. 20-3b). These resistant structures are dispersed through the air and can remain dormant until conditions are favorable for growth. Zygospores then undergo meiosis

and germinate into structures that bear haploid spores. The spores then give rise to new hyphae. These hyphae may reproduce asexually, by forming haploid spores in black spore cases called **sporangia** (Fig. 20-3c), or sexually, by fusing to produce more zygospores.

The Sac Fungi (Division Ascomycota) Form Spores in a Saclike Case Called an Ascus

The 30,000 species of *ascomycetes*, also called **sac fungi**, are named after the saclike case, or **ascus** (plural, **asci**), in which their spores form during sexual reproduction. Some ascomycetes live in decaying forest vegetation and form beautiful cup-shaped reproductive structures or corrugated, mushroomlike fruiting bodies called *morels* (Fig. 20-4, p. 382). This division also includes many of the colorful molds that attack stored food and destroy fruit and grain crops and other plants. Some ascomycetes secrete the enzymes cellulase and protease, which can cause significant damage to cotton and wool textiles, especially in warm, humid climates where molds flourish. Ascomycetes cause both Dutch elm disease and chestnut blight, but other ascomycetes are a boon to plants, forming mutually beneficial associations with plant roots. A gastronomic delicacy, the *truffle*, is also a member of this diverse division (see "Evolutionary Connections: Fungal Ingenuity—Pigs, Shotguns, and Nooses," p. 386).

Claviceps purpurea, an ascomycete that attacks rye plants, produces structures called ergots that release several toxins, one of which is the active ingredient in the drug LSD. If infected rye is made into flour and consumed, the toxins can lead to ergot poisoning—convulsions, hallucinations, and ultimately death. This happened frequently in northern Europe in the Middle Ages, but modern agricultural techniques have essentially eliminated the disease. Another *Claviceps* toxin has medicinal

Table 20-1 The Major Divisions of Fungi

Common Name (Division)	Reproductive Structures	Cellular Characteristics	Economic and Health Impacts	Representative Genera
Zygote fungi (Zygomycota)	Produce sexual diploid zygospores	Cell walls contain chitin; septa are absent	Cause soft fruit rot and black bread mold	*Rhizopus* (causes black bread mold); *Pilobolus* (dung fungus)
Sac fungi (Ascomycota)	Sexual spores formed in saclike ascus	Cell walls contain chitin; septa are present	Cause molds on fruit; can damage textiles; cause Dutch elm disease and chestnut blight; include yeasts and morels	*Saccharomyces* (yeast); *Ophiostoma* (causes Dutch elm disease)
Club fungi (Basidiomycota)	Sexual reproduction involves production of haploid basidiospores on club-shaped basidia	Cell walls contain chitin; septa are present	Cause smuts and rusts on crops; include some edible mushrooms	*Amanita* (poisonous mushroom); *Polyporus* (shelf fungus)
Imperfect fungi (Deuteromycota)	Not observed engaging in sexual reproduction	Cell walls contain chitin; septa are present	Cause athlete's foot, ringworm, histoplasmosis; source of penicillin	*Penicillium* (produces penicillin); *Arthrobotrys* (nematode predator)

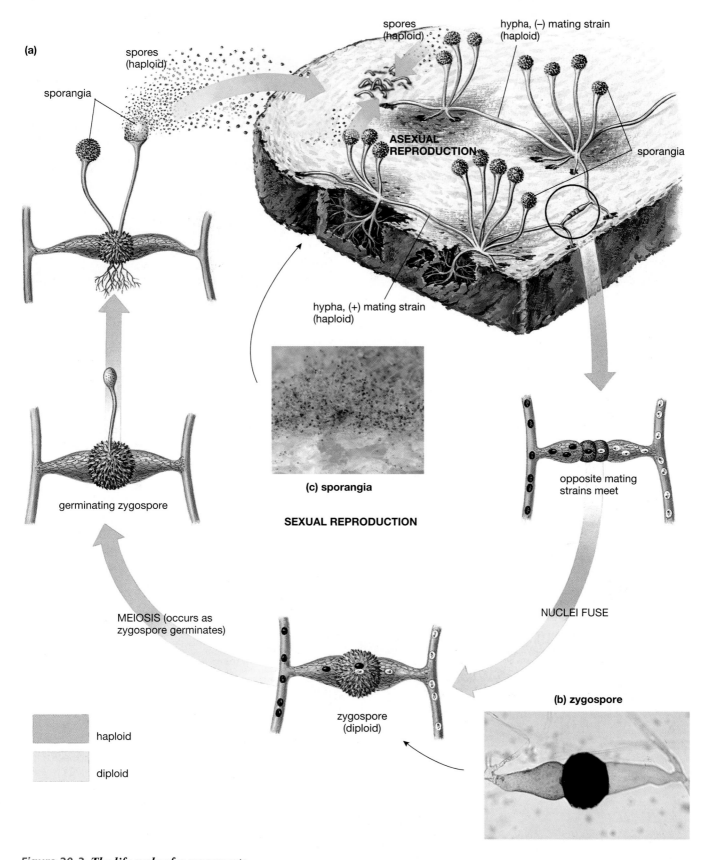

(a)

spores (haploid)

sporangia

spores (haploid)

hypha, (−) mating strain (haploid)

ASEXUAL REPRODUCTION

sporangia

hypha, (+) mating strain (haploid)

(c) sporangia

SEXUAL REPRODUCTION

germinating zygospore

opposite mating strains meet

MEIOSIS (occurs as zygospore germinates)

NUCLEI FUSE

(b) zygospore

haploid

diploid

zygospore (diploid)

Figure 20-3 The life cycle of a zygomycete
(a) Top: During asexual reproduction in the black bread mold (genus *Rhizopus*), haploid spores are produced within sporangia. The spores disperse and germinate on food such as bread, producing haploid hyphae. These in turn may produce additional spores within sporangia by mitosis. Bottom: During sexual reproduction, hyphae that appear identical but are actually of different mating types (designated + and − on the bread) contact one another and fuse, producing *(b)* a diploid zygospore. The zygospore undergoes meiosis and germinates, producing a structure on which *(c)* spore cases, or sporangia, are formed. The sporangia produce and liberate haploid spores, which begin the asexual part of the reproductive cycle.

(a)

(b)

Figure 20-4 *Diverse ascomycetes*
(a) The cup-shaped fruiting body of the scarlet cup fungus. *(b)* The morel, an edible delicacy. (But consult an expert before sampling any wild fungus—some are deadly!)

effects if administered in low doses; that toxin is currently used in drugs that induce labor and control hemorrhaging after childbirth.

Among the ascomycetes we also find the yeasts, some of the few unicellular fungi. The yeasts include both the parasitic yeast that is a common cause of vaginal infections (Fig. 20-5) and the baker's and brewer's yeasts that make possible the proverbial loaf of bread and jug of wine. Some yeasts form hyphae when nutrients are scarce; the hyphae can elongate and reach distant food sources.

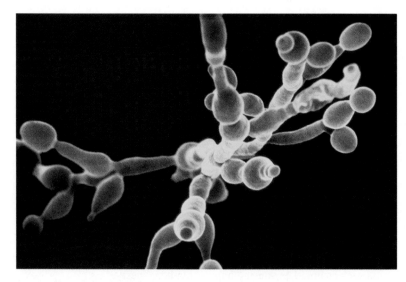

Figure 20-5 *The unusual yeast*
The yeast is an unusual, normally nonfilamentous ascomycete that reproduces most commonly by budding. The yeast shown here is *Candida*, a common cause of vaginal infections.

The Club Fungi (Division Basidiomycota) Produce Club-Shaped Reproductive Structures Called Basidia

Basidiomycetes are called the **club fungi** because they produce club-shaped reproductive structures. The division Basidiomycota consists of about 25,000 species, including the familiar mushrooms, puffballs (Fig. 20-6a), and shelf fungi, sometimes called monkey-stools (Fig. 20-6b). Although several mushroom species are considered delicacies, mushrooms can be deadly. Some members of the genus *Amanita* (Fig. 20-6c) contain potent toxins that are among the most deadly poisons ever found. Basidiomycetes can also be dangerous to plants; they include some devastating plant pests descriptively called *rusts* and *smuts*, which cause billions of dollars worth of damage to grain crops annually (Fig. 20-6d). Some members of this group, however, enter into mutually beneficial relationships with plants.

Basidiomycetes typically reproduce sexually; their life cycle is shown in Figure 20-7 (p. 384). Mushrooms and puffballs are actually reproductive structures: dense aggregations of hyphae that emerge under proper conditions from a massive underground mycelium. On the undersides of mushrooms are leaflike gills that produce specialized club-shaped diploid cells called **basidia**. Basidia give rise to haploid reproductive **basidiospores** by meiosis. These are released by the billions from the gills of mushrooms or the inner surface of puffballs and are dispersed by wind and water.

Figure 20-6 Diverse basidiomycetes
(a) The giant puffball *Lycopedon giganteum* may produce up to 5 trillion spores. *(b)* Shelf fungi, the size of dessert plates, are conspicuous on trees. *(c)* *Amanita verna*, a poisonous mushroom called the "angel of death." *(d)* Corn smut causes major losses of the corn crop each year.

Falling on fertile ground, a mushroom basidiospore may germinate and form haploid hyphae of two different mating types. When the two types meet, some of the cells fuse and produce an underground mycelium. These hyphae grow outward from the original spore in a roughly circular pattern as the older hyphae in the center die. The subterranean body periodically sends up numerous mushrooms, which emerge in a ringlike pattern called a **fairy ring** (Fig. 20-8, p. 385). The diameter of the fairy ring can reveal the approximate age of the fungus—the wider the ring diameter, the older the mushrooms, and their average rate of growth is known. Some fairy rings are estimated to be 700 years old. Recently, researchers have used genetic analysis to trace the extent of some basidiomycete mycelia. One, estimated to be at least 1500 years old, covers 38 acres in northern Michigan; another, on the forested slopes of Mt. Adams in Washington State, encompasses 1500 acres, although its age is "only" 400 to 1000 years.

The Imperfect Fungi (Division Deuteromycota) Seem to Reproduce Only by Asexual Means

Deuteromycetes are called the **imperfect fungi** because none have been observed to form sexual reproductive structures. In some species the sexual stage has been lost during evolution; in others it may exist but has not yet been observed. This large division includes about 25,000 described species of great diversity and considerable importance to humans. It was a member of this

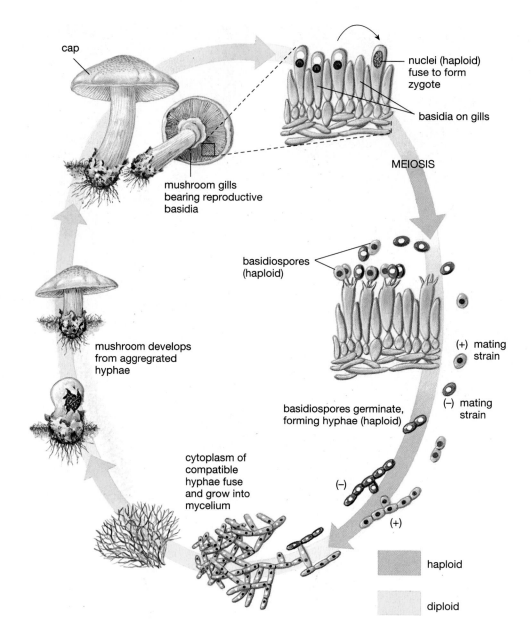

cap

nuclei (haploid)
fuse to form
zygote

basidia on gills

mushroom gills
bearing reproductive
basidia

MEIOSIS

basidiospores
(haploid)

(+) mating
strain

(−) mating
strain

mushroom develops
from aggregrated
hyphae

basidiospores germinate,
forming hyphae (haploid)

(−)

(+)

cytoplasm of
compatible
hyphae fuse
and grow into
mycelium

haploid

diploid

Figure 20-7 The life cycle of a "typical" basidiomycete
The mushroom (top left) is a reproductive structure, formed from aggregated hyphae made
up of cells that each contain two haploid nuclei. Within the cap, leaflike gills bear numerous
basidia (top right). Within each basidium, the two haploid nuclei fuse, producing a diploid
zygote. The zygote then undergoes meiosis, forming haploid basidiospores that are released
from the basidia (right). After dispersal by wind or water, the basidiospores germinate, forming haploid hyphae of two different mating types (+ and −). When hyphae of the different
mating types meet, some of the cells fuse. These cells, each containing two haploid nuclei, produce an extensive underground mycelium (bottom). In the presence of adequate water and
nutrients, portions of the mycelium aggregate, swell, and differentiate, poking up through the
soil as mushrooms and completing the cycle.

division that contaminated and killed the bacterial cultures of the microbiologist Alexander Fleming by accident. His keen observations led to the isolation of
penicillin, the first antibiotic, from the fungus *Penicillium* (Fig. 20-9). We also owe to deuteromycetes the in-

describable flavor and aroma of Roquefort and Camem-
bert cheeses. Other imperfect fungi are human parasites, causing diseases such as ringworm and athlete's
foot. Some are not content to live on dead organisms or
even to parasitize live ones—they act as predators, lay-

Figure 20-8 A mushroom fairy ring
Mushrooms emerge in a fairy ring from an underground fungal mycelium, growing outward from a central point where a single spore germinated, perhaps centuries ago.

ing deadly traps for unsuspecting roundworms (see "Evolutionary Connections: Fungal Ingenuity—Pigs, Shotguns, and Nooses" on p. 386).

Some Fungi Form Symbiotic Relationships with Certain Algae and Plants

Lichens Are Formed by Fungi That Live with Photosynthetic Algae or Bacteria

Lichens are a **symbiotic** (literally, "living together") association between fungi, normally ascomycetes, and unicellular green algae or cyanobacteria (Fig. 20-10). The fungus provides its partner with support and protection, and the photosynthetic algal or bacterial symbiont provides food for itself and for the fungus. Together, these organisms form a unit so tough and undemanding of nutrients that it is among the first living things to colonize newly formed volcanic islands. Brightly colored lichens have invaded other inhospitable habitats ranging from deserts to the Arctic and can even grow on bare rock. Their size and slow rate of growth suggest that some lichens in the Arctic are 4000 years old. About 20,000 species of lichen have been identified, each a combination of a different fungus and one of several types of green algae or cyanobacteria and each with a unique color and growth pattern (Fig. 20-11).

Figure 20-9 Penicillium, *an imperfect fungus*
Penicillium growing on an orange. Reproductive structures, which coat the fruit's surface, are visible, while hyphae beneath draw nourishment from inside. The spore-bearing structures of *Penicillium* are called conidiophores; the spores, termed conidia, are asexual. The antibiotic penicillin was first isolated from this fungus.

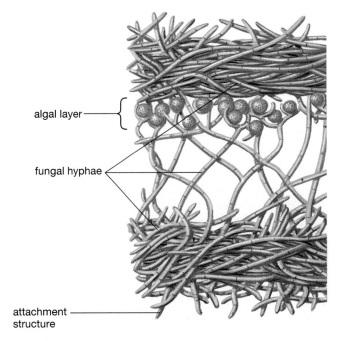

algal layer

fungal hyphae

attachment structure

Figure 20-10 The lichen: a symbiotic partnership
Most lichens have a layered structure bounded on the top and bottom by an outer layer formed from fungal hyphae. Attachments formed from fungal hyphae emerge from the lower layer and anchor the lichen to a surface, such as a rock or a tree. An algal layer in which the alga and fungus grow in close association lies beneath the upper layer of hyphae.

(a)

(b)

Figure 20-11 Diverse lichens
(a) A colorful encrusting lichen, growing on dry rock, illustrates the tough independence of this symbiotic combination of fungus and alga. *(b)* A leafy lichen grows from a dead tree branch.

Mycorrhizae Are Fungi Associated with the Roots of Many Plants

Mycorrhizae (singular, **mycorrhiza**) are important symbiotic associations between fungi and plant roots. Over 5000 species of mycorrhizal fungi (mostly basidiomycetes) are known to grow in intimate association with the roots of about 80% of all plants that have roots, including most trees. These associations benefit both the plant and its fungal partner. The hyphae of mycorrhizal fungi surround the plant root and commonly invade the root cells (Fig. 20-12). The fungus digests and absorbs

minerals and organic nutrients from the soil, passing some of them directly into the root cells. The fungus also absorbs water and passes it to the plant—an advantage in dry, sandy soils. In return, sugar produced photosynthetically by the plant is passed from the root to the fungus. Plants that participate in this unique relationship, especially those in poor soils, tend to grow larger and more vigorously than do those deprived of the fungus. Disturbing new data suggest that mycorrhizae and other types of fungi may be undergoing a dramatic decline that could threaten the health of forests and the communities of plants and animals that rely on them (see "Earth Watch: The Case of the Disappearing Mushrooms" on p. 388).

Some scientists believe that mycorrhizal associations may have been important in the invasion of land by plants more than 400 million years ago. Such a relationship between an aquatic fungus and a green alga (ancestral to terrestrial plants) could have helped the alga acquire the water and mineral nutrients it needed to survive out of water.

Figure 20-12 Mycorrhizae enhance plant growth
Hyphae of mycorrhizae entwining about the root of an aspen tree. Plants grow significantly better in a symbiotic association with these fungi, which help make nutrients and water available to the roots.

Evolutionary Connections
Fungal Ingenuity—Pigs, Shotguns, and Nooses

Natural selection, operating over millions of years on the diverse forms of fungi, has produced some remarkable adaptations by which fungi disperse their spores and obtain nutrients. A few of these adaptations are highlighted here.

Figure 20-13 The truffle
Truffles, rare ascomycetes (each about the size of a small apple), are a gastronomic delicacy.

Figure 20-14 An explosive zygomycete
The delicate, translucent reproductive structures of the zygomycete *Pilobolus* will literally blow their tops when ripe, dispersing the black caps with their payload of spores.

The Rare, Sexy Truffle

Although many fungi are prized as food, none are as avidly sought as the truffle (Fig. 20-13). A single specimen resembling a small, shriveled, blackened apple may sell for $100. Truffles are the spore-containing structures of an ascomycete that forms a mycorrhizal association with the roots of oak trees. Although it develops underground, the truffle has evolved a fascinating mechanism to entice animals, especially wild pigs, to dig it up. The truffle releases a chemical that closely resembles the pig's sex attractant. As aroused pigs dig up and devour the truffle, millions of spores are scattered to the winds. Truffle collectors use muzzled pigs to hunt their quarry; a good truffle-pig can smell an underground truffle 50 meters (about 150 feet) away!

The Shotgun Approach to Spore Dispersal

If you get close enough to a pile of horse manure to scrutinize it, you may be fortunate enough to observe the beautiful, delicate reproductive structures of the zygomycete *Pilobolus* (Fig. 20-14). Despite their daintiness, these are actually fungal shotguns. The clear bulbs, capped with sticky black spore cases, extend from hyphae that penetrate the dung. As the bulbs mature, the sugar concentration inside them increases, and water is drawn in by osmosis. Meanwhile, the bulb begins to weaken just below its cap. Suddenly, like an overinflated balloon, the bulb bursts and blows its spore-carrying top up to a meter (3 feet) away. *Pilobolus* bends toward the light, thereby increasing the chances that its spores will land on open pasture when the bulb bursts. There the spores adhere to grass until consumed by a grazing herbivore, perhaps a horse. Later (some distance away) the spores are deposited unharmed in a pile of their favorite food: manure. Growing hyphae penetrate this rich source of nutrients, sending up new projectiles to continue this ingenious cycle.

The Nematode Nemesis

Microscopic *nematodes* (roundworms) abound in rich soil, and fungi have evolved several fascinating forms of nematode-nabbing hyphae that allow them to exploit this rich source of protein. Some fungi produce sticky pods that adhere to passing nematodes, penetrate the worm body with hyphae, and then begin digesting it from within. One species shoots a microscopic harpoonlike spore into passing nematodes; the spore develops into a new fungus inside the worm. The fungal strangler *Arthrobotrys*, a member of the imperfect fungi, produces nooses formed from three hyphal cells. When a nematode wanders into the noose, its contact with the inner parts of the noose stimulates noose cells to swell with water (Fig. 20-15). In a fraction of a second the hole constricts, trapping the worm. Fungal hyphae then penetrate and feast on their prey.

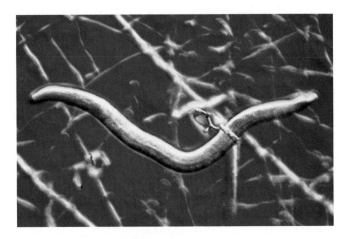

Figure 20-15 Nemesis of nematodes
Arthrobotrys, the nematode (roundworm) strangler, traps its nematode prey in a nooselike modified hypha that swells when the inside of the loop is contacted.

Earth Watch
The Case of the Disappearing Mushrooms

Mycologists (scientists who study fungi) and gourmet cooks may seem to have little in common, but recently they have become united in a common concern: European mushrooms are rapidly declining in numbers, in average size, and in species diversity. Although the problem is most easily recognized in Europe, where people have been gathering wild mushrooms for centuries, American mycologists are also alarmed; the same decline may be occurring here as well. Why are mushrooms disappearing? Overhunting of edible mushrooms is not the culprit, because poisonous forms are equally affected. The loss is evident in all types of mature forests, which eliminates changing forest management practices as the cause. The most likely cause is air pollution, because the loss of mushrooms is great-

est where the air contains the highest levels of ozone, sulfur, and nitrogen. Although mycologists have not yet determined exactly how air pollution harms mushrooms, the evidence is clear. In Holland, for example, over the past several decades the average number of fungal species per thousand square meters has dropped from 37 down to 12. Twenty out of 60 fungal species surveyed in England are declining. Concern is intensified by the fact that the mushrooms most affected are those whose hyphae form mycorrhizal associations with tree roots. Trees with diminished mycorrhizae may have less resistance to periodic droughts or extreme cold spells. Because air pollution is also harming forests directly, the additional loss of mycorrhizae could be devastating.

Summary of Key Concepts

1) What Are the Main Adaptations of Fungi?

Fungal bodies generally consist of filamentous hyphae, which are either multicellular or multinucleate and form large, intertwined networks called mycelia. Fungal nuclei are generally haploid. A cell wall of chitin surrounds fungal cells.

All fungi are heterotrophic, either parasitic or saprobic. They secrete digestive enzymes outside their bodies and absorb the liberated nutrients.

Fungal reproduction is varied and complex. Asexual reproduction can occur either through fragmentation of the mycelium or through asexual spore formation. Sexual spores form after compatible haploid nuclei fuse to form a diploid zygote, which undergoes meiosis to form haploid sexual spores. These spores produce haploid mycelia through mitosis.

2) How Do Fungi Affect Humans?

Fungi are extremely important decomposers in ecosystems. Their filamentous bodies penetrate rich soil and decaying

organic material, liberating nutrients through extracellular digestion.

The majority of plant diseases are caused by parasitic fungi. Some parasitic fungi can help control insect crop pests. Others can cause human diseases, including ringworm, athlete's foot, and common vaginal infections.

3) How Are Fungi Classified?

The major divisions of fungi and their characteristics are summarized in Table 22-1. A lichen is a symbiotic association between a fungus and unicellular algae or cyanobacteria. This self-sufficient combination can colonize bare rock. Mycorrhizae are associations between fungi and the roots of most vascular plants. The fungus derives photosynthetic nutrients from the plant roots and in return carries water and nutrients into the root from the surrounding soil.

Key Terms

ascus *p. 380*	**fairy ring** *p. 383*	**mycelium** *p. 378*	**septum** *p. 378*
basidiospore *p. 382*	**hypha** *p. 378*	**mycorrhiza** *p. 386*	**sporangium** *p. 380*
basidium *p. 382*	**imperfect fungus** *p. 383*	**sac fungus** *p. 380*	
club fungus *p. 382*	**lichen** *p. 385*	**saprobe** *p. 378*	

Thinking Through the Concepts

Multiple Choice

1. *There are no fungi that are*
   a. predators   b. photosynthetic
   c. decomposers   d. parasites
   e. symbiotic

2. *With what plant organ do mycorrhizae interact?*
   a. roots   b. leaves
   c. stems   d. fruits
   e. all parts of the plant

3. *What term refers to the mass of threads that forms the body of most fungi?*
   a. hyphae   b. mycelium   c. sporangia
   d. ascus   e. basidium

4. *Which of the following pairs is incorrect?*
   a. fruit rot: zygote fungus
   b. edible mushroom: club fungus
   c. black bread mold: sac fungus
   d. *Penicillium*: imperfect fungus
   e. yeast: sac fungus

5. *Which of the following is true both of fungi and of some types of bacteria?*
   a. Both produce gametes.
   b. Both engulf microscopic animals.
   c. Both absorb materials across cell walls.
   d. Both fix nitrogen.
   e. They interact to form lichens.

6. *Which of the following structures would you expect to find in the corn smut fungus?*
   a. ascospores   b. basidiospores
   c. asci   d. zygospores
   e. fairy rings

? Review Questions

1. Describe the structure of the fungal body. How do fungal cells differ from most plant and animal cells?

2. What portion of the fungal body is represented by mushrooms, puffballs, and similar structures? Why are these structures elevated above the ground?

3. What two plant diseases, caused by parasitic fungi, have had an enormous impact on forests in the United States? In which division are these fungi found?

4. List some fungi that attack crops. In which division is each?

5. Describe two modes of asexual reproduction in fungi.

6. What is the major structural ingredient in fungal cell walls?

7. List the major divisions of fungi, describe the feature that gives each its name, and give one example of each.

8. Describe how a fairy ring of mushrooms is produced. Why is the diameter related to its age?

9. Describe two symbiotic associations between plants and fungi. In each case, explain how each partner in these associations is affected.

Applying the Concepts

1. Dutch elm disease in the United States is caused by an *exotic*—that is, an organism (in this case a fungus) introduced from another part of the world. What damage has this introduction done? What other fungal pests fall into this category? Why are parasitic fungi particularly likely to be transported out of their natural habitat? What can governments do to limit this importation?

2. The discovery of penicillin revolutionized the treatment of bacterial diseases. However, penicillin is now rarely prescribed. Why is this? HINT: Refer back to Chapter 19.

3. The discovery of penicillin was the result of a chance observation by an observant microbiologist, Alexander Fleming. How would you search systematically for new antibiotics produced by fungi? Where would you look for these fungi?

4. Fossil evidence indicates that mycorrhizal associations between fungi and plant roots existed in the late Paleozoic era, when the invasion of land by plants began. This evidence suggests an important link between mycorrhizae and the successful invasion of land by plants. Why might mycorrhizae have been important fungi in the colonization of terrestrial habitats by plants?

5. General biology texts in the 1960s included fungi in the plant kingdom. Why do biologists no longer consider fungi as legitimate members of the plant kingdom?

6. What ecological consequences would occur if humans, using a new and deadly fungicide, destroyed all fungi on Earth?

Group Activity

Form a group of four students. Your group has been chosen to solve a major health problem, the evolution of pathogenic bacteria with resistance to antibiotics. Your group knows that fungi are the source of many antibiotics. Why do you think that fungi have evolved the ability to produce such substances? Use the answer to that question to discuss and develop a plan to find some new fungal antibiotics. Make sure your plan specifies how you will find the new fungal species and how you'll screen them for effectiveness.

For More Information

Angier, N. "A Stupid Cell with All the Answers." *Discover*, November 1986. Fascinating description of the uses of yeasts in molecular biology, including beautiful illustrations.

Barron, G. "Jekyll-Hyde Mushrooms." *Natural History*, March 1992. Fungi include a variety of forms; some absorb decaying plant material, and others prey on microscopic worms.

Dix, N. J., and Webster, J. *Fungal Ecology*. London: Chapman & Hall, 1995. A rather technical but comprehensive and readable account of fungal diversity that focuses on the roles of fungi in different ecological communities.

Kiester, E. "Prophets of Gloom." *Discover*, November 1991. Lichens can be used as indicators of air quality and provide evidence of deteriorating environmental conditions.

Radetsky, P. "The Yeast Within." *Discover*, March 1994. Describes the newly discovered ability of baker's yeast to form filaments, and the possible implications of this for the study of yeasts that cause vaginal infections.

Strobel, G. A., and Lanier, G. N. "Dutch Elm Disease." *Scientific American*, August 1981. A description of the life cycle of this parasitic fungus and of techniques that limit its spread.

Vogel, S. "Taming the Wild Morel." *Discover*, May 1988. Describes the research that has allowed these rare delicacies to be cultivated in the laboratory.

Answers to Multiple-Choice Questions
1. b   2. a   3. b   4. c   5. e   6. b

The bark of the Pacific yew is the source of a potent anticancer drug. Plants hold many such medicinal treasures, most of them still undiscovered.

The Plant Kingdom 21

At a Glance

Net Watch

On-line resources for this chapter are on the World Wide Web at:
http://www.prenhall.com/audesirk
(click on the Table of Contents link and then select Chapter 21).

We live in a green world, surrounded by members of the green kingdom, Plantae. The world as we know it could not exist without plants, because plants do the job of capturing the energy that animals, fungi, and other heterotrophs need for survival. Only photosynthetic organisms such as plants have the ability to convert solar energy to body tissue. Without photosynthesizers, the sunlight that bathes Earth would provide heat and light but would not provide the leaves, stems, fruits, and seeds that form the primary food source for so many ecosystems. Plants are the energetic foundation on which ecosystems are constructed. Our dependence on them, however, does not stop there. Plants also release into the atmosphere copious amounts of oxygen, the primary by-product of photosynthetic food production. Thus, the oxygen-rich atmosphere on which aerobic organisms such as humans depend is largely a result of the worldwide presence of plants.

In addition to making it possible for us to eat and breathe, plants help heal us when we're sick. The tissues of some plants contain chemicals that are potent medicines for humans. In fact, about a quarter of all prescription drugs used in the United States are derived from plants, even though only a tiny fraction of the world's plant species have been tested for medicinal properties. The list of medicines extracted from plants includes antibiotics, antihistamines, pain relievers, and anticancer drugs. One of the most effective of the anticancer drugs is Taxol®, which is used to treat ovarian cancer. Taxol is extracted from the bark of the Pacific yew, a tree that is native to the forests of northwestern North America. Unfortunately, it takes six 100-hundred-year-old yew trees to produce enough Taxol to treat one patient, so demand for the drug poses a threat to the old-growth forests in which yews live. Plant prospectors are scouring the world in search of other yew species that might also yield Taxol, even as chemists labor to devise a way to manufacture synthetic Taxol economically. As this example shows, our desire to exploit useful plants frequently conflicts with our simultaneous desire to preserve the diversity of plant species that doubtless holds a multitude of as-yet-undiscovered medicinal treasures.

Plants are *sessile*—that is, stationed in one spot throughout their lives—and their many adaptations reflect this fundamental constraint. Plants can't dash into a burrow to hide from a predator, migrate to escape a harsh winter, step down to the water hole for a drink, or roam the countryside in search of a mate. So their evolutionary strategies for survival and reproduction have followed different routes, with fascinating results. For example, the intimate relationships between flowering plants and their animal pollinators are among the most spectacular examples of the power of adaptive coevolution. Even the more mundane aspects of plant survival, such as gathering nutrients and surviving seasonal changes, have yielded adaptive structures and processes that are marvels of evolutionary design.

evolutionarily ancient plants, such as certain green algae whose ancestors may have given rise to terrestrial plants, the sporophyte generation remains tiny and short-lived, represented only by a diploid zygote. In *nonvascular plants*, such as mosses, which are believed to be similar to very early land plants, the sporophyte grows into a multicellular body but is smaller than the gametophyte and remains attached to it (see Fig. 21-6). In the seedless *vascular plants*, such as ferns, which represent the next stage in plant evolution, the sporophyte is dominant; the gametophyte is a much smaller, but still independent, plant. Finally, in the most recently evolved group of plants, the *seed plants*, the gametophytes are microscopic and barely recognizable as an alternate generation (see Figs. 21-11 and 21-13). These tiny gametophytes, however, still produce the eggs and sperm that unite and form the zygote that develops into the diploid sporophyte.

1) What Are the Key Features of Plants?

Most plants are multicellular and able to use photosynthesis to convert water and carbon dioxide to sugar. Neither multicelluarity nor the ability to photosynthesize is unique to plants, but the simultaneous appearance of those traits in a single organism is quite rare outside the green kingdom. The most distinctive feature of plants, however, is their reproductive cycle.

Plants Have Both a Sporophyte and a Gametophyte Generation

Plants generally have a "two-generation" life cycle. In contrast to animals, whose bodies are composed of diploid cells, plants produce separate diploid and haploid generations that alternate with one another. The diploid generation, in which the plant body consists of diploid cells (containing two sets of chromosomes), is called the **sporophyte** and produces haploid spores (containing one set of chromosomes) by meiosis. These haploid spores grow mitotically to produce the haploid generation, called the **gametophyte**. The gametophyte plant body is haploid and produces haploid gametes by mitosis. These gametes fuse to form a diploid **zygote**, a fertilized sex cell that then develops into the diploid sporophyte, continuing this complex life cycle, appropriately described as **alternation of generations** (as we shall see in Chapter 24).

Increasing prominence of the sporophyte generation, accompanied by decreasing size and duration of the gametophyte generation, appears to be a general trend in plant evolution (Table 21-1). In some of the most

2) What Is the Evolutionary Origin of Plants?

The ancestors of all plants were most likely photosynthetic protists that lived in water. Most modern **algae** (singular, **alga**) have remained in the ancestral watery habitat and retained many of the characteristics of their evolutionary predecessors. The algae may thus provide us with clues as to the nature of plants in the earliest phase of their evolution. The first algae appeared between 500 million and 600 million years ago. Algae lack true roots, stems, and leaves and also lack complex reproductive structures such as flowers and cones. Algal gametes are normally shed directly into the water, where they unite and develop.

Algal life cycles are complex and vary considerably between, and even within, the algal divisions. In some genera, such as the branching, pond-loving green alga *Chara*, the haploid gametophyte is dominant. In *Fucus*, a brown alga found at the waterline on rocky seashores, the diploid stage is dominant; the only haploid cells are gametes (see Fig. 21-3a). In other genera, the sporophyte and gametophyte may appear nearly identical, as in the marine green alga *Ulva* (Fig. 21-1, p. 394). Alternatively, sporophytes and gametophytes may be very different in size and appearance, as in the giant kelp, whose sporophyte can be longer than a football field (see Fig. 21-3c) but whose gametophyte is filamentous and microscopic.

Algae are colored by pigments that capture light energy for photosynthesis. These pigments are typically red or brown, absorbing the green, violet, and blue light that

Table 21-1 Features of the Major Plant Groups

Division	Relationship of Sporophyte and Gametophyte	Transfer of Reproductive Cells	Early Embryonic Development	Dispersal	Water and Nutrient Transport Structures	Typical Habitat
Red algae (Rhodophyta)	Highly variable	Gametes or asexual zoospores released into water	Occurs independently after zygote or zoospore settles to substrate	By water currents	Absent	Mostly marine, some freshwater
Brown algae (Phaeophyta)	Highly variable	Gametes or asexual zoospores released into water	Occurs independently after zygote or zoospore settles to substrate	By water currents	Absent (normally)	Almost entirely marine
Green algae (Chlorophyta)	Highly variable	Gametes or asexual zoospores released into water	Occurs independently after zygote or zoospore settles to substrate	By water currents	Absent	Mostly freshwater, some marine
Liverworts and mosses (Bryophyta)	Gametophyte dominant—sporophyte develops from zygote retained on gametophyte	Motile sperm swims to stationary egg retained on gametophyte	Occurs within archegonium of gametophyte	Haploid spores carried by wind	Present	Moist terrestrial
Ferns (Pteridophyta)	Sporophyte dominant—develops from zygote retained on gametophyte	Motile sperm swims to stationary egg retained on gametophyte	Occurs within archegonium of gametophyte	Haploid spores carried by wind	Present	Moist terrestrial
Conifers (Coniferophyta)	Sporophyte dominant—microscopic gametophyte develops within sporophyte	Wind-dispersed pollen carries sperm to stationary egg in cone	Occurs within a protective seed containing a food supply	Seeds containing diploid sporophyte embryo dispersed by wind or animals	Present	Varied terrestrial habitats—dominate in dry, cold climates
Flowering plants (Anthophyta)	Sporophyte dominant—microscopic gametophyte develops within sporophyte	Pollen, dispersed by wind or animals, carries sperm to stationary egg within flower	Occurs within a protective seed containing a food supply; seed encased in fruit	Fruit, carrying seeds, dispersed by animals, wind, or water	Present	Varied terrestrial habitats—dominant terrestrial plant

most readily penetrates deep water. The combination of these pigments with green chlorophyll lends algae their distinctive colors and gives algal divisions their names: division Rhodophyta—the red algae; division Phaeophyta—the brown algae; and division Chlorophyta—the green algae.

The Red Algae (Division Rhodophyta) Live Primarily in Clear Tropical Oceans

The 4000 species of red algae, ranging from bright red to nearly black, derive their color from red pigments that mask their green chlorophyll. Red algae are mostly

Figure 21-1 Life cycle of **Ulva,** *a large marine green alga*
In this case the sporophyte and gametophyte plants are equal in size and nearly indistinguishable. But the sporophyte is diploid and produces haploid spores through meiosis, whereas the male and female gametophytes are haploid and produce gametes by mitosis. The fusion of these gametes results in a diploid zygote that develops into the sporophyte plant.

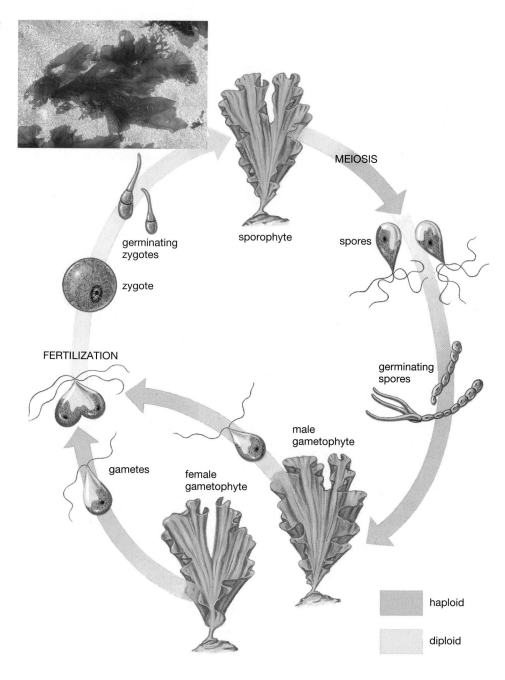

marine and strictly multicellular (Fig. 21-2). They dominate in deep, clear tropical waters, where their red pigments absorb the deeply penetrating blue-green light and transfer this light energy to chlorophyll, where it is used in photosynthesis.

Some species of red algae deposit calcium carbonate, which forms limestone, in their tissues and contribute to the formation of reefs (see Fig. 21-2). Other species are harvested for food in Asia. Red algae also provide certain gelatinous substances with commercial uses, including carrageenan (used as a stabilizing agent in products such as paints, cosmetics, and ice cream) and agar (a substrate for growing bacterial

colonies in laboratories). However, the major importance of these and other algae lies in their photosynthetic ability: The energy they capture and the food they synthesize help support the heterotrophs in marine ecosystems.

The Brown Algae (Division Phaeophyta) Dominate in Cool Coastal Waters

The brown algae increase their light-gathering ability by means of brownish yellow pigments that (in combination with green chlorophyll) produce the brown to olive-green color of these algae. Like the red algae,

Figure 21-2 *Red algae*
Red coralline algae from the Pacific Ocean off California
provide an anchoring site for bright yellow hydroids (phylum
Cnidaria). Coralline algae, which deposit calcium carbonate
within their bodies, contribute to coral reefs in tropical
waters.

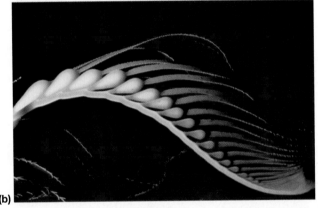

the 1500 species of brown algae are almost entirely
marine and strictly multicellular. This group includes
the dominant "seaweed" species that dwell along rocky
shores in the temperate (cooler) oceans of the world,
including the eastern and western coasts of the United States. Brown algae live in habitats ranging from
nearshore, where they cling to rocks that are exposed
at low tide, to far offshore. Several species use gas-
filled floats to support their bodies (Fig. 21-3a,b).
Some of the giant kelp plants found along the Pacific
coast reach heights of 100 meters (328 feet) and may
grow more than 15 centimeters (6 inches) in a single
day. With their dense growth and towering height (Fig.
21-3c), kelp plants form undersea forests that provide
food, shelter, and breeding areas for a variety of marine animals.

The Green Algae (Division Chlorophyta), Living Mostly in Ponds and Lakes, Probably Gave Rise to the Land Plants

The green algae, a large and diverse group of 7000
species, are mostly multicellular (although there are
a few unicellular forms). Some green algae, such as

Figure 21-3 *Diverse brown algae*
(a) Fucus, a genus found near shores, is shown here exposed
at low tide. Notice the gas-filled floats, which provide buoyancy in water. *(b)* The growing tip of the giant kelp. Gas-filled
floats are clearly visible at the base of leaflike blades. *(c)* The
sporophyte generation of the giant kelp plant *Macrocystis*
forms underwater forests off southern California.

Spirogyra, form thin filaments from long chains of cells (Fig. 21-4a). Other species of green algae form colonies containing clusters of cells that are somewhat interdependent and that constitute a structure intermediate between unicellular and multicellular forms. These colonies range from a few cells to a few thousand cells, as in species of *Volvox*. Most green algae are small, but some large species inhabit the sea. For example, the green alga *Ulva*, or sea lettuce, is similar in size to the leaves of its namesake (Fig. 21-4b).

The green algae are of special interest because green algal ancestors are believed to have given rise to the terrestrial plants. Three lines of evidence support the hypothesis that land plants evolved from green algal ancestors:

1. Green algae use the same type of chlorophyll and accessory pigments in photosynthesis as do land plants.
2. Green algae store food as starch, as do land plants, and have cell walls made of cellulose, similar in composition to those of land plants.
3. Most green algae live in fresh water, where they were originally subjected to environmental pressures that resulted in adaptations that helped the algae cope with the challenges of dry land, as described below.

In contrast to the nearly constant environmental conditions of the ocean, freshwater habitats (ponds, swamps, streams, and lakes) are highly variable. Water temperature can fluctuate seasonally or even daily, and changing levels of rainfall can lead to fluctuations in the concentration of chemicals in the water or even to periods in which the aquatic habitat dries up. Ancient freshwater algae must have developed adaptations that enabled

them to withstand extremes of temperature and periods of dryness. These adaptations served their descendants well as they invaded land.

3 How Did Plants Invade and Flourish on Land?

The terrestrial world may be green now, but it didn't start out that way. When plants first made the transition ashore, the land was barren and desolate, inhospitable to life. From a plant's evolutionary viewpoint, however, it was also a land of opportunity, free of competitors and predators, and full of carbon dioxide and sunlight (the raw materials for photosynthesis, which are present in far higher concentrations in the air than in water). So once natural selection had shaped the adaptations that helped plants overcome the obstacles to terrestrial living, plants prospered and diversified.

The Plant Body Increased in Complexity as Plants Made the Evolutionary Transition from Water to Dry Land

When plants pioneered the land, they underwent a series of evolutionary changes in response to the challenges posed by terrestrial environments. On land, the supportive buoyancy of water is missing, the plant body is no longer bathed in a nutrient solution, and the air tends to dry things out. These conditions favored the evolution of structures that support the body, vessels that transport water and nutrients to all parts of the plant, and structures that conserve water. The resulting adaptations to dry land include:

1. Roots or rootlike structures that anchor the plant and absorb water and nutrients from the soil
2. Conducting vessels that transport water and minerals upward from the roots and that move photosynthetic products from the leaves to the rest of the plant body
3. The stiffening substance **lignin**, which is a rigid polymer, impregnates the water-conducting and mineral-conducting vessels and supports the plant body, helping the plant expose maximum surface area to sunlight
4. A waxy **cuticle** covering the surfaces of leaves and stems that limits the evaporation of water
5. Pores called **stomata** (singular, **stoma**) in the leaves and stems that open to allow gas exchange but close when water is scarce, thus reducing the amount of water lost to evaporation

But as we shall see next, adaptations in the adult forms of plants were not enough to exploit dry land. To be fully successful on land, plants needed new methods of transporting sperm to eggs.

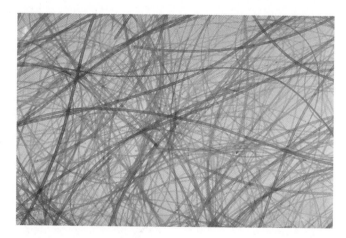

Figure 21-4 A form of green algae
(a) *Spirogyra* is a filamentous green alga composed of strands only one cell thick. **(b)** *Ulva*, also known as sea lettuce, is a green alga whose sporophyte and gametophyte generations are similar in appearance (see Fig. 21-1).

The Invasion of Land Required Protection and a Means of Dispersal for Sex Cells and Developing Plants

In aquatic and marine forms such as algae, the gametes (sex cells) and zygotes (fertilized sex cells) can be carried passively by water currents or can actively swim, like the flagellated gametes and spores (called **zoospores**) of some algae. But land plants cannot rely on water to carry their sex cells and disperse their fertilized eggs. Life on land demands a means of dispersal independent of water and also requires that developing embryos be protected from drying out. These challenges were met by the evolution of *pollen*, *seeds*, and later, *flowers* and *fruits*.

The dry, microscopic pollen grains that nonflowering seed plants began to produce allowed wind, instead of water, to carry the male gametes. Later came the evolution of attractive flowers, which entice animal pollinators who deliver pollen more precisely than does wind. The seed provided waterproof protection and nourishment for the developing embryos, and some modifications of the seed coat enhanced dispersal. Early fruits started to attract animal foragers, who consumed them and incidentally dispersed the indigestible seeds in their feces.

Two major groups of land plants arose from the ancient algal ancestors. One group, the *nonvascular plants*, or **bryophytes**, straddles the boundary between aquatic and terrestrial life, much like the amphibians of the animal kingdom. The other group, the *vascular plants*, or **tracheophytes**, became completely adapted to life on land.

The Liverworts and Mosses (Division Bryophyta) Are Only Partially Adapted to Dry Land

The 16,000 bryophyte species show some of the adaptations necessary for a completely terrestrial existence. The bryophytes, like the algae, lack true roots, leaves, and stems. Rootlike anchoring structures called **rhizoids** bring water and nutrients into the plant body, but bryophytes are *nonvascular*—they lack well-developed structures for conducting water and nutrients. They must rely on slow diffusion or poorly developed conducting tissues to distribute water and other nutrients, so their body size is limited. Most bryophytes are less than 2 centimeters (under 1 inch) tall.

Among the bryophyte features that do represent adaptations to terrestrial existence are their enclosed reproductive structures. The **archegonia** (singular, **archegonium**), in which eggs develop, and the **antheridia** (singular, **antheridium**), where sperm are formed, prevent the gametes from drying out. In some bryophyte species both archegonia and antheridia are located on the same plant; in other species each individual plant is either male or female. In all bryophytes the sperm must swim to the egg (which emits a chemical attractant) through a film of water. Bryophytes that live in dry areas must time their reproduction to coincide with the infrequent rains.

The major bryophyte representatives are the liverworts and mosses (Fig. 21-5). The liverworts and most

Figure 21-5 Bryophytes
(a) Liverworts grow inconspicuously in moist, shaded areas. This is the female gametophyte plant, bearing umbrella-like archegonia, which hold the eggs. Sperm must swim up the stalks through a film of water to fertilize the eggs. **(b)** Moss plants showing both stages in the life cycle. The short, leafy green plants are the haploid gametophytes; the stalks are the diploid sporophyte generation. The sporophytes are about 1 centimeter (less than half an inch) in height.

mosses are confined to moist areas. Some mosses, however, possess a waterproof cuticle that retains moisture and stomata that can be closed, preventing water loss. These mosses can survive in deserts, on bare rock, and in far northern and southern latitudes where humidity is low and liquid water is scarce for much of the year.

The life cycle of a moss is shown in Figure 21-6. As in many algae, the larger "leafy" plant body is the

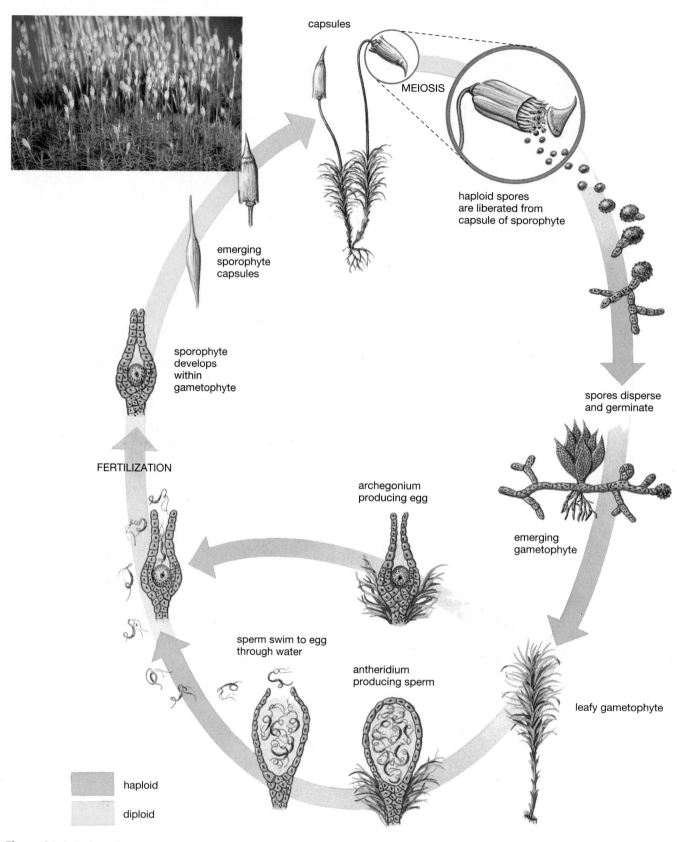

capsules

MEIOSIS

haploid spores
are liberated from
capsule of sporophyte

emerging
sporophyte
capsules

spores disperse
and germinate

sporophyte
develops
within
gametophyte

emerging
gametophyte

FERTILIZATION

archegonium
producing egg

sperm swim to egg
through water

antheridium
producing sperm

leafy gametophyte

haploid

diploid

Figure 21-6 Life cycle of a moss

The life cycle of a moss shows alternation of diploid and haploid generations. The leafy green gametophyte (lower right), which typically grows in patches resembling a cushion, is actually the haploid gametophyte generation that produces sperm and eggs. The sperm develop in the antheridium and must swim through a film of water to the egg (which remains in the archegonium where it is formed). The zygote develops into a stalked, diploid sporophyte that emerges from the gametophyte plant. The sporophyte is topped by a brown capsule in which haploid spores are produced by meiosis. These are dispersed and germinate, producing another green gametophyte generation. **(Inset)** Moss plants. The short, leafy green plants are haploid gametophytes; the reddish brown stalks are diploid sporophytes, about 1 centimeter (less than a half inch) tall.

haploid gametophyte, which forms sperm and eggs by mitosis. Bryophytes retain the fertilized egg in the archegonium. There the embryo grows and matures into a small diploid sporophyte that remains attached to the parent gametophyte plant. At maturity, the sporophyte produces haploid spores by meiosis within a capsule. When the capsule is opened, spores are released and dispersed by the wind. If a spore lands in a suitable environment, it may develop into another haploid gametophyte plant.

The Vascular Plants, or Tracheophytes, Have Conducting Vessels That Also Provide Support

Although the bryophytes were quite successful, they left vast areas of the land unoccupied. When the moister regions were covered with the short green fuzz of bryophytes, any plant that could stand taller than a few centimeters would benefit by basking in sunlight while shading its short competitors. Natural selection thus favored two types of adaptations that allowed some plants to become taller: (1) structures that provided support for the body and (2) vessels that conducted water and nutrients absorbed by the roots into the upper portions of the plant. In the **vascular** (from the Latin word for "vessel-bearing") plants, specialized groups of conducting cells (which we will call **vessels**) impregnated with the stiffening substance lignin served both supportive and conducting functions.

The Seedless Vascular Plants Include the Club Mosses, Horsetails, and Ferns

The seedless vascular plants reached treelike proportions and dominated the landscape during the Carboniferous period (from 286 million to 360 million years ago). Their bodies—transformed by heat, pressure, and time—are burned today as coal. Their modern representatives, the club mosses, horsetails, and ferns, have diminished in size and importance and have been largely replaced by the more-versatile seed plants.

The club mosses (division Lycophyta) are now limited to representatives a few centimeters in height (Fig. 21-7a). Their leaves are small and scalelike, resembling the leaflike structures of mosses. Club mosses of the genus *Lycopodium*, commonly known as ground pine, form a beautiful ground cover in some temperate coniferous and deciduous forests.

Modern horsetails (division Sphenophyta) form a single genus, *Equisetum*, with only 15 species, most less than 1 meter tall (Fig. 21-7b). The bushy branches of some species lend them the common name horsetails; the leaves are reduced to tiny scales on the branches. They are also called "scouring rushes," because they were used by early settlers to scour pots and floors. All species of *Equisetum* deposit large amounts of silica (glass) in their outer layer of cells, giving them an abrasive texture.

The ferns (division Pteridophyta), with 12,000 species, are far more successful (Fig. 21-7c). In the tropics, "tree ferns" still reach heights reminiscent of their ancestors in the Carboniferous period. Ferns are the only seedless vascular plants that have broad leaves. Broad leaves can capture more sunlight, and this advantage over the small-leaved club mosses and horsetails may account for the relative success of modern ferns.

The life cycle of a fern is depicted in Figure 21-8 (p. 401). A major difference between vascular plants and bryophytes is that the diploid sporophyte is dominant in vascular plants. On special leaves of club mosses and ferns, and on conelike structures of horsetails, haploid spores are produced in structures called *sporangia*. The spores are dispersed by the wind and give rise to tiny, haploid gametophyte plants, which produce sperm and eggs. The gametophyte generation retains two traits that are reminiscent of the bryophytes. First, the small gametophytes lack conducting vessels. Second, as in bryophytes, the sperm must swim through water to reach the egg, so these plants still depend on the presence of water for sexual reproduction.

The Seed Plants Dominate the Land, Aided by Two Important Adaptations: Pollen and Seeds

Plants are so distinct from animals that they occupy different kingdoms. There are more than a few parallels between them, though. For example, the domination of terrestrial habitats by both plants and animals hinged on evolutionary innovations that made it possible (1) to achieve fertilization without the need for gametes to travel through water and (2) to allow embryos to develop without being immersed in water. Animals accomplished these tasks by evolving internal fertilization and shelled eggs. In plants, natural selection accomplished the same tasks with a different set of adaptations: pollen and seeds. These key innovations have enabled seed plants to dominate the land for the past 250 million years.

Pollen (also called a *pollen grain*) is all that remains of the male gametophyte of seed plants. In seed plants, both male and female gametophytes (which produce the sex cells) are greatly reduced in size, whereas the sporophyte is large. The female gametophyte is a small group of haploid cells that produces the egg, and the male gametophyte is the pollen grain. Sperm-producing cells are carried within the pollen grains, which are dispersed by wind or by animal pollinators such as bees. Thus, seed plants are not limited in their distribution by the need for water in which sperm can swim to the egg; they are fully adapted to life on dry land.

The second reproductive adaptation is the seed itself (seed structure is discussed in detail in Chapter 24). Analogous to the eggs of birds and reptiles, **seeds** consist of an embryonic plant, a supply of food for the embryo, and a

Figure 21-7 Some seedless vascular plants
Seedless vascular plants are found in moist woodland habitats. *(a)* The club mosses (sometimes called ground pines) grow in temperate forests. This specimen is releasing spores. *(b)* The giant horsetail (genus *Equisetum*) extends long, narrow branches in a series of rosettes. Its leaves are reduced to insignificant scales. At right is a cone-shaped spore-forming structure. *(c)* The leaves of this deer fern are emerging from coiled fiddleheads.

protective outer coat (Fig. 21-9, p. 402). The *seed coat* maintains the embryo in a state of suspended animation or dormancy until conditions are proper for growth. The stored food helps sustain the emerging plant until it develops roots and leaves and can make its own food by photosynthesis. Some seeds possess elaborate adaptations that make possible dispersal by wind, water, and animals. These adaptations have helped them invade nearly every habitat on Earth.

Seed plants are grouped into two general types: (1) *gymnosperms*, which lack flowers, and (2) *angiosperms*, the flowering plants. Although these groupings are not official taxonomic categories, they are useful in organizing our discussion of the seed plants.

Gymnosperms Are Nonflowering Seed Plants

Gymnosperms (whose name means "naked seed" in Greek) evolved earlier than the flowering plants. One group, the **conifers** (division Coniferophyta), with 500 species, still dominates large areas of our planet. Other gymnosperms, such as the ginkgos and cycads (divisions Ginkgophyta and Cycadophyta, respectively), have declined to a small remnant of their former range and abundance.

The ginkgos were probably the first of the surviving seed plants to evolve, becoming widespread during the Jurassic period, which began 208 million years ago. Today they are represented by the single species *Ginkgo biloba*, the maidenhair tree. Ginkgo trees are either

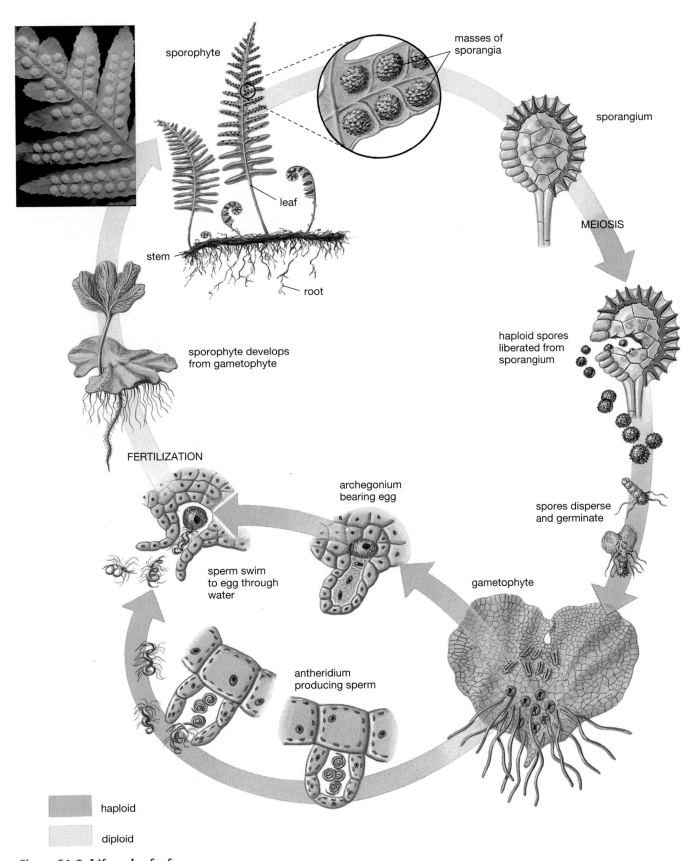

masses of sporangia

sporophyte

leaf

stem

root

sporophyte develops from gametophyte

FERTILIZATION

sperm swim to egg through water

archegonium bearing egg

antheridium producing sperm

sporangium

MEIOSIS

haploid spores liberated from sporangium

spores disperse and germinate

gametophyte

haploid

diploid

Figure 21-8 Life cycle of a fern
The fern's life cycle shows alternation of generations. The dominant plant body (upper left) is the diploid sporophyte. Haploid spores, formed in sporangia located on the underside of certain leaves, are dispersed by the wind to germinate on the moist forest floor into inconspicuous haploid gametophyte plants. On the lower surface of these small, sheetlike gametophytes, male antheridia and female archegonia produce sperm and eggs. The sperm must swim to the egg, which remains in the archegonium. The zygote develops into the large sporophyte plant. **(Inset)** Underside of a fern leaf, showing clusters of sporangia.

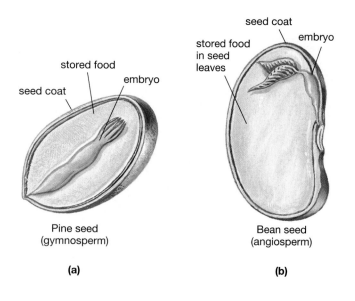

Figure 21-9 Seeds
Seeds from **(a)** a gymnosperm (pine) and **(b)** an angiosperm (bean). Both consist of an embryonic plant and stored food confined within a seed coat. In the angiosperm seed, food is stored within large seed leaves, which take up most of the volume of the seed. Seeds exhibit diverse adaptations for dispersal, including **(c)** the dandelion's tiny, tufted seeds that float in the air and **(d)** the massive, armored seeds (protected inside the fruit) of the coconut palm, which can survive prolonged immersion in seawater as they traverse oceans.

male or female; the female trees bear foul-smelling, fleshy seeds the size of cherries (Fig. 21-10a). Ginkgos have been maintained by cultivation, particularly in Asia; if not for this cultivation, they might be extinct today. Because they are more resistant to pollution than are most other trees, ginkgos (normally, the male trees) have been extensively planted in American cities. Recently, the leaves of the ginkgo have gained notoriety as an herbal remedy that purportedly improves memory.

Cycads look like large ferns, from which they probably evolved (Fig. 21-10b). Today there are approximately 160 species, most of which dwell in tropical or subtropical climates. Most cycads are about 1 meter (3 feet) in height, although some species can reach 20 meters (65 feet). Cycads grow slowly and live for a long time; one Australian specimen is estimated to be 5000 years old. The fleshy seeds of cycads were once a staple food in Guam, but they contain a toxin that may cause a neurological disorder that resembles Parkinson's disease.

Conifers spread widely as Earth became drier during the Permian period, which followed the Carboniferous period. Today they are most abundant in the cold latitudes of the far north and at high elevations where conditions are rather dry. Not only is rainfall limited in these areas, but soil water remains frozen and unavailable during the long winters. Conifers, including pines, firs, spruce, hemlocks, and cypresses, are adapted to dry, cold conditions in several ways. First, conifers retain green leaves throughout the year, enabling these plants to continue photosynthesizing and growing slowly during times when most other plants become dormant. Conifers are often called **evergreens** for this reason. Second, conifer leaves are actually thin needles covered with a thick cuticle whose small waterproof surface minimizes evaporation. Finally, conifers produce an "antifreeze" in their sap that enables them to continue transporting nutrients in subfreezing temperatures. This substance gives them their fragrant "piney" scent.

Reproduction is similar in all conifers, with pines serving as a good illustration. The tree itself is the diploid

(a)

(b)

Figure 21-10 Two uncommon gymnosperms
(a) The ginkgo, or maidenhair tree, has been kept alive by cultivation in China and Japan. Relatively resistant to pollution, these trees have become popular in American cities. This ginkgo is female and bears fleshy seeds the size of large cherries, which are noted for their foul smell when ripe. *(b)* A cycad. Common in the age of dinosaurs, these are now limited to about 160 species living in warm, moist climates. Like ginkgos, cycads have separate sexes.

(Fig. 21-11) sporophyte. It develops both male and female cones. The male cones are relatively small (normally 2 centimeters, about 3/4 of an inch, or less) and delicate structures. They release clouds of pollen during the reproductive season and then disintegrate (Fig. 21-11, top). Each pollen grain is a male gametophyte, consisting of several specialized haploid cells, some of which form tiny winglike structures that allow the pollen to be carried by the wind for long distances. Immense clouds of pollen are released by the male cones; inevitably some land by chance on the female cone. Each female cone consists of a series of woody scales arranged in a spiral around a central axis (Fig. 21-11, top). At the base of each scale are two **ovules** (immature seeds) within which diploid spore cells form and undergo meiosis to produce haploid female gametophytes. These gametophytes develop and produce egg cells. A pollen grain landing nearby sends out a pollen tube that slowly burrows into the female gametophyte. After nearly 14 months, the tube finally reaches the egg cell and releases sperm that fertilize it. The fertilized egg becomes enclosed in a seed as it develops into a tiny embryonic plant. The seed is liberated when the cone matures and its scales separate.

Angiosperms Are Flowering Seed Plants
Modern flowering plants, or **angiosperms** (division Anthophyta), are incredibly diverse, with more than 230,000 species. They range in size from the diminutive duckweed, a few millimeters in diameter, that floats on ponds (Fig. 21-12a, p. 405) to the mighty eucalyptus tree, over 100 meters (328 feet) tall (Fig. 21-12b). From desert cactus to

tropical orchids to grasses to parasitic mistletoe, angiosperms dominate the plant kingdom.

The angiosperm domination of Earth has persisted for more than 100 million years. The oldest fossils of flowering plants are estimated to be 127 million years old. Angiosperms may have evolved from gymnosperm ancestors that formed an association with animals (most likely insects), which carried their pollen from plant to plant. The insects benefited by eating some of the protein-rich pollen, whereas the plant no longer had to produce prodigious quantities of pollen and send it flying on the fickle winds to ensure fertilization. Animals can in some cases carry pollen farther than wind can and with greater accuracy and less waste. According to this scenario, the relationship between certain types of ancient gymnosperms and their animal pollinators was so beneficial that, through natural selection, plants evolved flowers that attracted insects and other animals.

Three major adaptations have contributed to the enormous success of angiosperms: (1) flowers, (2) fruits, and (3) broad leaves. In the angiosperm life cycle (Fig. 21-13, p. 406), the dominant sporophyte plant develops **flowers**, the structures in which both male and female gametophytes are formed. (We shall discuss this process in more detail in Chapter 24.) Male gametophytes (pollen) are formed inside a structure called the *anther*; the female gametophyte develops from an ovule within a part of the flower called the *ovary*. The egg in turn develops within the female gametophyte. Fertilization occurs when the pollen forms a tube through the *stigma*, a sticky pollen-catching structure of the flower, and bores into the ovule.

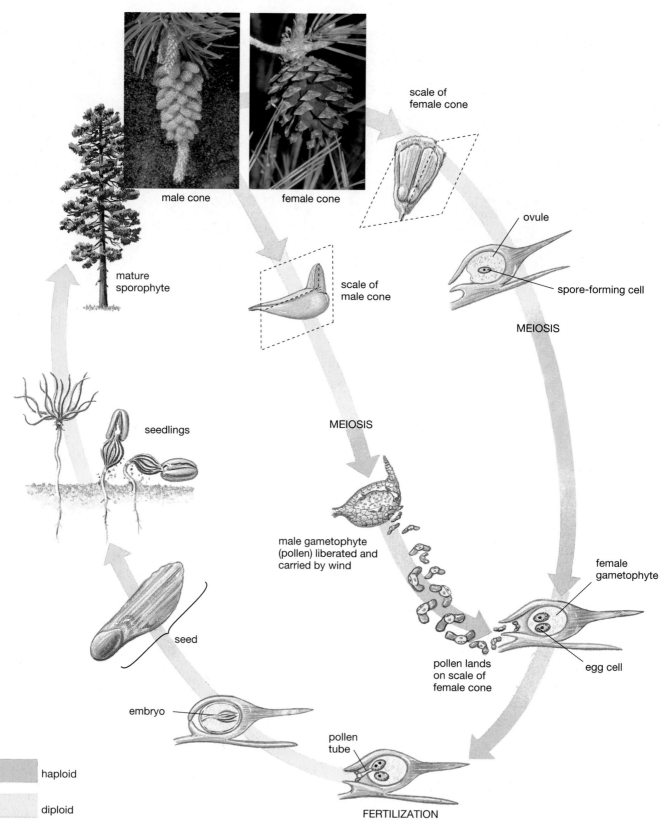

scale of
female cone

ovule

spore-forming cell

MEIOSIS

male cone

female cone

scale of
male cone

mature
sporophyte

MEIOSIS

seedlings

male gametophyte
(pollen) liberated and
carried by wind

female
gametophyte

seed

pollen lands
on scale of
female cone

egg cell

embryo

pollen
tube

FERTILIZATION

haploid

diploid

Figure 21-11 *Life cycle of the pine*

The pine tree is the sporophyte generation (upper left), bearing both male and female cones. Within the ovules at the base of each scale of the female cone, diploid spore-forming cells undergo meiosis, producing haploid spore cells, one of which develops into the female gametophyte. The female gametophyte in turn produces egg cells. In the male cones, meiosis produces the male gametophytes: the pollen. Pollen grains, carrying sperm, are dispersed by the wind and land on the scales of the female cone. The pollen grows a pollen tube that penetrates the female gametophyte and conducts the sperm to the egg. The fertilized egg then develops into an embryonic plant enclosed in a seed formed from the ovule. The seed is eventually released from the cone, germinates, and grows into a sporophyte tree.

(a)

(b)

(c)

(d)

(e)

Figure 21-12 Diverse angiosperms
(a) The smallest angiosperm is the duckweed, found floating on ponds. These specimens are about 3 millimeters (1/8 inch) in diameter. *(b)* The largest angiosperms are eucalyptus trees, which can reach 45 meters (150 feet) in height. Both *(c)* grasses and many trees, such as *(d)* this birch, in which flowers are shown as buds (green) and blossom (brown), have inconspicuous flowers and rely on wind for pollination. *(e)* Flowers, such as those shown on this butterfly weed and eucalyptus tree (inset in part b), entice insects and other animals that carry pollen between individual plants.

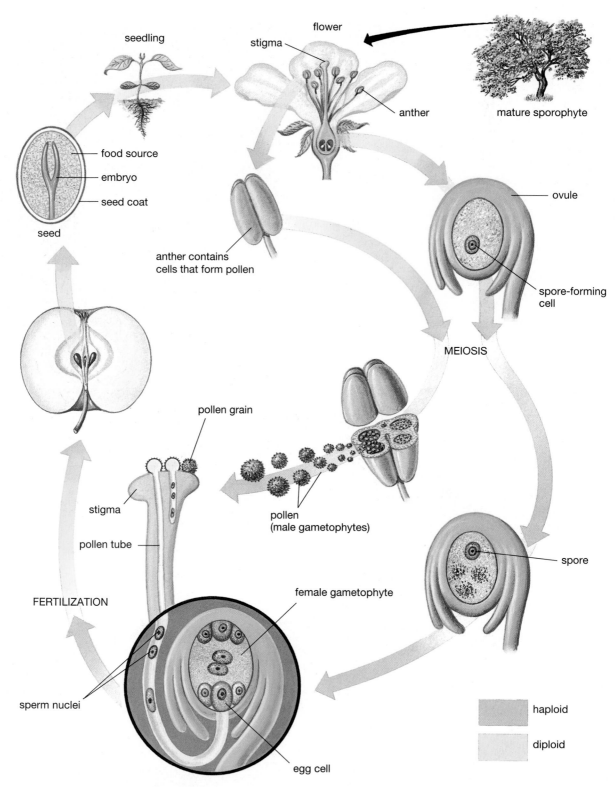

labels: seedling · flower · stigma · anther · mature sporophyte · food source · embryo · seed coat · seed · ovule · anther contains cells that form pollen · spore-forming cell · MEIOSIS · pollen grain · pollen (male gametophytes) · stigma · pollen tube · spore · female gametophyte · FERTILIZATION · sperm nuclei · egg cell · haploid · diploid

Figure 21-13 *Life cycle of a flowering plant*

The dominant plant (upper right) is the diploid sporophyte, whose flowers normally produce both male and female gameto-phytes. Male gametophytes (pollen grains) are produced within the anthers, where diploid spore-forming cells undergo meiosis, producing haploid spore cells. These spores divide mitotically to produce pollen, in which two sperm are formed. The female ga-metophyte develops within the ovule of the ovary. There diploid spore-forming cells undergo meiosis, producing a haploid spore cell. The spore divides mitotically to produce the female gametophyte, whose contents include one egg cell. After landing on the stigma, a pollen grain produces a pollen tube that burrows down to the ovary and into the female gametophyte. There it releases its sperm. One sperm fuses with the egg to form a zygote. The second eventually forms a source of food for the developing em-bryo. The ovule gives rise to the seed, which contains the endosperm and an embryo that develops from the zygote. The seed is dispersed, germinates, and develops into a mature sporophyte.

There the zygote develops into an embryo enclosed in a seed formed from the ovule.

The ovary surrounding the seed matures into a **fruit**, the second adaptation that has contributed to the success of angiosperms. The word *angiosperm* (from the Greek for "enclosed seed") refers to the enclosure of the seed within a fruit. Just as flowers encourage animals to transport pollen, so, too, many fruits entice animals to disperse seeds. These seeds pass through animal digestive tracts unharmed; examples can easily be observed in bird droppings. As dog owners are well aware, some fruits (called burs) disperse by clinging to animal fur. Others, such as the fruits of maples, form wings that carry the seed through the air. The variety of dispersal mechanisms made possible by the fruit has helped the angiosperms invade nearly all possible habitats.

A third feature that gives angiosperms an adaptive advantage in warmer, wetter climates is broad leaves. When water is plentiful, as it is during the warm growing season of temperate and tropical climates, broad leaves give trees an advantage by collecting more sunlight for photosynthesis. The extra energy gained during the spring and summer allows the trees to drop their leaves and enter a dormant period, which reduces evaporative water loss during periods when water is in short supply (Fig. 21-14). In temperate climates, such periods occur during the fall and winter, when virtually all temperate angiosperm trees and shrubs drop their leaves annually. In the tropics and subtropics, most angiosperms are evergreen, but species that inhabit certain tropical climates where periods of drought are common may drop their leaves to conserve water during the dry season.

The advantages of broad leaves are offset by some evolutionary costs. In particular, broad, tender leaves are much more appealing to herbivores than are the tough, waxy needles of conifers. As a result, angiosperms have developed a range of defenses against mammalian and insect herbivores. These adaptations include physical defenses such as thorns, spines, and resins that toughen the leaves. But the evolutionary struggle for survival has also led to a host of chemical defenses—compounds that make plant tissue poisonous or distasteful to potential predators. Many of the compounds responsible for chemical defense have properties that humans have exploited for medicinal and culinary uses. Medicines such as Taxol and aspirin, stimulants such as nicotine and caffeine, and

Figure 21-14 *Two ways of coping with the dryness of winter* The evergreen (a conifer) retains its needles throughout the year. The small surface area and heavy cuticle of the needles slow the loss of water through evaporation. In contrast, the aspen (an angiosperm) sheds its leaves each fall. The dying leaves turn brilliant shades of gold as pigments used to capture light energy for photosynthesis are exposed when the chlorophyll disintegrates.

spicy flavors such as mustard and peppermint are all derived from angiosperm plants.

The flowering plants can be grouped into two broad classes based on their internal and external structure, discussed in more detail in Chapter 23. The **monocots** (class Monocotyledoneae) are a group of about 65,000 species including grasses, corn and other grains, irises, lilies, and palms. The **dicots** (class Dicotyledoneae) are a considerably larger group with about 170,000 species, including most of the angiosperm trees, shrubs, and herbs.

Summary of Key Concepts

1) What Are the Key Features of Plants?

The kingdom Plantae consists of eukaryotic, photosynthetic, normally multicellular organisms. The ability of plants and other photosynthetic organisms to capture the energy of sunlight in high-energy molecules provides nearly all other forms of life on Earth with a source of usable energy. Plants exhibit alternation of generations in which a haploid gametophyte generation alternates with a diploid sporophyte generation. There has been a general evolutionary trend toward reduc-

tion of the haploid gametophyte, which is dominant in bryophytes but microscopic in seed plants.

2) What Is the Evolutionary Origin of Plants?

The first algae appeared between 500 million and 600 million years ago. The simplest plants alive today are the aquatic algae. There are three major divisions of algae. The red algae (Rhodophyta) dominate in clear tropical waters, and the brown algae (Phaeophyta) populate temperate oceans. The green algae (Chlorophyta) are primarily small, freshwater forms. Algal divisions are named for their predominant colors, which result from a combination of green chlorophyll and light-trapping pigments. Algae lack true roots, stems, and leaves and rely on the water to carry their sex cells.

③ How Did Plants Invade and Flourish on Land?

Green algal ancestors probably gave rise to the first land plants. As plants became increasingly adapted to a terrestrial existence, they developed (1) rootlike structures for anchorage and for absorption of water and nutrients; (2) conducting vessels to transport water and nutrients throughout the plant; (3) a stiffening substance, called lignin, to impregnate the vessels and support the plant body; (4) a waxy cuticle to slow the loss of water through evaporation; and (5) stomata that can open, allowing gas exchange, and that can also close, preventing water loss.

As plants invaded the land, new reproductive adaptations evolved. Reduction of the male gametophyte to pollen allowed wind to replace water in carrying sperm to eggs. Flowers attracted animals, who carry pollen more precisely and efficiently than wind can, and fruit enticed animals to disperse seeds. Seeds nourish, protect, and help disperse developing embryos.

Two major groups of plants, bryophytes and tracheophytes, arose from the ancient algal ancestors. Bryophytes, including the liverworts and mosses, are small, simple land plants that lack conducting vessels. Although some have adapted to dry areas, most live in moist habitats. Reproduction in bryophytes requires water through which the sperm swims to the egg.

In tracheophytes, or vascular plants, a system of vessels has evolved, stiffened by lignin, that conducts water and nutrients absorbed by the roots into the upper portions of the plant and supports the body as well. Owing to this support system, seedless vascular plants, including the club mosses (division Lycophyta), horsetails (division Sphenophyta), and ferns (division Pteridophyta), can grow larger than bryophytes. As in bryophytes, the sperm of tracheophytes must swim to the egg for sexual reproduction to occur, and the gametophyte lacks conducting vessels.

Vascular plants with seeds have two major new adaptive features: pollen and seeds. They are often classified into two categories: gymnosperms and angiosperms. Gymnosperms include ginkgos, cycads, and the highly successful conifers. These plants were the first fully terrestrial plants to evolve. Their success on dry land is partially due to the evolution of the male gametophyte into the pollen grain. Pollen protects and transports the male gamete, eliminating the need for the sperm to swim to the egg. The seed, a protective resting structure containing an embryo and a supply of food, is a second important adaptation contributing to the success of seed plants.

Angiosperms, the flowering plants, dominate much of the land today. In addition to pollen and seeds, angiosperms also produce flowers and fruits. The flower allows angiosperms to utilize animals as pollinators. In contrast to wind, animals can in some cases carry pollen farther and with greater accuracy and less waste. Fruits may attract animal consumers, which incidentally disperse the seeds in their feces.

Key Terms

alga *p. 392*	**conifer** *p. 400*	**gymnosperm** *p. 400*	**sporophyte** *p. 392*
alternation of generations *p. 392*	**cuticle** *p. 396*	**lignin** *p. 396*	**stoma** *p. 396*
	dicot *p. 407*	**monocot** *p. 407*	**tracheophyte** *p. 397*
angiosperm *p. 403*	**evergreen** *p. 402*	**ovule** *p. 403*	**vascular** *p. 399*
antheridium *p. 397*	**flower** *p. 403*	**pollen** *p. 399*	**vessel** *p. 399*
archegonium *p. 397*	**fruit** *p. 407*	**rhizoid** *p. 397*	**zoospore** *p. 397*
bryophyte *p. 397*	**gametophyte** *p. 392*	**seed** *p. 399*	**zygote** *p. 392*

Thinking Through the Concepts

Multiple Choice

1. *Which of the following groups of organisms live primarily in freshwater habitats?*
   a. red algae   b. brown algae   c. green algae
   d. liverworts   e. bryophytes

2. *In which of the following plants is the gametophyte the dominant generation?*
   a. mosses   b. ferns   c. pine trees   d. sunflowers
   e. The sporophyte is dominant in all of the above.

3. *What is the function of a fruit?*
   a. It attracts pollinators.
   b. It provides food for the developing embryo.
   c. It stores excess food produced by photosynthesis.
   d. It helps ensure seed dispersal from the parent plant.
   e. It evolved so that people would cultivate the plant, ensuring its survival.

4. *What is the function of lignin?*
   a. It provides support for the plant.
   b. It waterproofs plant surfaces.
   c. It stores food.

d. It promotes gas exchange in plant leaves.
   e. It transports dissolved nutrients.

5. *Which of the following plants produces sperm that swim to the egg?*
   a. sugar maple tree   b. Douglas fir tree, a conifer
   c. rattlesnake fern   d. common dandelion
   e. none of the above

6. *Which of the following is the correct sequence during alternation of generations?*
   a. sporophyte—diploid spores—gametophyte—haploid gametes
   b. sporophyte—haploid spores—gametophyte—haploid gametes
   c. sporophyte—haploid gametes—gametophyte—haploid spores
   d. sporophyte—haploid gametes—gametophyte—diploid spores
   e. sporophyte—diploid gametes—gametophyte—diploid spores

? Review Questions

1. What is meant by "alternation of generations"? What two generations are involved? How does each reproduce?

2. Explain the evolutionary changes in plant reproduction that adapted plants to increasingly dry environments.

3. Describe evolutionary trends in the life cycles of plants. Emphasize the relative sizes of the gametophyte and sporophyte.

4. Assuming that green algae have fewer accessory pigments than do red or brown algae, would you expect to find these algae in shallow or deep water? Where would you find the red algae, and why?

5. From which algal division did green plants probably arise? Explain the evidence that supports this hypothesis.

6. List the structural adaptations necessary for the invasion of dry land by plants. Which of these adaptations are possessed by bryophytes? By ferns? By gymnosperms and angiosperms?

7. The number of species of flowering plants is greater than the number of species in the rest of the plant kingdom. What feature(s) are responsible for the enormous success of angiosperms? Explain why.

8. List the adaptations of gymnosperms that have helped them become the dominant tree in dry, cold climates.

9. What is a pollen grain? What role has it played in helping plants colonize dry land?

10. The majority of all plants are seed plants. What is the advantage of a seed? How do plants that lack seeds meet the needs served by seeds?

Applying the Concepts

1. You are a geneticist working for a firm that specializes in plant biotechnology. Explain what *specific* parts (fruit, seeds, stems, roots, etc.) of the following plants you would try to alter by genetic engineering, what changes you would try to make, and why, on (a) corn, (b) tomatoes, (c) wheat, and (d) avocados.

2. Prior to the development of synthetic drugs, over 80% of all medicines were of plant origin. Even today, indigenous tribes in remote Amazonian rain forests can provide a plant product to treat virtually any ailment. Herbal medicine is also widely and successfully practiced in China. Most of these drugs are unknown to the Western world. But the forests from which much of this plant material is obtained are being converted to agriculture. We are in danger of losing many of these potential drugs before they can be discovered. What steps can you suggest to preserve these natural resources while also allowing nations to direct their own economic development?

3. Only a few hundred of the hundreds of thousands of species in the plant kingdom have been domesticated for human use. One example is the almond. The domestic almond is nutritious and harmless, but its wild precursor can cause cyanide poisoning. The oak makes potentially nutritious seeds (acorns) that contain very bitter-tasting tannins. If we could breed the tannin out of acorns, they might become a delicacy. Why do you suppose we have failed to domesticate oaks?

Group Activity

Work with a partner to examine the photographs in Figure 21-12; choose your two favorite ones. Then, working independently, identify and write down two distinctive traits in each photo that you believe are adaptations that helped individuals of the pictured species to survive and reproduce (for example, narrow, flat leaves in Fig. 21-12c). For each trait that you listed, describe a scenario that explains the steps by which the trait could have evolved by natural selection in populations in which the trait was not originally present. Your scenario should include the environmental conditions that must have prevailed during the evolutionary process.

Next, compare results with your partner. For each photo, discuss all the traits that either or both of you listed and the evolutionary scenarios that you developed. Agree on a single scenario that explains how all the traits evolved. Working together, summarize your agreed-upon scenarios in writing. Working independently, answer the following questions for each scenario. (1) Did you initially agree or disagree with your partner about how this trait probably evolved? (2) If you initially disagreed, did you eventually change your mind? (3) If you changed your mind, what argument convinced you?

For More Information

Cox, P. A., and Balick, M. J. "The Ethnobotanical Approach to Drug Discovery." *Scientific American*, June 1994. Biologists seek new pharmaceutical compounds by analyzing the plants used as drugs by indigenous cultures.

Diamond, J. "How to Tame a Wild Plant." *Discover*, September 1994. Cultivated plants have ecological and genetic properties that make them well suited for agriculture.

Doyle, J. "DNA, Phylogeny, and the Flowering of Plant Systematics." *BioScience*, June 1993. Chloroplast DNA is used in reconstructing evolutionary relationships, but there are difficulties with each of the methods in use.

Grant, M. C. "The Trembling Giant." *Discover* 1993. The largest living thing is a clone of 47,000 aspen trees covering 106 acres in the Wasatch Mountains of Utah.

Joyce, C. *Earthly Goods: Medicine-Hunting in the Rainforest*. Boston: Little, Brown, 1994. Science and adventure combine in this account of prospecting for new medicines and the people who do it.

Kaufman, P. B. *Plants—Their Biology and Importance*. New York: Harper & Row, 1989. Complete, readable coverage of all aspects of plant taxonomy, physiology, and evolution.

Milot, V. "Blueprint for Conserving Plant Diversity." *BioScience*, June 1989. Points out the importance of preserving genetic diversity in endangered plant species.

Nicholson, R. "Death and Taxus." *Natural History*, September 1992. The bark of the yew tree contains compounds that may help in the treatment of cancer.

Answers to Multiple-Choice Questions

1. c   2. a   3. d   4. a   5. c   6. b

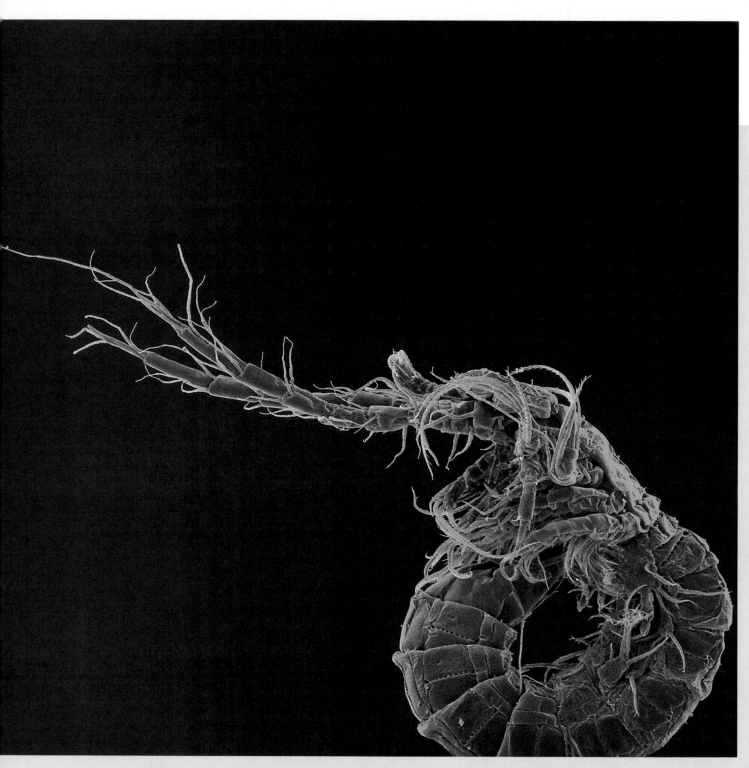

This microscopic crustacean of the genus Derocheilocaris *lives in the tiny spaces between grains of sand on the beach. Like other arthropods, it escapes the notice of most humans but is a member of Earth's most diverse and numerous group of animals.*

The Animal Kingdom 22

At a Glance

 Net Watch
On-line resources for this chapter are on the World Wide Web at:
http://www.prenhall.com/audesirk
(click on the Table of Contents link and then select Chapter 22).

What's your favorite animal? If you pose this question to a 6-year-old, it's highly unlikely that he or she will answer "the hissing cockroach" or "the whipworm." It's much more likely that the reply will contain references to doggies or horsies or birdies. Even if you discuss animals with an adult, the discussion is still likely to center on furry or feathered creatures. Humans are relatively large vertebrates; our perception of the animal world tends to focus on other relatively large vertebrates. But this is a distorted picture. By any measure, vertebrates represent just a tiny fraction of the animal kingdom. In fact, to a first approximation all animals are insects. Insects and other arthropods make up the vast majority of animal species and a huge proportion of the individual animals on Earth. Other animal phyla that we rarely think about are also far more diverse and numerous than vertebrates.

There are far more species of mollusks (clams, snails, slugs) than of mammals, and far more types of roundworms than of birds. Even the most diverse vertebrate group, the fishes, is outnumbered by the spiders in terms of diversity.

The animal world is dominated by small, inconspicuous, boneless creatures, many of which live in habitats that we rarely contemplate and cannot easily visit. Animal life abounds in pond muck, on sea bottoms, in treetops, beneath leaf litter, and between grains of beach sand. So our study of the animal kingdom will consist largely of an attempt to get up to speed on the multitude of body plans, modes of life, and behaviors that characterize the vast array of invertebrate life. (Don't worry, we'll also get to the vertebrates!) The evolutionary origin of animal diversity is something of a mystery, because the fossil record doesn't pick up the story until animals had already diverged into the phyla that we see today. Strangely, no major new animal body plans seem to have arisen in the 630 million years since these early fossils were deposited. But even though evolution has not altered the fundamental architecture of animals in quite some time, a multitude of successful design adjustments have been made. As a result of all this evolutionary innovation, animals today flourish on land, in the seas, and in the air. If you can imagine a way of making a living on this Earth, chances are that some animal species is already doing it.

1 What Characteristics Define an Animal?

If asked to answer the above question, you might be tempted to answer, "I know one when I see one." If you had to develop a precise definition, however, you might find it difficult to devise one that is simple and concise to describe the term *animal*. Animals are best defined by a list of characteristics, none of which is unique to animals but that, taken together, distinguish animals from members of the other kingdoms:

1. Animals are multicellular.
2. Animals are heterotrophic—that is, they obtain their energy by consuming the bodies of other organisms.
3. Animals typically reproduce sexually. Although animal species exhibit a tremendous diversity of reproductive styles, most are capable of sexual reproduction.
4. Animal cells lack a cell wall.
5. Animals are motile during some stage of their life. Even the stationary sponges have a free-swimming *larval* stage (a juvenile form).
6. Animals are usually able to make rapid responses to external stimuli as a result of the activity of nerve cells, muscle or contractile tissue, or both.

2 What Are the Major Evolutionary Trends in Animals?

Over Evolutionary Time, Animals Have Increased in Complexity

In our survey of the animal kingdom, as we progress from the sponges through the cnidarians (jellyfish and their relatives) and through the three phyla of worms, you will recognize a clear evolutionary trend toward increasing complexity. This trend can be seen by comparing the following traits among animal phyla: (1) the level of cellular organization and specialization; (2) the presence and type of symmetry in body plan (described below); (3) the degree of **cephalization** (the concentration of sensory organs and a brain in a defined head region); (4) the presence and type of body cavity; (5) the presence of **segmentation** (repeated, similar body parts); and (6) the structure of the digestive system. Because these traits differ among animal species and are tied to anatomical changes that occurred during the evolutionary history of animals, they are important in reconstructing animal phylogenies, such as the simple one pictured in Figure 22-1.

The trend toward greater complexity culminates in the arthropods, mollusks, echinoderms, and chordates (see Table 22-1). Although there are striking structural differences among these four phyla, the differences reflect adaptations to different environments and lifestyles rather than further increases in complexity. Within the chordates, however, another trend toward greater complexity is clear. The numbers of this phylum, which includes vertebrate animals such as fish and humans, exhibit a trend toward increasing size and sophistication of the brain.

Animal Phyla Show Trends toward Increasing Cellular Organization

As new types of animals emerged over the course of evolution, specialized cells in the newer forms tended to become increasingly organized into **tissues**—groups of similar cells integrated into a functional unit, such as a muscle. Different tissues tended to combine to form **organs**: two or more tissues integrated to perform a specialized function, illustrated by structures such as the kidney and the eye. Organs, in turn, formed **organ systems**: two or more organs that work together to perform a specific function, as illustrated by the digestive or excretory system.

In sponges, the most ancient phylum of animals, individual cells may have specialized functions but act more or less independently and are not organized into true tissues or organs. Cnidarians, the animal phylum most closely related to sponges, show well-defined tissues, such as their *nerve net*, which coordinates movement and sensory information, but cnidarians lack organs. Flatworms,

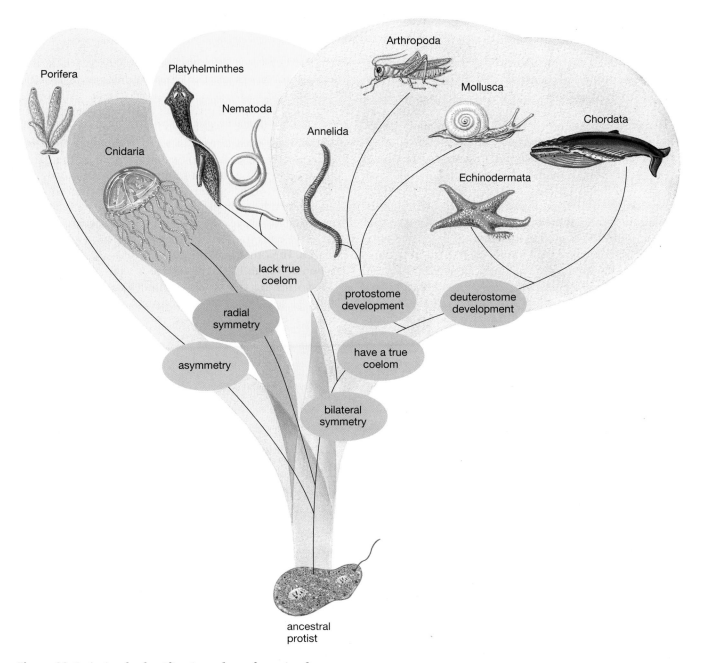

Figure 22-1 A simple classification scheme for animals
This classification scheme is based on anatomical features. Animals may be asymmetrical, radially symmetrical, or bilaterally symmetrical. Bilaterally symmetrical animals either lack or possess the body cavity called a coelom. Animals that have a coelom can be further subdivided into those with protostome embryological development and those with deuterostome development.

which were the next animal phylum to emerge, do have organs, including gonads and eyespots, and also have an organ system, the reproductive system. Organ systems are also found in all the remaining, more recently emerged animal phyla.

A second trend in cellular organization is in the number of tissue layers, called **germ layers**, that arise during embryonic development. The most primitive forms, sponges, have no true tissues. Cnidarians, the next most primitive animals, have two germ layers: an inner layer of **endoderm** (forming the lining of most hollow organs) and an outer layer of **ectoderm** (forming epithelial tissue that covers the body and lines its inner cavities and nervous tissue composed of nerve cells). Flatworms and all the more complex animals have three germ layers: In between the endoderm and ectoderm is a layer of **mesoderm** (forming muscle and, when present, the circulatory and skeletal systems).

Body Forms Became Symmetrical Early in the Evolutionary History of Animals

Animals that can be bisected along at least one plane so that the resulting halves are mirror images of one another are said to be symmetrical (the opposite of symmetry is asymmetry). Only the simplest animals, the sponges, have asymmetrical bodies. Many sponges are irregular and variable in shape, even within a single species. Other types of sponges show **radial symmetry**, in which any plane through a central axis divides the animal into roughly equal halves (Fig. 22-2). This was the first type of symmetry to evolve in animals; it is also seen in cnidarians and in some adult echinoderms. **Bilateral symmetry** evolved later and is characteristic of all more complex animals, including larval echinoderms. A bilaterally symmetrical animal can be divided into roughly mirror-image halves only along a particular, single plane through the central axis. A bilaterally symmetrical animal has an upper, or **dorsal**, surface and a lower, or **ventral**, surface.

Cephalization Increased over Evolutionary Time

Radially symmetrical animals tend either to be **sessile** (fixed to one spot, as in sea anemones) or to drift around on currents (as in jellyfish). Such animals may encounter food or threats from any direction, so a body that is essentially "facing" all directions at once is advantageous. In contrast, most bilaterally symmetrical animals move under their own power, in a particular direction. Resources are most likely to be encountered by the part of the animal that is closest to the direction of movement. The evolution of bilateral symmetry was therefore accompanied by cephalization, which produces an **anterior** (head) end, where sensory cells, sensory organs, clusters of nerve cells, and organs for ingesting food are concentrated. The other end of a cephalized animal is designated **posterior** and typically features a tail (Fig. 22-2). The flatworms are the simplest animals to show cephalization. Although they have a defined head with sensory organs, they still ingest food through a muscular tube

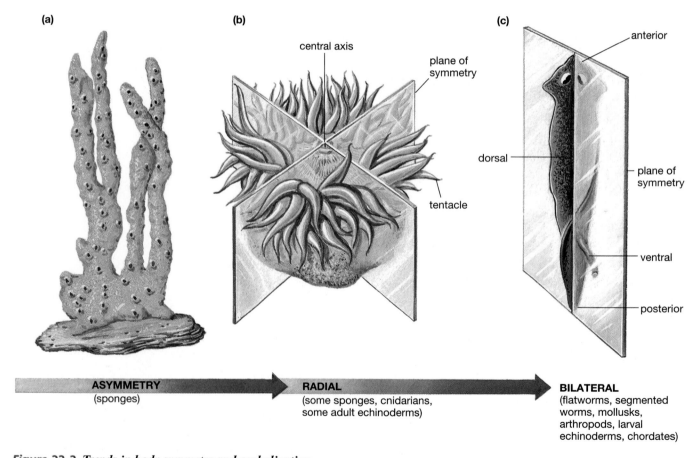

Figure 22-2 Trends in body symmetry and cephalization
(a) Sponges, the simplest animals, lack a head; most are asymmetrical. **(b)** The bodies of some sponges, cnidarians (anemones, hydra), and some adult echinoderms (sea urchins, sea stars) are radially symmetrical. Any plane that passes through the central axis divides the body into mirror-image halves. Animals in these groups lack a well-defined head. **(c)** Nearly all the more complex animals show bilateral symmetry. The body can be split into two mirror-image halves only along a particular plane that runs down the midline. Animals with bilateral symmetry have an anterior head end, a posterior tail end, a dorsal upper surface, and a ventral underside.

located near the middle of their bodies. More-complex animals have well-defined heads that contain a brain and bear important sensory structures.

Body Cavities Arose in More-Complex Animals

Another major trend in the evolution of animals has been the development of a fluid-filled cavity between the digestive tube (or gut, where food is digested and absorbed) and the outer body wall. Simple animals such as cnidarians and flatworms lack an internal cavity between their gut and body wall; the space is filled with solid tissue. In most other animal phyla, the gut and body wall are separated by a space, which creates a "tube-within-a-tube" body plan. In evolutionary terms, this new body plan (1) freed the gut from the constraints imposed by direct attachment to the body wall and (2) created a space in which new organs and organ systems could develop.

In many phyla, including our own, the fluid-filled body cavity is completely surrounded by a thin lining of tissue that develops from mesoderm. Such a cavity is known as a **coelom** (Fig. 22-3). A true coelom can serve a variety of functions. In the earthworm, it acts as a kind of skeleton, providing support for the body and a framework against which muscles can act. In other animals, internal organs are suspended in the coelomic fluid, which serves as a protective buffer between them and the outside world. The coelom has also allowed the internal organs such as the heart and digestive tract to move independently of the body wall. Thanks to your coelom, you may remain externally inactive after a meal even though your digestive tract is churning energetically.

A handful of animal phyla, including the roundworms, have a fluid-filled cavity between the body wall and gut, but that cavity is not completely surrounded by mesoderm-derived tissue. This kind of cavity is not considered be a true coelom. It is called a **pseudocoelom** (the prefix *pseudo* means "false").

Among the animal phyla that have a true coelom, embryological development follows a variety of pathways. These diverse developmental pathways, however, can be grouped into two categories on the basis of features such as the method by which the coelom arises. These two modes of embryological development are known as **protostome** and **deuterostome** development. In protostome development, the coelom forms within the space between the body wall and the digestive cavity. In deuterostome development, the coelom forms as outgrowths of the digestive cavity. The protostomes and the deuterostomes represent distinct evolutionary branches within the animals with a coelom. Annelids, arthropods, and mollusks exhibit protostome development; echinoderms and chordates are deuterostomes.

Segmentation First Arose in Annelid Worms

Segmentation, or the presence of similar repeated body segments, first arose in the annelids (earthworms and

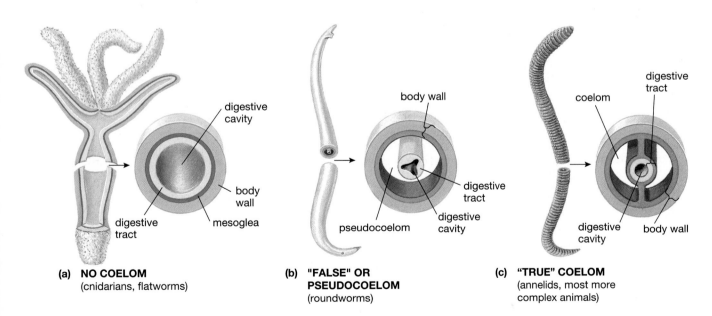

(a) **NO COELOM**
(cnidarians, flatworms)

(b) **"FALSE" OR PSEUDOCOELOM**
(roundworms)

(c) **"TRUE" COELOM**
(annelids, most more complex animals)

Figure 22-3 Trends in body cavities
(a) Cnidarians and flatworms have no cavity between the body wall and digestive tract. *(b)* Roundworms have a fluid-filled cavity, called a pseudocoelom, between the body wall and digestive tract. The pseudocoelom is partially, but not completely, lined with tissue derived from the embryo's mesoderm germ layer. *(c)* Annelids and other complex animals have a true coelom, a fluid-filled cavity between the body wall and the digestive tract that is completely lined with tissue derived from mesoderm.

their relatives; see Fig. 22-13). Segmentation appears to be an evolutionary device by which body size can increase, with a minimum of new genetic information required. Because the genetic "blueprint" for each segment is similar, new segments can arise from comparatively small genetic changes that cause existing segments to be duplicated. In more-complex animals, the repeated segments have become specialized for specific functions or are visible only in certain body parts, such as the series of similar vertebrae in the vertebrate backbone.

Digestive Systems Increased in Complexity

The simplest form of digestion is found in the sponges, which lack a specialized digestive tract. Digestion is entirely intracellular in sponges; individual cells trap, ingest, and digest smaller unicellular organisms. The cnidarians and flatworms have a digestive system that consists of a sac with a single opening (see Figs. 22-7 and 22-9). This opening, politely called a mouth, also serves as an anus through which undigested material is expelled.

In the roundworms and all of the more complex animals, an efficient, tubular, one-way digestive system has evolved. A mouth, located at the anterior end near the sensory structures, allows food to be ingested as it is detected. The food is then processed in stages as it passes through a series of specialized regions. First it is physically broken down, then enzymatically digested, then absorbed. Wastes are voided through a separate anus, normally located near the posterior end of the animal.

3) What Are the Major Animal Phyla?

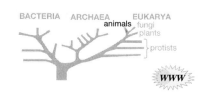

It's easy to overlook the differences among the multitude of small, boneless animals in the world. Even Carolus Linnaeus, the originator of modern taxonomy, recognized only two phyla of animals without backbones

Table 22-1 Comparison of the Major Animal Phyla

Common name (Phylum)		Sponges (Porifera)	Hydra, Anemones, Jellyfish (Cnidaria)	Flatworms (Platyhelminthes)	Roundworms (Nematoda)
Body Plan	**Level of organization**	Cellular—lack tissues and organs	Tissue—lack organs	Organ system	Organ system
	Germ layers	Absent	Two	Three	Three
	Symmetry	Absent	Radial	Bilateral	Bilateral
	Cephalization	Absent	Absent	Present	Present
	Body cavity	Absent	Absent	Absent	Pseudocoel
	Segmentation	Absent	Absent	Absent	Absent
Internal Systems	**Digestive system**	Intracellular	Gastrovascular cavity; some intracellular	Gastrovascular cavity	Separate mouth and anus
	Circulatory system	Absent	Absent	Absent	Absent
	Respiratory system	Absent	Absent	Absent	Absent
	Excretory system (fluid regulation)	Absent	Absent	Canals with ciliated cells	Excretory gland cells
	Nervous system	Absent	Nerve net	Head ganglia with longitudinal nerve cords	Head ganglia with dorsal and ventral nerve cords
	Reproduction	Sexual; asexual (budding)	Sexual; asexual (budding)	Sexual (some hermaphroditic); asexual (body splits)	Sexual (some hermaphroditic)
	Support	Endoskeleton of spicules	Hydrostatic skeleton	Absent	Hydrostatic skeleton

(insects and "worms"). Since the time of Linnaeus, however, systematists have come to appreciate that the animal lineage underwent many ancient splits. Today, we recognize about 27 distinct phyla of animals. The features of the major animal phyla are compared in Table 22-1. For convenience, biologists often place animals in one of two major categories: (1) **vertebrates**, those with a backbone, or vertebral column, and (2) **invertebrates**, those lacking a backbone. The vertebrates—fish, amphibians, reptiles, birds, and mammals—are perhaps the most conspicuous animals from a human point of view, but more than 97% of all known animal species on Earth are invertebrates. The vertebrates constitute only part of a single phylum, the phylum Chordata.

The earliest animals were invertebrates. They probably originated from colonies of protozoa whose members had become specialized to perform distinct roles within the colonial body. In our survey of the kingdom Animalia, we will begin with the sponges, whose body plan most closely resembles the probable ancestral protozoan colonies. Our discussion will proceed through a series of representative invertebrate phyla and will end with the vertebrates in the phylum Chordata.

4 The Sponges: Phylum Porifera

Sponges, members of the phylum Porifera, lack true tissues and organs; as such, they are the simplest multicellular animals. They resemble colonies in which single-celled organisms live together for mutual benefit. The simple cellular organization of sponges was discovered in an experiment by the embryologist H. V. Wilson in 1907. He mashed a sponge through a piece of silk, thereby dissociating it into single cells and cell clusters. He then placed them in seawater for 3 weeks, after which time the cells had reaggregated into a functional sponge. This experiment indicated that individual sponge cells were able to survive and function independently.

Segmented Worms (Annelida)	Insects, Arachnids, Crustaceans (Arthropoda)	Snails, Clams, Squid (Mollusca)	Sea Stars, Sea Urchins (Echinodermata)	Vertebrates (Chordata)
Organ system	Organ system	Organ system	Organ system	Organ system
Three	Three	Three	Three	Three
Bilateral	Bilateral	Bilateral	Bilateral larvae, radial adults	Bilateral
Present	Present	Present	Absent	Present
Coelom	Coelom	Coelom	Coelom	Coelom
Present	Present	Absent	Absent	Present (but reduced)
Separate mouth and anus	Separate mouth and anus	Separate mouth and anus	Separate mouth and anus (normally)	Separate mouth and anus
Closed	Open	Open	Absent	Closed
Absent	Tracheae, gills, or book lungs	Gills, lungs	Tube feet, skin gills, respiratory tree	Gills, lungs
Nephridia	Excretory glands resembling nephridia	Nephridia	Absent	Kidneys
Head ganglia with paired ventral cords; ganglia in each segment	Head ganglia with paired ventral nerve cords; ganglia in segments, some fused	Well-developed brain in some cephalopods; several paired ganglia, most in the head; nerve network in body wall	Head ganglia absent; nerve ring and radial nerves; nerve network in skin	Well-developed brain; dorsal nerve cord
Sexual (some hermaphroditic)	Normally sexual	Sexual (some hermaphroditic)	Sexual (some hermaphroditic); asexual by regeneration (rare)	Sexual
Hydrostatic skeleton	Exoskeleton	Hydrostatic skeleton	Endoskeleton of plates beneath outer skin	Endoskeleton of cartilage or bone

Although some sponges have a definite size and shape, others grow in free-form shape over rocks in their aquatic habitats (Fig. 22-4). All sponges have a similar general body plan (Fig. 22-5). The body is perforated by numerous tiny pores, through which water enters, and by fewer, large openings (called *oscula*), through which it is expelled. Within the sponge, water travels through canals. During its passage, oxygen is extracted, microorganisms are filtered out and taken into individual cells where they are digested, and wastes are released. Sponges have three major cell types (see Fig. 22-5), each with a specialized role. Flattened **epithelial cells** cover their outer body surfaces. Some epithelial cells are modified into *pore cells*, which surround pores, controlling their size and regulating the flow of water. The pores are closed when harm-

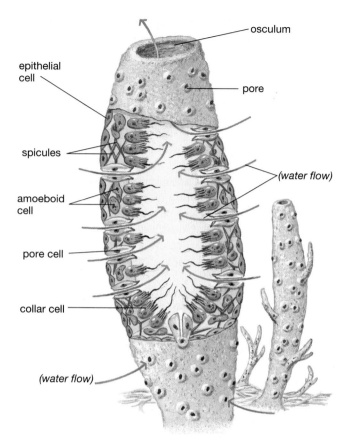

Figure 22-5 The body plan of sponges
All sponges have a similar body plan. Currents created by collar cells draw water in through numerous tiny pores. Microscopic food particles are filtered out by collar cells and shared among the various cell types. Water exits through larger pores, the oscula. Spicules form a supportive internal skeleton.

(a)

(b)

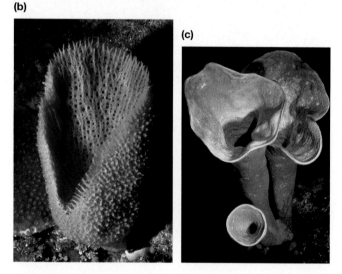

(c)

Figure 22-4 The diversity of sponges
Sponges come in a wide variety of sizes, shapes, and colors. Some are more than a meter tall; others, such as *(a)* this fire sponge, grow in free-form pattern over undersea rocks. *(b)* Tiny appendages attach this tubular sponge to rocks, whereas *(c)* this reef sponge with flared tubular openings attaches to a coral reef.

ful substances are present. **Collar cells** maintain a flow of water through the sponge by beating a flagellum that extends into the inner canal. The collar that surrounds the flagellum acts as a fine sieve, filtering out microorganisms that are then ingested by the cell. Some of the food is passed to the **amoeboid cells**. These cells roam freely between the epithelial and collar cells, digesting and distributing nutrients, producing reproductive cells, and secreting small skeletal projections called **spicules**.

Sponges can grow to more than a meter in height. An internal skeleton composed of spicules provides support for the body (see Fig. 22-5). The spicules may be formed from calcium carbonate (chalk), silica (glass), or protein. The natural bath sponge, now rarely used, is a proteinaceous sponge skeleton.

More than 5000 species of sponges have been identified; all are aquatic, and most are marine. All adult sponges are sessile, attaching themselves to rocks or other underwater surfaces. Sponges may reproduce asexually by **budding**, in which the adult produces miniature versions of itself that drop off and assume an independent existence, or sexually through the fusion of sperm and eggs. Fertilized eggs develop inside the adult into active larvae that escape through the oscula. Water currents disperse the larvae to new areas, where they settle and develop into adult sponges.

5 The Hydra, Anemones, and Jellyfish: Phylum Cnidaria

The *cnidarians* are clearly a step above the sponges in complexity. Their cells are organized into distinct tissues, including contractile tissue that acts like muscle. The nerve cells are organized into tissue called a **nerve net**, which branches through the body and controls the contractile tissue to bring about movement and feeding behavior. Most cnidarians lack true organs, however, and they have no brain.

Cnidarians come in a bewildering and beautiful variety of forms (Fig. 22-6), all of which are actually variations on two basic body plans: the **polyp** (Fig. 22-7a) and the **medusa** (Fig. 22-7b). The generally tubular polyp is adapted to a life spent quietly attached to rocks. The polyp has **tentacles**, extensions that reach upward for grasping, stinging, and immobilizing prey. Although the medusa ("jellyfish") swims weakly by contracting its bell-

Figure 22-6 Cnidarian diversity
(a) A red-spotted anemone spreads its tentacles to capture prey. *(b)* A close-up of coral reveals bright yellow polyps in various stages of tentacle extension. At the lower right, areas where the coral has died expose the calcium carbonate skeleton that supports the polyps and forms the reef. A strikingly patterned crab (phylum Arthropoda: class Crustacea) sits atop the coral, holding tiny white anemones in its claws. Their stinging tentacles help protect the crab. *(c)* A small medusa.
(d) The Portuguese man-of-war, a cnidarian whose stings are dangerous to humans. A stunned fish is trapped in its tentacles.

(a) POLYP

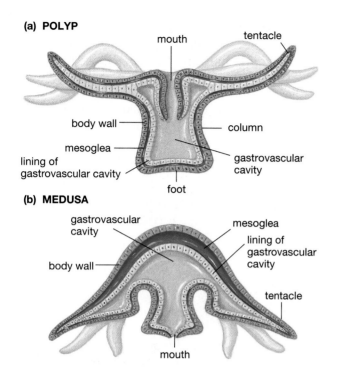

(b) MEDUSA

Figure 22-7 Polyp and medusa
The two basic body forms of cnidarians are actually variations on a single, simple theme. **(a)** The polyp form is seen in hydra (see Fig. 22-8), sea anemones (Fig. 22-6a), and the individual polyps within a coral (Fig. 22-6b). **(b)** The medusa form, seen in the jellyfish (Fig. 22-6c), resembles an inverted polyp. Both forms exhibit radial symmetry, with body parts arranged in a circle around a central axis.

shaped body, it is primarily carried by ocean currents, trailing its tentacles like multiple fishing lines. Both polyp and medusa develop from just two germ layers—the interior endoderm and the exterior mesoderm; between those layers is a jellylike **mesoglea**. Polyps and medusae are radially symmetrical, with body parts arranged in a circle around the mouth and digestive cavity (see Fig. 22-2). This arrangement of parts is well suited to these animals, which are either sessile or carried randomly by water currents, because they are prepared to capture prey or defend themselves from any direction.

Although all cnidarians are predatory, none hunt actively. Instead, they rely on their victims' blundering by chance into the grasp of their enveloping tentacles. Cnidarian tentacles are armed with **cnidocytes**, cells containing structures that, when stimulated by contact, explosively inject poisonous or sticky darts into prey (Fig. 22-8). Stung and firmly grasped, the prey is forced through an expansible mouth into a digestive sac, the **gastrovascular cavity**. Digestive enzymes secreted into this cavity break down some of the food, and further digestion occurs within the cells lining the cavity. Because the gastrovascular cavity has only a single opening, when digestion is completed, undigested material is expelled through the mouth. Although this two-way traffic prevents continuous feeding, it is adequate to support the low energy demands of these animals.

Cnidarians can reproduce both asexually and sexually. Some medusae and some polyps, such as hydras and sea anemones, bud off miniature replicas of themselves (see Fig. 35-1). Sexual reproduction in cnidarians involves the fusion of sperm and eggs that have been released into the water or are retained within the parent. The fertilized egg typically develops into a free-swimming, ciliated larval stage that settles to the ocean floor and becomes a tiny polyp.

Of the 9000 or more species of the phylum Cnidaria, all are aquatic and most are marine. One of these, the corals, are of particular ecological importance (see Fig. 22-6b). These polyps secrete a hard protective "house" of limestone, a rock consisting mainly of calcium carbonate. The limestone persists long after the corals' death, serving as a base to which others may attach themselves. The cycle continues until, after thousands of years, massive coral reefs are formed. Corals are restricted to the warm, clear waters of the tropics, where their reefs form undersea habitats that become

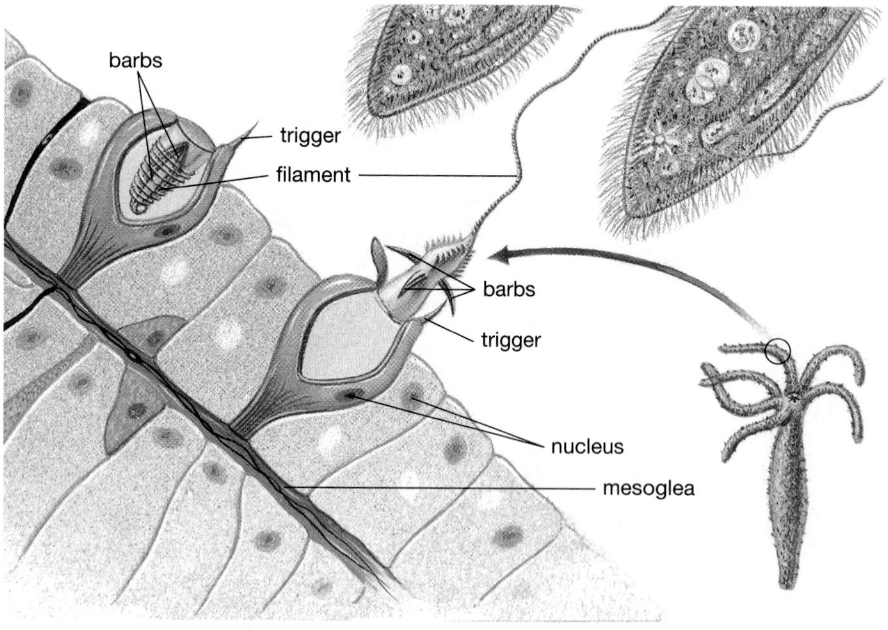

Figure 22-8 Cnidarian weaponry: the cnidocyte
At the slightest touch to the trigger of a special structure in their cnidocytes, cnidarians, such as this hydra, violently expel a poisoned or sticky dart. The barbs and hollow filament turn inside out, impaling the prey and injecting a paralyzing venom. These structures are microscopic. Only a few species inject enough venom to harm a human.

the basis of an ecosystem of stunning diversity and unparalleled beauty (see Chapter 41).

6 The Flatworms: Phylum Platyhelminthes

Flatworms—members of the phylum Platyhelminthes—do not look anything like cnidarians. Yet, the two share some features that have led biologists to speculate that they evolved from a common ancestor. Both have a gastrovascular cavity with a single opening. Certain flatworms also show striking similarities to the larval stage of cnidarians.

Flatworms, however, are clearly more complex than cnidarians. First, they are bilaterally symmetrical rather than radially symmetrical (see Fig. 22-2). This body plan, found in all the more complex animals, is an adaptation to active movement. The anterior end, where the sense organs are concentrated, first encounters the environment ahead. On the basis of what is encountered, these sense organs inform the organism to feed, forge onward, or retreat. In **free-living** (nonparasitic) flatworms such as the freshwater *planarians* (Fig. 22-9), the sense organs consist of eyespots for detecting light and dark and cells responsive to chemical and tactile stimuli. To process information, flatworms have clusters of nerve cells called **ganglia** (singular, **ganglion**) in the head that form a simple brain. Paired neural structures called **nerve cords** conduct nervous signals to and from the ganglia.

Flatworms are the simplest organisms with well-developed organs, in which tissues are grouped into functional units. When a free-living flatworm encounters food, normally smaller animals, it sucks up its prey by using a muscular tube called a **pharynx**, which is located in the middle of the ventral side of its body. The food is digested in an intricately branched gastrovascular cavity that distributes nutrients to all parts of the body (see Fig. 22-9a). Free-living flatworms have a simple system for regulating water levels in their bodies and for excreting some dissolved wastes. This system consists of a network of canals that end in bulbs containing beating cilia. The beating cilia drive liquids through the system, emptying excess fluids to the outside through numerous tiny pores.

Flatworms lack both respiratory and circulatory systems. Nutrients are distributed by the branched digestive tract, from which they readily diffuse into nearby cells. Gas exchange by diffusion between the cells and the environment is aided by the flatness of the body, which ensures that all body cells are relatively close to the outside environment.

Flatworms can reproduce both sexually and asexually. Free-living forms may reproduce by cinching themselves around the middle until they separate into two halves, each of which regenerates its missing parts. All forms can reproduce sexually; most are **hermaphroditic**—

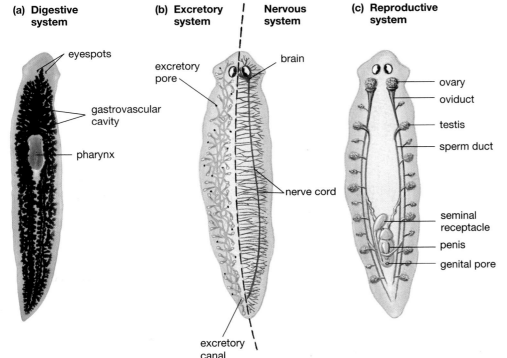

(a) Digestive system
- eyespots
- gastrovascular cavity
- pharynx

(b) Excretory system | **Nervous system**
- excretory pore
- brain
- nerve cord
- excretory canal

(c) Reproductive system
- ovary
- oviduct
- testis
- sperm duct
- seminal receptacle
- penis
- genital pore

Figure 22-9 Flatworm organ systems
Flatworms such as planarians have well-developed organ systems. *(a)* The elaborately branched digestive system, the centrally located ventral pharynx, and eyespots in the head are clearly visible. *(b)* (Left) The excretory system consists of branching tubes that conduct excess fluid to the outside through numerous pores. Cilia keep the fluid moving. (Right) The nervous system of flatworms shows clear cephalization, with eyes and a brain composed of ganglia cells in a well-defined head. Ladderlike nerve cords carry signals through the rest of the body. *(c)* Flatworm reproductive systems include both male and female reproductive organs, such as an ovary and testis.

that is, they possess both male and female sexual organs (Figs. 22-9c and 22-10). This feature is a great advantage to parasitic forms, because each worm can reproduce through self-fertilization, even if it is the only individual present in its host.

Although many flatworms, such as the planarians, are free-living, those of major importance to humans are parasites. **Parasites** are organisms that live in or on the body of another organism, called a *host*, which is harmed as a result of the relationship. The parasitic flatworms include the *tapeworms,* several of which can infect humans. In most cases, people become infected by eating improperly cooked beef, pork, or fish that has been infected by the worms. Worm larvae form encapsulated resting

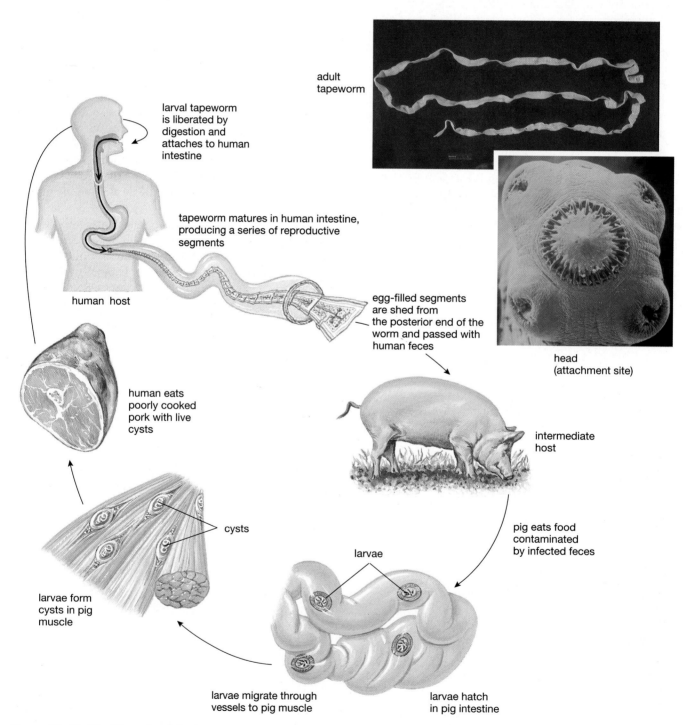

adult tapeworm

larval tapeworm is liberated by digestion and attaches to human intestine

tapeworm matures in human intestine, producing a series of reproductive segments

human host

egg-filled segments are shed from the posterior end of the worm and passed with human feces

head (attachment site)

human eats poorly cooked pork with live cysts

intermediate host

cysts

pig eats food contaminated by infected feces

larvae

larvae form cysts in pig muscle

larvae migrate through vessels to pig muscle

larvae hatch in pig intestine

Figure 22-10 The life cycle of the human pork tapeworm
Each reproductive unit, or proglottid, is a self-contained reproductive factory that includes both male and female sex organs.

structures, called **cysts**, in the muscles of these animals. The cysts hatch in the human digestive tract, where they attach to the intestine and mature. There they may grow to a length of more than 7 meters (20 feet), absorbing digested nutrients directly through their outer surface, fertilizing themselves, and releasing packets of eggs that are shed in the host's feces. If pigs eat grass contaminated with infected human feces, the eggs hatch in the pig's digestive tract, releasing larvae that burrow into its muscles and form cysts, thereby continuing the infective cycle (Fig. 22-10).

Another group of parasitic flatworms is the *flukes*. Of these, the most devastating are liver flukes (common in Asia) and blood flukes, such as those of the genus *Schistosoma*, which cause the disease schistosomiasis. Like most parasites, flukes have a complex life cycle that includes an intermediate host (a snail in the case of *Schistosoma*). Prevalent in Africa and parts of South America, schistosomiasis affects an estimated 200 million people worldwide. Its symptoms include diarrhea, anemia, and possible brain damage. Efforts to control the spread of schistosomiasis are sometimes impeded by the unintentional consequences of other human activities. For example, in Egypt, the irrigation ditches filled by the Aswan Dam, completed in 1968, have contributed to the spread of schistosomiasis by creating an extensive new habitat for the snail host.

7) The Roundworms: Phylum Nematoda

Nematodes, also called roundworms, have been enormously successful in colonizing nearly every habitat on Earth. Although only about 10,000 species in the phylum Nematoda have been named, there may be as many as 500,000. Most are microscopic, such as the one shown in Figure 22-11, but some parasitic forms reach a meter in length. Nematodes have a rather simple body plan, featuring a tubular gut that runs from mouth to anus. This one-way digestive tract represents a major evolutionary advance over the flatworm digestive system. A fluid-filled pseudocoelom surrounds the organs and forms a **hydrostatic skeleton**, a framework against which muscles can act. A tough, flexible, nonliving cuticle encloses and protects the thin, elongated body (Fig. 22-11). Sensory organs in the head transmit information to a simple "brain," composed of a ring of ganglia. Nematodes lack both circulatory and respiratory systems. Because most are extremely thin and all have low energy requirements, diffusion suffices for gas exchange and the distribution of nutrients. Most nematodes reproduce sexually, and the sexes are separate, with the male (who is normally smaller) fertilizing the female by placing sperm inside her body.

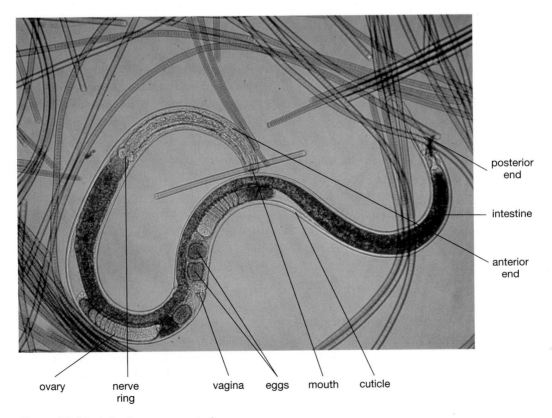

posterior end

intestine

anterior end

ovary   nerve ring   vagina   eggs   mouth   cuticle

Figure 22-11 A freshwater nematode
Eggs can be seen inside this female freshwater nematode, which feeds on algae.

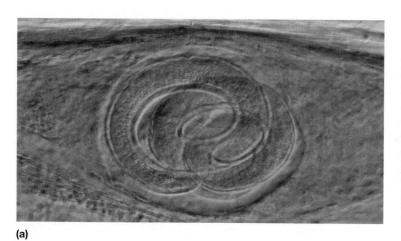

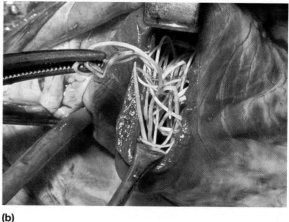

(a)

(b)

Figure 22-12 *Some parasitic nematodes*
(a) Encysted larva of the *Trichinella* worm in the muscle tissue of a pig, where it may live for up to 20 years.
(b) Adult heartworms in the heart of a dog. The juveniles are released into the bloodstream, where they may be ingested by mosquitoes and passed to another dog by the bite of an infected mosquito.

Although you may be blissfully unaware of their presence, roundworms are nearly everywhere outdoors, where they play an important role in breaking down organic matter. A single rotting apple may contain 90,000 worms. Billions thrive in each acre of topsoil. Nearly all plants and animals are host to several parasitic species.

During your life, you may be parasitized by one of the 50 species of roundworms that specialize on humans. Most such worms are relatively harmless, but there are important exceptions. For example, hookworm larvae in soil may bore into human feet, enter the bloodstream, and travel to the intestine, where they cause continuous bleeding. The *Trichinella* worm, which causes trichinosis, might be ingested by eating improperly cooked infected pork. Infected pork may contain up to 15,000 larval cysts per gram (Fig. 22-12a). The cysts hatch in the human digestive tract and invade blood vessels and muscles, causing bleeding and muscle damage. Another dangerous nematode parasite, the heartworm of dogs, is transmitted by mosquitoes (Fig. 22-12b). In the southern United States, and increasingly in other parts of the country, it poses a severe threat to the health of unprotected pets.

8 The Segmented Worms: Phylum Annelida

As the name suggests, a prominent feature of the segmented worms, or *annelids* (from the Latin word for "little ring"), is the division of the body into a series of repeating segments. Externally, these segments appear as ringlike depressions on the surface. Internally, many of the segments contain identical copies of nerve ganglia, excretory structures, and muscles. Segmentation is advantageous for locomotion, because the body compartments, each of which is controlled by separate muscles, collectively are capable of far greater complexity of movement than is seen in the nonsegmented worms. Another evolutionary innovation that distinguishes annelids from flatworms and roundworms is a fluid-filled true coelom between the body wall and the digestive tract (see Fig. 22-3c). The incompressible fluid in the coelom of many annelids is confined by the partitions between the segments and serves as a hydrostatic skeleton, making possible such feats as burrowing through soil.

In contrast to the nematodes, the annelids have a well-developed **closed circulatory system** that distributes gases and nutrients throughout the body (see Chapter 27). In closed circulatory systems (including yours), blood remains confined to the heart and blood vessels. In the earthworm, for example, blood with oxygen-carrying hemoglobin is pumped through well-developed vessels by five pairs of "hearts" (Fig. 22-13). These hearts are actually short, expanded segments of specialized blood vessels that contract rhythmically. The blood is filtered and wastes are removed by excretory organs called **nephridia** (singular, **nephridium**), which are found in many of the segments. Nephridia resemble the individual tubules of the vertebrate kidney (see Chapter 30). The annelid nervous system consists of a simple ganglionic brain in the head and a series of repeating paired segmental ganglia joined by a pair of ventral nerve cords along the length of the body.

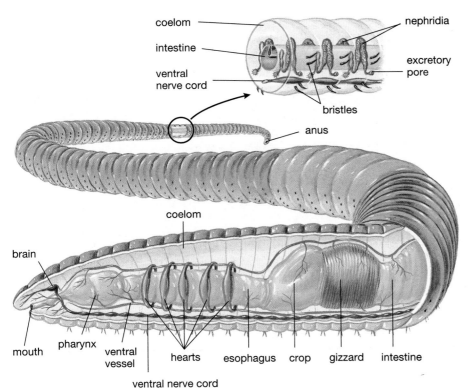

Figure 22-13 An annelid, the earthworm
This diagram shows an enlargement of segments, many of which are repeating similar units separated by partitions. The digestive system, which has both a mouth and an anus, is divided into a series of compartments specialized to process food in an orderly sequence.

Digestion in annelids occurs in a series of compartments, each specialized for a different phase of food processing (see Fig. 22-13). For example, in the earthworm, a muscular pharynx draws in food, consisting of bits of decaying plant and animal debris in soil. The food is conducted through the esophagus to a storage chamber, the *crop*, and then released slowly into the muscular *gizzard*, where it is ground into tiny particles by muscular contractions of the gizzard and the sharp-edged sand grains it contains. The food then passes into the intestine, where it is digested and nutrients are absorbed. Undigested food and soil exit through the anus.

The phylum Annelida includes about 9000 described species, including the familiar earthworm and its relatives, the *oligochaetes* (meaning "few bristles"). In general, these worms exchange gas by diffusion through moist skin. The largest group of annelids, the *polychaetes* (meaning "many bristles") live primarily in the ocean. Some have numerous bristles and paired fleshy paddles on most of their segments, used in locomotion. Others live in tubes from which they project feathery gills that both exchange gases and sift the water for microscopic food (Fig. 22-14a,b). A third group of annelids (class Hirudinea) consists of the *leeches* (Fig. 22-14c). These worms, living in freshwater or moist terrestrial habitats, are either carnivorous or parasitic—some preying on smaller invertebrates, others sucking the blood of larger animals.

9 The Insects, Arachnids, and Crustaceans: Phylum Arthropoda

Arthropods are the dominant animals on Earth. In terms of both number of individuals and number of species, no other phylum comes close to the phylum Arthropoda. About 1 million arthropod species have been discovered, and scientists estimate that up to 9 million remain undescribed. The phylum includes a huge array of forms, such as the insects (class Insecta), the spiders and their relatives (class Arachnida), and the crabs, shrimp, and their relatives (class Crustacea). Anyone who has spread a picnic beside a stream will recall shooing flies from the potato salad or cautiously observing a yellowjacket attack a sandwich and flying off with a small piece of meat in its grasp. The picnicker who is distracted by a spectacular butterfly may well turn back to find the cookies covered with ants and a spider spinning its web across the basket handle. Nearby, dragonflies hover at the water's edge, casting shadows on the crayfish in the water below. As dusk falls, the outing may well be brought to a rapid close by advancing swarms of mosquitoes.

The success of the arthropods can be attributed to several important adaptations that have allowed them to exploit nearly every habitat on Earth. These adaptations include an exoskeleton, segmentation, efficient

(a)

(b)

(c)

Figure 22-14 *Diverse annelids*
(a) A polychaete annelid projects brightly spiraling gills from a tube attached to rock. When the gills retract, the tube is covered by the trap door visible on the lower right. *(b)* The "fireworm" polychaete swims by using paddles on each segment. The bristles on each paddle can deliver a fiery sting. *(c)* This leech, a freshwater annelid, shows numerous segments. The sucker encircles its mouth, allowing it to attach to its prey. Doctors used leeches medicinally up until the 1800s to suck "tainted" blood from patients; even today leeches are used by some surgeons to prevent blood from building up around healing wounds.

gas-exchange mechanisms, and well-developed circulatory, sensory, and nervous systems.

The **exoskeleton** (*exo* meaning "outside" in Greek) is an external skeleton that encloses the arthropod body like a suit of armor. In places it is thin and flexible, allowing movement of the paired, *jointed appendages* from which the phylum Arthropoda (Greek for "jointed foot") derives its name. The exoskeleton is secreted by the *epidermis* (the outer layer of skin) and is composed chiefly of protein and a polysaccharide called **chitin**. This external skeleton protects against predators and is responsible for arthropods' greatly increased agility relative to their wormlike ancestors. By providing stiff but flexible appendages and rigid attachment sites for muscles, the exoskeleton makes possible the flight of the bumblebee and the intricate, delicate manipulations of the spider as it weaves its web (Fig. 22-15). The exoskeleton also contributed enormously to the arthropod invasion of dry terrestrial habitats by providing a watertight covering for delicate, moist tissues such as those used for gas exchange.

Like a suit of armor, the arthropod exoskeleton poses some unique problems. First, because it cannot expand as the animal grows, the exoskeleton must be shed, or

Figure 22-15 *The exoskeleton allows precision movements*
A garden spider, having immobilized its prey with a paralyzing venom, rapidly encases the prey in its web. Such dexterous manipulations are made possible by the exoskeleton and jointed appendages characteristic of arthropods.

molted, periodically and replaced with a larger one (Fig. 22-16). Molting uses energy and leaves the animal temporarily vulnerable until the new skeleton hardens. ("Soft-shelled" crabs are simply regular "hard-shelled" crabs that are caught and eaten during this delicate period.) The exoskeleton is also heavy; its weight increases exponentially as the animal grows. It is no coincidence that the largest arthropods are crustaceans (crabs and lobsters), whose watery habitat supports much of their weight.

Segmentation in arthropods is evidence that they shared a common ancestor with annelids. Arthropod segments, however, tend to be fewer, less distinct from one another, and specialized for different functions such as sensing the environment, feeding, and movement (Fig. 22-17). For example, in insects sensory and feeding structures are concentrated on the anterior-most segment, known as the **head**, and digestive structures are largely confined to the **abdomen**, which is the segment at the

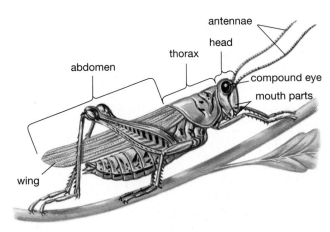

Figure 22-17 *Segments are fused and specialized in insects* Insects such as this grasshopper show fusion and specialization of body segments into a distinct head, thorax, and abdomen. Segments are visible beneath the wings on the abdomen.

Figure 22-16 *The exoskeleton must be molted periodically* A newly emerged praying mantis (a predatory insect) hangs beside its outgrown exoskeleton (left).

animal's posterior end. Between the head and the abdomen is the **thorax**, the segment to which structures used in locomotion, such as wings and walking legs, are attached.

Efficient gas exchange is required to supply adequate oxygen to the muscles that allow the rapid flight, swimming, or running displayed by many arthropods. Gas exchange is accomplished by **gills** (thin, external respiratory membranes) in aquatic forms such as the crustaceans, and by either **tracheae** (singular, **trachea**; a network of narrow, branching respiratory tubes) or **book lungs** (specialized respiratory structures of arachnids) in terrestrial forms. Arthropods also have well-developed circulatory systems with a feature not seen in annelids: the **hemocoel**, or blood cavity. Blood not only travels through vessels but also empties into the hemocoel, where it bathes the internal organs directly. This arrangement, known as an **open circulatory system**, is also present in most mollusks (as we shall see in Chapter 27).

Most arthropods possess a well-developed sensory system, including **compound eyes**, which have multiple light detectors (Fig. 22-18) and acute chemical and tactile senses. The arthropod nervous system is similar in plan to that of annelids, but more complex. It consists of a brain composed of fused ganglia in the head, connected to a series of ganglia along the length of the body and linked by a ventral nerve cord. The capacity for finely coordinated movement combined with sophisticated sensory abilities and a well-developed nervous system has allowed the evolution of complex behavior. In fact, the social behavior of some insect species, such as the honeybee, is as intricate and complex as any vertebrate behavior. Although many people associate communication and learning with

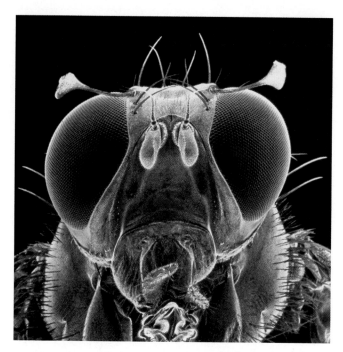

Figure 22-18 *Arthropods possess compound eyes*
This scanning electron micrograph shows the compound eye of a fruit fly. Compound eyes consist of an array of similar light-gathering and sensing elements whose orientation gives the arthropod a panoramic view of the world. Insects have reasonably good image-forming ability and good color discrimination.

vertebrates, both play important roles in insect societies (as we shall see in Chapter 37).

Insects Are the Most Diverse and Abundant Arthropods

The number of described *insect* species (members of the class Insecta; Fig. 22-19) is about 850,000, roughly three times the total number of known species in all other classes of animals combined. Insects have three pairs of legs, normally supplemented by two pairs of wings. Insects' capacity for flight distinguishes them from all other invertebrates and has contributed to their enormous success (Fig. 22-19c). As anyone who has unsuccessfully pursued a fly can testify, flight helps in escaping from predators. It also allows the insect to find widely dispersed food. Swarms of locusts (Fig. 22-19d) have been traced all the way from Saskatchewan, Canada, to Texas on the trail of food. Flight requires rapid and efficient gas exchange, which insects accomplish by means of tracheae. The network of tracheae conducts air to all parts of the body.

During their development, insects undergo **metamorphosis**, which commonly involves a radical change in body form from juvenile to adult. In insects with complete metamorphosis, the immature form, called a **larva**, is worm-shaped (for example, the maggot of a housefly or the caterpillar of a moth or butterfly; see Fig. 22-19e).

Figure 22-19 *The diversity of insects*
(a) The rose aphid sucks sugar-rich juice from plants. *(b)* A mating pair of Hercules beetles. The large "horns" are found only on the male. *(c)* A June beetle displays its two pairs of wings as it comes in for a landing. The outer wings protect the abdomen and the inner wings, which are relatively thin and fragile. *(d)* Insects such as this locust can cause devastation of both crops and natural vegetation. *(e)* Caterpillars are larval forms of moths or butterflies. This caterpillar larva of the Australian fruit-sucking moth displays large eyespot patterns that may frighten potential predators, who mistake them for eyes of a large animal.

The larva hatches from an egg, grows by eating voraciously and shedding its exoskeleton several times, then forms a nonfeeding form called a **pupa**. Encased in an outer covering, the pupa undergoes a radical change in body form, emerging in its adult winged form. The adults mate and lay eggs, continuing the cycle. Metamorphosis may include a change in diet as well as in shape, thereby eliminating competition for food between adults and juveniles and in some cases allowing the insect to exploit different foods when they are most available. For instance, a caterpillar that feeds on new green shoots in springtime metamorphoses into a butterfly that drinks nectar from the summer's blooming flowers. Some insects undergo a more gradual metamorphosis (called *incomplete metamorphosis*), hatching as young that bear some resemblance to the adult, then gradually acquiring more-adult features as they grow and molt.

Spiders, Scorpions, and Their Relatives Are Members of the Class Arachnida

The *arachnids* comprise about 50,000 species of terrestrial arthropods, including spiders, mites, ticks, and scorpions (Fig. 22-20). All members of the class Arachnida have eight walking legs, and most are carnivorous; many subsist on a liquid diet of blood or predigested prey. For example, spiders, the most numerous arachnids, first immobilize their prey with a paralyzing venom. They then inject digestive enzymes into the helpless victim (typically an insect) and suck in the resulting soup. Arachnids breathe by using either tracheae, book lungs, which are unique to arachnids, or both. Arachnids have simple eyes, each with a single lens, in contrast to the compound eyes of insects and crustaceans. The eyes are particularly sensitive to movement, and in some species they probably can form images. Most spiders have eight eyes placed in such a way as to give them a panoramic view of predators and prey.

Crabs, Shrimp, Crayfish, and Their Relatives Are Members of the Class Crustacea

The roughly 30,000 species of *crustaceans*, including crabs, crayfish, lobster, shrimp, and barnacles, make up the Crustacea, the only class of arthropods that is primarily aquatic (Fig. 22-21). Crustaceans range in size from the microscopic maxillopod *Derocheilocaris* (see the chapter-opening photograph), which lives in the spaces

Figure 22-20 The diversity of arachnids
(a) The tarantula is among the largest spiders but is relatively harmless. **(b)** Scorpions, found in warm climates including deserts of the American Southwest, paralyze their prey with venom from a stinger at the tip of the abdomen. A few species can harm humans. **(c)** Ticks before (left) and after feeding on blood. The uninflated exoskeleton is flexible and folded, allowing the animal to become grotesquely bloated while feeding.

Figure 22-21 *The diversity of crustaceans*
(a) The microscopic waterflea, *Daphnia*, is common in freshwater ponds. Notice the eggs developing within the body. *(b)* The sowbug, found in dark, moist places such as under rocks, leaves, and decaying logs, is one of the few crustaceans to invade the land successfully. *(c)* The hermit crab protects its soft abdomen by inhabiting an abandoned snail shell. *(d)* The goose-neck barnacle uses a tough, flexible stalk to anchor itself to rocks, boats, or even animals such as whales. Other types of barnacles attach with shells that resemble miniature volcanoes (see Fig. 22-24b). Early naturalists thought barnacles were mollusks until the jointed legs, seen extending into the water, were observed.

between grains of sand, to the largest of all arthropods, the Japanese crab, with legs spanning nearly 4 meters (12 feet). Crustaceans have two pairs of sensory *antennae*, but the rest of their appendages are highly variable in form and number, depending on the habitat and lifestyle of the species. Most crustaceans have compound eyes similar to those of insects, and nearly all respire by means of gills.

10 The Snails, Clams, and Squid: Phylum Mollusca

Mollusks share a common ancestry with their fellow protostomes, the annelids and arthropods. In number and variety (about 50,000 species have been described), they are second only to the arthropods. Members of the phylum Mollusca (whose name comes from the Latin *mollis*, meaning "soft") have a moist, muscular body supported by a hydrostatic skeleton. Some protect their body with a shell of calcium carbonate; others escape predation by moving swiftly or, if caught, by tasting terrible. Mollusks have a **mantle**, an extension of the body wall that forms a chamber for the gills and secretes the shell in shelled species. Most mollusks have an open circulatory system, like that of arthropods, with blood directly bathing the organs in a hemocoel. The nervous system, like that of annelids and arthropods, consists of ganglia connected by nerves, but many more of the ganglia are concentrated in the brain. Reproduction is sexual; some species have separate sexes, and others are hermaphroditic. Although mollusks are enormously diverse, a simplified diagram of the body plan of a mollusk is shown in Figure 22-22.

Among the many classes of mollusks, we will discuss three in more detail: snails and their relatives (the class Gastropoda), clams and their relatives (the class Bivalvia), and octopuses and their relatives (the class Cephalopoda).

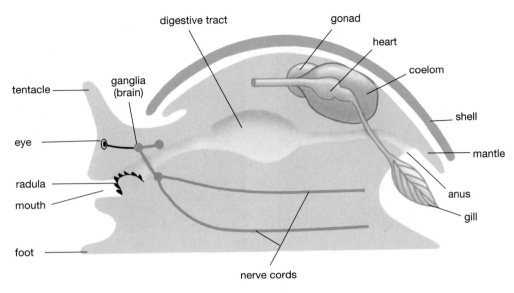

Figure 22-22 *A generalized mollusk*
The general body plan of a mollusk, showing the mantle, foot, gills, shell, radula. and other features that are seen in most (but not all) species of mollusk.

Snails and Their Relatives Are Members of the Class Gastropoda

Gastropoda is by far the largest class of mollusks, with about 35,000 known species. *Gastropods* crawl on a muscular *foot* (*gastropoda* is from the Greek for "stomach foot"), and many of these mollusks have shells that vary widely in form and color (Fig. 22-23a). One of the most beautiful varieties of gastropods, the sea slugs, lack shells, but their brilliant colors warn potential predators that they are poisonous or at least taste bad (Fig. 22-23b).

Gastropods feed with a **radula**, a flexible ribbon of tissue studded with spines that is used to scrape algae from rocks or grasp larger plants or prey (see Fig. 22-22). Most gastropods use gills for respiration in addition to their moist skin, through which dissolved gases readily diffuse. The gills of snails are enclosed in a cavity beneath the shell, whereas in many sea slugs gas exchange takes place through the skin. A few gastropods (including the destructive garden snails and slugs) live in moist terrestrial habitats. These terrestrial gastropods (and some freshwater forms that evolved from them) use a simple lung for breathing.

(a)

(b)

Figure 22-23 *The diversity of gastropod mollusks*
(a) A Florida tree snail displays a brightly striped shell, and eyes at the tip of stalks that are retracted instantly if touched. **(b)** Spanish shawl sea slugs prepare to mate. The brilliant colors of many sea slugs warn potential predators that they are distasteful.

(a)

(b)

Figure 22-24 *The diversity of bivalve mollusks*
(a) This swimming scallop parts its hinged shells. The upper shell is covered with an encrusting sponge.
(b) Mussels attach to rocks in dense aggregations exposed at low tide. White barnacles are attached to the mussel shells and surrounding rock.

Scallops, Clams, Oysters, and Their Relatives Are Members of the Class Bivalvia

Included among the *bivalves* are scallops, oysters, mussels, and clams (Fig. 22-24). Not only do members of the class Bivalvia lend exotic variety to the human diet, but they also are extremely important members of the nearshore marine community, where they attach to rocks that are alternately covered and exposed by the tides. Bivalves possess two shells connected by a flexible hinge (hence their name, meaning "two shells"). A strong muscle clamps the shells closed in response to danger; this muscle is what you are served when you order scallops in a restaurant.

Clams use a muscular foot for burrowing in sand or mud. In mussels, which are sessile, the foot is smaller and is used to help secrete threads that anchor the animal to rocks. Scallops lack a foot and move by a sort of whimsical jet propulsion achieved by flapping their shells together. Bivalves are filter feeders, using their gills as both respiratory and feeding structures. Water is circulated over the gills, which are covered with a thin layer of mucus that traps microscopic food particles. Food is conveyed to the mouth by the beating of cilia on the gills. Probably because they filter-feed and do not move extensively, bivalves have "lost their heads" during the course of their evolution.

Octopuses, Squid, and Their Relatives Are Members of the Class Cephalopoda

The *cephalapods* (whose name means "head-foot"), a fascinating group that includes octopuses, squid, nautiluses, and cuttlefish (Fig. 22-25), are among the largest, swiftest, and smartest of all invertebrates. All cephalopods are predatory carnivores, and all are marine. In these mollusks, the foot has evolved into tentacles with well-developed chemosensory abilities and suction disks for detecting and grasping prey. Prey grasped by tentacles may be immobilized by a paralyzing venom in the saliva before being torn apart by beaklike jaws. The cephalopod eye rivals our own in complexity and exceeds it in efficiency of design (as we shall see in Chapter 33).

Cephalopods move rapidly by jet propulsion, which is accomplished by the forceful expulsion of water from the mantle cavity. The octopus may also travel along the seafloor by using its tentacles like multiple, undulating legs. Members of the class Cephalopoda are the only mollusks with a closed circulatory system. It is probably an adaptation to the very active lifestyle of the cephalopods, as it allows more-efficient transport of oxygen and nutrients. The cephalopod brain, especially that of the octopus, is exceptionally large and complex for an invertebrate brain. It is enclosed in a skull-like case of cartilage and endows the octopus with highly developed capabilities to learn and remember. In the laboratory, octopuses can rapidly learn to associate certain symbols with food and to open a screw-cap jar to obtain food.

11) The Sea Stars, Sea Urchins, and Sea Cucumbers: Phylum Echinodermata

Although *echinoderms* have evolved a bewildering diversity of forms, they have never left their ancestral home on the ocean floor. Their descriptive common names re-

(a)

(b)

Figure 22-25 *The diversity of cephalopod mollusks*
(a) An octopus can crawl rapidly by using its eight suckered tentacles. It can alter its color and skin texture to blend with its surroundings. In emergencies this mollusk can jet backward by vigorously contracting its mantle. Octopuses and squid can emit clouds of dark purple ink to confuse pursuing predators. *(b)* The squid moves entirely by contracting its mantle to generate jet propulsion, which pushes the animal backward through the water. The giant squid is the largest invertebrate, reaching a length of 15 meters (almost 50 feet), including tentacles. *(c)* The chambered nautilus secretes a shell with internal, gas-filled chambers that provide buoyancy. Note the well-developed eyes and the tentacles used to capture prey.

(c)

flect this fact: sand dollar, sea urchin, sea star (or starfish), sea cucumber, and sea lily (Fig. 22-26).

The free-swimming embryos of members of the phylum Echinodermata are bilaterally symmetrical, but echinoderm adults have radial symmetry and lack a head. This absence of cephalization is consistent with the sluggish or (in some forms) sessile existence of echinoderms. Most echinoderms move very slowly as they feed on algae or small particles sifted from sand or water. Sea stars, though, are predators and can slowly pursue prey such as bivalve mollusks. Echinoderms move on numerous tiny **tube feet**, delicate cylindrical projections that extend from the lower surface of the body and terminate in a suction cup.

Tube feet are part of a unique echinoderm feature, the **water-vascular system**, which functions in locomotion,

(a)

(b)

(c)

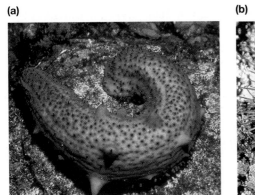

Figure 22-26 *The diversity of echinoderms*
(a) A sea cucumber feeds on debris in the sand. *(b)* The sea urchin's spines are actually projections of the internal skeleton. *(c)* The sea star has reduced spines and typically has five arms.

(a)

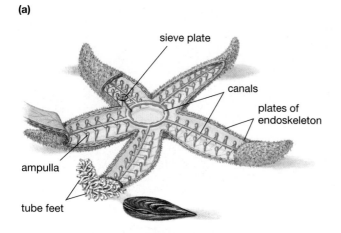

(b)

Figure 22-27 The water-vascular system of echinoderms
(a) Seawater enters through the sieve plate and is transported into the circular central canal, branches of which distribute water to each of the arms. The water inflates squeeze bulb-like ampullae, which expand and contract to extend or retract the tube feet. The plates of the endoskeleton are embedded in the body wall. **(b)** The sea star often feeds on mollusks such as this mussel. A feeding sea star attaches numerous tube feet to the mussel's shells, exerting a relentless pull. Then, the sea star turns the delicate tissue of its stomach inside out, extending it through its centrally located ventral mouth. The stomach can fit through an opening in the bivalve shells of less than 1 millimeter. Once insinuated between the shells, the stomach tissue secretes digestive enzymes that weaken the mollusk, causing it to open further. Partially digested food is transported to the upper portion of the stomach, where digestion is completed.

respiration, and food capture (Fig. 22-27). Seawater enters through an opening (the **sieve plate**) on the animal's upper surface and is conducted through a circular central canal, from which branch a number of radial canals. These canals conduct water to the tube feet, each of which is controlled by a muscular squeeze bulb (**ampulla**). Contraction of the bulb forces water into the tube foot, causing it to extend. The suction cup may be pressed against the seafloor or a food object, to which it adheres tightly until pressure is released.

Echinoderms have a relatively simple nervous system with no distinct brain. Movements are loosely coordinated by a system consisting of a nerve ring that encircles the esophagus, radial nerves to the rest of the body, and a nerve network through the epidermis. In sea stars, simple receptors for light and chemicals are concentrated on the arm tips, and sensory cells are scattered over the skin. The echinoderms lack a circulatory system, although movement of the fluid in their well-developed coelom serves this function. Gas exchange occurs through the tube feet, and in some forms, numerous tiny "skin gills" project through the epidermis. Sea cucumbers possess an internal system of canals called a *respiratory tree*.

Most species reproduce by shedding sperm and eggs into the water, where fertilization occurs; deuterostome development yields a free-swimming larva. The sexes are normally separate. Sea stars have the ability to regenerate lost parts; new individuals may form from a single arm, provided that part of the central body is attached to it. When mussel fisherman tried to rid mussel beds of predatory sea stars by hacking them into pieces and throwing the pieces back, needless to say, the strategy backfired!

Echinoderms possess an **endoskeleton** (*endo* meaning "inside" in Greek) composed of plates of calcium carbonate beneath the outer skin (see Fig. 22-27a). The name *echinoderm* (Greek, "hedgehog skin") comes from projections of the endoskeleton that extend as bumps or spines through the outer layer of skin. These spines are especially well developed in sea urchins and much reduced in sea stars and sea cucumbers.

12 The Tunicates, Lancelets, and Vertebrates: Phylum Chordata

It's hard to believe that humans have anything in common with a sea squirt. Nonetheless, we share the phylum Chordata not only with the birds and the apes but also with the tunicates (sea squirts) and little fishy-looking creatures called *lancelets*. What characteristics unite us with these creatures that seem so different? All chordates share deuterostome development (which is also characteristic of echinoderms) and are further united by four features that all possess at some stage of their lives:

1. A **notochord**. A stiff but flexible rod that extends the length of the body and provides an attachment site for muscles.
2. A **dorsal, hollow nerve cord**. Lying dorsal to the digestive tract, this hollow neural structure develops a thickening at its anterior end that becomes a brain.

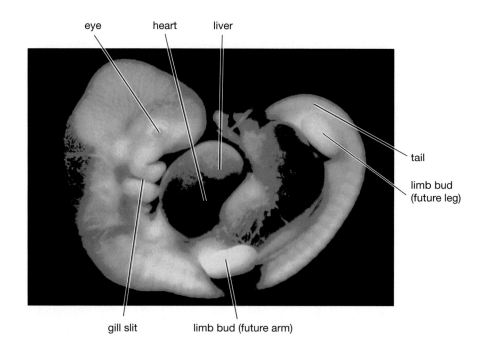

eye heart liver

tail

limb bud (future leg)

gill slit limb bud (future arm)

Figure 22-28 *Chordate features in the human embryo*
This 5-week-old human embryo is about 1 centimeter long and clearly shows external gill slits (more properly called grooves, since they do not penetrate the body wall) and a tail. Although the tail will disappear completely, the gill grooves contribute to the formation of the lower jaw.

3. **Pharyngeal gill slits**. Located in the pharynx (the cavity behind the mouth), these may form functional respiratory openings or may appear only as grooves during an early stage of development.
4. A **post-anal tail**. An extension of the body past the anus.

This list may seem puzzling because, although humans are chordates, at first glance we seem to lack every feature except the nerve cord. But evolutionary relationships are sometimes seen most clearly during early stages of development; it is during our embryonic life that we develop, and lose, our notochord, our gill slits, and our tails (Fig. 22-28). Humans share these chordate features with all other vertebrates and with two invertebrate chordate groups, the lancelets and the tunicates.

The Invertebrate Chordates Include Lancelets and Tunicates

The invertebrate chordates lack the backbone that is the defining feature of vertebrates. These chordates comprise two groups of organisms, the *lancelets* and the *tunicates*. The small (5 centimeters, or about 2 inches, long) fish-like lancelet spends most of its time half-buried in the sandy sea bottom, filtering tiny food particles from the water. As can be seen in Figure 22-29a, all the typical chordate features are present in the adult lancelet.

The tunicates form a larger group of marine invertebrate chordates, which includes the sea squirts. It is difficult to imagine a less likely relative of humans than the sessile, filter-feeding, vaselike sea squirt (Fig 22-29b). Its ability to move is limited to forceful contractions of its saclike body, which can send a jet of seawater into the face of anyone who plucks it from its undersea home;

hence the common name sea squirt. Adult sea squirts may be immobile, but their larvae swim actively and possess all the diagnostic chordate features (Fig. 22-29b).

The Vertebrates Have a Backbone and Other Adaptations That Have Contributed to Their Success

In vertebrates (members of the subphylum Vertebrata), the embryonic notochord is normally replaced during development by a backbone, or **vertebral column**. The vertebral column is composed of bone or **cartilage**, a tissue that resembles bone but is less brittle and more flexible. This column supports the body, offers attachment sites for muscles, and protects the delicate nerve cord and brain. It is also part of a living endoskeleton that can grow and repair itself. Because this internal skeleton provides support without the armorlike weight of the arthropod exoskeleton, it has allowed vertebrates to achieve great size and mobility and has contributed to their invasion of the land and the air.

Vertebrates show other adaptations that have contributed to their successful invasion of most habitats. One is the presence of paired appendages. These first appeared as fins in fish and served as stabilizers for swimming. Over millions of years, some fins were modified by natural selection into legs that allowed animals to crawl onto dry land, and later into wings that allowed some to take to the air. Another adaptation that has contributed to the success of vertebrates is an increase in the size and complexity of their brains and sensory structures. These parallel adaptations allow vertebrates to perceive their environment in detail and to respond to it in a great variety of ways.

(a) Lancelet

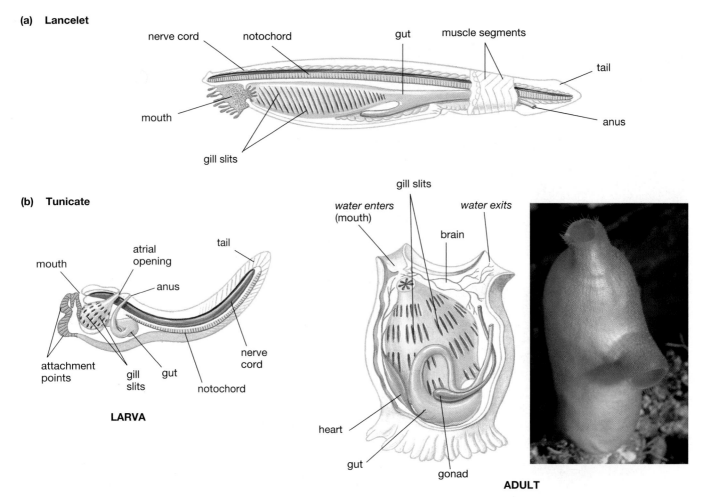

(b) Tunicate

Figure 22-29 Invertebrate chordates
(a) A lancelet, a fishlike invertebrate chordate. The adult organism exhibits all the diagnostic features of chordates. *(b)* The sea squirt larva (left) also exhibits all the chordate features. The adult sea squirt (a type of tunicate, middle) has lost its tail and notochord and has assumed a sedentary life, as shown in the photo (right).

The sequence of evolutionary events that led from invertebrate chordates to the first, fishlike vertebrates remains shrouded in mystery, because fossils of intermediate forms have never been discovered. Today, vertebrates are represented by seven major classes: jawless fishes (class Agnatha), cartilaginous fishes (class Chondrichthyes), bony fishes (class Osteichthyes), amphibians (class Amphibia), reptiles (class Reptilia), birds (class Aves), and mammals (class Mammalia).

Jawless Fishes (Class Agnatha) Were the First Vertebrates to Evolve

Vertebrates arose in the sea; the earliest vertebrate fossils are those of strange *jawless fishes* protected by bony armor plates. Today, two groups of jawless fishes (members of the class Agnatha) survive: the *hagfishes* and the *lampreys,* which are not closely related but nonetheless have many features in common. Both have skeletons of cartilage and are shaped like eels. Both hagfishes and

lampreys have unpaired fins along the midline of the body. Both lack scales, and their smooth, slimy skin is perforated by circular gill openings.

Hagfishes are exclusively marine (Fig. 22-30a). Purple to pink, they live in communal burrows in the mud, feeding primarily on polychaete worms. They eagerly attack dying fish, however, using pincerlike teeth that surround the tongue to burrow into the coelomic cavity and ingest the prey's soft internal organs. They are regarded with great disgust by fishermen because they produce large quantities of slime as a defense against predators.

Lampreys live in both fresh and salt waters; the marine forms return to fresh water to spawn. Lampreys include both parasitic and nonparasitic species. A parasitic lamprey has a suckerlike mouth lined with teeth, which it uses to attach itself to larger fish (Fig. 22-30b). Using rasping teeth on its tongue, the lamprey excavates a hole in the host's body wall, through which it sucks blood and body fluids. Beginning in the 1920s, lampreys spread into

major classes of jawed fishes that survive today: the cartilaginous fishes (class Chondrichthyes) and the bony fishes (class Osteichthyes).

Cartilaginous Fishes (Class Chondrichthyes) Have an Endoskeleton Composed of Cartilage

The class Chondrichthyes, whose name is derived from Greek words meaning "cartilage fishes," includes 625 marine species, among them the sharks, skates, and rays (Fig. 22-31). The *cartilaginous fishes* are graceful predators that lack any bone in their skeleton, which is formed entirely of cartilage. The body is protected by a leathery

(a)

(b)

(a)

(b)

Figure 22-30 Jawless fishes
(a) The colorful hagfishes live in communal burrows in mud, feeding on polychaete worms. **(b)** Some lampreys are parasitic, attaching to fish (such as this carp) with suckerlike mouths lined with rasping teeth **(inset)**.

Figure 22-31 Cartilaginous fishes
(a) A sand tiger shark displaying several rows of teeth. As outer teeth are lost, they are replaced by new ones formed behind them. Both sharks and rays lack a swim bladder and tend to sink toward the bottom when they stop swimming. **(b)** The tropical blue-spotted sting ray swims by graceful undulations of lateral extensions of the body.

the Great Lakes, where, lacking effective predators, they have multiplied prodigiously and greatly reduced commercial fish populations, including the lake trout. Vigorous measures to control the lamprey population have allowed some recovery of the other fish populations of the Great Lakes.

About 425 million years ago, in the mid-Silurian period, primitive jawless fishes that were ancestral to the lampreys and hagfishes gave rise to a group of fish that possessed an important new structure found in all the more-advanced vertebrates: jaws. Jaws allowed fish to grasp and chew their food, permitting them to exploit a much wider range of food sources than could jawless fish. Although the earliest forms of jawed fishes have been extinct for 230 million years, they gave rise to the two

skin roughened by tiny scales. Members of this group respire via gills. Although some must swim to circulate water through their gills, most can pump water across their gills. Like all fish, the cartilaginous fishes have a two-chambered heart (as we shall see in Chapter 27).

Many shark species sport several rows of razor-sharp teeth, the back rows moving forward as the front teeth are lost to age and use (Fig. 22-31). Although a few species consider us potential prey, most sharks are shy of humans. Sharks include the largest fishes, such as the gentle whale shark, which can grow to more than 15 meters (45 feet) in length. Skates and rays (Fig. 22-31b) are also retiring creatures, although some can inflict dangerous wounds with a spine near their tail, and others produce a powerful electrical shock that can stun their prey.

Bony Fishes (Class Osteichthyes) Have Invaded Nearly Every Aquatic Habitat

Just as our size bias makes us overlook the most-diverse invertebrate groups, out habitat bias makes us overlook the most-diverse vertebrates. The most diverse and abundant vertebrates are not the birds or predominantly terrestrial mammals. No, the vertebrate diversity crown belongs to the lords of the oceans and fresh water, the *bony fishes* of the class Osteichthyes. "Osteichthyes" (derived from Greek words for "bone fish") refers to their skeletons, which are composed of bone rather than cartilage. From the snakelike moray eel to bizarre, luminescent deep-sea forms to the streamlined tuna, this enormously successful group has spread to nearly every possible watery habitat, both freshwater and marine (Fig. 22-32). Although about 17,000 species have been identified, scientists predict that perhaps nearly twice this number may exist if the undescribed species from deep waters and remote areas are considered. For example, a bony fish called a *coelacanth*, believed to have been extinct for 75 million years, was caught in deep water off the coast of South Africa in 1939 (Fig. 22-32d).

A feature found in early representatives of the bony fishes and retained in a few modern species is the presence of lungs that supplement the gills. Lungs are advantageous in freshwater habitats that periodically become foul and stagnant or dry up entirely. The *swim bladder,* a sort of internal balloon that allows most bony fishes to float effortlessly at any level, probably evolved from the lungs of freshwater ancestors. Some groups of bony fishes developed modified fleshy fins that could be used (in an emergency) as legs, dragging the fish from a

(a)

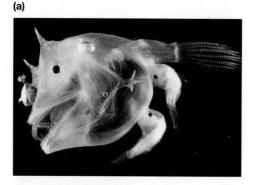

(b)

Figure 22-32 *The diversity of the bony fishes*

Bony fishes have colonized nearly every aquatic habitat. **(a)** This female deep-sea angler fish attracts prey with a living lure that projects just above her mouth. The fish is ghostly white because in the 2000-meter depth where anglers live, no light penetrates and thus colors are superfluous. Male deep-sea angler fish are extremely small and remain as permanent parasites attached to the female, always available to fertilize her eggs. Two parasitic males can be seen attached to this female. **(b)** This tropical green moray eel lives in rocky crevices. A small fish (a banded cleaner goby) on its lower jaw eats parasites that cling to the moray's skin. **(c)** The tropical seahorse may anchor itself with its prehensile tail (adapted for grasping) while the animal feeds on small crustaceans. **(d)** A rare photo of a coelacanth, a fish once believed to be long extinct, in its natural habitat off the coast of South Africa.

(c)

(d)

drying puddle to a deeper pool. From such ancestors arose the group that made the first tentative invasion of the land: the amphibians.

Success on Dry Land Required Numerous Adaptations

To those animals that first crawled from the water, life on land offered many advantages, including abundant food and freedom from aquatic predators. But the price was high. Deprived of water's support, the body was heavy and clumsy to drag along on modified fins or weak, poorly developed legs. Unsupported by water, gills collapsed and became useless. The dry air and relentless sun sucked vital water from unprotected skin and eggs; terrestrial temperatures fluctuated dramatically compared with the relatively steady environmental conditions of the sea. The terrestrial habitat thus favored the preservation of mutations that happened to be adaptive on land. These mutations led to changes in skeletal structure that provided better support for the body, coverings for skin and eggs that reduced water loss due to evaporation, protection of the respiratory membranes, control of body temperature, and more efficient circulation.

Amphibian (Class Amphibia) Means "Double Life"

The class Amphibia consists of about 2500 species that straddle the boundary between aquatic and terrestrial existence (Fig. 22-33). The limbs of amphibians show varying degrees of adaptation to movement on land, from the belly-dragging crawl of salamanders to the long leaps of frogs and toads. Lungs replace gills in most adult forms, and a three-chambered heart (in contrast to the two-chambered heart of fishes) circulates blood more efficiently (as we shall see in Chapter 27). Amphibian lungs, however, are poorly developed and must be supplemented by the skin, which serves as an additional respiratory organ. This respiratory function requires that the skin remain moist, a constraint that greatly restricts the range of amphibian habitats on land.

Amphibians are also tied to moist habitats by their breeding behavior, which requires water. Their fertilization is normally external and must therefore occur in water so that the sperm can swim to the eggs. The eggs must remain in water, as they are protected only by a jellylike coating that leaves them vulnerable to loss of water by evaporation. Fertilized eggs develop into aquatic larvae such as the tadpoles of frogs and toads. The dramatic transformation of these aquatic larvae into semiterrestrial adults gives the class Amphibia its name, which means "double life." Their double life and their thin, permeable skin have made amphibians particularly vulnerable to pollutants and to environmental degradation, as described in "Earth Watch: Frogs in Peril."

(a)

(b)

(c)

Figure 22-33 Amphibian means "double life"
The double life of amphibians is illustrated by the transition from *(a)* the completely aquatic larval tadpole to *(b)* the adult bullfrog leading a semiterrestrial life. *(c)* The red salamander is restricted to moist habitats in the eastern United States. Salamanders hatch in a form that closely resembles the adult.

Earth Watch
Frogs in Peril

Frogs and toads have frequented Earth's ponds and swamps for nearly 150 million years and somehow survived the Cretaceous catastrophe that extinguished the dinosaurs and so many other species about 65 million years ago. Their evolutionary longevity, however, appears to offer inadequate defense against the environmental changes wrought by human activities. During the 1990s, *herpetologists* (biologists who study reptiles and amphibians) from around the world have documented an alarming decline in amphibian populations. Thousands of species of frogs, toads, and salamanders are experiencing a dramatic decrease in numbers, and many have apparently gone extinct.

The phenomenon is not a local one; population crashes have been reported from every part of the globe. Yosemite toads and yellow-legged frogs are disappearing from the mountains of California; tiger salamanders have been nearly wiped out in the Colorado Rockies. Leopard frogs, eagerly chased by rural children, are becoming rare in the United States. Logging destroys the habitats of amphibians from the Pacific Northwest to the tropics (Fig. E22-1), but even amphibians in protected areas are dying. In the Monteverde Cloud Forest Preserve in Costa Rica, the golden toad (see Fig. 35-6) was common in the early 1980s but has not been seen since 1989. The gastric brooding frog of Australia fascinated biologists by swallowing its eggs, brooding them in its stomach, and later regurgitating fully formed offspring. This species was abundant and seemed safe in a national park. Suddenly, in 1980, the gastric brooding frog disappeared and hasn't been seen since.

The causes of the worldwide decline in amphibian diversity are poorly understood, but it seems unlikely that a single factor is responsible. All of the most likely causes stem from human modification of the biosphere—the portion of Earth that sustains life.

Habitat destruction, especially the draining of wetland habitats that are especially hospitable to amphibian life, is one major cause of the decline. The unique biology of amphibians also makes them especially vulnerable to toxic substances in the environment. In particular, amphibian bodies at all stages of life are protected only by a thin, permeable skin through which pollutants can easily penetrate. To make matters worse, the double life of amphibians exposes their permeable skin to a wide range of aquatic and terrestrial habitats and to a correspondingly wide range of environmental toxins. For example, many temperate amphibians' eggs and larvae develop in ponds and streams during the spring, a time when the melting of snow and ice formed from acid rain causes a pulse of intense acidity in freshwater ecosystems. In addition, many amphibians are herbivorous as larvae and insectivorous as adults, so they're exposed to both herbicides and insecticides in their food as their diet shifts.

Amphibian eggs can also be damaged by ultraviolet (UV) light, according to recent research by Andrew Blaustein, an ecologist at Oregon State University. Blaustein demonstrated that the eggs of some species of frogs in the Pacific Northwest are sensitive to damage from UV light and that the most sensitive species are experiencing the most drastic declines. Unfortunately, many parts of Earth are subject to increasingly intense UV radiation levels, because atmospheric pollutants have caused a thinning of the protective ozone layer (as we shall see in Chapter 40).

Increased UV radiation may also be tied to another disturbing trend among frogs and toads, the increasing incidence of grotesquely deformed individuals. The deformities seem to result from environmental rather than genetic causes, as they occur in a variety of species and in widely separated geographic areas. Some researchers think the deformities may be caused by newly developed pesticides that have been applied on a large scale in recent years. Other biologists, including Stanley Sessions of Hartwick College, cite evidence suggesting that some deformities, especially the most common one, extra limbs, are caused by parasitic infections during embryonic development. Sessions has found that many frogs with extra legs are infested with a parasitic flatworm, and he has shown how such infestations could cause deformities. What's the tie-in with UV light? Well, UV radiation may harm frogs by suppressing the immune response. In Sessions's view, a compromised immune system could leave a developing tadpole more vulnerable to parasitic infection.

Many scientists believe that the troubles of amphibians signal an overall deterioration of Earth's ability to support life. According to this line of reasoning, the highly sensitive amphibians are providing an early warning of environmental degradation that will eventually affect more-resistant organisms as well. Equally worrisome is the observation that amphibians are not just sensitive indicators of the health of the biosphere; they are also crucial components of many ecosystems. They may keep insect populations in check, in turn serving as food for larger carnivores. Their decline will further disrupt the balance of these delicate communities. Margaret Stewart, an ecologist at the State University of New York, Albany, aptly summarized the problem: "There's a famous saying among ecologists and environmentalists: 'Everything is related to everything else.' . . .You can't wipe out one large component of the system and not see dramatic changes in other parts of the system."

Figure E22-1 Amphibians in danger
The corroboree toad, shown here with its eggs, is rapidly declining in its native Australia. Tadpoles are developing within the eggs. The thin water-permeable and gas-permeable skin of the adult and the jellylike coating around the eggs make them vulnerable to both air and water pollutants.

Reptiles (Class Reptilia) Have Adaptations That Suit Them to Life on Dry Land

The approximately 7000 species of reptiles include the lizards and snakes (by far the most successful of the modern groups) and the turtles, alligators, and crocodiles, which have survived virtually unchanged from prehistoric times (Fig. 22-34). Reptiles (members of the class

(a)

(b)

(c)

Figure 22-34 The diversity of reptiles
(a) The mountain king snake has a color pattern very similar to that of the poisonous coral snake, which potential predators avoid. This mimicry helps the harmless king snake elude predation. *(b)* The outward appearance of the American alligator, found in swampy areas of the South, has changed little during the past 150 million years. *(c)* The tortoises of the Galapagos Islands, Ecuador, may live to be more than 100 years old.

Figure 22-35 The amniote egg
An anole lizard struggles free of its egg. The amniote egg encapsulates the developing embryo in a liquid-filled membrane (the amnion), ensuring that development occurs in a watery environment, even if the egg is far from water.

Reptilia) evolved from an amphibian ancestor about 250 million years ago. Early reptiles—the dinosaurs—ruled the land for nearly 150 million years.

Some reptiles, particularly desert dwellers such as tortoises and lizards, are completely independent of their aquatic origins. This independence was achieved through a series of adaptations, of which three are outstanding: (1) Reptiles evolved a tough, scaly skin that resists water loss and protects the body. (2) Reptiles evolved internal fertilization, in which the male deposits sperm within the female's body. (3) Reptiles evolved a shelled **amniote egg**, which can be buried in sand or dirt, far from water with its hungry predators. The shell prevents the egg from drying out on land. An internal membrane, the **amnion**, encloses the embryo in the watery environment that all developing animals require (Fig. 22-35).

In addition to these features, reptiles have more-efficient lungs than do earlier vertebrates, dispensing with the skin as a respiratory organ. The three-chambered heart became modified, allowing better separation of oxygenated and deoxygenated blood, and the limbs and skeleton evolved adaptations that provided better support and more-efficient movement on land.

Birds (Class Aves) Have Many Adaptations That Support the Demands of Flight

After spreading throughout the sea and over the land, vertebrates took to the air, where insect food was abundant and earthbound predators were absent. The 9000 or so species of the class Aves attest to the advantages of

Figure 22-36 *The diversity of birds*
(a) The delicate hummingbird beats its wings about 60 times per second and weighs about 0.15 ounce (4 grams). *(b)* This young frigate bird, a fish-eater from the Galapagos Islands, has nearly outgrown its nest. *(c)* The ostrich is the largest of all birds, weighing more than 136 kilograms (300 pounds); its eggs weigh more than 1500 grams (3 pounds).

Figure 22-37 **Archaeopteryx,** *the "missing link" between reptiles and birds*
The earliest known bird is *Archaeopteryx,* preserved in 150-million-year-old limestone. Feathers, a feature unique to birds, are clearly visible. But the reptilian ancestry of birds is also apparent: Like a modern reptile (but unlike a modern bird), *Archaeopteryx* had teeth, a tail, and claws.

flight (Fig. 22-36). The first birds appear in the fossil record roughly 150 million years ago (Fig. 22-37) and are distinguished from reptiles by the presence of feathers, which are derived from the body scales of birds' reptile ancestors. Modern birds retain scales on their legs—a testimony to their reptilian origin.

Bird anatomy and physiology are dominated by adaptations that help meet the rigorous demands of flight. In particular, birds are exceptionally light for their size. Hollow bones reduce the weight of the skeleton to a fraction of that of other vertebrates, and many bones present in reptiles are lost or fused with other bones in birds. Reproductive organs are considerably reduced in size during nonbreeding periods, and female birds possess only a single ovary, further minimizing weight. The shelled egg that contributed to the reptiles' success on land frees the mother bird from carrying her developing offspring. Feathers form lightweight extensions to the wings and the tail for the lift and control required for flight, and they also provide lightweight protection and insulation for the body. The nervous system of birds accommodates the special demands of flight with extraordinary coordination and balance combined with acute eyesight.

Birds are also able to maintain body temperatures high enough to allow their muscles and metabolic processes to operate at peak efficiency, supplying the power necessary to fly regardless of the outside temperature. This physiological ability to maintain an internal temperature that is usually higher than that of the surrounding environment is characteristic of both birds and mammals, which are therefore sometimes described as *warm blooded.* In contrast, the body temperature of invertebrates, fish, amphibians, and reptiles varies with the temperature of their environment, though these animals may exert some control of their body temperature by behavioral means (such as basking in the sun or seeking shade).

(a)

(b)

(c)

(d)

Figure 22-38 *The diversity of mammals*
(a) A humpback whale gives its offspring a boost. *(b)* A bat, the only mammal capable of true flight, navigates at night by using a kind of sonar. Large ears aid in detecting echoes as its high-pitched cries bounce off nearby objects. *(c)* Mammals are named after the mammary glands with which females nurse their young, as illustrated by this mother cheetah. *(d)* The male orangutan can reach 75 kilograms (165 pounds). These gentle, intelligent apes occupy swamp forests in limited areas of the tropics. They are endangered by hunting and habitat destruction.

Warm-blooded animals such as birds have a high metabolic rate, which increases their demand for energy and requires efficient oxygenation of tissues. Birds thus must eat frequently and possess circulatory and respiratory adaptations that help meet the need for efficiency. The heart of a bird has four chambers, serving to prevent the mixing of oxygenated and deoxygenated blood. The respiratory system of birds is supplemented by air sacs that provide a continuous supply of oxygenated air to the lungs, even while the bird exhales.

Mammals (Class Mammalia) Possess Hair and Produce Milk for Their Offspring

Whereas one branch of the reptile evolutionary tree gave rise to feathered birds, a different group developed hair and diverged to form the mammals. The mammals first appeared approximately 250 million years ago but did not diversify and assume terrestrial prominence until after the dinosaurs went extinct roughly 65 million years ago. Today the class Mammalia is represented by some 4500 species. Like birds, mammals are warm blooded and have high metabolic rates. In most mammals, fur protects and insulates the warm body. Like birds, mammals have four-chambered hearts that increase the amount of oxygen delivered to the tissues. Legs designed for running rather than crawling make many mammals fast and agile. In contrast to birds, whose bodies are almost uniformly molded to the requirements of flight, mammals have evolved a remarkable diversity of form. The bat, mole, impala, whale, seal, monkey, and cheetah exemplify the radiation of mammals into nearly all habitats, with bodies finely adapted to their varied lifestyles (Fig. 22-38).

(a)

(b)

Figure 22-39 Nonplacental mammals
(a) Monotremes, such as this platypus from Australia, lay leathery eggs resembling those of reptiles. The newly hatched young obtain milk from slitlike openings in the mother's abdomen. *(b)* Marsupials, such as the wallaby, give birth to extremely immature young who immediately grasp a nipple and develop within the mother's protective pouch **(inset)**.

Mammals are named for the milk-producing **mammary glands** used by all female members of this class to suckle their young (Fig. 22-38c). In addition to these unique glands, the mammalian body is arrayed with sweat, scent, and sebaceous (oil-producing) glands, none of which are found in reptiles. With the exception of the egg-laying **monotremes**, such as the platypus and spiny anteater (Fig. 22-39a), mammalian embryos develop in the *uterus*, a muscular organ in the female reproductive tract. In one group of mammals, the **marsupials** (including opossums, koalas, and kangaroos), the period of uterine development is short and the young are born at a very immature stage of development. Immediately after birth they crawl into a protective pouch (Fig. 22-39b), firmly grasp a nipple, and, nourished by milk, complete their development. Most mammals, called **placental** mammals, retain their young in the uterus for a much longer period. *Placental* refers to the **placenta**, the uterine structure that functions in gas, nutrient, and waste exchange between the circulatory systems of the mother and embryo.

The mammalian nervous system has contributed significantly to the success of the mammals by making possible behavioral adaptation to changing and varied environments. The brain is more highly developed than in any other class, endowing mammals with unparalleled curiosity and learning ability. The highly developed brain allows mammals to alter their behavior on the basis of experience and helps them survive in a changing environment. Relatively long periods of parental care after birth allow some mammals to learn extensively under parental guidance; humans and other primates are good examples. In fact, the intellectual development of humans has led to their domination of the environment, as explored in the following section.

Evolutionary Connections
Are Humans a Biological Success?

Physically, human beings are fairly unimpressive biological specimens. For such large animals, we are not very strong or very fast, and we lack the natural weapons of fang and claw. It is the human brain, with its tremendously developed cerebral cortex, that truly sets us apart from other animals. Our brains give rise to our minds, which, in bursts of solitary brilliance and in the collective pursuit of common goals, have created wonders. No other animal could sculpt the graceful columns of the Parthenon, much less reflect on the beauty of this ancient Greek temple. We alone can eradicate smallpox and polio, domesticate other life-forms, penetrate space with rockets, and fly to the stars in our imaginations.

And yet, are we, as it appears at first glance, the most successful of all living things? The duration of human existence is a mere instant in the 3.5-billion-year span of life on Earth. But during the last 300 years, the human population has increased from 0.5 billion to 5.5 billion and now grows by a million people every 4 days. Is this a measure of our success? As we have expanded our range over the globe, we have driven at least 300 major species to extinction. Within your lifetime, the rapid destruction of tropical rain forests and other diverse habitats may wipe out millions of species of plants, invertebrates, and vertebrates, most of which we will never know. Many of our activities have altered the environment in ways that are hostile to life, including our own. Acid from power plants and automobiles rains down on the land, threatening our forests and lakes—and eroding the marble of the Parthenon. Deserts spread as land is stripped by overgrazing and the demand for firewood. Our aggressive tendencies, spurred by pressures of expanding wants and needs, their scope magnified by our technological prowess, have given us the capacity to destroy ourselves and most other life-forms as well.

The human mind is the source of our most pressing problems—and our greatest hope for solving them. Will we now devote our mental powers to reducing our impact, controlling our numbers, and preserving the ecosystems that sustain us and other forms of life? Are we a phenomenal biological success—or a brilliant catastrophe? Perhaps the next few centuries will tell.

Summary of Key Concepts

1) What Characteristics Define an Animal?
Animals are multicellular, sexually reproducing, heterotrophic organisms. Most can perceive and react rapidly to environmental stimuli and are motile at some stage in their lives. Their cells lack a cell wall.

2) What Are the Major Evolutionary Trends in Animals?
Several trends are apparent in the evolution of animals. There was a trend toward greater overall complexity, culminating with the arthropods, mollusks, echinoderms, and chordates. Cellular organization increased from specialized cells through tissues to organs to organ systems. Three separate germ layers arose. Body forms evolved from asymmetry through radial symmetry to bilateral symmetry. Sense organs and clusters of neurons became increasingly concentrated in the head, a process called cephalization. Body cavities were originally absent until animals with pseudocoeloms and, later, true coeloms evolved. Segmentation, first seen in annelids, is present to some degree in most of the more complex animals. In the simple sponges, digestion is intracellular; later, a saclike gastrovascular cavity evolved, and finally, a one-way digestive tract.

3) What Are the Major Animal Phyla?
Vertebrates, animals with backbones, constitute a single subphylum within the phylum Chordata. All other animals lack a backbone and are called invertebrates. More than 97% of the animal species on Earth are invertebrates.

4) The Sponges: Phylum Porifera
The bodies of sponges are typically free-form in shape and are sessile. Sponges have relatively few types of cells. Despite the division of labor among the cell types, there is little coordination of activity. Sponges lack the muscles and nerves required for coordinated movement, and digestion occurs exclusively within the individual cells.

5) The Hydra, Anemones, and Jellyfish: Phylum Cnidaria
Far more coordination is evident among cells of cnidarians than occurs in sponges. A simple network of nerve cells directs the activity of contractile cells, allowing loosely coordinated movements. Digestion is extracellular, occurring in a central gastrovascular cavity with a single opening. Cnidarians exhibit radial symmetry, an adaptation to both the free-floating lifestyle of the medusa and the sedentary existence of the polyp.

6) The Flatworms: Phylum Platyhelminthes
Flatworms (phylum Platyhelminthes) are the first phylum to show a distinct head with sensory organs and a simple brain. A system of canals forming a network through the body aids in excretion.

7) The Roundworms: Phylum Nematoda
The roundworms (phylum Nematoda) are the first phylum to possess a separate mouth and anus.

8) The Segmented Worms: Phylum Annelida
The segmented worms (phylum Annelida) are the most complex of the worms, with a well-developed closed circulatory system and excretory organs that resemble the basic unit of the vertebrate kidney. The segmented worms have a compartmentalized digestive system, like that of vertebrates, which processes food in a sequence. Annelids also have a true coelom, a fluid-filled space between the body wall and the internal organs, found in most complex animals, including vertebrates.

9 **The Insects, Arachnids, and Crustaceans: Phylum Arthropoda**

Arthropods are the most diverse and abundant organisms on Earth. They have invaded nearly every available terrestrial and aquatic habitat. Jointed appendages and well-developed nervous systems make possible complex, finely coordinated behavior. The exoskeleton (which conserves water and provides support) and specialized respiratory structures (which remain moist and protected) enable the insects and arachnids to inhabit dry land. The diversification of insects has been enhanced by their ability to fly. Crustaceans, which include the largest arthropods, are restricted to moist, usually aquatic habitats and respire by using gills.

10 **The Snails, Clams, and Squid: Phylum Mollusca**

Mollusks lack a skeleton; some forms protect the soft, moist, muscular body with a single shell (many gastropods and a few cephalopods) or a pair of hinged shells (the bivalves). The lack of a waterproof external covering limits this phylum to aquatic and moist terrestrial habitats. Although the body plan of gastropods and bivalves limits the complexity of their behavior, the cephalopod's tentacles are capable of precisely controlled movements. The octopus has the most complex brain and the best-developed learning capacity of any invertebrate.

11 **The Sea Stars, Sea Urchins, and Sea Cucumbers: Phylum Echinodermata**

Echinoderms are an exclusively marine group. Like other complex invertebrates and chordates, echinoderm larvae are bilaterally symmetrical; however, the adults show radial symmetry. This, in addition to a primitive nervous system that lacks a definite brain, adapts them to a relatively sedentary existence. Echinoderm bodies are supported by a nonliving internal skeleton that sends projections through the skin. The water-vascular system, which functions in locomotion, feeding, and respiration, is a unique echinoderm feature.

12 **The Tunicates, Lancelets, and Vertebrates: Phylum Chordata**

The phylum Chordata includes two invertebrate groups, the lancelets and tunicates, as well as the familiar vertebrates. All chordates possess a notochord, a dorsal hollow nerve cord, pharyngeal gill slits, and a post-anal tail at some stage in their development. Vertebrates are a subphylum of chordates that have a backbone, which is part of a living endoskeleton. Vertebrate evolution is believed to have proceeded from the fishes to amphibians to reptiles, which gave rise to both birds and mammals. The heart increases in complexity from the two chambers in fishes, to three in amphibians and most reptiles, to four in the warm-blooded birds and mammals. The evolution of efficient transport of oxygen supported active lifestyles, high metabolic rates, and constant body temperatures.

During the progression from fishes to amphibians to reptiles, a series of adaptations evolved that helped vertebrates colonize dry land. All amphibians have legs, and most have simple lungs for breathing in air rather than in water. Most are confined to relatively damp terrestrial habitats by their need to keep their skin moist, use of external fertilization, and requirement that their eggs and larvae develop in water. Reptiles, with well-developed lungs, dry skin covered with relatively waterproof scales, internal fertilization, and the amniote egg with its own water supply, are well adapted to the driest terrestrial habitats. Birds and mammals are also fully terrestrial and have additional adaptations, such as an elevated body temperature, which allows the muscles to respond rapidly regardless of the temperature of the environment. The bird body is molded for flight, with feathers, hollow bones, efficient circulatory and respiratory systems, and well-developed eyes. Mammals have insulating hair and give birth to live young that are nourished with milk. The mammalian nervous system is the most complex in the animal kingdom, providing mammals with enhanced learning ability that helps them adapt to changing environments.

Key Terms

abdomen *p. 427*
amnion *p. 441*
amniote egg *p. 441*
amoeboid cell *p. 418*
ampulla *p. 434*
anterior *p. 414*
bilateral symmetry *p. 414*
book lung *p. 427*
budding *p. 418*
cartilage *p. 435*
cephalization *p. 412*
chitin *p. 426*
closed circulatory system *p. 424*
cnidocyte *p. 420*
coelom *p. 415*
collar cell *p. 418*
compound eye *p. 427*
cyst *p. 423*
deuterostome *p. 415*

dorsal *p. 414*
ectoderm *p. 413*
endoderm *p. 413*
endoskeleton *p. 434*
epithelial cell *p. 418*
exoskeleton *p. 426*
free-living *p. 421*
ganglion *p. 421*
gastrovascular cavity *p. 420*
germ layer *p. 413*
gill *p. 427*
head *p. 427*
hemocoel *p. 427*
hermaphroditic *p. 421*
hydrostatic skeleton *p. 423*
invertebrate *p. 417*
larva *p. 428*
mammary gland *p. 444*
mantle *p. 430*
marsupial *p. 444*

medusa *p. 419*
mesoderm *p. 413*
mesoglea *p. 420*
metamorphosis *p. 428*
molt *p. 427*
monotreme *p. 444*
nephridium *p. 424*
nerve cord *p. 421*
nerve net *p. 419*
notochord *p. 434*
open circulatory system *p. 427*
organ *p. 412*
organ system *p. 412*
parasite *p. 422*
pharyngeal gill slit *p. 435*
pharynx *p. 421*
placenta *p. 444*
placental *p. 444*
polyp *p. 419*

post-anal tail *p. 435*
posterior *p. 414*
protostome *p. 415*
pseudocoelom *p. 415*
pupa *p. 429*
radial symmetry *p. 414*
radula *p. 431*
segmentation *p. 412*
sessile *p. 414*
sieve plate *p. 434*
spicule *p. 418*
tentacle *p. 419*
thorax *p. 427*
tissue *p. 412*
trachea *p. 427*
tube foot *p. 433*
ventral *p. 414*
vertebral column *p. 435*
vertebrate *p. 417*
water-vascular system *p. 433*

Thinking Through the Concepts

Multiple Choice

1. *Which of the following animals have radial symmetry?*
a. jellyfish
b. sea star
c. sea anemone
d. sea urchin
e. all of the above

2. *Animals in which of the following phyla have collar cells?*
a. Porifera
b. Cnidaria
c. Annelida
d. Gastropoda
e. Chordata

3. *What is a radula?*
a. a flexible supportive rod on the dorsal surface of chordates
b. a stinging cell used by sea anemones to capture prey
c. a gas-exchange structure in insects
d. a spiny ribbon of tissue used for feeding in snails
e. a locomotory structure of sea stars

4. *Which of the following classes of animals includes the first vertebrates to appear on Earth?*
a. Agnatha, the jawless fishes
b. Chondrichthyes, the sharks
c. Osteichthyes, the bony fishes
d. Cephalopoda, the cephalopods
e. none of the above

5. *Which of the following animals molts its exoskeleton, allowing the animal to grow larger?*
a. the blue crab
b. the bat sea star
c. the scallop
d. the sea urchin
e. the Venus clam

6. *Which of the following has a gastrovascular cavity?*
a. sponges
b. sea stars
c. nematodes
d. flatworms
e. earthworms

? Review Questions

1. List the distinguishing characteristics of each of the phyla discussed in this chapter, and give an example of each.

2. Briefly describe each of the following adaptations, and explain the adaptive significance of each: amniote egg, bilateral symmetry, cephalization, closed circulatory system, coelom, placenta, radial symmetry, segmentation.

3. Describe and compare respiratory systems in the three major arthropod classes.

4. Describe the advantages and disadvantages of the arthropod exoskeleton.

5. State in which of the three major mollusk classes each of the following characteristics is found:
a. two hinged shells
b. a radula
c. tentacles
d. some sessile members
e. the best-developed brains
f. numerous eyes

6. Give three functions of the water-vascular system of echinoderms.

7. To what lifestyle is radial symmetry an adaptation? Bilateral symmetry?

8. List the vertebrate class (or classes) that have each of the following:
a. a skeleton of cartilage
b. a two-chambered heart
c. the amniote egg
d. warm-bloodedness
e. a four-chambered heart
f. a placenta
g. lungs supplemented by air sacs

9. Distinguish between vertebrates and invertebrates. List the major phyla found in each broad grouping.

10. List four distinguishing features of chordates.

11. Describe the ways in which amphibians are adapted to life on land. In what ways are amphibians still restricted to a watery or moist environment?

12. List the adaptations that distinguish reptiles from amphibians and help reptiles adapt to life in dry terrestrial environments.

13. List the adaptations of birds that contribute to their ability to fly.

14. How do mammals differ from birds, and what adaptations do they share?

15. How has the mammalian nervous system contributed to the success of mammals?

Applying the Concepts

1. The class Insecta is the largest taxon of animals on Earth. Its greatest diversity is in the tropics, where habitat destruction and species extinction are occurring at an alarming rate. What biological, economic, and ethical arguments can you advance to persuade people and governments to preserve this biological diversity?

2. Animals, particularly laboratory rats, are used extensively in medical research. What advantages and disadvantages do lab rats have as research subjects in medicine? What are the advantages and disadvantages of using animals more closely related to humans?

3. Discuss at least three ways in which the ability to fly has contributed to the success and diversity of insects.

4. Discuss and defend what attributes you would use to define biological success among animals. Are humans a biological success by these standards? Why or why not?

Group Activity

Form a group of four students. Your group is part of a team of scientists who have been studying a species of sea slug (a marine mollusk). You observed that any fish that bites the sea slug recoils as if stung. But you previously observed that sea slugs that live in areas devoid of jellyfish (the slugs' favorite food) *don't* seem to have this stinging ability. Your group's task is to (1) develop a hypothesis to explain your observations and (2) design an experiment to test your hypothesis.

For More Information

Blaustein, A. "Amphibians in a Bad Light." *Natural History*, October 1994. Recent declines in amphibian population size and overall diversity are linked to possible harm from ultraviolet light that is penetrating a depleted ozone layer.

Brusca, R. C., and Brusca, G. J. *Invertebrates*. Sunderland, MA: Sinauer, 1990. A thorough survey of the invertebrate animals; in textbook format but readable and filled with beautiful and informative drawings.

Chadwick, D. H. "Planet of the Beetles." *National Geographic*, March 1998. The beauty and diversity of beetles, which comprise one-third of the world's insects, are described in text and photographs.

Diamond, J. "Stinking Birds and Burning Books." *Natural History*, October 1994. A description of a recently described species of bird (the pitohuis) and its peculiar chemical ecology.

Duellman, W. E. "Reproductive Strategies of Frogs." *Scientific American*, July 1992. Free-living tadpoles are only one way in which these amphibians progress from egg to adult.

Hamner, W. "A Killer Down Under." *National Geographic*, August 1994. Among the most poisonous animals in the world is the box jellyfish, which lives off the coast of northern Australia.

Horridge, G. A. "The Compound Eye of Insects." *Scientific American*, July 1977. A close look at the multifaceted eyes of insects.

Luoma, J. R. "Vanishing Frogs." *Audubon*, May–June 1997. Brief summary of declining amphibian populations, accompanied by striking photographs of some of the affected frog species.

McMenamin, M. A. S., and McMenamin, D. L. S. *The Emergence of Animals: The Cambrian Breakthrough.* New York: Columbia University Press, 1990. A look at the adaptive radiation that led to an explosion of animal forms at the beginning of the Cambrian period.

Montgomery, S. "New Terror of the Deep." *International Wildlife*, July–August 1992. A description of the threat that overfishing by humans poses to shark populations.

Morell, V. "Life on a Grain of Sand." *Discover*, April 1995. The sand beneath shallow waters is home to an incredible range of microscopic creatures.

Rahn, H., Ar, A., and Paganelli, C. V. "How Bird Eggs Breathe." *Scientific American*, February 1979. (Offprint No. 1420). A description of the remarkable adaptations of the amniote egg for gas exchange.

Rennie, J. "Living Together." *Scientific American*, January 1992. Most parasites do not kill their hosts, and the interaction of parasite and host provides some fascinating insights into the evolution of life on Earth.

Answers to Multiple-Choice Questions
1. e 2. a 3. d 4. a 5. a 6. d

Unit Four
Plant Anatomy and Physiology

A prairie blanketed with Texas bluebonnets in the spring. Aside from providing a beautiful vista, flowers such as these are elaborately modified leaves that help attract pollinators.

Some of the diversity of plant forms is seen in these flowers and organ pipe cactus in Organ Pipe National Monument, Arizona. The thick stems of cacti store water and carry out photosynthesis. Their leaves, reduced to spines, conserve water while providing protection.

Plant Form and Function 23

 Net Watch

On-line resources for this chapter are on the World Wide Web at: http://www.prenhall.com/audesirk (click on the Table of Contents link and then select Chapter 23).

Many of us take plants for granted. We admire a field of wildflowers or a giant sequoia, but we seldom stop to think about the intricate structures and functions that allow even the most-common plants to survive and prosper. Plants cannot move to escape enemies, to find food or water, to avoid the arrival of winter, or to locate a mate. Yet spruces perch atop the permanently frozen terrain in Alaska, cypresses and mangroves stand immersed in swamps in Florida, and cacti bloom in the searing heat of Death Valley, California. Redwoods and eucalyptus trees tower 100 meters above the forest floor; grasses survive grazing by animals year after year; and bristlecone pines in the White Mountains of California may live to be 4000 years old. Clearly, plants have evolved many extremely successful adaptations that enable them to thrive in a wide variety of habitats.

If you have an opportunity, visit Organ Pipe National Monument, Arizona, in the spring. If the winter rains have come, you may be greeted with a spectacular array of wildflowers bloomings amidst the giant stems of the organ pipe cacti. How do plants obtain and store water in this harsh desert environment? Why are the stems of the cacti so fat and green, and where are their leaves? You already know that plants need carbon dioxide from the air to carry out photosynthesis. How do plants in dry environments allow air to enter their leaves yet prevent too much water from evaporating? How do plants transport water from their roots to the tops of their stems—in the case of giant redwoods, a trip 100 meters long?

In this unit, we explore the adaptations that make land plants so successful. We begin with the tissues and structures that constitute the plant body and how they help plants acquire nutrients from the air and soil and move the nutrients throughout their bodies. Because of the dominance of *angiosperms*—flowering plants—in most terrestrial habitats and their importance to humans, we will focus on them.

1) How Are Plant Bodies Organized?

Flowering plants consist of two major regions, the *root system* and the *shoot system* (Fig. 23-1). The **root system** consists of all the roots of a plant. **Roots** are branched portions of the plant body that are normally embedded in soil. Plant roots serve six major functions: They (1) anchor the plant in the ground; (2) absorb water and minerals from the soil; (3) store surplus sugars manufactured during photosynthesis; (4) transport water, minerals, sugars, and hormones to and from the shoot; (5) produce some hormones; and (6) interact with soil fungi and microorganisms that provide nutrients to the plant.

The rest of the plant is the **shoot system**, normally located above ground. The shoot system consists of *leaves*, *buds*, and (in season) *flowers* and *fruits*, all borne on **stems**, which are typically branched. The functions of shoots include (1) photosynthesis, mainly in leaves and young green stems; (2) transport of materials among leaves, flowers, fruits, and roots; (3) reproduction; and (4) hormone synthesis.

Figure 23-2 illustrates the two broad groups of flowering plants: the **monocots** (class Monocotyledoneae, which includes grasses, lilies, palms, and orchids) and the **dicots** (class Dicotyledoneae, which includes deciduous trees and bushes and many garden flowers). As we shall see in Chapter 24, the first leaflike structures an angiosperm embryo produces are called *cotyledons*, or *seed leaves*, for which these two broad groups are named. Monocots and dicots differ in the structure of their flowers, leaves, vascular tissue, root pattern, and seeds. Don't worry about terms that are not yet familiar to you; for now, look over Figure 23-2, and refer to it later as we examine the parts of the flowering plants in more detail.

2) How Do Plants Grow?

Animals and plants develop in dramatically different ways. One difference is the timing and distribution of growth. As you grew from a baby to an adult, all parts of

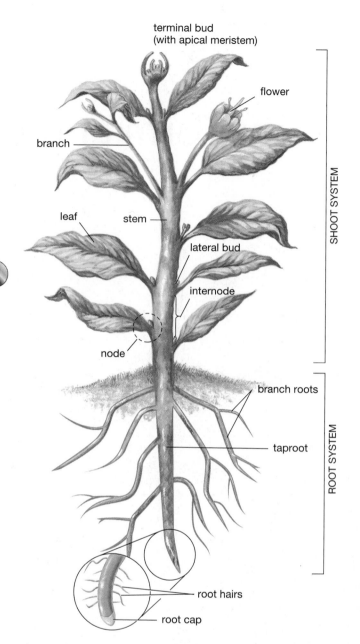

Figure 23-1 Flowering plant structure
A flowering plant consists of the root system and the shoot system. The root system is normally highly branched. Root hairs, typically microscopic, increase surface area for nutrient absorption. The shoot system includes stems (from which branches grow), with buds, and leaves. In the appropriate season, the shoot may bear flowers and fruit. Our model plant is a dicot.

your body became larger. When you reached your adult height, you stopped growing (upward, at least!). In contrast, flowering plants grow throughout their lives, never reaching a stable "adult" body form. What's more, most plants grow longer only at the tips of their branches and roots, and structures that developed earlier remain in exactly the same place; a swing tied to a tree branch does not move farther from the ground each year. Why do plants grow this way?

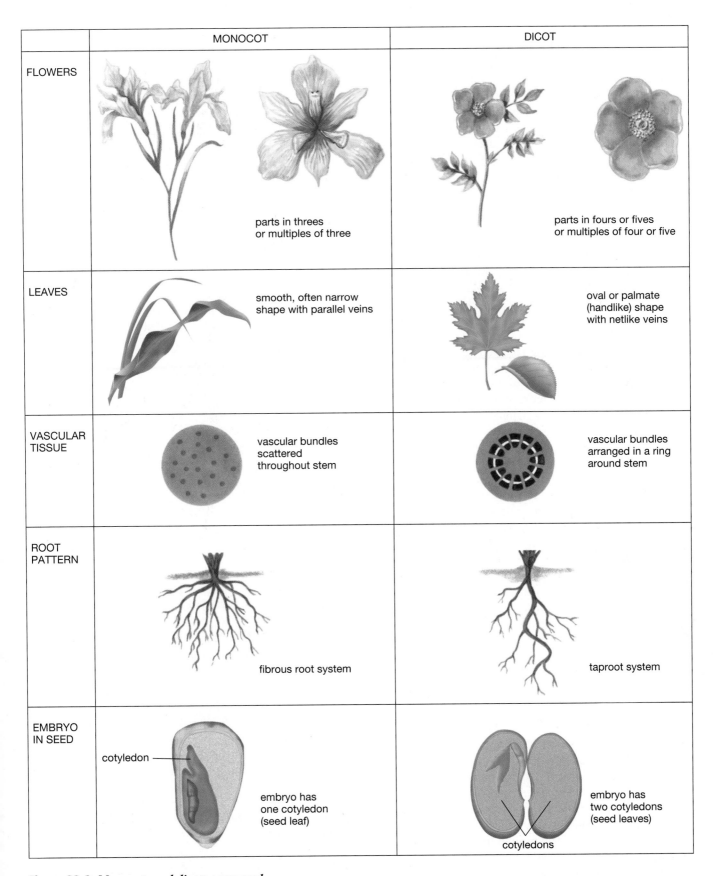

	MONOCOT	DICOT
FLOWERS	parts in threes or multiples of three	parts in fours or fives or multiples of four or five
LEAVES	smooth, often narrow shape with parallel veins	oval or palmate (handlike) shape with netlike veins
VASCULAR TISSUE	vascular bundles scattered throughout stem	vascular bundles arranged in a ring around stem
ROOT PATTERN	fibrous root system	taproot system
EMBRYO IN SEED	cotyledon — embryo has one cotyledon (seed leaf)	embryo has two cotyledons (seed leaves) cotyledons

Figure 23-2 Monocots and dicots compared
Distinguishing traits of the two major classes of flowering plants, the monocots and the dicots. The word *monocot* refers to the fact that the embryo in the seed has a single ("mono") cotyledon, or seed leaf; *dicot* means "two cotyledons."

From the moment they sprout, plants are composites of two fundamentally different categories of cells: meristem cells and differentiated cells. Embryonic, undifferentiated **meristem cells** are capable of mitotic cell division. Some of the offspring, or daughter cells, of the dividing meristem cells lose the ability to divide and become part of the nongrowing portions of the plant body. These **differentiated cells** are specialized in structure and function. Continued divisions of meristem cells, then, keep a plant growing throughout its life, whereas their differentiated daughter cells form more stable or permanent parts of the plant, such as mature leaves or the trunks of trees.

Plants grow via the division and differentiation of two major types of meristem cells: apical meristems and lateral meristems. **Apical meristems** ("tip meristems") are located at the ends of roots and shoots, including main stems and branches (see Figs. 23-9, 23-13, and 23-15). **Lateral meristems** ("side meristems"), also called **cambia** (singular, **cambium**), form cylinders that run parallel to the long axis of roots and stems (see Fig. 23-15).

Plant growth takes two forms: (1) *primary growth* and (2) *secondary growth*. **Primary growth** occurs by the mitotic cell division of apical meristem cells followed by differentiation of the resulting daughter cells. This type of growth takes place in the growing tips of roots and shoots of plants. Primary growth is responsible both for increase in length and for the development of the specialized plant structures. The elongation of roots and shoots through primary growth allows them to enter new space from which to collect light, nutrients, and water. It also explains why your swing never gets any higher off the ground.

Secondary growth, which is responsible for increases in diameter, occurs by the division of lateral meristem cells and differentiation of their daughter cells. Secondary growth causes the stems and roots of most conifers and dicots to become thicker and woodier as they age. Although we later discuss secondary growth only in stems, keep in mind that secondary growth also occurs in roots.

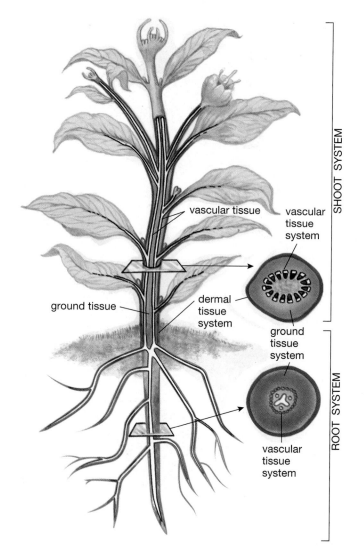

Figure 23-3 *The structure of the root and shoot*
Both the root and shoot of a flowering plant consist of three tissue systems: dermal, ground, and vascular tissue systems. (The structure of the leaves, vascular tissue system, and roots indicate that our sample plant is a dicot.)

3) What Are the Tissues and Cell Types of Plants?

The major structures of land plants, including roots, stems, and leaves, consist of three *tissue systems*—(1) dermal, (2) ground, and (3) vascular tissue systems—each consisting of more than one type of tissue. Each tissue, in turn, is composed of one or more specialized cell types (Fig. 23-3). The **dermal tissue system** covers the outer surfaces of the plant body. The **ground tissue system**, which consists of all nondermal and nonvascular tissues, makes up most of the body of young plants. Its functions include photosynthesis, support, and storage. The **vascular tissue system** transports water, minerals, sugars, and plant hormones throughout the plant. Some flowering plants, termed *herbaceous*, are soft-bodied plants with flexible

stems, such as lettuce, beans, and grasses. Such plants normally live only one year. Other plants, such as trees and bushes, are described as *woody*; most are perennial (living many years) and develop hard, thickened, woody stems as a result of secondary growth. As you will see, different types of tissue are present in herbaceous and woody plants.

The Dermal Tissue System Forms the Covering of the Plant Body

The dermal tissue system is the outer covering of the plant body. There are two types of dermal tissue: (1) *epidermal tissue* and (2) *periderm*.

Epidermal tissue forms the **epidermis**, the outermost cell layer that covers the leaves, stems, and roots of all young plants (Fig. 23-4a). Epidermal tissue also covers

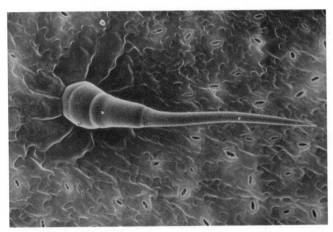

(a)

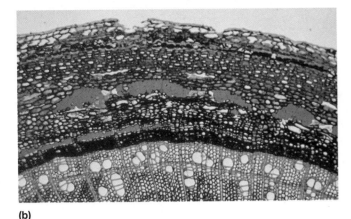

(b)

Figure 23-4 Dermal tissues cover plant surfaces
(*a*) The epidermis of a young root or shoot is a single layer of cells. In shoot epidermis, such as the epidermis of this zinnia leaf, the outer surfaces of the cells are covered with cuticle, a waxy, waterproof coating that reduces the evaporation of water. The "leaf hair" protruding from the epidermis also reduces evaporation by slowing the movement of air across the leaf surface. (*b*) Woody stems and roots develop a dense, tough, waterproof periderm (the darkest layer), which consists mostly of thick-walled cork cells.

flowers, seeds, and fruit. In herbaceous plants, the epidermis is retained as the outer covering of the entire plant body throughout its life. The epidermal tissue of the aboveground parts of a plant is generally composed of thin-walled cells packed tightly together and covered with a waterproof, waxy **cuticle**. Secreted by the epidermal cells, the cuticle reduces the evaporation of water from the plant and helps protect it from the invasion of disease microorganisms. The epidermal cells of roots, in contrast, are not covered with cuticle, because a waterproof cuticle would prevent the absorption of water and minerals.

Some epidermal cells produce fine extensions called *hairs*. Many root epidermal cells bear **root hairs**, long projections that greatly increase the absorptive surface area of the root. Epidermal hairs on the stems and leaves of desert plants reduce evaporative water loss by reflecting sunlight and producing an unstirred layer of air near the plant's surface. In contrast, some tropical plants use their hairy leaves to capture and hold water.

Periderm replaces epidermal tissue on the roots and stems of woody plants as they age. This dermal tissue is composed primarily of **cork cells**, which have thick, waterproof walls and are dead at maturity (Fig. 23-4b). Cork cells form the protective outer layers of the bark of trees and woody shrubs and the woody covering of their roots.

The Ground Tissue System Makes Up Most of the Young Plant Body

The ground tissue system, which makes up the bulk of a young plant, consists of all nondermal and nonvascular tissues. There are three types of ground tissue: (1) *parenchyma*, (2) *collenchyma*, and (3) *sclerenchyma*.

Parenchyma tissue is the most abundant of the ground tissues. Parenchyma cells are thin-walled cells, alive at maturity, that typically carry out most of the metabolic activities of the plant (Fig. 23-5a). Depending on their location within the plant body, parenchyma cells have such diverse functions as photosynthesis, storage of sugars and starches, or secretion of hormones. Under the proper conditions, many parenchyma cells are capable of mitotic cell division. Roots that are adapted for storage, including carrots and sweet potatoes, are packed with parenchyma cells that store carbohydrates, such as starch and sugar. Parenchyma cells full of starch pack the white potato as well.

Collenchyma tissue consists of elongated, polygonal (many-sided) cells with unevenly thickened cell walls (Fig. 23-5b). Collenchyma cells are alive at maturity but normally cannot divide. Although strong, the cell walls of collenchyma are still somewhat flexible. In herbaceous plants and in the leaf stalks and young growing stems of all plants, collenchyma tissue is an important source of support. The strings in celery stalks, for example, are mostly collenchyma cells in association with vascular tissue.

Sclerenchyma tissue consists of cells with thick, hardened secondary cell walls (Fig. 23-5c), reinforced with the stiffening substance *lignin*. Like collenchyma, sclerenchyma cells support and strengthen the plant body, but unlike collenchyma, they die after they differentiate. Their hardened cell walls then remain as a source of support. Sclerenchyma tissue can be found in many parts of the plant body, including xylem and phloem (described next). Sclerenchyma cells provide the fibers of hemp and jute, which are used for making rope. Other types of sclerenchyma cells form nut shells, the outer covering of peach pits, and the gritty texture of pears.

The Vascular Tissue System Consists of Xylem and Phloem

The vascular tissue system consists of two complex conducting tissues: (1) *xylem* and (2) *phloem*. The major role of each tissue is the transport of materials. Xylem

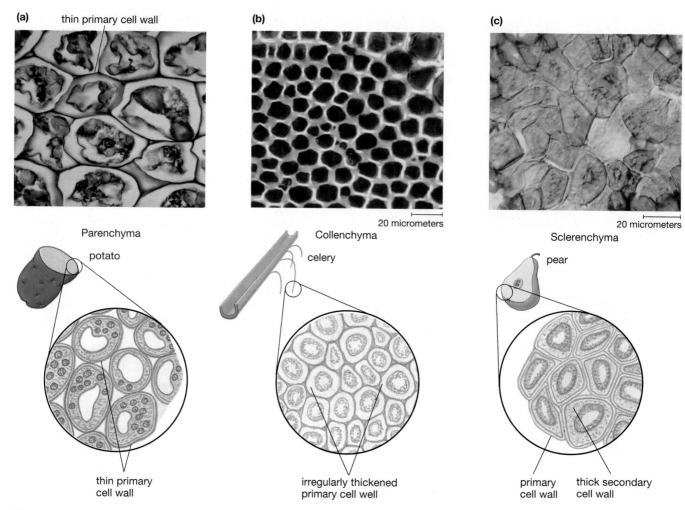

(a) thin primary cell wall

Parenchyma
potato

thin primary
cell wall

(b)

20 micrometers

Collenchyma
celery

irregularly thickened
primary cell well

(c)

20 micrometers

Sclerenchyma
pear

primary   thick secondary
cell wall   cell wall

Figure 23-5 *The structure of ground tissue*
Ground tissue consists of three cell types: parenchyma, collenchyma, and sclerenchyma. **(a)** Parenchyma cells are living cells with thin, flexible cell walls. These parenchyma cells are used for starch storage in a potato. **(b)** Collenchyma cells, such as these from a celery stalk, are living cells with irregularly thickened, but still somewhat flexible, walls. They help support the plant body. **(c)** Sclerenchyma cells, such as these "stone cells" that give the meat of a pear its slightly gritty texture, have thick, rigid secondary cell walls. Sclerenchyma cells also form the shells of nuts.

transports water and minerals up from the roots to the rest of the plant; phloem conveys water, sugars, amino acids, and hormones throughout the plant.

Xylem Conducts Water and Dissolved Minerals from the Roots to the Rest of the Plant

Xylem conducts water and minerals from roots to shoots in tubes that are made from *tracheids* and *vessel elements* (Fig. 23-6). **Tracheids** are thin cells with slanted ends resembling the tips of hypodermic needles. Tracheids are stacked atop one another with the slanted ends overlapping. The overlapping end walls contain **pits**, porous sections where secondary cell walls are absent. These pits allow water and minerals to pass from one tracheid to the next, or from a tracheid to an adjacent vessel element, by crossing only the thin, water-permeable primary cell wall.

Vessel elements also meet end to end, but they are larger in diameter than tracheids. Their ends may be either flat or overlapping and tapered. Perforations connect adjoining elements. In some cases, the ends of vessel elements completely disintegrate, leaving an open tube (see Fig. 23-6). Thus, vessel elements form wide-diameter, relatively unobstructed pipelines called **vessels** from root to leaf.

Most conifers have only tracheids; flowering plants normally have both tracheids and vessel elements. As these cells differentiate, they develop thick cell walls that help support the weight of the plant. In some trees (pines, for example), the bulk of the tree trunk consists of the thick cell walls of tracheids. The final step in the differentiation of both tracheids and vessel elements is death: The cytoplasm and plasma membrane disintegrate, leaving behind a hollow tube of cell wall.

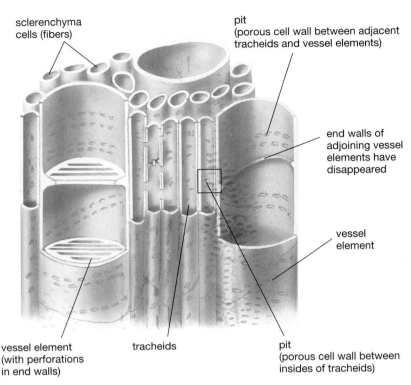

sclerenchyma cells (fibers)

pit (porous cell wall between adjacent tracheids and vessel elements)

end walls of adjoining vessel elements have disappeared

vessel element

vessel element (with perforations in end walls)

tracheids

pit (porous cell wall between insides of tracheids)

Figure 23-6 The structure of xylem
Xylem is a mixture of cell types, including parenchyma and sclerenchyma fibers (ground tissue) and two types of conducting cells, tracheids and vessel elements. Tracheids are thin, with tapered, overlapping ends joined by pits. The pits are not holes; they consist of a water-permeable primary cell wall that separates the interiors of adjoining cells. Vessels consist of vessel elements stacked atop one another. The end walls of some vessel elements are absent, and others have narrow openings that connect adjacent vessel elements. Both tracheids and vessel elements have pits in their side walls, allowing water and dissolved minerals to move sideways between adjacent conducting tubes.

Phloem Conducts Water, Sugars, Amino Acids, and Hormones throughout the Plant

Phloem carries water with dissolved substances synthesized by the plant, including sugars, amino acids, and hormones. Among the supportive sclerenchyma cells of phloem are **sieve tubes**, constructed of a single strand of cells called **sieve-tube elements** (Fig. 23-7). As sieve-tube elements mature, most of their internal contents disintegrate, leaving behind only a thin layer of cytoplasm that lines the plasma membrane. At the ends of sieve-tube elements, where adjacent cells meet, holes form in the primary cell walls, creating structures called **sieve plates**. The interiors of adjacent sieve-tube elements are connected through the openings in the sieve plate. A continuous conducting system is forged by the linking of many sieve-tube elements end to end in this way.

A sieve-tube element has a plasma membrane, a few small mitochondria, and some endoplasmic reticulum. It is therefore considered to be alive, but it normally lacks ribosomes, Golgi complexes, and a nucleus. How, then, can a sieve-tube element remain alive? Each sieve-tube element is nourished by a smaller, adjacent **companion cell**. Companion cells are connected to sieve-tube elements by cytoplasm-filled channels called *plasmodesmata*, described in Chapter 5. The companion cells maintain the integrity of the sieve-tube elements by donating high-energy compounds and perhaps even by repairing the sieve-tube plasma membrane. As we shall see later, companion cells also regulate the movements of sugars into and out of the sieve tubes.

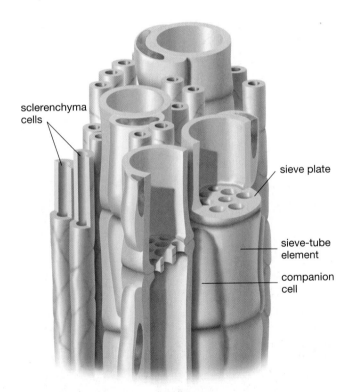

sclerenchyma cells

sieve plate

sieve-tube element

companion cell

Figure 23-7 The structure of phloem
Phloem is a mixture of cell types, including sclerenchyma, sieve-tube elements, and companion cells. Sieve-tube elements have primary cell walls and a thin layer of cytoplasm around a fluid-filled core. Sieve-tube elements, stacked end to end, form the conducting system of phloem. Where they join, sieve-tube elements form sieve plates, where membrane-lined pores allow fluid to pass from cell to cell. Each sieve-tube element has a companion cell that nourishes it and regulates its function.

4 Roots: Anchorage, Absorption, and Storage

As a seed sprouts, the **primary root**—the first root to develop—grows down into the soil. Many dicots, such as carrots and dandelions, develop a taproot system. A **taproot system** consists of the primary root, which normally becomes longer and stouter with time, and many smaller roots that grow out from the sides of the primary taproot (Fig. 23-8a). In contrast, in monocots such as grasses and palms, the primary root soon dies off, replaced by many new roots that emerge from the base of the stem. These secondary roots are nearly equal in size, forming a **fibrous root system** (Fig. 23-8b).

Primary Growth Causes Roots to Elongate

In young roots of both taproot and fibrous root systems, divisions of the apical meristem give rise to four distinct regions (Fig. 23-9). At the very tip of the root, daughter cells produced on the "soil side" of the apical meristem differentiate into the **root cap**. The root cap protects the apical meristem from being scraped off as the root pushes down between the rocky particles of the soil. Root-cap cells have thick cell walls and secrete a slimy lubricant that helps ease the way between soil particles. Nevertheless, root-cap cells wear away and must be continuously replaced by new cells from the meristem.

Daughter cells produced on the "shoot side" of the apical meristem differentiate into one of three parts: an outer envelope of epidermis; a *vascular cylinder* at the core of the root; and, between the two, a *cortex* (Fig. 23-9).

The Epidermis of the Root Is Very Permeable to Water

The root's outermost covering of cells is the epidermis, which is in contact with the soil and any air or water trapped among the soil particles. The cell walls of the epidermal cells are highly water permeable. Therefore, water can penetrate into the interior of the root either by passing through the membranes of the epidermal cells or by passing between cells of the epidermis, through the porous cell walls. Many epidermal cells grow root hairs into the surrounding soil (Fig. 23-10). By increasing the surface area, root hairs increase the root's ability to absorb water and minerals. Root hairs may add dozens of square meters of surface area to the roots of even small plants.

Cortex Makes Up Much of the Interior of a Young Root

Cortex occupies most of the inside of a young root. The cortex consists of an outer mass of large, loosely packed parenchyma cells just beneath the epidermis and an inner layer of smaller, close-fitting cells that form a ring around the vascular cylinder. That ring is called the **endodermis** (see Fig. 23-9). Sugars produced by photosynthesis in the shoot are transported down to the parenchyma cells of the cortex, where they are converted to starch and stored. These cells are particularly abundant in roots specialized for carbohydrate storage, such as the thick roots of carrots and dandelions.

The endodermis is a layer of cells with highly specialized cell walls. Where endodermal cells contact each other, their cell walls are filled with a waxy material, forming the **Casparian strip**. The Casparian strip resembles the mortar in a brick wall: The waxy waterproofing covers the top, bottom, and sides of the endodermal cells. However, it does not cover the cell surfaces that face the rest of the cortex or those that face the vascular cylinder. Water and dissolved minerals can travel *around* both epidermal and cortex parenchyma cells by moving through their porous cell walls. But the waxy Casparian strip seals off the vascular cylinder cell walls, forcing water and minerals that enter the vascular cylinder to pass through the living membranes of the endodermal cells (Fig. 23-11, p. 460). These membranes regulate the types and amounts of materials that the roots can absorb.

(a)   (b)

Figure 23-8 Typical root systems in dicots and monocots
(a) Dicots typically have a taproot system, consisting of a long central root with many smaller, secondary roots branching from it. **(b)** Monocots normally have a fibrous root system, with many roots of equal size.

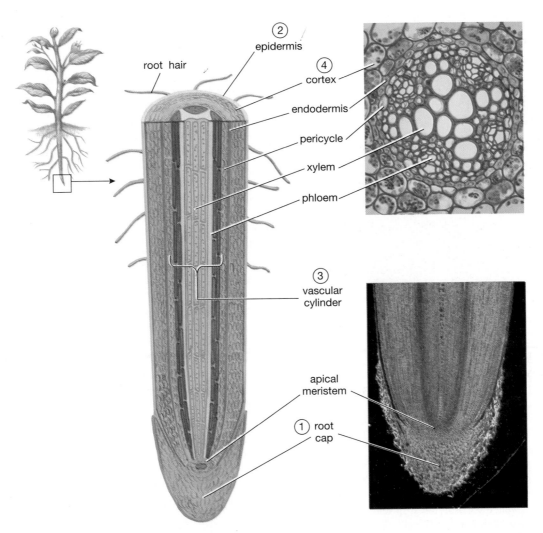

Figure 23-9 *Primary growth in roots*
Primary growth in roots results from cell divisions in the apical meristem, located near the tip of the root. Four regions are formed: the root cap, at the very tip of the root; and, via the subsequent differentiation of daughter cells, the epidermis, vascular cylinder, and cortex.

Figure 23-10 *Root hairs*
Root hairs, shown here in a sprouting radish, greatly increase a root's surface area for the absorption of water and minerals from the soil.

The Vascular Cylinder Contains Conducting Tissues

The **vascular cylinder** contains the conducting tissues of xylem and phloem, which transport water and dissolved materials within the plant. The outermost layer of the vascular cylinder, called the **pericycle**, is a remnant of meristem that retains the ability to divide. Under the influence of plant hormones, pericycle cells divide and form the apical meristem of a **branch root**, a root that forms as a branch of an existing root (Fig. 23-12). Branch root development is similar to primary root development except that the branch must first break out through the cortex and epidermis of the primary root. It does so partly by crushing the cells that lie in its path and partly by secreting enzymes that digest them away. The vascular tissues of the branch root connect with the vascular tissues of the primary root.

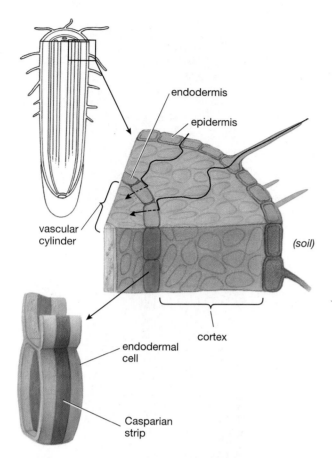

Figure 23-11 The role of the Casparian strip
The Casparian strip is a band of waterproof material in the walls between cells of the endodermis. Arranged like the mortar in a brick wall, the Casparian strip seals off the top, bottom, and sides of the endodermal cells, thereby forcing water to move through the cells rather than between them.

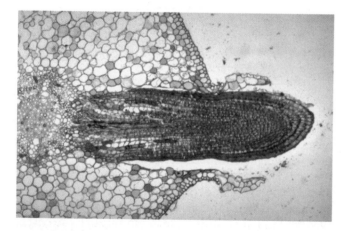

Figure 23-12 Branch roots
Branch roots emerge from the pericycle of a root. The central axis of this branch root is already differentiating into vascular tissue.

5 Stems: Reaching for the Light

The Stem Has a Complex Organization Including Four Types of Tissue

Like roots, stems develop from a small group of actively dividing cells, the apical meristem, that lies at the tip of the young shoot. The apical meristem lies within the **terminal bud**. The terminal bud consists of meristem tissue surrounded by developing leaves called **leaf primordia** (singular, **primordium**) at the tip of the shoot. The daughter cells of the apical meristem differentiate into the specialized cell types of stem, buds, leaves, and flowers.

As the shoot grows, small clusters of meristem cells are "left behind" at the surface of the stem. These meristem cells form the leaf primordia, which develop into the mature leaves unique to each species of plant. The meristem cells also produce **lateral buds**, which, under appropriate conditions, grow into branches. (We will discuss the growth of branches shortly.) Leaf primordia and lateral buds appear at characteristic locations, called **nodes**, on the stem; regions of stem between these nodes are called **internodes** (Fig. 23-13).

Most young stems are composed of four tissues: (1) epidermis (dermal tissue), (2) vascular tissues, (3) cortex, and (4) *pith*. (Both cortex and pith are ground tissues.) As Figure 23-2 illustrates, monocots and dicots differ somewhat in the arrangement of vascular tissues. We will discuss only dicot stems here.

The Epidermis of the Stem Is Specialized to Retard Water Loss While Allowing Carbon Dioxide to Enter

In the stem (and leaves), the epidermis is exposed to dry air, making it a potential pathway for water loss. Epidermal cells of the stem, unlike those of the root, secrete a waxy covering, the cuticle, that reduces evaporation of water. The cuticle also, however, reduces the diffusion of carbon dioxide and oxygen into and out of the plant. Hence the epidermis is commonly perforated with adjustable pores called **stomata** (singular, **stoma**) that regulate this exchange. Stomata are discussed in more detail later in this chapter.

The Cortex and Pith Support the Stem, Store Food, and May Photosynthesize

Cortex (located between the epidermis and vascular tissues) and **pith** (inside the vascular tissues at the center of the stem) are similar in most respects; in fact, in some stems it is difficult to tell where cortex ends and pith begins. Cortex and pith perform three major functions: support, storage, and, in some cases, photosynthesis.

1. *Support.* In very young stems, water filling the central vacuoles of cortex and pith cells causes turgor pres-

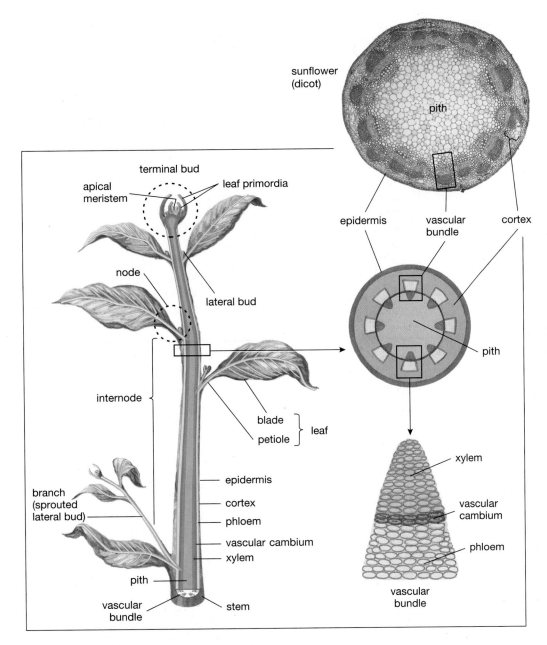

Figure 23-13 *The structure of a young dicot stem*
At the tip of the stem, the terminal bud includes the apical meristem and several leaf primordia, produced by the meristem. Other daughter cells of the apical meristem differentiate into epidermis, cortex, pith, and vascular tissues. As the young stem grows, the leaf primordia develop into mature leaves. Meanwhile, epidermis, cortex, pith, and vascular cells elongate between the points of attachment of leaf to stem, pushing the leaves apart. A lateral bud, a remnant of meristem tissue, remains between each leaf and the stem. Lateral buds may sprout into branches. Points on the shoot where leaves and lateral buds are located are nodes; the naked stem between nodes is an internode. In cross section, vascular tissue forms a ring of vascular bundles in dicots such as the sunflower shown in the photomicrograph.

sure (see Chapter 6). Turgor pressure stiffens the cells, much as air inflates a tire. If you forget to water your houseplants, their drooping tips show the importance of turgor pressure in keeping young stems erect. Somewhat older stems also have collenchyma or scle-

renchyma cells with thickened cell walls, which don't depend on turgor pressure for strength.

2. *Storage.* Parenchyma cells in both cortex and pith convert sugar into starch and store the starch as a food reserve.

3. *Photosynthesis.* In many stems, the outer layers of cortex cells contain chloroplasts and carry out photosynthesis. In some desert plants, such as cacti, the leaves are reduced or absent, and the cortex of the stem is the only green photosynthetic part of the plant.

Vascular Tissues in Stems Transport Water, Dissolved Nutrients, and Hormones

As in those of roots, the vascular tissues of stems transport water, minerals, sugars, and hormones. Vascular tissues are continuous in root, stem, and leaf, interconnecting all the parts of the plant. The **primary xylem** and **primary phloem** found in young stems arise from the apical meristem. In young dicot stems, the primary xylem, **vascular cambium** (meristematic tissue that produces *secondary xylem* and *secondary phloem*), and primary phloem may form concentric cylinders or may appear as a ring of bundles running up the stem, with each bundle containing both phloem and xylem (see Fig. 23-13). Sec-

ondary growth in dicot stems, as we shall discuss below, always results in concentric cylinders of xylem and phloem.

Stem Branches Form from Lateral Buds Consisting of Meristem Cells

Branches grow from lateral buds. A lateral bud is a cluster of dormant meristem cells left behind by the apical meristem as the stem grows. Lateral buds are located just above the attachment points of the leaves, at nodes (see Fig. 23-13). When stimulated by the appropriate hormones (as we shall see in Chapter 25), the meristem cells of a lateral bud are activated and the bud sprouts, growing into a branch (Fig. 23-14). As the meristem cells divide, they release hormones that change the developmental fate of the cells between the bud and the vascular tissues of the stem. Parenchyma cells of the cortex differentiate into xylem and phloem, ultimately con-

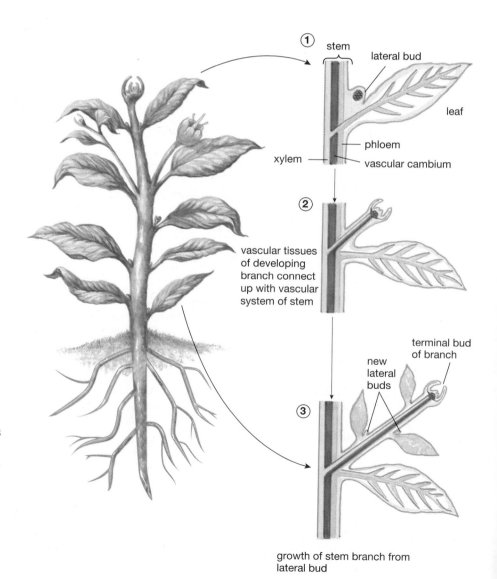

Figure 23-14 How branches form
Stem branches grow from lateral buds located at the outer surface of a stem. The bud apical meristem generates an outward-growing branch, replicating the pattern of nodes and internodes unique to each type of plant. Meanwhile, cortex cells beneath the sprouting bud differentiate into vascular tissues and connect with the vascular system of the stem.

growth of stem branch from lateral bud

necting with the main vascular systems in the stem. As the branch grows, it duplicates the development of the stem: It has an apical meristem at its tip and produces its own leaf primordia and lateral buds as it grows.

Secondary Growth Produces Thicker, Stronger Stems

In conifers and perennial dicots, stems may last up to hundreds of years, becoming thicker and stronger each year. This secondary growth in stem thickness results from cell division in two lateral meristems: the vascular cambium and *cork cambium* (Fig. 23-15).

Vascular Cambium Produces Secondary Xylem and Phloem

The vascular cambium is a cylinder of meristem cells located between the primary xylem and primary phloem. Daughter cells of the vascular cambium produced toward the inside of the stem differentiate into **secondary xylem**; those produced toward the outside of the stem differen-

tiate into **secondary phloem** (see Fig. 23-15). Because the center of the stem is already filled with pith and primary xylem, newly formed secondary xylem pushes the vascular cambium and all outer tissues farther out, increasing the diameter of the stem. This secondary xylem, with its thick cell walls, forms the wood that makes up most of the trunk of a tree. Young xylem, called **sapwood** (located just inside the vascular cambium), transports water and minerals; older xylem, the **heartwood**, contributes only to the strength of the trunk.

Phloem cells are much weaker than xylem cells. As they die over time, the sieve-tube elements and companion cells are crushed between the hard xylem on the inside of the trunk and the tough cork on the outside (see below). Only a thin strip of recently formed phloem remains alive and functioning.

In trees adapted to temperate latitudes, such as oaks and pines, cell division in the vascular cambium ceases during the cold of winter. In spring, the cambium cells divide, forming new xylem and phloem. The young cells grow by absorbing water and swelling while the

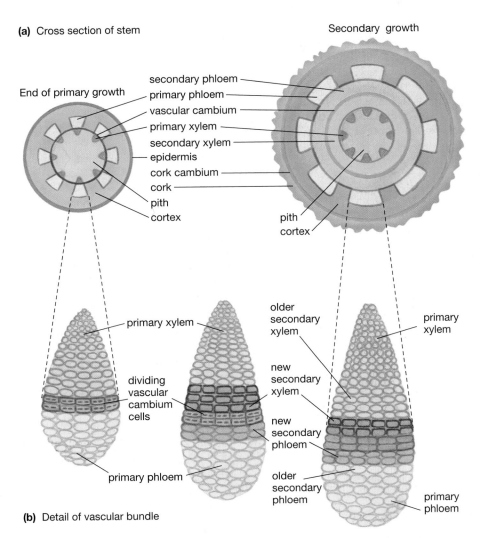

(a) Cross section of stem

Secondary growth

End of primary growth

- secondary phloem
- primary phloem
- vascular cambium
- primary xylem
- secondary xylem
- epidermis
- cork cambium
- cork
- pith
- cortex

pith
cortex

(b) Detail of vascular bundle

- primary xylem
- dividing vascular cambium cells
- primary phloem

- older secondary xylem
- new secondary xylem
- new secondary phloem
- older secondary phloem

- primary xylem
- primary phloem

Figure 23-15 Secondary growth in a dicot stem
(a) Cross section of a dicot stem at the end of primary growth (left) and during early secondary growth (right).
(b) Anatomical details of a vascular bundle during secondary growth. A vascular cambium forms between the primary xylem and primary phloem. When vascular cambium cells divide, daughter cells formed on the inside of the vascular cambium differentiate into secondary xylem; cells formed on the outside of the cambium differentiate into secondary phloem. Because xylem and pith already fill the inside of the stem, newly formed secondary xylem forces the cambium, phloem, and all outer tissues farther out, increasing the diameter of the stem. The cork cambium produces cork cells that cover the outside of the stem.

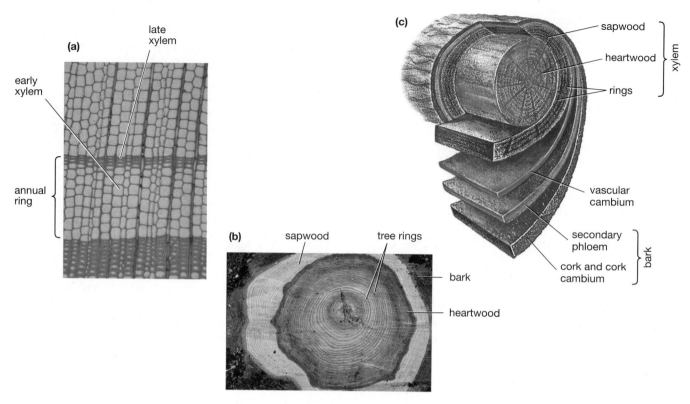

Figure 23-16 How tree rings are formed
Many trees form annual rings of xylem. *(a)* As this micrograph shows, secondary xylem cells formed during the wet spring are large, whereas secondary xylem cells formed during the hotter, drier summer are small. *(b)* Tree rings are clearly visibile in this section of trunk from a larch tree. The ratio of cell wall to "hole" (the now-empty interior of the cell) determines the color of the wood: Early wood, formed during the spring, with lots of hole, is pale; late wood, formed during the summer, with lots of wall, is dark. The water-transporting xylem of sapwood forms a lighter layer inside the bark. Xylem of the older heartwood, where the rings are most easily visible, no longer transports water and minerals. *(c)* The layers of a tree trunk.

newly formed cell walls are still soft. As the cells mature, the cell walls thicken and harden, preventing further growth. Water is readily available in spring; therefore, young xylem cells swell considerably and are large when mature. As summer progresses and water becomes scarcer, new xylem cells absorb less water and consequently are smaller when they mature. As a result, tree trunks in cross section show a pattern of alternating pale regions (large cells formed in spring) and dark regions (small cells formed in summer), as shown in Figure 23-16. This pattern forms the familiar **annual rings** of growth in temperate trees. You can determine the approximate age of a tree that has been cut by counting these growth rings.

Secondary Growth Causes the Epidermis to Be Replaced by Woody Cork

Recall that epidermal cells are mature, differentiated cells that can no longer divide. Therefore, as new secondary xylem and phloem are added each year, enlarging the stem, the epidermis can't expand to keep up with the increasing circumference. The epidermis splits off and dies.

Apparently prodded by hormones, some parenchyma cells in the cortex become rejuvenated and form a new lateral meristem, the **cork cambium** (see Fig. 23-15). These cells divide, forming daughter cells toward the outside of the stem. These daughter cells, called *cork cells*, or simply cork, develop tough, waterproof cell walls that protect the trunk both from drying out and from physical damage. Cork cells die as they mature and may form a protective layer half a meter thick in some tree species, such as the fire-resistant sequoia (Fig. 23-17). As the trunk expands from year to year, the outermost layers of cork split apart or peel off, accommodating the growth. Corks used to plug bottles are made from the outermost layer of cork from a certain type of oak, carefully peeled off by harvesters.

The common term **bark** includes all the tissues outside the vascular cambium: phloem, cork cambium, and cork cells. The complete removal of a strip of bark all the way around a tree, called *girdling*, is invariably fatal to a tree, because it severs the phloem. With the phloem gone, sugars synthesized in the leaves cannot reach the roots. Deprived of energy, the roots no longer take up water and minerals, causing the tree to die.

Figure 23-17 *Cork forms the outer layer of bark*
An ancient sequoia in the Sierra Nevada of California. The cork cambium of a sequoia produces new layers of cork each year, eventually producing a protective, fire-resistant outer covering half a meter or more thick. This massive cork layer contributes to the sequoia's great longevity; forest fires that kill lesser trees merely burn off a few inches of sequoia cork, leaving the living parts of the tree inside unharmed. The small blackened areas on this cork are from past fires.

6 Leaves: Nature's Solar Collectors

www

Only structures with chlorophyll are able to make valuable sugar out of widely available ingredients: sunlight, carbon dioxide, and water. **Leaves** are the major photosynthetic structures of most plants. Water is obtained from the soil and transported to the leaf through the xylem, and carbon dioxide (CO_2) must diffuse into the leaf from the air. You might believe that an ideal leaf should have a large surface area for gathering light and should be porous to permit CO_2 to enter from the air for photosynthesis. However, a large, porous leaf would also lose large amounts of water by evaporation. Waterproofing the entire surface would reduce the diffusion of CO_2 into the leaf. The leaves of flowering plants represent an elegant compromise among these conflicting demands (Fig. 23-18).

Leaves Have Two Major Parts: Blades and Petioles

A typical angiosperm leaf consists of a broad, flat portion, the **blade**, connected to the stem by a stalk called the **petiole** (see Fig. 23-13). The petiole positions the blade in space, normally orienting the leaf for maximum exposure to the sun. Inside the petiole are vascular tissues of xylem and phloem that are continuous with those in the stem, root, and blade. Within the blade, the vascular tissues branch into **vascular bundles**, or **veins**.

The leaf epidermis consists of a layer of nonphotosynthetic, transparent cells that secrete a waxy, waterproof cuticle on their outer surfaces. The epidermis and its cuticle are pierced by adjustable pores, the stomata, which regulate the diffusion of CO_2 and water into and out of the leaf. Each stoma is surrounded by two sausage-shaped **guard cells**, which regulate the size of the opening into the interior of the leaf (Fig. 23-18). Unlike the surrounding epidermal cells, guard cells contain chloroplasts and can carry out photosynthesis. As we shall see later, photosynthesis in the guard cells contributes to their ability to adjust the size of the pore.

Beneath the epidermis lies the loosely packed parenchyma cells of the **mesophyll** ("middle of the leaf"). In many leaves, mesophyll cells are of two types: a layer

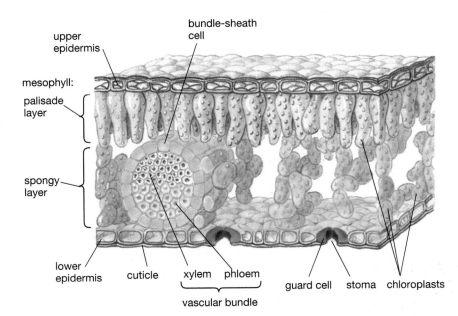

Figure 23-18 *The structure of a typical dicot leaf*
The cells of the epidermis lack chloroplasts and are transparent, allowing sunlight to penetrate to the chloroplast-containing mesophyll cells beneath. The stomata that pierce the epidermis and the loose, open arrangement of the mesophyll cells ensure that CO_2 can diffuse into the leaf from the air and reach all the photosynthetic cells.

of columnar **palisade cells** just beneath the upper epidermis, and a layer of irregularly shaped **spongy cells** above the lower epidermis. Both palisade and spongy cells contain chloroplasts; these cells perform most of the photosynthesis of the leaf. The openness of the leaf interior (Fig. 23-18) allows CO_2 to diffuse easily to all the mesophyll cells. Vascular bundles, each containing both xylem and phloem, are embedded within the mesophyll, with fine veins reaching very close to each photosynthetic cell. Thus, each mesophyll cell receives energy from sunlight transmitted through the clear epidermis; carbon dioxide from the air, diffusing through the stomata; and water from the xylem. The sugars it produces are carried away to the rest of the plant by the phloem.

7) What Are Some Special Adaptations of Roots, Stems, and Leaves?

Figure 23-19 An adaptation of roots
This *Cattleya* orchid is growing on a tree in the Amazon basin. Its roots hang down below the tree branch.

Not all roots are sinuous fibers, not all stems are smooth and upright, and not all leaves are flat and fanlike. Just as evolution has changed the basic shape of the vertebrate forelimb to suit the demands of running, swimming, and flying, so too, have plant parts become modified in response to environmental demands. You may be surprised to learn that many familiar structures are derived from unlikely parts of a plant.

Although we will highlight unusual adaptations, don't forget that *all plants are adapted to their environments*. The "typical" leaf of an oak or maple is just as much a special adaptation as is a cactus spine or daffodil bulb.

Some Specialized Roots Store Food; Others Photosynthesize

Roots have probably undergone fewer unusual modifications of their basic structure than have either stems or leaves. Some roots have extreme specializations for storage, such as the beet, carrot, or sweet potato. Some of the most bizarre root adaptations occur in certain orchids that grow perched on trees. A few of these aerial orchids have green, photosynthetic roots; in fact, for some orchids, the green roots are the only photosynthetic part of the plant (Fig. 23-19).

Some Specialized Stems Produce New Plants, Store Water or Food, or Produce Thorns or Climbing Tendrils

Many plants have modified stems that perform functions very different from the original one of raising leaves up to the light. Strawberries, for example, grow horizontal **runners** that snake out over the soil, sprouting new strawberry plants where nodes touch the soil (Fig. 23-20a). These new plants are connected with the "mother" plant, but once the plantlets form roots, they can live independently.

Some plants, such as the cactus in the chapter-opening photograph and the baobab tree (Fig. 23-20b), store water in aboveground stems. Many other plants store carbohydrates in underground stems. The common white potato is actually a storage stem; each eye is a lateral bud, ready to send up a branch next year, using the energy stored as starch in the potato to power the growth of the branch. Irises have horizontal underground stems called **rhizomes**, which store carbohydrates produced during the summer. Irises can be propagated by cutting up the rhizome; if it contains enough stored food, each piece with a node can generate a complete plant.

Many aboveground stems produce modified branches with special functions. One common branch adaptation is the **thorn**, generally growing from the normal branch location just above the site of attachment of a leaf. Hard, pointy thorns discourage animals from dining on the branches. Some of the branches of grapes and Boston ivy are modified into grasping **tendrils**, which coil around trees, trellises, or buildings, providing the otherwise prostrate plant better access to sunlight.

Specialized Leaves May Conserve and Store Water, Store Food, or Even Capture Insects

The most important environmental factors affecting the growth of leaves are temperature and availability of light and water. For example, plants growing on the floor of a tropical rain forest have plenty of water year-round but very little light, owing to the deep shade cast by several

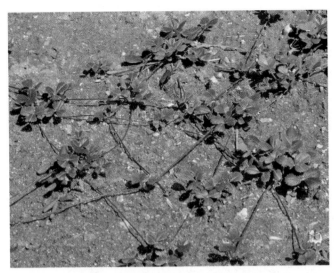

(a)

(b)

Figure 23-20 Some adaptations of stems
(a) The beach strawberry can reproduce via runners, horizontal stems that spread out over the surface of the sand. If a node of a runner touches the soil, it will sprout roots and develop into a complete plant. ***(b)*** The baobab tree develops an enormously fat, water-storing trunk. The baobab grows in dry regions; when it rains, it is to the tree's advantage to store all the water it can get. Some baobab trees have trunks so large that they have been hollowed out and used as small houses and, in one case, as a jail!

layers of trees above them. Consequently, their leaves tend to be extremely large—an adaptation demanded by the low light level and permitted by the abundant water.

At the other extreme, deserts receive bright sunlight virtually every day but have limited water and scorching temperatures. Desert plants have evolved two strikingly different adaptations to this situation. Some have very thick leaves with large cells that store water from the infrequent rains against the inevitable long droughts (Fig. 23-21a). Such leaves are covered with a thick cuticle that greatly reduces the evaporation of water. Cacti use the opposite strategy, reducing the leaves to thin spines that protect the plant from herbivores and reduce water loss (Fig. 23-21b). Photosynthesis in cacti occurs in cortex cells of the green, water-storing stems.

Modified leaves in other plants function in ways unrelated to photosynthesis or water conservation. The common edible pea, for example, grasps fences, mailbox posts, or other plants with clinging tendrils. Unlike the tendrils of grapes, which are derived from branches, pea tendrils are slender, supple leaflets. Some plants, such as onions, daffodils, and tulips, use thick, fleshy leaves as storage organs. A daffodil bulb consists of a short stem bearing thick, overlapping leaves that store nutrients during the winter (Fig. 23-21c).

Finally, a few plants have turned the table on the animals and have become predators. For example, both Venus flytraps and sundews (Fig. 23-21d) have leaves that are modified into snares for trapping unwary insects. These plants live in nitrogen-poor swamps and derive most of their nitrogen supply from the bodies of their prey.

As varied and in some cases bizarre as these leaf specializations are, the most extreme and most important leaf modification is the flower. As we will see in Chapter 24, these "reproductive leaves" enabled flowering plants to become the dominant plants on land.

8) How Do Plants Acquire Nutrients?

WWW

Nutrients are elements essential to normal life; they differ for different organisms. Plants require relatively large quantities of the following nutrients: carbon (obtained from CO_2), hydrogen (from water), oxygen (from air and water), phosphorus (from phosphate ions in soil), nitrogen (from nitrate and ammonium ions in soil), and magnesium, calcium, and potassium (as ions in soil). Plants also require very small quantities of nutrients such as iron, chlorine, copper, manganese, zinc, boron, and molybdenum. Carbon dioxide and oxygen normally enter a plant by diffusion from the air into leaves, stem, and roots. Roots extract water and all other nutrients, collectively called **minerals**, from the soil.

Figure 23-21 *Some adaptations of leaves*

(a) Desert plants receive plenty of light but very little water. Some desert plants have evolved fleshy leaves that store water from the occasional rains, just as the baobab tree does in its trunk. *(b)* Spines of desert cacti are leaves whose surface area has been minimized, reducing evaporation and protecting the plant from grazing animals. *(c)* Daffodil bulbs are enormous buds, with short central stems surrounded by thick, water- and food-storing leaves. Like other monocots, these bulbs form roots as outgrowths of the base of the stem. *(d)* Predatory leaves of the sundew bear glistening droplets that attract hungry insects. The droplets are incredibly sticky; the unsuspecting insect becomes not the eater but the eaten. Enzymes secreted by the leaf digest the insect, and the leaf absorbs the liberated nutrients.

Roots Acquire Minerals by a Four-Step Process

Soil consists of bits of pulverized rock, air, water, and organic matter (Fig. 23-22). Although both the rock particles and the organic matter contain many essential nutrients, only minerals dissolved in the soil water are accessible to the roots. The concentration of minerals in the soil water is very low, usually much lower than the concentration within plant cells and fluids. For example, the concentration of potassium (K^+) within root cells is at least 10 times greater than that in soil water, so diffusion cannot move potassium into the root. As a general principle, most minerals are moved into a root against their concentration gradients by active transport. (Re-

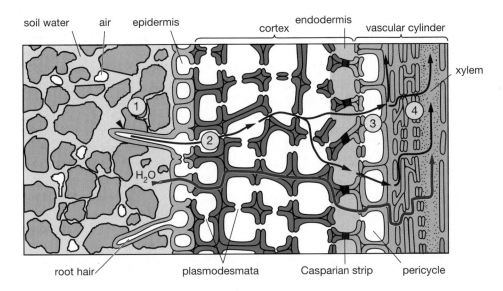

soil water air epidermis cortex endodermis vascular cylinder xylem

root hair plasmodesmata Casparian strip pericycle

H_2O

Figure 23-22 **Mineral and water uptake by roots**

The Casparian strip separates the extracellular space in the root into two compartments: an outer compartment (green) that is continuous with the soil water, and an inner compartment (blue) that is continuous with the inside of the conducting cells of the xylem. The black line is a pathway for both water and minerals; the blue line is an alternative pathway for water alone. ① Active-transport proteins in root hair membranes pump minerals into the root-hair cytoplasm; water follows by osmosis. ② Minerals diffuse inward from cell to cell through plasmodesmata that interconnect the cytoplasm of all living cells in the root; water follows by osmosis. ③ The outermost living cells at the root interior, the pericycle cells, actively transport minerals out of their cytoplasm into the extracellular space surrounding the xylem; water follows by osmosis. ④ Entry of the minerals raises the concentration of minerals in the extracellular space, so the minerals diffuse into the xylem cells through the pits in their walls, drawing in water by osmosis. Water can also move freely (blue path) through the porous cell walls of the epidermis and cortex until it reaches the Casparian strip. There it is forced to move through the interior of endodermal cells before it enters the vascular cylinder.

call from Chapter 5 that the movement of molecules from areas of low concentration to areas of high concentration requires energy.) Sugar synthesized in the leaves is transported through the phloem to the roots, where mitochondria in the root cells use it to produce ATP (adenosine triphosphate) by cellular respiration. Some of this ATP drives the active transport of minerals. Because ATP production by mitochondria requires oxygen, soil must have some air spaces within it. Flooding (or overwatering) can kill plants by depriving their roots of oxygen.

Most mineral absorption by roots occurs in a four-step process, as diagrammed in Figure 23-22:

① *Active transport into root hairs.* Root hairs projecting from the epidermal cells provide most of the surface area of the root and are in intimate contact with the soil water. The plasma membranes of the root hairs use the energy of ATP to transport minerals from the soil water, concentrating the minerals in the root hair cytoplasm.

② *Diffusion through cytoplasm to pericycle cells.* The cytoplasm of adjacent living plant cells is inter-connected by plasmodesmata. Minerals can diffuse through plasmodesmata from the epidermal cells into the cortex, endodermis, and pericycle cells.

③ *Active transport into the extracellular space of the vascular cylinder.* At the center of the vascular cylinder lies the xylem, into which the minerals must ultimately be transported. The tracheids and vessel elements of xylem are dead, without cytoplasm or plasma membrane—merely an outer skeleton of cell wall shot full of holes (see Fig. 23-6). Any minerals that enter the extracellular space surrounding the xylem can easily diffuse into the xylem cells through the holes in their walls. Therefore, pericycle cells actively transport minerals out of their own cytoplasm into the extracellular space around the xylem.

④ *Diffusion into the xylem.* The active transport of minerals into the extracellular space of the vascular cylinder increases the concentration of minerals in the extracellular space. This high concentration creates a gradient that promotes the diffusion of minerals from the extracellular space into the tracheids and vessel elements of the xylem.

You can now appreciate one of the functions of the waterproof Casparian strip that seals the spaces between the endodermal cells that surround the vascular cylinder. If water and dissolved minerals could flow through the extracellular space *between* endodermal cells, then minerals would leak back out of the extracellular space of the vascular cylinder as fast as they were pumped in. This leakage would waste the energy that was used in actively transporting the minerals into the root and would reduce the concentration gradient that allows the minerals to diffuse into the conducting cells of the xylem. The Casparian strip, however, effectively leakproofs the vascular cylinder, retaining the concentrated mineral solution within the vascular cylinder.

Symbiotic Relationships Help Plants Acquire Nutrients

Many minerals are too scarce in soil water to support plant growth, although plenty of minerals may be bound up in the surrounding rock particles. One nutrient—nitrogen—is almost always in short supply both in rock particles and in soil water. Most plants have evolved beneficial relationships with other organisms that help the plants acquire these scarce nutrients. Examples include root–fungus relationships, called *mycorrhizae*, and root–bacteria relationships formed in nodules of legumes.

Fungal Mycorrhizae Help Plants Acquire Minerals

Under normal conditions, water-soluble minerals are released very slowly from rock particles. Furthermore, the chemical forms of the minerals may not be suitable for uptake by the plasma membranes of plant root cells. Most land plants form symbiotic relationships with fungi to form root–fungus complexes called **mycorrhizae** (singular, **mycorrhiza**), which help the plant extract and absorb minerals. Fungal strands intertwine between the root cells and extend out into the soil (Fig. 23-23). In some way that is not yet understood, the fungus renders nutrients, such as nitrogen and phosphorus, accessible for uptake by the roots, perhaps by converting rock-bound minerals into simple soluble compounds that root plasma membranes can transport. The fungus, in return, receives sugars and amino acids from the plant. Both the fungus and the plant can thereby grow in places where neither could survive alone, including deserts and high-altitude, rocky soils that are low in nutrients.

Recent research has revealed that, in some forests, mycorrhizae form an immense underground web that interlinks trees—even trees of different species. This web of fungi transfers carbon compounds produced by one tree to another. Trees with access to abundant sunlight subsidize their shaded neighbors, with the mycorrhizae (like an underground Robin Hood) transferring photosynthetic products from the rich to the poor. Researchers hypothesize that nutrient transfer among trees by mycor-

(a) **(b)**

Figure 23-23 Mycorrhizae, a root–fungus symbiosis
(a) A tangled meshwork of fungal strands surrounds and penetrates into the root.
(b) Seedlings growing under identical conditions with (on the right) and without (left) mycorrhizal fungi show the importance of mycorrhizae in plant nutrition. Plants that participate in this unique relationship tend to grow larger and more vigorously than do those deprived of the fungus.

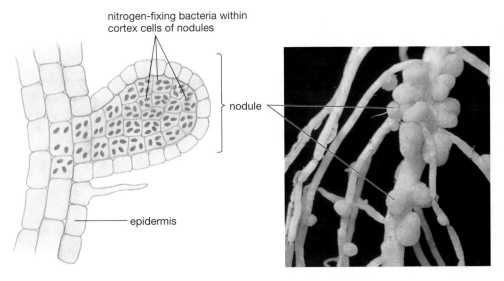

nitrogen-fixing bacteria within cortex cells of nodules

nodule

epidermis

Figure 23-24 *Nitrogen fixation in legumes*
A diagram and photograph of root nodules, containing nitrogen-fixing bacteria, in legumes.

rhizae may be an important factor in the overall health of the forest, which in turn benefits the mycorrhizae.

Bacteria-Filled Nodules on the Roots of Legumes Help Those Plants Acquire Nitrogen

Amino acids, nucleic acids, and chlorophyll all contain nitrogen, so plants need large amounts of this element. Although nitrogen gas (N_2) makes up about 79% of the atmosphere, plants can take up nitrogen only through their roots, in the form of ammonium ion (NH_4^+) or nitrate ion (NO_3^-).

Although N_2 diffuses from the atmosphere into the air spaces in the soil, it cannot be used by plants. Plants don't have the enzymes needed to carry out **nitrogen fixation**, the conversion of N_2 into ammonium or nitrate. A variety of **nitrogen-fixing bacteria** do have these enzymes. Some of these bacteria are free-living in the soil. However, nitrogen fixation is very costly, energetically speaking, using at least 12 ATPs per ammonium ion synthesized. Consequently, bacteria don't routinely manufacture a lot of extra ammonium and liberate it into the soil.

Some plants, particularly the **legumes** (peas, clover, and soybeans), enter into a mutually beneficial relationship with certain species of nitrogen-fixing bacteria. By secreting chemicals into the soil, legumes attract nitrogen-fixing bacteria to their roots. Once there, the bacteria enter the root hairs. The bacteria then digest channels through the cytoplasm of the epidermal cells and into underlying cortex cells. As both bacteria and their host cortex cells multiply, a **nodule**, or swelling that houses the root–bacteria complexes, is formed (Fig. 23-24). A cooperative relationship develops. The plant transports sugars from its leaves down to the cortex, just as it normally would for storage. The bacteria with-

in the cortex cells take up the sugar and use its energy for all of their metabolic processes, including nitrogen fixation. The bacteria obtain so much energy that they produce more ammonium than they need. The surplus ammonium diffuses into the cytoplasm of their host cells, providing the plant with a steady supply of usable nitrogen. Surplus ammonium also diffuses into the surrounding soil, making it better able to support other types of plants. Farmers plant legumes not only for their commercial value but also to enrich the soil with ammonium for future crops.

9 How Do Plants Acquire Water?

Once a high concentration of minerals builds up in the vascular cylinder, water absorption becomes very straightforward (see Fig. 23-22). Water moves by osmosis from regions of high water concentration to regions of low water concentration. Dissolved minerals bind water molecules and thus tie them up, lowering the concentration of free water molecules. Therefore, soil water (low in minerals) has a high free-water concentration, whereas the vascular cylinder water (high in minerals) has a low free-water concentration. This situation promotes the movement of water from the soil into the vascular cylinder.

Water can take two routes on its way to the vascular cylinder: (1) It follows the minerals, moving through cells by osmosis (black path in Fig. 23-22), and (2) it also moves readily through the highly porous cell walls of the epidermal and cortex cells (blue path in Fig. 23-22). At the endodermis, the waterproof Casparian strip blocks further movement of water through the spaces between cells, so water moves across the membranes of the endodermal cells by osmosis. After leaving the endodermal cells, water moves into the vascular cylinder, entering the tracheids and vessel elements of the xylem through porous pits in their cell walls.

As you will see when we discuss transport in xylem, water is pulled up the xylem by the force of the evaporation of water from the leaves. This movement further lowers the water concentration within the vascular cylinder and enhances the diffusion of water across the endodermis.

2. *Tension.* This "water chain" is pulled up the xylem; evaporation provides the necessary energy.

Let's briefly examine both factors.

Hydrogen Bonds between Water Molecules Produce Cohesion

You may recall from Chapter 2 that water is a polar molecule, with the oxygen carrying a slight negative charge and the hydrogens, a slight positive charge. As a result,

10 How Do Plants Transport Water and Minerals?

Once they enter the root xylem, water and minerals still must be moved to the uppermost reaches of the plant. (In redwood trees, the distance may be more than 100 meters!) The processes of active transport, diffusion, and osmosis, all of which suffice to move water and minerals from soil to root xylem, would be hopelessly slow for getting these substances to the top of a tree. Land plants accordingly move fluids up the xylem from root to stem and leaf by **bulk flow**. During bulk flow, molecules of fluids move together, rather than molecule by molecule, from areas of high pressure to areas of lower pressure. Because minerals are dissolved in water, they are passively carried along as the water flows upward. Why is water pressure sufficiently higher in the roots and lower in the leaves to overcome the force of gravity and make the water flow upward? The *cohesion–tension theory* provides an explanation.

Water Movement in Xylem Is Explained by the Cohesion–Tension Theory

According to the **cohesion–tension theory**, water is pulled up the xylem, powered by the evaporation of water from the leaves (Fig. 23-25). As its name suggests, this theory has two essential parts:

1. *Cohesion.* Attraction among water molecules holds water together, forming a solid chainlike column within the xylem tubes.

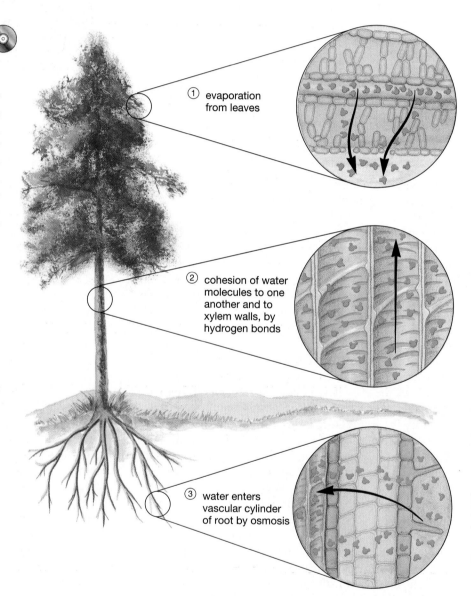

① evaporation from leaves

② cohesion of water molecules to one another and to xylem walls, by hydrogen bonds

③ water enters vascular cylinder of root by osmosis

Figure 23-25 *The cohesion–tension theory of water flow from root to leaf in xylem* ① Water molecules evaporate out of the leaves, and other water molecules replace them from the xylem of the leaf veins. ② Within the xylem, hydrogen bonding holds nearby water molecules together so firmly that the column of water behaves like a chain. The top of the "water chain" is pulled up by evaporation, and the rest of the chain, all the way down to the roots, comes along as well. ③ As the molecules of the water chain retreat up the xylem in the roots, the decreased water concentration within the root xylem and the surrounding extracellular space causes water to enter from the soil water by osmosis, thus steadily replenishing the bottom of the chain.

nearby water molecules attract one another, forming weak hydrogen bonds. Just as individually weak cotton threads together make the strong fabric of your jeans, the network of individually weak hydrogen bonds in water collectively produces a very high cohesion, or tendency to resist being separated. Experiments have demonstrated that the column of water within the xylem is at least as strong—and as unbreakable—as a steel wire of the same diameter. This is the "cohesion" part of the theory: Hydrogen bonds among water molecules provide the cohesion that holds together a chain of water that extends the entire height of the plant within the xylem. Supplementing the cohesion between water molecules is adhesion between water molecules and the walls of xylem. Attraction of water molecules to the cell walls of

the thin xylem tubes helps the water creep upward, just as water is pulled upward into a very narrow glass tube. This principle is called *capillary action.*

Transpiration Produces the Tension That Pulls Water Upward

Evaporation of water through the stomata of leaves and stems (though primarily leaves) is a process called **transpiration**, provides the force for water movement—the "tension" part of the theory. As a leaf transpires, the concentration of water in the mesophyll cells drops. This drop causes water to move by osmosis from the xylem into the dehydrating mesophyll cells. Water molecules leaving the xylem are attached to other water molecules in the same xylem tube by hydrogen bonds. Therefore, when one

Earth Watch
Plants Help Regulate the Distribution of Water

The land teems with a remarkable diversity of plant life. The distribution of plants on Earth is limited by environmental factors and the adaptations of the plants. Probably the most important environmental factor influencing plant distribution is water: Cacti inhabit deserts because they can withstand drought; orchids and mahogany trees need the frequent drenching rains of the rain forest. What people often overlook is the flip side of this plant–water relationship: Plants, through transpiration, help regulate the amount and distribution of rainfall, soil water, and even river flow.

Consider the Amazon rain forest (Fig. E23-1). An acre of soil supports hundreds of towering trees, each bearing millions of leaves. The surface area of the leaves dwarfs the surface area of the soil, so up to 75% of all the water evaporating from the acre of forest is due to leaf transpiration. This transpiration raises the humidity of the air and causes rain to fall. In fact, about

half of the water transpired from the leaves falls again as rain, with the overall result that about one-third of the total rainfall is water recycled by transpiration. Thus, in a very real sense, the high humidity and frequent showers that the rain forest needs to survive are partly *created by the forest itself!* If large areas of rain forest are cut down, less water evaporates in that area, so less rain falls, and new rain forest tree seedlings cannot grow. An entirely different plant community would probably become established on the disturbed land and might become a *permanent* new community.

Plant transpiration might even have a moderating influence on some aspects of the global warming that is being created by increasing atmospheric CO_2 levels, due mostly to the burning of fossil fuels (such as coal and oil) during industrial activity and to the cutting down of forests (as we shall see in Chapter 41). According to predictions about global warming, higher CO_2 levels will raise temperatures on Earth, and a warmer planet would increase water evaporation. Increased evaporation, in turn, should lead to drier soils and the expansion of deserts. Stomata, however, open partly in response to low CO_2 levels within the guard cells. Elevated atmospheric CO_2 levels also raise CO_2 within the guard cells and cause partial stomatal closing. Further, a recent study found that plants grown in an atmosphere with high CO_2 levels have fewer stomata per unit area of leaf than do plants grown at low, preindustrial CO_2 levels. If there are fewer stomata, and they are chronically partially closed, then we might expect less transpiration from plants that will live in a warmer future than from preindustrial plants. Some researchers suggest that soil moisture and river flow in the southwest United States might actually *increase* in response to increased atmospheric CO_2, despite the global warming effect.

These examples show that plants wield an enormous influence on what we often consider to be nonliving aspects of the biosphere, such as humidity, rainfall, soil water, and river flow. The responses of plants to human activities, neither simple nor easily predictable, can have major impacts on ecosystems.

Figure E23-1 *The Amazon rain forest*
The rain forest community helps mold its own environment.

water molecule leaves, it pulls adjacent water molecules up the xylem. As these water molecules move upward, other water molecules farther down the tube move up to replace them. This process continues all the way to the roots, where water in the extracellular space around the xylem is pulled in through the holes in the walls of vessel elements and tracheids. This upward and inward movement of water finally causes soil water to move through the endodermal cells into the vascular cylinder by osmosis. The force generated by the evaporation of water from the leaves, transmitted down the xylem to the roots, is so strong that water can be absorbed even from quite dry soils. Can this cohesion–tension theory explain the movement of water from soil to the topmost leaves of giant redwoods? Using a special apparatus, botanists have measured xylem water tensions strong enough to pull water up 200 meters (more than 600 feet).

Summing Up
Water Transport in Xylem

Transpiration from leaves removes water from the top of a xylem tube. This water is replaced by water from farther down the tube, so water moves up the xylem by bulk flow. This upward flow removes water from the root xylem and the extracellular space surrounding it, promoting the osmosis of water from the soil into the vascular cylinder of the root. *The flow of water is unidirectional, from root to shoot, because only the shoot can transpire.*

Adjustable Stomata Control the Rate of Transpiration

Although it provides the force that transports water and minerals to the leaves at the top of the plant, transpiration is by far the largest source of water loss—a loss that may threaten the very survival of the plant, especially in hot, dry weather. Most water is transpired through the stomata of the leaves and stem, so you might think that a plant could prevent water loss simply by closing its stomata. Don't forget, however, that photosynthesis requires carbon dioxide from the air, which diffuses into the leaf mainly through open stomata. *Therefore, a plant, by opening and closing its stomata, must achieve a balance between carbon dioxide uptake and water loss.*

A stoma consists of a central opening surrounded by two sausage-shaped, photosynthetic guard cells that regulate the size of the opening (Fig. 23-26). With some exceptions, stomata open during the day, when sunlight allows photosynthesis, and close at night, conserving water. They will also close in the sunlight if the plant is losing too much water. Plants whose leaves are oriented horizontally normally have more stomata on the shaded, lower surface than on the sunny, upper surface, reducing evaporation.

How do plants regulate stomatal opening and closing? Stomata open when the guard cells take up water and elongate, bowing outward and increasing the space between them. Stomata close when guard cells lose water and shrink, reducing the space between them. The entry of water, in turn, is regulated by changes in the potassium content of the guard cells. The plant opens its stomata by actively pumping potassium into the guard cells, causing water to follow by osmosis. When potassium leaves the guard cells, water leaves again by osmosis, and the stomata close.

Several factors regulate the potassium concentration inside guard cells. The three most important factors are availability of both light and carbon dioxide and water levels within the leaf. These factors help the plant achieve a balance between the need to photosynthesize and the need to conserve water:

(b)

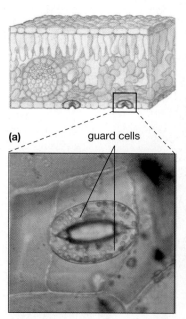

(a) guard cells

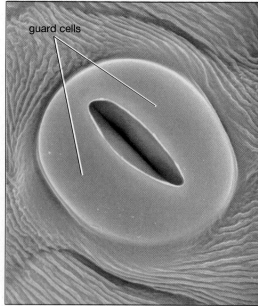

guard cells

Figure 23-26 Stomata
Stomata seen through *(a)* the light microscope and *(b)* scanning electron microscope. In the light micrograph, note that the guard cells contain chloroplasts (the green ovals within the cells) but that the other epidermal cells do not.

1. *Light reception.* When light strikes special pigments contained within the guard cells, it triggers a series of reactions that cause potassium to be actively transported into the guard cells. Water follows, and the stomata open. At night, when light is not present to activate the pigments, the potassium pumping stops. The "extra" potassium within the guard cells diffuses back out, and the stomata close, conserving water.

2. *Carbon dioxide concentration.* Low CO_2 concentrations (such as those that occur during the day when photosynthesis exceeds cellular respiration) stimulate the active transport of potassium into the guard cells. This transport causes stomata to open and allows CO_2 to diffuse in. At night, cellular respiration in the absence of photosynthesis raises CO_2 levels, halting the inward transport of potassium and allowing the guard cells to close.

3. *Water.* If a leaf loses water faster than it can replace water from the xylem, it begins to wilt, and the mesophyll cells release a hormone called **abscisic acid**. This hormone strongly inhibits the active transport of potassium into the guard cells, overriding the stimulatory effects of light and low CO_2 levels, so potassium pumping stops. As potassium diffuses out of the guard cells, water follows by osmosis, the guard cells shrink, and the stomata close. As you might guess, when your house or garden plants are wilted, they are unable to carry out normal levels of photosynthesis.

11 How Do Plants Transport Sugars?

Sugars synthesized in the leaves must be moved to other parts of the plant. There they nourish nonphotosynthetic structures such as roots or flowers and can be stored in the cortex cells of the root and stem. Sugar transport is the function of phloem.

Botanists studying phloem contents have employed a most unlikely lab assistant: the aphid. *Aphids* are insects that feed on the fluid contained in phloem sieve tubes. An aphid inserts its *stylet*, a pointed, hollow tube, through the epidermis and cortex of a young stem into a sieve tube (Fig. 23-27). The aphid can then relax and let the plant do the work. The fluid in the sieve tubes is under pressure and flows through the stylet into the aphid's digestive tract (sometimes with enough pressure to force its way out the other end!). By cutting off the aphid but leaving its stylet in place, botanists have collected sieve-tube fluid and found that it consists mostly of sucrose and water—as much as 25% sucrose by weight. How is this concentrated sugar solution moved about the plant?

The Pressure-Flow Theory Explains Sugar Movement in Phloem

The movement of fluid in phloem is directed by sugar production and use. Any structure that actively synthesizes sugar is said to be a **source** away from which

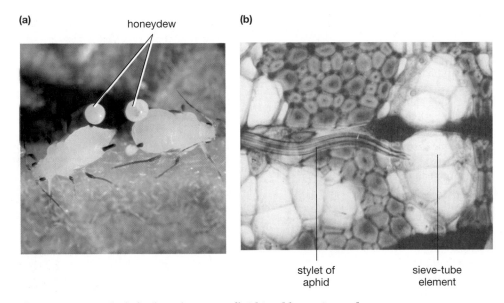

(a) honeydew

(b) stylet of aphid · sieve-tube element

Figure 23-27 Aphids feed on the sugary fluid in phloem sieve tubes
(a) When an aphid pierces a sieve tube, pressure in the tube forces the fluid out of the plant and into the aphid's digestive tract. The pressure can be so great that fluid is forced completely through the aphid and out its anus, as "honeydew." This fluid is collected by certain species of ants that act as "shepherds" to the aphids, defending them from predators in return for a diet of sweet honeydew. **(b)** The flexible stylet of an aphid, passing through many layers of cells to penetrate a sieve-tube element.

phloem fluid will be transported. Conversely, any structure that uses up sugar or converts sugar to starch is said to be a **sink** toward which phloem fluids will flow. A newly forming leaf will be a sink as it develops, with phloem flowing up into it from more-mature leaves. When the leaf matures, it will photosynthesize and produce sugar, becoming a source for phloem flow to other newly developing leaves, to flowers or fruits, or to the roots. Therefore, fluid in phloem can move either up or down the plant, depending on the metabolic demands of the various parts of the plant at any given time.

The most widely accepted mechanism of sugar transport in phloem is the **pressure-flow theory** (Fig. 23-28), which relies on differences in hydrostatic pressure (water pressure) to move fluid through sieve tubes. Let's illustrate this theory by following sucrose movements from a mature leaf to a developing fruit:

① *Sucrose source: photosynthesis.* When a leaf is photosynthesizing rapidly, it manufactures lots of glucose, much of which is converted to the larger molecule sucrose.

② *Phloem sieve-tube loading.* Much of this sucrose is actively transported into companion cells of the phloem in the leaf veins. This movement raises the concentration of sucrose within the companion cells, so sucrose then diffuses down its concentration gradient through plasmodesmata into adjacent sieve-tube elements. This diffusion in turn raises the sucrose concentration in the leaf sieve tube.

③ *Osmosis into the leaf sieve tube.* The high sucrose concentration in the leaf sieve tube lowers the water concentration in the sieve tube, causing water to enter the sieve tube by osmosis from nearby xylem. Hydrostatic pressure increases as more water molecules enter the tube.

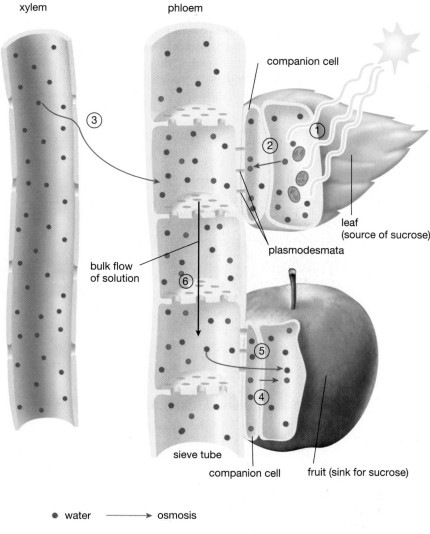

Figure 23-28 The pressure-flow theory This theory relies on differences in hydrostatic pressure to move fluid through phloem sieve tubes. ① A photosynthesizing leaf manufactures sucrose (red dots), which ② is actively transported (red arrow) into a nearby companion cell in phloem. The sucrose diffuses into the adjacent sieve-tube element through plasmodesmata, raising the concentration of sucrose in the sieve-tube element. ③ Water (blue dots) leaves nearby xylem and moves into the "leaf end" of the sieve tube by osmosis (blue arrow), raising the hydrostatic pressure as increasing numbers of water molecules enter the fixed volume of the tube. ④ The same sieve tube connects to a developing fruit. At the "fruit end" of the tube, sucrose enters the companion cells by diffusion through plasmodesmata. It is then actively transported out of the companion cells and into the fruit cells. ⑤ Water moves out of the tube by osmosis, lowering the hydrostatic pressure within the tube. ⑥ High pressure in the leaf end of the phloem and low pressure in the fruit end cause water, together with any dissolved solutes, to flow in bulk from leaf to fruit (black arrow).

④ *Sucrose sink: developing fruit.* Meanwhile, some distance away along the same sieve tube, sucrose is actively transported out of sieve-tube elements and companion cells into the cells of a fruit. The concentration of sugar in the fruit is raised, and the concentration of sugar in the sieve tubes is lowered.

⑤ *Osmosis out of the fruit sieve tube.* Water leaves the sieve tube by osmosis and follows the sugar into the fruit. Hydrostatic pressure drops within the tube.

⑥ *Bulk flow, driven by a hydrostatic pressure gradient.* Water that, by osmosis, follows the sucrose into the sieve tube near the leaf causes hydrostatic pressure to build up in the leaf portion of the sieve tube. Meanwhile, water entering the fruit by osmosis causes reduced hydrostatic pressure in the sieve tube near the fruit. In response to this pressure gradient, water moves by bulk flow from the leaf portion of the phloem into the fruit portion of the phloem, carrying the dissolved sugar.

Summary of Key Concepts

1) How Are Plant Bodies Organized?

The body of a land plant consists of root and shoot. Roots are normally underground and have six functions: They (1) anchor the plant in the soil; (2) absorb water and minerals from the soil; (3) store surplus photosynthetic products; (4) transport water, minerals, photosynthetic products, and hormones; (5) produce some hormones; and (6) interact with soil fungi and microorganisms that provide nutrients. Shoots are normally located above ground and consist of stem, leaves, buds, and (in season) flowers and fruit. Shoot functions include (1) photosynthesis, (2) transport of materials, (3) reproduction, and (4) hormone synthesis.

2) How Do Plants Grow?

Plant bodies are composed of two main classes of cells: meristem cells and differentiated cells. Meristem cells are undifferentiated cells that retain the capacity for mitotic cell division. Differentiated cells arise from divisions of meristem cells, become specialized for particular functions, and normally do not divide.

Most meristem cells are located in apical meristems at the tips of roots and shoots and in lateral meristems in the shafts of roots and shoots. Primary growth (growth in length and differentiation of parts) results from the division and differentiation of cells from apical meristems; secondary growth (growth in diameter) results from the division and differentiation of cells from lateral meristems.

3) What Are the Tissues and Cell Types of Plants?

Plant bodies consist of three tissue systems: the dermal, ground, and vascular systems. The dermal tissue system forms the outer covering of the plant body. The dermal tissue system of leaves and of primary roots and stems is normally a single cell layer of epidermis. Dermal tissue after secondary growth is a multilayered covering of cork.

The ground tissue system consists of a variety of cell types, including parenchyma, collenchyma, and sclerenchyma. Most are involved in photosynthesis, support, or storage. Ground tissue makes up most of a young plant during primary growth. During secondary growth of stems and roots, ground tissue becomes an increasingly small part of the plant body.

The vascular tissue system consists of xylem, which transports water and minerals from the roots to the shoots, and phloem, which transports water, sugars, amino acids, and hormones throughout the plant body.

4) Roots: Anchorage, Absorption, and Storage

Primary growth in roots results in a structure consisting of an outer epidermis, an inner vascular cylinder of conducting tissues, and cortex between the two. The apical meristem near the tip of the root is protected by the root cap. Cells of the root epidermis absorb water and minerals from the soil. Root hairs are projections of epidermal cells that increase the surface area for absorption. Most cortex cells store surplus sugars (normally in the form of starch) produced by photosynthesis. The innermost layer of cortex cells is the endodermis, which controls the movement of water and minerals from the soil into the vascular cylinder. The vascular cylinder contains the conducting tissues—xylem and phloem.

5) Stems: Reaching for the Light

Primary growth in dicot stems results in a structure consisting of an outer, waterproof epidermis; supporting and photosynthetic cells of cortex beneath the epidermis; vascular tissues of xylem and phloem; and supporting and storage cells of pith at the center. Leaves and lateral buds are found at nodes along the surface of the stem. Under the proper hormonal conditions, lateral buds may sprout into a branch.

Secondary growth in stems results from cell divisions in the vascular cambium and cork cambium. Vascular cambium produces secondary xylem and secondary phloem, increasing the stem's diameter. Cork cambium produces waterproof cork cells that cover the outside of the stem.

6) Leaves: Nature's Solar Collectors

Leaves are the main photosynthetic organs of plants. The blade of a leaf consists of a waterproof outer epidermis surrounding mesophyll cells, which have chloroplasts and which carry out photosynthesis, and vascular bundles of xylem and phloem, which carry water, minerals, and photosynthetic products to and from the leaf. The epidermis is perforated by adjustable pores called stomata that regulate the exchange of gases and water.

7) What Are Some Special Adaptations of Roots, Stems, and Leaves?

Through evolution, the roots, stems, and leaves of many plants have been modified into diverse structures such as spines, tendrils, thorns, and bulbs. These unusual structures are commonly involved in water or energy storage, support, or protection.

8 How Do Plants Acquire Nutrients?

Most minerals are taken up from the soil water by active transport into the root hairs. These minerals diffuse into the root through plasmodesmata to the pericycle, just inside the vascular cylinder. There they are actively transported into the extracellular space of the vascular cylinder. The minerals diffuse from the extracellular space into the tracheids and vessel elements of xylem.

Many plants have fungi called mycorrhizae associated with their roots that help absorb soil nutrients. Nitrogen can be absorbed only as ammonium or nitrate, both of which are scarce in most soils. Legumes have evolved a cooperative relationship with nitrogen-fixing bacteria that invade legume roots. The plant provides the bacteria with sugars, and the bacteria use some of the energy in those sugars to convert atmospheric nitrogen to ammonium, which is then absorbed by the plant.

9 How Do Plants Acquire Water?

Because the cells of the root epidermis and cortex are loosely packed and have porous walls, water in the soil has a continuous, uninterrupted pathway through the outer layers of the root, up to the waterproof layer of the Casparian strip between endodermal cells. Both mineral uptake and the upward movement of water in xylem contribute to a concentration gradient across the endodermal cells, with a higher concentration of free water molecules in the extracellular space outside the endodermis than in the extracellular space inside the endodermis. Therefore, water moves by osmosis across the plasma membranes of the endodermal cells into the extracellular space of the vascular cylinder.

10 How Do Plants Transport Water and Minerals?

The cohesion–tension theory explains xylem function: The cohesion of water molecules to one another by hydrogen bonds holds together the water within xylem tubes almost as if it were a solid chain. As water molecules evaporate from the leaves, the hydrogen bonds pull other water molecules up the xylem to replace them. This movement is transmitted down the xylem to the root, where water loss from the vascular cylinder promotes water movement across the endodermis from the soil water by osmosis.

11 How Do Plants Transport Sugars?

The pressure-flow theory explains sugar transport in phloem: Parts of the plant that synthesize sugar (for example, leaves) export sugar into the sieve tube. Increasing sugar concentrations attract water entry by osmosis, causing high hydrostatic pressure in that part of the phloem. Parts of the plant that consume sugar (for example, fruits) remove sugar from the sieve tube. The loss of sugar causes the loss of water by osmosis, resulting in low hydrostatic pressure. Water and dissolved sugar move by bulk flow in the sieve tube from areas of high to low pressure.

Key Terms

abscisic acid   *p. 475*
annual ring   *p. 464*
apical meristem   *p. 454*
bark   *p. 464*
blade   *p. 465*
branch root   *p. 459*
bulk flow   *p. 472*
cambium   *p. 454*
Casparian strip   *p. 458*
cohesion–tension theory   *p. 472*
collenchyma   *p. 455*
companion cell   *p. 457*
cork cambium   *p. 464*
cork cell   *p. 455*
cortex   *p. 458*
cuticle   *p. 455*
dermal tissue system   *p. 454*
dicot   *p. 452*
differentiated cell   *p. 454*
endodermis   *p. 458*
epidermal tissue   *p. 454*
epidermis   *p. 454*

fibrous root system   *p. 458*
ground tissue system   *p. 454*
guard cell   *p. 465*
heartwood   *p. 463*
internode   *p. 460*
lateral bud   *p. 460*
lateral meristem   *p. 454*
leaf   *p. 465*
leaf primordium   *p. 460*
legume   *p. 471*
meristem cell   *p. 454*
mesophyll   *p. 465*
mineral   *p. 467*
monocot   *p. 452*
mycorrhiza   *p. 470*
nitrogen fixation   *p. 471*
nitrogen-fixing bacterium   *p. 471*
node   *p. 460*
nodule   *p. 471*
nutrient   *p. 467*
palisade cell   *p. 466*
parenchyma   *p. 455*

pericycle   *p. 459*
periderm   *p. 455*
petiole   *p. 465*
phloem   *p. 457*
pit   *p. 456*
pith   *p. 460*
pressure-flow theory   *p. 476*
primary growth   *p. 454*
primary phloem   *p. 462*
primary root   *p. 458*
primary xylem   *p. 462*
rhizome   *p. 466*
root   *p. 452*
root cap   *p. 458*
root hair   *p. 455*
root system   *p. 452*
runner   *p. 466*
sapwood   *p. 463*
sclerenchyma   *p. 455*
secondary growth   *p. 454*
secondary phloem   *p. 463*
secondary xylem   *p. 463*
shoot system   *p. 452*

sieve plate   *p. 457*
sieve tube   *p. 457*
sieve-tube element   *p. 457*
sink   *p. 476*
source   *p. 475*
spongy cell   *p. 466*
stem   *p. 452*
stoma   *p. 460*
taproot system   *p. 458*
tendril   *p. 466*
terminal bud   *p. 460*
thorn   *p. 466*
tracheid   *p. 456*
transpiration   *p. 473*
vascular bundle   *p. 465*
vascular cambium   *p. 462*
vascular cylinder   *p. 459*
vascular tissue system   *p. 454*
vein   *p. 465*
vessel   *p. 456*
vessel element   *p. 456*
xylem   *p. 456*

Thinking Through the Concepts

Multiple Choice

1. *In a mycorrhizal relationship, a plant root has a symbiotic relationship with*
a. a fungus that helps obtain minerals from the soil
b. a fungus that helps in the fixation of nitrogen from the air into a form usable by the plant
c. bacteria that help obtain minerals from the soil
d. bacteria that help in the fixation of nitrogen from the air into a form usable by the plant
e. an alga that helps the plant photosynthesize

2. *Which of the following is involved in the cohesion–tension theory of water movement in plants?*
a. the presence of hydrogen bonds that hold water molecules together
b. the attraction of water molecules to the walls of the xylem
c. the diffusion of water from cells in the root to cells in the shoot
d. the evaporation of water through the stomata
e. All of the above are involved in this theory.

3. *You and a friend carve your initials 5 feet above the ground on a tree on campus. The tree is now 40 feet tall. When you come back for your twenty-fifth reunion, the tree will be 100 feet tall. How high above the ground should you look for your initials?*
a. 5 feet
b. 40 feet
c. 60 feet
d. 95 feet
e. 100 feet

4. *If you wanted to show a friend how much you had learned in biology class, which of the following would you NOT use to identify a plant as a dicot?*
a. leaves with veins arranged like a net
b. hand-shaped leaves
c. six petals
d. a taproot
e. a seed with two cotyledons

5. *When water enters a plant root, it is forced to travel through the endodermal cells by the waxy*
a. cuticle
b. epidermis
c. periderm
d. xylem
e. Casparian strip

6. *Which of the following is NOT a special adaptation of certain roots, stems, or leaves?*
a. roots–photosynthesis
b. stems–water storage
c. leaves–defense
d. roots–prey capture
e. stems–producing new plants

? Review Questions

1. Describe the locations and functions of the three tissue systems in land plants.

2. Distinguish between primary growth and secondary growth, and describe the cell types involved in each.

3. Distinguish between meristem cells and differentiated cells. Which meristems cause primary growth? Which ones form secondary growth? Where is each type located?

4. Diagram the internal structure of a root after primary growth, labeling and describing the function of epidermis, cortex, endodermis, pericycle, xylem, and phloem. What tissues are located in the vascular cylinder?

5. How do xylem and phloem differ?

6. What are the main functions of roots, stems, and leaves?

7. What types of cells form root hairs? What is the function of root hairs?

8. Diagram the internal structure of leaves. What structures regulate water loss and CO_2 absorption by a leaf?

9. What role does abscisic acid play in controlling the opening and closing of stomata? Describe the daily cycle of the opening and closing of guard cells. How are various environmental conditions involved in this process?

10. How are minerals and water taken up by roots? Diagram the structures involved, the pathways for water and minerals from soil water to xylem, and the transport processes at each step.

11. How does the pressure-flow theory explain the movement of sugars through a plant?

12. Describe the cohesion–tension theory of water movement in xylem. What supplies the cohesion, and what is the source of the tension? How do these two forces interact to move water through a plant?

13. What occurs during nitrogen fixation? Describe the formation of a root nodule in a legume. How do both the bacteria and the legume benefit from this relationship?

Applying the Concepts

1. A mutant form of aphid, the klutzphid, inserts its stylet into the vessel elements of xylem. What materials are found in the fluids of xylem? Could an aphid live on xylem fluid? Would xylem fluid flow into the aphid? Explain your answer.

2. One of the foremost goals of molecular botanists is to insert the genes for nitrogen fixation, or the ability to enter into symbiotic relationships with nitrogen-fixing bacteria, into crop plants such as corn or wheat (see Chapter 13). Why would the insertion of such genes be useful? What changes in farming practices would this technique allow?

3. We learned in Chapter 2 about the peculiar characteristics of water. Discuss several ways in which the evolution of vascular plants has been greatly influenced by water's special characteristics.

4. A major environmental problem is desertification, in which overgrazing by cattle or other animals results in too few plants in an area. Show how what you know about the movement of water through plants enables you to understand this process, in which there is less water in the atmosphere, less rain, and thus dry, desertlike conditions.

5. The tropical rain forest contains a large number of as-yet unidentified plants, many of which may have uses as medicines or food. If you were given the job of searching a particular portion of the rain forest for useful products, how would you use the information you gained from this chapter to help you narrow your search? What kinds of plant tissues or organs would be most likely to contain such products?

6. The desert tends to have two types of plants with respect to their root systems—small grasses or herbs, and shrubs or small trees. The grasses and herbs typically form fibrous root systems. The shrubs and trees form taproot systems. What advantages can you think of for each system? How does each type of root allow for survival in a desert environment?

7. Grasses (monocots) form their primary meristem near the ground surface rather than at the tips of branches the way dicots do. How does this feature allow you to grow a lawn and mow it every week in the summer? What would happen if you had a dicot lawn and tried to mow it?

8. Discuss the structures and adaptations that might occur in the leaves of plants living in (a) dry, sunny habitats; (b) wet, sunny habitats; (c) dry, shady habitats; and (d) wet, shady habitats. Which of these habitats do you think would be most inhospitable (for example, in which habitat would it be most difficult to design a functioning leaf)?

Group Activity

Like other large perennials, palms can make new leaves and roots if those organs get damaged—for example, by freezing. Nevertheless, palms are tropical plants precisely because they cannot recover from freezing-cold winters. Take 3 minutes to write your own explanation of the reason the structure of palms prevents them from living in areas with cold winters. Pass your explanation to the person on your right. Take another 3 minutes to arrive at an *alternative* explanation than the one suggested by the person on your left. Write your alternative beneath their suggested explanation. Discuss all the possible explanations with the rest of your class.

For More Information

Baskin, Y. "Forests in the Gas." *Discover*, October 1994. As carbon dioxide increases in the atmosphere, the relationships of plants in the natural environment will change.

Cochran, M. F. "Chestnuts: Back from the Brink." *National Geographic*, February 1990. Fungi girdle chestnuts, causing almost all chestnut trees in the United States to die. Recent discoveries are helping this tree come back.

Day, S. "A Shot in the Arm for Plants." *New Scientist*, January 9, 1993. Plants exposed to insects or disease may produce chemicals that are transported throughout their body to protect them from future attack.

Frits, H. C. "Tree Rings and Climate." *Scientific American,* May 1972. The width of the annual rings of trees reflects the length of the growing season and the amount of rainfall and can be used to determine prehistoric climate.

Line, L. "The Return of an American Classic." *Audubon*, September/October 1997. Dutch elm disease has nearly wiped out the majestic American elm tree. But a few hardy survivors with resistance to this fungus are being propagated for sale by the year 2000.

Mansfield, T. A., and Davies, W. J. "Mechanisms for Leaf Control of Gas Exchange." *BioScience,* March 1985. How stomata control gas exchange through the surface of a leaf.

Meyerowitz, E. M. "The Genetics of Flower Development." *Scientific American*, May 1994. An easy-to-read article about research in which flower parts are rearranged to determine how plant tissues develop.

Stuart, D. "Green Giants." *Discover*, April 1990. Sequoias, the most-massive living organisms on Earth, are among the oldest as well. Stuart describes how they can grow so large and live so long.

Vogel, S. "When Leaves Save the Tree." *Natural History*, September 1993. The shape and arrangement of leaves on a tree allow it to withstand wind and storm.

Weiss, R. "When Plants Act Like Animals." *National Wildlife*, December 1994–January 1995. A fun look at plant movements.

Zimmer, C. "The Processing Plant." *Discover*, September 1995. The purple pitcher plant supplements its diet by digesting flies, with the surprising help of insect larvae that thrive in the pitcher plant's "digestive chamber."

Zimmer, C. "The Web Below." *Discover*, November 1997. An underground web of mycorrhizae transfers nutrients between trees and helps maintain forest health.

Zimmerman, M. H. "How Sap Moves in Trees." *Scientific American,* March 1963. A delightful description of the use of aphids as research tools in botany.

Answers to Multiple-Choice Questions

1. a 2. e 3. a 4. c 5. e 6. d

"By the time I get to the wood, I am carrying all manner of seeds hooked in my coat or piercing my socks or sticking by ingenious devices to my shoestrings. I let them ride. After all, who am I to contend against such ingenuity? It is obvious that nature, or some part of it in the shape of these seeds, has intentions beyond this field, and has made plans to travel with me."

Loren Eiseley in The Immense Journey *(1957)*

Poppies blanket California's Antelope Valley after early spring rains. Desert wildflowers bloom and set seed rapidly, taking advantage of the brief spring moisture before the drought and heat of summer. The bright colors, shapes, and scents of flowers attract animal pollinators.

Plant Reproduction and Development

24

At a Glance

Net Watch
On-line resources for this chapter are on the World Wide Web at:
http://www.prenhall.com/audesirk
(click on the Table of Contents link and then select Chapter 24).

As you walk through a wildflower-strewn meadow or glimpse the dry, poppy-carpeted hillsides of the Antelope Valley in southern California, you may be tempted to think that the floral display was created just for human enjoyment. In fact, plants develop showy flowers for the birds and bees—and beetles, moths, and even bats. Perhaps the real wonder is that we too see flowers as beautiful, sometimes breathtakingly so.

Next time you walk through a meadow, look carefully at some flowers. Why are flowers so strikingly and diversely colored? Wouldn't it be simpler and more efficient for the plant if the flowers were green and could photosynthesize? The yellow structures nestled in the center of the flower are a decorative contrast in shape and color, but what do they do? Why do some plants produce large, showy flowers, whereas most trees and grasses have inconspicuous flowers that people often overlook? Do all plants produce some kind of flower?

If you get sneezy just imagining walking through a flowering or grassy field in late summer, you might have a special interest in pollen. What is pollen, and why is the summer air so full of it? Return to your meadow in the fall. Where poppy flowers swayed, you'll find elongated seed pods in various shades of green and brown—some filled with tiny black seeds, others split along the sides and empty. Your socks will be prickly with small burs that (to paraphrase Loren Eiseley) have made plans to travel with you. Although the plants are dying, you know that the field will bloom again in the spring. How does the cycle of life operate in a plant? Let's find out.

1 What Are the Features of Plant Life Cycles?

Many plants can reproduce either sexually or asexually. Asexual reproduction in plants normally involves part of a single plant (say, a stem) giving rise to a new plant. The cells from which offspring are produced asexually arise by mitosis from cells of the parent plant. Therefore, these offspring are genetically identical to the parent. In

Chapter 23, you encountered several methods of asexual reproduction, including the spreading of runners by strawberries, bulb production by daffodils, and the sprouting of rhizomes by irises. Asexual reproduction is often highly effective, allowing plants to colonize an entire area where the original parent found optimal conditions.

However, if an offspring is genetically identical to its parent, then the offspring is only as well adapted to the environment as its parent was. What if the environment changes? Most sexually produced offspring combine genes from both parents, and therefore they may be endowed with traits that differ from those of either parent. This new combination of traits may help the offspring cope with changing environments or survive in slightly different habitats. As a result, most organisms, including plants, reproduce sexually, at least some of the time.

During the familiar animal life cycle, animals with diploid cells produce haploid gametes (sperm or eggs) by the process of meiosis. The gametes fuse to form a new diploid cell that develops into the adult organism through repeated mitotic cell divisions. The plant life cycle, however, is a bit more complex. Plants have two distinct, multicellular "adult" forms, one diploid and one haploid, that give rise to each other. For this reason, the plant life cycle is called **alternation of generations**: Diploid adults (called *sporophytes*) alternate with haploid adults (called *gametophytes*).

Plants with the most easily visible alternating generations (mosses and ferns, for example) do not produce flowers. To illustrate the alternation of generations, let's examine the life cycle of a fern (Fig. 24-1), starting with the diploid adult form. This stage of the life cycle, the **sporophyte** ("spore plant" in Greek), bears reproductive cells. These cells undergo meiosis to produce haploid cells that are **spores**, not gametes. The difference between spores and gametes is that spores do not fuse together to form a diploid cell. Instead, wind blows a fern spore off the parent leaf, and the spore lands on the soil. There the spore **germinates** (begins to grow and develop), dividing repeatedly by mitosis to form a multicellular, haploid organism. This organism produces gametes and hence is called the **gametophyte** ("gamete plant" in Greek). Because its cells are haploid, the gametophyte can produce sperm and eggs without further meiosis. A single gametophyte normally produces both sperm and eggs, but typically at different times, thereby preventing self-fertilization. Sperm and egg fuse to form a zygote that develops into a new diploid sporophyte plant.

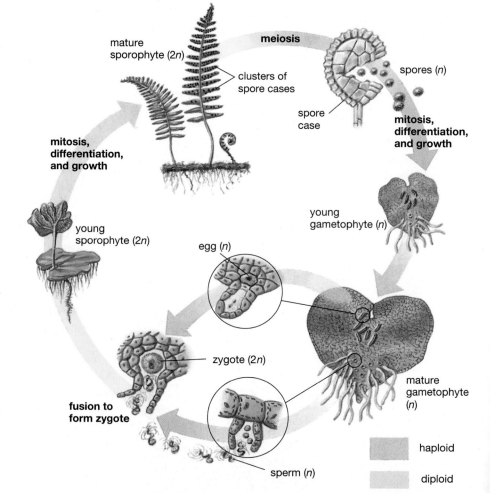

Figure 24-1 The life cycle of a fern—a nonflowering plant
Ferns typify the alternation-of-generations life cycle found in all plants, in which separate multicellular haploid and multicellular diploid "adult" organisms occur at different parts of the life cycle. (The text describes the stages of the life cycle.) The letter *n* refers to the haploid state, and *2n* to the diploid state. When you see ferns, look on the undersides of their leaves for clusters of brownish sporangia.

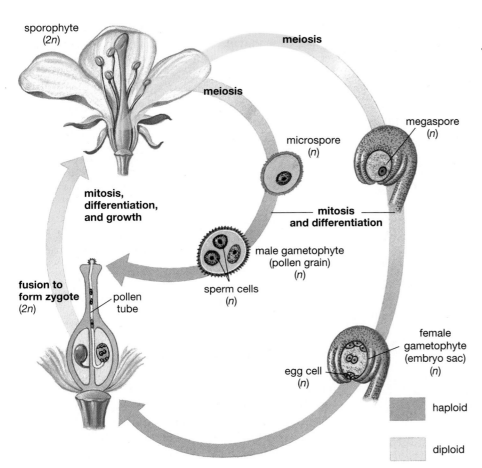

Figure 24-2 The life cycle of a flowering plant
Although this cycle shows the same basic stages as the life cycle of a fern (see Fig. 24-1), the haploid stages are much smaller and cannot live independently of the diploid plant.

Alternation of generations occurs in all plants. In primitive land plants, including mosses and ferns, the gametophyte is an independent, although normally small, plant. It liberates mobile sperm cells that reach an egg by swimming through thin films of water that cover adjacent gametophytes or by being splashed by raindrops from one plant to the next. Therefore, ferns and mosses can reproduce only in moist habitats.

But many terrestrial habitats are relatively dry. One way to reproduce in drier places is for the plant to surround its sperm in a watertight package that can be transported to another plant, where the sperm are liberated directly into the egg-bearing structures of the second plant. The seed plants (gymnosperms and flowering plants) do just that. In the flowering plants, two types of spores are formed by meiosis within the flowers borne by the sporophyte generation (Fig. 24-2). These haploid spores develop into gametophytes, not in the soil but *within the flower*. **Flowers** are the reproductive structures of flowering plants. The gametophytes are microscopic and do not live independently of the sporophyte. One type of spore, the *megaspore*, undergoes a few mitotic divisions and develops into the female gametophyte, a small cluster of cells permanently retained within the flower. The other type of spore, the *microspore*, develops into the male gametophyte: a tough, watertight **pollen grain** containing two sperm. The pollen grain drifts on the wind or is carried by an animal from one flower to another. On the recipient flower, the pollen grain elongates, burrowing through the flower's tissues to the female gametophyte within. This miniature male gametophyte liberates its sperm inside the female gametophyte, where fertilization occurs. The zygote becomes enclosed in a drought-resistant **seed**. The seed, including an embryonic plant and a food reserve within a protective outer coating, may lie *dormant* (it won't germinate) for months or years, waiting for conditions favorable for growth.

Here we examine sexual reproduction in flowering plants, from the evolution of the flower through the formation of the seed and the development of the new seedling.

2 How Did Flowers Evolve?

The flower is actually a sexual display that enhances a plant's reproductive success. By enticing animals to transfer pollen from one plant to another, flowers enable stationary plants to "court" distant members of their own species. This critical advantage has allowed the flowering plants to become the dominant plants on land.

The earliest seed plants were the gymnosperms, represented today mainly by conifers, a group that includes pines, firs, and spruces. As we described in Chapter 21, conifers do not produce flowers; instead, they bear male and female gametophytes on separate cones. During early spring, the small, male cones release millions of pollen grains that float about on breezes (Fig. 24-3). So many grains are floating around that some enter the pollen chambers located on the scales of the female cones, where they are captured by sticky coatings of sugars and resins. The pollen grains germinate and tunnel to the female gametophytes at the base of each scale. Sperm are liberated and fertilize the eggs within a female gametophyte, and a new generation begins.

Clearly, wind pollination is an inefficient operation, because most of the pollen grains are lost. In a world of stationary plants and mobile animals, if a gymnosperm could entice an animal to carry its pollen from male to female cone, it would greatly enhance its reproductive rate and hence its evolutionary success. As it happens, gymnosperms and insects were poised to establish just such a relationship about 150 million years ago.

Insects, especially beetles, are among the most abundant animals on Earth. They exploit nearly every possible food resource on land, including the reproductive parts of gymnosperms. About 150 million years ago, some beetles fed on both the protein-rich pollen of male cones and the sugar-rich secretions of female cones. Beetles can make quite a mess when they feed, and pollen feeders often wind up with pollen dusted all over their bodies. If the same beetle were to visit one plant, eat pollen, and then wander over to another plant of the same species to dine on the sugary secretions of a female cone, some of the loose pollen would quite likely rub off on the female cone.

The stage was set for the evolution of flowering plants. Efficient pollination by insects requires that a given insect visit several plants of the same species, pollinating them along the way. For the plants, two key adaptations were necessary. First, enough pollen or *nectar* (the sugary secretions) must be produced within the reproductive structures so that insects will regularly visit them to feed. Second, the location and richness of these storehouses of pollen and nectar must be advertised to the insects, both to show them where to go and to entice them to specialize on that particular plant species. Any mutation that contributed to these adaptations would enhance the reproductive success of the plant that carried the mutation and would be favored by natural selection. By about 130 million years ago, flowers had evolved with exactly these adaptations. The advantages of flowers are so great that in today's temperate and tropical zones, flowering plants are overwhelmingly dominant, and numerous animals, including bees, moths, butterflies, hummingbirds, and even some mammals, feed at and pollinate flowers.

Complete Flowers Have Four Major Parts

We have seen that flowers are the reproductive structures of flowering plants. Evolution commonly produces new structures by modifying old ones, and flower parts are actually highly modified leaves, shaped by mutation and natural selection into a form that enhances pollination. **Complete flowers**, such as those of crocuses, roses, and tomatoes, consist of a central axis on which four successive sets of modified leaves are attached (Fig. 24-4). These modified leaves form the *sepals*, *petals*, *stamens*, and *carpels*. The **sepals** are located at the base of the flower. In dicots, the sepals are typically green and leaflike; in monocots, most sepals resemble the petals (see Fig. 24-4). In either case, sepals surround and protect the flower bud as the remaining three structures develop. Just above the sepals are the **petals**, which are normally brightly colored and fragrant, advertising the location of the flower.

The male reproductive structures, the **stamens**, are attached just above the petals. Most stamens consist of a long and slender **filament**. At the filament's tip is an **anther**, the structure that produces pollen. The female reproductive structures, the **carpels**, occupy the uppermost position in the flower. An idealized carpel is somewhat vase-shaped, with a sticky **stigma** for catching pollen mounted atop an elongated **style**. The style connects the stigma with the bulbous **ovary**. Inside the ovary are one or more **ovules**, in which the female gametophytes develop. When mature, each ovule will become a seed, and the ovary will develop into a protective, adhesive, or edible enclosure, the **fruit**.

As you may know from your own gardening experience, not all flowers are complete. **Incomplete flowers** lack one or more of the four floral parts. For example, many plants, such as cucumbers and squashes, have separate male and female flowers, which may be borne on the same plant (Fig. 24-5). Alternatively, male and female flowers may be on different plants, as in the American holly: Male flowers lack carpels, and female flowers lack stamens. Incomplete flowers may also lack sepals or petals.

Figure 24-3 Conifers are wind-pollinated
Even slight breezes blow thick clouds of pollen from ripe male cones.

(a)

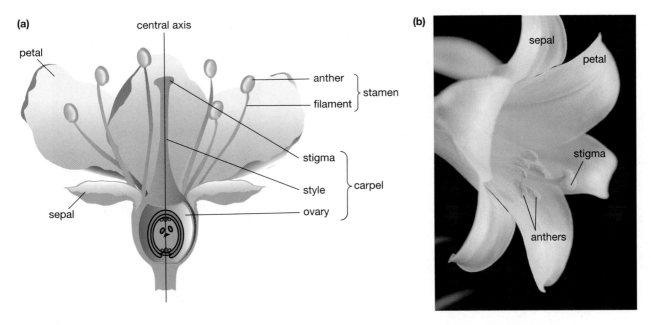

(b)

Figure 24-4 A complete flower
(a) A complete flower has four parts: sepals, petals, stamens (the male reproductive structures), and at least one carpel (the female reproductive structure). This drawing shows a complete dicot flower. **(b)** The lily is a complete monocot flower, with three sepals (virtually identical to the petals), three petals, six stamens, and three fused carpels. Each stamen consists of a filament bearing an anther at its tip. The carpels consist of an ovary hidden in the base of the flower, with a long style protruding out, ending in a sticky stigma. The anthers are well below the stigma, probably preventing self-pollination: Pollen cannot simply fall from the anther onto the stigma.

Figure 24-5 Some plants have separate male and female flowers
Plants of the squash family, such as these zucchinis, bear separate female (left) and male (right) flowers. Although individual flowers cannot be self-pollinated, the plant could still be self-pollinated if an insect carried pollen from a male flower to a female flower of the same plant. However, each plant initially produces only male flowers, so some cross-pollination between plants that flower at slightly different times is virtually ensured.

3 How Have Flowers and Pollinators Adapted to One Another?

Most wind-pollinated flowers, such as those of grasses and oaks, are inconspicuous and unscented, many of them scarcely more than naked stamens that liberate pollen to the wind (Fig. 24-6). Sufferers of "hay fever" (an allergic response to pollen) are especially aware of how much of this pollen ends up in the air. The beautiful flowers so much admired by humans, however, are pollinated by animals.

The distinctive shapes, colors, and odors of animal-pollinated flowers as well as the sensory capabilities and lifestyles of their pollinators are an example of **coevolution**—evolution in two species that interact extensively with one another, such that each acts as a major force of natural selection on the other (see Chapter 15). Animal-pollinated flowers must attract useful pollinators and frustrate undesirable visitors who might eat nectar or pollen without fertilizing the flower in return. The animals, in turn, have been under environmental pressures to locate flowers quickly, identify the flowers that can provide adequate nutrition, and extract the nectar or pollen with a minimum expenditure of energy.

Figure 24-6 *A wind-pollinated flower*
The flowers of grasses and many deciduous trees are wind-pollinated, with anthers (yellow structures hanging beneath the flowers) exposed to the wind. Petals are normally reduced or absent.

Animal-pollinated flowers can be loosely grouped into three categories, depending on the benefits (real or imagined) that they offer to potential pollinators: (1) food, (2) sex, or (3) a nursery.

Some Flowers Provide Food for Pollinators

Many flowers provide food for foraging animals such as beetles, bees, moths and butterflies, or hummingbirds. In return, the animals unwittingly distribute pollen from flower to flower.

Coevolution of Flowers and Insect Pollinators Has Produced Diverse Adaptations

Bees are culinary specialists, often feeding only on nectar and pollen. Bees use both scent and sight to locate and identify flowers from the air. We can thank the bees for most of the sweet-smelling flowers, because sweet "flowery" odors attract these pollinators. Bees also have good color vision, but they do not see exactly the same range of colors that humans do (Fig. 24-7a). Although bees cannot distinguish red from gray or black, their color vision extends into the ultraviolet (UV) light range, which

(a)

human vision

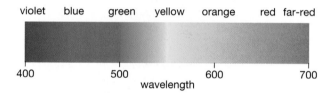

(b)

bee vision

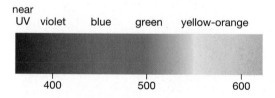

Figure 24-7 *Ultraviolet patterns guide bees to nectar*
The spectra of color vision for **(a)** humans and **(b)** bees overlap considerably in the blue, green, and yellow ranges but differ on the edges. Humans are sensitive to red, which bees do not perceive; bees can see UV light, which is invisible to the human eye. Many flowers photographed under ordinary daylight (in part *a*) and under UV light (in part *b*) show striking differences in color patterns. Bees can see the UV patterns that presumably lead them to the nectar- and pollen-containing centers of the flowers.

we cannot see. To attract a bee from afar, bee-pollinated flowers must look brightly colored *to a bee*. Typically, these flowers are white, blue, yellow, or orange, and many have other markings, such as central spots or lines pointing toward the center, that reflect UV light (Fig. 24-7b). Look carefully at the flower colors on the hills shown in our chapter-opening photograph. How are these flowers pollinated?

Bee-pollinated flowers have several structural adaptations that help ensure the transfer of pollen. Many bee-pollinated flowers, such as nasturtiums and foxgloves, produce nectar at the bottom of a tube; in the Scotch broom flower, nectar forms in a crevice between ingeniously evolved petals (Fig. 24-8). Either pollen-laden stamens (normally in newly opened flowers) or the sticky stigma of the carpel (in older flowers) sticks out of the top of the tube or emerges from confinement when the bee's

Figure 24-9 *The carrion flower is pollinated by flies*
Flies are attracted by its aroma of rotting meat, which gives this flower its name.

(a)

(b)

Figure 24-8 *"Pollinating" a pollinator*
(a) In the Scotch broom flowers, the bee finds nectar near the junction of top and bottom petals. *(b)* The bee's weight deflects the bottom petals downward, causing curved, pollen-laden stamens to pop up and cover the bee's hairy back with pollen. The bee will carry the pollen to other Scotch broom flowers, leaving some on ready stigmas.

weight deflects the petals downward. When a bee visits a young flower, the stamens brush pollen onto her back. She may then visit an older flower and repeat her foraging behavior. This time, she leaves pollen behind on the stigma.

Many flowers adapted for moth and butterfly pollinators are superficially similar to bee-pollinated flowers. However, their nectar tubes are normally deep and narrow, accommodating the long tongues of moths and butterflies. Flowers pollinated by night-flying moths open only in the evening; most are white and give off strong, musky odors that help the moth locate the flower in the dark.

Many beetles prefer to feed on animal material, which has a high nutrient content. Therefore, beetle-pollinated flowers typically smell like dung or rotting flesh, which attracts scavenging beetles. Flies have much the same taste in foods (or lack thereof!) as do beetles, and most fly-pollinated flowers, such as the carrion flower (Fig. 24-9), also emit a powerful stench.

Hummingbird-Pollinated Flowers Have Deep, Nectar-Filled Tubes

Hummingbirds (Fig. 24-10a) are one of the few vertebrates that are important pollinators, although several mammals also visit flowers (Fig. 24-10b,c). Birds have notoriously poor senses of smell, and hummingbird-pollinated flowers seldom synthesize fragrant chemicals. Instead, these flowers always produce much greater amounts of nectar than other flowers do, because hummingbirds need lots of energy. If the flowers didn't produce lots of nectar, the hummers would go elsewhere and the flower would remain unpollinated. But a large supply

(a)

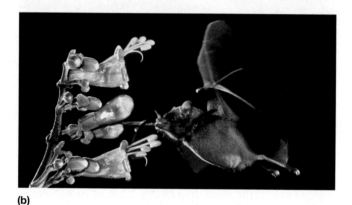

(b)

(c)

Figure 24-10 Some vertebrate pollinators
(a) A hummingbird hovers before a red flower that has elongated petals. Pollen-laden stamens protrude from the flower, ensuring that pollen will be deposited on the bird's head. *(b)* A tropical bat feeds at a cluster of tubular flowers with protruding stamens and stigma. As the bat hovers before the flower, the top of its head touches either the anthers or stigma or both, thus pollinating the flower. *(c)* As the honey possum stuffs its face into this flower, pollen adheres to its muzzle and whiskers. A visit to another flower may result in pollen transfer.

of nectar would also attract insects. A bee, for example, might return again and again to the same flower, never transferring pollen to another flower. Not surprisingly, hummingbird-pollinated flowers have evolved several adaptations that keep insects from drinking their nectar. These flowers are normally tubular, matching the long bills and tongues of hummers. The tube is much too deep for bees to reach the nectar at its base; the flower also lacks a lip on which bees could rest while dining. In addition, most hummingbird-pollinated flowers are red or orange (see Fig. 24-10a). Red is particularly attractive to a hummingbird but appears gray or black to a bee.

Sexy Deceptions Attract Pollinators

To pollinate their flowers, a few plants, most notably the orchids, take advantage of the mating drive and stereotyped behaviors of male wasps and flies. Some orchid flowers mimic female wasps both in scent and shape (Fig. 24-11). The males land atop these "females"

Figure 24-11 Sexual deception promotes pollination
This male wasp is actually trying to copulate with an orchid flower. The result is successful reproduction—not for the wasp but for the orchid.

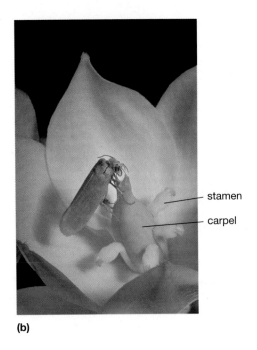

stamen

carpel

(a)

(b)

Figure 24-12 A mutually dependent relationship
(a) Yuccas bloom on the dry plains of eastern Colorado in early summer. *(b)* Within many of the yucca flowers, yucca moths carry out their part in one of nature's most unusual and most effective cooperative relationships between plant and animal.

and attempt to copulate, but they get only a packet of pollen for their efforts. As they repeat their attempts on other orchids of the same species, the pollen packet is transferred.

Some Plants Provide Nurseries for Pollinators

Perhaps the most-elaborate relationships between plants and pollinators occur in a few cases in which insects fertilize a flower and then lay their eggs in the flower's ovary. This arrangement occurs between milkweeds and milkweed bugs, figs and certain wasps, and yuccas and yucca moths (Fig. 24-12). The yucca moth's remarkable behavior results in the pollination of yuccas and a well-stocked pantry for its own offspring. A female moth visits a yucca flower, collects pollen, and rolls it into a compact ball. The moth flies off with the pollen ball to another yucca flower, drills a hole in the ovary wall, and lays its eggs inside the ovary. Then it smears pollen from the pollen ball all over the stigma of the flower, performing this genetically programmed behavior flaw-

lessly! By pollinating the yucca, the moth ensures that the plant will provide a supply of developing seeds for its caterpillar offspring. Because the caterpillars eat only a fraction of the seeds, the yucca also reproduces successfully. The mutual adaptation of yucca and yucca moth is so complete that neither can reproduce without the other.

 ## How Do Gametophytes Develop in Flowering Plants?

In the life cycle of flowering plants, as Figure 24-2 illustrates, the familiar plant of meadow, garden, and farm is the diploid sporophyte. The pollen grain (male) and the **embryo sac** (female) are the haploid gametophytes that develop within the flowers borne by the sporophytes. Both are much smaller than the gametophyte stages of ferns and mosses and cannot live independently of the sporophyte.

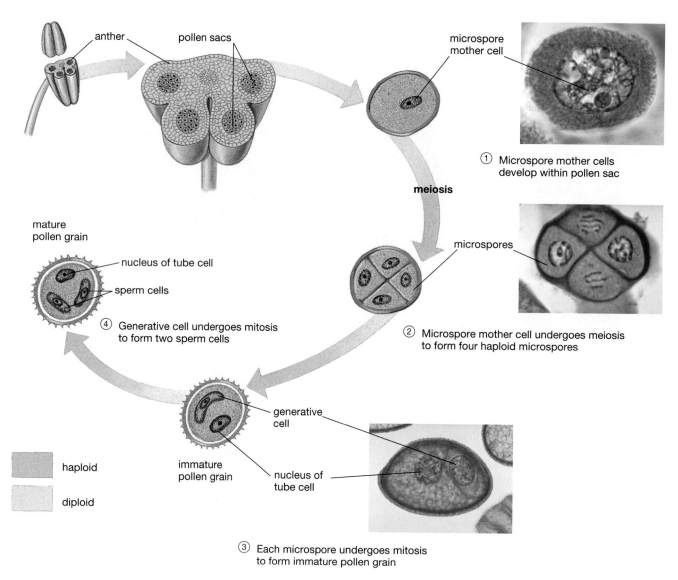

Figure 24-13 Pollen development in flowering plants
① Within the anthers, diploid microspore mother cells form. ② These cells undergo meiosis to produce four haploid microspores. ③ Through mitosis, each microspore divides into two (haploid) cells: the tube cell and the generative cell. The entire generative cell resides within the cytoplasm of the tube cell. These two cells form the juvenile pollen grain (the male gametophyte). ④ Later (in some species, after pollination), the generative cell divides to form two sperm cells, which remain in the tube cell cytoplasm. This three-celled "organism" is the mature male gametophyte.

Pollen Is the Male Gametophyte

Each anther consists of four chambers called *pollen sacs* (Fig. 24-13). Within each sac, hundreds to thousands of diploid **microspore mother cells** develop. Each microspore mother cell undergoes meiosis (see Chapter 11) to produce four haploid **microspores**. Each microspore divides once, by mitosis, to produce a male gametophyte, or pollen grain. Each pollen grain consists of only two cells: a large **tube cell** and a smaller **generative cell** that resides *within the cytoplasm of the tube cell* (Fig. 24-13). A tough surface coat develops around the pollen grain, protecting the cells within during their journey to the carpel (Fig. 24-14).

When the pollen has matured, the pollen sacs of the anther split open. In wind-pollinated flowers, such as those of grasses and oaks, the pollen grains spill out, a fortunate few to be carried by wind currents to other flowers of the same species. In animal-pollinated flowers, the pollen adheres weakly to the anther case until the pollinator comes along and brushes or picks it off.

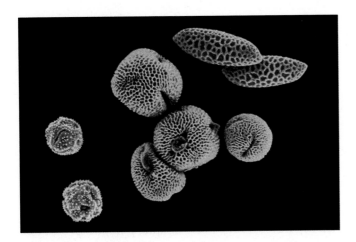

Figure 24-14 *Pollen grains*
The tough outer coverings of many pollen grains are elaborately sculptured in species-specific shapes and patterns. The pollen grains in this color-enhanced SEM photo are from a geranium (orange), a tiger lily (fuschia), and a dandelion (yellow).

The Embryo Sac Is the Female Gametophyte

In an ovary, one or more dome-shaped masses of cells differentiate into ovules. Each ovule consists of outer layers of cells called **integuments**, which surround a single, diploid **megaspore mother cell** (Fig. 24-15). That cell divides by meiosis to produce four large haploid **megaspores**. Three megaspores degenerate, and one survives. This remaining megaspore undergoes an unusual set of mitotic divisions. Three nuclear divisions produce a total of eight haploid nuclei. Plasma membranes then divide the cytoplasm into *seven*, not eight, cells: three small cells at each end, with one nucleus apiece, and one remaining large cell in the middle with two nuclei, called **polar nuclei**. This seven-celled organism, called the *embryo sac* (because it is where the embryo will develop), is the haploid female gametophyte. The large central cell with two polar nuclei is the **primary endosperm cell**. The **egg** is the central small cell at the bottom of the embryo sac, located near an opening in the integuments of the ovule.

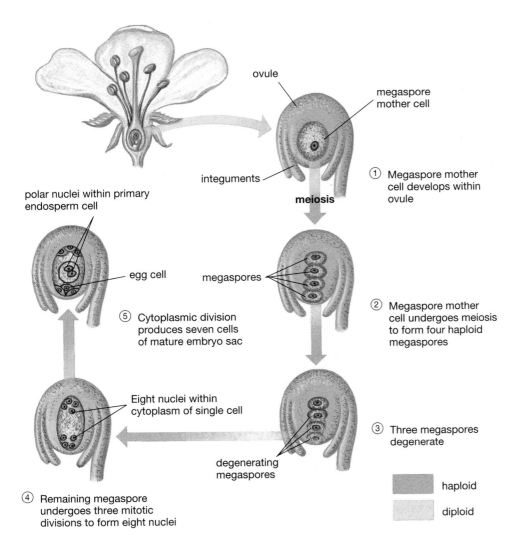

Figure 24-15 *Development of the female gametophyte*
① A single diploid megaspore mother cell matures within the integuments of an ovule.
② Through meiosis, it gives rise to four haploid megaspores.
③ Three of these degenerate.
④ The nucleus of the remaining megaspore then divides mitotically three times, producing eight haploid nuclei, four at each end of the as-yet undivided cell.
⑤ Cytoplasmic division then forms seven cells, six with one nucleus apiece, and one large primary endosperm cell with two nuclei, called polar nuclei. One of the small cells, near the opening in the integuments, is the egg cell. The other five cells degenerate soon after fertilization.

5 How Does Pollination Lead to Fertilization?

When a pollen grain lands on the stigma of a flower of the same species of plant, a remarkable chain of events occurs (Fig. 24-16). The pollen grain absorbs water from the stigma. The tube cell elongates, growing down the style toward an ovule in the ovary. Meanwhile, the generative cell divides mitotically to form two sperm cells.

If all goes well, the pollen tube reaches the pore in the integument of an ovule and breaks into the embryo sac. The tube's tip ruptures, releasing the two sperm. One sperm fertilizes the egg cell to form the diploid zygote that will develop into the embryo and eventually into a new sporophyte. The second sperm enters the primary endosperm cell. Its nucleus fuses with *both* polar nuclei,

forming a triploid nucleus (having three sets of chromosomes). Through repeated mitotic divisions, the primary endosperm cell will develop into the triploid **endosperm**, a food-storage tissue within the seed. The fusion of the egg with one sperm and the fusion of the polar nuclei of the primary endosperm cell with the second sperm is often called **double fertilization**, a process unique to flowering plants. The other five cells of the embryo sac degenerate soon after fertilization.

The distinction between pollination and fertilization is important. **Pollination** occurs when a pollen grain lands on a stigma; **fertilization** is the fusion of sperm and egg. Although pollination is necessary for fertilization, these are two separate events. For example, pollination will not lead to fertilization if the tube cell fails to grow properly, if the embryo sac has no egg, or if a sperm from another pollen grain has already reached the egg.

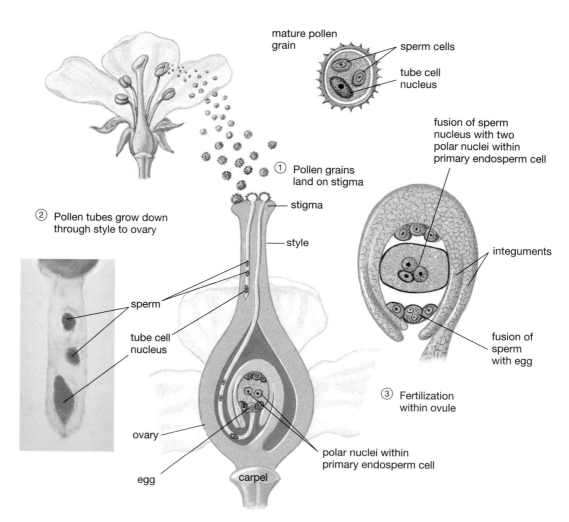

Figure 24-16 Pollination and fertilization of a flower
① The pollen grain lands on the sticky surface of the stigma and germinates. ② The tube cell elongates, burrowing down through the style. The sperm nuclei follow, inside the cytoplasm of the tube cell. ③ Upon reaching an ovule, the tube breaks into the embryo sac, and the tube nucleus degenerates. One sperm enters the primary endosperm cell; its nucleus fuses with both endosperm nuclei to form a triploid cell. The other sperm fertilizes the egg cell to form the diploid zygote that will develop into the new embryonic plant. The other cells of the embryo sac degenerate.

6 How Do Seeds and Fruits Develop?

Drawing on the resources of the parent plant, the embryo sac and the surrounding integuments of the ovule develop into a seed. The seed is surrounded by the ovary, which develops into a fruit (Fig. 24-17). Having already served their functions of attracting pollinator and producing pollen, petals and stamens shrivel and fall away as the fruit enlarges.

The Seed Develops from the Ovule and Embryo Sac

The integuments of the ovule develop into the **seed coat**, the outer covering of the seed. As we shall see, in many plants the characteristics of the seed coat play a role in regulating when the seed will germinate.

Meanwhile, within the integuments, two distinct developmental processes occur (Fig. 24-18a). First, the triploid endosperm cell divides rapidly. Its daughter cells absorb nutrients from the parent plant, forming a large, food-filled endosperm. Second, the zygote develops into the embryo. Both dicot and monocot embryos consist of three parts:

the shoot, the root, and the **cotyledons**, or *seed leaves*. The cotyledons absorb food molecules from the endosperm and transfer them to other parts of the embryo.

In dicots ("two cotyledons"), the cotyledons normally absorb most of the endosperm during seed development, so the mature seed is virtually filled with embryo (Fig. 24-18b, left). In monocots ("one cotyledon"), the cotyledon absorbs some of the endosperm during seed development, but most of the endosperm remains in the mature seed (Fig. 24-18b, right). Food grains, including wheat, corn, and rice, are monocots. We (like the developing plant) use the stored endosperm as food. In the case of wheat, we grind up the endosperm to make flour and sometimes consume the embryo of the wheat seed in the form of "wheat germ."

The future primary root develops at one end of the main embryonic axis. The future shoot, at the other end, is normally divided into two regions at the attachment site of the cotyledons. Below the cotyledons, but above the root, is the **hypocotyl** (*hypo* in Greek means "beneath" or "lower"); above the cotyledons the shoot is called the **epicotyl** (*epi* means "above"). At the tip of the epicotyl lies the apical meristem of the shoot, the daughter cells of which differentiate into the specialized cell

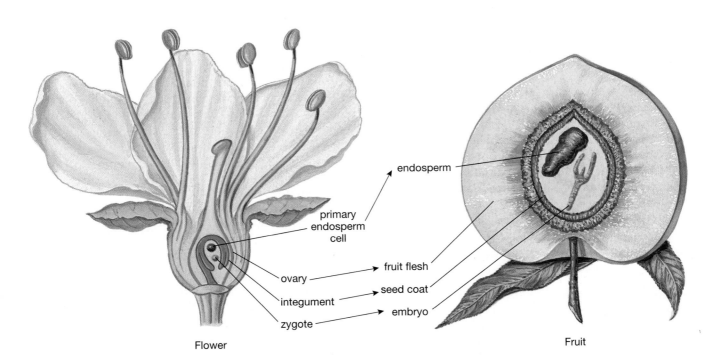

Flower

Fruit

Figure 24-17 Development of fruit and seeds
The fruit and the seed within it develop from various parts of the flower. Starting at the outside, the ovary wall ripens into the fruit flesh, which may be soft and tasty like a peach or variously hard, hooked, or tufted in other fruits. The integuments of the individual ovule, which surround the embryo sac, form the seed coat. The fruit and seed coat are derived from tissues of the parent sporophyte plant. Within the seed, the triploid endosperm cell divides repeatedly, absorbs nutrients from the parent plant, and becomes the endosperm, a food-storage structure within the seed. Finally, the zygote, also within the seed, develops into the embryo.

Figure 24-18 Seed development
(a) In a generalized seed, the endosperm develops first, absorbing nutrients from the parent plant. The embryo develops later, absorbing nutrients from the endosperm to fuel its growth. **(b)** Monocot and dicot seeds differ in the number of cotyledons and the fate of the endosperm. **(left)** Dicot seeds, such as the shepherd's purse, have two cotyledons, which normally absorb most of the endosperm as the seed develops; hence, the mature seed is mostly cotyledon. **(right)** Monocot seeds, such as the corn kernel, retain a large endosperm. (Cornmeal is the ground-up endosperm of corn seeds.) The embryo produces a single cotyledon. As the seed germinates, the cotyledon absorbs the food reserves of the endosperm and transfers them to the growing embryo.

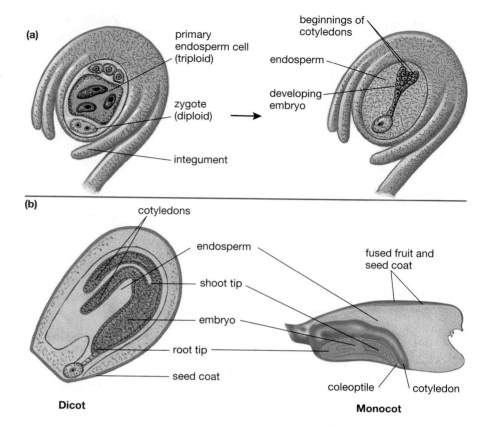

types of stem, leaves, and flowers. One or two developing leaves may already be present.

The Fruit Develops from the Ovary Wall

The wall of the ovary develops into a fruit (see Fig. 24-17). There are a bewildering variety of fruits, with outer layers that are variously fleshy, hard, winged, or even spiked like a medieval mace. The selective forces favoring the evolution of all fruits, however, are similar: Fruits help disperse the seeds to distant locations away from the parent plant. (See "Earth Watch: On Dodos, Bats, and Disrupted Ecosystems" and "Evolutionary Connections: Adaptations for Seed Dispersal.")

Seed Dormancy Helps Ensure Germination at an Appropriate Time

All seeds need warmth and moisture to germinate. But many newly matured seeds will not germinate immediately, even under ideal conditions. Instead, they enter a period of **dormancy**, during which they will not germinate. Dormancy is normally marked by lowered metabolic activity and resistance to adverse environmental conditions. Seed dormancy solves two problems, one intrinsic to the plant itself and one related to environmental factors. First, if a seed germinated while still enclosed in a fruit and hanging from a tree or vine, it might exhaust its food reserves before it ever touched the ground. Further, seedlings germinating within a fruit that con-

tains many seeds would grow in a dense cluster, competing with one another for nutrients and light. Second, environmental conditions suitable for seedling growth (such as adequate moisture and temperatures) may not coincide with seed maturation. Seeds that mature in late summer in temperate climates, for example, face the harsh winter to come. Spending the winter as a dormant seed is clearly preferable to death by freezing as a tender young sprout. In the warm, moist tropics, seed dormancy is much less common than in temperate regions because environmental conditions are suitable for germination throughout the year.

Mechanisms for Maintaining and Breaking Dormancy Are Adaptations to Differing Environments

Plants have evolved many mechanisms that produce seed dormancy, and the seeds of each species have their own set of requirements (in addition to adequate moisture and proper temperature) that must be met before germination can occur. Perhaps the three most common requirements are drying, exposure to cold, and disruption of the seed coat:

1. *Drying.* Many seeds must dry out before they are able to germinate. Drying prevents the seed from germinating while it is still within the fruit. Many such seeds are dispersed by animals that eat fruit but cannot digest the seeds, which are therefore excreted in their feces.
2. *Cold.* Seeds of many temperate and arctic plants will not germinate unless exposed to prolonged subfreez-

Earth Watch
On Dodos, Bats, and Disrupted Ecosystems

Flowering plants dominate terrestrial ecosystems largely because of the mutually beneficial relationships they forged with the animals that pollinate their flowers and disperse their seeds. Some plant and animal species, such as the yucca and yucca moths, have become totally dependent on one another. Within complex ecosystems, these mutually beneficial relationships sustain both plant and animal populations and ultimately the ecosystem itself.

On the island of Mauritius in the Indian Ocean, the tambalacoque tree, like most of the native plants on this island, is threatened. Tambalacoque trees produce a large, edible fruit something like a peach, with a pulpy outside surrounding a stone-hard pit. Today, the remaining trees produce healthy fruits that fall to the ground and rapidly rot in the tropical climate. The enclosed seeds are highly susceptible to destruction by fungal and bacterial infections, which are promoted by the rotting fruit.

Before humans arrived, the island was home to the dodo (Fig. E24-1a). Early sailors found the large, slow dodos to be easy prey, and by 1681 they had hunted the dodo to extinction. Other native animals, including giant tortoises, large-billed parrots, and the giant skink (a large reptile), were also driven to extinction as humans introduced monkeys, pigs, and deer to Mauritius and destroyed natural habitats by clearing the land for farming. Scientists believe that some of these extinct animals ate the tambalacoque fruit before it had a chance to rot, thoroughly cleaning the seeds and thus protecting them from attack by the fungi that now destroy the seeds. The animals also dispersed the seeds throughout the island, ensuring that some of them reached habitats favorable for germination. As native animals are destroyed, plants that coevolved with them are also threatened, and the natural ecosystem that evolved over millennia collapses.

In many tropical forests, bats are the most important agents of seed dispersal (Fig. E24-1b). Bats may fly more than 20 miles each night, consuming up to twice their weight in fruit and defecating the seeds in flight. Biologist Donald Thomas compared germination rates of seeds before and after the seeds' passage through the bats' digestive tract. After passing through a bat, nearly all the seeds germinated; seeds planted directly from fruit, however, had only a 10% germination rate. Today in the tropi-

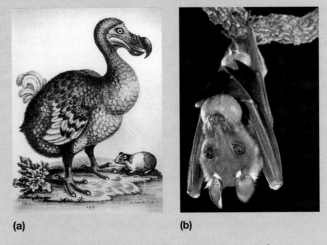

(a)   **(b)**

Figure E24-1 Animal seed dispersers are crucial to some ecosystems
(a) The dodo, driven to extinction in the late 1600s, probably helped disperse and promote germination of the tambalacoque trees of Mauritius. The trees are seriously threatened and are rarely germinating spontaneously in the wild. **(b)** A Wahlberg's epauleted bat in Kenya eats a ripe fig. Without bats and other seed-dispersing animals, some types of tropical forests may not survive.

cal forests of southern Mexico, fruit-eating animals such as monkeys, deer, and tapir have been overhunted. Tropical fruits are rotting on the forest floor or sending up doomed sprouts under the shade of their parents; dispersal has stopped. As Alejandro Estrada of the University of Mexico put it, "The continued existence of tropical forests whose primates and . . . birds and bats have been shot is just as precarious as if their trees had been chain-sawed and bulldozed." The web of interdependent lifeforms linked by interactions forged over millennia of coevolution is fragile and easily disrupted. Only by understanding and preserving the complex and crucial interactions among plants and animals can we hope to conserve diverse, functioning ecosystems.

ing temperatures, followed by warmth and moisture. This requirement ensures that the seeds stay dormant during mild days in autumn and sprout only after winter yields to spring.

3. *Disruption of the seed coat.* The seed coat itself can be a barrier to seed germination. Many seed coats are impermeable to water and oxygen, or they bind the developing embryo so tightly that growth simply cannot occur; others contain chemicals that inhibit germination (as we shall see in Chapter 25). In

deserts, for example, rainfall is spotty and scarce. Years may go by without enough water for plants to germinate, grow, flower, and set more seed. Therefore, the seed must not sprout unless a rainfall is heavy enough to allow the plant to complete its life cycle, because more rain may not fall in time. The seed coats of most desert plants have water-soluble chemicals that inhibit germination. Only a hard rainfall can wash away enough of the inhibitors to allow sprouting.

7 How Do Seeds Germinate and Grow?

www

During germination, growth and development of a seed, the formerly dormant embryo resumes growth and emerges from the seed. The embryo absorbs water, which makes it swell and burst its seed coat. The root is normally the first structure to emerge from the seed coat, growing rapidly and absorbing water and minerals from the soil. Much of the water is transported to cells in the shoot. As its cells elongate, the stem lengthens, pushing up through the soil.

The Shoot Tip Is Protected by the Coleoptile

The growing shoot faces a serious difficulty: It must push through the soil without scraping away the apical meristem and tender leaflets at its tip. A root, of course, must always contend with tip abrasion; its apical meristem is protected by a root cap (see Fig. 23-9). Shoots, however, spend most of their time in the air and do not develop permanent protective caps. Instead, germinating shoots have other mechanisms that cope with the abrasion of sprouting (Fig. 24-19). In monocots, the **coleoptile**, a tough sheath, encloses the shoot tip like a glove (Fig. 24-19a). The coleoptile "glove" pushes aside the soil particles as

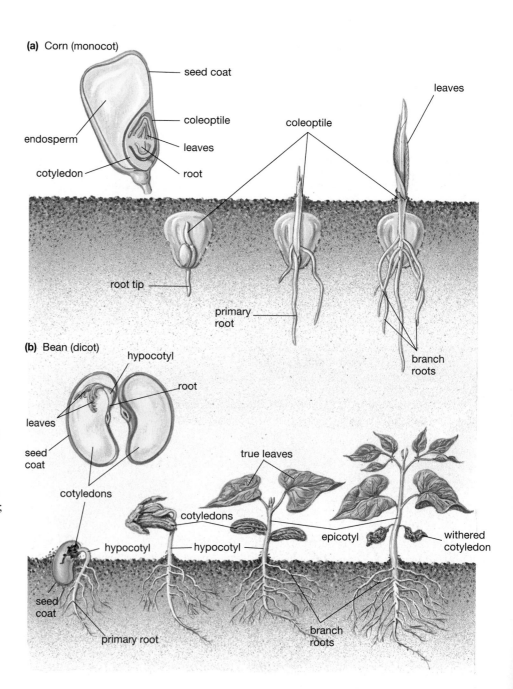

Figure 24-19 Seed germination Seed germination in (*a*) corn, a monocot, and (*b*) the common bean, a dicot, differs in detail but involves the same three principles: (1) use of stored food to provide energy for seedling growth until photosynthesis can take over; (2) rapid growth and branching of the root to absorb water and nutrients; and (3) protection of the delicate shoot tip as it moves upward through the soil. In monocots (a), the delicate shoot tip is protected within a tough sheath, the coleoptile, which pushes up through the soil when the seed sprouts. In dicots (b), the hypocotyl (shown here) or the epicotyl forms a hook that emerges from the soil first, protecting the shoot tip.

it grows. Once out in the air, the coleoptile tip degenerates, allowing the tender shoot "finger" to emerge. Dicots do not have coleoptiles. Rather, the dicot shoot forms a hook in the hypocotyl or epicotyl (Fig. 24-19b). The bend of the hook, encased in epidermal cells with tough cell walls, leads the way through the soil, clearing the path for the downward-pointing apical meristem with its delicate new leaves.

Cotyledons Nourish the Sprouting Seed

Food stored in the seed provides the energy for sprouting. Recall that the cotyledons of dicots had already absorbed the endosperm while the seed was developing and are now swollen and full of food. In dicots with hypocotyl hooks, the elongating shoot carries the cotyledons out of the soil into the air. These aboveground cotyledons typically become green and photosynthetic (Fig. 24-20) and transfer both previously stored food and newly synthesized sugars to the shoot. In dicots with epicotyl hooks, the cotyledons stay below the ground, shriveling up as the embryo absorbs their stored food. Monocots retain most of their food reserve in the endosperm until germination, when it is digested and absorbed by the cotyledon as the embryo grows. The cotyledon remains below ground in the remnants of the seed.

Controlling the Development of the Seedling

Once out in the air, the shoot rapidly spreads its leaves to the sun. Simultaneously, the root system delves into the soil. The apical meristem cells of shoot and root divide, giving rise to the mature structures discussed in Chapter 23. Eventually this plant, too, will mature, flower, and set seed, renewing the cycle of life. How this cycle is regulated—why shoots grow upward while roots grow downward, and how plants produce flowers at the proper time of year—is the subject of Chapter 25.

Evolutionary Connections
Adaptations for Seed Dispersal

Successful plant reproduction requires a suitable site for the seed to germinate and the young plant to grow. The growth of a seedling that sprouts right next to its parent will be inhibited by the larger plant's shade. If the offspring survives, it will compete with its parent for water and nutrients, to the detriment of both plants. Further, a plant species will become more widespread if its members send out some seeds to distant habitats. In flowering plants, seed dispersal is the function of fruits. A wide variety of fruits has evolved, each dispersing seeds in a different way.

Shotgun Dispersal

A few plants develop explosive fruits that eject their seeds meters away from the parent plant. Mistletoes, common parasites of trees, produce fruits that shoot out sticky seeds. If one seed strikes a nearby tree, it sticks to the bark and germinates, sending rootlike fibers into the vascular tissues of its host, from which it draws its nourishment. Because the proper germination site for a mistletoe seed is not the ground but a tree limb, shooting the seeds, not dropping them, is clearly useful.

Wind Dispersal

Dandelions and maples (Fig. 24-21) produce lightweight fruits with surfaces that catch the wind. (Yes, each individual hairy tuft on a dandelion ball is a separate fruit!) Each fruit typically contains a single small seed; having only one seed reduces weight and lets the fruit remain aloft longer. These featherweight fruits aid the seed in traveling away from the parent plant, from a few meters for maples to kilometers for a milkweed or dandelion on a windy day.

Water Dispersal

Many fruits can float on water for a time and may be dispersed by streams or rivers. The coconut fruit,

Figure 24-20 Cotyledons nourish the developing plant
In the squash family, a hypocotyl hook carries the cotyledons out of the soil. The cotyledons expand into photosynthetic leaves (the pair of smooth oval leaves). The first true leaf (crinkled single leaf) develops a little later. Eventually, the cotyledons shrivel up and die.

(a)   (b)

*Figure 24-21 **Wind-dispersed fruits***
These fruits normally contain only one or two lightweight seeds. Some, such as **(a)** dandelions, have filamentous tufts that catch the breezes. Others, such as **(b)** maple fruits, are miniature glider–helicopters, silently whirling away from the tree as they fall. To see how the wings aid in seed dispersal, take two maple fruits and pluck the wing off one. Hold both fruits over your head and drop them. The wingless fruit will fall at your feet; the winged one will glide some distance away.

however, is the ultimate floater. Round, buoyant, and watertight, the coconut drops off its parent palm, rolls to the sea, and floats for weeks or months until it washes ashore on some distant isle (Fig. 24-22). There it germinates, perhaps establishing a new coconut colony on a formerly barren island.

Animal Dispersal

Perhaps the majority of fruits use animals as agents of seed dispersal. Two quite distinct strategies have evolved for dispersal by animals: Grab an animal as it passes by, or entice it to eat the fruit but not digest the seeds.

Anyone who takes a long-haired dog on a walk through an abandoned field knows about fruits that hitchhike on animal fur. Burdocks, burr clover, foxtails, and sticktights all develop fruits with prongs, hooks, spines, or adhesive hairs (Fig. 24-23). The parent plants hold these fruits very loosely, so even slight contact with fur pulls the fruit free of the plant and leaves it stuck on the animal. Some of these fruits may fall off the next time the animal brushes against a tree or rock or come out when the animal grooms its fur.

Unlike hitchhiker fruits, edible fruits benefit both animal and plant. The plant stores sugars and tasty flavors in a fleshy fruit that surrounds the seeds, enticing hungry animals to eat the fruit (Fig. 24-24). Some fruits, such as peaches and plums, contain large, hard seeds that animals usually do not eat. After consuming the flesh of the fruit, the animals discard the seeds. Other fruits, including blackberries, raspberries, strawberries, and tomatoes, have small seeds that are swallowed along with the fruit flesh. The seeds then pass through the animal's digestive tract without harm. In some cases, passing through an animal's gut may even be essential to seed germination, by scraping or digesting away part of the seed coat. Besides transport away from its parent, a seed that is swallowed and excreted benefits in another way: It ends up with its own supply of fertilizer!

*Figure 24-22 **Dispersal by water***
After a long journey at sea, this coconut was washed high onto a beach by a storm. The large size and massive food reserves of coconuts are probably adaptations required for successful germination and seedling growth on barren, sandy beaches.

*Figure 24-23 **The cocklebur seed uses hooked spines to hitch a ride on furry animals.***

Figure 24-24 The colors of ripe fruits attract animals
A bright red raspberry fruit has attracted a resplendent quetzal in Costa Rica. Only ripe fruits with mature seeds inside are sweet and brightly colored. Most unripe fruits are unpalatable: green, hard, and bitter. This too is an evolutionary adaptation. The immature seeds within unripe fruit may not survive passage through an animal's gut.

Summary of Key Concepts

1) What Are the Features of Plant Life Cycles?

The sexual life cycle of plants, called alternation of generations, includes both a multicellular diploid form (the sporophyte generation) and a multicellular haploid form (the gametophyte generation).

In seed plants, the gametophyte stage is greatly reduced. The male gametophyte is the pollen grain, a drought-resistant structure that can be carried from plant to plant by wind or animals. The female gametophyte is also reduced and is retained within the body of the sporophyte stage. In this way, seed plants can reproduce independently of liquid water.

2) How Did Flowers Evolve?

Flowering plants evolved from gymnosperms. In gymnosperms, wind blows pollen from male cones to female cones. Wind pollination is inefficient, though. In many habitats, flowering plants enjoy a selective advantage over gymnosperms because many types of flowers attract insects that carry pollen from plant to plant.

Complete flowers consist of four parts: sepals, petals, stamens (male reproductive structures), and carpels (female reproductive structures). The sepals form the outer covering of the flower bud. Most petals (and in some cases, the sepals) are brightly colored and attract pollinators to the flower. The stamen consists of a filament that bears at its tip an anther, in which pollen (the male gametophyte) develops. The carpel consists of the ovary, in which one or more embryo sacs (the female gametophytes) develop, and a style. The style bears at its end a sticky stigma, to which pollen adheres during pollination.

Incomplete flowers lack one or more of the four floral parts.

3) How Have Flowers and Pollinators Adapted to One Another?

Most flowers are pollinated by the wind or by animals, normally insects or birds. Flowers show specific adaptations to pollination by particular pollinators, because flowering plants have coevolved with their pollinators.

4) How Do Gametophytes Develop in Flowering Plants?

Pollen develops in the anthers. The diploid microspore mother cell undergoes meiosis to produce four haploid microspores. Each of these divides mitotically to form pollen grains. An immature pollen grain consists of two cells: the tube cell and the generative cell. The generative cell divides once to produce two sperm cells.

The embryo sac develops within the ovules of the ovary. A diploid megaspore mother cell undergoes meiosis to form four haploid megaspores. Three of these degenerate; the fourth undergoes three sets of mitotic divisions to produce the eight nuclei of the embryo sac. These eight nuclei come to reside in only seven cells. One of these cells, with a single nucleus, is the egg cell; another, with two nuclei, is the primary endosperm cell. These two cells are involved in seed formation; the rest of the cells degenerate.

5) How Does Pollination Lead to Fertilization?

Pollination is the transfer of pollen from anther to stigma. When a pollen grain lands on a stigma, its tube cell grows through the style to the embryo sac. The generative cell divides to form two sperm cells that travel down the style within the tube cell, eventually entering the embryo sac. One sperm fuses with the egg to form a diploid zygote, which will give rise to the embryo. The other sperm fuses with the binucleate primary endosperm cell to produce a triploid cell. This cell will give rise to the endosperm, a food-storage tissue within the seed.

6) How Do Seeds and Fruits Develop?

The embryo develops a root, shoot, and cotyledons. Cotyledons digest and absorb food from the endosperm and transfer it to the growing embryo. Monocot embryos have one cotyledon, and dicot embryos have two. The seed is enclosed within a fruit, which develops from the ovary wall. The function of the fruit is to disperse the seeds away from the parent plant. Seeds typically remain dormant for some time after fruit ripening. Environmental conditions involved in the breaking of dormancy include an initial drying, exposure to cold, or disruption of the seed coat.

7) How Do Seeds Germinate and Grow?

Seed germination requires warmth and moisture. Energy for germination comes from food stored in the endosperm; that food is transferred to the embryo by the cotyledons.

Key Terms

alternation of generations *p. 484*	endosperm *p. 494*	megaspore *p. 493*	primary endosperm cell *p. 493*
anther *p. 486*	epicotyl *p. 495*	megaspore mother cell *p. 493*	seed *p. 485*
carpel *p. 486*	fertilization *p. 494*	microspore *p. 492*	seed coat *p. 495*
coevolution *p. 487*	filament *p. 486*	microspore mother cell *p. 492*	sepal *p. 486*
coleoptile *p. 498*	flower *p. 485*	ovary *p. 486*	spore *p. 484*
complete flower *p. 486*	fruit *p. 486*	ovule *p. 486*	sporophyte *p. 484*
cotyledon *p. 495*	gametophyte *p. 484*	petal *p. 486*	stamen *p. 486*
dormancy *p. 496*	generative cell *p. 492*	polar nucleus *p. 493*	stigma *p. 486*
double fertilization *p. 494*	germination *p. 484*	pollen grain *p. 485*	style *p. 486*
egg *p. 493*	hypocotyl *p. 495*	pollination *p. 494*	tube cell *p. 492*
embryo sac *p. 491*	incomplete flower *p. 486*		
	integument *p. 493*		

Thinking Through the Concepts

Multiple Choice

1. When a bee gets nectar from a flower, the bee picks up pollen from the _____ and carries it to another flower.
a. stigma   b. ovary   c. sepals
d. anthers   e. filament

2. In plants, the gametophyte produces eggs and sperm by
a. mitosis   b. meiosis
c. spore formation   d. fertilization
e. germination

3. If you found a dark, rotten-smelling flower near the ground, it would most likely be pollinated by
a. bats   b. bees   c. butterflies
d. moths   e. beetles

4. Reproduction in flowering plants is known as double fertilization because
a. one tube nucleus fuses with one egg, and one sperm nucleus fuses with another egg
b. two sperm nuclei fuse with one egg
c. one sperm fuses with two eggs
d. one sperm nucleus fuses with one egg, and another sperm nucleus fuses with the two haploid nuclei of the primary endosperm cell
e. two polar nuclei fuse with one egg nucleus

5. In a peach, the fruit is derived from the
a. wall of the ovary   b. endosperm
c. megaspores   d. sepals
e. petals

6. The tassel on top of a corn plant is made up of male flowers that lack female structures. This flower is
a. the sporophyte   b. the gametophyte
c. incomplete   d. complete
e. homologous

? Review Questions

1. Diagram the plant life cycle, comparing ferns with flowering plants. Which stages are haploid, and which are diploid? At which stage are gametes formed?

2. What are the advantages of the reduced gametophyte stages in flowering plants, compared with the more substantial gametophytes of ferns?

3. Diagram a complete flower. Where are the male and female gametophytes formed? What are these gametophytes called?

4. How does an egg develop within an embryo sac? How does this structure allow double fertilization to occur?

5. What does it mean when we say that pollen is the male gametophyte? How is pollen formed?

6. What are the parts of a seed, and how does each part help in the development of a seedling?

7. Describe the characteristics you would expect to find in flowers that are pollinated by the wind, beetles, bees, and hummingbirds, respectively. In each case, explain why.

8. What is the endosperm? From which cell of the embryo sac is it derived? Is endosperm more abundant in the mature seed of a dicot or of a monocot?

9. Describe three mechanisms whereby seed dormancy is broken in different types of seeds. How are these mechanisms related to the normal environment of the plant?

10. How do monocot and dicot seedlings protect the delicate shoot tip during seed germination?

11. Describe three types of fruits and the mechanisms whereby these fruit structures help disperse their seeds.

Applying the Concepts

1. A friend gives you some seeds for you to grow in your yard. When you plant some, nothing happens. What might you try to get the seed to germinate?

2. In areas where farms have been left uncultivated for several years, it is often possible to see certain kinds of trees growing in straight lines, which mark old fences where birds sat and deposited seeds they had eaten. Why are such seeds more likely to germinate than are those of the same species that have not passed through a bird's digestive tract? How might an anthropologist use such lines of trees to study past inhabitants of an area?

3. Charles Darwin once described a flower that produced nectar at the bottom of a tube 25 centimeters (10.5 inches) deep. He predicted that there must be a moth or other animal with a 25-centimeter-long tongue to match; he was right. Such specialization almost certainly means that this particular flower could be pollinated only by that specific moth. What are the advantages and disadvantages of such specialization?

4. Many plants that we call *weeds* were brought from another continent either accidentally or purposefully. In a new environment they have few competitors or animal predators, so they tend to grow in such large amounts that they come to be considered weeds. Think of all the ways you can in which humans become involved in plant dispersal. To what degree do you think humans have changed the distributions of plants? In what ways is this change helpful to humans? In what ways is it a disadvantage?

5. In the tropics there are a number of plant–animal coevolutionary relationships in which both are dependent on the relationship. In light of the rapid rate of destruction of tropical ecosystems, how does this type of relationship leave both organisms particularly vulnerable to extinction? What political and economic problems might this rapid rate of extinction create?

Group Activity

Break up into groups of four to design an idealized flower. Each member of the group should use one piece of paper to create (by folding, drawing, taping, or whatever) one floral structure that can be used by a specific type of pollinator—bees, butterflies, bats, birds, or wind. For example, consider a bee-pollinated flower: One person makes a pattern of flower structure (taking into account nectar placement that is appropriate for pollination by bees) for the other group members to follow; a second person makes a landing platform for the bee; a third person positions the nectar by folding the paper into an appropriate shape (such as a tube versus a dish); and the fourth person specifies the appropriate flower color for attracting the bee.

At the end of 7 minutes, assemble your floral design, and compare it with the work of other groups. Each group should then explain why each specific design was chosen.

For More Information

Eiseley, L. "How Flowers Changed the World." *National Wildlife*, April/May 1996. The late philosopher/naturalist Loren Eiseley eloquently explains how the evolution of flowers has changed the history of life on Earth. Beautifully written and lavishly illustrated.

Fleming, T. H. "Cardon and the Night Visitors." *Natural History*, October 1994. A beautifully illustrated look at the way bats pollinate the world's largest cactus and may help determine the numbers of each of the plant's three sexual types.

Handel, S. N., and Beattie, A. J. "Seed Dispersal by Ants." *Scientific American*, August 1990. Many plants depend on ants for seed dispersal, even producing fat deposits on the outside of the seed as a lure for the ants. Dispersal is not the only benefit to the seed: Being "planted" in an anthill seems to provide an ideal environment for germination and growth as well.

Heinrich, B. "Of Bedouins, Beetles, and Blooms." *Natural History*, May 1994. A colorful look at pollination strategies, especially at the large number of red flowers in Israel that use their color to attract beetles for sex.

Kearns, C., and Inouye, D. "Pistil-Packing Flies." *Natural History*, September 1993. A beautifully illustrated look at the wide variety of flowers that are pollinated by flies in alpine areas.

Moore, P. D. "The Buzz about Pollination." *Nature*, November 7, 1996. The buzzing of bees may actually shake loose the pollen of certain specially adapted flowers, a kind of "sonication pollination."

Murawski, D. A. "A Taste for Poison." *National Geographic*, December 1993. Certain passion-vine butterflies form cyanide apparently from eating pollen from the passion vine. Other butterflies mimic the poisonous types.

Nabhan, G. P. "The Parable of the Poppy and the Bee." *Nature Conservancy*, March/April 1996. An engaging account of the relationship between the poppy-loving bee and the dwarf bearclaw poppy.

Seymour, R. S., and Schultz-Motel, P. "Thermoregulating Lotus Flowers." *Nature*, September 26, 1996. The lotus flower generates a significant amount of heat and effectively regulates its own temperature. The heat may serve as an attractant to pollinators.

Simons, P. "An Explosive Start for Plants." *New Scientist*, January 2, 1993. Explosive movements in some plants spread seeds and spores and snap pollen onto bees.

Sunquist, F. "Blessed Are the Fruit Eaters." *International Wildlife*, May/June 1992. Especially in the tropics, fruit-eating mammals and birds are crucial to the dispersal and germination of seeds. Overhunting threatens not only the animals but also the plants that depend on them.

Answers to Multiple-Choice Questions
1. d   2. a   3. e   4. d   5. a   6. c

The brightly colored flesh of the saguaro cactus fruit attracts a lesser long-nosed bat. Nearly all of the activities of plants, from growth to fruit production and ripening, are under the control of plant hormones.

Plant Responses to the Environment 25

Have you ever bitten into a sweet, juicy red apple and seen the small brown seeds in the center, and wondered why the apple tree puts so much energy into encasing its seeds in fruit? When we eat fruit, we take advantage of an adaptation that has evolved through millions of years of interactions between plants and animals. In the wild, this mutually beneficial relationship is seen from forests to deserts.

The enticing ripe fruit of the giant saguaro cactus displays its bright red seeds to birds and bats. Many of the seeds will pass unharmed through the digestive tracts of these wide-ranging animals, then be deposited in bird and bat droppings throughout the desert. What stimulates the fruit to ripen and break open when the seeds are ready for dispersal? Why are so many fruits and berries red when they are ripe but green when they aren't?

Perhaps you live far from the desert, near deciduous forests. If so, you've probably seen masses of acorns on oak branches in late fall, and gray squirrels busily gathering and burying the nuts, storing them against the coming winter. With so many acorns "squirreled away," some are forgotten and others are not needed. Many, conveniently planted by the squirrels, will sprout next spring as the sun warms the forest soil.

The squirrels, of course, haven't planted the acorns so that new oaks will grow. The acorns are oriented randomly, some with the future stem end down and root end up. Buried beneath several inches of soil, how does the germinating seed "know" which way is up? If it sprouts and sends its shoot up into the air, how does it develop into a mature oak? How does it distinguish the seasons, flowering in spring and going dormant in fall?

Plants do not have specialized organs comparable to our eyes or ears for perceiving the environment. Nevertheless, plants *do* perceive many features of their world: the direction of gravity; the direction, intensity, and duration of sunlight; the strength of the wind; the munching of a caterpillar; and in the Venus flytrap, even the touch of a fly on a leaf. Plants also respond to these stimuli by regulating their own growth and development in appropriate ways. In most cases, plants control their bodily functions through the action of chemicals called *hormones*, or growth regulators. In this chapter we explore how plants detect external stimuli and, in response, produce and distribute hormones. We will also look at how these hormones govern the development of the plant body and the production of new life.

1) How Were Plant Hormones Discovered?

Everyone who keeps houseplants on a windowsill knows that the plants bend toward the window as they grow, in response to the sunlight streaming in. More than a hundred years ago, Charles Darwin and his son, Francis, studied this phenomenon of growth toward the light, or **phototropism.**

First, the Direction of Information Transfer Was Determined

The Darwins illuminated grass coleoptiles from various angles. (Recall that a *coleoptile* is the protective sheath surrounding a monocot seedling; see Fig. 24-19a.) They noted that a region of the coleoptile a few millimeters below the tip bent toward the light until the tip faced directly into the light source (Fig. 25-1). If they covered the very tip of the coleoptile with an opaque (lightproof) cap, the coleoptile didn't bend. A clear cap placed over the coleoptile, however, did not prevent bending. Further, bending toward the light still occurred if the Darwins covered the bending region with an opaque sleeve. The Darwins concluded that (1) the tip of the coleoptile perceives the direction of light and (2) bending occurs farther down the coleoptile; therefore, (3) the tip must transmit information about the light direction down to the bending region.

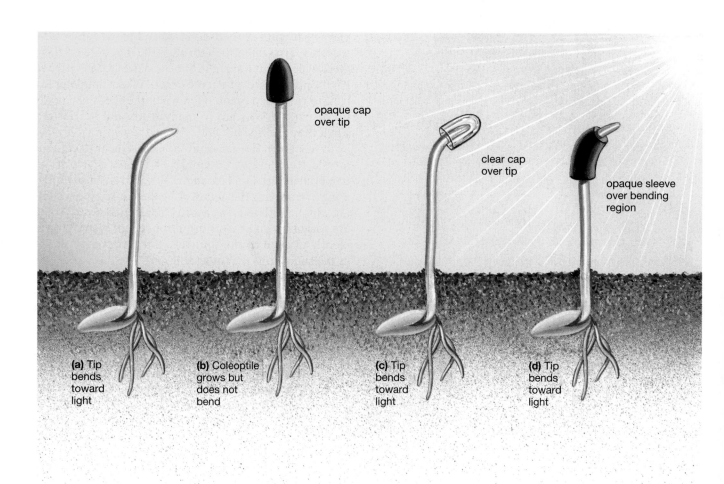

(a) Tip bends toward light   **(b)** Coleoptile grows but does not bend

opaque cap over tip

clear cap over tip

opaque sleeve over bending region

(c) Tip bends toward light   **(d)** Tip bends toward light

Figure 25-1 The phototropism experiments of Charles and Francis Darwin
(a) If light strikes a coleoptile from the side, a region a few millimeters below the tip bends as it grows, until the tip points toward the light source. **(b)** If the tip is covered with an opaque cap, the coleoptile does not bend. **(c)** If a clear cap is placed over the tip, bending still occurs, proving that the mere presence of a cap does not prevent bending. **(d)** If a flexible opaque collar is placed around the bending region, the coleoptile still bends toward the light.

How does the coleoptile bend? Although the Darwins didn't know this, growth in the bending region is due to the elongation of preexisting cells. Therefore, the coleoptile must bend as a result of unequal elongation of cells (Fig. 25-2). The cells on the outside of the bend (away from the light) elongate faster than do the cells on the inside of the bend (toward the light). If one side of the coleoptile becomes longer than the opposite side, then the entire shaft must bend away from the longer side. On the basis of this information, we can reinterpret the Darwins' results and say that the information transmitted from the tip to the bending region causes greater elongation of cells on the side of the coleoptile shaft away from the light.

Next, the Information Was Found to Be Chemical in Nature

About 30 years after the Darwins' experiments, Peter Boysen-Jensen showed that the information transferred down from the tip is chemical in nature (Fig. 25-3). He cut the tips off coleoptiles and found that the remaining stump neither elongated nor bent toward the light. If he replaced the tip and placed the patched-together coleoptile in the dark, it elongated straight up. In the light, it

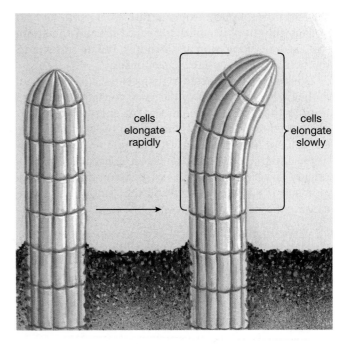

Figure 25-2 *The mechanism of bending by coleoptiles*
Coleoptiles bend by unequal elongation of cells on opposite sides of the coleoptile. If cells on one side elongate more rapidly than cells on the other side, the coleoptile bends away from the longer side.

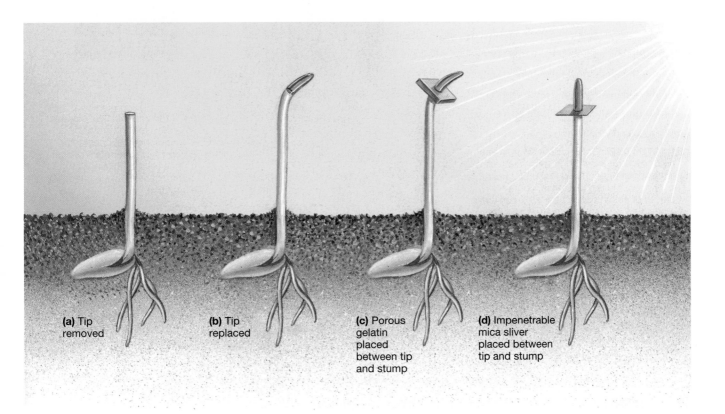

(a) Tip removed

(b) Tip replaced

(c) Porous gelatin placed between tip and stump

(d) Impenetrable mica sliver placed between tip and stump

Figure 25-3 Boysen-Jensen's experiments
These experiments demonstrated the chemical nature of the signal passing from the tip to the bending region of the coleoptile during phototropism. *(a)* If the tip is cut off a coleoptile, the remaining stump neither elongates nor bends toward the light. *(b)* Replacing the tip restores elongation and phototropism. *(c)* A permeable gelatin "filter" between the tip and stump does not prevent phototropism, but *(d)* an impenetrable slice of mica does.

showed normal phototropism. Inserting a thin layer of porous gelatin between the severed tip and the stump still allowed elongation and bending, but an impenetrable barrier eliminated these responses.

Boysen-Jensen concluded that a chemical is produced in the tip and moves down the shaft, causing cell elongation. In the dark, the chemical that causes the cells to elongate diffuses straight down from the tip and causes the coleoptile to elongate straight up. Presumably, light causes the chemical to become more concentrated on the "shady" far side of the shaft, so cells on the shady side elongate faster than do cells on the "sunny" near side, causing the shaft to bend toward the light.

Finally, the Chemical Auxin Was Identified

The next step was to isolate and identify the chemical. In the 1920s, Frits Went devised a way to collect the elon- gation-promoting chemical. He cut off the tips of oat coleoptiles and placed them on a block of agar (a porous, gelatinous material) for a few hours (Fig. 25-4a). Went hoped that the chemical would migrate out of the coleop- tiles into the agar. He then cut up the agar, now presum- ably loaded with the chemical, and placed small pieces on the tops of coleoptile stumps growing in darkness. When he put a piece of agar squarely atop a stump, the stump elongated straight up (Fig. 25-4b). All the stump cells received equal amounts of the chemical and elon- gated at the same rate. If he placed a piece on one side of a cut stump, the stump would invariably bend away from the side with the agar (Fig. 25-4c). It was apparent that cells on the side under the agar received more of the chemical and were more stimulated to elongate. Went called the chemical **auxin**, from a Greek word meaning "to increase." Kenneth Thimann later purified auxin and determined its molecular structure.

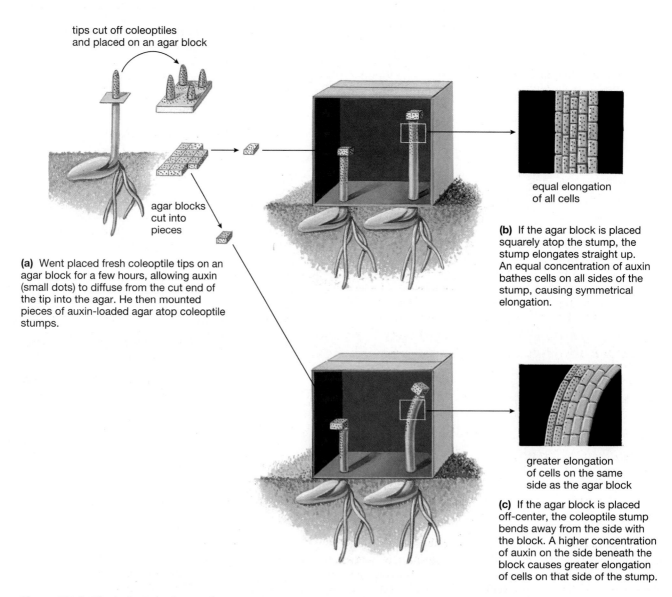

tips cut off coleoptiles and placed on an agar block

agar blocks cut into pieces

(a) Went placed fresh coleoptile tips on an agar block for a few hours, allowing auxin (small dots) to diffuse from the cut end of the tip into the agar. He then mounted pieces of auxin-loaded agar atop coleoptile stumps.

equal elongation of all cells

(b) If the agar block is placed squarely atop the stump, the stump elongates straight up. An equal concentration of auxin bathes cells on all sides of the stump, causing symmetrical elongation.

greater elongation of cells on the same side as the agar block

(c) If the agar block is placed off-center, the coleoptile stump bends away from the side with the block. A higher concentration of auxin on the side beneath the block causes greater elongation of cells on that side of the stump.

Figure 25-4 The isolation of auxin by Frits Went

2) What Are Plant Hormones, and How Do They Act?

Animal physiologists have long recognized that chemicals called **hormones** are produced by cells in one location and transported to other parts of the body, where they exert specific effects. By analogy, auxin and other plant-regulating chemicals are called **plant hormones**. So far, plant physiologists have identified five major classes of plant hormones: (1) auxins, (2) gibberellins, (3) cytokinins, (4) ethylene, and (5) abscisic acid (Table 25-1). Several other types of hormones are thought to exist but have not yet been isolated and identified. Each hormone can elicit a variety of responses from plant cells, depending on the type of cell, its physiological state, and the presence of other hormones. Further, the roles of some plant hormones vary among plant species. This should not be too surprising; after all, in animals a single hormone can be involved in actions as diverse as a salmon's transition from fresh water to salt water, a frog's metamorphosis from tadpole to adult, and a snake's shedding of its skin (see "Evolutionary Connections: The Evolution of Hormones," Chapter 35).

Auxin, as we have seen, promotes the elongation of cells in coleoptiles and other parts of the shoot. In roots, low concentrations of auxin stimulate elongation; slightly higher concentrations inhibit elongation. Both light and gravity affect the distribution of auxin in roots and shoots, so auxin plays a major role in both phototropism and **gravitropism** (directional growth with respect to gravity). Auxin also affects many other aspects of plant development. It stimulates root branching, the differentiation of conducting tissues (xylem and phloem), and the development of fruits. Auxin may also prevent sprouting by lateral buds.

Gibberellins are a group of chemically similar molecules that, like auxin, promote the elongation of cells in stems. In some plants, gibberellins stimulate flowering, fruit development, seed germination, and bud sprouting.

Cytokinins promote cell division in many plant tissues; consequently, they stimulate the sprouting of buds and the development of fruit, endosperm, and the embryo. Cytokinins also stimulate plant metabolism, preventing or at least delaying the aging of plant parts, especially leaves.

Ethylene is an unusual plant hormone in that it is a gas at normal environmental temperatures. Ethylene is best known, and most commercially valuable, for its ability to cause fruit to ripen. It also stimulates the separation of cell walls into *abscission layers*, which, we shall soon see, allow leaves, flowers, and fruit to drop off at the appropriate times without harming the plant.

Abscisic acid is a hormone that helps plants withstand unfavorable environmental conditions. As you learned in Chapter 23, it causes stomata to close when water availability is low. It inhibits the activity of gibberellins, thus helping maintain dormancy in buds and seeds at times when germination would be dangerous.

In the rest of this chapter, we describe a year in the life of a plant, illustrating how hormones regulate its growth and development.

3) How Do Hormones Regulate the Plant Life Cycle?

The life cycle of a plant results from a complex interplay between its genetic information and its environment. Hormones mediate many of the genetic determinants of growth and development as well as nearly all responses to environmental factors. At each stage in its life cycle, a plant produces a distinctive set of hormones that interact with one another in directing the growth of the plant body.

Abscisic Acid Maintains Seed Dormancy; Gibberellin Stimulates Germination

As we pointed out in Chapter 24, a seed maturing within a juicy fruit on a warm autumn day has ideal conditions for germination, yet it remains dormant until the following spring. In many seeds, *abscisic acid enforces dormancy*. Abscisic acid slows down the metabolism of the embryo within the seed, preventing its growth. The seeds of some desert plants contain high concentrations of abscisic acid; only a really hard rain can wash out the abscisic acid, freeing the embryo from its inhibitory effects and allowing the seed to germinate. Seeds of most high-latitude plants require a prolonged period of cold, such as occurs during winter, to break dormancy; in these seeds, chilling induces the destruction of abscisic acid.

Germination is stimulated by other hormones, especially gibberellin. The same environmental conditions that cause the breakdown of abscisic acid also promote the synthesis of gibberellin. In germinating seeds,

Table 25-1 Hormone Actions in Plants	
Hormone	**Functions**
Abscisic acid	Closing of stomata; seed dormancy; bud dormancy
Auxins	Elongation of cells in coleoptiles and shoots; phototropism; gravitropism in shoots and roots; root growth and branching; apical dominance; development of vascular tissue; fruit development; retarding senescence in leaves and fruit; ethylene production in fruit
Cytokinins	Promotion of sprouting of lateral buds; prevention of leaf senescence; promotion of cell division; stimulation of fruit, endosperm, and embryo development
Ethylene	Ripening of fruit; abscission of fruits, flowers, and leaves; inhibition of stem elongation; formation of hook in dicot seedlings
Gibberellins	Germination of seeds and sprouting of buds; elongation of stems; stimulation of flowering; development of fruit

gibberellin initiates the synthesis of enzymes that digest the food reserves of the endosperm and cotyledons, making sugars, lipids, and amino acids available to the growing embryo.

Auxin Controls the Orientation of the Sprouting Seedling

When the growing embryo breaks out of the seed coat, it immediately faces a crucial problem: Which way is up? The roots must burrow downward, while the shoot must grow upward to find the light. Whether the seed was buried by squirrels or fell randomly to the ground, it faces a high probability of being oriented "upside-down."

Auxin apparently controls the responses of both roots and shoots to light and gravity.

Auxin Stimulates Shoot Elongation away from Gravity and toward Light

Let's look at the growth of a shoot as it first emerges from the seed, buried underground. As we described earlier, auxin is synthesized in shoot tips, moves down the shaft of the stem, and stimulates cell elongation. If the stem is not exactly vertical, organelles in the cells of the stem detect the direction of gravity and cause auxin to accumulate on the stem's lower side (Fig. 25-5a). Therefore, the lower cells elongate rapidly, forcing the stem to

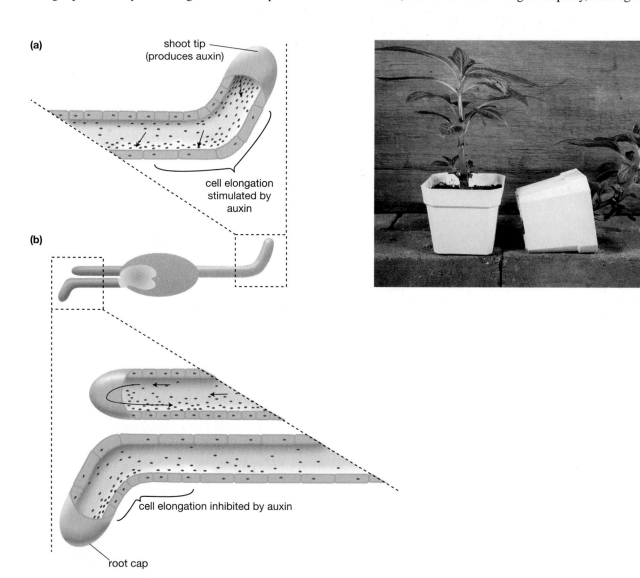

Figure 25-5 *The mechanism of gravitropism in shoots and roots*
(a) In shoots, auxin (blue) produced by the shoot tip accumulates on the lower side of the shoot. Auxin promotes rapid elongation of cells on the lower side, causing the shoot to bend upward. The photo shows two impatiens plants; the right-hand one was placed on its side in the dark for 16 hours. In just this short time, faster elongation of cells on the lower side of the stem has brought the plant to a nearly vertical orientation. *(b)* In roots, auxin (blue) is transported to the root from the shoot. The root cap senses the direction of gravity and redirects the flow of auxin, so auxin accumulates on the lower side of the root. Auxin inhibits cell elongation in roots. Because cells on the top elongate faster than do those on the bottom, the root bends downward.

bend upward. When the shoot tip is vertical, the auxin distribution becomes symmetrical. The stem then grows straight up, emerging from the soil into the light.

Auxin mediates phototropism in addition to gravitropism. Ordinarily, the distribution of auxin caused by light is the same as the distribution caused by gravity, because the direction of brightest light (the sun) is roughly opposite that of gravity. For example, if a young shoot still buried underground is close enough to the surface that some light penetrates down to it, both light and gravity cause auxin to be transported to the lower side of the shoot and promote upward bending. Thus, under normal conditions gravitropism and phototropism add to each other's effects.

Auxin May Control the Direction of Root Growth

Gravitropism in roots is less well understood than it is in stems. According to one model, auxin controls the direction of root growth (Fig. 25-5b). Auxin is transported from the shoot down to the root. If the root is not vertical, the root cap senses the direction of gravity and causes the auxin to accumulate on the lower side of the root. Unlike shoots, in which moderate concentrations of auxin *stimulate* cell elongation, in roots these same concentrations of auxin *inhibit* cell elongation. Therefore, cell elongation in the lower side of the root, where auxin accumulates, is inhibited, whereas cell elongation remains unaffected in the upper side of the root. As a result, the root bends downward. When the root tip points directly downward, the auxin distribution becomes equal on all sides, and the root continues to grow straight downward. Note that *auxin slows down but does not eliminate root cell elongation*: A vertical root continues to grow.

Plants May Sense Gravity by Means of Organelles Called Statoliths

Although auxin plays a major role in the unequal growth rates of cells that cause bending away from gravity (in shoots) and toward gravity (in roots), auxin itself does not detect gravity. Instead, specialized cells in stems and in root caps contain starch-filled organelles called **statoliths**. By staining these statoliths and observing them under the microscope, plant physiologists have discovered that they settle to the downward side of the cell within minutes (Fig. 25-6). When a plant is laid on its side, causing a formerly vertical root to be horizontal, the time it takes the statoliths to settle is similar to the time it takes the root to begin its unequal cell elongation so that it once more heads downward. Exactly how the falling statoliths initiate the response to gravity in stems and roots is still under investigation.

The Genetically Determined Shape of the Mature Plant Is the Result of Interactions among Hormones

As a plant grows, both its root and shoot develop branching patterns that are largely determined by its genetic heritage. For example, the stems of some plants, such as sunflowers, hardly branch at all; others, such as oaks and cottonwoods, branch profusely in seeming confusion; still others branch in a very regular pattern, producing the conical shapes of firs and spruces.

The amount of growth in shoot and root systems must also be kept in balance. The shoot must be large enough to supply the roots with sugars; the roots must be large enough to provide the shoot with water and minerals.

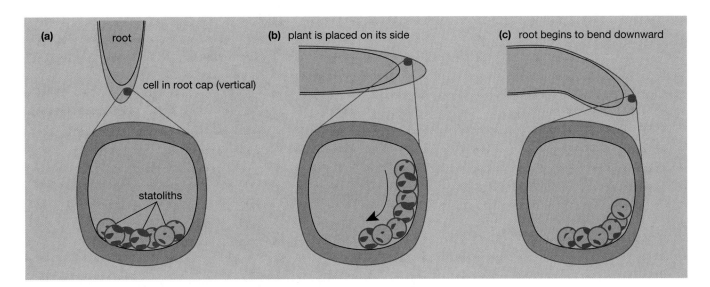

Figure 25-6 Statoliths allow plants to sense gravity
Statoliths are organelles that fall to the downward side of a cell. **(a)** Normal orientation of statoliths in a root cap cell. **(b)** When the plant is placed on its side (as in Figure 25-5), the statoliths begin tumbling downward, **(c)** coming to rest on the new lower surface. This change triggers the bending of the root downward.

Interactions between auxin and cytokinin regulate root and stem branching, thereby regulating the relative sizes of root and shoot systems.

Stem Branching Is Influenced by the Growing Tip of the Shoot

Gardeners know that pinching back the tip of a growing plant makes the tip become bushier. The botanical explanation for this practice is that the growing tip suppresses the sprouting of lateral buds to form branches, a phenomenon known as **apical dominance**. Although exactly how the sprouting of lateral buds is controlled remains a subject of research, some evidence shows that the proper levels of auxin and cytokinin must be present (Fig. 25-7). Auxin is produced by the shoot tip (its concentration is highest there) and transported down the stem, gradually decreasing in concentration. Cytokinin is produced by the roots (where its concentration is highest) and is transported up the stem. Therefore, the relative concentrations of these two hormones will vary along the length of the stem. Buds at different positions will experience differing hormonal influences.

Auxin by itself appears to inhibit the sprouting of lateral buds, whereas auxin and cytokinin together stimulate bud sprouting. The lateral buds closest to the shoot tip receive a great deal of auxin, probably enough to inhibit their growth, but receive very little cytokinin because they are so far from the roots. Therefore, they remain dormant. Lower buds receive less auxin while receiving much more cytokinin. They are stimulated by optimal concentrations of both hormones, so they sprout (Fig. 25-7). In many types of plants, this interaction between auxin and cytokinin produces an orderly progression of bud sprouting, from the bottom to the top of the shoot. The exact ratio of cytokinin to auxin that promotes sprouting varies among species.

Auxin Stimulates Root Branching

Even in extremely low concentrations, auxin stimulates the branching of roots. As we described in Chapter 23,

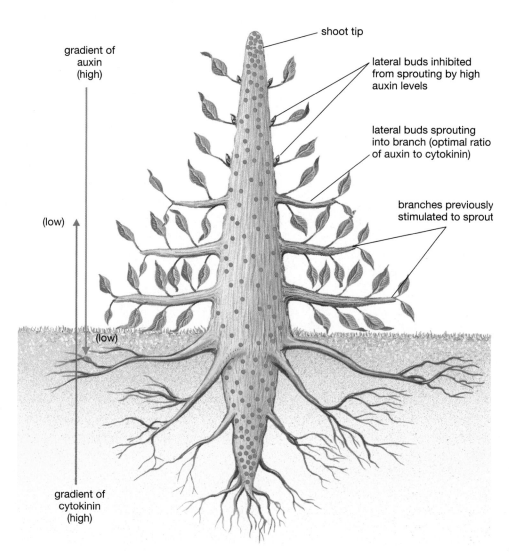

Figure 25-7 The role of auxin and cytokinin in lateral bud sprouting
A simplified diagram of the interplay of auxin (blue dots) and cytokinin (red dots) in the control of sprouting of lateral buds. Auxin is produced by shoot tips and moves downward; cytokinin is produced by root tips and moves upward. The auxin concentration decreases with increasing distance from the *shoot* tip, and the cytokinin concentration decreases with increasing distance from the *root* tip. The sprouting of lateral buds depends on the ratio of cytokinin to auxin, which varies among species as well as along the stem.

gradient of auxin (high)

(low)

(low)

gradient of cytokinin (high)

shoot tip

lateral buds inhibited from sprouting by high auxin levels

lateral buds sprouting into branch (optimal ratio of auxin to cytokinin)

branches previously stimulated to sprout

branch roots arise from the pericycle layer of the vascular cylinder. Auxin, transported down from the stem, stimulates pericycle cells to divide and form a branch root.

Hormone Gradients Create a Balance between the Root and Shoot Systems

Through the interaction of auxin and cytokinin, the root and shoot systems regulate each other's growth. This interaction is important, because those systems supply complementary nutrients. An enlarging root system synthesizes large amounts of cytokinin, which stimulates lateral buds to break dormancy and sprout. If the root system isn't keeping up with the growing shoot system, less cytokinin is produced. The sprouting of lateral buds is delayed, slowing the growth of the shoot system. Simultaneously, as the stem grows and branches, it produces lots of auxin, which stimulates root branching and growth. Thus, neither root nor shoot can outgrow the other, and the plant is adequately supplied with all its needs.

Daylength Controls Flowering

Ultimately, the plant matures enough to reproduce. The timing of flowering and seed production is finely tuned to the physiology of the plant and the rigors of its environment. In temperate climates, plants must flower early enough so that their seeds can mature before the killing frosts of autumn. Depending on how quickly the seed and fruit develop, flowering may occur in spring, as it does in oaks; in summer, as in lettuce; or even in autumn, as in asters.

What environmental cues do plants use to determine the season? Most cues, such as temperature or water availability, are quite variable: October can be warm, a late snow may fall in May, or the summer might be unusually cool and wet. *The only reliable cue is daylength*: Longer days always mean that spring and summer are coming; shorter days foretell the onset of autumn and winter.

With respect to flowering, plants are classified as day-neutral, long-day, or short-day (Fig. 25-8). A **day-neutral plant** flowers as soon as it has grown and developed

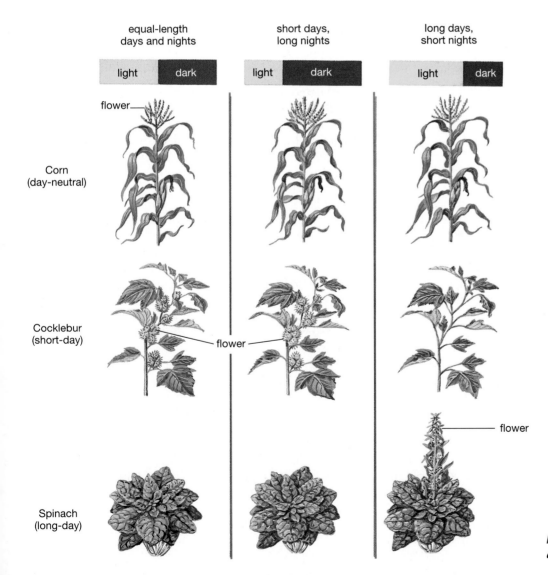

Figure 25-8 The effects of daylength on flowering

enough, regardless of the length of the day. Day-neutral plants include tomatoes, corn, and snapdragons. A **long-day plant** flowers when the daylength is *longer than some species-specific critical value*. A **short-day plant** flowers when the daylength is *shorter than some species-specific critical value*. Thus, spinach is classified as a long-day plant because it flowers only if the day is *longer than* 13 hours, and cockleburs are short-day plants because they flower only if the day is *shorter than* 15.5 hours. Note that both will flower with 14 hours of light. Spinach, however, will also flower in much longer daylengths (for example, 16 hours of light), but cocklebur will not. Conversely, cocklebur will flower in much shorter daylengths (for example, 12 hours), but spinach will not.

Hormones Called Florigens Both Stimulate and Inhibit Flowering

Grafting a branch from a (short-day) cocklebur plant maintained on a short-day schedule onto a cocklebur plant on a long-day schedule will induce flowering. In fact, plant physiologists have been able to induce flowering in the cocklebur by exposing a *single leaf* to short days in a special chamber while the rest of the plant was experiencing long days. Clearly, a signal must be traveling from leaf to bud. Studies on a variety of plants suggest that still-unidentified hormones can both trigger and inhibit flowering. These substances, which may differ among plant species, collectively are called **florigens** (literally, "flower makers"). Daylength is a crucial stimulus for these hormones, but how do plants detect daylength?

Pigments Called Phytochromes Measure Daylength by Resetting the Biological Clock

To measure daylength, a plant needs two things: (1) some sort of metabolic *clock* to measure time (how long it has been light or dark) and (2) a *light-detecting system* to set the clock. Virtually all organisms have an internal **biological clock** that measures time even without environmental cues. However, environmental cues, and particularly light, can reset the clock. In most cases, the operating method of the biological clock is poorly understood.

The light-detecting system of plants is a pigment, called **phytochrome** (meaning simply "plant color"), in the leaves. Phytochrome occurs in two interchangeable forms. One form strongly absorbs red light and is called P_r; the other form absorbs far-red light (almost infrared) and is accordingly called P_{fr} (Fig. 25-9). In most plants, P_{fr} is the active form of phytochrome; that is, a suitable concentration of P_{fr} stimulates or inhibits physiological processes, such as flowering or setting the biological clock. P_r, the inactive form, has no effect on these same processes.

Phytochrome flips back and forth from one form to the other when the pigment absorbs light of the appro-

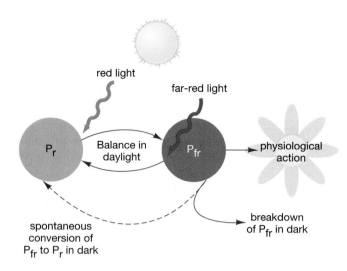

Figure 25-9 The light-sensitive pigment phytochrome Phytochrome exists in two forms, inactive (P_r) and active (P_{fr}). P_r is converted to P_{fr} by red light. P_{fr} may then participate in physiological responses, be converted to P_r by far-red light, revert spontaneously to P_r, or break down to other, inactive compounds.

priate color: Upon absorbing red light, P_r is converted into P_{fr}; upon absorbing far-red light, P_{fr} is transformed back into P_r. Daylight consists of all wavelengths of visible light, including both red and far-red. Therefore, during the day a leaf contains both forms of phytochrome. In the dark, P_{fr} rather rapidly breaks down or reverts to P_r.

Plants seem to use the phytochrome system and their internal biological clocks to detect daylength. Cockleburs, for example, flower under a lighting schedule of 8 hours of light and 16 hours of darkness. However, interrupting the middle of the dark period with just a minute or two of light prevents flowering. Thus, although cockleburs are usually classified as short-day plants, they might more accurately be called "long-night" plants, because what really matters is how long continuous darkness lasts.

The color of the light used for the night flash is also important. A midnight flash of red light inhibits flowering, but a far-red flash allows flowering. This observation, of course, implicates phytochrome in the control of flowering. Scientists are still researching, however, how the response of phytochrome to light determines whether a plant will flower. It seems likely that the biological clock measures the length of the night and that light reception by phytochrome "tells" the clock when sunrise and sunset have occurred, but this is not certain. How phytochromes influence the production of florigens is another area of active research.

Phytochromes Influence Other Responses of Plants to Their Environment

Phytochrome is involved in many plant responses. For example, P_{fr} inhibits the elongation of seedlings. Because P_{fr} breaks down or reverts to P_r in the dark, seedlings germinating in the darkness of the soil contain no P_{fr} and consequently elongate very rapidly, emerging from the soil. Seedlings growing beneath other plants will be exposed largely to far-red light, because the green chlorophyll of the leaves above them will absorb most of the red light but transmit the far-red. Far-red light converts P_{fr} to P_r, so shaded seedlings grow rapidly, which may bring them out of the shade. Once out in the sunlight, P_{fr} forms. The P_{fr} slows down elongation, which prevents the seedlings from becoming too spindly.

Other plant responses that are stimulated by P_{fr} include leaf growth, chlorophyll synthesis, and the straightening of the epicotyl or hypocotyl hook of dicot seedlings. As in the case of stem elongation, these responses are adaptations related to burial in the soil or shading by the leaves of other plants. For example, a newly germinating shoot needs to stay in its protective bend while still in the soil (that is, in the dark) and straighten out only in the open air, where sunlight converts P_r to P_{fr}.

Hormones Coordinate the Development of Seeds and Fruit

When a flower is pollinated, auxin or gibberellin released by the pollen stimulates the ovary to begin developing into a fruit. If fertilization also occurs, the developing seeds release still more auxin or gibberellin, or both, into the surrounding ovary tissues. Cells of the ovary multiply and grow larger, commonly storing starches and other food materials, forming a mature fruit. In this way, the plant coordinates the development of seeds and fruit.

Seeds and fruits acquire nutrients for growth and development from their parent plant. If the seed is separated from the parent too soon, it may not complete its development. Not surprisingly, seed maturation and fruit ripening are closely coordinated. Most unripe fruits are inconspicuously colored (normally green like the rest of the plant), hard, bitter, and in some cases even poisonous. As a result, animals seldom eat unripe fruit.

When the seeds mature, the fruit ripens: It becomes sweeter as starches are converted to sugar, softer, and more brightly colored, making it more noticeable and attractive to animals (Fig. 25-10; see also the chapter-opening photo). If you walk in a field in the fall, look around at the wild fruits on the dying plants. A few may be blue, and many are red, such as the "hips" on wild roses. Almost none are green when ripe. Or look around the produce section of your supermarket at all the brightly colored fruits—adapted to attract animal seed dispersers. Interestingly, gibberellin sprayed on fruit such as grapefruit causes the peel to remain tough and green, although the inside continues to ripen. Citrus growers in

Figure 25-10 Ripe fruit becomes attractive to animal seed dispersers
The prickly pear cactus fruit is green, hard, and bitter before it ripens, so animals are discouraged from eating it. After the seeds mature, the fruit becomes soft, red, and tasty, attracting animals such as this desert tortoise. The mature seeds are not harmed by the animal's digestive tract and are dispersed in the animal's feces.

Florida can now use this technique to discourage fruit flies, which are attracted to the yellow color and must penetrate the ripening peel to lay their eggs.

Ripening is stimulated by ethylene. Ethylene is synthesized by fruit cells in response to a surge of auxin that is released by the seeds (another mechanism by which both seed and fruit development are coordinated). Because ethylene is a gas, a ripe fruit continually leaks ethylene into the air. In nature this probably doesn't make much difference. When you store fruit in a closed container, however, ethylene released from one fruit will hasten ripening in the rest, which is why "one rotten apple spoils the barrel." The discovery of the role of ethylene in ripening revolutionized modern fruit and vegetable marketing. The gibberellin-sprayed green (but ripe) grapefruit described earlier will turn its normal yellow when exposed to ethylene. Bananas grown in Central America can be picked green and tough and shipped to North American markets. By exposing them to ethylene at their destination, grocers can market perfectly ripe fruit. Unfortunately, not all fruits seem to ripen properly when separated from the plant. The pinkish tomato with all the flavor of wet cardboard that we buy in supermarkets is an example. Because grapes and strawberries are not sensitive to ethylene, they must be allowed to ripen on the plant.

Senescence and Dormancy Prepare the Plant for Winter

The season is autumn. The time has come to let any un-eaten fruits drop to the ground. For perennial broadleaf plants, the leaves must be shed as well, because they would be a liability in winter, unable to photosynthesize but still allowing water to evaporate. Leaves, fruits, and flowers undergo a rapid aging called **senescence**. The culmination of senescence is the formation of the **abscission layer** at the base of the petiole (Fig. 25-11). Ethylene stimulates this layer of thin-walled cells to produce an enzyme that digests the cell wall holding a leaf, fruit, or flower to the stem, allowing the leaf or fruit to drop off. Cells interior to the abscission remain intact.

Senescence and abscission are complex processes controlled by several different hormones. In most plants, healthy leaves and developing seeds produce auxin, which in turn helps maintain the health of the leaf or fruit. Simultaneously, the roots synthesize cytokinin, which is transported up the stem and out the branches. Cytokinin also prevents senescence—a leaf plucked from a tree and floated in water in which cytokinin is dissolved stays green for weeks.

As winter approaches, cytokinin production in the roots slows down, and fruits and leaves produce less auxin. Perhaps driven by these hormonal changes, much of the organic material in leaves is broken down to simple molecules that are transported to the roots for winter storage. Meanwhile, ethylene is released by both aging leaves and ripening fruit. Ethylene stimulates the production of enzymes that destroy the cell walls holding together the abscission layers at the base of the petiole. When the abscission layers weaken, leaves and fruits fall from the branches.

Other changes also occur that prepare the plant for winter. Buds, which developed into new leaves and branches during spring and summer, now become dormant, waiting out the winter tightly wrapped up. Dormancy in buds, as in seeds, is enforced by abscisic acid. Metabolism slows to a crawl, and the plant enters its long winter sleep, waiting for the signals of warmth and longer days in spring before awakening once again.

4 Do Plants Communicate by Means of Chemicals?

The hundred-million-year war between plants and their parasites and predators has led to the evolution of sophisticated defenses that surprise people who are accustomed to thinking of plants as passive, helpless organisms. Researchers studying how plants respond to attack by predators or disease-causing viruses have recently made some fascinating discoveries. Plants under attack help protect themselves and neighboring plants by releasing volatile chemicals into the air around them. How does this "chemical cry for help" work?

Many plants produce *salicylic acid*, the compound from which aspirin is derived. Ilya Raskin and his colleagues at Rutgers University found that, when infected with a plant virus, tobacco plants produce salicylic acid in large quantities. The salicylic acid activates a kind of immune response in the plants, helping them fight off the viral attack. The plant also converts some of the salicylic acid to *methyl salicylate*, the substance used to flavor "wintergreen" candy. This highly volatile compound diffuses into the air from virus-infected plant tissues and is absorbed by other plants nearby. Within the healthy neighbors, the methyl salicylate is reconverted to salicylic acid, enhancing the neighbors' immune defenses and making them better able to resist the viral infection. These findings suggest that future farmers may treat their plants with aspirin instead of toxic chemicals to deter plant-disease organisms.

Researchers working with maize, a relative of corn, have discovered a sophisticated form of communication between plants and animals: The maize plant calls in a predator to attack a caterpillar that is munching on it. When a maize plant is chewed by a caterpillar, the plant responds by releasing a mixture of volatile chemicals. These chemicals attract a parasitic wasp that lays its eggs in the body of the caterpillar. This "biological time bomb" kills the caterpillar, as the larvae hatch and consume their

bud

petiole

abscission layer

Figure 25-11 The abscission layer
This cross section shows the abscission layer at the base of a maple leaf. A new leaf bud is visible above the dying leaf.

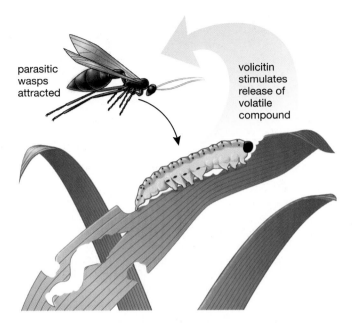

Figure 25-12 A chemical cry for help
Some plants that are attacked by caterpillars respond to a substance in the caterpillars' saliva by releasing volatile compounds that attract parasitic wasps. These wasps ultimately destroy the caterpillars by laying eggs inside the caterpillars' bodies.

host from the inside out. Studying this phenomenon, scientists discovered that merely tearing the leaves as a caterpillar does will not elicit the chemical alarm signal; the attack on the plant had to come from the caterpillar. Recently, the reason was discovered: Trace amounts of a chemical called *volicitin* in caterpillar saliva are required to cause the response (Fig. 25-12). As Ilya Raskin explains, "Plants can't run away and they can't make . . . noises. But they are wonderful chemists."

Evolutionary Connections
Rapid-Fire Plant Responses

All plants are alive, but some are more lively than others. Watch a fly brush against the sensory hairs in a Venus flytrap, and you will see a response that is almost animal-like in its purposefulness and speed of movement (Fig. 25-13). Why is it useful for a Venus flytrap to catch a fly, and how does the plant accomplish this task?

Many soils, particularly those of bogs, are nitrogen-deficient. As an evolutionary response to chronic nitrogen shortages, several bog plants, including the Venus flytrap, pitcher plant, and sundew, have resorted to eating animals. By snaring an insect now and then, the plant obtains nitrogen from the chitin (in the exoskeleton), proteins, and nucleic acids of its prey. Although the evolutionary advantage seems clear, the *mechanism* of move-

ment is much less obvious. How does the plant perceive the touch of a fly, and how does it move its leaves rapidly enough to catch the fly?

In this chapter you have seen how hormones such as auxin can trigger the expansion of cells, causing plants to move toward or away from light or gravity. But hormones don't carry signals fast enough to catch a fly. Venus flytraps have evolved a way to transmit signals that resemble the nerve impulse of animals, thus permitting their animal-like predatory behavior.

Each of the fringed trapping leaves of a Venus flytrap bears three sensory "hairs" on its inside surface (Fig. 25-13a). Insects are attracted to nectar secreted by the leaves. If an insect, in its foraging, touches one hair twice in rapid succession or touches two hairs, the hairs initiate a change in electrical potential analogous to the action potential of animal nerve cells (to be discussed in Chapter 33). The electrical potential sets off a rapid chain of events that causes the trap to close (Fig. 25-13b).

(a)

(b)

Figure 25-13 A Venus flytrap captures its prey

In a beautiful set of experiments, botanists Stephen Williams and Alan Bennett found that the flytrap leaf closes because it undergoes *irreversible, differential growth*. The flytrap leaves can be pictured most simply as two layers of cells, outer and inner (Fig. 25-14). The electrical potential triggered by hair movement stimulates the cells of the outer layer to pump hydrogen ions (H^+) extremely rapidly into their cell walls. Enzymes in the cell walls are activated by acidic conditions (created, as you'll remember from Chapter 2, by a high concentration of H^+) and loosen the cellulose fibers of the walls. As the walls weaken, the high osmotic pressure inside causes the cells to absorb water from extracellular fluids, swiftly growing by about 25%. Because the outer layer expands while the inner layer does not, the leaf is pushed closed. Although reopening the trap takes several hours, the fundamental mechanism is similar: The cells on the inside of the leaf expand, pushing apart the lobes of the trap.

So much energy is used up by the hydrogen ion pumps that *closing the trap consumes nearly one-third of all the ATP within the entire leaf.* It is therefore very important that something digestible actually be in the leaf before it closes the trap!

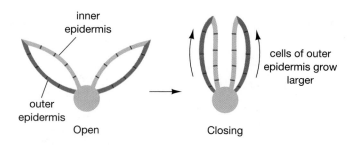

Figure 25-14 *How the flytrap springs its leafy trap*

Although much is known about the mechanisms that produce movement in the Venus flytrap, mysteries still remain. How is touch transformed into an electrical stimulus by the sensory hairs? What is the nature of the electrical potential change? How does the electrical signal cause the cells to begin pumping hydrogen ions? As so often happens in biology, the answer to one question immediately poses several new, usually tougher, questions.

Summary of Key Concepts

1) How Were Plant Hormones Discovered?

Most responses of plants to their environment are produced by the actions of plant hormones. Early experiments on the mechanism of phototropism, the growth of plants toward the light, led to the first discovery of a plant hormone, auxin.

2) What Are Plant Hormones, and How Do They Act?

Plant hormones are chemicals that are produced by cells in one part of a plant body and transported to other parts of the plant, where they exert specific effects. The five major classes of plant hormones are auxin, gibberellins, cytokinins, ethylene, and abscisic acid. The major functions of these hormones are summarized in Table 25-1.

3) How Do Hormones Regulate the Plant Life Cycle?

Dormancy in seeds is enforced by abscisic acid. Falling levels of abscisic acid, and rising levels of gibberellin, trigger germination. As the seedling grows, it shows differential growth with respect to the direction of light (phototropism) and gravity (gravitropism). Auxin mediates phototropism and gravitropism in shoots, and gravitropism in roots. In shoots, auxin stimulates the elongation of cells. In roots, similar concentrations of auxin inhibit elongation. Plants apparently detect gravity by means of organelles called statoliths.

Branching in stems results from the interplay of two hormones, auxin (produced in shoot tips and transported downward) and cytokinin (synthesized in roots and transported up the shoot). High concentrations of auxin inhibit the growth of lateral buds. An optimum concentration of both auxin and cytokinin stimulates the growth of lateral buds. Auxin also stimulates the growth of branch roots.

The timing of flowering is normally controlled by daylength. Flowering is both stimulated and inhibited by hormones called florigens. Plants appear to detect light and darkness by changes in phytochrome, a pigment in the leaves. Plant processes influenced by phytochrome responses to light include flowering, straightening the epicotyl or hypocotyl hook, seedling elongation, leaf growth, and chlorophyll synthesis.

Developing seeds produce auxin, which diffuses into the surrounding ovary tissues and causes the production of a fruit. A surge of auxin as the seed matures stimulates fruit cells to release another hormone, ethylene, which causes the fruit to ripen. Ripening includes the conversion of starches to sugars, softening of the fruit, development of bright colors, and commonly the formation of an abscission layer at the base of the petiole.

Several changes prepare perennial plants of temperate zones for winter. Leaves and fruits undergo a rapid aging process called senescence, including the formation of an abscission layer. Senescence occurs as a result of a fall in levels of auxin and cytokinin and, perhaps, a rise in ethylene concentrations. Other parts of the plant, including buds, become dormant. Dormancy in buds is enforced by high concentrations of abscisic acid.

4) Do Plants Communicate by Means of Chemicals?

When under attack, plants can release volatile chemicals into the air to help protect themselves and neighboring plants.

Key Terms

abscisic acid *p. 509*
abscission layer *p. 516*
apical dominance *p. 512*
auxin *p. 508*
biological clock *p. 514*

cytokinin *p. 509*
day-neutral plant *p. 513*
ethylene *p. 509*
florigen *p. 514*
gibberellin *p. 509*

gravitropism *p. 509*
hormone *p. 509*
long-day plant *p. 514*
phototropism *p. 506*
phytochrome *p. 514*

plant hormone *p. 509*
senescence *p. 516*
short-day plant *p. 514*
statolith *p. 511*

Thinking Through the Concepts

Multiple Choice

1. *If you put an underripe banana in a bag with an apple, the banana will quickly ripen because of the hormone _____ produced by the apple.*
   a. auxin
   b. cytokinin
   c. gibberellin
   d. abscisic acid
   e. ethylene

2. *Roots turn downward in a process known as*
   a. phototropism
   b. abscission
   c. apical dominance
   d. gravitropism
   e. dormancy

3. *If you grow coleus plants, you will need to cut or pinch off the top bud frequently to keep the plant from becoming tall and spindly. Pinching off this bud will slow the production of _____ by the apical bud and allow the plant to become bushy.*
   a. auxin
   b. cytokinin
   c. gibberellin
   d. abscisic acid
   e. ethylene

4. *In your warm house, poinsettias act as short-day plants that turn red when the daylength is less than 12 hours. If you want to get a poinsettia ready for Christmas, which of the following would work best?*
   a. Give it continuous light.
   b. Give it continuous darkness.
   c. Keep it in the dark, but shine a light on it for an hour every 13 hours.
   d. Keep it in the light except for turning off the light once a day for 1 hour.
   e. None of these would get it to turn red.

5. *When a seed is first formed in the fall, it typically will not germinate because _____ must first be washed from the seed by a hard rain or broken down by cycles of freezing and thawing.*
   a. auxin
   b. cytokinin
   c. gibberellin
   d. abscisic acid
   e. ethylene

6. *The seed of Question 5 germinates as levels of _____ increase.*
   a. auxin
   b. cytokinin
   c. gibberellin
   d. abscisic acid
   e. ethylene

? Review Questions

1. What did the Darwins, Boysen-Jensen, and Went each contribute to our understanding of phototropism? Do their experiments truly prove that auxin is the hormone controlling phototropism? What other experiments would you like to see?

2. How do hormones interact to cause apical dominance? To control seed dormancy?

3. How can one hormone, an auxin, cause shoots to grow up and roots to grow down?

4. What is the phytochrome system? Why does this chemical exist in two forms? How do the two forms interact to help control the plant life cycle?

5. Which hormones cause fruit development? Where do these hormones come from? Which hormone causes fruit ripening?

6. What is a biological clock?

7. Describe the role of phytochrome in stem elongation in seedlings that grow in the shade of other plants. What is the likely adaptive significance of this response?

8. What is apical dominance? How do auxin and cytokinin interact in determining the growth of lateral buds?

9. Which hormone(s) is (are) involved in leaf and fruit drop? In bud dormancy?

Applying the Concepts

1. Suppose you got a job in a greenhouse in which the owner was trying to start the flowering of chrysanthemums (a short-day plant) for Mother's Day. You accidentally turned on the light in the middle of the night. Would you be likely to lose your job? Why or why not? What would happen if you turned on the lights in the day?

2. A student reporting on a project said that one of her seeds did not grow properly because it was planted upside down so that it got confused and tried to grow down. Do you think the teacher accepted this explanation? Why or why not? Which plant hormone or hormones would be involved?

3. Agent Orange, a combination of two synthetic auxins, was used in Vietnam to defoliate the rain forest during the Vietnam War. When they are similar to natural growth hormones, how can synthetic auxins be used to harm or kill plants? What do you think would happen if natural auxins were used in excess quantities on plants?

4. Bean sprouts such as those you might eat in a salad have to be grown in the dark for them to form the long, yellowish stems that you see. We call such stems *etiolated*. If they are grown in the light, they will be short and green. Why do seedlings grow etiolated in the dark? Under what conditions does etiolation occur in nature? How do plant hormones enable these seedlings to form this shape?

5. Suppose that on July 4, you discover that both a short-day plant and a long-day plant have bloomed in your garden. Discuss how it is possible for both to bloom.

Group Activity

Form groups of three. Take three scraps of paper, and mark one "SDP," another "LDP," and the third "DNP" (for short-day, long-day, and day-neutral plant, respectively). Shuffle the papers; each of you should randomly choose one. Try to identify with the type of plant you chose. Take 3 minutes to write down the advantages and disadvantages of being that type of plant. Be creative but not frivolous. Take another 2 minutes to answer: If you could choose the flowering category you'd most like to become, which would you choose, and why? Then discuss your decisions with your group.

For More Information

Evans, M. L., Moore, R., and Hasenstein, K.-H. "How Roots Respond to Gravity." *Scientific American*, December 1986. Although botanists still dispute the mechanisms of root gravitropism, these authors convincingly argue that it is mediated by auxin.

Farmer, E. E. "New Fatty Acid–Based Signals: A Lesson from the Plant World." *Science*, May 9, 1997. Describes the research leading to the discovery of volicitin, which attracts parasitic wasps to plants under attack by caterpillars.

Horton, T. "Longleaf Pine: A Southern Revival." *Audubon*, March–April 1995. The entire life cycle of this plant is centered around its adaptations to fire.

Marchand, P. J. "Waves in the Forest." *Natural History*, February 1995. Compares the life rhythms of forests in Japan and New England in which entire areas of the forest live and die in cycles.

Mlot, C. "Where There's Smoke, There's Germination." *Science News*, May 31, 1997. Researchers have recently discovered that nitrogen dioxide produced by fires can induce germination in plants that live in ecosystems where fires are common.

Moffatt, A. S. "How Plants Cope with Stress." *Science*, November 1, 1994. A newly discovered hormone, "systemin," similar to animal hormones, enables plants to respond to stress.

Pennisi, E. "Plants Relay Signals Much as Animals Do." *Science News*, February 13, 1993. A short news article on the effects of ethylene that shows how plants respond to hormones.

Raloff, J. "When Tomatoes See Red." *Science News*, December 13, 1997. Researchers are discovering that different colors of mulch influence growth and disease resistance of tomatoes, by reflecting different wavelengths of light onto the plant's leaves.

Saunders, F. "Keep the Aspirin Flying." *Discover*, January 1998. A short article describing how plants use methyl salicylate to help nearby plants resist infection.

Answers to Multiple-Choice Questions
1. e   2. d   3. a   4. c   5. d   6. c

Unit Five
Animal Anatomy and Physiology

The animal body is an exquisite expression of the elegance with which evolution has linked form to function. All the animal body systems work in concert to maintain life.

"As a physiologist I am especially impressed by the common ignorance of bodily organs and their functions. It seems to me now, as it seemed to Robert Boyle nearly 300 years ago, that it is 'highly dishonorable for a Reasonable Soul to live in so Divinely built a Mansion as the Body [he or] she resides in, altogether unacquainted with the exquisite structure of it.'"

W. B. Cannon *in* **The Way of an Investigator** *(1945)*

Despite brutally drying wind, subzero temperatures, and an absence of shelter, the environment inside the bodies of these penguins remains moist, warm, and protected.

Homeostasis and the Organization of the Animal Body 26

At a Glance

What must an animal accomplish to survive and reproduce? Most people's initial response to this question focuses on the anatomical and behavioral characteristics that help animals secure food, protection, and mates. But animals' internal characteristics have also been honed by generations of competition for limited resources. Natural selection has favored body plans and processes that use resources efficiently and extract the most benefits from them at the lowest possible cost. The result of this selection for efficiency is a diverse array of elegant systems for crucial functions such as extracting energy and nutrients from food, exchanging gases with the environment, sensing and moving around in the environment, producing new individuals, and defending against microbial attack.

Many of these adaptive physiological systems function by regulating the exchange of materials between an animal's body and its external environment. Molecules are constantly entering and leaving the body, but physiological systems control this exchange such that the body's internal state is relatively constant. The body's cells are thus insulated from the vagaries of the external environment and can go about their biochemical business without interference from drastic fluctuations in their immediate surroundings. Whales can dive from sea level to depths at which the water pressure is tremendous; desert scorpions can survive the daily switch from hot days to cold nights; and penguins can stay active on the frigid ice floes of the Antarctic. The typically intricate mechanisms of regulation and control that provide animals with internal stability are among nature's true marvels.

The relatively constant, protected conditions that prevail inside animal bodies have provided the internal environment in which a mind-boggling array of structures has evolved. Each of these structures represents a response to an evolutionary problem, and each group of animals has evolved its own idiosyncratic set of solutions.

Net Watch
On-line resources for this chapter are on the World Wide Web at:
http://www.prenhall.com/audesirk
(click on the Table of Contents link and then select Chapter 26).

523

But all of these diverse structures share a common property: Their form is intimately related to their function. Natural selection has modified the shapes of cells, tissues, and organs in ways that reveal their purpose to the careful observer. Why do nerve cells have long, thin projections? Why is lung tissue so spongy? Why does our small intestine follow such a long, convoluted path? To answer these and a thousand other similar questions, just think for a moment about what those structures do.

1) How Do Animals Maintain Internal Constancy?

Each animal lives in a particular environment, its body surrounded by air or water. But only a few of the animal's cells are in direct contact with the external environment. The vast majority of cells are inside the animal's body, separated from the outside environment and bathed in liquid. The liquid is typically warm and slightly salty, and it forms a nurturing internal environment for the cells. This wet, salty internal environment is required for the cells' survival and must be maintained even if the animal's external environment consists of dry air or fresh, salt-free water. What's more, it's not enough that the internal environment simply be wet and salty. Instead, cells require a very specific, complex mixture of dissolved substances and a narrow temperature range. Because the body's cells cannot survive if the conditions of the internal environment deviate from a small range of acceptable states, the cells devote a large portion of their energy to actions that serve to keep the environment stable.

This "constancy of the interior milieu" was first recognized by the French physiologist Claude Bernard in the mid-nineteenth century. Later, in the 1920s, Walter B. Cannon coined the term **homeostasis** to describe the constancy of the body's internal environment. Although the word *homeostasis* (derived from Greek words meaning "to stay the same") implies a static, unchanging state, the internal environment in fact seethes with activity as the body continuously adjusts to internal and external changes.

The internal state of an animal body is thus better described as *dynamic equilibrium*. Many physical and chemical changes do occur, but the net result of all this activity is that physical and chemical parameters are kept within the narrow range that cells require to function. These equilibrium conditions are maintained by a host of mechanisms that are collectively known as *feedback systems*. *Negative feedback systems* counteract the effects of changes in the internal environment; less-common *positive feedback systems* reinforce changes when such reinforcement serves a physiological need.

Negative Feedback Reverses the Effects of Changes

The most important mechanism governing homeostasis is **negative feedback**, in which the response to a change is to counteract the change. In other words, an input stimulus causes an output response that "feeds back" to the initial input and decreases its effects. Because the initial change triggers a response that reverses its effects, the overall result is to return the system to its original condition. This kind of feedback is called "negative" because it reverses or negates the initial change.

A familiar example of negative feedback is your home thermostat (Fig. 26-1a). In a thermostat, an input stimulus (temperature drops below a set point) is detected by a thermometer, which activates an output response by switching on a heating device. That output restores the temperature to the *set point* (the thermostat setting— that is, the temperature you desire), and the heater is switched off. Continuously repeated on-off cycles keep your home's temperature near the set point. Note that the thermostat's negative feedback mechanism requires a *control center* with a set point, a *sensor* (the thermometer), and an *effector* (the furnace), which accomplishes the change.

Negative feedback mechanisms abound in physiological systems. For example, negative feedback keeps your body's temperature within about 1° of 98.6°F (37°C). The set point in this system is stored in a control center in your *hypothalamus*, a region of your brain that controls many homeostatic responses (Fig. 26-1b). Nerve endings in your hypothalamus, abdomen, spinal cord, skin, and large veins act as temperature sensors and transmit this information to the hypothalamus. When your body temperature drops, the hypothalamus activates various effector mechanisms that raise your body temperature, including shivering (which generates heat through muscular activity), constriction of blood supply to the skin (which reduces heat loss), and elevation of metabolic rate (which generates heat). When normal body temperature is restored, the hypothalamus switches off these temperature control mechanisms.

As you continue your study of animal physiology, you will find many additional examples of homeostatic control that operate by negative feedback. Negative feedback systems regulate blood oxygen content, water balance, blood sugar levels, and many other components of the "internal milieu."

Positive Feedback Drives Events to a Conclusion

In a **positive feedback** system, in contrast to a negative feedback system, a change produces a response that intensifies the original change (Fig. 26-1c). The end result is that change tends to proceed in the same direction as the initial stimulus (rather than reversing to return to a set point). Positive feedback, as you can imagine, tends

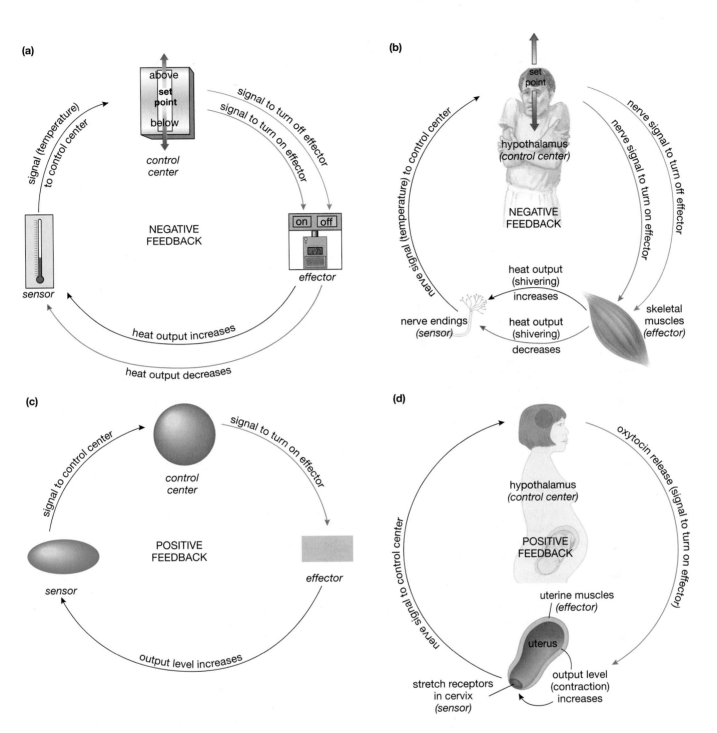

Figure 26-1 Negative and positive feedbacks
(a) In negative feedback systems, a sensor detects changes and relays information to a control center, which encodes a specific set point. The control center sends a signal to an effector, whose output returns the system to its set point. The sensor detects the restoration, relays the new information to the control center, which turns off the effector. **(b)** Body temperature is sensed by neurons in numerous locations throughout the body. These temperature readings are relayed to the hypothalamus (the control center with a set point). A drop in body temperature causes the hypothalamic neurons to send a signal to effectors such as muscles, which begin shivering. Shivering generates heat, which helps restore body temperature to its set point, shutting off the signal, and causing shivering to cease. **(c)** In positive feedback, a change in the system is sensed by a sensor and relayed to a control center with no set point. The control center then sends a signal to an effector. Rather than try to return the system to previous conditions, the effector's output causes an even greater change. **(d)** At the onset of labor, uterine contractions are sensed by stretch receptor neurons in the cervix, which send a signal to the hypothalamus that results in the secretion of a hormone (oxytocin) from the posterior pituitary gland. The hormone, in turn, stimulates more and stronger uterine contractions, ultimately leading to the expulsion of the baby and placenta from the uterus.

to create chain reactions that must somehow be controlled. For example, in nuclear fission, each particle that is split from an atom triggers the splitting of another atom, the pieces of which trigger the fission of other atoms, and so on. Controlled, the chain reaction supplies nuclear power. When deliberately set out of control, it produces an atomic explosion. A familiar biological example of positive feedback is population growth (as we shall see in Chapter 38); each offspring gives rise to still more offspring. The ecologist Paul Ehrlich coined the apt expression "population bomb" to describe unchecked population growth.

In physiological systems, events governed by positive feedback mechanisms are generally self-limiting and occur relatively infrequently. Positive feedback occurs, for example, during childbirth (Fig. 26-1d). The early contractions of labor begin to force the baby's head against the cervix, located at the base of the uterus; this pressure causes the cervix to dilate (open). Stretch-receptive neurons in the cervix respond to this expansion by signaling the hypothalamus, which responds by triggering the release of a hormone (oxytocin) that stimulates more and stronger uterine contractions. Stronger contractions create further pressure on the cervix, which in turn prompts the release of more hormones. The feedback cycle is finally terminated by the expulsion of the baby and its placenta.

The Body's Internal Systems Act in Concert

The constancy of an animal body's internal environment is maintained by a coordinated, integrated network of systems. The job of regulating a multitude of factors throughout the body cannot be accomplished by a few, independent feedback mechanisms. Instead, numerous mechanisms are constantly at work, responding to various stimuli that change continuously as the animal's activities and circumstances change. If all these control systems worked independently, without regard to the activities of all the other systems, it would not be possible to maintain homeostasis in the body as a whole.

Fortunately, evolution has ensured that the various systems work together. For example, the systems that take substances into the body act in concert with the systems responsible for removing substances from the body. This kind of coordinated action is possible because the body has mechanisms for moving substances and signals from one part of the body to another. Each cell in the body is indirectly connected to all of the others by an elaborate network of vessels and nerves that can carry molecules and messages to the appropriate locations. Molecules can be transported to the sites where they are needed and away from locations where their presence is harmful. Messages can be carried from sensors to effectors and back again.

2) How Is the Animal Body Organized?

The animal body is an engineering marvel. Using a single basic component, the cell, evolution has designed an astonishingly complex but completely self-regulating system that accomplishes hundreds of functions simultaneously. The parts fit together with a degree of precision and integration that human engineers can only dream of. This exquisite machine is made up of **tissues**, each tissue composed of dozens to billions of structurally similar cells that act in concert to perform a particular function (Fig. 26-2). Tissues are the building blocks of **organs**, such as the stomach, small intestine, kidneys, and urinary bladder. Organs in turn are organized into **organ systems**; for example, the digestive system includes the stomach, small intestine, large intestine, and other organs.

Animal Tissues Are Composed of Similar Cells That Perform a Specific Function

A tissue is composed of cells that are similar in structure and designed to perform a specialized function. Tissue may also include extracellular components produced by these cells, as in the case of cartilage and bone. We present a brief overview of the four major categories of animal tissue: (1) epithelial tissue; (2) connective tissue; (3) muscle tissue; and (4) nerve tissue.

Epithelial Tissue Forms Membranes That Cover the Body and Line Its Cavities

The cells of **epithelial tissues** (also called the *epithelium*) form continuous sheets known as **membranes**. Epithelial membranes cover the body and line body cavities. Thus, one surface of epithelial membranes is free; that is, it faces either the outside of the body or the inside of cavities within the body. Epithelial membranes create a barrier that either resists the movement of substances across it (such as the skin) or that allows the movement of specific substances across it (such as the lining of the small intestine). Epithelial membranes can serve as effective barriers because their cells are packed closely together and connected to one another by tight junctions (see Chapter 5). No blood vessels penetrate epithelial tissue; it is nourished by diffusion from capillaries within the connective tissue that lies beneath it.

Another important property of epithelial tissues is that they are continuously lost and replaced by mitotic cell division. Consider the abuse suffered, for example, by the epithelium that lines your mouth. Scalded by coffee and scraped by corn chips, it would be destroyed within a few days if it were not continuously replacing itself. The stomach lining, abraded by food and attacked by acids and protein-digesting enzymes, is completely replaced every 2 to 3 days. Your skin's outer layer, the epider-

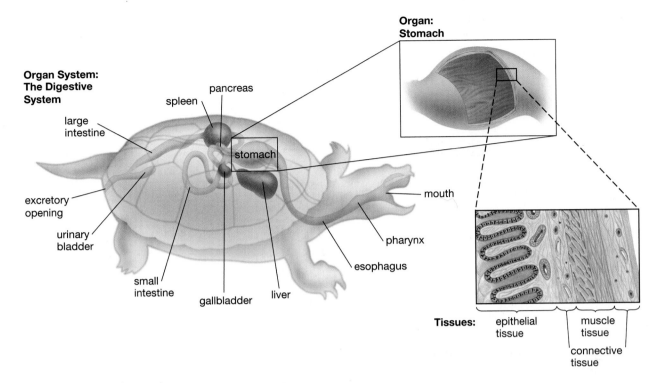

Figure 26-2 Organ systems, organs, and tissues
The animal body contains organ systems, each composed of several organs, which are in turn composed of several types of tissue.

mis, is renewed about twice a month. Epithelial tissues are classified according to the shape of their cells and the number of cell layers present, as summarized in Table 26-1.

Some Epithelial Tissues Also Form Glands

During development, some epithelial tissues fold inward, and their cells change shape and function and form **glands**, clusters of cells that are specialized to secrete (release) substances. Glands are classified into two broad categories: (1) exocrine glands and (2) endocrine glands. **Exocrine glands** remain connected to the epithelium by a passageway, or *duct*. Examples of exocrine glands are sweat glands and sebaceous (oil-secreting) glands; both types are found in the skin and are derived from skin epithelium (see Fig. 26-9). Exocrine glands called salivary glands release saliva into the mouth, and still other exocrine glands line the stomach, where they secrete a protective layer of mucus. **Endocrine glands** become separated from the epithelium that produced them. Most products of endocrine glands are hormones, which are secreted into the extracellular fluid that surrounds the glands and then diffuse into nearby capillaries. Endocrine glands and their hormones are covered in detail in Chapter 32.

Connective Tissues Have Diverse Structures and Functions

Connective tissues serve mainly to support and bind other tissues. Most connective tissues feature cells that are surrounded by large quantities of extracellular substances, typically secreted by the connective tissue cells themselves. Connective tissues can be placed into three main categories: (1) *loose connective tissue*, which include the layer that separates the skin from deeper structures; (2) *fibrous connective tissues*, which include *tendons* and *ligaments*; and (3) *specialized connective tissues*, which include *cartilage, bone, fat, blood*, and *lymph*. With the exception of blood and lymph, these connective tissues are interwoven with fibrous strands of an extracellular protein called **collagen**, secreted by the cells. Connective tissue underlies all epithelial tissue and contains capillaries and fluid-filled spaces that nourish the epithelium. Underlying the epidermis of the skin, for example, is connective tissue called the **dermis**, which is richly supplied with capillaries (see Fig. 26-9).

Loose connective tissue is characterized by a diffuse network of loosely woven fibers; it serves mainly to bind epithelial cells to underlying tissues and to cushion and support organs. In fibrous connective tissue, collagen fibers are densely packed in an orderly parallel arrange-

Table 26-1 Major Types of Epithelial Tissue

Tissue Type	Diagram	Properties and Function	Location in Body
Simple squamous epithelium	Thin, flattened cells, one layer thick	Thin layer that allows the movement of substances; allows diffusion and filtration to occur	Lines lung alveoli; forms capillary walls; lines blood vessels
Stratified squamous epithelium	Thin, flattened cells, many layers thick	Thick layer that is rapidly replaced by divisions of underlying cells; has protective function; may secrete mucus	Upper layer of skin; lines mouth, anal canal, vagina
Simple cuboidal epithelium	Cube-shaped cells, one layer thick	Thin layer; functions in absorption and secretion	Lines kidney tubules; has secretory role in salivary glands, thyroid, pancreas, and liver
Simple columnar epithelium	Elongated cells, one layer thick; nuclei lined up	Forms a thick layer that functions in secretion and absorption; may secrete mucus	Lines the esophagus, stomach, intestines, and uterus
Pseudostratified columnar epithelium	Elongated cells, one layer thick; nuclei at different levels	Typically possesses beating cilia; may secrete mucus; functions in trapping and transporting particles out of respiratory surfaces; moving sex cells	Lines the respiratory tract; lines the tubes of the reproductive system (oviducts, vas deferens)

ment—a design that gives **tendons** and **ligaments** the strength necessary for their respective functions of attaching muscles to bones and bones to bones. **Cartilage** is a flexible and resilient form of connective tissue that consists of widely spaced cells surrounded by a thick, nonliving matrix. The matrix is made of collagen secreted by the cartilage cells (Fig. 26-3). Cartilage covers the ends of bones at joints, provides the supporting framework for the respiratory passages, supports the ear and nose, and forms shock-absorbing pads between the vertebrae. **Bone** (Fig. 26-4) resembles cartilage but is hardened by deposits of calcium phosphate. Bone forms in concentric circles around a central canal, which contains a blood vessel. (We shall discuss cartilage and bone in depth in Chapter 34.) **Fat** cells, collectively called **adipose tissue** (Fig. 26-5), are specially modified for long-term energy storage (see Chapter 3). This tissue plays

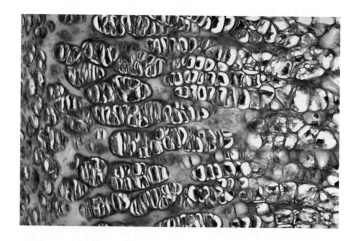

Figure 26-3 Cartilage
This specialized connective tissue supports the body and protects soft tissues. It is more flexible and less rigid than bone.

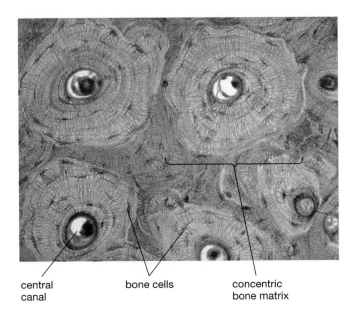

central canal   bone cells   concentric bone matrix

Figure 26-4 *Bone consists of cells embedded in a hard matrix*
Concentric circles of bone, deposited around a central canal that contains a blood vessel, are clearly visible in this micrograph. Bone cells appear as dark spots trapped in small chambers within the hard matrix the cells themselves deposit.

an especially important role in the physiology of animals adapted to cold environments, such as the penguins in the chapter-opening photo.

Although they are fluids, **blood** and **lymph** are considered connective tissues because the fluids of which they are composed is largely extracellular. The extracellular fluid of blood is *plasma,* and that of lymph is called lymph

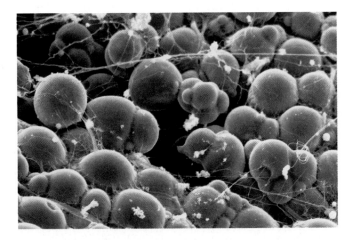

Figure 26-5 *Adipose tissue*
Adipose tissue, such as that shown here from a mouse, is made up almost exclusively of fat cells; very little matrix is present. Each fat cell contains a droplet of oil that takes up most of the volume of the cell.

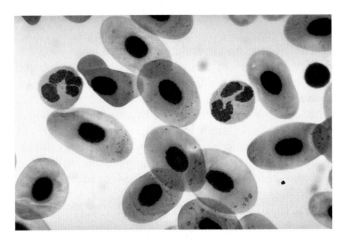

Figure 26-6 *Blood*
The most prominent cellular components of blood are red blood cells. Their primary function is to carry molecules, especially oxygen, to the appropriate body tissues. White blood cells are important in the immune response. The blood cells in this micrograph are from a frog. Note that the red cells have nuclei; mammalian red blood cells have no nuclei.

as well. The cellular portion of blood consists of red and white blood cells and cell fragments called *platelets* (Fig. 26-6). Blood carries dissolved oxygen and other nutrients to cells and carries away carbon dioxide and other cellular waste products. Blood also carries hormonal signals throughout the body. Lymph consists largely of fluid that has leaked out of blood capillaries and is carried back to the circulatory system within lymphatic vessels.

Muscle Tissue Has the Ability to Contract
The long, thin cells of muscle tissue (Fig. 26-7) contract (shorten) when stimulated, then relax passively. There are three types of muscle tissue: (1) skeletal; (2) cardiac;

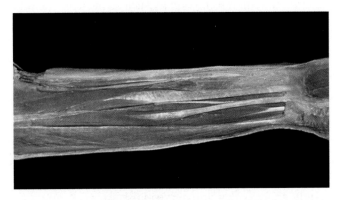

Figure 26-7 *Muscle tissue consists of contractile fibers*
Muscle fibers are composed of long, thin cells that shorten when stimulated. When the cells in a muscle fiber shorten, the fiber contracts and the muscle as a whole shortens. In animal bodies, muscle contractions accomplish work, such as moving limbs and pumping blood.

and (3) smooth. **Skeletal muscle** is generally under voluntary, or *conscious,* control. As its name implies, its main function is to move the skeleton, as occurs when you walk or turn the pages of this text. **Cardiac muscle** is located only in the heart. Unlike skeletal muscle, it is spontaneously active and not under conscious control. Cardiac muscle cells are interconnected by gap junctions, which allow electrical signals to spread rapidly through the heart. **Smooth muscle,** so named because it lacks the orderly arrangement of thick and thin filaments seen in cardiac and skeletal muscles, is embedded in the walls of the digestive tract, the uterus, the bladder, and large blood vessels. It produces slow, sustained contractions that are normally involuntary. Muscles and their contraction mechanism are covered in Chapter 34.

Nerve Tissue Is Specialized to Transmit Electrical Signals

Nerve tissue makes up the brain, the spinal cord, and the nerves that travel from them to all parts of the body. Nerve tissue is composed of two types of cells: (1) neurons and (2) glial cells. **Neurons** are specialized to generate electrical signals and to conduct these signals to other neurons, muscles, or glands. A neuron has four major parts, each with a specialized function (Fig. 26-8). The **dendrites** of a neuron receive signals from other neurons or from the external environment. The **cell body** directs the maintenance and repair of the cell. The **axon** conducts the electrical signal to its target cell, and the **synaptic terminals** transmit the signal to the target cell. **Glial cells** surround, support, and protect neurons and regulate the composition of the extracellular fluid, allowing neurons to function optimally. We shall discuss nerve tissue more fully in Chapter 33.

Organs Include Two or More Interacting Tissue Types

Organs are formed from at least two tissue types that function together; some organs, such as the skin, include all four of the tissue types described earlier. In this section, we examine the components and functions of a representative animal organ, the skin.

The structure of the skin is, in a general sense, representative of many organs. An outer epithelium is underlain by connective tissue that contains a blood supply, a nerve supply, in some cases muscle, and glandular structures derived from the epithelium. If an organ is hollow, such as the bladder or blood vessels, its interior is also lined with epithelium underlain by connective tissue. Different organs have different types and proportions of glandular, muscular, and nervous tissues.

The Skin Illustrates the Properties of Organs

The **epidermis,** or outer layer of the skin, is a specialized epithelial tissue (Fig. 26-9). It is covered by a protective layer of dead cells produced by underlying living epi-

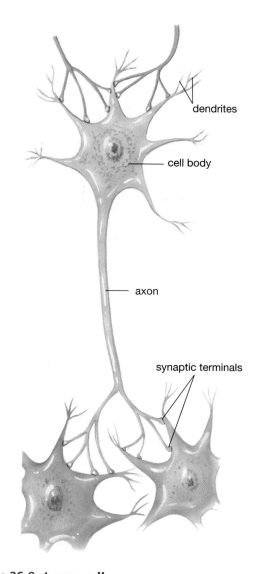

dendrites

cell body

axon

synaptic terminals

Figure 26-8 A nerve cell
A nerve cell, or neuron, has four major parts, each specialized for a specific function.

dermal cells. These dead cells are packed with the protein **keratin,** which helps keep the skin both airtight and relatively waterproof.

Immediately beneath the epidermis lies a layer of connective tissue, the dermis. The loosely packed cells of the dermis are permeated by *arterioles* (small arteries). Arterioles feed blood pumped from the heart into a dense meshwork of capillaries that nourish both the dermal and epidermal tissue and empty into a network of *venules* (small veins) in the dermis. Loss of heat through the skin is precisely regulated by neurons controlling the degree of dilation (expansion) of the arterioles. When cooling is required, the arterioles dilate and flood the capillary beds with blood, thus releasing excess heat; when heat conservation is required, the arterioles supplying the skin capillaries are constricted. Lymph vessels collect and

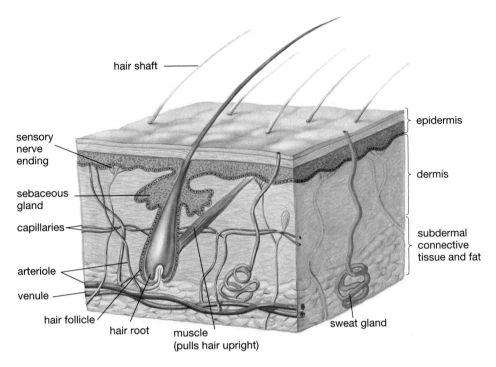

Figure 26-9 Skin
Mammalian skin, a representative organ, in cross section.

carry off extracellular fluid within the dermis. Various sensory nerve endings responsive to temperature, touch, pressure, vibration, and pain are scattered throughout the dermis and epidermis and provide feedback to the nervous system.

The dermis is also packed with glands derived from epithelial tissue. Glands called **hair follicles** produce hair from protein-containing secretions. Sweat glands produce watery secretions that cool the skin and excrete substances such as salts and urea. **Sebaceous glands** secrete an oily substance (*sebum*) that lubricates the epithelium.

In addition to the epithelial, connective, and nerve tissues already mentioned, the skin also contains muscle tissue. Tiny muscles attached to the hair follicles can cause the hairs of the skin to "stand on end" in response to signals from motor neurons. Although this reaction is useless for heat retention in humans, most mammals are able to increase the thickness of their insulating fur in cold weather by erecting individual hairs.

Organ Systems Consist of Two or More Interacting Organs

Organ systems consist of two or more individual organs (in some cases located in different regions of the body) that work together, performing a common function. An example is the digestive system, in which the mouth, esophagus, stomach, intestines, and other organs that supply digestive enzymes, such as the liver and pancreas, all function together to convert food into nutrient molecules (see Fig. 26-2). The major organ systems of the vertebrate body and their representative organs and functions are listed in Table 26-2. The structure and physiology of these organ systems are the subjects of the following nine chapters.

Table 26-2 Major Vertebrate Organ Systems

Organ System	Major Structures	Physiological Role	Organ System	Major Structures	Physiological Role
Circulatory system	Heart, blood vessels, blood	Transports nutrients, gases, hormones, metabolic wastes; also assists in temperature control	Endocrine system	A variety of hormone-secreting glands, including the hypothalamus, pituitary, throid, pancreas, and adrenals	Controls physiological processes, typically in conjunction with the nervous system
Lymphatic/immune lsystem	Lymph, lymph nodes and vessels, white blood cells	Carries fat and excess fluids to blood; destroysinvading microbes	Nervous system	Brain, spinal cord, peripheral nerves	Controls physiological processes in conjunction with the endocrine system; senses the environment, directs behavior
Digestive system	Mouth, esophagus, stomach, small and large intestines, glands producing digestive secretions	Supplies the body with nutrients that provide energy and materials for growth and maintenance	Muscular system	Skeletal muscle / Smooth muscle / Cardiac muscle	Moves the skeleton / Controls movement of substances through hollow organs (digestive tract, large blood vessels) / Initiates and implements heart contractions
Excretory system	Kidneys, ureters, bladder, urethra	Maintains homeostatic conditions within bloodstream; filters out cellular wastes, certain toxins, and excess water and nutrients	Skeletal system	Bones, cartilage, tendons, ligaments	Provides support for the body, attachment sites for muscles, and protection for internal organs
Respiratory system	Nose, trachea, lungs (mammals, birds, reptiles, amphibians), gills (fish and some amphibians)	Provides an area for gas exchange between the blood and the environment; allows oxygen acquisition and carbon dioxide elimination	Reproductive system	Males: testes, seminal vesicles, penis / Female (mammal): ovaries, oviducts, uterus, vagina, mammary glands	Male: produces sperm, inseminates female / Female (mammal): Produces egg cells, nurtures developing offspring

Summary of Key Concepts

1) How Do Animals Maintain Internal Constancy?

Homeostasis refers to the tendency of many physiological processes to maintain an organism's internal conditions within a narrow range that permits the continuation of life. These conditions are maintained through negative feedback, in which a change triggers a response that counteracts the change and restores conditions to a set point. Temperature control as well as many hormone systems use negative feedback to maintain homeostasis. Positive feedback, in which a change initiates events that intensify the change, occurs relatively rarely and is self-limiting. For example, the uterine contractions that lead to childbirth are driven by positive feedback. The animal body contains many feedback mechanisms that work in concert as a coordinated, integrated system.

2) How Is the Animal Body Organized?

The animal body is composed of organ systems consisting of two or more organs. Organs, in turn, are made up of tissues. A tissue is a group of cells and extracellular material that forms a structural and functional unit and is specialized for a specific task. Animal tissues include epithelial, connective, muscle, and nerve tissue.

Epithelial tissue forms membranous coverings over internal and external body surfaces and also gives rise to glands. Connective tissue normally contains considerable extracellular material and includes dermal tissue, bone, cartilage, tendons, ligaments, fat, and blood. Muscle tissue is specialized for movement. There are three types of muscle tissue: skeletal, cardiac, and smooth. Nerve tissue is specialized for the generation and conduction of electrical signals.

Organs include at least two tissue types that function together. Mammalian skin is a representative organ. The epidermis, an epithelial tissue, covers and protects the dermis beneath it. The dermis contains blood and lymph vessels, a variety of glands, and tiny muscles that erect the hairs. Animal organ systems include the digestive, excretory, immune, respiratory, circulatory/lymphatic, nervous, muscular, skeletal, endocrine, and reproductive systems, summarized in Table 26-2.

Key Terms

adipose tissue *p. 528*	dendrite *p. 530*	hair follicle *p. 531*	organ *p. 526*
axon *p. 530*	dermis *p. 527*	homeostasis *p. 524*	organ system *p. 526*
blood *p. 529*	endocrine gland *p. 527*	keratin *p. 530*	positive feedback *p. 524*
bone *p. 528*	epidermis *p. 530*	ligament *p. 528*	sebaceous gland *p. 530*
cardiac muscle *p. 530*	epithelial tissue *p. 526*	lymph *p. 529*	skeletal muscle *p. 530*
cartilage *p. 528*	exocrine gland *p. 527*	membrane *p. 526*	smooth muscle *p. 530*
cell body *p. 530*	fat *p. 528*	negative feedback *p. 524*	synaptic terminal *p. 530*
collagen *p. 527*	gland *p. 527*	nerve tissue *p. 530*	tendon *p. 528*
connective tissue *p. 527*	glial cell *p. 530*	neuron *p. 530*	tissue *p. 526*

Thinking Through the Concepts

Multiple Choice

1. *The skin contains*
   a. epithelial tissue
   b. connective tissue
   c. nerve tissue
   d. muscle tissue
   e. all of the above

2. *Glands that become separated from the epithelium that produced them are called _____ glands.*
   a. sebaceous
   b. sweat
   c. exocrine
   d. endocrine
   e. saliva

3. *Epithelial membranes*
   a. cover the body
   b. line body cavities
   c. may create barriers that alter the movement of certain substances
   d. are continuously replaced by cell division
   e. all of the above

4. *All of the following are examples of connective tissue EXCEPT*
   a. tendons
   b. ligaments
   c. blood
   d. muscle
   e. adipose tissue

5. *Which of the following statements about muscle is true?*
   a. Smooth muscle is important in locomotion.
   b. Skeletal muscle is not under conscious control.
   c. Cardiac muscle utilizes gap junctions.
   d. Smooth muscle is called voluntary muscle.
   e. Smooth muscle moves the skeleton.

6. *All of the following are found in the dermis EXCEPT*
   a. arteries
   b. sensory nerve endings
   c. hair follicles
   d. sebaceous glands
   e. cells packed with keratin

? Review Questions

1. Define *homeostasis,* and explain how negative feedback helps maintain it. Explain one example of homeostasis in the human body.

2. Explain positive feedback, and provide one physiological example. Explain why this type of feedback is relatively rare in physiological processes.

3. Explain why body temperature in humans cannot be maintained at *exactly* 37°C (98.6°F) at all times.

4. Describe the structure and functions of epithelial tissue.

5. What property distinguishes connective tissue from all other tissue types? List five types of connective tissue, and briefly describe the function of each type.

6. Describe the skin, a representative organ. Include the various tissues that compose it and the role of each tissue.

Applying the Concepts

1. Why does life on land present more difficulties in maintaining homeostasis than does life in water? What made it evolutionarily advantageous for organisms to colonize dry land?

2. The majority of homeostatic regulatory mechanisms in animals are "autonomic"—that is, not requiring conscious control. Discuss several reasons why this type of regulation is more advantageous to the animal than is conscious regulation of homeostatic controls.

3. Third-degree burns are usually painless. Skin regenerates only from the edges of these wounds. Second-degree burns regenerate from cells located at the burn edges, in hair follicles and in sweat glands. First-degree burns are painful but heal rapidly from undamaged epidermal cells. From this information, draw the depth of first-, second-, and third-degree burns on Figure 26-9.

4. A coroner dictates the following description during an autopsy: "The tissue I am looking at forms part of the fetal skeleton. The matrix appears transparent. Fibers of collagen are present but are small and evenly dispersed in the matrix. Chondrocytes appear in tiny spaces, *lacunae,* within the matrix. Blood vessels have not yet penetrated the matrix." What tissue is the coroner describing?

5. The pancreas is both an exocrine and an endocrine gland. During embryonic development, it appears as an outgrowth of the rudimentary digestive tube. Sketch the stages you might see as this compound organ develops.

6. Imagine you are a health-care professional teaching a prenatal class for fathers. Design a real-world analogy with sensors, electrical currents, motors, and so on to illustrate feedback relationships involved in the initiation of labor that a layperson could understand.

Group Activity

The physiological systems that arise as a result of natural selection often seem as if they have been designed to achieve their functions. This seemingly intentional design in natural systems is an illusion, but the human process of intentional design can nonetheless help us understand how physiological systems work. To try this for yourself, form a group of four students. Your group is part of a larger team that is trying to design a self-regulating robot vehicle. You've been given two tasks. First, design a system that will keep the temperature of the lubricant in the vehicle's wheel-bearings between 38°C and 43°C (100°F and 110°F), no matter how fast the vehicle is going. Second, design a system that will ensure that the vehicle swerves to avoid obstacles. Your designs may include any imaginable sensors, effectors, and control centers but must be totally self-contained. Submit a diagram and a brief written description of your designs.

For More Information

Degabriele, R. "The Physiology of the Koala." *Scientific American*, June 1980 (Offprint No. 1476). Koalas eat poisonous eucalyptus leaves and almost never drink water. How can they survive?

Nuland, S. *The Wisdom of the Body*. New York: Alfred A. Knopf, 1997. Human physiology as seen through a surgeon's eyes. A first-hand account of the beauty and power of the body's mechanisms for maintaining homeostasis.

Pool, R. "Saviors." *Discover*, May 1998. Many seriously ill individuals die while waiting for an organ transplant. Thanks to genetic engineering, organ donors of the future may be raised on a farm.

Schmidt-Nielson, K. "The Physiology of the Camel." *Scientific American*, December 1959. Do camels really store water in their humps? How long can one go without drinking?

Schmidt-Nielson, K., and Schmidt-Nielson, B. "The Desert Rat." *Scientific American*, July 1953 (Offprint No. 1050). The ultimate desert dweller, the kangaroo rat never needs to drink water.

Storey, K. B., and Storey, J. M. "Frozen and Alive." *Scientific American*, December 1990. Some animals have special adaptations that allow them to withstand freezing.

Wills, Christopher. "The Skin We're In." *Discover*, November 1994. A scientist looks at the controversial issue of skin color.

Answers to Multiple-Choice Questions
1. e 2. c 3. e 4. d 5. c 6. e

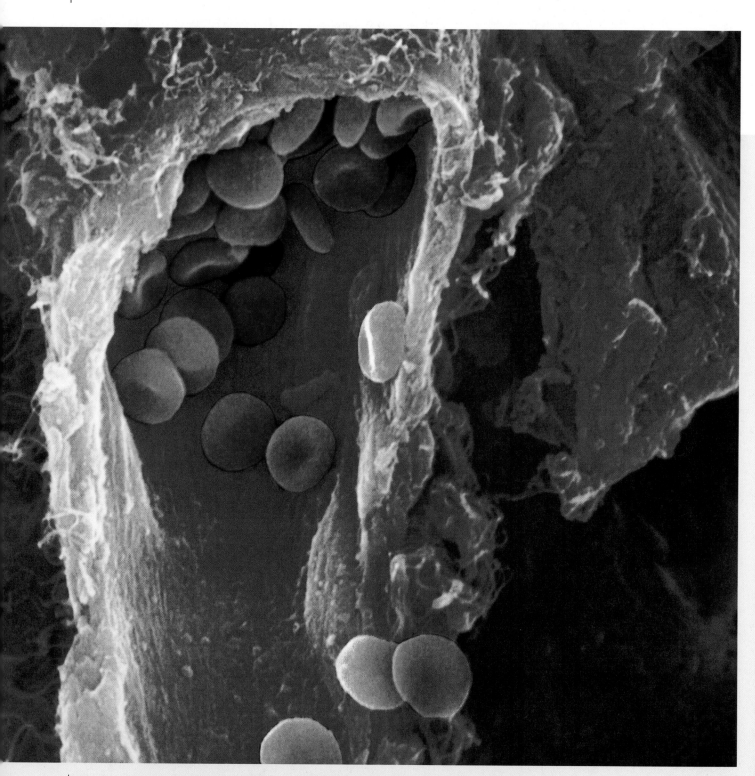

Red blood cells flow through a vein in this scanning electron micrograph. In its 4-month life span, a red blood cell will travel over 900 miles through the circulatory system, ferrying oxygen to cells throughout the body.

Circulation 27

Net Watch

On-line resources for this chapter are on the World Wide Web at:
http://www.prenhall.com/audesirk
(click on the Table of Contents link and then select Chapter 27).

Rest your hand on your desk, contract your upper arm muscle—hard—and then relax it. Repeat this sequence about once a second for a minute or so. Imagine continuing this all day and night, day after day, 2.5 billion times without stopping. Now you have a vague idea of how impressive a muscle your heart is. But your heart is impressive not only in its strength and endurance—it is marvelously complex in structure. Its four separate chambers alternately fill and empty, receiving blood from large vessels and delivering it first to the lungs, then throughout the rest of the body. The contractions of the chambers are precisely timed, and blood flow is maintained in the proper direction. Further, the rate of contractions is finely coordinated with your body's demands for oxygen: speeding up during exercise or emergencies, slowing during rest.

What causes your heart to beat steadily, and how is the rate of contraction changed? What keeps the chambers contracting in proper order? Why doesn't the blood flow back into the vessels that delivered it? How is your heart designed so that blood low in oxygen goes to the lungs, and blood high in oxygen goes to the rest of the body? If you read this introduction again after you've read this chapter, the answers will be a lot clearer. But first, let's look at some more fundamental questions: Why do animals have circulatory systems, and what are some specific functions of the vertebrate circulatory system? Why has it aptly been called "the river of life"?

1) What Are the Major Features and Functions of Circulatory Systems?

Billions of years ago, the first cells were nurtured by the sea in which they evolved. The sea brought them nutrients, which diffused into the cells, and washed away the wastes, which had diffused out. Today microorganisms and some simple multicellular animals still rely almost exclusively on diffusion for the exchange of nutrients and wastes with the environment. Sponges, for example, circulate seawater through pores in their bodies, bringing the environment close to each cell. As larger, more-complex animals evolved, individual cells became increasingly distant from the outside world. The constant demands of a cell require, however, that diffusion distances be kept short so that adequate nutrients reach the cell and that the cell isn't poisoned by its own wastes. With the evolution of the circulatory system, a sort of "internal sea" was created, serving the same purpose as the sea did for the first cells. This internal sea transports a fluid (blood) rich in food and oxygen close to each cell and carries away wastes produced by the cells.

All circulatory systems have three major parts:

1. A fluid, **blood**, that serves as a medium of transport.
2. A system of channels, or **blood vessels**, that conduct the blood throughout the body.
3. A pump, the **heart**, that keeps the blood circulating.

Animals Have One of Two Types of Circulatory Systems

Two major types of circulatory systems are found in animals: (1) open and (2) closed. **Open circulatory systems** are present in many invertebrates, including arthropods such as crustaceans, spiders, and insects and mollusks such as snails and clams. Animals with an open circulatory system have a heart, blood vessels, and a large open space within the body called a **hemocoel**. The heart pumps blood through vessels that release the blood into the hemocoel and pick it up from the hemocoel for delivery back to the heart (Fig. 27-1a). Within the hemocoel (which may occupy 20% to 40% of the body volume) tissues and internal organs are directly bathed in blood.

Closed circulatory systems are present in invertebrates such as the earthworm and very active mollusks such as squid and octopuses. Closed circulatory systems are also a characteristic of all vertebrates, including humans. In closed circulatory systems, the blood (whose volume is only 5% to 10% of body volume) is confined to the heart and a continuous series of blood vessels (Fig. 27-1b). Closed circulatory systems allow more-rapid blood flow, more-efficient transport of wastes and nutrients, and higher blood pressure than is possible in open systems.

In this chapter we look at the different parts and functions of the vertebrate circulatory system. We shall focus on the human as a representative vertebrate.

The Vertebrate Circulatory System Has Many Diverse Functions

The circulatory system has many roles and is most highly developed in the vertebrates. Among the most important functions of the vertebrate circulatory system are the following:

• Transport of oxygen from the lungs or gills to the tissues and of carbon dioxide from the tissues to the lungs or gills.

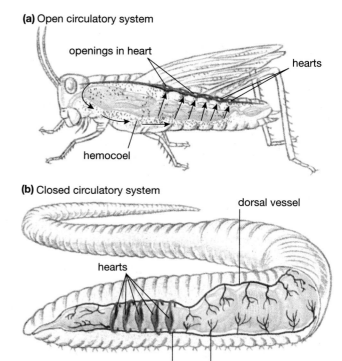

(a) Open circulatory system

openings in heart

hearts

hemocoel

(b) Closed circulatory system

dorsal vessel

hearts

ventral vessel   capillary network

Figure 27-1 Open and closed circulatory systems
(a) In the open circulatory system of insects and other arthropods, a series of hearts pumps blood through vessels into the hemocoel, where blood directly bathes the other organs. When the hearts relax, blood is sucked back into them through openings guarded by one-way valves. When the hearts contract, the valves are pressed shut, forcing the blood to travel out through the vessels and back to the hemocoel. **(b)** In a closed circulatory system, blood remains confined to the heart and the blood vessels. In the earthworm, five contractile vessels serve as hearts and pump blood through major ventral and dorsal vessels from which smaller vessels branch.

• Distribution of nutrients from the digestive system to all body cells.
• Transport of waste products and toxic substances to the liver, where many of them are detoxified, and to the kidney for excretion.
• Distribution of hormones from the glands and organs that produce them to the tissues on which they act.
• Regulation of body temperature, which is achieved partly by adjustments in blood flow.
• Prevention of blood loss by means of the clotting mechanism.
• Protection of the body from bacteria and viruses by circulating antibodies and white blood cells.

In the following sections we examine the three parts of the circulatory system: (1) the heart, (2) the blood, and (3) the vessels, with an emphasis on the human system. Finally, we describe the lymphatic system, which works closely with the circulatory system.

2) What Are the Features and Functions of the Vertebrate Heart?

Increasingly Complex and Efficient Hearts Have Arisen during Vertebrate Evolution

No circulatory system can be constructed without a dependable pump. Blood must be moved through the body continuously throughout an animal's life. The vertebrate heart consists of muscular chambers capable of strong contractions. Chambers called **atria** (singular, **atrium**) collect blood. Atrial contractions send blood into the **ventricles**, chambers whose contractions circulate blood through the body. During the course of vertebrate evolution, the heart has become increasingly complex, with more separation between oxygenated blood (which has picked up oxygen from the lungs or gills) and deoxygenated blood (which, in passing through body tissues, has lost oxygen).

The hearts of fishes, the first vertebrates to evolve, consist of sequential contractile chambers, with a single atrium that empties into a single ventricle. Blood pumped from the ventricle passes first through the gill *capillaries*, thin-walled vessels where it picks up oxygen, and then to the rest of the body (Fig. 27-2a).

Over evolutionary time, as fish gave rise to amphibians and amphibians to reptiles, a three-chambered heart evolved, consisting of two atria and one ventricle. In the three-chambered hearts of amphibians and most reptiles, deoxygenated blood from the body is delivered into the right atrium, and blood from the lungs into the left atrium. Both atria empty into the single ventricle. Although some mixing occurs there, the deoxygenated blood tends to remain in the right portion of the ventricle and be pumped into vessels that enter the lungs, while most of the oxgenated blood remains in the left portion of the ventricle and is pumped to the rest of the body (Fig. 27-2b). The separation is enhanced by a partial wall between the right and left portions of the ventricle in reptiles.

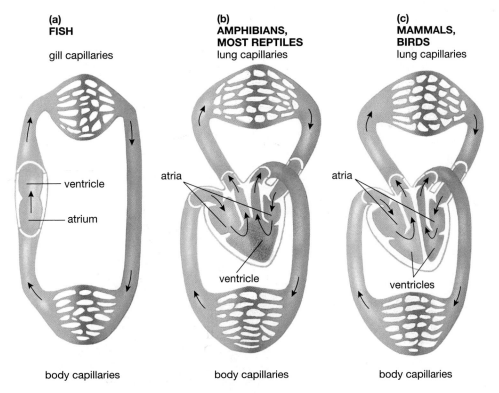

(a) FISH

gill capillaries

ventricle

atrium

body capillaries

(b) AMPHIBIANS, MOST REPTILES

lung capillaries

atria

ventricle

body capillaries

(c) MAMMALS, BIRDS

lung capillaries

atria

ventricles

body capillaries

Figure 27-2 The evolution of the vertebrate heart
(a) The earliest vertebrate heart is represented by the two-chambered heart of fishes. Blood from the body tissues is collected in the atrium and transferred to the ventricle. Contraction of the ventricle sends blood through the gill capillaries, where it picks up oxygen and gives off carbon dioxide, and then to the body capillaries, where it delivers oxygen to the tissues and picks up carbon dioxide. *(b)* In amphibians and most reptiles, the heart has two atria—the left receiving oxygenated blood from the lungs, the right receiving deoxygenated blood from body tissues. Both atria empty into a single ventricle. Many reptiles have a partial wall down the middle of the ventricle. *(c)* The hearts of birds and mammals are actually two separate pumps, so mixing of oxygenated and deoxygenated blood is prevented.

The warm-blooded birds and mammals have high metabolic demands and require more-efficient delivery of oxygen to their tissues than do cold-blooded animals. That demand is met by the four-chambered heart (Fig. 27-2c). The complete separation of oxygenated and deoxygenated blood, which is made possible by separate right and left ventricles, ensures that blood that reaches the tissues has the highest possible oxygen content.

The Vertebrate Heart Consists of Muscular Chambers Whose Contraction Is Controlled by Electrical Impulses

Human hearts (and those of other mammals and birds) can be considered as two separate pumps, each with two chambers. In each pump, an atrium receives and briefly stores the blood, passing it to a ventricle that propels it through the body (Fig. 27-3). One pump is for **pulmonary circulation** and consists of the right atrium and ventricle. Oxygen-depleted blood from the body empties into the right atrium through a large **vein** (a vessel that carries blood *toward* the heart) called the *superior vena cava*. The right atrium contracts, transferring the blood to the right ventricle. Contraction of the right ventricle sends the oxygen-depleted blood to the lungs via pulmonary **arteries** (vessels that carry blood *away from* the heart). The other pump, consisting of the left atrium and ventricle, powers **systemic circulation**. Newly oxygenated blood

from the lungs enters the left atrium through pulmonary veins and is passed to the left ventricle. Strong contractions of the left ventricle, the heart's most muscular chamber, send the oxygenated blood coursing out through a major artery, the *aorta*, to the rest of the body.

The Coordinated Contractions of Atria and Ventricles Produce the Cardiac Cycle

The alternating contraction and relaxation of the heart chambers is called the **cardiac cycle**. The two atria contract in synchrony, emptying their contents into the ventricles. A fraction of a second later, the two ventricles contract simultaneously, forcing blood into arteries that exit the heart. Both atria and ventricles then relax briefly before the cycle is repeated (Fig. 27-4). At normal resting heart rate, the cardiac cycle lasts just under 1 second. In determining blood pressure (Fig. 27-5), the higher of the two readings (called *systolic pressure*) is measured during ventricular contraction and the lower reading (called *diastolic pressure*) is measured between contractions.

Valves Maintain Directionality of Blood Flow, Whereas Electrical Impulses Coordinate the Sequence of Contractions

Coordinating the activity of the four chambers to maintain proper blood flow through the heart and its vessels presents some challenges. First, when the ventricles contract, the blood must be directed out through the arter-

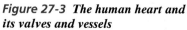

Figure 27-3 The human heart and its valves and vessels
The heart is drawn as if it were in a body facing you, so that right and left appear reversed. The right atrium receives deoxygenated blood and passes it to the right ventricle, which pumps it to the lungs. Blood returning from the lungs enters the left atrium, which passes it to the left ventricle, which pumps oxygenated blood through the rest of the body. Note the thickened walls of the left ventricle, which must pump blood much farther than does the right ventricle. One-way valves are located between the aorta and the left ventricle, between the pulmonary artery and the right ventricle, and between the atria and ventricles.

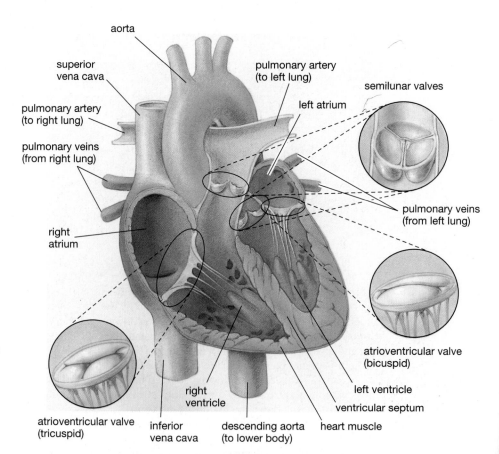

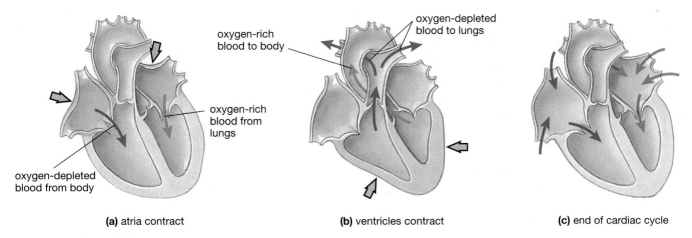

(a) atria contract   (b) ventricles contract   (c) end of cardiac cycle

Figure 27-4 The cardiac cycle
(a) First the atria contract, forcing blood into the ventricles. **(b)** Next, the ventricles contract, forcing blood out through arteries to the lungs and the rest of the body. **(c)** The cardiac cycle ends as all chambers of the heart relax and as blood refills the atria and begins to flow passively into the ventricles.

ies and not back up into the atria. Then, once blood has entered the arteries, it must be prevented from flowing back as the heart relaxes. These problems are solved by four simple one-way *valves* (see Fig. 27-3). Pressure in one direction opens them readily, but reverse pressure forces them tightly closed. **Atrioventricular valves** separate the atria from the ventricles; a **tricuspid valve** (from Latin, meaning "three-pointed" valve) separates the right

ventricle and right atrium; and a **bicuspid valve** (Latin, "two-pointed") lies between the left atrium and the left ventricle. Two **semilunar valves** (Latin, "half-moon") allow blood to enter the pulmonary artery and the aorta when the ventricles contract but prevent it from returning as the ventricles relax.

A second challenge is to create smooth, coordinated contractions of the muscle cells that make up each

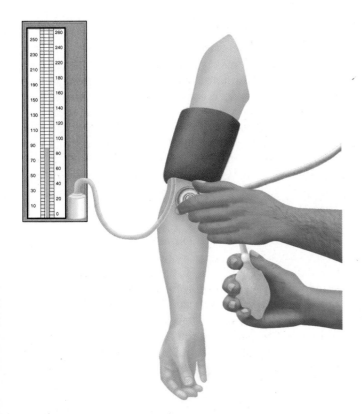

Figure 27-5 Measuring blood pressure
Blood pressure is measured with an inflatable blood pressure cuff and a stethoscope. The cuff is placed around the upper arm, the stethoscope over the artery just below the cuff. The cuff is inflated until its pressure closes off the arm's main artery, blood ceases to flow, and no pulse can be detected below the cuff. Then the pressure is gradually reduced. When the pulse is first audible in the artery, the pressure pulses created by the contracting left ventricle are just overcoming the pressure in the cuff and blood is flowing. This is the upper reading: the systolic pressure. Cuff pressure is then further reduced until no pulse is audible, indicating that blood is flowing continuously through the artery and that the pressure between ventricular contractions is just overcoming the cuff pressure. This is the lower reading: the diastolic pressure. A blood pressure reading of 130 (systolic) over 80 (diastolic) is within the normal range. The numbers are in millimeters of mercury, a standard measure of pressure also used in barometers.

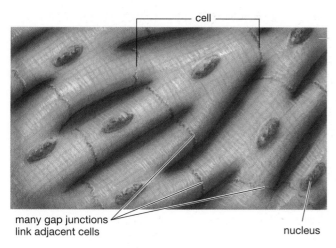

cell

many gap junctions
link adjacent cells

nucleus

Figure 27-6 The structure of cardiac muscle
Cardiac muscle cells are branched. Adjacent plasma membranes meet in folded areas that are densely packed with gap junctions (pores), which connect the interiors of adjacent cells. This arrangement allows direct transmission of electrical signals between the cells, coordinating their contractions.

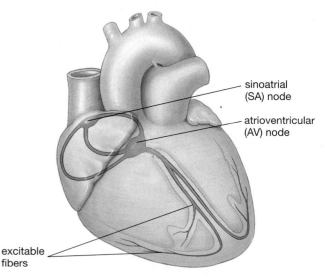

sinoatrial
(SA) node

atrioventricular
(AV) node

excitable
fibers

Figure 27-7 The heart's pacemaker and its connections
The sinoatrial (SA) node, a spontaneously active mass of modified muscle fibers in the right atrium, serves as the heart's pacemaker. The signal to contract spreads from the SA node through the muscle fibers of both atria, finally exciting the atrioventricular (AV) node in the lower right atrium. The AV node then transmits the signal to contract through bundles of excitable fibers that stimulate the ventricular muscle.

chamber. Muscle cells produce electrical signals that cause contraction. Individual heart muscle cells communicate directly with one another through gap junctions in their adjacent plasma membranes (Fig. 27-6). These connecting pores, introduced in Chapter 5, allow the electrical signals that cause contraction to pass freely and rapidly between heart cells. The contraction of the heart is initiated and coordinated by a **pacemaker**, a cluster of specialized heart muscle cells that produce spontaneous electrical signals at a regular rate. Although the nervous system can alter the rate of these signals, they are initiated by the pacemaker muscle cells themselves. The heart's primary pacemaker is the **sinoatrial (SA) node**, located in the wall of the right atrium (Fig. 27-7). Signals from the SA node spread rapidly through both the right and left atria, causing the atria to contract in smooth synchrony.

A final challenge is to coordinate contractions of all four chambers. The atria must contract first and empty their contents into the ventricles so that the atria can refill while the ventricles contract. Thus, there must be a delay between the contractions of the atria and those of the ventricles. From the SA node, an electrical impulse creates a wave of contraction that sweeps through the muscles of the atria until it reaches a barrier of unexcitable tissue between the atria and the ventricles. There the excitation is channeled through the **atrioventricular (AV) node**, a small mass of specialized muscle cells located in the floor of the right atrium (see Fig. 27-7). The impulse is delayed at the AV node, postponing the ventricular contraction for about 0.1 second after the atria contract. This delay gives the atria time to complete the

transfer of blood into the ventricles before ventricular contraction begins. From the AV node, the signal to contract spreads to the base of the two ventricles along tracts of excitable fibers. The impulse then travels rapidly from these fibers through the communicating muscle fibers, causing the ventricles to contract in unison.

When the pacemaker fails to coordinate muscle contractions, uncoordinated, irregular contractions impair the heart's function, a condition known as **fibrillation**. Fibrillation of the ventricles can be fatal, because blood is not pumped out of the heart to the brain and other organs but is merely sloshed around. A defibrillating machine applies a jolt of electricity to the heart, synchronizing the contraction of the ventricular muscle and sometimes allowing the pacemaker to resume its normal coordinating function.

The Nervous System and Hormones Influence Heart Rate

Left on its own, the SA node pacemaker would maintain a steady rhythm of about 100 beats per minute. However, the heart rate is significantly altered by the influence of nerve impulses and hormones. In a resting individual, activity of the parasympathetic nervous system (which controls body functions during periods of rest; see Chapter 33) slows the heart rate to about 70 beats per minute. When exercise or stress creates a demand for greater blood flow to the muscles, the parasympathetic influence is reduced and the sympathetic nervous system (which prepares the body for emergency action) accelerates the heart rate. Likewise, the hormone epinephrine (also known as adrenaline) increases the heart rate while mobilizing the entire body for response to threatening or unfamiliar events. When astronauts were landing on the Moon, their heart rates were over 170 beats per minute, even though they were sitting still!

3) What Are the Features and Functions of Blood?

Blood is the medium in which dissolved nutrients, gases, hormones, and wastes are transported through the body. It has two major components: (1) a fluid called **plasma** and (2) specialized cells (*red blood cells*, *white blood cells*, and *platelets*) that are suspended in the plasma (Table 27-1). On the average, the cellular components of blood account for 40% to 45% of its volume; the other 55% to 60% is plasma. The average human has 5 to 6 liters of blood, constituting about 8% of the total body weight.

Plasma Is Primarily Water in Which Proteins, Salts, Nutrients, and Wastes Are Dissolved

About 90% of the straw-colored plasma is water in which a number of substances are dissolved. Those substances include proteins, hormones, nutrients (glucose, vitamins, amino acids, lipids), gases (carbon dioxide, oxygen), salts (sodium, calcium, potassium, magnesium), and wastes such as urea. Plasma proteins are the most abundant of the dissolved substances. The three major plasma proteins are (1) *albumins*, which help maintain the blood's osmotic pressure (which controls the flow of water across plasma membranes); (2) *globulins*, which transport nutrients and play a role in the immune system; and (3) *fibrinogen*, important in blood clotting, which is discussed later in this chapter.

Table 27-1 Human Blood Cells

Cell Type		Description	Average Number Present	Major Function
Red blood cell (erythrocyte)		Biconcave disk without nucleus, about one-third hemoglobin; approximately 8 μm[b] in diameter	5,000,000 per mm³[a]	Transports oxygen and a small amount of carbon dioxide
White blood cells (leukocytes)			7500 per mm³	
1. Neutrophil		About twice the size of red blood cells; nucleus with two to five lobes	62% of white blood cells	Destroys relatively small particles by phagocytosis
2. Eosinophil		About twice the size of red blood cells; nucleus with two lobes	2% of white blood cells	Inactivates inflammation-producing substances; attacks parasites
3. Basophil		About twice the size of red blood cells; nucleus with two lobes	Less than 1% of white blood cells	Releases anticoagulant to prevent blood clots; releases histamine, which causes inflammation
4. Monocyte		Two to three times larger than red blood cells; nuclear shape varies from round to lobed	3% of white blood cells	Gives rise to macrophage, which destroys relatively large particles by phagocytosis
5. Lymphocyte		Only slightly larger than red blood cell; nucleus nearly fills cell	32% of white blood cells	Functions in the immune response
Platelet		Cytoplasmic fragment of cells in bone marrow called megakaryocytes	250,000 per mm³	Important in blood clotting

[a]mm³ = cubic millimeter

[b]μm = micrometer

Red Blood Cells Carry Oxygen from the Lungs to the Tissues

The oxygen-carrying red blood cells, also called **erythrocytes**, make up about 99% of all blood cells. They constitute about 40% of the total blood volume in females and 45% in males. Each cubic millimeter of blood contains about 5 million erythrocytes. A red blood cell resembles a ball of clay squeezed between thumb and forefinger (Fig. 27-8). Its biconcave shape (dented inward on both sides) provides a larger surface area than would a spherical cell of the same volume, thus increasing the cell's ability to absorb and release oxygen through its plasma membrane.

The red color of erythrocytes is caused by the pigment **hemoglobin** (Fig. 27-9). This large, iron-containing protein accounts for about one-third the weight of each red blood cell and carries about 97% of the blood's oxygen. Each hemoglobin molecule carries a pigment complex known as a *heme* group, which includes an atom of iron. One hemoglobin molecule can bind and carry up to four molecules of oxygen, which allows blood to hold far more oxygen than would be possible if all the gas had to be dissolved in plasma. Furthermore, the chemical bond between oxygen and hemoglobin is finely tuned. Hemoglobin binds loosely to oxygen, picking up oxygen in the capillaries of the lungs, where oxygen concentration is high, and releasing it in other tissues of the body, where the oxygen concentration is lower. After releasing its oxygen, some of the hemoglobin picks up carbon dioxide from the tissues for transport back to the lungs. The role of blood in gas exchange is discussed in more detail in Chapter 28.

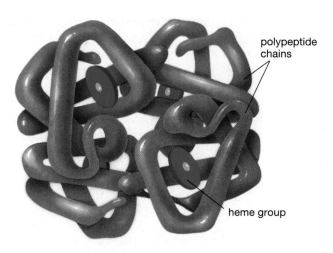

polypeptide chains

heme group

Figure 27-9 Hemoglobin
A molecule of hemoglobin is composed of four polypeptide chains (two pairs of similar chains), each surrounding a heme group. The heme group contains an iron atom and is the site of oxygen binding. When saturated, each hemoglobin molecule can carry four oxygen molecules (eight oxygen atoms).

Red Blood Cells Have a Relatively Short Life Span

Red blood cells are formed in the *bone marrow*, the soft interior portion of certain bones, including those of the chest, upper arms, upper legs, and hips. During their development, mammalian red blood cells lose their nuclei and, with them, their ability to divide. Without the ability to synthesize cellular materials, their lives are necessarily short; each cell lives only about 120 days. Every second, more than 2 million red blood cells die and are replaced by new ones from the bone marrow. Dead or damaged red blood cells are removed from circulation, primarily in the liver and spleen, and broken down to release their iron. The salvaged iron is carried in the blood to the bone marrow, where it is used to make more hemoglobin and packaged into new red blood cells. Although the recycling process is efficient, small amounts of iron are excreted daily and must be replenished by the diet. Bleeding from injury or menstruation also tends to deplete iron stores.

Negative Feedback Regulates Red Blood Cell Numbers

The number of red blood cells in the blood is maintained at an adequate level by a negative feedback system that involves the hormone **erythropoietin**. Erythropoietin is produced by the kidneys in response to oxygen deficiency. This lack of oxygen may be caused by a loss of blood, insufficient production of hemoglobin, high altitude (where less oxygen is available), or lung disease that interferes with gas exchange in the lungs. The hormone stimulates the rapid production of new red blood cells by the bone marrow. When adequate oxygen levels are restored, erythropoietin production is inhibited and the rate of red blood cell production returns to normal (Fig. 27-10).

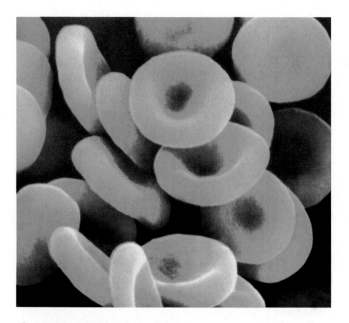

Figure 27-8 Red blood cells
This false-color scanning electron micrograph clearly shows the biconcave disk shape of red blood cells.

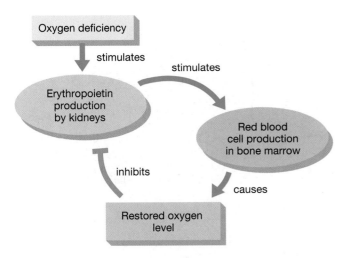

Figure 27-10 Red blood cell regulation by negative feedback

Blood Type Is Determined by Specific Proteins on Red Blood Cell Membranes

Blood is classified as type A, B, AB, or O depending on the presence or absence of specific proteins (designated A and B) on the plasma membranes of red blood cells. Blood type is inherited; the genetics and properties of these major blood types are discussed in detail in Chapter 12 (see Table 12-1).

Another type of protein on red blood cells is the **Rh factor**. If it is present, blood is described as *Rh-positive*; if the protein is absent, the blood is *Rh-negative*. Transfusion of Rh-positive blood into an Rh-negative individual triggers the production of antibodies to the Rh-positive protein. On further exposure, the antibodies attack and destroy the Rh-positive red blood cells. If

an Rh-negative woman mates with an Rh-positive man, their children are likely to be Rh-positive, because Rh-positive blood is a dominant genetic trait. Their first Rh-positive child will trigger antibody production in the mother's blood, commonly without noticeable ill effects. Subsequent Rh-positive children, however, could be born with **erythroblastosis fetalis**. In this condition, the mother's antibodies invade the fetus and attack its red blood cells, causing severe anemia in the newborn. Fortunately, this condition can now be avoided by injections of a substance that prevents formation of Rh antibodies by the pregnant woman.

White Blood Cells Help Defend the Body against Disease

There are five common types of white blood cells, or **leukocytes**, which together make up less than 1% of the total cellular component of the blood. These cells, described and illustrated in Table 27-1, can be distinguished from one another visually by their staining characteristics, size, and the shape of their nuclei. All white blood cells are derived from cells that originate in bone marrow. Most white blood cells function in some way to protect the body against foreign invaders and use the circulatory system to travel to the site of invasion. For example, **monocytes** and **neutrophils** travel through capillaries to wounds where bacteria have gained entry, then ooze out through narrow openings in the capillary walls. After leaving the capillaries, monocytes differentiate into **macrophages** ("big-eaters"), amoebalike cells that engulf foreign particles. Macrophages and neutrophils feed on bacterial invaders or other "foreign" cells, including cancer cells (Fig. 27-11). They typically die in the process, and their dead bodies accumulate and contribute to the white substance called *pus*, seen at infection sites.

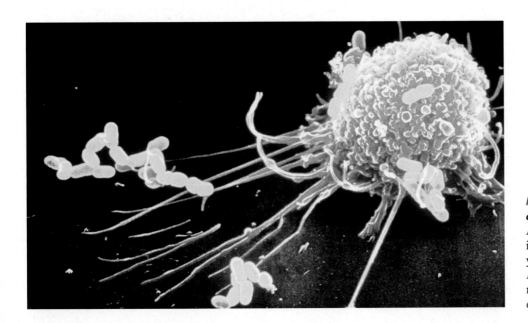

Figure 27-11 A white blood cell attacks bacteria
An amoebalike white blood cell is seen capturing bacteria (in yellow). These bacteria are *Escherichia coli*, intestinal bacteria that can cause disease if they enter the bloodstream.

Lymphocytes, described in detail in Chapter 31, are white blood cells responsible for the production of antibodies that help provide immunity against disease. Cells that give rise to lymphocytes migrate from the bone marrow to tissues of the lymphatic system, such as the thymus, spleen, and lymph nodes, described later in this chapter. Least abundant are the **eosinophils** and **basophils**. Eosinophil production is stimulated by parasitic infections. Eosinophils converge on the invaders, releasing substances that kill the parasite. Basophils release substances that inhibit blood clotting as well as chemicals, such as histamine, that participate in allergic reactions and in responses to tissue damage and microbial invasion.

Platelets Are Cell Fragments That Aid in Blood Clotting

Platelets are not complete cells but fragments of large cells called **megakaryocytes**. Megakaryocytes remain in the bone marrow, where they pinch off membrane-enclosed pieces of their cytoplasm that we call platelets (Fig. 27-12). The platelets then enter the blood and play a central role in blood clotting. Like red blood cells, platelets lack a nucleus, and their life span is even shorter—about 10 to 12 days.

Blood clotting is a complex process without which we would bleed to death from normal wear and tear on the body. Clotting starts when platelets and other factors in the plasma contact an irregular surface, such as a damaged blood vessel. The ruptured surface of an injured blood vessel causes platelets to adhere and partially block the opening. The adhering platelets and the injured tissues initiate a complex sequence of events among circulating plasma proteins. These events result in the pro-

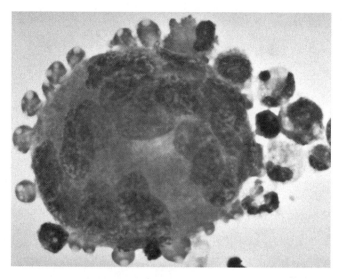

Figure 27-12 The production of platelets
Megakaryocytes are large cells in bone marrow from which membrane-enclosed pieces of cytoplasm, called platelets, constantly bud off and enter the blood, helping initiate and control blood clotting. Here a single megakaryocyte is producing dozens of platelets.

duction of the enzyme **thrombin**. Thrombin catalyzes the conversion of the plasma protein *fibrinogen* into insoluble, stringlike molecules called **fibrin**. Fibrin molecules adhere to one another, forming a fibrous network (Fig. 27-13a). This protein web immobilizes the fluid portion of the blood, causing it to solidify in much the same way that gelatin does as it cools. The web traps red blood cells, further increasing the density of the clot (Fig. 27-13b).

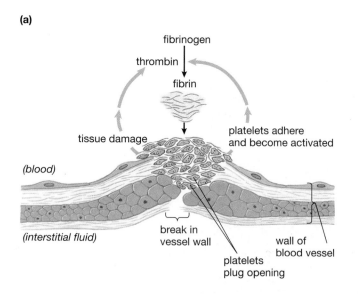

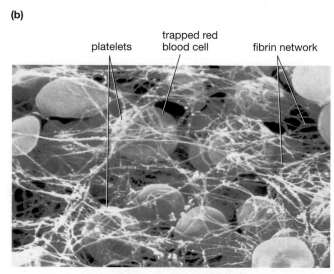

Figure 27-13 Blood clotting
(a) Injured tissue and adhering platelets cause a complex series of biochemical reactions among blood proteins. These reactions produce thrombin, which catalyzes the conversion of fibrinogen to insoluble fibrin strands. **(b)** Threadlike fibrin proteins produce a tangled sticky mass that traps red blood cells and eventually forms a clot.

Platelets adhere to the fibrous mass and send out sticky projections that attach to one another. Within half an hour, the platelets contract, pulling the mesh tighter and forcing liquid out. This action creates a denser, stronger clot (on the skin it is called a *scab*) and also constricts the wound, pulling the damaged surfaces closer together in a way that promotes healing.

4) What Are the Structures and Functions of Blood Vessels?

www

Some of the major blood vessels of the human circulatory system are diagrammed in Figure 27-14. In this section we examine each type of blood vessel in detail. As it leaves the heart, blood travels from arteries to *arterioles* to capillaries to *venules* to veins, which return it finally to the heart. These vessels are shown in sequence from left to right in Figure 27-15.

Arteries and Arterioles Are Thick-Walled Vessels That Carry Blood Away from the Heart

After leaving the heart, blood first enters large vessels called arteries. These have thick walls containing smooth muscle and elastic tissue (Fig. 27-15). With each surge of blood from the ventricles, the arteries expand slightly, like thick-walled balloons. They recoil between heartbeats, helping pump the blood and maintain a steady flow through the smaller vessels. Arteries branch into vessels of smaller diameter called **arterioles**, which play a major role in determining how blood is distributed within the body, as described later.

Capillaries Are Microscopic Vessels That Allow the Blood and Body Cells to Exchange Nutrients and Wastes

The entire circulatory system is an elaborate device for providing each cell of a complex, multicellular organism with the same type of exchange—diffusion—that is used

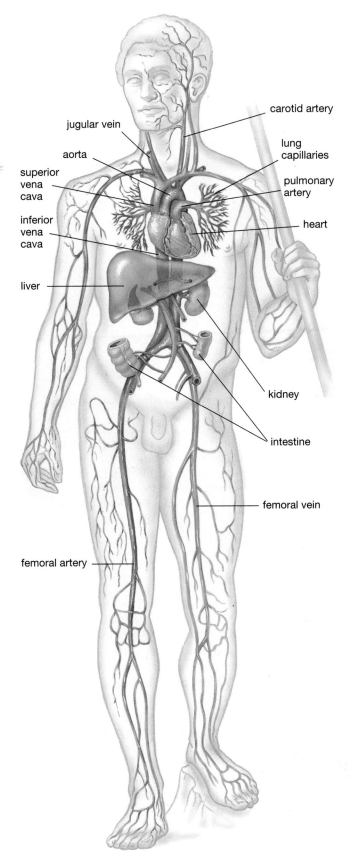

*Figure 27-14 **The human circulatory system***
Oxygenated blood is shown in red, deoxygenated blood in blue. Most veins (illustrated on the right) carry deoxygenated blood to the heart. Most arteries (shown on the left) conduct oxygenated blood away from the heart. The pulmonary arteries and veins are exceptions: Pulmonary arteries carry deoxygenated blood; pulmonary veins, oxygenated blood. The liver and some other organs are illustrated to show their location and approximate size. All organs receive blood from arteries, send it back via veins, and are nourished by exchange through beds of capillaries. For simplicity, only lung capillaries are illustrated.

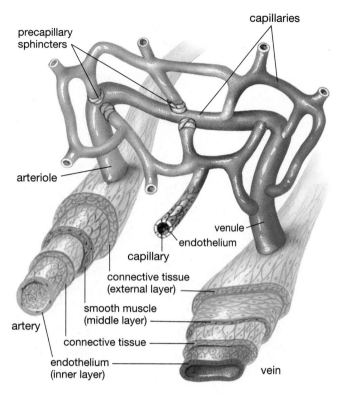

Figure 27-15 Structures and interconnections of blood vessels
Arteries and arterioles are more muscular than are veins and venules. Capillaries have walls only one cell thick. Oxygenated blood moves from arteries to arterioles to capillaries. Capillaries empty deoxygenated blood into venules, which empty into veins. The movement of blood from arterioles into capillaries is regulated by muscular rings called precapillary sphincters.

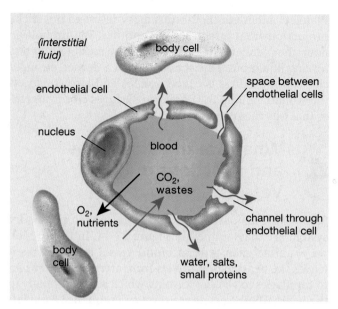

Figure 27-16 Cross section of capillary wall structure
Capillary wall structure facilitates the exchange of dissolved gases, wastes, and nutrients. Narrow spaces separate adjacent endothelial cells, and channels also penetrate the plasma membranes. Blood pressure forces water that contains salts and small proteins out through these two types of openings into the interstitial fluid. Gases and most wastes and nutrients diffuse through the endothelial cell's plasma membranes along their concentration gradients.

by the simplest unicellular organisms. Diffusion is accomplished at the level of the **capillaries**, the tiniest of all vessels. There wastes, nutrients, gases, and hormones are exchanged between the blood and the body cells. Capillaries are finely adapted to their role of exchange. Their walls are only a single cell thick. Most nutrients, oxygen, and carbon dioxide diffuse readily through capillary plasma membranes. Salts and small charged molecules (including some small proteins) move through fluid-filled spaces within the capillary plasma membrane or between adjacent capillary cells (Fig. 27-16). The pressure within capillaries causes a continuous leakage of fluid from the blood plasma into the spaces that surround the capillaries and tissues. This fluid, known as the **interstitial fluid**, consists primarily of water in which are dissolved nutrients, hormones, gases, wastes, and small proteins from the blood. Because they are too large to fit through plasma membrane channels, large plasma proteins, red blood cells, and platelets are unable to leave the capillaries, but white blood cells can ooze through the openings between

capillary cells. The exchange of materials between capillary blood and nearby cells occurs through this interstitial fluid, which bathes nearly all the body's cells.

Capillaries are so narrow that red blood cells must pass through them in single file (Fig. 27-17). Consequently, all the blood is sure to pass very close to the capillary walls, where exchange occurs. In addition, capillaries are so numerous that no body cell is more than 100 micrometers (0.004 inch; about as thick as four pages of this book) from a capillary. These factors facilitate the exchange of materials by diffusion. It is estimated that the total length of capillaries in a human is more than 80,600 kilometers (50,000 miles), enough to encircle the globe twice! The speed of blood flow drops very quickly as blood is forced through this narrow, almost endless network of capillaries, allowing more time for diffusion to occur.

Veins and Venules Carry Blood Back to the Heart

Blood from the capillaries, now carrying carbon dioxide and other cellular wastes, drains into larger vessels called **venules**, which empty into still larger *veins* (see Fig. 27-15). Veins provide a low-resistance pathway by which

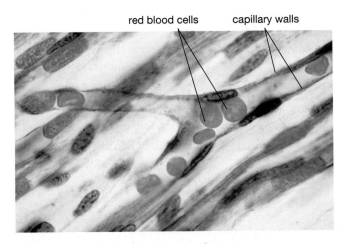

red blood cells   capillary walls

Figure 27-17 Red blood cells flow through a capillary
Capillaries are so narrow that red blood cells must pass
through them single file, thereby facilitating the exchange of
gases by diffusion.

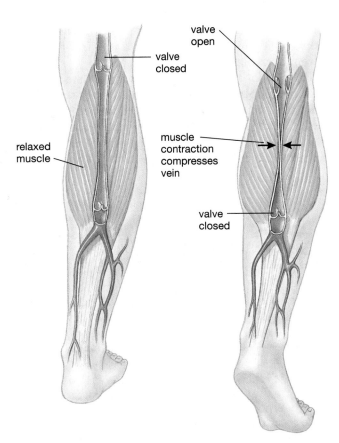

valve
open

valve
closed

relaxed
muscle

muscle
contraction
compresses
vein

valve
closed

Figure 27-18 Valves direct the flow of blood in veins
Veins and venules have one-way valves that maintain blood
flow in the proper direction. When the vein is compressed by
nearby muscles, the valves allow blood to flow toward the
heart but clamp shut to prevent backflow.

blood can return to the heart. The walls of veins are thinner, less muscular, and more expandable than those of
arteries, although both contain a layer of smooth muscle. Because blood pressure in the veins is low, the contractions of skeletal muscle during exercise and breathing
must assist in the return of blood to the heart. These muscular movements squeeze the veins, forcing blood
through them.

When veins are compressed, you might predict that
blood would be forced away from the heart as well as toward it. To prevent this, veins are equipped with one-way
valves that allow blood to flow only toward the heart
(Fig. 27-18). When you sit or stand for long periods, the
lack of muscular activity allows blood to accumulate in
the veins of the lower legs. This accumulation accounts for
the swollen feet often experienced by airplane passengers. Long periods of inactivity can also contribute to
varicose veins, in which the valves become stretched and
weakened.

If blood pressure should fall—for instance, after extensive bleeding—veins can help restore it. The sympathetic nervous system (which prepares the body for
emergency action) automatically stimulates the contraction of the smooth muscles in the vein walls. This action
decreases the internal volume of the veins and raises
blood pressure, speeding up the return of blood to
the heart.

Arterioles Control the Distribution of Blood Flow

The muscular walls of arterioles are under the influence
of nerves, hormones, and chemicals produced by nearby
tissues. Arterioles can therefore contract and relax in re-

sponse to the changing needs of the tissues and organs
they supply. For instance, as you read in your paperback
thriller ". . . the blood drained from her face as she be-
held the gruesome sight . . . ," keep in mind that the hero-
ine is experiencing the constriction of the arterioles that
supply her skin with blood. In such threatening situations,
the sympathetic nervous system is activated and stimulates the smooth muscle of the arterioles to contract. This
contraction raises blood pressure overall, but selective
constriction also redirects blood to the heart and muscles, where it may be needed for vigorous action, and
away from the skin, where it is less essential.

On a hot summer day, however, you become flushed
as the arterioles in your skin expand and bring more
blood to the skin capillaries. Bringing this blood closer
to the surface enables your body to dissipate excess heat
to the outside and to maintain a proper internal temperature. In contrast, in extremely cold weather, your fingers and toes can become frostbitten because the
arterioles that supply the extremities constrict. The blood
is shunted to vital organs, such as the heart and brain,

which cannot function properly if their temperature drops. By minimizing blood flow to the heat-radiating extremities, your body conserves heat.

The flow of blood in capillaries is regulated by tiny rings of smooth muscle called **precapillary sphincters**, which surround the junctions between arterioles and capillaries (see Fig. 27-15). These sphincters open and close in response to local changes that signal the needs of nearby tissues. For example, the accumulation of carbon dioxide, lactic acid, or other cellular wastes signals the need for increased blood flow to the tissues. These signals cause the precapillary sphincters as well as the muscles in nearby arterioles to relax, thus increasing blood flow through the capillaries.

What happens when blood vessels rupture, are narrowed by deposits of cholesterol, or are blocked by clots, damming the "river of life"? We explore these questions in "Health Watch: Matters of the Heart."

Health Watch
Matters of the Heart

Cardiovascular disorders, or disorders of the heart and blood vessels, are the leading cause of death in the United States, killing nearly 1 million Americans annually. Consider the stresses under which your circulatory system constantly operates. Your heart is expected to contract vigorously more than 2.5 billion times during your lifetime without once stopping to rest. It is also expected to force blood through a series of vessels whose total length would encircle the globe twice. Add the possibility that the complex network of vessels may become constricted, weakened, or clogged for a number of reasons, and it's easy to see why the cardiovascular system is a prime candidate for malfunction.

Hidden Killers: High Blood Pressure and Atherosclerosis

High blood pressure, also called **hypertension**, is normally caused by the constriction of arterioles, which results in increased resistance to blood flow. In the majority of the 50 million Americans with this condition, the cause of the constriction is unknown. Heredity seems to play a role. For some individuals who are predisposed to hypertension, high salt intake in the diet, as well as obesity, can be aggravating factors. Although normal blood pressure tends to increase somewhat with age, an approximate borderline reading for high blood pressure is 140/90.

High blood pressure commonly gives few warning signals, but it undermines the cardiovascular system in several insidious ways. First, it causes a strain on the heart by increasing resistance to blood flow. Although the heart may enlarge in response to this added demand, its own blood supply may not increase proportionately. The heart muscle is then inadequately supplied with blood, especially during exercise. Lack of sufficient oxygen to the heart can cause chest pain called **angina**. Second, high blood pressure contributes to "hardening of the arteries," or *atherosclerosis*, described below. Third, high blood pressure, in conjunction with hardened arteries, can lead to the rupture of an artery and internal bleeding. The rupture of vessels supplying the brain causes **stroke**, in which brain function is lost in the area deprived of blood and of the vital oxygen and nutrients the blood carries.

Mild hypertension is in some cases alleviated by weight reduction, exercise, stress management, and (for some individuals) reduction of dietary salt. For more-severe cases, drugs might be prescribed: *Diuretics* increase urination and reduce blood volume; other drugs reduce the heart rate or cause dilation (expansion) of the arteries and arterioles.

Atherosclerosis (derived from the Greek *athero*, meaning "gruel" or "paste," and *scleros*, "hard") causes a loss of elasticity in the large arteries and a thickening of the arterial walls. The thickening results from deposits called **plaques**, which are composed of cholesterol and other fatty substances as well as calcium and fibrin. Plaques are deposited within the wall of the artery between the smooth muscle and the inner lining (Fig. E27-1). A plaque may rupture through the lining into the interior of the vessel. This rupture stimulates the blood platelets to adhere to the vessel wall and initiate blood clots. These clots further obstruct the artery and may completely block it. Arterial clots are responsible for the most-serious consequences of atherosclerosis: heart attacks and strokes.

About 1.5 million Americans suffer **heart attacks** each year, and about half a million people die from them. A heart attack occurs when one of the coronary arteries (arteries that supply the heart muscle itself; see Fig. E27-1a) is blocked. If a blood clot suddenly breaks loose, it may be carried to a narrower part of the artery, obstructing blood flow. Deprived of nutrients and oxygen, the heart muscle supplied by the blocked artery rapidly and painfully dies. If the damaged area is small, the individual may recover, but the death of large areas of heart muscle is almost instantly fatal. Although heart attacks are the major cause of death from atherosclerosis, this disease causes plaques and clots to form in arteries throughout the body. If a clot or plaque obstructs an artery that supplies the brain, it can cause a stroke, with results similar to those caused by a ruptured artery.

As with hypertension, the exact cause of atherosclerosis is unclear, but it is promoted by hypertension, cigarette smoking, genetic predisposition, obesity, diabetes, lack of exercise, and high blood levels of a certain type of cholesterol bound to a carrier molecule called *low-density lipoprotein* (LDL). If LDL-bound cholesterol levels are too high, cholesterol can be deposited in arterial walls. In contrast, cholesterol bound to *high-density lipoprotein* (HDL) is metabolized or excreted and hence is often called "good" cholesterol.

Traditional treatment of atherosclerosis includes the use of drugs or changes in diet and lifestyle to lower blood pressure and blood cholesterol levels. For angina, nitroglycerin is used to

5) What Are the Structures and Functions of the Lymphatic System?

www

The **lymphatic system** consists of a network of lymph capillaries and larger vessels that empty into the circulatory system, numerous small *lymph nodes*, patches of lymphocyte-rich connective tissue (including the *tonsils*), and two additional organs: the *thymus* and the *spleen*

(Fig. 27-19). Although not strictly part of the circulatory system, the lymphatic system is closely associated with it. The lymphatic system has several important functions:

1. Removal of excess fluid and dissolved substances that leak from the capillaries.
2. Transport of fats from the small intestine to the bloodstream.
3. Defense of the body by exposing bacteria and viruses to white blood cells.

dilate blood vessels and ease the pain caused by partial blockage of the coronary arteries. A tube can be surgically inserted into a partially blocked artery. At the end of the tube is a tiny balloon, which is inflated to squash the plaque flat, a drill that shaves the plaque off the artery wall, or a laser that vaporizes it. A new use for a laser beam is *laser revascularization*, in which a laser is used to shoot 15 to 30 small (1-mm-diameter) channels through the ventricular walls. Blood clots form on the outside of the ventricle, keeping blood from leaking out. As the heart beats, blood from inside the ventricles flows in and out of the channels, supplying oxygen to the ventricular muscle and partially replacing the function of coronary arteries clogged by plaques. *Coronary bypass surgery*, performed on more than 350,000 individuals each year in the United States, consists of bypassing one or more obstructed coronary arteries with a piece of vein,

normally obtained from the patient's leg. Researchers are working to perfect artificial veins, either formed from human tissue or from synthetic material so the patient's own veins can be spared.

If a heart attack occurs, rapid treatment can minimize the damage and significantly increase the chances of survival. Blood clots in coronary arteries are commonly dissolved by injecting substances (streptokinase or TPA—tissue plasminogen activator) into the coronary artery. These substances stimulate the production of an enzyme that breaks down fibrin, the protein that binds the clot together. Although heart disease is still the leading cause of death in the United States, steady progress in treatment has significantly reduced the rate of early deaths from atherosclerosis.

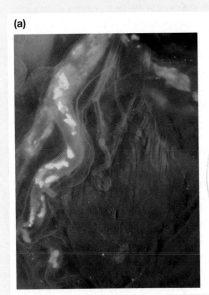
(a)

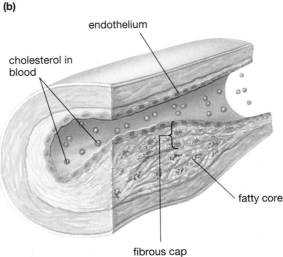
(b)

endothelium

cholesterol in blood

fatty core

fibrous cap

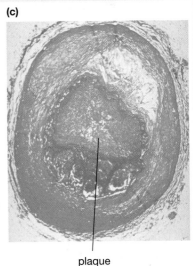
(c)

plaque

Figure E27-1 Plaques clog arteries
(a) In this remarkable photo of coronary arteries, plaques are seen in glowing yellow. If they block a coronary artery, a heart attack will occur. ***(b)*** Diagrammatic cross section of an artery with a plaque. If the fibrous cap ruptures, a clot will form that can completely obstruct the artery, or the clot can break loose and clog a narrower artery "downstream." ***(c)*** Cross section of a coronary artery with a large plaque that has completely obstructed it.

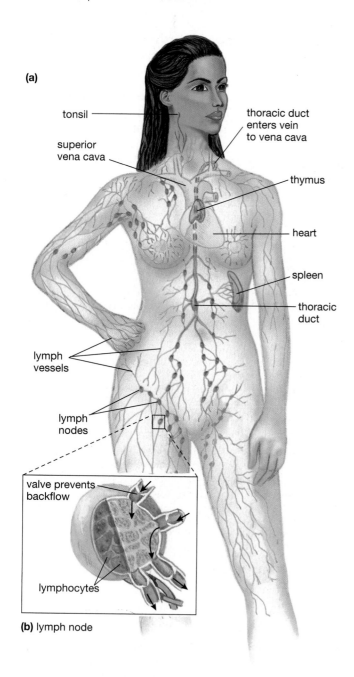

(a)

tonsil

superior vena cava

thoracic duct enters vein to vena cava

thymus

heart

spleen

thoracic duct

lymph vessels

lymph nodes

valve prevents backflow

lymphocytes

(b) lymph node

Figure 27-19 The human lymphatic system
(a) Lymph vessels, lymph nodes, and two auxiliary lymph organs, the thymus and spleen. Lymph is returned to the circulatory system by way of the thoracic duct, which empties into the vena cava, a large vein. *(b)* A cross section of a lymph node. The node is filled with channels lined with white blood cells (lymphocytes) that attack foreign matter in the lymph.

Lymphatic Vessels Resemble the Veins and Capillaries of the Circulatory System

Like blood capillaries, *lymph capillaries* form a complex network of microscopically narrow, thin-walled vessels into which substances can move readily. In contrast to those of blood capillaries, lymph capillary walls are composed of cells with openings between them that act as one-way

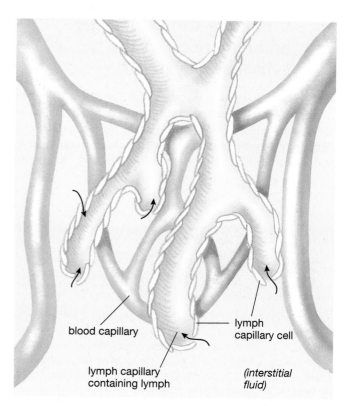

blood capillary

lymph capillary cell

lymph capillary containing lymph

(interstitial fluid)

Figure 27-20 Lymph capillary structure
Lymph capillaries end blindly in the body tissues, where pressure from the accumulation of interstitial fluid forces the fluid into the lymph capillaries through valvelike openings between lymph capillary cells.

valves. These openings allow relatively large particles, along with fluid, to be carried into the lymph capillary. Also unlike blood capillaries, which form a continuous connected network, lymph capillaries dead-end in the body's tissues (Fig. 27-20). Materials collected by the lymph capillaries flow into larger lymph vessels. Large lymph vessels have somewhat muscular walls, but, as in blood veins, most of the impetus for lymph flow comes from the contraction of nearby muscles, such as those used in breathing and walking. As in blood veins, the direction of flow is regulated by one-way valves (Fig. 27-21).

The Lymphatic System Returns Fluids to the Blood

As described earlier, dissolved substances are exchanged between the capillaries and body cells by means of interstitial fluid (derived from blood plasma), which bathes nearly all the body's cells. In an average person, about 3 liters more fluid leaves the blood capillaries than is reabsorbed by them each day. One function of the lymphatic system is to return this excess fluid and its dissolved proteins and other substances to the blood. As interstitial fluid accumulates, its pressure forces the fluid through the openings in the lymph capillaries (see Fig. 27-20). The lymphatic system transports this fluid, now called **lymph**, back to the circulatory system.

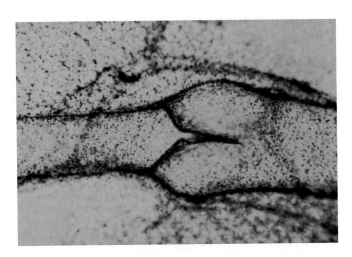

Figure 27-21 *A valve in a lymph vessel*
Like blood-carrying veins, lymph vessels have internal one-way valves that direct the flow of lymph toward the large veins into which they empty.

The Lymphatic System Transports Fats from the Small Intestine to the Blood

You will learn in Chapter 29 that the small intestine is richly supplied with lymph capillaries. After absorbing digested fats, intestinal cells release fat globules into the interstitial fluid. These globules are too large to diffuse into blood capillaries but can easily move through the openings between lymph capillary cells. Once in the lymph, they are dumped into the vena cava, a large vein that enters the heart. After a fatty meal, these fat globules may make up 1% of the lymphatic fluid.

The Lymphatic System Helps Defend the Body against Disease

In addition to its other roles, the lymphatic system helps defend the body against foreign invaders, such as bacteria and viruses. In the linings of the respiratory, digestive, and urinary tracts are patches of connective tissue that contain large numbers of lymphocytes. The largest of these patches are the **tonsils**, located in the cavity behind the mouth. The large lymph vessels are interrupted periodically by kidney bean–shaped structures about 2.5 centimeters (1 inch) long called **lymph nodes** (see Fig. 27-19). Lymph is forced through channels within the nodes, which are lined with masses of macrophages. Lymphocytes are also produced in the lymph nodes. Both macrophages and lymphocytes recognize and destroy foreign particles, such as bacteria and viruses, and are killed in the process. The painful swelling of lymph nodes that accompanies certain diseases (mumps is an extreme example) is largely a result of the accumulation of dead lymphocytes and macrophages and dead virus-infested cells they have engulfed.

The thymus and the spleen are often considered part of the lymphatic system (see Fig. 27-19a). The **thymus**, an organ that produces lymphocytes, is located beneath the breastbone slightly above the heart. The thymus is particularly active in infants and young children but decreases in size and importance in early adulthood. The **spleen**, another lymphocyte-producing organ, is located in the left side of the abdominal cavity, between the stomach and diaphragm. Just as the lymph nodes filter lymph, the spleen filters blood, exposing it to macrophages and lymphocytes that destroy foreign particles and aged red blood cells.

Summary of Key Concepts

1) What Are the Major Features and Functions of Circulatory Systems?
Circulatory systems transport blood rich in dissolved nutrients and oxygen close to each cell, where nutrients can be released and wastes absorbed by diffusion. All circulatory systems have three major parts: (1) blood, a fluid; (2) vessels, a system of channels to conduct the blood; and (3) a heart, a pump to circulate the blood. Invertebrates have open or closed circulatory systems; nearly all vertebrates have closed systems. In open systems, blood is pumped by a heart into a hemocoel, where it directly bathes internal organs. In closed systems, the blood is confined to the heart and blood vessels.

2) What Are the Features and Functions of the Vertebrate Heart?
Vertebrate circulatory systems transport gases, hormones, and wastes; distribute nutrients; help regulate body temperature; and defend the body against disease.

The vertebrate heart evolved from two chambers in fishes, to three in amphibians and most reptiles, to four in birds and mammals. In the four-chambered heart, blood is pumped

separately to the lungs and through the body, maintaining complete separation of oxygenated and deoxygenated blood. Deoxygenated blood is collected from the body in the right atrium and passed to the right ventricle, which pumps it to the lungs. Oxygenated blood from the lungs enters the left atrium, is passed to the left ventricle, and pumped to the rest of the body.

The cardiac cycle consists of two stages: (1) atrial contraction, followed by (2) ventricular contraction. The direction of blood flow is maintained by valves within the heart. The contractions of the heart are initiated and coordinated by the sinoatrial node, the heart's pacemaker. Heart rate can be modified by the nervous system and by hormones such as epinephrine.

3) What Are the Features and Functions of Blood?
Blood is composed of both fluid and cellular materials. The fluid plasma consists of water that contains proteins, hormones, nutrients, gases, and wastes. Red blood cells, or erythrocytes, are packed with a large iron-containing protein called hemoglobin, which carries oxygen. Their numbers are regulated by the hormone erythropoietin. Proteins on

their plasma membrane determine blood type. There are five types of white blood cells, or leukocytes, that fight infection (see Table 27-1). Platelets, which are fragments of megakaryocytes, are important for blood clotting.

4) What Are the Structures and Functions of Blood Vessels?

Blood leaving the heart travels (in sequence) through arteries, arterioles, capillaries, venules, veins, and then back to the heart. Each vessel is specialized for its role. Elastic, muscular arteries help pump the blood. The thin-walled capillaries are the sites of exchange of materials between the body cells and the blood. Veins provide a low-resistance path back to the heart, with one-way valves that maintain the direction of blood flow. The distribution of blood is regulated by the constriction and dilation of arterioles under the influence of the sympathetic nervous system and local factors such as the amount of carbon dioxide in the tissues. Local factors also regulate precapillary sphincters, which control blood flow in the capillaries.

5) What Are the Structures and Functions of the Lymphatic System?

The human lymphatic system consists of lymphatic vessels, which resemble blood veins and capillaries; tonsils; lymph nodes; and the thymus and spleen. The lymphatic system removes excess interstitial fluid that leaks through blood capillary walls. It transports fats to the bloodstream from the small intestine and fights infection by filtering the lymph through lymph nodes, where white blood cells ingest foreign invaders, such as viruses and bacteria. The thymus, which is most active in young children, produces lymphocytes that function in immunity. The spleen filters blood past macrophages and lymphocytes, which remove bacteria and damaged blood cells.

Key Terms

angina *p. 550*	closed circulatory system *p. 538*	lymph *p. 552*	pulmonary circulation *p. 540*
arteriole *p. 547*	eosinophil *p. 546*	lymphatic system *p. 551*	Rh factor *p. 545*
artery *p. 540*	erythroblastosis fetalis *p. 545*	lymph node *p. 553*	semilunar valve *p. 541*
atherosclerosis *p. 550*	erythrocyte *p. 544*	lymphocyte *p. 546*	sinoatrial (SA) node *p. 542*
atrioventricular (AV) node *p. 542*	erythropoietin *p. 544*	macrophage *p. 545*	spleen *p. 553*
atrioventricular valve *p. 541*	fibrillation *p. 542*	megakaryocyte *p. 546*	stroke *p. 550*
atrium *p. 539*	fibrin *p. 546*	monocyte *p. 545*	systemic circulation *p. 540*
basophil *p. 546*	heart *p. 538*	neutrophil *p. 545*	thrombin *p. 546*
bicuspid valve *p. 541*	heart attack *p. 550*	open circulatory system *p. 538*	thymus *p. 553*
blood *p. 538*	hemocoel *p. 538*	pacemaker *p. 542*	tonsil *p. 553*
blood clotting *p. 546*	hemoglobin *p. 544*	plaque *p. 550*	tricuspid valve *p. 541*
blood vessel *p. 538*	hypertension *p. 550*	plasma *p. 543*	vein *p. 540*
capillary *p. 548*	interstitial fluid *p. 548*	platelet *p. 546*	ventricle *p. 539*
cardiac cycle *p. 540*	leukocyte *p. 545*	precapillary sphincter *p. 550*	venule *p. 548*

Thinking Through the Concepts

Multiple Choice

1. *Which of the following is NOT an important function of the vertebrate circulatory system?*
   a. transport of nutrients and respiratory gases
   b. regulation of body temperature
   c. protection of the body by circulating antibodies
   d. removal of waste products for excretion from the body
   e. defense against blood loss, through clotting

2. *Which event initiates blood clotting?*
   a. contact with an irregular surface by platelets and other factors in plasma
   b. production of the enzyme thrombin
   c. conversion of fibrinogen into fibrin
   d. conversion of fibrin into fibrinogen
   e. excess flow of blood through a capillary

3. *The sites of exchange of wastes, nutrients, gases, and hormones between the blood and body cells are the*
   a. arteries   b. arterioles
   c. capillaries   d. veins
   e. all blood vessels

4. *What produces systolic blood pressure?*
   a. contraction of the right atrium
   b. contraction of the right ventricle
   c. contraction of the left atrium
   d. contraction of the left ventricle
   e. the pause between heartbeats

5. *Which of the following is NOT a component of plasma?*
   a. water   b. globulins
   c. fibrinogen   d. albumins
   e. platelets

6. *Lymph most closely resembles which of the following?*
   a. blood   b. urine
   c. plasma   d. interstitial fluid
   e. water

? Review Questions

1. Trace the flow of blood through the circulatory system, starting and ending with the right atrium.

2. List three types of blood cells, and describe their principal functions.

3. What are five functions of the vertebrate circulatory system?

4. In what way do veins and lymph vessels resemble one another? Describe how fluid is transported in each of these vessels.

5. Describe three important functions of the lymphatic system.

6. Distinguish among plasma, interstitial fluid, and lymph.

7. Describe veins, capillaries, and arteries, noting their similarities and differences.

8. Trace the evolution of the vertebrate heart from two chambers to four chambers.

9. Explain in detail what causes the vertebrate heart to beat.

10. Describe the cardiac cycle, and relate the contractions of the atria and ventricles to the two readings taken during the measurement of blood pressure.

11. Describe how red blood cell number is regulated by a negative feedback system.

12. Describe the formation of an atherosclerotic plaque. What are the risks associated with atherosclerosis?

Applying the Concepts

1. Discuss the steps you can take now and in the future to reduce your risks of developing heart disease.

2. Discuss why a four-chambered heart is much more efficient than a two-chambered heart in delivering oxygenated blood to the various body parts. What evolutionary changes in the lifestyles of organisms selected for the evolution of the four-chambered heart?

3. Heart surgeons have attempted to transplant baboon hearts into humans whose hearts were failing. Discuss the implications of this operation from as many angles as you can think of.

4. Considering the prevalence of cardiovascular disease and the high, increasing costs of treating it, certain treatments may not be available to all who might benefit from them. What factors would you take into account in rationing cardiovascular procedures, such as heart transplants?

5. Joe, a 45-year-old executive of a major corporation, has been diagnosed with mild hypertension. What treatments or lifestyle changes might Joe's physician recommend? If Joe's hypertension becomes more severe, what treatments might Joe's physician use? Should Joe be concerned about mild hypertension? Explain your answer.

Group Activity

Form groups of three or four to discuss the following questions: What cardiovascular changes—circulatory system and heart—would you expect to occur in a mammal as the result of engaging in daily endurance-type activities (such as running or swimming) for a period of several months that greatly increase maximum oxygen consumption capability? Think in terms of all the cardiovascular parameters that would promote oxygen transport in the blood. Which of those changes are more beneficial and thus more likely to occur? What evolutionary changes in the mammalian cardiovascular system have occurred to facilitate endurance-type activities?

For More Information

Hajjar, D. P., and Nicholson, A. C. "Atherosclerosis." *American Scientist*, September/October 1995. Scientists are learning more about the biochemical reactions that occur during plaque formation, an understanding that will lead to more-effective treatments in the future.

Katzir, A. "Optical Fibers in Medicine." *Scientific American*, May 1989. Describes the use of fiber optics in medicine, with emphasis on their role in cardiovascular surgery.

Lawn, R. M. "Lipoprotein(a) in Heart Disease" *Scientific American*, June 1992. Describes the interaction of this protein with blood fats and its role in heart disease.

Lillywhite, H. B. "Snakes, Blood Circulation, and Gravity." *Scientific American*, December 1988. Describes how the cardiovascular systems of snakes are adapted to their diverse lifestyles.

Nucci, M. L., and Abuchowski, A. "The Search for Blood Substitutes." *Scientific American*, February 1998. As the authors summarize: "The threat of blood shortages and fears about contamination have hastened attempts to find life-sustaining alternatives."

Perutz, M. "Hemoglobin Structure and Respiratory Transport." *Scientific American*, December 1978. How hemoglobin plays the dual role of transporting oxygen to the tissues and carbon dioxide back to the lungs.

Radetsky, P. "The Mother of All Blood Cells." *Discover*, March 1995. Discusses the discovery by Irving Weissman of a "stem cell" that gives rise to all other types of blood cells.

Seppa, N. "Secondary Smoke Carries High Price." *Science News*, January 17, 1998. New research suggests that smoking causes atherosclerosis, that the damage persists after a smoker quits, and that exposure to secondhand smoke significantly increases the buildup of plaque in the carotid artery of nonsmokers.

Wu, C. "Engineered Blood Vessel Is Only Human." *Science News*, January 17, 1998. Researchers have constructed an experimental blood vessel composed entirely of human tissue.

Zimmer, C. "The Body Electric." *Discover*, February 1993. Describes computer imaging of the heart's electrical activity.

Zucker, M. "The Functioning of Blood Platelets." *Scientific American*, June 1980. Describes the complex role of platelets in blood clotting.

Respiratory structures require large, moist surfaces for gas exchange. This Spanish shawl nudibranch (literally, "naked gill"), a mollusk, extends numerous threadlike gills into the surrounding seawater. Gases are exchanged through capillaries just below the gill surface.

Respiration 28

Late again! Sprinting up two flights of stairs to your classroom, you feel your calves "burning." Remembering Chapter 8, you think "Aha! That's lactic acid building up—my muscle cells are fermenting glucose, because they can't get enough oxygen for cellular respiration." As you slip into your seat, quietly panting and feeling your heart pounding, the discomfort eases. With less exertion, adequate oxygen is now available; the lactic acid is being reconverted to pyruvate and then broken down into carbon dioxide (CO_2) and water, while providing additional energy.

You're experiencing firsthand the relationship between cellular respiration and breathing. Each cell in your body demands a continuous influx of energy to maintain itself. When you call on your muscles to carry you upstairs quickly, the demands are extreme. As cellular respiration converts the energy in nutrients such as sugar into ATP that can be used by body cells, the process requires a steady supply of oxygen and generates carbon dioxide as a waste product. The rapid beating of your heart as you relax after your sprint upstairs is a reminder that the circulatory system works in close harmony with the respiratory system, extracting oxygen from air in your lungs, carrying it within diffusing distance of each cell, then picking up carbon dioxide for release in the lungs.

How, exactly, does breathing support cellular respiration? What does the inside of a lung look like, and how is it adapted for gas exchange? Why are lungs inside our bodies, instead of on the outside, which is already exposed to the air? How do aquatic animals, such as the nudibranch, "breathe"? In this chapter we explore the specialized structures of respiratory systems. We start with a variety of adaptations for gas exchange that don't utilize lungs, then move to lung-breathing vertebrates, with an emphasis on the human respiratory system.

Net Watch

On-line resources for this chapter are on the World Wide Web at:

http://www.prenhall.com/audesirk

(click on the Table of Contents link and then select Chapter 28).

1) What Are Some Evolutionary Adaptations for Gas Exchange?

www

Gas exchange in all organisms ultimately relies on diffusion. Cellular respiration depletes O_2 and increases CO_2 levels, creating concentration gradients that favor the diffusion of carbon dioxide out of cells and the diffusion of oxygen into them. Although animal respiratory systems are amazingly diverse, they all share two features that facilitate diffusion: (1) The respiratory surface must remain moist, because gases must be dissolved in water when they diffuse into or out of cells. (2) The respiratory system must have a large surface area in contact with the environment to allow adequate gas exchange.

In the following sections we examine some diverse respiratory systems, each shaped by the environment in which it evolved. Notice how each type of system meets the demand for a large, moist surface that allows gas exchange.

Some Animals in Moist Environments Lack Specialized Respiratory Structures

Some animals that live in moist environments are able to exchange gases without specialized respiratory structures. The outside of their bodies provide an adequate surface area for the diffusion of gases. If the body is extremely small and elongated, as in microscopic nematode worms, gases need to diffuse only a short distance to reach all cells of the body. Alternatively, an animal's body may be thin and flattened, producing a large surface area for diffusion. In flatworms, most cells are close to the moist skin through which gases can diffuse (Fig. 28-1a).

If energy demands are sufficiently low, the relatively slow rate of gas exchange by diffusion may suffice even for a larger, thicker body. For example, jellyfish can be quite large, but their cells that are far from the surface are relatively inert and require little oxygen (Fig. 28-1b).

Another adaptation for gas exchange is to bring the environment (normally, water) close to all the body cells, allowing direct exchange of gases between the body cells

(a)

(b)

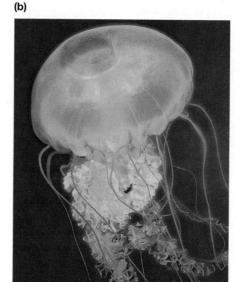

(c)

Figure 28-1 Some animals lack specialized respiratory structures
Most animals that lack a respiratory system have low metabolic demands and a large, moist body surface. **(a)** The flattened body of this marine flatworm exchanges gases with the water. **(b)** The cells in the bell-shaped body of a jellyfish have a low metabolic rate, and seawater flowing in and out of the bell during swimming allows adequate gas exchange. **(c)** Flagellated cells draw currents of water through numerous openings in the body of the sponge. These currents carry microscopic food particles and allow the cells to exchange gases with the water.

and water. Sponges, for example, circulate seawater through channels within their bodies, where it comes close to all of their cells (Fig. 28-1c).

Some animals combine a large skin surface through which diffusion occurs with a well-developed circulatory system (see Fig. 27-1b). In the earthworm, for example, gases diffuse through the moist skin and are distributed throughout the body by an efficient circulatory system. Blood in skin capillaries rapidly carries off oxygen that has diffused through the skin, maintaining a concentration gradient that favors the diffusion of oxygen inward. The worm's elongated shape ensures a large skin surface relative to its internal volume, and the worm's sluggish metabolism demands relatively little oxygen. The skin must stay moist to remain effective as a gas-exchange organ; a dry earthworm will suffocate.

Respiratory Systems Facilitate Gas Exchange by Diffusion

Most animals have evolved specialized respiratory systems that interface closely with their circulatory systems to exchange gases between the cells and the environment. The transfer of gases from the environment to the blood and then to the cells and back again normally occurs in stages that alternate bulk flow with diffusion. During **bulk flow**, fluids or gases move in bulk through relatively large spaces, from areas of higher pressure to areas of lower pressure. Bulk flow contrasts with diffusion, in which molecules move individually from areas of higher concentration to areas of lower concentration (see Chapter 5). In general, gas exchange in respiratory systems occurs in the following stages, illustrated for the mammalian system in Figure 28-2:

① Air or water, containing oxygen, is moved past a respiratory surface by bulk flow, commonly facilitated by muscular breathing movements.

② Oxygen and carbon dioxide are exchanged through the respiratory surface by diffusion; oxygen is carried into the capillaries of the circulatory system, and carbon dioxide is removed.

③ Gases are transported between the respiratory system and the tissues by bulk flow of blood as it is pumped throughout the body by the heart.

④ Gases are exchanged between the tissues and the circulatory system by diffusion. At the tissues, oxygen diffuses out of the capillaries and carbon dioxide diffuses into them along their concentration gradients.

Gills Facilitate Gas Exchange in Aquatic Environments

Gills are the respiratory structures of many aquatic animals. The simplest type of gill, found in certain mollusks and amphibians, consists of numerous projections of the body surface into the surrounding water (as shown by the nudibranch in the chapter-opening photo). In gener-

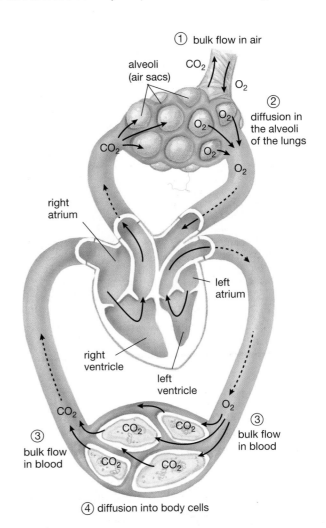

Figure 28-2 An overview of gas exchange
Oxygen is transported first by bulk flow as air is pumped into and out of the lungs. It moves by diffusion through capillary walls in the alveoli of the lungs and then is transported again by bulk flow in the bloodstream. Finally, diffusion carries oxygen into body cells. Carbon dioxide follows the opposite path. Red blood is oxygenated; blue blood is deoxygenated.

al, gills are elaborately branched or folded, maximizing their surface area. In some animals, gill size is determined by the availability of oxygen in the surrounding water. For example, salamanders living in stagnant water (which has little opportunity to mix with air) have larger gills than do those living in well-aerated water. Gills have a dense profusion of capillaries just beneath their delicate outer membrane. These capillaries bring blood close to the surface, where gas exchange occurs.

Fish gills are covered by a protective bony flap, the **operculum**, which keeps the delicate gill membranes from being nibbled off by predators and also streamlines the body, allowing the fish to swim faster. Fish create a continuous current over their gills by pumping water into their mouths and ejecting it through the opercular

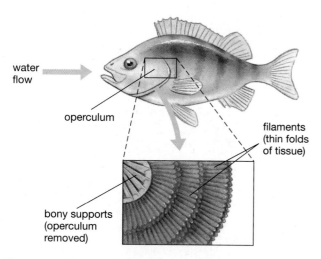

Figure 28-3 Gills exchange gases with water
The gill reaches its greatest complexity in the fish, where it is made of thin folds of tissue called filaments and is protected under the operculum, a bony flap. A one-way flow of water is maintained over the gill by the pumping of water through the mouth and out the opercular opening.

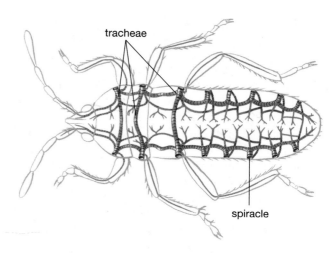

Figure 28-4 Insects breathe via tracheae
The tracheae of insects, such as this beetle, branch intricately throughout the body and open to the air through spiracles in the abdominal wall.

openings (Fig. 28-3). They can increase the flow of water by swimming with their mouths open; some fast swimmers, such as tuna and certain sharks, rely exclusively on swimming to ventilate their gills. Gills are useless out of water, because they collapse and dry out in air. As terrestrial animals evolved, therefore, their respiratory organs required both support and protection from desiccation.

Terrestrial Animals Have Internal Respiratory Structures

Terrestrial (land-dwelling) animals live in an atmosphere far higher in oxygen than aquatic environments, but extracting oxygen from dry air presents special challenges. All respiratory surfaces must remain moist, because gases must be dissolved in water to diffuse across membranes. But land-dwelling animals cannot afford the continuous loss of water that would occur by evaporation if their moist respiratory surfaces were on the outside of the body. Thus, land animals have evolved structures in which respiratory surfaces are moistened, supported, and protected from drying. Natural selection has produced a variety of terrestrial respiratory structures, such as tracheae in insects and lungs in vertebrates.

Insects Respire by Means of Tracheae

Insects use a system of elaborately branching internal tubes called **tracheae** (singular, **trachea**), which convey air directly to the body cells. Reinforced with chitin (which also supports the insect's external skeleton), tracheae subdivide and branch into smaller channels

(*tracheoles*) that penetrate the body tissues and allow gas exchange (Fig. 28-4). Each body cell is close to a tracheole, minimizing diffusion distances. Air enters the tracheae through a series of openings called **spiracles**, located along each side of the abdomen. Spiracles have valves that allow them to be opened or closed. Some large insects use muscular pumping movements of the abdomen to enhance air movement through the tracheae.

Most Terrestrial Vertebrates Respire by Means of Lungs

Lungs are chambers containing moist, delicate respiratory surfaces that are protected within the body, where water loss is minimized and the body wall provides support. The first vertebrate lung probably appeared in a freshwater fish and consisted of an outpocketing of the digestive tract. Gas exchange in this simple lung helped the fish survive in stagnant water, in which oxygen is scarce. Amphibians, which straddle the boundary between aquatic and terrestrial life, may use gills in the larval stage and lungs in the more terrestrial adult form. For example, the purely aquatic tadpole exchanges its gills for lungs as it develops into a more terrestrial frog (Fig. 28-5a,b). Frogs and salamanders use their moist skin as a supplemental respiratory surface.

The scales of reptiles (Fig. 28-5c) reduce the loss of water through the skin and allow reptiles to survive in dry environments. But scales also reduce the diffusion of gases through the skin, so the lungs of reptiles are better developed than are those of amphibians.

Birds and mammals are exclusively lung breathers. The bird lung has evolved special adaptations that allow extremely efficient gas exchange, which is necessary to

(a) (b) (c)

Figure 28-5 Amphibians and reptiles have different respiratory adaptations
(a) The bullfrog, an amphibian, begins life as a fully aquatic tadpole with feathery external gills that will later become enclosed in a protective chamber. *(b)* During metamorphosis into an air-breathing adult frog, the gills are lost and replaced by simple saclike lungs. In both tadpole and adult, gas exchange also occurs by diffusion through the skin, which must be kept moist to function as a respiratory surface. *(c)* The fully terrestrial reptile, such as this mangrove snake, is covered with dry scales that restrict gas exchange through the skin. Reptilian lungs are more efficient than are those of amphibians.

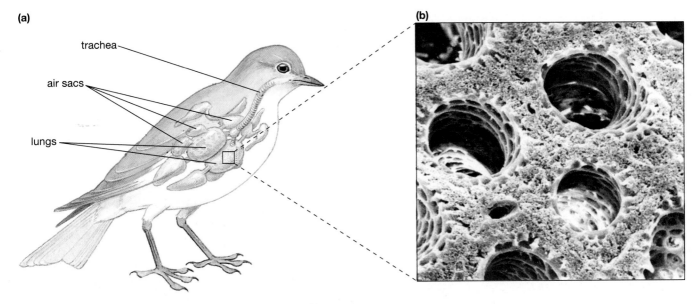

(a) (b)

trachea
air sacs
lungs

Figure 28-6 The bird respiratory system is extremely efficient
(a) Birds have air sacs in addition to their lungs. When air is inhaled, some air fills the lungs; the rest travels past the lungs to fill air sacs. As air is exhaled, the fresh air that has been temporarily stored in the air sacs fills the lungs on its way out. *(b)* Birds use tubular gas-exchange organs called parabronchi rather than saclike alveoli. The parabronchi allow the air to flow entirely through the lung to and from the air sacs.

support the enormous energy demands of flight. As a bird breathes in, it draws air through its lungs, where oxygen is extracted, and simultaneously pulls air into air sacs, some of which are located beyond the lungs (Fig. 28-6a). As the bird breathes out, oxygenated air from the air sacs is forced back through the lungs, allowing the bird to extract oxygen even as it exhales. In contrast to other vertebrate lungs, which are saclike, bird lungs are filled with hollow thin-walled tubes called *parabronchi*, which allow air to pass through them in both directions (Fig. 28-6b).

2) What Are the Features and Functions of the Human Respiratory System?

The respiratory system in humans and other lung-breathing vertebrates can be divided into two parts: (1) the **conducting portion** and (2) the **gas-exchange portion.** The conducting portion consists of a series of passageways that carry air into the gas-exchange portion, where gas is exchanged with the blood in tiny sacs in the lungs.

The Conducting Portion of the Respiratory System Carries Air to the Lungs

Air enters through the nose or the mouth, passes through the nasal cavity or oral cavity into a common chamber, the **pharynx**, and then travels through the **larynx** (Fig. 28-7). The opening to the larynx is guarded by the *epiglottis*, a flap of tissue supported by cartilage. During normal breathing, the epiglottis is tilted upward, allowing air to flow freely into the larynx. During swallowing, the epiglottis tilts downward and covers the larynx, directing substances into the esophagus instead. If an individual attempts to inhale and swallow at the same time, this reflex may fail and food can become lodged in the larynx, blocking air from entering the lungs. What should you do if you see this happen? The *Heimlich maneuver* (Fig. 28-8) is easy to perform and has saved countless lives.

Within the larynx are the **vocal cords**, bands of elastic tissue controlled by muscles. Muscular contractions can cause the vocal cords to obstruct the opening within the larynx partially, so exhaled air causes them to vibrate, giving rise to the tones of speech or song. The tones can be varied in pitch by stretching the cords and articulated into words by movements of the tongue and lips.

Inhaled air continues past the larynx into the **trachea**, a flexible tube whose walls are reinforced with semicircular bands of stiff cartilage. Within the chest, the trachea splits into two large branches called **bronchi** (singular, **bronchus**), one leading to each lung. Inside the lung, each bronchus branches repeatedly into ever-smaller tubes called **bronchioles**. Bronchioles lead finally to the microscopic **alveoli** (singular, **alveolus**), tiny air sacs where gas exchange occurs (see Fig. 28-7).

During its passage through the conducting system, air is warmed and moistened. Much of the dust and bacteria

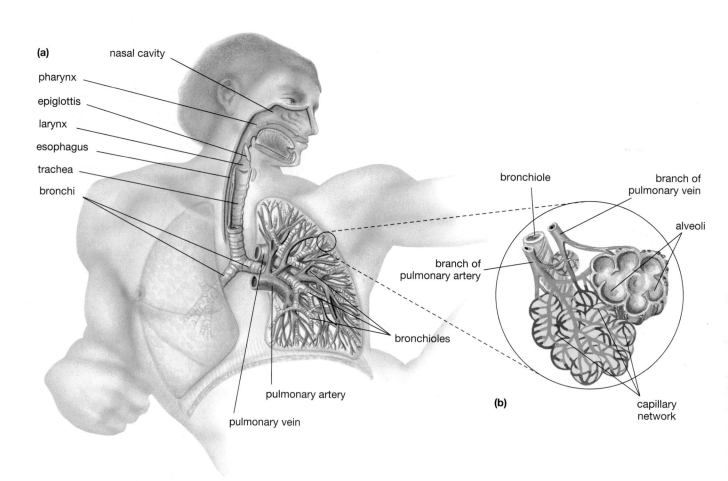

Figure 28-7 The human respiratory system
(*a*) Air enters mainly through the nasal cavity and mouth and passes through the pharynx and the larynx into the trachea. The epiglottis prevents food from going down the trachea rather than the esophagus. The trachea splits into two large branches, the bronchi, which lead into the two lungs. The smaller branches of the bronchi, the bronchioles, lead to the microscopic alveoli, which are enmeshed in capillaries, where gas exchange occurs. The pulmonary artery carries deoxygenated blood (in blue) to the lungs; the pulmonary vein carries oxygenated blood (in red) back to the heart. (*b*) Close-up of alveoli and their surrounding capillaries.

semble tiny bubbles and provide an enormous surface area for diffusion—about 75 square meters (800 square feet), 80 times the total skin surface area of an adult human. The alveoli, which cluster about the end of each bronchiole like a bunch of grapes, are entirely enmeshed in capillaries (see Fig. 28-7). Because both the alveolar wall and the adjacent capillary walls are only one cell thick, the air is extremely close to the blood in the capillaries. The lung cells remain moist because they are coated by a thin layer of watery fluid, lining each alveolus. Gases dissolve in this fluid and diffuse through the alveolar and capillary membranes (Fig. 28-9).

After the blood circulates through the body tissues, it is pumped to the lungs by the heart. The incoming blood surrounding the alveoli is low in oxygen (because the body cells have used it up) and high in carbon dioxide (which is released by the cells). This is especially true after heavy exertion, such as when you have sprinted up two flights of stairs on your way to class. In the alveoli, oxygen diffuses from the air, where its concentration is high, into the blood, where its concentration is low. Conversely, carbon dioxide diffuses out of the blood, where its concentration is high, into the air in the alveoli, where

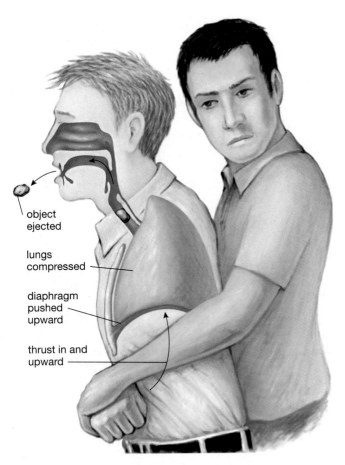

Figure 28-8 *The Heimlich maneuver can save lives*
If a person has choked on food or other object and is unable to breathe, stand behind him and grasp your hands together just above his navel but below the sternum (the "breastbone" to which the ribs are attached). Quickly and forcefully pull your linked hands upward and inward toward your body. This movement will push upward on the victim's diaphragm and force air out of his lungs, possibly dislodging the object. Repeat this if necessary.

it carries is trapped in mucus secreted by cells that line the respiratory passages. The mucus, with its trapped debris, is continuously swept upward toward the pharynx by cilia that line the bronchioles, bronchi, and trachea. On reaching the pharynx, the mucus is coughed up or swallowed. Smoking interferes with this cleansing process by paralyzing the cilia. (See "Health Watch: Smoking—A Life and Breath Decision.")

Gas Exchange Occurs in the Alveoli

The lung is clearly designed to maximize respiratory surface area. The dense network of bronchioles conducts air to tiny structures, the alveoli, which are nearly *all* surface. Each lung is packed with 1.5 million to 2.5 million alveoli. These microscopic (0.2-millimeter-diameter) chambers give magnified lung tissue the appearance of sponge cake (see Fig. E28-2). The thin-walled alveoli re-

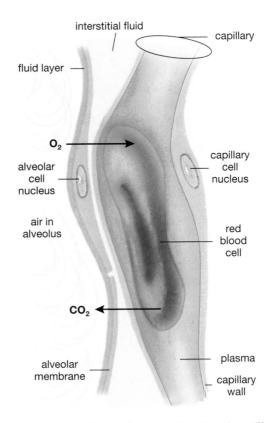

Figure 28-9 *Gas exchange between alveoli and capillaries*
The relationship between the alveolar membranes and the capillary walls facilitates the exchange of oxygen (O_2) and carbon dioxide (CO_2) by diffusion. Because the alveoli and capillaries are only one-cell thick and the cells are coated in a thin layer of fluid, gases dissolve and diffuse easily.

Health Watch
Smoking—A Life and Breath Decision

An estimated 419,000 Americans die of smoking-related diseases each year. Further, the cost of health care and loss of productivity due to smoking-related ailments costs the United States about $97 billion annually. The American Cancer Society estimates that nearly 159,000 individuals died of lung cancer in 1996 and that 87% of those deaths were attributable to smoking. The rest of the deaths due to smoking result from a combination of smoking-induced emphysema, chronic bronchitis, heart disease, and cancers other than lung cancer, including cancers of the mouth, larynx, esophagus, pancreas, bladder, and kidney.

Tobacco smoke has a dramatic impact on the human respiratory tract. As smoke is inhaled through the nose, trachea, and bronchi, toxic substances such as nicotine and sulfur dioxide paralyze the cilia that line the respiratory tract; a single cigarette can inactivate them for a full hour. Because the function of these ciliary sweepers is to remove inhaled particles, smoking inhibits them just when they are most needed. The visible portion of cigarette smoke consists of billions of microscopic carbon particles. Adhering to these particles are toxic substances, of which a dozen or more are *carcinogenic* (cancer causing). With the cilia out of action, the particles stick to the walls of the respiratory tract or enter the lungs. Thus, smokers encounter a higher-than-normal risk of cancer in all areas of the respiratory tract touched by smoke.

Cigarette smoke also impairs the amoebalike white blood cells (macrophages) that defend the respiratory tract by engulfing foreign particles and bacteria. Consequently, still more bacteria, dust, and smoke particles enter the lungs. In response to the irritation of cigarette smoke, the respiratory tract increases the production of mucus, a third method of trapping foreign particles. But without the cilia to sweep it along, the mucus builds up and can obstruct the airways; the familiar "smoker's cough" is an attempt to expel the mucus. Microscopic smoke particles find a secure lodging place in the tiny, moist alveoli deep within the lungs. There they accumulate over the years until the lungs of a heavy smoker are literally blackened. The longer the delicate tissues of the lungs are exposed to the carcinogens on the trapped particles, the greater the chance that cancer will develop (Fig. E28-1).

Chronic bronchitis and *emphysema* together kill about 96,000 people each year in the United States, and smoking is the leading contributing factor to each of these diseases. **Chronic bronchitis** is a persistent lung infection characterized by coughing, swelling of the lining of the respiratory tract, an increase in

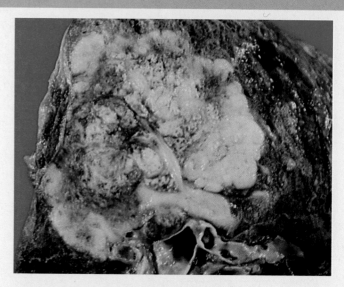

Figure E28-1 Smoking causes lung cancer
A tumor of lung cancer is visible as a large, pale mass; the lung tissue surrounding it is blackened by trapped smoke particles.

mucus production, and a decrease in the number and activity of cilia. The result: a decrease in air flow to the alveoli. **Emphysema** occurs when toxic substances in cigarette smoke, such as nitrogen oxides and sulfur dioxide, cause the body to produce substances that reduce the elasticity and increase the brittleness of lung tissue. As the brittle alveoli rupture, the lung gradually loses its normal sponge-cake appearance (Fig. E28-2a) and more closely resembles blackened Swiss cheese (Fig. E28-2b). The loss of the alveoli, where gas exchange occurs, leads to oxygen deprivation of all body tissues. In an individual with emphysema, breathing becomes labored and grows increasingly worse until death.

Carbon monoxide, present in high levels in cigarette smoke, binds tenaciously to red blood cells in place of oxygen. This binding reduces the blood's oxygen-carrying capacity and thereby increases the work the heart must do. Chronic bronchitis and emphysema compound this problem. Smoking also causes *atherosclerosis*, or thickening of the arterial walls by fatty deposits that can lead to heart attacks (see Chapter 27). As a result, smok-

its concentration is lower (see Fig. 28-9). Blood from the lungs, oxygenated and purged of carbon dioxide, returns to the heart, which pumps it to the body tissues. In the tissues, the concentration of oxygen is lower than in the blood, so oxygen diffuses into the cells. Carbon dioxide, which has built up in the cells, diffuses into the blood (see Fig. 28-2).

Oxygen and Carbon Dioxide Are Transported via Different Mechanisms

In the blood, oxygen binds loosely and reversibly with **hemoglobin**, a large, iron-containing protein in the red blood cells, as described in Chapter 27 (see Fig. 27-9). Each hemoglobin molecule can bind up to four oxygen

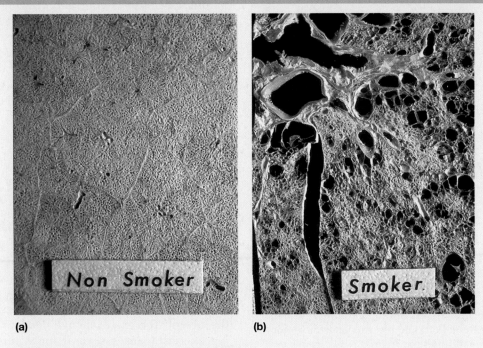

Figure E28-2 Smoking causes emphysema
(a) Normal lung tissue from a nonsmoker, seen in cross section, has nearly invisibly small openings, the alveoli, surrounded by healthy tissue. *(b)* The lung of a smoker suffering from emphysema is full of large holes, each caused by the rupture of hundreds of alveoli.

ers are 70% more likely than nonsmokers to die of heart disease. The carbon monoxide in cigarette smoke may also contribute to the reproductive problems experienced by pregnant women who smoke, including increased incidence of miscarriage, lower birth weight of their babies, and learning impairment of their young children.

"Passive smoking," or breathing secondhand smoke, poses real health hazards for both children and adults. Researchers have concluded that children whose parents smoke are more likely to contract bronchitis, pneumonia, ear infections, coughs, and colds and to have decreased lung capacity. Children who grow up with smokers are more likely to develop asthma and allergies; for children with asthma, the number and severity of asthma attacks are increased by secondhand smoke.

Among adults, angina sufferers are likely to experience chest pain after being in a smoke-filled room. Studies have concluded that nonsmoking spouses of smokers face a 30% higher risk of both heart attack and lung cancer than do spouses of nonsmokers. A recent study links even relatively infrequent exposure to secondhand smoke with atherosclerosis. Government agencies report that secondhand smoke is responsible for an estimated 3000 lung-cancer deaths and 37,000 deaths from heart disease in nonsmokers each year.

For smokers who quit, healing begins immediately and the chances of heart attack, lung cancer, and numerous other smoking-related illnesses gradually diminish. Atherosclerosis may, however, continue to progress.

molecules (eight oxygen atoms). Nearly all the oxygen carried by the blood is bound to hemoglobin. By removing oxygen from solution in the plasma, hemoglobin maintains a concentration gradient that favors the diffusion of oxygen from the air into the blood. Thanks to hemoglobin, our blood can carry about 70 times as much oxygen as it could if the oxygen were simply dissolved in the plasma.

As hemoglobin binds oxygen, the protein undergoes a slight change in shape, which alters its color. Deoxygenated blood is dark maroon-red and appears bluish through the skin; oxygenated blood is a bright cherry-red. Carbon monoxide gas (CO) is highly toxic, because it "fools" hemoglobin, binding in place of oxygen and more than 200 times as tenaciously. Hemoglobin

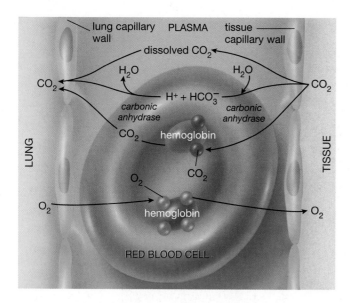

Figure 28-10 *The chemistry of gas exchange*
Carbon dioxide from tissues diffuses into blood through the capillary wall (right). Most CO_2 combines with water in the presence of carbonic anhydrase to form bicarbonate. Some CO_2 is carried by hemoglobin; a small amount is dissolved in plasma. At the lungs, CO_2 diffuses into the alveoli along its concentration gradient (left). Meanwhile, O_2 diffuses into the capillary from the alveoli (left) and is carried in hemoglobin to the tissues, where it diffuses out (right).

containing CO is also bright red, but it is incapable of transporting oxygen. Most victims of asphyxiation have bluish lips and nail beds, because their hemoglobin is deoxygenated; the lips and nail beds of victims of carbon monoxide poisoning are brighter red than normal.

Carbon dioxide is transported in three different ways. In the presence of an enzyme found in red blood cells (*carbonic anhydrase*), about 70% of the CO_2 reacts with water to form bicarbonate ion (HCO_3^-), which then diffuses into the plasma. About 20% of the CO_2 binds to hemoglobin (which has released its O_2 to the tissues) for its return trip to the lungs; the remaining 10% stays dissolved in the plasma as CO_2 (Fig. 28-10). Both the production of bicarbonate ion and the binding of CO_2 to hemoglobin reduce the concentration of dissolved CO_2 in the blood and increase the gradient for CO_2 to flow from the body cells into the blood.

Air Is Inhaled Actively and Exhaled Passively

Outside the lungs, the chest cavity is airtight—bounded by neck muscles and connective tissue on top and by the dome-shaped, muscular **diaphragm** on the bottom. Surrounding and protecting the lungs is the *rib cage* within the wall of the chest. Lining the rib cage and surrounding the lungs is a double layer of **pleural membranes**. These membranes contribute to the airtight seal between the lungs and the chest wall.

Breathing occurs in two stages: (1) **inhalation**, when air is actively drawn into the lungs, and (2) **exhalation**, when it is passively expelled from the lungs. Inhalation is accomplished by enlarging the chest cavity. To do so, the diaphragm muscles contract, drawing the diaphragm downward. The rib muscles also contract, lifting the ribs up and outward (Fig. 28-11). When the chest cavity is expanded, the lungs expand with it, because a vacuum holds them tightly against the inner wall of the chest. (If the chest is punctured and air leaks in between the pleural membranes and the lung, the lung will collapse.) As the lungs expand, their increased volume creates a partial vacuum that draws air into the lungs.

Exhalation occurs automatically when the muscles that cause inhalation are relaxed. The relaxed diaphragm domes upward, and the ribs fall down and inward, decreasing the size of the chest cavity and forcing air out of the lungs. More air can be forced out by contracting the abdominal muscles. After exhalation, the lungs still contain air. This air prevents the thin alveoli from collapsing and fills the space within the conducting portion of the respiratory system. A normal breath moves only about 500 milliliters (1 pint) of air into the respiratory system. Of this, only about 350 milliliters reaches the alveoli for gas exchange. Deeper breathing during exercise causes several times this volume to be exchanged.

Breathing Rate Is Controlled by the Respiratory Center of the Brain

Breathing occurs rhythmically and automatically without conscious thought. But, unlike the heart muscle, the muscles used in breathing are not self-activating; each contraction is stimulated by impulses from nerve cells. These impulses originate in the **respiratory center**, which is located in the medulla just above the spinal cord. Nerve cells in the respiratory center generate cyclic bursts of impulses that cause the alternating contraction and relaxation of the respiratory muscles.

The respiratory center receives input from several sources and adjusts breathing rate and volume to meet the body's changing needs. The respiratory rate is regulated to maintain a constant level of carbon dioxide in the blood, as monitored by carbon dioxide receptors in the medulla. An elevated level of carbon dioxide—caused, for example, by an increase in cellular activity, such as that experienced by your muscle cells when you run up stairs—signals a need for more oxygen and causes the receptors to stimulate an increase in the rate and depth of breathing. These receptors are extremely sensitive; an increase in carbon dioxide of only 0.3% can double the breathing rate.

The respiratory rate is much less sensitive to changes in oxygen concentration, because normal breathing supplies an overabundance of oxygen. But if blood oxygen levels fall drastically, receptors in the aorta and carotid arteries stimulate the respiratory center. Surprisingly,

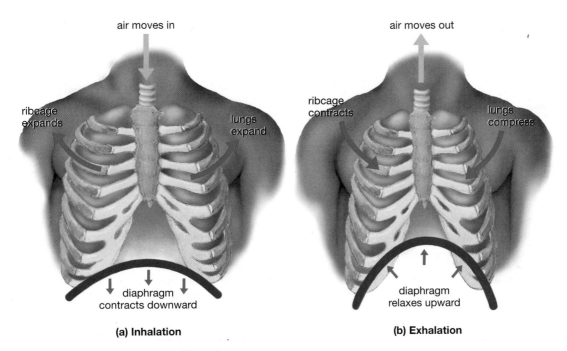

air moves in

ribcage expands

lungs expand

diaphragm contracts downward

(a) Inhalation

air moves out

ribcage contracts

lungs compress

diaphragm relaxes upward

(b) Exhalation

Figure 28-11 *The mechanics of breathing*
During inhalation, rhythmic nerve impulses from the brain stimulate the diaphragm to contract (pulling it downward) and the muscles surrounding the ribs to contract (moving them up and outward). The result is an increase in the size of the chest cavity, causing air to rush in. Relaxation of these muscles (exhalation) allows the diaphragm to dome upward and the rib cage to collapse, forcing air out of the lungs.

when you begin strenuous activity, such as running, an increase in breathing rate actually *precedes* any changes in blood gas levels. Apparently, when higher brain centers activate muscles during heavy exercise, they simultaneously stimulate the respiratory center to increase breathing rate. Breathing activity is then "fine-tuned" by the receptors that monitor carbon dioxide concentrations.

Summary of Key Concepts

1) What Are Some Evolutionary Adaptations for Gas Exchange?

Respiration makes possible the exchange of oxygen and carbon dioxide between the body and the environment by diffusion of these gases across a moist surface. In moist environments, animals whose bodies are very small or flattened may rely exclusively on diffusion through the body surface. Animals with low metabolic demands and/or well-developed circulatory systems may also lack specialized respiratory structures. Larger, more active animals have evolved specialized respiratory systems. Animals in aquatic environments have evolved gills, such as those of fish and many amphibians. On land, moist respiratory surfaces mustbe protected internally. This need has selected for the evolution of tracheae in insects and lungs in terrestrial vertebrates.

The transfer of gases between respiratory systems and tissues occurs in a series of stages that alternate bulk flow with diffusion. Air or water moves by bulk flow past the respiratory surface, and gases in blood are carried by bulk flow. Gases move by diffusion across membranes between the respiratory system and the capillaries and between the capillaries and the tissues.

2) What Are the Features and Functions of the Human Respiratory System?

The human respiratory system consists of a conducting portion and a gas-exchange portion. Air passes first through the conducting portion, consisting of the nose and mouth, pharynx, larynx, trachea, bronchi, and bronchioles, and then into the gas-exchange portion, composed of alveoli (microscopic sacs). Blood within a dense capillary network surrounding the alveoli releases carbon dioxide and absorbs oxygen from the air.

Most of the oxygen in the blood is bound to hemoglobin within red blood cells. Hemoglobin binds oxygen at concentrations typical of those in alveolar capillaries and releases it at the lower oxygen concentrations of the body tissues. Carbon dioxide diffuses into the blood from the tissues and is transported as bicarbonate, bound to hemoglobin, or dissolved in blood plasma.

Breathing involves actively drawing air into the lungs by contracting the diaphragm and the rib muscles, which expand the chest cavity. Relaxing these muscles causes the chest cavity to collapse, expelling the air. Respiration is controlled by nerve impulses that originate in the medulla's respiratory center. The respiration rate is modified by receptors, such as those in the medulla that monitor carbon dioxide levels in the blood.

Key Terms

alveolus   *p. 562*
bronchiole   *p. 562*
bronchus   *p. 562*
bulk flow   *p. 559*
chronic bronchitis   *p. 564*
conducting portion   *p. 561*

diaphragm   *p. 566*
emphysema   *p. 564*
exhalation   *p. 566*
gas-exchange portion   *p. 561*
gill   *p. 559*
hemoglobin   *p. 564*

inhalation   *p. 566*
larynx   *p. 562*
lung   *p. 560*
operculum   *p. 559*
pharynx   *p. 562*
pleural membrane   *p. 566*

respiratory center   *p. 566*
spiracle   *p. 560*
trachea (in birds/mammals)
   p. 562
trachea (in insects)   *p. 560*
vocal cord   *p. 562*

Thinking Through the Concepts

Multiple Choice

1. *With which other system do specialized respiratory systems most closely interface in exchanging gases between the cells and the environment?*
   a. the skin
   b. the excretory system
   c. the circulatory system
   d. the muscular system
   e. the nervous system

2. *The gas-exchange portion of the human respiratory system is the*
   a. larynx   b. trachea
   c. bronchi   d. pharynx
   e. alveoli

3. *Which of the following animals use tracheae for respiration?*
   a. mollusks   b. snails
   c. insects   d. fish
   e. bookworms

4. *How is most of the oxygen transported in the blood?*
   a. dissolved in plasma
   b. bound to hemoglobin
   c. in the form of CO_2
   d. as bicarbonate
   e. dissolved in water

5. *Which of the following statements regarding cigarette smoking is true?*
   a. Cigarette smoke damages the alveoli.
   b. Cigarette smoke decreases the amount of oxygen in the blood.
   c. Cigarette smoke causes many types of cancer.
   d. Cigarette smoke can lead to heart damage.
   e. all of the above

6. *Which of the following pairs of respiratory adaptations and animals are NOT correct?*
   a. gills: fish
   b. parabronchi: birds
   c. lungs: mammals
   d. moist skin: snakes
   e. spiracles: insects

? Review Questions

1. Describe three arthropod respiratory systems and two vertebrate respiratory systems.

2. Trace the route taken by air in the vertebrate respiratory system, listing the structures through which it flows and the point at which gas exchange occurs.

3. Explain some characteristics of animals in moist environments that may supplement respiratory systems or make them unnecessary.

4. How are human respiratory movements initiated? How are they modified, and why are these controls adaptive?

5. What events occur during human inhalation? Exhalation? Which of these is always an active process?

6. Trace the pathway of an oxygen molecule in the human body, starting with the nose and ending with a body cell.

7. Describe the effects of smoking on the human respiratory system.

8. Explain how bulk flow and diffusion interact to promote gas exchange between air and blood and between blood and tissues.

9. Compare carbon dioxide and oxygen transport in the blood. Include the source and destination of each.

10. Explain how the structure and arrangement of alveoli make them well suited for their role in gas exchange.

Applying the Concepts

1. Heart–lung transplants are performed in some cases, but donors are scarce. On the basis of your knowledge of the respiratory and circulatory systems and of lifestyle factors that might damage them, what criteria would you use in selecting a recipient for such a transplant?

2. Nicotine is a drug in tobacco that is responsible for several of the effects that smokers crave. Discuss the advantages and disadvantages of low-nicotine cigarettes.

3. Discuss why a brief exposure to carbon monoxide is much more dangerous than a brief exposure to carbon dioxide.

4. Describe several adaptations that might evolve to help members of a species of mammal respire better if the population began living continuously for many generations at very high altitudes.

5. Mary, a strong-willed 3-year-old, threatens to hold her breath until she dies if she doesn't get her way. Can she carry out her threat? Explain.

Group Activity

Form groups of three or four to address the following issue: The level of oxygen in the atmosphere is currently 21%, but evidence from gases trapped in amber 25 million to 40 million years old indicates that the level was once as high as 30%. If the level were to drop slowly over the next few centuries due to pollution of the oceans, where much of Earth's oxygen is produced, what might happen? What short-term adaptations would humans and other mammals have to make to adjust to this change? What sorts of evolutionary changes in mammalian respiratory systems would be favored in this new environment? As the level of oxygen in the atmosphere dropped, which types of present-day organisms would you expect to be affected most, which least, and why?

For More Information

Adler, T. "Fish Dishes May Catch on among Smokers." *Science News*, July 30, 1994. Discusses chronic obstructive pulmonary disease.

Feder, M. E., and Burggren, W. W. "Skin Breathing in Vertebrates." *Scientific American*, May 1985. Describes how some vertebrates supplement the action of their lungs or gills to perform gas exchange better.

Harding, C. "Going to Extremes." *National Wildlife*, August/September 1993. Describes adaptations that allow diving animals to plunge to enormous depths without running out of oxygen.

Houston, C. "Mountain Sickness." *Scientific American*, October 1992. The mechanisms of potentially fatal altitude sickness are explained.

Schmidt-Nielsen, K. "How Birds Breathe." *Scientific American*, December 1971. Specializations of the bird respiratory system include additional air sacs and even hollow bones.

Seppa, N. "Secondary Smoke Carries High Price." *Science News*, January 17, 1998. New research suggests that smoking causes atherosclerosis, that the damage continues after a smoker quits, and that exposure to secondhand smoke significantly increases the buildup of plaque in the carotid artery of nonsmokers.

Answers to Multiple-Choice Questions
1. c   2. e   3. c   4. b   5. e   6. d

An Atlantic puffin on the Shetland Islands, Scotland, pauses with a beakful of sand eels, the nutritious payoff of a foraging flight.

Nutrition and Digestion 29

At a Glance

Net Watch

On-line resources for this chapter are on the World Wide Web at:
http://www.prenhall.com/audesirk
(click on the Table of Contents link and then select Chapter 29).

Animals are machines for converting food into more animals. Natural selection has favored the animals that accomplish the conversion most efficiently. Food is an animal's only source of energy, but some of that energy must be spent on finding, securing, and digesting the food. Only the "leftover" energy is available for body maintenance, growth, and reproduction. So the most-successful individuals will be the ones with the largest difference between energy intake and energy expended for feeding and digestion. This evolutionary drive for efficiency has fostered a range of animal behaviors and physiological systems that exploit every conceivable food source.

Many different strategies have proved to be successful, because natural selection can favor any profitable strategy. For example, compare a sponge with the Atlantic puffin (a seabird). The sponge waits passively for bits of detritus to float by, but the Atlantic puffin actively flies long distances in search of fish. Both strategies persist because both are efficient. The sponge's diet may consist of foods with low energy content, but its investment in getting and digesting the food is likewise low. The puffin, in contrast, invests a great deal of energy in finding and digesting food, but its payoff is highly nutritious, energy-packed food. In evolutionary terms, the bottom-line profit (not the initial investment) is the most important thing.

The human digestive system is a relatively unspecialized one that has been designed by evolution to handle a wide variety of food sources. The manner in which this system processes food is something akin to a highly automated factory assembly line, except that the goal is *dis*assembly. Food travels from mouth to anus; in the course of this circuitous route, it is subjected to a carefully orchestrated succession of digestive operations. By the time its passage is complete, the food has been chopped, mashed, mixed, churned, and bathed in a series of powerful chemicals. Everything of value has been extracted, and the residue is ejected. This stepwise deconstruction of foodstuffs requires coordinated action from the integrated array of structures that make up the digestive system.

1) What Nutrients Do Animals Need?

Nutrition is the process of acquiring and, if necessary, processing nutrients into a usable form. Animal nutrients fall into five major categories: (1) lipids, (2) carbohydrates, (3) proteins, (4) minerals, and (5) vitamins. These substances provide the body with its basic needs:

- energy to fuel cellular metabolism and activities;
- the chemical building blocks, such as amino acids, to construct complex molecules unique to each animal; and
- minerals and vitamins that participate in a variety of metabolic reactions.

The Primary Sources of Energy Are Carbohydrates and Fats

Each cell in an animal's body relies on a continuous expenditure of energy to maintain its incredible complexity and perform its specific functions. Three nutrients provide energy for animals: fats, carbohydrates, and proteins. In a "typical" American diet, fats provide about 38%, carbohydrates about 46%, and protein about 16% of the energy. These molecules are broken down during cellular respiration, and the energy derived from them is used to produce adenosine triphosphate (ATP) (see Chapter 8).

The energy in nutrients is measured in calories. A **calorie** is the amount of energy required to raise the temperature of 1 gram of water by 1 degree Celsius. The calorie content of foods is measured in units of 1000 calories (*kilocalories*), also known as **Calories** (with a capital *C*). The human body at rest burns about 1550 Calories per day (somewhat more for males and less for females). Exercise significantly boosts caloric requirements: Well-trained athletes can temporarily raise their calorie

consumption from a resting rate of about 1 Calorie per minute to nearly 20 Calories per minute during vigorous exercise (Table 29-1).

Lipids Include Fats, Phospholipids, and Cholesterol

Lipids are a diverse group of molecules that generally contain long chains of carbon atoms and are insoluble in water. The principal types of lipids are *fats*, or *triglycerides; phospholipids;* and *cholesterol* (see Chapter 3). Fats are used primarily as a source of energy. Phospholipids are important components of cellular membranes and also provide the insulating covering of neurons (nerve cells). Cholesterol is used in the synthesis of cellular membranes, bile (which aids in fat breakdown), and certain hormones. Some animal species can synthesize all the specialized lipids they need. Others must acquire specific types of lipid building blocks, called **essential fatty acids**, from their food. For example, humans are unable to synthesize linoleic acid, which is required for the synthesis of certain phospholipids, so we need to obtain this essential fatty acid from our diet.

Humans and most other animals store energy primarily as fat. When an animal's diet provides more energy than is expended through metabolic activities, most of the excess carbohydrate, fat, or protein is converted to fat for storage. About 3600 Calories are stored in each pound of fat. Fats have two major advantages as energy-storage molecules. First, they are the most concentrated energy source, containing more than twice the energy per unit weight of either carbohydrates or protein (about 9 Calories per gram for fats compared with about 4 per gram for proteins and carbohydrates). Second, lipids are *hydrophobic*—that is, they do not mix with water. Fat deposits, therefore, do not cause any extra accumulation of water in the body. For both these reasons, fats store more calories with less weight

Activity	Calories/ min	500 Calories Cheeseburger	300 Calories Ice Cream Cone	70 Calories Apple
Cross-country skiing (5 mph)	12	42 min	25 min	6 min
Rowing (machine, vigorous)	11	45 min	27 min	6 min
Jogging (5.5 mph)	11	45 min	27 min	6 min
Bicycling (10 mph)	7	1 hr 10 min	43 min	10 min
Dancing	7	1 hr 10 min	43 min	10 min
Gardening	6.5	1 hr 17 min	46 min	11 min
Swimming (crawl, 20 yd/min)	5	1 hr 40 min	1 hr	14 min
Walking (3 mph)	3.8	2 hr 12 min	1 hr 19 min	18 min
Sitting quietly	1.7	4 hr 54 min	2 hr 56 min	41 min

Table 29-1 Approximate Energy Consumed by a 150-Pound Person for Different Activities

Time to "Work Off"

Figure 29-1 Fat provides insulation
These walruses can withstand the icy waters of the polar seas because they are insulated with a thick layer of fat beneath the skin.

than do other molecules. Minimizing weight allows an animal to move faster (important for escaping predators and hunting prey) and to use less energy for movement (important when food supplies are limited).

Mammals, which maintain an elevated body temperature, commonly make their fat deposits do double duty by providing insulation as well as storing energy. Fat is typically stored in a layer beneath the skin. There it insulates the body, because fat conducts heat at only one-third the rate of other body tissues. Mammals who live near the North or South Pole or in cold ocean waters are particularly dependent on this insulating layer, which reduces the amount of energy they must expend to keep warm (Fig. 29-1).

Carbohydrates, Including Sugars and Starches, Are a Source of Quick Energy

Carbohydrates consist of monosaccharide and disaccharide sugars as well as longer chains of sugars called *polysaccharides* (see Chapter 3). Polysaccharides include starches, the principal energy-storage material of plants; *glycogen,* a short-term energy-storage molecule in animals; and *cellulose,* the major structural component of plant cell walls. During digestion, carbohydrates are broken down into sugars and absorbed. For practical purposes, body cells obtain their energy from a single sugar: glucose. Glucose can be derived from fats, amino acids, and the carbohydrates consumed in the diet.

Animals, including humans, store the carbohydrate **glycogen** (a large, highly branched chain of glucose molecules) in the liver and muscles. Although humans can potentially store hundreds of pounds of fat, we store less than half a pound of glycogen. During exercise, such as running, the body draws on this store of glycogen as a source of quick energy. When the activity is prolonged, as in the case of a marathon runner, the stored glycogen can be totally depleted. The expression "hitting the wall" describes the extreme fatigue that long-distance runners may experience after exhausting their glycogen supply.

Proteins, Composed of Amino Acids, Perform a Wide Range of Functions within the Body

Each day, the human body breaks down 20 to 30 grams of its own protein, which is metabolized for energy. This protein is replaced by protein taken in through the diet, and any excess amino acids are either broken down and used for energy or stored as fat. The breakdown of protein produces the waste product **urea**, which is filtered from the blood by the kidneys. Specialized diets in which protein is the major energy source place extra stress on the kidneys. The major role of dietary protein is as a source of amino acids to make new molecules (see also Chapter 3).

Amino acids are used to synthesize certain hormones, other amino acids, some neurotransmitters (chemicals used in communication between neurons), and new proteins. These proteins have diverse roles in the body, acting as enzymes, receptors on cell membranes, oxygen transport molecules (hemoglobin), structural components (hair and nails), hormones, antibodies, and muscle proteins.

In the digestive tract, dietary protein is broken down into its amino acid subunits. Then, in the body cells, the amino acids are linked in specific sequences to form new proteins. The human liver can synthesize (from other amino acids) 9 of the 20 different amino acids used in proteins. Those that cannot be synthesized, called **essential amino acids**, must be supplied by the diet in foods such as meat, milk, eggs, corn, beans, and soybeans. Because many plant proteins are deficient in some of the essential amino acids, individuals on a vegetarian diet must include a variety of plants whose proteins together will provide all nine, or they risk protein deficiency. The nine essential amino acids and examples of how different plant proteins complement one another in providing them are illustrated in Table 29-2.

Table 29-2 The Essential Amino Acids		
Present in:		
Beans and Other Legumes	Both Legumes and Grains	Corn and Other Grains
Isoleucine Lysine	Valine Histidine Threonine Phenylalanine Leucine	Tryptophan Methionine

Minerals Are Elements and Small Inorganic Molecules Required by the Body

Animals require a wide variety of **minerals**, which are elements and small inorganic molecules (Table 29-3). Minerals must be obtained through the diet, either from food or dissolved in drinking water, because the body cannot manufacture them. Required minerals include calcium, magnesium, and phosphorus, which are major constituents of bones and teeth. Others, such as sodium and potassium, are essential for muscle contraction and the conduction of nerve impulses. Iron is used in the production of hemoglobin, and iodine is found in hormones produced by the thyroid gland. In addition, trace amounts of several other minerals, including zinc, copper, and selenium, are required, typically as parts of enzymes.

Vitamins Are Required in Small Amounts and Play Many Roles in Metabolism

Vitamins are a diverse group of organic compounds that animals require in small amounts. In general, the body cannot synthesize vitamins (or cannot do so in adequate

Table 29-3 Human Water and Mineral Requirements[a]

Mineral	RDA[b] for Healthy Adult Male (milligrams)	Dietary Sources	Major Functions in Body	Deficiency Symptoms
Water	1.5 liters/day	Solid foods, liquids, drinking water	Transport of nutrients Temperature regulation Metabolic reactions	Thirst, dehydration
Calcium	800	Milk, cheese, green vegetables, legumes	Bone and tooth formation Blood clotting Nerve impulse transmission	Stunted growth Rickets, osteoporosis Convulsions
Phosphorus	800	Milk, cheese, meat, poultry, grains	Bone and tooth formation Acid–base balance	Weakness Demineralization of bone Loss of calcium
Potassium	2500	Meats, milk, fruits	Acid–base balance Body water balance Nerve function	Muscular weakness Paralysis
Chlorine	2000	Table salt	Formation of gastric juice Acid–base balance	Muscle cramps Apathy Reduced appetite
Sodium	2500	Table salt	Acid–base balance Body water balance Nerve function	Muscle cramps Apathy Reduced appetite
Magnesium	350	Whole grains, green leafy vegetables	Activation of enzymes in protein synthesis	Growth failure Behavioral disturbances Weakness, spasms
Iron	10	Eggs, meats, legumes, whole grains, green vegetables	Constituent of hemoglobin and enzymes involved in energy metabolism	Iron-deficiency anemia (weakness, reduced resistance to infection)
Fluorine	2	Fluoridated water, tea, seafood	May be important in maintenance of bone structure	High frequency of tooth decay
Zinc	15	Widely distributed in foods	Constituent of enzymes involved in digestion	Growth failure Small sex glands
Iodine	0.14	Seafish and shellfish, dairy products, many vegetables, iodized salt	Constituent of thyroid hormones	Goiter
Copper Silicon Vanadium Tin Nickel Selenium Manganese	Not established (trace amounts)	Widely distributed in foods	Some unknown; some work in conjunction with enzymes	Unknown

[a]Modified from N. S. Scrimshaw and V. R. Vernon, "The Requirements of Human Nutrition," *Scientific American*, September 1976.

[b]Recommended Daily Allowance.

amounts), so they must be obtained from food. When it is exposed to sunlight, our skin can manufacture some vitamin D, but most of us spend so much time indoors that we do not synthesize enough and must supplement it through our diet. The vitamins considered essential in human nutrition are listed in Table 29-4.

Human vitamins are often grouped into two categories, water soluble and fat soluble. *Water-soluble vitamins* include vitamin C and the 11 compounds that make up the B-vitamin complex. These substances dissolve in the water of the blood plasma and are excreted by the kidneys. They are not stored in the body in any appreciable amounts. Water-soluble vitamins generally work in conjunction with enzymes to promote chemical reactions that supply energy or synthesize materials. Because each vitamin participates in several metabolic processes, a deficiency of a single vitamin can have wide-ranging effects (see Table 29-4).

Table 29-4 Human Vitamin Requirements

Vitamin	RDA for Adult (milligrams)	Dietary Sources	Major Functions in Body	Deficiency Symptoms
Water soluble				
Vitamin B$_1$ (thiamin)	1.9	Milk, meat, bread	Coenzyme in metabolic reactions	Beriberi (muscle weakness, peripheral nerve changes, edema, heart failure)
Vitamin B$_2$ (riboflavin)	1.8	Widely distributed in foods	Constituent of coenzymes in energy metabolism	Reddened lips, cracks at corner of mouth, lesions of eye
Niacin	20	Liver, lean meats, grains, legumes	Constituent of two coenzymes in energy metabolism	Pellagra (skin and gastrointestinal lesions; nervous, mental disorders)
Vitamin B$_6$ (pyridoxine)	2	Meats, vegetables, whole-grain cereals	Coenzyme in amino acid metabolism	Irritability, convulsions, muscular twitching, dermatitis, kidney stones
Pantothenic acid	5–10	Milk, meat	Constituent of coenzyme A with a role in energy metabolism	Fatigue, sleep disturbances, impaired coordination
Folacin	0.4	Legumes, green vegetables, whole wheat	Coenzyme involved in nucleic and amino acid metabolism	Anemia, gastrointestinal disturbances, diarrhea, red tongue
Vitamin B$_{12}$	0.003	Meats, eggs, dairy products	Coenzyme in nucleic acid metabolism	Pernicious anemia, neurological disorders
Biotin	0.15–0.3	Legumes, vegetables, meats	Coenzymes required for fat synthesis, amino acid metabolism, and glycogen formation	Fatigue, depression, nausea, dermatitis, muscular pains
Choline	Not established Usual diet provides 500–900	Egg yolk, liver, grains, legumes	Constituent of phospholipids, precursor of the neurotransmitter acetylcholine	None reported in humans
Vitamin C (ascorbic acid)	45	Citrus fruits, tomatoes, green peppers	Maintenance of cartilage, bone, and dentin (hard tissue of teeth); collagen synthesis	Scurvy (degeneration of skin, teeth, blood vessels; epithelial hemorrhages)
Fat soluble				
Vitamin A (retinol)	1	Beta-carotene in green, yellow, red vegetables Retinol added to dairy products	Constituent of visual pigment Maintenance of epithelial tissues	Night blindness, permanent blindness
Vitamin D	0.01	Cod-liver oil, eggs, dairy products	Promotes bone growth and mineralization Increases calcium absorption	Rickets (bone deformities) in children; skeletal deterioration
Vitamin E (tocopherol)	15	Seeds, green leafy vegetables, margarines, shortenings	Antioxidant, prevents cellular damage	Possibly anemia
Vitamin K	0.03	Green leafy vegetables Product of intestinal bacteria	Important in blood clotting	Bleeding, internal hemorrhages

The *fat-soluble* vitamins A, D, E, and K have even more varied roles. Vitamin K, for example, helps regulate blood clotting, and vitamin A is used to produce visual pigment (the light-capturing molecule in the retina of the eye). Fat-soluble vitamins can be stored in body fat and may accumulate in the body over time. For this reason, fat-soluble vitamins may be toxic if consumed in excessively high doses.

Nutritional Guidelines Help People Obtain a Balanced Diet

Most Americans are fortunate to live amid an abundance of food. However, the overwhelming diversity of food available in an American supermarket and the easy availability of "fast food" can lead to poor nutritional choices. To help people make informed choices, the U.S. government has developed recommendations and goals for the "average American," which are summarized in Table 29-5.

Further help is provided by the food pyramid, designed by the U.S. Department of Agriculture, which illustrates the relative abundances of different food groups in an optimal diet (Fig. 29-2). A third source of information is the nutritional labeling now required on commercially packaged foods. These labels provide complete information about calorie, fiber, fat, sugar, and vitamin content (Fig. 29-3). Some fast-food chains also make fliers available that list nutritional information about their products.

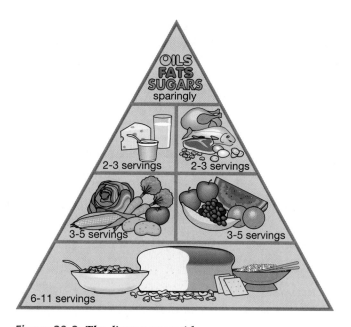

Figure 29-2 The dietary pyramid
Recommended by the U.S. Department of Agriculture, this chart shows suggested daily servings.

NUTRITION FACTS	
Serving Size	1 Cup (55g)
Servings Per Package	8
Amount Per Serving	
Calories 210	Calories from Fat 0
	% Daily Value*
Total Fat 0g	0%
Saturated Fat 0g	0%
Cholesterol 0mg	0%
Sodium 20mg	1%
Total Carbohydrate 46g	15%
Dietary Fiber 6g	24%
Sugars 12g	
Protein 6g	
Vitamin A* •	Vitamin C 2%
Calcium 4% •	Iron 18%
Thiamin 38% •	

*Percent Daily Values are based on a 2,000 calorie diet. Your Daily Values may be higher or lower depending on your calorie needs:

	Calories	2,000	2,500
Total Fat	Less than	65g	80g
Saturated Fat	Less than	20g	25g
Cholesterol	Less than	300mg	300mg
Sodium	Less than	2,400mg	2,400mg
Total Carbohydrate		300g	375g
Dietary Fiber		25g	30g

Calories per gram:
Fat 9 • Carbohydrate 4 • Protein 4

Figure 29-3 New food labels
The U.S. government recently required more-complete nutritional labeling of foods, such as in this sample. The weight in grams of various nutrients (such as fat, cholesterol, and sodium) is converted into a percentage of the recommended daily allotment, assuming a 2000-Calorie diet. At the bottom of the label, the total recommended number of grams of these nutrients is listed for a 2000- and a 2500-Calorie diet.

Table 29-5 Dietary Changes Advocated by the U.S. Government

	Percentage of Total Daily Energy Intake	
Dietary Component	Average American Diet	Dietary Goals
Carbohydrates	46	58
Lipids	38	30
Proteins	16	12

Summary of Recommendations:

1. Increase consumption of fruits, vegetables, and whole grains.
2. Decrease consumption of refined sugars.
3. Decrease consumption of fats; replace saturated with unsaturated fats.
4. Decrease consumption of animal fats by selecting lean meats, poultry, and fish.
5. Decrease consumption of high-cholesterol foods, such as butter and eggs.
6. Decrease consumption of salt and foods high in salt content.
7. Decrease caloric intake to maintain desirable weight.

Health Watch
Eating Disorders—Betrayal of the Body

Food is a fundamental requirement of all animals; no animal can survive without eating. Natural selection has therefore provided animals with mechanisms to ensure that the impulse to eat arises when nutrients are needed. In humans, however, these natural impulses, so crucial to our health and well-being, can go terribly awry. In recent decades we have seen an increase in the occurrence of *eating disorders*, ailments characterized by the disruption of normal eating behavior.

Eating disorders include two particularly debilitating illnesses, *anorexia nervosa* and *bulimia nervosa*. People with anorexia nervosa experience an intense fear of gaining weight, and they achieve extreme weight loss by eating very little food. This drastically reduced eating is commonly accompanied by other measures, including self-induced vomiting, laxative intake, and excessive exercise. The consequences of the behaviors associated with anorexia nervosa are disastrous. Anorexics become emaciated, losing both fat and muscle mass. This emaciation can in turn disrupt cardiac, digestive, endocrine, and reproductive functions. Up to 18% of anorexics die as a result of their disorder.

The main symptom of bulimia nervosa is recurrent bouts of binge eating, in which extraordinarily large amounts of food are consumed in a short period. The binge eating of bulimics is typically followed by the same kinds of purging behaviors common in anorexics, such as vomiting and laxative consumption. In most cases, the health consequences of bulimia nervosa are not as dire as are those of anorexia nervosa, because there is no drastic weight loss.

Girls and young women are most at risk for eating disorders (Fig. E29-1). More than 90% of diagnosed eating disorders occur in females; in the large majority of these cases, onset of the illness takes place between 13 and 20 years of age. The causes of eating disorders are not well understood. In a body wracked by the near-starvation of anorexia nervosa, the effects of the illness are hard to distinguish from a defect that might be a cause. No clear genetic link to the disorders has been found thus far, so suspicion has fallen on environmental factors, especially on societal pressures to be thin.

Unfortunately, existing treatments for eating disorders are not terribly effective. The most common treatment, hospitalization (to restore nutritional health) and counseling, does not typ-

ically lead to full recovery. Antidepressant drugs are helpful in some cases but are more likely to be effective for bulimia nervosa than for the more dire anorexia nervosa. Hope may be on the horizon, however, owing to a recent explosion of research findings about the hormones that control appetite. For example, researchers have discovered a class of hormones known as *orexins* (from the Greek word *orexis*, meaning "appetite"), which, when injected into mice, bind to receptors in the brain and cause food consumption to increase dramatically. Such new discoveries raise hopes that as our understanding of the physiological control of appetite grows, we may yet be able to devise chemical treatments for eating disorders.

Figure E29-1 Eating disorders
The late Princess Diana made headlines when she announced that she had been suffering from bulimia.

2) How Is Digestion Accomplished?

www

Digestion, the physical grinding up and chemical breakdown of food, occurs in the digestive system. The digestive systems of animals take in and then digest the complex molecules of their food into simpler molecules that can be absorbed. Material that cannot be used or broken down is then expelled from the body.

Animals eat the bodies of other organisms, bodies that may resist becoming food. The plant body, for example,

armors each cell with a wall of indigestible cellulose. Animal bodies may be covered with equally indigestible fur, scales, or feathers. In addition, the complex lipids, proteins, and carbohydrates in food do not occur in a form that can be used directly. They must be broken down before they can be absorbed and distributed to the cell of the animal that has consumed them, where they are recombined in unique ways. Different types of animals meet the challenge of acquiring nutrients with various types of digestive tracts, each finely tuned to a unique diet and lifestyle. Amid this diversity, however, are certain tasks that all digestive systems must accomplish:

1. *Ingestion.* The food must be brought into the digestive tract through an opening, usually called a **mouth**.
2. *Mechanical breakdown.* The food must be physically broken down into smaller pieces. This is accomplished by gizzards or teeth as well as by the churning action of the digestive cavity itself. The particles produced by mechanical breakdown provide a large surface area for attack by digestive enzymes.
3. *Chemical breakdown.* The particles of food must be exposed to digestive enzymes and other digestive fluids that break down large molecules into smaller subunits.
4. *Absorption.* The small molecules must be transported out of the digestive cavity and into cells.
5. *Elimination.* Indigestible materials must be expelled from the body.

In the following section, we will explore briefly some of the diverse mechanisms by which animal digestive systems accomplish these functions.

Digestive Systems Are Adapted to the Lifestyle of Each Animal

Digestion within Single Cells Occurs in the Sponges

Simple digestive systems can be as efficient as more-complex ones if a comparatively small amount of energy is used to acquire and digest food. Sponges, for example, are sedentary feeders. They do not have a large investment in specialized digestive structures; thus, they prosper with a system in which digestion occurs *within* individual cells. Such **intracellular digestion** occurs after a cell has engulfed microscopic food particles. Once engulfed by a cell, the food is enclosed in a **food vacuole**, a space surrounded by membrane that serves as a temporary stomach. The vacuole fuses with small packets of digestive enzymes called **lysosomes**, and food is broken down within the vacuole into smaller molecules that can be absorbed into the cell cytoplasm. Undigested remnants remain in the vacuole, which eventually expels its contents outside the cell. Intracellular digestion is seen in the single-celled protists and in the simplest animals. Sponges, for example, rely entirely on intracellular digestion (Fig. 29-4). This process limits their menu to mi-

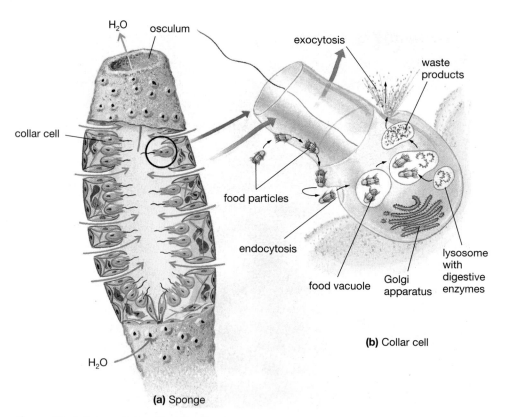

(a) Sponge

(b) Collar cell

Figure 29-4 Intracellular digestion in a sponge
(a) Internal anatomy of a simple sponge, showing the direction of water flow and the location of the collar cells. *(b)* Enlargement of a single collar cell. Water is filtered through the collar, and food particles (single-celled organisms) are trapped on the outside of the collar and conducted down to the plasma membrane. There the food is engulfed by endocytosis, then digested in lysosomes, and wastes are expelled.

croscopic food particles, such as protists, that they filter from the surrounding sea by means of the sievelike "collars" of special collar cells (see Chapter 22).

A Sac with One Opening Forms the Simplest Digestive System

Larger, more-complex organisms evolved a chamber within the body where chunks of food could be broken down by enzymes that act outside the cells. This process is called **extracellular digestion**. One of the simplest of these chambers is found in cnidarians, such as sea anemones, hydra, and jellyfish. As we learned in Chapter 22, these animals possess a digestive sac called a **gastrovascular cavity**, with a single opening through which food is ingested and wastes are ejected (Fig. 29-5). Although it is generally referred to as the mouth, this opening also serves as an anus. Food captured by stinging tentacles is escorted into the gastrovascular cavity, where enzymes break it down. Cells lining the cavity absorb the nutrients and engulf small food particles. Further digestion occurs via intracellular digestion. The undigested remains are eventually expelled through the same opening by which they entered.

While one meal is being digested, a second cannot be processed efficiently, because the same chamber is used. Thus, this type of digestive system is unsuited to active animals, which require frequent meals, or to animals whose food offers so little nutrition that they must feed continually. The needs of such animals are met by a digestive system that consists of a one-way tube with an opening at each end.

Digestion in a Tube Allows More-Frequent Meals

Most animals, from nematode worms to earthworms, mollusks, arthropods, echinoderms, and vertebrates, have a digestive system that is basically a tube through the body. A tubular digestive tract allows the animal to eat frequently. Further, it consists of a series of specialized regions that process the food in an orderly sequence: first physically grinding it up, then enzymatically breaking it down, then absorbing the small nutrient molecules into body cells. Specialized tubular digestive tracts adapt different types of animals to eat a wide range of foods and to extract the maximum amount of nutrients from them.

In its simplest form, as seen in the threadlike nematode worm, the tube is relatively unspecialized along its length (Fig. 29-6a). In more-complex organisms, such as the earthworm, the tube consists of a series of compartments, each with a specific role in the breakdown of food (Fig. 29-6b). The earthworm extracts nutrients from decaying organic material in soil. A tubular digestive system is essential to the earthworm, which continuously ingests soil as it burrows through the earth, passing the soil out at one end while taking it in at the other. The worm's muscular **pharynx** is a cavity that connects the mouth with the **esophagus**, a muscular passageway. Ingested soil and bits of vegetation are passed through the pharynx and the esophagus to the **crop**, a thin-walled storage organ. The crop collects the food and gradually passes it to the **gizzard**. There, bits of sand and the contraction of muscles grind the food into smaller particles. The food then travels to the intestine, where enzymes break it down into simple molecules that can be absorbed by the cells lining the intestine.

Like the earthworm, humans and other vertebrates have tubular digestive tracts with several compartments in which food is first physically, then chemically, broken down before being absorbed by individual cells. Animals with tubular digestive systems thus use extracellular digestion to break down their food. Vertebrate digestive tracts are specialized for the particular diet of the animal. Birds, for example, that eat seeds, shelled invertebrates, or bony mammalian prey typically possess a large,

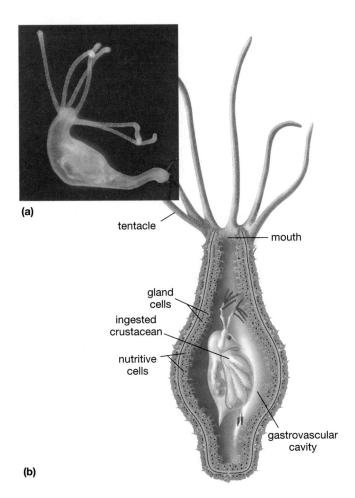

(a)

tentacle — mouth

gland cells

ingested crustacean

nutritive cells

gastrovascular cavity

(b)

Figure 29-5 Digestion in a sac

(a) A *Hydra* has just captured and ingested a waterflea (*Daphnia,* a microscopic crustacean). *(b)* Within the gastrovascular cavity of *Hydra,* gland cells secrete enzymes that digest the prey into smaller particles and nutrients. Elongated cells lining the cavity ingest these particles by intracellular digestion (see Fig. 29-4), and digestion is completed intracellularly. Undigested waste is then expelled through the single opening.

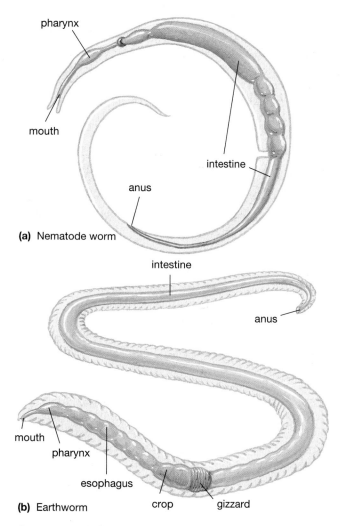

(a) Nematode worm

(b) Earthworm

Figure 29-6 Tubular digestive tracts

(a) The nematode worm, an abundant, normally microscopic animal, is among the simplest animals to have a one-way digestive system, but that system is relatively unspecialized. Nematodes typically live within their food source (several hundred thousand may be found in a rotting apple) and may eat almost continuously. *(b)* The earthworm has a one-way digestive system in which food is passed through a series of compartments. Each compartment is specialized to play a specific role in breaking down food and absorbing it.

cell, it is potentially one of the most abundant food energy sources on Earth. *Ruminant* animals—cows, sheep, goats, camels, and hippos, to name a few—have evolved elaborate digestive systems that can break down cellulose. *Rumination,* or *cud-chewing,* is the process of regurgitating food and rechewing it, one of several adaptations that enable these animals to digest tough plant material. Ruminant stomachs consist of four chambers (Fig. 29-7b). The first chamber, the *rumen,* has evolved into a massive fermentation vat. There, microorganisms—including many species of bacteria and ciliates—thrive in a mutually beneficial relationship with the ruminant. These microorganisms produce **cellulase**, the enzyme that breaks

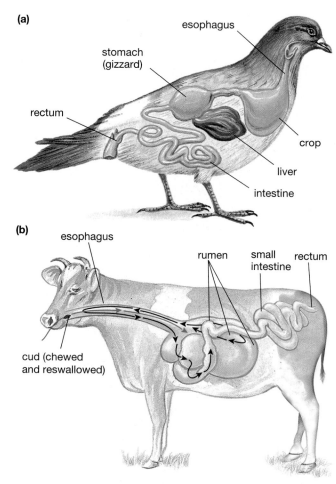

Figure 29-7 Adaptations of vertebrate digestive systems

(a) The digestive system of birds is adapted to the demands of flight. The expandable crop serves as a storage organ, allowing the bird to store food to meet the enormous caloric demands of flight. The gizzard replaces the teeth, using muscular action and small stones that are stored in this organ to break down the hard seeds and insect exoskeletons prevalent in the diet of many birds. Undigested food is expelled.
(b) The stomach of the cow has four chambers. The largest is the rumen, which houses a flourishing population of microorganisms that digest the cellulose in the cow's vegetarian diet. Arrows trace the path of food through the digestive tract.

highly muscular gizzard (Fig. 29-7a) in which these resistant foods are ground up before passing to the intestines for further processing. Lacking teeth, birds have developed other means to accomplish the grinding component of digestion. Let's look at the digestive tracts of the cow and the human as well.

Adaptations Help Ruminants Digest Cellulose

Cellulose, like starch, consists of long chains of glucose molecules, but these molecules are linked together in a way that resists the attack of animal digestive enzymes (see Chapter 3). Because cellulose surrounds each plant

down cellulose into its component sugars. After it is processed in the rumen, the plant material, now called *cud,* is regurgitated, chewed, and reswallowed to the rumen. The extra mechanical breakdown exposes more of the cellulose and cell contents to the enzymes of the microorganisms for further digestion. Gradually the cud is then released into the remaining three chambers for further breakdown before it enters the intestine.

3) How Do Humans Digest Food?

The human digestive system (Fig. 29-8) is comparatively unspecialized, adapted for processing a wide variety of different foods. *Herbivores* are animals, such as the ru-

minants discussed above, that eat only plants; *carnivores* are animals that eat only other animals. In contrast, humans are *omnivores*, able to consume foods from both plant and animal (as well as, for that matter, fungal and protist) sources.

The Mechanical and Chemical Breakdown of Food Begins in the Mouth

Both the mechanical and the chemical breakdown of food begin in the mouth. In mammals, the mechanical work is done in large part by teeth. *Incisors* at the front of the mouth snip off pieces of food; the pointed *canine* teeth beside them are useful for tearing the pieces apart; and the *premolars* and *molars* at the back of the mouth have flat surfaces for grinding food to a paste. In different species of mammals, teeth are specialized for the diet of the

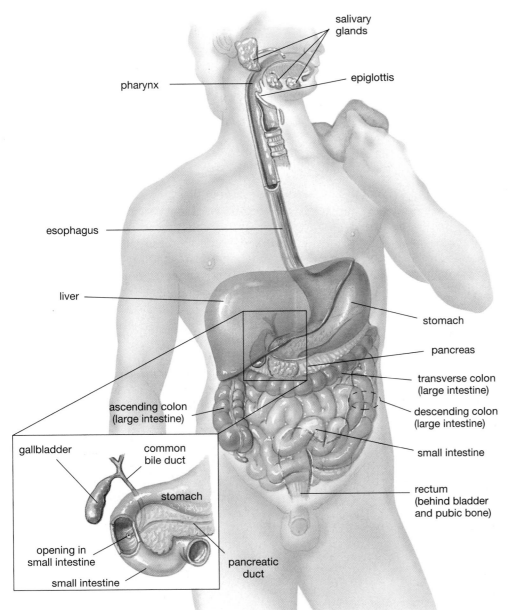

Figure 29-8 The human digestive tract
The digestive system includes both the digestive tube and organs such as the salivary glands, liver, gallbladder, and pancreas, all of which produce and store digestive secretions.

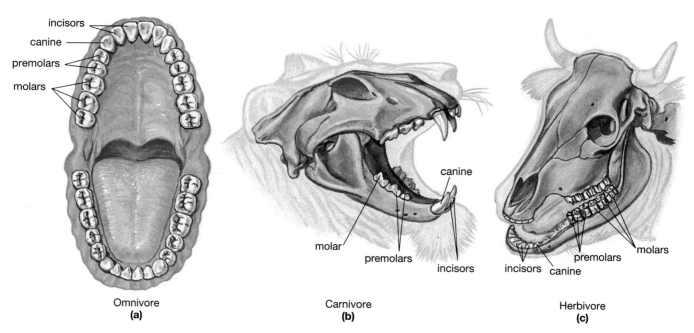

Figure 29-9 Teeth are adapted to different diets
Mammalian teeth have been modified in various ways by evolution. *(a)* The varied, omnivorous diet of humans has fostered the evolution of an unspecialized dentition that includes flat incisors for biting, pointed canines for tearing, premolars for grinding, and molars for crushing and chewing. *(b)* Carnivores (such as the lion) have modest incisors but greatly enlarged canines for stabbing and tearing flesh. Carnivores also have a reduced set of cheek teeth (molars and premolars) that are specialized for shearing through tendon and bone. *(c)* Herbivores (such as the cow) have incisors that are specialized for snipping leaves, and their canines have been reduced in size and moved forward to help with that job. The cheek is filled with a full set of wide, flat premolars and molars that grind up the tough, cellulose-containing plant material.

animal (Fig. 29-9). In adult humans, 32 teeth of varying sizes and shapes cut and grind food into small pieces.

While the food is pulverized by the teeth, the first phase of chemical digestion occurs as three pairs of salivary glands pour out saliva in response to the smell, feel, taste, and (if you're hungry) even the thought of food.

Saliva contains the digestive enzyme **amylase**, which begins the breakdown of starches into sugar (Table 29-6). Saliva has other functions as well. It contains a bacteria-killing enzyme and antibodies that help guard against infection. It also lubricates the food to facilitate swallowing and dissolves some food molecules such as acids

Table 29-6 Digestive Structures and Secretions

Site of Digestion	Secretion	Source of Secretion	Role in Digestion
Mouth	Amylase	Salivary glands	Breaks down starch into disaccharides
	Mucus, water	Salivary glands	Lubricates, dissolves food
Stomach	Hydrochloric acid	Cells lining stomach	Allows pepsin to work, kills bacteria, solubilizes minerals
	Pepsin	Cells lining stomach	Breaks down proteins into large peptides
	Mucus	Cells lining stomach	Protects stomach
Small intestine	Sodium bicarbonate	Pancreas	Neutralizes acidic chyme from stomach
	Amylase	Pancreas	Breaks down starch into disaccharides
	Peptidases	Pancreas	Split large peptides into small peptides
	Trypsin	Pancreas	Breaks down proteins into large peptides
	Chymotrypsin	Pancreas	Breaks down proteins into large peptides
	Lipase	Pancreas	Breaks down lipids into fatty acids and glycerol
	Bile	Liver	Emulsifies lipids
	Peptidases	Cells lining small intestine	Split small peptides into amino acids
	Disaccharidases	Cells lining small intestine	Split disaccharides into monosaccharides

(a)

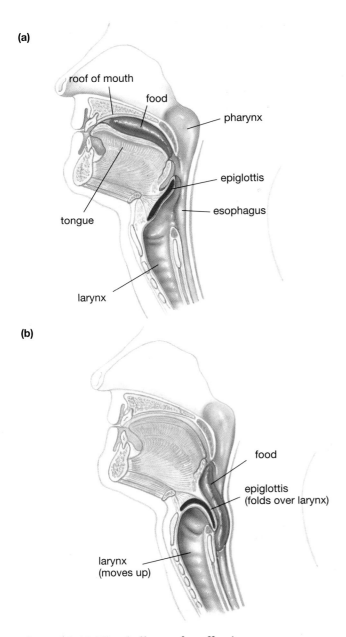

(b)

Figure 29-10 The challenge of swallowing
(a) Swallowing is complicated by the fact that both the esophagus (part of the digestive system) and the larynx (part of the respiratory system) open into the pharynx. *(b)* During swallowing, the larynx moves upward beneath a small flap of cartilage, the epiglottis. The epiglottis folds down over the larynx, sealing off the opening to the respiratory system and directing food down the esophagus instead.

and sugars, carrying them to *taste buds* on the tongue. The taste buds are sensory receptors that help identify the type and quality of the food.

With the help of the muscular tongue, the food is manipulated into a mass and pressed backward into the **pharynx**, a muscular cavity connecting the mouth with the esophagus (Fig. 29-10a). The pharynx also connects, via the *larynx*, the nose and mouth with the *trachea*, which conducts air to the lungs. This arrangement occasionally

causes problems, as anyone who has ever choked on a piece of food can attest. Normally, however, the swallowing reflex elevates the larynx so it meets the **epiglottis**, a flap of tissue that blocks off the respiratory passages. Food is thus directed into the esophagus rather than into the trachea (Fig. 29-10b).

The Esophagus Conducts Food to the Stomach, Where Digestion Continues

The esophagus is a muscular tube that propels food from the mouth to the stomach. Circular muscles surrounding the esophagus contract in sequence above the swallowed food mass, squeezing it down toward the stomach. This muscular action, called **peristalsis**, also occurs in the stomach and intestines, where it helps move food along the digestive tract (Fig. 29-11). Peristalsis is so effective that a person can actually swallow when upside down. Mucus secreted by cells that line the esophagus helps

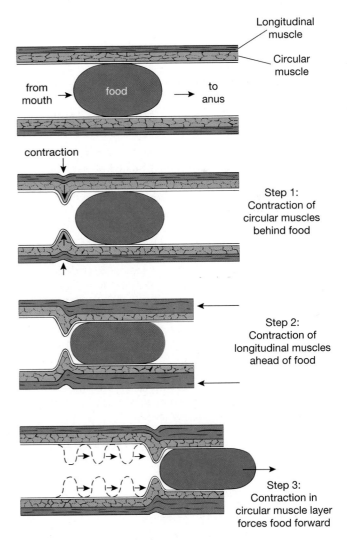

Figure 29-11 Muscular movements of the digestive tract
Food is propelled through the digestive system by the rhythmic contractions of circular and longitudinal muscles.

protect it from abrasion and lubricates the food during its passage.

The **stomach** in humans is an expandable muscular sac capable of holding from 2 to 4 liters (as much as a gallon) of food and liquids. Food is retained in the stomach by a ring of circular muscle that separates the lower portion of the stomach from the upper *small intestine*. This muscle, called the **pyloric sphincter**, regulates the passage of food into the small intestine, as described later. The stomach has three major functions. First, the stomach stores food and releases it gradually into the small intestine at a rate suitable for proper digestion and absorption. Thus, the stomach enables us to eat large, infrequent meals. Carnivores carry this ability to an extreme. A lion, for instance, may consume about 18 kilograms (40 pounds) of meat at one meal, then spend the next few days quietly digesting it. A second function of the stomach is to assist in the mechanical breakdown of food. In addition to peristalsis, its muscular walls undergo a variety of contracting, churning movements that help break apart large pieces of food.

Finally, glands in the lining of the stomach secrete enzymes and other substances that facilitate the chemical breakdown of food. These include gastrin, hydrochloric acid (HCl), pepsinogen, and mucus. **Gastrin** (a hormone) stimulates the secretion of hydrochloric acid by specialized stomach cells. Other cells release *pepsinogen*, an inactive form of the protein-digesting enzyme *pepsin*. Pepsin is a **protease**, an enzyme that helps break proteins into shorter chains of amino acids called *peptides* (see Table 29-6). Pepsin must be secreted in an inactive form to prevent it from digesting the very cells that produce it. The highly acidic conditions in the stomach (pH 1 to 3), though, convert pepsinogen into pepsin, which functions best in an acidic environment.

As you may have noticed, the stomach produces all the ingredients necessary to digest itself. Indeed, this is what happens when a person develops ulcers (see "Health Watch: Are You an Ulcer Candidate?"). However, cells lining the stomach normally produce a large quantity of thick mucus that coats the stomach lining and serves as a barrier to self-digestion. The protection is not

Health Watch
Are You an Ulcer Candidate?

About 1 out of every 10 Americans eventually develops an ulcer. *Ulcers* occur when the mucus barriers of the stomach and upper small intestine break down, and the inner lining and deeper layers of the digestive tract are eroded by acid and the protein-digesting enzyme pepsin. The upper small intestine is the most common site for ulcers (Fig. E29-2), because it receives the highly acidic chyme but is less protected by mucus than the stomach is.

The causes of ulcers are complex and poorly understood. A tendency to develop ulcers can be inherited. Some, but not all, ulcer sufferers secrete excess stomach acid. In some cases, decreased secretion of sodium bicarbonate or mucus by cells that line the stomach and small intestine may contribute to the disease. There is a correlation between the development of ulcers and stress, which may stimulate acid and pepsin secretion. Smoking and chronic consumption of alcohol and aspirin all may reduce the resistance of the digestive tract lining to the effects of pepsin and acids and can aggravate ulcers. In 1983, B. J. Marshall, an Australian physician, identified a bacterium in stomach tissue from ulcer patients. When his claims that the bacterium caused the ulcer were met with skepticism, he swallowed a batch of them and soon developed ulcer symptoms. The bacterium (*Helicobacter pylori*) lives in the digestive tracts of nearly all ulcer patients, and antibiotics that kill this bacterium have been very effective in reducing the recurrence of ulcers. Although an estimated half of the world's population harbors *H. pylori*, the bacterium causes ulcers in only 10% to 20% of those infected with it.

Ulcers can sometimes be relieved by stress-reduction programs, the elimination of smoking and drinking, and dietary changes. Medications for ulcers include antibiotics, antacids, and drugs that decrease stomach acid production.

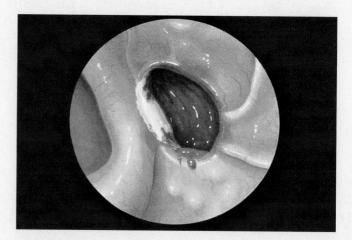

Figure E29-2 *A bleeding ulcer in the stomach*
An illustration of a bleeding ulcer in the stomach as it would be seen through a fiber-optic viewing device called an endoscope. The stomach lining has been digested away, and blood seeps through the opening.

perfect, however, and the cells lining the stomach are digested to some extent and must be replaced every few days.

Food in the stomach is gradually converted to a thick, acidic liquid called **chyme**, which consists of partially digested food and digestive secretions. Peristaltic waves then propel the chyme toward the small intestine. The pyloric sphincter allows only about a teaspoon of chyme to be expelled with each contraction, which occurs about every 20 seconds. It takes 2 to 6 hours, depending on the size of the meal, to empty the stomach completely. The continued churning movements of an empty stomach are felt as "hunger pangs."

Only a few substances, including water, some drugs, and alcohol, can enter the bloodstream through the stomach wall. Alcohol that is consumed when the stomach is empty is immediately absorbed into the bloodstream, with strong and rapid effects. Because food in the stomach slows alcohol absorption, the advice "never drink on an empty stomach" is based on sound physiological principles.

Most Digestion Occurs in the Small Intestine

The **small intestine** is a coiled, narrow tube approximately 3 meters (about 10 feet) long. Its diameter ranges from about 2 to 5 centimeters (1 to 2 inches) in an adult human. The small intestine has two major functions: to digest food into small molecules and to absorb these molecules, passing them to the bloodstream. The first role of the small intestine—digestion—is accomplished with the aid of digestive secretions from three sources: (1) the liver, (2) the pancreas, and (3) the cells of the small intestine itself.

The Liver and Gallbladder Provide Bile, Important in Fat Breakdown

The **liver** is perhaps the most versatile organ in the body. Its many functions include storage of fats and carbohydrates for energy, regulation of blood glucose levels, synthesis of blood proteins, storage of iron and certain vitamins, conversion of toxic ammonia (released by the breakdown of amino acids) into urea, and detoxification of other harmful substances such as nicotine and alcohol. The role of the liver in digestion is to produce *bile*, a liquid stored and concentrated in the **gallbladder** and released into the small intestine through a tube called the *bile duct* (see Fig. 29-8).

Bile is a complex mixture composed of **bile salts**, water, other salts, and cholesterol. Bile salts are synthesized in the liver from cholesterol and amino acids. Although they assist in the breakdown of lipids, bile salts are not enzymes. Rather, they act as detergents or emulsifying agents, dispersing globs of fat in the chyme into microscopic particles. These particles expose a large surface area for attack by **lipases**, lipid-digesting enzymes produced by the pancreas.

The Pancreas Supplies Several Digestive Secretions to the Small Intestine

The **pancreas** lies in the loop between the stomach and small intestine (see Fig. 29-8). It consists of two major types of cells. One type produces hormones involved in blood sugar regulation (as we shall see in Chapter 32), and the other produces a digestive secretion called **pancreatic juice**, which is released into the small intestine. Pancreatic juice neutralizes the acidic chyme and digests carbohydrates, lipids, and proteins. About 1 liter (1.06 quarts) of pancreatic juice is released into the small intestine each day. This secretion contains water, sodium bicarbonate, and several digestive enzymes (see Table 29-6). Sodium bicarbonate (the active ingredient in baking soda) neutralizes the acidic chyme in the small intestine, producing a slightly basic pH. Pancreatic digestive enzymes require this basic pH for proper functioning, in contrast to the stomach's digestive enzymes, which require an acidic pH.

The pancreatic digestive enzymes break down three major types of food: (1) Amylase breaks down carbohydrates, (2) lipases digest lipids, and (3) several proteases separate proteins and peptides. The pancreatic proteases include *trypsin, chymotrypsin,* and *carboxypeptidase.* Both trypsin and chymotrypsin break proteins and peptides into shorter peptide chains. Carboxypeptidase then completes protein digestion by separating the individual amino acids from the ends of the peptide chains. These proteases are secreted in an inactive form and become activated after they reach the small intestine.

The Intestinal Wall Completes the Digestive Process

The wall of the small intestine is studded with cells that are specialized to complete the digestive process and absorb the small molecules that result. These cells have various enzymes on their external membranes, which form the lining of the small intestine. These enzymes include proteases, which complete the breakdown of peptides into amino acids, and sucrase, lactase, and maltase, which break down disaccharides into monosaccharides (see Chapter 3). Small amounts of lipase digest lipids. Because these enzymes are actually embedded in the membranes of the cells that line the small intestine, this final phase of digestion occurs *as* the nutrient is being absorbed into the cell. Like the stomach, the small intestine is protected from digesting itself by large amounts of mucous secretions from specialized cells in its lining.

Most Absorption Occurs in the Small Intestine

The small intestine is not only the principal site of chemical digestion; it is also the major site of nutrient **absorption** into the blood. The small intestine has numerous foldings and projections that give it an internal surface area that is 600 times that of a smooth tube of the same length (Fig. 29-12). Covering the entire folded surface of

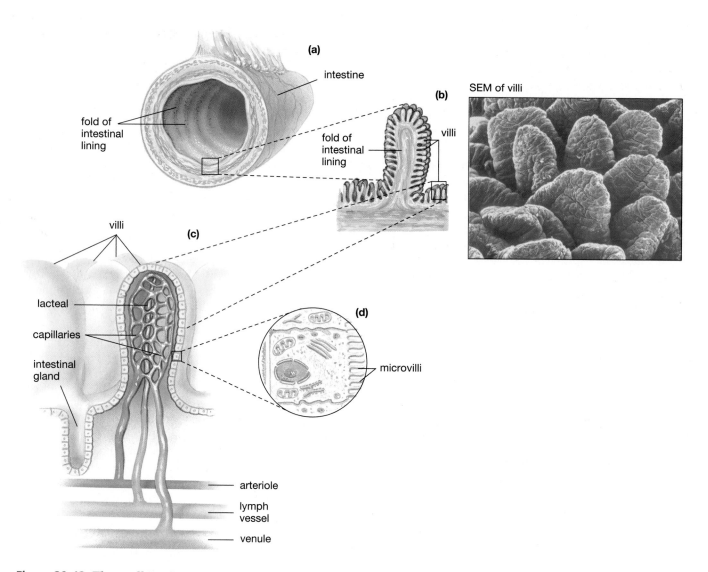

Figure 29-12 The small intestine
The folds of the small intestine maximize the surface area available to absorb nutrients. *(a)* Macroscopic folds in the intestinal lining are themselves carpeted in *(b)* tiny, fingerlike projections called villi, which enclose *(c)* a network of capillaries and lymph vessels. *(d)* If we use a microscope to zoom in on one villus, we see that the epithelial cells on its surface are sheathed in plasma membranes that have yet another level of microscopic projections, microvilli.

the wall are minute, fingerlike projections called **villi** (literally, "shaggy hairs"; singular, **villus**). Villi, which range from 0.5 to 1.5 millimeters in length, give the intestinal lining a velvety appearance to the naked eye. They move gently back and forth in the chyme that passes through the intestine. This movement increases their exposure to the molecules to be digested and absorbed. Further, each individual cell of the villi bears a fringe of microscopic projections called **microvilli**. Taken together, these specializations of the small intestine wall give it a surface area of about 250 square meters (more than 2200 square feet).

Segmentation movements, rhythmic, unsynchronized contractions of the circular muscles of the intestine, slosh

the chyme back and forth, bringing nutrients into contact with the absorptive surface of the small intestine. When absorption is complete, coordinated peristaltic waves conduct the leftovers into the *large intestine*.

Nutrients absorbed by the small intestine include water, monosaccharides, amino acids and short peptides, fatty acids produced by lipid digestion, vitamins, and minerals. The mechanisms by which this absorption occurs are varied and complex. In most cases, energy is expended to transport nutrients into the intestinal cells. (Water follows by osmosis as described in Chapter 6.) The nutrients then diffuse out of the intestinal cells into the interstitial fluid, where they then enter the bloodstream.

Each villus of the small intestine is provided with a rich supply of blood capillaries and a single lymph capillary, called a **lacteal**, to carry off the absorbed nutrients and distribute them throughout the body (see Fig. 29-12). Most of the nutrients enter the bloodstream through the capillaries, but fat subunits take a different route. After diffusing into the epithelial cells, they are resynthesized into fats, combined with other molecules, and then released as droplets into the interstitial fluid. By this means, they enter the lymph capillary and are eventually delivered to the bloodstream when the lymph vessels empty into the veins.

Water Is Absorbed and Feces Are Formed in the Large Intestine

The **large intestine** in an adult human is about 1.5 meters (5 feet) long and about 7.5 centimeters (3 inches) in diameter, which is considerably wider than the small intestine. The large intestine has two parts: For most of its length it is called the **colon**, but its final 15 centimeters (6 inches) is called the **rectum**. Into the large intestine flow the leftovers of digestion: a mixture of water, undigested fats and proteins, and indigestible fibers, such as the cell walls of vegetables and fruits. The large intestine contains a flourishing population of bacteria that live on unabsorbed nutrients. These bacteria earn their keep by synthesizing vitamin B_{12}, thiamin, riboflavin, and, most important, vitamin K, which would otherwise be deficient in a normal diet. Cells lining the large intestine absorb these vitamins as well as leftover water and salts.

After absorption is complete, the result is the semisolid *feces*. Feces consist of indigestible wastes and the dead bodies of bacteria, which account for about one-third of the dry weight of feces. The feces are transported by peristaltic movements until they reach the rectum. Expansion of this chamber stimulates the desire to defecate. Although defecation is a reflex (as any new parents can attest), it is initiated voluntarily after about the age of two.

Digestion Is Controlled by the Nervous System and Hormones

Considerable coordination is required to break down a chef salad into amino acids and peptides, sugars, fatty acids, vitamins, minerals, and indigestible cellulose. As your mouth responds to the first bite, your stomach must be warned that food is on the way. In addition, the enzymes of the stomach and small intestine require different environments for proper functioning (highly acidic in the stomach, slightly basic in the small intestine), and secretions into various parts of the digestive tract must be coordinated with the arrival of food. Not surprisingly, the secretions and activity of the digestive tract are coordinated by both nerves and hormones (Table 29-7).

The initial phase of digestion is under the control of the nervous system and involves responses to signals that originate in the head. These signals include the sight, smell, taste, and sometimes the thought of food, as well as the muscular activity of chewing. In response to these stimuli, saliva is secreted to the mouth while nervous signals to the stomach walls initiate the secretion of acid and the hormone gastrin, which stimulates further acid secretion. The concentration of acid is regulated by a negative feedback mechanism (see Chapter 26). When acid levels reach a certain point, they inhibit gastrin secretion, thereby inhibiting further acid production.

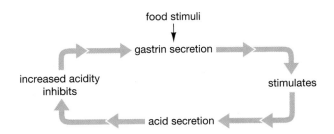

The arrival of food in the stomach triggers the second phase of digestion. Stimulation of the stomach wall causes the production of a large quantity of mucus, which

Table 29-7 Some Important Digestive Hormones			
Hormone	**Site of Production**	**Stimulus for Production**	**Effect**
Gastrin	Stomach	Food in mouth Distension of stomach Peptides in stomach	Stimulates acid secretion by cells in stomach
Secretin	Small intestine	Acid in small intestine	Stimulates bicarbonate production by pancreas and liver; increases bile output by liver
Cholecystokinin	Small intestine	Amino acids, fatty acids in small intestine	Stimulates secretion of pancreatic enzymes and release of bile by gallbladder
Gastric inhibitory peptide	Small intestine	Fatty acids and sugars in small intestine	Inhibits stomach movements and release of stomach acid

protects the stomach against self-digestion. The acidity of the stomach converts pepsinogen to its active form, pepsin, which begins protein digestion. However, the presence of protein in food tends to reduce the concentration of stomach acid. Thus, as protein is broken down, the acidity drops and the release of gastrin is no longer inhibited; gastrin is released again and stimulates further acid production. The cells secreting stomach acid are also activated by the expansion of the stomach and by the presence of peptides produced by protein digestion.

As the liquid chyme is gradually released into the small intestine, its acidity stimulates the release of a second hormone, **secretin**, into the bloodstream by cells of the upper small intestine. Secretin causes the pancreas to pour sodium bicarbonate into the small intestine. Sodium bicarbonate neutralizes the acidity of the incoming chyme, creating an environment in which the pancreatic enzymes can function. A third hormone, **cholecystokinin**, is also produced by cells of the upper small intestine in response to the presence of chyme. This hormone stimulates the pancreas to release various digestive enzymes into the small intestine. It also stimulates the gallbladder to contract, squeezing bile through the bile duct to the small intestine. Bile assists in fat breakdown, as described earlier.

Gastric inhibitory peptide, a hormone secreted by the small intestine in response to the presence of fatty acids and sugars in chyme, inhibits acid production and peristalsis in the stomach. This inhibition slows the rate at which chyme is dumped into the small intestine, providing additional time for digestion and absorption to occur.

Summary of Key Concepts

1) What Nutrients Do Animals Need?

Each type of animal has specific nutritional requirements. These requirements include molecules that can be broken down to liberate energy, such as lipids, carbohydrates, and protein; chemical building blocks used to construct complex molecules, such as amino acids that can be linked together to form proteins; and minerals and vitamins that facilitate the diverse chemical reactions of metabolism.

2) How Is Digestion Accomplished?

Digestive systems must accomplish five tasks: ingestion, mechanical followed by chemical breakdown of food, absorption, and elimination of wastes. Digestive systems convert the complex molecules of the bodies of other animals or plants that have been eaten into simpler molecules that can be utilized. Animal digestion at its simplest is intracellular, as occurs within the individual cells of a sponge. Extracellular digestion, utilized by all more-complex animals, occurs in a body cavity. The simplest form is a saclike gastrovascular cavity in organisms such as flatworms and hydra. Still more-complex animals utilize a tubular compartment with specialized chambers where food is processed in a well-defined sequence.

3) How Do Humans Digest Food?

In humans, digestion begins in the mouth, where food is physically broken down by chewing and chemical digestion is initiated by saliva. Food is then conducted to the stomach by peristaltic waves of the esophagus. In the acidic environment of the stomach, food is churned into smaller particles, and protein digestion begins. Gradually, the liquefied food, now called chyme, is released to the small intestine. There it is neutralized by sodium bicarbonate from the pancreas. Secretions from the pancreas, liver, and the cells of the intestine complete the breakdown of proteins, fats, and carbohydrates. In the small intestine, the simple molecular products of digestion are absorbed into the bloodstream for distribution to the body cells. The large intestine absorbs the remaining water and converts indigestible material to feces.

Digestion is regulated by the nervous system and hormones. The smell and taste of food and the action of chewing trigger the secretion of saliva in the mouth and the production of gastrin by the stomach. Gastrin stimulates stomach acid production. As chyme enters the small intestine, three additional hormones are produced by intestinal cells: secretin, which causes sodium bicarbonate production to neutralize the acidic chyme; cholecystokinin, which stimulates bile release and causes the pancreas to secrete digestive enzymes into the small intestine; and gastric inhibitory peptide, which inhibits acid production and peristalsis by the stomach. This inhibition slows the movement of food into the intestine.

Key Terms

Thinking Through the Concepts

Multiple Choice

1. *An acidic mixture of partially digested food that moves from the stomach into the small intestine is called*
a. cholecystokinin
b. bile
c. lymph
d. secretin
e. chyme

2. *The hormone responsible for stimulating the secretion of hydrochloric acid by stomach cells is*
a. pepsin
b. gastrin
c. cholecystokinin
d. insulin
e. secretin

3. *A sudden increase in the amount of secretin circulating in the blood is an indication that food has recently been introduced to the*
a. mouth
b. pharynx
c. stomach
d. small intestine
e. large intestine

4. *Humans lack digestive enzymes to attack chitin, a complex polysaccharide in the exoskeleton of lobster and crayfish. We also lack enzymes that degrade*
a. peptides
b. plant starch
c. cellulose
d. sucrose
e. lipids

5. *Which of the following is NOT characteristic of bile?*
a. It is produced in the gallbladder.
b. It is a mixture of special salts, water, and cholesterol.
c. It acts as a detergent or emulsifying agent.
d. It helps expose a large surface area of lipid for attack by lipases.
e. It works in the small intestine.

6. *"Water-soluble compound that works primarily as an enzyme helper" would be a good definition of*
a. vitamin C
b. vitamin A
c. vitamin B
d. vitamin E
e. vitamin D

? Review Questions

1. List four general types of nutrients, and describe the role of each in nutrition.

2. List and describe the function of the three principal secretions of the stomach.

3. List the substances secreted into the small intestine, and describe the origin and function of each.

4. Name and describe the muscular movements that usher food through the human digestive tract.

5. Vitamin C is a vitamin for humans but not for dogs. Certain amino acids are essential for humans but not for plants. Explain.

6. Name four structural or functional adaptations of the human small intestine that ensure good digestion and absorption.

7. Describe protein digestion in the stomach and small intestine.

Applying the Concepts

1. The food label on a soup can shows that the product contains 10 grams of protein, 4 grams of carbohydrate, and 3 grams of fat. How many Calories are in this soup?

2. Stomach ulcers that resist antibiotic therapy are treated with several kinds of drugs. Anticholinergic drugs decrease nerve signals to the stomach walls that are produced by the sight, smell, and taste of food. Antacids neutralize stomach acid. Why would it be inadvisable to take anticholinergics and antacids together?

3. Small birds have high metabolic rates, efficient digestive tracts, and high-calorie diets. Some birds consume an amount of food equivalent to 30% of their body weight every day. They rarely eat leaves or grass but often eat small stones. The bird's small intestine has an attached pancreas and liver. Using this information, and Figure 29-7a, explain how a bird's digestive tract is adapted to its lifestyle (foods consumed, flight, habitat, and so on).

4. Control of the human digestive tract involves several feedback loops and messages that coordinate activity in one chamber with those taking place in subsequent chambers. List the coordinating events you discovered in this chapter, in order, beginning with tasting, chewing, and swallowing a piece of meat and ending with residue that enters the large intestine. What turns on and what shuts off each process?

5. Symbiotic protozoa in the digestive tracts of termites produce cellulase that the hosts use. In return, termites provide protozoa with food and shelter. Imagine that the human species is gradually invaded, over many generations, by symbiotic protozoa capable of producing cellulase. What evolutionary adaptive changes in body structure and function might occur simultaneously?

6. Trace a ham and cheese with lettuce sandwich through the human digestive system, discussing what happens to each part of the sandwich as it passes through each region of the digestive tract.

7. One of the common remedies for constipation (difficulty eliminating feces) is a laxative solution that contains magnesium salts. In the large intestine, magnesium salts are absorbed very slowly by the intestinal wall, remaining in the intestinal tract for long periods of time. Thus, the salts affect water movement in the large intestine. On the basis of this information, explain the laxative action of magnesium salts.

Group Activity

While rummaging around in the basement of a natural history museum, a curator came across a case of preserved animals that were collected in the early 1900s by a British expedition to Brazil. None of the museum's experts can determine the species-level identity of any of the specimens; the labels have been lost, and many of the animals are now extinct or very rare. As part of the effort to study the specimens, the museum director has asked you to form teams of three to examine the digestive systems of the mysterious animals. Your first group of three specimens includes:

- An invertebrate, 7.6 cm (3 inches) long, with a very flat body; no appendages (limbs); no eyes; and a long, narrow digestive tube with a single opening to the outside world.

- A legged vertebrate, 0.6 m (2 feet) long, with 30 teeth (all of which are either dagger-shaped or long and narrow with sharp edges); a single stomach that seems large for the animal's size; a short, straight small intestine; a prominent gallbladder; and very thick, muscular walls of the digestive tube, especially the stomach.

- A legless vertebrate, 0.9 m (3 feet) long, with sharp incisor teeth at the very front of its mouth; teeth at the back of its mouth that have flat, wide, bumpy surfaces; a stomach divided into three separate chambers; a long, highly coiled small intestine; and no gallbladder.

Formulate a specific, well-reasoned speculation about the food and feeding habits of each of these long-dead animals. Be sure to account for *all* of the information provided above for each specimen.

For More Information

Blaser, M. "The Bacteria Behind Ulcers." *Scientific American*, February 1996. New discoveries about the causes of ulcers are leading to new treatment strategies.

Davenport, H. "Why the Stomach Does Not Digest Itself." *Scientific American*, January 1972. How the stomach protects itself from its own strongly acidic secretions, and what happens when the defenses fail.

Diamond, J. "Dining with the Snakes." *Discover*, April 1994. Follow food as it is consumed and digested by a snake.

Kretchmer, N. "Lactose and Lactase." *Scientific American*, April 1972. Discusses differences in tolerance to milk sugar among humans from different populations.

Martin, R. J., White, B. D., and Hulsey, M. G. "The Regulation of Body Weight." *American Scientist*, November–December 1991. Describes some of the complex mechanisms that control eating, obesity, and the body's energy balance.

Mason, M. "Why Ulcers Run in Families." *Health*, September 1994. The story of the bacterium that causes most ulcers.

Mayfield, E. "A Consumer's Guide to Fats." *FDA Consumer*, May 1994. Commonsense information about fat, cholesterol, diet, and health.

Moog, F. "The Lining of the Small Intestine." *Scientific American*, November 1981. A description of this intricate tissue, which is responsible for absorbing nutrients into the body.

Nuland, S. B. "The Beast in the Belly." *Discover*, February 1995. An interesting medical tale of bacterial disease, digestive enzymes, and food.

Willett, W. C. "Diet and Health: What Should We Eat?" *Science*, April 22, 1994. Summarizes studies that suggest that diet can play a major role in the prevention of disease.

Answers to Multiple-Choice Questions 1. e 2. b 3. d 4. c 5. a 6. c

Perpetually bathed in fresh water, this brook trout faces a physiological challenge. The freshwater fish must somehow cope with an unending flow of water from its surroundings into its body.

The Urinary System 30

The typical trout urinates more or less constantly, excreting an amount of urine equivalent to its total blood volume every 2 or 3 hours. A kangaroo rat, conversely, might excrete a few milliliters of urine over the course of a day. Both the trout's need to eliminate water and the rat's need to conserve it have the same root cause: The volume of water inside an animal's body must be kept relatively constant. To accomplish this constancy, each species has evolved a regulatory mechanism that is effective in its particular environment. For the trout, which spends its life bathed in fresh water, the main task is getting rid of the water that is driven into its body by osmosis. For the kangaroo rat, which spends its life bathed in dry desert air, the main task is to prevent water from escaping its body. In both cases, the end result is maintenance of body water volume within a limited range.

Why do animals make large investments in the structures, energy, and control mechanisms required for water balance? If the volume of water inside body cells fluctuates too much and the materials dissolved in this intracellular fluid become too concentrated or too diluted, the chemical equilibrium of the cells will be disrupted, with disastrous consequences for the animal. The only way to keep intracellular water volume stable is to make sure that the cells are bathed in a precisely controlled volume of extracellular fluid.

As with many physiological systems, the hardware that evolution has installed for water regulation typically serves other crucial purposes as well. The mammalian kidney, for example, is the key component of a physiological system that not only regulates water exchange but also eliminates digestive waste products and ensures that the chemical composition of the blood and extracellular fluid remains stable and within the bounds required for cell metabolism.

Net Watch
On-line resources for this chapter are on the World Wide Web at:
http://www.prenhall.com/audesirk
(click on the Table of Contents link and then select Chapter 30).

1) What Are the Functions of Vertebrate Urinary Systems?

The urinary system plays a crucial role in **homeostasis**, the maintenance of a stable internal environment (as we learned in Chapter 26). All of the homeostatic functions of the urinary system in vertebrates are performed simultaneously as blood is filtered through the kidneys. The **kidneys** are organs in which the fluid portion of the blood is collected. From this fluid, water and important nutrients are reabsorbed into the blood, while toxic substances, cellular waste products, and excess vitamins, salts, hormones, and water are left behind to be excreted as urine. The rest of the urinary system channels and stores urine until it is released from the body, a process called **excretion**. The mammalian urinary system helps maintain homeostasis in several ways. These include:

1. Regulation of the blood levels of ions such as sodium, potassium, chloride, and calcium
2. Regulation of the water content of the blood
3. Maintenance of proper pH of the blood
4. Retention of important nutrients such as glucose and amino acids in the blood
5. Secretion of hormones, such as *erythropoietin,* which stimulates red blood cell production (see Chapter 27)
6. Elimination of cellular waste products such as *urea*

Urea is a product of amino acid metabolism. As you may recall from Chapter 29, the digestive system breaks proteins into their amino acid building blocks, which are then absorbed. When amino acids are taken into cells, some are used directly to synthesize new proteins. Others have their amino ($-NH_2$) groups removed and are then used either as a source of energy or in the synthesis of new molecules. The amino groups are released as **ammonia** (NH_3), which is very toxic. In mammals, ammonia is carried in the blood to the liver, where it is converted to urea, a far less toxic substance (Fig. 30-1). Urea is filtered from the blood by the kidneys and excreted in **urine**, a fluid consisting of water and dissolved wastes and some excess nutrients.

By excreting waste in the form of urea, mammals avoid damage from ammonia's acute toxicity. Still, mammals pay a cost for this evolutionary strategy. Urea production consumes energy and metabolically useful molecules, such as carbon and oxygen; after the urea is excreted, these useful commodities are lost to the animal. Furthermore, water must be excreted along with urea, even if water loss is disadvantageous. This problem is avoided by birds and reptiles, which excrete the by-products of protein digestion in the form of **uric acid**. Uric acid is not very soluble and can be excreted in crystalline form with very little loss of water.

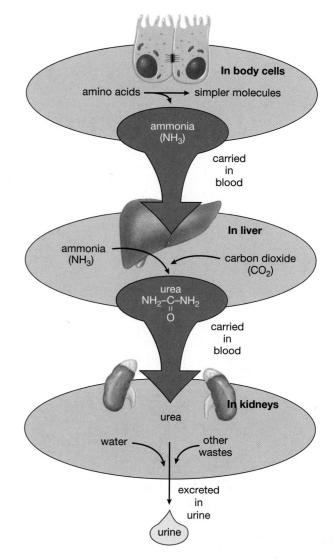

Figure 30-1 A flow diagram showing the formation and excretion of urea

2) How Do Simpler Animals Regulate Excretion?

Even in relatively simple invertebrate animals, urinary systems perform functions similar to those of vertebrates: They collect and filter body fluids, retaining nutrients and releasing wastes.

Flame Cells Filter Fluids in Flatworms

The first specialized excretory structures to arise in animal evolution were most likely **protonephridia**, which consist of a single cell that opens to the environment through a tube. One animal that still retains a protonephridial system is the flatworm, which is found under rocks in streams. The flatworm's urinary system (often called an *excretory system*) consists of a network of tubes that branch throughout the body (Fig. 30-2). At intervals

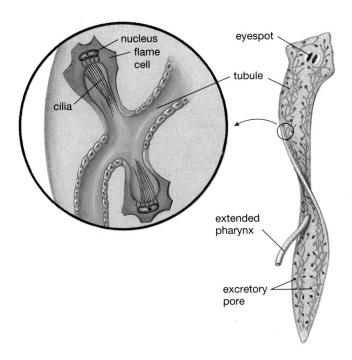

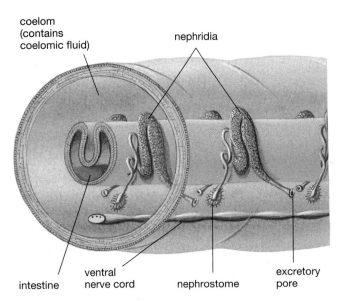

Figure 30-3 ***The excretory system of the earthworm***
This system consists of structures called nephridia, one pair per segment. Coelomic fluid is drawn into the nephrostome, and urine is released through the excretory pore. Each nephridium resembles a vertebrate nephron.

Figure 30-2 ***The simple excretory system of a flatworm***
Hollow flame cells direct excess water and dissolved wastes into a network of tubes. The beating cilia of the flame cells help circulate the fluid to excretory pores.

the tubes end blindly in single-celled bulbs called **flame cells**, which derive their name from the tuft of beating cilia that extends into the hollow bulb. Under the microscope, the beating of the cilia resembles the flickering of a flame. Water and some dissolved wastes are filtered into the bulbs, where the beating cilia produce a current that conducts the fluid through the tubular network. There, more waste products are added and nutrients withdrawn. Eventually, the waste liquid reaches one of numerous pores that release it to the outside. Flatworms also have a large skin surface through which wastes leave by diffusion.

Nephridia in Earthworms Resemble Parts of the Vertebrate Kidney

Earthworms, mollusks, and several other invertebrates have simple kidney-like structures called **nephridia** (singular, **nephridium**). In the earthworm, fluid fills the body cavity, or coelom, that surrounds the internal organs. This coelomic fluid collects both wastes and nutrients from the blood and tissues. The fluid is conducted into a funnel-shaped opening called the **nephrostome** and swept by cilia along a narrow, twisted tube (Fig. 30-3). There, salts and other dissolved nutrients are absorbed back into the blood, leaving behind water and wastes. The resulting urine is stored in an enlarged bladderlike portion of the nephridium and is then excreted through an opening, the **excretory pore**, in the body wall. The earthworm body is composed of repeating segments, nearly every one of which contains its own pair of nephridia.

3 How Does the Human Urinary System Function?

The problems of water regulation, solute regulation, and elimination of dissolved wastes are intertwined and often in conflict with one another. The excretion of soluble wastes requires the simultaneous excretion of water, but the need to maintain water balance may demand that the water be retained. Furthermore, the water in which the wastes are dissolved is very likely also to contain dissolved nutrients and salts that the body cannot afford to lose. How can wastes be excreted without undue loss of water and desirable molecules? One elegantly designed solution is found in the mammalian kidneys, complex organs that in some ways resemble dense collections of nephridia. The kidneys are part of a larger group of structures collectively called the *urinary system* (Fig. 30-4). The kidneys produce the urine; other portions of the system transport, store, and eliminate it. In the following sections we examine the major structures of the urinary system, tracing the pathway taken by waste products as they travel through the system. Then we explain kidney function in greater detail; finally, we describe the role of the kidneys in homeostasis.

Human kidneys are paired, kidney-bean-shaped organs located on either side of the spinal column and extending slightly above the waist. Each is approximately 13 centimeters (5 inches) long, 8 centimeters (3 inches) wide, and 2.5 centimeters (1 inch) thick. Blood carrying dissolved cellular wastes enters each kidney through a **renal artery**. After the blood has been filtered, it exits through

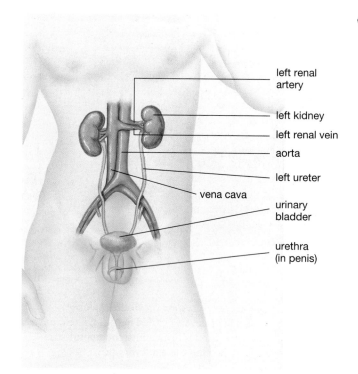

the **renal vein** (Fig. 30-5). Urine leaves each kidney through a narrow, muscular tube called the **ureter**. Using peristaltic contractions, the ureters transport urine to the *urinary bladder,* or simply **bladder**. This hollow, muscular chamber collects and stores the urine.

The walls of the bladder, which contain smooth muscle, are capable of considerable expansion. Urine is retained in the bladder by two sphincter muscles located at its base, just above the junction with the urethra. When the bladder becomes distended, receptors in the walls signal its condition and trigger reflexive contractions. The sphincter nearest the bladder, the *internal sphincter*, is opened during this reflex. However, the lower, or *external, sphincter*, is under voluntary control, so the brain can suppress the reflex unless bladder distension becomes acute. The average adult bladder will hold about 500 milliliters (approximately a pint) of urine, but the desire to urinate is triggered by considerably smaller accumulations. Urine completes its journey to the outside through the **urethra**, a single narrow tube about 3.8 centimeters

(a)

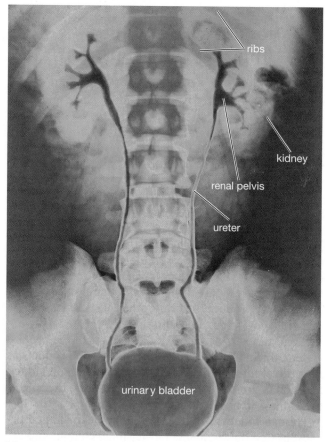

(b)

Figure 30-4 *The human urinary system*
(a) Diagrammatic view of the human urinary system and its blood supply. ***(b)*** Color-enhanced radiographic view (called a pyelogram).

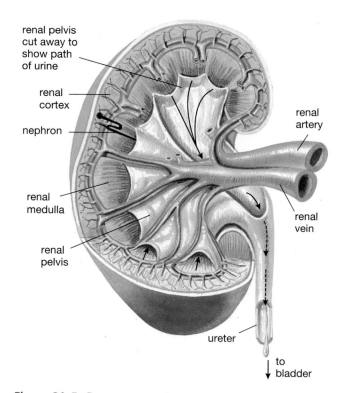

Figure 30-5 *Cross section of a kidney*
The cross section shows the blood supply and internal structure of a kidney. The renal artery, which brings blood to the kidney, and the renal vein, which carries the filtered blood away, branch extensively within the kidney. The two are joined by a highly permeable capillary network through which substances are exchanged between the blood and the nephrons. A nephron, considerably enlarged, is drawn to show its orientation in the kidney. The renal pelvis is the branched collecting chamber that funnels urine out of the kidney.

(1.5 inches) long in the female and about 20 centimeters (8 inches) long in the male.

Urine Is Formed in the Kidneys

Each kidney contains a solid outer layer where urine is formed and a hollow inner chamber called the **renal pelvis** (see Fig. 30-4b). The renal pelvis is a branched collecting chamber that funnels urine into the ureter (see Fig. 30-5). The outer layer of the kidney is divided into a fan-shaped inner **renal medulla** and an overlying **renal cortex**. Microscopic examination of these structures reveals an array of tiny individual filters, or **nephrons**. More than 1 million nephrons are packed into the cortex of each kidney, with many extending into the renal medulla.

The nephron has three major parts: (1) the **glomerulus**, a dense knot of capillaries from which fluid from the blood is filtered into (2) a surrounding cuplike structure, called **Bowman's capsule**, and (3), a long, twisted **tubule** (from the Latin, "little tube"). The tubule is further subdivided into first the **proximal tubule**, then the **loop of Henle** (which extends into the renal medulla), and finally the **distal tubule**, which leads to the **collecting duct** (Fig. 30-6; see also Fig. E30-1). In the tubule, nutrients are selectively reabsorbed from the filtrate back into the blood,

and wastes and some of the water are left behind to form urine. Additional wastes are also secreted into the tubule from the blood. Different portions of the tubule selectively modify the filtrate as it travels through. These processes are examined in the following sections and covered in more detail in "A Closer Look: The Nephron and Urine Formation" (pp. 600–601).

Blood Is Filtered by the Glomerulus

Blood is conducted to each nephron by an arteriole that branches from the renal artery. Within a cup-shaped portion of the nephron—Bowman's capsule—the arteriole branches into numerous microscopic capillaries that form the glomerulus, an intertwined mass (see Fig. 30-6). The walls of the glomerular capillaries are extremely permeable to water and small dissolved molecules, but they prevent the movement of most large proteins, such as the albumin found in blood. Beyond the glomerulus, the capillaries reunite to form an arteriole whose diameter is smaller than that of the incoming arteriole. The differences in diameter between the incoming and outgoing arterioles create pressure within the glomerulus, driving water and many of the dissolved substances from the blood through the capillary walls. This process is called **filtration** (Fig. 30-7, step ①), and the resulting fluid is called the **filtrate**. The watery filtrate, resembling blood plasma minus its plasma proteins, is collected in Bowman's capsule for transport through the nephron.

With the filtrate removed, the blood in the arteriole leaving the glomerulus is now very concentrated, having had much of its water removed but retaining substances too large to pass through the glomerular capillary walls, such as blood cells, large proteins, and fat droplets. Beyond the glomerulus, the arteriole branches into smaller, highly porous capillaries. These capillaries surround the tubule, forming intimate contacts with it. At these points of contact, water and nutrients remaining in the filtrate after filtration are reabsorbed as the filtrate passes through the nephron and are returned to the blood; in addition, wastes remaining in the blood after filtration are passed into the filtrate for disposal.

The Filtrate Is Converted to Urine in the Nephron

The filtrate collected in Bowman's capsule contains a mixture of both wastes and essential nutrients, including most of the blood's vital water. The nephron restores the nutrients and most of the water to the blood, while retaining wastes for elimination. This is accomplished by two processes: (1) tubular reabsorption and (2) tubular secretion.

Tubular reabsorption is the process by which cells of the proximal tubule remove water and nutrients from the filtrate within the tubule and pass them back into the blood (Fig. 30-7, step ②). The reabsorption of salts and other nutrients, such as amino acids and glucose, generally occurs by active transport (that is, cells of the tubule

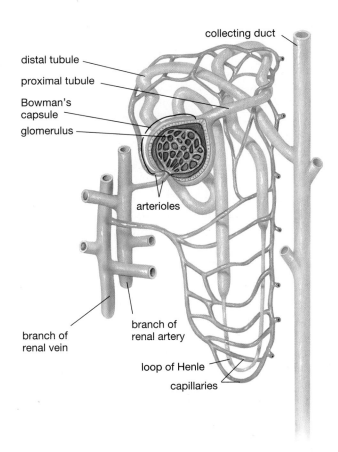

Figure 30-6 An individual nephron and its blood supply

collecting duct

distal tubule

proximal tubule

Bowman's capsule

glomerulus

arterioles

branch of renal vein

branch of renal artery

loop of Henle

capillaries

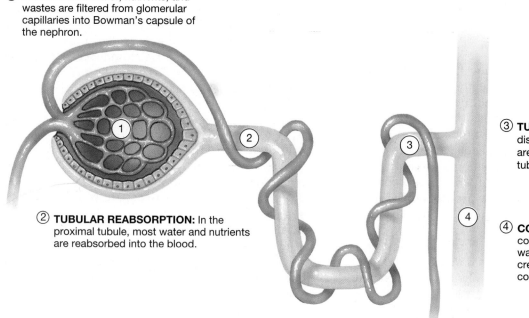

① **FILTRATION:** Water, nutrients, and wastes are filtered from glomerular capillaries into Bowman's capsule of the nephron.

② **TUBULAR REABSORPTION:** In the proximal tubule, most water and nutrients are reabsorbed into the blood.

③ **TUBULAR SECRETION:** In the distal tubule, additional wastes are actively secreted into the tubule from the blood.

④ **CONCENTRATION:** In the collecting duct, additional water may leave the blood, creating urine that is more concentrated than the blood.

Figure 30-7 Urine formation in the nephron
A summary of events that occur during the formation of urine in the nephron.

expend energy to transport these substances out of the tubule). These nutrients then enter adjacent capillaries by diffusion. Water follows the nutrients out of the tubule by osmosis. Wastes such as urea remain in the tubule and become more concentrated as water leaves.

Tubular secretion is the process by which wastes and excess substances that were not initially filtered out into Bowman's capsule are removed from the blood for excretion (Fig. 30-7, step ③). These wastes are actively secreted into the distal tubule by tubule cells. Secreted substances include hydrogen and potassium ions, ammonia, and many drugs.

The Loop of Henle Allows Urine to Become Concentrated

The kidneys of mammals and birds are able to produce urine that is more concentrated (has a higher osmotic pressure) than their blood. The ability to concentrate urine is a result of the structure of both the nephron and the collecting duct into which several nephrons empty (Fig. 30-7, step ④). Urine can become concentrated because there is an osmotic concentration gradient of salts and urea in the interstitial fluid that surrounds the loop of Henle. This gradient is produced by the loop of Henle within the renal medulla; the longer the loop, the greater the concentration gradient. The most concentrated fluid (with the greatest amount of dissolved substances and the least amount of water), which is far more concentrated than blood, surrounds the bottom of the loop. The

collecting duct passes through this osmotic gradient. As the filtrate passes through the portion of the collecting duct surrounded by the osmotically concentrated fluid, additional water leaves the filtrate by osmosis, but wastes do not. Therefore, as it moves through the collecting duct, the filtrate, now called urine, can reach an osmotic equilibrium, becoming as concentrated as the surrounding fluid. Because the rest of the excretory system does not allow water to enter or urea to escape, the urine remains concentrated.

It is important to produce concentrated urine when water is scarce and to produce dilute, watery urine when there is excess water in the blood. The degree of concentration of the urine is controlled by the amount of *antidiuretic hormone*, to be described shortly.

The Kidneys Are Important Organs of Homeostasis

Each drop of blood in your body passes through a kidney about 350 times a day; thus, the kidney is able to fine-tune the composition of the blood and thereby helps maintain homeostasis. The importance of this task is illustrated by the fact that kidney failure results in death within a short time (see "Health Watch: When the Kidneys Collapse," p. 602).

The Kidneys Regulate the Water Content of the Blood
One of the most important functions of the kidney is to regulate the water content of the blood (see Fig. 30-7).

Human kidneys filter 125 milliliters (about half a cup) of fluid from the blood each minute. Thus, without reabsorption of water, you would produce more than 180 liters (45 gallons) of urine daily! Water reabsorption occurs passively by osmosis as the filtrate travels through the tubule and the collecting duct.

The amount of water reabsorbed into the blood is controlled by a negative feedback mechanism (see Chapter 26). That feedback involves the amount of **antidiuretic hormone** (**ADH**; also called *vasopressin*) that circulates in the blood. This hormone increases the permeability of the distal tubule and the collecting duct to water, thereby allowing more water to be reabsorbed from the urine. ADH is produced by cells in the hypothalamus and is released by the posterior pituitary gland (as we shall see in Chapter 32).

The release of ADH is regulated by receptor cells in the hypothalamus that monitor the osmotic concentration of the blood and by receptors in the heart that monitor blood volume. For example, as the lost traveler staggers through the searing desert sun, dehydration occurs. With the loss of water, the osmotic concentration of his blood rises and his blood volume falls, triggering the release of more ADH (Fig. 30-8). The release of ADH increases water reabsorption and produces urine more concentrated than the blood.

In contrast, a partygoer overindulging in beer will experience a decrease in blood concentration and an increase in blood volume, and her receptors will cause a decrease in ADH output. Reduced ADH concentration will make the distal tubule and collecting duct less permeable to water. When the ADH level is very low, little water is reabsorbed after the urine leaves the loop of Henle, and the urine produced will be more dilute than the blood. In extreme cases, urine flow may exceed 1 liter (about 1 quart) per hour. As the proper water level is restored, the increased osmotic concentration of the blood and decreased blood volume will stimulate increased ADH production, which will in turn stimulate increased water reabsorption.

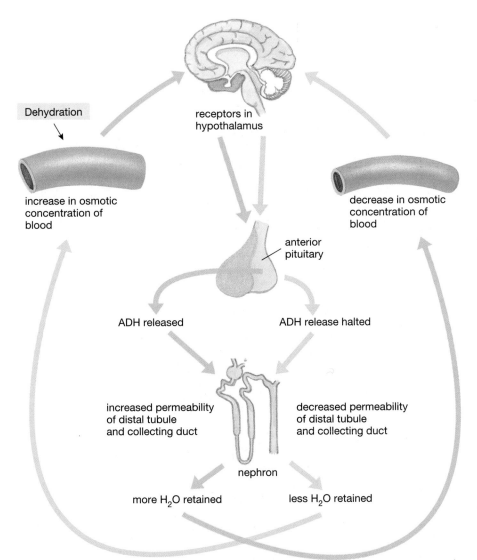

Dehydration

receptors in hypothalamus

increase in osmotic concentration of blood

decrease in osmotic concentration of blood

anterior pituitary

ADH released

ADH release halted

increased permeability of distal tubule and collecting duct

decreased permeability of distal tubule and collecting duct

nephron

more H$_2$O retained

less H$_2$O retained

Figure 30-8 Regulation of the water content of the blood
The water content of the blood is regulated in part by a negative feedback mechanism. Sensors in the hypothalamus gauge the concentration of solutes in the blood; deviations from a set point cause the anterior pituitary gland to turn on or turn off a release of the hormone ADH. Any perturbation of the system (for example, by dehydration) triggers a change that will restore blood solute concentration to the appropriate level.

A Closer Look
The Nephron and Urine Formation

The complex structure of the nephron is finely adapted to its function. In Figure E30-1, the nephron is presented in diagram form to illustrate the processing that occurs in each part. Circled numbers refer to the following descriptions:

① *Filtration.* Water and dissolved substances are forced out of the glomerular capillaries into Bowman's capsule, from which they are funneled into the proximal tubule.

② *Tubular reabsorption.* In the proximal tubule, most of the important nutrients remaining in the filtrate are actively pumped out through the walls of the tubule and are reabsorbed into the blood. These nutrients include about 75% of the salts as well as amino acids, sugars, and vitamins. The proximal tubule is highly permeable to water, so water follows the nutrients, moving out of the tubule and back into the blood by osmosis.

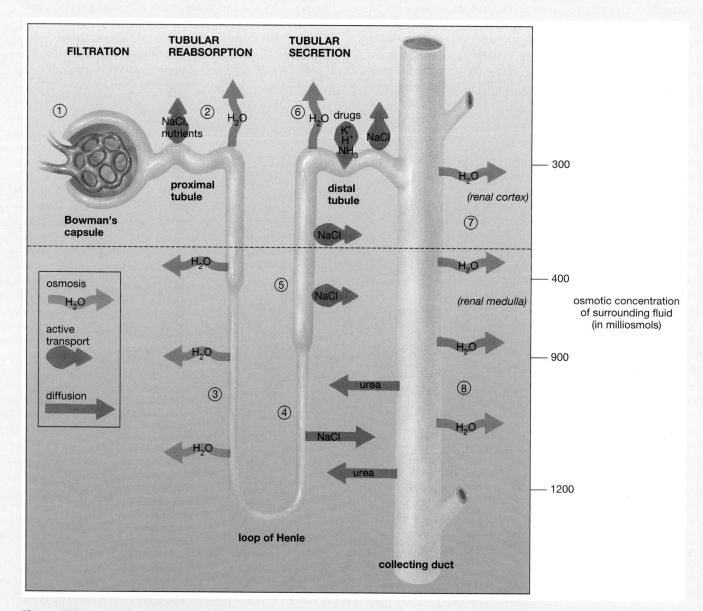

Figure E30-1 *Details of urine formation*
A single nephron, showing the movement of materials through different regions. The osmotic concentration of the filtrate inside the nephron increases (so the concentration of dissolved substances rises) from top to bottom in this diagram. Outside the nephron, darker shades of blue represent higher osmotic concentrations of the surrounding fluid.

③ The loop of Henle, which is unique to birds and mammals, is essential for urine concentration. It maintains a salt concentration gradient in the extracellular fluid that surrounds it, with the highest concentration at the bottom of the loop. The descending portion of the loop of Henle is very permeable to water but not to salt or other dissolved substances. As the filtrate passes through the descending portion, water leaves by osmosis as the concentration of the surrounding fluid increases.

④ The thin portion of the ascending loop of Henle is relatively impermeable to water and urea but is permeable to salt, which moves out of the filtrate by diffusion. Why? Although the osmotic concentrations inside and outside the tubule are about equal, at this stage the level of urea is higher outside and that of salt is higher inside. Thus, the concentration gradient favors the movement of salt outward. Because water cannot follow it, the filtrate now becomes less concentrated than its surroundings.

⑤ The thick portion of the ascending loop of Henle is also impermeable to water and urea. There, salt is actively pumped out of the filtrate, leaving water and wastes behind.

⑥ The watery filtrate, low in salt but retaining wastes such as urea, now arrives at the distal portion of the tubule, where more salt is pumped out. Because this portion is permeable to water, water follows by osmosis. Tubular secretion occurs throughout the tubule but is especially active in the distal portion. There, substances such as K^+, H^+, NH_3, and some drugs and toxins are actively pumped into the tubule.

⑦ By the time the filtrate reaches the collecting duct, very little salt is left and about 99% of the water has been reabsorbed into the bloodstream. The collecting duct conducts the fluid, now called urine, down through the increasing concentration gradient created by the loop of Henle. The collecting duct is very permeable to water when antidiuretic hormone (ADH) is present, so water moves out by osmosis as the concentration of the external fluid increases. If ADH is absent, the collecting duct remains impermeable to water, and the urine stays dilute and watery.

⑧ The lower portion of the collecting duct is also permeable to urea. Therefore, as the filtrate moves farther down the collecting duct, some urea diffuses out, contributing to the osmotic concentration of the surrounding fluid. As water (when ADH is present) and urea move out, the concentration of dissolved wastes in the urine in the collecting duct approaches equilibrium with the high osmotic concentration of the external fluid.

Mammalian Kidneys Are Adapted to Diverse Environments

In addition to regulating water balance by using hormones such as ADH, different types of mammals have kidneys whose structures are adapted to the availability of water in their natural habitats. Mammals that must conserve water do so by producing urine that is more concentrated than their blood. The degree of concentration that can be achieved is determined by the length of the loop of Henle. The longer the loop, the higher the salt concentration produced in the fluid that surrounds it. The higher the salt concentration, the more water moves out of the urine by osmosis as urine passes through the collecting duct, and the more concentrated the urine becomes. As you might predict, animals living in very dry climates have the longest loops of Henle, whereas those living in watery environments have relatively short loops.

The beaver, for example, has only short-looped nephrons and is unable to concentrate its urine to more than twice its plasma concentration. Human kidneys have a mixture of long-looped and short-looped nephrons and can concentrate urine to about four times the plasma concentration. The masters of urine concentration are desert rodents such as kangaroo rats, which can produce urine 14 times their plasma concentration (Fig. 30-9).

Figure 30-9 An adaptation to a dry environment
The desert kangaroo rat of the southwestern United States can dispense with drinking partly because its long loops of Henle allow it to produce very concentrated urine.

Health Watch
When the Kidneys Collapse

The kidney is a relatively rugged organ. Normally the need to urinate is our only evidence of its industrious activity. But if the kidneys fail, death occurs rapidly. The kidneys are vulnerable to attack from several sources. Some toxic substances can destroy kidney tubules. These substances include heavy metals (mercury, arsenic), organic compounds such as solvents (carbon tetrachloride), insecticides, and overdoses of some antibiotics and pain relievers. A second source of injury is through infection by bacteria, particularly intestinal bacteria that reach the kidney via the urethra. Still another source of injury is *glomerulonephritis.* In this disease, an abnormal immune response to an infection causes insoluble particles to form. These particles become trapped in the glomeruli, resulting in inflammation and death of the tubules. Kidney damage is typically treated by using a remarkable machine: the artificial kidney (Fig. E30-2).

First used in 1945, the artificial kidney operates on the simple principle of diffusion of substances from areas of high concentration to areas of low concentration across an artificial permeable membrane. The passive diffusion of substances across an artificial semipermeable membrane is called **dialysis**; when blood is filtered according to this principle, the process is called **hemodialysis**. During hemodialysis, the patient's blood is diverted from the body and pumped through narrow tubing made of a cellophane membrane suspended in dialyzing fluid. This membrane has pores too small to permit the passage of blood cells and large proteins but large enough to pass small molecules such as water, sugar, salts, amino acids, and urea. The composition of the dialyzing fluid is adjusted to have normal blood levels of these substances so that important nutrients in the blood will be retained. As a result, only molecules whose concentration is in excess of normal will diffuse into the dialyzing fluid. Urea, for example, is present in a relatively high concentration in the blood and absent in the dialyzing fluid, so it diffuses out of the blood and into the dialyzing fluid.

One challenge of dialysis is to eliminate excess water. Because the blood has a higher osmolarity than does the dialyzing fluid, water tends to diffuse *into* the blood from the dialyzing fluid. This process is counteracted by applying pressure to the blood, which forces some of the small water molecules out of the blood against their concentration gradient.

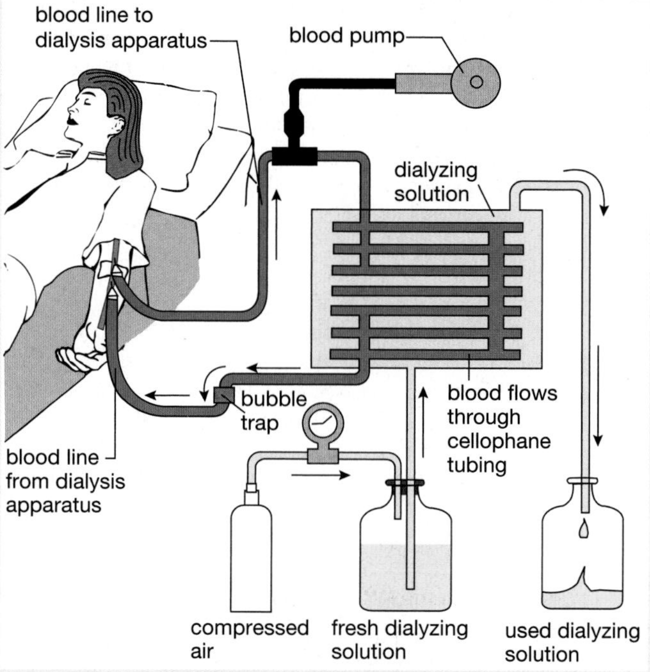

Figure E30-2 A diagram of the artificial kidney

Roughly 500 milliliters (1 pint) of blood is in the machine at any one time, flowing at a rate of several hundred milliliters per minute. Still, the patient must remain attached to the dialysis machine for 4 to 6 hours three times a week. Although patients have remained alive on dialysis for up to 20 years, their lives are far from normal. Their diet and fluid intake must be carefully regulated. Even so, their blood composition fluctuates and many more toxic substances accumulate than is normal. Despite its drawbacks, dialysis keeps about 200,000 Americans alive. For those who are fortunate enough to find a compatible donor, kidney transplants are a far better solution; about 20,000 of these are performed yearly in the United States.

Kangaroo rats (as you might predict) have only very long-looped nephrons. Because they have an extraordinary ability to conserve water, they can completely dispense with drinking, relying entirely on water derived from their food.

Kidney structure is not the only factor that affects water balance in the mammalian body. The body's water content is also affected by water inputs from drinking and eating, by evaporative water loss through the skin (for example, sweating), and by an animal's behavior. For instance, kangaroo rats have developed their extraordinary ability to conserve water mainly because their en-

vironment affords them little opportunity to drink. They must rely entirely on water derived from their food, and they have developed behaviors that help them both avoid unnecessary water loss and acquire the greatest possible amount of water from their food. Kangaroo rats live in underground burrows and venture above ground only at night. This behavior minimizes their exposure to hot desert temperatures by day, which in turn minimizes the amount of water they lose by sweating. During their nightly sojourns to the surface, kangaroo rats gather seeds to eat. But instead of eating them right away, the rats carry the seeds back to their burrows. Why? Because the air in the underground burrows is far moister than the parched desert air on the surface, and a seed stored underground will absorb precious water before it is eaten. For the kangaroo rat, behaviors that reduce exposure to dry air and dry food are key factors in maintaining homeostasis.

Kidneys Help Regulate the Blood Pressure and Oxygen Content of the Blood

Two hormones produced by the kidneys are extremely important in regulating blood pressure and the blood's oxygen-carrying capacity. When blood pressure falls, the kidneys release **renin** into the bloodstream. Renin acts as an enzyme, catalyzing the formation of a second hormone, **angiotensin**, from a protein that circulates in the blood. Angiotensin in turn causes arterioles to constrict, elevating blood pressure. The constriction of the arterioles that carry blood to the kidneys also reduces the rate of blood filtration, causing less water to be removed from the blood. Water retention causes an increase in blood volume and, consequently, an increase in blood pressure.

In response to low blood oxygen levels, the kidneys release a second hormone, **erythropoietin**, described in Chapter 27. Erythropoietin travels in the blood to the bone marrow, where it stimulates more-rapid production of red blood cells, whose role is to transport oxygen.

Kidneys Monitor and Regulate Dissolved Substances in the Blood

As the kidney filters the blood, it monitors and regulates blood composition in order to maintain a constant internal environment. Substances the kidney regulates, in addition to water, include nutrients such as glucose, amino acids, vitamins, urea, and a variety of ions, including sodium, potassium, chloride, and sulfate. The kidney maintains a constant blood pH by regulating the amount of hydrogen and sodium bicarbonate ions. This remarkable organ also eliminates potentially harmful substances, including some drugs, food additives, pesticides, and toxic substances from cigarette smoke, such as nicotine.

Summary of Key Concepts

1) What Are the Functions of Vertebrate Urinary Systems?

The urinary system plays a crucial role in homeostasis, the maintenance of a stable internal environment. The kidneys of mammalian urinary systems regulate the water and ion contents of the blood and blood pH. These organs help retain nutrients and eliminate cellular wastes and toxic substances, and they secrete the hormone erythropoietin.

2) How Do Simpler Animals Regulate Excretion?

The simple excretory system of the flatworm consists of a network of tubules that branch through the body. Flame cells circulate body fluid through the tubules, where nutrients are reabsorbed. Wastes, including excess water, are excreted through numerous excretory pores. Many of the more complex invertebrates, including earthworms and mollusks, use nephridia. In the earthworm, these are paired structures, resembling vertebrate nephrons, that are found in most of the earthworm's segments. Coelomic fluid is drawn into a ciliated opening, the nephrostome, and nutrients and water are reabsorbed. Wastes and excess water are released through the excretory pore.

3) How Does the Human Urinary System Function?

The human urinary system consists of kidneys, ureters, bladder, and urethra. Kidneys produce urine, which is conducted by the ureters to the bladder, a storage organ. Distension of the muscular bladder wall triggers urination, during which urine passes out of the body through the urethra.

Each kidney consists of more than a million individual nephrons in an outer renal cortex, with many extending into an inner renal medulla. Urine formed in the nephrons enters collecting ducts that empty into the renal pelvis, from which it is funneled into the ureter.

Each nephron is served by an arteriole that branches from the renal artery. The arteriole further branches into a mass of porous-walled capillaries called the glomerulus. There water and dissolved substances are filtered from the blood by pressure. The filtrate is collected in the cup-shaped Bowman's capsule and conducted along the tubular portion of the nephron. During tubular reabsorption, nutrients are actively pumped out of the filtrate through the walls of the tubule. Nutrients then enter capillaries that surround the tubule, and water follows by osmosis. Some wastes remain in the filtrate; others are actively pumped into the tubule by tubular secretion. The tubule forms the loop of Henle, which creates a salt concentration gradient surrounding it. After completing its passage through the tubule, the filtrate enters the collecting duct, which passes through the concentration gradient. Final passage of the filtrate through this gradient via the collecting duct allows the concentration of the urine.

The kidneys are important organs of homeostasis. The water content of the blood is regulated by antidiuretic hormone (ADH), produced in the hypothalamus and released by the posterior pituitary gland. Low blood volume and high osmotic concentration of the blood signal dehydration and stimulate the release of ADH into the bloodstream. The ADH increases the permeability to water of the distal tubule and the collecting duct, allowing more water to be reabsorbed into the blood. In addition, the kidneys control blood pH, remove toxic substances, and regulate ions such as sodium, chloride, potassium, and sulfate. Excess glucose, vitamins, and amino acids are also excreted by the kidneys.

Key Terms

ammonia p. 594
angiotensin p. 603
antidiuretic hormone (ADH)
  p. 599
bladder p. 596
Bowman's capsule p. 597
collecting duct p. 597
dialysis p. 602
distal tubule p. 597
erythropoietin p. 603

excretion p. 594
excretory pore p. 595
filtrate p. 597
filtration p. 597
flame cell p. 595
glomerulus p. 597
hemodialysis p. 602
homeostasis p. 594
kidney p. 594
loop of Henle p. 597

nephridium p. 595
nephron p. 597
nephrostome p. 595
protonephridium p. 594
proximal tubule p. 597
renal artery p. 595
renal cortex p. 597
renal medulla p. 597
renal pelvis p. 597
renal vein p. 596

renin p. 603
tubular reabsorption p. 597
tubular secretion p. 598
tubule p. 597
urea p. 594
ureter p. 596
urethra p. 596
uric acid p. 594
urine p. 594

Thinking Through the Concepts

Multiple Choice

1. *Which of the following is false?*
   a. Urea is more toxic than ammonia.
   b. Ammonia is converted to urea in the liver.
   c. Ammonia is produced in body cells.
   d. The fluid collected in Bowman's capsule is called the filtrate.
   e. *Tubule* means "little tube."

2. *The walls of the _____ are made more or less permeable to water, depending on the need to conserve water.*
   a. ureter   b. urethra
   c. proximal tubule   d. collecting duct
   e. glomerulus

3. *Which of the following conditions will cause a decrease in ADH production?*
   a. dehydration
   b. drinking beer
   c. an increase in osmotic pressure of blood
   d. abnormally high blood sugar
   e. strenuous exercise

4. *The function of the glomerulus and Bowman's capsule of the nephron is to*
   a. reabsorb water into the blood
   b. eliminate ammonia from the body
   c. reabsorb salts and amino acids
   d. filter the blood and capture the filtrate
   e. concentrate the urine

5. *What determines the ability of a mammal to concentrate its urine?*
   a. the number of nephrons
   b. the length of the tubules
   c. the length of the collecting duct
   d. the size of the glomerulus
   e. the length of the loop of Henle

6. *Which of the following processes does NOT occur in the nephron and collecting duct?*
   a. filtration
   b. elimination of urea from the body
   c. reabsorption of nutrients
   d. tubular secretion
   e. concentration of urine

? Review Questions

1. Explain the two major functions of excretory systems.
2. Trace a urea molecule from the bloodstream to the external environment.
3. What is the function of the loop of Henle? The collecting duct? Antidiuretic hormone?
4. Describe and compare the processes of filtration, tubular reabsorption, and tubular secretion.
5. Describe the role of the kidneys as organs of homeostasis.
6. Compare and contrast the excretory systems of humans, earthworms, and flatworms. In what general ways are they similar? Different?

Applying the Concepts

1. Discuss the differences in function of the two major capillary beds in the kidneys: the glomerular capillaries and those surrounding the tubules.
2. Desert animals need to conserve water. These animals have larger kidneys than do animals that live in moist environments and thus need not conserve water. The larger kidneys allow for a greater distance between the glomerulus and the bottom of the loop of Henle. Discuss why this anatomical difference assists water conservation in the desert animals.
3. Some "quick weight loss" diets require the ingestion of much protein-rich food and the elimination of carbohydrates. Two side effects of such diets are increased thirst and increased urination. Explain the connections between the diets and the side effects.
4. Your teenage son is dying from injuries sustained in a motorcycle accident. You have been asked to allow your son's kidneys to be donated to people who currently rely on dialysis. Explain what benefits these kidneys would confer on a dialysis patient. How could two dialysis patients be helped by your son?
5. Some employers require their employees to submit to urine tests before they can be employed and at random intervals during their employment. Refusal to take the test or failure to "pass" the test could be grounds for termination. What is the purpose of the urine test? What types of employers might find such tests necessary? How do you feel about urine tests for obtaining or keeping a job? Explain your answers.

Group Activity

Mammals excrete nitrogen-containing wastes as urea, but urea is not the only molecule that can serve this purpose. One possible alternative is ammonia, which (compared with urea) is very soluble and requires much less energy to synthesize. Ammonia is also a small molecule that can easily diffuse through membranes. Unfortunately, ammonia is relatively toxic to animals. Another molecule that can be used to excrete nitrogen is uric acid. Uric acid is not toxic, but it is nearly insoluble and typically forms crystals. Also, uric acid takes a great deal of energy to manufacture. How might evolution have proceeded if there were no urea? NASA would like to know, because the astronomers there have discovered two planets that lack urea. Each planet has habitats just like those on Earth, but when it comes to nitrogen-bearing biological chemicals, Planet A has only ammonia and Planet B has only uric acid. NASA has asked you to predict the habitats in which animals will live and the design of their physiological systems for excreting nitrogenous waste and for maintaining water balance.

To accomplish this task, form a team of four students, and then divide your team in half. One pair should make a written prediction for Planet A, the other pair for Planet B. Then exchange predictions and (in your original pair) evaluate and critique the other pair's prediction. Finally, gather the entire team and agree on two final predictions (one for each planet) to submit. Include annotated diagrams of the animals' excretory systems.

For More Information

Greenberg, A. (ed.). *Primer of Kidney Disease.* San Diego: Academic Press, 1998. An up-to-date, comprehensive review of kidney diseases, written by a variety of experts.

O'Brien, C. "Lucky Break for Kidney Disease Gene." *Science*, June 24, 1994. Discusses research on a dominant gene that causes kidney disease.

Plafrey, C., and Cossins, A. "Fishy Tales of Kidney Function." *Nature*, September 24, 1994. A description of research findings that have increased our understanding of how fish cope with the physiological challenges posed by watery and/or salty environments.

Taylor, C. R. "The Eland and the Oryx." *Scientific American*, January 1969. An account of the physiological adaptations by which two species of African antelopes cope with the hot, dry environments in which they live.

Answers to Multiple-Choice Questions
1. a   2. d   3. b   4. d   5. e   6. b

"If you wish to be well and keep well, take Braggs Vegetable Charcoal and Charcoal Biscuits. Absorbs all impurities in the stomach and bowels, effectively warding off cholera, smallpox, typhoid, and all malignant fevers. Eradicate worms in children. Sweeten the breath."

From an early 1900s newspaper ad

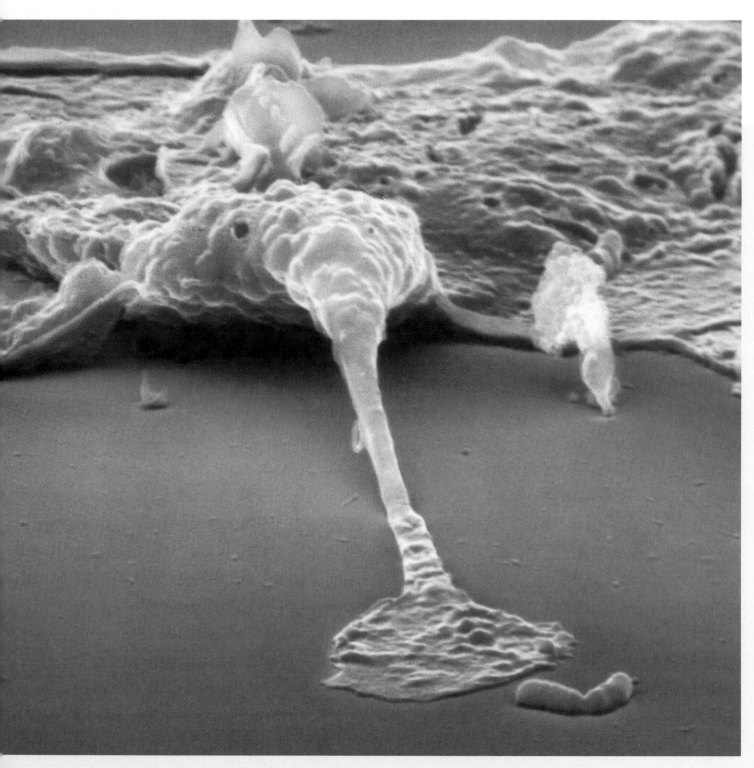

A macrophage, one of the key cellular warriors in the never-ending battle against invaders of the body, attacks a rod-shaped bacterium.

Defenses Against Disease: The Immune Response

31

At a Glance

 Net Watch

On-line resources for this chapter are on the World Wide Web at:

http://www.prenhall.com/audesirk

(click on the Table of Contents link and then select Chapter 31).

Why do we get sick? Because the animal body is under constant attack. Our environment is teeming with viruses, bacteria, fungi, and protists that, should they have the opportunity to take up residence, can harm or even kill an animal. The environment is likewise home to a huge array of toxic substances that can pose a threat to health. Given the diversity and ubiquity of potential threats, it may make more sense to ask, "Why don't we get sick more often?" In fact, we owe our health to our evolutionary heritage. Among our ancestors, those who best survived and reproduced were those whose bodies best resisted attack by parasites and toxins. These resistant individuals passed on their methods of resistance to subsequent generations. However, the most evolutionarily successful parasites are those whose attack strategies are most capable of overcoming resistance. As a result, animals and their parasites have engaged in a never-ending, constantly escalating battle in which ever more sophisticated defense systems are challenged by ever more effective tactics for penetrating those defenses.

This evolutionary arms race has honed our defenses and endowed us with a multifaceted system of resistance to parasites. This system of defense combines a variety of different mechanisms into a stunningly complex system for resisting parasitic invasion. You're probably familiar with some of the components of the system, though you may not have recognized them as such. Fever, inflammation, pain, and coughing may seem more like problems than solutions, but all are part of the body's fight against invasion. Much of the battle, however, takes place invisibly, at the cellular level. The body mobilizes an army of cells to patrol its territory and seek out and attack invaders. This cellular army, however, faces a complex challenge as it combats diverse and resourceful opponents whose survival depends on their gaining access to the stable environment and abundant resources inside the host's body. How can the body deny entry to invaders? If intruders somehow gain entry, how can the immune system discriminate between invaders and the body's own cells and molecules? Once an invader is identified, how can it be killed or rendered harmless? How are some invaders able to evade or overcome the immune system defenses? Just like military defense systems, the immune system fights its battles on many fronts and makes use of many kinds of specialized weapons. This chapter will introduce you to some of the key specialists and to their strategies for detecting and attacking invaders.

1 How Does the Body Defend Against Invasion?

The human body has three lines of defense against microbial attack: (1) external barriers that keep microbes out of the body; (2) nonspecific internal defenses that combat all invading microbes; and (3) the immune system, which directs its assault—an *immune response*—against specific microbes (Fig. 31-1).

The Skin and Mucous Membranes Form Barriers to Invasion

The best defense strategy is to prevent invaders from entering the body in the first place. In animal bodies, this first line of defense is performed by the two surfaces that are exposed to the environment: (1) the **skin** and (2) the **mucous membranes** of the digestive and respiratory tracts. These surfaces are barriers to invasion.

Intact Skin Is Both a Barrier to Entry and an Inhospitable Environment for Microbial Growth

The outer surface of the skin consists of dry, dead cells filled with horny proteins similar to those in hair and nails. Consequently, most microbes that land on the skin cannot obtain the water and nutrients they need. Even those few bacteria and fungi that gain a foothold will most likely be ejected before they can do harm, because skin cells are constantly sloughed off and replaced from below. The skin is further protected by secretions from sweat glands and sebaceous glands. These secretions contain natural antibiotics, such as lactic acid, that inhibit the growth of bacteria and fungi. These multiple defenses make the unbroken skin an extremely effective barrier against microbial invasion.

Antimicrobial Secretions, Mucus, and Ciliary Action Defend the Mucous Membranes Against Microbes

The warm, moist mucous membranes that line the digestive and respiratory tracts are much more hospitable to microbes than is the dry, oily skin, but these membranes nonetheless incorporate effective defense mechanisms. In particular, the membranes secrete mucus that contains antibacterial enzymes such as lysozyme, which destroys bacterial cell walls. The mucus also physically traps microbes that enter the body through the nose or mouth (Fig. 31-2). Cilia on the membranes sweep up the mucus, microbes and all, until it is either coughed or sneezed out of the body, or swallowed. If microbes are swallowed, they enter the stomach, where they encounter a combination of extreme acidity (pH 1 to 3) and protein-digesting enzymes, which can kill many types of microbes. Farther along in the digestive tract, the intestine is inhabited by bacteria that are harmless to the human body but secrete substances that destroy invading foreign bacteria or fungi. Despite these defenses, many disease organisms manage to enter the body through the mucous membranes.

Nonspecific Internal Defenses Combat Microbes

Invading parasites that penetrate the first line of defense at the skin and mucous membranes encounter an array of internal defenses. Some of these defenses are nonspecific—that is, they attack a wide variety of microbes rather than targeting specific invaders as the immune response does. Nonspecifc defenses fall into three main categories. (1) The body has a standing army of **phagocytic cells**, which destroy microbes, and **natural killer cells**, which destroy cells of the body that have been infected by viruses. The steady trickle of microbes that pass through the body's external barriers are mostly mopped up by these cells. (2) An injury, with its combination of tissue damage and relatively massive invasion of microbes, provokes an **inflammatory response**. The inflammatory

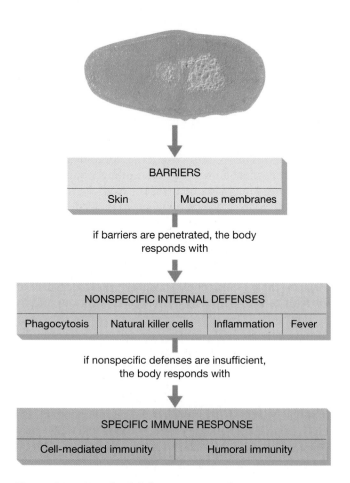

Figure 31-1 *Levels of defense against infection*

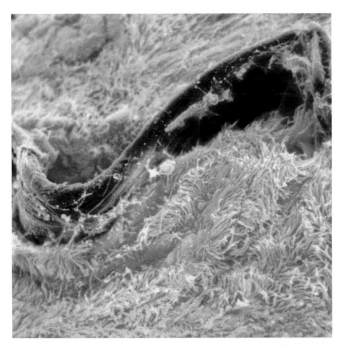

0.5 micrometer

Figure 31-2 ***The protective function of mucus***
Mucus traps microbes and debris in the respiratory tract (a brownish-green strand of dirt is shown caught in mucus atop the orange cilia). The cilia lining the walls of the respiratory tract then sweep both mucus and foreign matter out of the body.

response simultaneously recruits new members of the army of phagocytic cells and killer cells and walls off the injured area, isolating the infected tissue from the rest of the body. (3) If a population of microbes succeeds in establishing a major infection, the body commonly produces a **fever**, which both slows down microbial reproduction and enhances the body's own fighting abilities.

Phagocytic Cells and Natural Killer Cells Destroy Invading Microbes

The body contains several types of amoeboid white blood cells that can engulf and digest microbes. The most important of these are the **macrophages** (literally, "big eaters"), white blood cells that crawl around in the extracellular fluid. Macrophages ingest microbes by phagocytosis. In addition to their immediate effect of destroying an individual microbe, macrophages also play a crucial role in the immune response by acting as *antigen-presenting cells*—"presenting" parts of the microbe to other cells of the immune system.

Natural killer cells are another class of white blood cells. In general, natural killer cells do not directly attack invading microbes. Instead, natural killer cells strike at the body's own cells that have been invaded by viruses. Once infected by a virus, cells normally bear some viral proteins on their surfaces. Natural killer cells seek out

and recognize these proteins and kill the infected cell. Natural killer cells also recognize and kill cancerous cells. Rather than eat their victims, natural killer cells strike from the outside. They secrete proteins onto the plasma membrane of the infected or cancerous cell. Some of the proteins insert themselves into the target plasma membrane and form a ring (much like making a barrel out of individual staves of wood), thereby opening up a large pore in the membrane (Fig. 31-3). Killer cells also secrete enzymes that break up some of the molecules of the target cell. Shot full of holes and chewed on by enzymes, the target cell soon dies.

The Inflammatory Response Defends Against Localized Injury

Large-scale breaches of the skin or mucous membranes, such as a cut, elicit an inflammatory response. Damaged cells release the chemical **histamine** into the wounded area. Histamine makes capillary walls leaky and relaxes the smooth muscle that surrounds arterioles, leading to increased blood flow. With extra blood flowing through leaky capillaries, fluid seeps from the capillaries into the tissues around the wound. The wound becomes red, swollen, and warm. (*Inflammation* literally means "to set on fire.") Meanwhile, other chemicals released by injured cells initiate blood clotting (see Chapter 27), which blocks damaged blood vessels, preventing microbes from entering the bloodstream, and also seals off the wound from the outside world, limiting the entry of more microbes.

Still other chemicals, some released by wounded cells and others produced by the microbes themselves, attract macrophages and other phagocytic cells to the wound. Some of the phagocytic cells emerge from local tissues; others arrive via the circulatory system and squeeze out through the leaky capillary walls. After entering the wound, the phagocytic cells engulf microbes, dirt, and damaged cells (Fig. 31-4, p. 611). Unfortunately, each phagocyte can eat just so many microbes, and then it dies. If tissue damage is too severe or the wound is too dirty, then the phagocytes may be unable to complete the clean-up job. In that case, the fluid around the wound turns into pus, a thick mixture of microbes, tissue debris, and living and dead white blood cells.

Fever Combats Large-Scale Infections

If invaders survive the inflammatory response in sufficient numbers, they may infect larger areas of the body and trigger a fever. Advertisements and TV dramas have encouraged us to regard fevers as dangerous and debilitating, but a fever is in fact part of the body's defense against infection. Fever has both beneficial effects for the body's normal defenses and detrimental effects on the invading microbes. For example, fever increases the activity of the phagocytic white blood cells that attack bacteria, thereby producing a shorter and less serious infection. At the same time, feverish body temperatures of about 39°C (102°F) force many bacteria to use more

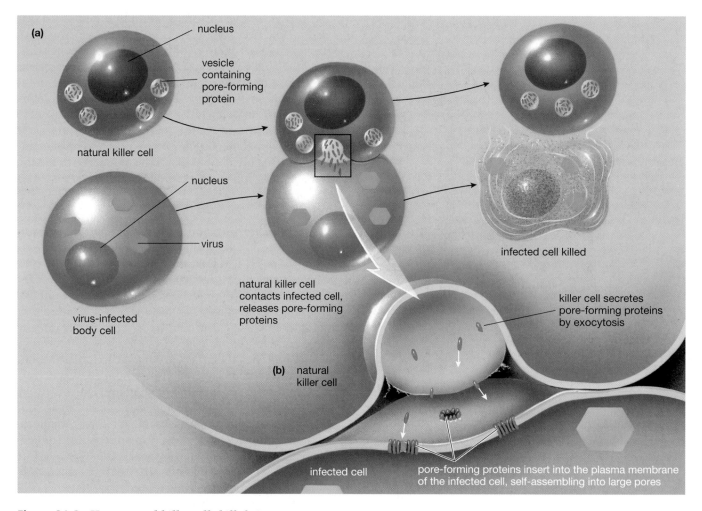

Figure 31-3 *How natural killer cells kill their targets*
(a) A natural killer cell contains vesicles full of pore-forming proteins. When it encounters a suitable target cell (here, a virus-infected body cell), the killer cell briefly contacts the target and releases the pore-forming proteins by exocytosis. The pore-forming proteins assemble into large holes in the target cell's plasma membrane. The cell's contents leak out, and the cell dies. *(b)* A close-up view of the killing process. Pore-forming proteins leave the killer cell and insert themselves into the target cell's plasma membrane. The proteins line up next to one another like the staves in a barrel, forming large pores. The membrane of the killer cell is not affected; apparently it contains other proteins that bind to and inactivate any pore-forming proteins that try to insert themselves into the killer's membrane.

iron for reproduction than is used at the human body's normal temperature of 37°C (98.6°C), so fever and reduced iron in the blood combine to slow down bacterial reproduction.

Fever also helps fight viral infections, by increasing the production of the protein **interferon**. Some types of cells synthesize and release interferon after invasion by viruses, and the interferon travels to other cells and increases their resistance to viral attack. Owing to this benefit of fever, it can be a mistake to attempt to control or reduce fevers. In one study, individuals with colds were treated with aspirin (to reduce fever) or a *placebo* (an inactive substance that looks like the real drug, so the subjects don't know whether they have been given the drug or not). The subjects given aspirin had far more viruses in their noses and throats—and, consequently,

sneezed and coughed out far more viruses—than did the subjects given the placebo. The immune systems of subjects with fevers lowered by aspirin weren't as effective at controlling infections, and these subjects were much more infectious to other people.

The onset of fever is controlled by the hypothalamus, the part of the brain that contains the temperature-sensing nerve cells that are the body's thermostat. Normally, the thermostat is set at about 37°C (98.6°F). When disease organisms invade, however, the thermostat is turned up. Certain macrophages, in responding to the infection, release hormones collectively called **endogenous pyrogens** ("self-produced fire-makers"). Pyrogens travel in the bloodstream to the hypothalamus and raise the thermostat's set point, triggering responses that increase body temperature: shivering, increased fat me-

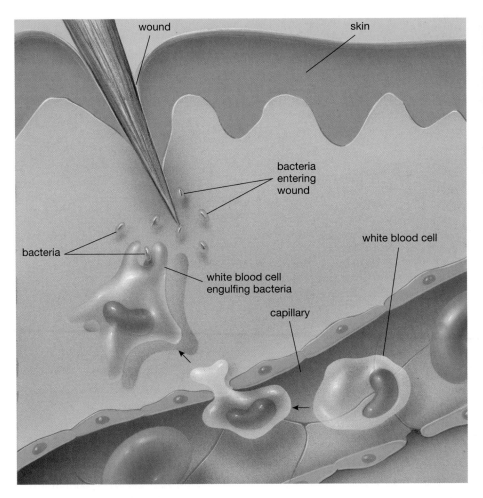

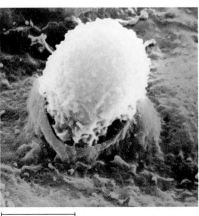

5 micrometers

Figure 31-4 *The inflammatory response*
During an inflammatory response triggered by a cut in the skin, the cells of the capillary walls separate slightly. Macrophages and other phagocytic cells squeeze through the gaps and enter the wounded tissue, where they engulf microbes, dirt, and debris from damaged cells. The micrograph shows a white blood cell leaving a capillary to join the fray against bacteria that have entered a cut.

tabolism, and the constriction of blood vessels. Pyrogens also cause other cells to reduce the concentration of iron in the blood.

2) What Are the Key Characteristics of the Immune Response?

Phagocytic cells, natural killer cells, the inflammatory response, and fever are all *nonspecific* defenses; their role is to prevent or overcome *any* microbial invasion of the body. Unfortunately, however, these nonspecific defenses are not impregnable. When they fail to do the job, the body mounts a highly specific **immune response** directed against the particular organism that has successfully invaded the body.

The essential features of the immune response to infection were recognized more than 2000 years ago by the Greek historian Thucydides. He observed that occasion-

ally someone would contract a disease, recover, and never catch that particular disease again—the person had become immune. With rare exceptions, however, immunity to one disease confers no protection against other diseases. Thus, the immune system attacks one type of microbe, overcomes it, and provides future protection against that microbe but no others. This is why we refer to the immune response as a *specific* defense against invasion.

The immune system consists of about 2 trillion *lymphocytes*, a kind of white blood cell (see Chapter 27). Lymphocytes are distributed throughout the body in the blood and lymph, though many are clustered in specific organs, particularly the thymus, lymph nodes, and spleen. The immune response arises from interactions among the various types of lymphocytes and the molecules that they produce. The theater of the immune response has a large cast of characters and is difficult to follow without a program. Table 31-1 provides a brief overview of the major actors and their roles.

Table 31-1 The Major Molecules and Cells of the Immune Response

Molecules

Antigens	Large organic molecules, normally proteins, polysaccharides, or glycoproteins, that can trigger an immune response; typically located on the surface of cells.
Antibodies	Proteins produced by the cells of the immune system that bind to antigens and either neutralize the antigenic molecule itself or mark cells that bear the antigens for destruction.
Major histocompatibility complex (MHC)	A set of proteins found on the surface of cells that "label" the cell as belonging to a unique individual organism.
Effector molecules	A diverse group of molecules, including histamine and the cell-destroying proteins of killer cells and complement (soluble proteins found in blood).
Regulatory molecules	Hormonelike molecules produced by cells of the immune system that regulate the immune response.

Cells

Macrophages	Phagocytic white blood cells that both destroy invading microbes and help alert other immune cells to the invasion.
B cells	Lymphocytes that produce antibodies; when stimulated, certain of their daughter cells (*plasma cells*) secrete large quantities of antibodies into the bloodstream.
T cells	A set of lymphocytes that regulate the immune response or kill certain types of cells.
Cytotoxic T cells	Destroy specific targeted cells, normally either foreign eukaryotic cells, infected body cells, or cancerous body cells.
Helper T cells	Stimulate immune responses by both B cells and killer T cells.
Suppressor T cells	Inhibit immune responses by other lymphocytes.
Memory cells	A subset of the offspring of B and T cells that are long-lived and provide future immunity against a second invasion by the same antigen.

A Successful Immune Response Recognizes, Overcomes, and Remembers

The key actors in the immune response are two types of lymphocytes, called **B cells** and **T cells** (Fig. 31-5). Like all white blood cells, B lymphocytes and T lymphocytes arise from precursor cells in the bone marrow (see Chapter 27). Some of these lymphocyte precursors are released into the bloodstream and come to rest in the thymus, where they complete their differentiation into T (for thymus) cells. In contrast, B cells differentiate in the bone marrow itself. The two cell types play quite different roles in the

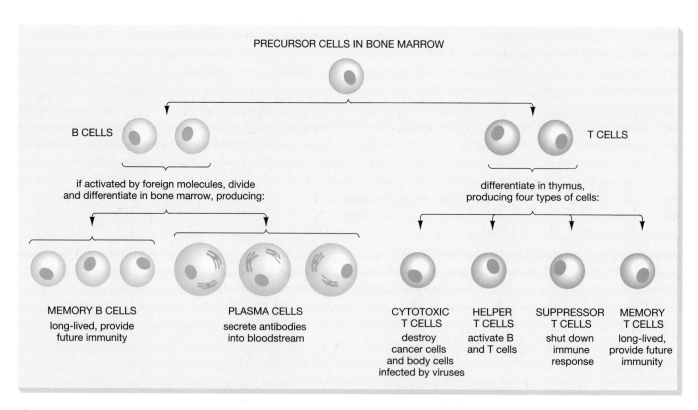

Figure 31-5   The major cells of the immune system and their roles in the immune response

immune response, but immune responses produced by both B cells and T cells consist of the same three fundamental steps: (1) recognizing the invader, (2) launching a successful attack to overcome the invader, and (3) retaining a memory of the invader to ward off future infections.

3) How Are Threats Recognized?

www

To understand how the immune system recognizes invading microbes and initiates a response, we must answer three related questions: (1) How do the immune cells recognize foreign molecules? (2) How can immune cells recognize and produce specific responses to so many different types of molecules? (3) How do immune cells avoid mistaking the body's own cells and molecules for invaders?

We can begin to answer these questions by examining the structure and function of two kinds of large proteins: **antibodies** and **T-cell receptors**. Antibodies are either attached to the surfaces of B cells or dissolved in the blood plasma, where they are called *immunoglobulins* (often abbreviated Ig). T-cell receptors, as their name implies, are attached to the surfaces of T cells. They are never secreted from the cells into the bloodstream.

Antibodies Bind to Foreign Molecules, Triggering the Immune Response

Antibodies are, or are built from, Y-shaped molecules composed of two pairs of peptide chains: one pair of identical large (heavy) chains and one pair of identical small (light) chains (Fig. 31-6). Both heavy and light chains consist of a **constant region**, which is similar in all antibodies of the same class (see Table 31-2), and a **variable region**, which differs among antibody classes. The combination of light and heavy chains results in an antibody with two functional parts: the "arms" and the "stem" of the Y.

Antibodies Are Both Receptors for Foreign Molecules and Effectors That Help Destroy Invading Molecules and Microbes

Antibodies perform two distinct roles in the immune response: They act as both receptors and effectors. In an antibody's role as a receptor, the stem attaches the antibody to the plasma membrane of a B cell (Fig. 31-7a); the two arms of the antibody protrude outward, sampling the blood and lymph for molecules called **antigens** (short for "*anti*body response *gen*erating"). Antigens either are attached to the surfaces of cells (for example, the body's own cells or invading microbes) or are dissolved in the blood or extracellular fluid (snake venom and bacterial toxins are dissolved antigens). In general, only large, complex molecules such as proteins, polysaccharides, and glycoproteins can act as antigens. The binding of an antigen to "receptor antibodies" triggers responses in the cells that bear the antibodies. A cell has that never bound an antigen is said to be *virgin*.

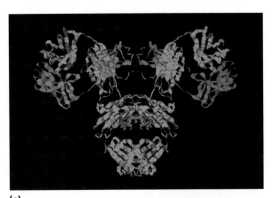
(a)

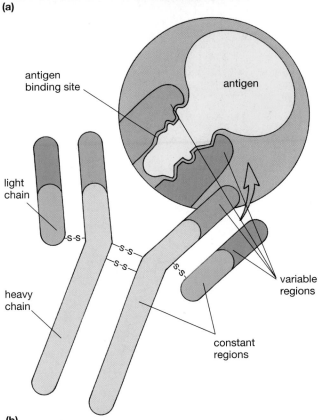
(b)

Figure 31-6 Antibody structure
(a) Computer-generated model of antibody structure. *(b)* Antibodies are proteins composed of two pairs of peptide chains (light chains and heavy chains) arranged like the letter Y. Constant regions on both chains form the stem of the Y; variable regions on the two chains form a specific binding site at the end of each arm of the Y. Different antibodies have different variable regions, forming unique binding sites. The human body synthesizes millions of distinct antibodies, each binding a different antigen (see the enlargement).

Antibodies also act as effectors (Fig. 31-7b). In their effector role, antibodies circulate in the bloodstream, where they neutralize poisonous antigens or destroy microbes that bear antigens. There are five classes of true antibodies (not including T-cell receptors): IgM, IgG, IgA, IgE, and IgD (Table 31-2). Each of the five classes serves a different function in the defense of the body. The

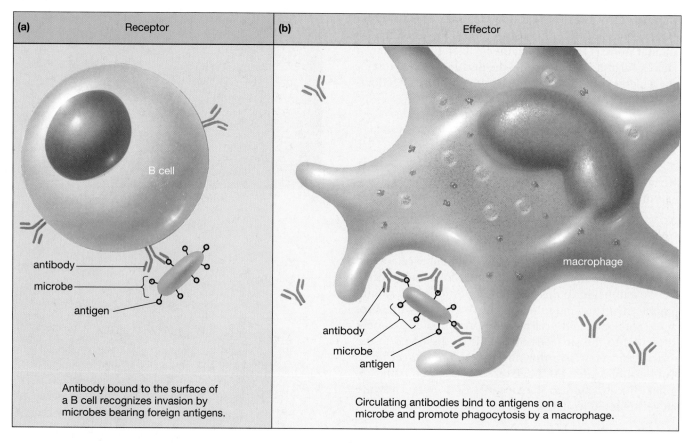

Figure 31-7 *Examples of receptor and effector functions of antibodies*
There are many other effector functions that are described in the text but not illustrated here.

Table 31-2 Antibody Classes and Their Roles

IgM		IgM is composed of five Y-shaped antibody "monomers," or units, held together at their stems. IgM is usually the first antibody secreted during an immune response. In the bloodstream it agglutinates—clumps together—antigens (because it has so many binding sites for antigens), activates complement proteins, and stimulates phagocytosis of bound microbes by macrophages.
IgG		IgG, the most common antibody in the blood, is composed of a single antibody monomer. It activates both complement and macrophages. Special transport processes carry IgG across the placenta, where it protects the developing fetus against disease.
IgA		IgA is normally a "dimer" of two antibody monomers held together by their stems. An additional protein wound around the antibody stems helps IgA to be secreted from the bloodstream into saliva, tears, and mucus. Therefore, IgA is abundant on the surfaces of the respiratory and digestive tracts. It binds microbes on these surfaces, prevents them from entering the body, and allows them to be swept out of the body with mucus or other secretions. Thus, IgA provides a front line of defense for mucous membranes.
IgE		IgE, the "allergy antibody," is composed of a single antibody monomer. The IgE stem binds to mast cells in connective tissue and to certain white blood cells, including basophils and eosinophils. Its normal function appears to be protection against parasites, which are expelled by the sneezing and coughing induced by mast cell activation and weakened or killed by eosinophil activation. Allergic responses are produced when harmless substances bind to IgE. Allergy "shots" work by stimulating the synthesis of IgG antibodies that bind the same antigen the IgE antibodies bind. If enough IgG antibodies are present, they bind up most of the allergy antigen and keep the allergy antigens from contacting IgE.
IgD		IgD is a single antibody monomer, normally bound to the plasma membranes of B cells. It can help activate the B cell to which it is bound.

different types of antibody are distinguished by small differences in the structure of the constant region—that is, the stem. The stems of IgG antibodies form a coating on microbes that targets the microbes for ingestion by macrophages; IgA stems promote the secretion of the antibody into the respiratory and digestive tracts. IgE stems bind to certain cells that line the respiratory and digestive tracts and release histamine during allergic responses. Five Y-shaped segments are joined at their stems in IgM, forming large immune complexes; and IgD stems can help activate the B cells to which they are bound.

The Variable Regions of an Antibody Form Binding Sites That Recognize Specific Antigens

The variable regions at the tips of an antibody molecule's arms form highly specific binding sites for antigens. These binding sites are a lot like the active sites of enzymes (Chapter 4): Each binding site has a particular shape and electrical charge, so only certain molecules can fit in and bind. The binding sites are so specific that each antibody can bind at most a few types of antigen molecules—perhaps only one.

The Constant Regions of an Antibody Determine Its Effector Mechanism

The constant regions of a Y-shaped antibody molecule determine the mechanism by which the antibody will act against invaders. As mentioned above, the constant regions are *similar* in different antibody classes, but they are not identical. For example, the heavy-chain constant regions that make up the stem of one type of antibody may serve to attach the antibody to the plasma membrane of a cell; the constant regions of another type of antibody may bind to certain proteins in the blood, called **complement** proteins, to promote the destruction of microbes.

T-Cell Receptors Bind Antigens and Trigger Responses

T-cell receptors are somewhat different from antibodies in structure and function. First, they are found only on the surfaces of T cells, not in the bloodstream. Second, they consist of two peptide chains of about equal size. However, the ends of the two chains protrude out from the T cell, forming highly specific binding sites for an antigen, just as the ends of the arms of an antibody molecule do. Third, T-cell receptors serve only a receptor function, recognizing antigen molecules and triggering a response in the T cell. They do not directly contribute to the destruction of invading microbes or toxic molecules.

The Immune System Can Recognize Millions of Molecules

During your lifetime, your body will be challenged by a multitude of different potential invaders. Your friends and family members will sneeze cold and flu viruses into the air you breathe. Roadside weeds and trees will release pollen that will find its way to your lungs. Your food may play host to a bacterial population or even to botulinus toxin (which causes botulism, a kind of food poisoning). Your drinking water could contain the amoebae that cause dysentery. As you relax outdoors, a tick that carries the bacteria responsible for Lyme disease might bite you. There is no escape from the pervasive and persistent assaults of the invaders. Fortunately, your immune system recognizes and responds to virtually all of the millions of antigens that you might encounter (both natural and artificial), because your immune cells produce millions of different antibodies and T-cell receptors, each capable of binding a different antigen. These amazing recognition abilities are one of the main reasons you don't get sick more often.

The recognition abilities of the immune system are fortunate, but they are not easy to explain. Antibodies and T-cell receptors are, after all, proteins, and proteins are encoded by genes. It would seem that, to recognize millions of different antigens, each human would have to possess millions of different genes for antibodies and T-cell receptors. But there are probably fewer than 100,000 genes in the entire human genome. How, then, can a subset of these genes code for millions of antibodies and T-cell receptors? The answer lies in two distinct but complementary mechanisms that join forces to produce an enormous diversity of antibodies and T-cell receptors from a comparatively small number of genes. We describe next how antibody diversity develops in B cells; similar phenomena produce a large diversity of T-cell receptors.

1. There are no genes for entire antibody molecules in cells of the body. Instead, the genome includes genes that encode a variety of different antibody fragments that can be pieced together in many combinations. Developing B cells retain not all of the different genes for antibody fragments but instead only a few of the fragment genes; the rest are excised (cut out) from the DNA as the cell matures (Fig. 31-8). The particular group of antibody genes that ends up in each B cell is a randomly selected subset of the original, larger group of genes. The selection process, however, ensures that each B cell retains genes for both a light chain and a heavy chain, and that the genes for each chain include both a constant region and a variable region. (The light and heavy chains are synthesized separately and then assembled into a complete antibody molecule.)

   In the original body cells that give rise to B cells, antibody-fragment genes are situated on the chromosome such that all of the various fragments for constant regions are near one another and all of the fragments for variable regions are also near another. Thus, when the genes for unneeded fragments are excised, the retained fragment genes end up physically very close to one another, forming a single, continuous gene for a particular light or heavy chain.

Figure 31-8 Recombination during the construction of antibody genes *(a)* Each precursor cell of the immune system contains one or perhaps a few genes for the constant regions (C) of the light and heavy chains of antibodies and many genes for the variable regions (V). *(b)* During the development of each B cell, these genes are rearranged, moving one of the variable-region genes next to a constant-region gene. For each chain, each cell thus generates a "recombined antibody gene" that differs from the recombined antibody genes generated by other B cells. *(c)* The different antibodies synthesized by each B cell in (b).

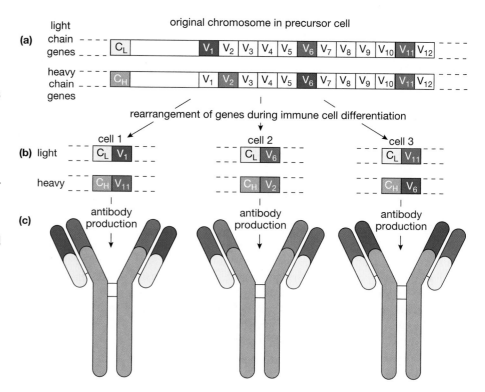

Therefore, each B cell produces a single type of antibody, specified by the chance recombination of variable- and constant-region genes, that is different from the antibody produced by other B cells (except its own daughter cells).

The random selection of antibody-fragments genes from sizable pools of choices yields a vast number of possible unique combinations. It may help you to think of antibody gene formation in terms of card playing. Each B cell is dealt a "hand" of two variable-region genes: one for the light chain and one for the heavy chain, randomly chosen from two large "decks" of genes. With each deck containing hundreds of "cards" (genes), virtually every cell will synthesize its own unique antibody.

2. The genes for certain antibody fragments contain "hot spots" for mutations. The extremely high mutation rates at these genetic locations rapidly generate new antibody genes. As B cells reproduce, some of their daughter cells accumulate mutations in their antibody genes. Thus, two sister cells can produce different antibodies.

Kmart Versus Custom Tailoring: The Immune System Does Not Design Antibodies or T-Cell Receptors Expressly to Bind Invading Antigens

The end result of mutation and gene recombination is that each B cell has its own particular antibody genes, different from those of most other B cells. At any time, the human body contains an army of perhaps 100 million different antibodies, so antigens almost always encounter antibodies that can bind them. It is important to recognize that the immune system does not "design" antibodies to fit invading antigens. Instead, the immune system randomly synthesizes millions of different antibodies. The fact that virtually every possible invading antigen binds to at least a few antibodies is due purely to the immense numbers of different antibodies present in the body.

A more familiar example may help make this clear. When the Queen of England wants a dress, tailors go to Buckingham Palace, obtain the Queen's exact measurements, and custom-sew a dress to the Queen's specifications. Most of the rest of us go to a department store and look through the racks of ready-made clothes. If a store has a large enough selection, we will probably find something that fits reasonably well and is more or less the style we want.

The immune system is like a department store: The array of antibodies is simply there, waiting. The antibodies have not been designed to bind to any particular antigen. An antigen, on entering the body, may randomly bump into the variable regions of many antibodies. It cannot bind to most antibodies (in our clothing analogy, these antibodies don't "fit"), but, with millions of different antibodies around, the antigen inevitably encounters one that binds it pretty well. Antigen–antibody binding triggers changes in the B cells, normally leading to the destruction of microbes that bear that antigen. We will examine these changes in more detail in a moment.

The Immune System Distinguishes "Self" from "Non-Self"

The surfaces of the body's own cells bear large proteins and polysaccharides, just as microbes do. Some of these proteins, collectively called the **major histocompatibility complex (MHC)**, are unique to each individual (except identical twins, who have the same genes and hence the same MHC proteins). Because your MHC proteins are different from those of everyone else, they act as antigens in other people's bodies. (That is why transplants are rejected; the recipient's immune system recognizes MHC proteins on the donor's cells as foreign and destroys the transplanted tissue.)

Why doesn't your own immune system respond to these "self" antigens and destroy your own cells? The key seems to be the continuous presence of the body's antigens while the immune cells mature. As an embryo develops, some differentiating immune cells do indeed produce antibodies or T-cell receptors that can bind the body's own proteins and polysaccharides, not only the MHC proteins, but lots of other molecules as well. However, if these *immature* immune cells contact molecules that bind to their antibodies or T-cell receptors, the immune cells are destroyed. In this way, potentially dangerous immune cells, which respond to the body's own cells as they would to foreign cells, are eliminated during immune system development. Thus, the immune system distinguishes "self" from "nonself" by retaining only those immune cells that do not respond to the body's own molecules.

4) How Are Threats Overcome?

If the body is invaded, the immune system mounts two types of attack: (1) **Humoral immunity** is provided by B cells and the circulating antibodies they secrete into the bloodstream; invaders are attacked before they can enter body cells. (2) **Cell-mediated immunity** is produced by T cells, which attack invaders that have made their way into body cells. These two types of immune responses are not completely independent, but the entire system is more easily understood if the two components are considered separately.

Humoral Immunity Is Produced by Antibodies in Blood

Each B cell bears a specific antibody on its surface. When an infection occurs, the antibodies borne by a few B cells bind to antigens on the invader. (The process is actually more complex than this, but binding is the end result.) Antigen–antibody binding causes these B cells to divide rapidly. This process is called **clonal selection**, because the resulting population of cells is a clone (genetically identical to the parent B cells) that has been "selected" to multiply by the presence of particular invading antigens (Fig. 31-9). The daughter cells differentiate into two cell types: (1) **memory cells** and (2) **plasma cells**. Memory cells

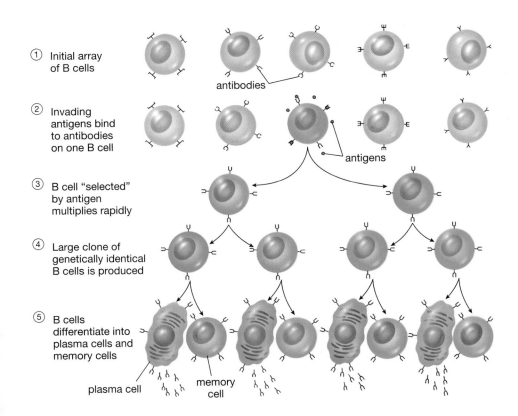

① Initial array of B cells

antibodies

② Invading antigens bind to antibodies on one B cell

antigens

③ B cell "selected" by antigen multiplies rapidly

④ Large clone of genetically identical B cells is produced

⑤ B cells differentiate into plasma cells and memory cells

plasma cell

memory cell

Figure 31-9 Clonal selection among B cells by invading antigens
① The body contains a large array of B cells, each of which synthesizes and bears on its surface a specific antibody, different from the antibodies borne by other B cells. ② On entering the bloodstream, an antigen binds to the antibodies on one (or a few) B cells. ③ Antigen–antibody binding causes the selected B cell to multiply rapidly, resulting in ④ a clone of B cells, all synthesizing the same antibody. ⑤ The daughter B cells differentiate into plasma cells, which secrete antibodies directed against the specific activating antigen, or into long-lived memory cells.

(which can be B cells or T cells) do not release antibodies but do play an important role in future immunity to the particular invader that stimulated their production (as we shall soon see). Plasma cells become enlarged and packed with endoplasmic reticulum (Fig. 31-10); these

(a)

2 micrometers

(b)

4 micrometers

endoplasmic
reticulum

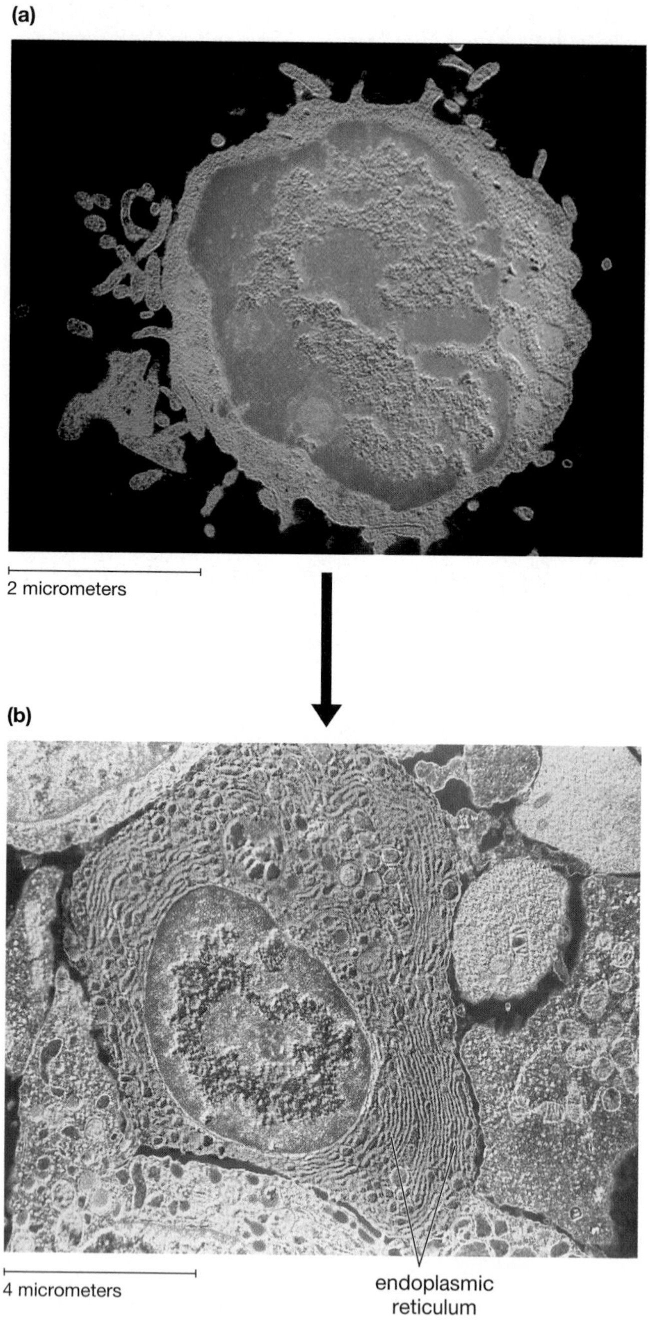

Figure 31-10 A B cell becomes a plasma cell
Colorized micrographs of B cells *(a)* before and *(b)* after conversion to plasma cells. The plasma cell is much larger than the B cell (note the difference in scale) and is virtually filled with rough endoplasmic reticulum that synthesizes antibodies.

cells churn out huge quantities of their own specific antibodies. These antibodies are released into the bloodstream (hence the name "humoral" immunity; to the ancient Greeks, blood was one of the four "humors," or body fluids).

Because antibodies circulate in the bloodstream, humoral immunity can defend against only invaders that are in the blood or extracellular fluid. Bacteria (most of which never enter the body's cells), bacterial toxins, and some fungi and protists are therefore vulnerable to the humoral immune response. Invaders that penetrate into the body's cells, as viruses do, are safe from antibody attack as long as they are within the cytoplasm of a body cell. Antibodies can attack viruses only when the viruses are outside the body's cells. That occurs during the first stage of an infection or when the viruses have finished replicating in one host cell, ruptured it, and have been released into the body fluids. As we will see shortly, however, cell-mediated immune responses—and not humoral responses—are the primary defense against viral infections.

Monoclonal Antibodies Are Pure Antibodies Produced by Cloning

In addition to their normal role in defending the body, antibodies are valuable for many medical and scientific purposes. Because a given antibody binds only to one specific type of antigen, antibodies are invaluable tools for finding and/or marking molecules in living animals or in dissected tissues. For example, pregnancy tests typically use an antibody that binds to a hormone released by a developing embryo. The medical and scientific usefulness of antibodies, however, depends on the availability of pure samples of a single antibody, such as a sample produced by a single clone of plasma cells.

A good way to acquire a supply of pure antibody would be to inject an animal with the antigen that induces B cells to produce the desired antibody and then to collect the appropriate clone of plasma cells. This clone of plasma cells could then be grown in the laboratory, and the antibody produced by the cells could be collected. In fact, normal plasma cells cannot be grown outside the body, but cancerous tumors of plasma cells, called *myelomas*, are fairly easy to grow. So, to produce pure antibodies, B cells from antigen-injected animals are induced to merge with myeloma cells to form hybrid cells called **hybridomas**. The hybridomas can proliferate under laboratory conditions but still produce the desired antibody. Antibodies produced by this method are known as **monoclonal antibodies**.

Antibodies Destroy Extracellular Microbes and Molecules by Several Mechanisms

Antibodies in the blood may affect antigenic molecules, microbes that bear antigens, and infected body cells in four ways:

1. **Neutralization**. The antibody may combine with or cover up the binding site of a toxic antigen such as a bacterial toxin, thereby preventing the toxin from harming the body.

2. **Promotion of phagocytosis**. The antibody may coat the surface of a microbe. The protruding stems of the antibody seem to identify the microbe as a specific target for circulating phagocytic white blood cells to engulf.

3. **Agglutination**. Each antibody has two binding sites for antigen, one on each arm (the exceptions being IgM antibodies, which have 5 pairs of arms and thus 10 binding sites, and IgA antibodies, which can have 2 pairs of arms). These binding sites may attach to antigens on two different microbes, holding them together. As more and more antibodies link up with antigens on different microbes, the microbes clump together, or *agglutinate*. Agglutination seems to enhance phagocytosis.

4. **Complement reactions**. The antibody–antigen complex on the surface of an invading cell may trigger a series of reactions with blood proteins called the **complement system**. When these complement proteins bind to the antibody stems, the proteins attract phagocytic white blood cells to the site, promote phagocytosis of the foreign cells, or in some instances directly destroy the invaders by creating holes in their plasma membranes, much as natural killer cells do (see Fig. 31-3).

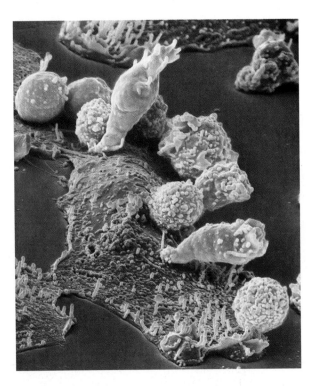

Figure 31-11 Cell-mediated immunity at work
Cytotoxic T cells (white) contact a large cancer cell in this false-color SEM photo.

Cell-Mediated Immunity Is Produced by T Cells

Cell-mediated immunity is the primary defense against the body's own cells when they have become cancerous or have been infected by viruses. Cell-mediated immunity is also important in overcoming infection by fungi or by protists. Three types of T cells contribute to cell-mediated immunity: (1) *cytotoxic T cells*, (2) *helper T cells*, and (3) *suppressor T cells*.

Three Types of T Cells Contribute to the Immune Response

Each of the three different types of T cells performs different functions in the cell-mediated immune response:

1. **Cytotoxic T cells** release proteins that disrupt the infected cell's plasma membrane (Fig. 31-11). This attack is activated when receptors on the cytotoxic T cell's membrane bind to antigens on the surface of an infected cell. In at least some instances, the proteins released by cytotoxic T cells are similar or identical to those that natural killer cells use to create giant holes in the target cell's membrane.

2. When the surface receptors of **helper T cells** bind an antigen, the cells release chemicals that assist other immune cells in their defense of the body. These hormonelike chemicals stimulate cell division and differentiation in both B cells and cytotoxic T cells that respond to the same microbial invasion. In fact, very little immune response, either cell-mediated or humoral, can occur without the boost provided by helper T cells. That is why AIDS, which destroys helper T cells, is such a deadly disease.

3. **Suppressor T cells** act after an infection has been conquered. They help shut off the immune response in both B and cytotoxic T cells.

After an infection is over, some suppressor T cells and helper T cells persist and function as memory T cells. Like memory B cells, these help protect the body against future infection.

Figure 31-12 compares the humoral immune response with the cell-mediated immune response.

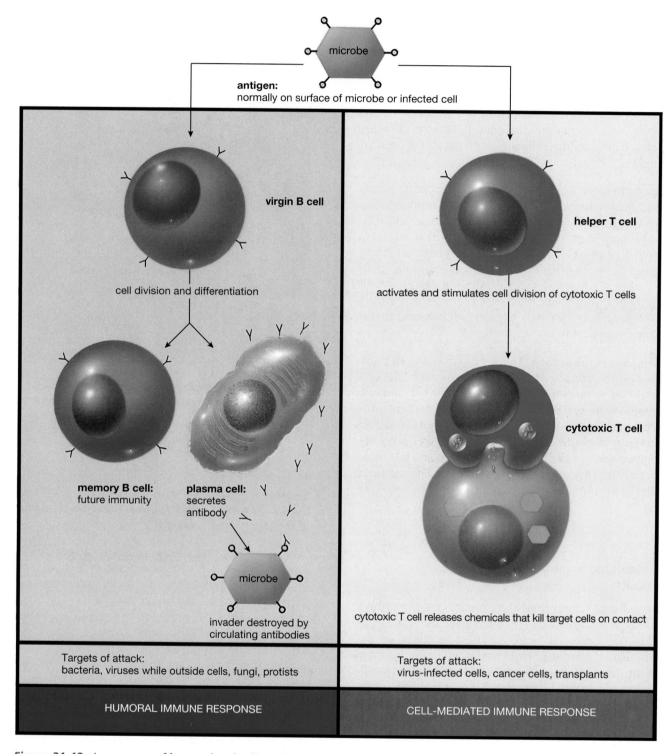

Figure 31-12 A summary of humoral and cell-mediated immune responses

5) How Are Invaders "Remembered"?

As Thucydides observed 2000 years ago, a person who overcomes a disease typically remains immune to future encounters with that disease for many years. Retaining immunity is the function of memory cells. Plasma cells and cytotoxic T cells may do the immediate job of fighting disease organisms, but they normally live only a few days. Conversely, B and T memory cells may survive for many years. Memory cells are a major factor in explaining why we don't get sick more often. If the body is invaded by foreign cells that bear antigens to which the

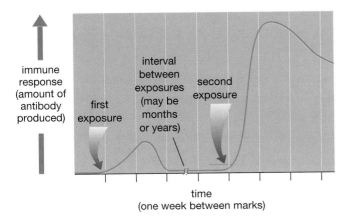

immune response (amount of antibody produced)

first exposure

interval between exposures (may be months or years)

second exposure

time
(one week between marks)

Figure 31-13 *Memory cells and the immune response*
The immune response to the first exposure to a disease organism is fairly slow and not very large, as B and T cells are selected and multiply. A second exposure activates memory cells formed during the first response, so the second response is both faster and larger.

immune system has previously mounted a response, the appropriate memory cells will recognize the invaders. These memory cells will then multiply rapidly and produce a second immune response by generating huge populations of plasma cells and cytotoxic T cells.

In the first encounter with a disease microbe, only a few B and T cells respond. Each of these, however, leaves behind hundreds or thousands of memory cells. Further, memory cells respond to antigens much more rapidly than could their parent B and T cells. Therefore, the second immune response is very rapid (Fig. 31-13). In most instances, second or subsequent invasions by the same microbe are overcome so quickly that there are no noticeable symptoms of infection. Cold and flu viruses may appear to be exceptions to this rule, because most people are subject to repeated bouts of these maladies. In fact, however, the immune system does provide lasting protection against cold and flu viruses. The problem, as we describe in "Health Watch: Flu—The Unbeatable Bug" (pp. 626–627) is that this year's colds and flus aren't caused by the same viruses that you fought off last year.

6) How Does Medical Care Augment the Immune Response?

www

Antibiotics Slow Down Microbial Reproduction

Our description of the body's responses to infection may seem to suggest that nothing should harm us. As everyone knows, however, that is unfortunately not the case. If

untreated, many diseases kill their victims, normally for a simple reason: The body provides ideal conditions for the growth and reproduction of disease microbes. Such microbes can multiply rapidly, sometimes dividing as fast as once an hour. The infection thus becomes a race between the invading microbes and the immune response. If the initial infection is massive or if the microbes produce particularly toxic products, the full activation of the immune response may be too late.

Antibiotics are drugs that help combat infection by slowing down the growth and multiplication of many invaders, including bacteria, fungi, and protists (but not viruses). Although antibiotics usually don't destroy every single microbe, they give the immune system enough time to finish the job. One problem with antibiotics, however, is that they are potent agents of natural selection. The occasional mutant microbe that is resistant to an antibiotic will pass on the gene(s) for resistance to its offspring. The result: Resistant mutants proliferate, whereas susceptible microbes die off. Eventually, many antibiotics become ineffective in treating diseases against which they were formerly effective. The phenomenon of antibiotic resistance is discussed more fully in Chapter 19 in "Health Watch: Antibiotics—Miracle Drugs Rendered Useless?"

Vaccinations Stimulate the Development of Memory Cells

As early as A.D. 1000, people in India, China, and Africa deliberately exposed themselves to mild cases of smallpox to acquire immunity to the disease. In 1798, Edward Jenner discovered that infection with cowpox conferred immunity to smallpox. This discovery initiated the modern practice of *immunization*. In the late 1800s, Louis Pasteur extended the use of immunization to several other diseases by injecting weakened or dead microbes into healthy individuals. The weakened microbes do not cause disease (or at least not a severe case) but bear antigens that elicit vigorous immune responses. As it would to a real infection, the immune system produces an army of memory cells, conferring immunity against subsequent exposure to the living, dangerous microbes. These injections of weakened or killed microbes to confer immunity are called **vaccinations**, from the Latin word for "cow," in honor of Jenner's pioneering efforts with cowpox. Today, many diseases, including polio, diphtheria, typhoid fever, and measles, can be controlled through vaccination. Smallpox, one of the deadliest diseases of all, has been completely eradicated as a result of a vaccination program sponsored by the World Health Organization.

Through genetic engineering (see Chapter 13), we now enjoy the prospect of manufacturing tailor-made vaccines. One method is to synthesize the antigenic proteins of disease-causing microbes. These antigens can then be used as vaccines without having to raise,

A Closer Look
Cellular Communication during the Immune Response

The immune system is a strange "system." Unlike the nervous system, for example, it is not composed of physically attached structures. Instead, as befits its mission of patrolling the entire body for microbial invaders, the immune system consists of an army of separate cells. Nevertheless, the army is highly coordinated. This coordination requires complex communications involving antigens, antibodies, hormones, receptors, and cells. For example, when a virus invades the body (Fig. E31-1, step ①), it sets off a cascade of events that can be loosely divided into three components.

Activation of Helper T Cells One component of the immune response begins when macrophages ingest the virus (Fig. E31-1, step ②) and digest it. Antigens that have been "chewed off" the virus become attached to certain proteins of the macrophage's major histocompatibility complex (MHC) and are displayed, or *presented,* on the surface of the macrophage. These antigen–MHC complexes are recognized by virgin helper T cells (step ③). Next, receptors on helper T cells bind to the antigen–MHC complex, and the helper T cells release a hormone called *interleukin-2* (step ④). This hormone stimulates cell division and differentiation (step ⑤) in both the releasing cell and in any other T cells that have bound to an antigen–MHC complex. Some of the resulting daughter helper T cells become memory cells that provide future immunity (step ⑥); other daughter cells become mature T cells that assist in activating—that is, stimulating the immune response of—cytotoxic T cells and B cells (step ⑦).

Activation of Cytotoxic T Cells: Cell-Mediated Immunity Meanwhile, other copies of the virus are infecting ordinary body cells, such as those lining the respiratory tract (step ⑧). Infected body cells display viral antigens on their surfaces, bound to another set of MHC molecules. Virgin cytotoxic T cells bind to the antigen–MHC complex on the body cells (step ⑨) and are simultaneously activated by interleukin-2 released by the activated helper T cells. This combination of binding and stimulation causes the cytotoxic T cells to multiply and become activated (step ⑩). When activated cytotoxic T cells then encounter in-fected cells presenting the antigen–MHC complex, the T cells release toxic proteins that kill the infected cell by lysis (step ⑪).

Activation of B Cells: Humoral Immunity Some B cells bear antibodies on their surfaces that bind antigens on the surface of free viruses that have not yet invaded a body cell (step ⑫). This antigen–antibody binding stimulates some B cell division and maturation, but full activation of B cells requires a boost from helper T cells. This boost is provided when B cells that have bound antigen ingest that antigen (by receptor-mediated endocytosis; see Chapter 5), attach the antigen to MHC molecules, and present the antigen–MHC complex on their surfaces. The antigen–MHC complex is recognized by activated helper T cells (step ⑬), which then release several types of interleukin hormones that stimulate the division and differentiation of antigen-binding B cells (step ⑭). Some of their progeny become memory cells (step ⑮); others become plasma cells that secrete antibodies into the bloodstream (step ⑯).

Summary of Immune Cell Communication This interlocking communication network is quite complex. We can summarize its essentials, however, in five generalizations:

1. Macrophages bind to antigens, engulf them, and present them on their surfaces, along with "self-identification" MHC molecules, to helper T cells.
2. When they recognize the antigen–MHC complex, helper T cells multiply rapidly.
3. Meanwhile, cytotoxic T cells and B cells recognize the same antigen.
4. Hormones released by helper T cells stimulate cell division and maturation of only those cytotoxic T cells and B cells that have also been activated by antigen binding.
5. The activated cytotoxic T cells and B cells then provide cell-mediated and humoral immunities, respectively.

As you can see, helper T cells are essential in turning on both phases of the immune response. A loss of helper T cells, such as that caused by the virus that causes AIDS, virtually eliminates the immune response to many diseases.

---

isolate, and weaken the disease microbes themselves or even to inject people with microbes. A vaccine against anthrax, a severe disease of livestock, has been manufactured by this procedure. A second technique is to insert genes that encode the antigens of, for example, herpes into the genome of harmless microbes such as the cowpox virus. These "designer" microbes produce herpes antigens without being able to cause the disease and have been shown to be effective in experiments with animals.

7 Can the Immune System Malfunction?

Allergies Are Inappropriately Directed Immune Responses

More than 35 million Americans suffer from **allergies**, adverse reactions to substances that are not harmful in themselves and to which many other people do not respond. Common allergies include those to pollen, dust, mold

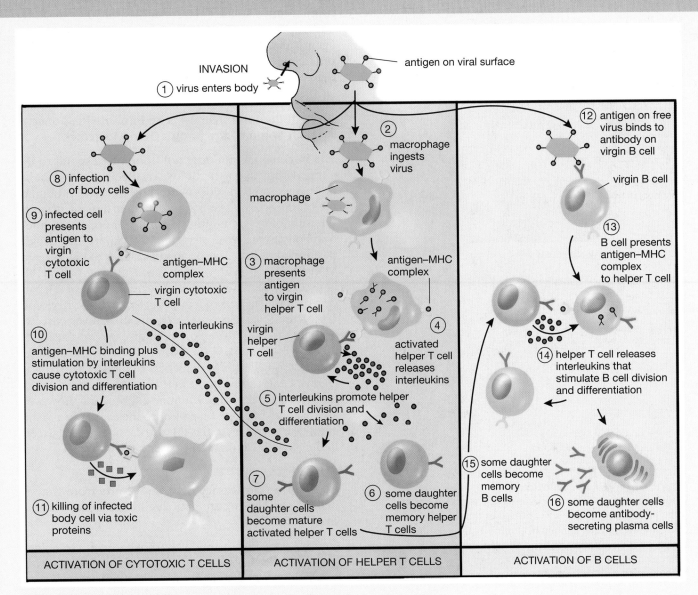

Figure E31-1 *The interactions among immune cells during the immune response* The numbers are keyed to those used in the essay.

spores, and bee stings. Allergies are actually a form of immune response. A foreign substance, such as a pollen grain, enters the bloodstream and is recognized as an antigen by a particular type of B cell. This B cell proliferates, producing plasma cells that pour out IgE antibodies against the pollen antigens (Fig. 31-14). The stems of the IgE antibodies attach to the plasma membranes of **mast cells**, histamine-containing cells located in the respiratory and digestive tracts. Pollen antigens that later encounter and bind to the attached IgE antibodies in the respiratory

tract trigger the release of histamine, which causes increased mucus secretion, leaky capillaries, and other symptoms of inflammation. Because airborne substances such as pollen grains, dust, and mold spores typically enter the nose and throat, the resulting allergic reactions often include the runny nose, sneezing, and congestion typical of "hay fever." Food allergies cause analogous symptoms, including cramps and diarrhea, in the digestive tract.

Why are only some individuals allergic to a given substance? We are all exposed to the same antigens,

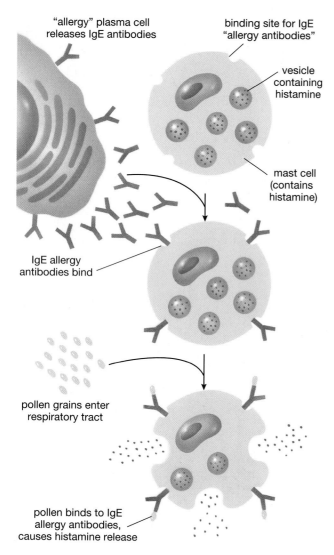

"allergy" plasma cell releases IgE antibodies

binding site for IgE "allergy antibodies"

vesicle containing histamine

mast cell (contains histamine)

IgE allergy antibodies bind

pollen grains enter respiratory tract

pollen binds to IgE allergy antibodies, causes histamine release

Figure 31-14 Allergic reactions
Upon exposure to certain antigens, such as pollen grains, some plasma cells synthesize antibodies that bind to mast cells (middle), which contain histamine. If these bound antibodies later encounter the same antigen (bottom), the mast cells release histamine into their surroundings, causing local inflammation and the symptoms of allergy.

so people without allergies must lack the genes for the allergy-causing antibodies or must for some other reason fail to produce as much antibody as do allergic individuals. From an evolutionary perspective, we might also ask, Why are *any* people allergic? Do IgE antibodies serve some useful function that prevents them from being eliminated by natural selection? The immune system evolved to protect the body from parasitic invasion; thus, we might expect that IgE antibodies were selected for such a function. Some scientists speculate that IgE does indeed play a role in the defense against various parasites that invade the body through mouth, nose, or anus and attach to the lining of the nasal

passages, throat, or intestine. If these parasites attempt to burrow into the body tissues (see Chapter 22), then the typical symptoms induced by IgE antibodies—increased mucus secretions, sneezing, coughing, intestinal convulsions, and diarrhea—would help dislodge and expel the parasites. That IgE antibodies may act against pollen as well as parasites is perhaps an unfortunate side effect of an otherwise useful aspect of our defense system.

An Autoimmune Disease Is an Immune Response Against Some of the Body's Own Molecules

The immune system does not normally respond to the antigens borne on the body's own cells. Occasionally, however, something goes awry, and "anti-self" antibodies are produced. The result is **autoimmune disease**, in which the immune system attacks some component of one's own body. Some types of anemia, for example, are caused by antibodies that destroy an individual's red blood cells. Many cases of insulin-dependent (juvenile-onset) diabetes occur because the insulin-secreting cells of the pancreas are the victims of a misdirected immune response. Unfortunately, at present there is no known cure for autoimmune diseases. For some diseases, replacement therapy can alleviate the symptoms—for instance, by administering insulin to diabetics or blood transfusions to anemics. Alternatively, the autoimmune response can be suppressed with drugs. Immune suppression, however, also reduces immune responses to the everyday assaults of disease microbes, so this therapy cannot be used except in the most life-threatening cases.

An Immune Deficiency Disease Results from the Inability to Mount an Effective Immune Response to Infection

On rare occasions, a child is born with **severe combined immune deficiency (SCID)**, a defect in which few or no immune cells are formed. A child with SCID may survive the first few months of postnatal life, protected by antibodies acquired from the mother during pregnancy or in her milk. Once these antibodies are lost, however, common bacterial infections can prove fatal. Some immune-deficient children must live in a germproof "bubble," isolated from contact with every unsterilized object, including other people. One form of therapy is to transplant bone marrow (from which immune cells arise) from a healthy donor into the child. In some children, marrow transplants have resulted in some antibody production, occasionally enough to confer normal immune responses. In 1990, researchers at the National Institutes of Health began clinical trials of injecting genetically engineered bone marrow cells into children with SCID (see Chapter 13). The therapy has been somewhat successful but is not yet in widespread use.

AIDS Is a Devastating Immune Deficiency Disease

The most common and widespread immune deficiency disease is **acquired immune deficiency syndrome**, or **AIDS**. Two viruses, named **human immunodeficiency viruses** 1 and 2 (HIV-1 and HIV-2), cause AIDS. These viruses undermine the immune system by infecting and destroying helper T cells, which normally play a crucial role in the body's defense by stimulating both the cell-mediated and humoral immune responses.

AIDS does not directly kill its victims. Instead, an individual with AIDS becomes increasingly susceptible to other diseases as the helper T-cell population declines. In fact, it was unexpected occurrences of unusual diseases that first led to the recognition of AIDS in 1981. In that year, a man entered the UCLA Medical Center with a fungal infection in his throat. A few weeks later, he developed a rare form of pneumonia (*Pneumocystis carinii*) that is normally seen only in cancer patients and people whose immune system has been intentionally suppressed to prevent the rejection of organ transplants. The UCLA patient soon died; physicians across the United States began encountering similar cases of individuals with debilitating effects of diseases that were rare or not ordinarily serious. Although their particular diseases varied, all the affected individuals had one feature in common: Their immune systems lacked helper T cells and therefore failed to ward off invading microbes.

The Human Immunodeficiency Virus Is a Retrovirus That Infects and Destroys Helper T Cells

Both HIV-1 and HIV-2 are **retroviruses**—viruses that have RNA as their genetic material and that reproduce by transcribing that RNA into DNA and then inserting the DNA into the chromosomes of a host cell. As illustrated in Figure 31-15, HIV consists of an outer envelope, taken from an infected cell's plasma membrane as the virus leaves the cell, and two *capsules,* or protein coatings, the innermost of which contains RNA and an enzyme called **reverse transcriptase**. When HIV attacks a helper T cell, the virus's outer envelope binds to the plasma membrane of the cell and allows the virus to enter the cell. Once the virus is inside, its reverse transcriptase catalyzes the transcription of the virus's RNA genome into DNA (a process known as *reverse transcription,* because it reverses the "normal" DNA-to-RNA direction of transcription in living cells). The DNA product of the viral reverse transcription then travels to the nucleus and is inserted into the T cell's genome, where it may lie dormant for years. Eventually, however, the infected cell's own metabolic machinery will begin to transcribe and translate the viral "DNA copy" hidden in the cell's chromosomes. In effect, more viruses are manufactured, bud off of the cell, and disperse into the bloodstream. Ultimately, the proliferating viruses kill the infected helper T cell.

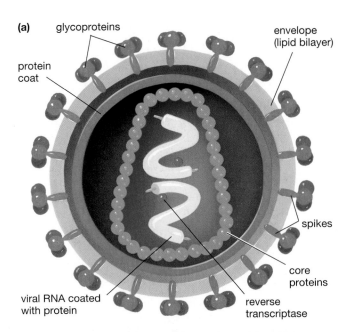

(a)

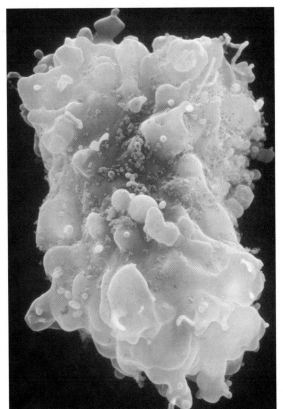

(b)

Figure 31-15 The HIV virus causes AIDS
(a) HIV, the virus that causes AIDS, consists of an outer envelope, taken from the cells it infects, and an inner protein capsule, which contains RNA (the genetic material of HIV) and the enzyme reverse transcriptase (which copies the RNA over into DNA when the virus infects a cell). The proteins protruding through the envelope attach to the plasma membranes of helper T cells. These proteins are potential targets for AIDS vaccines. *(b)* The blue specks in the colorized scanning electron micrograph are HIV viruses that have just emerged from the large helper T cell.

Health Watch
Flu—The Unbeatable Bug

Every winter, a wave of *influenza*, or flu, sweeps across the world. Thousands of the elderly, the newborn, and those already suffering from illness succumb, while hundreds of millions more suffer the respiratory distress, fever, and muscle aches of milder cases. Occasionally, devastating flu varieties appear. In the great flu pandemic of 1918, the worldwide toll was 20 million dead in one winter. In 1968, the Hong Kong flu infected 50 million Americans, causing 70,000 deaths in 6 weeks.

Flu Viruses   Flu is caused by several viruses that invade the cells of the respiratory tract, turning each cell into a factory for manufacturing new viruses. The outer surface of a flu virus is studded with proteins, some of which are recognized by the immune system as antigens. This recognition ensures that most people survive the flu because their immune systems inactivate the viruses or kill off virus-infected body cells before the viruses finish reproducing. This is the same mechanism by which other viruses, such as those that cause mumps or measles, are conquered. So why can't we become immune to the flu, as we can to measles?

The answer lies in a flu virus's amazing ability to change. Flu virus genes are made up of RNA, which lacks the proofreading mechanisms that reduce mutations in genes made of DNA (see Chapter 10). Therefore, flu RNA genes mutate rapidly: On average, 10 mutations will appear in every million newly synthesized viruses. Most single mutations don't change the properties of the viral antigens very much. Four or five mutations in the same virus, however, may alter the surface antigens enough that the immune system doesn't fully recognize the virus as the same old flu that was beaten off last year. Some of the memory cells don't recognize it at all, and the immune response produced by the rest of the memory cells doesn't work as well as it should. The virus, although slowed down somewhat, gets a foothold in the body and multiplies until a new set of immune cells recognizes the mutated antigens and starts up a new immune response. So you get the flu again this year.

Deadly New Strains   Far more serious are the dramatically new flu viruses that occasionally appear, as in the epidemic of 1918, the Asian flu of 1957, and the Hong Kong flu of 1968. In these viruses, entirely new antigens seem to appear suddenly. The novel antigens are not just slight variations of the old set, but they have distinctive structures that the human immune system has never before encountered. Where do the genes that encode these new antigens come from? Believe it or not, they come from viruses that infect birds and pigs. The intestinal tracts of birds, especially ducks, may host viruses strikingly similar to human flu viruses, though infected birds suffer from no noticeable disease. The bird viruses can't be transmitted to people, and human flu viruses likewise don't infect birds. But both human and bird viruses can infect pigs, so both viruses can in some cases simultaneously infect the same pig cell. Once in a great while (perhaps only three times during the twentieth century), new viruses that spring from a double-infected pig cell end up with a mixture of genes from human and bird viruses (Fig. E31-2). Some of these hybrid viruses combine the worst genes (at least from our perspective) of each: From the human virus, the deadly new viruses pick up the genes needed to subvert human cellular metabolism to produce new viruses; from the bird virus, they pick up genes for new surface antigens. The hybrid viruses can move easily from pigs to humans, because pigs live near humans and, like us, pigs cough when they have the flu.

Have you ever wondered why flu strains are called "Asian" or "Hong Kong"? The reason is that Southeast Asia is usually the place where new strains crop up. Many farmers in Asia, especially in southern China, have "integrated" farms. Crops are grown to feed pigs and ducks, and the feces from the pigs and ducks are used to fertilize fish ponds. This is a very efficient farming practice, but, unfortunately, it also places ideal mixing vessels for flu viruses (pigs) in close proximity to humans and ducks.

The HIV Virus Is Transmitted by the Exchange of Body Fluids

HIV cannot survive for very long outside the body. The virus can be transmitted only by the direct exchange of body fluids—including blood, semen, vaginal secretions, and breast milk—by activities that causes these body fluids to come into contact with a break in the skin or a mucous membrane. HIV infection can thus spread through the population by sexual activity, by sharing of needles among intravenous drug users, by blood transfusions (more common before it became standard practice to screen all donated blood for anti-HIV antibodies), or

when maternal and fetal blood are mixed during pregnancy or childbirth. In many individuals, HIV infection progresses into AIDS.

In the early days of the AIDS epidemic in the United States, almost all individuals with AIDS were homosexual men or intravenous drug users. Although these groups are still the main avenues for infection in North America and Europe, women and heterosexual men who are not drug users have also contracted AIDS. In fact, in the developed world the number of AIDS cases among homosexuals has leveled off in the past few years, while the number of infected heterosexuals has continued to in-

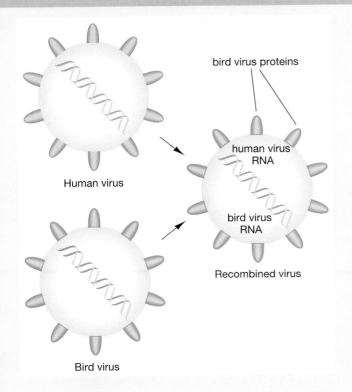

Human virus

Bird virus

bird virus proteins

human virus RNA

bird virus RNA

Recombined virus

Figure E31-2 Formation of deadly new strains of influenza
Rarely occurring recombination of genes from bird and human flu viruses can result in deadly new strains of flu. Note the protein "spikes" projecting from the virus coats in the colorized EM photo of flu viruses. These projections attach to plasma membranes of cells in the human respiratory system, helping the virus gain entry into the cells.

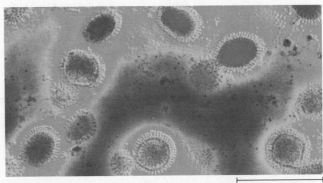

0.5 micrometer

Effects of New Strains If a human is infected by a hybrid virus, the immune system must start from scratch, selecting out entirely new lines of B cells and T cells to attack the intruder. But in the meantime, the virus multiplies so rapidly that many individuals die or become so weakened that they contract some other fatal disease. Other individuals recover, with immune systems now primed to resist any further assault from the new virus. In subsequent years, a few point mutations might allow a slightly altered strain of the new virus to infect millions of people, but with a partial immune response ready, few fatalities occur. Once again, for most of us, the flu becomes a routine annoyance. At least until the next time the improbable happens again.

crease. It is now obvious that AIDS can be transmitted through heterosexual contact. Heterosexual intercourse is the most common means of infection in Africa and Asia, where the largest number of individuals with AIDS live.

There Are Partially Effective Treatments, but No Cures, for AIDS

For persons already infected with AIDS, there are two categories of therapy. First, infections that are not directly caused by the HIV virus—such as *Kaposi's sarcoma* (a deadly form of cancer affecting the skin) or *Pneumocys-*

tis carinii pneumonia—can be treated as they would be in any patient. More-effective treatments for these diseases, developed in recent years, have improved the quality of life for AIDS patients and helped them live longer. Second, the progress of AIDS can be slowed, but not stopped, by drugs that target retroviruses. These drugs fall into two general classes: (1) *reverse transcriptase inhibitors* and the more recently developed (2) *protease inhibitors*.

Reverse transcriptase inhibiting drugs such as ziduvidine (AZT) or dideoxyinosine (ddI) neutralize viral reverse transcriptase and thus inhibit viral reproduction by slowing the production of DNA from RNA. Ideally, the viral RNA is thus never copied completely over into DNA, so new virus cannot be synthesized. Unfortunately, these drugs are not completely successful in stopping reverse transcription. They also interfere to some extent with normal DNA replication, and in some patients they have very severe side effects.

Protease inhibitors also attack a viral enzyme, but not reverse transcriptase; instead, they neutralize an enzyme that helps assemble viruses from the components produced by the host cell's pirated genetic machinery. The new protease inhibitor drugs are especially effective when used in combination with reverse transcriptase inhibitors. By attacking simultaneously at two points in the HIV life cycle, a "cocktail" of multiple drugs is more effective than any of the drugs alone at staving off the effects of AIDS.

Drug treatments are expensive, have negative side effects, and do not cure AIDS. Clearly, the best solution would be one that prevents HIV infections. No such solution now exists, but an army of researchers is feverishly trying to find one. One promising approach is to block the particular plasma-membrane surface receptors that HIV needs to gain entry to a cell. This approach received a boost from the discovery, in 1996, that some individuals carry a mutation that confers resistance to HIV-1: They lack a gene that codes for one of the surface receptors that HIV-1 requires. These resistant individuals suffered no ill effects from their mutation, which encourages researchers to try to develop treatments that would block the receptors in other individuals. Several research teams are exploring the possibility of flooding the body with other compounds that would bind to these receptors and prevent HIV from binding to them.

The ultimate goal of AIDS prevention research is to develop a vaccine. This is a tricky business, in large part because, for unknown reasons, the anti-HIV antibodies that individuals with AIDS produce naturally do little to prevent the progress of the HIV infection. Therefore, vaccines would have to evoke a very different, and more effective, immune response than does normal HIV infection. Further, HIV has an incredible mutation rate, perhaps a thousand times faster than that of flu viruses, so different infected individuals can have remarkably different strains of HIV. Even more perplexing, HIV can be very different even when isolated at different times from the same person. Nevertheless, trials of AIDS vaccines are under way in several countries, though none yet appear to be particularly promising.

AIDS Is a Widespread, Lethal Disease

We are currently in the midst of a worldwide AIDS epidemic in which almost 12 million people had died by the beginning of 1998. UNAIDS, the United Nations organization responsible for tracking AIDS cases, estimates that 30 million people are infected with HIV and that

16,000 new infections occur each day. Most of those infected live in the developing world (Fig. 31-16), especially in sub-Saharan Africa, which currently has the highest infection rate. Southern Africa has been especially hard hit. For example, the government of Zimbabwe estimates that one of every five adults there is infected with HIV.

According to UNAIDS, AIDS is now one the world's most deadly infectious diseases, rivaling other notorious killers such as malaria and tuberculosis in terms of annual deaths. However, those other infectious diseases receive less attention and research funding than does AIDS. Some observers fear that excessive focus on AIDS may lead to comparative neglect of other serious public health problems.

How concerned should you be about AIDS, personally? People in the high-risk categories should certainly take precautions against infection. Because the disease is so deadly and the incidences of other sexually transmitted diseases are on the rise, "safe-sex" practices are advisable for everyone. Each year, thousands of health-care workers suffer "accidental needle sticks"; these workers should be extremely careful when handling blood products and needles. At the same time, the odds that a patient will acquire AIDS from a physician or dentist are astronomically small, and no one should forego medical care for fear of contracting AIDS.

Cancer Can Overwhelm the Immune Response

The usual development of any organ begins with rapid growth during embryonic life, slower growth in childhood, and finally maintenance of a constant size during adulthood. Individual cells may die and be replaced (as happens continuously in the stomach lining), but most organs remain about the same size throughout adult life. A **tumor** is a population of cells that has escaped from normal regulatory processes and grows at an abnormal rate. The cells in a *benign tumor*, or *polyp*, normally remain constrained in one area, but the cells in a *malignant tumor* grow uncontrollably. As a malignant tumor grows, it uses increasing amounts of the body's energy and nutrient supplies and literally squeezes out vital organs nearby. **Cancer** is a disease characterized by the unchecked growth of malignant tumor cells.

Cancer is one of the most dreaded words in the English language, and with good reason. The disease kills more than 500,000 Americans each year and about 6 million people worldwide. A sobering 40% of Americans will eventually contract cancer. Advances in detection and treatment have reduced the death rates for some forms of cancer, but the overall cancer death rate has increased since the mid-1970s. Despite decades of intensive research, the quest for a cancer cure remains unfulfilled. Why can't we cure or prevent cancer? We can, after all, prevent smallpox and polio and can cure dozens of other diseases. What makes cancer different?

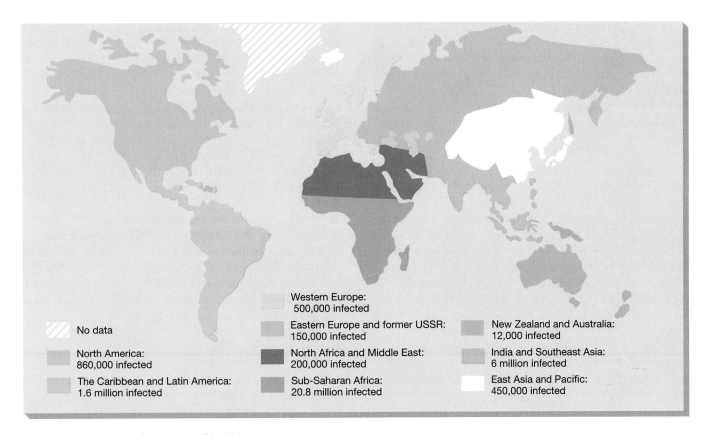

Western Europe:
500,000 infected

No data

North America:
860,000 infected

The Caribbean and Latin America:
1.6 million infected

Eastern Europe and former USSR:
150,000 infected

North Africa and Middle East:
200,000 infected

Sub-Saharan Africa:
20.8 million infected

New Zealand and Australia:
12,000 infected

India and Southeast Asia:
6 million infected

East Asia and Pacific:
450,000 infected

Figure 31-16 HIV infections worldwide
The worldwide incidence of infection with HIV, estimated by UNAIDS and the World Health Organization in 1997.

Unlike most other diseases, cancer is not a straightforward invasion of the body by a foreign organism. Although some cancers are triggered by viruses, cancer is essentially a failure of the mechanisms that control the growth of the body's own cells—a disease in which the body destroys itself.

Cancer Is Caused by Mutations in Genes That Control Cell Division

To learn the causes of cancer, we must answer two questions: (1) What changes occur in a cancerous cell that allow it to escape normal growth controls? (2) What agents can initiate these cellular changes?

The genes that have been implicated in cancer formation fall into two main classes: (1) **proto-oncogenes** and (2) **tumor-suppressor genes**. In normal cells, these genes produce proteins that function in the complex biochemical pathways that control cell proliferation. Normally, cells divide only when they receive a chemical signal that activates the genes responsible for initiating growth. Cells stop dividing when they receive signals that stimulate the production of proteins that shut down the replication machinery. Appropriate growth in any organ or tissue requires the proper balance between the growth-stimulating pathways, which use the products of proto-oncogenes, and growth-inhibiting pathways, which use the products of tumor-suppressor genes. In the delicate balance that maintains the proper rate of cell growth, proto-oncogenes are the "accelerators" that respond to "go" signals, and tumor-suppressor genes are the "brakes" that respond to "stop" signals.

Cancerous tumors can develop when cells fail to respond to the body's "stop" and "go" signals. Such failures are typically the result of mutations. Mutations in proto-oncogenes can convert them to oncogenes, which are activated even if they don't receive the appropriate cellular signal. The inappropriately activated **oncogenes** produce a continuous supply of growth-stimulating proteins, even when they should be inactive. The accelerator of cell growth is perpetually engaged. In contrast, tumor-suppressor genes become dangerous when they experience mutations that *deactivate* the genes and prevent them from playing their role in halting cell division. The inappropriately deactivated tumor-suppressor genes fail to produce key proteins needed to inhibit cell growth. The brakes of cell growth are never applied.

In general, any single mutation in a cell cannot induce cancer development. Instead, several mutations of proto-oncogenes and/or tumor-suppressor genes must occur in the same cell. It can take years for the necessary number of mutations to accumulate, which is why cancers generally tend to appear later in life. What causes the mutations that ultimately lead to cancer? Probably the most common causes of cancer are environmental insults, chiefly chemicals and radiation. We are besieged by cancer-causing chemicals (*carcinogens*), including avoidable ones such as cigarette smoke, less avoidable ones such as pesticides, diesel exhaust, and asbestos, and virtually unavoidable ones that occur naturally in foods and even some that are synthesized in our own digestive tracts. The most dangerous from of carcinogenic radiation is sunlight, which causes a majority of skin cancers. Some types of cancer, including cervical cancer, can be caused by viral infections that disrupt cellular DNA. Proto-oncogene and tumor-suppressor gene mutations can also be inherited. Individuals who inherit such genes have an elevated risk of developing cancer, because they begin life with each body cell containing part of the set of mutations required for tumor development.

A Cancer Develops: Oncogenes and Tumor-Suppressor Genes Interact in Colon Cancer

For most types of cancer, the exact sequence of causal events is not known. Some cancers, however, are better understood. For example, the development of colon cancer apparently involves mutations in at least one oncogene and no fewer than three tumor-suppressor genes in a single cell or its progeny (Fig. 31-17). The final step, damage to the *p53* tumor-suppressor gene on chromosome 17, also seems to be involved in dozens of other types of cancer.

The Immune System Defends Against Cancerous Cells

Cancer cells form in our bodies every day. Even the best of preventive measures cannot eliminate cancer completely. We cannot avoid some carcinogens, such as gamma rays from the sun, radioactivity from the rocks beneath our feet, and naturally produced carcinogens in our food. Fortunately, natural killer cells and cytotoxic T cells screen the body for cancer cells and destroy nearly all of them before they have a chance to proliferate and spread. Cancer cells are, of course, "self" cells (the body's own cells), and the immune system does not respond to "self." How are cancer cells weeded out? It seems likely that the very processes that cause cancer also cause new and slightly different proteins to appear on the surfaces of cancer cells. Cytotoxic T cells encounter these new proteins, recognize them as "non-self" antigens, and destroy the cancer cells.

Without constant surveillance by the immune system, it is unlikely that any of us would survive more than a few years. However, we can reduce our own chances of developing cancer. First and foremost, avoid cigarette smoke; smoking causes an estimated 150,000 lung cancer deaths each year in the United States alone. Second, eat a diet high in fruits and vegetables and low in red meat and saturated fats. Other behaviors that lower our risk of cancer include avoiding excessive alcohol consumption, staying out of the midday sun, and getting regular exercise. If possible, limit your exposure to known workplace carcinogens, such as benzene, formaldehyde, arsenic, and X-ray radiation.

Medical Treatments for Cancer Depend on Distinguishing and Selectively Killing Cancerous Cells

Some cancer cells may not be recognized as threats by the immune system. Even tumors that are attacked by the immune system may develop variant cell types that are resistant to immune attack (a process analogous to the one by which microbes develop antibiotic resistance). In other cases, tumors can actively suppress the immune system or can simply grow so fast that the immune response can't keep up. If the immune system is thwarted by any of these means, the tumor grows and spreads. At this point, the individual's health depends on medical treatment. Unfortunately, the available treatments are only partially effective and have serious drawbacks. The rate of cure is increasing, but it is still scarcely a third of all cancers. (The suggested reading "A War Not Won" by Tim Beardsley is instructive but depressing.)

The three main forms of treatment are (1) surgery, (2) radiation, and (3) chemotherapy. The surgical removal of the tumor is the first step in the treatment of many cancers. Ideally, surgery will remove all cancerous tissue and the patient will be cured. Unfortunately, the surgeon may not be able to see and remove very small patches of cancer cells that may extend from the main tumor. And surgery cannot be used to treat cancer that has begun to spread throughout the body. Alternatively, cancerous cells can be bombarded with radiation, in an effort to kill them. Unlike surgery, radiation may destroy even microscopic clusters of cancer cells. Like surgery, however, radiation cannot be used to treat widespread cancers, because to irradiate the whole body would damage a great deal of healthy tissue. Both surgery and radiation therapy can be traumatic and dangerous.

Chemotherapy, or drug treatment, is commonly used to supplement surgery and/or radiation or to treat cancers that cannot be treated with surgery or radiation. Various chemotherapeutic drugs offer different mechanisms of action. All of these drugs, however, aim to kill cancer cells without also killing normal cells. Most of the available drugs attack the machinery of cell division. Because cancer cells divide much more frequently than do normal cells, the hope is that attacks on dividing cells will selectively kill cancer cells. Unfortunately, other cells of the body divide too, and chemotherapy inevitably kills some healthy cells. It is damage to dividing cells in the

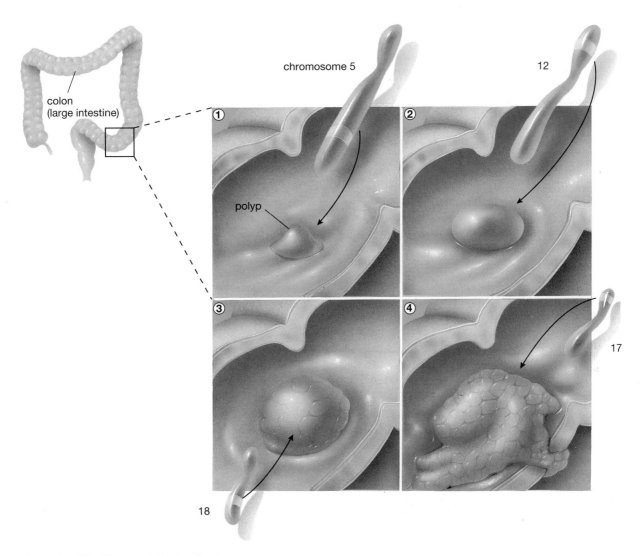

Figure 31-17 *The genetic basis of colon cancer*
The development of cancer of the colon involves four genetic changes: ① A tumor-suppressor gene on chromosome 5 is mutated in one cell of the colon. The cell multiplies, forming a polyp, a small, projecting growth. ② In one of the cells of the polyp, a proto-oncogene on chromosome 12 mutates into an oncogene. This cell multiplies more rapidly, and the polyp enlarges. ③ A second tumor-suppressor gene, on chromosome 18, mutates in one of the cells that already has the first two mutations. This cell's division rate increases, and the polyp becomes a tumor. ④ Finally, the *p53* tumor-suppressor gene on chromosome 17 mutates in a cell with all three previous mutations. Its daughter cells divide wildly, invading the wall of the colon and spreading cancer to other parts of the body.

hair follicles and intestinal lining that produces the well-known side effects of nausea, vomiting, and hair loss.

A tremendous amount of research has been devoted to the search for cancer treatments that are more effective and have fewer unpleasant side effects. One approach with some potential is *gene therapy*, in which properly functioning genes would be introduced into the genomes of cancerous cells to replace the damaged (mutated) genes that were causing uncontrolled growth. Similarly, many researchers are working on developing "designer molecules" that would interfere with the growth-stimulating proteins generated by out-of-control oncogenes. Other approaches include therapies that would stimulate the immune system to attack tumors, and developing antibody molecules that recognize tumor cells and could be used to deliver drugs and other treatments directly to tumor cells without affecting healthy cells. Any of these ideas might one day be translated into new treatments, but none, so far, has reached that stage. Research and clinical trials continue, however, and cancer patients may one day soon reap the benefits of innovative new treatments.

Summary of Key Concepts

1) How Does the Body Defend Against Invasion?

The human body has three lines of defense against invasion by microbes: (1) the barriers of skin and mucous membranes; (2) nonspecific internal defenses, including phagocytosis, killing by natural killer cells, inflammation, and fever; and (3) the immune response.

The skin physically blocks the entry of microbes into the body. It is also covered with secretions from sweat and sebaceous glands that inhibit bacterial and fungal growth. The mucous membranes of the respiratory and digestive tracts secrete antibiotic substances, IgA antibodies, and mucus. Microbes are trapped in the mucus, swept up to the throat by cilia, and expelled or swallowed.

If microbes do enter the body, white blood cells travel to the site of entry and engulf the invading cells. Natural killer cells secrete proteins that kill infected body cells or cancerous cells. Injuries stimulate the inflammatory response, in which chemicals are released that attract phagocytic white blood cells, increase blood flow, and make capillaries leaky. Later, blood clots wall off the injury site, preventing further invasion and spread of the microbes to other parts of the body. Fever is caused by endogenous pyrogens, chemicals released by white blood cells in response to infection. High temperatures inhibit bacterial growth and accelerate the immune response.

2) What Are the Key Characteristics of the Immune Response?

The immune response involves two types of lymphocytes: B cells and T cells. Plasma cells, which are descendants of B cells, secrete antibodies into the bloodstream, causing humoral immunity. Cytotoxic T cells destroy some microbes, cancer cells, and virus-infected cells on contact, causing cell-mediated immunity. Helper T cells stimulate both the humoral and cell-mediated immune responses. Immune responses have three steps: recognition, attack, and memory.

3) How Are Threats Recognized?

Antibodies (on B cells) and T-cell receptors (on T cells) recognize foreign molecules and trigger the immune response. Antibodies are Y-shaped proteins composed of a constant region and a variable region. Antigens are molecules that generate an antibody response. Antibodies function both as receptors, detecting the presence of antigens, and as effectors, actively working to destroy antigens. Each B cell synthesizes only one type of antibody, unique to that particular cell and its progeny. The diversity of antibodies arises from gene shuffling and the mutation of antibody genes during B-cell development. Each antibody has specific sites that bind one or a few types of antigen. Normally, only foreign antigens are recognized by the B cells.

4) How Are Threats Overcome?

Antigens bind to and activate only those B and T cells with the complementary antibodies or T-cell receptors. In humoral immunity, B cells with the proper antibodies, stimulated by the presence of particular antigens, divide rapidly to produce plasma cells that synthesize massive quantities of the antibody. The circulating antibodies destroy antigens and antigen-bearing microbes by four mechanisms: (1) neutralization, (2) promotion of phagocytosis by white blood cells, (3) agglutination, and (4) complement reactions. In cell-mediated immunity, T cells with the proper receptors bind antigens and divide rapidly. Cytotoxic T cells bind to antigens on microbes, infected cells, or cancer cells and then kill the cells. Helper T cells stimulate, and suppressor T cells turn off, both the B-cell and cytotoxic-T-cell responses.

5) How Are Invaders "Remembered"?

Some progeny cells of both B and T cells are long-lived memory cells. If the same antigen reappears in the bloodstream, these memory cells are immediately activated, divide rapidly, and cause an immune response that is much faster and more effective than the original response.

6) How Does Medical Care Augment the Immune Response?

Antibiotics kill microbes or slow down their reproduction, thus allowing the immune system more time to respond and exterminate the invaders. Vaccinations are injections of antigens from disease organisms, typically the weakened or dead microbes themselves. An immune response is evoked by the antigens, providing memory and a rapid response should a real infection occur.

7) Can the Immune System Malfunction?

Allergies are immune responses to normally harmless foreign substances, such as pollen or dust. Mast cells respond to the presence of these substances by releasing histamine, which causes a local inflammatory response. Some diseases are caused by defective immune responses. Autoimmune diseases arise when the immune system destroys some of the body's own cells. Immune deficiency diseases occur when the immune system cannot respond strongly enough to ward off normally minor diseases.

Infection with one of two viruses, called human immunodeficiency viruses (HIV) 1 and 2, can lead to AIDS (acquired immune deficiency syndrome). These viruses invade and destroy helper T cells. Without helper T cells to stimulate the immune responses of B cells and cytotoxic T cells, an individual with AIDS is extremely susceptible to a wide assortment of infections, which are eventually fatal.

Cancer is a population of the body's cells that divides without control. Some cancers may be caused by the activation of growth genes, called oncogenes, that cause cells to grow and multiply. These genes may be activated by viral infection or by mutations caused by chemicals or radiation. Other cancers may be caused by the loss or inactivation of tumor-suppressor genes, which turn off or suppress cell division in normal cells.

Key Terms

Thinking Through the Concepts

Multiple Choice

1. *In addition to the immune system, we are protected from disease by*
 a. the skin
 b. mucous membranes
 c. natural secretions such as acids, protein-digesting
  enzymes, and antibiotics
 d. cilia
 e. all of the above

2. *Body cells infected by viruses are destroyed mostly by*
 a. IgA   b. phagocytes
 c. natural killer cells   d. histamines
 e. natural antibiotics

3. *Fevers*
 a. decrease interferon production
 b. decrease the concentration of iron in the blood
 c. decrease the activity of phagocytes
 d. increase the reproduction rate of invading bacteria
 e. do all of the above

4. *T cells and B cells are*
 a. lymphocytes   b. macrophages
 c. natural killer cells   d. red blood cells
 e. phagocytes

5. *During allergic responses, IgE*
 a. covers invaders and marks them for phagocytosis
 b. pokes holes in plasma membranes of invading parasites
 c. binds to mast cells, causing them to release histamine
 d. triggers the production of antigens by lymphocytes
 e. promotes the secretion of antibodies into the digestive
  and respiratory tracts

6. *What shuts off the immune response in T cells and B cells
 after an infection has been conquered?*
 a. IgA   b. histamine
 c. pyrogens   d. natural killer cells
 e. suppressor T cells

? Review Questions

1. List the human body's three lines of defense against invading microbes. Which are nonspecific (that is, act against all types of invaders), and which are specific (act only against a particular type of invader)? Explain your answer.

2. How do natural killer cells and cytotoxic T cells destroy their targets?

3. Describe humoral immunity and cell-mediated immunity. Include in your answer the types of immune cells involved in each, the location of antibodies and receptors that bind foreign antigens, and the mechanisms by which invading cells are destroyed.

4. How does the immune system construct so many different antibodies?

5. How does the body distinguish "self" from "non-self"?

6. Diagram the structure of an antibody. What parts bind to antigens? Why does each antibody bind only to a specific antigen?

7. What are memory cells? How do they contribute to long-lasting immunity to specific diseases?

8. What is a vaccine? How does it confer immunity to a disease?

9. Compare and contrast an inflammatory response with an allergic reaction from the standpoint of cells involved, substances produced, and symptoms experienced.

10. Distinguish between autoimmune diseases and immune deficiency diseases, and give one example of each.

11. Describe the causes, progression, and eventual outcome of AIDS. How do AIDS treatments work?

12. What is cancer? How do oncogenes cause cancer? How do tumor-suppressor genes prevent cancer? How can environmental factors "turn on" oncogenes to cause cancer?

Applying the Concepts

1. Some types of cancer are partly hereditary; that is, members of certain families are more likely than the general population to develop certain cancers, but not all family members develop the cancer. Given the multistep mechanism for colon cancer outlined in Figure 31-17, could you propose a genetic model that would account for predisposition to cancer while not requiring that all people who carry, for example, a defective tumor-suppressor gene necessarily develop cancer?

2. Why is it essential that antibodies and T-cell receptors bind only relatively large molecules (like proteins) and not relatively small molecules (such as amino acids)?

3. The text states that smallpox has been eradicated from Earth. As of early 1998, this is not quite true: There are smallpox virus stocks in two laboratories—one in the United States, one in Russia. A debate is raging about whether these stocks should be destroyed (see *The Scientist*, December 9, 1996, pp. 11–14, and *Science,* January 27, 1995, pp. 5197–5198). In brief, one side argues that having smallpox around is too dangerous, particularly with the proliferation of terrorist groups. The other side argues that we may be able to learn things from smallpox, answers to questions we don't know enough to ask yet, that may help us conquer future diseases. Both stocks are tentatively scheduled to be destroyed in June 1999. Do you believe the smallpox stocks should be destroyed? Why or why not?

4. The essay "Health Watch: Flu—The Unbeatable Bug" states that the flu virus is different each year. If that is true, what good is it to get a "flu shot" each winter?

5. Pollen, house dust, and poison ivy provoke allergic reactions in some individuals. When these allergens cannot be avoided, a "hyposensitization" procedure is usually prescribed. Hyposensitization is most likely due to an opposite process—immunization. Explain how giving a patient gradually increasing doses of poison-ivy extract over a period of time helps subdue allergic dermatitis.

6. Transplant patients typically receive the drug cyclosporine. This drug inhibits the production of interleukin-2, a regulatory molecule that stimulates helper T cells to proliferate. How does cyclosporine prevent the rejection of transplanted organs? Some patients who received successful transplants many years ago are now developing various kinds of cancers. Propose a hypothesis to explain this phenomenon.

7. Throughout pregnancy a mother may pass IgG antibodies across the placenta. After birth she may pass IgAs on through breast milk. The immunity provided to her offspring lasts only a few months. Explain.

Group Activity

This chapter has introduced you to some of the mechanisms that have evolved to defend the vertebrate body against attack. The attackers, however, have not stood idly by in the face of evolutionary improvements in our defenses. Many viruses, bacteria, fungi, and invertebrate animals have themselves developed sophisticated strategies for evading and resisting the immune response. Can you guess what some of these strategies might be? Form a group of four students, and try to design the structures and behaviors that might have emerged in these invaders through natural selection. Assign two members of your group to design a structural solution and a behavioral solution that an invader might use to avoid being recognized by the immune system. The other two members of the group should design one structure and one behavior that might resist immune attack even after recognition. Next, bring your entire group together and discuss the strengths and drawbacks of your proposed solutions. Your final product should be a "designer organism" that has multiple ways of evading immune recognition and attack. Be sure to specify your organism's size, life cycle, structure, and food source.

For More Information

Beardsley, T. "A War Not Won." *Scientific American*, January 1994. The war on cancer remains a standoff.

Bloom, F. (ed.). "The New Face of AIDS." *Science*, June 28, 1996. A collection of short articles on the main scientific, economic, and policy issues engendered by the AIDS epidemic.

Caldwell, M. "The Dream Vaccine." *Discover*, September 1997. An account of the discoveries that led to the development of DNA vaccines and a discussion of the potential of this new technique to lead to vaccinations for AIDS, tuberculosis, and other killers.

Hoffman, M. "AIDS: A Howdonit Story." *American Scientist*, January–February 1997. The mechanisms by which HIV invades cells and causes disease.

Janeway, C. A., Jr. "How the Immune System Recognizes Invaders." *Scientific American*, September 1993. All about antibodies, major histocompatibility complexes, antigen presentation, and the immune response.

Leder, P. "The Genetics of Antibody Diversity." *Scientific American*, May 1982. How only a few hundred genes can be used to make millions of antibodies.

Levins, R., Awerbuch, T., Brinkmann, U., Eckardt, I., Epstein, P., Makhoul, N., de Possas, C. A., Puccia, C., Spielman, A., and Wilson, M. E. "The Emergence of New Diseases." *American Scientist*, January–February 1994. Diseases that appear to be new continue to challenge the human body and physicians.

Lichtenstein, L. M. "Allergy and the Immune System." *Scientific American*, September 1993. Allergic responses, evolved as protection against parasites, turn against us when we respond violently to harmless pollen and foods.

O'Brien, S., and Dean, M. "In Search of AIDS-Resistance Genes." *Scientific American*, September 1997. A newly discovered genetic trait has turned the search for an AIDS cure in a new direction.

Marrack, P., and Kappler, J. W. "How the Immune System Recognizes the Body." *Scientific American*, September 1993. How the immune system's potential weapons against the body are disarmed.

Nesse, R., and Williams, G. *Why We Get Sick*. New York: Vintage Books, 1994. An evolutionary perspective on disease and immunity.

Nossal, G. J. V. "Life, Death, and the Immune System." *Scientific American*, September 1993. A nice introduction to the major players of the immune system.

Paul, W. E. "Infectious Diseases and the Immune System." *Scientific American*, September 1993. Microbes and the immune system engage in constant evolutionary warfare.

Radetsky, P. "Immune to a Plague." *Discover*, June 1977. The story of the scientific detectives who are unraveling the mysteries of genetic resistance to AIDS.

Steinman, L. "Autoimmune Disease." *Scientific American*, September 1993. An estimated 5% of Americans suffer from some type of autoimmune disorder.

Young, J. D.-E., and Cohn, Z. A. "How Killer Cells Kill." *Scientific American*, January 1988. Cytotoxic T cells strike by secreting proteins that form large holes in the plasma membrane of their targets.

Answers to Multiple-Choice Questions

1. e   2. c   3. b   4. a   5. c   6. e

The transition from childhood to adulthood is marked by profound physical and behavioral changes. These changes are under the control of hormones, the chemical messengers of the animal body.

Chemical Control of the Animal Body: The Endocrine System

32

At a Glance

Parents of teenagers have been known to say, "What ever became of that sweet little boy who used to live here?" The physical and behavioral changes associated with puberty are so extreme that it can seem like a new person has been created. And the situation is even more drastic in other types of organisms. Think of the difference between a caterpillar and a butterfly or between a tadpole and a frog. What causes these dramatic rearrangements of body tissue? Hormones.

Hormones are secreted chemical messages that initiate changes in the cells that receive them. Other, specialized cells send these messages via the circulatory system, normally after the sending cells have been activated by some environmental or physiological stimulus. The changes induced by hormonal messages may be prolonged and irreversible, as in puberty or metamorphosis. More typically, the induced changes are transient and reversible and function in systems for physiological control and regulation. In fact, the **endocrine system** (consisting of hormones and the various cells that secrete and receive them) can be viewed as the postal system of physiology, moving information and instructions between cells that may be some distance apart. Regulation requires communication; in animal bodies, hormones provide much of that communication.

1 What Are the Characteristics of Animal Hormones?

A **hormone** is a chemical that is secreted by specialized cells in one part of the body and is transported in the bloodstream to another part of the body, where it affects particular cells called **target cells**. Hormones are released by the cells of major endocrine glands and endocrine organs located throughout the body (Fig. 32-1).

Hormones Have a Variety of Chemical Structures

A chemical earns the right to be called a hormone on the basis of the kind of function it performs, but hormone molecules assume a wide variety of forms. The structural

Net Watch

On-line resources for this chapter are on the World Wide Web at: http://www.prenhall.com/audesirk

(click on the Table of Contents link and then select Chapter 32).

637

Figure 32-1 Major mammalian endocrine glands
The major mammalian endocrine glands discussed in the text are the hypothalamus–pituitary complex, the thyroid and parathyroid glands, the pancreas, the sex organs (ovaries in females, testes in males), and the adrenal glands. Other organs that secrete hormones include the pineal gland, thymus, kidneys, heart, and digestive tract.

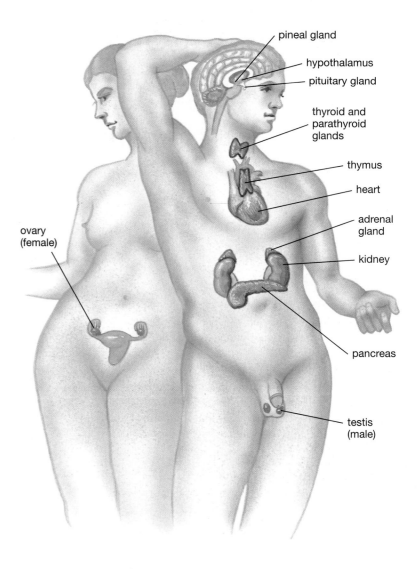

diversity of vertebrate hormones can, however, be grouped into four general classes (Table 32-1):

1. *Peptide hormones.* Most animal hormones are composed of peptides, which are chains of amino acids. The term *peptide* technically refers to comparatively short amino acid chains, whereas longer chains are called *proteins.* Both peptides and proteins can serve as hormones, but, for convenience, all hormones composed of amino acid chains are described as **peptide hormones.**

2. *Amino-acid derivatives.* A few hormones consist of comparatively simple molecules derived from single amino acids. For example, the amino acid tyrosine forms the basis for the hormones *epinephrine* (also called *adrenaline*) and *norepinephrine.*

3. *Steroid hormones.* **Steroid hormones,** sometimes known simply as *steroids,* have a chemical structure resembling cholesterol, from which most of them are synthesized. Steroid hormones are secreted by the ovaries, placenta, testes, and adrenal cortex.

4. *Prostaglandins.* Molecules in this category consist of two fatty acid carbon chains attached to a five-carbon ring. **Prostaglandins** don't fit neatly into our definition of hormones, because they are produced by nearly every type of cell in the body (instead of by specialized cells in endocrine glands), and they act mainly on nearby cells (instead of traveling in the circulatory system to act on more-distant cells). Nonetheless, the effects of prostaglandins can be similar to those of hormones, and they often act in concert with the endocrine system. Hence, prostaglandins are frequently considered to be a special type of hormone.

Hormones Function by Binding to Specific Receptors on Target Cells

Because nearly all cells have a blood supply, a hormone that has been released into the bloodstream will reach nearly every cell of the body. But to exert their precise control, hormones must act only on certain target cells. Target cells have receptors for particular hormone molecules; cells that lack the appropriate receptor will not respond to the hormonal message. In addition, a given hormone may have several different effects depending

Table 32-1 The Chemical Diversity of Vertebrate Hormones

Chemical Type	Examples
Peptides and proteins (synthesized from multiple amino acids)	Oxytocin
Amino-acid derivatives (synthesized from single amino acids)	Norepinephrine
	Thyroxine
Steroids (synthesized from cholesterol)	Testosterone
	Estradiol
Prostaglandins (synthesized from fatty acids)	Prostaglandin E_1

on the nature of the target cell it contacts. Receptors for hormones are found in two general locations on target cells: (1) on the plasma membrane and (2) inside the cell, normally within the nucleus.

Some Hormones Bind to Surface Receptors

Most peptide and amino acid-based hormones are soluble in water but not in lipid. Hence, these hormones cannot penetrate plasma membranes, which are largely composed of phospholipids. Instead, the hormones react with protein receptors that protrude from the outside surface of the target cell's plasma membrane (see Chapter 5). Some of these surface receptors are linked to plasma membrane channels that open in response to the binding of the hormone. For example, when epinephrine binds to receptors on heart muscle cells, calcium channels open, allowing more calcium to flow into the muscle cells, which in turn increases the strength of contraction of the heart.

Direct mechanisms such as that involved in epinephrine's effect on heart muscles are relatively uncommon among hormones that bind to surface receptors in the endocrine system. More commonly, the mechanism involves a **second messenger**, so called because it transfers the signal from the first messenger—the hormone—to molecules within the cell. In second-messenger systems, a hormone binding to a receptor on the outside cell surface triggers the release *inside* the cell of a chemical (the second messenger) that initiates a cascade of biochemical reactions (Fig. 32-2a). In many cases, the binding of the receptor activates an enzyme that catalyzes the conversion of ATP to **cyclic AMP** (cAMP), a nucleotide that regulates many cellular activities (see Chapter 3). Cyclic AMP acts as a second messenger and initiates a chain of reactions inside the cell. Each reaction in the chain involves an increasing number of molecules, amplifying the original signal. The end result varies with the target cell. For example, channels may be opened in the plasma membrane, or substances may be synthesized or secreted.

Other Hormones Bind to Intracellular Receptors

Steroid hormones (and the nonsteroidal hormones that are produced in the thyroid gland) are lipid soluble and are therefore able to pass through the plasma membrane. Once inside the cell, these hormones bind to receptors inside the cell—typically, to protein receptors in the nucleus (Fig. 32-2b). The receptor–hormone complex binds to DNA and stimulates particular genes to transcribe messenger RNA, which moves to the cytoplasm and directs the synthesis of proteins that alter the cell's activity. Steroid and thyroid hormones thus act by stimulating the expression of genes that are otherwise switched off. These hormones may therefore take minutes or even days to exert their full effects (in contrast to the relatively rapid and short-lived effects of hormones that bind to cell-surface receptors).

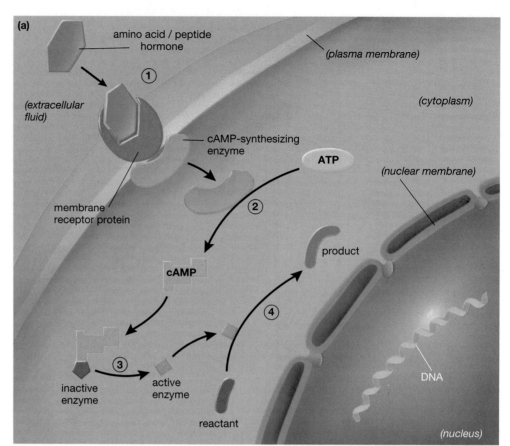

Figure 32-2 Modes of action of hormones

(a) ① Peptide and amino-acid hormones, which are not soluble in lipid, bind to a receptor on the outside the target cell's plasma membrane. ② Hormone-receptor binding triggers the synthesis of cyclic AMP (cAMP). ③ Cyclic AMP in turn activates enzymes that ④ promote specific cellular reactions that produce new products. This cAMP "cascade" may generate a variety of responses. Examples include an increase in glucose synthesis induced by epinephrine and an increase in estrogen synthesis induced by luteinizing hormone.

(b) ① Lipid-soluble steroid hormones diffuse readily through the plasma membrane into the target cell and then ② into the nucleus, where they combine with a protein receptor molecule. ③ The hormone–receptor complex binds to DNA and facilitates the binding of RNA polymerase to promoter sites on specific genes, accelerating ④ the transcription of DNA into messenger RNA (mRNA). ⑤ The mRNA then directs protein synthesis. In hens, for example, estrogen promotes the transcription of the albumin gene, causing the synthesis of albumin (egg white), which is packaged in the egg as a food supply for the developing chick.

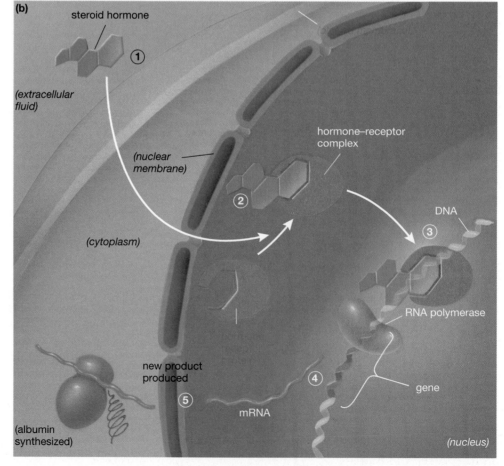

Hormones Are Regulated by Feedback Mechanisms

A signal that's always turned on is incapable of carrying an instruction for change. If a hormone is to provide a useful signal for physiological control, there must be a way to turn its message on and off. In animals, the switching mechanism normally involves negative feedback: The secretion of a hormone stimulates a response in target cells that inhibits further secretion of the hormone.

Control of hormone release by negative feedback is especially important, because most hormones exert such powerful effects on the body that it would be harmful to have too much hormone working for too long. For example, suppose you have jogged several miles on a hot, sunny day and have lost a pint of water through perspiration. In response to the loss of water from your bloodstream, your pituitary gland releases *antidiuretic hormone* (ADH), which causes your kidneys to reabsorb water and to produce a very concentrated urine (see Chapter 30). However, if you arrive home and drink a quart of Gatorade®, you will more than replace the water you lost in sweat. Continued retention of this excess water could raise blood pressure and possibly damage your heart. Negative feedback ensures that ADH secretion is turned off when your blood water content returns to normal, and your kidneys begin eliminating the excess water (see Fig. 30-8).

In a few cases, positive feedback controls hormone release. For example, as mentioned in Chapter 26, contractions of the uterus early in childbirth cause the release of the hormone *oxytocin* by the posterior pituitary. The oxytocin stimulates stronger contractions of the uterus, which cause more oxytocin release, creating a positive feedback cycle. Simultaneously, oxytocin causes uterine cells to release prostaglandins, which further enhance uterine contractions, another example of positive feedback.

2) What Are the Structures and Functions of the Mammalian Endocrine System?

www

Endocrinologists (biologists who study the endocrine system) are far from fully understanding hormonal control in mammals. New hormones and new roles for known hormones are discovered nearly every year. The key functions of the major endocrine glands and endocrine organs, however, have been known for many years. We shall soon discuss the endocrine functions of the hypothalamus–pituitary complex, the thyroid and parathyroid glands, the pancreas, the sex organs, and the adrenal glands (see Fig. 32-1). Table 32-2 (pp. 642–643) lists these and other glands, their major hormones, and their principal functions.

Mammals Have Both Exocrine and Endocrine Glands

Mammals have two types of glands: (1) exocrine glands and (2) endocrine glands (Fig. 32-3). **Exocrine glands** produce secretions that are released outside the body

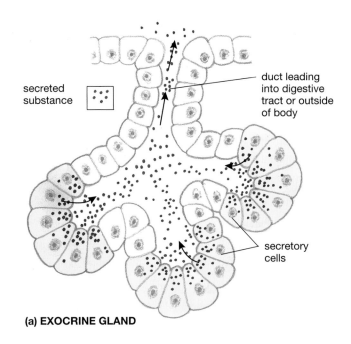

(a) EXOCRINE GLAND

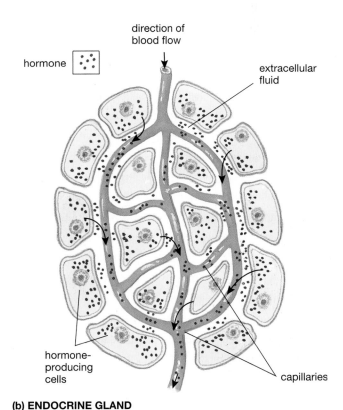

(b) ENDOCRINE GLAND

Figure 32-3 The structure of exocrine and endocrine glands
(a) The secretory cells of exocrine glands secrete substances into ducts that open outside the body (sweat glands, mammary glands) or into the digestive tract (pancreas, salivary glands). *(b)* Endocrine glands consist of hormone-producing cells embedded in a network of capillaries. The cells secrete hormones into the extracellular fluid, from which they diffuse into the capillaries.

Table 32-2 Mammalian Endocrine Glands and Hormones

Endocrine Gland	Hormone	Type of Chemical	Principal Function
Hypothalamus (via posterior pituitary)	Antidiuretic hormone (ADH)	Peptide	Promotes reabsorption of water from kidneys; constricts arterioles
	Oxytocin	Peptide	In females, stimulates contraction of uterine muscles during childbirth, milk ejection, and maternal behaviors; in males, causes sperm ejection
Hypothalamus (to anterior pituitary)	Releasing and inhibiting hormones	Peptides	At least nine hormones; releasing hormones stimulate release of hormones from anterior pituitary; inhibiting hormones inhibit release of hormones from anterior pituitary
Anterior pituitary	Follicle-stimulating hormone (FSH)	Peptide	In females, stimulates growth of follicle, secretion of estrogen, and perhaps ovulation; in males, stimulates spermatogenesis
	Luteinizing hormone (LH)	Peptide	In females, stimulates ovulation, growth of corpus luteum, and secretion of estrogen and progesterone; in males, stimulates secretion of testosterone
	Thyroid-stimulating hormone (TSH)	Peptide	Stimulates thyroid to release thyroxine
	Growth hormone	Peptide	Stimulates growth, protein synthesis, and fat metabolism; inhibits sugar metabolism
	Adrenocorticotropic hormone (ACTH)	Peptide	Stimulates adrenal cortex to release hormones, especially glucocorticoids
	Prolactin	Peptide	Stimulates milk synthesis in and secretion from mammary glands
Thyroid	Thyroxine	Amino acid derivative	Increases metabolic rate of most body cells; increases body temperature; regulates growth and development
	Calcitonin	Peptide	Inhibits release of calcium from bones
Parathyroid	Parathormone	Peptide	Stimulates release of calcium from bone; promotes absorption of calcium by intestines; promotes reabsorption of calcium by kidneys
Pancreas	Insulin	Peptide	Decreases blood glucose levels by increasing uptake of glucose into cells and converting glucose to glycogen, especially in liver; regulates fat metabolism
	Glucagon	Peptide	Converts glycogen to glucose, raising blood glucose levels

(*exo* means "out of" in Greek) or into the digestive tract (which is actually a hollow tube continuous with the outside world). Exocrine gland secretions are released through tubes or openings called **ducts**. The exocrine glands include the sweat glands and sebaceous (oil-producing) glands of the skin, the lacrimal (tear-producing) glands of the eye, the mammary (milk-producing) glands, as well as glands (including salivary glands and the pancreas) that produce digestive secretions.

Endocrine glands, sometimes called *ductless glands,* release their hormones within the body (*endo* means "inside of"). An endocrine gland generally consists of clusters of hormone-producing cells embedded within a network of capillaries. The cells secrete their hormones into the extracellular fluid surrounding the capillaries. The hormones then enter the capillaries by diffusion and are distributed throughout the body by the bloodstream.

The Hypothalamus Controls the Secretions of the Pituitary Gland

If the endocrine system is the body's postal service, then the hypothalamus is the main post office and the pituitary gland is the administrative headquarters. Together, these structures coordinate the action of many key hormonal messaging systems. The **hypothalamus** is a part of the brain that contains clusters of specialized nerve cells, called **neurosecretory cells**. Neurosecretory cells synthesize peptide hormones, store them, and release them when stimulated. The **pituitary gland** is a pea-sized gland that dangles from the hypothalamus by a stalk (Fig. 32-4, p. 644). Anatomically, the pituitary consists of two distinct lobes, or parts: (1) the **anterior pituitary** and (2) the **posterior pituitary**. The hypothalamus controls the release of hormones from both parts. The anterior pituitary is a true endocrine gland, composed of several types of hormone-secreting cells enmeshed in a network of cap-

Table 32-2 (*Continued*)

Endocrine Gland	Hormone	Type of Chemical	Principal Function
Ovaries[a]	Estrogen	Steroid	Causes development of female secondary sexual characteristics and maturation of eggs; promotes growth of uterine lining
	Progesterone	Steroid	Stimulates development of uterine lining and formation of placenta
Testes[a]	Testosterone	Steroid	Stimulates development of genitalia and male secondary sexual characteristics; stimulates spermatogenesis
Adrenal medulla	Epinephrine (adrenaline) and norepinephrine (noradrenaline)	Amino acid derivative	Increase levels of sugar and fatty acids in blood; increase metabolic rate; increase rate and force of contractions of the heart; constrict some blood vessels
Adrenal cortex	Glucocorticoids	Steroid	Increase blood sugar; regulate sugar, lipid, and fat metabolism; anti-inflammatory effects
	Aldosterone	Steroid	Increases reabsorption of salt in kidney
	Testosterone	Steroid	Causes masculinization of body features, growth

Other Sources of Hormones

Pineal gland	Melatonin	Amino acid derivative	Regulates seasonal reproductive cycles and sleep–wake cycles; may regulate onset of puberty
Thymus	Thymosin	Peptide	Stimulates maturation of cells of immune system
Kidney	Renin	Peptide	Acts on blood proteins to produce hormone (angiotensin) that regulates blood pressure
	Erythropoietin	Peptide	Stimulates red blood cell synthesis in bone marrow
Heart	Atrial natriuretic peptide (ANP)	Peptide	Increases salt and water excretion by kidneys; lowers blood pressure
Digestive tract[b]	Secretin, gastrin, cholecystokinin, and others	Peptides	Control secretion of mucus, enzymes, and salts in digestive tract; regulate peristalsis

[a]See Chapters 35 and 36.

[b]See Chapter 29.

illaries. The posterior pituitary, however, is derived from an outgrowth of the hypothalamus.

The Anterior Pituitary Produces and Releases a Variety of Hormones

The anterior pituitary (see Fig. 32-4) produces several peptide hormones. Four of these regulate hormone production in other endocrine glands. **Follicle-stimulating hormone (FSH)** and **luteinizing hormone (LH)** stimulate the production of sperm and testosterone in males, and of eggs, estrogen, and progesterone in females. We shall discuss the roles of FSH and LH in more detail in Chapter 35. **Thyroid-stimulating hormone (TSH)** stimulates the thyroid gland to release its hormones, and **ACTH**, or **adrenocorticotropic hormone** ("hormone that stimulates the adrenal cortex"), causes the release of hormones from the adrenal cortex. We discuss the effects of thyroid and adrenal cortical hormones later in this chapter.

The remaining hormones of the anterior pituitary do not act on other endocrine glands. **Prolactin**, in conjunction with other hormones, stimulates the development of the mammary glands (which are exocrine glands) during pregnancy. **Endorphins** are anterior pituitary hormones that inhibit the perception of pain by binding to certain receptors in the brain. Some narcotic drugs, including morphine and heroin, mimic endorphins and bind to the same sites in the brain. **Melanocyte-stimulating hormone (MSH)** stimulates the synthesis of the skin pigment melanin. **Growth hormone** regulates the body's growth. This hormone acts on all the body's cells—increasing protein synthesis, fat utilization, and the storage of carbohydrates. During maturation, it has a stimulatory effect on bone growth, which influences the ultimate size of the adult organism. Much of the normal variation in human height is due to differences in the secretion of growth hormone from the anterior pituitary. Too little growth hormone causes some cases of *dwarfism;* too

Figure 32-4 The hypothalamus controls the pituitary

The pituitary gland consists of secretory cells enmeshed in capillary beds. Neurosecretory cells of the hypothalamus control hormone release in the anterior lobe of the pituitary by producing releasing hormones (left). These cells release their hormones into a capillary network directly "upstream" from the anterior pituitary and travel downstream to capillaries in the anterior pituitary. There the releasing hormones contact the various endocrine cells of the anterior pituitary. Only endocrine cells with matching plasma membrane receptors respond to a given releasing hormone, so each releasing hormone stimulates a particular type of endocrine cell to release its hormone while leaving other types unaffected. The posterior lobe of the pituitary (right) is an extension of the hypothalamus. Neurosecretory cells in the hypothalamus have cell endings on a capillary bed in the posterior lobe, where they release oxytocin or antidiuretic hormone (ADH).

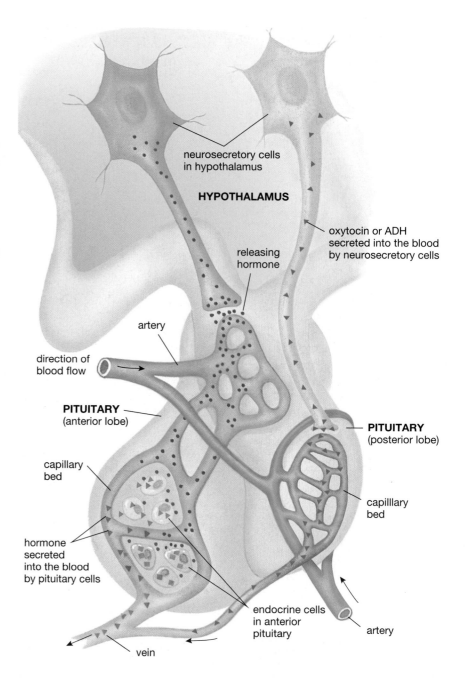

much can cause *gigantism* (Fig. 32-5). Although in adulthood many bones lose their ability to grow, growth hormone continues to be secreted throughout life, helping regulate protein, fat, and sugar metabolism.

A major advance in the treatment of pituitary dwarfism occurred when molecular biologists successfully inserted the gene for human growth hormone into bacteria, which churn out large quantities of the substance (see Chapter 13). Previously, the main commercial source of growth hormone was human cadavers, from which tiny amounts were extracted at great cost. Thanks to the new, cheaper source, many more children with underactive pituitary glands who would previously have been dwarfs can achieve normal height.

Hypothalamic Hormones Exert Control over the Anterior Pituitary

Neurosecretory cells of the hypothalamus produce at least nine peptide hormones that regulate the release of hormones from the anterior pituitary. These peptides are called **releasing hormones** or **inhibiting hormones** depending on whether they stimulate or prevent the release of pituitary hormone, respectively (see Fig. 32-4). Releasing and inhibiting hormones are synthesized in nerve cells in the hypothalamus, secreted into a capillary bed in the lower portion of the hypothalamus, and travel a short distance through blood vessels to a second capillary bed that surrounds the endocrine cells of the anterior pituitary. There, the releasers and inhibitors

Figure 32-5 When the anterior pituitary malfunctions
An improperly functioning anterior pituitary can produce either too much or too little growth hormone. Too little can result in dwarfism; too much causes gigantism.

diffuse out of the capillaries and influence pituitary hormone secretion.

Because the releasing and inhibiting hormones are secreted very close to the anterior pituitary, they are produced only in minute amounts. Not surprisingly, they were extremely difficult to isolate and study. Andrew Schally and Roger Guillemin, American endocrinologists who shared the Nobel Prize in Medicine in 1977 for characterizing several of these hormones, used the brains of millions of sheep and pigs (obtained from slaughterhouses) to extract enough releasing hormone to analyze.

The Posterior Pituitary Releases Hormones Produced by Cells in the Hypothalamus

The posterior pituitary contains the endings of two types of neurosecretory cells whose cell bodies are located in the hypothalamus. These neurosecretory cell endings are enmeshed in a capillary bed into which they release hormones to be carried into the bloodstream (see Fig. 32-4). Two peptide hormones are synthesized in the hypothalamus and released from the posterior pituitary: **antidiuretic hormone (ADH)** and **oxytocin.**

Antidiuretic hormone, which literally means "hormone that prevents urination," helps prevent dehydration. As you learned in Chapter 30, by increasing the permeability to water of the collecting ducts of nephrons

in the kidney, ADH causes water to be reabsorbed from the urine and retained in the body. Interestingly, alcohol inhibits the release of ADH, greatly increasing urination, so a beer drinker may temporarily lose more fluid than he or she has taken in.

Oxytocin triggers the "milk letdown reflex" in nursing mothers by causing muscle tissue within the breasts to contract during lactation (breastfeeding). This reflex ejects milk from the saclike milk glands into the nipples (Fig. 32-6). Oxytocin also causes contractions of the muscles of the uterus during childbirth, helping expel the fetus from the womb.

Recent studies using laboratory animals indicate that oxytocin also has behavioral effects. In rats, for example, oxytocin injections cause virgin females to exhibit maternal behavior such as building a nest, licking pups, and retrieving pups that have strayed. Oxytocin may also have a role in male reproductive behavior. In several types of animals, oxytocin stimulates the contraction of muscles that surround the tubes that conduct sperm from the testes to the penis, causing ejaculation.

The Thyroid and Parathyroid Glands Affect Metabolism and Calcium Levels

The **thyroid gland** lies at the front of the neck, nestled around the larynx (Fig. 32-7a, p. 647). The thyroid produces two major hormones: (1) **thyroxine**, an iodine-containing modified amino acid that raises the metabolic rate of most body cells, and (2) **calcitonin**, a peptide important in calcium metabolism.

Thyroxine influences most of the cells in the body, elevating their metabolic rate and stimulating the synthesis of enzymes that break down glucose and provide energy. These cellular effects cause the body to increase oxygen consumption and heart rate. Thyroxine's influence on metabolic rate seems to be at least in part connected to regulating body temperature and response to stress. Exposure to cold, for example, greatly increases thyroid hormone production.

In juvenile animals, including humans, thyroxine helps regulate growth by stimulating both metabolic rate and the development of the nervous system. Undersecretion of thyroid hormone early in life causes *cretinism,* a condition characterized by mental retardation and dwarfism. Fortunately, early diagnosis and thyroxine supplementation can prevent this condition. Conversely, oversecretion of thyroxine in developing vertebrates can trigger precocious development. In 1912, in one of the first demonstrations of hormone action, a physiologist discovered that thyroxine can induce early metamorphosis in tadpoles (see "Evolutionary Connections: The Evolution of Hormones").

Levels of thyroxine in the bloodstream are finely tuned by negative feedback loops. Thyroxine release is stimulated by thyroid-stimulating hormone (TSH) from the anterior pituitary, which in turn is stimulated by a releasing hormone from the hypothalamus. The amount of TSH released from the pituitary is regulated by thyroxine levels

Figure 32-6 Hormones and breastfeeding

The control of milk letdown by oxytocin during breastfeeding is regulated by feedback between a baby and its mother. The breast, or mammary gland, is an exocrine gland. There, clusters of milk-producing cells surround hollow bulbs, where milk collects in lactating women. The bulbs are surrounded by muscle that can expel the milk through the nipple. Milk is expelled when suckling by the baby stimulates nerve endings that send a signal to the mother's hypothalamus, causing the secretion of oxytocin into the bloodstream by the posterior pituitary. When oxytocin reaches the muscles that surround the milk ducts, it causes them to contract and expel milk through the nipple. This cycle continues until the infant is full and stops suckling. With the nipple no longer being stimulated, oxytocin release stops, the muscles relax, and milk flow ceases.

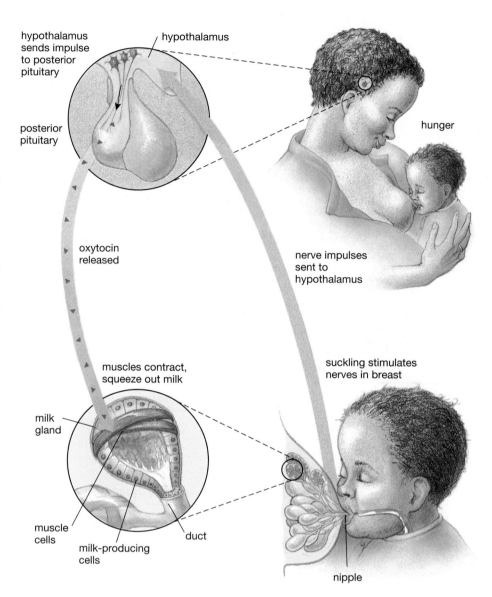

hypothalamus sends impulse to posterior pituitary

hypothalamus

posterior pituitary

oxytocin released

hunger

nerve impulses sent to hypothalamus

muscles contract, squeeze out milk

suckling stimulates nerves in breast

milk gland

muscle cells

milk-producing cells

duct

nipple

in the blood (Fig. 32-8): High levels of thyroxine inhibit the secretion of both the releasing hormone and TSH, thus inhibiting further release of thyroxine from the thyroid.

A diet deficient in iodine can reduce the production of thyroxine and trigger a feedback mechanism that acts to restore normal hormone levels by dramatically increasing the number of thyroxine-producing cells. This compensating mechanism leads to excessive growth of the thyroid gland; the enlarged gland may bulge from the neck, producing a condition called **goiter** (Fig. 32-7b). Goiter was once common in some regions of the United States, but widespread use of iodized salt has now all but eliminated this condition in developed countries.

The four small disks of the **parathyroid glands** are embedded in the back of the thyroid gland (see Fig. 32-7a). The parathyroids secrete **parathormone**, which, along with calcitonin, controls the concentration of calcium in the blood and other body fluids. Calcium is essential for many processes, including nerve and muscle function, so the cal-

cium concentration in body fluids must be kept within narrow limits. Parathormone and calcitonin regulate calcium absorption and release by the bones, which serve as a bank into which calcium can be deposited or withdrawn as necessary. In response to low blood calcium, the parathyroids release parathormone, which causes release of calcium from bones. The parathyroids increase in size in pregnant and lactating women, thereby enhancing parathormone output and allowing the mother's body to meet the extra demands for calcium imposed by the developing fetus and, later, milk production. If blood calcium levels become too high, the thyroid releases calcitonin, which inhibits the release of calcium from bones.

The Pancreas Is Both an Exocrine and an Endocrine Gland

The **pancreas** is a gland that produces both exocrine and endocrine secretions. The exocrine portion synthesizes digestive secretions that are released into the *pancreatic*

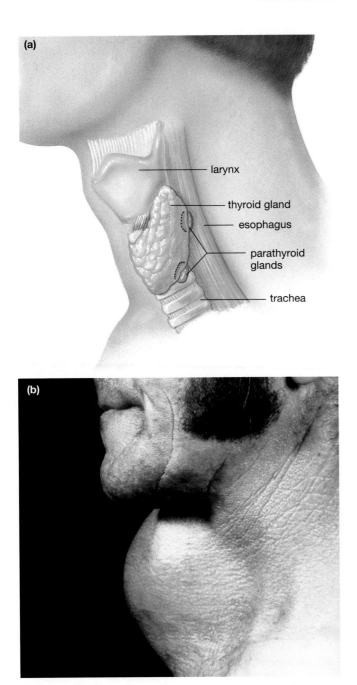

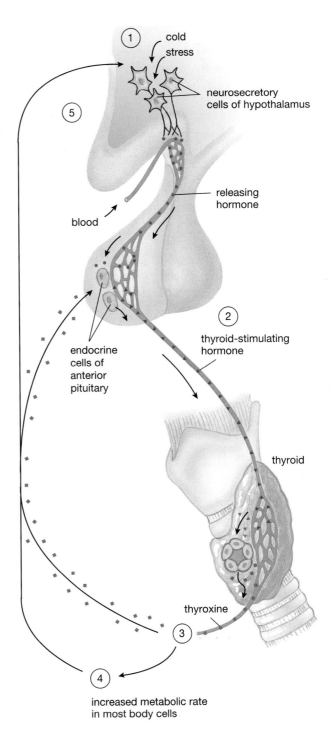

Figure 32-7 The thyroid and parathyroid glands
(a) The thyroid and parathyroid glands are located around the front of the larynx in the neck. *(b)* Individuals with iodine-deficient diets may have goiter, a condition in which the thyroid becomes greatly enlarged.

Figure 32-8 Negative feedback in thyroid gland function
① Low body temperature or stress stimulates neurosecretory cells of the hypothalamus, whose releasing hormones trigger the release of ② thyroid-stimulating hormone (TSH) in the anterior pituitary. ③ The TSH then stimulates the thyroid gland to release thyroxine. ④ Thyroxine causes an increase in the metabolic activity of most body cells, generating ATP energy and heat. ⑤ Both the raised body temperature and higher thyroxine levels in the blood inhibit the releasing-hormone cells and the TSH-producing cells.

duct and flow into the small intestine (see Chapter 29). The endocrine portion consists of clusters of cells called **islet cells**, which produce peptide hormones. One type of islet cell produces the hormone **insulin**; another type produces the hormone **glucagon**.

Insulin and glucagon work in opposition to regulate carbohydrate and fat metabolism: Insulin reduces the blood glucose level; glucagon increases it (Fig. 32-9). Together the two hormones help keep the blood glucose level nearly

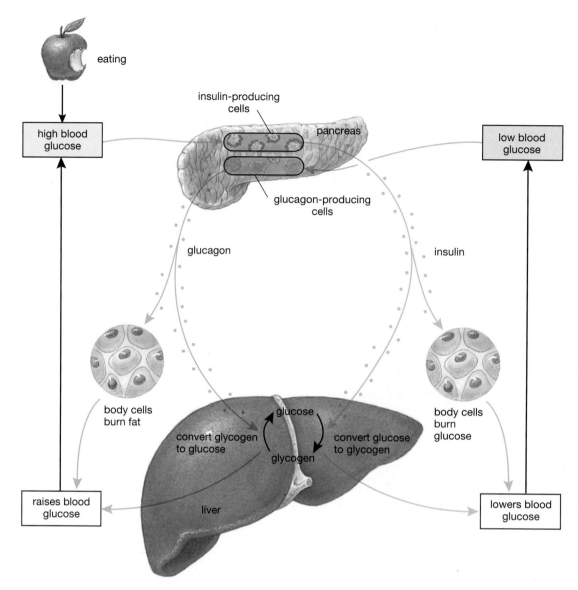

eating

high blood glucose

insulin-producing cells

pancreas

low blood glucose

glucagon-producing cells

glucagon

insulin

body cells burn fat

glucose

body cells burn glucose

convert glycogen to glucose

glycogen

convert glucose to glycogen

raises blood glucose

liver

lowers blood glucose

Figure 32-9 *The pancreas controls blood glucose levels*
The pancreatic islet cells contain two populations of hormone-producing cells: one producing insulin, and the other glucagon. These two hormones cooperate in a two-part negative feedback loop to control blood glucose concentrations. High blood glucose stimulates the insulin-producing cells and inhibits the glucagon-producing cells; low blood glucose stimulates the glucagon-producing cells and inhibits the insulin-producing cells. This dual control quickly corrects either high or low blood glucose levels.

constant. When blood glucose rises (for example, after you've eaten a meal), insulin is released. Insulin causes body cells to take up glucose and either metabolize it for energy or convert it to fat or *glycogen* (a starchlike molecule) for storage. When blood glucose levels drop (for example, after you've skipped breakfast or have run a 10-kilometer race), glucagon is released. Glucagon activates an enzyme in the liver that breaks down glycogen (which is stored primarily in the liver), releasing glucose into the blood. Glucagon also promotes lipid breakdown, which releases fatty acids that can be metabolized for energy.

Defects in insulin production, release, or reception by target cells result in **diabetes mellitus**, a condition in which

blood glucose levels are high and fluctuate wildly with sugar intake. The lack of functional insulin in diabetics causes the body to rely much more heavily on fats as an energy source, leading to high circulating levels of lipids, including cholesterol. Severe diabetes causes fat deposits in the blood vessels, resulting in high blood pressure and heart disease; diabetes is an important cause of heart attacks in the United States. The fatty deposits in small vessels can also damage the retina of the eye, leading to blindness, and the kidneys, leading to kidney failure. Until recently, the insulin supplements required to treat diabetes were formulated from insulin extracted from the pancreas of cows and pigs, obtained from slaughterhouses. Now, however, large quantities of human insulin are produced by bacteria into which the gene for human insulin has been inserted.

The Sex Organs Secrete Steroid Hormones

The **testes**, in males, and **ovaries**, in females, are important endocrine organs. The testes secrete several steroid hormones, collectively called **androgens**. The most important of these is **testosterone**. The ovary secretes two types of steroid hormones: (1) **estrogen** and (2) **progesterone**. The role of the sex hormones in development, the menstrual cycle, and pregnancy is discussed in Chapters 35 and 36.

The sex hormones also play a key role in *puberty*, the phase of life that links maturity and immaturity. Mature mammals are capable of reproduction, and puberty encompasses the physiological changes that endow immature individuals with this reproductive capacity. In humans, the development of full reproductive competence is accompanied by a growth spurt and by the development of *secondary sexual characteristics*, including pubic hair, breasts (in females), and facial hair, larger muscles, and a prominent Adam's apple (in males).

Puberty begins when, for reasons not fully understood, the hypothalamus starts to secrete increasing amounts of releasing hormones that in turn stimulate the anterior pituitary to secrete more lutenizing hormone (LH) and follicle-stimulating hormone (FSH) into the bloodstream. Both LH and FSH stimulate target cells in the testes and ovaries to produce higher levels of sex hormones. The resulting increase in circulating sex hormones ultimately affects tissues throughout the body. It also leads to activation of the adult reproductive system, to the new growth underlying secondary sexual characteristics, and to the behavioral changes that make puberty such an interesting time for parents and teenagers.

The Adrenal Glands Have Two Parts That Secrete Different Hormones

Like the pituitary gland and pancreas, the **adrenal glands** (Latin for "on the kidney") are two glands in one: (1) the *adrenal medulla* and (2) the *adrenal cortex* (Fig. 32-10). The **adrenal medulla** is located in the center of each gland (*medulla* means "marrow" in Latin). It consists of secretory cells derived during development from nervous tissue, and its hormone secretion is controlled directly by the nervous system. The adrenal medulla produces two hormones—**epinephrine** and **norepinephrine** (also called *adrenaline* and *noradrenaline,* respectively)—in response to stress. These hormones, which are amino-acid derivatives, prepare the body for emergency action. They increase the heart and respiratory rates, cause blood glucose levels to rise, and direct blood flow away from the digestive tract and toward the brain and muscles. The adrenal medulla is activated by the sympathetic nervous system, which prepares the body to respond to emergencies, as described in Chapter 33.

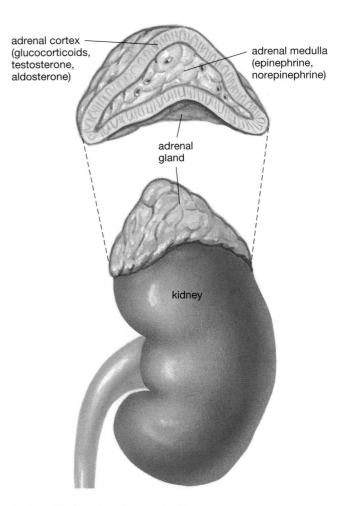

Figure 32-10 *The adrenal glands*
Atop each kidney sits an adrenal gland, which is a two-part gland composed of very dissimilar cells. The outer cortex consists of ordinary endocrine cells that secrete steroid homones. The inner medulla is derived from nervous tissue during development and secretes epinephrine and norepinephrine.

The outer layer of the adrenal gland forms the **adrenal cortex** (*cortex* is Latin for "bark"). The cortex secretes three types of steroid hormones, called **glucocorticoids**, synthesized from cholesterol. These hormones help control glucose metabolism. Glucocorticoid release is stimulated by ACTH from the anterior pituitary; ACTH release in turn is stimulated by releasing hormones produced by the hypothalamus in response to stressful stimuli, including trauma, infection, or exposure to temperature extremes. In some respects, the glucocorticoids act similarly to glucagon, raising blood glucose levels by stimulating glucose production and promoting the use of fats instead of glucose for energy production.

You may have noticed that many different hormones are involved in glucose metabolism: thyroxine, insulin, glucagon, epinephrine, and the glucocorticoids. Why? The reason can probably be traced to a metabolic requirement of the brain. Although most body cells can produce energy from fats and proteins as well as from carbohydrates, brain cells can burn only glucose. Thus, blood glucose levels cannot be allowed to fall too low, or brain cells rapidly starve, leading to unconsciousness and death.

The adrenal cortex also secretes **aldosterone**, which regulates the sodium content of the blood. Sodium ions, derived from salt in the diet, are the most abundant positive ions in blood and extracellular fluid. The sodium ion gradient across plasma membranes (high outside, low inside) is a factor in many cellular events, including the production of electrical signals by nerve cells. If blood sodium falls, the adrenal cortex releases aldosterone, which causes the kidneys and sweat glands to retain sodium. Then salt and other sources of dietary sodium, combined with aldosterone-induced sodium conservation, raise blood sodium levels again, shutting off further aldosterone secretion.

Finally, the adrenal cortex produces the male sex hormone testosterone, although normally in much smaller amounts than the testes produce. The adrenal cortex produces this hormone in women as well as in men. Tumors of the adrenal medulla can lead to excessive testosterone release, causing masculinization of women. Many of the "bearded ladies" who once appeared in circus sideshows probably had this condition.

Many Types of Cells Produce Prostaglandins

Unlike most other hormones, which are synthesized by a limited number of cells, prostaglandins are produced by many, perhaps all, cells of the body. These hormones are modified fatty acids synthesized by the cell from membrane phospholipids. Several prostaglandins are known, and probably a great many more await discovery. One prostaglandin causes arteries to constrict and stops bleeding from the umbilical cords of newborn infants. Another prostaglandin works in conjunction with oxytocin during labor, stimulating uterine contractions. Prostaglandin-soaked vaginal suppositories are used to induce labor. Menstrual cramps are caused by the overproduction of uterine prostaglandins, stimulating uterine contractions.

Some prostaglandins cause inflammation (such as occurs in arthritic joints) and stimulate pain receptors. Drugs such as aspirin and ibuprofen, which inhibit prostaglandin synthesis, can provide relief. Other prostaglandins expand the air passages of the lungs; one day, asthma sufferers may benefit from research into this effect. Still others stimulate the production of the protective mucus that lines the stomach, a potential boon for ulcer patients. Research on this diverse and potent family of compounds is still in its infancy, but it promises many health benefits in the future.

Other Endocrine Organs Include the Pineal Gland, Thymus, Kidneys, Heart, and Digestive Tract

The **pineal gland** is located between the two hemispheres of the brain, just above and behind the hypothalamus (see Fig. 32-1). Named for its resemblance to a pine cone, the pineal gland is smaller than a pea. In 1646, the philosopher

René Descartes described it as "the seat of the rational soul." Since then, scientists have learned more about this organ, but many of its functions are still poorly understood. The pineal produces the hormone **melatonin**, an amino-acid derivative. Melatonin is secreted in a daily rhythm, which in mammals is regulated by the eyes. In some vertebrates, such as the frog, the pineal itself contains photoreceptive cells, and the skull above it is thin, so the pineal can detect sunlight and thus daylength. By responding to daylengths characteristic of different seasons, the pineal appears to regulate the seasonal reproductive cycles of many mammals. Despite years of research, the function of the pineal gland in humans and of melatonin is still unclear. One hypothesis is that the pineal gland and melatonin secretion influence sleep–wake cycles. Improper pineal function may contribute to the depression that some people experience during the short days of winter.

The **thymus** is located in the chest cavity behind the breastbone, or sternum (see Fig. 32-1). In addition to producing white blood cells, the thymus produces the hormone **thymosin**, which stimulates the development of specialized white blood cells (T cells) that play an important role in the immune system (see Chapter 31). The thymus is extremely large in infants but, under the influence of sex hormones, begins decreasing in size after puberty.

The kidney, which plays a central role in maintaining body fluid homeostasis, has recently been recognized as an important endocrine organ as well. When the oxygen content of the blood drops, the kidney produces the hormone **erythropoietin**, which increases red blood cell production (see Chapter 27). The kidney also produces a second hormone, **renin**, in response to low blood pressure, such as may be caused by bleeding. Renin is an enzyme that catalyzes the production of the hormone **angiotensin** from proteins in the blood. Angiotensin raises blood pressure by constricting arterioles. It also stimulates aldosterone release by the adrenal cortex, causing the kidneys to retain sodium, which in turn increases blood volume.

The heart seems an unlikely endocrine organ, but in 1981 a substance that was extracted from heart atrial tissue and injected into rats was found to cause an increase in the output of salt and water by the kidneys. Two years later, the active substance, **atrial natriuretic peptide (ANP)**, was described and its amino acid sequence determined. This peptide is released by cells in the atria of the heart when blood volume increases. Increased blood volume causes extra distension of the heart. Atrial natriuretic peptide causes a reduction of blood volume by decreasing the release of both ADH and aldosterone. The pace of modern biomedical research is astonishing: Only 5 years after the factor was first detected in heart extract, clinical trials of synthetic ANP were started to test its value in treating high blood pressure and related problems.

The stomach and small intestine produce a variety of peptide hormones that help regulate digestion. These hormones include **gastrin**, **secretin**, and **cholecystokinin**, discussed in Chapter 29.

Evolutionary Connections

The Evolution of Hormones

Not long ago, vertebrate endocrine systems were considered unique to our phylum, and the endocrine chemicals were thought to have evolved expressly for their role in vertebrate physiology. In recent years, however, physiologists have discovered that hormones are evolutionarily ancient. Insulin, for example, is found not only in vertebrates but also in protists, fungi, and bacteria, although research has not yet determined the function of insulin in most of those organisms. Protists also manufacture ACTH, even though they have no adrenal glands to stimulate. Yeasts have receptors for estrogen but no ovaries. Thyroid hormones have been found in certain invertebrates, such as worms, insects, and mollusks, as well as in vertebrates. Even among vertebrates, the effects of chemically identical hormones, secreted by the same glands, may vary dramatically from organism to organism. Let's look briefly at the diverse effects that the thyroid hormone thyroxine has on several different organisms.

Some fish undergo radical physiological changes during their lifetimes. A salmon, for example, begins life in fresh water, migrates to the ocean, and returns to fresh water to spawn. In the stream where the salmon hatched, fresh water tends to enter the fish's tissues by osmosis; in salt water, the fish tends to lose water, becoming dehydrated. The salmon's migrations, therefore, require complete revamping of salt and water control. In salmon, one of the functions of thyroxine is to produce the metabolic changes necessary to go from life in streams to life in the ocean and back.

In amphibians, thyroxine has the dramatic effect of triggering metamorphosis. In 1912, in one of the first demonstrations of the action of any hormone, tadpoles were fed minced horse thyroid. As a result, the tadpoles metamorphosed prematurely into miniature adult frogs (Fig. 32-11). In high mountain lakes in Mexico, where the water is deficient in the iodine needed to synthesize thyroxine, natural selection has produced one species of salamander that has the ability to reproduce while still in its juvenile form.

Thyroxine regulates the seasonal molting of most vertebrates. From snakes to birds to the family dog, surges of thyroxine stimulate the shedding of skin, feathers, or hair. In humans (who neither migrate regularly, metamorphose, nor molt), thyroxine regulates growth and metabolism.

The use of chemicals to regulate cellular activity is extremely ancient. The diversity of life on Earth rests upon a conservative foundation: A relative handful of chemicals coordinate activities within single cells and among groups of cells. Life's diversity originated in part by changing the systems used to deliver the chemicals and by evolving new types of responses. Early in their evolution, animals developed a complement to hormonal communication that provides faster, more precise delivery of chemical messages: the nervous system. As we explain in the next chapter, the nervous system permits rapid responses to environmental stimuli, flexibility in response options, and ultimately consciousness itself.

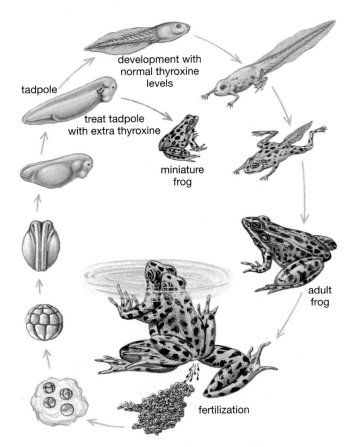

Figure 32-11 *Thyroxine controls metamorphosis in amphibians*
The life cycle of the frog begins with fertilization of the eggs (bottom); then development into an aquatic, fishlike tadpole; growth of the tadpole; and ultimately metamorphosis into an adult frog. Metamorphosis is triggered by a surge of thyroxine from the tadpole's thyroid gland. If injected with extra thyroxine, a young tadpole will metamorphose ahead of schedule into a miniature adult frog.

Summary of Key Concepts

1) What Are the Characteristics of Animal Hormones?

A hormone is a chemical secreted by cells in one part of the body that is transported in the bloodstream to another part of the body, where it affects the activity of specific target cells. Four types of molecules are known to act as hormones: peptides, amino-acid derivatives, steroids, and prostaglandins.

Most hormones act on their target cells in one of two ways: (1) Peptide hormones and amino-acid derivatives bind to receptors on the surfaces of target cells and activate intracellular second messengers, such as cyclic AMP. The second messengers then alter the metabolism of the cell. (2) Steroid hormones diffuse through the plasma membranes of the target cells and bind with receptor proteins in the cytoplasm. The hormone–receptor complex travels to the nucleus and promotes the transcription of specific genes. Thyroid hormones also penetrate the plasma membrane but diffuse into the nucleus, where they bind to receptors associated with the chromosomes and influence gene transcription.

Hormone action is commonly regulated through negative feedback, a process in which a hormone causes changes that inhibit further secretion of that hormone.

2) What Are the Structures and Functions of the Mammalian Endocrine System?

Hormones are produced by endocrine glands, which are clusters of cells embedded within a network of capillaries.

Hormones are secreted into the extracellular fluid and diffuse into the capillaries. The major endocrine glands of the human body are the hypothalamus–pituitary complex, the thyroid and parathyroid glands, the pancreas, the sex organs, and the adrenal glands. The hormones released by these glands and their actions are summarized in Table 32-2. Prostaglandins, unlike other hormones, are not secreted by discrete glands but are synthesized and released by many cells of the body. Other endocrine organs include the pineal gland, thymus, kidneys, heart, and the stomach and small intestine.

Key Terms

adrenal cortex *p. 649*
adrenal gland *p. 649*
adrenal medulla *p. 649*
adrenocorticotropic hormone (ACTH) *p. 643*
aldosterone *p. 650*
androgen *p. 649*
angiotensin *p. 650*
anterior pituitary *p. 642*
antidiuretic hormone (ADH) *p. 645*
atrial natriuretic peptide (ANP) *p. 650*
calcitonin *p. 645*
cholecystokinin *p. 650*
cyclic AMP *p. 639*
diabetes mellitus *p. 648*
duct *p. 642*

endocrine gland *p. 642*
endocrine system *p. 637*
endorphin *p. 643*
epinephrine *p. 649*
erythropoietin *p. 650*
estrogen *p. 649*
exocrine gland *p. 641*
follicle-stimulating hormone (FSH) *p. 643*
gastrin *p. 650*
glucagon *p. 647*
glucocorticoid *p. 649*
goiter *p. 646*
growth hormone *p. 643*
hormone *p. 637*
hypothalamus *p. 642*
inhibiting hormone *p. 644*
insulin *p. 647*

islet cell *p. 647*
luteinizing hormone (LH) *p. 643*
melanocyte-stimulating hormone (MSH) *p. 643*
melatonin *p. 650*
neurosecretory cell *p. 642*
norepinephrine *p. 649*
ovary *p. 649*
oxytocin *p. 645*
pancreas *p. 646*
parathormone *p. 646*
parathyroid gland *p. 646*
peptide hormone *p. 638*
pineal gland *p. 650*
pituitary gland *p. 642*
posterior pituitary *p. 642*
progesterone *p. 649*

prolactin *p. 643*
prostaglandin *p. 638*
releasing hormone *p. 644*
renin *p. 650*
second messenger *p. 639*
secretin *p. 650*
steroid hormone *p. 638*
target cell *p. 637*
testis *p. 649*
testosterone *p. 649*
thymosin *p. 650*
thymus *p. 650*
thyroid gland *p. 645*
thyroid-stimulating hormone (TSH) *p. 643*
thyroxine *p. 645*

Thinking Through the Concepts

Multiple Choice

1. *Steroid hormones*
   a. alter the activity of genes
   b. trigger rapid, short-term responses in cells
   c. work via second messengers
   d. initiate open channels in plasma membranes
   e. bind to cell-surface receptors

2. *Examples of posterior pituitary hormones are*
   a. FSH and LH
   b. prolactin and parathormone
   c. secretin and cholecystokinin
   d. melatonin and prostaglandin
   e. ADH and oxytocin

3. *Negative feedback to the hypothalamus controls the level of _____ in the blood.*
   a. thyroxine   b. estrogen
   c. glucocorticoids   d. insulin
   e. all of the above

4. *The primary targets for FSH are cells in the*
   a. hypothalamus
   b. ovary
   c. thyroid gland
   d. adrenal medulla
   e. pituitary gland

5. *The kidney is a source of*
   a. thyroxine and parathormone
   b. calcitonin and oxytocin
   c. renin and erythropoietin
   d. ANP and epinephrine
   e. glucagon and glucocorticoids

6. *Hormones that are produced by many different body cells and cause a variety of localized effects are known as*
   a. peptide hormones   b. parathormones
   c. releasing hormones   d. prostaglandins
   e. exocrine hormones

? Review Questions

1. What are the four types of molecules used as hormones in vertebrates? Give an example of each.

2. What is the difference between an endocrine gland and an exocrine gland? Which type releases hormones?

3. When peptide hormones attach to target cell receptors, what cellular events follow? How do steroid hormones behave?

4. Diagram the process of negative feedback, and give an example of negative feedback in the control of hormone action.

5. What are the major endocrine glands in the human body, and where are they located?

6. Describe the structure of the hypothalamus–pituitary complex. Which pituitary hormones are neurosecretory? What are their functions?

7. Describe how releasing hormones regulate the secretion of hormones by cells of the anterior pituitary. Name the hormones of the anterior pituitary, and give one function of each.

8. Describe how the hormones of the pancreas act together to regulate the concentration of glucose in the blood.

9. Compare the adrenal cortex and adrenal medulla by answering the following questions: Where are they located within the adrenal gland? What are their embryological origins? Which hormones do they produce? Which organs do their hormones target? What homeostatic processes regulate blood levels of the respective hormones?

Applying the Concepts

1. A student decides to do a science project on the effect of the thyroid gland on frog metamorphosis. She sets up three aquaria with tadpoles. She adds thyroxine to the water of one, the drug thiouracil to a second, and nothing to the third. Thiouracil reacts with thyroxine inside tadpoles to produce an ineffective compound. Assuming that the student uses appropriate physiological concentrations, predict what will happen.

2. Diabetes mellitus is common diabetes. Two other forms, adrenal diabetes and pituitary diabetes, are also characterized by high levels of blood glucose. What specific parts of regulatory cycles are disrupted in these latter forms to cause hyperglycemia (abnormally high sugar content in the blood)?

3. Suggest a hypothesis about the endocrine system to explain why many birds lay their eggs in the spring and why poultry farmers keep lights on at night in their egg-laying operations.

4. Anabolic steroids, used by risk-taking athletes and bodybuilders, are chemically related to testosterone. They increase bone mass and muscle mass and seem to improve athletic performance. But anabolic steroids can cause liver problems, heart attacks, strokes, testicular atrophy, and personality changes in males. Females on anabolic steroids also have liver and circulatory problems. In addition, their voices deepen, their bodies develop more hair, and their menstrual cycles are disturbed. Explain how the same compound can produce different effects in males and females.

5. Some parents who are interested in college sports scholarships for their children are asking physicians to prescribe growth hormone treatments. Farmers also have an economic incentive to treat cows with growth hormone, which can now be produced in large quantities by genetic-engineering techniques. What biological and ethical problems do you foresee for parents, children, physicians, coaches, college scholarship boards, food consumers, farmers, the U.S. Food and Drug Administration, and biotechnology companies?

Group Activity

In 1994, researchers discovered that if young mice are injected with a protein called *leptin*, they begin puberty earlier than do mice that received no leptin. On the basis of this finding, the researchers believe that the activation of genes that encode leptin may be the first step in the onset of puberty. Receptors for leptin have been found on cells in the hypothalamus, but they've also been found on cells in the reproductive organs of both males and females. Which receptors are responsible for leptin's role in initiating puberty? Form a group of four students to begin answering this question. Two members of your group should develop a hypothetical mechanism by which leptin could initiate puberty by binding with receptors on the hypothalamus. Your proposed mechanism should be as detailed as possible (specify all glands, organs, and hormones that are involved, and make a drawing that shows all pathways and regulatory feedback mechanisms). The other two members of your group should develop an equally detailed hypothesis of how leptin initiates puberty by binding to cells in the reproductive system. After your two competing hypotheses are complete, reconvene the group. Discuss the strengths and weaknesses of each hypothesis, and agree on one key experiment that, if done, would enable you to determine which hypothesis is most likely to be correct.

For More Information

Atkinson, M., and MacLaren, N. "What Causes Diabetes?" *Scientific American*, July 1991. An in-depth look at what's behind a common but serious disease.

Beardsley, T. "Melatonin Mania." *Scientific American*, April 1996. The hormone melatonin has been touted as a treatment for ailments ranging from insomnia to cancer. What do scientists think of the evidence?

Berridge, M. J. "The Molecular Basis of Communication within the Cell." *Scientific American*, October 1985. Both the "classical" cyclic AMP and more-recently discovered second messengers convey information from cell-surface receptors to DNA and cellular metabolism.

Gold, C. "Hormone Hell." *Discover*, September 1996. Environmental pollutants that mimic the chemical activity of hormones may be playing havoc with animal (including human) physiology.

McLachlan, J., and Arnold, S. "Environmental Estrogens." *American Scientist*, September–October 1996. A sober evaluation of the effects of estrogenlike substances in the environment, written by two of the leading researchers in the field.

Sapolsky, R. "Testosterone Rules." *Discover*, March 1997. An engaging inquiry into the question of whether testosterone causes aggression.

Snyder, S. H. "The Molecular Basis of Communication between Cells." *Scientific American*, October 1985. The similarities and differences between neural and hormonal control systems in the body.

Vogel, S. "The Mouse on the Left Needs Leptin." *Discover*, January 1996. The hormone leptin regulates appetite and fat deposition in mice. Can it become a weight-loss treatment for humans?

Answers to Multiple-Choice Questions
1. a   2. e   3. e   4. b   5. c   6. d

"Know thyself."

—Inscribed above the entrance to the Temple of Apollo, home of the Oracle of Apollo at Delphi

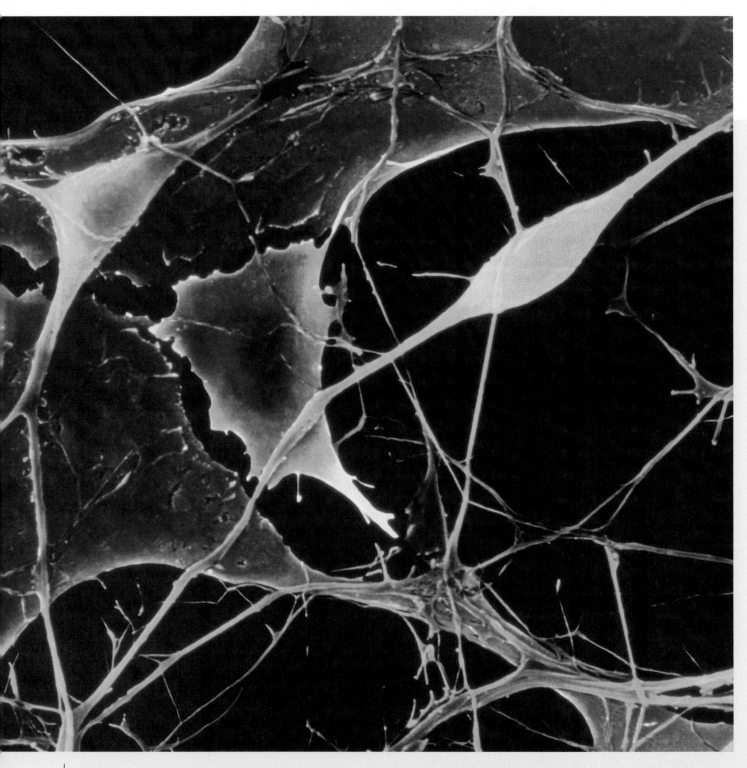

Color-enhanced scanning electron micrograph of neurons (yellow), from the human cerebral cortex of the brain, that are growing on glial cells (green) in a culture dish. The threadlike structures projecting from the neurons are dendrites and axons, which receive and carry electrical signals.

The Nervous System and the Senses 33

At a Glance

 Net Watch
On-line resources for this chapter are on the World Wide Web at: http://www.prenhall.com/audesirk
(click on the Table of Contents link and then select Chapter 33).

To read this page, your eyes must process the light reflected from it in complex ways. Your brain must decode the signals it receives from your eyes into distinct images, then translate these images into words and interpret the words' meanings. Sound waves from your stereo enter your ears and are changed to neural signals and interpreted as music. Your senses of smell and taste are simultaneously stimulated as you drink coffee. You breathe, swallow, and shift positions in your chair. Untaxed by these marvelous but ordinary activities, your mind directs dexterous manipulations of a pen as you take notes, stores memories of what you have read, and still has the capacity to appreciate the music. Suddenly, a piercing screech from the hall outside your dorm room galvanizes your entire body. You feel your muscles tense and your heart and breathing rate increase as you open the door to an intensified noise and the blinking of the fire-alarm lights. Tense and annoyed (because it is probably another false alarm), you join your dorm mates for the trek outside.

What type of specialized cells and signals does your nervous system use to communicate information within your body? How do your ears, eyes, nose, and tongue respond to the various sensory stimuli of music and fire alarm, light, and coffee? What in your brain produced the momentary sense of fear when you were startled by the loud noise? What part of your nervous system caused your heart and breathing rate to accelerate when you heard the alarm? How do your ears and brain distinguish a loud noise from soft music? Before the fire alarm rudely interrupted you, why were you able to concentrate on your reading with so many other stimuli surrounding you?

In this chapter we explore the fascinating complexity of the nervous system and the senses that supply it with information.

1) How Do Nervous and Endocrine Communication Differ?

Although it is convenient to discuss hormonal control (Chapter 32) separately from nervous control, in some ways the two are quite similar. Both hormone-producing cells and nerve cells synthesize "messenger" chemicals that they release into extracellular spaces (Fig. 33-1). However, there are four main differences in how the nervous and endocrine systems use chemical messages. (1) Whereas hormone-producing and neurosecretory cells (such as those found in the hypothalamus; see Chapter 32) release hormones into the bloodstream, nerve cells usually release their chemical messengers (called *neurotransmitters*) very close to the cells they influence, commonly from less than a micrometer away. (2) Whereas blood-borne hormones bathe millions of cells indiscriminately, a nerve cell releases its neurotransmitter onto one or a small group of specific target cells. (3) A nerve cell speeds information from one part of the body to another by means of electrical signals that travel within the cell itself, and it releases its neurotransmitter only when it reaches its target. Hormones move much more slowly and are released at varying distances from the target cell. (4) The effects of messages sent by nerve cells called *neurons* tend to be of much shorter duration than are the effects of hormones.

Yet even these differences blur as we learn more about the hormonal and nervous systems of animals. Some hormones are in fact produced and released into the bloodstream by nerve cells, such as the neurosecretory cells of the hypothalamus (described in Chapter 32). Other chemicals that biologists formerly believed were strictly hormones, such as insulin, have recently been discovered in the brain, where they are synthesized and released by neurons and act more like neurotransmitters. As you read

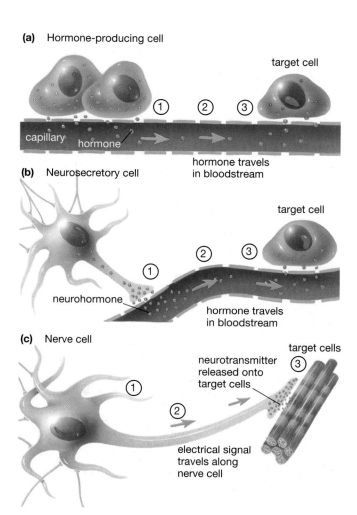

(a) Hormone-producing cell

target cell

capillary   hormone

hormone travels in bloodstream

(b) Neurosecretory cell

target cell

neurohormone

hormone travels in bloodstream

(c) Nerve cell

target cells

neurotransmitter released onto target cells

electrical signal travels along nerve cell

Figure 33-1 Three major types of "control cells"
All release chemicals that influence the activity of other cells of the body: **(a)** hormone-producing cells, **(b)** neurosecretory cells, and **(c)** nerve cells. Each type involves three common steps: ① The chemical-releasing cell is stimulated. ② It sends a message to distant cells. ③ Selected target cells respond. Hormone-producing cells and neurosecretory cells release their chemical messages into the bloodstream, which transports the chemicals to distant target cells. Nerve cells grow long cell extensions to their target cells and release their chemical messages directly onto the target cell.

about the nervous system in this chapter, remember that no system in your body works alone. The endocrine and nervous systems are closely coordinated in their control of bodily functions.

2) What Are the Structures and Functions of Neurons?

Our study of the nervous system begins with the individual nerve cell, or **neuron**. As the fundamental unit of the nervous system, each neuron must perform four specialized functions:

1. Receive information from the internal or external environment or from other neurons.
2. Integrate the information it receives and produce an appropriate output signal.
3. Conduct the signal to its terminal ending, which may be some distance away.
4. Transmit the signal to other nerve cells, glands, or muscles.

Each neuron also carries on normal metabolic activities.

Although neurons vary enormously in structure, a "typical" vertebrate neuron has four distinct structural regions that carry out the functions above. These regions are the *dendrites*, the *cell body*, the *axon*, and the *synaptic terminals* (Fig. 33-2).

Dendrites, branched tendrils that extend outward from the nerve cell body, are specialized to respond to signals from other neurons or from the external environment. In other words, they perform the first function listed at left. Their branched form provides a large surface area to receive these signals. Dendrites of *sensory neurons* (neurons that respond to a stimulus) have special membrane adaptations that allow them to respond to specific stimuli from the environment, such as pressure, odorous molecules, light, or heat. In neurons of the brain and spinal cord, dendrites respond to the chemical neurotransmitters released by other neurons. These dendrites have protein receptors in their membranes that bind specific neurotransmitters and produce electrical signals as a result of this binding. These signals are weak, do not travel very far, and may be positive or negative.

Electrical signals travel down the dendrites and converge on the neuron's **cell body**, which serves as an integration center. In its integrating role, the cell body adds up the various positive and negative signals from the dendrites (fulfilling the second function listed at left). If the sum of all these signals is sufficiently positive, the neuron will produce an **action potential**, the electrical output signal of the neuron. The cell body, containing the usual assortment of organelles, also carries on activities common to most other body cells. These activities include the synthesis of complex molecules such as proteins, lipids, and carbohydrates and the coordination of the metabolic activities of the cell.

In a typical neuron, a long, thin fiber called an **axon** extends outward from the cell body. The action potential begins where the axon leaves the cell body, at a site called the **spike initiation zone**. (Action potentials are often called "spikes" because of their shape on a recording device.) Single axons may stretch from your spinal cord to your toes, a distance of about a meter (about 3 feet), making neurons the longest cells in the body. Axons are distribution lines, carrying action potentials from the cell body to the *synaptic terminals*, located at the far end of each axon—the third function listed at left.

Axons are normally bundled together into **nerves**, much like bundles of wires in an electrical cable. However, unlike electrical power distribution cables (in which energy is lost along the way from power station to customer), the plasma membranes of axons conduct action potentials undiminished from the cell body to their synaptic terminals. Some axons are wrapped with insulation called **myelin**, which allows more-rapid conduction of the electrical signal. Myelin is formed from non-neuronal

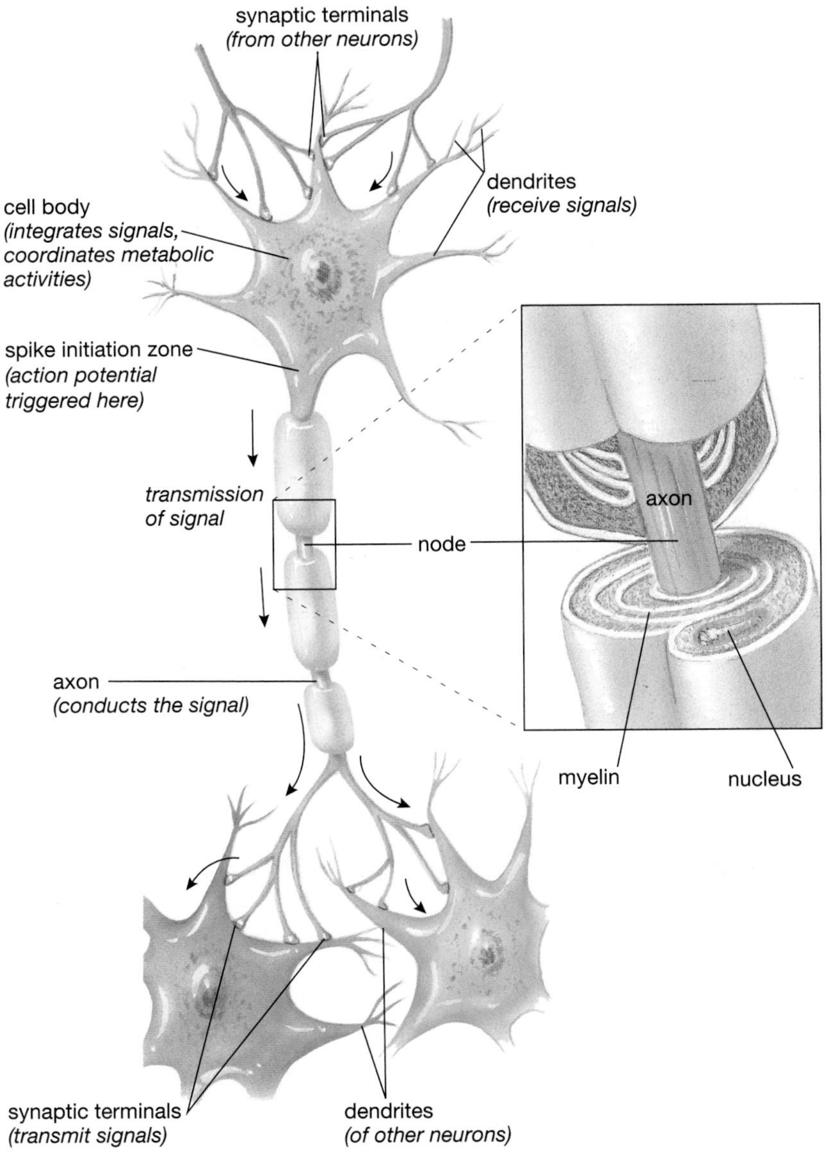

cell body
(integrates signals,
coordinates metabolic
activities)

synaptic terminals
(from other neurons)

dendrites
(receive signals)

spike initiation zone
(action potential
triggered here)

transmission
of signal

node

axon

axon
(conducts the signal)

myelin

nucleus

synaptic terminals
(transmit signals)

dendrites
(of other neurons)

Figure 33-2 A nerve cell, showing its specialized parts and their functions

cells that wrap around the axon (Fig. 33-2). In vertebrates, axons bundled into nerves emerge from the brain and spinal cord and extend out to all regions of the body.

Signals are transmitted to other cells at **synaptic terminals**, swellings at the branched endings of axons (see Fig. 33-2). Most synaptic terminals contain a **neurotransmitter**, a specific type of chemical that they release in response to the passage of an action potential down the axon. The synaptic terminals of one neuron may communicate with a gland, a muscle, or the dendrites or cell body of a second neuron—the fourth function listed on p. 657. The output of the first cell becomes the input to the second. The site at which synaptic terminals communicate with other cells is called the **synapse**.

3) How Is Neural Activity Generated and Communicated?

WWW

Neurons Create Electrical Signals across Their Membranes

In the early 1950s, biologists using the giant axon of a squid—a mollusk—developed ways to record electrical events inside individual neurons (Fig. 33-3). The researchers found that unstimulated, inactive neurons maintain a constant electrical difference, or *potential*, across their plasma membranes, similar to that across the poles of a battery. As in a battery, the electrical potential across a neuron membrane stores energy. This potential, called the **resting potential**, is always negative inside the cell and ranges from −40 to −90 millivolts (thousandths of a volt).

If the neuron is stimulated, either naturally or with an electrical current, the negative potential inside the neuron can be altered. Depending on the nature of the stimulus, the inside potential can be made either more or less negative (Fig. 33-4). If the potential is made sufficiently less negative, it reaches a level called **threshold** (roughly 15 millivolts less negative than the resting potential). At threshold, an action potential is triggered at the spike initiation zone. During the action potential, the neuron's potential rapidly rises to become about +50 millivolts inside. Action potentials last a few milliseconds (thousandths of a second) before the cell's negative resting potential is restored (see Fig. 33-4). The positive charge of the action potential flows rapidly down the axon to the synaptic terminal, where the signal is communicated to another cell at a synapse. In "A Closer Look: Ions and Electrical Signals" (pp. 660–661), we examine these electrical potentials, the language of the nervous system.

Neurons Communicate at Synapses

Once an action potential has been conducted to the synaptic terminal of the neuron, the signal must be transmitted to another cell, normally another neuron. This transmission occurs at synapses. The signals transmitted at synapses are called **postsynaptic potentials** (PSPs).

(a)

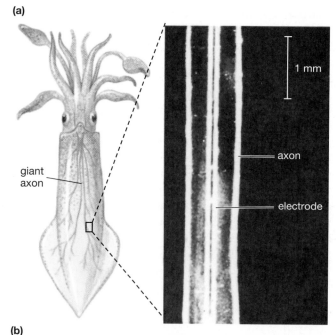

(b)

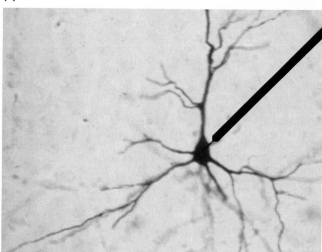

Figure 33-3 Some ways of studying neurons
(a) Using the giant axon of the squid, the British physiologists Bernard Katz, Alan Hodgkin, and Andrew Huxley pushed electrodes (thin wires or narrow tubes) down the inside of the axon. These electrodes were connected to voltmeters to record the electrical potential difference between the inside and outside of the axon, about −70 millivolts.
(b) Modern neurobiologists use hollow glass electrodes drawn to a needlelike tip less than 1 micrometer in diameter and filled with a salt solution. The sharp tip penetrates the neuron without damaging it.

At a synaptic terminal, an action potential encounters a synapse, where parts of two neurons are close together and are specialized to communicate with one another. The **synaptic cleft**, a tiny gap, separates the synaptic terminal of the first neuron, the **presynaptic neuron**, from the second, or **postsynaptic neuron** (Fig. 33-5). Both the

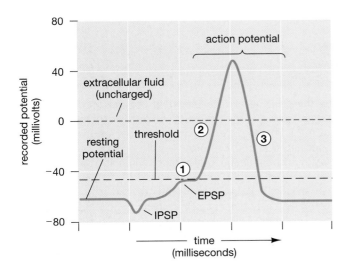

Figure 33-4 *The electrical events during an action potential*
A recording (by oscilloscope) of the electrical events in a nerve cell. The resting potential is about 60 millivolts negative with respect to the outside. ① When the cell is stimulated to reach threshold by a postsynaptic potential, membrane channels permeable to sodium open up and sodium enters the cell, powered by diffusion and by electrical attraction; ② the inside of the cell becomes positively charged. ③ Shortly thereafter, other membrane channels permeable to potassium open and potassium leaves, driven by diffusion and electrical repulsion from the now-positive inside of the cell, until the resting potential is reestablished.

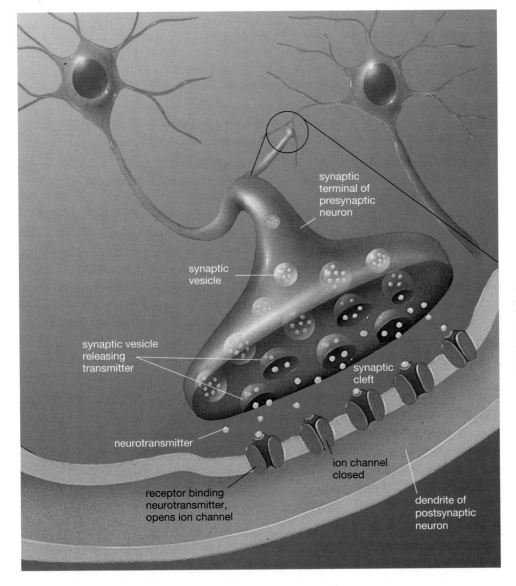

Figure 33-5 *The structure and operation of the synapse*
The synaptic terminal contains numerous vesicles that enclose neurotransmitter, for which the postsynaptic neuron has membrane receptors. When an action potential enters the synaptic terminal of the presynaptic neuron, the vesicles dump their neurotransmitter into the synaptic cleft—the space between the neurons. The neurotransmitter diffuses rapidly across the space, binds to the postsynaptic receptors, and causes ion channels to open. Ions flow through these open channels, causing a postsynaptic potential in the postsynaptic cell.

A Closer Look
Ions and Electrical Signals

How Is the Resting Potential Generated?

The resting potential is based on a balance between chemical and electrical gradients, maintained by active transport and by a membrane that is selectively permeable to specific ions. The ions of the cytoplasm consist mainly of positively charged potassium ions (K^+) and large, negatively charged organic molecules such as proteins, which cannot leave the cell (Fig. E33-1). Out-side the cell, the extracellular fluid contains more positively charged sodium ions (Na^+) and negatively charged chloride ions (Cl^-). These concentration differences are maintained by a specialized membrane protein called a *sodium–potassium pump*, which simultaneously pumps K^+ into and Na^+ out of the cell.

In an unstimulated neuron, only K^+ can cross the plasma membrane. The ion does so by traveling through specific membrane proteins called *potassium channels* (shown in yellow). Although *sodium channels* (shown in purple) are also present, they remain closed in unstimulated neurons. Because the K^+ concentration is higher inside the cell than outside, that ion tends to diffuse out, leaving the negatively charged organic ions behind:

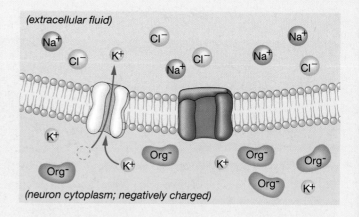

(extracellular fluid)

(neuron cytoplasm; negatively charged)

Figure E33-1 The neuron maintains ionic gradients
The ionic composition of a neuron's cytoplasm is significantly different from that of the extracellular fluid. The neuron maintains high concentrations of K^+ and large organic ions (Org^-); the extracellular fluid is high in Na^+ and Cl^-.

The outward diffusion of K^+ stops when the negative charge (which tends to pull the K^+ back inside the cell) becomes sufficiently large to counteract the diffusion gradient across the membrane. This negative charge at rest is the neuron's resting potential.

dendrites and the cell bodies of neurons are typically covered with synapses.

When an action potential reaches a synaptic terminal, the inside of the terminal becomes positively charged. This charge triggers storage vesicles in the synaptic terminal to release a chemical neurotransmitter into the synaptic cleft. The neurotransmitter molecules rapidly diffuse across the gap and bind briefly to receptors in the membrane of the postsynaptic neuron before diffusing away or being taken back up into the pre-synaptic neuron (see Fig. 33-5). The synapse includes the synaptic terminal of the presynaptic cell, the synaptic cleft, and the specialized membrane of the postsynaptic cell just across the cleft that contains receptors for neurotransmitters.

Excitatory or Inhibitory Potentials Are Produced at Synapses and Integrated in the Cell Body

Receptor proteins in the postsynaptic membrane bind to a specific type of neurotransmitter. This binding causes specific types of ion channels in the postsynaptic mem-

Action Potentials Can Carry Messages Rapidly over Long Distances

Neuron signals are carried long distances by action potentials. Action potentials occur if electrical signals from other neurons bring the potential at the spike initiation zone to threshold. At threshold, Na⁺ channels (purple) are triggered to open, allowing a rapid influx of Na⁺ (① in the figure below). Soon after the Na⁺ channels are opened, they spontaneously close and additional K⁺ channels (orange) are triggered to open by the positive charge inside the axon, allowing more K⁺ to flow out of the cell and restore the negative resting potential (②):

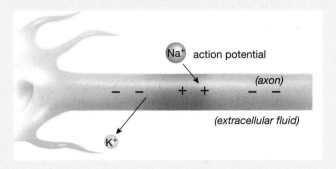

As the wave of positive charges passes a given point along the axon, the resting potential is restored as K⁺ flows out:

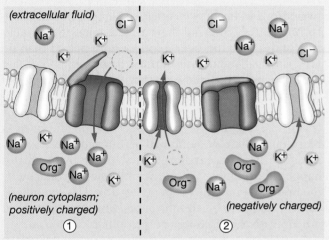

An action potential resembles a fast-moving wave of positive charge that travels, undiminished in size, along the axon to the synaptic terminal. The positive charge carried into the axon by Na⁺ at the spike initiation zone causes Na⁺ channels farther along the axon to open. More Na⁺ can then flow in, and the process continues:

Action potentials are "all-or-none"—that is, if the neuron does not reach threshold, there will be no action potential, but if threshold is reached, a full-sized action potential will occur and travel the entire length of the axon.

A tiny fraction of the total potassium and sodium in and around each neuron is exchanged during each action potential. The gradients of Na⁺ and K⁺ are maintained by the sodium–potassium pump.

brane to open, allowing ions to flow across the plasma membrane along their concentration gradients. The flow of ions in the postsynaptic neuron causes a small, brief change in electrical charge, the postsynaptic potential. Depending on what type of channels are opened and what type of ions flow, postsynaptic potentials can be either *excitatory* (EPSPs), making the neuron less negative inside and more likely to produce an action potential, or *inhibitory* (IPSPs), making it more negative and less likely to produce an action potential (see Fig. 33-4). Postsynaptic potentials are small, rapidly fading signals, but they travel far enough to

reach the cell body. There they determine whether an action potential will be produced, as described below.

The dendrites and cell body of a single neuron can receive EPSPs and IPSPs from the synaptic terminals of thousands of presynaptic neurons. The postsynaptic potentials produced by different presynaptic neurons are then "added up," or **integrated**, in the cell body of the postsynaptic neuron. The postsynaptic cell will produce an action potential only if the excitatory and inhibitory potentials, when added together, raise the electrical potential inside the neuron above threshold.

The Nervous System Uses Many Neurotransmitters

Over the past few decades, researchers have become increasingly aware that the brain is a teeming cauldron; its neurons synthesize and respond to a vast array of chemicals, including many of the hormones once thought unique to the endocrine system. For example, hormones that control digestive-tract secretions are now known to be synthesized in the brain, where they influence appetite. At least 50 neurotransmitters and **neuropeptides** (small protein molecules with neurotransmitter-like actions) have been identified, and more are added to the list annually. In the following sections, we first discuss a few "classical" neurotransmitters that have been recognized for many years and whose roles are partially understood. Then we describe a few recently recognized neuropeptides released by neurons that alter the activity of groups of neurons over longer periods of time.

Acetylcholine

The neurotransmitter **acetylcholine** is found in many areas of the brain and is the only transmitter released at the synapses between neurons and skeletal muscles, where it is always excitatory. The drug curare, found in the skin of certain small South American frogs, blocks acetylcholine receptors. Natives of South America used the drug on arrow tips. By preventing muscle contraction, curare triggers paralysis and may cause death. Specific groups of acetylcholine-producing neurons have been shown to break down in the central nervous system of individuals with Alzheimer's disease.

Dopamine

The degeneration of neurons that produce the neurotransmitter **dopamine** in a specific brain region causes Parkinson's disease, characterized by difficulty in initiating conscious movements and by uncontrolled tremors. The drug *levodopa* (L-dopa) is used by unharmed neurons in the brain to synthesize dopamine, partially replacing dopamine that has been lost. Schizophrenics are commonly treated successfully with drugs that block dopamine receptors. Cocaine and amphetamines block the re-uptake of dopamine into the presynaptic neuron, causing the neurotransmitter to remain in the synapse and prolonging its effects. Abuse of these drugs can produce symptoms resembling schizophrenia.

Serotonin

The neurotransmitter **serotonin** acts in the brain and spinal cord. It can inhibit pain sensory neurons in the spinal cord; electrical devices that stimulate these neurons may be implanted in individuals with chronic pain. Serotonin is also believed to affect sleep and mood. Animals in which serotonin production is blocked are unable to sleep normally. Too little serotonin may cause depression. The antidepressant Prozac®, for example, selectively blocks re-uptake of serotonin into the presynaptic neuron, enhancing the neurotransmitter's effects.

Norepinephrine

Norepinephrine (also called *noradrenaline*) is chemically very similar to the hormone epinephrine (adrenaline), secreted by the adrenal glands. Noradrenaline is released onto many organs, such as the heart, digestive system, and lungs (by neurons of the sympathetic nervous system, which we shall discuss soon). When the fire alarm shrilled, norepinephrine was released onto nearly all your body organs, because the effects of norepinephrine prepare the body to respond to stressful situations. Both amphetamines and cocaine block norepinephrine uptake into the presynaptic neuron, prolonging the neurotransmitter's effects. Some sudden deaths among cocaine users are caused by overstimulation by excess norepinephrine.

Neuropeptides

Dozens of neuropeptides are synthesized and released by neurons. Many may be released along with other neurotransmitters, altering their effectiveness. Some neuropeptides function over a longer time period than do the "classical" neurotransmitters described above, and they may influence many neurons at once. Examples of neuropeptides are the **opioids** (literally, "opiate-like substances"), such as **endorphin**. For centuries the analgesic (pain-relieving) effects of plant-derived opiates, such as morphine, opium, codeine, and heroin, have been recognized. Perhaps, researchers reasoned, the plant opiates happen to resemble unknown substances produced by the brain that diminish pain perception. This hypothesis led to a search for such substances, which was rewarded in 1975 with the discovery of the opioids.

Certain opioids are responsible for the suppression of pain in times of extreme stress, such as on a battlefield (or a football field). Opioids released during strenuous exercise may account for the well-known "runner's high." The analgesic effects of acupuncture are apparently caused by its ability to stimulate the release of opioids. Opioids and other neuropeptides participate in the maintenance of body temperature and blood pressure. They have also been implicated in the regulation of behavioral states such as hunger, sexual excitement, anger, and depression, and they may even be involved in learning.

4) How Is the Nervous System Designed?

The individual neuron uses a language of action potentials. Yet somehow this basic language allows even simple animals to perform a variety of complex behaviors. One key to the versatility of the nervous system is the presence of complex networks of neurons. These neural networks range from thousands to billions of cells. As in computers, small, simple elements can perform amazing feats when connected properly.

Information Processing Requires Four Basic Operations

Before we delve into the basic anatomy of nervous systems, we should first examine the operating principles. At a minimum, a nervous system must be able to perform four operations:

1. Determine the type of stimulus.
2. Signal the intensity of a stimulus.
3. Integrate information from many sources.
4. Initiate and direct the response.

Let's examine each of these operations.

The Type of Stimulus Is Distinguished by Wiring Patterns in the Brain

The nervous system must be able to identify the type of stimulus—for example, light, touch, or sound. All action potentials are pretty much alike; their properties tell us nothing about the kind of stimulus that elicited them. Instead, the nervous system monitors *which* neurons are firing action potentials. Thus, your brain interprets action potentials that occur in the axons of your optic nerves (originating in the eye and traveling to a specific area of the brain) as the sensation of light, action potentials in olfactory nerves (originating in receptors in the nose and traveling to a different region of the brain) as odors, and so on. In this way you have no trouble distinguishing the action potentials caused by the music from your stereo from those caused by the bitter taste of coffee on your tongue.

However, this genetic wiring may occasionally yield false information. Being poked in the eye may trigger action potentials in the optic nerve, because of slight trauma. Even though the stimulus is mechanical, your brain interprets all action potentials that occur in the optic nerve as light, so you "see stars." For the same reason, a blow to the head can make your ears "ring."

The Intensity of a Stimulus Is Coded by the Frequency of Action Potentials

Because all action potentials are of roughly the same magnitude and duration, no information about the strength, or **intensity**, of a stimulus (for example, the loudness of a sound) can be encoded in a single action potential. Instead, intensity is coded in two other ways. First, intensity can be signaled by the frequency of action potentials in a single neuron. The more intense the stimulus, the faster the neuron produces action potentials, or *fires* (Fig. 33-6). Second, most nervous systems have many neurons that can respond to the same input. Stronger stimuli tend to excite more of these neurons, whereas weaker stimuli excite fewer. Thus, intensity can also be signaled by the number of similar neurons that fire at the same time (see Fig. 33-6). The loud wail of the fire alarm activated many of your neurons involved in hearing and caused them to fire action potentials very rapidly.

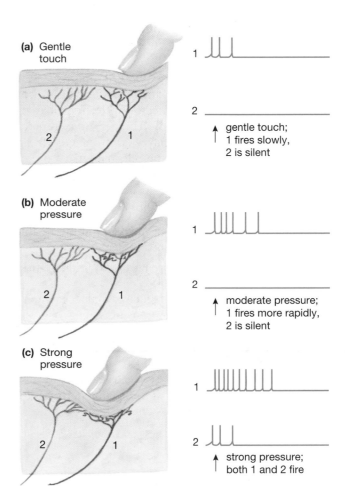

(a) Gentle touch

1

2

↑ gentle touch; 1 fires slowly, 2 is silent

(b) Moderate pressure

1

2

↑ moderate pressure; 1 fires more rapidly, 2 is silent

(c) Strong pressure

1

2

↑ strong pressure; both 1 and 2 fire

Figure 33-6 Signaling stimulus intensity
The intensity of a stimulus is signaled by the rate at which individual neurons produce action potentials and by the number of neurons that do so. Consider two touch receptors that have endings in adjacent patches of skin. *(a)* A gentle touch elicits only a few action potentials and from only one of the sensory neurons. *(b)* Moderate pressure stimulates only one receptor, but this receptor now fires faster, informing the brain that the touch is more intense than before. *(c)* Strong pressure activates both receptors, firing one very fast and the other more slowly, thus signaling to the brain that the pressure is very intense and localized over the fastest-firing receptor.

The Nervous System Processes Information from Many Sources through Convergence

Your brain is continuously bombarded by sensory stimuli that originate both inside and outside the body. The brain must filter all these inputs, determine which ones are important, and decide how to respond. Nervous systems integrate information much as do individual neurons, through **convergence**. In this process, many neurons funnel their signals to fewer neurons. For example, many sensory neurons may converge onto a smaller number of brain cells. The brain cells that might be considered to be "decision-making cells" add up the postsynaptic potentials that result from the synaptic activity of these sensory neurons; depending on their relative strengths (and other internal factors such as hormones or metabolic activity), they produce appropriate outputs.

Divergence of Signals Allows Complex Responses

The output of the decision-making cells is responsible for initiating activity. The actions directed by the brain may involve many parts of the body. These actions require **divergence**, the flow of electrical signals from a relatively small number of decision-making cells onto many different neurons that control muscle or glandular activity.

Neural Pathways Direct Behavior

Most behaviors are controlled by neuron-to-muscle pathways composed of four elements:

1. **Sensory neurons**, which respond to a stimulus, either internal or external to the body.
2. **Association neurons**, which receive signals from many sources, including sensory neurons, hormones, neurons that store memories, and many others. On the basis of this input, association neurons activate motor neurons.
3. **Motor neurons**, which receive instructions from association neurons and activate muscles or glands.
4. **Effectors**, normally muscles or glands that perform the response directed by the nervous system.

The Simplest Behavior Is the Reflex

The simplest type of behavior in animals is the **reflex**, a largely involuntary movement of a body part in response to a stimulus. Examples of human reflexes include the familiar knee-jerk and pain-withdrawal reflexes, both of which are produced by neurons in the spinal cord. The pain-withdrawal reflex (which moves a body part away from a painful stimulus, such as a hot stovetop) uses one of each of the three types of neurons and an effector (Fig. 33-7). Reflexes of this sort do not require the brain, although, as we know, other pathways inform the brain of pricked fingers and may in fact trigger other, more-complex behaviors (cursing, for example!).

Nearly all animals are capable of much more subtle and varied behavior than can be accounted for by simple reflexes. In principle, these more complex behaviors can be organized by *interconnected neural pathways* in which several types of sensory input (along with memories, hormones, and other factors) converge on a set of association neurons. By integrating the postsynaptic potentials from several sources, the association neurons can "decide" what to do and can stimulate the motor neurons to direct the appropriate activity in muscles and glands.

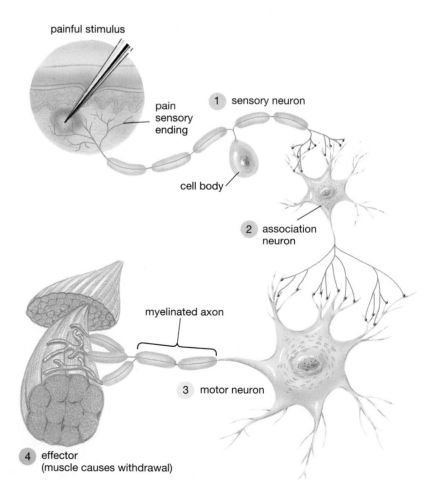

Figure 33-7 The neuron-to-muscle pathway for a simple behavior
The pain-withdrawal reflex illustrates the three general types of neurons and an effector, the muscle. The neurons vary in size and shape, a reflection of their varied functions.

Increasingly Complex Nervous Systems Are Increasingly Centralized

In all the animal kingdom, there are really only two nervous-system designs: (1) a diffuse nervous system, such as that of cnidarians (*Hydra*, jellyfish, and their relatives; Fig. 33-8a), and a centralized nervous system, found to varying degrees in more-complex organisms. Not surprisingly, nervous-system design is highly correlated with the animal's lifestyle. The radially symmetrical cnidarians have no "front end," so there has been no evolutionary pressure to concentrate the senses in one place. A *Hydra* sits anchored to the seafloor, and prey or other dangers are equally likely to come from any direction. Cnidarian nervous systems are composed of a network of neurons, often called a **nerve net**, woven through the animal's tissues. Here and there we find a cluster of neurons, called a **ganglion** (plural, **ganglia**), but nothing resembling a real brain.

Almost all other animals are bilaterally symmetrical, with definite head and tail ends. Because the head is the first part of the body to encounter food, danger, and potential mates, it is advantageous to have sense organs concentrated here. Sizable ganglia evolved that integrate the information gathered by the senses and initiate appropriate action. Over evolutionary time, the sense organs gathered in the head and the ganglia became centralized into a brain. This trend, called **cephalization**, is clearly seen in the mollusks (Fig. 33-8b). Cephalization reaches its peak in the vertebrates, in which nearly all the cell bodies of the nervous system are localized in the brain and spinal cord.

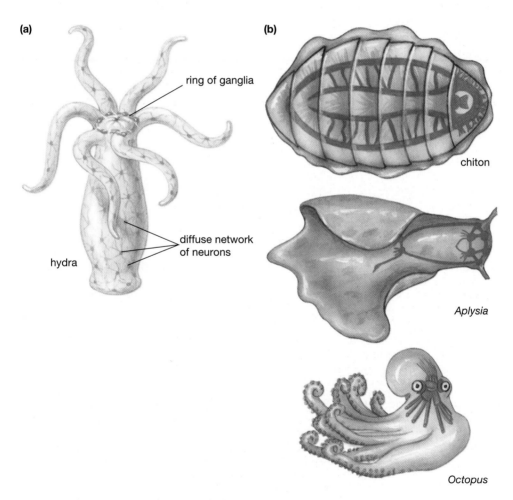

(a)

ring of ganglia

diffuse network of neurons

hydra

(b)

chiton

Aplysia

Octopus

Figure 33-8 Nerve net and cephalization
(a) The diffuse nervous system of *Hydra* contains a few concentrations of neurons, particularly at the bases of the tentacles, but no brain. Neural signals are conducted in virtually any direction throughout the body. *(b)* The trend toward increasing concentration of the nervous system in the head is illustrated by various mollusks. (top) The slow-moving chiton has had little environmental pressure for concentrating its sense organs and brain in the head. (middle) Some marine snails, such as the shell-less *Aplysia*, can crawl quite rapidly or even swim. Many of their neurons are aggregated into a brain. (bottom) Mollusk mobility and intelligence culminate in *Octopus*, with its large, complex brain and learning capabilities, which rival those of some mammals.

5) How Is the Human Nervous System Organized?

www

The human nervous system can be divided into two parts: (1) central and (2) peripheral. The **central nervous system** consists of a **brain** and a **spinal cord** that extends down the dorsal part of the torso (Fig. 33-9). The **peripheral nervous system** consists of nerves that connect the central nervous system to the rest of the body.

The Peripheral Nervous System Links the Central Nervous System to the Body

The peripheral nervous system consists of **peripheral nerves**, which link the brain and spinal cord to the rest of the body, including the muscles, the sensory organs, and the organs of the digestive, respiratory, excretory, and circulatory systems. Within the peripheral nerves are axons of sensory neurons that bring sensory information *to the* central nervous system from all parts of the body. Peripheral nerves also contain the axons of motor neurons that carry signals *from* the central nervous system to the organs and muscles.

The motor portion of the peripheral nervous system can be subdivided into two parts: (1) the **somatic nervous system** and (2) the **autonomic nervous system**. Motor neurons of the somatic nervous system form synapses on skeletal muscles and control voluntary movement. As you take notes, lift a coffee cup, and adjust your stereo, your somatic nervous system is in charge. The cell bodies of somatic motor neurons are located in the *gray matter* of the spinal cord (to be described shortly), and their axons go directly to the muscles they control. (Muscles and their control are discussed in Chapter 34.)

Motor neurons of the autonomic nervous system control involuntary responses. They form synapses on the heart, smooth muscle, and glands. The autonomic nervous system is controlled both by the *medulla* and the *hypothalamus* of the brain, described later in this chapter. It consists of two divisions: (1) the **sympathetic division** and (2) the **parasympathetic division** (Fig. 33-10). The two divisions of the autonomic nervous system generally make synaptic contacts with the same organs but normally produce opposite effects.

The sympathetic division acts on organs in ways that prepare the body for stressful or highly energetic activity, such as fighting, escaping, or giving a speech. During

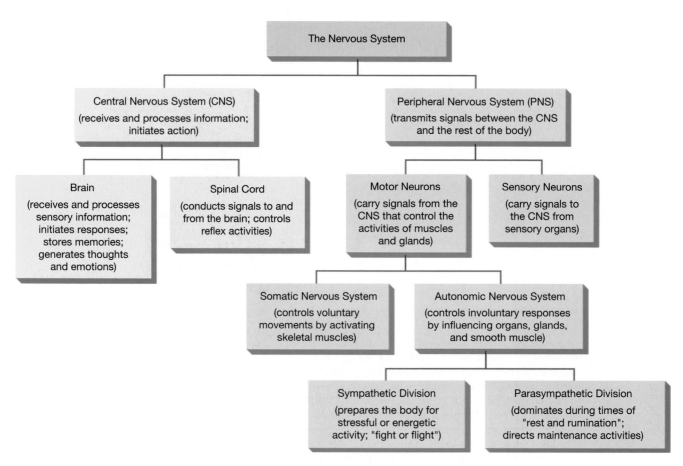

Figure 33-9 *The organization of the vertebrate nervous system*
This diagram shows the major parts of the vertebrate nervous system, their subdivisions, and functions.

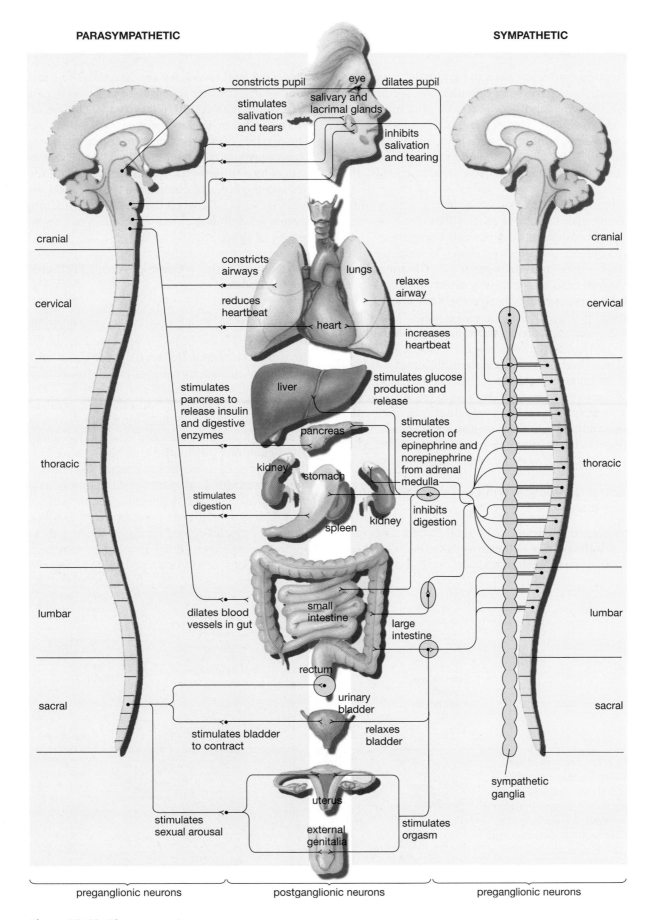

PARASYMPATHETIC

SYMPATHETIC

constricts pupil — eye — dilates pupil

stimulates salivation and tears

salivary and lacrimal glands

inhibits salivation and tearing

cranial

cervical

constricts airways

lungs

relaxes airway

reduces heartbeat

heart

increases heartbeat

stimulates glucose production and release

stimulates pancreas to release insulin and digestive enzymes

liver

pancreas

stimulates secretion of epinephrine and norepinephrine from adrenal medulla

kidney

stomach

stimulates digestion

spleen

kidney

inhibits digestion

thoracic

cervical

cranial

thoracic

dilates blood vessels in gut

small intestine

large intestine

lumbar

lumbar

sacral

rectum

urinary bladder

stimulates bladder to contract

relaxes bladder

sacral

sympathetic ganglia

stimulates sexual arousal

uterus

external genitalia

stimulates orgasm

preganglionic neurons — postganglionic neurons — preganglionic neurons

Figure 33-10 The autonomic nervous system
The autonomic nervous system has two divisions, the sympathetic and parasympathetic. The two divisions supply nerves to many of the same organs but produce opposite effects. Activation of the autonomic nervous system is involuntarily commanded by signals from the hypothalamus, part of the interior of the brain above the spinal cord (see Fig. 33-14).

such "fight-or-flight" activities, the sympathetic nervous system curtails activity of the digestive tract, redirecting some of its blood supply to be used by the muscles of arms and legs. Heart rate accelerates. The pupils of the eyes open wider, admitting more light, and the air passages in the lungs expand, accommodating more air. These things happened when you were startled by the fire alarm.

The parasympathetic division, in contrast, dominates during maintenance activities that can be carried on at leisure, often called "rest and rumination." Under its control, the digestive tract becomes active, heart rate slows, and air passages in the lungs constrict. As you read this text, your parasympathetic nervous system is probably dominating your unconscious functions.

Two differences in the organization of the sympathetic and parasympathetic divisions are evident in Figure 33-10. First, their nerves originate at different levels of the central nervous system. Second, although both divisions use a two-neuron pathway to carry messages to each target organ, in the sympathetic division, the synapse occurs in ganglia near the spinal cord. In the parasympathetic division, the synapse occurs in smaller ganglia located at or very near each target organ.

The Central Nervous System Consists of the Spinal Cord and Brain

The spinal cord and brain make up the central nervous system. It is the portion of the nervous system where sensory information is received and processed, thoughts are generated, and responses are directed. The central nervous system consists primarily of association neurons—somewhere between 10 billion and 100 billion of them!

The brain and spinal cord are protected in three ways. The first line of defense is a bony armor, consisting of the *skull,* which surrounds the brain, and the *vertebral column,* which protects the spinal cord. Beneath the bones lies a triple layer of connective tissue called **meninges** (see Fig. 33-14). Between the layers of the meninges, the **cerebrospinal fluid,** a clear lymphlike liquid, cushions the brain and spinal cord.

The Spinal Cord Is a Cable of Axons Protected by the Backbone

The spinal cord is a neural cable about as thick as your little finger that extends from the base of the brain to the lower back (Fig. 33-11). It is protected by the bones of the vertebral column. Between the vertebrae, nerves carrying axons of sensory neurons and motor neurons arise from the dorsal and ventral portions of the spinal cord, respectively, and merge to form the peripheral nerves of the spinal cord (part of the peripheral nervous system). In the center of the spinal cord is a butterfly-shaped area of **gray matter**. Gray matter consists of the cell bodies of several types of neurons that control voluntary muscles and the autonomic nervous system, and neurons that communicate with the brain and other parts of the spinal cord. The gray matter is surrounded by **white matter**, formed of myelin-coated axons of neurons that extend up or down the spinal cord. In addition to relaying neural signals between the brain and the rest of the body, the

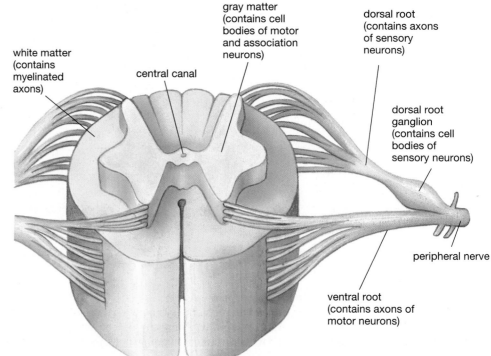

Figure 33-11 The spinal cord The spinal cord runs from the base of the brain to the hips, protected by the vertebrae. Peripheral nerves emerge from between the vertebrae. In cross section, the spinal cord has an outer region of myelinated axons (white matter) that travel to and from the brain, surrounding an inner, butterfly-shaped region of dendrites and the cell bodies of association and motor neurons (gray matter). The cell bodies of the sensory neurons are located outside the cord in the dorsal root ganglion.

gray matter
(contains cell
bodies of motor
and association
neurons)

white matter
(contains
myelinated
axons)

central canal

dorsal root
(contains axons
of sensory
neurons)

dorsal root
ganglion
(contains cell
bodies of
sensory neurons)

peripheral nerve

ventral root
(contains axons of
motor neurons)

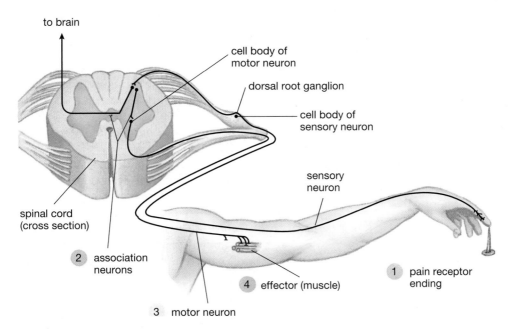

Figure 33-12 *The pain-withdrawal reflex*
This simple reflex circuit includes one each of the four elements of a neural pathway. (1) The sensory neuron has pain-sensitive endings in the skin and a long fiber leading to the spinal cord. That neuron stimulates (2) an association neuron in the spinal cord, which in turn stimulates (3) a motor neuron, also in the cord. The axon of the motor neuron carries action potentials to (4) muscles, causing them to contract and withdraw the body part from the damaging stimulus. The sensory neuron also makes a synapse on association neurons not involved in the reflex that carry signals to the brain, informing it of the danger.

spinal cord contains the neural pathways for certain simple behaviors, including reflexes.

To illustrate some of the functions of the parts of the spinal cord, let's examine a simple spinal reflex, the pain-withdrawal reflex, which involves neurons of both the central and the peripheral nervous systems (Fig. 33-12). The cell bodies of the sensory neurons from the skin (in this case, signaling pain) are located in **dorsal root ganglia** on spinal nerves just outside the spinal cord. Both association neuron and motor neuron cell bodies are found in the gray matter in the center of the spinal cord. Association neurons for the pain-withdrawal reflex, for example, not only form synapses on motor neurons but also have axons that extend up to the brain. Signals carried along these axons alert the brain to the painful event. The brain, in turn, sends impulses down axons in the white matter to cells in the gray matter. These signals can modify spinal reflexes. With sufficient motivation, you can suppress the pain-withdrawal reflex; to rescue a child from a burning building, for example, you could reach into the flames.

In addition to simple reflexes, the entire program for operating some fairly complex activities also resides within the spinal cord. All the neurons and interconnections needed for walking and running, for example, are contained in the spinal cord. In these cases, the brain's role is to initiate and guide the activity of spinal neurons. The advantage of this semi-independent arrangement is probably an increase in speed and coordination, because messages do not have to travel all the way up the cord to the brain and back down again (in the case of walking) merely to swing forward one of your legs. Moreover, the motor neurons of the spinal cord also control the muscles involved in conscious, voluntary activities, such as eating, writing, or playing tennis. Axons of the brain cells that direct these activities carry signals down the cord and stimulate the appropriate motor cells.

The Brain Consists of Many Parts Specialized for Specific Functions

All vertebrate brains have the same general structure, with major modifications corresponding to lifestyle and intelligence. Embryologically, the vertebrate brain begins as a simple tube that soon develops into three parts: (1) the *hindbrain*, (2) *midbrain*, and (3) *forebrain* (Fig. 33-13a). Scientists believe that in the earliest vertebrates, these three anatomical divisions were also functional divisions: The **hindbrain** governed automatic behaviors such as breathing and heart rate, the **midbrain** controlled vision, and the **forebrain** dealt largely with the sense of smell. In nonmammalian vertebrates, these three divisions remain prominent. However, in mammals, and

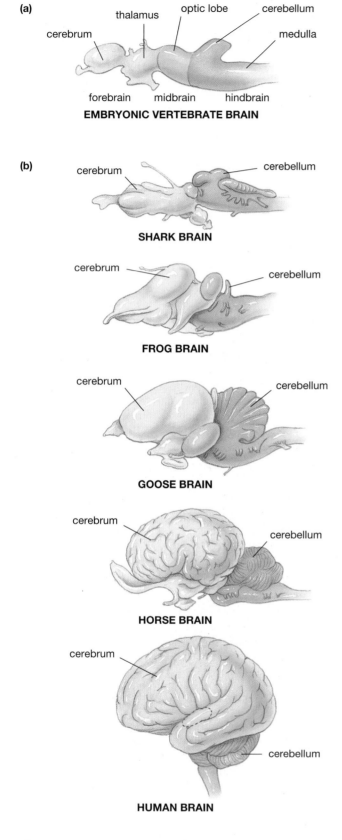

(a)

thalamus — optic lobe — cerebellum

cerebrum — medulla

forebrain — midbrain — hindbrain

EMBRYONIC VERTEBRATE BRAIN

(b)

cerebrum — cerebellum

SHARK BRAIN

cerebrum — cerebellum

FROG BRAIN

cerebrum — cerebellum

GOOSE BRAIN

cerebrum — cerebellum

HORSE BRAIN

cerebrum — cerebellum

HUMAN BRAIN

Figure 33-13 A comparison of vertebrate brains
(a) The embryonic vertebrate brain shows three distinct regions: the forebrain, midbrain, and hindbrain. *(b)* This basic structure persists in all adult brains, but the relative size and importance of the parts vary enormously.

particularly in humans, the brain regions are significantly modified in adults. Some have been reduced in size, and others, especially the forebrain, greatly enlarged (Fig. 33-13b). A section through the midline of the human brain reveals many of its structural features, as shown in Figure 33-14.

The Hindbrain Includes the Medulla, Pons, and Cerebellum

In humans, the hindbrain is represented by the *medulla*, the *pons,* and the *cerebellum* (see Fig. 33-14). In both structure and function, the **medulla** is very much like an enlarged extension of the spinal cord. Like the spinal cord, the medulla has neuron cell bodies at its center, surrounded by a layer of myelin-covered axons. The medulla controls several automatic functions, such as breathing, heart rate, blood pressure, and swallowing. Certain neurons in the **pons**, located above the medulla, appear to influence transitions between sleep and wakefulness and between stages of sleep. Other neurons influence the rate and pattern of breathing. The **cerebellum** is crucially important in coordinating movements of the body. It receives information from command centers in the conscious areas of the brain that control movement and also from position sensors in muscles and joints. By comparing what the command centers ordered with information from the position sensors, the cerebellum guides smooth, accurate motions and body position. The cerebellum is also involved in learning and memory storage for behaviors. As you take notes, your cerebellum is instructing your brain about the order and timing of muscle movements in your hand. Not surprisingly, the cerebellum is largest in animals whose activities require fine coordination. It is best developed in birds (see Fig. 33-13b), which engage in the complex activity of flight.

The Midbrain Contains the Reticular Formation

The midbrain is extremely reduced in humans (see Figs. 33-13b and 33-14). It contains an auditory relay center and a center that controls reflex movements of the eyes. Another important relay center, the **reticular formation**, passes through it. The neurons of the reticular formation extend all the way from the central core of the medulla up into lower regions of the forebrain. It receives input from virtually every sense, from every part of the body, and from many areas of the brain as well. The reticular formation plays a role in sleep and arousal, emotion, muscle tone, and certain movements and reflexes. It filters sensory inputs before they reach the conscious regions of the brain, although the selectivity of the filtering seems to be set by higher brain centers. Activities of the reticular formation allow you to read and concentrate in the presence of a variety of distracting stimuli, such as the music from your stereo and the smell of coffee. The fact that a mother wakens upon hearing the faint cry of her infant but sleeps through loud traffic noise outside her

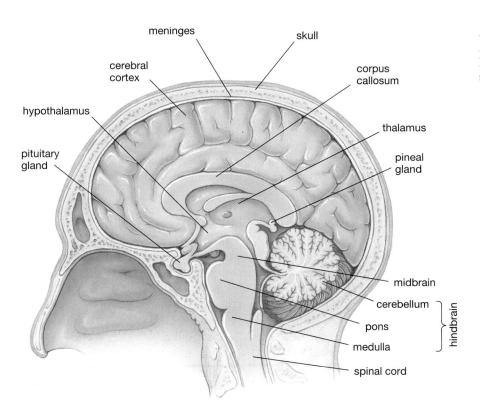

meninges, skull
cerebral cortex
corpus callosum
hypothalamus
thalamus
pituitary gland
pineal gland
midbrain
cerebellum
pons } hindbrain
medulla
spinal cord

Figure 33-14 *The human brain*
A section through the midline of the
human brain reveals some of its major
structures.

window testifies to the effectiveness of the reticular formation in screening inputs to the brain.

The Forebrain Includes the Thalamus, Limbic System, and Cerebral Cortex

The forebrain, also called the **cerebrum**, can be divided
into three functional parts: (1) the *thalamus*, (2) the *limbic system*, and (3) the *cerebral cortex*. In mammals, the
cerebral cortex is much enlarged compared with that of
fish, amphibians, and reptiles. This trend culminates in
the human cerebral cortex (see Fig. 33-13b).

The Thalamus   The **thalamus** (Fig. 33-15) carries sensory information to the limbic system and cerebral cortex. This information includes sensory input from
auditory and visual pathways, from the skin, and from
within the body. Signals from the cerebellum and limbic
system back to the cerebral cortex are also channeled
through this busy thoroughfare.

The Limbic System   Anatomically, the **limbic system** is
a diverse group of structures located in an arc between
the thalamus and cerebral cortex (see Fig. 33-15). These
structures work together to produce our most basic and
primitive emotions, drives, and behaviors, including fear,
rage, tranquility, hunger, thirst, pleasure, and sexual responses. Portions of the limbic system are also important

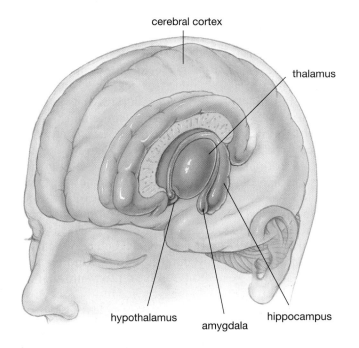

cerebral cortex
thalamus
hypothalamus
amygdala
hippocampus

Figure 33-15 *The limbic system and thalamus*
The limbic system extends through several brain regions. It
seems to be the center of most unconscious emotional behaviors, such as love, hate, hunger, sexual responses, and fear. The
thalamus is a crucial relay center among the senses, the limbic
system, and the cerebral cortex.

in the formation of memories. The limbic system includes the *hypothalamus*, the *amygdala*, and the *hippocampus* as well as nearby regions of the cerebral cortex.

The **hypothalamus** (literally, "under the thalamus") contains many different clusters of neurons. Some of these neurons are neurosecretory cells that release hormones into the blood; others control the release of a variety of hormones from the pituitary gland (see Chapter 32). Other regions of the hypothalamus direct the activities of the autonomic nervous system. The hypothalamus, through its hormone production and neural connections, acts as a major coordinating center, controlling body temperature, hunger, the menstrual cycle, water balance, and the sleep–wake cycle.

Clusters of neurons in the **amygdala** produce sensations of pleasure, punishment, or sexual arousal when stimulated. Conscious humans whose amygdalas are electrically stimulated have reported feelings of rage or fear. Recent studies have revealed that damage to the human amygdala early in life eliminates the ability both to feel fear and to recognize fearful facial expressions in other people. Your amygdala was probably activated when you were startled by the fire alarm.

The shape of the **hippocampus** as it curves around the thalamus inspired its name, which is derived from the Greek word meaning "sea horse." As in the amygdala and hypothalamus, behaviors that reflect a variety of emotions, including rage and sexual arousal, can be elicited by stimulating portions of the hippocampus. The hippocampus also plays an important role in the formation of long-term memory and is thus required for learning, discussed in more detail later in this chapter.

The Cerebral Cortex In humans, by far the largest part of the brain is the **cerebral cortex**, the outer layer of the forebrain. The cerebral cortex and underlying parts of the forebrain are divided into two halves, called **cerebral hemispheres**. These halves communicate with each other by means of a large band of axons, the **corpus callosum**. The cerebral cortex is the most sophisticated information-processing center known, but it is also the area of the brain that scientists know the least about. Tens of billions of neurons are packed into this thin surface layer. To accommodate this profusion of cells, the cortex forms folds called **convolutions**, which greatly increase its area. In the cortex, cell bodies of neurons predominate, giving this outer layer of the brain a gray appearance. These neurons receive sensory information, process it, store some in memory for future use, direct voluntary movements, and are responsible for the poorly understood processes that we call "thinking."

The cerebral cortex is divided into four regions on the basis of anatomical criteria: the (1) *frontal*, (2) *parietal*, (3) *occipital*, and (4) *temporal lobes* (Fig. 33-16). Functionally, the cortex contains primary sensory areas where signals originating in sensory organs such as the eyes and ears are received and converted into subjective impres-

sions—for instance, light and sound. Nearby association areas interpret the sounds as speech or music, for example, and the visual stimuli as recognizable objects or words on this page. Association areas also link the stimuli with previous memories stored in the cortex and generate commands to produce speech. Primary sensory areas in the parietal lobe interpret sensations of touch that originate in all parts of the body, which are "mapped" in an orderly sequence.

In an adjacent region of the frontal lobe, primary motor areas generate commands for movements in corresponding areas of the body. These commands control the motor neurons that form synapses with muscles, allowing you to take notes or to walk out of the dorm (see Fig. 33-16). The association area of the frontal lobe behind the bones of the forehead seems to be involved in complex reasoning, such as decision making ("Shall I leave the dorm or assume it's a false alarm?"), predicting the consequences of actions ("If I don't leave and it's *not* a false alarm . . . !"), controlling aggression, and planning for the future.

Damage to the cortex from trauma, stroke, or a tumor results in specific deficits, such as problems with speech, difficulty reading, or the inability to sense or move specific parts of the body. Most brain cells of adults cannot reproduce, so if a brain region is destroyed, it cannot repair itself. Hence, these deficits may be permanent. Fortunately, however, training can sometimes allow undamaged regions of the cortex to take control over and restore some of the lost functions.

6 How Does the Brain Produce the Mind?

Historically, people have always had difficulty reconciling the physical presence of a few pounds of grayish material in the skull with the range of thoughts, emotions, and memories of the human mind. This "mind–brain problem" has occupied generations of philosophers and, more recently, neurobiologists. Beginning with observations of individuals with head injuries and progressing to sophisticated surgical, physiological, and biochemical experiments, the outlines of how the brain creates the mind are beginning to emerge. Here, we will be able to touch on only a few of the most fascinating features.

The "Left Brain" and "Right Brain" Are Specialized for Different Functions

Although the cerebrum consists of two extremely similar-looking hemispheres, it has been known since the early 1900s that this symmetry does not extend to brain function. Much of what is known of the differences in hemisphere function comes from two sources: studies of accident or stroke victims with localized damage to one

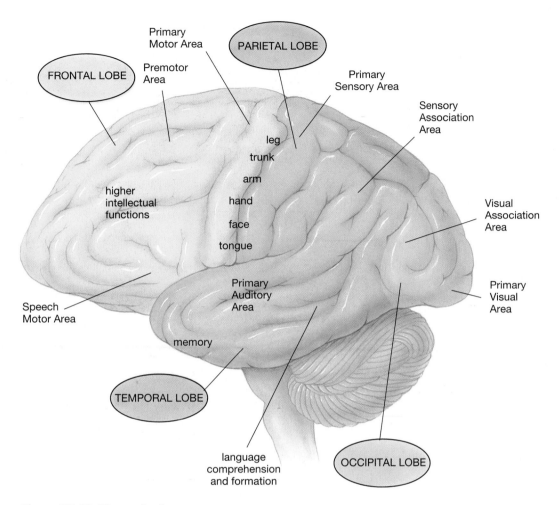

Figure 33-16 ***The cerebral cortex***
Structural (colored) and functional (labeled) regions of the human left cerebral cortex. A map of the right cerebral cortex would be similar, except that speech and language are less well developed there.

hemisphere and studies of patients who have had their corpus callosum (which connects the two hemispheres) severed. This surgical procedure is still performed in certain cases of uncontrollable epilepsy to prevent the spread of seizures through the brain.

Roger Sperry, of the California Institute of Technology, worked with individuals whose hemispheres had been surgically separated by cutting the corpus callosum. In his studies, Sperry made use of the knowledge that axons within each optic tract (not severed by the surgery) follow a pathway that causes the left half of each visual field to be projected on the right cerebral hemisphere and the right half to be projected onto the left hemisphere (Fig. 33-17). Through an ingenious device that projected different images onto the left and right visual fields (thus sending different signals to each hemisphere), Sperry and other investigators have gained more insight into the roles of the two hemispheres.

If Sperry projected an image of a nude figure onto just the left visual field, the patients would blush and smile but would claim to have seen nothing, because the image had reached only the nonverbal right side of the brain! The same figure projected onto the right visual field was readily described verbally. These experiments, begun in the 1950s and refined since then, have revealed that in right-handed people, the left hemisphere is dominant in speech, reading, writing, language comprehension, mathematical ability, and logical problem solving. The right side of the brain is superior to the left in musical skills, artistic ability, recognition of faces, spatial visualization, and the ability to recognize and express emotions. For his pioneering work, Sperry was awarded the Nobel Prize for Physiology in 1981.

Recent experiments indicate that the left-right dichotomy is not as rigid as was once believed. Patients who have suffered a stroke that disrupted the blood

Figure 33-17 The visual pathway and specializations of the two cerebral hemispheres
In general, each hemisphere controls sensory and motor functions of the opposite side of the body. The left side seems to predominate in rational and computational activities; the right side governs creative and spatial abilities. Images on the left half of the visual field are projected onto the right half of the retina and from there reach only the right visual cortex (orange), whereas images on the right half of the visual field reach the left visual cortex (blue).

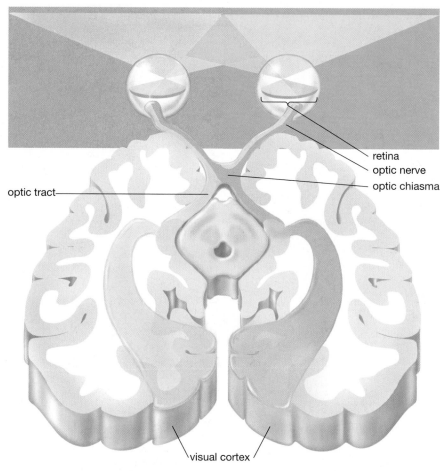

retina
optic nerve
optic chiasma
optic tract
visual cortex

LEFT HEMISPHERE
1. Controls right side of body
2. Input from right visual field, right ear, left nostril
3. Centers for language, mathematics

RIGHT HEMISPHERE
1. Controls left side of body
2. Input from left visual field, left ear, right nostril
3. Centers for spatial perception, music, creativity

supply to their left hemisphere typically show symptoms such as loss of speaking ability. Commonly, however, training can partially overcome these speech or reading deficits, even though the left hemisphere itself has not recovered. This fact suggests that the right hemisphere has some latent language capabilities. Interestingly, female stroke victims are more likely to recover some lost abilities than are males, and females also have a slightly larger corpus callosum. These findings suggest a gender difference in the degree of specialization of the two hemispheres and in the extent of their interconnections. Further evidence of this difference has recently been provided by sensitive techniques that allow imaging of neural activity in the brains of normal subjects while they perform various mental tasks. When subjects were asked to compare word lists for rhyming words, a specific region of the left cortex of male subjects became active, but in females, similar areas in *both* the left and right hemispheres were activated. (Further brain-imaging stud-

ies are described in "Scientific Inquiry: Peering Into the 'Black Box,'" pp. 676–677.)

The Mechanisms of Learning and Memory Are Poorly Understood

Although hypotheses abound as to the cellular mechanisms of learning and memory, we are a long way from understanding these phenomena. In mammals, and particularly in humans, however, we do know a fair amount about two other aspects of learning and memory: the time course of learning and some of the brain sites involved in learning, memory storage, and recall.

Memory May Be Brief or Long Lasting
Experiments show that learning occurs in two phases: an initial **working memory** followed by **long-term memory**. For example, if you look up a number in the phone book, you will probably remember the number long enough to

dial but then forget it promptly. This is working memory. But if you call the number frequently, eventually you will remember the number more or less permanently. This is long-term memory.

Some working memory seems to be electrical in nature, involving the repeated activity of a particular neural circuit in the brain. As long as the circuit is active, the memory stays. In other cases, working memory involves temporary biochemical changes within neurons of a circuit, with the result that synaptic connections between them are strengthened.

In contrast, long-term memory seems to be structural—the result, perhaps, of persistent changes in the expression of certain genes. It may require the formation of new, long-lasting synaptic connections between specific neurons or the long-term strengthening of existing but weak synaptic connections. Working memory can be converted into long-term memory. This process seems to involve the hippocampus, which processes new memories and then transfers them to the cerebral cortex for permanent storage.

The Temporal Lobes Are Important for Memory

The mechanisms of learning, memory storage, and memory retrieval are subjects of extensive research. There is strong evidence that the hippocampus, deep within the brain's temporal lobe, is involved in learning. For example, intense electrical activity occurs in the hippocampus during learning. Even more striking are the results of hippocampal damage. An individual in whom both hippocampi (one in each hemisphere) are destroyed retains much of his or her former memories but is unable to learn new information after the loss. A patient whose hippocampi and associated brain structures were removed surgically in 1953 in an attempt to control seizures remains unable to recall his address or find his way home after many years at the same residence. He can entertain himself indefinitely by reading the same magazine over and over, and people whom he sees daily require reintroduction at each encounter. This example and others suggest that the hippocampus is responsible for transferring information from working memory to long-term memory.

The temporal lobes of the cerebral hemispheres seem to be important in the *retrieval,* or recall, of long-term memories. In a famous series of experiments in the 1940s, neurosurgeon Wilder Penfield electrically stimulated the temporal lobes of conscious patients undergoing brain surgery. The patients did not merely recall memories but felt that they were experiencing the past events right there in the operating room!

Insights on How the Brain Creates the Mind Come from Diverse Sources

Until about 100 years ago, the mind was more appropriately a subject for philosophers than for scientists, because tools for studying the brain did not yet exist.

Through the first half of the twentieth century, the mind was treated by psychologists as a "black box" whose internal workings could be deduced only through the investigation of how past and present experiences were interpreted and influenced behavior. New discoveries, however, are rapidly changing our views of the workings of the brain.

During recent decades, we have begun to understand the neural bases of at least some psychological phenomena. Many forms of mental illness, such as schizophrenia, manic depression, and autism, once thought to be due to childhood trauma or inept parenting, are now recognized as the results of biochemical imbalances in the brain. Studies are revealing a strong heritability factor (and hence, a biological basis) for traits that were once considered entirely learned, such as shyness and alcoholism.

A striking illustration of how the physical structure of the brain is related to personality was unwittingly provided by Phineas Gage in 1848. An explosive charge he was setting was triggered prematurely. The blast blew a metal rod that weighed 13 pounds and was more than a yard long through his skull, damaging both of his frontal lobes (Fig. 33-18). Amazingly, he got up, took a cart into town, and informed the local doctor "here is business enough for you." Although Gage survived for many years

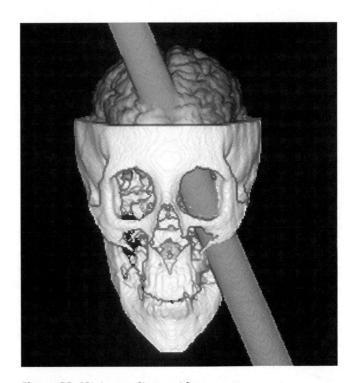

Figure 33-18 A revealing accident
On the basis of studies of the skull of Phineas Gage, scientists have created this computer-generated reconstruction of the path taken by the steel rod that was blown through his head by an explosion. He survived, but the frontal lobe of both hemispheres of his cerebral cortex was extensively damaged.

Scientific Inquiry
Peering Into the "Black Box"

For most of human history, the brain has been seen as a "black box" whose inputs and outputs were observable but whose internal workings were inherently unknowable. New techniques, however, are providing exciting insights into brain function. Two new imaging techniques, the *PET scan* (positron emission tomography, described in Chapter 2) and *fMRI* (functional magnetic resonance imaging), allow researchers to observe the brain in action. Regions of the brain that are most active have higher energy demands; these regions utilize more glucose and attract a greater flow of oxygenated blood than do inactive areas.

In PET scans, scientists inject the subject with radioactive glucose and then monitor levels of radioactivity (which reflect glucose consumption) throughout the brain. Different levels of glucose utilization are translated into colors on cross-sectional images of the brain. By monitoring radioactivity while a specific task is performed, scientists can identify parts of the brain that are involved in that task.

In contrast, fMRI detects differences in the way both oxygenated and deoxygenated blood respond to a powerful magnetic field applied in pulses by an enormous electromagnet that surrounds the body. Active brain regions, which receive larger quantities of oxygenated blood, can be distinguished from inactive regions via fMRI without the use of radioactivity and over much shorter time spans than through PET.

Using fMRI or PET, researchers can observe changes as the brain responds to an odor or to a visual or auditory stimulus or performs a specific reasoning task. Through brain scans, scientists have confirmed that different aspects of the processing of language occur in distinct areas of the cerebral cortex (Fig. E33-2). Using fMRI, researchers analyzed the frontal lobe areas used in generating words in individuals who spoke two languages. In subjects who had grown up speaking two languages, the same region of the frontal lobe was used in speaking each language. In subjects who had learned a second language later

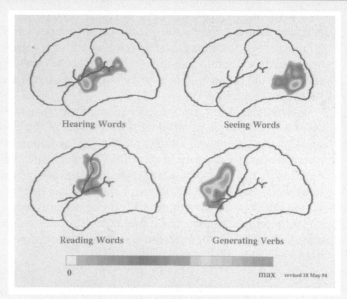

Figure E33-2 Localization of language tasks
Changes in glucose utilization, measured by PET, reveal different cortical regions involved in different language-related tasks, on the basis of research by Dr. Marcus Raichle of the Washington University School of Medicine in St. Louis. The scale ranges from white (lowest) to red (highest).

in life, different but adjacent frontal lobe areas were activated for the two languages. PET or fMRI scans can also precisely localize damaged portions of the brain, such as the result of a stroke (Fig. E33-3a). We can also observe the contrast between disturbed brain functioning, as in Alzheimer's disease, and normal brain functioning (Fig. E33-3b).

after his accident, his personality changed radically. Before the accident, Gage was conscientious, industrious, and well-liked. After his recovery, he became impetuous, profane, and incapable of working toward a goal. Subsequent research has implicated the frontal lobe in emotional expression, control of aggression, and the ability to work for delayed rewards.

Other sites of damage have revealed additional anatomical specializations. One patient with very localized damage to the left frontal lobe of the cerebral cortex was unable to name fruits and vegetables (although he could name everything else). Describing this patient, one science writer quipped, "Does the brain have a produce section?" Other victims of brain damage have developed a selective inability to recognize faces or, in a recent case, to recognize any object that is *not* a face, sug-

gesting that the brain has a region specialized to recognize faces that is separate from the region that allows it to recognize objects in general.

In the past, much of our understanding of the human mind–brain connection came from the study of victims of brain damage such as that caused by a stroke, trauma, tumor, or surgical procedure. Typically, the exact extent of the damage remained unknown until revealed by autopsy. New techniques, such as the PET and fMRI scans, now permit insight into the functioning of normal, as well as diseased, brains (see "Scientific Inquiry: Peering Into the 'Black Box'"). These and increasingly sophisticated techniques of the future will create ever-larger windows into the "black box" that is the human brain and a clearer understanding of how the human brain generates the human mind.

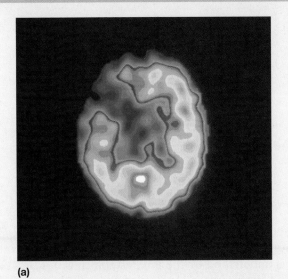

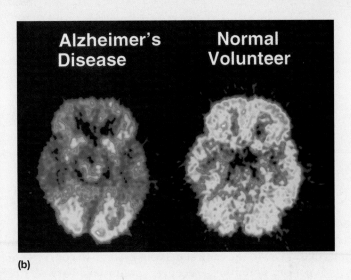

(a)

(b)

Figure E33-3 *PET scans reveal malfunctioning brains*
(a) As a result of stroke, part of the brain dies from lack of blood flow. The damaged region can be precisely localized by the lack of neural activity to the upper-left region of this brain scan. **(b)** The brain of an individual with Alzheimer's disease is compared with that of a healthy elderly person. The top of each image corresponds to the front of the brain, just behind the forehead, and the bottom to the back of the brain. Metabolic activity decreases from red areas to yellow to green and to blue; black areas exhibit little or no metabolic activity.

7) How Do Sensory Receptors Work?

The word *receptor* is used in several contexts in biology. In the most general sense, a **receptor** is a structure that changes when it is acted on by a stimulus from its surroundings, causing a signal to be produced. All receptors are **transducers**—structures that convert signals from one form to another. A receptor may be a membrane protein that changes configuration when it binds a specific hormone or neurotransmitter, as discussed in previous chapters. Alternatively, as described in this chapter, a **sensory receptor** may be an entire cell (typically a neuron) that is specialized to produce an electrical response to particular stimuli—that is, it translates sensory stimuli into the

language of the nervous system. All sensory receptors produce electrical signals, but each receptor type is specialized to produce its signal only in response to a particular type of environmental stimulus. Some receptors, called *free nerve endings*, consist of branching dendrites of sensory neurons; other receptors have specialized structures that help them respond to a specific stimulus. Many sensory receptors are clustered into sensory organs, such as the eye, ear, skin, or tongue. Their electrical activity gives rise to the subjective perceptions of light, sound, touch, and taste that we describe as our "senses."

The stimulation of a sensory receptor causes a **receptor potential**, an electrical signal whose size is proportional to the strength of the stimulus. Receptor potentials may give rise to action potentials in the receptor neurons themselves. Alternatively, receptor potentials in small

Table 33-1 Vertebrate Receptor Types			
Type of Receptor	**Sensory Cell Type**	**Stimulus**	**Location**
Thermoreceptor	Free nerve ending	Heat, cold	Skin
Mechanoreceptor	Hair cell	Vibration, motion, gravity	Inner ear
	Specialized nerve endings and free nerve endings in skin (Pacinian corpuscle, Merkel's disc)	Vibration, pressure, touch	Skin
	Specialized nerve endings in muscles or joints (muscle spindle, Golgi tendon organ)	Stretch	Muscles, tendons
Photoreceptor	Rod, cone	Light	Retina of eye
Chemoreceptor	Olfactory receptor	Odor (airborne molecules)	Nasal cavity
	Taste receptor	Taste (waterborne molecules)	Tongue
Pain receptor	Free nerve ending	Chemicals released by tissue injury	Widespread in body

receptor cells cause the release of neurotransmitter onto postsynaptic neurons, which in turn produce action potentials that travel to the central nervous system. In contrast to action potentials, receptor potentials vary according to the intensity of the stimuli: The stronger the stimulus, the larger the receptor potential. Intensity is then conveyed to the nervous system by the frequency, not the size, of action potentials.

Sensory receptor cells are named after the stimulus to which they respond, as summarized in Table 33-1. **Thermoreceptors** ("heat" receptors) are free nerve endings that respond to fluctuations in temperature. Most **mechanoreceptors** produce a receptor potential in response to stretching of their plasma membranes. Embedded in and directly beneath human skin are several types of mechanoreceptors, making the skin exquisitely sensitive to touch, pressure, and vibration. Mechanoreceptors located in the walls of the stomach, rectum, and bladder signal fullness when they are stretched. Position-sensing mechanoreceptors, located in joints and muscles, sense the orientation and direction of movement of various body parts. These position sensors allow you to walk without watching your feet or to eat without watching the fork on its way to your mouth. Another type of mechanoreceptor, **hair cells**, are located in the inner ear. These receptor cells for sound, motion, and gravity bear hairlike structures; bending of these hairs produces a receptor potential.

Many other sensory receptor cells produce receptor potentials in response to stimuli that alter specific receptor proteins embedded in their plasma membranes. **Photoreceptors** (receptors for light), **chemoreceptors** (receptors for chemicals, which we perceive as tastes or odors), and **pain receptors** possess extensive areas of membrane studded with specialized receptor proteins. When light or a chemical contacts a receptor protein specialized to respond to that stimulus, the protein changes shape. This change alters the permeability of the plasma membrane, generating a receptor potential.

Sensory receptors are our links with the world around us. As Aristotle observed in the fourth century B.C. "Nothing is understood by the intellect which is not first perceived by the senses." In the following sections we focus on some senses that are central to human perception of the world.

8   How Is Sound Sensed?

Sound is produced by any vibrating object—a drum, a motor, vocal cords, or the speaker of your stereo. These vibrations are transmitted through air in the form of sound waves and are intercepted by our ears. The ears of humans and other animals are elaborate and remarkable structures that detect the direction, loudness, and pitch of sound.

The Structure of the Ear Helps Capture, Transmit, and Transduce Sound

The ear of humans and most other vertebrates consists of three parts: the outer, middle, and inner ear (Fig. 33-19a). The **outer ear** consists of the **external ear** and the **auditory canal**. The external ear, with its fleshy folds, modifies sound waves. These modifications help the brain determine the location of the sound source. The air-filled auditory canal conducts the sound waves to the **middle ear**, consisting of the **tympanic membrane**, or *eardrum*, three tiny bones called the *hammer*, *anvil*, and *stirrup*, and the **Eustachian tube**. The Eustachian tube connects the middle ear to the pharynx and equalizes the air pressure between the middle ear and the atmosphere.

Within the middle ear, sound vibrates the tympanic membrane, which in turn vibrates the hammer, the anvil, and the stirrup. These bones transmit vibrations to the **inner ear**. The fluid-filled hollow bones of the inner ear form the spiral-shaped **cochlea** ("snail" in Latin) and other structures that detect head movement and the pull of grav-

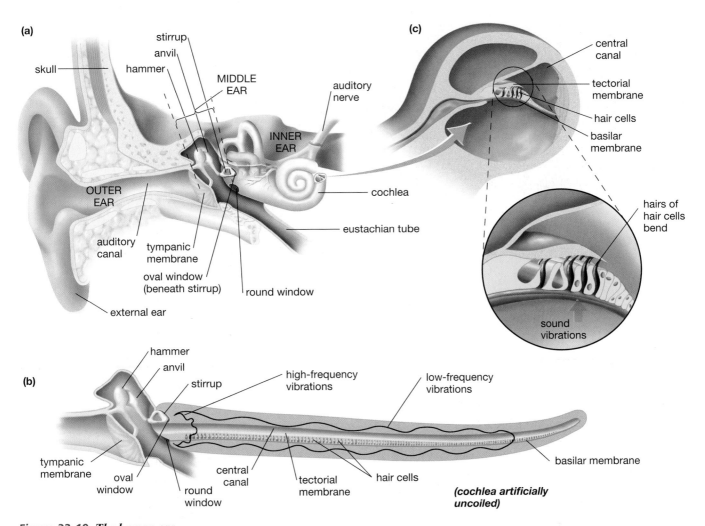

Figure 33-19 The human ear
(a) Overall anatomy of the ear. *(b)* Uncoiled, the cochlea consists of an outer tube surrounding the central canal. Different frequencies of sound waves cause the basilar membrane to vibrate at different locations. *(c)* The hairs of hair cells span the gap between the basilar and tectorial membranes in the central canal. Sound vibrations move the membranes relative to one another, bending the hairs and producing a receptor potential in the hair cells.

ity. Sound enters the cochlea when the stirrup vibrates the **oval window**, a thin membrane that covers a hole in the cochlea. The *round window*, a membrane below the oval window, allows the incompressible fluid in the cochlea to shift back and forth as the oval window vibrates.

Sound Transduction Is Aided by the Structure of the Cochlea

If we were to unroll the cochlea, in lengthwise cross section it would consist of two fluid-filled tubes, an outer U-shaped canal, and a straight central canal (Fig. 33-19b). The central canal contains the **basilar membrane**, on top of which are the receptors, or hair cells. Protruding into the central canal is the **tectorial membrane**, a gelatinous structure in which the "hairs" of the hair cells are embedded.

How do these structures allow the perception of sound? The oval window passes vibrations to the fluid in the cochlea, which in turn vibrates the basilar membrane, causing it to move up and down. This movement bends the hairs embedded in the tectorial membrane (Fig. 33-19c), producing receptor potentials in the hair cells. The hair cells then release transmitter onto neurons of the **auditory nerve**. Action potentials triggered in axons of the auditory nerve travel to the brain.

The cochlea also allows us to perceive *loudness* (the magnitude of sound vibrations) and *pitch* (the frequency of sound vibrations). A weak sound causes small vibrations, which bend the hairs only slightly, and results in a low rate of action potential propagation in the auditory nerve axons. A loud sound, such as the fire alarm, causes large vibrations, which cause greater bending of the hairs and a larger receptor potential, producing a

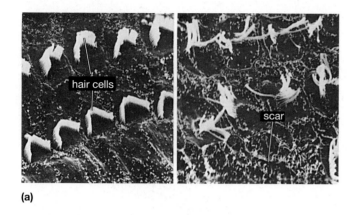

(a)

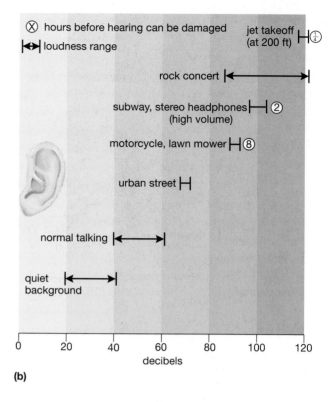

(X) hours before hearing can be damaged

|←→| loudness range

jet takeoff (at 200 ft)

rock concert

subway, stereo headphones (high volume) ②

motorcycle, lawn mower ⑧

urban street

normal talking

quiet background

0   20   40   60   80   100   120
decibels

(b)

Figure 33-20 Loud sound can damage hair cells
(a) Scanning electron micrographs show the effect of intense sound on the hair cells of the inner ear. (left) Top view of the hairs of the hair cells in a normal guinea pig; hairs emerge from each receptor in a V-shaped pattern. (right) After 24-hour exposure to a sound level approached by loud rock music (2000 vibrations per second at 120 decibels), many of the hairs are damaged or missing, leaving "scars." Hair cells in humans do not regenerate, so such hearing loss is permanent. (Scanning electron micrographs by Robert S. Preston, courtesy of Professor J. E. Hawkins, Kresge Hearing Research Institute, University of Michigan Medical School.)
(b) Sound levels of everyday noises and their potential to damage hearing. Sound intensity is measured in *decibels* on a logarithmic scale; a 10-decibel sound is 10 times as loud as a 1-decibel sound, and a 20-decibel sound is 100 times as loud. You will feel pain at sound intensities above 120 decibels. [Source: Deafness Research Foundation, National Institute on Deafness and Other Communication Disorders]

high rate of action potential propagation in the auditory nerve. Loud sounds sustained for a long time can actually damage the hair cells (Fig. 33-20a), resulting in hearing loss, a fate suffered by many prominent rock musicians and their fans. In fact, many sounds in our everyday environment have the potential to damage hearing (Fig. 33-20b).

The perception of pitch is a little more complex. Humans can detect vibration frequencies from about 30 vibrations per second (very low pitched) to about 20,000 vibrations per second (very high pitched). The basilar membrane is stiff and narrow at the end near the oval window but more flexible and wider near the tip of the cochlea. This progressive change in structure causes each portion of the membrane to vibrate in synchrony with a particular frequency of sound (see Fig. 33-19b). The brain interprets signals from receptors near the oval window as high-pitched sound, whereas signals from receptors farther along the uncoiled cochlea are interpreted as lower in pitch.

9 How Is Light Sensed?

Animal vision varies in its ability to provide a sharp, accurate representation of the real world, and several types of eyes have evolved independently. All forms of vision, however, use photoreceptors. These sensory cells contain receptor molecules called **photopigments** (because they are colored), which absorb light and chemically change in the process. This chemical change alters ion channels in the receptor cell membrane, producing a receptor potential.

The Compound Eyes of Arthropods Produce a Mosaic Image

The arthropods (insects, spiders, and crustaceans) evolved **compound eyes**, which consist of a mosaic of many individual light-sensitive subunits called **ommatidia** (singular, **ommatidium**; Fig. 33-21). Each ommatidium functions as an on/off, bright/dim detector. Using a large number of

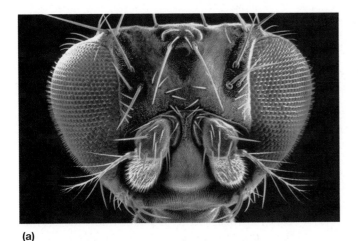

(a)

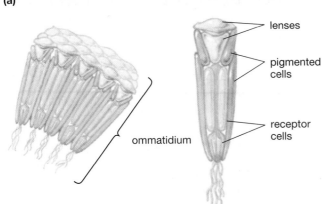

(b)

Figure 33-21 Compound eyes
(a) Scanning electron micrograph of the head of a fruit fly, showing a compound eye on each side of the head. *(b)* Each eye is made up of numerous individual light-receptive ommatidia. Within each ommatidium are several receptor cells, capped by a lens. Pigmented cells surrounding each ommatidium prevent light from passing through to adjacent receptors.

individual units (up to 36,000 per eye in a dragonfly), most arthropods probably obtain a reasonably faithful, although grainy, image of the world. Compound eyes are excellent at detecting movement, an advantage in avoiding predators and in hunting. In addition, many arthropods, such as bees and butterflies, have good color perception.

The Mammalian Eye Collects, Focuses, and Transduces Light Waves

In the eyes of mammals, incoming light, such as that reflected from this page, first encounters the **cornea**, a transparent covering over the front of the eyeball. Behind the cornea is a chamber filled with a watery fluid called **aqueous humor**, which provides nourishment for both the lens and cornea. The amount of light entering the eye is adjusted by the **iris**, pigmented muscular tissue whose circu-

lar opening, the **pupil**, can be expanded or contracted. Light passing through the pupil encounters the **lens**, a structure that resembles a flattened sphere and is composed of transparent protein fibers. The lens is suspended behind the pupil by ligaments and muscles that regulate its shape. Behind the lens is another, much larger chamber filled with the **vitreous humor**, a clear jellylike substance that helps maintain the shape of the eye (Fig. 33-22a).

After passing through the vitreous humor, light reaches the **retina**, a multilayered sheet of photoreceptors and neurons. There the light energy is converted into electrical nerve impulses that are transmitted to the brain (Fig. 33-22b). The retina is richly supplied with blood vessels and contains a layer of pigment that absorbs stray light rays that escape the photoreceptors. Behind the retina is the **choroid**, a darkly pigmented tissue. The choroid's rich blood supply helps nourish the cells of the retina. Its dark pigment also absorbs stray light whose reflection inside the eyeball would interfere with clear vision. Surrounding the outer portion of the eyeball is the **sclera**, a tough connective tissue layer that is visible as the white of the eye and is continuous with the cornea.

In vertebrates (such as deer) that are most active at dusk, the choroid may be modified to reflect light rather than to absorb it. By reflecting light that escaped the photoreceptors during its initial passage, the choroid gives the receptors a second chance to detect it, maximizing the animal's ability to see in dim light. Reflective choroids give the eyes of these animals an eerie red or blue glow when bright light (such as from car headlights) is reflected back through the wide-open pupil.

The Adjustable Lens Allows Focusing of Both Distant and Nearby Objects

The visual image is focused most sharply on a small area of the retina called the **fovea**. Although focusing is begun by the cornea, whose rounded contour bends light rays, the lens is responsible for final, sharp focusing. The shape of the lens can be adjusted such that, when viewed from the side, it is either rounded to focus on nearby objects or flattened to focus on distant objects (Fig. 33-23a, p. 683).

If your eyeball is even a millimeter too long, you will be *nearsighted*: unable to focus on distant objects. *Farsighted* people, whose eyeballs are slightly too short, cannot focus on nearby objects. These conditions can be corrected by external lenses of the appropriate shape (Fig. 33-23b,c). Nearsightedness can also be corrected by surgery that slightly flattens the cornea. The lens stiffens as humans age, causing farsightedness, because the lens can no longer round up enough to focus on nearby objects. By their mid-forties, most people require glasses for close work such as reading.

Light Striking the Retina Is Captured by Photoreceptors; the Signal Is Processed by Layers of Overlying Neurons

The vertebrate eye provides the sharpest vision in the animal kingdom, even though the complex, multilayered retina is "built backward" from an engineering perspective (see Fig. 33-22b). The photoreceptors, called **rods** and **cones** after their shapes, have their light-gathering parts farthest from the light, at the rear of the retina. Between the receptors and incoming light are several layers of neurons that process the signals from the photoreceptors. The retinal layer nearest the vitreous humor consists of **ganglion cells**, whose axons make up the **optic nerve**. Ganglion cell axons must pass back through the retina to reach the brain at a location called the **optic** disc, or **blind spot** (Fig. 33-24; see also Fig. 33-22a). This area lacks receptors; objects focused there seem to disappear. You can locate your blind spot by closing your left eye and focusing steadily on the star below with your right eye. Start with the book about a foot away and gradually move it closer. The spot will disappear when the image falls on your optic disc.

The receptor potential from the photoreceptors is processed by other retinal neurons in ways that enhance our ability to detect edges, movement, dim light, and changes in light intensity. The much-modified signal from the photoreceptors is finally converted to action potentials carried by the optic nerve to the brain. In the brain, further processing ultimately results in the sensation of vision.

Rods and Cones Differ in Distribution and Light Sensitivity

Photoreception in both rods and cones begins with the absorption of light by photopigment molecules that are embedded in the plasma membranes of the photoreceptor cells (see Fig. 33-22b). Light hitting a photopigment molecule causes a change in the permeability of the receptor's plasma membrane to ions, producing a receptor potential in the photoreceptor cell.

Although cones are located throughout the retina, they are concentrated in the fovea, where the lens focuses images most sharply (see Fig. 33-22a). The fovea appears as a depression near the center of the retina, because the layers of signal-processing neurons are pushed aside there while still retaining their synaptic connections. This arrangement allows light to reach the cones of the fovea with relatively little interference.

Human eyes have three varieties of cones, each containing a slightly different photopigment. Each type of photopigment is most strongly stimulated by a particular wavelength of light, corresponding roughly to red, green, or blue. The brain distinguishes color according to the relative intensity of stimulation of different cones. For example, the sensation of yellow is caused by fairly equal stimulation of red and green cones. About 3% of males have difficulty distinguishing red from green, because they possess a defective gene for the red or green photopigment on the X chromosome (see Chapter 13).

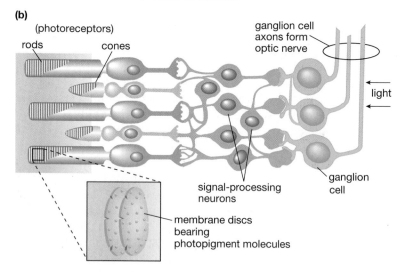

Figure 33-22 The human eye
(a) The anatomy of the human eye. **(b)** The human retina has rods and cones (photoreceptors), signal-processing neurons, and ganglion cells. Each rod and cone bears a long extension packed with membranes in which the light-sensitive molecules are embedded.

(a) Normal eye

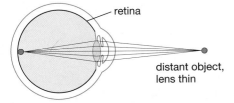

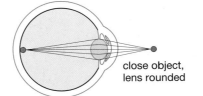

distant object, lens thin

close object, lens rounded

(b) Nearsighted eye (long eyeball)

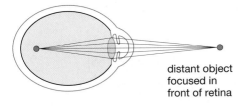

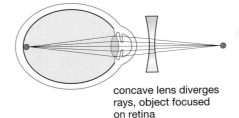

distant object focused in front of retina

concave lens diverges rays, object focused on retina

(c) Farsighted eye (short eyeball)

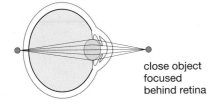

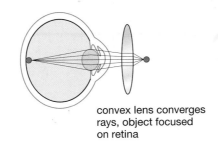

close object focused behind retina

convex lens converges rays, object focused on retina

Figure 33-23 Focusing in the human eye
(a) (left) To focus on a distant object, the lens is made thinner, causing relatively little bending of the light rays. (right) To focus on a nearby object, the lens assumes a more nearly spherical shape, bending the light rays more sharply. *(b)* (left) Most nearsighted people have eyeballs that are too long. Light rays focus in front of the retina and are out of focus again by the time they strike the retina. (right) Eyeglasses with a concave lens cause the light rays to diverge slightly so that the focal point falls on the retina. *(c)* (left) Farsighted people have eyeballs that are too short. The lens cannot bend incoming light rays enough to focus on the retina, so the focal point falls behind the retina. (right) Eyeglasses with a convex lens converge light rays, causing the focal point to fall on the retina.

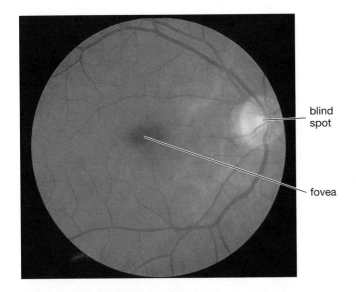

blind spot

fovea

Figure 33-24 The human retina
A photograph of a portion of the human retina, taken through the cornea and lens of a living person. The blind spot (optic disc) and fovea are visible. Blood vessels supply oxygen and nutrients; notice that they are dense over the blind spot (where they won't interfere with vision) and scarcer near the fovea.

Rods dominate in the peripheral portions of the retina. Rods are far more sensitive to light than cones are and so are largely responsible for our vision in dim light. Unlike cones, rods do not distinguish colors; in moonlight, which is too dim to activate the cones, the world appears in shades of gray.

Not all animals have both rods and cones. Animals that are active almost entirely during the day (certain lizards, for example) may have all-cone retinas, whereas many nocturnal animals (such as the ferret) or those dwelling in dimly lit habitats (such as deep-sea fishes) have mostly rods.

Binocular Vision Allows Depth Perception

Among vertebrates, we find two basically different eye placements: Most herbivores have one eye on each side of the head, but predators and omnivores such as humans have both eyes facing forward (Fig. 33-25a). The forward-facing eyes of predators and omnivores have slightly different but extensively overlapping visual fields. This **binocular vision** allows depth perception and accurate judgment of the size and distance of an object from the eyes. These abilities are important to a cat about to

(a)

(b)

Figure 33-25 Eye position differs in predators and prey
There are two normal placements for the eyes—one on each side of the head or both in front. Eye placement is related to lifestyle. *(a)* Most predators, such as this owl, and primates have eyes in front; both eyes can be focused on a target, providing binocular vision. Each eye gets a slightly different view of the target, so size and distance can be judged fairly accurately. *(b)* Most herbivorous prey animals, such as rabbits, mice, horses, and deer, have eyes placed at the sides, the better to scan all around for possible predators.

pounce on a mouse or to a monkey leaping from branch to branch.

In contrast, the widely spaced eyes of herbivores (Fig. 33-25b) have little overlap in their visual fields; accurate depth perception is sacrificed in favor of a nearly 360-degree field of view. This view allows these animals, who are frequently preyed upon, to spot a predator approaching from any direction.

10 How Are Chemicals Sensed?

Through chemical senses, animals find food, avoid poisonous materials, and may locate homes or find mates. Terrestrial vertebrates have two separate chemical senses: one for airborne molecules, called *smell*, or **olfaction**, and one for chemicals dissolved in water or saliva, the sense of **taste**.

The Ability to Smell Arises from Olfactory Receptors

In humans and most other vertebrates, receptors for smell, or olfaction, are nerve cells located in a patch of mucus-covered epithelial tissue in the upper portion of each nasal cavity (Fig. 33-26). The human olfactory epithelium is small compared with that of many other mammals (dogs, for example) whose sense of smell is

hundreds of times more acute than ours. Olfactory receptors have hairlike dendrites that protrude into the cavity and lie embedded in a layer of mucus. Odorous molecules in the air, such as those from your coffee, diffuse into the mucus layer and bind with receptors on the dendrites.

Recent research suggests that there may be 1000 types of receptor proteins embedded in the olfactory dendrites. Each receptor protein is specialized to bind a particular type of molecule and to stimulate the olfactory receptor to send a message to the brain.

Taste Receptors Are Located in Clusters on the Tongue

The human tongue bears about 10,000 **taste buds**, structures embedded in small bumps that cover the surface of the tongue. Each taste bud consists of a cluster of 60 to 80 taste receptor cells surrounded by supporting cells in a small pit. The cells in the pit communicate with the mouth through a pore (Fig. 33-27). Microvilli (thin membrane projections) of taste receptor cells protrude through the pore. Dissolved chemicals enter the pore and bind to receptor molecules on the microvilli, producing a receptor potential.

There are four major types of taste receptors: (1) sweet, (2) sour, (3) salty, and (4) bitter. We perceive a great variety of tastes as a result of two mechanisms. First, a particular substance may stimulate two or more recep-

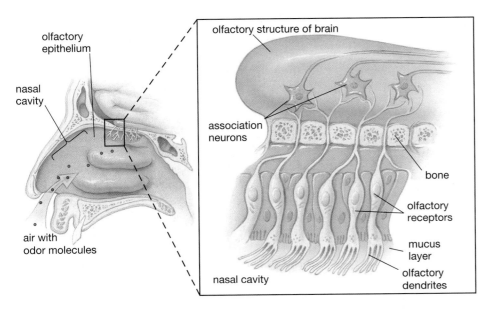

Figure 33-26 Human olfactory receptors
The receptors for olfaction in humans are neurons bearing hairlike projections that protrude into the nasal cavity. The projections are embedded in a mucus layer in which odor molecules dissolve before contacting the receptors.

(a) The human tongue

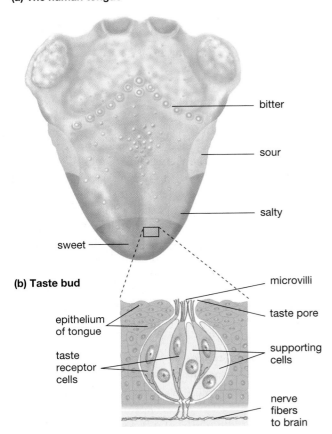

(b) Taste bud

Figure 33-27 Human taste receptors
(a) The human tongue is covered with bumps, each containing many taste buds. Taste receptor cells within the taste buds are specialized for each of the four tastes and are concentrated in specific areas on the tongue. Other areas are less sensitive to these tastes. **(b)** Each taste bud consists of supporting cells surrounding 60 to 80 taste receptor cells, whose microvilli have protein receptors that bind tasty molecules, producing a receptor potential.

tor types to different degrees, making it taste "salty-sweet," for example. Second and more important, material being tasted normally releases molecules into the air inside the mouth. These odor molecules diffuse to the olfactory receptors, which contribute an odor component to the basic flavor. (Remember, the mouth and nasal passages are connected.)

To prove that what we call taste is really mostly smell, try holding your nose (and closing your eyes) while you eat different flavors of jelly beans. The tastes of grape, lime, and cherry jelly beans will be indistinguishable, like sweet, sticky pastes. Likewise, when you have a bad cold, notice how normally tasty foods seem bland and lose much of their appeal and how your coffee tastes merely bitter.

Pain Is Actually a Specialized Chemical Sense

Whether you burn, cut, or crush a fingertip, you will feel the same sensation: pain. Most pain is produced by tissue damage. Researchers have found that pain perception is actually a special kind of chemical sense (Fig. 33-28).

When cells and capillaries are damaged by a cut or a burn, for example, their contents flow into the extracellular fluid. The cell contents include potassium ions, which stimulate pain receptors. Damaged cells also release enzymes that convert certain blood proteins into a chemical called **bradykinin**, another stimulus that activates pain receptors. Each part of the body has a separate set of pain receptor neurons that provide input to particular brain cells. Hence, the brain can identify the location of the pain.

Drugs that provide pain relief, such as morphine or Demerol®, block synapses in the pain pathways of the brain or spinal cord. In ways that we are just beginning to understand, the brain can modulate its perception of pain through its own narcotic-like endorphins. In critical

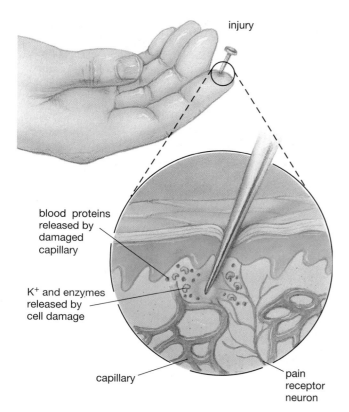

Figure 33-28 Pain perception
Pain perception is a specialized chemical sense. An injury, such as a tack stab, damages both cells and blood vessels. The damaged cells release K⁺, which activates pain receptor neurons, and also release enzymes. Those enzymes convert certain blood proteins into bradykinin, which also stimulates pain-sensitive neurons.

situations, such as combat or during escape from a fire, endorphins may allow us to function by blocking our perception of pain until the emergency is over.

Evolutionary Connections
Uncommon Senses

We have reviewed the "common" senses of sound, sight, smell, taste, and pain. But had this text been written for bats, it undoubtedly would have included a large section on *echolocation* and nearly omitted the coverage of vision! Here we review a few of the "uncommon" senses that have evolved in response to different environments.

Echolocation
Some animals who hunt in darkness or murky water have evolved a type of sonar called **echolocation**, similar to the navigational system used by ships. Using echolocation, bats can navigate and hunt insect prey in total darkness. An echolocating bat emits pulses of noise at ultrasonic frequencies (higher than can be detected by the human ear)

that bounce off nearby objects. The intensity of this sound would make it unpleasantly loud if we could detect it. The patterns of returning sound convey accurate information about the size, shape, surface texture, and location of objects in the environment. Little brown bats can detect wires only 1 millimeter thick from a distance of 2 meters (over 6 feet). Several adaptations contribute to this remarkable sensitivity. The bat's enormous, elaborately folded outer ears collect the returning echoes and help the bat locate their source (Fig. 33-29a). As the bat emits its cry, muscles attached to the bones of the middle ear contract briefly, reducing the bones' vibrations and preventing the bat from being deafened by its own calls. The tympanic membrane and bones of the middle ear are exceptionally light and easily vibrated by the faint returning echoes.

Porpoises produce ultrasonic clicks within their nasal passages and emit them through the front of their head (Fig. 33-29b). There, a large, flexible, oil-filled sac acts like an acoustic lens, directing the sound forward in a broad beam (for navigation) or a narrow beam (for prey location). An echolocating porpoise can locate a pea-sized object on the floor of its tank and can distinguish among species of fish. Porpoises may also use the narrowly focused beam to stun fish with a blast of sound, making them easier to capture.

Detecting Electrical Fields
Some fish, called weak electric fish, use electrical fields for **electrolocation** in much the same way that bats and porpoises use sound waves for echolocation. These fish produce high-frequency electrical signals from an electric organ just in front of their tails; they detect these signals via *electroreceptor* cells located along each side of their body (Fig. 33-30). Objects near the fish distort the electrical field that surrounds the fish. This distortion is detected by the electroreceptors, which send an altered pattern of action potentials to the brain. The fish uses this information to detect and localize nearby objects.

The platypus, whose bill is covered with electroreceptors (as well as sensitive mechanoreceptors), hunts for crayfish and tadpoles at night in murky freshwater ponds and streams.

Detecting Magnetic Fields
Homing pigeons are famous for their ability to fly home after being released some distance away. They can accurately locate their home roost even under cloudy skies in terrain with few landmarks. If a researcher straps a small magnet to a pigeon's back, the bird will find its way home successfully in sunny weather but will lose its way under overcast skies. Apparently, pigeons can navigate either by the sun or, if the sun is hidden, by magnetic fields. In cloudy weather, the experimental magnets confuse the pigeon's magnetic compass. How do pigeons detect magnetic fields? Pigeons have deposits of magnetite (a magnetic iron compound) located just beneath their skull. These deposits may act as a built-in magnet that is used to tell direction.

(a)

(b)

Figure 33-29 Echolocation
Echolocation by bats and porpoises is based on very different anatomical structures. *(a)* A long-eared bat, showing the enormous size and elaborate folds of the external ears that help it localize returning echoes. *(b)* The bottlenose porpoise focuses ultrasonic clicks by using the oil-filled sac in the front of its head.

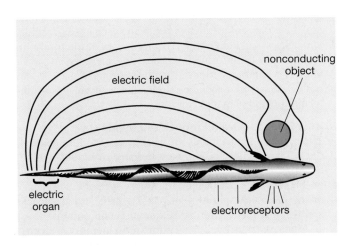

Figure 33-30 Electrolocation
Weak electric fish locate nearby objects by sensing distortions the objects produce in the animal's own electrical field. That field is generated by electric organs near the tail and detected by electroreceptors along the sides of the body.

Eels of eastern North America and western Europe swim out of streams and rivers into the Atlantic Ocean and migrate to the Sargasso Sea (near the West Indies) to spawn. Eels probably also use magnetic fields for navigation. You may recall from high school physics that moving an electrical conductor through a magnetic field induces an electrical current in the conductor. (That is how we generate electricity commercially.) Seawater, with its high salt concentration, is a fairly good conductor. The currents of the Gulf Stream, which flow from the Gulf of Mexico along the east coast of the United States, provide movement through Earth's magnetic field. The Gulf Stream generates an extremely weak electrical field, roughly equivalent to the field produced by a 1-volt battery with its poles more than 7 kilometers (12 miles) apart. At first, this field seemed far too weak to be detected. But researchers discovered that eels can detect electrical fields as weak as a 1-volt battery with poles *over 5000 kilometers* (3000 miles) apart. For an eel, finding the Gulf Stream must be a piece of cake!

Summary of Key Concepts

1) How Do Nervous and Endocrine Communication Differ?

There are four main differences between hormonal and neural controls: (1) Hormones are released into the blood, but nerve cells release neurotransmitters directly onto their target cells; (2) a nerve cell releases neurotransmitter onto one or a few target cells rather than millions of cells; (3) neurons use rapid electrical signals at short range, whereas hormones may travel slowly over longer distances; and (4) neural signals have a much shorter duration than do most hormonal messages. These distinctions, however, are not absolute, and the hormonal and nervous systems are closely coordinated.

2) What Are the Structures and Functions of Neurons?

Nervous systems are composed of billions of individual cells called neurons. A neuron has four major specialized functions, which are reflected in its structure: (1) Dendrites receive information from the environment or from other neurons. (2) The cell body adds together electrical signals from the dendrites and from synapses on the cell body itself and "decides" whether to produce an action potential. (The cell body also coordinates the cell's metabolic activities.) (3) The axon conducts the action potential to its output terminal: the synapse. (4) Synaptic terminals transmit the signal to other nerve cells, to glands, or to muscles.

3) How Is Neural Activity Generated and Communicated?

An unstimulated neuron maintains a negative resting potential inside the cell. Signals received from other neurons are small, rapidly fading changes in potential called postsynaptic potentials. Inhibitory and excitatory postsynaptic potentials (IPSPs and EPSPs) make the neuron less likely or more likely, respectively, to produce an action potential. If postsynaptic potentials, added together within the cell body, bring the neuron to threshold, an action potential will be triggered. The action potential is a wave of positive charge that travels along the axon to the synaptic terminals, undiminished in magnitude.

A synapse, where two neurons communicate, consists of the synaptic terminal of the presynaptic neuron and a specialized region of the postsynaptic neuron, separated by the synaptic cleft. Neurotransmitter from the presynaptic neuron, released in response to an action potential, binds to receptors, opening ion channels in the postsynaptic cell's plasma membrane. Ions flow, producing either an EPSP or an IPSP, depending on the type of ion channels opened.

Neurotransmitters include acetylcholine, dopamine, serotonin, and norepinephrine. Neuropeptides such as opioids modify synaptic transmission and may influence many neurons over a longer time span than do other neurotransmitters.

4) How Is the Nervous System Designed?

Information processing in the nervous system requires four operations. The nervous system must (1) determine the type of stimulus, (2) signal the intensity of the stimulus, (3) integrate information from many sources, and (4) initiate and direct the response. The nervous system collects and processes sensory information from many sources. These sensory stimuli may come together (converge) on fewer neurons whose activity in response to the stimuli determines the action taken. The "decision" to act may then be transmitted to many more neurons (divergence), which direct the activity.

Neural pathways normally have four elements: (1) sensory neurons, (2) association neurons, (3) motor neurons, and (4) effectors. Overall, nervous systems consist of numerous interconnected neural pathways, which may be either diffuse or centralized.

5) How Is the Human Nervous System Organized?

The nervous system of humans and other vertebrates consists of the central nervous system and the peripheral nervous system.

The peripheral nervous system is further subdivided into sensory and motor portions. The motor portions consist of the somatic nervous system (which controls voluntary movement) and the autonomic nervous system (directing involuntary responses).

Within the central nervous system, the spinal cord contains (1) neurons controlling voluntary muscles and the autonomic nervous system and neurons communicating with the brain and other parts of the spinal cord; (2) axons leading to and from the brain; and (3) neural pathways for reflexes and certain simple behaviors. The brain consists of three parts—the hindbrain, midbrain, and forebrain, each further subdivided into distinct regions.

The hindbrain in humans consists of the medulla and pons, which control involuntary functions (such as breathing), and the cerebellum, which coordinates complex motor activities (such as typing). In humans, the small midbrain contains the reticular formation, a filter and relay for sensory stimuli. The forebrain includes the thalamus, a sensory relay station that shuttles information to and from conscious centers in the forebrain. The diverse structures of the limbic system of the forebrain are involved in emotion, learning, and the control of instinctive behaviors such as sex, feeding, and aggression. Finally, the cerebral cortex of the forebrain is the center for information processing, memory, and initiation of voluntary actions. It includes primary sensory and motor areas and association areas that analyze sensory information and plan movements.

6) How Does the Brain Produce the Mind?

The cerebral hemispheres are each specialized. In general, the left hemisphere is dominant in speech, reading, writing, language comprehension, mathematical ability, and logical problem solving. The right hemisphere specializes in recognizing faces and spatial relationships, artistic and musical abilities, and recognition and expression of emotions.

Memory takes two forms. Short-term memory is electrical or chemical. Long-term memory probably involves structural changes that increase the effectiveness of synapses. The hippocampus is an important site for learning and for the transfer of information into long-term memory. The temporal lobes are important for memory.

7) How Do Sensory Receptors Work?

Receptors are transducers, converting signals from one form to another. Receptor cells are named after the stimulus to which they respond.

8 **How Is Sound Sensed?**

In the vertebrate ear, air vibrates the tympanic membrane, which transmits vibrations to the bones of the middle ear and then to the oval window of the fluid-filled cochlea. Within the cochlea, vibrations bend the hairs of hair cells, which are receptors located between the basilar and tectorial membranes. This bending produces receptor potentials in the hair cells that cause action potentials in the axons of the auditory nerve, which leads to the brain.

9 **How Is Light Sensed?**

In the vertebrate eye, light enters the cornea and passes through the pupil to the lens, which focuses an image on the fovea of the retina. Two types of photoreceptor, rods and cones, are located deep in the retina. They produce receptor potentials in response to light. These signals are processed through several layers of neurons in the retina and finally are translated into action potentials in the optic nerve, which leads to the brain. Rods are more abundant and more light-sensitive than are cones and provide vision in dim light. Cones, which are concentrated in the fovea, provide color vision.

10 **How Are Chemicals Sensed?**

Terrestrial vertebrates detect chemicals in the external environment either by olfaction or by taste. Each olfactory or taste receptor cell type responds to only one or a few specific types of molecules, allowing discrimination among tastes and odors. Olfactory neurons of vertebrates are located in a tissue that lines the nasal cavity. Taste receptors are located in clusters called taste buds on the tongue. Pain is a special type of chemical sense in which sensory neurons respond to chemicals released by damaged cells.

Key Terms

Thinking Through the Concepts

Multiple Choice

1. _____ *are integration centers in neurons.*
   a. dendrites
   b. axons
   c. cell bodies
   d. ion channels
   e. synapses

2. *The role of the axon is to*
   a. integrate signals from the dendrites
   b. release neurotransmitter
   c. conduct the action potential to the synaptic terminal
   d. synthesize cellular components
   e. stimulate a muscle, gland, or another neuron

3. *Which of the following statements is FALSE?*
   a. The hindbrain contains the medulla, pons, and cerebellum.
   b. The thalamus is an important sensory relay structure.
   c. The cerebral cortex in humans is folded into convolutions.
   d. The hippocampus has an important role in recognizing and experiencing fear.
   e. The hypothalamus coordinates the activities of the autonomic nervous system.

4. *The fovea is*
   a. the blind spot
   b. a clear area in front of the pupil and iris
   c. the tough outer covering of the eyeball
   d. the substance that gives the eyeball its shape
   e. the central focal region of the vertebrate retina

5. *Light entering the eye and striking the choroid would travel, in order, through the*
   a. lens, vitreous humor, cornea, aqueous humor, and retina
   b. retina, aqueous humor, lens, vitreous humor, and cornea
   c. cornea, aqueous humor, retina, vitreous humor, and lens
   d. cornea, aqueous humor, lens, vitreous humor, and retina
   e. lens, aqueous humor, cornea, vitreous humor, and retina

6. *Pain perception is carried out by a special type of*
   a. mechanoreceptor
   b. photoreceptor
   c. chemoreceptor
   d. magnetoreceptor
   e. thermoreceptor

? Review Questions

1. List four major parts of a neuron, and explain the specialized function of each part.

2. Diagram a synapse. How are signals transmitted from one neuron to another at a synapse?

3. How does the brain perceive the intensity of a stimulus? The type of stimulus?

4. What are the four elements of a simple nervous pathway? Describe how these elements function in the human pain-withdrawal reflex.

5. Draw a cross section of the spinal cord. What types of neurons are located in the spinal cord? Explain why severing the cord paralyzes the body below the level where it is severed.

6. Describe the functions of the following parts of the human brain: medulla, cerebellum, reticular formation, thalamus, limbic system, and cerebrum.

7. What structure connects the two cerebral hemispheres? Describe the evidence that the two hemispheres are specialized for distinct intellectual functions.

8. Distinguish between long-term and working memory.

9. What are the names of the specific transducers used for taste, vision, hearing, smell, and touch?

10. Why are we apparently able to distinguish hundreds of different flavors, when we have only four types of taste receptors? How are we able to distinguish so many different odors?

11. Describe the structure and function of the various parts of the human ear by tracing a sound wave from the air outside the ear to the cells that cause action potentials in the auditory nerve.

12. How does the structure of the inner ear allow for the perception of pitch? Of sound intensity?

13. Diagram the overall structure of the human eye. Label the cornea, iris, lens, sclera, retina, and choroid. Describe the function of each structure.

14. How does the lens change shape to allow focusing of distant objects? What defect makes focusing on distant objects impossible, and what is this condition called? What type of lens can be used to correct it, and how does it do so?

15. List the similarities and differences between rods and cones.

16. Distinguish between taste and olfaction.

17. Describe how pain is signaled by tissue damage.

Applying the Concepts

1. Argue for or against the statement "Consciousness by its nature is incomprehensible; the brain will never understand the mind."

2. In Parkinson's disease, which afflicts several million Americans, the cells that produce the neurotransmitter dopamine degenerate in a small part of the brain that is important in the control of movement. Some physicians have reported improvement in patients after injecting cells taken from the same general brain region of an aborted fetus into appropriate parts of the brain of a Parkinson patient. Discuss this type of surgery from as many viewpoints as possible: ethical, financial, practical, and so on. On the basis of your responses, is fetal transplant surgery the answer to Parkinson's disease?

3. If the axons of human spinal cord neurons were unmyelinated, would the spinal cord be larger or smaller? Would you move faster or slower? Explain.

4. Explain the statement "Your sensory perceptions are purely a creation of your brain." Discuss the implications for communicating with other humans, with other animals, and with intelligent life from another universe.

5. Discuss the biological, ethical, and social implications of regulating noise as a form of pollution in the following places: rock concerts, construction sites, a neighborhood that was built near an existing airport, and factories.

6. Corneal transplants can help restore vision and greatly improve the recipient's quality of life. What properties of the cornea make it an excellent candidate for transplantation? Suggest some ways in which society could improve the availability of corneal and other tissues for transplantation.

Group Activity

Form groups of three or four for the following exercise: Devise and carry out a simple experiment to demonstrate that there are fewer cones and less-acute vision in the peripheral portions of the human retina than in and near the fovea. (Recall the different properties and abilities of rods and cones.) Try this experiment on all members of the group. Estimate the percentage of your field of vision that has relatively few cones in it.

For More Information

Angier, N. "Storming the Wall." *Discover*, May 1990. Describes the blood–brain barrier and new methods of penetrating it with drugs.

Axel, R. "The Molecular Logic of Smell." *Scientific American*, October 1995. Describes research that uncovers some of the mechanisms by which the nose and brain decipher scents.

Beardsley, T. "The Machinery of Thought." *Scientific American*, August 1997. Using PET and MRI on monkeys and humans, researchers are learning more about where working memory resides.

Bower, B. "Creatures in the Brain." *Science News*, April 13, 1996. Using imaging techniques, scientists have discovered clues as to how different regions of the brain are specialized for different concepts.

Dolnick, E. "Obsessed." *Health*, September 1994. Researchers have linked obsessive–compulsive disorders to imbalances in neurotransmitters.

Freedman, D. H. "In the Realm of the Chemical." *Discover*, June 1993. Describes how smell and taste help us experience our world.

Gutin, J. C. "Good Vibrations." *Discover*, June 1993. Research uncovers mechanisms of hearing.

Holloway, M. "Rx for Addiction." *Scientific American*, March 1991. By studying drug addiction, neurobiologists are learning more about the brain and new ways to counteract addictions.

Kimura, D. "Sex Differences in the Brain." *Scientific American*, September 1992. Hormonal differences between the sexes influence brain development.

Koretz, J. F., and Handelman, G. H. "How the Human Eye Focuses." *Scientific American*, July 1988. Describes focusing in the human eye with an emphasis on the loss of focusing ability that occurs with age.

McKean, K. "Pain." *Discover*, October 1986. Describes the discovery of bradykinin and its role in pain perception.

Nathans, J. "The Genes for Color Vision." *Scientific American*, February 1989. Good discussion of the physiology and genetics of color blindness.

Raichle, M. E. "Visualizing the Mind." *Scientific American*, April 1994. Brain imaging techniques partially open the "black box" of the mind.

Scientific American special issue on Mind and Brain, September 1992. Articles describe the development of the brain, the biochemical basis of learning, major neurological disorders, and aging effects on the brain.

Answers to Multiple-Choice Questions
1. c   2. c   3. d   4. e   5. d   6. c

The grace and beauty of the acrobatic movements of Michael Jordan (number 23) are made possible by a coordinated interaction between his skeletal and muscular systems.

Action and Support: The Muscles and Skeleton 34

At a Glance

Net Watch

On-line resources for this chapter are on the World Wide Web at:
http://www.prenhall.com/audesirk
(click on the Table of Contents link and then select Chapter 34).

The flight of a bumblebee, the pounce of a cat, the soaring leap of Michael Jordan, and virtually all other instances of animal movement depend on the same humble yet elegant mechanism. That mechanism is the action of muscles. Muscle tissue is composed of muscle cells, which perform only one trick: They exert a force by contracting. Under the influence of natural selection, however, this simple unidirectional force has been applied to complex structural elements such as wings, hands, and fins, and its action has come to be coordinated by the nervous system. The resulting capacity for movement gives animals the ability to search for food, seek out new habitats, flee from danger—and, on occasion, to move in ways that we find awe-inspiring.

Muscles and skeletons also perform more-mundane, but still crucial, functions. Pumping blood through the circulatory system, moving food through the digestive system, and breathing are some of the essential processes that depend on muscle contraction. Skeletons play an equally essential role by opposing the force of gravity. Terrestrial organisms in particular depend on skeletal support to maintain their shapes. Without your skeleton, you'd be nothing more than a formless, quivering mound of tissue.

1) How Do Muscles Work?

All muscular work requires that muscles alternately contract and lengthen, but muscles are active only during the contraction phase. The lengthening that follows contraction is passive; it occurs when muscles relax and are stretched out by other forces. For example, a relaxed muscle may be lengthened by contractions of opposing muscles, the weight of a limb, or pressure from food on the muscular walls of the stomach.

Animals have a variety of muscle types, each specialized to perform a particular function. Mammals have evolved three types of muscle: (1) skeletal, (2) cardiac, and (3) smooth. All work on the same basic principles but differ in function, appearance, and control (Table 34-1).

693

Table 34-1 Location, Characteristics, and Functions of the Three Muscle Types

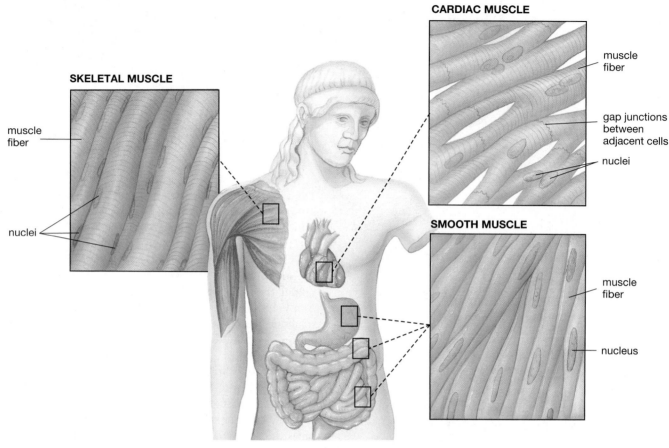

Property	Smooth	Cardiac	Skeletal
		Type of Muscle	
Muscle appearance	Unstriped	Irregular stripes	Regular stripes
Cell shape	Spindle	Branched	Spindle or cylindrical
Number of nuclei	One per cell	Many per cell	Many per cell
Speed of contraction	Slow	Intermediate	Slow to rapid
Contraction caused by	Spontaneous, stretch, nervous system, hormones	Spontaneous	Nervous system
Function	Controls movement of substances through hollow organs	Pumps blood	Moves the skeleton
Voluntary control	Normally no[a]	Normally no[a]	Yes

[a]Smooth and cardiac muscles normally contract without conscious control. In some cases, however, their contractions may be initiated or modified voluntarily. For example, heart rate can be voluntarily slowed after biofeedback training, and bladder contractions are initiated consciously.

Skeletal muscle, so named because it is used to move the skeleton, is also called **striated muscle**, because it has a striped appearance under the microscope (*striated* means "striped"). Most skeletal muscle is under voluntary, or conscious, control. It can produce contractions ranging from quick twitches (as in blinking) to powerful, sustained tension (as in carrying an armload of textbooks). **Cardiac muscle** is so named because it is located only in the heart. It is spontaneously active, initiating its own contractions, but it is influenced by nerves and hormones. Like skeletal muscle, cardiac muscle has a striped appearance under the microscope. **Smooth muscle**, as its name suggests, lacks the orderly striped appearance of skeletal and cardiac muscles. Smooth muscle lines the walls of the digestive tract and large blood vessels and produces slow, sustained contractions. These contractions are primarily involuntary; that is, they are not under conscious control.

Our discussion of muscles begins with and emphasizes skeletal muscle, the most abundant form of muscle in the body. We then survey cardiac and smooth muscle.

The Structure of Skeletal Muscle Cells Is Closely Tied to Their Function

In most eukaryotic cells, motions such as shape changes, locomotion, and movement of organelles within cells depend on microfilaments constructed of the protein **actin** (see Chapter 6). When microfilaments are involved in cellular motion, the movement typically depends on interactions between the actin microfilaments and strands of another protein known as **myosin**. During movements, the strand of actin and myosin slide past one another and change the shape of the cell. This evolutionarily ancient mechanism forms the basis of animal muscle cell function.

Individual muscle cells, called **muscle fibers**, are among the largest cells in the human body. Ranging from 10 to 100 micrometers in diameter (a bit smaller than the period at the end of this sentence), each muscle fiber runs the entire length of the muscle, which may be as long as 35 centimeters (about 14 inches) in a human thigh (Fig. 34-1a,b). Each muscle fiber, in turn, contains many **myofibrils**, individual contractile subunits extending from

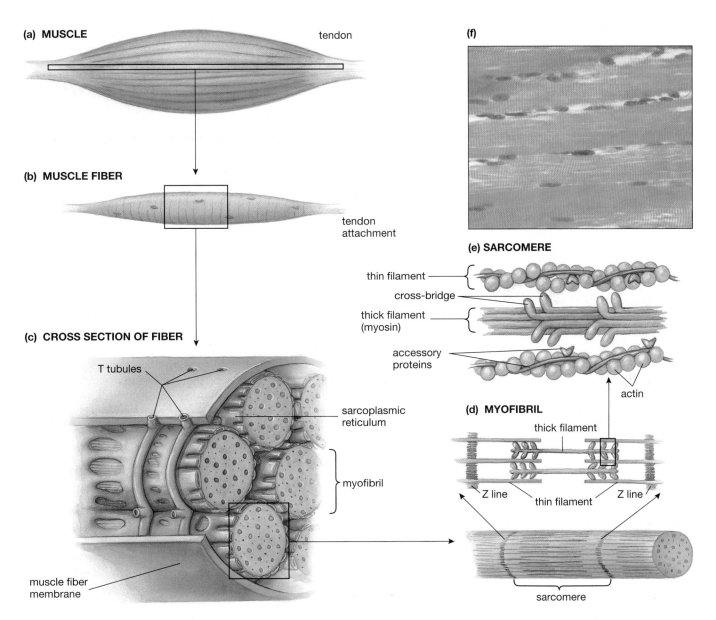

Figure 34-1 *Muscle structure*
(*a*) A muscle is made up of (*b*) individual muscle cells, called fibers. (*c*) The sarcoplasmic reticulum subdivides each muscle fiber into smaller cylinders called myofibrils. (*d*) The myofibril consists of a series of subunits called sarcomeres, attached end to end. Within each sarcomere are (*e*) alternating thick and thin filaments, which can be connected by cross-bridges (projections of the myosin molecules that make up the thick filaments). (*f*) In vertebrate skeletal muscles, the actin and myosin filaments are regularly arranged in the muscle fiber, giving it a striped appearance.

one end of the fiber to the other. Each cylindrical myofibril is surrounded by **sarcoplasmic reticulum**. Like the endoplasmic reticulum from which it is derived, the sarcoplasmic reticulum is a series of double sheets of membrane forming interconnected hollow tubes (Fig. 34-1c). The fluid within the sarcoplasmic reticulum stores high concentrations of calcium ions. Deep indentations of the muscle fiber membrane, called **transverse tubules** or **T tubules**, extend down into the muscle fiber, passing very close to portions of the sarcoplasmic reticulum. This arrangement of T tubules and sarcoplasmic reticulum is crucial to the control of muscle contraction, as described later.

Within each myofibril is a beautifully precise arrangement of actin and myosin filaments. The myofibrils are organized into subunits called **sarcomeres** (Fig. 34-1d), which are aligned end to end throughout the length of the myofibril. Within each sarcomere, actin molecules (in association with two accessory proteins) form the **thin filaments**. Suspended between the thin filaments are **thick filaments**. Thick filaments are composed of myosin in contact with the thin filaments by way of small branches of the myosin molecules called **cross-bridges** (Fig. 34-1e). The thin filaments are attached to structures called **Z lines**, which lie at the junction points between adjacent sarcomeres. The thick and thin filaments are lined up in all the myofibrils, giving the cell its striped appearance (Fig. 34-1f).

Muscle Contraction Results from Thick and Thin Filaments Sliding Past One Another

Muscle contraction is controlled by a process that depends on the molecular structure of the thin filament. The actin protein that makes up most of the thin filament is formed from a double chain of subunits, resembling a twisted double strand of pearls. Each of the subunits has a binding site for a myosin cross-bridge. In a relaxed muscle cell, however, these binding sites are blocked by molecules of accessory proteins (see Fig. 34-1e). These accessory proteins prevent the myosin cross-bridges from attaching to the actin of the thin filament.

When a muscle contracts, the accessory proteins of the thin filament are moved aside, exposing the binding sites on the actin. As soon as the sites are exposed, myosin cross-bridges attach. Using energy from the splitting of adenosine triphosphate (ATP), the cross-bridges repeatedly bend, release, and reattach farther along, much like a sailor pulling in an anchor line hand over hand (Fig. 34-2a). The thin filaments are pulled past the thick filaments, shortening the sarcomere and contracting the muscle (Fig. 34-2b). Because the thick and thin filaments slide past one another during contraction, muscle contraction is described as using a *sliding-filament mechanism*.

How Is Muscle Contraction Controlled?

Ultimately, muscle contraction is controlled by the nervous system, which exercises control by regulating the concentration of calcium ions around the filaments of muscle cells. First, a muscle cell is stimulated by a *motor neuron* (a neuron that controls muscles; see Chapter 33). An action potential in the muscle cell travels into the cell's interior by passing down the T tubules (see Fig. 34-1c). On reaching the sarcoplasmic reticulum, the action potential causes calcium ions to be released from the sarcoplasmic reticulum and to flow into the cytoplasm of the muscle cell. Once in the cytoplasm, calcium ions bind to the accessory proteins of the thin filament, changing their shape such that the myosin binding sites are exposed. With its binding sites exposed, the fiber contracts. As soon as the action potential fades away, active transport proteins in the sarcoplasmic reticulum membrane pump the calcium ions back inside their intracellular "holding area" in the sarcoplasmic reticulum. As a result, the accessory proteins return to a shape that blocks the myosin binding sites, and the fiber relaxes.

The points at which motor neurons form synapses with muscle cells are known as **neuromuscular junctions** (Fig. 34-3, p. 698). Neuromuscular junctions differ from most other synapses in two important ways: (1) They are always excitatory, never inhibitory. (2) In contrast to the summation of multiple synaptic inputs that is required to generate an action potential in a typical neuron, every action potential in a motor neuron elicits an action potential in a muscle fiber, causing all its sarcomeres to contract. So how does the nervous system control the strength and degree of muscle contraction? Namely, by controlling the number of muscle fibers stimulated and the frequency of action potentials in each fiber. A single action potential doesn't cause a muscle cell to contract fully, because calcium is pumped away too rapidly for many cross-bridge movement cycles to occur. Thus, many action potentials in rapid succession are required to contract a muscle fully. If rapid firing is prolonged, the muscle produces a sustained maximal contraction called **tetany**, such as you experience when you carry an armful of books like this one.

Most motor neurons innervate (form synapses onto) more than one muscle fiber. The group of fibers on which a single motor neuron forms synapses is called a **motor unit**. The number of muscle fibers in a motor unit varies from muscle to muscle. Muscles used for large-scale movement, such as those of the thigh and buttocks that help Michael Jordan leap into the air for a slam dunk, may have hundreds of muscle fibers in each motor unit. In muscles used for fine control of small body parts, such as those of the lips, eyes, and tongue, only a few muscle cells may be stimulated by each motor neuron. Therefore, when a single motor neuron fires an action potential, that action potential can cause the contraction of a few muscle cells or of many, depending on the size of the motor unit.

(a)

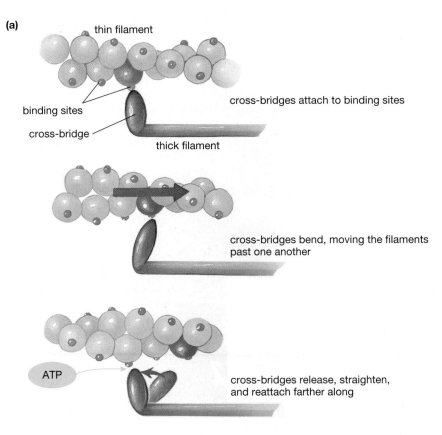

Figure 34-2 Muscle movement
(a) The cross-bridges connecting myosin to actin swivel as if on hinges, pulling the actin filaments (which are attached to the ends of the sarcomere) toward the center of the sarcomere. Repeated cycles of attachment, swiveling, release (which requires ATP energy to break the attachment), and reattachment result in muscle contraction.
(b) Muscle contraction causes the thick and thin filaments to slide past one another, shortening the individual sarcomeres and hence the muscle cell.

(b)

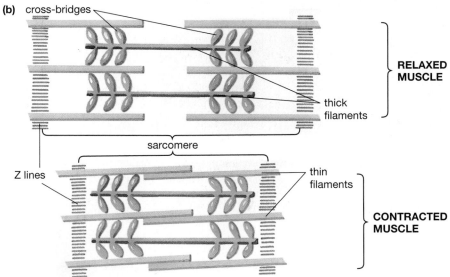

Cardiac Muscle Powers the Heart

Cardiac muscle is located only in the heart. Cardiac muscle, like skeletal muscle, is striated owing to the regular arrangement of sarcomeres with their alternating thick and thin filaments. As with skeletal muscle, the contraction of cardiac muscle is induced when action potentials spread into the cell through the T tubules and cause a release of calcium from the sarcoplasmic reticulum. In contrast to skeletal muscle, calcium also enters the cytoplasm from the extracellular fluid. Unlike skeletal muscle fibers, which contract in response to action potentials that originate in motor neurons, cardiac muscle fibers can initiate their own contractions. This quality is particularly well developed in the specialized cardiac muscle fibers of the *sinoatrial* (SA) *node*, which serves as the heart's pacemaker (see Chapter 27). Action potentials originating in the pacemaker are spread rapidly throughout the heart by specialized areas in which numerous gap junctions connect the membranes of adjacent muscle cells. Gap junctions allow action poten-

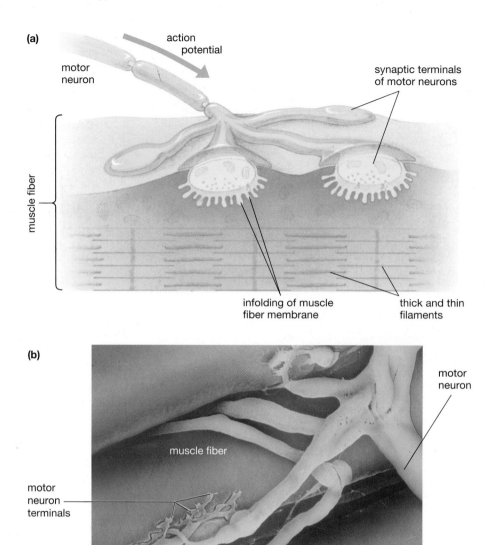

(a)

action potential

motor neuron

synaptic terminals of motor neurons

muscle fiber

infolding of muscle fiber membrane

thick and thin filaments

(b)

motor neuron

muscle fiber

motor neuron terminals

Figure 34-3 The neuromuscular junction
(a) Diagram of a neuromuscular junction in cross section. Action potentials in the motor neuron stimulate the muscle fiber membrane, which is folded beneath the terminal to allow more surface area for receptors. **(b)** A scanning electron micrograph of motor neuron terminals that form synapses on muscle fibers.

tials to travel from one cell to the next, synchronizing their contractions (see Table 34-1). These connecting areas are absent in skeletal muscle, where each cell is individually stimulated by a branch of a motor neuron.

Smooth Muscle Produces Slow, Involuntary Contractions

Smooth muscle surrounds blood vessels and most hollow organs, including the uterus, bladder, and digestive tract. As its name suggests, smooth muscle lacks the regular arrangement of sarcomeres that characterizes skeletal and cardiac muscles (see Table 34-1). Smooth muscle generally produces either slow, sustained contractions

(such as the constriction of arteries to elevate blood pressure during times of stress) or slow, wavelike contractions (such as the peristaltic waves that move food through the digestive tract). Like cardiac muscle cells, most smooth muscle cells are directly connected to one another by gap junctions, allowing synchronized contraction. Smooth muscle lacks sarcoplasmic reticulum; all the calcium needed for contraction flows in from the extracellular fluid during the action potential. Smooth muscle contraction may be initiated by stretching, by hormones, by nervous signals, or by some combination of these stimuli. Although contractions of the bladder can be initiated voluntarily, most smooth muscle contraction is under involuntary control.

2) What Does the Skeleton Do?

www

For most of us, the word *skeleton* conjures up the image of our own collection of bones, the skull smiling like a Halloween decoration. But skeletons are in fact as diverse as any other structure in the animal kingdom and need not even be made of bone. A **skeleton** can be broadly defined as a supporting framework for the body. Within the animal kingdom, skeletons come in three radically different forms: (1) *hydrostatic skeletons* (made of fluid), (2) *exoskeletons* (on the outside of the animal), and (3) *endoskeletons* (internal).

The **hydrostatic skeletons** of worms, mollusks, and cnidarians are the simplest, consisting of a fluid-filled sac (Fig. 34-4a). Fluid, which cannot be compressed, provides excellent support; but because fluid is formless, these animals rely on two layers of surrounding muscles in the body wall—one circular, the other longitudinal—to determine their shape. The wavelike movements of a burrowing earthworm, alternately extending to stringlike thinness and then fattening as it contracts, provide an excellent illustration of the flexibility of hydrostatic skeletons.

Exoskeletons (literally, "outside skeletons") encase the bodies of arthropods (such as spiders, crustaceans, and insects). Exoskeletons vary tremendously in thickness

(a)

circular muscle

longitudinal muscle

hydrostatic skeleton

LONGITUDINAL MUSCLES CONTRACTED

CIRCULAR MUSCLES CONTRACTED

(b)

flexible hinge material

rigid exoskeleton

pair of antagonistic muscles

Figure 34-4 *Not all skeletons are made of bone*
(a) Hydrostatic skeletons. The skeleton of worms, many mollusks, and cnidarians such as this sea anemone is essentially a fluid-filled tube with soft walls. The tube is surrounded by two layers of muscles that are perpendicular to one another: The fibers of circular muscles form bands around the circumference of the tube, and the fibers of longitudinal muscles run lengthwise. Because fluids are incompressible and can take any shape, if the circular muscles contract, the animal will become long and thin; if the longitudinal muscles contract, it will become short and fat.
(b) Exoskeletons. Arthropods, such as this Sally lightfoot crab, have armorlike skeletons on the outside of their bodies. Joints allow movement, produced by pairs of muscles (antagonistic muscles—flexor and extensor) that span the joint.

and rigidity, from the thin flexible covering of many insects and spiders to the armorlike covering of many crustaceans (Fig. 34-4b). All exoskeletons are thin and flexible at the *joints* (the juncture of two bones), allowing complex and skillful movements such as those of a web-spinning spider. Exoskeletons and hydrostatic skeletons are discussed further in Chapter 22.

Endoskeletons, the internal skeletons of humans and other vertebrates, are found only in echinoderms and chordates (see Chapter 22). These are actually the least common type of skeleton.

The Vertebrate Skeleton Serves Many Functions

The bony endoskeleton of humans and most vertebrates serves a wide variety of functions:

1. The skeleton provides a rigid framework that supports the body and protects the internal organs. The central nervous system (CNS), for example, is almost completely enclosed within the skull and vertebral column; the rib cage protects the lungs and the heart with its major blood vessels.
2. Vertebrates depend on bones for locomotion. Although muscle contractions provide the power, skeletal elements provide the structures that actually move the animal. Natural selection for efficient locomotion has produced the wonderfully designed wings, limbs, fins, and other complex skeletal structures that allow vertebrates to fly, run, swim, and even slam dunk.
3. Bones produce red blood cells, white blood cells, and platelets (see Chapter 27). In adults, these cells of the circulatory system are produced by *red bone marrow,* located in porous areas of bone in the sternum (breastbone), ribs, upper arms and legs, and hips.
4. Bone serves as a storage site for calcium and phosphorus. Bone contains 99% of the calcium and 90% of the phosphorus in the human body. It absorbs and releases these minerals as needed, maintaining a constant concentration in the blood. *Yellow bone marrow*, dominated by fat cells, also stores energy reserves.
5. The skeleton even participates in sensory transduction. As you may recall from Chapter 33, three tiny bones of the middle ear (hammer, anvil, and stirrup) transmit the sound vibrations between the eardrum and the cochlea.

The 206 bones of the human skeleton can be placed in two categories: (1) the axial skeleton and (2) the appendicular skeleton. The **axial skeleton**, whose bones form the axis of the body, includes the bones of the head, vertebral column, and rib cage. The **appendicular skeleton**, whose bones form the appendages (extremities) and their attachments to the axial skeleton, includes the pectoral (shoulder) and pelvic (hip) girdles and the bones of the arms, legs, hands, and feet (Fig. 34-5).

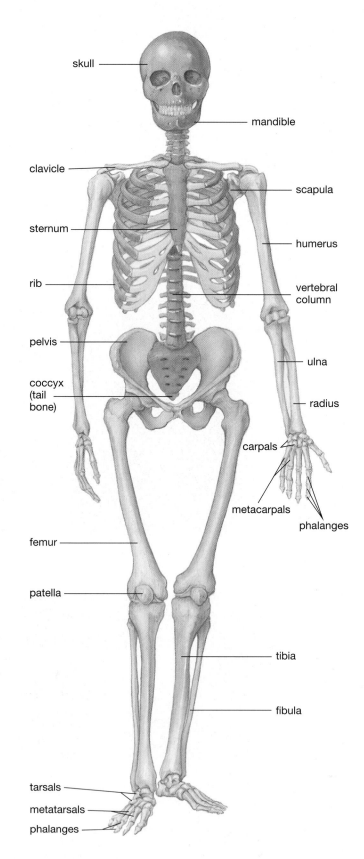

Figure 34-5 *The human skeleton*
The human skeleton, showing the axial skeleton (tinged in blue-gray) and the appendicular skeleton (bone color).

3) Which Tissues Compose the Vertebrate Skeleton?

The vertebrate skeleton is composed primarily of two types of tissue: cartilage and bone. Both (1) *cartilage* and (2) *bone* are rigid tissues that consist of living cells embedded in a matrix of a protein called **collagen** (see Chapter 26).

Cartilage Provides Flexible Support and Connections

Cartilage plays many roles in the human skeleton. For example, during the development of the embryo, the skeleton is first formed from cartilage that is only later replaced by bone (Fig. 34-6). Cartilage also covers the ends of bones at joints, supports the flexible portion of the nose and external ears, connects the ribs to the sternum (breastbone), and provides the framework for the

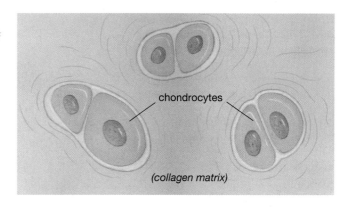

Figure 34-7 Cartilage
In cartilage, the chondrocytes, or cartilage cells, are embedded in an extracellular matrix of the protein collagen, which they secrete.

larynx, trachea, and bronchi of the respiratory system. In addition, it forms tough, shock-absorbing pads that are in the knee joints and that form the **intervertebral discs** between the vertebrae of the backbone.

The living cells of cartilage are called **chondrocytes**. These cells secrete a flexible, elastic, nonliving matrix of collagen that surrounds them and forms the bulk of the cartilage (Fig. 34-7). No blood vessels penetrate cartilage; to exchange wastes and nutrients, chondrocytes must rely on the gradual diffusion of materials through the collagen matrix. As you might predict, cartilage cells have a very slow metabolic rate; damaged cartilage repairs itself very slowly, if at all.

Bone Provides a Strong, Rigid Framework for the Body

Bone is the most rigid form of connective tissue. Although bone resembles cartilage, the collagen fibers of bone are hardened by deposits of the mineral *calcium phosphate*. Bones, such as those supporting your arms and legs, consist of a hard outer shell of **compact bone**, with **spongy bone** in the interior (Fig. 34-8). Compact bone is dense and strong and provides an attachment site for muscle. Spongy bone is lightweight, rich in blood vessels, and highly porous. Bone marrow, where blood cells form, is found in cavities of spongy bone. In contrast to cartilage, bone is well supplied with blood capillaries.

Three types of cells are associated with bone: (1) **osteoblasts** (bone-forming cells), (2) **osteocytes** (mature bone cells), and (3) **osteoclasts** (bone-dissolving cells). Early in development, when bone is replacing cartilage, osteoclasts invade and dissolve the cartilage; then osteoblasts replace it with bone.

As bones grow, osteoblasts form a thin layer covering the outside of the bone. The osteoblasts secrete a hardened matrix of bone and gradually become entrapped within it. They then stop secreting matrix and become mature osteocytes. Osteocytes are nourished by nearby

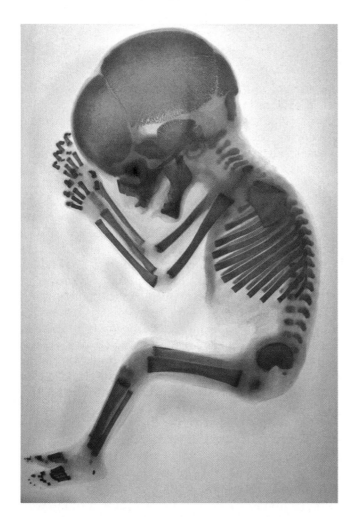

Figure 34-6 Bone replaces cartilage during development
In this 16-week-old human fetus, bone is stained magenta. The clear areas at the wrists, knees, ankles, elbows, and breastbone show where cartilage will later be replaced by bone.

Figure 34-8 The fine structure of bone
(a) A typical bone such as those in the arms and legs is made of an outer layer of compact bone and spongy bone inside. For simplicity, blood vessels are not shown. *(b)* Osteons are clearly visible in this micrograph. Each includes a central canal containing a capillary. The capillary nourishes the osteocytes, embedded in the concentric rings of bone material.

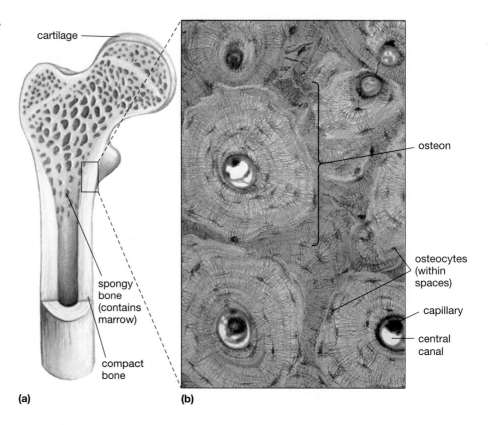

(a)   (b)

capillaries and are connected to other osteocytes by thin extensions that the bone cells send out through narrow channels in the bone. Although unable to produce more bone, osteocytes may secrete substances that control the continuous remodeling of bone.

Bone Remodeling Allows Skeletal Repair and Adaptation to Stresses

Each year 5% to 10% of all the bone in your body is dissolved away and replaced, a process called *bone remodeling*. This process allows your skeleton to alter its shape subtly in response to the demands placed on it. For example, bones that carry heavy loads or are subjected to extra stress, like those in Michael Jordan's legs, become thicker to provide more strength and support. Archeologists excavating the skeletons of individuals buried in volcanic ash at Pompeii were able to identify archers because the bones of their right and left arms differed considerably in thickness. Normal stresses are actually a major factor in maintaining bone strength. The bones of an arm or leg that is immobilized in a cast rapidly lose significant amounts of calcium.

Bone remodeling is the result of the coordinated activity of the osteoclasts that dissolve bone and the osteoblasts that rebuild it. Osteoclasts cling to the bone surface, secreting acids and enzymes that dissolve the hard matrix. Working in small groups, osteoclasts tunnel into the bone, creating channels. These channels are invaded by capillaries and by osteoblasts. The osteoblasts fill the channel with concentric deposits of new bone matrix, leaving only a small opening for the capillary. As a result of this process, hard bone is made up of tightly packed units called **osteons** (also known as *Haversian systems*), each consisting of concentric layers of bone with embedded osteocytes. The concentric deposits surround a *central canal*, through which a capillary passes (see Fig. 34-8). Osteoclasts and osteocytes also play a crucial role in the repair of bone fractures, as described in "A Closer Look: Sticks and Stones—The Repair of Broken Bones."

Bone remodeling replaces bone as it ages and becomes brittle. As the body gets older, however, the remodeling process slows, and bones tend to become more fragile as a result. (See "Health Watch: Osteoporosis—The Hidden Crippler," pp. 704–705.) The continuous turnover of bone also allows the body to maintain constant levels of calcium in the blood: Calcium from bones is retained in the blood if blood calcium levels drop but is returned to bone if blood calcium is adequate or high. This process is regulated by hormones; calcitonin and parathormone cause bones to absorb calcium from the blood and release calcium into the blood, respectively (see Chapter 32).

4) How Does the Body Move?

In addition to providing support for the body, the skeleton facilitates movement by providing a framework that muscles can move. Movement of the skeleton is accomplished by the action of pairs of **antagonistic muscles**: One muscle actively contracts, causing the other to be passively extended (Fig. 34-9; see also Fig.

A Closer Look
Sticks and Stones—The Repair of Broken Bones

Bones are strong, but they're not indestructible. All too many of us have experienced the trauma and pain of a broken bone. We cringe at each recollection of the ill-fated mishap, the audible cracking sound, the dawning realization that the painful swelling and oddly bent-looking limb necessitated a trip to the emergency room. There, the attending physician coaxes the damaged bone back into its proper orientation and immobilizes it with a cast or splint. The rest of the healing process is up to the body's own repair mechanisms. Over the next 6 weeks or so, the body orchestrates a systematic restoration of the bone's strength and integrity.

The healing process begins when the break is surrounded by a large blood clot from vessels ruptured during the injury (Fig. E34-1a). Phagocytic cells and osteoclasts in the blood ingest and dissolve cellular debris and bone fragments. The fracture ruptures the *periosteum*, a thin layer of connective tissue that normally surrounds the bone and that is rich in capillaries, osteoblasts, and osteoblast-forming cells. The osteoblasts, in conjunction with cartilage-forming cells, then secrete a *callus*, a porous mass of bone and cartilage that surrounds the break (Fig. E34-1b). The callus replaces the original blood clot and holds the ends of the bones together while remodeling processes reform the original shape of the bone. Once the callus is in place, osteoclasts, osteoblasts, and capillaries invade it. Nourished by the capillaries, osteoclasts break down cartilage while osteoblasts add new bone (Fig. E34-1c). Finally, osteoclasts remove excess bone, restoring the bone's original shape (Fig. E34-1d).

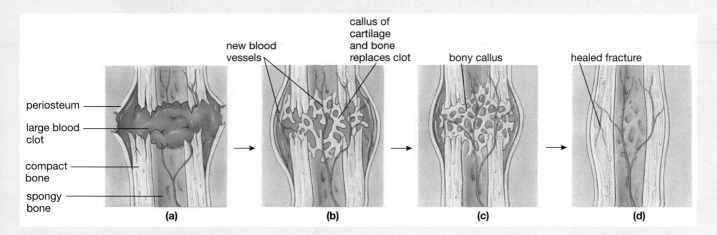

Figure E34-1 *The steps in bone repair*

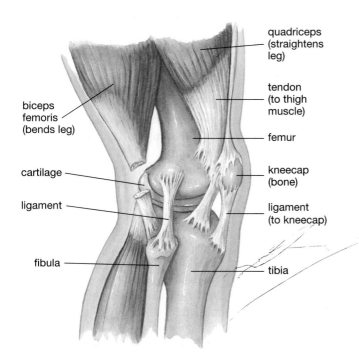

Figure 34-9 A hinge joint
The human knee—a hinge joint—showing antagonistic muscles (here, the biceps femoris and the quadriceps of the thigh), tendons, and ligaments. (The tendon of the biceps femoris has been cut for viewing purposes.) The complexity of this joint, coupled with the extreme stresses placed on it during activities such as jumping, running, or skiing, make it very susceptible to injury.

Health Watch
Osteoporosis—The Hidden Crippler

As a human grows and matures, bone density increases steadily, reaching a peak at about age 35. After this point, however, the activity of osteoclasts exceeds that of osteoblasts and bone density begins a slow, natural decline. Some bone loss is normal, but in people with the insidious (though preventable) condition known as **osteoporosis** (literally, "porous bones"; Fig. E34-2a), the loss is sufficient to weaken the bones, making them vulnerable to fractures and deformities. As many as 28 million Americans, most of them women past menopause, have

osteoporosis. In many cases, the vertebrae of individuals with osteoporosis compress, causing a hunch-backed appearance (Fig. E34-2b). In extreme cases, simple activities such as lifting a shopping bag, opening a window, or sneezing can break a bone. Of women living to age 85, nearly one-third will fracture a hip weakened by osteoporosis. As a result of hip fractures, the elderly can become much less self-sufficient and inactive, and one-quarter of them die within 6 months of complications such as pneumonia.

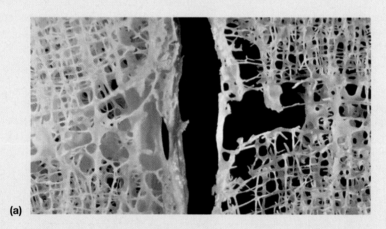
(a)

(b)

Figure E34-2 Osteoporosis
(a) Cross section of (left) a normal bone compared with (right) a bone from a woman with osteoporosis. **(b)** The devastating effects of osteoporosis extend beyond the obvious deformities, such as a hunch-backed appearance. Its victims are also at high risk for bone fractures.

34-4). Antagonistic muscles alter the configuration of the skeleton by causing movement around joints (in exoskeletons and vertebrate endoskeletons) or by altering the shape of the internal fluid (in hydrostatic skeletons; see Fig. 34-4).

Muscles Move the Skeleton around Flexible Joints

In vertebrates, bones act as levers that can be moved by skeletal muscles to which they are attached. The bones typically move around **joints**, the points at which two bones meet. Not all joints are movable, but in those that are designed to move, the portion of each bone that forms the joint is coated with a layer of cartilage, whose smooth, resilient surface allows the bone surfaces to slide past one another during movement. On either side of a joint, skeletal muscles are attached to bones by bands of tough, fibrous connective tissue called **tendons**. The bones them-

selves are joined at joints by bands of fibrous connective tissue called **ligaments** (see Fig. 34-9).

Most skeletal muscles are arranged in antagonistic pairs on opposite sides of a joint (see Figs. 34-4b and 34-9). When one of the muscles in a pair contracts, it moves a bone and simultaneously stretches the opposing muscle. In the most common types of joints, skeletal muscles span the joint; their contraction moves the bone on one side of the joint while the bone on the other side of the joint remains in a fixed position. These joints, including the elbows, knees, and finger joints, are called **hinge joints**. Like a hinged door, these joints are movable in only two dimensions.

In hinge joints, pairs of muscles lie in roughly the same plane as the joint, and the members of the muscle pair are known as the **flexor** and the **extensor**. (For instance, in the diagram in Fig. 34-4b, the red muscle is the flexor; when it contracts, the joint—and therefore the leg—flexes. The blue muscle is the extensor; when it con-

Women are eight times as likely to suffer from osteoporosis as men. Why? One reason is that the bones of women are about 30% less massive than those of men to start with, so they can less afford to lose bone. Another factor is dietary calcium, which tends to be lower in women's diets than in men's. Two-thirds of women between the ages of 18 and 30 get less than the RDA (Recommended Daily Allowance) of calcium, and women tend to consume even less calcium as they become older. As a result, when women reach the age when bone loss begins naturally, their bones may already be more fragile than they should be. Another factor unique to women is the role of the hormone estrogen. In women, estrogen stimulates osteoblasts and helps maintain bone density. After menopause, when estrogen production drops dramatically, women may lose 3% to 5% of their bone mass each year for several years. Half of all women over age 65 are estimated to have some degree of osteoporosis.

Another important factor contributing to osteoporosis is lack of weight-bearing exercise, such as walking, dancing, or running. Bones thrive on moderate stress, but as people age they tend to be less active. Being inactive or bedridden (or being weightless, as astronauts have discovered) results in rapid loss of bone calcium. In contrast, exercise, even in elderly people, can reverse bone loss and even increase bone mass. Alcoholism and smoking also contribute to bone loss and osteoporosis.

Fortunately, much of the pain, incapacitation, and expense (estimated at $14 billion per year) caused by fractures due to osteoporosis can be prevented. Clearly, the best way to prevent osteoporosis is a combination of regular exercise and adequate dietary or supplemental calcium and vitamin D (which is crucial to the proper metabolism of calcium). These steps will ensure that bone mass is as high as possible before natural, age-related losses begin and will also minimize such losses in old age. For women approaching menopause, estrogen supplements can help protect bone during the period of rapid bone loss that immediately follows menopause. Many women, however, choose not to undergo estrogen-replacement therapy, owing to its adverse effects on breast and uterine tissues, and turn to other treatments for osteoporosis. Another hormone, calcitonin, has fewer undesirable side effects and also has beneficial effects on bone deposition. Calcitonin, however, is a protein hormone that would be broken down by the digestive system if administered orally; until recently, it could be administered only by daily injection. Not long ago, the Federal Drug Administration (FDA) approved a calcitonin formulation that can be administered as a nasal spray.

Effective nonhormonal treatments for osteoporosis are also becoming increasingly available. For example, the mineral sodium fluoride, in combination with supplements of calcium and vitamin D, is sometimes prescribed to increase bone deposition. In addition, two classes of newly available drugs have been shown to increase bone density in individuals with osteoporosis. One, *biosophonates,* includes compounds that bind to bone and render the bone resistant to digestion by osteoclasts. The other, an even newer class of drugs (approved by the FDA in late 1997) is "selective estrogen receptor modulators," which are molecules that mimic estrogen's effects on the skeletal system while blocking estrogen's effects on breast and uterine tissues. None of these treatments is without potentially negative side effects, however, and none of them can actually cure osteoporosis, so the best remedy for the disease remains prevention.

tracts, the joint extends.) One end of each muscle, called the **origin**, is fixed to a relatively immovable bone on one side of the joint; the other end, the **insertion**, is attached to a mobile bone on the far side of the joint. When the flexor muscle contracts, it bends the joint; when the extensor muscle contracts, it straightens the joint. Thus, alternate contractions of flexor and extensor muscles cause the movable bone to pivot back and forth at the joint.

Other joints, such as those of the hip and shoulder, are **ball-and-socket joints**, in which the rounded end of one bone fits into a hollow depression in another (Fig. 34-10). Ball-and-socket joints allow movement in several directions: Simply compare the wide-ranging swinging of your arm or upper leg with the limited bending of your knee. The range of motion in ball-and-socket joints is made possible by at least two pairs of muscles, oriented at right angles to each other, that provide flexibility of movement.

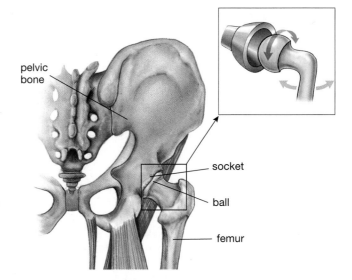

Figure 34-10 A ball-and-socket joint
The human hip—a ball-and-socket joint—consists of a rounded end (the ball; seen at the end of the femur) that fits into a cuplike depression (the socket; seen in the bones of the pelvis). This arrangement permits rotational movement.

Summary of Key Concepts

1) How Do Muscles Work?

Muscle can only actively contract. Skeletal muscle fibers (muscle cells) consist of numerous subunits—myofibrils—surrounded by the sarcoplasmic reticulum. Sarcomeres within each myofibril are formed by alternating thick filaments of myosin and thin filaments of actin and two accessory proteins.

During skeletal muscle contraction, the thick filaments form cross-bridges to exposed sites on the thin filaments. Using ATP, the cross-bridges bend, release, and reattach in a way that slides the filaments past each other, shortening the muscle fiber. When stimulated by motor neurons at neuromuscular junctions, muscles produce action potentials. Calcium, stored in the sarcoplasmic reticulum, is released when an action potential invades the muscle fiber. Calcium binds to one accessory protein, which pulls the other off the binding sites on the thin filaments, allowing contraction to occur. Both the strength and the degree of muscle contractions are determined by the number of muscle fibers stimulated and the frequency of action potentials in each fiber. Rapid firing can result in maximum contraction, or tetany, if sustained.

Cardiac, or heart, muscle also consists of sarcomeres that contain alternating thick and thin filaments. Its cells tend to contract rhythmically and spontaneously, but these contractions are synchronized by electrical signals produced by specialized muscle fibers in the sinoatrial node. Cardiac muscle fibers are interconnected electrically by gap junctions located between cells, allowing coordinated contraction.

Smooth muscle lacks organized sarcomeres, but like cardiac muscle, its cells are electrically coupled by gap junctions. Smooth muscle surrounds hollow organs (uterus, digestive tract, bladder) and blood vessels, producing slow sustained or rhythmic contractions, which are normally involuntary.

2) What Does the Skeleton Do?

Three types of skeletons are found in animals: Hydrostatic skeletons (in cnidarians, mollusks, and worms) use fluid confined in a chamber and surrounded by muscle. Exoskeletons, found in arthropods, are outer coverings with flexible joints.

Endoskeletons, including the bony vertebrate skeleton, are found in both echinoderms and chordates.

The vertebrate skeleton provides support for the body, attachment sites for muscles, and protection for internal organs. Red blood cells, white blood cells, and platelets form in the marrow of bones. Bone acts as a storage site for calcium and phosphorus. The axial skeleton includes the skull, vertebral column, and rib cage. The appendicular skeleton consists of the pectoral and pelvic girdles and the bones of the arms, legs, hands, and feet.

3) Which Tissues Compose the Vertebrate Skeleton?

Cartilage is located at the ends of bones and forms pads in the knee joints and the intervertebral discs. It also supports the nose, ears, and respiratory passages. During embryological development, cartilage is the precursor of bone. Cartilage is formed by chondrocytes, which surround themselves with a matrix of fibrous collagen.

Bone is formed by osteoblasts, which secrete a collagen matrix that becomes hardened by calcium phosphate. A typical bone consists of an outer shell of compact, hard bone, to which muscles are attached, and inner spongy bone, which provides a space for bone marrow. Remodeling of bone occurs continuously. Osteoclasts tunnel through the bone by means of acids and enzymes. Nourishing capillaries invade the tunnels, and osteoblasts fill the space with concentric layers of new bone, leaving a small central canal for the capillaries. This process produces osteons. Osteoblasts trapped within the bone are called osteocytes.

4) How Does the Body Move?

Skeletal muscles form antagonistic pairs that move the skeleton. In the vertebrate skeleton, movement occurs around joints, where bones are joined by ligaments. Muscles attach to bones on either side of the joint by tendons. The contraction of one muscle bends the joint and straightens its antagonistic muscle. At hinge joints, muscles are attached to the immovable bone at their origins; their insertions attach to the mobile bone. The contraction of the flexor muscle bends the joint; the contraction of its antagonistic extensor straightens it.

Key Terms

actin   *p. 695*
antagonistic muscles   *p. 702*
appendicular skeleton   *p. 700*
axial skeleton   *p. 700*
ball-and-socket joint   *p. 705*
cardiac muscle   *p. 694*
cartilage   *p. 701*
chondrocyte   *p. 701*
collagen   *p. 701*
compact bone   *p. 701*
cross-bridge   *p. 696*

endoskeleton   *p. 700*
exoskeleton   *p. 699*
extensor   *p. 704*
flexor   *p. 704*
hinge joint   *p. 704*
hydrostatic skeleton   *p. 699*
insertion   *p. 705*
intervertebral disc   *p. 701*
joint   *p. 704*
ligament   *p. 704*
motor unit   *p. 696*
muscle fiber   *p. 695*

myofibril   *p. 695*
myosin   *p. 695*
neuromuscular junction   *p. 696*
origin   *p. 705*
osteoblast   *p. 701*
osteoclast   *p. 701*
osteocyte   *p. 701*
osteon   *p. 702*
osteoporosis   *p. 704*
sarcomere   *p. 696*
sarcoplasmic reticulum   *p. 696*

skeletal muscle   *p. 694*
skeleton   *p. 699*
smooth muscle   *p. 694*
spongy bone   *p. 701*
striated muscle   *p. 694*
tendon   *p. 704*
tetany   *p. 696*
thick filament   *p. 696*
thin filament   *p. 696*
transverse (T) tubule   *p. 696*
Z line   *p. 696*

Thinking Through the Concepts

Multiple Choice

1. *Thin filaments in myofibrils consist of*
a. actin and accessory proteins
b. sarcomeres
c. cross-bridges
d. Z lines
e. myosin

2. *The deep infoldings of muscle fiber membranes that conduct action potentials are called*
a. sarcoplasmic reticula
b. Z lines
c. myofilaments
d. T tubules
e. sarcomeres

3. *The force of muscle contraction depends on the*
a. number of muscle fibers stimulated
b. number of motor units stimulated
c. frequency of action potentials in each motor unit
d. frequency of action potentials in each muscle fiber
e. all of the above

4. *Smooth muscle fibers can be distinguished from striated ones because smooth fibers*
a. contract more rapidly
b. lack regular arrangements of sarcomeres
c. lack gap junctions
d. contain only actin filaments
e. contain sarcoplasmic reticulum

5. *Bone-dissolving cells are called*
a. chondrocytes
b. osteoblasts
c. osteoclasts
d. osteocytes
e. erythroblasts

6. *Osteons contain all of the following EXCEPT*
a. blood vessels
b. intervertebral discs
c. osteocytes
d. calcium phosphate crystals
e. concentric layers of bone

? Review Questions

1. Sketch a relaxed muscle fiber containing a myofibril, sarcomeres, and thick and thin filaments. How would a contracted muscle fiber look by comparison?

2. Describe the process of skeletal muscle contraction, beginning with an action potential in a motor neuron and ending with the relaxation of the muscle. Your answer should include the following words: *neuromuscular junction, T tubule, sarcoplasmic reticulum, calcium, thin filaments, binding sites, thick filaments, sarcomere, Z line,* and *active transport.*

3. Explain the following two statements: Muscles can only actively contract; muscle fibers lengthen passively.

4. What are the three types of skeletons found in animals? For one of these, describe how the muscles are arranged around the skeleton and how contractions of the muscles result in movement of the skeleton.

5. Compare the structure and function of the following pairs: spongy and compact bone, smooth and striated muscle, and cartilage and bone.

6. Explain the functions of osteoblasts, osteoclasts, and osteocytes.

7. How is cartilage converted to bone during embryonic development? Where is cartilage located in the body, and what functions does it serve?

8. Describe a hinge joint and how it is moved by antagonistic muscles.

Applying the Concepts

1. Discuss some of the problems that would result if the human heart were made of skeletal muscle instead of cardiac muscle.

2. Muscle fibers in individuals with Duchenne muscular dystrophy (DMD) lack a protein called dystrophin, which normally helps control calcium release from sarcoplasmic reticulum. Lack of dystrophin leads to a constant leaking of calcium ions, which activates an enzyme that dissolves muscle fibers. The gene that causes DMD is inherited as a sex-linked recessive gene. Women with this gene have a 50% chance of passing the disease to their male children and a 50% chance of passing the gene to their female children. Afflicted children gradually become unable to walk and die of respiratory problems as young adults. Recently tests have been developed that allow a woman to determine if she is a carrier and if her fetus has inherited the gene. What factor would make a woman a candidate for this test? If a woman discovers she carries this gene, what are her options with regard to having children? Discuss the ethical implications of these various options.

3. Myasthenia gravis is caused by the abnormal production of antibodies that bind to acetylcholine receptors on muscle cells and that eventually destroy the receptors. The disease causes muscles to become flaccid, weak, or paralyzed. Drugs, such as neostigmine, that inhibit the action of acetylcholinesterase (an enzyme that breaks down acetylcholine) are used to treat myasthenia gravis. How does neostigmine restore muscle activity?

4. Some insects would have a tough time flying if one nerve impulse was required for each muscle contraction. Gnats, for example, may beat their wings 1000 times a second. At such high frequencies, contraction is "myogenic," originating from the stretching caused by the contraction of antagonistic muscles. Also, insect flight muscle cells are filled with giant mitochondria. Suggest a mechanism to explain how myogenic contraction works inside cells. Explain the significance of giant mitochondria.

5. Human muscle cells contain a mixture of three types of muscle fibers: slow-twitch oxidative, fast-twitch oxidative, and fast-twitch glycolytic. Slow-twitch muscle cells break down ATP slowly; they contain many mitochondria and large amounts of myoglobin, a dark pigment that acts as a reservoir for oxygen. All fast-twitch muscle cells break down ATP rapidly. Fast-twitch oxidative fibers have moderate amounts of myoglobin and glycogen. Fast-twitch glycolytic fibers have little myoglobin and lots of glycogen. The relative numbers of these fibers in different muscles is under genetic control. Use this information to explain the location of dark and white meat in birds. Predict the relative numbers of slow-twitch and fast-twitch fibers in muscle samples from the thighs of world-class marathon runners and world-class sprinters.

6. Some hours after death, skeletal muscle fibers run out of ATP and become rigid, although they are not contracted, a condition called rigor mortis. On the basis of your knowledge of the mechanisms of muscle contraction, including the role of calcium and ATP, explain what must be happening at the cellular level to cause this muscle stiffness.

Group Activity

Among vertebrate species, the structure of skeletal and muscle systems is closely tied to an animal's mode of locomotion. Natural selection favors the arrangements of bones and muscles that are most effective in achieving the type of movement that an animal needs to pursue its way of life. It is therefore frequently possible to infer an animal's lifestyle by examining its muscular and skeletal anatomies. Form a group of three students; examine the accompanying drawings, and explain the differences between the two pictured hind limbs. First, list all the differences that you can find between the two limbs (include at least 10 specific differences). Then discuss your list of differences and agree on the advantages and disadvantages of each feature on the list. For example, what would be the costs and benefits of the different muscle insertion points on the two femurs? Write down your conclusions. Finally, referring to your list of advantages and disadvantages, make a detailed prediction about the locomotion and way of life of each animal. Does each animal fly, run, swim, dig, or climb? Is each fast or slow? Given their modes of locomotion, what does each eat and how does each acquire its food? Be prepared to defend your hypothesis.

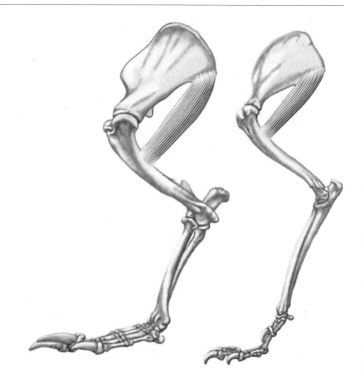

For More Information

Alexander, R. "Muscles Fit for the Job." *New Scientist*, April 15, 1989. Diverse muscles are adapted for different functions.

Huyghe, P. "No Bone Unturned." *Discover*, December 1988. The story of a remarkable forensic anthropologist who uncovers the secrets hidden in unidentified skeletons.

Merton, P. A. "How We Control the Contraction of Our Muscles." *Scientific American*, May 1972. Useful muscular activity requires precise simultaneous adjustments of the contractions of many muscles.

Moore, K. "Little Is beyond His Scope." *Discover*, March 1985. A description of intricate arthroscopic surgery on joints.

Smith, K. K., and Kier, W. M. "Trunks, Tongues, and Tentacles: Moving with Skeletons of Muscle." *American Scientist*, January–February 1989. In certain organs of many animals, muscles provide support as well as movement.

Stossel, T. P. "The Machinery of Cell Crawling." *Scientific American*, September 1994. Cell movement relies on the orderly assembly and disassembly of scaffold proteins.

Answers to Multiple-Choice Questions
1. a   2. d   3. e   4. b   5. c   6. b

After emerging from their winter hibernation, red-sided garter snakes gather in massive aggregations devoted to reproduction. Despite the large numbers of snakes present, only a few males will succeed in the competition for copulations.

Animal Reproduction 35

At a Glance

Evolutionary success is determined by reproductive success. Natural selection is the result of a contest to transmit genes to future generations, and the genes that are most likely to be transmitted are those that encode traits that help organisms reproduce. Individuals that are best at reproducing will pass on the most copies of their genes, so natural selection preserves adaptations that help an animal find a mate, secure access to a mate, produce viable gametes, nourish developing embryos, and/or protect offspring. Among these adaptations are such wonders as the massive mating aggregations of the red-sided garter snake, the song of the mockingbird, the huge tusks of male African elephants, the kangaroo's pouch, and the human placenta.

The evolutionary success of human reproductive mechanisms is evident in our explosively expanding population. The human reproductive system has long been a major preoccupation of biological science, and our ever-increasing understanding of this well-designed and elegantly regulated system for making new humans has enabled us to exert increasing control over the reproductive process. Paradoxically, the development of reproductive technology has focused in equal measure on techniques for preventing unwanted pregnancies and on methods for enabling reproduction in those who lack normal reproductive capability. In any case, progress on either end of the spectrum would have been impossible had scientists not devoted tremendous effort toward answering basic questions about human reproductive biology. How are sperm and eggs formed? How is the formation of these gametes regulated? What happens when sperm and egg meet? We must understand these processes in order to develop safe and effective ways to disrupt them or to enhance them in accordance with our needs and desires.

Net Watch

On-line resources for this chapter are on the World Wide Web at:
http://www.prenhall.com/audesirk
(click on the Table of Contents link and then select Chapter 35).

711

1) How Do Animals Reproduce?

Animals reproduce either sexually or asexually. As you learned in Chapter 11, in **sexual reproduction** an animal produces haploid gametes through meiosis. Two gametes, normally from separate parents, fuse to form a diploid offspring. Because an offspring receives some genes from each of its two parents, its genome is not identical to either parent's genome. In contrast, **asexual reproduction** involves only a single animal that produces offspring through repeated mitoses of cells in some part of its body. The offspring are, therefore, genetically identical to the parent.

Humans reproduce sexually, and we tend to regard sexual reproduction as the normal, best method. In reality, sexual reproduction is a rather inefficient method of producing offspring. Asexual reproduction is far more efficient, because there is no need to find a mate, court, and fend off rivals, and there is no waste of sperm and eggs that never unite to form an offspring. Not surprisingly, many animals reproduce asexually, at least some of the time. Let's begin, then, with a brief survey of asexual reproduction among animals before we move on to sexual reproduction.

Asexual Reproduction Does Not Involve the Fusion of Sperm and Egg

Budding Produces a Miniature Version of the Adult

Many sponges and cnidarians, such as *Hydra* and some sea anemones, reproduce by **budding** (Fig. 35-1). A miniature version of the animal—a **bud**—grows directly on the body of the adult, drawing nourishment from its par-

ent. When it has grown large enough, the bud breaks off and becomes independent.

Fission Followed by Regeneration Can Produce a New Individual

Many animals are capable of **regeneration**, the ability to regrow lost body parts. For example, sea stars will quickly regenerate an arm that is lost to an accident or predator. Even if multiple arms are lost, they can all be replaced as long as most of the central disc remains intact. The regenerative abilities of sea stars do not, however, amount to a reproductive strategy, because in most cases no new individuals are formed (notwithstanding the tales told by oyster "ranchers" who tried to rid their oyster beds of predatory sea stars and inaccurately claimed that each fragment of a hacked-up sea star would regenerate an entire animal).

Nonetheless, some species do use regeneration for true reproduction; they reproduce by **fission**. Several annelid and flatworm species can reproduce by spontaneously disintegrating into two or more pieces, each of which regenerates an entire body (Fig. 35-2). A few brittle star species routinely reproduce in a similar fashion; they split apart, and each half regenerates a complete animal. A few corals can divide lengthwise to produce two smaller but complete individuals. Some cnidarian medusae (that is, cnidarians with bell-shaped bodies; see Chapter 22) reproduce by a fission process in which the entire bell folds in half and splits the stomach and other organs into two parts in preparation for division into two individuals.

During Parthenogenesis, Eggs Develop without Fertilization

The females of some animal species can reproduce by a process known as **parthenogenesis**, in which haploid egg cells develop into adults without being fertilized. **Fertilization** is the union of haploid gametes to form a diploid

Figure 35-1 Budding
The offspring of some cnidarians, such as the frilled sea anemone shown here, grow as buds that appear as miniature adults sprouting from the body of the parent. When sufficiently developed, the buds break off and assume independent existence.

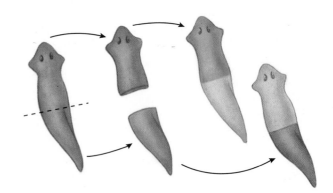

Figure 35-2 Fission followed by regeneration
Some flatworm species reproduce by dividing across the middle. Then, each offspring regenerates the missing half of its body.

Figure 35-3 *A female aphid gives live birth*
In spring and early summer, when food is abundant, aphid females reproduce parthenogenetically. In fact, the development of the reproductive tract proceeds so rapidly that females are born pregnant! In fall, they reproduce sexually, as the females mate with males. Aphids have thus evolved the ability to exploit the advantages of both asexual reproduction (rapid population growth during times of abundant food, no energy spent in seeking a mate, no wasted gametes) and sexual reproduction (genetic recombination).

zygote. Parthenogenetically produced offspring of some species remain haploid. Male honeybees, for example, are haploid, developing from unfertilized eggs; their diploid sisters develop from fertilized eggs. Conversely, some fish, amphibians, and reptiles regain the diploid number of chromosomes in parthenogenetically produced offspring by duplicating all their chromosomes either before or after meiosis. All the resulting offspring are females.

Some species of fish, including relatives of the mollies and platies that are popular in tropical fish stores, and some lizards, such as the whiptail, have done away with males completely. Their populations consist entirely of parthenogenetically reproducing females. Still other animals, such as the aphid, can reproduce either sexually or parthenogenetically, depending on environmental factors such as the season of the year or the availability of food (Fig. 35-3).

Sexual Reproduction Requires the Union of Sperm and Egg

Given the obvious efficiency of asexual reproduction, no one is sure why sex arose. But it did, and sexual reproduction is now a prominent feature in the lives of most animals. It also has an important side effect. The genetic recombination that results from sexual reproduction tends to create novel genotypes—and therefore novel

phenotypes—that are an important source variation on which natural selection may act.

In animals, sexual reproduction occurs when a haploid sperm fertilizes a haploid egg, generating a diploid offspring. In most animal species, an individual is either male or female. These species are termed **dioecious** (Greek for "two houses"). The sexes are defined by the type of gamete that each produces. Females produce **eggs**, which are large, nonmotile cells containing substantial food reserves. Males produce small, motile **sperm**, which have almost no cytoplasm and hence no food reserves.

In **monoecious** ("one house") species, such as earthworms and many snails, single individuals produce both sperm and eggs. Such individuals are commonly called **hermaphrodites** (after Hermaphroditos, a male Greek god whose body was merged with that of a female water nymph, producing a half-male and half-female being). In most hermaphroditic species, reproduction involves a mutual exchange of sperm between individuals. In some hermaphroditic species, however, individuals can fertilize their own eggs if necessary. These animals, including tapeworms and many pond snails, are relatively immobile and may find themselves isolated from other members of their species. Obviously, the ability to fertilize oneself is advantageous under these circumstances.

For dioecious species and for hermaphrodites that cannot self-fertilize, successful reproduction requires that sperm and eggs from different animals be brought together for fertilization. The union of sperm and egg is accomplished in a variety of ways, depending on the mobility of the animals and on whether they breed in water or on land.

External Fertilization Occurs outside the Parents' Bodies

In **external fertilization**, the union of the sperm and egg takes place outside the bodies of the parents. When animals breed in water, the parents release sperm and eggs into the water, through which the sperm swim to reach an egg. This procedure is called **spawning**. Because sperm and egg are relatively short-lived, spawning animals must synchronize their reproductive behaviors, both *temporally* (male and female spawn at the same time) and *spatially* (male and female spawn in the same place). Synchronization may be achieved by way of signals, behaviors, environmental cues, or some combination of these factors.

Most spawning animals rely on environmental cues to some extent. Breeding usually occurs only during certain seasons of the year, and cues such as seasonal changes in daylength typically stimulate the physiological changes that lead to readiness for breeding. More-precise synchrony, however, is required to coordinate the actual release of sperm and egg. Grunion, fish that inhabit coastal waters off southern California, regulate their unusual reproductive rituals by the season, time of day, and phase of the moon (Fig. 35-4a). On fall nights during the highest tides (which occur during a full moon), grunion swim

(a)

(b)

(c)

Figure 35-4 Environmental cues may synchronize spawning
(a) At the highest tides of fall, grunion swarm ashore on the few undeveloped beaches left in southern California. The fish burrow slightly into the sand and release sperm and eggs. When the next-highest tide comes, the young fish wash back into the ocean. The eggs hatch in the warm sand and develop within 2 weeks. *(b)* Along the Great Barrier Reef of Australia, thousands of corals spawn simultaneously, creating this "blizzard" effect. Spawning in these corals is linked to the phase of the moon. *(c)* Close-up of a package of sperm and eggs erupting from a spawning hermaphroditic coral.

up onto sandy beaches. Writhing masses of males and females release their gametes into the wet sand and then swim back out to sea on the next wave. Many corals of Australia's Great Barrier Reef also synchronize spawning by the phase of the moon. On the fourth or fifth night after the full moons of November and December, all the corals of a particular species on an entire reef release a blizzard of sperm and eggs into the water (Fig. 35-4b).

Other animals communicate their sexual readiness to one another by sending visual, acoustic, or chemical signals. Chemical signals are especially common among immobile or sluggish invertebrates, such as mussels and sea stars. These animals release chemicals called **pheromones** into the water, where the pheromone can affect the behavior of other animals. Normally, when a female is ready to spawn, she releases eggs and a pheromone into the water. Nearby males, detecting the mating pheromone, quickly release millions of sperm. The sperm themselves are lured by a chemical attractant released by the eggs in some, if not most, animals. Such "egg pheromones," which have been detected in animals as diverse as sea stars and humans, help ensure fertilization.

Synchronized timing alone does not guarantee efficient reproduction. Corals, sea stars, and mussels all waste enormous quantities of sperm and eggs because the gametes are released too far apart. In species of mobile animals, both temporal *and* spatial synchrony can be ensured by mating behaviors. Most fish, for example, have some sort of courtship ritual in which the male and female come very close together and release their gametes in the same place and at the same time (Fig. 35-5). Frogs

Figure 35-5 Courtship rituals synchronize release of sperm and eggs
Violent courtship rituals among Siamese fighting fish *(Betta splendens)* ensure fertilization of the female's eggs, as male and female curl about one another, releasing sperm and eggs together. The male retrieves the eggs as they fall, spits them into his bubble nest (seen here as bubbles floating on the surface above him), and cares for the offspring during their first few weeks of life.

Figure 35-6 Golden toads in amplexus
The smaller male rides atop the female and stimulates her to release eggs.

carry this ritual one step further, by assuming a characteristic mating pose called **amplexus** (Fig. 35-6). At the edges of ponds and lakes, the male frog mounts the female and prods her in the side. This prodding stimulates her to release eggs, which he immediately fertilizes by releasing a cloud of sperm above them.

Internal Fertilization Occurs within the Female Body

In **internal fertilization**, sperm are taken into the body of the female, where fertilization occurs. Internal fertilization is an important adaptation to terrestrial life: Sperm must be bathed in fluid until they reach the eggs; on land, this liquid passage is best achieved inside the female's body. Even in aquatic environments, internal fertilization increases the likelihood of successful fertilization, because the sperm and eggs are confined together in a small space rather than relying on encounters within a large volume of water.

Internal fertilization normally occurs by **copulation**, the behavior by which the male deposits sperm directly into the reproductive tract of the female (Fig. 35-7). In a variation of internal fertilization, males of some species package their sperm in a container called a **spermatophore** (Greek for "sperm carrier"). In many spermatophore-producing species, including some scorpions, grasshoppers, and salamanders, no copulation occurs. The male simply drops a spermatophore on the ground, and if a female finds it, she fertilizes herself by inserting it into her reproductive cavity, where the enclosed sperm are liberated. Other species have evolved more sophisticated ways to transfer a spermatophore. For example, in some scorpion species the male attaches his spermatophore to the ground, then engages his mate in a "push-pull dance," during which he maneuvers her back

Figure 35-7 Internal fertilization is essential for reproduction on land
(a) Ladybugs mate on a dandelion flower. **(b)** South American tortoises must cope with confining shells. **(c)** King penguins mate comfortably in the snow.

and forth until she is positioned over the spermatophore. Male octopuses of the genus *Argonauta* use a special tentacle to transfer the spermatophore to the female's mantle cavity. After effecting the transfer, the tentacle breaks off and remains inside the cavity. Before this procedure was first observed, biologists believed that the detached arms inside the females were parasitic worms, which were even given their own Latin name.

When animals must copulate to reproduce (or when a mating ritual is required for spawning), males may compete for access to copulations with females. This competition has driven the evolution of a wide variety of sexually selected structures and reproductive behaviors (see Chapter 15). One spectacular example of competition for access to mates occurs in the early spring of each year in the woods of western Canada. As the snows melt and the ground warms, red-sided garter snakes begin to emerge from the underground dens where they have hibernated in groups of tens of thousands of individuals. The males emerge first and linger near the den's opening. Later, when the females emerge, a mating frenzy begins. In a sea of thousands of writhing snake bodies, each female attracts a crowd of dozens or even hundreds of males. Only one will copulate successfully. The ensuing contest appears to a human observer to be total chaos, with each female at the center of a ball of undulating males and the den transformed into a seemingly deranged snarl of snake bodies.

Simply depositing sperm into the body of the female, however, does not guarantee fertilization. Fertilization can occur only if a mature egg is released into the female reproductive tract during the limited time when sperm are present. Just as with spawning animals, the mating behavior of males and females must be synchronized. Among mammals, for example, copulation usually occurs only at certain seasons of the year or when the female

signals readiness to mate. The season or signal typically coincides with **ovulation**, the release of the egg cell from the ovary. Copulation itself triggers ovulation in a few mammals, such as rabbits. An alternative strategy, employed by many female snails and insects, is to store sperm for days, weeks, or even months, thus ensuring a supply of sperm whenever eggs are ready.

2 How Does the Human Reproductive System Work?

Like other mammals, humans have separate sexes and reproduce sexually, with internal fertilization. The **gonads** of mammals are paired organs that produce sex cells—sperm and eggs. Although most mammal species reproduce only during certain seasons of the year and consequently produce sperm and eggs only at that time, human reproduction is not restricted by season. Men produce sperm more or less continuously, and women ovulate about once a month.

The Male Reproductive Tract Includes the Testes and Accessory Structures

The central structures of the male reproductive tract are the **testes** (singular, **testis**), which are the gonads that produce sperm. The male reproductive system (Table 35-1 and Fig. 35-8) also includes accessory structures that secrete substances that activate and nourish the sperm, store it, and conduct it to the female reproductive tract.

The Testes Are the Site of Sperm Production
The testes, which produce both sperm and male sex hormones, are located in the **scrotum**, a pouch that hangs outside the main body cavity. This location keeps the

Table 35-1 Structures and Functions of the Human Male Reproductive Tract

Structure	Type of Organ	Function
Testis	Gonad	Produces sperm and testosterone
Epididymis and vas deferens	Ducts	Store sperm; conduct sperm from testes to penis
Urethra	Duct	Conducts semen from vas deferens and urine from urinary bladder to the tip of the penis
Penis	External "appendage"	Deposits sperm in female reproductive tract
Seminal vesicles	Glands	Secrete fluids that contain fructose (energy source) and prostaglandins (possibly cause "upward" contractions of vagina, uterus, and oviducts, assisting sperm transport to oviducts); fluids may wash sperm out of ducts of male reproductive tract into vagina
Prostate	Gland	Secretes fluids that are basic (neutralize acidity of vagina) and contain factors that enhance sperm motility
Bulbourethral glands	Glands	Secrete mucus (may lubricate penis in vagina)

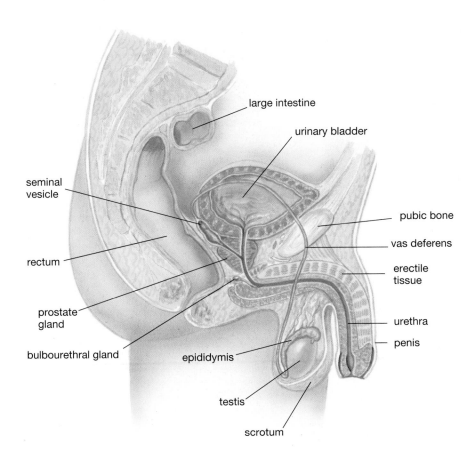

large intestine

urinary bladder

seminal vesicle

rectum

prostate gland

bulbourethral gland

epididymis

testis

scrotum

pubic bone

vas deferens

erectile tissue

urethra

penis

Figure 35-8 *The human male reproductive tract*
The male gonads, the testes, hang beneath the abdominal cavity in the scrotum. Sperm pass from the seminiferous tubules of a testis to the epididymis, and from there through the vas deferens and urethra to the tip of the penis. Along the way, fluids are added from three sets of glands: the seminal vesicles, the bulbourethral glands, and the prostate gland.

testes about 4°C cooler than the core of the body and provides the optimal temperature for sperm development. (Tight jeans may look sexy, but they push the scrotum up against the body, raising the temperature of the testes. Some researchers believe that wearing tight pants can reduce sperm counts and decrease fertility. This is not, however, a reliable means of birth control!) Coiled, hollow **seminiferous tubules**, in which sperm are produced, nearly fill each testis (Fig. 35-9a). In the spaces between the tubules are the **interstitial cells**, which synthesize the male hormone *testosterone*.

Just inside the wall of each seminiferous tubule lie **spermatogonia** (singular, **spermatogonium**), the diploid cells from which the sperm eventually will arise, and the much larger **Sertoli cells** (Fig. 35-9b,c). Each time a spermatogonium divides, it can take one of two developmental paths. First, it may undergo mitosis. Mitosis ensures that the male has a steady supply of new spermatogonia throughout his life. Alternatively, it may undergo **spermatogenesis**—that is, the production of sperm by the process of meiosis followed by differentiation (Fig. 35-9d).

Spermatogenesis begins with the growth and differentiation of spermatogonia into **primary spermatocytes**, large diploid cells that will develop into sperm. The primary spermatocytes then undergo meiosis (see Chapter 11). At the end of meiosis I, each primary spermatocyte

gives rise to two haploid **secondary spermatocytes**. Each secondary spermatocyte divides again, during meiosis II, to produce two **spermatids**, for a total of four spermatids per primary spermatocyte. Spermatids undergo radical rearrangements of their cellular components as they differentiate into sperm.

Sertoli cells regulate the process of spermatogenesis and nourish the developing sperm. The spermatogonia, spermatocytes, and spermatids are embedded in infoldings of the Sertoli cells. As spermatogenesis proceeds, they migrate up from the outermost edge of the seminiferous tubule to the central cavity of the tubule (Fig. 35-9c). The mature sperm, several hundred million a day, are then liberated into the central cavity.

A human sperm (Fig. 35-10, p. 719) is unlike any other cell of the body. Most of the cytoplasm disappears, leaving a haploid nucleus nearly filling the head. Atop the nucleus lies a specialized lysosome called the **acrosome**. The acrosome contains enzymes that will be needed to dissolve protective layers around the egg, enabling the sperm to enter and fertilize it. Behind the head is the *midpiece*, which is packed with mitochondria. These organelles provide the energy needed to move the *tail*, which protrudes out the back. Whiplike movements of the tail, which is really a long flagellum, propel the sperm through the female reproductive tract.

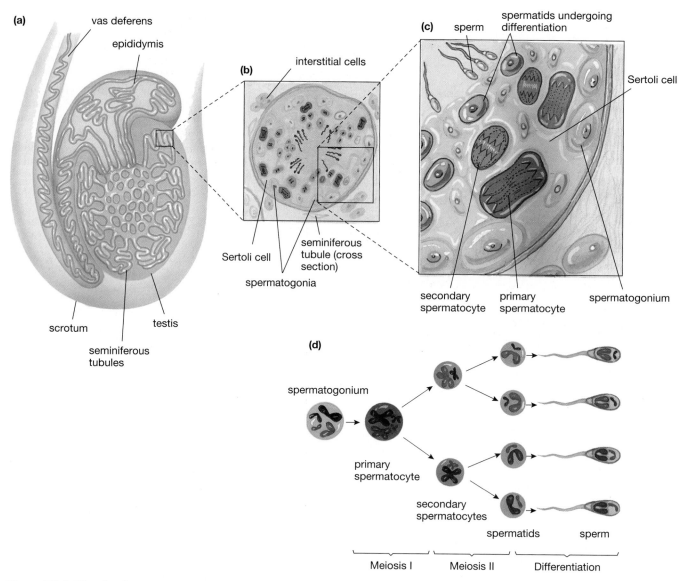

Figure 35-9 *The development of sperm cells*
(a) A lengthwise section of the testis, showing the location of the seminiferous tubules, epididymis, and vas deferens. *(b)* Cross section of a seminiferous tubule. The walls of the seminiferous tubules are lined with Sertoli cells and spermatogonia, protruding into the central cavity of the tubule. *(c)* As spermatogonia undergo meiosis, the daughter cells move inward, embedded in infoldings of the Sertoli cells. There they differentiate into sperm, drawing on the Sertoli cells for nourishment. Mature sperm are freed into the central cavity of the tubules for transport to the penis. Testosterone is produced by the interstitial cells in the spaces between tubules. *(d)* Spermatogenesis is accomplished by meiotic divisions that produce haploid sperm (compare with the actual locations shown in part [c]). Although four chromosomes are shown for clarity, in humans the diploid number is 46 and the haploid number is 23.

In humans and other mammals, immature males do not produce sperm. Spermatogenesis does not begin until *puberty,* a period of rapid growth and transition to sexual maturity. At puberty, the hypothalamus releases **gonadotropin-releasing hormone (GnRH),** which stimulates the anterior pituitary to produce **luteinizing hormone (LH)** and **follicle-stimulating hormone (FSH).** Luteinizing hormone stimulates the interstitial cells of the testes to produce the hormone **testosterone** (Fig. 35-11). This

testosterone, in combination with FSH, stimulates the Sertoli cells and spermatogonia, causing spermatogenesis.

Testosterone also stimulates the development of *secondary sexual characteristics* (such as the growth of facial hair in males and breast development in females), maintains sexual drive, and is required for successful *intercourse* (a term we will use for human copulation). Sperm, however, are not involved in these functions. Therefore, if FSH release could be suppressed, blocking

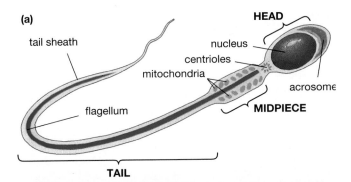

(a)

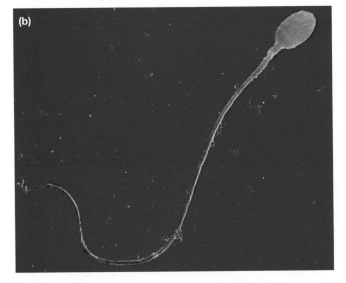

(b)

Figure 35-10 A human sperm cell
(**a**) A mature sperm is a stripped-down cell equipped with only the essentials: a haploid nucleus containing the male genetic contribution to the future zygote, a lysosome (called the acrosome) containing enzymes that will digest the barriers surrounding the egg, mitochondria for energy production, and a tail (a long flagellum) for locomotion. (**b**) False-color electron micrograph of a human sperm.

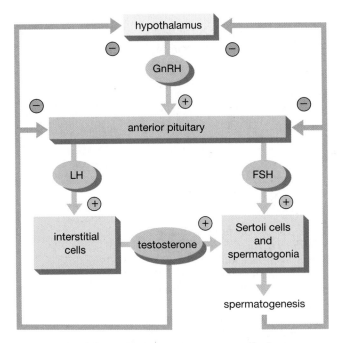

Figure 35-11 Hormonal control of spermatogenesis
Gonadotropin-releasing hormone (GnRH) from the hypothalamus stimulates the anterior pituitary to release LH and FSH. LH stimulates the interstitial cells to produce testosterone. Testosterone and FSH stimulate the Sertoli cells and the spermatogonia, causing spermatogenesis. Testosterone and chemicals produced during spermatogenesis inhibit further release of FSH and LH, forming a negative feedback loop that keeps the rate of spermatogenesis and the concentration of testosterone in the blood nearly constant.
(+ stimulates; − inhibits)

spermatogenesis, while LH release continues, thereby allowing continued testosterone production, a man would be *infertile* but not *impotent*; in other words, he could not produce sperm but could maintain an erection of the **penis**, the male organ of copulation. Efforts are under way to develop a drug to do just that, as a form of male birth control.

Accessory Structures Produce Semen and Conduct the Sperm outside the Body

The seminiferous tubules merge to form the **epididymis**, a single continuous, folded tube (see Fig. 35-9). The epididymis leads into the **vas deferens**, which leaves the scrotum and enters the abdominal cavity. Most of the hundreds of millions of sperm produced each day are stored in the vas deferens and epididymis. The vas def-

erens joins the **urethra**, which connects the bladder to the tip of the penis. This final common path is shared, at different times, by sperm (during ejaculation) and urine (during urination).

The fluid ejaculated from the penis, called **semen**, consists of sperm mixed with secretions from three glands that empty into the vas deferens or urethra: (1) the **seminal vesicles**, (2) the **prostate gland**, and (3) the **bulbourethral glands**. The secretions activate swimming by the sperm, provide energy for swimming, and neutralize the acidic fluids of the vagina that normally inhibit bacterial growth (see Table 35-1).

The Female Reproductive Tract Includes the Ovaries and Accessory Structures

The female reproductive tract is almost entirely contained within the abdominal cavity (Table 35-2 and Fig. 35-12). It consists of paired gonads—the **ovaries** (Fig. 35-13a, p. 721)—and accessory structures that accept sperm, conduct the sperm to the egg, and nourish the developing embryo.

Table 35-2 Structures and Functions of the Human Female Reproductive Tract		
Structure	**Type of Organ**	**Function**
Ovary	Gonad	Produces eggs, estrogen, and progesterone
Fimbria	Mouth of duct	Cilia sweep egg into oviduct
Oviduct	Duct	Conducts egg to uterus; site of fertilization
Uterus	Muscular chamber	Site of development of fetus
Cervix	Connective tissue ring	Closes off lower end of uterus, supports fetus, and prevents foreign material from entering uterus
Vagina	Large "duct"	Receptacle for semen; birth canal

Figure 35-12 The human female reproductive tract
Eggs are produced in the ovaries and swept by cilia into the oviduct. A male deposits sperm in the vagina, from which they move up through the cervix and uterus into the oviduct. Sperm and egg normally meet in the oviduct, where fertilization occurs. The fertilized egg attaches to the lining of the uterus, where the embryo develops.

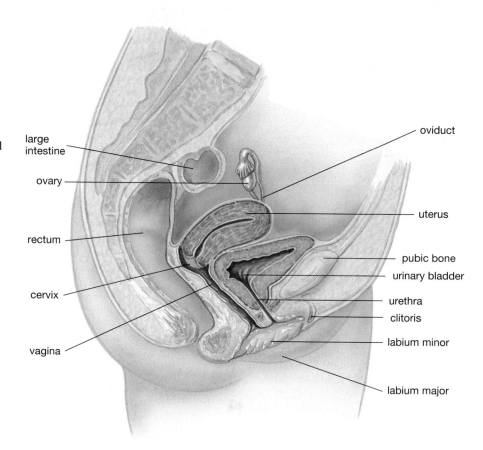

The Ovaries Are the Site of Egg Production

Oogenesis, the formation of egg cells, begins in the developing ovaries of a female **fetus** (an embryo that is sufficiently developed to be recognizably human). This process starts with the formation of precursor egg cells called **oogonia** (singular, **oogonium**). By the end of the third month of fetal development, no oogonia remain, as they have all divided by mitosis and grown into **primary oocytes**. As fetal development continues, meiosis begins in all primary oocytes but is halted at prophase of meiosis I. By birth, a lifetime supply of primary oocytes is already in place, and no new ones will be generated later in life. The ovaries start out with about 2 million primary oocytes. Many primary oocytes die each day, until at puberty (normally 11 to 14 years of age) only about 400,000 remain. That number is more than enough, because only a few oocytes resume

meiosis during each month of a woman's reproductive span (from puberty to *menopause* at age 45 to 55).

Surrounding each oocyte is a layer of much smaller cells that both nourish the developing oocyte and secrete female sex hormones. Together, the oocyte and these accessory cells make up a **follicle** (Fig. 35-13b). Approximately once a month during a woman's reproductive years, she undergoes a *menstrual cycle*, which is described later in this chapter. During the menstrual cycle, pituitary hormones stimulate the development of a dozen or more follicles, although normally only one follicle matures completely. At this time, the primary oocyte completes its first meiotic division (which was halted during development), dividing into a single **secondary oocyte** and a **polar body**, which is little more than a discarded set of chromosomes (Fig. 35-13c). Meanwhile, the small acces-

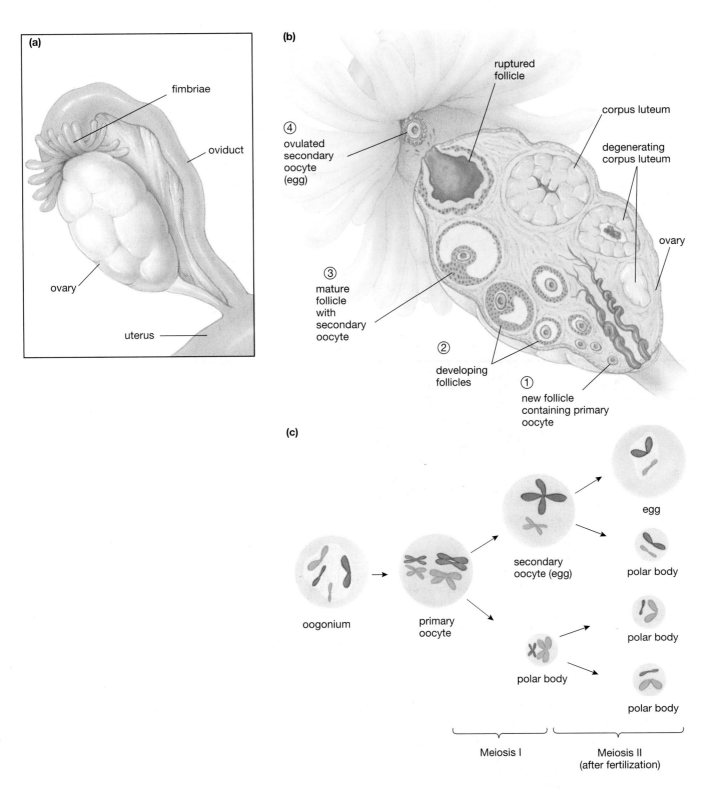

Figure 35-13 Oogenesis in the human female
(a) External view of the ovary and oviduct. *(b)* The development of follicles in an ovary, portrayed in a time sequence (clockwise from the lower right). ① A primary oocyte begins development within a follicle. ②,③ The follicle grows, providing both hormones and nourishment for the enlarging oocyte. ④ At ovulation, the secondary oocyte, or egg, bursts through the ovary wall, surrounded by some follicle cells. The remaining follicle cells develop into the corpus luteum, which secretes hormones. If fertilization does not occur, the corpus luteum breaks down after a few days. *(c)* The cellular stages of oogenesis. The oogonium enlarges to form the primary oocyte. At meiosis I, almost all the cytoplasm is included in one daughter cell, the secondary oocyte. The other daughter cell is a small polar body that contains chromosomes but little cytoplasm. At meiosis II, almost all the cytoplasm of the secondary oocyte is included in the egg, and a second small polar body discards the remaining "extra" chromosomes. The first polar body may also undergo the second meiotic division. In humans, meiosis II does not occur unless the egg is fertilized.

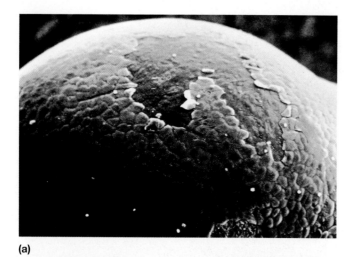

(a)

(b)

Figure 35-14 A follicle erupts from the ovary
The mature follicle grows so large and is filled with so much fluid that it moves to *(a)* the surface of the ovary and *(b)* literally bursts through the ovary wall like a miniature volcano. It then releases the secondary oocyte into the oviduct.

sory cells of the follicle multiply and secrete the hormone **estrogen**. As the follicle matures, it grows, eventually erupting through the surface of the ovary and releasing the secondary oocyte (Fig. 35-14). The second meiotic division may then occur in the **oviduct** (the tube leading out of the ovary), but only if the secondary oocyte is fertilized. For convenience, we will refer to the ovulated secondary oocyte as the *egg.*

Some of the follicle cells accompany the egg, but most remain behind in the ovary. These cells enlarge and become glandular, forming the **corpus luteum** (see Fig. 35-13b). The corpus luteum secretes both estrogen and a second hormone, **progesterone**. If fertilization does not occur, the corpus luteum breaks down a few days later.

Accessory Structures Include the Oviducts, Uterus, and Vagina

Each ovary is adjacent to, but not continuous with, an oviduct (sometimes called the **fallopian tube** in humans; see Fig. 35-13a). The open end of the oviduct is fringed with ciliated "fingers" called **fimbriae** (singular **fimbria**), which nearly surround the ovary. The cilia create a current that sweeps the egg into the mouth of the oviduct. Fertilization normally occurs in the oviduct. The **zygote**, as the fertilized egg is now called, is swept down the oviduct by beating cilia and released into the pear-shaped **uterus**, or *womb.* There it will develop for 9 months. The wall of the uterus has two layers that correspond to its dual functions of (1) nourishment and (2) childbirth. The inner lining, or **endometrium**, is richly supplied with blood vessels. (This lining will form the mother's contribution to the **placenta**, the structure that transfers oxygen, carbon dioxide, nutrients, and wastes between fetus and mother, as we shall see in Chapter 36.) The outer muscular wall of the uterus, the **myometrium**, contracts strongly during delivery, expelling the infant out into the world.

Developing follicles secrete estrogen, which stimulates the uterine lining to grow an extensive network of blood vessels and nutrient-producing glands. After ovulation, estrogen and progesterone released by the corpus luteum promote continued growth of the endometrium. Thus, if an egg is fertilized, it encounters a rich environment for growth. If the egg is not fertilized, however, the corpus luteum disintegrates, estrogen and progesterone levels fall, and the overgrown endometrium disintegrates as well. The uterus contracts, squeezing out the excess endometrial tissue (perhaps causing menstrual cramps in the process). The resulting flow of tissue and blood, as a result of the erosion of blood vessels in the endometrium, is called **menstruation** (from the Latin *mensis,* meaning "month").

The outer end of the uterus is nearly closed off by the **cervix**, a ring of connective tissue. The cervix holds a developing baby in the uterus, expanding only at the onset of labor to permit passage of the child. Beyond the cervix is the **vagina**, which opens to the outside of the body. The vagina serves both as the receptacle for the penis during intercourse and as the birth canal.

The Menstrual Cycle Is Controlled by Complex Hormonal Interactions

A human male is able to produce large numbers of sperm continuously and can therefore contribute gametes to a fertilization at any time. In contrast, a woman does not produce mature gametes (ovulate) unless her reproductive tract is properly prepared for pregnancy. Ovulation can occur only when the uterus is prepared to receive and nourish a fertilized egg. The necessary coordination of ovulation and uterine preparation is accomplished by a complex **menstrual cycle**, which is regulated by hormonal interactions among the hypothalamus, pituitary gland, and ovary.

We will begin our discussion of the menstrual cycle with the spontaneous release of gonadotropin-releasing hormone (GnRH) by cells in the hypothalamus. You may recall from Chapter 32 that neurosecretory cells in the hypothalamus control hormone release by the anterior pituitary. Some of the neurosecretory cells produce GnRH. A key to understanding the menstrual cycle is understanding that these cells spontaneously release GnRH all the time, unless actively prevented from doing so by other hormones—notably, progesterone.

Gonadotropin-releasing hormone stimulates the anterior pituitary to release FSH and LH (step ① in Fig. 35-15). Both FSH and LH circulate in the bloodstream and initiate the development of several follicles within the ovaries, and the cells of these developing follicles secrete estrogen. Under the combined influences of FSH, LH, and estrogen, the follicles grow during the next 2 weeks (step ②). Simultaneously, the primary oocyte within each follicle enlarges, storing both food and regulatory substances (mostly proteins and messenger RNA) that will be needed by the fertilized egg during early development. For reasons that are not completely understood, only one or, rarely, two follicles complete development each month.

As the maturing follicle enlarges, it secretes ever greater amounts of estrogen (step ③). This estrogen has three effects. First, it promotes the continued development of the follicle and of the primary oocyte within it. Second, it stimulates the growth of the endometrium of the uterus. Third, high levels of estrogen stimulate both the hypothalamus and pituitary, resulting in a surge of LH and FSH at about the 12th day of the cycle (step ④).

The function of the peak in FSH concentration is not well understood, but the surge of LH has three important consequences: (1) It triggers the resumption of meiosis I in the oocyte, resulting in the formation of the secondary oocyte and the first polar body; (2) it causes the final explosive growth of the follicle, culminating in ovulation (step ⑤); and (3) it transforms the remnants of the follicle that remain in the ovary into the corpus luteum.

The corpus luteum secretes both estrogen and progesterone (step ⑥). The combination of these hormones inhibits the hypothalamus and pituitary, shutting down the release of FSH and LH (step ⑦), thereby preventing the development of more follicles. Simultaneously, estrogen and progesterone stimulate further growth of the endometrium, which eventually becomes about 5 millimeters thick. (Note that the effects of progesterone and estrogen depend on the target organ. Progesterone *stimulates* the endometrium but *inhibits* hormone release from the hypothalamus and pituitary.)

In menstrual cycles in which pregnancy does not occur, the corpus luteum starts to disintegrate about 1 week after ovulation (step ⑧). This disintegration is precipitated

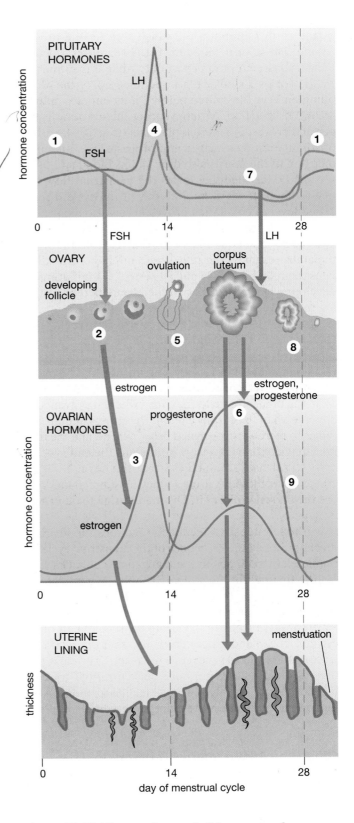

Figure 35-15 *Hormonal control of the menstrual cycle*
The menstrual cycle is generated by interactions among the hormones of the hypothalamus, of the anterior pituitary, and of the ovaries. The hormonal changes drive cyclic changes in the uterine lining. The circled numbers on the graphs refer to the hormonal interactions discussed in the accompanying text.

by the corpus luteum itself, which secretes the progesterone that in turn shuts down LH secretion. Because the corpus luteum can persist only while it is stimulated by LH (or by a similar hormone released by the developing embryo, described below), it essentially induces its own destruction. With the corpus luteum gone, estrogen and progesterone levels plummet (step ⑨). Deprived of stimulation by estrogen and progesterone, the endometrium of the uterus also dies, and its blood and tissue are shed. This shedding forms the menstrual flow that begins about the 27th or 28th day of the cycle. The reduced level of circulating progesterone also means that it no longer inhibits the hypothalamus and pituitary, and the spontaneous release of GnRH from the hypothalamus resumes. The release of GnRH in turn stimulates the release of FSH and LH (step ①), initiating the development of a new set of follicles and thereby restarting the cycle.

During pregnancy, the embryo itself prevents these changes from occurring. Shortly after the ball of cells formed by the dividing fertilized egg embeds itself in the endometrium, it starts secreting an LH-like hormone called **chorionic gonadotropin (CG)**. This hormone travels in the bloodstream to the ovary, where it prevents the breakdown of the corpus luteum. The corpus luteum continues to secrete estrogen and progesterone, and the uterine lining continues to grow, nourishing the embryo. The embryo releases so much CG that the hormone is excreted in the mother's urine. In fact, most pregnancy tests use the presence of CG in a woman's urine to determine pregnancy.

Although negative feedback regulates the levels of most hormones, the hormones of the menstrual cycle are regulated by both positive and negative feedbacks. During the first half of the cycle, FSH and LH stimulate estrogen production by the follicles. High levels of estrogen then *stimulate* the mid-cycle surge of FSH and LH release (positive feedback). During the second half of the cycle, estrogen and progesterone together *inhibit* the release of FSH and LH (negative feedback). The early positive feedback causes hormone concentrations to reach high levels, and the later negative feedback shuts the system down again unless pregnancy intervenes.

Copulation Allows Internal Fertilization

As terrestrial mammals, humans use internal fertilization to deposit sperm into the moist environment of the female's reproductive tract. The penis is inserted into the vagina, where sperm are released during ejaculation. The sperm swim upward in the female reproductive tract, from the vagina through the opening of the cervix into the uterus, and up into the oviducts. If the female has ovulated within the past day or so, the sperm will meet an egg in one of the oviducts. Only one sperm can succeed in fertilizing the egg and thus begin the development of a new human being.

During Copulation, Sperm Are Deposited in the Female's Vagina

The male role in copulation begins with erection of the penis. Before erection, the penis is relaxed (flaccid), because the arterioles that supply it are constricted, allowing little blood flow (Fig. 35-16a). Under the dual influences of psychological and physical stimulation, the arterioles dilate and blood flows into spaces in the tissue within the penis. As these tissues swell, they squeeze off the veins that drain the penis (Fig. 35-16b). Pressure builds up, causing an erection. After the penis is inserted into the vagina, movements further stimulate touch receptors on the penis, triggering ejaculation. *Ejaculation*

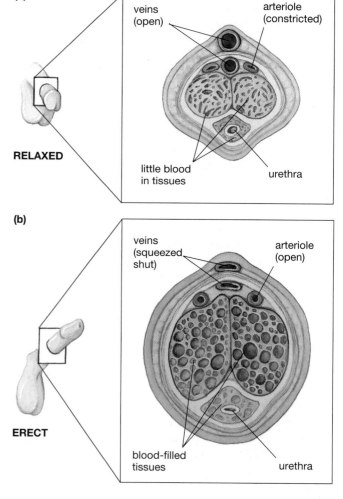

Figure 35-16 Changes in blood flow within the penis cause erection
(a) Normally, smooth muscles encircling the arterioles leading into the penis are contracted, limiting blood flow. *(b)* During sexual excitement, these muscles relax, and blood flows into spaces within the penis. The swelling penis squeezes off the veins leaving the penis, thereby increasing the pressure produced by fluids within the penis and causing it to become elongated and firm.

occurs when muscles encircling the epididymis, vas deferens, and urethra contract, forcing semen out of the penis and into the vagina. On average, 3 or 4 milliliters of semen, containing 300 million to 400 million sperm, is ejaculated. *Male orgasm* causes both ejaculation and a feeling of intense pleasure and release.

Similar changes occur in the female. Sexual excitement causes increased blood flow to the vagina and to the external parts of the reproductive tract. These external parts consist of the **labia** (singular, **labium**)—paired folds of skin known as the *labia minor* and *labia major*—and the **clitoris**, a rounded projection (see Fig. 35-12). The clitoris, which is derived from the same embryological tissue as the tip of the penis, becomes erect. Stimulation by the penis of the male typically, but not always, results in *female orgasm*, a series of rhythmic contractions of the vagina and uterus accompanied by sensations of pleasure and release. Female orgasm is not necessary for fertilization.

The intimate contact involved in copulation creates a situation in which disease organisms can readily be transmitted. Ever since the "sexual revolution," which began in the 1960s, many individuals have had multiple sexual partners, and the incidence of *sexually transmitted diseases (STDs)* has greatly increased (see "Health Watch: Sexually Transmitted Diseases").

During Fertilization, the Sperm and Egg Nuclei Unite

Neither sperm nor egg lives very long. An unfertilized egg may remain viable for a day, and sperm, under ideal conditions, may live for two. Therefore, fertilization can succeed only if copulation occurs within a couple of days before or after ovulation. You will recall that when it leaves the ovary, the egg is surrounded by follicle cells. These cells, now called the **corona radiata**, form a barrier between the sperm and the egg (Fig. 35-17a). A second barrier, the jellylike **zona pellucida** ("clear area"), lies between the corona radiata and the egg. Recent research suggests that the human egg releases a chemical attractant that lures the sperm toward it.

In the oviduct, hundreds of sperm reach the egg and encircle the corona radiata, each sperm releasing enzymes from its acrosome (Fig. 35-17b). These enzymes weaken both the corona radiata and the zona pellucida, allowing the sperm to wriggle through to the egg. If there aren't enough sperm, an insufficient amount of enzyme is released, and none of the sperm will reach the egg. This may be the reason that natural selection has led to the ejaculation of so many sperm. Perhaps 1 in 100,000 reach the oviduct, and 1 in 20 of those find the egg, so only a few hundred of the 300 million sperm that were ejaculated join the attack on the barriers around the egg.

When the first sperm finally contacts the surface of the egg, the plasma membranes of egg and sperm fuse, and the sperm's head is drawn into the egg's cytoplasm. As the sperm enters, it triggers two vital changes in the egg: (1) Vesicles near the surface of the egg release chemicals into the zona pellucida, reinforcing it and preventing further sperm from entering the egg. (2) The egg

(a)

(b)

Figure 35-17 *The secondary oocyte and fertilization*
(a) A human secondary oocyte shortly after ovulation. Sperm must digest their way through the small follicular cells of the corona radiata and the clear zona pellucida to reach the oocyte itself. *(b)* Sperm surround the oocyte, attacking its defensive barriers.

undergoes its second meiotic division, producing a haploid gamete at last. Fertilization occurs as the haploid nuclei of sperm and egg fuse, forming a diploid nucleus that contains all the genes of a new human being.

Anomalies in the male or female reproductive systems can prevent fertilization. For example, a blocked oviduct can prevent sperm from reaching the egg. Likewise, men who produce fewer than 20 million sperm per milliliter of semen (about one-fifth the normal amount)

Health Watch
Sexually Transmitted Diseases

Sexually transmitted diseases (STDs), caused by viruses, bacteria, protists, or arthropods that infect the sexual organs and reproductive tract, are a serious and growing health problem worldwide. The World Health Organization estimates that there are 250 million new STD cases each year. As the name implies, these diseases are transmitted either exclusively or primarily through sexual contact. Here we discuss some of the more common of these diseases.

Bacterial Infections **Gonorrhea**, an infection of the genital and urinary tract, is one of the most common of *all* infectious diseases in the United States, estimated to infect about 325,000 people annually. The causative bacterium, *Neisseria gonorrhoeae*, cannot survive outside the body and is transmitted almost exclusively by intimate contact. It penetrates the membranes that line the urethra, anus, cervix, uterus, oviducts, and throat. In males, inflammation of the urethra results in painful urination and a discharge of pus from the penis. About 10% of infected males and 50% of infected females do not seek treatment, because they have symptoms that are mild or absent. They become carriers who can readily spread the disease. Gonorrhea can lead to infertility by blocking the oviducts with scar tissue. Treatment by penicillin was formerly highly successful, but penicillin-resistant strains now require the use of other antibiotics. Infants born to infected mothers can acquire the bacterium during delivery. The bacterium attacks the eyes of newborns and was once a major cause of blindness. Today, most newborns are immediately given antibiotic eyedrops preventively to kill the bacterium.

Syphilis is a far more dangerous, though less prevalent, disease than gonorrhea. It is caused by a spiral-shaped bacterium *Treponema pallidum,* which enters the mucous membranes of the genitals, lips, anus, or breasts. Like gonorrhea, it is readily killed by exposure to air and is spread only by intimate contact. Syphilis begins with a sore at the site of infection. Syphilis can be cured with antibiotics. If untreated, syphilis bacteria spread through the body, multiplying and damaging many organs, including the skin, kidneys, heart, and brain, in some cases with fatal results. About 4 out of every 1000 newborns in the United States have been infected with syphilis before birth. The skin, teeth, bones, liver, and central nervous system of such infants may be damaged.

Chlamydia causes inflammation of the urethra in males and the urethra and cervix in females. In many cases there are no obvious symptoms, so the infection goes untreated and is spread. Like the gonorrhea bacterium, the chlamydia bacterium, *Chlamydia trachomatis,* can infect and block the oviducts, resulting in sterility. Chlamydial infection can cause eye inflammations in infants born to infected mothers.

Viral Infections **Acquired immune deficiency syndrome**, or **AIDS**, is caused by the HIV virus, discussed in detail in Chapter 31. Because the virus does not survive exposure to air, it is spread primarily by sexual activity and by contaminated blood and needles. The HIV virus attacks the immune system, leaving the victim vulnerable to a variety of infections, which almost invariably prove fatal. Children born to mothers with AIDS can become infected before or during birth. There is no cure, although certain drugs, such as AZT, can prolong life.

Genital herpes reached epidemic proportions during the 1970s and continues to spread to more than a million new victims yearly. It causes painful blisters on the genitals and surrounding skin and is transmitted primarily when blisters are present. The herpes virus never leaves the body but resides in certain nerve cells, emerging unpredictably, possibly in response to stress. The first outbreak is the most serious; subsequent outbreaks produce fewer blisters and can be quite infrequent. The drug acyclovir, which inhibits viral DNA replication, can reduce the severity of outbreaks. Pregnant women with an active case of genital herpes can transmit the virus to the developing fetus, causing severe mental or physical disability or stillbirth. Herpes can also be transmitted from mother to infant if the infant contacts blisters during childbirth.

Genital warts are growths or bumps that appear on the external genitalia, in or around the vagina or anus, or on the cervix in females, and on the penis, scrotum, groin, or thigh in males. The warts, which are normally painless, are caused by the *human papillomavirus (HPV)*, which is transmitted by skin-to-skin contact during sex. The most common treatment is removal of the warts, typically by freezing the warts with liquid nitrogen. Antiviral drugs can also be effective. Like the herpes virus, the human papillomavirus resides inside cells and is therefore difficult to cure permanently. Repeated treatments are typically required, and subsequent outbreaks are possible even after treatment. In women, genital warts are considered to be a risk factor for certain types of cervical cancer.

Protists and Arthropod Infections **Trichomoniasis** is caused by *Trichomonas*, a flagellated protist that colonizes the mucous membranes that line the urinary tract and genitals of both males and females. The symptoms are a discharge caused by inflammation in response to the parasite. The protist is spread by intercourse but can also be acquired through contaminated clothing and toilet articles. Lengthy untreated infections can result in sterility.

Crab lice, also called *pubic lice*, are microscopic arachnids that live and lay their eggs in pubic hair. Their mouth parts are adapted for penetrating the skin and sucking blood and body fluids, a process that causes severe itching. "Crabs" are not only irritating; they can also spread infectious diseases. They can be controlled through careful hygiene and chemical treatments.

normally cannot fertilize a woman's egg during intercourse because too few sperm reach the egg. If the sperm are otherwise normal, such men can father children by *artificial insemination*, in which a large quantity of their semen is injected directly into the oviduct. Today, some couples seek high-technology help in the form of *in vitro* fertilization (see "Scientific Inquiry: *In Vitro* Fertilization").

During Pregnancy, the Developing Embryo Grows within the Uterus

The zygote begins to divide while being carried by cilia down the oviduct to the uterus, a journey that takes about 4 days (Fig. 35-18a,b). By about 1 week after fertilization, the zygote has developed into a hollow ball of cells, the blastocyst (Fig. 35-18b). The **inner cell mass**, a thickened region of the blastocyst, will become the embryo itself. The sticky outer ball will adhere to the uterus and burrow into the endometrium, a process called **implantation**. The growing embryo is nourished by blood from ruptured uterine vessels plus glycogen secreted by glands in the endometrium.

Obtaining nutrients directly from the nearby cells of the endometrium suffices only for the first week or two of embryonic growth. During this time, the placenta, composed of interlocking tissues of the embryo and the endometrium, begins to form. Through the placenta, the embryo will receive nutrients and oxygen and will dispose of wastes into the maternal circulation. The details of embryonic development are presented in Chapter 36.

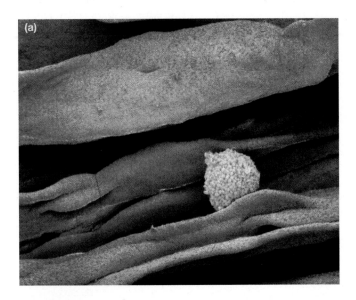

Figure 35-18 The journey of the egg
(a) The egg, surrounded by the corona radiata, travels down the oviduct toward the uterus. It emits chemicals that attract sperm, increasing its chances of being fertilized. *(b)* The egg is fertilized in the oviduct and slowly travels down to the uterus. Along the way, the zygote divides a few times, until a hollow blastocyst is formed. The inner cell mass forms the embryo; the surrounding cells adhere to the uterine endometrium, burrow in, and begin forming the placenta.

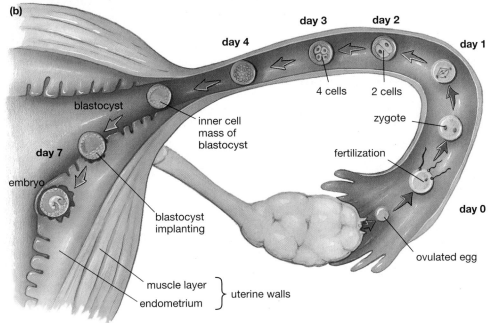

Scientific Inquiry
In Vitro Fertilization

Louise Brown, the first "test-tube baby," was born in England in 1978. Since that well-publicized event, almost 300 centers for *in vitro* fertilization (IVF) have been established in the United States; these have produced more than 30,000 healthy babies. The demand for IVF is fueled by an epidemic of infertility. In the 1970s, one out of every six couples in the United States was unable to conceive after trying for 1 year or more; that rate has now tripled. One reason for the increase in infertility is that many modern couples delay childbearing, and fertility declines with age. A second reason is a higher incidence of sexually transmitted diseases such as chlamydia and gonorrhea, which can scar and block the oviducts or sperm ducts. Both blocked ducts and low sperm counts can be overcome with IVF, because oocytes are removed directly from the ovaries, and their meeting with sperm is guaranteed within the confines of a small glass dish (*in vitro* is Latin for "in glass"). The popularity of IVF, which can cost as much as $10,000 per attempt, is a testimony to the strong biological drive to have children. With an average success rate below 20%, a typical conception via IVF costs more than $40,000.

Although the technique is simple in concept, the procedure is complex and delicate. First, the woman is given daily injections of drugs, hormones, or both to stimulate multiple ovulation. Using blood tests and ultrasound imaging of the ovaries, physicians determine when the time is ripe for ovulation. Then the woman is injected with human chorionic gonadotropin, which begins the process of expelling the oocytes from the follicles. Just before the oocytes are ejected, surgeons insert a thin fiber-optic viewing device through a small abdominal incision to locate the mature follicles. Next, they insert a long, hollow needle into each ripe follicle and suck out the follicular cells and fluid, which are examined under a microscope. With luck, an oocyte is present. Normally, at least four oocytes are harvested and incubated in a glass dish to which freshly collected sperm are then added. In 48 hours, about two-thirds of the oocytes will have been fertilized and have reached the stage at which eight cells have formed. A few of these early embryos are sucked into a tube and expelled very gently into the uterus. (Extra embryos can be frozen for later use in case the first attempt at implantation is unsuccessful.) Transplanting multiple embryos increases the success rate for implantation, but it also increases the probability of multiple births, which carry considerably higher risks than do single births.

The IVF technique has become a weapon in the fight to save endangered species. The National Zoo in Washington, D.C., has been working since 1984 to adapt IVF technology to help endangered species reproduce. Zoo officials have developed a mobile IVF laboratory that can travel to zoos where endangered species are housed. A tremendous advantage of IVF is that it will allow sperm from a male of an endangered species to be transported between continents, if necessary, to fertilize an appropriate female. This method eliminates the danger and trauma of transporting the animals themselves. It also overcomes the very real probability that, once together, the animals will refuse to mate. In April 1990, the first "test-tube" tiger was born (Fig. E35-1). Only 200 individuals of this rare Siberian subspecies remain in the wild. The use of IVF increases the chances that we may save this and other precious forms of life on Earth.

The standard technique outlined above can overcome many common causes of infertility but is of little use to couples in which the man produces no live sperm cells. Until recently, men

Milk Secretion Is Stimulated by Hormones of Pregnancy

As the fetus grows, nourished by nutrients diffusing through the placenta, changes in the mother's breasts prepare her to continue nourishing her child after it is born. When pregnancy occurs, large quantities of estrogen and progesterone (acting together with several other hormones) stimulate **mammary glands**, milk-producing glands in the breasts, to grow, branch, and develop the capacity to secrete milk (Fig. 35-19). The mammary glands are arranged in a circle around the nipple; each gland has a milk duct that leads to the *nipple*, a projection of epithelial tissue. The actual secretion of milk, called **lactation**, is promoted by the pituitary hormone prolactin (see Chapter 32).

The level of prolactin rises steadily from about the fifth week of pregnancy until birth. Immediately after birth, estrogen and progesterone levels plummet, and prolactin takes over, stimulating the production of milk. Milk is released when the infant's suckling stimulates nerve endings in the nipples. The stimulated nerves send a signal to the hypothalamus, triggering an extra surge of prolactin and oxytocin from the pituitary. Oxytocin causes muscles surrounding the mammary glands to contract, ejecting the milk into the ducts that lead to the nipples (see Chapter 32).

During the first few days after birth, the mammary glands secrete a thin, yellowish fluid called **colostrum**. Colostrum is high in protein and contains antibodies that are absorbed directly through the infant's intestine and help protect the newborn against some diseases.

membrane of an egg cell and inject a single sperm directly into the egg's cytoplasm (Fig. E35-2). In almost all cases, this procedure leads to a successful fertilization. Apparently, sperm are capable of fertilization very early in their developmental process, and the normal process of maturation serves mainly to provide them with the features required to reach the egg and penetrate its protective layers.

Figure E35-1 *The world's first test-tube tiger*
Born in 1990, this test-tube tiger is the first successful use of IVF in the fight to save endangered species.

with this kind of sterility could not be biological fathers, even with the aid of IVF. Now, however, even men who are unable to produce viable sperm may be able to father their own children by means of *intracytoplasmic sperm injection* (ICSI). In ICSI, immature sperm (or even sperm precursor cells) are extracted from the testes. Then, a micropipette is used to pierce the plasma

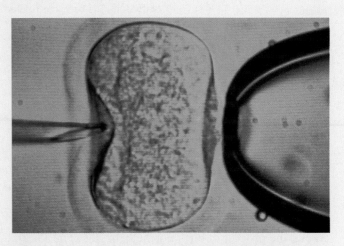

Figure E35-2 *Piercing an egg to inject a sperm cell*
An egg, stripped of its protective layers and held in place with a pipette, is injected with a single sperm cell that is placed directly into the egg's cytoplasm.

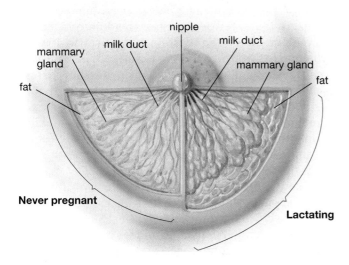

Figure 35-19 *The structure of the mammary glands*
During pregnancy, both fatty tissue and the milk-secreting glands and ducts increase in size.

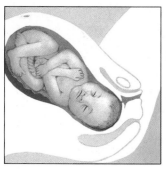

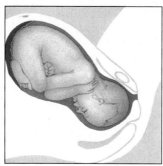

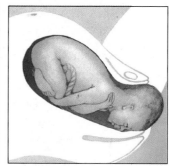

(a) The baby is oriented head downward, facing the mother's side. The cervix is beginning to thin and expand in diameter (dilate).

(b) The cervix is completely dilated to 10 centimeters (almost 4 inches wide), and the baby's head has entered the vagina, or birth canal. The baby has rotated to face the mother's back.

(c) The baby's head is emerging.

(d) The baby has rotated to the side once again as the shoulders emerge.

Figure 35-20 Delivery

Colostrum is gradually replaced by mature milk, which is higher in fat and milk sugar (lactose) and lower in protein.

Reproduction Culminates in Labor and Delivery

Near the end of the ninth month, give or take a few weeks, the process of birth normally begins (Fig. 35-20). Birth is the result of a complex interplay between (1) uterine stretching caused by the growing fetus and (2) fetal and maternal hormones that finally trigger **labor** (contractions of the uterus that result in the birth).

Unlike skeletal muscles, uterine muscles can contract spontaneously, and stretching enhances these contractions. As the baby grows, it stretches the uterine muscles, which contract occasionally weeks before delivery. The final trigger for labor is probably provided by the fetus. The near-term fetus produces steroid hormones that cause increased estrogen and prostaglandin production by the placenta and uterus. These hormones make the uterus even more likely to contract. When the combination of hormones and stretching activate the uterus beyond some critical point, strong contractions begin, signaling the onset of labor. As the contractions proceed, the baby's head pushes against the mother's cervix, making it expand in diameter (dilate). Stretch receptors in the walls of the cervix send signals to the hypothalamus, triggering oxytocin release. Under the dual stimulation of prostaglandin and oxytocin, the uterus contracts even more strongly. This positive feedback cycle is finally halted when the baby emerges from the vagina, or *birth canal.*

After a brief rest, uterine contractions resume, causing the uterus to shrink remarkably. During these contractions, the placenta is sheared from the uterus and expelled through the vagina as the *afterbirth.* Further release of prostaglandins in the *umbilical cord,* the stalk that connects the fetus and the placenta, causes the mus-

cles that surround fetal blood vessels in the umbilical cord to contract. This contraction shuts off blood flow. (Tying off the cord is standard practice but is not usually necessary; if it were, other mammals would not survive birth!) A new human being has been born.

3 How Can Fertility Be Limited?

Natural selection favors individuals that reproduce successfully, but reproduction alone is not sufficient for evolutionary success. The persistence of an animal's genes is ensured only if the animal's offspring survive long enough to reproduce. During most of human evolution, child mortality was high, so natural selection tended to favor people who produced enough children to offset the likelihood that most would not survive to adulthood. Today, however, most human populations have infant mortality rates that are far lower than those of ancient peoples. Humans no longer need to have many children to ensure that a few will survive to adulthood, but we nonetheless retain the reproductive drives with which evolution has endowed us. As a result, each passing week sees almost 2 million new people added to our increasingly overcrowded planet, and the control of birth rates has become an environmental necessity. On the individual level, techniques for limiting reproduction allow people to plan their families and provide the best opportunities for themselves and their children.

Historically, limiting fertility has not been easy. In the past, some cultures have tried such inventive, if bizarre, techniques as swallowing froth from the mouth of a camel or placing crocodile dung in the vagina. Since the 1970s, however, several effective techniques have been developed for **contraception,** or the prevention of pregnancy. These techniques are described next, and their reliability and some possible side effects are summarized in Table 35-3. Of course, the choice of a contraceptive should be made only in consultation with a physician, who can provide more-complete information.

Permanent Contraception Can Be Achieved through Abstinence or Sterilization

Abstaining from sexual intercourse is the only completely effective means (other than *sterilization*) of preventing sperm and egg from meeting. Abstinence also has the advantage of complete safety from possible contraceptive side effects and from sexually transmitted diseases.

In the long run, the most effortless method of contraception is **sterilization**, in which the pathways through which sperm or egg must travel are interrupted. In men, the vas deferens leading from each testis may be severed in an operation called a **vasectomy**. Sperm are still produced, but they cannot reach the penis during ejaculation. The surgery is performed under a local anesthetic, and vasectomy has no known physical side effects on health or sexual performance. In the United States, more than 500,000 vasectomy operations are performed annually and about 20% of all men over the age of 35 have had vasectomies. The slightly more complex operation of **tubal ligation** renders a woman infertile by cutting her oviducts. About 37% of women of childbearing age in the United States have chosen this form of birth control. Ovulation still occurs, but sperm cannot travel to the egg; nor can the egg reach the uterus. Sterilization is generally permanent. Sometimes, however, in a delicate and expensive operation, a surgeon can reconnect the vas deferens or oviducts.

There Are Three Major Approaches to Temporary Contraception

Temporary contraception techniques fall into three general categories: (1) preventing ovulation, (2) preventing sperm and egg from meeting when ovulation does occur, and (3) preventing the implantation of a fertilized egg in the uterus.

Synthetic Hormones Can Prevent Ovulation

As you learned earlier in this chapter, during a normal menstrual cycle, ovulation is triggered by a mid-cycle surge of LH. An obvious way to prevent ovulation is to suppress LH release by providing a continuous supply of estrogen and progesterone. Estrogen and progesterone (generally in synthetic form) are the components of **birth control pills**. "The Pill" is an extremely effective form of birth control, but it must be taken daily, normally for 21 days each menstrual period.

Two new long-term contraceptives, Norplant® and Depo-Provera®, have been approved for use in the United States since 1990. Both contain synthetic hormones, resembling progesterone, that prevent ovulation. Norplant consists of six slim, 1.3-inch-long silicone rub-

ber rods inserted under the skin of the upper arm. The rods provide a gradual, steady diffusion of hormone into the bloodstream for 5 years. In extensive tests, Norplant has proved slightly more effective than the birth control pill. Most women using Norplant become fertile within months after the capsule is removed. Depo-Provera is injected once every 3 months.

Barrier Methods Prevent Sperm from Reaching the Egg

There are several effective *barrier methods,* which prevent the encounter of sperm and egg. The **diaphragm** and **cervical cap** are rubber caps that fit snugly over the cervix, preventing sperm from entering the uterus. In conjunction with a **spermicide**, diaphragms and cervical caps are very effective and have no serious known side effects. Alternatively, the male can wear a **condom** over the penis, preventing sperm from being deposited in the vagina. A female condom is now available that completely lines the vagina. Diaphragms and condoms must be applied shortly before intercourse, typically at a time when the participants would rather be thinking about something else. Furthermore, if the diaphragm or condom has even a small hole, sperm can still enter the uterus and fertilize the egg.

Other, less effective procedures include the use of spermicides alone, **withdrawal** (the removal of the penis from the vagina just before ejaculation), and **douching** (washing sperm out of the vagina before, it is hoped, they have entered the uterus). Spermicides have some contraceptive effect, but withdrawal and douching are essentially useless. A final method of preventing fertilization is the **rhythm method**: abstinence from intercourse during the ovulatory period of the menstrual cycle. In practice, rhythm normally has a high failure rate, because of lack of discipline on the part of the users or inaccuracies in determining the menstrual cycle, which varies somewhat from month to month. Because a slight rise in body temperature and changes in the discharge of mucus from the cervix can be used to predict ovulation, the rhythm method can be made more effective if these indicators are monitored and sexual activity is regulated accordingly.

Intrauterine Devices Prevent Implantation

Even if an egg is fertilized, pregnancy will not occur unless the blastocyst implants in the uterus, so contraception can be achieved by preventing implantation. One approach to this kind of contraception is an **intrauterine device (IUD)**, a small copper or plastic loop, squiggle, or shield that is inserted into, and removed from, the uterus by a physician. Highly effective (if it stays in place), the IUD seems to work by irritating the uterine lining so that it cannot receive the embryo. A second method of preventing implantation is the "morning after" pill, which contains a massive dose of estrogen. For some women, preventing implantation has the major drawback that it is, in effect, an extremely early form of abortion.

Table 35-3 Nonpermanent Birth Control Techniques*

	Technique	Mechanism
Abstinence	Deciding not to be sexually active; Saying No.	Avoids any chance of contracting a sexually transmitted disease (STD).
The Pill	A pill made of a combination of synthetic hormones similar to those produced by the ovaries. Taken at the same time every day, regardless of intercourse.	The day-by-day action of the pills protects from pregnancy by preventing ovary from releasing an egg. With no egg present for a sperm to fertilize, pregnancy cannot occur. A back-up method should be used along with the Pill in first couple months of use.
Norplant	Long-lasting. Physician inserts 6 tiny flexible capsules filled with hormone (progestogen) under skin of upper arm to provide protection up to 5 years. Fertility returns when Norplant is removed.	Progestogen is released in steady low doses, preventing pregnancy by blocking ovulation.
Depo-Provera (injectable)	An injectable form of contraception (given as a shot) that protects from pregnancy for 3 months. Contains a chemical similar to the hormone (progesterone) that ovaries produce during the second half of the menstrual cycle.	Works by preventing egg cells from ripening. If an egg is not released from ovaries during a menstrual cycle, it can't be fertilized by sperm, so pregnancy can't occur. But injections must be given every 3 months—no more, no less.
Diaphragm/ Cervical Cap	Fit over the cervix and prevent pregnancy by preventing sperm from entering uterus. Both must be individually fitted by a medical professional. Both are made of rubber-like materials. Cap is half the size of diaphragm and can be left in place twice as long.	One teaspoonful of spermicide containing non-oxynol-9 should be placed inside the dome and another around the edge of diaphragm or cap; spermicide seals barrier and kills sperm.
Foam (spermicides)	Sperm-killing foam inserted into vagina before intercourse. Foam should contain non-oxynol-9 for best protection against pregnancy and STDs.	Inserted deep into vagina with plastic applicator, forms chemical barrier over uterine entrance. Sperm die when they hit foam.
Condom (for men)/ Condom (for women)	For men: Thin rubber or latex (latex is strongest) disposable sheath (normally lubricated with non-oxynol-9, not with Vaseline®) worn over penis during intercourse. For women: Soft, loose-fitting prelubricated polyurethane pouch with two rings: one inserted deep in vagina, the other, when in place, remains just outside the vagina.	For both: Placed correctly, catches sperm so they can't enter vagina. Also shields both partners from exposure to AIDS and other STDs. ("Natural" condoms don't protect against STDs.)
Rhythm Method	Combination of cervical mucus (Billings) and basal body temperature (BBT) methods to determine fertile period when pregnancy can occur. Billings: based on recognizing changes in mucus discharge that occur just *before* ovulation; BBT: based on body temperature changes just *after* ovulation.	Professional nurse practitioner or physician can indicate how changes in cervical mucus and body temperature can be used to predict the fertile period.
IUD (intrauterine device)	Small plastic device treated with copper or hormones. Nylon thread attached for easy checking. Fitted by medical professional.	Prevents egg from being implanted in uterine wall. Copper IUD is replaced every 6 years; hormonal IUD, every year.

*Modified from *"No!" and Other Methods of Birth Control*, Private Line, P.O. Box 31, Kenilworth, Illinois 60043.

Abortion Removes the Embryo from the Uterus

When contraception fails, pregnancy can be terminated by **abortion**. Abortion commonly involves dilating the cervix and removing the embryo and placenta by suction. Alternatively, abortion can be induced by drugs such as RU-486, which binds to progesterone receptors and blocks the actions of progesterone, which are essential to the maintenance of pregnancy. RU-486 is taken as a pill and is routinely used in France, Britain, Sweden, and China to terminate pregnancy up to about a month after conception. The drug is available on a very limited basis in the United States, but prospects for wider availability are poor, because no major pharmaceutical company is willing to manufacture or distribute such a controversial drug. In fact, European pharmaceutical companies have also halted production of RU-486, and it is unclear if

Table 35-3 *(Continued)*

Side Effects	Dangers	Reliability
None	None	100% effective
Regular periods, less anemia, less cramping, less of benign breast disease. May inhibit some forms cancer. Negative effects (normally disappear within 3 months) may include nausea, spotting, missed periods, headaches, mood changes, dark skin areas. Major but rare: blood clots, high blood pressure, gallbladder disease, heart attacks, liver tumors.	Women with certain health problems should not take the Pill. A health professional should choose the appropriate pill and offer advice. Smoking increases the chance of blood clots or stroke, even in young people on the Pill.	1 woman out of 300 becomes pregnant in a given year when using the Pill correctly. Physician must be consulted about possible interactions between the Pill and any other medications taken.
Irregular menstrual cycle for first 6 months with longer periods, spotting between periods or skipped periods. Possible headaches, weight gain.	Method is too new for accurate statistics. Counseling to understand method is extremely important. If pregnancy does occur, Norplant must be removed at once.	1 woman out of 300 becomes pregnant in a given year when using Norplant for the first 2 years. After 2 years, less protection for overweight women.
Irregular menstrual bleeding at first, then little or no menstruation, minor weight gain. Possibly headache, nervousness, abdominal pain, dizziness, fatigue. Physician should be consulted if unusual symptoms occur.	Important that injection be given during first 5 days of menstrual period to avoid possibility of receiving the shot during a pregnancy—which could result in ectopic pregnancy (outside uterus). If pregnancy occurs, physician should be consulted immediately.	1 woman out of 300 in a given year becomes pregnant while receiving Depo-Provera injections as prescribed.
None, unless the spermicide used causes bladder or yeast infection.	Slight danger of toxic shock syndrome (TSS). Should be used for birth control only, not to control vaginal secretions. Must be inserted correctly and used with spermicidal foam or jelly.	5 to 10 women out of 100 in a given year become pregnant when using a diaphragm or cap consistently and correctly. (In case of weight gain or loss, a diaphragm may no longer fit correctly and should be checked by a medical professional.)
Occasionally the foam can cause a mild vaginal irritation, cause bladder or yeast infection, or irritate partner.	None, except a condom should be used as well to give the better protection against STDs and pregnancy.	4 to 29 women out of 100 become pregnant in a given year when using foam alone.
For both: Next to avoiding intercourse, condoms used along with spermicidal foam containing non-oxynol-9 are the best available way to protect against AIDS and other STDs—or pregnancy. (Allergy to spermicidal foam or lubricant is possible.)	None. Condoms should always be used to protect against disease, regardless of other techniques used— even sterilization. For women: A female condom, properly used, provides even better protection than a male condom against STDs.	2 to 10 women out of 100 become pregnant in a given year when the man uses a condom correctly every time. The time period for clinical testing of the female condom has been too short for accurate statistics. The manufacturer reports that 2 to 3 women out of 100 become pregnant in a given year when the woman uses a condom every time as directed.
None, but offers no protection against STDs.	Pregnancy can occur unless intercourse is avoided during the fertile period. Some women have trouble identifying mucus changes.	10 to 15 women out of 100 become pregnant in a given year while using the Billings or BTT method. Relying on the calendar date alone is not reliable.
Possible cramps and heavy menstrual flow caused by the body's effort to push out IUD. Heavy menstrual periods during the first few months.	Risk of pelvic inflammatory disease (PID) or tubal pregnancy. Physician should be consulted if fever or stomach pain occurs.	1 to 5 women out of 100 in a given year become pregnant while using an IUD.

new sources will develop before current inventories are depleted.

Because abortions are normally surgical procedures, they are potentially more dangerous to a woman's health than most of the contraceptive techniques described above. Although science can describe fetal development during pregnancy, it cannot provide judgments about when a fetus becomes a "person" or about the relative merits of fetal rights versus maternal rights. Therefore, abortion remains controversial.

Additional Contraceptive Methods Are Under Development

Further advances in contraception are under development. For example, a vaginal ring that continuously releases synthetic progesterone has been tested in several countries but is not yet on the market. A pill that prevents ovulation by blocking receptors for gonadotropin-releasing hormone is in the testing stage. Once-a-month contraceptive pills, most of which prevent the uterus from becoming fully prepared for implantation, are also under development. Another possibility is a contraceptive vaccine that would remain effective for one to several years. One such vaccine induces antibodies against human chorionic gonadotropin, which is essential for the implantation of the embryo in the uterus. Another vaccine induces the formation of antibodies to a protein called SP-10, which is unique to sperm.

Most contraceptive techniques are directed at the woman, not the man, because it is much easier to interfere with ovulation than with sperm formation. A woman ovulates only once a month, and ovulation itself does not influence a woman's sexual drives. In contrast, testosterone is essential for both sperm formation and sexual performance; early "male pills" caused impotence as well as infertility. Nevertheless, a major research effort is under way to develop male contraceptives equivalent to the Pill. A promising contraceptive undergoing clinical trials is a daily dose of testosterone and a modified form of gonadotropin-releasing hormone, which together seem to block sperm production without affecting sexual performance.

Summary of Key Concepts

1) How Do Animals Reproduce?

Animals reproduce either sexually or asexually. In sexual reproduction, haploid gametes, normally from two separate parents, unite and produce an offspring that is genetically different from either parent. In asexual reproduction, offspring tend to be genetically identical to the parent. Asexual reproduction can occur by budding, fission, or parthenogenesis.

Among animals that engage in sexual reproduction, the female produces large, nonmotile eggs and the male produces small, motile sperm. Animals are either monoecious (a single animal produces both sperm and eggs) or dioecious (a single animal produces one type of gamete). Fertilization, the union of sperm and egg, can occur outside the bodies of the animals (external fertilization) or inside the body of the female (internal fertilization). External fertilization must occur in water so that the sperm can swim to meet the egg. Internal fertilization normally occurs by copulation, in which the male deposits sperm directly into the female's reproductive tract.

2) How Does the Human Reproductive System Work?

The human male reproductive tract consists of paired testes, which produce sperm and testosterone, and accessory structures that conduct the sperm to the female's reproductive tract and secrete fluids that activate swimming by the sperm and provide energy. In human males, spermatogenesis and testosterone production are stimulated by FSH and LH, secreted by the anterior pituitary. Spermatogenesis and testosterone production are nearly continuous, beginning at puberty and lasting until death.

The human female reproductive tract consists of paired ovaries, which produce eggs as well as the hormones estrogen and progesterone, and accessory structures that conduct sperm to the egg and receive and nourish the embryo during prenatal development. In human females, oogenesis, hormone production, and development of the lining of the uterus vary in a monthly menstrual cycle. The cycle is controlled by hormones from the hypothalamus (gonadotropin-releasing), anterior pituitary (FSH and LH), and ovaries (estrogen and progesterone).

During copulation, the male inserts his penis into the female's vagina and ejaculates semen. The sperm move through the vagina and uterus into the oviduct, where fertilization normally takes place. The unfertilized egg is surrounded by two barriers, the corona radiata and the zona pellucida. Enzymes released from the acrosomes in the head of sperm digest these layers, permitting sperm to reach the egg. Only one sperm enters the egg and fertilizes it.

The fertilized egg undergoes a few cell divisions in the oviduct and then implants in the uterine lining. Implantation and subsequent release of chorionic gonadotropin by the embryo maintain the integrity of the corpus luteum and the endometrium during early pregnancy, preventing further menstrual cycles.

During pregnancy, mammary glands in the mother's breasts enlarge under the influence of estrogen, progesterone, and other hormones. After birth, milk secretion is triggered by prolactin and oxytocin, whose release is triggered by suckling.

After about 9 months, uterine contractions are triggered by a complex interplay of uterine stretch and prostaglandin and oxytocin release. As a result, the uterus expels the baby and then the placenta.

3) How Can Fertility Be Limited?

Permanent contraception can be achieved by abstinence or by sterilization: severing the vas deferens in males (vasectomy) or the oviducts in females (tubal ligation). Temporary contraception techniques include those that prevent ovulation: birth control pills, Norplant, and Depo-Provera. Barrier methods, which prevent sperm and egg from meeting, include the diaphragm, the cervical cap, and the condom, accompanied by spermicide. Spermicide alone is less effective. Withdrawal and douching are ineffective techniques. The rhythm method involves abstinence around the time of ovulation. Intrauterine devices prevent implantation of the blastocyst. Abortion causes the expulsion of the developing embryo.

Key Terms

abortion *p. 732*	**endometrium** *p. 722*	**lactation** *p. 728*	**semen** *p. 719*
acquired immune deficiency	**epididymis** *p. 719*	**luteinizing hormone (LH)**	**seminal vesicle** *p. 719*
syndrome (AIDS) *p. 726*	**estrogen** *p. 722*	*p. 718*	**seminiferous tubule** *p. 717*
acrosome *p. 717*	**external fertilization**	**mammary gland** *p. 728*	**Sertoli cell** *p. 717*
amplexus *p. 715*	*p. 713*	**menstrual cycle** *p. 722*	**sexually transmitted disease**
asexual reproduction	**fallopian tube** *p. 722*	**menstruation** *p. 722*	**(STD)** *p. 726*
p. 712	**fertilization** *p. 712*	**monoecious** *p. 713*	**sexual reproduction** *p. 712*
birth control pill *p. 731*	**fetus** *p. 720*	**myometrium** *p. 722*	**spawning** *p. 713*
blastocyst *p. 727*	**fimbria** *p. 722*	**oogenesis** *p. 720*	**sperm** *p. 713*
bud *p. 712*	**fission** *p. 712*	**oogonium** *p. 720*	**spermatid** *p. 717*
budding *p. 712*	**follicle** *p. 720*	**ovary** *p. 719*	**spermatogenesis** *p. 717*
bulbourethral gland *p. 719*	**follicle-stimulating hormone**	**oviduct** *p. 722*	**spermatogonium** *p. 717*
cervical cap *p. 731*	**(FSH)** *p. 718*	**ovulation** *p. 716*	**spermatophore** *p. 715*
cervix *p. 722*	**genital herpes** *p. 726*	**parthenogenesis** *p. 712*	**spermicide** *p. 731*
chlamydia *p. 726*	**genital wart** *p. 726*	**penis** *p. 719*	**sterilization** *p. 731*
chorionic gonadotropin (CG)	**gonad** *p. 716*	**pheromone** *p. 714*	**syphilis** *p. 726*
p. 724	**gonadotropin-releasing**	**placenta** *p. 722*	**testis** *p. 716*
clitoris *p. 725*	**hormone (GnRH)**	**polar body** *p. 720*	**testosterone** *p. 718*
colostrum *p. 728*	*p. 718*	**primary oocyte** *p. 720*	**trichomoniasis** *p. 726*
condom *p. 731*	**gonorrhea** *p. 726*	**primary spermatocyte**	**tubal ligation** *p. 731*
contraception *p. 730*	**hermaphrodite** *p. 713*	*p. 717*	**urethra** *p. 719*
copulation *p. 715*	**implantation** *p. 727*	**progesterone** *p. 722*	**uterus** *p. 722*
corona radiata *p. 725*	**inner cell mass** *p. 727*	**prostate gland** *p. 719*	**vagina** *p. 722*
corpus luteum *p. 722*	**internal fertilization** *p. 715*	**regeneration** *p. 712*	**vas deferens** *p. 719*
crab lice *p. 726*	**interstitial cell** *p. 717*	**rhythm method** *p. 731*	**vasectomy** *p. 731*
diaphragm *p. 731*	**intrauterine device (IUD)**	**scrotum** *p. 716*	**withdrawal** *p. 731*
dioecious *p. 713*	*p. 731*	**secondary oocyte** *p. 720*	**zona pellucida** *p. 725*
douching *p. 731*	**labium** *p. 725*	**secondary spermatocyte**	**zygote** *p. 722*
egg *p. 713*	**labor** *p. 730*	*p. 717*	

Thinking Through the Concepts

Multiple Choice

1. *Budding and fission are processes used by*
a. dioecious species
b. hermaphroditic organisms
c. organisms requiring new gene combinations for each generation
d. sexually reproducing species
e. asexually reproducing species

2. *All of the following are barrier contraceptive devices EXCEPT the*
a. IUD
b. female condom
c. cervical cap
d. diaphragm
e. male condom

3. *In humans, spermatogenesis yields _____ sperm for each diploid sex cell, and oogenesis yields _____ secondary oocyte(s) for each sex cell.*
a. one, four   b. two, one
c. one, two   d. four, one
e. four, two

4. *Which structure adds the* final *secretions to semen as it moves out of the male reproductive tract?*
a. epididymis
b. bulbourethral gland
c. seminal vesicle
d. prostate gland
e. interstitial cells

5. *During fetal development, oogonia in human females halt meiosis at*
a. the secondary oocyte stage
b. the polar body stage
c. metaphase of meiosis II
d. telophase of meiosis II
e. prophase of meiosis I

6. *The primary hormone that inhibits GnRH is*
a. FSH
b. LH
c. progesterone
d. estrogen
e. a hypothalamic releasing factor

? **Review Questions**

1. List the advantages and disadvantages of asexual reproduction, sexual reproduction, external fertilization, and internal fertilization, including an example of an animal that uses each type.

2. Compare the structures of the egg and sperm. What structural modifications do sperm have that facilitate movement, energy use, and digestion?

3. What is the role of the corpus luteum in a menstrual cycle? In early pregnancy? What determines its survival after ovulation?

4. Construct a chart of common sexually transmitted diseases. List the disease's name, the cause (organism), symptoms, and treatment.

5. List the structures, in order, through which a sperm passes on its way from the seminiferous tubules of the testis to the oviduct of the female.

6. Name the three accessory glands of the male reproductive tract. What are the functions of the secretions they produce?

7. Diagram the menstrual cycle, and describe the interactions among hormones secreted by the pituitary gland and ovaries that produce the cycle.

8. Describe the changes in the breast that prepare a mother to nurse her child. How do hormones influence these changes and stimulate milk production?

9. Describe the events that lead to the expulsion of the baby and placenta from the uterus. Explain why this is an example of positive feedback.

Applying the Concepts

1. Identify and discuss some of the ethical issues involved in *in vitro* fertilization.

2. Discuss the most effective or appropriate method of birth control for each of the following couples: Couple A, which has intercourse three times a week but never wants to have children; Couple B, which has intercourse once a month and may want to have children someday; and couple C, which has intercourse three times a week and wants to have children someday.

3. Female condoms were recently introduced in the United States. What advantages and disadvantages can you think of for this form of birth control?

4. *Pelvic endometriosis* is a relatively common disease of women in which bits of the endometrial lining find their way onto abdominal organs and respond in typical ways to hormones during a menstrual cycle. When the uterine lining bleeds during menstruation, so do these implants. Common treatments are oral contraceptives, Danazol® (a compound that inhibits gonadotropins), and synthetic GnRH analogues that are, paradoxically, powerful inhibitors of FSH and LH. How does each of these compounds provide relief?

5. Would contraceptive drugs that block cell receptors for FSH be useful in males and/or females? Explain. What side effects would such drugs have?

6. Think of all the choices a couple has to obtain a child, including: *in vitro* fertilization using the couple's eggs and sperm, *in vitro* fertilization using a donor's sperm or egg, and insemination of a surrogate mother with sperm from the couple's husband. Think of some more. What ethical issues do these various options present?

Group Activity

The president of the National Academy of Sciences has chosen you to serve on a panel that will try to answer the question "Why is there sex?" The panel will meet next week, so form a group of four students to develop a working hypothesis. The problem you face is this: Sexual reproduction is inefficient. To illustrate this, have your group imagine a population of 20 insects (10 male, 10 female) in which each female can produce two offspring in her lifetime. The females in this population, however, are of two types. Five females reproduce asexually, giving birth only to more females. The other five reproduce sexually and give birth to males and females in equal proportions. After five generations, how many females of each type will be in the population? Clearly, sexual reproduction puts individuals at a severe disadvantage in the contest to reproduce. Although it's true that sexual recombination of genes can have advantages over the very long term, the theory of natural selection tells us that a trait can persist only if it helps individuals survive and reproduce in the very short term. So we're faced with a mystery. What direct advantage was gained by the first sexually reproducing organisms? Was a series of intermediate steps required? Were particular environmental conditions required? Did sex originally arise for a purpose other than reproduction? The Academy is expecting you to produce a detailed, plausible scenario for the origin of sex, so gather your group for critical discussion, and submit your hypothesis.

For More Information

Alexander, N. "Future Contraceptives." *Scientific American*, September 1995. A review of the new techniques that may be used for contraception in the twenty-first century.

Crews, D. "Animal Sexuality." *Scientific American*, January 1994. Explores the wide range of mechanisms that control male and female sexual development among different types of animals.

Eberhard, W. G. "Runaway Sexual Selection." *Natural History*, December 1987. Interesting article on the evolution of male genitalia.

Morell, V. "A Clone of One's Own." *Discover*, May 1998. Someday, someone will probably clone a person. What will be the outcome?

Riddle, J. M., and Estes, J. W. "Oral Contraceptives in Ancient and Medieval Times." *American Scientist*, May–June 1992. How did women control their fertility before modern medicine stepped in?

Ulmann, A., Teutsch, G., and Philibert, D. "RU-486." *Scientific American*, June 1990. Describes the abortion pill now widely used in France and other applications for this drug.

Wassarman, P. W. "Fertilization in Mammals." *Scientific American*, December 1988. Describes the events leading to the penetration of the egg by the sperm, with emphasis on the role and composition of the zona pellucida.

Wright, K. "Human in the Age of Mechanical Reproduction." *Discover*, May 1998. The child's question "Where do babies come from?" has become far more difficult to answer as technology has provided more than a dozen ways to have a child.

Wright, K. "The Sniff of Legend." *Discover*, April 1994. This article explores human pheromones, sex attractants, and a sixth sense organ in the nose.

Answers to Multiple-Choice Questions
1. e   2. a   3. d   4. b   5. e   6. c

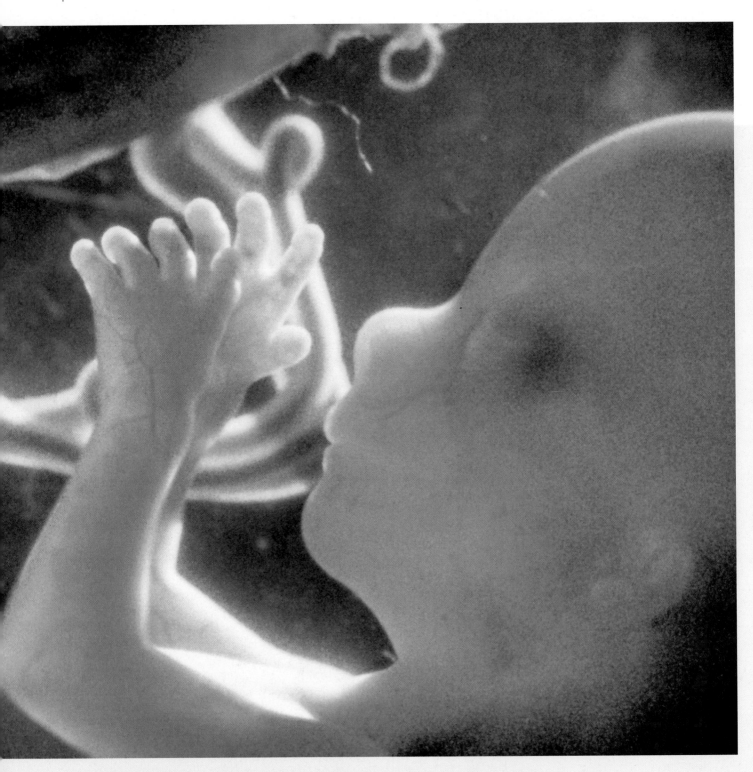

This 4-month-old human fetus must continue to develop inside its mother for another 5 months, obtaining oxygen and nutrients from the mother's bloodstream. These vital resources pass to the fetus by way of the placenta and the umbilical cord, which is visible in the background.

Animal Development 36

At a Glance

Net Watch

On-line resources for this chapter are on the World Wide Web at:
http://www.prenhall.com/audesirk
(click on the Table of Contents link and then select Chapter 36).

How does form arise from formlessness? This question represents one of the central mysteries of biology and is the basis of the burgeoning field of developmental biology. Developmental biologists want to know how a single cell—the zygote formed from the fusion of sperm and egg—transforms itself into a complex organism. As cell division proceeds, some daughter cells become nerves, some become muscles, some become sensory organs, and so on. Because the cells of the embryo proliferate by mitosis, each cell has an identical genome. So, what mechanism directs cells to become different specialized types? Perhaps even more puzzling is the question of how animal form arises. How do the cells of, say, a developing dog arrange themselves into the shape of a dog? How do they "know" to arrange themselves as a leg with five toes on a paw and not as a fin, or a wing, or a leg with a hoof, or a tail, or an eye? And once all that is settled, how do the cells age? As you've probably guessed, developmental biology is a field with more questions than answers. Yet, answers are beginning to emerge, and exciting new findings are published almost daily.

1) How Do Cells Differentiate during Development?

Development is the process by which an organism proceeds from fertilized egg through adulthood. **Differentiation** is the specialization of embryonic cells into different cell types, such as muscle cells, brain cells, and so on. How do cells become differentiated from one another during development? One possibility might be that differentiation results from a progressive loss of genes. By this scenario, the zygote would contain all the genes needed to direct the construction of the entire organism. Each differentiated cell would retain only those genes needed for its particular function in the body; unnecessary genes would be lost during the process of specialization. This idea is appealing but has proved to be inaccurate.

A clever experiment by the British molecular biologist J. B. Gurdon showed that gene loss cannot be the mechanism of differentiation. Gurdon transplanted nuclei from intestinal cells of tadpoles of the African clawed frog, *Xenopus,* into unfertilized eggs whose own nuclei had been destroyed (Fig. 36-1). Although the operation was very difficult and most of the eggs died, some of the eggs with intestinal nuclei, when fertilized, developed into normal adult frogs. This experiment showed that the nucleus of an intestinal cell provided all the genetic information necessary to form a normal tadpole. More generally, each differentiated cell in an animal contains all the genetic information needed for the development of the entire or-

ganism. Different types of cells differ not in which genes they contain but in which genes they use. In other words, cell types differ because different genes are activated, transcribed to messenger RNA, and translated into proteins.

Gene Transcription Is Precisely Regulated during Development

In any cell at any time, only a portion of the cell's genes are transcribed. (Recall from Chapter 10 that transcription is the production of messenger RNA, using the gene as a blueprint.) The particular combination of genes that is transcribed in a cell largely determines the shape, structure, and activity of that cell. Thus, differentiation during development is accomplished by the selective activation of transcription in different sets of genes. In general, the mechanism by which transcription is controlled involves regulatory molecules, typically proteins or proteins combined with activating substances such as steroid hormones, that travel to the nucleus and bind to the chromosomes (see Chapter 10). By binding, these substances can either block or promote the transcription of particular genes.

During the development of the egg (*oogenesis*) in many animal species (but not in mammals), various gene-regulating substances become concentrated in different specific places in the egg cytoplasm. As the fertilized egg divides, each of its daughter cells receives different gene-regulating substances (Fig. 36-2). Hence, the developmental fate of a daughter cell is determined by the gene-regulating substances that are present in the part of the egg's cytoplasm it inherits. Experiments that uncovered this principle are described later in the chapter.

Differentiation is also regulated by cell–cell interactions. During later embryonic development (and continuing throughout adult life), cells constantly receive chemical messages, including nutrients, hormones, and neurotransmitters, from other cells of the body. These chemical messages can alter the developmental fate of a cell by altering the transcription of genes and the activity of enzymes within a cell, as described in Chapter 32.

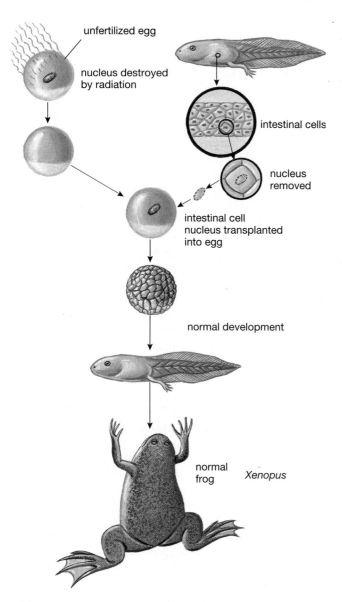

Figure 36-1 *Cells retain all of their genes during differentiation*
J. B. Gurdon's experiment proving that cells do not lose genes as they differentiate. Gurdon destroyed the nuclei of unfertilized frog eggs and then transplanted nuclei of intestinal cells from a tadpole into the egg. The resulting egg cells developed into normal tadpoles and eventually adult frogs, demonstrating that the intestinal cells retained all the genes necessary for normal development of an entire organism.

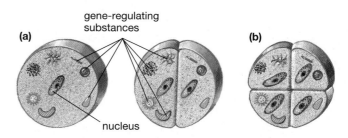

Figure 36-2 *Distribution of gene-regulating substances*
Gene-regulating substances are not distributed evenly—nor randomly—during the division of a fertilized egg. *(a)* Different gene-regulating substances (represented by the various symbols) are positioned in particular places in the egg's cytoplasm during oogenesis. *(b)* As the egg divides, these materials remain in about the same positions, so the daughter cells inherit different substances.

2) How Do Indirect and Direct Development Differ?

Animal development begins with an egg that contains food reserves of **yolk**, rich in lipids and protein. This food reserve is crucial to the **embryo**, as the organism is called in its early stages of development. The yolk provides nourishment for the early embryo until it develops into a form that can obtain food from outside sources. The amount of yolk in an egg corresponds closely with the way in which an animal develops. Animal development normally proceeds down one of two paths: (1) indirect development or (2) direct development.

During Indirect Development, Animals Undergo a Radical Change in Body Form

In **indirect development**, the juvenile animal that hatches from the egg differs significantly from the adult. As it matures, the offspring undergoes radical changes in body form, such as the transformation of a caterpillar into a butterfly. Indirect development occurs in most invertebrates, including insects and echinoderms, and in a few vertebrates—notably, the amphibians. Animals with indirect development typically produce huge numbers of eggs, and each egg has only a small amount of yolk. The yolk nourishes the developing embryo during a rapid transformation into a small, sexually immature form called a **larva** (Fig. 36-3a,b). Because the yolk is small and the time spent as an embryo is relatively short, indirect development does not place great demands on the mother, and many offspring can be produced.

Some larval animals not only look very different from adult animals but also occupy entirely different habitats. In addition, most larvae feed on different organisms than they will as adults. For instance, the aquatic larva of the dragonfly feeds on aquatic organisms such as tadpoles, but the adult dragonfly, which is terrestrial, feeds on insects (Fig. 36-3b). Eventually, the larvae undergo a revolution in body form, or **metamorphosis**, and become sexually mature adults.

Although we tend to regard the adult form as the "real animal" and larvae as "preparatory stages," most of the life span of some animals, especially insects, is spent as a larva. The adult may live for only a few days, reproducing frantically and in some cases not even eating. The mayfly, for example, metamorphoses from an aquatic larva that may have spent a year or more feeding and growing. Emerging in huge swarms from freshwater streams, ponds, and lakes, adult mayflies live a few hours or, at most, a few days. The sole occupation of the adults is to mate and lay eggs; their fragile dead bodies then accumulate in piles to be swept away by the wind.

Newborn Animals That Undergo Direct Development Resemble Miniature Adults

Other animals, including such diverse groups as land snails, reptiles, birds, and mammals, show **direct development**, in which the newborn animal is a miniature, but sexually immature, version of the adult (Fig. 36-4). As the young animal matures, it may grow much bigger, but it does not radically change its body form.

Juveniles of directly developing species are typically much larger than larvae and consequently need much

Figure 36-3 Indirect development Many animals, including *(a)* many marine mollusks such as this common whelk, undergo indirect development. The larval stage tends to be very different from the adult in size, appearance, and lifestyle. The molluscan larva is barely visible to the naked eye. *(b)* The larval dragonfly is aquatic and feeds on tadpoles and small fish, whereas the adult form is terrestrial and eats other insects.

(a)

(b)

(c)

(d)

Figure 36-4 *Direct development*
The offspring of animals with direct development closely resemble their parents from the moment of birth, except in size. *(a)* Land snails, *(b)* lizards, and *(c)* birds hatch from large, yolk-filled eggs. *(d)* Mammalian mothers nourish their young within their bodies for weeks or months.

more nourishment before emerging into the world. Two strategies have evolved that meet the embryo's food requirement. Snails, reptiles, and birds produce large eggs that contain large amounts of yolk. An ostrich egg, for example, weighs several pounds. Mammals, some snakes, and a few fish have relatively little yolk in their eggs but instead nourish the developing embryo within the body of the mother (Fig. 36-4d). Either way (in contrast to indirect development), providing food for directly devel-

oping embryos places great demands on the mother, and relatively few offspring are produced.

Reptiles, Birds, and Mammals Produce Similar Extraembryonic Membranes

Amphibians were the first vertebrates to live on land, but their reproduction remained tied to water, and even now they don't venture very far from the standing water where their eggs must be deposited. Fully terrestrial life

Table 36-1 Vertebrate Embryonic Membranes

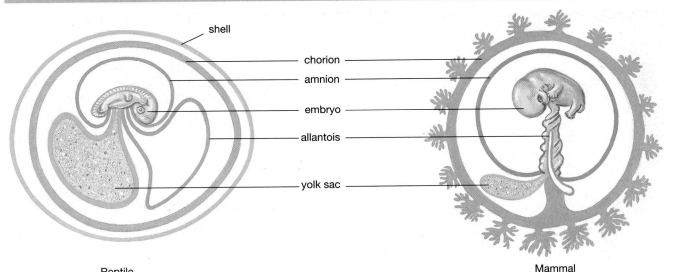

Reptile

Mammal

Membrane	Reptilian Embryo		Mammalian Embryo	
	Structure	Function	Structure	Function
Chorion	Membrane lining inside shell	Acts as respiratory surface; regulates exchange of gases and water between embryo and air	Fetal contribution to placenta	Provides surface for exchange of gases, nutrients, and wastes between embryo and mother
Amnion	Sac surrounding embryo	Encloses embryo in fluid	Sac surrounding embryo	Encloses embryo in fluid
Allantois	Sac connected to embryonic urinary tract; capillary-rich membrane lining inside of chorion, with blood vessels connecting to embryonic circulation	Stores wastes (especially urine); acts as respiratory surface	Provides blood vessels of umbilical cord	Carries blood between embryo and placenta
Yolk sac	Membrane surrounding yolk	Contains yolk as food; digests yolk and transfers nutrients to embryo; forms part of digestive tract	"Empty" membranous sac	Forms part of digestive tract

was not possible until a final piece of the puzzle evolved: the shelled **amniote egg**. This innovation, which encases the embryo in a protected, liquid-filled space, arose first in the reptiles and persists today in that group and in its descendants, birds and mammals. The amniote egg is characterized by four membranes, called **extraembryonic membranes**: (1) the *chorion*; (2) the *amnion*; (3) the *allantois*; and (4) the *yolk sac*. The **chorion** lines the shell and exchanges oxygen and carbon dioxide through the shell. The **amnion** encloses the embryo in a watery environment; the **allantois** surrounds wastes; and the **yolk sac** contains the stored food. Although mammalian eggs contain almost no yolk, much of the reptilian genetic program for development still persists, including the four extraembryonic membranes. Table 36-1 compares the structures and functions of these extraembryonic membranes in reptiles and mammals.

3) How Does Animal Development Proceed?

The transformation from fertilized egg—a single cell—to a multicellular, differentiated embryo is a beautiful, nearly magical process that has been carefully described for a number of animals. In textbooks, the changes in the developing embryo are generally depicted in stages, but the stages are just convenient "snapshots" for the purpose of illustration. The actual development is a smoothly continuous process. The initial stages of *cleavage, gastrulation, organogenesis,* and *growth* occur during embryonic life (Fig. 36-5), in which nearly all the organs that will be present in the adult are formed. After birth, the animal typically undergoes further growth, achieves *sexual maturity* (at which time the animal may reproduce), *ages,* and finally dies. Let's examine each of these events.

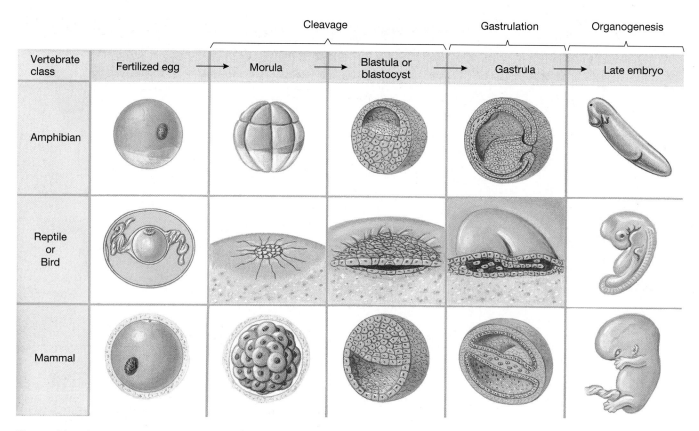

Vertebrate class	Fertilized egg	Morula	Blastula or blastocyst	Gastrula	Late embryo
		Cleavage		Gastrulation	Organogenesis
Amphibian					
Reptile or Bird					
Mammal					

Figure 36-5 Stages in the development of amphibians, reptiles (or birds), and mammals

Cleavage Distributes Gene-Regulating Substances

Development begins with **cleavage**, a series of mitotic divisions of the fertilized egg without an increase in overall size. Cleavage reduces cell size and distributes gene-regulating substances to the newly formed cells. As you know, an egg is a very large cell. Unlike most mitotic cell divisions—which proceed through a cycle in which they divide, grow, duplicate genetic material, then divide again—embryonic cell divisions during cleavage skip the growth phase. Consequently, as cleavage progresses, the available cytoplasm is split up into ever-smaller cells whose sizes approach those of cells in the adult organism. Finally, a solid ball of small cells, the **morula**, is formed. The morula as a whole is about the same size as the zygote. As cleavage continues, a cavity opens within the morula. The cells of the morula then become the outer covering of a hollow (typically spherical) structure, the **blastula**. The space inside the blastula is called the *blastocoel*.

The details of cleavage differ by species. The pattern of cleavage is largely determined by the amount of yolk present, because yolk hinders cytoplasmic division (cytokinesis). The almost yolkless eggs of sea urchins divide symmetrically, but eggs with extremely large yolks, such as a hen's egg, don't divide all the way through. Nevertheless, a hollow blastula is always produced, though in reptiles and birds the blastula is flattened rather than spherical.

During cleavage, different gene-regulating materials become incorporated into different daughter cells (see Fig. 36-2). This process was discovered decades ago in experiments on frog eggs (Fig. 36-6). The unfertilized frog egg has pale yolk on the "bottom," or *vegetal pole,* and pigmented cytoplasm on the "top," or *animal pole.* At fertilization, some of the pigment shifts toward the animal pole, leaving behind a **gray crescent** of intermediate coloring. The first cleavage division of frog eggs normally passes through the center of the gray crescent, so each daughter cell receives roughly half the crescent (Fig. 36-6a). If the two daughter cells are gently separated, each will develop into a normal tadpole. But scientists can experimentally force the first division to miss the gray crescent, so that one daughter cell receives the entire gray crescent, whereas the other receives none (Fig. 36-6b). If the cells are then separated, the cell with the gray crescent develops into a tadpole; the one with no crescent material merely forms a lump of cells that soon dies. Clearly, gene-regulating substances in the gray crescent region are required for the normal development of the tadpole.

Gastrulation Forms Three Tissue Layers

In the next step of development, an indentation called the **blastopore** forms on one side of the blastula. Blastula cells migrate in a continuous sheet in through the blastopore, much as if you punched in an underinflated basketball (Fig. 36-7a,b), to form three embryonic tissue layers. The enlarging dimple is destined to become the digestive tract; the cells that line its cavity are now called

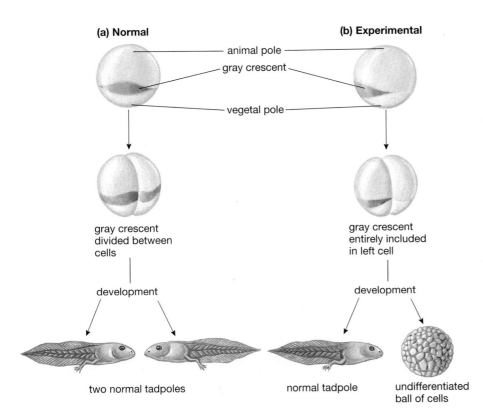

(a) Normal

animal pole

gray crescent

vegetal pole

gray crescent divided between cells

development

two normal tadpoles

(b) Experimental

gray crescent entirely included in left cell

development

normal tadpole

undifferentiated ball of cells

Figure 36-6 *Distribution of gene-regulating substances during cleavage*
Different gene-regulating substances are distributed to different daughter cells during cleavage. In frog eggs, a pigmented region known as the gray crescent contains substances needed for normal embryonic growth and differentiation. **(a)** Normally, the first cleavage division cuts neatly through the center of the gray crescent. If the two daughter cells are separated in the lab, each cell contains gray crescent material, and each can develop into a normal tadpole. **(b)** If the first cleavage division is experimentally forced to miss the gray crescent, one daughter cell receives the entire gray crescent and the other receives none. The cell with the crescent material develops normally, but the cell lacking crescent substance cannot.

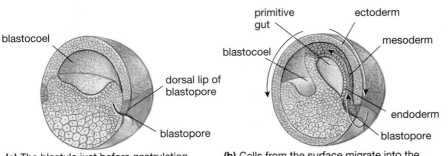

blastocoel

dorsal lip of blastopore

blastopore

(a) The blastula just before gastrulation. The opening inside is the blastocoel; the blastopore is the site at which gastrulation will begin.

primitive gut

ectoderm

mesoderm

blastocoel

endoderm

blastopore

(b) Cells from the surface migrate into the interior of the blastula through the blastopore. These cells will form the endoderm and mesoderm layers of the gastrula; the cells remaining on the surface will form ectoderm. The endoderm encloses the primitive gut.

primitive gut

developing notochord and muscle

remnant of blastocoel

yolk plug in blastopore

(c) Mesoderm cells differentiate into the notochord and muscle masses.

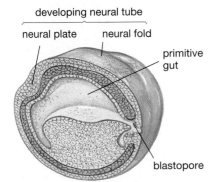

developing neural tube

neural plate   neural fold

primitive gut

blastopore

(d) The notochord induces ectoderm cells lying directly above it to form the neural tube.

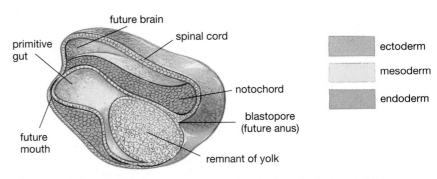

future brain

spinal cord

primitive gut

notochord

future mouth

blastopore (future anus)

remnant of yolk

(e) The neural tube enlarges and differentiates into brain and spinal cord. A future mouth is produced when the opening formed by the primitive gut breaks through at the end of the embryo opposite the blastopore. The blastopore is the future anus.

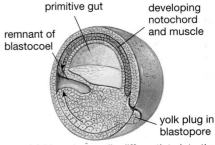

ectoderm

mesoderm

endoderm

*Figure 36-7 **Gastrulation in the frog***

Table 36-2 Derivation of Adult Tissues from Embryonic Cell Layers

Embryonic Layer	Adult Tissue
Ectoderm	Epidermis of skin; lining of mouth and nose; hair; glands of skin (sweat, sebaceous, and mammary glands); nervous system; lens of eye; inner ear
Mesoderm	Dermis of skin; muscle, skeleton; circulatory system; gonads; kidneys; outer layers of digestive and respiratory tracts
Endoderm	Lining of digestive and respiratory tracts; liver; pancreas

endoderm (Greek for "inner skin"). The cells remaining on the outside will form the epidermis of the skin and the nervous system and are called **ectoderm** ("outer skin"). Meanwhile, some cells migrate between the endoderm and ectoderm, forming a third and final layer, the **mesoderm** ("middle skin"). Mesoderm gives rise to muscles, the skeleton (including the **notochord**, a supporting rod found at some stage in all chordates; Fig. 36-7c), and the circulatory system (Table 36-2). This

process of cell movement is called **gastrulation**, and the three-layered embryo that results is the **gastrula**.

The Cellular Environment Influences the Developmental Fate of Cells

During gastrulation, the developmental fate of most of the embryo's cells is determined by chemical messages received from other cells, a process called **induction**. In amphibian embryos, cells derived from the gray crescent region of the zygote form the site of dimpling as the blastula is transformed into the gastrula. This area, called the *dorsal lip* of the blastopore, controls the developmental fate of the cells around it, as Hans Spemann and Hilde Mangold demonstrated in the 1920s (Fig. 36-8a). They transplanted the dorsal lip of the blastopore from one embryo to another. The transplanted dorsal lip then induced the cells of the host to form a second embryo, showing that dorsal lip tissue controls differentiation in the surrounding cells. A control experiment can also be performed, transplanting cells from regions of the gastrula other than the dorsal lip of the blastopore (Fig. 36-8b). These cells give rise to tissues appropriate to the region into which they were transplanted rather than the region from which they were taken. These experi-

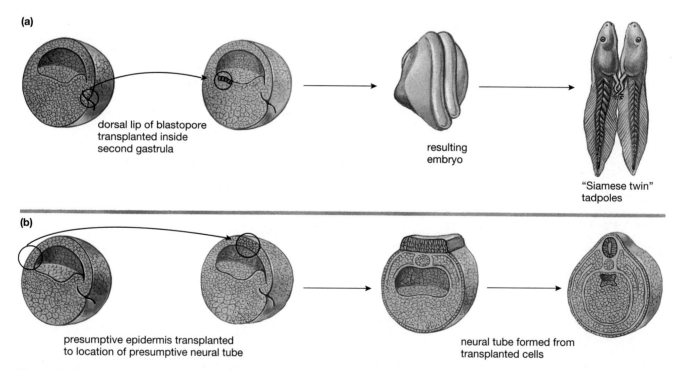

Figure 36-8 Induction and its role in differentiation
(a) The dorsal lip of the blastopore is removed from the gastrula of a heavily pigmented strain of newt and transplanted into a different region of the gastrula from a lightly pigmented strain of newt. The transplanted dorsal lip of the blastopore induced the adjacent regions of the "host" gastrula to form a separate, nearly complete tadpole with the dark pigmentation of the transplanted tissue. This experiment suggests that many cells are "followers," guided along certain developmental paths by their proximity to inducing tissues such as the dorsal lip of the blastopore. *(b)* Cells that would normally become skin from the gastrula of a darkly pigmented strain of newt are transplanted to another site on a "host" gastrula of the lighter strain. The host gastrula has induced cells that would normally have formed skin to produce a neural tube. Thus, the developmental fate of follower cells is determined by the region into which they are transplanted, not by the region from which they were removed.

ments clearly demonstrate that gene expression within differentiating cells (in this case, nonblastopore cells) can be controlled by substances produced by other cells (in this case, cells of the dorsal lip of the blastopore).

Adult Structures Develop during Organogenesis

Gradually, the ectoderm, mesoderm, and endoderm rearrange themselves into the organs characteristic of the animal species (see Table 36-2). This process, called **organogenesis**, also normally occurs by induction.

In some cases, adult structures are, in effect, "sculpted" by the death of excess cells produced during embryonic development. Death of some cells is programmed to occur at a precise time during development. At least two mechanisms seem to be at work in different tissues. Some cells die during development unless they receive a "survival signal." Embryonic vertebrates, for example, have

A Closer Look
Homeobox Genes and the Control of Body Form

In 1948, Edward Lewis of the California Institute of Technology began a systematic analysis of mutations in the fruit fly *Drosophila* that caused one body part to be replaced by another; for example, legs growing where antennae should be (Fig. E36-1). Lewis discovered that these organ substitutions could be caused by mutations in single genes, which he called "master genes," that control the activity of the many other genes necessary to produce the misplaced organ.

In 1983, researchers found that the different master genes in *Drosophila* shared a nearly identical DNA sequence, now called a **homeobox**. Each homeobox segment of DNA codes for a similar 60-amino-acid protein. The proteins encoded by each homeobox have, however, small crucial differences that allow them to specify which body regions will develop into which organs. These proteins are found in the cell nucleus, where they bind to specific genes, turning them on or off. Homeoboxes, by coding for proteins that activate or inactivate other genes, seem to specify directly the course of differentiation of embryonic cells. Homeoboxes determine the head-to-tail axis of the embryo and establish the appropriate sequence of structures along its length, thereby determining the overall shape of the body and the location and shape of its parts. Consequently, mutations in the homeobox can cause legs to replace antennae in the fruit fly.

Researchers studying diverse developmental questions, from the formation of color patterns on butterfly wings to the development of the forelimb in chickens and frogs, have found that chemical gradients commonly specify the fate of cells. In other words, during development, cells may take different forms in response to different concentrations of a regulatory substance. As a result, a gradient (such as occurs by diffusion from a concentrated source) of a single substance can produce an entire sequence of structural characteristics (such as the sequence of structures from the shoulder to the fingers of a forelimb). One source of such chemical gradients is the homeobox. For example, during the development of the clawed frog *Xenopus*, a forelimb bud forms from mesoderm that expresses an identified homeobox gene. The homeobox codes for a protein that forms a concentration gradient: high on the "thumb side" of the bud and low on the "pinkie side." As the limb lengthens, this protein forms a second gradient: high at the shoulder and lower toward the hand. Gradients of identical proteins have been found in the forelimbs of developing chick and mouse embryos as well.

Homeobox gene segments have a common evolutionary origin, and their discovery is merging the interests of evolutionary and developmental biologists. Researchers have discovered homeoboxes in humans and mice that are nearly identical to those of the fruit fly and that can be transplanted into the fruit fly and function normally. Because evolutionary lines leading to insects and mammals diverged 500 million years ago, the conservation of these short segments of DNA suggests that homeoboxes play a fundamental role in organizing the body plan of all multicellular animals. Recently, homeoboxes resembling those that specify the head end of the fruit fly have been discovered in the cnidarians (jellyfish), which are among the simplest of all multicellular animals.

The development of a single cell, the fertilized egg, into the fantastic complexity of a tadpole or a human infant is simultaneously one of life's greatest mysteries and one of biology's most exciting fields of research. Discovery of the homeobox has brought us one step closer to understanding this incredible journey.

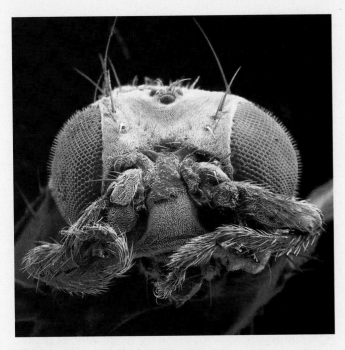

Figure E36-1 A homeobox mutant
A mutation in a homeobox segment of a gene in this *Drosophila* has caused legs to grow where antennae should be.

far more motor neurons in their spinal cords than do adult animals. Motor neurons are programmed to die unless they successfully form synapses with a skeletal muscle, releasing a chemical that prevents the death of its own motor neuron.

For other cells, the situation is just the reverse: Some cells live unless they receive a "death signal" from other cells of the developing animal. Many embryonic structures disappear during development. For example, all vertebrates pass through developmental stages with tails and webbed hands and feet. In humans, these stages can be seen clearly in the 5-week-old human embryo (see Fig. 36-12a). Two weeks later, the webbing cells have died, revealing separate fingers, and the tail cells are dying, causing the tail to regress (see Fig. 36-12b). In frogs, the tail is lost during metamorphosis from its tadpole larva. Thyroid hormone, which triggers metamorphosis, stimulates cells in the tail to synthesize enzymes that digest the tail away. If the thyroid gland is surgically removed, the frog retains its tail.

Sexual Maturation Is Controlled by Genes and the Environment

Development does not stop at birth; animals continue to change throughout their lives. For example, animals are not born with full reproductive capability but become sexually mature at an age that is determined by interactions between genetic and environmental factors. Most animals must undergo months to years of genetically regulated growth and development before they are physiologically capable of producing sperm or eggs.

Once the animal has reached the appropriate age, however, the precise onset of sexual maturity must typically await specific environmental stimuli. For example, most songbirds that breed in temperate regions become sexually mature in the spring, stimulated by the increasing daylength. Social factors can also influence maturation in some species. The average age of puberty among women, for instance, has dropped substantially during the past few centuries. Puberty at a younger age is due in part to improved nutrition, but social stimulation by early close contact between the sexes may also be involved.

Aging Seems to Be Genetically Programmed

Most of your cells will function less efficiently, or divide more slowly, as you age. Is death, for cells and for entire organisms, a programmed part of life? From an evolutionary standpoint, natural selection promotes only those mechanisms that keep an organism alive and healthy during the time that it is producing and nurturing its young. Repair mechanisms that extend longevity past this time are not favored and may even be harmful to the population as a whole. For example, older, nonreproductive individuals might compete with younger ones for limited resources, such as food.

Evidence of programmed cell death is seen in cells grown in dishes in the laboratory. These cells divide a relatively fixed number of times, then stop and eventually die. Recent research has suggested that this maximum life span varies from species to species and that longevity depends on the ability of the cell to repair damage to its DNA. Long-lived cells and long-lived animals have more natural enzymes that protect against DNA damage and are better at repairing damaged DNA than are short-lived cells and short-lived animals. Nevertheless, all normal cells stop dividing and die eventually.

Not all cells, however, are "normal." Cancer cells survive and reproduce indefinitely in cell culture or in the body, which is what makes cancer so devastating. Some lineages of cancer cells have been reproducing in culture for decades. Although scientists are still working to uncover the mechanism that regulates the life span of a cell, cancer cells provide evidence that this mechanism can be bypassed. However, cultured cancer cells frequently mutate and change their characteristics. Can we discover a method of defusing the self-destruct mechanism while retaining proper controls over the cells in our bodies? No one knows, but we can be certain it will not happen soon.

4) How Do Humans Develop?

Human development is controlled by the same mechanisms that control the development of other animals. In fact, our development strongly reflects our evolutionary heritage, something we shall emphasize in the brief discussion to follow. Figure 36-9 summarizes the stages of human embryonic development. You may want to refer to this figure as we go along.

During the First 2 Months, Rapid Differentiation and Growth Occur

A human egg is normally fertilized in a woman's oviduct and undergoes a few cleavage divisions on its way to the uterus. Just before being implanted in the uterine wall, the embryo, now called a **blastocyst** (the mammalian version of a blastula), consists of a thin-walled, hollow ball with a thicker **inner cell mass** on one side (Fig. 36-10, p. 750; see also Fig. 35-18b). The thin outer wall becomes the chorion and will form the embryonic contribution to the placenta; the inner cell mass develops into the embryo and the three other extraembryonic membranes.

After implantation, the inner cell mass grows and splits, forming two fluid-filled sacs that are separated by

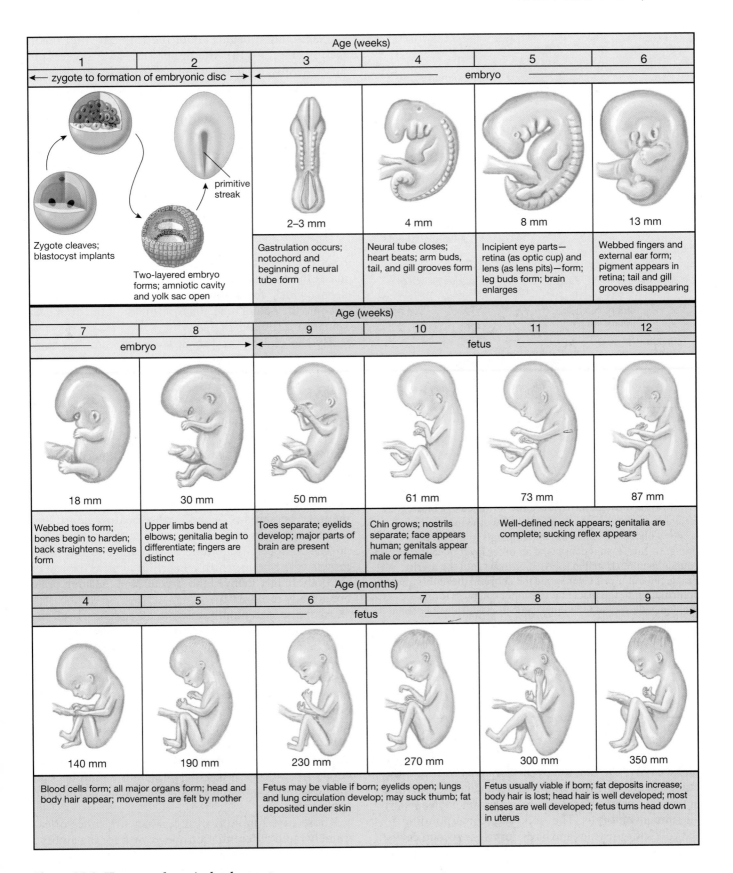

Figure 36-9 Human embryonic development
A calendar of human embryonic development, from blastula to birth.

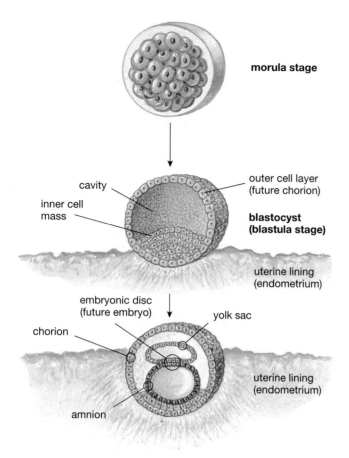

morula stage

cavity

inner cell mass

outer cell layer (future chorion)

blastocyst (blastula stage)

inner cell mass

uterine lining (endometrium)

embryonic disc (future embryo)

chorion

yolk sac

amnion

uterine lining (endometrium)

Figure 36-10 Human development during the first and second weeks
As it travels through the oviduct, the fertilized egg undergoes cleavage, forming a morula. In the uterus, the morula becomes a blastocyst (blastula) and implants in the uterine lining. At this stage it consists of an outer layer of cells surrounding an inner cell mass. As it burrows into the uterine lining, the outer cell layer forms the chorion, the embryonic contribution to the placenta. The inner cell mass forms the amnion, yolk sac, and the embryonic disc, which will become the embryo.

a double layer of cells called the **embryonic disc** (Fig. 36-11). One sac, bounded by the amnion, forms the amniotic cavity. The amnion eventually grows around the embryo, enclosing it in what has been called its "private aquarium," providing the watery environment needed by all animal embryos. The yolk sac, corresponding to the yolk sac of reptiles and birds, forms the second cavity, although in humans it contains no yolk. At this stage, the embryonic disc consists of an upper layer of future ectoderm cells (on the side facing the amniotic cavity) and a lower layer of future endoderm cells (on the side facing the yolk sac).

Gastrulation begins about the 15th day after fertilization (Fig. 36-11a). The upper and lower layers split apart slightly; a slit, the **primitive streak** (corresponding to the blastopore), appears in the center of the upper layer. Cells of the upper layer migrate through the primitive streak into the interior of the embryo, forming mesoderm. The upper layer is now called ectoderm and the lower layer is called endoderm. One of the earliest mesoderm structures to develop is the notochord (Fig. 36-11b).

During the third week of development, the embryo and its amniotic sac begin to curl toward the yolk sac (Fig. 36-11c). As the embryo grows, the endoderm pinches, forming a tube that will become the digestive tract (Fig. 36-11d). Simultaneously, the notochord induces the formation of a groove in the overlying ectoderm, which folds inward and then closes over to become the **neural tube**, the forerunner of the brain and spinal cord (Fig. 36-7d,e).

By the end of the fourth week, the amnion completely surrounds the embryo. The amnion is punctured only by the *umbilical cord*, which connects the embryo to the placenta.

By the end of the sixth week, the embryo clearly displays its chordate ancestry (see Chapter 22), having developed a notochord, a prominent tail, and *gill grooves* (indentations behind the head that are homologous to the gills that many chordates, such as fish, retain as adults; Fig. 36-12a, p. 752). These structures disappear as development continues. The embryo already has the rudimentary beginnings of the eyes, a beating heart, and separating fingers and toes on its tiny hands and feet (Fig. 36-12b). Especially notable at this stage is the rapid growth of the brain, which is nearly as large as the rest of the body. In fact, many of the structures of the adult brain are already recognizable.

As the second month draws to an end, nearly all the major organs have formed, and the embryo begins to look human (Fig. 36-12c). The gonads appear and develop into testes or ovaries, depending on the presence or absence of the Y chromosome. Sex hormones—either testosterone from the testes or estrogen from the ovaries—are secreted. These hormones will affect the future development of the embryonic organs, including not only the reproductive organs but also certain regions of the brain. After the second month of development, the embryo is called a **fetus**; this designation denotes that the developing being has taken on a generally human appearance.

These first 2 months of pregnancy are times of extremely rapid differentiation and growth for the embryo and also times of considerable danger. Although the fetus is vulnerable throughout development, rapidly developing organs are the most sensitive to environmental insults, such as drugs or certain medications taken by the mother.

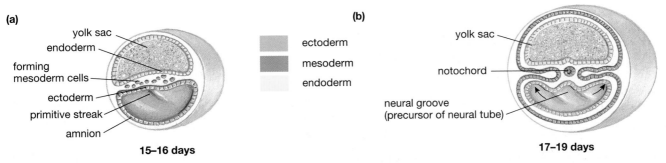

(a)

yolk sac
endoderm
forming
mesoderm cells
ectoderm
primitive streak
amnion

ectoderm
mesoderm
endoderm

15–16 days

Shortly after implantation, gastrulation occurs. The two-layered embryonic disc soon splits open, and the primitive streak develops in the ectoderm. Ectoderm cells migrate in, forming mesoderm.

(b)

yolk sac

notochord

neural groove
(precursor of neural tube)

17–19 days

Some mesoderm cells form the notochord, which induces development of the neural tube, forerunner of the brain and spinal cord.

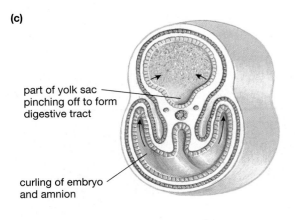

(c)

part of yolk sac
pinching off to form
digestive tract

curling of embryo
and amnion

20–21 days

During the third and fourth weeks of development, the embryo curls toward the yolk sac, forming a tubelike embryo typical of vertebrates.

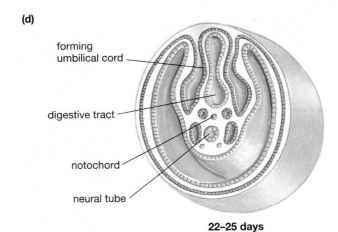

(d)

forming
umbilical cord

digestive tract

notochord

neural tube

22–25 days

As the embryo curls, part of the (empty) yolk sac pinches to form the digestive tract. The amnion curls with the embryo, eventually completely enclosing it, except where the umbilical cord extends through.

Figure 36-11 Human development during the third and fourth weeks

The Placenta Secretes Hormones and Exchanges Materials between Mother and Embryo

During the first few weeks of pregnancy, embryonic cells burrow into the thickened lining of the uterus (the endometrium). The outer cells of the embryo form the chorion, which penetrates the endometrium with finger-like projections called **chorionic villi** (singular, **villus**). From this complex interweaving of tissues arises the **placenta**, a marvelously intricate organ. The placenta has two major functions: (1) It secretes hormones, and (2) it allows the selective exchange of materials between the mother and the fetus.

The placenta, as it develops during the first 2 months of pregnancy, begins secreting estrogen and progesterone. Estrogen stimulates the growth of the mother's uterus and mammary glands, whereas progesterone also stimu-lates the mammary glands and inhibits premature contractions of the uterus.

The placenta also regulates the exchange of materials between the blood of the mother and the blood of the fetus without allowing the two to mix. The chorionic villi contain a dense network of fetal capillaries and are bathed in pools of maternal blood (Fig. 36-13, p. 753). This arrangement permits many small molecules to diffuse between fetal blood and maternal blood. Oxygen diffuses from maternal blood to fetal blood, and carbon dioxide from fetal blood to maternal blood. Nutrients, some aided by active transport, travel from mother to fetus. Fetal urea diffuses into the mother's blood, to be filtered out by the mother's kidneys.

While allowing exchange by diffusion, the membranes of the capillaries and chorionic villi act as barriers to the passage of some large proteins and most cells. Despite

(a)

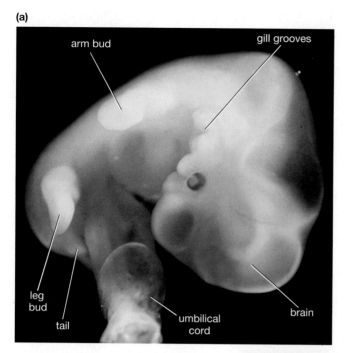

arm bud

gill grooves

leg bud

tail

umbilical cord

brain

(b)

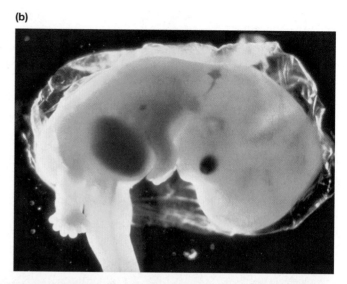

(c)

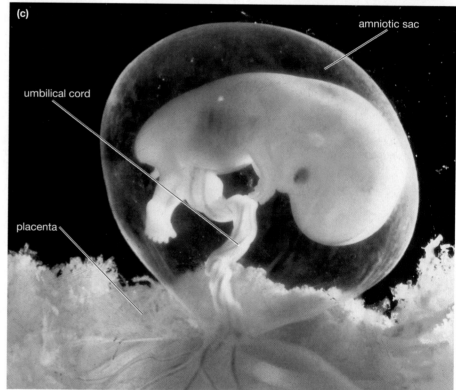

amniotic sac

umbilical cord

placenta

Figure 36-12 Human development during the fifth through eighth weeks (a) At the end of the fifth week, the human embryo is about half head. The feet and hands have begun to develop digits. A tail and gill grooves are clearly visible, evidence of our evolutionary relationship to other vertebrates. **(b)** By the seventh week, the human form has been more clearly defined by the selective death of cells that form the tail and connect the fingers and toes. The tail has nearly disappeared, and fingers and toes are separated from one another. **(c)** At the end of the eighth week, the embryo is clearly human in appearance and is now called a fetus. Most of the major organs of the adult body have begun to develop.

this barrier, some disease-causing organisms and many harmful chemicals can penetrate the placental barrier, as described in "Health Watch: The Placenta Provides Only Partial Protection" (p. 754).

Growth and Development Continue during the Last 7 Months

The fetus continues to grow and develop for another 7 months. (See the chapter-opening photo for a view at

about 4 months.) Although the rest of its body is "catching up" with the head in size, the brain continues to develop rapidly and the head remains disproportionately large. Nearly every nerve cell ever formed during the entire human life span develops during embryonic life. As the brain and spinal cord grow, they begin to generate noticeable behaviors. As early as the third month of pregnancy, the fetus begins to move and respond to stimuli. Some instinctive behaviors appear, such as sucking, which

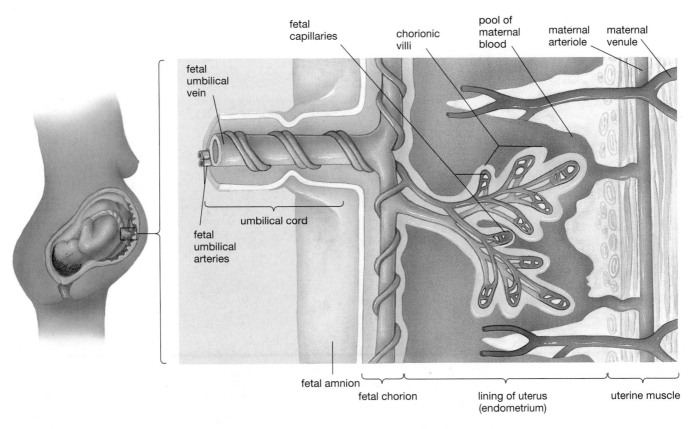

Figure 36-13 *The placenta*
The placenta is formed from both the chorion of the embryo and the endometrium of the mother. Capillaries of the endometrium break down, releasing blood to form pools within the placenta. Meanwhile, the chorion develops projections (the chorionic villi) that extend into these pools of maternal blood. Blood vessels from the umbilical cord branch extensively within the villi. The resulting structure separates the maternal and fetal blood supplies and generates a large surface area for the diffusion of oxygen, carbon dioxide, nutrients, and wastes between the fetal capillaries and the maternal blood pools. Umbilical arteries carry deoxygenated blood from the fetus to the placenta, and umbilical veins carry oxygenated blood back to the fetus.

will have obvious importance soon after birth. Structures that the fetus will need when it emerges from the womb (the mother's uterus), such as the lungs, stomach, intestine, and kidneys, enlarge and become functional, though they will not be used until after birth. Most fetuses 7 months or older can survive outside the womb, but larger and more-mature fetuses have a much greater chance of survival.

Development Culminates in Birth

Normally, during the last months of pregnancy the fetus becomes positioned head downward in the uterus, with the crown of the skull resting against (and being held up by) the cervix (see Chapter 35). After about 9 months of development, the head is so large that it can barely fit through the mother's pelvis. The skull is compressed into a slightly conical shape as it passes through the birth canal (the mother's vagina). Generally delivered head first, the infant is in for a rude awakening. The womb was soft, fluid-cushioned, and warm. All of a sudden, the baby must obtain oxygen and eliminate carbon dioxide by breathing. It must regulate its own body temperature, and it must suck-

le to obtain food. Considering that they have recently experienced what by any standard is a change for the worse, babies are often remarkably cheerful (Fig. 36-14)!

Figure 36-14 *On her own*

Health Watch
The Placenta Provides Only Partial Protection

When your grandparents were bearing children, physicians assumed that the placenta protected the developing fetus from most of the substances in maternal blood that could harm it. We know now that this is far from true. In fact, most medications and drugs and even some disease-producing organisms readily penetrate the placental barrier and affect the fetus.

Infections Can Cross the Placenta The German measles virus can cross the placenta and attack the fetus, causing potentially severe retardation and other defects. As mentioned in Chapter 35, the virus causing genital herpes (during active outbreaks) and the bacterium causing syphilis can cause mental or physical defects in the developing fetus. The virus causing AIDS can also cross the placenta, so some infants are born with this deadly disease.

Drugs Readily Cross the Placenta A tragic example of a drug that crosses the placenta is the tranquilizer *thalidomide,* commonly prescribed in Europe in the early 1960s (Fig. E36-2). Thalidomide's devastating effects on embryos were discovered only when many babies were born with missing or extremely abnormal limbs. In the late 1980s, the anti-acne drug Accutane® was found to cause gross deformities in babies born to women using it. (Accutane contains retinoic acid, which has been shown to activate homeobox genes; see "A Closer Look: Homeobox Genes and the Control of Body Form," p. 747.)

Although these are extreme examples, *any drug, including aspirin, has the potential to harm the fetus, and any woman who thinks she may be pregnant should seek medical advice about any drugs she takes.* The use of so-called recreational drugs, such as heroin and cocaine, has devastating effects on a developing fetus. Many babies of heroin addicts are born addicted. In the United States, hundreds of thousands of infants have been born in recent years to users of "crack" cocaine. Research suggests that these children may be impaired, both behaviorally and emotionally.

Figure E36-2 Drugs interfere with development
Children born to mothers who took the tranquilizer thalidomide during pregnancy cope heroically with missing and deformed limbs. Drugs with no obvious ill effects on the mother can have devastating impacts on her developing fetus.

Summary of Key Concepts

1) How Do Cells Differentiate During Development?
All the cells of an animal body contain a full set of genetic information, yet each cell is specialized for a particular function. During development, cells differentiate by stimulating and repressing the transcription of specific genes. Gene transcription is regulated in two ways: (1) The egg cytoplasm contains gene-regulating substances that are incorporated into specific daughter cells during the first few cleavage divisions. The developmental fate of each daughter cell is determined by which substances it receives. (2) Later in development, certain cells produce chemical messages that induce other cells to differentiate into particular cell types.

2) How Do Indirect and Direct Development Differ?
Animals undergo either direct or indirect development. In indirect development, eggs (normally those with relatively little yolk) hatch into larval feeding stages that later undergo metamorphosis to become adults with body forms notably different from that of the larva. In direct development, the newborn animal is sexually immature but otherwise resembles a small adult. Animals with direct development tend to

The Effects of Smoking Probably the most common toxic substances to which fetuses are exposed are those in cigarette smoke. Because many women who smoke are so addicted that they don't stop smoking during pregnancy, nearly a million human embryos are exposed to the poisons (including nicotine, carbon monoxide, and a host of carcinogens) each year in the United States alone. Women who smoke have a higher incidence of miscarriages than do nonsmokers and give birth to smaller infants who are more likely to die shortly after birth than are infants of nonsmokers. There is evidence that some children born to heavy smokers also suffer behavioral and intellectual impairment.

Fetal Alcohol Syndrome The effects of alcohol on a developing fetus can be devastating. When a pregnant woman drinks, alcohol in the blood of her unborn child reaches a level as high as that in her own blood. Many children born to alcoholic mothers exhibit **fetal alcohol syndrome (FAS)**. Such children are mentally retarded and can be hyperactive and irritable. FAS children have small heads and abnormally small, improperly developed brains (Fig. E36-3), facial abnormalities, inhibited growth, and a higher-than-normal incidence of defects of the heart and other organs.

FAS is the third most common cause of mental retardation in the United States, with one out of every 500 to 750 children being affected. It is likely that far higher numbers of children suffer from *fetal alcohol effect,* a milder form of FAS caused by alcohol consumption below the seven drinks per week that can cause severe FAS. A woman who takes one or two alcoholic drinks a day during the first 3 months of pregnancy also significantly increases her chances of having a miscarriage. *Researchers have established no safe level of alcohol consumption during any phase of pregnancy. The U.S. Surgeon General advises preg-*

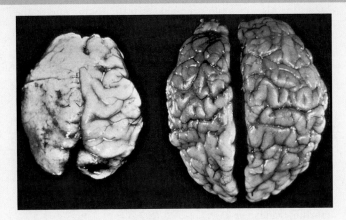

Figure E36-3 Alcohol impairs brain development
(left) The brain of a child with fetal alcohol syndrome and *(right)* the brain of a normal child of the same age show the devastating effects of alcohol on the developing brain.

nant women and those who are likely to become pregnant to avoid all alcohol consumption.

In summary, a pregnant woman should assume that whatever chemicals she ingests will find their way into the bloodstream of her developing infant. Women who are likely to become pregnant need to consider that crucial stages of development occur before most women even realize that they are pregnant. The mother's choices during this critical 9-month period can strongly influence her child's future well-being.

have either have large, yolk-filled eggs or nourish the developing embryo within the mother's body.

3) How Does Animal Development Proceed?
Animal development occurs in several stages. *Cleavage:* The zygote undergoes cell divisions with little intervening growth, so the egg cytoplasm is partitioned into smaller cells. Cleavage divisions result in the formation of the morula, a solid ball of cells. A cavity then opens up within the morula, forming the blastula, a hollow ball of cells. *Gastrulation:* A dimple forms in the blastula, and cells migrate from the surface into the interior of the ball, eventually forming a three-layered gastrula. The three cell layers of ectoderm, mesoderm, and endoderm give rise to all the adult tissues (see

Table 36-2). *Organogenesis:* The cell layers of the gastrula form organs characteristic of the animal species. *Sexual maturation:* The juvenile animal increases in size and achieves sexual maturity. *Aging:* Cells begin to function less efficiently, at least partly as a result of failure to repair their DNA, and eventually the animal dies.

4) How Do Humans Develop?
The human developmental program clearly reflects our reptilian ancestry. Reptilian embryos develop four extraembryonic membranes—the chorion, amnion, allantois, and yolk sac—that function in gas exchange, provision of the watery environment needed for development, waste storage, and yolk storage, respectively. These membranes are retained by

the human embryo but are modified to suit development inside the mother's womb (see Table 36-1). The chorion invades the uterine lining (endometrium); together, these tissues form the placenta. There the exchange of materials occurs by diffusion between maternal blood and fetal blood, but the blood supplies remain separate. By the end of the second month, the major organs have formed, and the embryo appears human and is called a fetus. In the next 7 months, until birth, the fetus continues to grow, and the lungs, stomach, intestine, and kidneys enlarge and become functional. Human embryonic development follows the same principles as the development of other mammals. The stages of human development are summarized in Figure 36-9.

Key Terms

allantois *p. 743*
amnion *p. 743*
amniote egg *p. 743*
blastocyst *p. 748*
blastopore *p. 744*
blastula *p. 744*
chorion *p. 743*
chorionic villus *p. 751*
cleavage *p. 744*
development *p. 739*

differentiation *p. 739*
direct development *p. 741*
ectoderm *p. 746*
embryo *p. 741*
embryonic disc *p. 750*
endoderm *p. 746*
extraembryonic membrane *p. 743*
fetal alcohol syndrome (FAS) *p. 755*

fetus *p. 750*
gastrula *p. 746*
gastrulation *p. 746*
gray crescent *p. 744*
homeobox *p. 747*
indirect development *p. 741*
induction *p. 746*
inner cell mass *p. 748*
larva *p. 741*
mesoderm *p. 746*

metamorphosis *p. 741*
morula *p. 744*
neural tube *p. 750*
notochord *p. 746*
organogenesis *p. 747*
placenta *p. 751*
primitive streak *p. 750*
yolk *p. 741*
yolk sac *p. 743*

Thinking Through the Concepts

Multiple Choice

1. *Cells become differentiated through all of the following events EXCEPT*
   a. binding of regulatory molecules to chromosomes
   b. chemical messages received from other cells
   c. unequal distribution of gene-regulating substances during cleavage
   d. transcription of different genes
   e. progressive loss of genes as cells divide

2. *Indirect development is characteristic of animals that normally produce*
   a. few eggs
   b. eggs with large amounts of yolk
   c. young that are sexually immature versions of adults
   d. all of the above
   e. none of the above

3. *In bird eggs, the allantois*
   a. exchanges oxygen and carbon dioxide
   b. produces the shell
   c. stores wastes
   d. encloses the embryo in a watery environment
   e. contains stored food

4. *The endoderm gives rise to the*
   a. lining of the digestive tract
   b. epidermis of skin
   c. skeletal system
   d. muscles
   e. nervous system

5. *In human development, ectoderm cells migrate through the primitive streak to form*
   a. endoderm   b. mesoderm
   c. the chorion   d. the yolk sac
   e. the amnion

6. *The process by which a tissue causes another tissue to differentiate is called*
   a. gastrulation   b. metamorphosis
   c. cleavage   d. induction
   e. indirect development

? Review Questions

1. Define *differentiation*. How do cells differentiate; that is, how is it that adult cells express some but not all the genes of the fertilized egg?

2. Describe the two aspects of differentiation that occur during development: the influence of the egg cytoplasm and induction by other cells.

3. Distinguish between indirect and direct development, and give examples of each.

4. What is yolk? How does it influence cleavage?

5. Name two structures derived from each of the three embryonic tissue layers—endoderm, ectoderm, and mesoderm.

6. What is gastrulation? Describe gastrulation in frogs and in humans.

7. Describe the process of induction, and give two examples.

8. How does cell death contribute to development?

9. Describe the structure and function of four extraembryonic membranes found in reptiles and birds. Are these four present in placental mammals? Explain.

10. Explain how the structure of the placenta prevents mixing of fetal and maternal blood while allowing the exchange of substances between the mother and the fetus.

11. List and explain two very different functions of the placenta.

12. Is the placenta an effective barrier against substances that can harm the fetus? Describe two types of harmful agents that can cross the placenta and their effects on the fetus.

Applying the Concepts

1. When fetal ectoderm tissue from the mouth of an embryo is transplanted into a region of mesoderm of another embryo of a different species, the mesoderm induces the ectoderm to form mouth structures of the type of organism from which the ectoderm tissue was taken. When ectoderm from a chick is implanted into a mouse embryo, the chicken tissue produces teeth. What does this experiment tell us about why the adage "scarce as hen's teeth" is true? Also, what does this experiment tell us about the evolutionary origin of chickens?

2. On the basis of your knowledge of genetics (Unit II) and evolution (Unit III), explain why the human embryo passes through a developmental stage in which it has gill grooves and a tail.

3. Embryologists have used embryo fusion to produce *tetraparental* (four-parent) mice and have also produced "geeps" from goat and sheep embryos. The resulting bodies are patchworks of cells from both animals. Why does fusion succeed with very early embryos (4-cell to 8-cell stages) and fail when much older embryos are used?

4. If the nuclei of frog intestinal cells can be transplanted into eggs from which the nucleus has been removed to produce clones of the parent, is it theoretically possible to produce human clones? Would such clones yield offspring that are *exactly* identical to the parents who supplied the nuclei? Explain.

5. Estimates of the number of women in the United States who self-medicate with over-the-counter (OTC) drugs during pregnancy range from 65% to 95%. Evidently the public doesn't perceive nonprescription substances as drugs. Many subtle disorders, especially of the nervous system, will probably be connected to over-the-counter use in coming years. If you were engaged in pharmaceutical research, which OTC drugs would you focus on as possible culprits? Which months of pregnancy would you study closely? Who would you chose as your study population?

6. At one stage of development, human embryos contain two ribbons of tissue capable of developing into breasts. The tissues stretch from the armpits down to the lower abdomen. Some children are born with *supernumerary* (extra) breasts located along these tissue ribbons. What normally happens to the process or timing of induction to limit the location of breasts? What must go wrong in supernumerary breast development? Why do dogs or cats have rows of mammary glands?

Group Activity

Your research team at the Institute for Animal Development has been presented with some fascinating new information. Other researchers working on insect development discovered that a gene called *Dll* is active during the development of legs. If *Dll* fails to turn on during development, the insect will develop no legs, only stumps. Intriguingly, *Dll* is also present in vertebrates, and it switches on during vertebrate leg development. What's more, *Dll* is also present in a diverse array of invertebrates, including sea urchins, sea squirts, and velvet worms. Your team has been asked to answer some key questions about these findings: (1) How can the same gene control the development of structures as different as an insect leg and a vertebrate leg? (2) What is the most likely function of the protein encoded by *Dll* (that is, what does the protein do)? (3) Why has *Dll* remained, virtually unchanged, in the genomes of all these different animals whose common ancestor lived hundreds of millions of years ago? Form a group of four students to propose logical, detailed, written answers to these questions. In formulating your answers, be sure to recall what you've learned about animal development.

For More Information

Beaconsfield, P., Birdwood, G., and Beaconsfield, R. "The Placenta." *Scientific American*, August 1980 (Offprint No. 1478). The placenta is one of the most remarkable of mammalian structures, allowing internal development of offspring within the mother.

Caldwell, M. "How Does a Single Cell Become a Whole Body?" *Discover*, November 1992. Explores the miracles of development in layperson's terms.

DeRobertis, E. M., Oliver, G., and Wright, C. V. E. "Homeobox Genes and the Vertebrate Body Plan." *Scientific American*, July 1990. Genes can be injected into amphibian eggs, allowing direct observation of the effects of specific gene products on development.

Dewitt, P. E. "Cloning: Where Do We Draw the Line?" *Time*, November 8, 1993. Does cloning of animals suggest that this technique will be applied to humans?

McGinnis, W., and Kuziora, M. "The Molecular Architects of Body Design." *Scientific American*, February 1994. A summary of the experimental evidence that development of body shape is controlled by similar genetic and molecular processes in all animals.

Nathanielsz, P. "The Timing of Birth." *American Scientist*, November–December 1996. An account of the interactions between fetal and maternal physiology that affect the timing of birth in mammals.

Nüsslein-Volhard, C. "Gradients That Organize Embryo Development." *Scientific American*, August 1996. A Nobel laureate describes developmental experiments that have revealed a great deal about how chemical gradients in embryos help guide cell differentiation.

Rusting, R. "Why Do We Age?" *Scientific American*, December 1992. The underlying causes of human aging are being explored, using experimental animals as diverse as fruit flies and roundworms.

Taubes, G. "Ontongeny Recapitulated." *Discover*, May 1998. Will biotechnology allow us to turn on genes that normally become silent after development has been completed and that will provide the opportunity to regrow limbs or to replace dying brain cells?

Answers to Multiple-Choice Questions
1. e   2. e   3. c   4. a   5. b   6. d

"As I watched the geese, it appeared to me as little short of a miracle that a hard, matter-of-fact scientist should have been able to establish a real friendship with the wild, free-living animals, and the realization of this fact made me strangely happy. It made me feel as though man's expulsion from the Garden of Eden had thereby lost some of its bitterness."

Konrad Lorenz in King Solomon's Ring *(1952)*

Familiar animals can exhibit puzzling and intriguing behaviors. These honeybees have left the safety of their hive to form a noisy, pulsating swarm on an exposed branch of a mesquite tree.

Animal Behavior 37

Net Watch

On-line resources for this chapter are on the World Wide Web at: http://www.prenhall.com/audesirk
(click on the Table of Contents link and then select Chapter 37).

The evolutionary contest to reproduce has shaped not only the structure and appearance of animals but also what animals do. Natural selection has favored behaviors that help animals find food, find shelter, find mates, avoid being eaten, and perform a host of other tasks that help them survive and reproduce. The range of behaviors by which these functions are accomplished is as wondrously diverse as animals themselves. You can encounter fascinating examples of animal behavior virtually anywhere. There's no need to travel to exotic locales to witness exotic behaviors. Consider, for example, the humble honeybee, a common, widespread, and familiar insect.

Perhaps the most obvious example of honeybee behavior is the species' defensive behavior. Bees respond to potential threats with a series of movements that concludes by inflicting a painful sting on an intruder. This behavior is apparent to even the most casual observer, but closer and more-patient observation reveals subtler behaviors. For example, honeybees must search for food over a large area to find enough nectar to support a whole colony. Individual bees frequently venture into unfamiliar territory in search of food, but somehow the bees find their way back to the hive. What's more, if one bee finds a good food source, in minutes a large crowd of bees is busily gathering food at the same spot. How does a honeybee find its way back home? And how do the other bees find out where the food is? Careful observation and experimentation have shown that honeybees are able to navigate by using an internal clock and celestial cues (such as the position of the sun) and that they communicate to other bees the direction and distance to a newly discovered food source.

That honeybees, with only a rudimentary nervous system, are able to engage in sophisticated defensive, searching, navigating, and communication behaviors is fascinating, perhaps even astonishing. And the behaviors mentioned above represent only a fraction of the behavioral repertoire of the honeybee. The list of behaviors by which honeybees reproduce, raise offspring, keep the hive functioning, and maintain their complex social organization is long, and each aspect of honeybee behavior raises interesting biological questions. There are hundreds of thousands of other animal species about which we might ask similar questions. The sheer number of species and behaviors may seem overwhelming, but behavioral biologists have uncovered some common themes that underlie behavioral diversity and that help us develop a framework for investigating and understanding behavior. What questions pop into *your* head when you notice what an animal is doing?

1) How Do Innate and Learned Behaviors Differ?

Innate Behaviors Can Be Performed without Prior Experience

Behavior is any observable response to external or internal stimuli. A fruit fly placed in a tunnel with light at one end and darkness at the other will fly toward the light; a hungry newborn human infant, touched on the side of her mouth, will turn her head and attempt to suckle. These are examples of **innate**, or **instinctive**, behaviors. Innate behavior is performed in reasonably complete form even the first time an animal of the right age and motivational state encounters a particular stimulus. (The proper motivational state for feeding, for example, would be hunger and not fear.)

Scientists can demonstrate that a behavior is innate by depriving the animal of the opportunity to learn it. For example, wild red squirrels bury extra nuts in the fall for retrieval during the winter. Red squirrels can be raised from birth in a bare cage on a liquid diet, providing them with no experience with nuts, digging, or burying. Presented with nuts for the first time, such a squirrel (after eating several) will carry one to the corner of its cage, then make covering and patting motions with its forefeet. Nut burying is therefore an innate behavior.

Innate behaviors can also be recognized by their occurrence immediately after birth, before there is any opportunity for learning. The cuckoo, for example, lays its eggs in the nest of another bird species, to be raised by the unwitting adoptive parent. Immediately after hatching, the cuckoo chick shoves the nestowner's eggs (or baby birds) out of the nest, eliminating its competitors for food (Fig. 37-1).

Many Simple Orientation Behaviors Are Innate

Many animals exhibit relatively simple innate behaviors that serve to ensure that the animal ends up in a suitable location. For example, the pillbug (a land-dwelling crustacean) moves faster, though in no particular direction, if the air around it becomes drier. Eventually, the pillbug encounters a damper area, where it slows and eventually stops. This simple **kinesis**—a behavior in which an organism changes the speed of random movements—enables pillbugs to reach the moist areas they need to survive.

In contrast to a kinesis, a **taxis** (plural, **taxes**) is a *directed* movement either toward (a positive taxis) or away from (a negative taxis) a stimulus. A moth flying toward a light is one example. Much of the behavior of very simple organisms, including single-celled protists, consists of taxes. For example, the protist *Euglena* (see Chapter 19)

(a)

(b)

Figure 37-1 *Innate behavior*
(a) The cuckoo chick, just hours after it hatches, and before its eyes have opened, evicts the eggs of its foster parents from the nest. **(b)** The parents, responding to the stimulus of the cuckoo chick's wide-gaping mouth, feed the chick, unaware that it is not related to them.

is photosynthetic, but its chlorophyll can be damaged by intense light. *Euglena* shows a positive taxis toward dim light but a negative taxis toward intense light. Mosquito behavior also involves several taxes. Males orient toward the high-pitched whine of the female. Female mosquitoes (only females suck blood) exhibit taxes to the warmth, humidity, and carbon dioxide exuded by their prey. Mosquito repellents probably act by blocking the receptors for these stimuli, rendering the insect incapable of sensing her victim.

Fixed Action Patterns Are Complex Innate Behaviors

Because innate behavior is performed without learning or prior experience, it tends to be highly *stereotyped*— that is, it is performed similarly each time. Such stereotyped behaviors are sometimes called **fixed action patterns**, especially if they are performed automatically in response to a particular stimulus. If the animal is at the right developmental stage and properly motivated, a fixed action pattern will be performed correctly the first time the appropriate stimulus, also called a **releaser** (because it "releases" specific stereotyped behavior), is presented.

The releaser of a given fixed action pattern is typically a very specific characteristic of the animal toward which the behavior is directed. For example, the female red-winged blackbird signals her readiness to mate by raising her tail to a particular angle, which serves as a releaser for copulatory behavior by the male. This releaser is so potent that males will even attempt to copulate with the disembodied tail of a stuffed female, provided the feathers are angled appropriately (Fig. 37-2). To determine the particular feature that acts as a releaser, *ethologists* (scientists who study behavior) may selectively exaggerate specific features of a stimulus. Niko Tinbergen, for instance, studied the instinctive fixed action pattern in which a herring gull chick pecks at a red spot on its parent's beak (Fig. 37-3a). This stereotyped behavior, which chicks can perform immediately after hatching, causes the parent to regurgitate food, which the chicks then eat. Tinbergen found that the releasing features of the bill are its long, thin shape, the red color, and the presence of color contrasts. When he offered chicks a thin red rod with white stripes painted on it, they pecked at it more often than at a real beak (Fig. 37-3b).

Learned Behaviors Are Modified by Experience

Natural selection may favor innate behaviors in many circumstances. For instance, it is clearly to the advantage of a herring gull chick to be able to peck at its parent's bill as soon as possible after hatching. But in other circumstances, rigidly fixed behavior patterns may be less useful. For example, a male red-winged blackbird that copulates with a stuffed female obviously will produce no offspring as a result of that behavior. In many situations, a greater degree of behavioral flexibility will be advantageous. The capacity to make *changes* in behavior

Figure 37-2 Releasers
A male red-winged blackbird is stimulated to mate by the angle of the female's tail, whether she is attached to it or not.

on the basis of experience is called **learning**. This deceptively simple definition encompasses a vast array of different phenomena. A frog learns to avoid distasteful insects, a baby shrew learns which adult is its mother, a human learns to speak a language, a blackbird learns to use the stars for navigation. Each of the many examples of animal learning represents the outcome of a unique evolutionary history and set of ecological needs, so learning processes are as diverse as animals themselves. Nonetheless, it can be useful to categorize types of learning, as long as we keep in mind that the categories are only rough guides, and that many examples of learning will not fit neatly into any category.

Habituation Is a Decline in Response to a Repeated Stimulus

A common form of simple learning is **habituation**, defined as a decline in response to a repeated stimulus. The ability to habituate prevents an animal from wasting its energy and attention on irrelevant stimuli. This form of learning is shown by the simplest animals and has even been demonstrated in protists, which are single-celled. For example, the protist *Stentor* (Fig. 37-4a) retracts when touched but gradually stops retracting if touching is continued. The sea anemone, which also lacks a brain, shows a similar response to touch (Fig. 37-4b).

The ability to habituate is clearly adaptive. If a sea anemone withdrew every time it was brushed by a strand of waving seaweed, the animal would waste a great deal of energy, and its retracted posture would prevent it from

(a)

Model   Response
(number of pecks)

(b)

Figure 37-3 Experiments define releasers
(a) A herring gull chick pecks at the red spot on its mother's bill, causing her to regurgitate food for it. Niko Tinbergen and others have shown that the thin shape of the bill and the contrasting spot serve as releasers for pecking, which occurs immediately after hatching. The pecking then serves as a releaser for food regurgitation by the parent. **(b)** Models are used to determine which features of the bill are releasers for pecking behavior. The length of the bar is proportional to the number of pecks delivered to the corresponding model. A brightly contrasting spot improves performance; the color red further increases pecking. But the red rod with contrasting stripes is most effective; the long, thin shape, the red color, and the contrast produce an exaggerated stimulus.

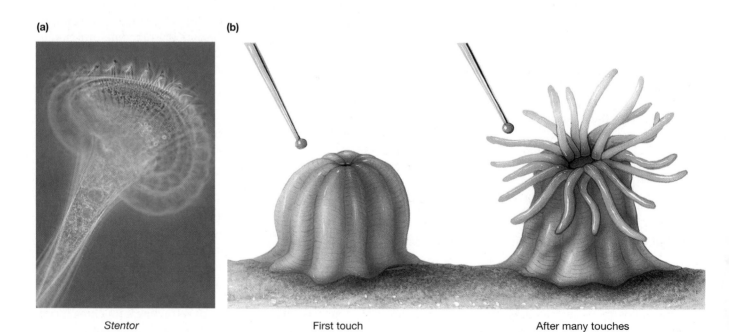

Stentor   First touch   After many touches

Figure 37-4 Habituation
(a) The protist *Stentor* is a complex single cell that can show habituation, a simple form of learning. **(b)** Habituation in the sea anemone. Touched for the first time, the animal withdraws. After many touches, it habituates to this harmless stimulus. Learning occurs even though the anemone possesses only a simple network of neurons and lacks a brain.

snaring food. Humans habituate to many stimuli: city dwellers to nighttime traffic sounds, and country dwellers to choruses of crickets and tree frogs. Each may initially find the other's habitat unbearably noisy at first but habituate after a time.

Conditioning Is a Learned Association between a Stimulus and a Response

A more complex form of learning, most commonly seen in the laboratory, is called *conditioning*. During **classical conditioning** an animal learns to perform a response (normally caused by one stimulus) to a new stimulus. In the early 1900s, Ivan Pavlov, a Russian physiologist who has been called the "father of conditioning," performed his now-famous experiment on dogs. Pavlov first placed dried meat powder into a dog's mouth, causing reflex salivation. Then he rang a bell just before the meat powder was presented. After several repetitions, the dog would salivate to the bell alone. The dog that leaps up at the sight of her leash or drools on the floor at the sound of a can opener has been classically conditioned, however unintentionally. This form of learning can also be demonstrated in simple organisms. For example, some investigators have reported that flatworms can learn to associate a flash of light with an electrical shock, which causes them to contract. After several exposures to light paired with shock, they contract in response to the light alone.

During **operant conditioning** an animal learns to perform a behavior (such as pushing a lever or pecking a button) to receive a reward or to avoid punishment. This technique is most closely associated with the American comparative psychologist B. F. Skinner. Skinner designed the "Skinner box," in which an animal (typically a white rat or pigeon) is isolated and allowed to train itself. The box might contain a lever that ejects a food pellet when pressed. The animal in its explorations inevitably bumps the lever and is rewarded. Soon the animal presses the lever repeatedly for food. Skinner boxes can be quite complex. A pigeon might learn that one lever will provide food only when a green light is shining, but when a red light is on, a specific spot on the wall must be pecked to avoid a shock.

In the natural environment, animals are faced with naturally occurring rewards and punishments analogous to those provided by experimental operant conditioning. By **trial-and-error learning**, animals acquire new and appropriate responses to stimuli through experience. For example, a hungry toad occasionally captures a mouthful of trouble: a bee. Having its tongue stung results in trial-and-error learning that takes only a single experience (Fig. 37-5). The toad modifies it response to flying insects

(a) A naive toad is presented with a bee.

(b) While trying to eat the bee, the toad is stung painfully on the tongue.

(c) Presented with a harmless robber fly, which resembles a bee, the toad cringes.

(d) The toad is presented with a dragonfly.

(e) The toad immediately eats the dragonfly, demonstrating that the learned aversion is specific to bees and insects resembling them.

Figure 37-5 Trial-and-error learning in the toad

to exclude bees and even other insects that resemble them. Much learning of this type occurs during play and exploratory behavior in animals with complex nervous systems (see "Evolutionary Connections: Why Do Animals Play?"). Trial-and-error learning makes a major contribution to the behavior of young children and of adults as well. It is how a child learns which foods taste good or bad, that a stove can be hot, and not to pull the cat's tail.

Insight Is Problem Solving without Trial and Error

In certain situations, animals seem able to solve problems suddenly, without the benefit of prior experience. This kind of sudden problem solving is sometimes called **insight learning**, because it seems at least superficially similar to the process by which humans mentally manipulate concepts to arrive at a solution. We cannot, of course, know for sure if nonhuman animals experience similar mental states when they solve problems.

In 1917, the animal behaviorist Wolfgang Kohler showed that a hungry chimpanzee, without any training, would stack boxes to reach a banana suspended from the ceiling (Fig. 37-6). This type of mental problem solving was once believed to be limited to very intelligent types of animals such as primates. In 1984, however, R. Epstein and associates at Harvard University observed pigeons under controlled conditions exhibiting similar insight: The pigeons moved a box beneath a suspended banana, then stood on the box to reach the banana. As investigators learn more about how to design appropriate experiments, we may find that insight among animals is much more common than was originally suspected.

There Is No Sharp Distinction between Innate and Learned Behaviors

Although the terms *innate* and *learned* can be useful tools in describing and understanding behaviors, these words also have the potential to lull us into an oversimplified view of animal behavior. In practice, few behaviors are unambiguously instinctive or unequivocally learned; most behavior is an intimate mixture of the two.

Seemingly Innate Behavior Can Be Modified By Experience

Behaviors that seem to be performed correctly on the first attempt without prior experience can later be modified by experience. Recall the example of the herring gull chicks' pecking their parent's beak for food. Immediately after hatching, the chicks prefer a striped rod to the real beak. Within a few days, however, the chicks learn enough about the appearance of their parents that they begin pecking more frequently at models more closely resembling the parents. After 1 week, young gulls have learned enough about how their parents look to prefer models of their own species to models of a closely related species. Eventually, the young birds learn to beg only from their own parents.

Habituation can also fine-tune an organism's innate responses to environmental stimuli. For example, young birds crouch when a hawk flies over but ignore harmless birds such as geese. Early observers hypothesized that only the very specific shape of predatory birds released crouching. Using an ingenious model, Niko Tinbergen and Konrad Lorenz (two of the founding fathers of **ethology**, the study of animal behavior) tested this hypothesis (Fig. 37-7). When moved in one direction, the

Figure 37-6 Insight
(a) A dog, lacking insight in this situation, fails a detour problem on the first try. It will eventually learn by trial and error. *(b)* Unable to reach the bananas, the chimpanzee exhibits insight by stacking boxes beneath the bananas to extend its reach.

(a)

(b)

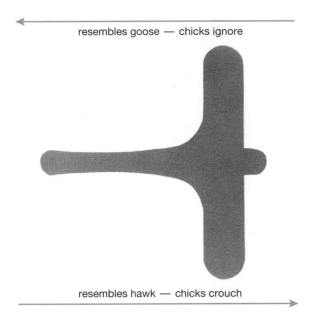

resembles goose — chicks ignore

resembles hawk — chicks crouch

Figure 37-7 Experience modifies innate responses
The model used by Konrad Lorenz and his student Niko Tin-bergen to investigate the response of chicks to the shape of objects flying overhead. The chicks' response depended on which direction the model moved. Moving toward the right, the model resembles a predatory hawk, but moving left it resembles a harmless goose.

model resembled a goose, which the chicks ignored. When its movement was reversed, it resembled a hawk and elicited crouching. Further research revealed that naive chicks instinctively crouch when *any* object moves over their heads. Over time, their response habituates to things that soar by harmlessly and frequently, such as leaves, songbirds, and geese. Predators are much less common, and the novel shape of a hawk still elicits instinctive crouching. Thus, learning modifies the innate response, making it more adaptive.

Learning May Be Governed by Innate Constraints

Learning always occurs within boundaries that help increase the chances that only the appropriate behavior is acquired. For example, young robins routinely learn their songs from the adult robins that they hear but almost never copy the songs of the sparrows, warblers, finches, or other species that they also hear. The innate constraints on learning are perhaps most strikingly illustrated by **imprinting**, a special form of learning in which learning is rigidly programmed to occur only at a certain critical period of development. A strong association is formed during a particular stage, called a **sensitive period**, in the animal's life. During this stage, the animal is primed to learn a specific type of information, which is then incorporated into a behavior that is not easily altered by further experience. Imprinting is best known in birds such as geese, ducks, and chickens. These birds learn to follow

the animal or object that they most frequently encounter during a sensitive period. (For mallard ducks this period occurs about 13 to 16 hours after hatching.) In nature, the mother is the object most likely to be nearby during the sensitive period, so the young birds imprint on her. In the laboratory, however, these birds can be forced to imprint on a toy train or other moving object (Fig. 37-8); if given a choice, they select a duck.

All Behavior Arises Out of Interaction between Genes and Environment

Many early ethologists saw innate behaviors as rigidly controlled by genetic factors and viewed learned behaviors as determined exclusively by an animal's environment. Today, however, this false dichotomy has given way to the realization that, just as no behavior is wholly innate or wholly learned, no behavior can be caused strictly by genes or strictly by the environment. Instead, all behavior develops out of an interaction between genes and their environment. The relative contributions of heredity and learning vary among animal species and among behaviors within an individual.

At the most basic level, we know that behavior cannot be determined by genes alone, because behavior is produced by the brain and nervous system, which are themselves products of a prolonged interaction between gene products and the environment in which they are expressed.

Figure 37-8 Konrad Lorenz and imprinting
Konrad Lorenz, known as the "father of ethology," is followed by goslings that imprinted on him shortly after they hatched. They now follow him as they would their mother.

Whether a behavior is produced at birth (as is an infant's suckling) or requires years of learning (as does human language), a functioning nervous system is required. The complexity of that nervous system, and its specific neural connections, determines which behaviors and which learning processes are possible. Similarly, specific kinds of behavior are greatly influenced by particular sensory systems that arise out of gene–environment interactions during development. As we saw in Chapter 33, echolocating bats produce and respond to sounds far higher in pitch than the human ear can detect, and they use those sounds to produce some type of mental "image" of objects around them. A bee is guided toward nectar by patterns of reflected ultraviolet light on flowers that are invisible to the human eye (see Fig. 24-7). A tick is exquisitely sensitive to the odor of butyric acid, produced by the skin of mammals. The scent will cause the tick to drop off its perch, normally landing on the passing animal.

The precise nature of the link among genes, environments, and behaviors is, in most cases, not well understood. The chain of events between the transcription of genes and the performance of a complex behavior may well be so complex that we will never be able to decipher it fully. Nonetheless, a great deal of evidence demonstrates the existence of both genetic and environmental components in the development of even the most complex behaviors. For example, consider the case of bird migration. Even though it's well known that migratory birds must learn by experience how to navigate via celestial cues, this environmental determinant is not the only process at work.

At the close of the summer, many birds disappear from the habitats where they've spent the breeding season and head for winter quarters hundreds or even thousands of miles to the south. Many of the migrating birds are traveling for the first time, because they were hatched only a few months earlier. Amazingly, these naïve birds depart at the proper time and in the proper direction and locate the proper wintering location, even though they cannot simply follow more-experienced birds (which typically depart a few weeks in advance of the first-year birds). Somehow, these young birds are able to execute a very difficult task the first time they try it. It seems that birds must be born with the ability to migrate; it must be "in their genes." And indeed, birds that are experimentally hatched and raised in isolation indoors still orient in the proper migratory direction when autumn comes, apparently without the need for any learning or experience. The conclusion that birds must have a genetically controlled ability to migrate in the right direction was further supported by hybridization experiments with blackcap warblers. This species breeds in Europe and migrates to Africa, but populations from different areas travel by different routes. Blackcaps from western Europe travel in a southwesterly direction to reach Africa, whereas birds from eastern Europe must travel to the southeast (Fig. 37-9). If birds from the two populations are cross-

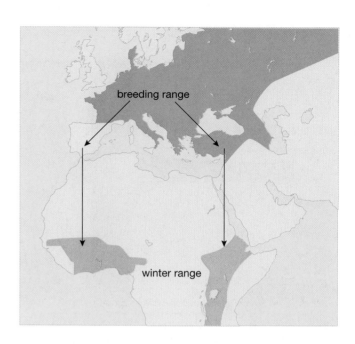

Figure 37-9 Genes influence migratory behavior
Blackcap warblers from western Europe begin their fall migration by flying in a southwesterly direction, but those from eastern Europe fly to the southeast when they begin migrating. If members of the two populations are crossbred in captivity, the hybrid offspring orient in a direction intermediate between the migratory directions of the parents. This result suggests that migratory behavior has a heritable component.

bred in captivity, however, the hybrid offspring will exhibit migratory orientation toward due south, intermediate between that of the two parents. This result suggests that the parents had genes that influenced migratory direction and that the hybrids inherited a mixture of these genes.

2) How Do Animals Communicate?

Animals frequently make information available for sharing. The sounds uttered, movements made, and chemicals emitted by animals can reveal the animals' physical location, level of aggression, readiness to mate, and so on. If revealing such information evokes a response from other individuals, and if that response tends to benefit the sender and the receiver, then a communication channel can form. **Communication** is defined as the production of a signal by one organism that causes another organism to change its behavior in a way beneficial to one or both.

Although animals of different species may communicate (picture a cat, its tail erect and bushy, hissing at a strange dog), most communication occurs between members of the same species. Potential mates must communicate, as must parents and offspring. At the same time, members of the same species compete most directly with one another for food, space, and mates. Communication is often used to resolve such conflicts with minimal damage.

The mechanisms by which animals communicate are astonishingly diverse and use all the senses. In the following sections, we look at communication by visual displays, sound, chemicals, and touch.

Visual Communication Is Most Effective over Short Distances

Animals with well-developed eyes, from insects to mammals, use visual signals to communicate. Visual signals can be *active,* in which a specific movement (such as baring fangs) or posture (such as lowering the head) conveys a message (Fig. 37-10). Alternatively, visual signals may be *passive,* in which case the size, shape, or color of the animal conveys important information, commonly about its sex and reproductive state. For example, when female mandrills become sexually receptive, they develop a large, brightly colored swelling on their buttocks (Fig. 37-11). Active and passive signals can be combined, as illustrated by the lizard in Figure 37-12 and the courtship behavior of the three-spined stickleback fish (see Fig. 37-25).

Like all forms of communication, visual signals have both advantages and disadvantages. On the plus side, they are instantaneous and can be rapidly revised to convey a variety of messages in a short period. Visual communication is quiet and unlikely to alert distant predators, although the signaler does make itself conspicuous to those

Figure 37-11 *A passive visual signal*
The female mandrill's colorfully swollen buttocks serve as a passive visual signal that she is fertile and ready to mate.

nearby. On the negative side, visual signals are generally ineffective in darkness and in dense vegetation, though female fireflies signal potential mates by using species-specific patterns of flashes. Finally, visual signals are limited to close-range communication.

Figure 37-10 *An active visual signal*
The wolf signals aggression by lowering its head, ruffling the fur on its neck and along its back, facing its opponent with a direct stare, and exposing its fangs. These signals can vary in intensity, communicating different levels of aggression.

Figure 37-12 *Active and passive visual signals combined*
The South American *Anolis* lizard raises his head high in the air (an active visual signal), revealing a brilliantly colored throat pouch (a passive visual signal) that warns others to keep their distance.

Communication by Sound Is Effective over Longer Distances

The use of sound overcomes many of the shortcomings of visual displays. Like visual displays, sound signals reach receivers almost instantaneously. But unlike visual signals, sound can be transmitted through darkness, dense forests, and even water, as the intricate songs of humpback whales testify. If the animal is sufficiently energetic, its call can carry much farther than the eye can see. The low, resonant song of the humpback whale can be heard by other whales up to hundreds of miles away. When the small kangaroo rat drums the Arizona desert floor with its hind feet, the sound can be heard more than 45 meters (150 feet) away. And the howls of a wolf pack carry for miles on a still night.

Auditory signals are similar to visual displays in that they can be varied to convey rapidly changing messages quickly. (Think of words and of the emotional nuances conveyed by the human voice during a conversation.) Changes in motivation can be signaled by a change in the loudness or pitch of the sound. An individual can convey different messages by variations in the pattern, volume, and pitch of the sound produced. Ethologist Thomas Struhsaker studied vervet monkeys in Kenya in the 1960s and found that they produced different calls in response to threats from each of their major predators: snakes, leopards, and eagles. In 1980, other researchers reported that the response of other vervet monkeys to each of these calls is appropriate to the particular predator. The "bark" that warns of a leopard or other four-legged carnivore causes monkeys on the ground to take to trees and those in trees to climb higher. The "rraup" call signaling an eagle or other hunting bird causes monkeys on the ground to look upward and take cover, whereas monkeys already in trees drop to the shelter of lower, denser branches. The "chutter" call indicates a snake and causes the monkeys to stand up and search the ground for this slower, lower predator.

The use of sound is by no means limited to birds and mammals. Male crickets produce species-specific songs that attract female crickets of the same species (see "Scientific Inquiry: Robot Cricket Finds Her Mate" on p. 770). The annoying whine of the female mosquito as she prepares to bite alerts nearby males that she will soon have the blood meal necessary for laying eggs. Male water striders vibrate their legs, sending species-specific patterns of vibrations through the water that attract mates and repel other males (Fig. 37-13). From these rather simple signals to the virtuoso performance of human language, sound is one of the most important forms of communication.

Chemical Messages Persist Longer But Are Hard to Vary

Chemical substances produced by an individual that influence the behavior of others of its species are called

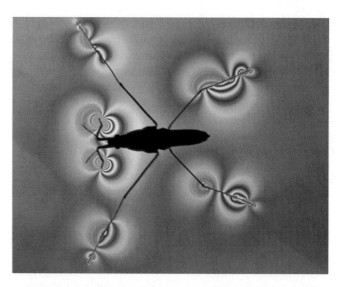

Figure 37-13 Communication by vibration
The light-footed water strider relies on the surface tension of water to support its weight. By vibrating its legs, the water strider sends signals that radiate out over the surface of the water. These vibrations advertise the striders' species and sex to others nearby.

pheromones. Chemicals may carry messages over long distances, and, unlike sound, take very little energy to produce. Pheromones may not even be detected by other species, including predators who might be attracted to visual or auditory displays. Like a signpost, a pheromone persists over time and can convey a message after the animal has departed. Wolf packs, hunting over areas up to 1000 square kilometers (386 square miles), mark the boundaries of their travels with pheromones in urine to warn other packs of their presence. As anyone who has walked a dog can attest, the domesticated dog reveals its wolf ancestry by staking out its neighborhood with urine that carries a chemical message: "I live in this area."

This type of communication requires that an animal be able to synthesize as well as respond to a different chemical for each message. As a result, in general, fewer messages are communicated with chemicals than with sight or sound. In addition, pheromone signals lack the diversity and gradation of auditory or visual signals. Nonetheless, chemicals powerfully convey a few simple but critical messages.

Pheromones act in one of two ways. (1) **Releaser pheromones** cause an immediate, observable behavior in the animal that detects them. They convey messages such as "This area is mine" or "I am ready to mate." Foraging ants and termites that discover food lay a trail of pheromones, secreted by abdominal glands, from the food back to the nest (Fig. 37-14). Its message: "Follow to find food." (2) **Primer pheromones** stimulate a physiological change in the animal that detects them, normally in its reproductive state. The queen honeybee produces a primer pheromone called **queen substance,**

Figure 37-14 *Communication by chemical messages*
To indicate a source of food to other termites, foraging termites have laid a trail of pheromones, secreted by glands in the abdomen, from the food to the nest. The others follow.

which is eaten by her hivemates and prevents other females in the hive from becoming sexually mature. The urine of mature males of certain species of mice contains a primer pheromone that influences female reproductive hormones. This pheromone causes the newly mature female to become sexually receptive and fertile for the first time. It will also cause a female mouse that is newly pregnant by another male to abort her litter and become sexually receptive to the new male. There is indirect evidence (discussed later in the chapter) that primer pheromones may even influence human reproductive cycles.

The sex attractant pheromones of some agricultural pests, such as the Japanese beetle and gypsy moth, have been synthesized. These synthetic pheromones can be used to confuse and disrupt mating or to lure these insects into traps. Pest control using pheromones has major environmental advantages. Pesticides kill beneficial as well as harmful insects and select for resistant insect strains. In contrast, pheromones are specific to one species, and insects resistant to the attraction of their own pheromones would not reproduce.

Communication by Touch Helps Establish Social Bonds

Physical contact between individuals is used in several ways, particularly to establish and maintain social bonds among group members. Primates, including humans, are "contact species" in which a variety of gestures—including kissing, nuzzling, patting, petting, and grooming—play an important social function (Fig. 37-15a). Many species of ants appear to communicate by touching other individuals with their antennae. The bond between parent and offspring is often cemented by close physical contact, and sexual activity is frequently preceded by ritualized contact (Fig. 37-15b).

Touch can also influence human well-being. Recent research shows that when the limbs of premature human infants are stroked and moved for 45 minutes daily, the infants are more active, responsive, and emotionally stable and gain weight more rapidly than do premature infants who receive the standard hospital treatment.

3) How Do Animals Interact? *www*

Sociality is a widespread feature of animal life. Most animals have at least a small measure of interaction with other individuals of their species; many spend the bulk of their lives in the company of others; and a few species have developed complex, highly structured societies. Social interaction can be cooperative or competitive and is typically a mixture of the two. Interactions of all types typically involve communication via one or more of the methods described in the previous section.

Figure 37-15 *Communication by touch*
(a) An adult olive baboon grooms a juvenile. Grooming not only reinforces social relationships but also removes debris and parasites from the fur.
(b) Touch is also important in sexual communication. These land snails (*Helix*) engage in courtship behavior that will culminate in mating.

Scientific Inquiry
Robot Cricket Finds Her Mate

The cheerful chirping of a cricket is actually the "call song" of the male as he attempts to attract a female. The female follows the song unerringly, deftly detouring around obstacles and ignoring other sounds en route to her prospective mate. How intelligent is this apparently purposeful behavior? Barbara Webb, a psychologist at the University of Edinburgh, Scotland, attacked this problem in a novel way; she built a robot female cricket (Fig. E37-1). Webb's goal was to find out whether mate-finding behavior could be distilled down to taxes, relatively simple responses to stimuli, such as responses that could be wired into an electronic robot (and thus wired into genetically predetermined neural connections). Although its tangle of wires appears bewildering, the circuitry of the robot is trivial when compared with the potential complexity of neural connections—even in a cricket's brain.

On the laboratory bench, a loudspeaker "male cricket" broadcasts its species-specific call song: short, regularly repeating tones. As the robot rolls forward, microphonic ears conduct the song to electronic circuitry that filters it from other sounds and adds together the repeating syllables of the song that reach each ear. The summed sounds in the ear closest to the loudspeaker reach a critical threshold level first, activating a mechanism that turns the robot toward the sound. The turning halts when an equal intensity of sound hits both ears. Sensory "bumpers" help the robot detour around obstacles. The success of "robocricket" surpassed Webb's expectations; it not only found its "mate" but it unexpectedly mimicked other cricket-searching behaviors. Placed between two loudspeakers that broadcast at equal volume, the robot, like a real cricket, arbitrarily chose one speaker. If the repeating syllables of the song were alternated between the two speakers, the robot (again, like a real cricket) first positioned itself exactly between them, then made an arbitrary choice. The electronic circuitry provides insights into mechanisms that could be used by a simple nervous system to produce complex adaptive behavior.

(a)   **(b)**

Figure E37-1 A robot model mimics cricket taxes
(a) The wiring in this model allows it to mimic several aspects of the mate-finding behavior of *(b)* a real cricket.

Competition for Resources Underlies Many Forms of Social Interaction

Aggressive Behavior Helps Secure Resources
One of the most obvious manifestations of competition for resources such as food, space, or mates is **aggression**, or antagonistic behavior, normally between members of the same species. Although the expression "survival of the fittest" evokes images of the strongest animal emerg-ing triumphantly from among the dead bodies of its competitors, in reality most aggressive encounters between members of the same species are rather harmless. Natural selection has favored the evolution of symbolic displays or rituals for resolving conflicts. During fighting, even the victorious animal can be injured, so serious fighters might not survive to pass on their genes. Aggressive *displays*, in contrast, allow the competitors to assess each other and acknowledge a winner on the basis of its size,

(a)

(b)

Figure 37-16 *Aggressive displays*
(a) Threat display of the male baboon. Despite the potentially lethal fangs so prominently displayed, aggressive encounters between baboons rarely cause injury. *(b)* The aggressive display of many male fish (here, striped grunts) includes elevating the fins and flaring the gill covers, thus making the body appear larger.

strength, and motivation rather than on the wounds it can inflict.

During aggressive displays, animals may exhibit weapons, such as claws and fangs (Fig. 37-16a), and often behave in ways that make them appear larger (Fig. 37-16b). Competitors might stand upright and erect their fur, feathers, ears, or fins (see Fig. 37-10). The displays might be accompanied by intimidating sounds (growls, croaks, roars, chirps) whose loudness can help decide the winner. Fighting tends to be a last resort when displays fail to resolve the dispute.

In addition to visual and vocal displays of strength, many animal species engage in ritualized combat. Deadly weapons may clash harmlessly (Fig. 37-17) or may not be used at all. In many cases these encounters involve shoving rather than slashing. Again, both the strength and the motivation of the combatants are determined, and the loser slinks away in a submissive posture that minimizes the size of its body.

Dominance Hierarchies Reduce Aggressive Interactions
Aggressive interactions use a great deal of energy, can cause injury, and can disrupt other important tasks such as finding food, watching for predators, courting a mate, or raising young. Thus, there are adaptive advantages to resolving conflicts with minimal aggression. In a **dominance hierarchy**, each animal establishes a rank that determines its access to resources. Domestic chickens, after a period of squabbling, sort themselves into a reasonably stable "pecking order." Thereafter, when competition for

food occurs, all hens defer to the dominant bird, all but the dominant bird give way to the second, and so on. Conflict is minimized because each bird knows its place. Dominance among male bighorn sheep is reflected in the size of their horns (Fig. 37-18). In wolf packs, one member of each sex is the dominant, or "alpha," individual to whom all others are subordinate. Although aggressive encounters occur frequently while the dominance

Figure 37-17 *Displays of strength*
Ritualized combat of fiddler crabs. The oversized claws, which could severely injure another animal, grasp harmlessly. Eventually one crab, sensing greater vigor in his opponent, will typically retreat unharmed.

Figure 37-18 A dominance hierarchy
The dominance hierarchy of the male bighorn sheep is signaled by the size of the horns; these rams increase in status from right to left. These backward-curving horns, clearly not designed to inflict injury, are used in ritualized combat.

hierarchy is being established, after each animal learns its place, disputes are minimized. The dominant individuals obtain most access to the resources needed for reproduction, including food, space, and mates.

Perhaps the most thoroughly studied dominance hierarchy is that of chimpanzees. The ethologist Jane Goodall (Fig. 37-19) has devoted more than 30 years to making meticulous observations of chimpanzee behavior in the field at Gombe National Park in Tanzania, and she has described and documented the animals' complex social organization. Chimpanzees live in groups, and dominance hierarchies among males are a key aspect of their social life. Many males devote a significant amount of time to maintaining their position in the hierarchy, largely by way of an aggressive *charging display* in which a male rushes forward, throws rocks, leaps up to shake vegetation, and otherwise seeks to intimidate rival males. The advantages that accrues to dominant males, however, are not entirely clear. According to Goodall, low-ranking males are still able to secure access to food and successful copulations

(albeit not quite as easily as higher-ranking males). In Goodall's view, little evolutionary advantage accrues to a dominant male, and the function of chimpanzee dominance hierarchies remains unexplained.

Animals May Defend Territories That Contain Resources

In many animal species, competition for resources takes the form of **territoriality**, the defense of an area where important resources are located. The defended resources may include places to mate, raise young, feed, or store food. Territorial animals generally restrict most or all of their activities to the defended area and advertise their presence there. Territories may be defended by males, females, a mated pair, or entire social groups (as in the case of the defense of their nest by social insects). However, territorial behavior is most commonly seen in adult males, and territories are normally defended against members of the same species, who compete most directly for the resources being protected. Territories are as diverse as the animals defending them. For example, they can be a tree where a woodpecker stores acorns (Fig. 37-20), small

Figure 37-20 A feeding territory
Acorn woodpeckers live in communal groups that excavate corn-sized holes in dead trees, stuffing the holes with green acorns for dining during the lean winter months. The group defends the trees vigorously against other groups of acorn woodpeckers and against acorn-eating birds of other species, such as jays.

Figure 37-19 Jane Goodall observing chimps at play

depressions in the sand used as nesting sites by cichlid fish, a hole in the sand used as a home by a crab, or an area of forest providing food for a squirrel.

Acquiring and defending a territory require considerable time and energy, yet territoriality is seen in animals as diverse as worms, arthropods, fish, birds, and mammals. The fact that organisms as unrelated as worms and humans independently evolved similar behavior suggests that territoriality provides some important adaptive advantages. Although the particular benefits depend on the species and the type of territory it defends, some broad generalizations are possible. First (as with dominance hierarchies), once a territory is established through aggressive interactions, relative peace prevails as boundaries are recognized and respected. The saying "good fences make good neighbors" also applies to nonhuman territories. One reason for this respect is that an animal is highly motivated to defend its territory and will often defeat even larger, stronger animals that attempt to invade it. Conversely, an animal outside its territory is much less secure and more easily defeated. This principle was demonstrated by Niko Tinbergen in an experiment using stickleback fish (Fig. 37-21).

For males of many species, successful territorial defense has a direct impact on reproductive success. In these species, females are attracted to a high-quality breeding territory, which might have features such as large size, abundant food, and secure nesting areas. Males who successfully defend the most-desirable territories have the greatest chance of mating and passing on their genes. For example, experiments have shown that male sticklebacks that defend large territories are more successful in attracting mates than are males who defend small territories. Females who select males with the best territories increase their own reproductive success and pass their

genetic traits (typically including their mate-selection preferences) to their offspring.

Territoriality limits the population size of some species, helping keep it within the limits set by the available resources. For instance, the Scottish red grouse defends feeding and nesting territories. Scientists found that variations in territory size were correlated with the availability of food. In years when food was abundant, territories in the study area were smaller and the number of mating pairs was greater than in lean years.

Territories are advertised through sight, sound, and smell. If the territory is small enough, the owner's mere presence, reinforced by aggressive displays at intruders, can be a sufficient defense. A mammal that owns a territory but cannot always be present may use pheromones to scent-mark its terrestrial boundaries. Male rabbits use pheromones secreted by glands in their chin and by anal glands to mark their territories. Hamsters rub the areas around their dens with secretions from special glands in their flanks.

Vocal displays are a common form of territorial advertisement. Male sea lions defend a strip of beach by swimming up and down in front of it, calling continuously. Male crickets produce a specific pattern of chirps to warn other males away from their burrows. Birdsong is a striking example of territorial defense. The husky trill of the male seaside sparrow is part of an aggressive display, warning other males to steer clear of his territory (Fig. 37-22). In fact, males that are unable to sing are unable to defend territories. The importance of singing to seaside sparrows' territorial defense was elegantly demonstrated by the ornithologist M. Victoria McDonald, who captured territorial males and performed an operation that left them temporarily unable to sing but still able to utter the other, shorter and quieter signals in their

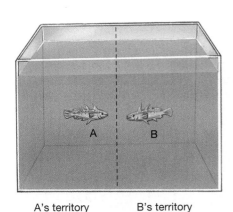

A's territory B's territory

(a) Two sticklebacks, A and B, established territories in an aquarium, then each was placed in a glass tube that could be moved between the two territories.

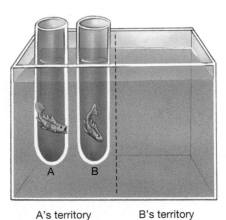

A's territory B's territory

(b) When both fish were moved into A's territory, A attempted to attack B, who assumed a submissive posture.

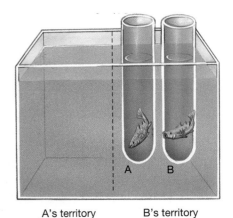

A's territory B's territory

(c) The reverse occurred when both were transported into B's territory.

Figure 37-21 *Territory ownership and aggression*
Niko Tinbergen's experiment demonstrating the effect of territory ownership on aggressive motivation.

Figure 37-22 *Defense of a territory by song*
A male seaside sparrow announces ownership of his territory.

vocal repertoires. The songless males were unable to defend territories or attract mates but regained their lost territories when their songs returned.

Sexual Reproduction Commonly Involves Social Interactions between Mates

In many sexually reproducing animal species, mating involves copulation or other close contact between male and female. Before such mating can occur, animals must identify one another as members of the same species, as members of the opposite sex, and as being sexually receptive. Many animals resist the close approach of another individual; this resistance must be overcome before mating can occur. Some animals, such as frogs and many species of fish, must release eggs and sperm at precisely the same moment for fertilization to occur. The need to fulfill all these requirements has resulted in exceedingly complex, diverse, and fascinating courtship behavior.

Vocal and Visual Signals Encode Sex, Species, and Individual Quality

Individuals that waste energy and gametes by mating with members of the wrong sex or wrong species will be at a disadvantage in the contest to reproduce, so natural selection has favored behaviors by which animals communicate their own sex and species. One result of this selection is that many campers have been kept awake by a raucous nighttime chorus of male tree frogs, each singing a species-specific song. Male grasshoppers and crickets also advertise sex and species by their calls, as does the female mosquito with her high-pitched whine.

The signals that encode sex and species may also be used by potential mates as a basis for comparison among rival suitors. For example, the male bellbird uses its deafening call to defend large territories and attract females from great distances. The females fly from one territory to another, alighting near the male in his tree. The male, beak gaping, leans directly over the flinching female and utters an earsplitting note. The female apparently endures this noise to compare the volumes of the various males, choosing the loudest (who would also be the best defender of a territory) as a mate.

Many species use visual displays for courting. The firefly, for example, flashes a message that identifies its sex and species. Male fence lizards bob their heads in a species-specific rhythm, and females distinguish and prefer the rhythm of their own species. The elaborate construction projects of the male satin bowerbird and the scarlet throat of the male frigate bird serve as flashy advertisements of sex, species, and male quality (Fig. 37-23).

(a)

(b)

Figure 37-23 *Sexual displays*
(a) During courtship, the male satin bowerbird builds a bower out of twigs and decorates it with colorful items that he gathers. *(b)* The male frigate bird of the Galapagos Islands inflates a scarlet throat pouch to attract passing females.

Figure 37-24 Sexual dimorphism in guppies
As is true of many animal species, in guppies the male (left) is normally much brighter in color than the female.

Sending these extravagant signals must be risky, as they make it much easier for predators to locate the sender. For males, the added risk is an evolutionary necessity, as females won't mate with males that lack the appropriate signal. Females, in contrast, typically do not need to attract males or assume the risk associated with a conspicuous signal, so in many species females are drab in comparison to males (Fig. 37-24).

The intertwined functions of sex recognition and species recognition, advertisement of individual quality, and synchronization of reproductive behavior commonly require a complex series of signals, both active and passive, by both sexes. Such signals are beautifully illustrated by the complex underwater "ballet" executed by the male and female three-spined stickleback fish (Fig. 37-25).

Chemical Signals Bring Mates Together

Pheromones can play an important role in reproductive behavior. The sexually receptive female silk moth, for example, sits quietly and releases a chemical message so powerful that it can be detected by males 4 to 5 kilometers (2.5 to 3 miles) away. The exquisitely sensitive and selective receptors on the antennae of the male silk moth respond to just a few molecules of the substance, allowing him to travel upwind along a concentration gradient to find the female (Fig. 37-26a, p. 777).

Water is an excellent medium for dispersing chemical signals, and fish commonly use a combination of pheromones and elaborate courtship movements to ensure the synchronous release of gametes. Mammals, with their highly developed sense of smell, often rely on pheromones released by the female during her fertile periods to attract males (Fig. 37-26b). The irresistible attraction of a female dog in heat to nearby males is one example; the primer pheromone in male mouse urine is another.

Aggression Can Be Suppressed during Reproductive Interactions

Most nonbreeding encounters between individuals of solitary species are competitive and aggressive. During the brief mating season, their reluctance to allow others to approach closely must be overcome for their mate but retained toward others. These conflicting needs introduce an element of tension into sexual encounters that may be overcome by submissive signals. The female Siamese fighting fish appeases the aggressive male with a submissive, head-down posture. Either the male or female of several species of birds defuses aggressive impulses by mimicking juvenile behavior such as begging. Courting male hamsters emit high-pitched cries like those of baby hamsters, eliciting a maternal response from the female.

Presenting "gifts" seems to inhibit aggression in some species. Finches present their mates with nesting material, terns give fish, and flightless cormorants offer one another gifts of seaweed (Fig. 37-27, p. 777). In some cases, the female of one species of fly devours the male when he makes sexual advances. The males of a closely related species avoid this fate by presenting the female with a gift: a dead insect. While she eats, the male mates and runs. The male of another species gains extra time to mate by "gift-wrapping" the insect in a silken web that she must remove before she can consume the gift.

Social Behavior within Animal Societies Requires Cooperative Interactions

Social groupings of animals are conspicuous but by no means universal. Group living has both advantages and disadvantages, and the relative weight of the positive and negative factors varies considerably among animal species.

On the negative side, social animals may encounter:

- Increased competition within the group for limited resources
- Increased risk of infection from contagious diseases
- Increased risk that offspring will be killed by other members of the group
- Increased risk of being spotted by predators

For a species to have developed social behavior, the benefits must have outweighed the costs. Benefits to social animals include:

- Increased abilities to detect, repel, and confuse predators
- Increased hunting efficiency or increased ability to spot localized food resources
- Advantages resulting from the potential for division of labor within the group
- Conservation of energy
- Increased likelihood of finding mates

(a) A male, inconspicuously colored, leaves the school of males and females to establish a breeding territory.

(b) As his belly takes on the red color of the breeding male, he displays aggressively at other red-bellied males, exposing his red underside.

(c) Having established a territory, the male begins nest construction by digging a shallow pit that he will fill with bits of algae cemented together by a sticky secretion from his kidneys.

(d) After he tunnels through the nest to make a hole, his back begins to take on the blue courting color that makes him attractive to females.

(e) An egg-carrying female displays her enlarged belly to him by assuming a head-up posture. Her swollen belly and his courting colors are passive visual displays.

(f) Using a zigzag dance, he leads her to the nest.

(g) After she enters, he stimulates her to release eggs by prodding at the base of her tail.

(h) He enters the nest as she leaves and deposits sperm, which fertilize the eggs.

Figure 37-25 Courtship of the three-spined stickleback

The degree to which animals of the same species cooperate varies significantly from one species to the next. Some types of animals, such as the mountain lion, are basically solitary; interactions between adults consist of brief aggressive encounters and mating. Other types of animals cooperate on the basis of changing needs. For example, the coyote is solitary when food is abundant but hunts in packs in the winter when food becomes scarce. Loose social groupings, such as pods of dolphins, schools of fish, flocks of birds, and herds of

(a)

(b)

Figure 37-26 Pheromone detectors
(a) Male moths find females not by sight but by following airborne pheromones released by females. These odors are sensed by receptors on the male's huge antennae, whose enormous surface area maximizes the chances of detecting the female scent. *(b)* When dogs meet, they typically sniff each other near the base of the tail. Scent glands there broadcast information about sex (both type and interest in) and status.

Figure 37-27 Courtship gifts defuse aggression
Returning to its nest, a flightless cormorant from the Galapagos Islands bears a "gift" of seaweed for its mate, who will aggressively snatch away the gift. A bird that fails to bring a gift will be driven away by its mate. The gift appears to allow the nesting bird to take out its aggressive impulses harmlessly.

musk oxen (Fig. 37-28), provide a variety of benefits. For example, the characteristic spacing of fish in schools or the V-pattern of geese in flight provides a hydrodynamic or aerodynamic advantage for the individuals, reducing the energy required to swim or fly. Zoologists hypothesize that schooling fish confuse predators—their myriad flashing bodies make it difficult for the predator to focus on and pursue a single individual.

At the far end of the social spectrum are a few highly integrated cooperative societies, found primarily among the insects and mammals. As you read the following section, you may notice that some cooperative societies are based on behavior that seems to sacrifice the individual for the good of the group. There are many examples: Young, mature Florida scrub jays may remain at their parents' nest and help them raise subsequent broods instead of breeding; worker ants die in defense of their nest; Belding ground squirrels may sacrifice their lives to warn the group of an approaching predator. These behaviors characterize **altruism**—behavior that decreases the reproductive success of one individual to the benefit of another.

How could such behavior evolve? Why aren't the individuals that perform such self-sacrificing deeds

Figure 37-28 *Cooperation in loosely organized social groups* A herd of musk oxen functions as a unit when threatened by predators such as wolves. Males form a circle, horns pointed outward, around the females and young.

eliminated from the population, taking with them the genes that contribute to this behavior? One possibility is that because closely related individuals are likely to share some genes, an individual can promote the survival of its own genes through behaviors that maximize the survival of its close relatives. This concept is called **kin selection**. Kin selection helps explain a variety of altruistic behaviors. Please refer to "Evolutionary Connections: Kin Selection and the Evolution of Altruism" (Chapter 15) for an exploration of the evolutionary basis of the self-sacrificing behaviors that contribute to the success of cooperative societies. In the following section, we present examples of complex societies in insect, fish, and mammal species that vividly illustrate cooperative behavior.

Honeybees Form Complex Insect Societies

Perhaps the most perplexing of all animal societies are those of the bees, ants, and termites. Scientists have long struggled to explain the evolution of a social structure in which most individuals never breed but instead labor slavishly to feed and protect the offspring of a different individual. Whatever its evolutionary explanation, the intricate organization of a social insect colony makes a compelling story. In these communities, the individual is a mere cog in an intricate, smoothly running machine and could not function by itself. Individual social insects are born into one of several castes within the society. These castes are groups of similar individuals that perform a specific function.

Honeybees emerge from their larval stage into one of three major preordained roles. One role is that of *queen.* Only one queen is tolerated in a hive at any time. Her functions are to produce eggs (up to 1000 per day for a lifetime of 5 to 10 years) and to regulate the lives of the workers. Male bees, called *drones,* serve merely as mates for the queen. Lured by her sex pheromones, drones mate with the queen, perhaps as many as 15 times, during her

first week of life. This relatively brief "orgy" supplies her with sperm that will last a lifetime, enough to fertilize more than 3 million eggs. Their sexual chore accomplished, the drones become superfluous and are eventually driven out of the hive or killed.

The hive is run by the third class of bees, sterile female *workers.* The worker's tasks are determined by her age and conditions in the colony. The newly emerged worker starts as a waitress, carrying food such as honey and pollen to the queen, to other workers, and to developing larvae. As she matures, special glands begin wax production, and she becomes a builder, constructing perfectly hexagonal cells of wax where the queen will deposit her eggs and the larvae will develop. She will take a shift as maid, cleaning the hive and removing the dead, and as a guard, protecting the hive against intruders. Her final role in life is that of a forager, gathering pollen and nectar, food for the hive. She will spend nearly half of her 2-month life in this role. Acting as a forager scout, she will seek new and rich sources of nectar and, having found one, will return to the hive and communicate its location to other foragers. She communicates by means of the **waggle dance**, an elegant form of symbolic expression (Fig. 37-29). Much of the meaning of the waggle dance was deciphered by Karl von Frisch during 35 years of research beginning in 1915.

Pheromones play a major role in regulating the lives of social insects. Honeybee drones are drawn irresistibly to the queen's sex pheromone (*queen substance*), which she releases during her mating flights. Back at the hive, she uses the same substance (now acting as a primer pheromone) to maintain her position as the only fertile female. The queen substance is licked off her body and passed among all the workers, rendering them sterile. The queen's presence and health are signaled by her continuing production of queen substance; a decrease in production (which occurs normally in the spring) alters the

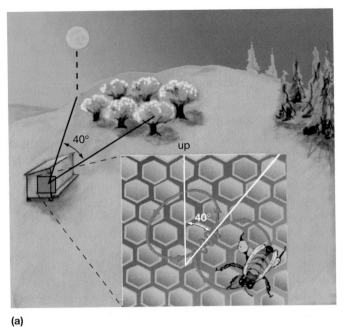

(a)

(b)

Figure 37-29 *Bee language: the waggle dance*
A forager, returning from a rich source of nectar, performs a waggle dance that communicates the distance and direction of the food source, as other foragers crowd around her, touching her with their antennae. The bee moves in a straight line while shaking her abdomen back and forth ("waggling") and buzzing her wings; the direction of the waggling indicates the direction of the food source. She repeats this over and over in the same location, circling back in alternating directions. The richness of the food source is communicated by the vigor of the dance and how long it continues. The rate at which she circles communicates the distance of the nectar source from the hive; the higher the rate, the closer the food source. The smell of the flowers on her body tells the other foragers what scent to search for. *(a)* If the dance is performed on a wall inside the hive, the angle that the straight run deviates from the vertical represents the angle between the sun and the flowers. *(b)* On a horizontal surface outside, the waggling run is aimed directly toward the flowers.

behavior of the workers. Almost immediately they begin building extra-large "royal cells" and feeding the larvae that develop in them a special glandular secretion known as "royal jelly." This unique food alters the development of the growing larvae so that, instead of a worker, a new queen emerges from the royal cell. The old queen will then leave the hive, taking a swarm of workers with her to establish residence elsewhere. If more than one new queen emerges, a battle to the death ensues, with the victorious queen taking over the hive.

Bullhead Catfish Illustrate a Simple Vertebrate Society
The nervous systems of vertebrates are far more complex than those of insects, and we might therefore expect vertebrate societies to be proportionately more complex. With the exception of human society, however, they are not. Perhaps because the vertebrate brain *is* more complex, vertebrate societies tend to be simpler than are those of the social insects such as honeybees, army ants, and termites. The robotic precision that makes complex insect societies possible is rarely found among verte-

brates, which generally exhibit greater behavioral flexibility than do insects and therefore greater behavioral variation among individuals.

The social interactions of bullhead catfish, described by John Todd of the Woods Hole Oceanographic Institution in Massachusetts, provide a fascinating illustration of a relatively simple vertebrate in which complex social interactions are based almost entirely on pheromones. Todd observed these nocturnal fish in large aquariums in the laboratory. He discovered that when a group was housed together, territories were staked out and a dominance hierarchy was established, with the dominant fish defending the largest and best-protected area of the tank. Contests between tankmates consisted of open-mouthed aggressive displays. Once a fish became dominant, its aggressive displays caused the subordinate fish to flee. Actual violence occurred only when a stranger was introduced into a tank with an established hierarchy. In this case, the established group exhibited cooperative behavior. The dominant fish allowed others to take refuge in his protected territory, then fought the

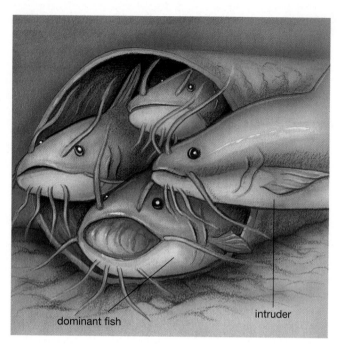

Figure 37-30 Cooperation among bullhead catfish
Three bullhead catfish occupy a section of pipe in a laboratory tank. The dominant fish is normally the exclusive occupant of the pipe, which is part of his territory. When an intruder is introduced into the tank, however, he allows two subordinate fish to seek refuge in the pipe and reacts aggressively to the intruder.

intruder (Fig. 37-30). When the newcomer was defeated, the dominant fish chased the others back out of his territory.

Todd discovered that blinding the bullheads did not cause any appreciable change in their social interactions. When their sense of smell was temporarily blocked, however, the fish acted like permanent strangers, interacting aggressively for weeks until their sense of smell returned. Both the status of an individual and a change in status are communicated by scent. If a dominant fish was removed from his tank and later returned, in most cases both his territory and his status were remembered and respected by his tankmates. But if he was removed and subjected to defeat in the tank of a more aggressive fish, his pheromones were somehow altered. Upon return to his home tank, he was attacked by his former subordinates.

When many newly caught fish are placed in the same tank, bullheads may form a dense and peaceful community that lacks territories or dominance hierarchies. Todd established such a community in one tank and placed a pair of aggressive rival fish in an adjacent tank. When water was pumped continuously from the "community tank" into the adjacent tank, the aggressive bullheads became peaceful, resuming their fighting only when the water

flow was stopped. Under the crowded conditions of the community tank, an "anti-aggression pheromone" is apparently produced, minimizing conflict.

Naked Mole Rats Form a Complex Vertebrate Society

Perhaps the most bizarre society among nonhuman mammals is that of the naked mole rat. These nearly blind, nearly hairless relatives of guinea pigs live in large underground colonies in southern Africa and exhibit a form of social organization not unlike that of an ant or termite colony. The colony is dominated by the queen, a single reproducing female to whom all other members are subordinate. The queen is the largest individual in the colony and maintains her status by aggressive behavior, particularly shoving. She prods and shoves lazy workers, stimulating them to become more active (Fig. 37-31). As in honeybee hives, there is a division of labor among the workers, in this case based on size. Small young rats clean the tunnels, gather food, and tunnel. Tunnelers line up head to tail and pass excavated dirt along the completed tunnel to an opening. Just below the opening, a larger mole rat flings the dirt into the air, adding it to a cone-shaped mound. Biologists observing this behavior from the surface dubbed it "volcanoing." In addition to volcanoing, large mole rats defend the colony against predators and members of other colonies.

If another female begins to become fertile, the queen apparently senses changes in the estrogen levels of the subordinate female's urine. The queen then selectively shoves the would-be breeder, causing stress that prevents her rival from ovulating. Large males are more likely to mate with the queen than are small ones, although all adult males are fertile. When the queen dies, a few of the

Figure 37-31 The naked mole rat queen
Encountering a lazy worker in one of the colony's underground tunnels, the queen shoves it and knocks it over to stimulate the worker to greater efforts.

females gain weight and begin shoving one another. The aggression may escalate until a rival is killed. Finally, a single female becomes dominant. Her body lengthens; she assumes the queenship and begins to breed. Litters averaging 14 pups are produced about four times a year. During the first month, the queen nurses her pups, and the workers feed the queen. Then the workers begin feeding the pups solid food.

It is far from clear why naked mole rats, alone among the mammals, have developed a social order in which almost all individuals sacrifice their reproductive potential. One likely contributing factor is that the queen produces the pups that grow up to form her colony, so all colony members are quite closely related. (Helping close relatives to reproduce can be adaptive; see "Evolutionary Connections: Kin Selection and the Evolution of Altruism" in Chapter 15). Another factor that may have favored the evolution of this communal social structure is the high cost of leaving the colony. It's unlikely that a single departing naked mole rat would be able to construct the elaborate underground burrow that the species inhabits.

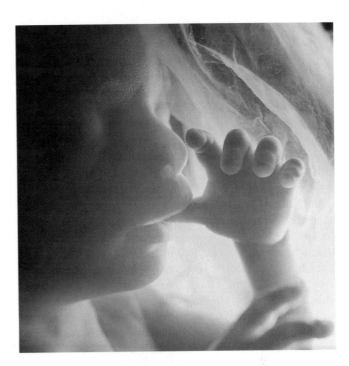

Figure 37-32 A human instinct
Thumb sucking is a difficult habit to discourage in young children, because sucking on appropriately sized objects is an instinctive, food-seeking behavior. This fetus sucks its thumb at about 4 1/2 months of development.

4) Can Biology Explain Human Behavior? *www*

Some scientists hypothesize that because humans are animals whose behaviors have an evolutionary history, the techniques and concepts of ethology can be used to help understand and explain human behavior. Human ethology, however, is, and will remain, a less rigorous science than animal ethology. We cannot treat people as laboratory animals; nor can we control all the variables that influence our attitudes and actions. Some observers argue that human culture has been freed from the constraints of our evolutionary past for so long that we cannot explain our behavior in terms of biological evolution. Nevertheless, some scientists have attempted to take an ethological, evolutionary approach to human behavior—for example, by trying to isolate the genetic components of human behavior. In the following sections we review some of these studies and their findings.

The Behavior of Newborn Infants Has a Large Innate Component

Much of the behavior of very young infants presumably has a large innate component, because there has been little time for learning to occur. The rhythmic movement of a baby's head in search of the mother's breast is a fixed action pattern that is expressed in the first days after birth. Sucking, which can be observed even in the human fetus, is also innate (Fig. 37-32). Other behaviors seen in newborns and even premature infants include walking movements when the body is supported and grasping with the hands and feet. Another example is smiling,

which may occur soon after birth. Initially, smiling can be released by almost any object looming over the newborn. This initial indiscriminate response, however, is soon modified by experience. Infants up to 2 months old will smile in response to a stimulus consisting of two dark, eye-sized spots on a light background, which at that stage of development is a more potent releaser of smiling than it is an accurate representation of a human face. But as the child's development continues, learning and further development of the nervous system interact to limit the response to more-correct representations of a face.

One of the most important insights to have emerged from studies of animal learning is that animals of a given species tend to have an inborn predilection for specific types of learning that are important to that species' mode of life. In humans, one such inborn predilection seems to be for the acquisition of language. Young children are able to acquire language rapidly and nearly effortlessly; they typically acquire a vocabulary of 28,000 words before the age of eight. A large body of research suggests that we are born with a brain that is already primed for this early facility with language. For example, the human fetus begins responding to sounds during the third trimester of pregnancy, and researchers have demonstrated that infants are able to distinguish a variety of consonant sounds by 6 weeks after birth. The latter finding stems from an ingenious experiment that took advantage of an infant's ability to make sucking movements in order to gauge the infant's ability to discriminate sounds. In the experiment, a baby responded to the presentation of various consonant sounds by sucking on a pacifier that contained a force transducer to record the sucking rate. When one sound (such as "ba") was presented repeatedly, the infant became habituated and decreased her sucking rate. But when a new sound (such as "pa") was presented, sucking rate increased, revealing that the infant perceived the new sound as different.

The sucking-measurement technique has also been used to demonstrate that newborns in their first 3 days of life can be conditioned to produce certain rhythms of sucking by using their mother's voice as reinforcement. In an experiment performed by William Fifer of the New York State Psychiatric Institute, infants clearly preferred their own mothers' voices over other female voices, as indicated by their sucking rhythm (Fig. 37-33). The infant's ability to learn his or her mother's voice and respond positively to it within days of birth has strong parallels to imprinting and may help initiate bonding with the mother.

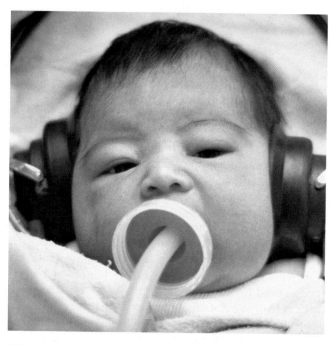

Figure 37-33 *Newborns prefer their mother's voices*
Using a nipple connected to a computer that plays audio tapes, researcher William Fifer demonstrated that newborns can be conditioned to suck at specific rates for the privilege of listening to their own mothers' voices through headphones. For example, if the infant sucks faster than normal, her mother's voice is played; if she sucks more slowly, another woman's voice is played. Researchers found that infants easily learned and were willing to work hard at this task just to listen to their own mothers' voices, presumably because they had become used to her voice in the womb. This response may be a human version of imprinting.

Innate Tendencies Can Be Revealed by Exaggerating Human Releasers

A behavior probably has an instinctive component if the stimulus that causes it (the releaser) can be exaggerated beyond the bounds of reality and elicit an even stronger response. The eye-spots that cause young infants to smile are one example. In turn, the smile of an infant, along with certain characteristic baby features, may release protective feelings in adults. These features include a relatively large head with a domed forehead, chubby cheeks, a small nose, short arms and legs, and a small, rounded body. Even 3-year-old children respond to these features with "mothering" behavior. The marketplace has exploited the releasing aspects of these features by exaggerating them in representations of baby animals and people and by using their innate appeal to sell dolls, posters, calendars, and cards.

One human signal with an innate, physiological basis is the involuntary enlargement of the pupil of the eye when viewing something pleasant, be it a loved one or a hot fudge sundae. We also react to this signal in others. To test this reaction, researchers showed male subjects photographs of smiling women; the photographs were identical except they had been retouched to enlarge or contract the pupils. The subjects overwhelmingly preferred the women with large pupils, although none were consciously aware of the pupil size. We react positively to someone who gazes at us with dilated pupils (although recognition of this feature is subconscious), because it implies interest and attraction. The positive response to enlarged pupils was recognized as long ago as the Middle Ages, when some women artificially enlarged their pupils with eyedrops that contained the drug belladonna (meaning "beautiful woman" in Italian).

Simple Behaviors Shared by Diverse Cultures May Be Innate

Another way to study the instinctive bases of adult human behavior is to compare simple acts performed by people from isolated and diverse cultures. This comparative approach, pioneered by the ethologist Irenaus Eibl-Eibesfeldt, has revealed several gestures that seem to form a universal, and therefore probably innate, human language. Such gestures include a variety of facial expressions for pleasure, rage, and disdain and movements, such as the "eye flash" greeting (in which the eyes are widely opened and the eyebrows rapidly elevated) and a hand upraised in greeting.

People May Respond to Pheromones

Humans may have unconscious responses to pheromones. Our sense of smell is relatively poor, and the role of odor as a means of human communication is largely unknown. However, an interesting study by Martha McClintock of Harvard University provided indirect evidence that primer pheromones may influence female reproductive physiology. Using dorm-living college women as subjects, she found that the menstrual cycles of roommates and close friends became significantly more synchronous over a 6-month period. The cycles of women randomly chosen from the dormitory did not. Further, she found that college women who spent time with men typically had significantly shorter menstrual cycles than did women who spent little time around men. Additional support comes from another study in which the timing of the menstrual cycle was altered by placing extract of male underarm sweat on the upper lip of women. Anatomist David Berliner believes he has isolated an odorless human pheromone from skin cells that produces a sense of well-being; he is currently marketing this discovery in the form of a perfume.

In most vertebrates, pheromones are detected by a small pitlike structure inside the nose called the *vomeronasal organ* (VNO), but the VNO was believed to be nonexistent in humans. In the mid-1980s, researchers finally took a closer look and found a VNO, small but unmistakable, in nearly every human nose they examined. Preliminary studies suggest that certain odorless chemicals cause responses in the sensory cells that line the VNO. In other vertebrates, nerve fibers from the VNO travel to the hypothalamus and amygdala, brain structures important in producing unconscious and emotional responses (see Chapter 33). Although far from conclusive, these findings are tantalizing and call for further investigation of the role of both primer and releaser pheromones in human behavior.

Comparisons of Identical and Fraternal Twins Reveal Genetic Components of Behavior

Twins present perhaps the best opportunity to examine the hypothesis that differences in human behavior are related to genetic differences. When twins are reared together, environmental influences on behavior are very similar for each member of the pair, so behavioral differences between twins must have a large genetic component. If a particular behavior is heavily influenced by genetic factors, we would expect to find similar expression of that behavior in *identical twins* (which arise from a single fertilized egg and have identical genes) but not in *fraternal twins* (which arise from two individual eggs and are no more similar genetically than are other siblings). Data from twin studies, and from other investigations of the genealogy of behaviors, have tended to confirm the heritability of many human behavioral traits. These studies have documented a significant genetic component for traits such as activity level, alcoholism, sociability, anxiety, intelligence, dominance, and even political attitudes. On the basis of tests designed to measure many aspects of personality, identical twins are about twice as similar in personality as are fraternal twins.

The most fascinating, if perhaps less rigorous, twin findings are based on anecdotal observations of identical twins separated soon after birth, reared in different environments, and reunited for the first time as adults. Identical twins reared apart were found to be as similar in personality as those reared together, indicating that the differences in their environments had little influence on their personality development. They have been found to share nearly identical taste in jewelry, clothing, humor, food, and names for children and pets. Personal idiosyncrasies such as giggling, nail biting, drinking patterns, hypochondria, and mild phobias are in some cases shared by these unacquainted twins.

The field of human behavioral genetics is controversial, because it challenges the long-held belief that environment is the most important determinant of human behavior. As discussed earlier in this chapter, we now recognize that all behavior has some genetic basis and that complex behavior in nonhuman animals typically combines elements of both learned and innate behavior. In the case of our own behavior, the debate over the relative importance of heredity and environment continues and is unlikely ever to be fully resolved. Human ethology is not yet recognized as a rigorous science, and it will always be hampered because we can neither view ourselves with detached objectivity nor treat each other as though we were laboratory rats. Despite these limitations, there is much to be learned about the interaction of learning and innate tendencies in people.

Evolutionary Connections
Why Do Animals Play?

Pigface, a giant 50-year-old African softshell turtle, spends hours each day batting a ball around his aquatic home in the National Zoo in Washington, D.C., to the delight of thousands of visitors and the puzzlement of behavioral biologists. Play has always been somewhat of a mystery. It has been observed in many birds and in most mammals, but, until zookeepers tossed Pigface a ball a few years ago, it had never been seen in animals as evolutionarily ancient as reptiles.

Animals at play are fascinating. Pygmy hippopotamuses push one another, shake and toss their heads, splash in the water, and pirouette on their hind legs. Otters delight in elaborate acrobatics. Bottlenose dolphins balance fish on their snouts, throw objects, and carry them in their mouths while swimming. Even baby vampire bats have been observed chasing, wrestling, and slapping each other with their wings. Solitary play typically involves a single animal manipulating an object, such as a cat with a ball of yarn, or the dolphin with its fish, or a macaque monkey making and playing with a snowball. Play can also be social. Often young of the same species play together, but parents may join them (Fig. 37-34a). Social play typically includes chasing, fleeing, wrestling, kicking, and gentle biting (Fig. 37-34b,c).

What are the features of play? (1) Play seems to lack any clear immediate function. (2) Play is abandoned in

(a)

(b)

(c)

Figure 37-34 _Young animals at play_
(a) A bonobo mother plays with her young. *(b)* Young polar bears wrestle on the ice of Hudson Bay, Canada. *(c)* Young arctic foxes gently biting each other.

favor of feeding, courtship, and escaping from danger. (3) Young animals play more frequently than do adults. (4) Play typically involves movements borrowed from other behaviors (attacking, fleeing, stalking, and so on). (5) Play uses considerable energy. (6) Play is potentially dangerous. Many young humans and other animals are injured, and some are killed, during play. In addition, play can distract the animal from the presence of danger while making it conspicuous to predators. So, why do animals play?

The only logical conclusion is that play must have survival value and that natural selection has favored those individuals who engage in playful activities. One of the best explanations for the survival value of play is the "practice theory," first proposed by K. Groos in 1898. He suggested that play allows young animals to gain experience in a variety of behaviors that they will use as adults. By performing these acts repeatedly in a lighthearted context, the animal practices skills that will later be important in hunting, fleeing, or social interactions.

More-recent research supports and extends Groos's proposal. Play occurs most intensely early in life when the brain is developing and crucial neural connections are being formed. John Byers, a zoologist at the University of Idaho, has observed that animals with larger brains tend to be more playful than are animals with smaller brains. Because larger brains are generally linked to increased learning ability, this relationship supports the idea that adult skills are learned during juvenile play. Watch children roughhousing or playing tag, and you will see how strength and coordination are fostered by play and how skills are developed that might have helped our hunting ancestors survive. Quiet play with other children, with dolls, blocks and other toys helps children prepare to interact socially, nurture their own children, and deal with the physical world.

Although Shakespeare tells us "play needs no excuse," there is good evidence that the tendency to play has evolved as an adaptive behavior in animals capable of learning. Play is quite literally "serious fun"!

Summary of Key Concepts

1) How Do Innate and Learned Behaviors Differ?

Although all animal behavior is influenced by both genetic and environmental factors, it can be useful to distinguish between behaviors whose development is not highly dependent on external factors and behaviors that require more-extensive environmental stimuli in order to develop. Behaviors in the first category are sometimes designated as innate and can be performed properly the first time an animal encounters the appropriate stimulus. Innate behaviors include kineses, in which animals orient by varying the speed of essentially random movements, stopping when they encounter favorable conditions. In contrast, taxes are directed movements toward or away from specific stimuli. A fixed action pattern is a complex innate behavior elicited by a specific stimulus called a releaser. Learning can in some cases modify the releasers for fixed action patterns. Fixed action patterns can be distinguished from learned behaviors (1) if they are performed without prior experience, (2) if manipulating the releaser can result in inappropriate behavior, and (3) if a genetic basis can be determined through breeding experiments.

Behavior that changes in response to input from an animal's social and physical environment is said to be learned. Learning is especially adaptive in environments that are changing and unpredictable, and learning can modify innate behavior to make it more appropriate. Learning assumes a wide variety of forms, depending on the evolutionary history and the ecological needs of animal species involved.

Among this diverse array of learning methods are imprinting, habituation, conditioning, trial and error, and insight. Imprinting is a special kind of learning that occurs during a limited sensitive period early in life. This form of simple learning typically involves attachment between parent and offspring or learning the features of a future mate.

Habituation is the decline in response to a harmless stimulus that is repeated frequently. It commonly modifies innate escape responses or defensive responses.

During classical conditioning, an animal learns to make a reflexive response, such as withdrawal or salivation, to a stimulus that did not originally elicit that response. During operant conditioning, an animal learns to make a new response, such as pressing a button, to obtain a reward or to avoid punishment.

Trial-and-error learning can modify innate behavior or can produce new behavior as a result of rewards and punishments provided by the environment.

Insight, the most complex form of learning, can be considered a form of mental trial-and-error learning. An animal showing insight makes a new and adaptive response to an unfamiliar situation.

Although the distinction between innate and learned behavior is conceptually useful, the distinction is not sharp in naturally occurring behaviors. In virtually all behaviors, learning and instinct interact to produce adaptive behavior. Certain types of learning, such as imprinting, occur instinctively, during a rigidly defined time span. Instinctive responses are typically modified by experience. Learning allows animals to modify these innate responses so that they occur only with appropriate stimuli.

2) How Do Animals Communicate?

Communication, an action by one animal that alters the behavior of another, is the basis of all social behavior. It allows animals of the same species to interact effectively in their quest for mates, food, shelter, and other resources. Animals communicate through visual signals, sound, chemicals (pheromones), and touch.

Visual communication is quiet and can convey subtle, rapidly changing information. Visual signals are active (body movements) or passive (body shape and color). Sound communication can also carry a wide range of rapidly changing information and is effective where vision is impossible. Although sound can attract predators, the animal may remain hidden while communicating.

Chemical signals take the form of releaser or primer pheromones. Releaser pheromones cause an immediate, observable behavior in the animal that detects them. Primer pheromones alter the physiological state of the recipient; releaser pheromones influence the recipient's behavior. Pheromones can be detected after the sender has departed, conveying a message over time. Physical contact reinforces social bonds and helps synchronize mating in a variety of animals, from mammals to snails.

3) How Do Animals Interact?

Although many competitive interactions are resolved through aggression, serious injuries are rare. Most aggressive encounters are settled by means of displays that communicate the motivation, size, and strength of the combatants.

Some species establish dominance hierarchies that minimize aggression and regulate access to resources. On the basis of initial aggressive encounters, each animal acquires a status in which it defers to more dominant individuals and dominates subordinates. When resources are limited, dominant animals obtain the largest share and are most likely to reproduce.

Territoriality, a behavior in which animals defend areas where important resources are located, also allocates resources and minimizes aggressive encounters. In general, territorial boundaries are respected, and the best-adapted individuals defend the richest territories and produce the most offspring.

Successful reproduction requires that animals recognize the species, sex, and sexual receptivity of potential mates. In some cases, they must also overcome a resistance to close approach by another individual. These requirements have resulted in the evolution of sexual displays that use all possible forms of communication.

Social living has both advantages and disadvantages, and species show a wide variation in the degree to which their members cooperate. Some species form cooperative societies. The most rigid and highly organized are those of the social insects such as the honeybee, in which the members follow rigidly defined roles throughout life. These roles are maintained through both genetic programming and the influence of primer pheromones. Nonhuman vertebrates also form complex, but normally less rigid, societies, such as those of bullhead catfish. Naked mole rats exhibit the most complex and rigid vertebrate social interactions, resembling insect societies.

4) Can Biology Explain Human Behavior?

The degree to which human behavior is genetically influenced is highly controversial. Because we cannot freely experiment on humans, and because learning plays a major role in nearly all human behavior, investigators must rely on studies of newborn infants, observation of responses to exaggerated stimuli, comparative cultural studies, correlations between certain behaviors and physiology (which suggest a role for pheromones), and studies of identical and fraternal twins. Evidence is mounting that our genetic heritage plays a role in personality, intelligence, simple universal gestures, our responses to certain stimuli, and our tendency to learn specific things such as language at particular stages of development.

Key Terms

aggression *p. 770*
altruism *p. 777*
behavior *p. 760*
classical conditioning *p. 763*
communication *p. 766*
dominance hierarchy *p. 771*
ethology *p. 764*

fixed action pattern *p. 761*
habituation *p. 761*
imprinting *p. 765*
innate *p. 760*
insight learning *p. 764*
instinctive *p. 760*
kinesis *p. 760*

kin selection *p. 778*
learning *p. 761*
operant conditioning *p. 763*
pheromone *p. 768*
primer pheromone *p. 768*
queen substance *p. 768*
releaser *p. 761*

releaser pheromone *p. 768*
sensitive period *p. 765*
taxis *p. 760*
territoriality *p. 772*
trial-and-error learning *p. 763*
waggle dance *p. 778*

Thinking Through the Concepts

Multiple Choice

1. *What type of behavioral process is represented when an animal presses a lever to receive a food reward?*
   a. operant conditioning   b. classical conditioning
   c. habituation   d. insight
   e. imprinting

2. *Which of the following is NOT a feature of play?*
   a. Young animals play more often than adults.
   b. Play commonly involves movements borrowed from other behaviors.
   c. Play uses considerable energy.
   d. Play always has a clear, immediate function.
   e. Play is potentially dangerous.

3. *A taxis is*
   a. a learned behavior
   b. a change in the rate of movement in response to a stimulus
   c. a stereotyped series of movements
   d. an innate movement toward or away from a stimulus
   e. a result of operant conditioning

4. *In which of the following ways does an individual benefit from a social grouping with other animals?*
   a. increased ability to detect, repel, or confuse predators
   b. increased hunting efficiency
   c. increased likelihood of finding mates
   d. energy conservation
   e. all of the above

5. *Which of the following pairs of communication forms and advantages are NOT accurate?*
   a. pheromones, long-lasting
   b. visual displays, instantaneous
   c. sound communication, effective at night
   d. pheromones, convey rapidly changing information
   e. touch, maintains social bonds

6. *In an insect society, such as the honeybee society,*
   a. the division of labor is based on biologically determined castes
   b. all adult members share labor equally
   c. all adult members have the opportunity to reproduce
   d. reproduction is altered seasonally among the adults
   e. the organization of the society is flexible and adaptable

? Review Questions

1. Distinguish between taxes and kineses.

2. What is the name for a stimulus that elicits a fixed action pattern? Give three examples of this type of stimulus.

3. Compare classic conditioning, operant conditioning, and trial-and-error learning. Give an example of each.

4. Compare insight with trial-and-error learning.

5. Explain why neither "innate" nor "learned" adequately describes the behavior of any given organism.

6. Explain why animals play. Include the features of play in your answer.

7. List four senses through which animals communicate, and give one example of each form of communication. After each, present both advantages and disadvantages of that form of communication.

8. Define and distinguish between primer and releaser pheromones. Give an example of each.

9. A bird will ignore a squirrel in its territory but will act aggressively toward a member of its own species. Explain why.

10. Why are most aggressive encounters among members of the same species relatively harmless?

11. Discuss advantages and disadvantages of group living.

12. In what ways do naked mole rat societies resemble those of the honeybee?

Applying the Concepts

1. You are a police officer assigned to the K-9 squad. Your new partner is a German shepherd, which you are assigned to train as a narcotics dog. You plan to use motivational techniques rather than negative reinforcement. Which behavior principles (processes) would you use in your training? What are the advantages of positive over negative reinforcement?

2. Your 3-year-old child persistently approaches the hot stove even though you have told her not to touch it. Which type of learning is she attempting to use? Is there any other learning process that might work that would not require her to touch the hot stove?

3. On the basis of the information you learned about mosquito taxes, design a mosquito trap or killer that exploits a mosquito's innate behaviors. Now design one for moths.

4. You raise honeybees but are new at the job. Trying to increase honey production, you introduce several queens into the hive. What is the likely outcome? What different things could you do to increase production?

5. Describe and give an example of a dominance hierarchy. What role does it play in social behavior? Give a human parallel, and describe its role in human society. Are the two roles similar? Why or why not? Now repeat this exercise for territorial behavior, both in humans and in another species of animal.

6. You are manager of an airport. Planes are being endangered by large numbers of flying birds, which can be sucked into engines, disabling the engines. Without harming the birds, what might you do to discourage them from nesting and flying near the airport and its planes?

Group Activity

To see how playing games can help us understand the evolution of behavior, try playing the following game. Imagine a population of animals in which two behavior patterns are present. The first pattern is called "cooperate," and cooperators always behave helpfully during encounters with another individual of the same species. The second behavior pattern is called "defect," and defectors always harm other individuals to gain an advantage. Whenever two individuals meet and interact, each individual gets a payoff, which can be positive, negative, or neutral. These payoffs are summarized in the accompanying "payoff matrix." If two cooperators meet, each gets one point, and so on according to the matrix.

To play the game, form a group of four students. Choose one member of the group to be a cooperator (always cooperates) and a second member to be a defector (always defects). In each round, the remaining two members will either cooperate or defect, depending on the results of a coin flip (heads cooperate, tails defect; each person flips separately). Next, on each of four slips of paper, write the name of a group member, and place

the slips of paper in a hat. To play a round of the game, each member of the group chooses a slip of paper and "interacts" with the member whose name is chosen. (Try again if you choose your own name, and be sure that both coin-flippers have selected a new role before each round.) Each group will thus have two interactions per round, with points generated according to the payoff matrix. Play 20 rounds, with each player keeping track of his or her own score. Which strategy scored the highest number of points in your group? Tabulate the results of all the groups in your class to determine the best overall strategy.

Next, play another 20 rounds of the game, but this time each member of the group should use the same strategy: "If my game partner cooperated last round, I'll cooperate; if my partner defected last round, I'll defect." (For the first round, players should choose a role by coin flip.) What happens to the scores now? Gather your group and, on the basis of the results of your two games, develop a hypothesis about the conditions under which cooperative behavior could evolve.

	Cooperate	Defect
Cooperate	Mutual help; each earns 2 points	Defector earns 1 point, Cooperator loses 1 point
Defect	Defector earns 1 point, Cooperator loses 1 point	Stand-off; no points earned or lost

For More Information

Benzer, S. "Genetic Dissection of Behavior." *Scientific American*, December 1973. Mutant fruit flies are used to study the genetic bases of behavior.

Brown, S. L. "Animals at Play." *National Geographic*, December 1994. Clearly written description of why animals play accompanied by beautiful photographs that capture wild animals at play.

Dawkins, R. *The Selfish Gene*, 2nd ed. New York: Oxford University Press, 1989. Exceptionally clear explanation of the evolutionary basis of altruistic behavior, written for both laypeople and scientists.

Dugatkin, L., and Godin, J. "How Females Choose Their Mates." *Scientific American*, April 1998. An entertaining look at how biologists study the processes by which females choose among available males.

Gould, J. L., and Gould, C. G. "The Instinct to Learn." *Science 81*, May 1981. Birds, bees, and perhaps even humans are genetically programmed to learn specific things at particular stages in life.

Horgan, J. "Eugenics Revisited: Trends in Behavioral Genetics." *Scientific American*, June 1993. Explores the fascinating interface between genetics and behavior.

Kirchner, W. H., and Towne, W. F. "The Sensory Basis of the Honeybee Dance Language." *Scientific American*, June 1994. Combines a historical look at the work of Karl von Frisch with a modern update on recent research that has almost fully elucidated the mechanisms of honeybee communication during the waggle dance.

Lorenz, K. *King Solomon's Ring: New Light on Animal Ways*. New York: Thomas Y. Cromwell, 1952. Beautifully written and filled with interesting anecdotes; provides important insights into early ethology.

Macdonald, D., and Brown, R. "The Smell of Success." *New Scientist*, May 1985. Describes the amazing diversity of mammalian pheromones.

Phillips, K. "School Riddles." *International Wildlife*, March/April 1995. Explores research on fish schooling behavior and how it helps avoid predation.

Pinker, S. *The Language Instinct*. New York: William Morrow, 1994. An entertaining account of linguists' current understanding of how we develop the ability to use language and how that ability may have evolved.

Seeley, T. D. "The Honeybee Colony as a Superorganism." *American Scientist*, November–December 1989. The honeybee society exemplifies the concept of kin selection in which the society, rather than the individual, serves as a "vehicle for the survival of genes."

Sherman, P., and Alcock, J. *Exploring Animal Behavior*. Sunderland, MA: Sinauer, 1998. A collection of articles in which behavioral biologists describe their research and their conclusions about the mechanisms, function, and evolution of behavior.

Sherman, P. W., Jarvis, J. U. M., and Braude, S. H. "Naked Mole Rats." *Scientific American*, August 1992. Describes the newly investigated society of the naked mole rat, a vertebrate whose social behavior resembles that of some social insects.

Stevens, J. "The Biology of Violence." *BioScience*, May 1994. Discusses the genetic aspects of violence.

Wilson, E. O. "Empire of the Ants." *Discover*, March 1990. This entertaining article examines the diversity of ant behavior that has contributed to their enormous success.

Answers to Multiple-Choice Questions

1. a   2. d   3. d   4. e   5. d   6. a

Unit Six
Ecology

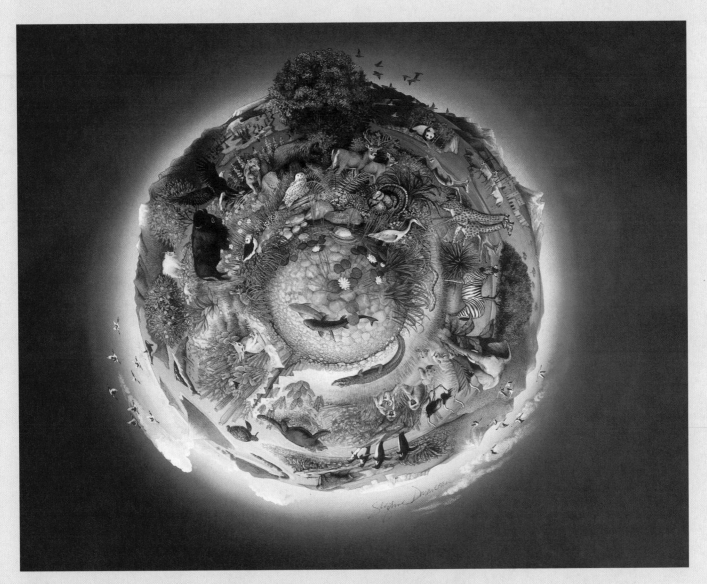

The beauty and interdependence of Earth's biosphere is illustrated in "Paradise," Part One of "The Trilogy of the Earth" by Suzanne Duranceau/Illustratrice, Inc.

> *"There is no exception to the rule that every organic being naturally increases at so high a rate that, if not destroyed, the Earth would soon be covered by the progeny of a single pair."*

Charles Darwin in On the Origin of Species (1859)

Lesser flamingos form a pink carpet in Kenya. Although this population appears to be fulfilling Darwin's prediction of covering Earth (see the quote above), these flamingos actually illustrate a "clumped" population distribution (that is, they live in groups), as do many social species.

Population Growth and Regulation

38

At a Glance

Net Watch

On-line resources for this chapter are on the World Wide Web at:
http://www.prenhall.com/audesirk
(click on the Table of Contents link and then select Chapter 38).

In eastern Colorado lies a remnant of shortgrass prairie, dominated by buffalo grass and blue grama grass. In spring and summer it is ablaze with wildflowers: paintbrush, vetch, sunflower, and bladderpod (Fig. 38-1). Prairie dogs stand upright by their burrows, alerting their "town" with vigorous high-pitched cries as a single hawk soars lazily overhead. The prairie ecosystem is home to a complex community of living organisms, including wildflowers, prairie dogs, the grass on which the prairie dogs feed, the hawk that preys on them, and the myriad microscopic organisms that keep the soil fertile.

Just to the north, a farm was abandoned a few years ago. Its fields were plowed but not planted, and several different types of plants—some introduced from other continents—are rapidly invading the rich open space. In a field where the farmer's cattle grazed, unpalatable sagebrush has replaced the lush prairie grasses, so the land can no longer support livestock. A "For Sale" sign in the field foreshadows high-density housing developments that will soon displace both farm and prairie to house a tiny fraction of the growing human population. The first new residents will be delighted to see a hawk, a magnificent bird of prey, practically in their backyards. But as the prairie dog town (closely spaced burrows that house families of the rodents) is bulldozed, this rare predator will also disappear.

How did overgrazing by cattle cause a change in vegetation that decreased the land's ability to support cattle? What keeps natural populations such as those of prairie dogs from overpopulating their habitat and starving? What effect does the hawk have on the prairie dog population? What happens when different organisms compete for the same type of food or other resources? Why has the human population continued to expand while other populations have fluctuated, remained stable, or declined? Look for answers to these questions in this chapter as we explore how populations grow and how population growth is controlled.

This chapter begins our unit on **ecology**, the study of interrelationships among living things and their nonliving environment. The environment consists of an *abiotic*

Figure 38-1 *The prairie ecosystem*
This montage shows representatives of *(a)* the natural prairie community. Insets show *(b)* a prairie wildflower, the Indian paintbrush; *(c)* a prairie dog; *(d)* a hawk; and *(e)* an overgrazed prairie in which sagebrush has replaced native grasses.

(nonliving) component, including soil, water, and weather, and a *biotic* (living) component, including all forms of life. The term **ecosystem** refers to all the organisms and their nonliving environment within a defined area, such as our prairie. Within an ecosystem, all the interacting populations of organisms are described as a **community**. Ecology (whose name is derived from the Greek word *oikos*, "a place to live") is a tremendously diverse, complex, and relatively young scientific discipline. In Chapters 39 through 41, we will proceed from *populations* to levels of increasing complexity: first to *communities* and the interactions within them, then to the organization of ecosystems, and finally to an exploration of the diversity of ecosystems that make up the *biosphere*.

1 How Do Populations Grow?

A **population** consists of all the members of a particular species who live within an ecosystem and can potentially interbreed. Members of our prairie dog town belong to the same population; those in a different town too far away to allow interbreeding are members of the same species but not part of the same population.

Studies of undisturbed ecosystems show that some populations tend to remain relatively stable in size over time, others fluctuate in a cyclical pattern, and still others vary sporadically in response to complex environmental variables. In contrast, the human population has exhibited steady growth for centuries. Let's examine how and why populations grow, then look at the forces that normally control this growth.

Three factors determine whether and how much the size of a population changes: (1) births, (2) deaths, and (3) migration. Organisms join a population through birth or **immigration** (migration in) and leave it through death or **emigration** (migration out). A population remains stable if, on the average, as many individuals leave as join. Population growth occurs when the number of births plus immigrants exceeds the number of deaths plus emigrants. Population growth drops when the reverse occurs. A simple equation for the change in population size within a given time period is as follows:

(births − deaths) + (immigrants − emigrants) = change in population size

In many natural populations, organisms moving in and out contribute relatively little to population change, leaving birth and death rates as the primary factors that influence population growth.

The ultimate size of any population (leaving out migration) is the result of a balance between two major opposing factors. The first is **biotic potential**, or the maximum rate at which the population could increase, assuming ideal conditions that allow a maximum birth rate and minimum death rate. Opposing this potential for growth are limits set by the living and nonliving environments. These limits include the availability of food and space, competition with other organisms, and certain interactions among species, such as predation and parasitism. Collectively, these limits are called **environmental resistance**. Environmental resistance can both decrease the birth rate and increase the death rate. The interaction between biotic potential and environmental resistance usually results in a balance between population size and available resources. To understand how populations grow and how their size is regulated, we must examine each of these forces in more detail.

Biotic Potential Can Produce Exponential Growth

Changes in population size (ignoring migration) are functions of the birth rate, the death rate, and the number of individuals in the original population. Rates of change in population size are often measured as the changes in these variables per individual during a given unit of time. For example, the birth rate may be expressed as the number of births per person per year.

The **growth rate** (r) of a population is a measure of the change in population size per individual per unit of time. This value is determined by subtracting the death rate (d) from the birth rate (b):

$$\underset{\text{(birth rate)}}{b} - \underset{\text{(death rate)}}{d} = \underset{\text{(growth rate)}}{r}$$

(If the death rate should exceed the birth rate, the growth rate will be negative, and the population will decline, but here we will focus on growing populations.) If we wish to calculate the annual growth rate of a population of 10,000 in which 1500 births and 500 deaths occur each year, we can use the simple equation

$$\text{birth rate } b - \text{death rate } d = \text{growth rate } r$$

$$\tfrac{1500}{10,000} - \tfrac{500}{10,000} = 0.10, \text{ or } 10\%$$

or

0.15 births per person per year
− 0.05 deaths per person per year

$= 0.10$, or 10% increase per person per year $= r$

To determine the number of individuals added to a population in a given time period, the growth rate (r) is multiplied by the original population size (N):

$$\text{population growth} = rN$$

In this example, population growth (rN) equals $0.10 \times 10,000 = 1000$ people per year. If this growth rate is constant, then the following year, r must be multiplied by an even larger population size ($N + rN = 11,000$). In other words, in the second year 1100 more individuals are added to the population size, which further increases the number of individuals added in the third year, and so on.

This pattern of a continuously accelerating increase in population size is **exponential growth**. During exponential growth, the population (over a given time period) grows by a fixed percentage of its size at the beginning of that time period. Thus, an increasing number of individuals is added to the population during each succeeding time period, causing population size to grow at an ever-accelerating pace. The graph of exponential population growth is often called a J-shaped growth curve, or a **J-curve**, after its shape (Fig. 38-2).

Such growth occurs whenever births consistently exceed deaths. This occurs if, on the average, each individual produces more than one surviving offspring during its lifetime. Although the number of offspring an individual produces each year varies from millions for an oyster to one or fewer for a human, each organism—whether working alone or as part of a sexually reproducing pair—has the potential to replace itself many times over during its lifetime. This capacity, biotic potential, has evolved because it helps ensure that at least one offspring survives to bear its own young. Several factors influence biotic potential, including the following:

- The age at which the organism first reproduces.
- The frequency with which reproduction occurs.
- The average number of offspring produced each time.
- The length of the organism's reproductive life span.
- The death rate of individuals under ideal conditions.

We will use examples in which these factors differ to illustrate the concept of exponential growth.

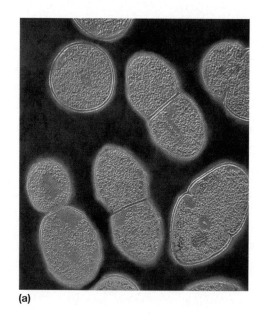

(a)

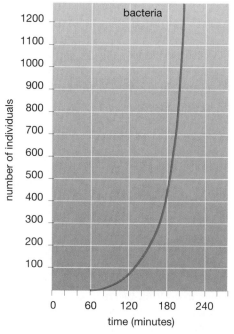

time (minutes)	number of bacteria
0	1
20	2
40	4
60	8
80	16
100	32
120	64
140	128
160	256
180	512
200	1024
220	2048

(b)

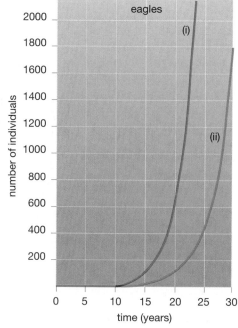

time (years)	number of eagles (i)	number of eagles (ii)
0	2	2
2	2	2
4	4	2
6	8	4
8	14	8
10	28	12
12	52	18
14	100	32
16	190	54
18	362	86
20	630	142
22	1314	238
24	2504	392
26	4770	644
28	9088	1066
30	17314	1764

Figure 38-2 *Exponential growth curves*
All such curves share a similar J shape; the major difference is the time scale. **(a)** Growth of a population of bacteria, starting with a single individual and with a doubling time of 20 minutes. **(b)** Growth of a population of eagles, starting with a single pair of hatchlings, for ages at first reproduction of 4 years (red line) and 6 years (green line). Notice in the table that after 26 years, the population of eagles that began reproducing at age 4 is nearly seven times that of the eagles that began reproducing at age 6.

The bacterium *Streptococcus* (Fig. 38-2a) is normally a harmless resident in and on the human body, where its population growth is restricted by environmental resistance. But in an ideal culture medium, such as warm custard, where *Streptococcus* may accidentally be intro-duced, each bacterial cell can divide every 20 minutes, doubling the population three times each hour. (The by-products of bacterial metabolism can result in food poi-soning under these conditions.) The larger the population grows, the more cells there are to divide. The biotic potential of bacteria is so great that, were nutrients unlimited, the offspring of a single bacterium would swamp Earth in a layer 7 feet deep within 48 hours!

In contrast, the golden eagle is a relatively long-lived, rather slowly reproducing species (Fig. 38-2b). Let's assume that the golden eagle can live 30 years and reaches sexual maturity at age 4 years, and that each pair of eagles produces two offspring annually for the remaining 26 years. Figure 38-2 compares the potential population growth of eagles to that of bacteria, assuming no deaths occur in either population during the time graphed. Notice that the shapes of the curves are virtually identical—both populations exhibit unrestricted exponential growth. Although the time scale differs, both population sizes eventually become infinitely large. Figure 38-2b also shows what happens if eagle reproduction begins at age 6 years instead of 4. Exponential growth still occurs, but the time required to reach a particular size increases considerably. This result has important implications for the human population: Delayed childbearing significantly slows population growth. If each woman bears three children in her early teens, the population will grow much faster than if each woman bears five children but begins at age 30.

So far we have looked only at birth rates. Even under ideal conditions, however, some deaths occur. To illustrate the effect of differing death rates, three bacterial populations are compared in Figure 38-3. Again, the shapes of the curves are the same. In each case the population eventually approaches infinite size, but the time

required to reach any given population size is increased by increasing mortality.

2 How Is Population Growth Regulated?

Exponential Growth Cannot Continue Indefinitely

In nature, exponential growth occurs only under special circumstances and only for a limited time. For example, exponential growth is seen in populations that undergo regular cycles, in which rapid population growth is followed by a sudden massive die-off. These **boom-and-bust cycles** occur in a variety of organisms for complex and varied reasons. Many short-lived, rapidly reproducing species, from algae to insects, have seasonal population cycles that are linked to predictable changes in rainfall, temperature, or nutrient availability (Fig. 38-4). In temperate climates, insect populations grow rapidly during the spring and summer, then crash with the killing hard frosts of winter. More-complex factors produce roughly 4-year cycles for small rodents such as voles and lemmings and much longer population cycles in hares, muskrats, and grouse.

Lemming populations, for instance, may grow until the rodents overgraze their fragile arctic tundra ecosystem. Lack of food, increasing populations of predators, and social stress caused by overpopulation may all contribute to a sudden high mortality. Many deaths occur as waves of lemmings emigrate from regions of high population density. During these dramatic mass movements, lemmings are easy targets for predators. Many lemmings accidentally drown; they begin swimming when they encounter a body of water, including the ocean, but cannot make it all the way across. The reduced lemming population eventually contributes to a decline in preda-

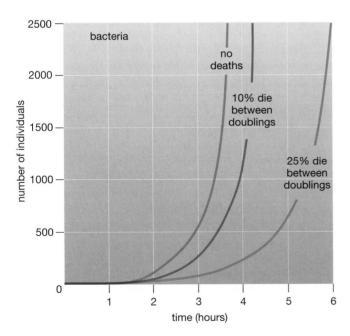

Figure 38-3 *The effect of death rates on population growth*
The graphs assume that a bacterial population doubles every 20 minutes. Notice that the population in which a quarter of the bacteria die every 20 minutes reaches 2500 only 2 hours and 20 minutes later than one in which no deaths occur.

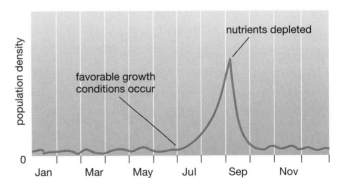

Figure 38-4 *A boom-and-bust cycle of population*
Population density of cyanobacteria (blue-green algae) in an annual boom-and-bust cycle in a lake. Algae survive at a low level through the fall, winter, and spring. Early in July, conditions become favorable for growth, and exponential growth occurs through August. Nutrients soon become depleted, and the population "goes bust."

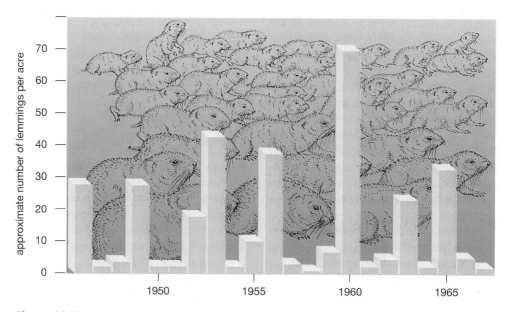

Figure 38-5 *Lemming population cycles*
Lemming population density follows roughly a 4-year cycle (data from Point Barrow, Alaska).

tor numbers (see "Scientific Inquiry: Cycles in Predator and Prey Populations," p. 800) and a recovery of the plant community on which the lemmings normally feed. These responses, in turn, set the stage for another round of exponential growth in the lemming population (Fig. 38-5).

In populations that do not show boom-and-bust cycles, exponential growth may occur temporarily under special circumstances—for example, if population-controlling factors, such as predators or parasites, are eliminated or if the food supply is increased. Exponential growth can also occur when individuals invade a new habitat where conditions are favorable and competition is scarce, such as the farm field that is plowed but then abandoned. Invasion of new habitats and exponential growth of the invaders commonly occur when people introduce foreign, or **exotic**, species into ecosystems, in many cases with tragic results (see "Earth Watch: The Invasion of Fire Ants and Kudzu," Chapter 39). One of the first invaders into the abandoned farm will probably be cheatgrass, an exotic species. As you will learn in the next section, all populations showing exponential growth must eventually stabilize or crash (fall precipitously).

Environmental Resistance Limits Population Growth

Exponential growth carries with it the seeds of its own demise. As individuals join the population, competition for resources intensifies. As plants invade the abandoned farm field and their populations grow, competition for space, water, sunlight, and soil nutrients increases until further expansion is impossible. In response to increases in populations of animals such as prairie dogs, predators

such as hawks may increase in number or may make this newly abundant prey a larger part of their diet. Parasites and diseases spread more readily owing to weakness and crowding, resulting from lack of food or from stress caused by adverse social interactions. Animals might emigrate to establish new populations or die. Consequently, after a period of exponential growth, populations tend to stabilize at or below the maximum number the environment can sustain. The rate of growth gradually declines and reaches a long-term state of equilibrium, fluctuating around a growth rate of zero. In this equilibrium, the birth rate is balanced by the death rate and population size is stabilized. This type of population growth, which is typical of long-lived organisms colonizing a new area, is represented graphically by an S-shaped growth curve, or **S-curve** (Fig. 38-6).

Populations may stabilize at a level at or below the **carrying capacity** of the ecosystem. The carrying capacity is the maximum population size that an ecosystem can support indefinitely. It is determined primarily by the sustained availability of two types of resources. The first type is renewable resources, which are replenished by natural processes. Renewable resources include nutrients, water, and light (the energy source for plants). The second type of resource is nonrenewable: space.

Organisms will starve if demands on renewable resources are too high. If space requirements are exceeded, animals may emigrate, but commonly to less-suitable areas where their death rate will be higher. Reproduction will decline, because animals may not find adequate breeding sites or the seeds of plants may not reach a suitable place to germinate. If a population exceeds its carrying capacity, excess demands on resources may

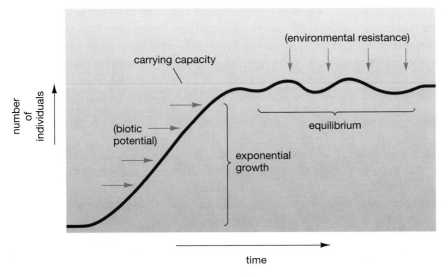

Figure 38-6 *The S-curve of population growth*
The population grows exponentially at first, then fluctuates around the carrying capacity. The growth is driven by biotic potential but levels off owing to environmental resistance.

(which cattle will not eat) to thrive. Once established, sagebrush replaces edible grasses and reduces the land's carrying capacity for cattle. Other dramatic cases of overgrazing have occurred when herbivores, such as reindeer, have been introduced onto islands without large predators (Fig. 38-7).

In nature, environmental resistance maintains populations at or *below* the carrying capacity of their environment. Factors of environmental resistance can be classified into two broad categories. (1) **Density-independent** factors limit population size regardless of the population density (number of individuals per given area). (2) **Density-dependent** factors increase in effectiveness as the population density increases. In the following sections, we will look more closely at these factors and how they control population growth.

damage ecosystems, reducing *their* carrying capacity. The result is either a population decline until the ecosystem recovers or a permanently reduced population. For example, overgrazing by cattle on dry western grasslands has reduced the grass cover and allowed sagebrush

Density-Independent Factors Limit Populations Regardless of Their Density

Perhaps the most important natural density-independent factor is weather. For example, many insects and annual plant populations are limited in size by the number of

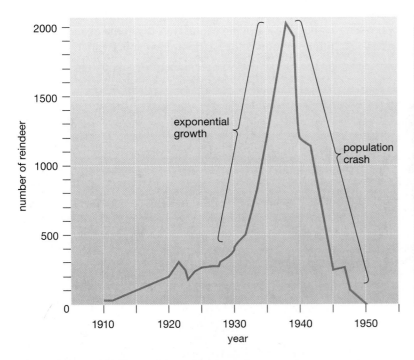

Figure 38-7 *The effects of exceeding carrying capacity*
Exceeding carrying capacity can damage an ecosystem, reducing its ability to support the population. In 1911, 25 reindeer were introduced onto one of the Pribilof Islands (St. Paul) in the Bering Sea off Alaska. Food was plentiful, and no predators of reindeer lived on the island. The herd grew exponentially (note the initial J shape) until it reached 2000 reindeer in 1938. At this point, the small island was seriously overgrazed, food was scarce, and the population declined dramatically. By 1950, only eight reindeer survived.

individuals that can be produced before the first hard freeze. Such populations typically do not reach their carrying capacity, because density-independent factors intervene first. Weather is largely responsible for the boom-and-bust population cycles described earlier. Human activities can also limit the growth of natural populations in ways that are independent of population density. Pesticides and pollutants can cause drastic declines in natural populations. Before it was banned in the 1970s, the pesticide DDT had drastically reduced populations of predatory birds, such as eagles, ospreys, and pelicans. Habitat destruction caused by the construction of farms, roads, and housing developments has eliminated most of the large prairie dog towns and pushed such diverse species as the American condor and (in China) the giant panda to the brink of extinction.

Organisms that live several years have evolved various mechanisms to compensate for seasonal changes in weather, thereby circumventing this form of density-independent population control. Many mammals, for example, develop thick coats and store fat for the winter; some also hibernate. Other animals, including many birds, migrate long distances to find food and a hospitable climate. Plants might survive the rigors of winter by entering a period of dormancy, dropping their leaves and drastically slowing their metabolic activities.

Density-Dependent Factors Become More Effective as Population Density Increases

For long-lived species, by far the most important elements of environmental resistance are density-dependent factors. Because they become increasingly effective as population density increases, density-dependent factors exert a negative feedback effect on population size. The larger the population grows, the more changes are triggered that counteract this growth. Density-dependent factors include community interactions, such as predation and parasitism, as well as competition within the species or with members of other species. These factors are discussed below and in Chapter 39.

Predators Help Control Their Prey Populations

Predation is the act of killing and eating another living organism. This broad definition includes the swift and violent capture of elk by wolves (Fig. 38-8) as well as the less dramatic munching on a cactus by a moth. Organisms that perform predation are called **predators**, and the organisms they kill and eat are called **prey**. Predation controls prey populations in a density-dependent manner. It becomes an increasingly important factor in population control as prey populations increase, simply because the more prey, the more often predators encounter them. Many predators will eat a variety of prey depending on what is most abundant and easiest to find. Coyotes might eat more mice when the mouse population is high but switch to eating more ground squirrels as the mouse population declines, incidentally allowing the mouse population to recover.

Predators can also exert density-dependent effects by increasing in number as their prey population grows. For instance, predators that feed heavily on lemmings, such as the arctic fox and snowy owl, regulate the number of offspring they produce according to the abundance of lemmings. The snowy owl might produce up to 13 chicks when lemmings are abundant but not reproduce at all in years when lemmings are scarce. In some cases, an increase in predators might cause a crash of the prey population, and a decline in the predator population follows. This pattern can result in out-of-phase **population cycles** of both predators and prey (see "Scientific Inquiry: Cycles in Predator and Prey Populations").

The influence of predators on prey populations varies considerably. Some predators feed primarily on prey made vulnerable because their populations have exceeded their carrying capacity. Such prey may be weakened by lack of food or may be exposed owing to lack of appropriate shelter. In such cases, predation may maintain prey populations near a density that can be sustained by the resources of the ecosystem.

In other cases, predators may maintain their prey at well below carrying capacity. A dramatic example of this phenomenon is the prickly pear cactus, which was intro-

Figure 38-8 Predators help control prey populations
Gray wolves have brought down an elk, who may have been weakened by old age or parasites.

(a) **(b)**

Figure 38-9 *An introduced predator controls an introduced pest*
(a) A pasture in Queensland, Australia, in 1926 is blanketed by the imported prickly pear cactus, whose spread was unchecked by predators. **(b)** The same site in 1929, after the introduction of a predatory cactus moth appropriately named *Cactoblastis cactorum*. (Courtesy of the Department of Lands, Queensland, Australia.)

duced into Australia from Latin America. Lacking natural predators, it spread uncontrollably, destroying millions of acres of valuable pasture and range land. Finally, in the 1920s, a cactus moth (a predator of the prickly pear) was imported from Argentina and released to feed on the cacti. Within a few years, the cacti were virtually eliminated. Today the moth continues to maintain its prey at very low population densities (Fig. 38-9).

Parasites Can Reduce Their Host Populations

In contrast to predation, **parasitism** is the process of feeding on a larger organism without killing it immediately or directly. **Parasites** live on or in their larger prey, or **host**. Although they weaken their hosts, most parasites do not kill them outright. They may, however, dramatically increase the death rate in a population where individuals are already weakened by the effects of overcrowding. Further, infestation by parasites makes organisms more vulnerable to predators.

Parasitism is density-dependent. Most parasites have limited motility and spread more readily from host to host at high host-population densities. For example, plant diseases spread readily through acres of densely planted crops, and childhood diseases spread rapidly through schools and day-care centers.

Both predators and parasites can have beneficial effects on the prey population as a whole. Predators, parasites, and their prey evolve together—they *coevolve*. Parasites and predators tend to destroy the least fit of the prey, leaving the better-adapted prey to reproduce. The result tends to be a balance in which the prey population is regulated but not eliminated. The population balance in ecosystems can be destroyed when species are introduced into new areas where they have no natural predators or parasites (see "Earth Watch: The Invasion of Fire Ants and Kudzu," Chapter 39).

Predators and parasites can be devastating when introduced into regions where they did not evolve and local prey species have had no opportunity to evolve defenses against them through natural selection. The smallpox virus, inadvertently carried by traveling Europeans, ravaged the native population of Hawaii, the Amerindians of Argentina, and the Aborigines of Australia. Introduced rats, snakes, and mongooses have exterminated many of Hawaii's native bird populations.

Competition for Resources Helps Control Populations

The resources that determine carrying capacity (space, nutrients, water, light) are often limited relative to the demand for them. In other words, there may not be enough resources to support all the organisms that are produced. Further, use by one individual limits their availability to another. For this reason, **competition**, interaction among individuals who attempt to utilize a limited resource, limits population size in a density-dependent manner. There are two major forms of competition: (1) **interspecific competition** (among individuals of different species; see Chapter 39), and (2) **intraspecific competition** (among individuals of the same species). Because the needs of members of the same species for water and nutrients, shelter, breeding sites, light, and other resources are almost identical, intraspecific competition is more intense than interspecific competition.

Organisms have evolved several ways to deal with intraspecific competition. Some organisms, including most plants and many insects, engage in **scramble competition**, a kind of free-for-all with resources as the prize. For

Scientific Inquiry
Cycles in Predator and Prey Populations

If we assume that prey, such as hares, are eaten exclusively by a particular predator, such as the lynx, it seems logical that both populations might show cyclic changes, with changes in the predator population size lagging behind changes in the prey population size. For example, a large hare population would provide abundant food for lynx and their offspring, which would then survive in large numbers. The increased lynx population would eat more hares, reducing the hare population. With fewer prey, fewer lynx would survive and reproduce, so the lynx population would decline a short time later.

Does this out-of-phase population cycle of predators and prey actually occur in nature? A classic example of such a cycle was demonstrated by using the ingenious method of counting all the pelts of northern Canada lynx and snowshoe hares purchased from trappers by the Hudson Bay Company between 1845 and 1935. The availability of pelts (which presumably reflects population size) showed dramatic, closely linked population cycles of these predators and their prey (Fig. E38-1). Unfortunately, many uncontrolled variables could have influenced the relationship between hares and lynx. One problem is that hare populations sometimes fluctuate even without lynx present, possibly because in the absence of predators hares overshoot their carrying capacity and reduce their food supply. Further, lynx do not feed exclusively on hares but can eat a variety of small mammals. Environmental variables such as exceptionally severe winters could have adversely affected both populations and produced similar cycles. Recently, researchers tested the hare/predator relationship more rigorously by fencing off 1-kilometer-square areas in northern Canada to exclude predators. The crash of the hare population was lessened both by providing extra food and by excluding predators, but by far the greatest success in preventing the crash of the hare population was achieved when the researchers excluded predators *and* provided extra food.

To test the predator/prey cycle hypothesis in an even more controlled manner, investigators turned to laboratory studies on

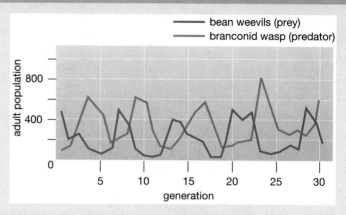

Figure E38-2 *Experimental predator–prey cycles*
Out-of-phase fluctuations in laboratory populations of the bean weevil and its braconid wasp predator.

populations of small predators and their prey. The study illustrated in Figure E38-2 involved braconid wasp predators and their bean weevil prey. As predicted, the two populations showed regular cycles, with the predator population rising and falling slightly later than the prey population. The wasps lay their eggs on weevil larvae, which provide food for the newly hatched wasps. A large weevil population assures a high survival rate for wasp offspring, increasing the predator population. Then, under intense predation pressure, the weevil population plummets, reducing the population size of the next generation of wasps. Reduced predation then allows the weevil population to increase rapidly, and so on.

It is highly unlikely that such a straightforward example will ever be found in nature, but this type of predator–prey interaction clearly can contribute to the fluctuations observed in many natural populations.

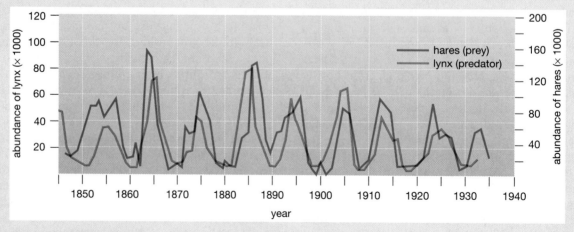

Figure E38-1 *Population cycles in predators and prey*
Snowshoe hares and their lynx predators are graphed on the basis of the number of pelts received by the Hudson Bay Company.

example, when a plant disperses its seeds in a small area, hundreds may germinate. However, as they grow, the larger plants begin to shade the smaller ones; those with the most extensive root systems absorb most of the water, and the weaker individuals eventually wither and die.

Many animals (and even a few plants) have evolved **contest competition**, which helps regulate population size and reduce direct competition. Contest competition consists of social or chemical interactions used to limit access to important resources (also see Chapter 37). Territorial species—such as wolves, many fish, rabbits, and songbirds—defend an area that contains important resources such as food or nesting sites. When the population begins to exceed the available resources, only the best-adapted individuals are able to defend adequate territories. Those without territories may not reproduce and are easy prey.

As population densities increase and competition becomes more intense, some animals react by emigrating. Large numbers leave their homes to colonize new areas, and many, sometimes most, die in the quest. The mass movements of lemmings, which in some instances end in marches into the sea, are apparently in response to overcrowding. Migrating swarms of locusts plague the African continent, stripping all vegetation in their path (Fig. 38-10).

Population size is a result of complex interactions between density-independent and density-dependent forms of environmental resistance. For example, a stand of pines weakened by drought (a density-independent factor) may more readily fall victim to the pine bark beetle (a density-dependent parasite). Likewise, a caribou weakened by parasites (density-dependent) is more likely to be killed by an exceptionally cold winter (a density-independent factor).

Figure 38-10 Emigration
In response to overcrowding, emigration can reduce local populations. Locust swarms provide a dramatic example.

 How Are Populations Distributed?

Organisms might live in flocks, herds, pairs, or as solitary individuals, or they might cluster around resources such as water holes. The spatial pattern in which members of a population are dispersed within a given area is that population's *distribution*. Distribution may vary with time, changing with the breeding season, for example. Ecologists recognize three major types of spatial distribution: *clumped*, *uniform*, and *random* (Fig. 38-11).

Individuals in Many Populations Clump Together into Groups

There are many populations whose members live in groups and whose distribution can be described as **clumped** (Fig. 38-11a). Examples include family or social groupings, such as elephant herds, wolf packs, prides of lions, flocks of flamingos, as in the chapter-opening photograph, and schools of fish. What are the advantages of clumping? Flocks provide many eyes that can search for localized food, such as a tree full of fruit or a lake with fish. Schooling fish and flocks of birds may confuse predators with their sheer numbers. Some species form temporary groups for mating and caring for their young. Other plant or animal populations cluster, not for social reasons, but because resources are localized. Cottonwood trees, for example, cluster along streams and rivers in grasslands. Animals also congregate around water holes in the dry savanna of Africa.

Some Individuals Disperse Themselves Almost Uniformly

Organisms with a **uniform** distribution maintain a relatively constant distance between individuals. This distribution is most common among animals that defend territories and exhibit territorial behaviors designed to protect scarce resources. Male Galapagos iguanas establish regularly spaced breeding territories. Shorebirds are also often found in evenly spaced nests, just out of reach of one another. Other territorial species, such as the tawny owl, mate for life and continuously occupy well-defined, relatively uniformly spaced territories. (For breeding animals, the spacing refers to pairs, not individuals.) Competition for water may cause some plants, such as sage and creosote bushes (Fig. 38-11b), to space themselves relatively evenly. This distribution helps ensure adequate resources for each individual.

In a Few Populations, Organisms Are Randomly Distributed

Organisms with a **random** distribution are relatively rare. Such individuals do not form social groups. The resources they need are more or less equally available throughout

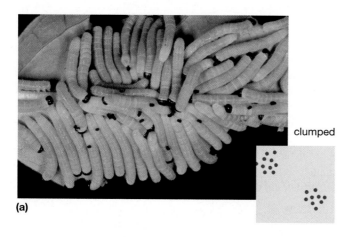

Figure 38-11 Population distributions
(a) Clumped: a gathering of caterpillars. *(b)* Uniform: creosote bushes in the desert. *(c)* Random: trees and plants in a rain forest.

the area they inhabit, and resources are not scarce enough to require territorial spacing. Trees and other plants in rain forests come close to being randomly distributed (Fig. 38-11c). There are probably no vertebrate species that maintain random distribution throughout the year, because they must breed—a behavior that makes social interaction inevitable.

4 What Survivorship Patterns Do Populations Exhibit?

Population patterns can be considered from the perspective of time as well as space. Over time, populations show characteristic patterns of deaths or (more optimistically) survivorship. These patterns, called **survivorship curves**, are revealed when the number of individuals of each age is graphed against their age. Three types of survivorship curve, which can be described as "late loss," "constant loss," and "early loss," according to the part of the life cycle during which most deaths occur, are shown in Figure 38-12.

Populations with *convex* survivorship curves have relatively low infant death rates, and most individuals survive to old age. This late-loss curve is characteristic of humans and many other large animals, such as Dall mountain sheep. These species produce relatively few offspring, which are protected by the parents. Populations with constant-loss survivorship curves have a fairly constant death rate; individuals have an equal chance of dying at any time during their life span. This phenomenon is seen in the American robin, the gull, and laboratory populations of organisms that reproduce asexually, such as hydra and bacteria. The *concave* curve is characteristic of organisms that produce large numbers of offspring that receive little parental care, being largely left to compete on their own. The death rate is very high among the offspring, but those that reach adulthood have a good chance of surviving to old age. Most invertebrates, most plants, and many fish exhibit such early-loss survivorship curves. In some populations of black-tailed deer, 75% of the population dies within the first 10% of its life span, giving this mammalian population an early-loss curve as well.

5 How Is the Human Population Changing?

Your first child received free schooling and free medical care, and you were offered longer maternity leave and a larger pension if you signed a promise to have no more children. Pregnant with your second child, you face fines equivalent to all the benefits provided to you for the first child, up to a full-year's pay. Each night, delegates come to your home to tell you of the harm you are doing to society. Your neighbors shun you, and the collective pressures finally force you to the hospital for a state-

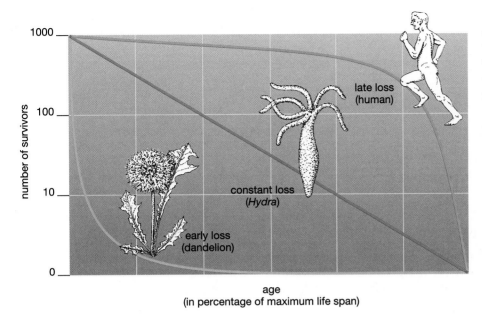

Figure 38-12 Population patterns in time
Three types of survivorship curve are shown. Because the life spans of these organisms differ so much, to compare them on the same graph, the percentages of survivors rather than the ages are used.

sponsored abortion. A distant future scenario from a pessimistic science fiction novel? Hardly. Similar policies were instituted in China in the 1980s. With a population of over 1 billion and an increase of more than 15 million people in 1983 alone, the government took drastic measures. Although China's coercive policies have been eased somewhat, women in China now average 1.8 children, less than 2.0 average for women in the United States. Nonetheless, China's current population of 1.2 billion still increases by over 12 million annually. What is China's demographic future? Because couples in China, on the average, are not replacing themselves, why does China's population continue to grow? Is the U.S. population increasing, and if so, is the increase a cause for concern? Let's examine the human population in light of what we know of exponential growth and carrying capacity.

The Human Population Is Growing Exponentially

Compare the graph of human population growth in Figure 38-13 with the exponential growth curves in Figure 38-2. The time spans are different, but each has the J-shape characteristic of exponential growth. It took over 1 million years for the human population to reach 1 billion, but the second billion was added in just 100 years, the third billion in 30 years, the fourth billion in 15 years, and the fifth billion in 12 years. The 1998 world population of more than 5.9 billion is projected to reach 6 billion in 1999. World population currently grows by about 86 million yearly—over 235,000 people every day, or 1.6 million per week! Why hasn't environmental resistance put an end to our exponential growth? What is the carrying capacity of the world for humans?

Like all populations, humans have encountered environmental resistance, but unlike other populations, we have responded to environmental resistance by overcoming it. As a result, the human population has grown exponentially for an unprecedented time span. To accommodate our growing numbers, we have altered the face of the globe. Human population growth has been spurred by a series of "revolutions" that conquered various aspects of environmental resistance and increased Earth's carrying capacity for people.

Technological Advances Have Increased Earth's Carrying Capacity for Humans

Primitive peoples produced a *cultural revolution* when they discovered fire, invented tools and weapons, built shelters, and designed protective clothing. Tools and weapons allowed more effective hunting and an increased food supply; shelter and clothing increased the habitable areas of the globe.

Domesticated crops and animals supplanted hunting and gathering by about 8000 B.C. This *agricultural revolution* provided people with a larger, more dependable food supply and further increased Earth's carrying capacity for humans. Increased food resulted in a longer life span and more childbearing years, but a high death rate from disease still restricted the population.

Human population growth continued slowly for thousands of years until the *industrial–medical revolution* began in England in the mid-eighteenth century, spreading through Europe and North America in the nineteenth century. Medical advances dramatically decreased the death rate by reducing environmental resistance from disease. These advances included the discovery of bacteria and their role in infection, leading to the control of bacterial diseases through improved sanitation and the use of antibiotics. Another advance was the discovery of viruses, leading to the development of vaccines for diseases such as smallpox. The revolution continues today as research proceeds on vaccines against major killers such as malaria and AIDS, and sophisticated medical procedures, such as coronary bypass operations and organ transplants, are improved and new procedures are developed.

In developed countries, such as those of Western Europe, the industrial–medical revolution resulted in an initial rise in population due to decreased deaths, followed by a decline in birth rates. This decline can be attributed to many factors, including better education, increased

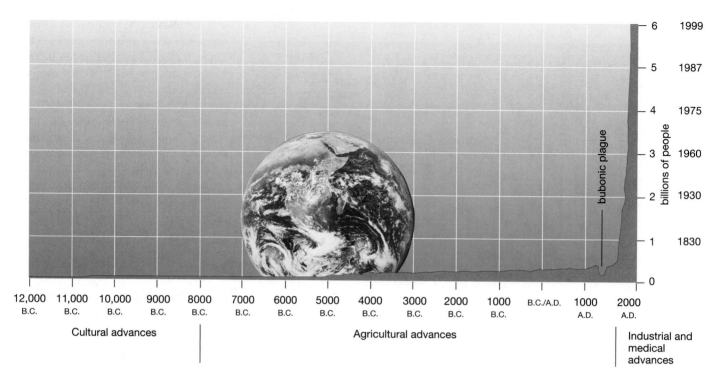

Figure 38-13 Human population growth
The human population from the Stone Age to the present has shown continued exponential growth. Note the dip in the fourteenth century caused by the bubonic plague. Note also the steadily decreasing time intervals at which additional billions are added. Inset: Earth is an island of life in a sea of emptiness; its space and resources are limited.

availability of contraceptives, a shift to a primarily urban lifestyle, and more career options for women. In developed countries, on the average, populations have more or less stabilized.

In developing countries, such as in most of Central and South America, Africa, and Asia, medical advances have decreased death rates and increased the life span, but birth rates remain high. These countries have not experienced the increase in wealth that was partly responsible for the birth rate decline in developed countries. Children serve as a form of social security in developing nations, because children may be the only support for parents in their old age. In agricultural societies, children are an important source of labor. Social traditions may offer prestige to the man who fathers and to the woman who bears many children. In Nigeria, the most crowded country in Africa, many men refuse to allow any form of birth control, many women desire large families, and the average woman bears six children. Nigeria is already suffering from loss of forests, the spread of deserts, soil erosion, and water pollution. With nearly half of its population of 104 million under the age of 15, continued population growth is a certainty. Although a majority of African nations have concluded that their growth rates are too high and are working toward reducing them, lack of education and lack of access to contraceptives impede progress in curbing population growth. Of the projected 6.3 billion people on Earth in the year 2000, 5 billion will

reside in developing countries. The prospects for population stabilization—*zero population growth*—in the near future are nil, barring major catastrophes that might dramatically increase deaths (as AIDS is doing in some parts of Africa). The reason can be seen clearly by looking at the age structures of the developing countries and comparing them with countries with stable populations, as we describe next.

The Age Structure of a Population Predicts Its Future Growth

The **age structure** of a population is the distribution of males and females of each age group, which can be shown graphically in an age-structure diagram. All age-structure diagrams rise to a peak at the top, because few people live into their nineties. The shape of the rest of the diagram, however, shows whether the population is expanding, stable, or shrinking (Fig. 38-14). If the number of children (age 0 to 14) exceeds the number of reproductive-age individuals (age 15 to 45), the population is expanding and the diagram resembles a pyramid. In the case of stable populations, the number of children is about equal to the number of reproducing adults, forming a nearly straight-sided figure. In shrinking populations, there are fewer children than reproducing adults and the age-structure diagram is narrow at the base.

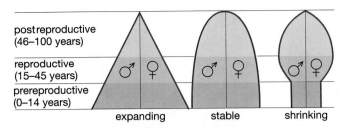

Figure 38-14 Generalized age-structure diagrams
Expanding, stable, and shrinking populations. The width of each figure is proportional to the number of individuals, with the male population on the left and the female population on the right. The vertical axis shows increasing age. The number of children (prereproductive, age 0 to 14 years) produced by the reproductive-age population determines whether the population is expanding (more children than reproductive-age adults), stable (children equal to reproductive-age adults), or shrinking (fewer children than reproductive-age adults).

Figure 38-15 shows the average age structures of the populations of developed and developing countries. The outermost boundaries represent the projected population structure for the year 2025; the inner ones are actual values for 1995. Each graph has been divided into three parts to show individuals who are prereproductive (age 0 to 14), reproductive (15 to 45), and postreproductive (46 and older). In 1997, the developing countries (in Asia, Africa, India, and South and Central America) had an average annual growth rate of 2.1%, and the developed countries (in North America, Europe, Australia, Japan, and New Zealand) showed an average annual growth rate of 0.1%. In the developed countries, projections for the year 2025 are only slightly higher than present levels (see Fig. 38-15a).

In contrast, each year in the developing countries, increasing numbers of people enter their reproductive years and give birth to an ever-increasing base of infants.

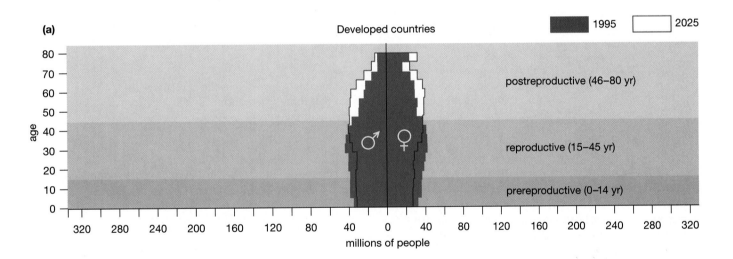

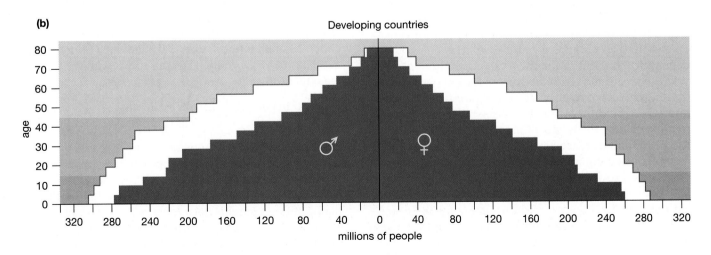

Figure 38-15 Actual age-structure diagrams
(*a*) Developed countries and (*b*) developing countries. Compare these with the idealized diagrams shown in Figure 38-14. The outer lines are projections to the year 2025.

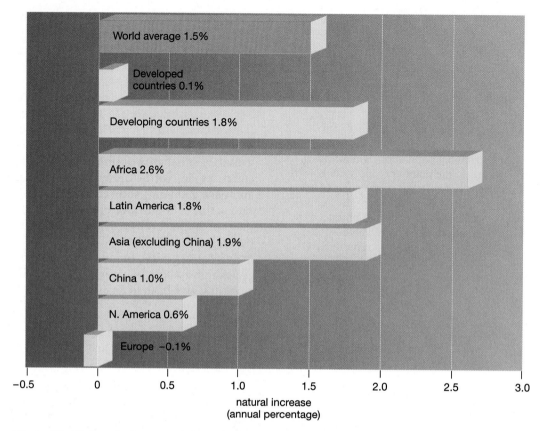

Figure 38-16 Population growth by world regions
Growth rates shown are due to natural increase (births – deaths) expressed as the percentage increase per year for various regions of the world. The bar that represents Europe includes the former Soviet Union. These figures do not include immigration or emigration.

Even if these countries were immediately to reach **replacement-level fertility (RLF)**—that is, if people of reproductive age had only enough children to replace themselves—population growth would continue for decades because of the built-in momentum produced by increasing numbers of people reaching reproductive age. This momentum fuels China's population growth, even though its fertility rate is below replacement level.

Study the predicted age-structure in 2025 for developing countries. Can you detect a trend that indicates the beginnings of stabilization for these populations? Compare the size of the reproductive population (parents) to the prereproductive population (children). The excess number of children over parents is expected to become smaller than in 1990s, indicating that these populations are beginning to reach RLF. Nevertheless, as the huge numbers of young people advance up the age pyramid, population growth will continue. The percentage growth per year caused by natural increase (births – deaths) for various parts of the world is shown in Figure 38-16.

Ironically, population growth in developing countries helps perpetuate the poverty and ignorance that in turn tend to sustain high birth rates. The relationship of income and education to birth rate has been documented in the United States. Here, on the average, women who do not complete high school have twice as many children as do those with more than 4 years of college.

The U.S Population Is Growing Rapidly

As shown in Figure 38-17, the U.S. population is growing exponentially. In fact, at nearly 1% annually, our growth rate is several times the average in the developed world. During 1997, for example, the U.S. population rose by about 2.3 million (more than 250 people per hour), bringing the total to more than 268 million. Recall the following equation:

population change = (births – deaths)
+ (immigrants – emigrants)

Let's examine each component of the equation to determine why the United States is growing so rapidly.

If each American woman had 2.1 children, we would have replacement-level fertility. RLF is slightly higher than 2 because parents must replace both themselves and children who die before reaching maturity. The U.S. birth rate in 1997 was just below replacement-level fertility, so why does the population continue to grow? Two factors are contributing to the rapid growth of the U.S. population: the "baby boom" and immigration.

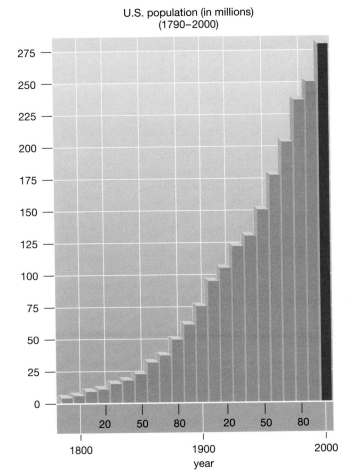

U.S. population (in millions)
(1790–2000)

Figure 38-17 *U.S. population growth*
Since 1790, U.S. population growth has shown the J-shaped curve characteristic of exponential growth. The estimate for the year 2000 is a conservative projection based on a 1.0% annual growth rate.

Part of the current U.S. growth rate is a legacy of the recent past. Parents of the late 1940s through the 1960s had families that were larger than replacement-level, resulting in a "baby boom" and a momentum in population growth that has not yet subsided. Even though women are averaging only 2 children each, because more women are having children, the U.S. population swells.

A second crucial component of the U.S. population equation is immigration, which contributes more to population growth here than in any other nation in the world. Legal immigration is well over 800,000 per year, causing about 30% of the total growth. Illegal immigration adds an estimated 275,000 people each year. According to Census Bureau projections, the U.S. population will not stabilize in the foreseeable future. The Census Bureau projects that by the year 2050, there will be 392 million Americans (about 125 million more than today) and the population will still be growing.

The rapid growth of the U.S. population has major environmental implications both for us and for the world. The average American uses five times as much energy as the global average, so with less than 5% of the world's population, we account for 25% of world energy use. Because people in developing countries use considerably less energy than the global average, the 2.3 million Americans added each year contribute more to climate change through the release of CO_2 and other greenhouse gases than do the 34 million people added annually by China and India combined.

When and how will human numbers ever stabilize? How many people can Earth support? There are no certain answers to these questions, but in "Earth Watch: Have We Exceeded Earth's Carrying Capacity?" we explore them in more detail.

Summary of Key Concepts

1) How Do Populations Grow?

Individuals join populations through birth or immigration and leave through death or emigration. The ultimate size of a stable population is the result of interactions among biotic potential, the maximum possible growth rate, and environmental resistance, which limits population growth.

All organisms have the biotic potential to more than replace themselves over their lifetime, resulting in population growth. Populations tend to grow exponentially, with increasing numbers of individuals added during each successive time period. Populations cannot indefinitely grow exponentially; they either stabilize or undergo periodic boom-and-bust cycles as a result of environmental resistance.

2) How Is Population Growth Regulated?

Environmental resistance restrains population growth by increasing the death rate or decreasing the birth rate. The maximum size at which a population may be sustained indefinitely by an ecosystem is termed the carrying capacity, determined by limited resources such as space, nutrients, and light. Environmental resistance generally maintains populations at or below the carrying capacity.

Population growth is restrained by density-independent forms of environmental resistance (such as weather) and density-dependent forms of resistance (including predation, parasitism, and competition).

3) How Are Populations Distributed?

Populations can be classified into three major types of distribution: clumped, uniform, and random. Clumped distribution may occur for social reasons or around limited resources. Uniform distribution is normally the result of territorial spacing. Random distribution is rare, occurring only when individuals do not interact socially and when resources are abundant and evenly distributed.

4) What Survivorship Patterns Do Populations Exhibit?

Populations show specific survivorship curves that describe the likelihood of survival at any given age. Convex (late-loss) curves are characteristic of long-lived species with few offspring, which receive parental care. Species with constant-

Earth Watch
Have We Exceeded Earth's Carrying Capacity?

A glance at the age structure of developing countries, where most of the world's population resides, shows a tremendous momentum for continued growth. World population in the year 2010 is predicted to be about 7 billion and growing. A modest United Nations projection is that the human population may stabilize in the year 2150 at 11.5 billion (Fig. E38-3). Can Earth support nearly twice its current population?

No one knows. Estimates of Earth's carrying capacity for humans have ranged from 3 billion to 44 billion people. Each of these estimates is based on major assumptions regarding technological developments and lifestyles. Earlier we defined carrying capacity as the maximum population that could be indefinitely sustained. This sustainability requires that the ecosystem not be damaged in ways that lower its ability to provide necessary resources. By this definition, we may have already exceeded Earth's carrying capacity for humans.

The upper limit of the planet's carrying capacity is determined by the ability of its plant life to harvest the energy from sunlight and produce high-energy molecules that other organisms can use as food. Stanford University biologist Peter Vitousek estimates that human activities have already reduced the productivity of Earth's forests and grasslands by 12%. Each year, millions of acres of once-productive land are turned into desert through overgrazing and deforestation, especially in developing countries (Fig. E38-4a). In a world where over 800 million people are chronically undernourished (Fig. E38-4b),

80% of the world's agricultural land is suffering moderate to severe erosion. The quest for more agricultural land is leading to deforestation and attempts to farm land that is poorly suited for agriculture. These actions contribute to the destruction of 40 million acres of rain forest annually and the extinction of enormous numbers of undescribed species. Each year the United States loses nearly half a million acres of prime farmland to development for homes, shopping malls, and roads. Already, most countries must import the grain they need from countries such as the United States. As our own growing population spreads onto our farmland, our ability to export grain to help sustain other nations will diminish, while their needs increase.

The demand for wood in developing countries is far outstripping production, and large areas are being deforested annually. Deforestation in turn causes erosion of precious topsoil, runoff of much-needed fresh water, and the spread of deserts. The total world fish harvest peaked in 1989 and has been declining since, despite increased investments in fishing equipment and fish farming. Almost two-thirds of commercial fish populations in both the Atlantic and Pacific Oceans have been fully exploited or overfished, and many fisheries, such as the New England cod harvest, have collapsed because of overfishing. These are clear indications that our present population, at its present level of technology, is already "overgrazing" the world ecosystem.

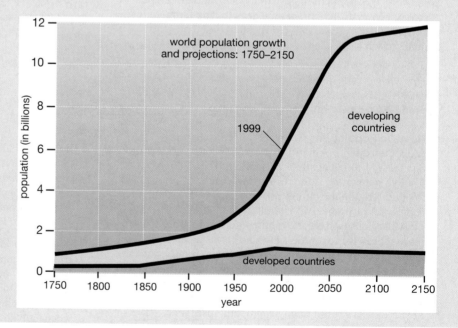

Figure E38-3 World population projected to the year 2150
The disparity in numbers between developed and developing countries is projected to increase even further than at present. (Modified from the Population Reference Bureau's data sheet: "World Population and the Environment," 1997.)

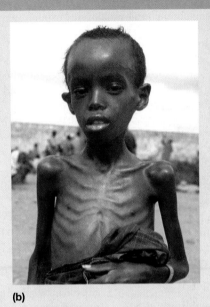

(a)   (b)

Figure E38-4 **Desertification**
(a) Human activities, including overgrazing livestock, deforestation, and poor agricultural practices, convert once-productive land into barren desert. *(b)* The loss of productive land, when combined with an expanding human population, can lead to tragedy.

There are some who predict that technological advances will continue to increase Earth's carrying capacity for humans almost indefinitely. In evaluating this forecast, keep in mind that technology takes time, wealth, and education to develop and implement. Rapid population growth in developing countries increases poverty, overloads the educational system, and hampers technological development while making its need more urgent. Ironically, the countries that so desperately need these predicted technological advances are often unable even to take advantage of today's level of technology. The high-yield crops developed in the 1960s have helped considerably but have had less than full impact, because they require expensive equipment, fertilizer, pesticides, and irrigation water to cultivate. Some developing countries cannot afford these items, and most will not be able to afford the new bioengineered crops currently under development.

In estimating how many people Earth can—or should—support, we must keep in mind that humans desire more out of life than a minimum caloric intake each day. Do we want to be able to eat meat, drive a car, live in a single-family dwelling, walk in a wilderness, and know that somewhere eagles, pandas, and elephants are living free? For people living at Earth's carrying capacity, these will probably be unattainable luxuries. On the basis of the consumption of resources by the average American, William Rees of the University of British Columbia has estimated that for all of Earth's nearly 6 billion people to live as many Americans do would require the resources of three additional Earth-sized planets.

Hope for the future lies in using the intelligence that has allowed us to overcome environmental resistance to see signs of overgrazing and to act before we have irrevocably damaged our biosphere. The human population *will* stop its exponential growth. Either we will voluntarily reduce our birth rate, or various forces of environmental resistance such as disease and starvation will increase our death rate; the choice is ours. Facing the problem of how to limit births is politically and emotionally difficult, but continued failure to do so will be disastrous. Our dignity, our intelligence, and our role as self-appointed stewards of life on Earth demand that we make the decision to halt population growth before we have permanently reduced Earth's ability to support all life, including our own.

loss curves have an equal chance of dying at any age. Concave (early-loss) curves are typical of organisms that produce numerous offspring, most of which die.

5) How Is the Human Population Changing?

The human population has exhibited exponential growth for an unprecedented time by overcoming certain aspects of environmental resistance and increasing Earth's carrying capacity for humans. This has been accomplished by the use of tools, agriculture, industry, and medical advances. Age-structure diagrams depict numbers of males and females in various age groups that comprise a population. Expanding populations have pyramidal age structures, stable populations show rather straight-sided age structure, and shrinking populations are illustrated by age structures that are constricted at the base.

Today, most of the world's people live in developing countries with rapidly expanding populations, where a variety of social and cultural conditions encourage large families. The United States is the fastest growing of the developed countries, owing to high immigration rates and the baby boom of the 1940s through the 1960s. Earth's carrying capacity for humans is unknown, but with a population of over 5.8 billion, resources are already too limited for all to be supported at a high standard of living. A steady decline in productive land and in wood and fish harvests suggests that we are already damaging our world ecosystem and decreasing its ability to sustain us.

Key Terms

age structure *p. 804*	**ecology** *p. 791*	**interspecific competition** *p. 799*	**predator** *p. 798*
biotic potential *p. 793*	**ecosystem** *p. 792*	**intraspecific competition** *p. 799*	**prey** *p. 798*
boom-and-bust cycle *p. 795*	**emigration** *p. 793*	**J-curve** *p. 793*	**random distribution** *p. 801*
carrying capacity *p. 796*	**environmental resistance** *p. 793*	**parasite** *p. 799*	**replacement-level fertility (RLF)** *p. 806*
clumped distribution *p. 801*	**exotic** *p. 796*	**parasitism** *p. 799*	**scramble competition** *p. 799*
community *p. 792*	**exponential growth** *p. 793*	**population** *p. 792*	**S-curve** *p. 796*
competition *p. 799*	**growth rate** *p. 793*	**population cycle** *p. 798*	**survivorship curve** *p. 802*
contest competition *p. 801*	**host** *p. 799*	**predation** *p. 798*	**uniform distribution** *p. 801*
density-dependent *p. 797*	**immigration** *p. 793*		
density-independent *p. 797*			

Thinking Through the Concepts

Multiple Choice

1. *Which of the following factors is* not *an example of density-dependent environmental resistance?*
a. weather b. competition
c. predation d. parasitism
e. lack of food

2. *For exponential growth to occur, it is necessary that*
a. there is no mortality
b. there are no density-independent limits
c. the birth rate consistently exceed the death rate
d. a species is very fast-reproducing
e. the species is an exotic invader in an ecosystem

3. *Which of the following currently contributes most to human population growth within the United States?*
a. the consequences of the baby boom
b. immigration
c. a birth rate above RLF
d. both a and b
e. all of the above

4. *Which is the most common type of spatial distribution?*
a. logistic b. uniform
c. random d. exponential
e. clumped

5. *Which continent has the highest rate of human population growth?*
a. North America b. Africa c. Asia
d. Australia e. South America

6. *If a population exceeds its carrying capacity,*
a. it must immediately crash
b. it can remain stable at this level indefinitely
c. it will continue to increase for the indefinite future
d. it must decline sooner or later
e. the food supply will increase to support it

? Review Questions

1. Define *biotic potential* and *environmental resistance*.

2. Draw the growth curve of a population before it encounters significant environmental resistance. What is the name of this type of growth, and what is its distinguishing characteristic?

3. Distinguish between density-independent and density-dependent forms of environmental resistance.

4. Describe (or draw a graph illustrating) what is likely to happen to a population that far exceeds the carrying capacity of its ecosystem. Explain your answer.

5. List three density-dependent forms of environmental resistance, and explain why each is density-dependent.

6. Distinguish between populations showing concave and convex survivorship curves. Which is characteristic of Americans, and why?

7. Given that the U.S. birth rate is currently at replacement-level fertility, why is our population growing?

Applying the Concepts

1. Explain natural selection in terms of biotic potential and environmental resistance.
2. The United States has a long history of accepting large numbers of immigrants. Discuss the implications of immigration for population stabilization.
3. What factors encourage rapid population growth in developing countries? What will it take to change this growth?
4. Contrast age structure in rapidly growing versus stable human populations. Why is there a momentum in population growth built into a population that is above RLF?
5. Why is the concept of carrying capacity difficult to apply to human populations? In reference to human population, should the concept be modified to include quality of life?

Group Activity

Work in groups of three or four. Prepare a statement to be read and defended to the other groups concerning U.S. population. Included in the statement must be current information on population statistics and patterns, using the textbook, the Internet, or other sources. Include also a statement on their policy concerning immigration (how much, how little, how come, . . .) and government or voluntary family planning and education. Address the following questions:

- Should the United States limit its population size; why or why not?
- Should there be mandatory or voluntary birth control, and why?
- How would you convince people to limit births?
- What is the importance of the education of girls and women on population growth?
- Should immigration be limited; why or why not? (If yes, explain how you would do this.)
- Will your plan address issues such as limited resources, quality of life, environmental disasters, and pollution? Explain your plan's implications, on the basis of the projections noted in the textbook, in terms of age structure and numbers of people expected by 2025.

Although there will undoubtedly be disagreement in your group, come to a consensus to prepare your statement. (You can agree to disagree for now.) The accompanying graph gives projections for the U.S. population for both a two-child and a three-child family growth rate.

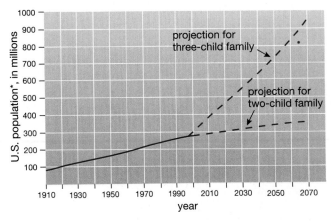

*Data based on "Population: the U.S. Problem, the World Crisis," Supplement to *The New York Times*, August 14, 1974.

When you are finished, meet with another group. For your group presentation, each group first reads its statement while the other group listens and takes down any notes. After each group has read its statement, questions can be asked of the other group and clarification sought. Compare areas where there is agreement or disagreement. After discussion, prepare a brief report of the similarities and differences between your group and the other, and submit that report with your own report to your instructor. Be prepared to be called on to share your group's ideas for class discussion.

For More Information

Bender, W., and Smith, M. "Population, Food, and Nutrition." *Population Bulletin*, February 1997. A comprehensive and up-to date summary of issues regarding food supply and the growing human population.

Bongaarts, J. "Can the Growing Human Population Feed Itself?" *Scientific American*, March 1994. A balanced account of the arguments between optimists and pessimists concerning the future world food supply.

Ehrlich, P. R., and Erlich, A. *Extinction*. New York: Random House, 1981. The causes and consequences of the disappearance of species.

Kates, R. W. "Sustaining Life on the Earth." *Scientific American*, October 1994. Provides a prescription for an environmentally sustainable future.

Kunzig, R. "Twilight of the Cod." *Discover*, April 1995. Documents the collapse of the cod fishery from Cape Cod to Newfoundland.

Korpimaki, E., and Krebs, C. J. "Predation and Population Cycles of Small Mammals." *Bioscience*, November 1996. A review of recent studies designed to evaluate cycles of predators and their prey.

Myers, J. H., and Krebs, C. J. "Population Cycles in Rodents." *Scientific American*, June 1974. Natural populations of small rodents show periodic 3- to 4-year cycles in population size.

Raloff, J. "The Human Numbers Crunch." *Science News*, June 22, 1996. Reviews the challenges for the next 50 years as humans grow in number.

Safina, C. "The World's Imperiled Fish." *Scientific American*, November 1995. The collapse of many fisheries clearly illustrates that wild fish populations cannot sustain themselves against modern fishing techniques.

Answers to Multiple-Choice Questions
1. a   2. c   3. d   4. e   5. b   6. d

Predation is an important community interaction that selects for many specialized adaptations, such as the sharp eyes of the red-tailed hawk.

Community Interactions 39

Summer vacation—a backpacking trip in the Rocky Mountains! You start your journey in low shrubland. Grasshoppers leap out from hiding spots and startle you with a loud clacking sound as they fly. A beautiful monarch butterfly alights on a milkweed plant near the trail, and a yellow-jacket wasp buzzes past your face. Your companion suddenly shouts and points upward. A red-tailed hawk has begun a dive-bombing descent. You hear a rustling as it lands and, a few seconds later, see it rise again with a small animal in its talons. A half hour later you reach the forest—pines surround you, and the cool shade is welcome. The grasses and sagebrush disappear, and the ground becomes more open. A mother deer and two spotted fawns watch you warily. As you reach a beautiful mountain lake later in the day, you are pleased to find it surrounded by open meadow—a great place to bask in the sun and have lunch. The high point of your trip is a 13,000-foot peak. Long before you reach the top, the forest gives way to stunted trees, then grasses, flowers, and low-growing shrubs. The views are panoramic and spectacular. In the far distance, you see a large plume of smoke—a forest fire! You watch, waiting to see helicopters dropping fire-extinguishing chemicals, but none appear. Later in the afternoon the sky darkens, and distant thunder rumbles. As lightning flashes in the distance and you hurry down to a lower elevation among the trees, you wonder about the fire. Did lightning start it? Will the thunderstorm put it out? As you finally leave the wilderness, exhausted but refreshed, you resolve to come back sometime in the future—perhaps with your children.

If you do return to your mountain lake in 20 years, will it be the same? Why did you find different communities of plants as you hiked up the mountain? Why do monarch butterflies seem especially attracted to milkweed? Why are yellow jackets brightly colored but the grasshoppers green and brown? What environmental pressures have, by natural selection, produced hawks that can spot prey from hundreds of feet in the air, and produced fawns with dappled coats? And why didn't someone do something about that forest fire? Let's find out.

Net Watch
www
On-line resources for this chapter are on the World Wide Web at:
http://www.prenhall.com/audesirk
(click on the Table of Contents link and then select Chapter 39).

1) Why Are Community Interactions Important?

An ecological **community** consists of all the interacting populations within an ecosystem. In the previous chapter, you learned that community interactions such as predation, parasitism, and competition help limit the size of populations. The interacting web of life that forms a community tends to maintain a balance between resources and the numbers of individuals consuming them. When populations interact with one another, influencing each other's ability to survive and reproduce, they serve as agents of natural selection. For example, in killing prey that are easiest to catch, predators leave behind those individuals with better defenses against predation. These individuals leave the most offspring, and over time their inherited characteristics increase within the prey population. Thus, as community interactions limit population size, they simultaneously shape the bodies and behaviors of the interacting populations. This process by which two interacting species act as agents of natural selection on one another over evolutionary time is called **coevolution.**

You have probably heard the expression "the balance of nature." This balance is the result of community interactions finely tuned to one another over evolutionary time. The sometimes fragile balance can be overturned when organisms are introduced into an ecosystem where they did not evolve, as described in "Earth Watch: The Invasion of Fire Ants and Kudzu" (p. 828).

The most important community interactions are competition, predation, parasitism, and mutualism, whose effects on participating species are shown in Table 39-1. Their importance is illustrated by the adaptations that have evolved under environmental pressures exerted by these interactions over evolutionary time, as described in the following sections.

2) What Are the Effects of Competition among Species?

Community interactions may be harmful or beneficial to the participating organisms (Table 39-1). During competition among members of different species, called **interspecific competition,** two or more species attempt to use the same limited resources, particularly food and/or space. In interspecific competition, each species involved is harmed, because access to resources is reduced. The intensity of interspecific competition depends on how similar the requirements of the species are. In other words, the degree of competition is proportional to the amount of overlap in the *ecological niches* of the competing species.

The Ecological Niche Defines the Place and Role of Each Species in Its Ecosystem

Although the word *niche* may call to mind a small cubbyhole, in ecology it means much more. Each species occupies a unique **ecological niche** that encompasses all aspects of its way of life. An ecological niche includes the organism's physical home or habitat. The habitat of a white-tailed deer, for example, is the eastern deciduous forest. In addition, the niche includes all the physical environmental factors necessary for survival, such as the range of temperatures under which the organism can survive, the amount of moisture it requires, the pH of the water or soil it may inhabit, the type of soil nutrients required, and the degree of shade it can tolerate. Although different types of organisms share many aspects of their niche with others, no two species ever occupy exactly the same ecological niche.

The ecological niche extends well beyond habitat. It also specifies how the organism gets it supply of energy and materials—what might be called the organism's role or "occupation" within its ecosystem. An organism's predators, prey, and competitors as well as its behaviors and interactions with other organisms are considered elements of its niche.

Adaptations Reduce the Overlap of Ecological Niches among Coexisting Species

Just as no two organisms can occupy exactly the same physical space at the same time, no two species can inhabit exactly the same ecological niche simultaneously and continuously. This important concept, often called the **competitive exclusion principle,** was formulated in 1934 by the Russian microbiologist G. F. Gause. If two species with the same niche were placed together and forced to compete for limited resources, inevitably one would outcompete the other, and the less well-adapted of the two would die out. Gause used two species of the protist *Paramecium, P. aurelia* and *P. caudatum,* to demon-

---

Table 39-1 Interactions among Organisms

Type of Interaction	Effect on Organism A	Effect on Organism B
Competition between A and B	Harms	Harms
Predation by A on B	Benefits	Harms
Symbiosis		
Parasitism by A on B	Benefits	Harms
Commensalism of A with B	Benefits	No Effect
Mutualism between A and B	Benefits	Benefits

strate this principle. In laboratory flasks, both species thrived on bacteria and fed in the same parts of the flask. Grown separately, each population flourished (Fig. 39-1a). But when Gause placed the two species together in a flask, one always eliminated, or "competitively excluded," the other (Fig. 39-1b). Gause then repeated the experiment, replacing *P. caudatum* with a different species, *P. bursaria*, which tended to feed in a different part of the flask. In this case, the two species of *Paramecium* were able to coexist indefinitely, because they occupied slightly different niches.

The ecologist R. MacArthur tested Gause's laboratory findings under natural conditions by investigating five species of North American warbler. All these birds hunt for insects and nest in the same type of spruce tree. Although the niches of these birds appear to overlap considerably, MacArthur found that each species concentrates its search in specific areas of the tree, employs different hunting tactics, and nests at slightly different times. By partitioning the resource, the warblers minimize the overlap of their niches and reduce competition among the different species (Fig. 39-2).

As MacArthur discovered, when two species with similar requirements coexist, they typically occupy a smaller niche than either would if it were by itself. This phenomenon, called **resource partitioning**, is an evolutionary adaptation that reduces the harmful effects of interspecific competition. Resource partitioning is the

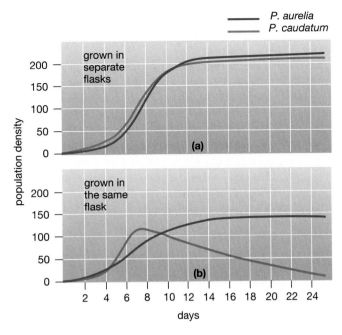

Figure 39-1 Competitive exclusion
(a) Raised separately with a constant food supply, both *Paramecium aurelia* and *P. caudatum* show the S-curve typical of a population that initially grows rapidly, then stabilizes.
(b) Raised together and forced to occupy the same niche, *P. aurelia* consistently outcompetes *P. caudatum* and causes that population to die off. (Modified from G. F. Gause, *The Struggle for Existence*. Baltimore: Williams & Wilkins, 1934.)

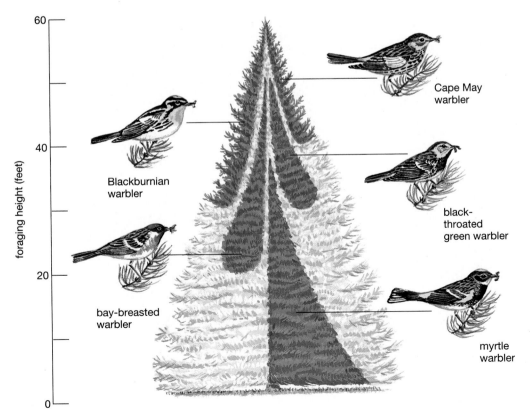

Figure 39-2 Resource partitioning
Each of these five insect-eating species of North American warblers searches for food in different regions of spruce trees.

outcome of the coevolution of species with extensive but not total niche overlap. Because natural selection favors individuals with fewer competitors, over evolutionary time the competing species develop physical and behavioral adaptations that minimize their competitive interactions. A dramatic example of resource partitioning was discovered by Charles Darwin among finches of the Galapagos Islands. The finches had evolved different bill sizes and shapes and different feeding behaviors that reduced the competition among them (see Chapter 16).

Competition Helps Control Population Size and Distribution

Individuals of the same species have essentially identical requirements for resources and occupy the same ecological niche. For this reason, *intra*specific competition is a major factor in controlling population size. Although natural selection helps minimize niche overlap among species, the species that are closely related or have similar needs still compete directly for limited resources. This interspecific competition may restrict the size and distribution of the competing populations.

A classic study of the effects of *inter*specific competition was performed by the ecologist J. Connell, using barnacles (crustaceans that attach permanently to rocks and other surfaces). Barnacles of the genus *Chthamalus* share the rocky shores of Scotland with another genus, *Balanus*, and their niches overlap considerably. Both genera live in the **intertidal zone**, an area of the shore that is alternately covered and exposed by the tides. Connell found that *Chthamalus* dominates the upper shore and *Balanus* dominates the lower. When he scraped off *Balanus*, the *Chthamalus* population increased, spreading downward into the area its competitor had once inhabited. Where the habitat is appropriate for both genera, *Balanus* conquers, because it is larger and grows faster. But *Chthamalus* tolerates drier conditions, so on the upper shore, where only high tides submerge the barnacles, it has a competitive advantage. As this example illustrates, interspecific competition limits both the size and the distribution of the competing populations.

3 What Are the Results of Interactions between Predators and Their Prey?

www

You probably think of a predator as an organism that kills and eats other organisms. This is accurate, but ecologists may include herbivorous animals in this general category because herbivores can have a major influence on the size and distribution of plant populations. We will define predation in its broadest sense, to include the cow grazing on prairie grass and the bat pursuing a moth (Fig. 39-3), the moth feeding on cactus (see Fig. 38-9), the

(a)

(b)

Figure 39-3 *Forms of predation*
(a) A cow grazes on prairie grass. The tough stems of grass have evolved under predation pressure by herbivores. *(b)* A bat uses a sophisticated echolocation system to hunt this moth, which has evolved special sound detectors and behaviors to avoid it.

jumping spider stalking its prey (see Fig. 39-12a), and the hawk as it swoops down on a small mammal (see the chapter-opening photograph). Most predators are either larger than their prey or hunt collectively, as wolves do when bringing down a moose (see Fig. 38-8). Predators are generally less abundant than their prey; in the next chapter, we will explain why.

Predator–Prey Interactions Shape Evolutionary Adaptations

To survive, predators must feed and prey must avoid becoming food. Therefore, predator and prey populations exert intense environmental pressure on one another, resulting in coevolution. As prey become more difficult to catch, predators must become more adept at hunting. Environmental pressure among predators and prey has produced the dappled color of the fawn. It has endowed the cheetah with speed and camouflage spots (see Fig. 39-7a) and its zebra prey with speed and camouflage stripes. It has produced the keen eyesight of the hawk and the warning call of the Belding ground squirrel, the stealth of the jumping spider and the remarkable spider mimicry of the fly it stalks. In the following sections, we examine some of the evolutionary results of predator–prey interactions.

Bats and Moths Have Evolved Counteracting Strategies

Most bats are nighttime hunters that navigate and locate prey by echolocation. They emit extremely high-frequency and high-intensity pulses of sound and, by analyzing the returning echoes, create an "image" of their surroundings and nearby objects. Under environmental pressure from this specialized prey-location system, certain moths (a favored prey of bats; Fig. 39-3b) have evolved simple ears that are particularly sensitive to the frequencies used by echolocating bats. When they hear a bat, these moths take evasive action, flying erratically or dropping to the ground. The bats may counter this defense by switching the frequency of their sound pulses away from the moth's sensitivity range. Some moths have evolved a way to jam the bats' echolocation mechanism by producing their own high-frequency clicks. In response, when hunting a clicking moth, a bat may turn off its own sound pulses temporarily and zero in on the moth by following the moth's clicks. These interactions dramatically illustrate the complexity of coevolutionary adaptations.

Camouflage Conceals Both Predators and Their Prey

An old saying of detective novels is that the best hiding place may be right out in plain sight! Both predators and prey (such as the dappled fawns and grasshoppers you encountered on your backpacking trip) have evolved colors, patterns, and shapes that resemble their surroundings. Such disguises, called **camouflage**, render

(a)

(b)

Figure 39-4 Camouflage by blending in
(a) The sand dab is a flattened, bottom-dwelling ocean fish with a mottled color that closely resembles the sand on which it rests. *(b)* This nightjar on its nest in Belize is barely visible among the surrounding leaf litter.

animals inconspicuous even when they are in plain sight (Fig. 39-4).

Some animals closely resemble specific objects such as leaves, twigs, bark, thorns, or even bird droppings (Fig. 39-5). Camouflaged animals tend to remain motionless rather than to flee their predators; a fleeing bird dropping would be quite conspicuous! Whereas many camouflaged animals resemble plants, a few types of plants have evolved to resemble rocks, which are ignored by their herbivorous predators (Fig. 39-6).

Predators who ambush are also aided by camouflage. For example, a spotted cheetah becomes inconspicuous in the grass as it watches for grazing mammals. The frogfish closely resembles the algae-covered rocks and algae on which it sits motionless, dangling a small lure from its upper lip (Fig. 39-7). Small fish notice only the lure and are engulfed as they approach it.

Figure 39-6 *A plant that mimics a rock*
This cactus of the American Southwest is appropriately called the "living rock cactus."

Figure 39-5 *Camouflage by resembling specific objects*
Resembling uninteresting parts of the environment allows some animals to avoid predation. *(a)* A moth whose color and shape resemble a bird dropping sits motionless on a leaf. *(b)* The leafy sea dragon (an Australian "seahorse" fish) has evolved extensions of its body that precisely duplicate the algae in which it normally hides. *(c)* These Florida treehopper insects avoid detection by resembling thorns on a branch.

Figure 39-7 *Camouflage assists predators*
(a) As it waits for prey, a cheetah blends into the background of the grass. *(b)* Combining camouflage and aggressive mimicry, a frogfish waits in ambush, its camouflaged body matching the algae-encrusted rock on which it normally rests. Above its mouth dangles a lure, closely resembling a small fish. The lure attracts small would-be predators, who will find themselves prey.

Bright Colors Often Warn of Danger

Some animals have evolved very differently, exhibiting bright **warning coloration** (Fig. 39-8; see also Figs. 39-9, 39-13). These animals are usually distasteful, and many are poisonous, such as the yellow jacket with its bright yellow and black stripes. Because poisoning your predator is small consolation if you have already been eaten, the bright colors declare, "Eat me at your own risk!" A single unpleasant experience is enough to teach predators to avoid these conspicuous prey.

Some Organisms Gain Protection Through Mimicry

Mimicry refers to a situation in which a species evolves to resemble something else, typically another type of organism. For example, once warning coloration evolved, there arose a selective advantage for tasty, harmless animals to resemble poisonous ones. The deadly coral snake has brilliant warning coloration, and the harmless mountain king snake avoids predation by resembling it (Fig. 39-9a,b). By resembling each other, two distasteful species

Figure 39-8 Warning coloration
The South American poison arrow frog, with its poisonous skin, advertises its unpleasant taste with bright and contrasting color patterns.

Figure 39-9 Warning mimicry
The warning coloration of **(a)** the poisonous coral snake is mimicked by **(b)** the harmless mountain king snake.
(c) Mimicry of the warning coloration of the distasteful monarch butterfly by the equally distasteful viceroy.
(d) After attempting to eat either of these species, a bird will avoid them both.

(a)

(b)

Figure 39-10 Aggressive mimicry
(a) Some fish, such as this cleaner wrasse, obtain food by eating parasites off the bodies of larger fish, a mutualistic relationship called a *cleaning symbiosis*. Fish welcome the approach of this predator of parasites. *(b)* The saber-toothed blenny has evolved to resemble the cleaner wrasse closely. Fish that allow the blenny to approach usually get a chunk bitten out of them—hence the term *aggressive mimicry*.

may each benefit from predators' painful experience with the other. For example, predators rapidly learn to avoid the conspicuous stripes on bees, hornets, and yellow jackets. The monarch butterfly is poisonous and distasteful, and viceroy butterflies (which researchers have recently discovered are equally distasteful) have wing patterns strikingly similar to those of monarchs (Fig. 39-9c,d). A common color pattern results in faster learning by predators—and less predation on all similarly colored species.

Some predators have evolved **aggressive mimicry**, a "wolf-in-sheep's-clothing" approach to predation, in which they entice their prey to come close by resembling a harmless animal or part of the environment. For example, although the body of the frogfish resembles the algae-covered rocks where it lurks, it dangles a wriggling

lure that resembles a small fish just above its mouth (see Fig. 39-7). A curious fish attracted to the lure is quickly swallowed. The saber-toothed blenny fish imitates a harmless fish, called a cleaner wrasse, that picks parasites from the skin of larger fish. Parasite-ridden fish welcome the attention of the cleaner wrasse and are fooled by the saber-tooth blenny, which bites them instead of their parasites (Fig. 39-10).

Certain prey species use still another form of mimicry called **startle coloration**. Several insects, including certain moths and caterpillars, have evolved patterns of color that closely resemble the eyes of a much larger, and possibly dangerous, animal (Fig. 39-11). If a predator gets too close, the prey suddenly flashes its eye-spots, startling the predator enough to make an escape.

(a)

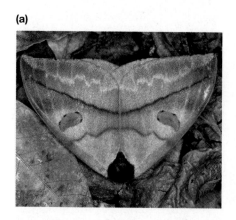

(b)

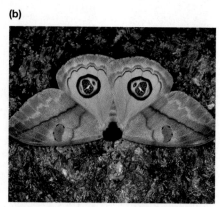

(c)

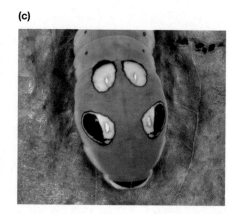

Figure 39-11 Startle coloration
(a) The peacock moth from Trinidad is well camouflaged, but should a predator approach too closely, *(b)* it suddenly opens its wings to reveal spots resembling large eyes. This startles the predator, giving the moth a chance to flee. *(c)* Would-be predators of this caterpillar larva of the swallowtail butterfly are deterred by its close resemblance to a snake. Note that the caterpillar's head is the "snake's nose."

(a)

(b)

Figure 39-12 Behavioral mimicry
(a) In response to the approach of a jumping spider, **(b)** the snowberry fly spreads its wings, revealing a pattern that resembles spider legs. The fly enhances the effect by performing a jerky, side-to-side dance that resembles the leg-waving display of a jumping spider defending its territory.

A sophisticated variation on the theme of prey who mimic predators is seen in snowberry flies, who are hunted by territorial jumping spiders. When a fly spots an approaching spider, it spreads its wings, moving them back and forth in a jerky dance. Seeing this display, chances are good that the spider will flee from the harmless fly. Why? Researchers recently observed that the markings on the fly's wings closely resemble the legs of another jumping spider. The jerky movements of the fly mimic those made by a jumping spider when it drives another spider from its territory (Fig. 39-12). Natural selection has finely tuned both the behavior and the appearance of the fly to avoid predation by jumping spiders.

Some Predators and Prey Engage in Chemical Warfare

Both predators and prey have evolved a variety of toxic chemicals for attack and defense. The venom of spiders and poisonous snakes such as the coral snake (see Fig. 39-9) serves both to paralyze prey and to deter its predators. Many plants produce defensive toxins. For example, lupines produce chemicals called alkaloids, which deter attack by the blue butterfly, whose larvae feed on the lupine's buds. In fact, different individuals of the same species of lupine produce different forms of alkaloids, thus making the evolution of resistance to it more difficult.

Certain mollusks, including squid, octopus, and some sea slugs, emit clouds of ink when attacked. These colorful chemical "smoke screens" confuse their predators and mask their own escape. A dramatic example of chemical defense is seen in the bombardier beetle. In response to the bite of an ant, the beetle releases secretions from special glands into an abdominal chamber. There, enzymes catalyze an explosive chemical reaction that shoots a toxic, boiling-hot spray onto the attacking ant (Fig. 39-13a).

Plants and Herbivores Have Many Coevolutionary Adaptations

Plants have evolved a variety of chemical adaptations that deter their herbivorous "predators." Many, such as the milkweed, synthesize toxic and distasteful chemicals. Animals rapidly learn not to eat foods that make them sick, and so milkweeds and other toxic plants suffer little nibbling. Consequently, such plants are often very abundant; any animal immune to the plant poisons enjoys a bountiful food supply. As plants evolved toxic chemicals for defense, certain insects evolved increasingly efficient ways to detoxify or even make use of the chemicals. The result is that nearly every toxic plant is eaten by at least one type of insect. For example, monarch butterflies lay their eggs on milkweed; when their larvae hatch, they consume the toxic plant (Fig. 39-13b). The caterpillars not only tolerate the milkweed poison but also store it in their tissues as a defense against their own predators. The stored toxin is even retained in the metamorphosed monarch butterfly (see Fig. 39-9).

Grasses have evolved tough silicon (glassy) substances in their blades, discouraging all herbivorous predators except those with strong, grinding teeth and powerful jaws, such as the cow in Figure 39-3a. Thus, grazing animals have come under environmental pressure for longer, harder teeth. An example is the coevolution of horses and the grasses they eat. On an evolutionary time scale, grasses evolved tougher blades that reduce predation, and horses evolved longer teeth with thicker enamel coatings that resist wear.

(a)

Figure 39-13 Chemical warfare
(a) The bombardier beetle sprays a hot toxic brew in response to a leg pinch. *(b)* A monarch caterpillar feeds on milkweed that contains a powerful toxin. Why do you think the caterpillar is colored with bright stripes instead of being green?

(b)

4) What Is Symbiosis?

Symbiosis, which literally means "living together," is defined as a close interaction between organisms of different species for an extended time. Considered in its broadest sense, symbiosis includes parasitism, mutualism, and commensalism. Although one species always benefits in symbiotic relationships, the second species may be unaffected, harmed, or helped (see Table 39-1). **Commensalism** is a relationship in which one species benefits while the other is relatively unaffected. Barnacles that attach themselves to the skin of a whale, for example, get a free ride through nutrient-rich waters without harming the whale. In parasitism and mutualism, in which both organisms are affected, the participants act on each other as strong agents of natural selection. Here we will discuss these two forms of community interaction.

Parasitism Harms, But Does Not Immediately Kill, the Host

In **parasitism**, one organism benefits by feeding on another. **Parasites** live in or on their prey, which are called *hosts*, normally harming or weakening them but not immediately killing them. Although it is sometimes difficult to distinguish clearly between a predator and a parasite, parasites are generally much smaller and more numerous than their hosts. Familiar parasites include tapeworms, fleas, and numerous disease-causing protozoa, bacteria, and viruses. Many parasites, particularly worms and protozoa, have complex life cycles involving two or more hosts (see Chapter 22). There are few parasitic vertebrates; the lamprey eel, which attaches itself to a host fish and sucks its blood, is a rare example.

The variety of infectious bacteria and viruses and the precision of the immune system that counters their attacks are evidence of the powerful forces of coevolution between parasitic microorganisms and their hosts. Consider the malaria parasite, which has provided strong environmental pressure for the defective hemoglobin gene in humans that causes sickle-cell anemia. The parasite can't survive in the affected red blood cells. In certain areas of Africa where malaria is common, up to 20% of the human population carries the sickle-cell gene. Another example is *Trypanosoma*, a parasitic protozoan that causes both human sleeping sickness and a disease in cattle called *nagana*. African antelope, which coevolved with this parasite, are relatively unaffected by it. Most infected cattle, a more recently introduced species, suffer but survive infection if they have been bred in an infested area for many generations. Newly imported cattle, however, generally die if not treated.

In Mutualistic Interactions, Both Species Benefit

When two species interact in a way that benefits both, the relationship is called **mutualism**. The mutualistic interactions between flowering plants and their pollinators are discussed in Chapter 25. Mutualistic associations occur in the digestive tracts of cows and termites, where protists and bacteria find food and shelter while helping their hosts extract nutrients, and in our own intestines, where bacteria synthesize certain vitamins. The nitrogen-fixing bacteria inhabiting special chambers on the roots of legume plants are another important example. The bacteria obtain food and shelter from the plant and in return trap nitrogen in a form the plant can utilize. Some mutualistic partners have coevolved to the extent that neither can survive alone. An interesting example is the ant–acacia mutualism described in "Scientific Inquiry: Ants and Acacias—An Unlikely Match."

Scientific Inquiry
Ants and Acacias—An Unlikely Match

Daniel Janzen, a doctoral student at the University of Pennsylvania, was walking down a road in Veracruz, Mexico, when he saw a flying beetle alight on a thorny tree, only to be driven off by an ant. Looking more closely, he saw that the tree, a bull's-horn acacia, was covered with ants. A large ant colony of the genus *Pseudomyrmex* made its home inside the enlarged thorns of the plant, whose soft pulpy interiors are easily excavated to provide shelter (Fig. E39-1).

To determine how important the ants are to the tree, Janzen began stripping the thorns by hand until he found and removed the thorn that housed the ant queen, thus destroying the colony. Later, he turned to more-efficient but dangerous methods, using an insecticide to eliminate all the ants on a large stand of acacias. The acacias were unharmed by the poison, Janzen became ill from it, and all the ants were killed. Within a year of the spraying, Janzen found nearly all the acacia trees dead, consumed by insects and other herbivores and shaded out by competing plants. The ground surrounding the trees, which the ants normally kept trimmed, was overgrown with vegetation. The trees were apparently dependent on their resident ants for survival.

Wondering if the ants could survive off the tree, Janzen painstakingly peeled the ant-inhabited thorns off 100 acacia trees; he suffered multiple stings in the process. Janzen housed each ant colony in a jar provided with local non-acacia vegetation and insects for food. All the ant colonies starved. Carefully examining the acacia, he found swollen structures filled with sweet syrup at the base of the leaves and protein-rich capsules on the leaf tips (Fig. E39-1, inset). Together, these materials provide a balanced diet for the ants.

Janzen's experiments strongly suggest that these species of ant and acacia have an obligatory mutualistic relationship: Neither can survive without the other. Of course, further observations were required to confirm this hypothesis. The fact that the ants starved in Janzen's jars did not rule out that they might survive successfully elsewhere, but, in fact, this species of ant is never found living independently. Similarly, the bull's-horn acacia is never found without its resident ant colony. Thus, a chance observation followed by careful research led to the discovery of an important mutualistic association.

Figure E39-1 A mutualistic relationship
Yellow, protein-rich capsules are produced at the tips of certain acacia leaves. These capsules provide food for the resident ants. **Inset:** A hole in the enlarged thorn of the bull's-horn acacia provides shelter for members of the ant colony. The ant entering the thorn is carrying a food capsule produced by the acacia. As the ant colony grows, more thorns are invaded.

Figure 39-14 Mutualism
The clown fish snuggles unharmed among the stinging tentacles of the anemone. The fish obtains protection and sometimes brings food to the anemone in this mutualistic relationship.

Mutualistic relationships involving vertebrates are rare and typically less intimate and extended, as in the relationship of the cleaner wrasse (see Fig. 39-10a) and the fish it cleans. The clown fish takes shelter among the venomous tentacles of the anemone, which are harmless to it. The fish derives shelter and protection and, at least occasionally, brings bits of food to its anemone host (Fig. 39-14).

5 How Do Keystone Species Influence Community Structure?

In some communities, a certain key species, called a **keystone species**, plays a major role in determining community structure—a role that is out of proportion to its abundance in the community. Removal of the keystone species dramatically alters the community. For example, R. Paine, an ecologist at the University of Washington, removed predatory starfish *Pisaster* (Fig. 39-15a) from sections of the rocky intertidal coast of Washington in 1969. Mussels (bivalve mollusks that are a favored prey of *Pisaster*) became so abundant that they outcompeted algae and other invertebrates that normally coexist in intertidal communities. Another marine invertebrate, the lobster, may be a keystone species off the east coast of Canada. Overfishing of the lobster allowed sea urchins, a prey of lobsters, to increase in numbers. The population explosion of sea urchins nearly eliminated certain types of algae on which the urchins prey, leaving large expanses of bare rock where a diverse community once existed. The African elephant is a keystone predator in

(a)

(b)

Figure 39-15 Keystone species
(a) The starfish *Pisaster* is a keystone species along the rocky coast of the Pacific Northwest. *(b)* The elephant is a keystone species on the African savanna.

the African savanna. By grazing on small trees and bushes, elephants prevent the encroachment of forests and help maintain the grassland community (Fig. 39-15b).

It is difficult to identify keystone species, because ideally the species should be selectively removed and the community studied for several years before and after its removal. However, many ecological studies performed since the concept was first introduced by ecologist Paine provide evidence that keystone species play important roles in a wide variety of communities. Why is it important to study keystone species? As humans increasingly infringe on natural ecosystems, it becomes increasingly urgent to understand community interactions and to preserve those species that are crucial to maintaining the natural community.

6 Succession: How Does a Community Change over Time?

In a mature terrestrial ecosystem, the populations comprising the community interact with one another and with their nonliving environment in intricate ways. But this tangled web of life did not spring fully formed from bare rock or naked soil; rather, it emerged in stages over a long period, a process called succession. **Succession** is a structural change in a community and its nonliving environment over time. It is a kind of "community relay" in which assemblages of plants and animals replace one another in a sequence that is somewhat predictable.

Succession occurs under a variety of circumstances but is most easily observed in freshwater and terrestrial ecosystems. Freshwater ponds and lakes tend to undergo a series of changes that, through time, transform them first into marshes and eventually to dry land. Shifting sand dunes are stabilized by creeping plants and may eventually support a forest. Volcanic eruptions may, as in the case of Mount St. Helens, wipe out previously existing communities, or they may create new islands ripe for colonization. Forest fires may destroy a community but leave behind a nutrient-rich environment that encourages rapid invasion of new life (Fig. 39-16).

The precise changes that occur during succession are as diverse as the environments in which succession occurs, but we can recognize certain general stages. In each case, succession is begun by a few hardy invaders called **pioneers**. If undisturbed, succession progresses to a diverse and relatively stable **climax community**, the endpoint of succession. Our discussion of succession will focus on plant communities.

There Are Two Major Forms of Succession: Primary and Secondary

Succession takes two major forms: primary and secondary. During **primary succession**, a community gradually colonizes bare rock, sand, or a clear glacial pool where there is no trace of a previous community. The generation of a community "from scratch" is a process that typically requires thousands or even tens of thousands of years. During **secondary succession**, a new community develops after an existing ecosystem is disturbed,

as in the case of a forest fire or an abandoned farm field. Secondary succession happens much more rapidly than does primary succession, because the previous community has left its mark in the form of soil and seeds. Succession in an abandoned farm field in the southeastern United States can reach its climax after just two centuries.

As species gradually replace one another during succession, they interact in various ways. Lichens, pioneer species that grow on bare rock during primary succession, actually change the environment in ways that favor their competitors by liberating nutrients. During secondary succession, most pioneer plants are annuals (grasses, weeds, or wildflowers that live for a single growing season), and they generally produce large numbers of easily dispersed seeds that help them colonize open spaces. These pioneer species are sun-loving and grow rapidly, but they don't compete well against longer-lived (perennial) species that gradually grow larger and shade out the pioneers. In the following examples, we examine these processes in more detail.

Primary Succession Can Begin on Bare Rock

Figure 39-17 (p. 827) illustrates primary succession as studied on Isle Royale, Michigan, an island in Lake Superior. Bare rock, such as that exposed by a retreating glacier or cooled from molten lava, begins to liberate nutrients such as minerals by *weathering*. In weathering, cracks form as the rock alternately freezes and thaws, contracting and expanding, and chemical action such as acid rain further breaks down the rock surface.

For lichens, which are symbiotic associations of fungi and algae, weathered rock provides a place to attach where there are no competitors and plenty of sunlight. Lichens can photosynthesize, and they obtain minerals by dissolving some of the rock with an acid they secrete. As the pioneering lichens spread over the rock, drought-resistant, sun-loving mosses begin growing in the cracks. Fortified by nutrients liberated by the lichens, the moss forms a dense mat that traps dust, tiny rock particles, and bits of organic debris. The death of some of the moss adds to a growing nutrient base, and the moss mat acts like a sponge, trapping moisture. Within the moss, seeds of larger plants, such as bluebell and yarrow, germinate. Eventually, their bodies contribute to a growing layer of soil. As woody shrubs such as blueberry and juniper take advantage of the newly formed soil, the moss and lichens may be shaded out and buried by decaying leaves and vegetation. Eventually, trees such as jack pine, blue spruce, and aspen take root in the deeper crevices, and the sun-loving shrubs are shaded out. Within the forest, shade-tolerant seedlings of taller or faster-growing trees, such as balsam fir, paper birch, and white spruce, thrive. In time they tower over and replace the original trees, which are intolerant of shade. After a thousand years or more, a tall climax forest thrives on what was once bare rock.

(a)

(b)

Figure 39-16 Succession in progress
A photographic comparison to demonstrate (primary) succession. **(a)** Left: On May 18, 1980, the explosion of Mount St. Helens in Washington State devastated the pine forest ecosystem on its sides. Right: Just 9 years later, pioneering evergreens were 4 to 6 inches tall, and shrubs carpeted the nutrient-rich ash. **(b)** Left: The Hawaiian volcano Mount Kilauea has erupted repeatedly since 1983, sending rivers of lava over the surrounding countryside. Right: A pioneer fern takes root on hardened lava. **(c)** Left: In the summer of 1988, extensive fires swept through the forests of Yellowstone National Park in Wyoming. Right: Within 2 years, flowering plants had sprung up in the sunlight and wildlife populations had started to rebound.

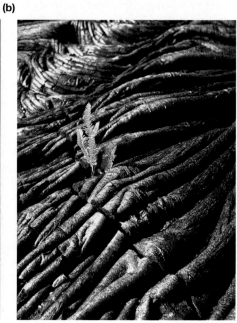

(c)

lichens and moss on bare rock | bluebell, yarrow | blueberry, juniper | jack pine, black spruce, aspen | balsam fir, paper birch, white spruce, climax forest

time (years)

0 → 1000

Figure 39-17 Primary succession
Primary succession as it occurs on bare rock in upper Michigan.

An Abandoned Farm Field Will Undergo Secondary Succession

Figure 39-18 illustrates succession on an abandoned southeastern farm. The pioneers are fast-growing annual weeds such as crabgrass, ragweed, and sorrel, which root in the rich soil already present and thrive in direct sunlight. A few years later, perennial plants such as asters, goldenrod, broom sedge grass, and woody shrubs, such as blackberry, invade. These plants multiply rapidly and dominate for the next few decades. Eventually, they are replaced by pines and fast-growing deciduous trees, such as tulip poplar and sweet gum, which sprout from wind-blown seeds. These trees become prominent after about 25 years, and a pine forest dominates the field for the rest of the first century. Meanwhile, shade-resistant, slow-growing hardwoods such as oak and hickory take root beneath the pines. After the first century, these begin to tower over and shade the pines, which eventually die from lack of sun. A relatively stable climax forest dominated by oak and hickory is present by the end of the second century.

Succession Also Occurs in Ponds and Lakes

In freshwater ponds or lakes, succession occurs both from changes within the pond or lake and as a result of an influx of nutrients from outside the ecosystem. Sediments and nutrients carried in by runoff from the surrounding land have a particularly large impact on small freshwater lakes, ponds, and bogs, which gradually undergo succes-

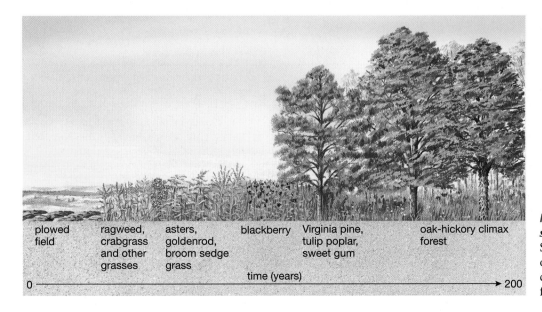

plowed field | ragweed, crabgrass and other grasses | asters, goldenrod, broom sedge grass | blackberry | Virginia pine, tulip poplar, sweet gum | oak-hickory climax forest

time (years)

0 → 200

Figure 39-18 Secondary succession
Secondary succession as it occurs on a plowed, abandoned southeastern farm field.

sion to dry land (Fig. 39-19). In forests, meadows may be produced by lakes undergoing succession. As the lake fills in from the edges, grasses colonize the newly exposed soil. As the lake shrinks and the meadow grows, trees will be encroaching around the meadow's edges. If you visit a forest lake 20 years after your first visit, it probably will be a bit smaller.

Succession Culminates in the Climax Community

Succession ends with a relatively stable climax community, which perpetuates itself if it is not disturbed by external forces (such as fire, invasion of an introduced species, or human activities). The populations within a climax community have ecological niches that allow them to coexist without replacing one another. In general, climax communities have more species and more types of community interactions than do early stages of succession. The plant species that dominate climax communities are generally longer-living and tend to be larger than pioneer species; this trend is particularly evident in ecosystems where forest is the climax community.

In your travels, you have undoubtedly noticed that the type of climax community varies dramatically from one area to the next. For example, if you drive through Colorado, you will see a shortgrass prairie climax community on the eastern plains (in those rare areas where it has not been replaced by farms), pine-spruce forests in the mountains, tundra on their uppermost reaches, and sagebrush-dominated climax community in the western valleys. The exact nature of the climax community is determined by numerous geological and climatic variables, including temperature, rainfall, elevation, latitude, type of rock (which determines the type of nutrients available), exposure to sun and wind, and many more.

Earth Watch
The Invasion of Fire Ants and Kudzu

Exotic species, or species introduced into an ecosystem where they did not evolve, sometimes find no predators or parasites in their new environment but find new prey that have few defenses against them. The unchecked population growth of these invaders may seriously damage the ecosystem as they displace, outcompete, and prey on native species. Both starlings and English sparrows have spread dramatically since their deliberate introduction into the eastern United States in the 1890s. Their success threatens the native bluebird. Red fire ants from South America were accidentally introduced into Alabama on shiploads of lumber in the 1930s and have since spread throughout the South, with its hospitable warm, moist climate. Fire ants displace and may kill other insects, birds, and small mammals. Their mounds can ruin farm fields, and their fiery stings and aggressive temperament can make backyards uninhabitable.

In 1986, a trading vessel bringing cargo from Europe up the St. Lawrence Seaway discharged fresh water that contained millions of larvae of the zebra mussel into the St. Clair River near Detroit. By 1989, the invading mussels had spread to Lake Erie; now they inhabit all the Great Lakes and the upper Mississippi River and are spreading throughout eastern waterways. They cover piers, boats, machinery, and beaches. At one water treatment plant on Lake Erie, zebra mussel populations reached 600,000 per square yard, clogging pipes and reducing water flow to a trickle (Fig. E39-2a). The mussels cover and suffocate other forms of shellfish, threatening many rare varieties with extinction.

A Japanese vine called kudzu was introduced to the South as a decorative plant in 1876 and then planted extensively in the 1940s to control erosion. Today kudzu is a major pest, overgrowing and killing trees and underbrush and occasionally engulfing small houses (Fig. E39-2b). The water hyacinth, introduced from South America as an ornamental plant, now clogs about 2 million acres of southern lakes and waterways, slowing boat traffic and displacing natural vegetation (Fig. E39-2c).

By evading the checks and balances imposed by millennia of coevolution, exotic species are wreaking havoc on natural ecosystems throughout the world. Recently, wildlife officials have made cautious attempts to reestablish these checks and balances by importing predators or parasites to attack selected exotic species. In Australia, a dozen European rabbits introduced in the 1840s had exploded in number to 300 million by 1996, wiping out native vegetation and competing for food with native species such as kangaroos. In 1996, officials released a virus deadly to rabbits at hundreds of sites across the continent. This controversial experiment has worked well so far. The virus has not spread to other species, rabbit populations have been reduced by 95% in some areas, and native plants and animals are rebounding. This type of control is fraught with danger, however, because introducing even more predators or parasites into an ecosystem can have unpredicted and possibly disastrous consequences for native species.

Figure 39-19 *Succession in a freshwater pond*
In small ponds, succession is speeded by an influx of materials from the surroundings. *(a)* In this small pond, dissolved minerals carried by runoff from the surroundings support aquatic plants, whose seeds or spores were carried in by the winds or by birds and other animals. *(b)* Over time, the decaying bodies of aquatic plants build up soil that provides anchorage for more-terrestrial plants. *(c)* Finally, the pond is entirely converted to dry land.

Figure E39-2 *Exotic species*
(a) Workers blast jets of hot water at zebra mussels coating the interior of a Michigan water treatment plant.
(b) The Japanese vine kudzu will rapidly cover entire trees and houses. *(c)* The beauty that became a beast—water hyacinths, originally from South America, today clog waterways in our southern states.

Natural events such as hurricanes, avalanches, and fires started by lightning may destroy sections of climax forest, reinitiating secondary succession and producing a patchwork of various successional stages within an ecosystem.

In many forests throughout the United States, rangers are allowing fires set by lightning to run their course, recognizing that this natural process is important for the maintenance of the entire ecosystem. Fires liberate nutrients that are used by plants. Fires also kill some (but normally not all) of the trees they engulf. Sunlight can then reach the forest floor, encouraging the growth of **subclimax** plants, which belong to a successional stage earlier than the climax stage. The combination of climax and subclimax regions within the ecosystem provides habitats for a larger number of species.

Human activities may dramatically alter the climax vegetation. Large stretches of grasslands in the west, for example, are now dominated by sagebrush due to overgrazing. The grass that normally outcompetes sagebrush is selectively eaten by cattle, allowing the sagebrush to prosper.

Many Ecosystems Are Maintained in a Subclimax State

Some ecosystems are not allowed to reach the climax stage but are maintained in a subclimax stage. The tallgrass prairie that once covered northern Missouri and Illinois is a subclimax of an ecosystem whose climax community is deciduous forest. The prairie was maintained by periodic fires, some set by lightning and others deliberately set by Native Americans to increase grazing land for buffalo. Forest now encroaches, and limited prairie preserves are maintained by carefully managed burning.

Agriculture also depends on the artificial maintenance of carefully selected subclimax communities. Grains are specialized grasses characteristic of the early stages of succession, and much energy goes into preventing competitors (weeds and shrubs) from taking over. The suburban lawn is a painstakingly maintained subclimax ecosystem. Mowing destroys woody invaders, and selective herbicides kill pioneers such as crabgrass and dandelions.

To study succession is to study the variations in communities over time. The climax communities that form during succession are strongly influenced by climate and geography—the distribution of ecosystems in *space*. Deserts, grasslands, and deciduous forests are climax communities formed over broad geographical regions with similar environmental conditions. These extensive areas of characteristic plant communities are called **biomes**. In Chapter 41 we shall explore some of the great biomes of the world. Although the communities comprising the various biomes differ radically in the types of populations they support, communities worldwide are structured according to general rules. These principles of ecosystem structure are described in Chapter 40.

Summary of Key Concepts

1) Why Are Community Interactions Important?
Community interactions influence population size, and the interacting populations within communities act on one another as agents of natural selection. Thus, community interactions also shape the bodies and behaviors of the interacting populations.

2) What Are the Effects of Competition among Species?
The ecological niche defines all aspects of a species' habitat and interactions with its living and nonliving environments. Each species occupies a unique ecological niche. Interspecific competition occurs when the niches of two populations within a community overlap. When two species with the same niche are forced under laboratory conditions to occupy the same ecological niche, one species always outcompetes the other. Species within natural communities have evolved in ways that avoid excessive niche overlap, with behavioral and physical adaptations that allow resource partitioning. Interspecific competition limits both the size and the distribution of competing populations.

3) What Are the Results of Interactions between Predators and Their Prey?
Predators eat other organisms and are generally both larger and less abundant than their prey. Predators and prey act as strong agents of selection on one another. Prey animals have evolved a variety of protective colorations that render them either inconspicuous (camouflage) or startling (startle coloration) to their predators. Some prey are poisonous and exhibit warning coloration by which they are readily recognized and avoided. The situation in which an animal has evolved to resemble another is called mimicry. Both predators and prey have evolved a variety of toxic chemicals for attack and defense. Plants that are preyed on have evolved elaborate defenses, ranging from poisons to thorns to overall toughness. These defenses, in turn, have selected for predators that can detoxify poisons, ignore thorns, and grind down tough tissues.

4) What Is Symbiosis?
Symbiotic relationships involve two species that interact closely over an extended time. Symbiosis includes parasitism, in which the parasite feeds on a larger, less abundant host, normally harming it but not killing it immediately. In commensal symbiotic relationships, one species benefits, typically by finding food more easily in the presence of the other species, which is not affected by the association. Mutualism benefits both symbiotic species.

5) How Do Keystone Species Influence Community Structure?

Keystone species have a greater influence on community structure than can be predicted by their numbers. Removal of a keystone species will radically alter community structure to an extent that would not be predicted on the basis of the species' abundance.

6) Succession: How Does a Community Change over Time?

Succession is a progressive change over time in the types of populations that comprise a community. Primary succession, which may take thousands of years, occurs where no remnant of a previous community existed (such as on rock scraped bare by a glacier or cooled from molten lava, a sand dune, or in a newly formed glacial lake). Secondary succession occurs much more rapidly, because it builds on the remains of a disrupted community, such as an abandoned field or the aftermath of a fire. Secondary succession on land is initiated by fast-growing, readily dispersing pioneer plants, which are eventually replaced by longer-lived, generally larger and more shade-tolerant species. Uninterrupted succession ends with a climax community, which tends to be self-perpetuating unless acted on by outside forces, such as fire or human activities. Some ecosystems, including tallgrass prairie and farm fields, are maintained in relatively early stages of succession by periodic disruptions.

Key Terms

aggressive mimicry *p. 820*
biome *p. 830*
camouflage *p. 817*
climax community *p. 825*
coevolution *p. 814*
commensalism *p. 822*
community *p. 814*

competitive exclusion
   principle *p. 814*
ecological niche *p. 814*
exotic species *p. 828*
interspecific competition
   p. 814
intertidal zone *p. 816*

keystone species *p. 824*
mimicry *p. 819*
mutualism *p. 823*
parasite *p. 822*
parasitism *p. 822*
pioneer *p. 825*
primary succession *p. 825*

resource partitioning *p. 815*
secondary succession *p. 825*
startle coloration *p. 820*
subclimax *p. 830*
succession *p. 825*
symbiosis *p. 822*
warning coloration *p. 819*

Thinking Through the Concepts

Multiple Choice:

1. *Which of the following is usually NOT true of climax communities?*
   a. They have more species than do pioneer communities.
   b. They have longer-lived species than do early successional stages.
   c. Climax communities are relatively stable.
   d. Some climax communities are maintained by fire.
   e. Climax communities vary with the location of the ecosystem.

2. *What were the differences in the niches of the* Paramecium *in Gause's second experiment that allowed them to coexist?*
   a. food eaten
   b. body size
   c. feeding area
   d. preferred water temperature
   e. preferred pH

3. *Which of the following is an example of a mutualistic relationship?*
   a. flowering plants and their pollinators
   b. a caterpillar eating a tomato plant
   c. bats and moths
   d. monarch and viceroy butterflies
   e. lupines and blue butterflies

4. *What is the function of aggressive mimicry?*
   a. to hide a prey from a predator
   b. to warn a predator that a prey is dangerous
   c. to warn a predator that a prey is distasteful
   d. to keep prey from recognizing a predator
   e. to startle a prey when it sees a predator

5. *What is coevolution?*
   a. two species selecting for traits in each other
   b. individuals of two species living together
   c. the presence of two species in the same community
   d. two species evolving separately through time
   e. individuals of two species learning how to coexist with or to hunt with each other

6. *The competitive exclusion principle implies that coexisting species*
   a. can use the same resources
   b. cannot eat exactly the same things
   c. cannot have identical ecological interactions
   d. cannot be exactly the same size
   e. cannot be closely related to one another

? Review Questions

1. Define ecological *community*, and list three important types of community interactions.

2. Describe four very different ways in which specific plants and animals protect themselves against being eaten. In each case, describe an adaptation that might evolve in predators of these species that would overcome their defenses.

3. List two important types of symbiosis; define and provide an example of each.

4. Which type of succession would occur on a clear-cut (a region in which all the trees have been removed by logging) in a national forest, and why?

5. List two subclimax and two climax communities. How do they differ?

6. Define *succession*, and explain why it occurs.

Applying the Concepts

1. Herbivorous animals that eat seeds are considered by some ecologists to be predators of plants, and herbivorous animals that eat leaves are considered to be parasites of plants. Discuss the validity of this classification scheme.

2. An interesting interspecific relationship exists between the tarantula spider and the tarantula hawk wasp. This wasp attacks tarantulas, paralyzing them with venom from their stingers. The wasp then lays eggs on the paralyzed spider. The eggs hatch, and the young eat the living, immobilized tissues of the spider. Discuss whether this relationship between spider and wasp exemplifies parasitism or predation.

3. An ecologist visiting an island finds two very closely related species of birds, one of which has a slightly larger bill than the other. Interpret this finding with respect to the competitive exclusion principle and the ecological niche, and explain both concepts.

4. Think about the case of the camouflaged frogfish and its prey. As the frogfish sits camouflaged on the ocean floor, wiggling its lure, a small fish approaches the lure and is eaten, while a very large predatory fish fails to notice the frogfish. Describe all the possible types of community interactions and adaptations these organisms have selected for. Remember that predators can also be prey and that community interactions are complex!

5. Design an experiment to determine whether the kangaroo is a keystone species in the Australian outback.

6. Why is it difficult to study succession? Suggest some ways you would approach this challenge for a few different ecosystems.

Group Activity

In groups of three, divide up the accompanying case studies. Each student should focus on one case study and present an overview of the case and the answers to the questions to the group. As a group, collectively provide intelligent responses to each of the cases presented in class discussions.

Case 1

A plateau in Arizona is covered mostly by evergreen forests and grasslands. The plateau is surrounded by rugged canyons, including the Grand Canyon, which form natural barriers to the movement of most animals. The animals on the plateau include deer, mountain lions, coyotes, bobcats, rabbits, ground squirrels, chipmunks, and wood rats. Mountain lions mainly eat deer. Coyotes and bobcats eat some deer but depend on rabbits and small rodents for most of their food. The rodents, rabbits, and deer eat grasses and low shrubs. The deer also eat the buds and tender branches of trees they can reach.

Before the arrival of Caucasians in the late 1800s, Native Americans lived on the plateau. They killed about 800 deer each year for food and hides. They seldom killed the predators. During this time the deer population did not change greatly from year to year.

A study of the plateau animals was made in 1906. The deer population was estimated to be about 4000. About this time, the plateau was declared a game reserve, and the killing of deer was prohibited. A campaign was begun to kill the predators. As a result, about 10,000 predators were killed between 1906 and 1939. By 1924 there were about 100,000 deer. However, much of the plant life had been eaten or destroyed. Thousands of deer died of starvation or disease during the next few years. By 1930, the number of deer was down to about 30,000.

Suppose that you have been called in by government officials, who need your advice on the management of the plateau's wildlife. How many deer would you suggest the plateau could reasonably support? What actions would you suggest to keep the deer population at the size you suggested? Give reasons for your answers to both questions.

Case 2

High mountain lakes look good but seldom support many large fish. An ecologist decided to add more fish food so that there would be more and larger fish. One summer, she dumped fertilizer into the lake. The plants grew and supported more insects, which caused fish populations to increase. By summer's end, there were so many plants in the lake that from above it looked green rather than the blue it used to be.

The ecologist seemed to think that she had done the right thing, so she decided to fertilize more mountain lakes the next summer. However, when she flew over the lake country early the following summer, she could not find the green lake, because it had turned blue again. In fact, the plants, insects, and fish were scarcer than they had been before the lake was fertilized. Explain why the plants died and why there were fewer insects and fish than before.

Case 3

To save a forest that was being destroyed by insects, a governmental agency used sprays to kill them. Because the area was a nature preserve, the agency thought it should explain to the public why it used the sprays. A special video was made telling about the harmful effects of insects, the need for the use of sprays, and how carefully the sprays had been applied. Furthermore, the video narrator said that the sprays had killed only the insects and that no other wildlife had been affected.

Explain why spraying would affect the animals other than insects even if the sprays did directly harm only the insects. Then explain how the sprays might also affect natural selection of the insects being sprayed.

For More Information

Amos, W. H. "Hawaii's Volcanic Cradle of Life." *National Geographic*, July 1990. A naturalist explores succession on lava flows.

Batten, M. "The Ant and the Acacia." *Science 84*, April 1984. Tells the story of Daniel Janzen's discovery of this mutualistic relationship.

Daniels, P. "How Flowers Seduce the Bugs and the Bees." *International Wildlife*, November/December 1984. Beautifully illustrated description of plants and their mutualistic pollinators.

Drollette, D. "Wide Use of Rabbit Virus Is Good News for Native Species." *Science*, January 10, 1997. Describes recent efforts that use an introduced virus to control introduced rabbits in Australia.

Horn, H. S. "Forest Succession." *Scientific American*, May 1975 (Offprint No. 1321). Describes succession in eastern deciduous forests.

Power, M., et al. "Challenges in the Quest for Keystones." *Bioscience*, September 1996. A comprehensive review of the importance of keystone species and the challenges of studying them.

Rinderer, T. E., Oldroyd, B. P., and Sheppard, W. S. "Africanized Bees in the U.S." *Scientific American*, December 1993. What are the likely impacts of these aggressive invaders?

Vitousek, P. M., D'Antonio, C. M., Loope, L. L., and Westbrooks, R. "Biological Invasions as Global Environmental Change." *American Scientist*, September/October 1996. The introduction of exotic species is having a major negative impact on ecosystems, human health, and local economies.

Answers to Multiple-Choice Questions

1. d   2. c   3. a   4. d   5. a   6. c

> "A thing is right, when it tends to preserve the integrity, stability, and beauty of the biotic community. It is wrong when it tends otherwise."
>
> **Aldo Leopold in** A Sand County Almanac **(1949)**

A garden can be a lesson in feeding relationships. Garden plants derive energy from sunlight. Plant-eaters such as the tomato hornworm (caterpillar) can be controlled naturally, such as by the Venturia *wasp. The wasp injects its eggs inside the caterpillar, which is consumed from within by wasp larvae.*

How Do Ecosystems Work? 40

Net Watch
On-line resources for this chapter are on the World Wide Web at:
http://www.prenhall.com/audesirk
(click on the Table of Contents link and then select Chapter 40).

What can you learn from a garden? Gardening is good for the mind and body, and it provides hands-on lessons in the way ecosystems function. A good garden starts with a compost heap, nourished year-round with corn husks, eggshells, lawn clippings, and leaves. If conditions are right, upon digging into the bottom of the heap you will find a minor miracle: rich, black soil, possibly home to a flourishing population of earthworms. Dig this into your garden, and your tomatoes, radishes, lettuce, and anything else you plant is likely to flourish as long as it gets enough sun and water. Lacking a compost heap, you may sprinkle your garden with commercial fertilizer. As the vegetables grow, you may discover, to your dismay, that your miniature farm has attracted some pests: A caterpillar called a tomato hornworm is gnawing voraciously at your tomatoes, systematically stripping the leaves. You may be lucky enough to see some small birds making a meal of these plant-eaters. In the fall, sated with healthy vegetables, you pull up the old tomato plants and throw them onto the compost heap—food for next year's garden.

What magic turns a heap of vegetable matter into nutrient-rich soil? Why did the corner of your garden that was shaded not produce as much as the sunny parts? What role do plant-eaters such as the caterpillar and meat-eaters such as the bird play in ecosystems? Which is better from a biological standpoint—using compost or commercial fertilizer?

In Chapters 7 and 8, you learned how energy is trapped by photosynthesis, released by cellular respiration, and used to construct the complex molecules of life. In Chapter 39, you learned some of the complex ways in which organisms interact within ecological communities. In this chapter, we relate some of these basic principles to the workings of ecosystems, from a garden to a prairie.

1) What Are the Pathways of Energy and Nutrients through Communities?

The activities of life, from the leaping of the jackrabbit to the active transport of molecules through a cell membrane, are powered by the energy of sunlight. Each time this energy is used, some of it is lost as heat. But while solar energy continuously bombards Earth and is continuously lost as heat, nutrients remain. They may change in form and distribution, but nutrients do not leave Earth and are constantly recycled. Two basic laws thus underlie ecosystem function. (1) Energy moves *through* the communities within ecosystems in a continuous one-way flow. Energy needs constant replenishment from an outside source, the sun. (2) In contrast to energy, nutrients constantly cycle and recycle in a circular flow *within* ecosystems (Fig. 40-1). These laws shape the complex interactions within ecosystems.

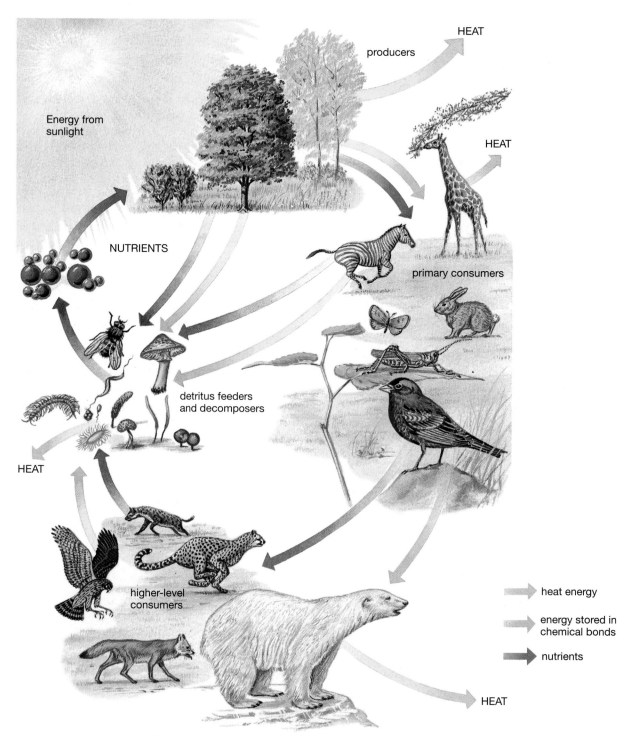

Figure 40-1 *Energy flow, nutrient cycling, and feeding relationships in ecosystems*
Note that nutrients (purple) neither enter nor leave the cycle. Energy (yellow), continuously supplied to producers as sunlight, is captured in chemical bonds and transferred through various levels of organisms (magenta). At each level, some energy is lost as heat (orange).

2) How Does Energy Flow through Communities?

Energy Enters Communities through the Process of Photosynthesis

Ninety-three million miles away, the sun fuses hydrogen molecules into helium molecules, releasing tremendous quantities of energy. A tiny fraction of this energy reaches Earth in the form of electromagnetic waves, including heat, light, and ultraviolet energy. Of the energy that reaches Earth, much is reflected by the atmosphere, clouds, and Earth's surface. Still more is absorbed as heat by Earth and its atmosphere, leaving only about 1% to power all life. Of this 1%, green plants capture 3% or less. The teeming life on this planet is thus supported by less than 0.03% of the energy reaching Earth from the sun. But how does this energy enter the biological community? The energy that powers life enters through the process of photosynthesis.

In photosynthesis (performed by plants, plantlike protists, and cyanobacteria), pigments such as chlorophyll absorb certain wavelengths of sunlight. This solar energy is then used in reactions that store energy in chemical bonds, producing sugar and ultimately the other complex molecules that form the bodies of photosynthetic organisms (Fig. 40-2). Photosynthetic organisms, from mighty oaks to the zucchini and tomato plants in your garden to single-celled diatoms in the ocean, are called **autotrophs** (Greek, "self-feeders") or **producers**, because they produce food for themselves. They directly or indirectly produce food for nearly all other forms of life as well. Organisms that cannot photosynthesize, called **heterotrophs** (Greek, "other-feeders") or **consumers**, must acquire energy prepackaged in the molecules of the bodies of other organisms.

The amount of life an ecosystem can support is determined by the energy captured by the producers in the ecosystem. The energy that photosynthetic organisms store and make available to other members of the community over a given period is called **net primary productivity**. Net primary productivity can be measured in units of energy (calories) stored per unit area in a given period. It is also measured as the **biomass**, or dry weight of organic material, in producers that is added to the ecosystem per unit area in a given time span. The productivity of an ecosystem is influenced by many environmental variables, including the amount of nutrients available to the producers, the amount of sunlight reaching them, the availability of water, and the temperature. In the desert, for example, lack of water limits productivity; in the open ocean, light is limited in deep water and nutrients are limited in surface water. When resources are abundant, as in estuaries (where rivers meet the ocean, carrying nutrients washed from the land) and in tropical rain forests, productivity is high. Some average productivities for a variety of ecosystems are shown in Figure 40-3.

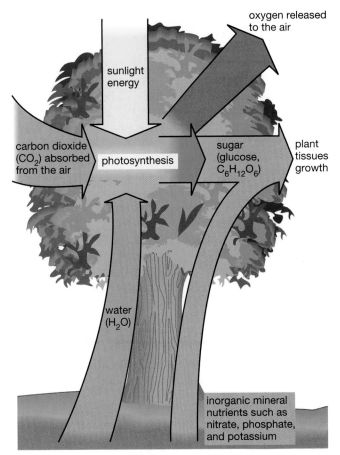

Figure 40-2 Primary productivity: photosynthesis
During photosynthesis, plants capture the energy of sunlight and store it in sugar synthesized from carbon dioxide and water. Using building blocks of sugar and nutrients taken up with water from the soil, photosynthetic organisms synthesize all the complex molecules that comprise their bodies.

Within the Community, Energy Is Passed from One Trophic Level to Another

Energy flows through communities from photosynthetic producers through several levels of consumers. Each category of organism is called a **trophic level** (literally, "feeding level"). The producers, from redwood trees to cyanobacteria, form the first trophic level, obtaining their energy directly from sunlight (see Fig. 40-1). Consumers form several trophic levels. Certain consumers occupy more than one trophic level as they eat organisms from different levels. Some consumers have evolved to feed directly and exclusively on producers, the most abundant living energy source in any ecosystem. These **herbivores** (meaning "plant eaters"), ranging from grasshoppers to giraffes to tomato hornworms, are also called **primary consumers**; they form the second trophic level. **Carnivores** (meaning "meat eaters"), such as the spider, hawk, wolf, and the bird that helpfully consumed your hornworm, are predatory flesh-eaters, feeding primarily

Figure 40-3 *Ecosystem productivity compared*
Average net primary productivity, in grams of organic material per square meter per year, of some terrestrial and aquatic ecosystems.

on primary consumers. Carnivores are also called **secondary consumers**; they form the third trophic level. Some carnivores occasionally eat other carnivores. When doing so, they form the fourth trophic level, **tertiary consumers**.

Food Chains and Food Webs Describe the Feeding Relationships within Communities

To illustrate who feeds on whom in a community, it is common to identify a representative of each trophic level that eats the representative of the level below it. This linear feeding relationship is called a **food chain**. As illustrated in Figure 40-4, different ecosystems have radically different food chains.

Natural communities rarely contain well-defined groups of primary, secondary, and tertiary consumers, however. A **food web** shows the many interconnecting food chains in a community, describing the actual feeding relationships within a given community much more accurately than does a food chain (Fig. 40-5, p. 840). Some animals, such as raccoons, bears, rats, and humans, are *omnivores* (Latin, "eating all")—that is, at different times they act as primary, secondary, and occasionally tertiary (third-level) consumers. Many carnivores will eat either herbivores or other carnivores, thus acting as secondary or tertiary consumers, respectively. An owl, for instance, is a secondary consumer when it eats a mouse, which feeds on plants, but a tertiary consumer when it eats a shrew, which feeds on insects. A shrew that eats a carnivorous insect is a tertiary consumer, and the owl that

fed on the shrew is then a quaternary (fourth-level) consumer. When digesting a spider, a carnivorous plant, such as the sundew, can tangle the web hopelessly by serving simultaneously as a photosynthetic producer and a secondary consumer!

Detritus Feeders and Decomposers Release Nutrients for Reuse

Among the most important strands in the food web are the *detritus feeders* and *decomposers*. The **detritus feeders** are an army of small and often unnoticed animals and protists that live on the refuse of life: molted exoskeletons, fallen leaves, wastes, and dead bodies. (*Detritus* means "debris.") The network of detritus feeders is extremely complex, including earthworms, mites, protists, centipedes, some insects and crustaceans, nematode worms, and even a few large vertebrates such as vultures. With the exception of vultures, these organisms probably abound in your garden and compost heap, including a unique land-dwelling crustacean called a pillbug (or "rolypoly"). These organisms consume dead organic matter, extract some of the energy stored within it, and excrete it in a further decomposed state. Their excretory products serve as food for other detritus feeders and decomposers. The **decomposers** are primarily fungi and bacteria that digest food outside their bodies by secreting digestive enzymes into the environment. They absorb the nutrients they need, and release the remaining nutrients. The black coating or gray fuzz you may notice on the decaying tomatoes and bread crusts in your compost are

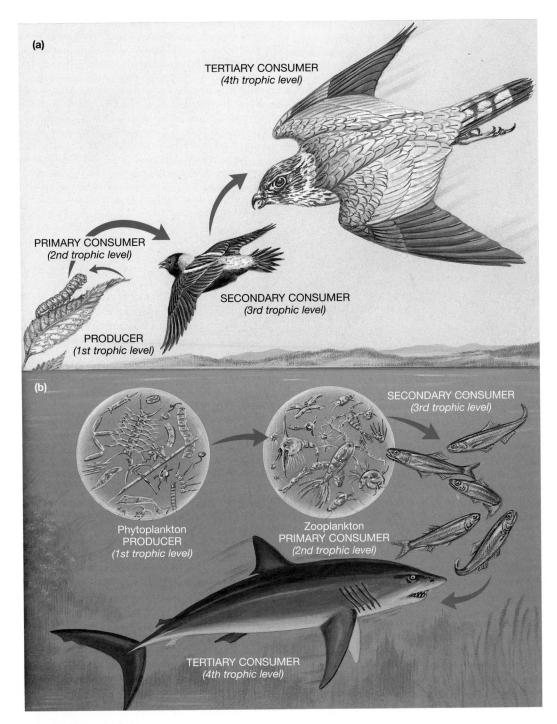

Figure 40-4 Food chains
(a) A simple terrestrial food chain. *(b)* A simple marine food chain.

fungal decomposers hard at work. Thanks to detritus feeders and decomposers, most of the stored energy is eventually utilized.

Through the activities of detritus feeders and decomposers, the bodies and wastes of living organisms are reduced to simple molecules, such as carbon dioxide, water, minerals, and organic molecules, that return to the atmosphere, soil, and water. By liberating nutrients

for reuse, detritus feeders and decomposers form a vital link in the nutrient cycles of ecosystems. In some ecosystems, such as deciduous forests, more energy passes through the detritus feeders and decomposers than through the primary, secondary, or tertiary consumers.

What would happen if detritus feeders and decomposers were to disappear? This portion of the food web, although inconspicuous, is absolutely essential to life on

Figure 40-5 A food web
A simple terrestrial food web on a short-grass prairie.

Earth. Without them, communities would gradually be smothered by accumulated wastes and dead bodies. The nutrients stored in these bodies would be unavailable to enrich the soil. The quality of the soil would become poorer and poorer until plant life could no longer be sustained. With plants eliminated, energy would cease to enter the community; the higher trophic levels, including humans, would disappear as well.

Energy Transfer through Trophic Levels Is Inefficient

As we discussed in Chapter 4, a basic law of thermodynamics is that energy use is never completely efficient. For example, as your car converts the energy stored in gasoline to the energy of movement, about 75% is immediately lost as heat. So too in living systems; the splitting of the chemical bonds of ATP to power a muscular contraction also produces heat as a by-product, as anyone who has exercised vigorously is well aware. Small amounts of waste heat are also produced by the germination of a seed and the thrashing of the tail of a sperm.

The transfer of energy from one trophic level to the next is also quite inefficient. When a caterpillar (a primary consumer) eats the leaves of a tomato plant (a producer), only some of the solar energy originally trapped by the plant is available to the insect. Some energy was used by the plant for growth and maintenance, and more was lost as heat during these processes. Some energy was converted into the chemical bonds of molecules such as cellulose, which the caterpillar cannot digest. Therefore, only a fraction of the energy captured by the first trophic level is available to organisms in the second trophic level. The energy consumed by the caterpillar is in turn partially used to power crawling and the gnashing of mouthparts. Some is used to construct the indigestible exoskeleton, and much is given off as heat. All this energy is unavailable to the songbird in the third trophic level when it eats the caterpillar. The bird loses energy as body heat, uses more in flight, and converts a considerable amount into indigestible feathers, beak, and bone. All this energy will be unavailable to the hawk that catches it. A simplified model of energy flow through the trophic levels in a deciduous forest ecosystem is illustrated in Figure 40-6.

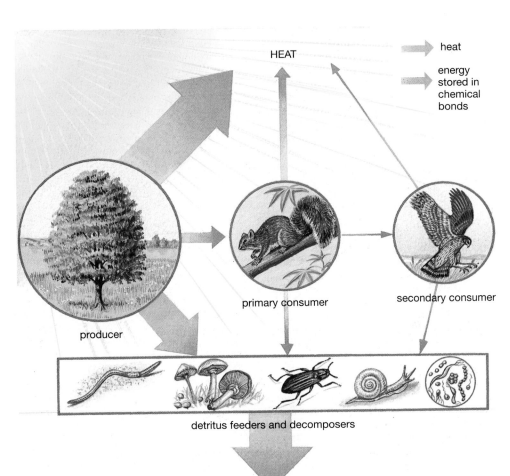

heat

energy stored in chemical bonds

HEAT

producer

primary consumer

secondary consumer

detritus feeders and decomposers

HEAT

Figure 40-6 Energy transfer and loss
This diagram shows approximate amounts of energy transferred between trophic levels in the form of chemical energy (magenta) and lost in the form of heat from each trophic level (orange) in a forest community. The width of the arrows is roughly proportional to the quantity of energy transferred or lost.

Energy Pyramids Illustrate Energy Transfer between Trophic Levels

Studies of a variety of communities indicate that the net transfer of energy transfer between trophic levels is roughly 10% efficient, although transfer among levels within different communities varies significantly. This means that, in general, the energy stored in primary consumers (herbivores) is only about 10% of the energy stored in the bodies of producers. In turn, the bodies of secondary consumers possess roughly 10% of the energy stored in primary consumers. In other words, for every 100 calories of solar energy captured by grass, only about 10 calories are converted into herbivores, and only 1 calorie into carnivores. This inefficient energy transfer between trophic levels is called the "10% law." An **energy pyramid**, which shows maximum energy at the base and steadily diminishing amounts at higher levels, illustrates the energy relationships between trophic levels graphically (Fig. 40-7). Ecologists sometimes use biomass as a measure of the energy stored at each trophic level. Because the dry weight of the bodies of organisms at each trophic is roughly proportional to the amount of energy stored in them, a *biomass pyramid* for a given community normally has the same general shape as its energy pyramid.

What does this mean for communities? If you wander through an undisturbed ecosystem, you will notice that the predominant organisms are plants. Plants have the most energy available to them, because they trap it directly from sunlight. The most abundant animals will be those feeding on plants, and carnivores will always be relatively rare. The inefficiency of energy transfer also has important implications for human food production. The lower the trophic level we utilize, the more food energy will be available to us; far more people can be fed on grain than on meat.

An unfortunate side effect of the inefficiency of energy transfer, coupled with human production and release of toxic chemicals, is *biological magnification*. This phenomenon is the concentration of certain persistent toxic chemicals in the bodies of carnivores, including ourselves, as described next.

Biological Magnification Occurs as Toxic Substances Are Passed through Trophic Levels

In the 1940s, the properties of the new insecticide DDT seemed close to miraculous. In the tropics, DDT saved millions of lives by killing the mosquitos that spread malaria. Increased crop yields resulting from DDT's destruction of insect pests saved millions more from starvation. DDT is long-lasting, so a single application keeps on killing. Swiss chemist Paul Müller, the discoverer of its properties as a pesticide, was awarded the 1948 Nobel Prize for Medicine and Physiology, and people looked forward to a new age of freedom from insect pests. Lit-

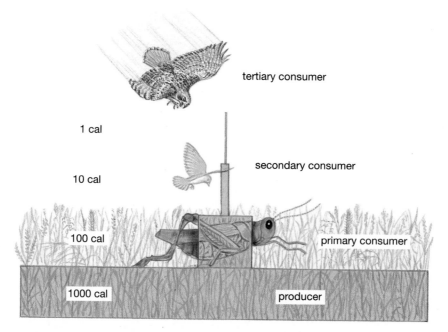

Figure 40-7 An energy pyramid for a prairie ecosystem
Each trophic level from producer to tertiary consumer has less energy stored in it. The width of each rectangle is proportional to the energy stored at that trophic level. A pyramid of biomass for this ecosystem would look quite similar.

tle did they realize that the indiscriminate use of this pesticide was unraveling the complex web of life.

For example, in the mid-1950s, the World Health Organization sprayed DDT on the island of Borneo to control malaria. A caterpillar that fed on the thatched roofs of houses was relatively unaffected, but a predatory wasp that fed on the caterpillar was destroyed. Thatched roofs collapsed, eaten by the uncontrolled caterpillars. Gecko lizards that ate the poisoned insects accumulated high concentrations of DDT in their bodies. Both they, and the village cats that ate the geckos, died of DDT poisoning. With the cats eliminated, the rat population exploded. The village was then threatened with an outbreak of plague, carried by the uncontrolled rats. The outbreak was avoided by the delivery of new cats to the villages.

In the United States, wildlife biologists during the 1950s and 1960s witnessed an alarming decline in populations of several predatory birds, especially fish-eaters, such as bald eagles, cormorants, ospreys, and brown pelicans. These top predators are never abundant, and the decline pushed some, including the brown pelican and the bald eagle, close to extinction. The aquatic ecosystems supporting these birds had been sprayed with relatively low amounts of DDT to control insects. Scientists were amazed to find concentrations of DDT in the bodies of predatory birds up to *one million* times the concentration present in the water. This led to the discovery of **biological magnification**, the process by which toxic substances accumulate in increasingly high concentrations in progressively higher trophic levels. In 1973, DDT was banned in the United States, but it is still used in some developing countries. In 1994, the bald eagle was removed from the endangered species list.

Both DDT and other substances that undergo biological magnification have two properties that make them dangerous: (1) Decomposer organisms cannot readily break them down into harmless substances—that is, they are not **biodegradable**; and (2) they are fat soluble but not water soluble. Thus, they accumulate in the bodies of animals, particularly in the fat, rather than being broken down and excreted in the watery urine. Because the transfer of energy from lower to higher trophic levels is extremely inefficient, herbivores must eat large quantities of plant material (which may have been sprayed with DDT), carnivores must eat many herbivores, and so on. Because DDT is not excreted, the predator accumulates the substance from all its prey over many years. Thus, DDT reaches highest levels in top predators, such as in predatory birds, where it interferes with the deposition of calcium into eggshells by mimicking and disrupting the functions of the natural sex hormone estrogen.

Today there is growing concern about the effects of a number of chlorinated compounds that are chemically related to DDT and share its tendencies to persist, accumulate, and interfere with normal sex hormone functions. These chemicals, which include dioxin and PCBs, are produced by a variety of manufacturing processes. These pollutants have been described as "environmental estrogens," because they appear to mimic the effects of the hormone estrogen or to enhance estrogen's potency.

Studies have focused on the Great Lakes, which until recently had relatively high levels of these chlorinated compounds. There, populations of fish-eating river otters have declined sharply, and a variety of fish-eating birds, including the newly returned bald eagles, produce deformed offspring or eggs that never hatch (Fig. 40-8). Minute amounts of dioxin given to pregnant rats cause lowered sperm counts and genital deformities in their male offspring. In Florida's Lake Apopka, a spill of chlorinated chemicals, including DDT, in 1980 produced a 90% decline in the birth rate of the lake's alligators and a reduction of penis size in alligators who reached maturity in the lake. Several possible effects in humans are currently being investigated and debated by researchers. These effects include increased breast cancer rates in

*Figure 40-8 **The price of pollution***
Deformities such as the twisted beak of this double-crested cormorant from Lake Michigan have been linked to dioxin and related chemicals. Abnormalities of the reproductive and immune systems are also common in many types of organisms exposed to these pollutants. Predatory animals are especially vulnerable owing to biological magnification.

women and lowered sperm counts and increased testicular cancer rates in men.

Understanding the properties of pollutants and the workings of food webs is crucial if we are to prevent human health hazards as well as widespread loss of wildlife. Humans have considerable cause for concern, because we often feed high on the food chain. When you eat a tuna sandwich, for example, you are acting as a tertiary or even quaternary consumer. In addition, our long life span provides more time for substances stored in the body to accumulate to toxic levels.

3 How Do Nutrients Move within Ecosystems?

www

In contrast to the energy of sunlight, nutrients do not flow down onto Earth in a steady stream from above. For practical purposes, the same pool of nutrients has been supporting life for more than 3 billion years. Nutrients are elements and small molecules that form all the chemical building blocks of life. Some, called **macronutrients**, are required by organisms in large quantities. These include water, carbon, hydrogen, oxygen, nitrogen, phosphorus, sulfur, and calcium. **Micronutrients**, including zinc, molybdenum, iron, selenium, and iodine, are required only in trace quantities. **Nutrient cycles**, also called *biogeochemical cycles*, describe the pathways these substances follow as they move from the communities to the nonliving portions of ecosystems and then back again to communities.

The sources and storage sites of nutrients are called **reservoirs**. The major reservoirs are generally in the nonliving, or **abiotic**, environment. For example, carbon has several major reservoirs: It is stored as carbon dioxide gas in the atmosphere, in dissolved form in oceans, and as fossil fuels underground. In the following section we briefly describe the cycles of carbon, nitrogen, phosphorus, and water.

Carbon Cycles through the Atmosphere, Oceans, and Communities

Chains of carbon atoms form the framework of all organic molecules, including those that form hair, muscle, blood, bone, and enzymes. Carbon enters the living community through capture of carbon dioxide (CO_2) during photosynthesis by producers. On land, producers (such as your garden plants) acquire CO_2 from the atmosphere, where it comprises 0.036% of the total gases. Aquatic producers in the ocean, such as seaweeds and diatoms, find abundant CO_2 for photosynthesis dissolved in the water. In fact, far more CO_2 is stored in the oceans than in the atmosphere. Producers return some CO_2 to the atmosphere and ocean during cellular respiration and incorporate the rest into their bodies. Primary consumers,

such as cows, shrimp, or tomato hornworms, eat the producers and acquire the carbon stored in their tissues. These herbivores also release some carbon through respiration and store the rest—and so on through the trophic levels. All living things eventually die, and their bodies are broken down by detritus feeders and decomposers. Cellular respiration by these organisms, which abound in your compost pile, returns CO_2 to the atmosphere and oceans. Carbon dioxide passes freely between these two great reservoirs (Fig. 40-9).

Some carbon cycles much more slowly. For example, mollusks and marine protists extract CO_2 dissolved in water and combine it with calcium to form calcium carbonate ($CaCO_3$), from which they construct their shells. After death, the shells of these organisms collect in undersea deposits, are buried, and may eventually be converted to limestone. Geological events may expose the limestone, which will dissolve gradually as water runs over it, making the carbon available to living organisms once more.

Another long-term part of the carbon cycle is the production of fossil fuels. **Fossil fuels** are formed from the remains of ancient plants and animals and serve as a third major reservoir for carbon. The carbon found in the organic molecules of the ancient organisms in these deposits is transformed by high temperatures and pressures over millions of years into coal, oil, or natural gas. The energy of prehistoric sunlight is also trapped in fossil fuels. When people burn fossil fuels to utilize this stored energy, CO_2 is released into the atmosphere. In addition to the burning of fossil fuels, human activities such as cutting and burning Earth's great forests (where much carbon is stored) are increasing the amount of CO_2 in the atmosphere, as we shall describe later in this chapter.

The Major Reservoir for Nitrogen Is the Atmosphere

The atmosphere contains about 79% nitrogen gas (N_2) and is thus the major reservoir for this important nutrient. Nitrogen is a crucial component of proteins, many vitamins, and the nucleic acids DNA and RNA. Interestingly, neither plants nor animals can extract this gas from the atmosphere. Instead, plants must be supplied with nitrate (NO_3^-) or ammonia (NH_3). But how is atmospheric nitrogen converted to these molecules? Ammonia is synthesized by certain bacteria and cyanobacteria that engage in **nitrogen fixation**, a process that combines nitrogen with hydrogen. Some of these bacteria live in water and soil. Others have entered a symbiotic association with plants called **legumes** (including alfalfa, soybeans, clover, and peas), where they live in special swellings on the roots (see Chapter 19). Decomposer bacteria can also produce ammonia from amino acids and urea found in dead bodies and wastes. Still other bacteria convert ammonia to nitrate.

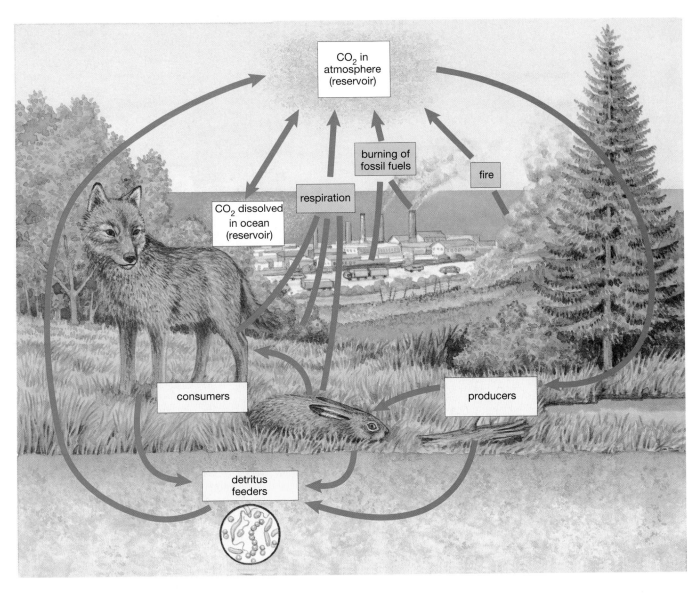

Figure 40-9 *The carbon cycle*
Carbon is captured from the atmosphere during photosynthesis and passed up through the trophic levels. It is
released during respiration from all trophic levels and by the burning of forests and fossil fuels.

Nitrogen is combined with oxygen by electrical storms
and by burning forests and fossil fuels. Plants incorporate the nitrogen from ammonia and nitrate into amino
acids, proteins, nucleic acids, and vitamins. These nitrogen-containing molecules from the plant are eventually
consumed, either by primary consumers, detritus feeders, or decomposers. As it is passed through the food web,
some of the nitrogen is released in wastes and dead bodies, which decomposer bacteria convert back to nitrate
and ammonia. The cycle is balanced by a continuous return of nitrogen to the atmosphere by **denitrifying bacteria**. These residents of wet soil, swamps, and estuaries

break down nitrate, releasing nitrogen gas back to the atmosphere (Fig. 40-10).

In human-dominated ecosystems, such as farm fields,
gardens, and suburban lawns, ammonia and nitrate are
supplied by chemical fertilizers. These fertilizers—like
those you sprinkled over your garden before you planted vegetables—are produced by using the energy in fossil fuels to artificially "fix" atmospheric nitrogen. In fact,
human activities now dominate the nitrogen cycle, a
cause for serious environmental concerns. By burning
fossil fuels, by burning forests and draining wetlands that
store nitrogen, by cultivating legumes, and by producing

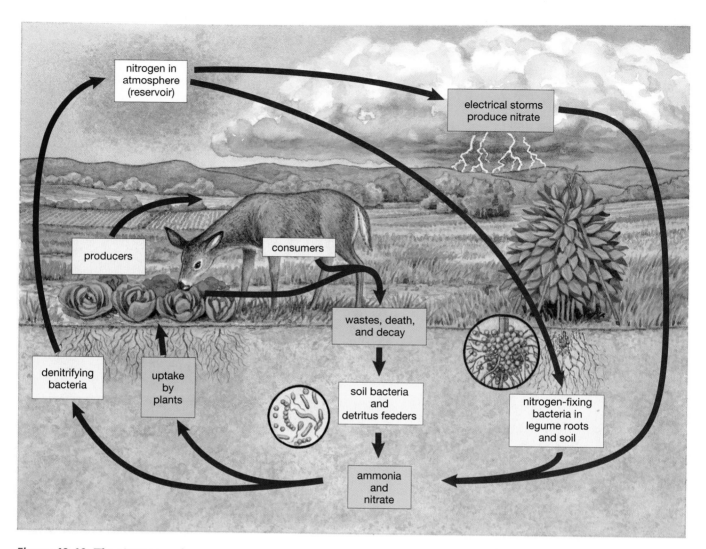

Figure 40-10 *The nitrogen cycle*
From its reservoir in the atmosphere, nitrogen gas is combined with oxygen to form nitrates by lightning or by the burning of forests or fossil fuels and carried to Earth dissolved in rain. Nitrogen-fixing bacteria produce ammonia. Nitrate and ammonia are also synthesized by humans for use in fertilizers. These are absorbed by plants and other producers and incorporated into biological molecules that are passed up through the trophic levels. Nitrate and ammonia are released by excretion or by decomposer bacteria. Other bacteria convert these molecules back to atmospheric nitrogen, completing the cycle.

fertilizers, humans each year are converting an estimated 140 million metric tons of nitrogen from its gaseous form to ammonia and nitrogen oxides (nitrogen combined with oxygen). Natural processes utilize 90 million to 140 million metric tons of nitrogen yearly. Various types of nitrogen oxides contribute to global warming, destroy the protective ozone layer, and acidify precipitation. Runoff from fertilized farm fields into the Mississippi River is carried to the Gulf of Mexico, where scientists believe it is causing a 7000-square-mile "dead zone" that appears there each summer. When they enter ecosystems, nitrogen oxides may change the composition of plant communities or destroy forests by acid rain, which we shall discuss later in this chapter.

The Major Reservoir for Phosphorus Is Rock

Phosphorus is a crucial component of biological molecules, including the energy transfer molecules (ATP and NADP), nucleic acids, and the phospholipids of cell membranes. It is also a major component of vertebrate teeth and bones. In contrast to the cycles of carbon and nitrogen, the phosphorus cycle has no atmospheric component. The reservoir of phosphorus in ecosystems is rock, where it is bound to oxygen in the form of phosphate. As phosphate-rich rocks are exposed and eroded, rainwater dissolves the phosphate. Dissolved phosphate is readily absorbed through the roots of plants and by other autotrophs, such as photosynthetic protists and cyanobacteria, where it is incorporated into biological molecules. From these producers, phosphorus is passed through food webs (Fig. 40-11). At each level, excess phosphate is excreted. Ultimately, decomposers return the phosphorus that remains in dead bodies back to the soil and water in the form of phosphate. It may then be reabsorbed by autotrophs, or it may become bound to sediment and eventually reincorporated into rock.

Some of the phosphate dissolved in fresh water is carried to the oceans. Although much of this phosphate ends up in marine sediments, some is absorbed by marine producers and eventually incorporated into the bodies of invertebrates and fish. Some of these, in turn, are consumed by seabirds, which excrete large quantities of phosphorus back onto the land. At one time, the guano (droppings) deposited by seabirds along the western coast of South America was mined, and it provided a major source of the world's phosphorus. Phosphate-rich rock is also mined, and the phosphate is incorporated into fertilizer. Soil that erodes from fertilized fields carries large quantities of phosphates into lakes, streams, and the ocean, where it stimulates the growth of producers. In lakes, phosphorus-rich runoff from land can stimulate such a rich growth of algae and bacteria that the natural community interactions of the lake are disrupted. You'll read more about this in the next chapter.

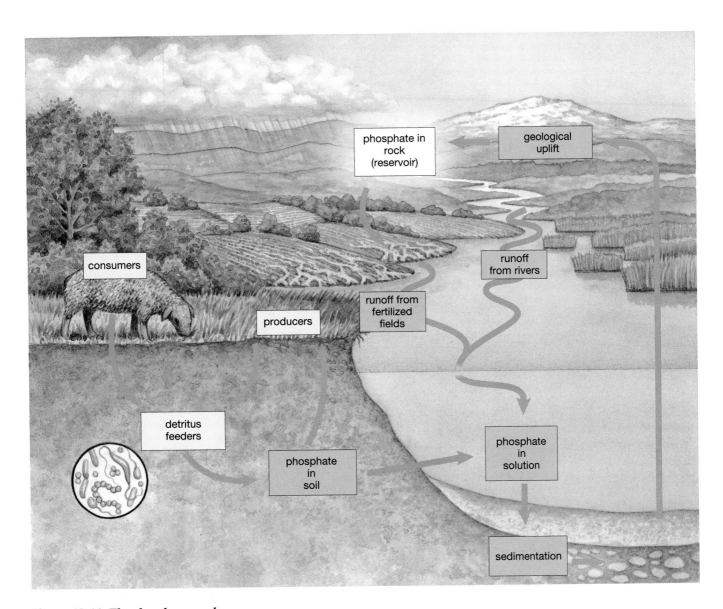

Figure 40-11 The phosphorus cycle
Phosphate dissolves from phosphate-rich rocks or from fertilizers and enters plants and other producers, where it is incorporated into biological molecules. These molecules are passed up through the trophic levels. Phosphate is excreted or returned to the soil and water by decomposer bacteria. It may then be reused by producers or eventually incorporated into rock.

Most Water Remains Chemically Unchanged during the Water Cycle

The water cycle, or **hydrologic cycle** (Fig. 40-12), differs from most other nutrient cycles in that most water remains in the form of water throughout the cycle and is not used in the synthesis of new molecules. The major reservoir of water is the ocean, which covers about three-quarters of Earth's surface and contains over 97% of the available water. The hydrologic cycle is driven by solar energy, which evaporates water, and by gravity, which draws the water back to Earth in the form of precipitation (rain, snow, sleet, and dew). Most evaporation occurs from the oceans, and much water returns directly to them by precipitation. Water falling on land takes various paths. Some water is evaporated from the soil, lakes, and streams. A portion runs off the land back to the oceans, and a small amount enters underground reservoirs. Because the bodies of living things are roughly 70% water, some of the water in the hydrologic cycle enters the living communities of ecosystems. It is absorbed by the roots of plants, and much of this is evaporated back to the atmosphere from their leaves. A small amount is combined with carbon dioxide during photosynthesis to produce high-energy molecules. Eventually these molecules are broken down during cellular respiration, releasing water back to the environment. Consumers get water from their food or by drinking.

As the human population has grown, fresh water has become scarce in many regions of the world where the human demand for it is high. This scarcity limits the abil-

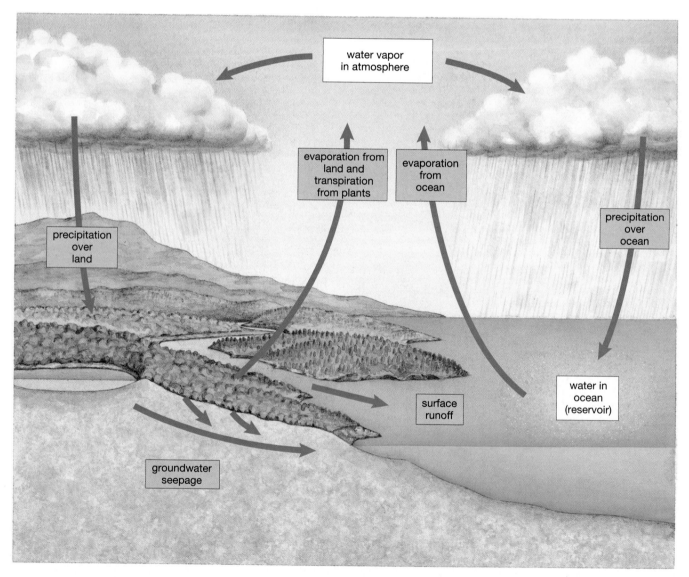

Figure 40-12 The hydrologic cycle
The hydrologic cycle is the simplest of the nutrient cycles.

ity to grow crops; currently, about 16% of cropland relies at least partially on irrigation. Lack of clean fresh water threatens human health where people rely on unsafe water for drinking. Even in the United States, to irrigate their crops, farmers are "mining" large underground pools of water that are not being replenished. Currently, about 500 major dams are being completed each year worldwide. Although damming rivers allows storage of fresh water, it disrupts communities associated with the river ecosystem and its natural floodplain.

4 What Is Causing Acid Rain and Global Warming?

Many of the environmental problems that beset modern society are the result of human interference in the way ecosystems function. Primitive peoples were sustained solely by the energy flowing from the sun, and they produced wastes that were readily taken back into the nutrient cycles. But as the population grew and technology increased, humans began to act more and more independently of these natural processes. We have mined substances, such as lead, arsenic, cadmium, mercury, oil, and uranium, that are foreign to natural ecosystems and toxic to many of the organisms in them (Fig. 40-13). In our factories, we synthesize substances never before found on Earth: pesticides, solvents, and a wide array of other industrial chemicals harmful to many forms of life. The Industrial Revolution, which began in earnest in the mid-nineteenth century, resulted in a tremendous increase in our reliance on energy from fossil fuels (rather than from sunlight) for heat, light, transportation, industry, and even agriculture. And industrial processes produce more nutrients than nutrient cycles can efficiently process.

In the following sections, we shall describe two environmental problems of global proportions that are a direct result of human reliance on fossil fuels: acid rain and global warming.

Overloading the Nitrogen and Sulfur Cycles Has Caused Acid Rain

Each year, the United States discharges about 30 million tons of sulfur dioxide into the atmosphere, two-thirds of it from power plants burning coal or oil (Fig. 40-14). The

Figure 40-13 A natural substance out of place
An oil-soaked sea bird is cleaned by rescuers after an oil spill off the coast of Alaska.

Figure 40-14 A source of acid deposition
Power plants burning high-sulfur coal with inadequate emission controls are a major source of atmospheric sulfur dioxide, the prime contributor to acid deposition.

rest is largely a by-product of industrial boilers, smelters, and refineries. Although volcanoes and hot springs also release sulfur dioxide, human industrial activities account for 90% of the sulfur dioxide in the atmosphere. Twenty-five million tons of nitrogen oxides are also released from vehicles, power plants, and industry in the United States annually.

Excess production of these substances was identified in the late 1960s as the cause of a growing environmental threat: *acid rain*, more accurately called **acid deposition**. Combined with water vapor in the atmosphere, nitrogen oxides are converted to nitric acid and sulfur dioxide to sulfuric acid. Days later, and commonly hundreds or thousands of miles from the source, the acids fall, either dissolved in rain or as microscopic dry particles. The acids are corrosive, eating away at statues and buildings (Fig. 40-15), damaging trees and crops, and turning lakes lifeless. Although both sulfuric and nitric acids form solutions in water vapor, sulfuric acid may also form particles under dry conditions. These particles, which may visibly cloud the air, are deposited even in the absence of rain, snow, or fog. The term *acid deposition* is used to describe both the wet and dry acid assaults on the environment.

Although sources of sulfur dioxide are widely distributed, in the United States they tend to be concentrated in the upper Ohio Valley, Indiana, and Illinois, where high-sulfur coal is burned in old power plants with few emission controls. As complex chemical reactions in the atmosphere produce sulfuric and nitric acids, winds carry them toward New England, where lakes and forests are particularly vulnerable. Some areas of the country have soft rocks that are rich in calcium carbonate, a natural buffer for acidity. However, in New England, hard rock, such as granite and basalt, predominate, and they cannot

Figure 40-15 Acid deposition is corrosive
This limestone statue at Rheims Cathedral in France is being dissolved away by acid deposition.

neutralize acid rain. Although the Northeast has been the most seriously damaged, it is not the only sensitive area; about 75% of the United States is at least moderately vulnerable to acid rain (Fig. 40-16).

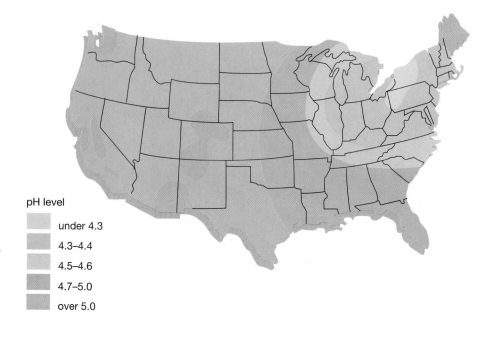

Figure 40-16 Acid rain distribution
The most acidic rain is concentrated in the Northeast, but thousands of lakes in the West are also vulnerable. (Source: National Acid Precipitation Assessment Program.)

pH level

under 4.3

4.3–4.4

4.5–4.6

4.7–5.0

over 5.0

Acid Deposition Damages Life in Lakes, Farms, and Forests

In New York's Adirondack Mountains, acid rain has made about 25% of all the lakes and ponds too acidic to support fish. But before the fish die, much of the food web that sustains them is destroyed. Clams, snails, crayfish, and insect larvae die first, then amphibians, and finally fish. The result is a crystal-clear lake—beautiful but dead. The impact is not limited to aquatic organisms. The loss of insect larvae and crustaceans has probably contributed to a dramatic decline in the population of black ducks, which feed on them.

Acid rain is a threat not only to lakes. Recent studies under laboratory conditions have shown that it also interferes with the growth and yield of many farm crops. Acidic water draining through the soil washes out essential nutrients such as calcium and potassium. Acid deposition may also kill decomposer microorganisms, thus preventing the return of nutrients to the soil. Plants, poisoned and deprived of nutrients, become weak and vulnerable to infection and insect attack. High in the Green Mountains of Vermont, scientists have witnessed the death of about half of the red spruce and beech trees and one third of the sugar maples since 1965. The snow, rain, and heavy fog that commonly cloak these eastern mountaintops are highly acidic. At a monitoring station atop Mount Mitchell in North Carolina, the pH of fog even on good days is 2.9, more acidic than vinegar. In this fragile and stressful environment, acid precipitation may well be tipping the balance against these forests (Fig. 40-17).

Figure 40-17 Acid deposition can destroy forests
Acid rain and fog have destroyed this forest atop Mount Mitchell in North Carolina.

Acid deposition increases the exposure of organisms to toxic metals, including aluminum, lead, mercury, and cadmium, which are far more soluble in acidified water than in water of neutral pH. Aluminum may inhibit plant growth. In acidic lakes and streams, aluminum dissolved from the surrounding rock and soil causes mucus to accumulate in the gills of fish, suffocating the fish. Drinking water in some households has been found dangerously contaminated with lead dissolved by acidic water from lead solder in old pipes. Boston's water supply must be chemically treated to counteract the acidity. Fish in acidified water have been found to have dangerous levels of mercury in their bodies, which is subject to biological magnification as it is passed through trophic levels.

Interference with the Carbon Cycle Is Causing Global Warming

Between 345 million and 280 million years ago, huge quantities of carbon were diverted from the carbon cycle when, under the warm, wet conditions of the Carboniferous period, the bodies of plants and animals were buried in sediments, escaping decomposition. Over time, heat and pressure converted their bodies into fossil fuels, such as coal, oil, and natural gas. Without human intervention, this carbon would have remained essentially untouched for millions of years. Ever since the Industrial Revolution, however, we have increasingly relied on the energy stored in these fuels. As we burn them in our power plants, factories, and cars, we release CO_2 into the atmosphere. Since 1850, the CO_2 content of the atmosphere has increased from 280 parts per million (ppm) to 360 ppm, an increase of more than 25%. That increase is continuing at the rate of 1.5 ppm yearly. The burning of fossil fuels is the single greatest cause of this increase.

A second important source of additional atmospheric CO_2 is global **deforestation**, the cutting of tens of millions of acres of forests each year. Deforestation is occurring principally in the tropics, where rain forests are rapidly being eliminated to increase the availability of agricultural land. The carbon stored in the massive trees in these forests returns to the atmosphere after they are cut, primarily through burning. During the past 200 years, human activities such as deforestation and the burning of fossil fuels have released as much excess CO_2 as is currently stored in all life on Earth.

Greenhouse Gases Trap Heat in the Atmosphere

Carbon dioxide still comprises only a tiny fraction of Earth's atmosphere, about 0.036%. However, CO_2 has an important property that makes its build-up a cause for concern: It traps heat. Atmospheric CO_2 acts something like the glass in a greenhouse, allowing energy in the form of sunlight to enter but absorbing and holding that

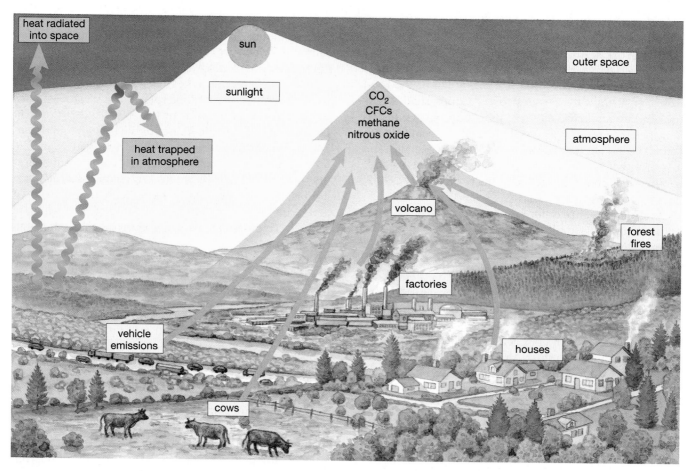

Figure 40-18 Increases in greenhouse gas emissions are causing global warming
Incoming sunlight warms Earth's surface and is radiated back to the atmosphere. Greenhouse gases absorb some of this heat, trapping it in the atmosphere. Human activities have greatly increased levels of greenhouse gases, resulting in a gradual rise in average global temperatures.

energy once it has been converted to heat (Fig. 40-18). Several other **greenhouse gases** share this property, including methane, chlorofluorocarbons (CFCs), water vapor, and nitrous oxide. The **greenhouse effect**, the ability of greenhouse gases to trap the sun's energy in a planet's atmosphere as heat, is a natural process. By keeping our atmosphere relatively warm, it allows life on Earth as we know it. However, human activities have amplified the natural greenhouse effect, producing a phenomenon called **global warming.**

Historical temperature records have revealed a global temperature increase of 0.5 to 0.7°C since 1860, paralleling the rise in atmospheric CO_2 and methane levels (Fig. 40-19). The 1980s were the warmest decade on record; the 11 years between 1987 and 1997 included 9 years of record-setting high temperatures, with 1997 posting the highest average global temperatures ever recorded. As a decade, the 1990s will undoubtedly set a new record.

Many interacting factors, such as changes in cloud cover caused by increased evaporation, possible increases in net primary productivity due to increased CO_2, and an uncertain ability of the ocean reservoir to absorb CO_2, make future climate prediction difficult and imprecise. However, many scientists believe that global warming is likely to cause a rise of 1.5° to 4.5°C in average world air temperatures by the end of the twenty-first century. This range may not seem like much, but average air temperatures some 20,000 years ago, during the peak of the last Ice Age, were only about 5°C lower than at present. The temperature increase since the last Ice Age radically changed the species composition of forests throughout North America and caused ocean levels to rise so that the shore moved inland by about 100 miles. As a result of global warming, within the next century Earth may be hotter than it has been at any time in the past million years, with the change occurring more rapidly than has any climate change in Earth's history.

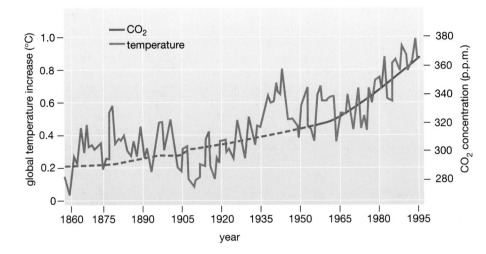

Figure 40-19 Global warming parallels CO₂ increases
The CO_2 concentration of the atmosphere (blue line) has shown a steady increase since 1860. The dashed portion of that curve represents measurements made from air trapped in ice cores; the solid portion reflects direct measurements made at Mauna Loa, Hawaii. Average global temperatures (red line) have also shown a gradual increase, paralleling the increasing atmospheric CO_2.

Global Warming Could Have Severe Consequences

As the geochemist James White of the Institute of Arctic and Alpine Research at the University of Colorado puts it, "If the Earth had an operating manual, the chapter on climate might begin with the caveat that the system has been adjusted at the factory for optimum comfort, so don't touch the dials." As polar ice caps and glaciers melt and ocean waters expand in response to atmospheric warming, sea levels rise. A dramatic sea-level rise would threaten coastal cities and flood coastal wetlands. These threatened ecosystems are the breeding grounds for many species of birds, fish, shrimp, and crabs, whose populations could be severely diminished.

Another serious consequence of the warming trend is a shift in the global distribution of temperature and rainfall. Even small temperature changes can dramatically alter the paths of major air and ocean currents, which would in turn change precipitation patterns in unpredictable ways. Some land that is currently cultivated only with the help of irrigation would become too dry for agriculture. Other areas would receive more rain than at present. Agricultural disruption could be disastrous for human populations already on the brink of starvation.

The impact of global warming on forests could be profound. The distribution of tree species is exquisitely sensitive to average annual temperature, and small changes could dramatically alter the extent and species composition of our forests. The United Nations-sponsored Intergovernmental Panel on Climate Change (IPCC) concluded that global warming at the high end of current estimates would wipe out much of the hardwood forests of the eastern United States; those forests would be replaced by grass and shrubland. The IPCC panel also noted that global warming will increase the range of tropical disease-carrying organisms, such as malaria-transmitting mosquitos, with direct consequences for human health.

Although the scenario seems bleak, global Earth Summit conferences, such as that held in Kyoto, Japan, in December 1997, offer a ray of hope. There, representatives from 160 nations proposed a treaty whose goal is to curb CO_2 emissions. If ratified, the treaty would commit industrialized nations to reducing CO_2 emissions to about 5% below 1990 levels by the year 2012. Developing countries, such as China and India, are not bound by the Kyoto agreement. Many observers are concerned that the benefits of the treaty could be overwhelmed by emissions from such nations as they strive to industrialize and increase their standard of living. Although the obstacles to a global reduction of CO_2 emissions are considerable, global conferences such as that in Kyoto are evidence of a worldwide recognition of global warming and a concern for its consequences, the first steps toward change.

What can we as individuals do to help reduce both acid rain and global warming? A car with a fuel efficiency of 20 miles per gallon emits 1 pound of CO_2 per mile, so we can substantially reduce emissions of CO_2 by using fuel-efficient vehicles, car-pools, and public transportation. As electricity is generated in fossil fuel-fired power plants, so are tremendous quantities of CO_2, sulfur dioxide, and nitrogen oxides. To conserve electricity, we can use more-efficient appliances, replace incandescent light bulbs with fluorescent light bulbs, and turn off unused lights. Insulating and weatherproofing our homes can significantly reduce fuel consumption, as will designing solar energy features into new homes. Recycling is also a tremendous energy saver—for example, 95% of the energy used to produce an aluminum can is conserved when the can is recycled. And we can support *reforestation* efforts to replace trees both in tropical rain forests and locally.

Summary of Key Concepts

1) What Are the Pathways of Energy and Nutrients through Communities?

Ecosystems are sustained by a continuous flow of energy from sunlight and a constant recycling of nutrients.

2) How Does Energy Flow through Communities?

Energy enters the biotic portion of ecosystems when it is harnessed by autotrophs during photosynthesis. Net primary productivity is the amount of energy that autotrophs store in a given unit of area over a given period of time.

Trophic levels describe feeding relationships in ecosystems. Autotrophs are the producers, the lowest trophic level. Herbivores occupy the second level as primary consumers. Carnivores act as secondary consumers when they prey on herbivores and as tertiary or higher-level consumers when they eat other carnivores.

Feeding relationships in which each trophic level is represented by one organism are called food chains. In natural ecosystems, feeding relationships are far more complex and are described as food webs. Detritus feeders and decomposers, which digest dead bodies and wastes, use and release the energy stored in these substances and free up nutrients for recycling. In general, only about 10% of the energy captured by organisms at one trophic level is converted to the bodies of organisms in the next highest level. The higher the trophic level, the less energy is available to sustain it. As a result, plants are more abundant than herbivores, and herbivores are more common than carnivores. The storage of energy at each trophic level is illustrated graphically as an energy pyramid. The energy pyramid explains biological magnification, the process by which toxic substances accumulate in increasingly high concentrations in progressively higher trophic levels.

3) How Do Nutrients Move within Ecosystems?

A nutrient cycle depicts the movement of a particular nutrient from its reservoir (normally in the abiotic, or nonliving, portion of the ecosystem) through the biotic, or living, portion of the ecosystem and back to its reservoir, where it is again available to producers. Carbon reservoirs include the oceans, atmosphere, and fossil fuels. Carbon enters producers through photosynthesis. From autotrophs it is passed through the food web and released to the atmosphere as CO_2 during cellular respiration.

The major reservoir of nitrogen is the atmosphere. Nitrogen gas is converted by bacteria and human industrial processes into ammonia and nitrate, which can be used by plants. Nitrogen passes from producers to consumers and is returned to the environment through excretion and the activities of detritus feeders and decomposers.

The reservoir of phosphorus is in rocks as phosphate, which dissolves in rainwater. Phosphate is absorbed by photosynthetic organisms, then passed through food webs. Some is excreted, and the rest is returned to the soil and water by decomposers. Some is carried to the oceans, where it is deposited in marine sediments. Humans mine phosphate-rich rock to produce fertilizer.

The major reservoir of water is the oceans. Water is evaporated by solar energy and returned to Earth as precipitation. The water flows into lakes, underground reservoirs, and in rivers to the oceans. Water is absorbed directly by plants and animals and is also passed through food webs. A small amount is combined with CO_2 during photosynthesis to form high-energy molecules.

4) What Is Causing Acid Rain and Global Warming?

Environmental problems occur when people interfere with the natural functioning of ecosystems. Human industrial processes release toxic substances and produce more nutrients than nutrient cycles can efficiently process. Through massive consumption of fossil fuels, we have increased the flow of energy and disrupted the natural cycles of carbon, sulfur, and nitrogen, causing acid rain and global warming (an amplification of the greenhouse effect). Solutions to both acid rain and global warming include reducing our use of fossil fuels and undertaking a major reforestation effort.

Key Terms

abiotic   *p. 844*
acid deposition   *p. 850*
autotroph   *p. 837*
biodegradable   *p. 843*
biological magnification   *p. 843*
biomass   *p. 837*
carnivore   *p. 837*
consumer   *p. 837*
decomposer   *p. 838*

deforestation   *p. 851*
denitrifying bacterium   *p. 845*
detritus feeder   *p. 838*
energy pyramid   *p. 842*
food chain   *p. 838*
food web   *p. 838*
fossil fuel   *p. 844*
global warming   *p. 852*
greenhouse effect   *p. 852*

greenhouse gas   *p. 852*
herbivore   *p. 837*
heterotroph   *p. 837*
hydrologic cycle   *p. 848*
legume   *p. 844*
macronutrient   *p. 844*
micronutrient   *p. 844*
net primary productivity   *p. 837*
nitrogen fixation   *p. 844*

nutrient cycle   *p. 844*
primary consumer   *p. 837*
producer   *p. 837*
reservoir   *p. 844*
secondary consumer   *p. 838*
tertiary consumer   *p. 838*
trophic level   *p. 837*

Thinking Through the Concepts

Multiple Choice

1. *Why do scientists think that human-induced global warming will be more harmful to plants and animals than were past, natural climate fluctuations?*
   a. because temperatures will change faster
   b. because the temperature changes will be larger
   c. because species now are less adaptable than species in the past
   d. because ecosystems are now more complicated than they used to be
   e. because the temperature changes will last longer

2. *The major source(s) of carbon for living things is (are)*
   a. coal, oil, and natural gas
   b. plants
   c. CO_2 in the atmosphere and oceans
   d. methane in the atmosphere
   e. carbon in animal bodies

3. *Why would you expect there to be a smaller biomass of big predators (lions, leopards, hunting dogs, etc.) than of grazing mammals (gazelles, zebra, elephants, etc.) in the African savanna?*
   a. too little cover for the predators to hide in
   b. the inefficiency of energy transfer between trophic levels
   c. like domestic cats, large predators are susceptible to hair balls
   d. many predators occur in social groups, thus limiting their numbers
   e. because grazers are better adapted to moving long distances, they can better follow the rains

4. *What is the one group of organisms that is able to fix atmospheric nitrogen into forms usable by living organisms?*
   a. plants
   b. fungi
   c. insects
   d. bacteria
   e. viruses

5. *The biological process by which carbon is returned to its reservoir is*
   a. photosynthesis
   b. denitrification
   c. deamination
   d. glycolysis
   e. cellular respiration

6. *As a black widow spider consumers her mate, what is the lowest trophic level she could be occupying?*
   a. 3rd   b. 1st
   c. 2nd   d. 4th
   e. 5th

? Review Questions

1. What makes the flow of energy through ecosystems fundamentally different from the flow of nutrients?

2. What is an autotroph? What trophic level does it occupy, and what is its importance in ecosystems?

3. Define *primary productivity*. Would you predict higher productivity in a farm pond or an alpine lake? Defend your answer.

4. List the first three trophic levels. Among the consumers, which are most abundant? Why would you predict that there will be a greater biomass of plants than herbivores in any ecosystem? Relate your answer to the "10% law."

5. How do food chains and food webs differ? Which is the more accurate representation of actual feeding relationships in ecosystems?

6. Define *detritus feeders* and *decomposers*, and explain their importance in ecosystems.

7. Trace the movement of carbon from its reservoir through the biotic community and back to the reservoir. How have human activities altered the carbon cycle, and what are the implications for future climate?

8. Explain how nitrogen gets from the air to a plant.

9. Trace a phosphorus molecule from a phosphate-rich rock into the DNA of a carnivore. What makes the phosphorus cycle fundamentally different from the carbon and nitrogen cycles?

10. Trace the movement of a water molecule from the moment it leaves the ocean until it eventually reaches a plant root, then a plant stoma, and then makes its way back to the ocean.

Applying the Concepts

1. What could your college or university do to reduce its contribution to acid rain and global warming? Be specific and, if possible, offer practical alternatives to current practices.

2. Relate fossil fuel consumption to (a) the decline of forests in the Northeast and (b) the loss of aquatic life in lakes in the Northeast and Canada. Trace each step from the burning of gasoline in a car or power plant to the death of organisms.

3. Define and give an example of *biological magnification*. What qualities are present in materials that undergo biological magnification? In which trophic level are the problems worst, and why?

4. Discuss the contribution of population growth to (a) acid rain and to (b) the greenhouse effect.

5. Describe what would happen to a population of deer if all predators were removed and hunting was banned. Include effects on vegetation as well as on the deer population itself. Relate your answer to carrying capacity as discussed in Chapter 38.

Group Activity

How do humans affect the food web? In the 1950s, the island country Borneo had a severe problem with malaria, which is carried by mosquitoes. The World Health Organization (WHO) decided to spray a pesticide called DDT to rid the island of mosquitoes. The accompanying diagram (adapted from The Center for Applied Research in Education, 1991) shows a simplified food chain of the ecosystem of those affected by the DDT. Additional facts:

- DDT causes nerve damage, slow reflexes, and eventual death in animals.

- When an animal consumes another animal that has consumed DDT, the consumer is also affected by the DDT.

- DDT kills mosquitoes; it also affects cockroaches but does not kill them.

In groups of three, predict the effects DDT will have on each member of the food web. Your group should be prepared to share with the class and to compare your findings to the actual events.

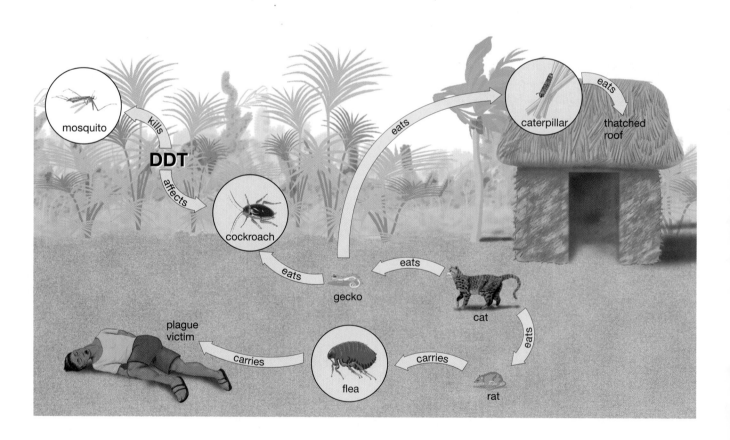

For More Information

Baskin, Y. "Ecologists Dare to Ask: How Much Does Diversity Matter?" *Science*, April 8, 1994. What is the scientific evidence for the importance of biodiversity to ecosystem function?

Bazzaz, F. A, and Fajer, E. D. "Plant Life in a CO_2-Rich World." *Scientific American*, January 1992. How increasing levels of CO_2 may alter the structure and function of ecosystems.

Earthworks Group. "Fifty Simple Things You Can Do to Save the Earth." Earthworks Press, Berkeley, 1989. Provides practical, simple advice on how individuals can make a difference.

Karl, T. R., Nicholls, N., and Gregory, J. "The Coming Climate." *Scientific American*, May 1997. What changes in weather patterns can we expect in a warmer world?

Matthews, S. W. "Under the Sun." *National Geographic*, October 1990. A comprehensive, readable exploration of the greenhouse effect, its causes, and its consequences.

McLachlan, J. A., and Arnold, S. F. "Environmental Estrogens." *American Scientist*, September/October 1996. Reviews the evidence that a variety of environmental pollutants are mimicking the effects of estrogen or other hormones.

Mlot, C. "Tallying Nitrogen's Increasing Toll." *Science News*, February 15, 1997. Human activities now dominate the nitrogen cycle, with increasingly disruptive effects.

Monastersky, R. "Health in the Hot Zone." *Science News*, April 6, 1996. Reviews the predicted effects on humans of global warming.

Monastersky, R. "Beyond Hot Air." *Science News*, May 24, 1997. This informative article asks the question, Will the world adopt strict limits on greenhouse gas emissions?

Post, W. M., and others. "The Global Carbon Cycle." *American Scientist*, July/August 1990. A readable and comprehensive look at the natural carbon cycle and how humans have influenced it.

Answers to Multiple-Choice Questions

1. a   2. c   3. b   4. d   5. e   6. d

"We must consider our planet to be on loan from our children, rather than being a gift from our ancestors."

Gro Harlem Brundtland, former Prime Minister of Norway

The stunning diversity of a coral reef: Reef-building corals and brightly colored fish and invertebrates characterize the now-dwindling coral reefs throughout the world.

Earth's Diverse Ecosystems 41

Net Watch
On-line resources for this chapter are on the World Wide Web at:
http://www.prenhall.com/audesirk
(click on the Table of Contents link and then select Chapter 41).

Looking back on your summer backpacking trip, you recall the series of different communities you traversed on your trek from open grassland to mountain peak. You might be surprised to learn that, if you had driven north for about 2000 miles starting in the grasslands of Colorado, you would have passed through a series of communities remarkably similar to those you traversed in your hike. Grasslands extending into Canada eventually give way to coniferous forests, then to arctic tundra. Some of the species of plants differ, but the communities are clearly recognizable. How can a single mountainside duplicate the conditions found in a trip north over thousands of miles?

A world traveler will notice that although the plant species may differ from region to region, very similar groups of plants are found wherever a particular climate exists. Deciduous forests, deserts, grasslands, tropical rain forests, and coniferous forests all have distinguishing features that identify them no matter where they are located or which species inhabit them. Marine ecosystems, such as coral reefs, are scattered in particular latitudes worldwide. For the individual fortunate and adventurous enough to dive among them, the characteristic groupings of reef-building corals with brightly colored and diverse fish and invertebrates make these ecosystems unmistakable—and unforgettable.

What are the characteristics of different communities on land and in water? What factors determine where these communities are located? In this, our final chapter, we explore the spatial distribution of life on Earth.

1) What Factors Influence Earth's Climate?

The distribution of life, particularly on land, is dramatically affected by both weather and climate. **Weather** refers to short-term fluctuations in temperature, humidity, cloud cover, wind, and precipitation in a region over periods of hours or days. **Climate**, in contrast, refers to patterns of weather that prevail from year to year and even century to century in a particular region. The amount of sunlight and water and the range of temperatures determine the climate of a given region. Whereas weather affects individual organisms, climate influences and limits the overall distribution of entire species.

Both Climate and Weather Are Driven by the Sun

Both climate and weather are driven by a great thermonuclear engine: the sun. Solar energy reaches Earth in a range of wavelengths, from short, high-energy ultraviolet (UV) rays, through visible light, to the longer infrared wavelengths that produce heat. The solar energy that reaches Earth drives the wind, the ocean currents, and the global water cycle. Before it reaches Earth's surface, however, sunlight is modified by the atmosphere. A layer relatively rich in ozone (O_3) is located in the middle atmosphere. This *ozone layer* absorbs much of the sun's high-energy UV radiation, which can damage biological molecules (see Earth Watch: The Ozone Hole—A Puncture in Our Protective Shield). Dust, water vapor, and clouds scatter light, reflecting some of the energy back into space. Carbon dioxide, water vapor, ozone, methane, and other *greenhouse gases* selectively absorb energy at infrared wavelengths, trapping heat in the atmosphere. Human activities have intensified this natural *greenhouse effect*, as described in Chapter 40.

Only about half the solar energy reaching the atmosphere actually strikes Earth's surface. Of this, small fractions are immediately reflected directly back into space or used in photosynthesis; the rest is absorbed as heat. Eventually, nearly all the incoming solar energy is returned to space, either as light or as infrared radiation (heat). The solar energy temporarily absorbed and stored as heat by the atmosphere and by Earth's surface maintains Earth's relative warmth.

Many Physical Factors Also Influence Climate

Many physical factors influence climate. Among the most important are latitude, air currents, ocean currents, and the presence of mountains and irregularly shaped continents.

Latitude Influences the Angle at Which Sunlight Strikes Earth

Latitude, measured in degrees, is the distance north or south from the equator, which is located at 0° latitude. The amount of sunlight that strikes a given area of Earth's surface has a major effect on average yearly temperatures. At the equator, sunlight hits Earth's surface nearly at a right angle, making the weather there consistently warm. Farther north or south, the sun's rays strike Earth's surface at a greater slant. This angle spreads the same amount of sunlight over a larger area, producing lower overall temperatures (Fig. 41-1).

Earth is tilted on its axis as it makes its yearly trip around the sun. The higher latitudes therefore experience considerable variation in the directness of sunlight throughout the year, resulting in pronounced seasons. For example, when the Northern Hemisphere is tilted toward the sun, it receives sunlight most directly and experiences summer; when that hemisphere is in winter, the Southern Hemisphere is closest to the sun (Fig. 41-1). The tilting hardly affects the angle of the equator to the sun's rays, so there is little seasonal variation there.

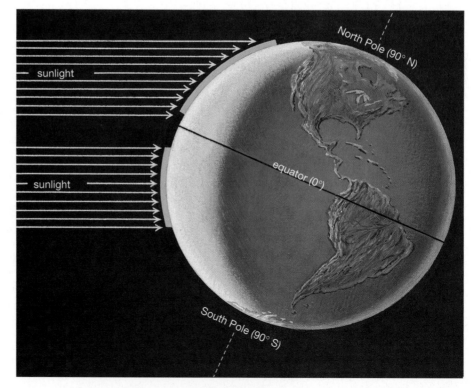

Figure 41-1 Earth's curvature and tilt produce seasons and climate
Seasons and climate are the result of an uneven distribution of solar energy. Near the equator, sunlight falls nearly perpendicularly on Earth's surface, so its warmth is concentrated onto a relatively small area. Toward the poles, the same amount of sunlight falls over a much larger surface area. Temperatures are thus highest at the equator and lowest at the poles. The tilt of Earth on its axis causes seasonal variations in the directness of sunlight. The Southern Hemisphere, when tilted toward the sun as shown here, receives more direct sunlight than does the Northern Hemisphere, which is tilted away from the sun. Thus, it is winter in the Northern Hemisphere and summer in the Southern Hemisphere.

Earth Watch www
The Ozone Hole—A Puncture in Our Protective Shield

A small fraction of the radiant energy produced by the sun, called *ultraviolet,* or UV, radiation, is so highly energetic that it can damage biological molecules. In small quantities, UV radiation helps human skin produce vitamin D and causes tanning in fair-skinned people. But in larger doses, UV causes sunburn and premature aging of skin, skin cancer, and cataracts, a condition in which the lens of the eye becomes cloudy.

Fortunately, most of the UV radiation is filtered out by ozone in the stratosphere, a layer of atmosphere extending from 10 to 50 kilometers (6 to 30 miles) above Earth. In pure form, *ozone* (O_3) is a bluish, explosive, and highly poisonous gas. In the stratosphere, the normal concentration of ozone is about 0.1 part per million (ppm), compared with 0.02 ppm in the lower atmosphere. This ozone-enriched layer is called the **ozone layer.** Ultraviolet light striking ozone and oxygen causes reactions that break down as well as regenerate ozone. In the process, the UV radiation is converted to heat, and the overall level of ozone remains reasonably constant—or so it did until humans intervened.

In 1985, British atmospheric scientists published a startling discovery: The springtime levels of stratospheric ozone over Antarctica had declined by more than 40% since 1977. A hole had been punctured in Earth's protective shield. In the localized *ozone hole* over Antarctica, ozone now dips to about one-third of its pre-depletion levels (Fig. E41-1). Although depletion of the ozone layer is most severe over Antarctica, the ozone layer is somewhat reduced over most of the world, including nearly all of the continental United States. Recently, springtime ozone levels over Greenland, Scandinavia, and western Siberia were reduced up to 40%. Evidence of ozone thinning over the Arctic has led researchers to predict that we will soon have an Arctic, as well as an Antarctic, ozone hole.

Satellite data reveal that, since the early 1970s, UV radiation per decade has risen by nearly 7% in the Northern Hemisphere and by almost 10% in the Southern Hemisphere. Skin cancer rates are predicted to rise as a result. Studies using mice show that exposure to UV light in dosages similar to a day at the beach without sunscreen can depress their immune systems, making them more vulnerable to a variety of diseases. Human health effects are only one cause for concern. Photosynthesis by phytoplankton, the producers for marine ecosystems, is reduced under the ozone hole above Antarctica. Some types of trees and farm crops are also harmed by increased UV radiation.

The decline in the thickness of the ozone layer is caused by increasing levels of chlorofluorocarbons (CFCs), compounds that contain chlorine, fluorine, and carbon. Developed in 1928, these gases were widely used as coolants in refrigerators and air conditioners, aerosol spray propellants, agents for producing Styrofoam®, and cleansers for electronic parts. These chemicals are very stable and were considered safe. Their stability, however, proved to be a major problem, because they remain chemically unchanged as they gradually rise into the stratosphere. There, under intense bombardment by UV light, the CFCs break down, releasing chlorine atoms. Chlorine catalyzes the breakdown of ozone to oxygen gas (O_2) while remaining unchanged itself. Clouds over the Arctic and Antarctic regions are composed of ice particles that provide a surface on which the reaction can occur.

But we have taken the first steps toward "plugging" the ozone hole. In an almost unprecedented example of global cooperation and concern, talks beginning in 1985 led to treaties in 1990 and 1992 in which industrialized nations throughout the world agreed to phase out ozone-depleting chemicals rapidly, eliminating CFCs by 1996. These compounds are no longer used in Styrofoam production, and CFC substitutes have been found for use in spray cans, refrigerators, and car air-conditioners as well. Ground-level global atmospheric chlorine levels (an indicator of CFC use) peaked in 1994, but since then chlorine levels near the ground have been slightly reduced. Because the highly stable CFCs can persist 50 to 100 years and take a decade or more to ascend into the stratosphere, the millions of tons already released will continue to erode our protective ozone shield, with possibly severe consequences. Scientists are optimistic, however, that the ozone hole may begin to get smaller by the year 2010 and will possibly be eliminated by 2050 as a direct result of global cooperation.

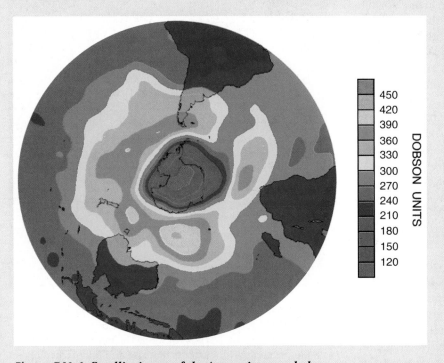

DOBSON UNITS

450
420
390
360
330
300
270
240
210
180
150
120

Figure E41-1 *Satellite image of the Antarctic ozone hole*
The ozone hole is seen in gray and pink on this image recorded by satellite in October 1997, when ozone levels dipped to less than one-third of normal. Dobson units measure the thickness of the ozone (in hundredths of millimeters) in a given layer of atmosphere if it were purified and held at standard temperature and pressure.

Air Currents Produce Broad Climatic Regions

Air currents are generated by Earth's rotation and by differences in temperature between different air masses. Because warm air is less dense than cold air, as the sun's direct rays fall on the equator, heated air rises there. The warm air of the tropics is also laden with water evaporated by solar heat (Fig. 41-2a). As the water-saturated air rises, it cools somewhat. Cool air cannot retain as much moisture as can warm air, so water condenses from the rising air and falls as rain. The direct rays of the sun and the rainfall produced when warm, moist air rises and is cooled create a band around the equator, called the *tropics*. This region is both the warmest and the wettest on Earth. The cooler dry air then flows north and south from the equator. Near 30° N and 30° S, the cooled air is dense enough to sink. As it sinks, the air is warmed by heat radiated from Earth. By the time it reaches the surface, it is both warm and very dry. Not surprisingly, the major deserts of the world are found at these latitudes (Fig. 41-2a,b). This air then flows back toward the equator. Farther north and south, this general circulation pattern is repeated, dropping moisture at around 60° N and 60° S and creating extremely dry conditions at the North and South Poles.

Ocean Currents Moderate Near-Shore Climates

Ocean currents are driven by Earth's rotation, winds, and the direct heating of the water by the sun. Continents interrupt the currents, breaking them into roughly circular patterns called **gyres**. Gyres rotate clockwise in the Northern Hemisphere and counterclockwise in the Southern Hemisphere (Fig. 41-3). Because water both heats and cools more slowly than does land or air, ocean currents tend to moderate temperature extremes. Coastal areas, then, generally have less-variable climates than do areas near the center of continents. For example, a gyre in the Atlantic Ocean brings warm water (the Gulf Stream) from equatorial regions north along the eastern coast of North America, creating a warmer, moister climate than is found farther inland. It then carries the still-warm water farther north and east, warming the western coast of Europe before returning south.

Continents and Mountains Complicate Weather and Climate

If Earth's surface were uniform, climate zones would occur in bands corresponding to latitude. These zones would result from the interaction of temperature and rainfall, which are determined by the rise and fall of air masses (see Fig. 41-2a). The presence of irregularly shaped continents (which heat and cool relatively quickly) amidst oceans (which heat and cool more slowly) alters the flow of wind and water and contributes to the irregular distribution of ecosystems.

Variations in elevation within continents further complicate the situation. As elevation increases, the atmosphere becomes thinner and retains less heat. The

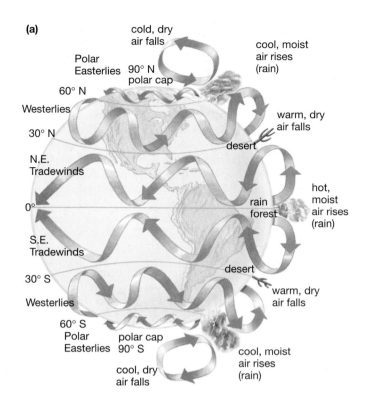

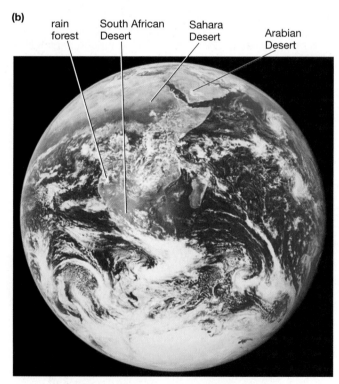

Figure 41-2 Distribution of air currents and climatic regions *(a)* Rainfall is determined primarily by the distribution of temperatures and by Earth's rotation. The interaction of these two factors creates air currents that rise and fall predictably with latitude, producing broad climatic regions. *(b)* Some of these regions are visible in this photograph of the African continent taken from *Apollo 11*. Along the equator are heavy clouds that drop moisture on the central African rain forests. Note the lack of clouds over the Sahara and Arabian Deserts near 30° N and the South African Desert near 30° S.

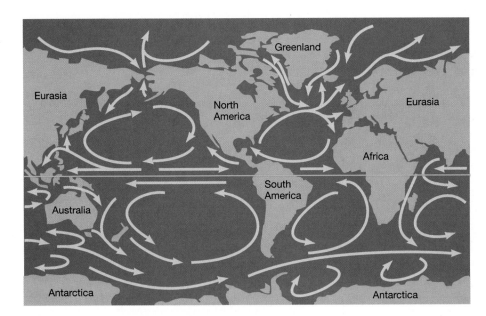

Figure 41-3 *Ocean circulation patterns are called gyres*
Gyres travel clockwise in the Northern Hemisphere and counterclockwise in the Southern Hemisphere. These currents tend to distribute warmth from the equator to northern and southern coastal areas.

temperature drops approximately 2°C (3.5°F) for every 305 meters (1000 feet) elevation gain. This characteristic explains why snow-capped mountains are found even in the tropics (Fig. 41-4).

Mountains also modify rainfall patterns. When water-laden air is forced to rise as it meets a mountain, it cools. Cooling reduces the air's ability to retain water, and the water condenses as rain or snow on the windward (near) side of the mountain. The cool, dry air is warmed again as it travels down the far side of the mountain and absorbs water from the land, creating a local dry area called a **rain shadow**. For example, mountain ranges such as the Sierra Nevada of the western United States wring the moisture from the westerly winds that come off the Pacific Ocean, leaving deserts in the rain shadow on their eastern sides (Fig. 41-5).

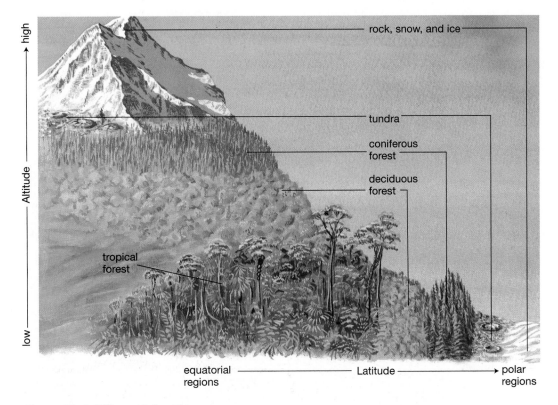

Figure 41-4 *Effects of elevation on temperature*
Climbing a mountain in some ways is like going north; in both cases, increasingly cool temperatures produce a similar series of biomes.

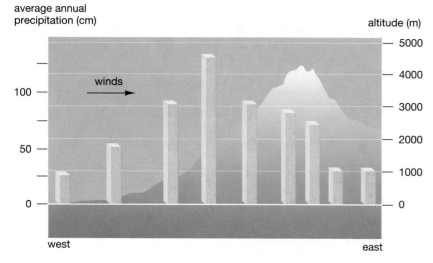

Figure 41-5 *The rain shadow of the Sierra Nevada*
Here, rainfall is graphed against altitude. Westerly (eastward-moving) winds deposit their moisture on the western slope, leaving a desert to the east.

2) What Are the Requirements of Life?

From the lichens on bare rock to the thermophilic (Greek for "heat-loving") algae in the hot springs of Yellowstone National Park to the bacteria thriving under the pressure-cooker conditions of a deep-sea vent, Earth teems with life. Underlying the diversity of habitats is the common ability to provide to varying degrees the four fun-

damental resources required for life. These resources are:

1. Nutrients from which to construct living tissue
2. Energy to power that construction
3. Liquid water to serve as a medium in which metabolic reactions occur
4. Appropriate temperatures in which to carry out these processes

As we shall see in the following sections, these resources are very unevenly distributed. Their availability limits the types of organisms that can exist within the various terrestrial and aquatic ecosystems on Earth.

Ecosystems are extraordinarily diverse, yet clear patterns exist. The community that is characteristic of each ecosystem is dominated by organisms specifically adapted to particular environmental conditions. Variations in temperature and in the availability of light, water, and nutrients shape the adaptations of organisms that inhabit an ecosystem. The desert community, for example, is dominated by plants adapted to heat and drought. The cacti of the Mojave Desert of the American Southwest are strikingly similar to the euphorbia of the deserts of Africa and of the nearby Canary Islands, although these plants are in separate families and are only distantly related genetically. Their spinelike leaves and thick, green, water-storing stems are adaptations for water conservation (Fig. 41-6). Likewise, the plants of the arctic

(a)

(b)

Figure 41-6 *Environmental demands mold physical characteristics*
Evolution in response to similar environments has molded the bodies of **(a)** American cacti and **(b)** Canary Islands euphorbia into nearly identical shapes, although they are in different families. Both of these desert plants have thick, water-conserving stems and leaves reduced to defensive spines that minimize evaporation.

tundra and those of the alpine tundra of the Rocky Mountains show growth patterns clearly recognizable as adaptations to a cold, dry, windy climate. Thus, in regions where the environmental conditions are similar, similar types of organisms are organized into similar types of communities.

3 How Is Life on Land Distributed?

Terrestrial organisms are restricted in their distribution largely by the availability of water and by temperature. Terrestrial ecosystems receive plenty of light, even on an overcast day, and the soil provides abundant nutrients. Water, however, is limited and very unevenly distributed, both in place and in time. Terrestrial organisms must be adapted to obtain water when it is available and to conserve it when it is scarce.

Like water, favorable temperatures are very unevenly distributed in place and time. At the South Pole, even in summer, the average temperature is usually well below freezing; not surprisingly, life is scarce there. Places such as central Alaska have favorable temperatures during only the summer, whereas the tropics have a uniformly warm, moist climate, and life abounds there.

Terrestrial Biomes Support Characteristic Plant Communities

Terrestrial communities are dominated and defined by their plant life. Because plants can't escape from drought, sun, or winter weather, they tend to be precisely adapted to the climate of a particular region. Large land areas with similar environmental conditions and characteristic plant communities are called **biomes** (Fig. 41-7). Biomes are generally named after the major type of vegetation found there. The predominant vegetation of each biome is determined by the complex interplay of rainfall and temperature (Fig. 41-8). These factors determine the available soil moisture needed for plant growth as well as for compensation for water losses through evaporation. In addition to the total amount of rainfall and the overall yearly average temperature, the variability of rain and temperature over the year determines which plants can grow in an area. Arctic tundra plants, for example, must be adapted to marshy conditions in the early summer but cold and desertlike conditions for much of the rest of the year, when water is solidly frozen and unavailable. In the following sections we discuss the major biomes, beginning at the equator and working our way poleward. We also discuss some of the effects human activities have on these biomes.

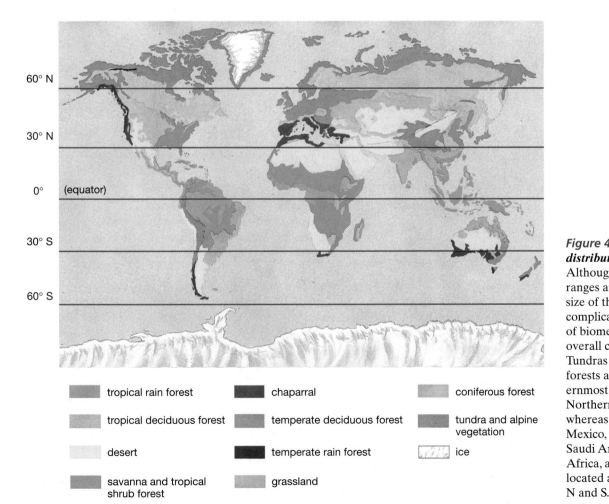

Figure 41-7 The distribution of biomes Although mountain ranges and the sheer size of the continents complicate the pattern of biomes, note the overall consistencies. Tundras and coniferous forests are in the northernmost parts of the Northern Hemisphere, whereas the deserts of Mexico, the Sahara, Saudi Arabia, South Africa, and Australia are located about 20° to 30° N and S.

Legend:
- tropical rain forest
- tropical deciduous forest
- desert
- savanna and tropical shrub forest
- chaparral
- temperate deciduous forest
- temperate rain forest
- grassland
- coniferous forest
- tundra and alpine vegetation
- ice

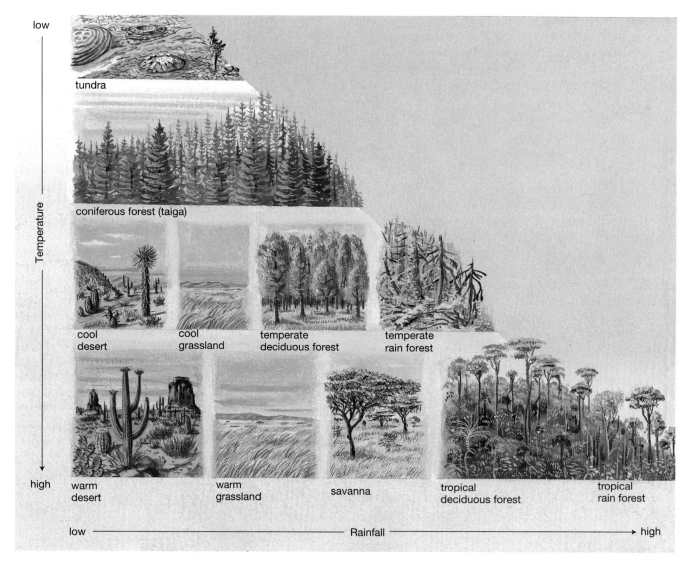

Figure 41-8 Rainfall and temperature influence biome distribution
Together, these factors determine the available soil moisture needed for plant growth.

Tropical Rain Forests

Large areas of South America and Africa lie along the equator. There the temperature averages between 25° and 30°C  (77° and 86°F) with little variation, and rainfall ranges from 250 to 400 centimeters (100 to 160 inches) annually. These evenly warm, evenly moist conditions combine to create the most diverse biome on Earth, the **tropical rain forest**, dominated by huge broadleaf evergreen trees (Fig. 41-9).

Although rain forests cover only 6% of Earth's total land area, they are believed to be home to between 5 million and 8 million species, representing half to two-thirds of the world's total. In Chapter 1, we introduced the term *biodiversity*. **Biodiversity** is the total number of species within an ecosystem and the resulting complexity of interactions among them. In short, it defines the biological "richness" of an ecosystem. Rain forests have the highest biodiversity of any ecosystem on Earth. For example, a recent survey of a 2.5-acre site in the upper Amazon basin revealed 283 species of trees, most of which were represented by a single individual. In a 5-square-kilometer (about 3-square-mile) tract of rain forest in Peru, scientists counted more than 1300 species of butterfly and 600 species of bird. In comparison, the entire United States is home to only 400 butterfly species and 700 bird species.

Tropical rain forests typically have several layers of vegetation. The tallest trees reach 50 meters (150 feet) and tower above the rest of the forest. Below is a fairly continuous canopy of treetops at about 30 to 40 meters (90 to 120 feet). Another layer of shorter trees typically stands below the canopy. Huge woody vines, commonly 100 meters (328 feet) or more in length, grow up the trees, reaching the sunlight far above. These layers of vegetation block out most of the sunlight. Many of the plants that live in the dim green light that filters through to the forest floor have enormous leaves to trap the little available energy.

Figure 41-9 The tropical rain forest biome
(a) Towering trees draped with vines reach for the light in the dense tropical rain forest. Because food and light energy are found high in trees, amid their branches dwells the most diverse assortment of animals on Earth, including *(b)* a fruit-eating toucan, *(c)* a shimmering butterfly, *(d)* a tree-climbing orchid, and *(e)* a howler monkey.

Because edible plant material close to the ground in tropical rain forests is relatively scarce, much of the animal life, including numerous birds, monkeys, and insects, is *arboreal* (living in the trees). Competition for the nutrients that do reach the ground is intense among both animals and plants. Even such unlikely sources of food as the droppings of monkeys are in great demand. For example, dung beetles feed and lay their eggs on monkey droppings. When ecologists attempted to collect droppings of the South American howler monkeys to find out what the monkeys had been eating, they found themselves in a race with the beetles, hundreds of which would arrive within minutes after a dropping hit the ground!

Almost as soon as bacteria or fungi release any nutrients from dead plants or animals into the soil, rainforest trees and vines absorb the nutrients. This is one of the reasons why, despite the teeming vegetation, agriculture is very risky and normally destructive in rain forests. Virtually all the nutrients in a rain forest are tied up in the vegetation, so the soil is very infertile. If the trees are carried away for lumber, few nutrients remain to support crops. Further, even if the nutrients are released by burning the vegetation, the heavy year-round rainfall quickly dissolves and erodes them away, leaving the soil infertile after a few seasons of cultivation. The exposed soil, which is rich in iron and aluminum, then takes on an impenetrable, bricklike quality as it bakes in the tropical sun. As a result, secondary succession on cleared rainforest land is slowed significantly. Even small forest cuttings take about 70 years to regenerate.

Human Impact Despite their unsuitability for agriculture, rain forests are being felled for lumber or burned down for ranching or agriculture at a rate of roughly 38 million acres each year, or more than 70 acres each minute (Fig. 41-10). At least 40% of the world's rain forests are now gone. Harvard University biologist Edward O. Wilson estimates that the destruction of tropical rain forests may be driving 27,000 species to extinction annually. With the loss of biodiversity, humanity loses access to a wealth of potential drugs and raw materials. For example, in Malaysia, the sap of a rare species of tree was discovered to yield a compound called calanolide A, which, in test tubes, completely blocks replication of the HIV virus. The removal of these forests is also contributing significantly to the build-up of carbon dioxide in the atmosphere, making the greenhouse effect more severe. It is estimated that one-quarter of the carbon released into the atmosphere during the past decade came from the loss of tropical rain forests. In West Africa, burning the forests is also causing acid rain, which is damaging the remaining trees.

Although disastrous losses are continuing, recognition of the problem is growing. Some areas have been set aside as protected preserves, and some reforestation efforts are under way. Local residents are becoming more involved in conservation efforts. In Brazil, a union of people who collect natural rubber from tree sap is fighting to preserve large tracts of land open only to rubber production and to the harvesting of fruits and nuts. This approach suggests the ultimate solution, which is tragically slow in coming—sustainable use, the harvesting of products without permanently damaging the trees or the ecosystem they support.

Figure 41-10 *Fire engulfs a tropical rain forest in Brazil* The cleared area will be converted to ranching or agriculture, doomed to failure because the soil quality is so poor.

Tropical Deciduous Forests

Slightly farther from the equator, the rainfall is not nearly as constant, and there are pronounced wet and dry seasons. In these areas, which include India, much of Southeast Asia, and parts of South and Central America, **tropical deciduous forests** grow. During the dry season, the trees cannot get enough water from the soil to compensate for evaporation from their leaves. As a result, the plants have adapted to the dry season by shedding their leaves, thereby minimizing water loss. If the rains fail to return on schedule, the trees delay the formation of new leaves until the drought passes.

Savanna

Along the edges of the tropical deciduous forest, the trees gradually become more widely spaced, with grasses growing between them. Eventually, grasses become the dominant vegetation, with only scattered trees and thorny scrub forests here and there; this biome is the **savanna** (Fig. 41-11). Savanna grasslands typically have a rainy season during which virtually all of the year's precipitation falls—30 centimeters (12 inches) or less. When the dry season arrives, it comes with a vengeance. Rain might not fall for months, and the soil becomes hard, dry, and dusty. Grasses are well adapted to this type of climate, growing very rapidly during the rainy season and dying back to drought-resistant roots during the arid times. Only a few specialized trees, such as the thorny acacia or the water-storing baobab, can survive the devastating savanna dry seasons. In areas in which the dry season becomes even more pronounced, virtually no trees can grow, and the savanna imperceptibly grades into tropical grassland.

Figure 41-11 The African savanna
(a) Elephants roam while *(b)* a cheetah feasts on its prey. *(c)* Large herds of grazing animals, such as zebras, can still be seen on African preserves. The herds of herbivores provide food for the greatest assortment of large carnivores on the planet, including *(d)* lions, leopards, cheetahs, hyenas, and wild dogs.

The African savanna has probably the most diverse and impressive array of large mammals on Earth. These mammals include numerous herbivores such as antelope, wildebeest, water buffalo, elephants, and giraffes and such carnivores as the lion, leopard, hyena, and wild dog.

Human Impact Africa's rapidly expanding human population threatens the wildlife of the savanna. Poaching has driven the black rhinoceros to the brink of extinction (Fig. 41-12) and endangers the African elephant. The abundant grasses that make the savanna a suitable habitat for so much wildlife also make it suitable for grazing domestic cattle. As the human population of East Africa increases, so does the pressure of cattle grazing on the savanna. Fences increasingly disrupt the migration of the great herds of herbivores in search of food and water. Ecologists have discovered that the native herbivores are much more efficient at converting grass into meat than are cattle. Perhaps the future African savanna may support herds of domesticated antelope and other large native grazers in place of cattle.

Figure 41-12 Poaching threatens African wildlife
Rhinoceros horns, believed by some to have aphrodisiac properties, fetch staggering prices and encourage poaching. The black rhino is now nearly extinct.

Deserts

Even drought-resistant grasses need at least 25 to 50 centimeters (10 to 20 inches) of rain a year, depending on its seasonal

distribution and the average temperature. When less rain than this falls, many grasses fail and **desert** biomes result. These biomes are found on every continent, typically around 20° to 30° N and S latitude, and also in the rain shadows of major mountain ranges. As with all biomes, deserts include a variety of environments. At one extreme are certain areas of the Sahara Desert or Chile, where it virtually never rains and no vegetation grows (Fig.

41-13a). The more common deserts, however, are characterized by widely spaced vegetation and large areas of bare ground. The plants tend to be spaced evenly, as if planted by hand (Fig. 41-13b). In many cases, the perennial plants are bushes or cacti with large, shallow root systems. The shallow roots quickly soak up the soil moisture after the infrequent desert storms. The rest of the plant is typically covered with a waterproof, waxy coating to prevent evaporation of precious water. Water is stored in the thick stems of cacti and other succulents (Fig. 41-13c). The spines of cacti are leaves modified for protection and water conservation, presenting almost no surface area for evaporation. In many deserts, all the rain

Figure 41-13 The desert biome

(a) Under the most extreme conditions of heat and drought, deserts can be almost devoid of life, such as these sand dunes of the Sahara Desert in Africa. *(b)* Throughout much of Utah and Nevada, the Great Basin Desert presents a monotonous landscape of widely spaced shrubs, such as sagebrush and greasewood. These shrubs often secrete a growth inhibitor from their roots, preventing germination of nearby plants and thus reducing competition for water. *(c)* The Anza Borrego Desert in California is typified by the prickly pear cactus, the brittle bush, and the spindly ocotillo, whose red flowers are pollinated by hummingbirds. *(d)* The kangaroo rat is an elusive inhabitant of the deserts of North America.

falls in just a few storms, and specialized annual wildflowers take advantage of the brief period of moisture to race through germination, growth, flowering, and seed production in a month or less (Fig. 41-14).

The animals of the deserts, like the plants, are specially adapted to survive on little water. Most deserts appear to be almost completely devoid of animal life during the day, because the animals seek relief from the sun and heat in cool, underground burrows. After dark, when deserts cool down considerably, horned lizards, snakes, and other reptiles emerge to feed, as do mammals such as the kangaroo rat (Fig. 41-13d) and birds such as the burrowing owl. Most of the smaller animals survive without ever drinking, getting all the water they need from their food and from that produced during cellular respiration in their tissues. Larger animals, such as desert bighorn sheep, are dependent on permanent water holes during the driest times of the year.

Human Impact   Although human activities are reducing the extent of many biomes, they are causing the spread of deserts, a process call **desertification**. A dramatic example is occurring in the Sahel, which borders the southern edge of the Sahara Desert in Africa. Twenty-five years of below-average rainfall, coupled with rapid growth of the human population, have caused a steady southward spread of the desert. Western technology allowed the Sahelians to drill deep wells, abandon their nomadic way of life, and increase their livestock herds. The natural vegetation was destroyed as livestock competed for forage during the drought (Fig. 41-15). Native shrubs were uprooted by goats, preventing regeneration. The growing human population began cultivating increasingly fragile land without allowing time for the soil to recover between crops. As the desert spread southward, so did the people, with their livestock and crops,

Figure 41-14 The Sonoran Desert
In spring this Arizona desert is carpeted with wildflowers. Through much of the year, annual wildflower seeds lie dormant, waiting for the spring rains to fall.

Figure 41-15 Overgrazing
In the Sahel region of Africa, overgrazing is degrading the already limited productivity of the land and contributing to the southward spread of the Sahara Desert.

causing further deforestation, soil destruction, and desertification of once-productive land. The Sahel has a human population exceeding the carrying capacity of the land. The loss of productivity of the ecosystem is nearly irreversible, and massive famines, such as occurred in Ethiopia in the mid-1980s, are a tragic result.

Chaparral

In many coastal regions that border on deserts, such as southern California and much of the Mediterranean, a unique type of vegetation grows, called **chaparral**. The annual rainfall in these regions is similar to that of a desert, but the proximity of the sea provides a slightly longer rainy season in the winter and frequent fogs during the spring and fall, which reduce evaporation. Chaparral consists of small trees or large bushes with thick waxy or fuzzy evergreen leaves that conserve water. These shrubs are also able to withstand the frequent summer fires started by lightning (Fig. 41-16).

Human Impact   About 400 B.C., the Greek philosopher Plato wrote that "there are mountains in Attica [southern Greece] which can now keep nothing but bees, but which were clothed, not so very long ago, with timber suitable for roofing very large buildings." The original

Greek forests were cut down not only for ceiling beams but also for the great Athenian naval fleets. Subsequently, heavy grazing by goats prevented regrowth of the forests. Only the chaparral plants, which could survive in the hot sun and were distasteful or poisonous to goats, regrew. The forests have never returned.

Grasslands

In the temperate regions of North America, deserts occur in the rain shadows east of the mountain ranges, such as the Sierra Nevada and Rocky Mountains. Eastward, as the rainfall gradually increases, the land begins to support more and more grasses, giving rise to the prairies of the Midwest. Most such **grassland**, or **prairie**, biomes are concentrated in the centers of continents. You can see a large grassland in the center of the Eurasian continent in Figure 41-7. In general, they have a continuous cover of grass and virtually no trees except along the rivers. From the tallgrass prairies of Iowa, Missouri, and Illinois (Fig. 41-17) to the shortgrass prairies of eastern Colorado, Wyoming, and Montana (Fig. 41-18), the North American grassland once stretched across almost half the continent.

Water and fire are the crucial factors in the competition between grasses and trees. The hot, dry summers and frequent droughts of the shortgrass prairies can be tolerated by grass but are fatal to trees. Trees can grow in the more eastern prairies, but historically the trees were destroyed by frequent fires, often set by Native Americans to maintain grazing land for the bison. Although the tops

Figure 41-16 *The chaparral biome*
This biome is limited primarily to coastal mountains in dry regions, such as the San Gabriel Mountains in southern California. Chaparral is maintained by frequent fires set by summer lightning. Although the tops of the plants may be burned off, the roots send up new sprouts the following spring.

Figure 41-17 *Tallgrass prairie in Missouri*
In the central United States, moisture-bearing winds out of the Gulf of Mexico produce summer rains, allowing a lush growth of tall grasses and wildflowers such as these coneflowers. Periodic fires, now carefully managed, prevent encroachment of forest.

Figure 41-18 Shortgrass prairie
The lands east of the Rocky Mountains receive relatively little rainfall, and *(a)* shortgrass prairie results,
characterized by low-growing bunch grasses such as buffalo grass and grama grass. *(b)* Pronghorn antelope,
(c) prairie dogs, and *(d)* protected bison herds occupy this biome, in which *(e)* wildflowers such as this
coneflower abound.

of the grasses are destroyed by fire, their root systems
normally survive; trees, however, are killed outright. The
resulting grasslands of North America once supported
huge herds of bison—as many as 60 million in the early
nineteenth century. Grasses growing and decomposing
for thousands of years produced what may be the most
fertile soil in the world.

Human Impact With the elimination of the bison and
the development of plows that could break the dense
turf, the former prairie has become the "breadbasket" of
North America, so named because enormous quantities
of grain are cultivated in its fertile soil. The tallgrass
prairie has been converted to agricultural land, except
for tiny protected remnants.

On the western shortgrass prairie, cattle and sheep
have replaced the bison and pronghorn antelope. As a
result of their overgrazing the grasses, the boundary be-

tween the cool deserts and the grassland has commonly
been altered in favor of desert plants. Much of the sagebrush desert of the American West is actually the result
of overgrazing shortgrass prairie (Fig. 41-19). Cattle prefer grass to sagebrush, so heavy grazing destroys the grass.
Consequently, moisture that the grass would have absorbed is left in the soil, encouraging the growth of the
woody sagebrush. Sagebrush soon becomes the dominant vegetation; thus, the prairie grasses are replaced by
plants characteristic of the cool desert.

Temperate Deciduous Forests
At their eastern edge, the
North American grasslands
merge into the **temperate de-
ciduous forest** biome, also
found in Western Europe and East Asia (Fig. 41-20).

Figure 41-19 Sagebrush desert or shortgrass prairie?
Biomes are influenced by human activities as well as by temperature, rainfall, and soil. The shortgrass prairie field on the right has been overgrazed by cattle, causing the grasses to be replaced by sagebrush.

Figure 41-20 The temperate deciduous forest biome
(a) In temperate deciduous forests of the eastern United States, *(b)* the white-tailed deer is the largest herbivore, and *(c)* birds such as the blue jay are abundant. *(d)* In spring, a profusion of woodland wildflowers (such as these hepaticas) blooms briefly before the trees produce leaves.

Higher precipitation (75 to 150 centimeters, or 30 to 60 inches) occurs there than in the grasslands, and, in particular, more summer rainfall occurs. The soil therefore retains enough moisture for trees to grow, and the resulting forest shades out grasses. In contrast to the tropical forests, the temperate deciduous forest biome has cold winters, usually with at least several hard frosts and often long periods of below-freezing weather. Winter in this biome has an effect on the trees similar to that of the dry season in the tropical deciduous forests: During periods of subfreezing temperatures, liquid water is not available to the trees. To reduce evaporation when water is in short supply, the trees drop their leaves in the fall. They produce leaves again in the spring, when liquid water becomes available. During the brief time in spring when the ground has thawed but the trees have not yet blocked off all the sunlight, numerous wildflowers grace the forest floor.

Insects and other arthropods are numerous and conspicuous in deciduous forests. The decaying leaf litter on the forest floor also provides food and habitat for bacteria, earthworms, fungi, and small plants. Many arthropods feed on these or on each other. A variety of vertebrates, including mice, shrews, squirrels, raccoons, and many species of birds, dwell in the deciduous forests.

Human Impact Large mammals such as black bear, wolves, bobcats, and mountain lions were formerly abundant. In many deciduous forests, deer are plentiful because their predators have been severely reduced and, in some cases, eliminated by humans. Clearing for lumber, agriculture, and housing has dramatically reduced America's deciduous forests, and virgin deciduous forests are now almost nonexistent.

Temperate Rain Forests

On the U.S. Pacific coast, from the lowlands of the Olympic Peninsula in Washington State to southeast Alaska, lies the **temperate rain forest** biome. Temperate rain forests, which are relatively rare, are also located along the southeastern coast of Australia and the southwestern coast of New Zealand (Fig. 41-21). As in the tropical rain forest, there is no shortage of liquid water year-round. This

Figure 41-21 The temperate rain forest biome
(a) The Hoh River temperate rain forest in Olympic National Park. The coniferous trees do not block the light as effectively as do broadleaf trees, so ferns, mosses, and wildflowers grow in the pale green light of the forest floor. *(b)* The dead feed the living, as new trees grow from the decay of this fallen giant, called a "nurse log," and *(c)* flowering foxglove and *(d)* fungi find ideal conditions amid the moist, decaying vegetation.

abundance of water is due to two factors. First, there is a tremendous amount of rain. The Hoh River rain forest in Olympic National Park receives more than 400 centimeters (160 inches) of rain annually, over 60 centimeters (24 inches) in the month of December alone. Second, the moderating influence of the Pacific Ocean prevents severe frost from occurring along the coast, so the ground seldom freezes.

The abundance of water means that the trees have no need to shed their leaves in the fall, and almost all the trees are evergreens. In contrast to the broadleaf evergreen trees of the tropics, temperate rain forests are dominated by conifers. The ground and typically the trunks of the trees are covered with mosses and ferns. As in tropical rain forests, so little light reaches the forest floor that tree seedlings usually cannot become established. Whenever one of the forest giants falls, however, it opens up a patch of light, and new seedlings quickly sprout, commonly right atop the fallen log. This event produces a "nurse log" (Fig. 41-21b).

Taiga

North of the grasslands and temperate forests, the **taiga**, also called the **northern coniferous forest** (Fig. 41-22), stretches horizontally across all of North America and Eurasia, including parts of the northern United States and much of southern Canada. Conditions in the taiga are harsher than those in the temperate deciduous forest. In the taiga, the winters are longer and colder, and the growing season is shorter. The few months of warm weather are too short to allow trees the luxury of regrowing leaves in the spring. As a result, the taiga is populated almost entirely by evergreen coniferous trees with small, waxy needles. The waxy coating and small surface area of the needles reduce water loss by evaporation during the cold months, and the leaves remain on the trees year-round. Thus, the trees are instantly ready to take advantage of good growing conditions when spring arrives and can continue slow growth late into the fall.

Figure 41-22 The taiga (or northern coniferous forest) biome
(a) The small needles and pyramidal shape of conifers allow them to shed heavy snows. *(b)* Winter is a challenge not only for the trees but also for animals such as this snowshoe hare and the bobcat that preys on it. The hare is also prey for *(c)* the great horned owl. Taiga animals face diminished food supply but increased energy requirements during subfreezing weather.

Because of the harsh climate in the taiga, the diversity of life there is much lower than in many other biomes. Vast stretches of central Alaska, for example, are covered by a somber forest that consists almost exclusively of black spruce and an occasional birch. Large mammals such as the wood bison, grizzly bear, moose, and wolf, which have mostly been eradicated in the southern regions of their original range, still roam the taiga, as do smaller animals such as the wolverine, fox, snowshoe hare, and deer.

Human Impact Owing to the remoteness of the northernmost taiga and the severity of its climate, a greater percentage of the taiga remains in undisturbed condition than any other North American biome except the tundra. Nonetheless, the taiga is a major source of lumber for construction; *clear-cutting*, the removal of all the trees in a given area, has destroyed huge expanses of forest (Fig. 41-23).

Tundra

The last biome encountered before we reach the polar ice caps is the arctic **tundra**, a vast treeless region bordering the Arctic Ocean (Fig. 41-24). Conditions in the tundra are severe. Winter temperatures in the arctic tundra often reach −55°C (−40°F) or below, winds howl at 50 to 100 kilometers (30 to 60 miles) an hour, and precipitation averages 25 centimeters (10 inches) or less each year, mak-

Figure 41-23 Clear-cutting
Clear-cutting, as seen in this Oregon forest, is relatively simple and cheap—but its environmental costs are high. Erosion will diminish the fertility of the soil, slowing new growth. Further, the dense stands of same-age trees that typically regrow are more vulnerable to attack by parasites than a natural stand of trees of various ages would be.

ing this a "freezing desert." Even during the summer, the temperatures can drop to freezing, and the growing season may last only a few weeks before a hard frost occurs. Somewhat less cold but similar conditions produce alpine tundra on mountaintops above the altitude where trees can grow.

Figure 41-24 The tundra biome
(a) Life on the tundra is adapted to cold. *(b)* Plants such as dwarf willows and perennial wildflowers (such as this dwarf clover) grow low to the ground, escaping the chilling tundra wind. Tundra animals, such as *(c)* caribou and *(d)* arctic foxes, can regulate blood flow in their legs, keeping them just warm enough to prevent frostbite, while preserving precious body heat for the brain and vital organs.

The cold climate of the arctic tundra results in **permafrost**, a permanently frozen layer of soil typically no more than half a meter (about 1.5 feet) below the surface. As a result, when summer thaws come, the water from melted snow and ice cannot soak into the ground and the tundra becomes a huge marsh. Trees cannot survive in the tundra; the permafrost limits root growth to the topmost meter or so of soil.

Nevertheless, the tundra supports a surprising abundance and variety of life. The ground is carpeted with small perennial flowers and dwarf willows no more than a few centimeters tall and commonly with a large lichen called "reindeer moss," a favorite food of caribou. The standing water provides a superb habitat for mosquitoes. The mosquitoes and other insects provide food for numerous birds, most of which migrate long distances to nest and raise their young during the brief summer feast. The tundra vegetation supports lemmings, which are eaten by wolves, snowy owls, arctic foxes, and even grizzly bears.

Human Impact The tundra is perhaps the most fragile of all the biomes because of its short growing season. A willow 10 centimeters (4 inches) high may have a trunk 7 centimeters (3 inches) in diameter and be 50 years old. Human activities in the tundra leave scars that persist for centuries. Fortunately for the tundra inhabitants, the impact of civilization is localized around oil drilling sites, pipelines, mines, and military bases.

Rainfall and Temperature Determine the Vegetation a Biome Can Support

Terrestrial biomes are greatly influenced by both temperature and rainfall, whose effects interact. Temperature strongly influences the effectiveness of rainfall in providing soil moisture for plants and standing water for animals to drink. The hotter it is, the more rapidly water evaporates, both from the ground and from plants. As a result of this interaction of temperature with rainfall (and to a lesser extent, the distribution of rain throughout the year), areas that receive almost exactly the same rainfall can have startlingly different vegetation, all the way from desert to taiga. Take a trip with us from southern Arizona to northern Alaska as we visit ecosystems that each receive about 28 centimeters (12 inches) of rain annually.

The Sonoran Desert near Tucson, Arizona (see Fig. 41-13c), has an average annual temperature of 20°C (68°F) and receives about 28 centimeters (1 foot) of rain each year. The landscape is dominated by giant saguaro cactus and low-growing, drought-resistant bushes. Fifteen hundred kilometers (931 miles) north, in eastern Montana, rainfall is about the same, but we have passed into the shortgrass prairie biome (see Fig. 41-18), largely because the average temperature is much lower, about 7°C (45°F). Central Alaska receives about the same annual rainfall (28 centimeters), yet is covered with taiga forest

(see Fig. 41-22). As a result of the low average annual temperature (about −4°C, or 25°F), permafrost underlies much of the ground. During the summer thaw, the taiga earns its Russian name "swamp forest," although its rainfall is about the same as that of the Sonoran Desert.

4) How Is Life in Water Distributed?

Although thus far this chapter has emphasized terrestrial biomes, the saltwater oceans and seas are the largest ecosystems on Earth, covering about 71% of its surface. Freshwater ecosystems, in contrast, cover less than 1%.

The unique properties of water lend some common features to aquatic ecosystems. *First*, because water is slower to heat and cool than air, temperatures in aquatic ecosystems are more moderate than are those in terrestrial ecosystems. *Second*, although water may appear quite transparent, it absorbs a considerable amount of the light energy that sustains life. Even in the clearest water, the intensity of light decreases rapidly with depth. At depths of 200 meters (600 feet) or more, little light is left to power photosynthesis. If the water is at all cloudy—for example, because of suspended sediment or microorganisms—the depth to which light can penetrate is greatly reduced. *Third*, nutrients in aquatic ecosystems tend to be concentrated near the bottom sediments where light levels tend to be too low to support photosynthesis. This separation of energy and nutrients limits aquatic life. Of the four requirements for life, aquatic ecosystems provide abundant water and appropriate temperatures. Thus, the major factors that determine the quantity and type of life in aquatic ecosystems are energy and nutrients.

Although they share some common features, aquatic ecosystems are extremely diverse. Freshwater ecosystems include rivers, streams, ponds, lakes, and marshes; marine (saltwater) ecosystems include estuaries, tide pools, the open ocean, and coral reefs. In the following sections, we look more closely at a few of these important aquatic ecosystems.

Freshwater Lakes Have Distinct Regions of Life

Freshwater lakes vary tremendously in size, depth, and nutrient content. Although each lake is unique, moderate to large lakes in temperate climates share some common features, including distinct zones of life.

Life Zones Are Determined by Access to Light and Nutrients

The distribution of life in lakes depends largely on access to light, to nutrients, and in some cases to a place for attachment (the bottom). The life zones of lakes, then, correspond to specific locations within the lake. We rec-

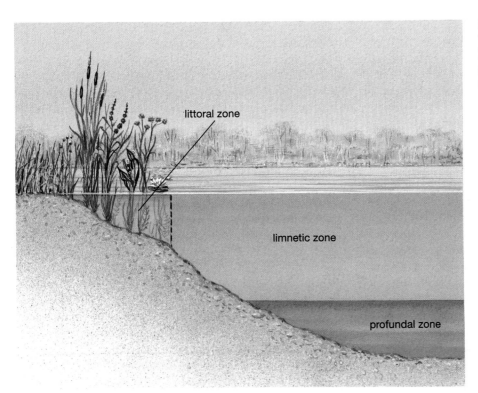

Figure 41-25 Lake life zones
There are three life zones in a "typical" lake: a near-shore littoral zone with rooted plants, an open-water limnetic zone, and a deep, dark profundal zone.

ognize three such zones: (1) the *littoral zone*, (2) the *limnetic zone*, and (3) *the profundal zone* (Fig. 41-25).

Near the shore is the **littoral zone**. In this zone, the water is shallow, and plants find abundant light, anchorage, and adequate nutrients from the bottom sediments. Not surprisingly, littoral-zone communities are the most diverse. Cattails and bulrushes abound nearest the shore, and water lilies and entirely submerged vascular plants and algae flourish at the deepest reaches of the littoral zone. The plants of the littoral zone trap sediments carried in by streams and by runoff from the surrounding land, increasing the nutrient content in this region. Living among the anchored plants are microscopic organisms called **plankton**. There are two forms of plankton: **phytoplankton** (Greek, "drifting plants"), which includes photosynthetic protists, bacteria, and algae, and **zooplankton** (Greek, "drifting animals"), such as protozoa and tiny crustaceans. The greatest diversity of animals in the lake is also found in this zone. Littoral invertebrate animals include small crustaceans, insect larvae, snails, flatworms, and hydra; littoral vertebrates include frogs, minnows, and aquatic snakes and turtles.

As the water increases in depth farther from shore, plants are unable to anchor to the bottom and still collect enough light for photosynthesis. This open-water area is divided into two regions: the upper limnetic zone and the lower profundal zone (Fig. 41-25). In the **limnetic zone**, enough light penetrates to support photosynthesis. Here phytoplankton including cyanobacteria (also called blue-green algae) serve as producers. These are eaten by protozoa and small crustaceans, which in turn are con-

sumed by fish. Below the limnetic zone lies the **profundal zone**, where light is insufficient to support photosynthesis. This area is nourished mainly by detritus that falls from the littoral and limnetic zones and by incoming sediment. It is inhabited primarily by decomposers and detritus feeders, such as bacteria, snails and insect larvae, and fish that swim freely among the different zones.

Freshwater Lakes Are Classified According to Their Nutrient Content

Although each lake is unique, freshwater lakes can be classified on the basis of their nutrient content as either *oligotrophic* or *eutrophic*.

Oligotrophic (Greek, "poorly fed") **lakes** are very low in nutrients. Many are formed by glaciers that scrape depressions in bare rock, and they are fed by mountain streams carrying little sediment. Because there is little sediment or microscopic life to cloud the water, oligotrophic lakes are clear, and light penetrates deeply. Therefore, photosynthesis (and by extension oxygenation, as oxygen is produced as a by-product of photosynthesis) in deeper water is possible, and the limnetic zone may extend all the way to the bottom. Oxygen-loving fish, such as trout, thrive in oligotrophic lakes.

Eutrophic (Greek, "well fed") **lakes** receive larger inputs of sediments, organic material, and inorganic nutrients (such as phosphorus) from their surroundings, allowing them to support dense communities. They are murkier, both from suspended sediment and dense phytoplankton populations, so the lighted limnetic zone is shallower. Dense "blooms" of algae occur seasonally in the limnetic zone. Their dead bodies fall into the profundal zone, where they are used as food by decomposer organisms. The metabolic activities of many of these decomposers use oxygen, depleting the oxygen content of the profundal zone in eutrophic lakes.

Lakes are transient ecosystems, although very large lakes may persist for millions of years. Over time, they gradually fill with sediment, undergoing succession to dry land (see Chapter 39). As nutrient-rich sediment accumulates, oligotrophic lakes tend to become eutrophic, a process called *eutrophication*.

Human Impact Human activities can greatly accelerate the process of eutrophication, because nutrients are carried into lakes from farms, feedlots, and sewage. Even if solid wastes are removed, water discharged from sewage-treatment plants is often rich in phosphates and nitrates dissolved from wastes and detergents. Excessively enriched water also washes off fertilized farm fields and feedlots, where the manure of thousands of cattle accumulates. The added nutrients support excessive growth of phytoplankton. These producers, particularly cyanobacteria, form a scum on the lake surface, depriving the submerged plants of sunlight. The plant bodies are decomposed by bacteria, depleting the dissolved oxygen. Deprived of oxygen, the fish, snails, and insect larvae die,

and their bodies fuel more bacterial growth, further depleting oxygen. Even without oxygen, certain bacteria that produce foul-smelling gases thrive. Although it is full of life, the nutrient-polluted lake smells dead. Most of the trophic levels, including the fish, have been eliminated, and the community is dominated by bacteria and microscopic algae.

Acid rain, caused by the burning of fossil fuels, poses a very different threat to freshwater ecosystems. Few organisms can withstand the low pH of acidified lakes, which typically appear clear and oligotrophic because they lack life. Acid rain is discussed in Chapter 40.

Marine Ecosystems Cover Much of Earth

In the oceans, the upper layer of water to a depth of about 200 meters (650 feet), where the light is strong enough to support photosynthesis, is called the **photic zone**. Below the photic zone lies the **aphotic zone**, where the only energy comes from the excrement and bodies of organisms that sink or swim down there (Fig. 41-26).

As in lakes, most of the nutrients in the oceans are at or near the bottom, where there is not enough light for photosynthesis. Nutrients dissolved in the water of the photic zone are constantly being incorporated into the bodies of living organisms. When these organisms die,

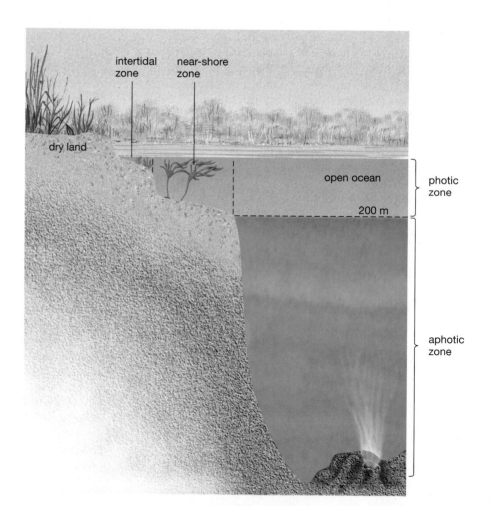

Figure 41-26 Ocean life zones
Photosynthesis can occur only in the upper photic zone, which includes the intertidal and near-shore zones and the upper waters of the open ocean. Life in the aphotic zone relies on energy-rich material that drifts down from the photic zone or (in the unique case of hydrothermal vent communities) on energy that is stored in hydrogen sulfide and trapped by chemosynthesis.

some sink into the aphotic zone, providing its organisms with nutrients. If no additional nutrients entered the photic zone, life there would eventually cease.

Fortunately, there are two sources of nutrients to the photic zone: the land, from which rivers constantly remove nutrients and carry them to the oceans, and **upwelling**, an upward flow that brings cold, nutrient-laden water from the ocean depths to the surface. Upwelling occurs along western coastlines, as in California, Peru, and West Africa, where prevailing winds displace surface water, causing it to be replaced by water from below. Upwelling also occurs around Antarctica. Not surprisingly, the major concentrations of life in the oceans are found where abundant light is combined with a source of nutrients, which occurs most commonly in regions of upwelling and in shallow coastal waters, including coral reefs.

Coastal Waters Support the Most Abundant Marine Life

The most-abundant life in the oceans is found in a narrow strip surrounding Earth's landmasses, where the water is shallow and a steady flow of nutrients washes off the land. Coastal waters consist of (1) the **intertidal zone**, the area that is alternately covered and uncovered by water with the rising and falling of the tides, and (2) the **near-shore zone**, relatively shallow but constantly submerged areas, including bays and coastal wetlands such as salt marshes and estuaries (Figs. 41-26, 41-27). (**Estuaries** are wetlands formed where river meet the ocean.) The near-shore zone is the only part of the ocean where large plants or seaweeds can grow, anchored to the bottom. In addition, the abundance of nutrients and

Figure 41-27 Near-shore ecosystems
(a) A salt marsh in the eastern United States. Expanses of shallow water fringed by marsh grass (*Spartina*) provide excellent habitat and breeding grounds for many marine organisms and shorebirds. *(b)* Although the shifting sands present a challenge to life, grasses stabilize them, and animals such as **(inset)** this *Emerita* crab burrow in the sandy intertidal zone. *(c)* A rocky intertidal shore in Oregon, where animals and algae grip the rock against the pounding waves and resist drying during low tide. **(Inset)** Colorful sea stars cling to the rocks surrounded by a seaweed, fucus. *(d)* Towering kelp sway through the clear water off southern California, providing the basis for a diverse community of invertebrates, fishes, and **(inset)** an occasional sea otter.

Earth Watch
Humans and Ecosystems

The expanding human population has left relatively few ecosystems undisturbed. Human impacts on natural ecosystems are so diverse and wide ranging that they far exceed the scope of this book. However, we can identify some general characteristics of ecosystems dominated by humans and contrast them to the characteristics of undisturbed ecosystems. Below are listed six characteristic differences between these two types of ecosystems and a few ideas for minimizing them.

First, ecosystems dominated by humans tend to be simpler—that is, to have fewer species and fewer community interactions—than do undisturbed ecosystems. Although a city street bustles with life and apparent complexity, count the number of species you encounter in an average block and compare it with the number you encounter while hiking in a wilderness for a similar distance. As humans enter an ecosystem, animals in the highest trophic levels are the first to go. Carnivores are always relatively rare, and their specialized needs are most easily disrupted. Many big carnivores, such as the wolf and mountain lion, require large, undisturbed hunting territories. Humans often selectively destroy large predators, believing them a threat to us or our livestock. As a result, even in relatively undisturbed areas many large carnivores have been eliminated. Grizzly bears and wolves no longer roam the Rocky Mountains. On the western shortgrass prairie, prairie dog towns fall as human towns and ranches rise (Fig. E41-2). The black-footed ferret, a predator of prairie dogs, faces possible extinction. Farm fields have been deliberately simplified from the original prairie biome to eliminate competition and predation and allow the maximum productivity of a single crop species. Nowhere is the contrast between human and natural ecosystems greater than in the tropical rain forests, whose unparalleled diversity is replaced by failed attempts at farming.

There is probably no practical way to restore to human ecosystems the great diversity found in undisturbed areas, nor is doing so always desirable from a human standpoint. Agriculture, for example, demands a simplified ecosystem (but not all biomes are suited to it). Cities concentrate human activities and culture and may lessen the human impact on the surrounding countryside. However, as we recognize the benefits of artificially simplified ecosystems, we must be aware of the need to preserve intact as many natural communities as possible. The undisturbed forest traps and purifies water and can reduce air pollution. Swamps and estuaries contain a wealth of detritus feeders and decomposers that purify water. Coastal wetlands are breeding places for millions of birds and spawning sites for the majority of our commercially important fish and crustacean species. In undisturbed, diverse ecosystems we find aesthetic pleasure as well as a storehouse of species whose commercial or medicinal values are not yet recognized. For example, nearly half the medicines in use today were originally discovered in plants, and we

Figure E41-2 Habitat destruction
Loss of habitat due to human activities is a major threat to most of Earth's wildlife.

have examined only a small fraction of existing plants for possible medical uses.

Second, whereas natural ecosystems run on sunlight, human ecosystems have become dependent on nonrenewable energy from fossil fuels. From the suburbanite pouring gas into a lawn mower to the farmer driving a tractor, managing a simplified ecosystem is an energy-intensive proposition. Energy must be expended to oppose the tendency of the natural system to restore complexity. Fertilizers also require large amounts of energy to produce, and farms must be heavily fertilized because nutrient cycles have been disrupted.

To counteract this trend, some farmers are returning to organic farming. Plant and animal waste and natural nutrient cycles are used to maintain soil fertility. Alternating legume crops, such as soybeans and alfalfa, with other crops helps maintain nitrogen in the soil. Mulching can help retain fertility, reduce water loss, and control weeds. The use of natural insect predators and limited pesticide spraying at critical times during a pest's life cycle can dramatically reduce the need for poisons. Organic farms can be as productive as more conventional farms while using 15% to 50% less energy from fossil fuels per quantity of food produced.

In our homes and commercial buildings, better insulation and increased use of solar heat can result in dramatic energy savings. By increasing our use of renewable energy sources, such as wind and especially sunlight, we can conserve fossil fuels, dramatically reduce pollution, and move human ecosystems a step closer to those that occur naturally.

Third, natural ecosystems recycle nutrients, whereas human ecosystems tend to lose nutrients. Walk around some suburban neighborhoods on trash-pickup day. Grass clippings and leaves are packed in plastic bags to be hauled away. To compensate for this loss of natural nutrients, many suburbanites heavily fertilize their lawns and gardens. A similar trend has occurred in modern farming. The exposed soil is eroded away by wind and water, removing crucial nutrients and requiring large inputs of fertilizer to replace them. Runoff from the field, carrying fertile topsoil and artificial fertilizers, may pollute nearby rivers, streams, and lakes. Pesticides may kill detritus feeders and decomposers, further disrupting natural nutrient cycles. Thus, whereas fertile soil accumulates in many natural ecosystems, it tends to be lost in those dominated by humans. Some 3 billion tons of topsoil are eroded annually from farms in the United States. The Mississippi River alone carries off 40 tons each hour.

Organic farming can help reverse this trend too. We can apply the same principles to our lawns and gardens by composting organic wastes either in our own gardens or in community compost facilities. Farmers can counteract erosion by using contour planting, in which row crops are oriented so that they slow the flow of water instead of funneling it downslope. Row crops can also be alternated in strips with dense, soil-catching crops such as wheat. Planting rows of trees as windbreaks helps prevent soil loss from blowing wind, and it creates a more diverse ecosystem and a nesting place for insect-eating birds. Farm fields need not be plowed in the fall and left unplanted to erode during the winter; in fact, an increasing number of farmers are planting crops in the stubble from the previous season, reducing erosion.

Fourth, natural ecosystems tend to store water and purify it through biological processes, whereas human ecosystems tend to pollute water and shed it rapidly. A thundershower strikes a forest, an adjacent city, and a farm. The rich soil of the forest sponges up the water, which gradually drains into the ground, filtered by the soil and purified by decomposers that break down organic contaminants. In the nearby city, water pours from sidewalks, rooftops, and streets, picking up soot, silt, oil, heavy metals, and garbage. It races down gutters into storm sewers, and a weakly toxic soup gushes into the nearest stream or river. Farm runoff carries priceless topsoil, expensive fertilizer, and animal manure into rivers and lakes, where these potential resources become pollutants.

Preventing erosion will simultaneously conserve water and reduce water pollution from farm runoff. Manure from cattle feedlots, which is a significant source of both groundwater and surface-water pollution, could be placed on fields, where it would restore needed nutrients.

Although water will continue to run off our cities, the pollutants it carries can be reduced by minimizing our reliance on fossil fuels. We must eliminate leaded gasoline worldwide; U.S. gas is now lead-free, but leaded gas remains available throughout most of the world. We must also tighten standards for emissions from diesel and gasoline engines and smokestacks. Efficient public transportation systems will reduce pollution (and the massive frustration) caused by congested traffic. Insulation will reduce power consumption in our homes and offices, as will increased reliance on solar heat.

Fifth, simple human ecosystems such as farms tend to be unstable, whereas natural ecosystems have many species and tend to remain stable over time. Herbivorous insects are controlled by natural predators, including other insects, birds, and shrews. Insect populations are also limited because their preferred plants are scattered among many other plants rather than growing in a pure stand, as on a farm. On farms, both pest insects and their natural predators are exposed to pesticides. Unfortunately, the pests may develop resistance to the poison, but their predators are killed. In these simplified communities, unfavorable weather conditions or the introduction of an exotic species can be disastrous.

Farmers can counteract this trend by planting smaller fields with a wider variety of crops. Alternating crops helps maintain soil fertility and also helps prevent the proliferation of disease and insect pests that are specialized for a particular crop. Populations of corn borers, for example, will die off during years when the field is planted in alfalfa. The use of biological controls such as natural insect predators and insect diseases can reduce reliance on pesticides.

Finally, human ecosystems are characterized by continuously growing populations, whereas nonhuman populations in natural ecosystems are relatively stable. As our population expands, the spread of human-dominated ecosystems presents a growing threat to the diversity of species and to the delicate balance that has evolved over the 3-billion-year history of life on Earth.

In summary, natural ecosystems tend to be complex, stable, and self-sustaining, powered by solar energy and nourished by recycled nutrients. They provide diverse habitats for wildlife, purify contaminants through the action of decomposers, and build up nutrient-rich soil. Modern human ecosystems are relatively simple and are sustained by large inputs of energy from fossil fuels. They tend to minimize wildlife habitat, to contaminate soil and water, and to lose nutrients and fertile soil. These problems are compounded by continued population growth, which is causing the expansion of human-dominated ecosystems at the expense of undisturbed ones.

As we have pointed out, human ecosystems do not have to be as disruptive and alien to the operation of natural ecosystems as we have allowed them to become. Through understanding, education, commitment, appropriate use of technology, and stabilization of our population, we can reverse many of these destructive trends.

sunlight in this zone promotes the growth of a veritable soup of photosynthetic phytoplankton. Associated with these plants and protists are animals from nearly every phylum: annelid worms, sea anemones, jellyfish, sea urchins, sea stars, mussels, snails, fish, and sea otters, to name just a few. A large number and variety of organisms live permanently in coastal waters, but many that spend most of their lives in the open ocean come into the coastal waters to reproduce. Bays, salt marshes, and estuaries in particular are the breeding grounds for a wide variety of organisms, such as crabs, shrimp, and an array of fish, including most of our commercially important species.

Human Impact Coastal regions are of great importance not only to the organisms that live or breed there but also to humans who use them for food sources, recreation, mineral and oil extraction, or living places. Like rain forests, wetlands of all types once covered 6% of Earth's surface, but nearly half of them have been dredged or filled in. In the United States, only about 100 million acres remain out of an original 215 million acres of wetlands. As populations increase in coastal states and resources such as oil become increasingly scarce, the conflict between preservation of coastal wetlands as wildlife and animal habitat and development of these areas for housing, harbors and marinas, and energy extraction will become increasingly intense. Although conservation efforts have slowed the loss of wetlands, the United States still loses well over 100,000 acres yearly. Because much of the life of the entire ocean is dependent on the well-being of the coastal waters, it is essential that we protect these fragile, vital areas.

Coral Reefs

Coral reefs are created by animals and plants. In warm tropical waters, with just the right combination of bottom depth, wave action, and nutrients, specialized algae and corals (types of cnidarians) build reefs from their own calcium carbonate skeletons (Fig. 41-28). Coral reefs

Figure 41-28 Coral reefs
(a) Coral reefs, composed of the bodies of corals and algae, provide habitat for an extremely diverse community of extravagantly colored animals. *(b)* Many fish, including this blue tang, feed on coral (note the bright yellow corals in background). A vast array of invertebrates such as *(c)* this sponge and *(d)* blue-ringed octopus live among the corals of Australia's Great Barrier Reef. This tiny octopus (15 centimeters, or 6 inches, fully extended) is one of the world's most venomous creatures, possessing enough venom to kill 10 adult humans.

are most abundant in tropical waters of the Pacific and Indian Oceans, the Caribbean, and the Gulf of Mexico as far north as southern Florida, where the maximum water temperatures range between 22°C and 28°C (72°F and 82°F).

Reef-building corals are involved in a mutualistic relationship with unicellular algae called dinoflagellates, which live embedded in the coral tissue. These corals grow best within the photic zone at depths of less than 40 meters (130 feet), where light can penetrate and allow their algal partners to photosynthesize. The algae benefit from the high nitrogen, phosphorus, and carbon dioxide levels in the coral tissues. In return, algae provide food for the coral and help produce calcium carbonate, which forms the coral skeleton. Coral reefs provide an anchoring place for many other algae, a home for bottom-dwelling animals, and shelter and food for the most diverse collection of invertebrates and fish in the oceans (Fig. 41-28). The Great Barrier Reef in Australia is home to more than 200 species of coral alone, and a single reef may harbor 3000 species of fish, invertebrates, and algae.

Human Impact Coral reefs are extremely sensitive to certain types of disturbance, especially silt caused by soil eroding from nearby land. As silt clouds the water, light is diminished and photosynthesis reduced, hampering growth of the corals. The reef may eventually become buried in mud, the corals smothered, and the entire marvelous community of diverse organisms destroyed. Erosion from construction, roadways, and poor land management has produced enough silt to ruin several reefs near Honolulu, Hawaii. The reef inhabitants have been replaced by large numbers of sediment-feeding invertebrates such as sea cucumbers. In the Philippines, logging has dramatically increased erosion, so coral reefs (as well as rain forests) are being destroyed. Another hazard is sewage and runoff from agriculture, which fertilizes near-shore ocean water, causing eutrophication and a dense growth of algae. The algae block sunlight from the coral's dinoflagellates and thereby deprive the corals of nutrients. Decaying algal bodies also deplete the water of oxygen, killing the coral.

Still another threat to the reef communities is overfishing. In at least 80 countries, a variety of species, including mollusks, turtles, fish, crustaceans, and even corals, are being harvested from reefs faster than they can replace themselves. Many of these species are sold to shell collectors and aquarium owners in developed countries. In some tropical countries, dynamite is used to kill coral reef fish, destroying entire sections of the coral reef community in the process. Tropical fish collectors often use poison to stun the fish before collecting them, leaving most dead. The removal of predators from reefs may disrupt the ecological balance of the community, allowing an explosion in populations of coral-eating sea urchins. For example, such sea urchins are destroying reefs along the Kenyan coast in Africa.

As with rain forests, both protection and sustainable use are crucial to the survival of these fragile and diverse underwater ecosystems. Carefully regulated harvesting and tourism produce far more economic benefits than do activities that destroy the reefs. Some countries have recognized this, and there are now many areas where reefs are protected.

The Open Ocean

Beyond the coastal regions lie vast areas of the ocean in which the bottom is too deep to allow plants to anchor and still receive enough light to grow. Most life in the open ocean (Fig. 41-29a) is limited to the upper photic zone, in which the life-forms are **pelagic**—that is, free-swimming or floating—for their entire lives (Fig. 41-29b,c,d). The food web of the open ocean is dependent on phytoplankton consisting of microscopic photosynthetic protists, mainly diatoms and dinoflagellates (Fig. 41-29e). These organisms are consumed by zooplankton, such as tiny crustaceans that are relatives of crabs and lobsters (Fig. 41-29f). Zooplankton in turn serve as food for larger invertebrates, small fish, and even marine mammals such as the humpback whale (Fig. 41-29d).

One challenge faced by inhabitants of the open ocean is to remain afloat in the photic zone, where sunlight and food are abundant. Many members of the planktonic community have elaborate flotation devices, such as oil droplets in their cells or long projections, to slow their rate of sinking (see Fig. 41-29e). Most fish have swim bladders that can be filled with gas to regulate their buoyancy. Some animals, and even some of the phytoplankton, actively swim to stay in the photic zone. Many small crustaceans migrate to the surface at night to feed, then sink into the dark depths during daylight, thus avoiding visual predators such as fish. The amount of pelagic life varies tremendously from place to place. The blue clarity of tropical waters is a result of a lack of nutrients, which limits the concentration of plankton in the water. Nutrient-rich waters that support a large plankton community are greenish and relatively murky.

Below the photic zone, the only available energy in most cases comes from the excrement and dead bodies that drift down from above. Nevertheless, a surprising quantity and variety of life exist in the aphotic zone, including fishes of bizarre shapes, worms, sea cucumbers, sea stars, and mollusks.

Human Impact Two major threats to the open ocean are pollution and overfishing. Open-ocean pollution takes several forms. Oceangoing vessels dump millions of plastic containers overboard daily, and plastic six-pack holders, foam cups, and packing material wash and blow off the land, collecting on parts of the ocean surface. The plastic looks like food to unsuspecting sea turtles, gulls, porpoises, seals, and whales, many of which die after trying to consume it. Until 1992, New York City placed its refuse and sewage sludge on barges and towed them out

Figure 41-29 *The open ocean*
(a) The open ocean supports abundant life in the photic zone, where light is available. *(b)* Porpoises skim the surface, *(c)* fish such as the blue jack swim, and *(d)* rare humpback whales leap clear of the water. *(e)* The photosynthetic phytoplankton are the producers on which most other marine life ultimately depends. Phytoplankton are eaten by *(f)* zooplankton, represented by this microscopic crustacean, a copepod. The spiny projections on these planktonic creatures help keep them from sinking below the photic zone.

to sea, creating a heavily contaminated area covering 40 square miles of open-ocean floor. The open ocean has served as a dumping ground for radioactive wastes. Oil contaminates the open ocean from many sources, including tanker spills, runoff from improper disposal on land, leakage from offshore oil wells, and natural seepage.

The increasing demand for fish to feed a growing human population, coupled with increasingly efficient fishing technologies, has resulted in depletion of many important fisheries. The cod fishery of the northeastern United States has collapsed from overfishing. Populations of lobsters, salmon, haddock, king crab, and many other types of seafood have also declined dramatically because of overfishing. Many scientists hope for the designation of more marine preserves where fishing is banned, allowing fish in these areas to reproduce and replenish depleted populations. Establishment of no-fishing zones is a difficult undertaking due to intense pressure from fishermen, and this concept is still in its infancy.

Hydrothermal Vent Communities

In 1977, a new and unusual source of nutrients, forming the basis of a spectacular undersea community, was discovered in the deep ocean. Geologists exploring the Galapagos Rift (an area of the Pacific floor where plates that form Earth's crust are separating) found vents spewing superheated water, black with sulfur and minerals. Surrounding these vents was a rich community of pink fish, blind white crabs, enormous mussels, giant white clams, sea anemones, and giant tube worms (Fig. 41-30). Twenty-two previously undescribed families and 284 new species of animals have been found in these novel **hydrothermal vent communities**.

In this unique ecosystem, sulfur bacteria serve as the primary producers, harvesting the energy stored in hydrogen sulfide spewed from vents in Earth's crust. This process, called *chemosynthesis*, replaces photosynthesis at these depths, over 2500 meters (7500 feet) below the surface. The bacteria proliferate in the warm water sur-

*Figure 41-30 **Hydrothermal vent communities***
Located in the ocean depths, vent communities include giant tube worms nearly 4 meters (12 feet) long. Parts of these worms are red with oxygen-trapping hemoglobin. They have no digestive tract but are host to sulfur bacteria, which provide the worms with energy by oxidizing hydrogen sulfide.

rounding the vents, covering nearby rocks with thick, matlike colonies. These colonies provide the food on which the animals of the vent community thrive. Many vent animals consume the bacteria directly. Others, such as the giant tube worm (which lacks a digestive tract), harbor the bacteria within cells in their bodies (see Fig. 41-30). In this mutualistic association, the bacteria pro-vide high-energy carbon compounds, and the tube worm provides hydrogen sulfide. The worm uses a unique form of hemoglobin to transport the hydrogen sulfide to the bacteria in its bloodstream.

The world still holds wonders and mysteries for those who seek them. We have only begun to explore the versatility and diversity of life on Earth.

Summary of Key Concepts

1) What Factors Influence Earth's Climate?

The availability of sunlight, water, and appropriate temper-atures determines the climate of a given region. Sunlight maintains Earth's temperature. Equal amounts of solar en-ergy are spread over a smaller surface at the equator than farther north and south, making the equator relatively warm, whereas higher latitudes have lower overall temperatures. Earth's tilt on its axis causes dramatic seasonal variation at northern and southern latitudes.

The rising of warm air and sinking of cool air in regular patterns from north to south produce areas of low and high moisture. These patterns are modified by the topography of continents and by ocean currents.

2) What Are the Requirements of Life?

The requirements for life on Earth include nutrients, energy, liquid water, and a reasonable temperature. The differences in the form and abundance of living things in various loca-tions on Earth are largely attributable to differences in the interplay of these four factors.

3) How Is Life on Land Distributed?

On land, the crucial limiting factors are temperature and liq-uid water. Large regions of the continents that have similar climates will have similar vegetation, determined by the in-teraction of temperature and rainfall or the availability of water. These regions are called biomes.

Tropical forest biomes, located near the equator, are warm and wet, dominated by huge broadleaf evergreen trees. Most nutrients are tied up in vegetation, and most animal life is arboreal. Rain forests, home to at least 50% of all species, are rapidly being cut for agriculture, although the soil is ex-tremely poor.

The African savanna is an extensive grassland with pro-nounced wet and dry seasons. It is home to the world's most diverse and extensive herds of large mammals.

Most deserts, hot and dry, are located between 20° and 30° of latitude and in the rain shadows of mountain ranges. In deserts, plants are widely spaced and have adaptations to conserve water. Animals tend to be small and nocturnal, also adapted to drought.

Chaparral exists in desertlike conditions, which are mod-erated by their proximity to a coastline, allowing small trees and bushes to thrive.

Grasslands, concentrated in the centers of continents, have a continuous grass cover and few trees. They produce the world's richest soils and have largely been converted to agriculture.

Temperate deciduous forests, whose broadleaf trees drop their leaves in winter to conserve moisture, dominate the eastern half of the United States and are also found in West-ern Europe and East Asia. Higher precipitation occurs there than in the grasslands. The wet temperate rain forests, dominated by evergreens, are on the northern Pacific coast of the United States.

The taiga, or northern coniferous forest, covers much of the northern United States, southern Canada, and northern Eurasia. It is dominated by conifers whose small waxy needles are adapted for water conservation and year-round photosynthesis.

The tundra is a frozen desert where permafrost prevents the growth of trees, and bushes remain stunted. Nonetheless, diverse arrays of animal life and perennial plants flourish in this fragile biome, which is found on mountain peaks and the Arctic.

4 How Is Life in Water Distributed?

Energy and nutrients are the major limiting factors in the distribution and abundance of life in aquatic ecosystems. Nutrients are found in bottom sediments and are washed in from surrounding land, concentrating them near shore and in deep water.

Freshwater lakes have three life zones. The littoral zone, near shore, is rich in energy and nutrients and supports the most diverse community. The limnetic zone is the lighted region of open water where photosynthesis can occur. The profundal zone is the deep water, where light is inadequate for

photosynthesis and the community is dominated by heterotrophic organisms.

Oligotrophic lakes are clear and low in nutrients, and they support sparse communities. Eutrophic lakes are rich in nutrients and support dense communities. During succession to dry land, lakes tend to go from an oligotrophic to a eutrophic condition.

Most life in the oceans is found in shallow water, where sunlight can penetrate, and is concentrated near the continents and in areas of upwelling, where nutrients are most plentiful.

Coastal waters, consisting of the intertidal zone and the near-shore zone, contain the most abundant life. Producers consist of aquatic plants anchored to the bottom and photosynthetic protists called phytoplankton.

Coral reefs are confined to warm, shallow seas. The calcium carbonate reefs form a complex habitat supporting the most diverse undersea ecosystem.

In the open ocean, most life is found in the photic zone, where light supports phytoplankton. In the lower aphotic zone, life is supported by nutrients that drift down from the photic zone. Specialized vent communities, supported by bacteria, thrive at great depths in the superheated waters where Earth's crustal plates are separating.

Key Terms

aphotic zone   p. 880
biodiversity   p. 866
biome   p. 865
chaparral   p. 872
climate   p. 859
coral reef   p. 884
desert   p. 870
desertification   p. 871
estuary   p. 881
eutrophic lake   p. 879
grassland   p. 872

gyre   p. 862
hydrothermal vent
   community   p. 886
intertidal zone   p. 881
limnetic zone   p. 879
littoral zone   p. 879
near-shore zone   p. 881
northern coniferous forest
   p. 876
oligotrophic lake   p. 879
ozone layer   p. 861

pelagic   p. 885
permafrost   p. 878
photic zone   p. 880
phytoplankton   p. 879
plankton   p. 879
prairie   p. 872
profundal zone   p. 879
rain shadow   p. 863
savanna   p. 868
taiga   p. 876

temperate deciduous forest
   p. 873
temperate rain forest   p. 875
tropical deciduous forest
   p. 868
tropical rain forest   p. 866
tundra   p. 877
upwelling   p. 881
weather   p. 859
zooplankton   p. 879

Thinking Through the Concepts

Multiple Choice

1. *Most plant species fit into only a few morphological types. Which type would you think is most restricted by temperature and rainfall?*
   a. trees
   b. shrubs
   c. grasses
   d. perennial herbs
   e. annual weeds

2. *In which biome is the smallest fraction of carbon and nutrients present in the soil?*
   a. tropical rain forest
   b. savanna
   c. tundra
   d. grassland
   e. coniferous forest

3. *How do mountain ranges create deserts?*
   a. by lifting land up into colder, drier air
   b. by completely blocking the flow of air into desert areas, thus preventing clouds from getting there
   c. by forcing air to first rise and then fall, thus causing rain on one side of the mountains and desert on the other
   d. by causing the global wind patterns that make certain latitudes very dry
   e. by causing very steep slopes that are subject to erosion

4. *What is the primary reason that plants from distant, but climatically similar, places commonly look the same?*
   a. common ancestry
   b. adaptation to the same physical conditions
   c. adaptation to similar herbivores
   d. continental drift
   e. effects of past climate change

5. *Which of these biomes has been increased in area by human activities?*
a. savanna
b. temperate rain forest
c. grassland
d. coniferous forest
e. desert

6. *What biome has the richest soil and has largely been converted to agriculture?*
a. tundra
b. coniferous forest
c. grassland
d. tropical rain forest
e. deciduous forest

? Review Questions

1. Explain how air currents contribute to the formation of the tropics and the large deserts.

2. What are large, roughly circular ocean currents called? What effect do they have on climate, and where is that effect strongest?

3. What are the four major requirements for life? Which two are most often limiting in terrestrial ecosystems? In ocean ecosystems?

4. Explain why traveling up a mountain takes you through biomes similar to those you would encounter traveling north for a long distance.

5. Where are the nutrients of the tropical forest biome concentrated? Why is life in the tropical rain forest concentrated high above the ground?

6. Explain two undesirable effects of agriculture in the tropical rain forest biome.

7. List some adaptations of (a) desert plants and (b) desert animals to heat and drought.

8. How and where does desertification occur?

9. How are trees of the taiga adapted to a lack of water and a short growing season?

10. How do deciduous and coniferous biomes differ?

11. What single environmental factor best explains why there is shortgrass prairie in Colorado, tallgrass prairie in Illinois, and deciduous forest in Ohio?

12. Where are the world's largest populations of large herbivores and carnivores located?

13. Where is life in the oceans most abundant, and why?

14. Why is the diversity of life so high in coral reefs? What human impacts threaten them?

15. Distinguish among the limnetic, littoral, and profundal zones of lakes in terms of their location and the communities they support.

16. Distinguish between oligotrophic and eutrophic lakes. Describe (a) a natural scenario and (b) a human-created scenario under which an oligotrophic lake might be converted to a eutrophic lake.

17. What is the reason and importance of the spring and fall overturn in temperate lakes?

18. Distinguish between the photic and aphotic zones. How do organisms in the photic zone obtain nutrients? How are nutrients obtained in the aphotic zone?

19. What unusual primary producer forms the basis for hydrothermal vent communities?

20. On the basis of the location of the worst atmospheric ozone depletion, which biomes are likely to be most affected by increased UV penetration?

Applying the Concepts

1. List at least six differences between human-dominated and undisturbed ecosystems, and discuss in some detail how these differences can be minimized.

2. In which terrestrial biome is your college or university located? Discuss similarities and differences between your location and the general description of that biome in the text. If you are living in a city, how has the urban environment modified your interaction with the biome?

3. During the 1960s and 1970s, many parts of the United States and Canada banned the use of detergents containing phosphates. Until that time, almost all laundry detergents and many soaps and shampoos had high concentrations of phosphates. What environmental concern do you think prompted these bans, and what ecosystem has benefited most from the bans?

4. Because ozone depletion is expected to get worse, not better, for decades to come, biologists have tried to assess which types of species will be most susceptible to increased UV penetration. Two of the groups that may be most vulnerable—ocean plankton and long-lived birds and mammals—are quite different from each other. Try to think of what makes each of these groups so susceptible to increased UV radiation.

5. Understanding the ways in which the four basic requirements for life determine where different biomes occur can help us predict the consequences of global warming. Global warming is expected to make most areas warmer, but it is also expected to change rainfall in ways that are hard to predict—some areas will get wetter, and others drier. Our ignorance of how rainfall will change is not very important in understanding shifts in more northerly biomes, but it is very important for our understanding of changes in tropical areas. Look at Figure 41-8 and explain why this is true.

6. More-northerly forests are far better able to regenerate after logging than are tropical rain forests. Try to explain why this is true. HINT: The cold soils of northern climates greatly slow down decomposition rates.

Group Activity

Work in groups of three or four. Your group is presented with the question, How can humans become more fully integrated into the biosphere? Use "Earth Watch: Humans and Ecosystems" on page 882 and "Evolutionary Connections: Are Humans a Biological Success?" on page 444 (in Chapter 22) to present a realistic answer to this question. Include the following components:

- Awareness and Assessment—What problems in ecosystems have been influenced by humans?

- Analysis—What effects (both positive and negative) might the actions or plan have?
- Education—How will you inform others of the problem and possible solutions?
- Politics—How can you get others involved?
- Action—How will the actions taken be monitored and evaluated to ensure success?

For More Information

Bell, R. H. V. "A Grazing Ecosystem in the Serengeti." *Scientific American*, July 1971. Describes the great migrations of the herbivores of the African savanna.

Chadwick, D. H. "Blue Refuges." *National Geographic*, March 1998. Beautiful photographs highlight 12 national marine sanctuaries set aside to preserve a variety of marine ecosystems along the coastal United States.

Dodd, J. L. "Desertification and Degradation in sub-Saharan Africa: The Role of Livestock." *BioScience* Vol. 44, 1994. An argument that the process of desert formation is not as clearly linked to human-caused degradation as is usually portrayed.

Goreau, T. F., Goreau, N. I., and Goreau, T. J. "Corals and Coral Reefs." *Scientific American*, August 1979. Discusses the ecology of the great reefs built by tiny polyps harboring symbiotic protists.

Hinrichsen, D. "Requiem for Reefs?" *International Wildlife*, March/April 1997. Stunning photographs and a compelling text describe the beauty of Earth's imperiled coral reefs.

Holloway, M. "Sustaining the Amazon." *Scientific American*, July 1993. Describes the threats to the Amazon rain forest and innovative ways to preserve tropical forests and their biodiversity.

Kusler, J. A., Mitsch, W. J., and Larson, J. S. "Wetlands." *Scientific American*, January 1994. Discusses the characteristics and values of these threatened but highly important ecosystems.

Mares, M. A. "Desert Rodents, Seed Consumption, and Convergence: Evolutionary Shuffling of Adaptations." *BioScience*, Vol. 43, 1993. An example of how animal as well as plant species and communities living in similar climates can show convergence.

Milton, S. J., Dean, R. J., du Plessis, M. A., and Siegfried, W. R. "A Conceptual Model of Arid Rangeland Degradation." *BioScience*, Vol. 44, 1994. Rangeland degradation proceeds in steps, each of them increasingly difficult and costly to reverse.

Mitchell, J. G. "Our National Forests." *National Geographic*, March 1997. Stunning photographs and engaging text describe the threats to forests in the United States.

Prather, M., Midgley, P., Rowland, F. S., and Storlarski, R. "The Ozone Layer: The Road Not Taken." *Nature*, June 13, 1996. Reviews the progress toward slowing ozone depletion and explores what would have happened had the ozone-protection treaties not been passed.

Toon, B., and Turco, R. P. "Polar Stratospheric Clouds and Ozone Depletion." *Scientific American*, June 1991. Describes the unique chemistry of the Antarctic clouds that causes ozone breakdown.

Tunnicliffe, V. "Hydrothermal-Vent Communities of the Deep Sea." *American Scientist*, July–August 1992. A comprehensive look at the structure and function of undersea vent communities, with excellent illustrations.

Williams, T. "What Good Is a Wetland?" *Audubon*, November/December 1996. In addition to supporting a diversity of wildlife, purifying drinking water, and protecting people from floods, wetlands are beautiful.

Wilson, E. O. "Threats to Biodiversity." *Scientific American*, September 1989. The elimination of habitats, particularly in the rain forest, is driving plants and animals to extinction in record numbers.

Answers to Multiple-Choice Questions

1. a   2. a   3. c   4. b   5. e   6. c

Metric System Conversions

To Convert Metric Units:	Multiply by:	To Get English Equivalent:
Length		
Centimeters (cm)	0.3937	Inches (in.)
Meters (m)	3.2808	Feet (ft)
Meters (m)	1.0936	Yards (yd)
Kilometers (km)	0.6214	Miles (mi)
Area		
Square centimeters (cm²)	0.155	Square inches (in.²)
Square meters (m²)	10.7639	Square feet (ft²)
Square meters (m²)	1.1960	Square yards (yd²)
Square kilometers (km²)	0.3831	Square miles (mi²)
Hectare (ha) (10,000 m²)	2.4710	Acres (a)
Volume		
Cubic centimeters (cm³)	0.06	Cubic inches (in.³)
Cubic meters (m³)	35.30	Cubic feet (ft³)
Cubic meters (m³)	1.3079	Cubic yards (yd³)
Cubic kilometers (km³)	0.24	Cubic miles (mi³)
Liters (L)	1.0567	Quarts (qt), U.S.
Liters (L)	0.26	Gallons (gal), U.S.
Mass		
Grams (g)	0.03527	Ounces (oz)
Kilograms (kg)	2.2046	Pounds (lb)
Metric ton (tonne) (t)	1.10	Ton (tn), U.S.
Speed		
Meters/second (mps)	2.24	Miles/hour (mph)
Kilometers/hour (kmph)	0.62	Miles/hour (mph)

To Convert English Units:	Multiply by:	To Get Metric Equivalent:
Length		
Inches (in.)	2.54	Centimeters (cm)
Feet (ft)	0.3048	Meters (m)
Yards (yd)	0.9144	Meters (m)
Miles (mi)	1.6094	Kilometers (km)
Area		
Square inches (in.²)	6.45	Square centimeters (cm²)
Square feet (ft²)	0.0929	Square meters (m²)
Square yards (yd²)	0.8361	Square meters (m²)
Square miles (mi²)	2.5900	Square kilometers (km²)
Acres (a)	0.4047	Hectare (ha) (10,000 m²)
Volume		
Cubic inches (in.³)	16.39	Cubic centimeters (cm³)
Cubic feet (ft³)	0.028	Cubic meters (m³)
Cubic yards (yd³)	0.765	Cubic meters (m³)
Cubic miles (mi³)	4.17	Cubic kilometers (km³)
Quarts (qt), U.S.	0.9463	Liters (L)
Gallons (gal), U.S.	3.8	Liters (L)
Mass		
Ounces (oz)	28.3495	Grams (g)
Pounds (lb)	0.4536	Kilograms (kg)
Ton (tn), U.S.	0.91	Metric ton (tonne) (t)
Speed		
Miles/hour (mph)	0.448	Meters/second (mps)
Miles/hour (mph)	1.6094	Kilometers/hour (kmph)

Metric Prefixes

Prefix			Meaning	
giga-	G	10^9	=	1,000,000,000
mega-	M	10^6	=	1,000,000
kilo-	k	10^3	=	1000
hecto-	h	10^2	=	100
deka-	da	10^1	=	10
		10^0	=	1
deci-	d	10^{-1}	=	0.1
centi-	c	10^{-2}	=	0.01
milli-	m	10^{-3}	=	0.001
micro-	μ	10^{-6}	=	0.000001

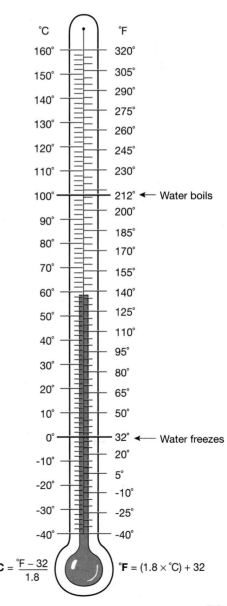

°C = $\dfrac{°F - 32}{1.8}$ °F = (1.8 × °C) + 32

Classification of Major Groups of Organisms*

Domain	Kingdom	Division or Phylum	Common Name
Bacteria (prokaryotic, peptidoglycan in cell wall)			bacteria
Archaea (prokaryotic, no peptidoglycan in cell wall)			archaeans
Eukarya (eukaryotic)			
	Protista (unicellular)		protists
		Division Pyrrophyta	dinoflagellates
		Division Chrysophyta	diatoms
		Division Euglenophyta	euglenoids
		Division Myxomycota	plasmodial slime molds
		Division Acrasiomycota	cellular slime molds
		Division Oomycota	egg fungi
		Phylum Sarcomastigophora	zooflagellates, amoebae
		Phylum Apicomplexa	sporozoans
		Phylum Ciliophora	ciliates
	Fungi (multicellular, heterotrophic, absorb nutrients)		fungi
		Division Zygomycota	zygote fungi
		Division Ascomycota	sac fungi
		Division Deuteromycota	imperfect fungi
		Division Basidiomycota	club fungi
	Plantae (multicellular, photosynthetic)		plants
		Division Rhodophyta	red algae
		Division Phaeophyta	brown algae
		Division Chlorophyta	green algae
		Division Bryophyta	liverworts, mosses
		Division Pteridophyta	ferns
		Division Coniferophyta	evergreens
		Division Anthophyta	flowering plants
	Animalia (multicellular, heterotrophic, ingest nutrients)		animals
		Phylum Porifera	sponges
		Phylum Cnidaria	hydras, sea anemones, jellyfish, corals
		Phylum Platyhelminthes	flatworms
		Phylum Nematoda	roundworms
		Phylum Annelida	segmented worms
		Class Oligochaeta	earthworms
		Class Polychaeta	tube worms
		Class Hirudinea	leeches
		Phylum Arthropoda	arthropods ("jointed legs")
		Class Insecta	insects
		Class Arachnida	spiders, ticks
		Class Crustacea	crabs, lobsters
		Phylum Mollusca	mollusks ("soft-bodied")
		Class Gastropoda	snails
		Class Pelecypoda	mussels, clams
		Class Cephalopoda	squid, octopuses
		Phylum Echinodermata	sea stars, sea urchins, sea cucumbers
		Phylum Chordata	chordates
		Subphylum Tunicata	tunicates
		Subphylum Cephalochordata	lancelets
		Subphylum Vertebrata	vertebrates
		Class Agnatha	lampreys, hagfish
		Class Chondrichthyes	sharks, rays
		Class Osteichthyes	bony fishes
		Class Amphibia	frogs, salamanders
		Class Reptilia	turtles, snakes, lizards
		Class Aves	birds
		Class Mammalia	mammals

*This table lists only those taxonomic categories described in the textbook.

Glossary

abdomen: the body segment at the posterior end of an animal with segmentation; contains most of the digestive structures.

abiotic (ā-bī-ah´-tik): nonliving; the abiotic portion of an ecosystem includes soil, rock, water, and the atmosphere.

abortion: the procedure for terminating pregnancy; the cervix is dilated, and the embryo and placenta are removed.

abscisic acid (ab-sis´-ik): a plant hormone that generally inhibits the action of other hormones, enforcing dormancy in seeds and buds and causing the closing of stomata.

abscission layer: a layer of thin-walled cells, located at the base of the petiole of a leaf, that produces an enzyme that digests the cell wall holding leaf to stem, allowing the leaf to fall off.

absorption: the process by which nutrients are taken into cells.

accessory pigments: colored molecules other than chlorophyll that absorb light energy and pass it to chlorophyll.

acellular slime mold: a type of funguslike protist that forms a multinucleate structure that crawls in amoeboid fashion and ingests decaying organic matter; also called plasmodial slime mold.

acetylcholine (ah-sēt´-il-kō´-lēn): a neurotransmitter in the brain and in synapses of motor neurons that innervate skeletal muscles.

acid: a substance that releases hydrogen ions (H^+) into solution; a solution with a pH of less than 7.

acid deposition: the deposition of nitric or sulfuric acid, either dissolved in rain (acid rain) or in the form of dry particles, as a result of the production of nitrogen oxides or sulfur dioxide through burning, primarily of fossil fuels.

acidic: with an H^+ concentration exceeding that of OH^-; releasing H^+.

acquired immune deficiency syndrome (AIDS): an infectious disease caused by the human immunodeficiency virus (HIV); attacks and destroys T cells, thus weakening the immune system.

acrosome (ak´-rō-sōm): a vesicle, located at the tip of an animal sperm, that contains enzymes needed to dissolve protective layers around the egg.

actin (ak´-tin): a major muscle protein whose interactions with myosin produce contraction; found in the thin filaments of the muscle fiber; see also *myosin*.

action potential: a rapid change from a negative to a positive electrical potential in a nerve cell. This signal travels along an axon without a change in intensity.

activation energy: in a chemical reaction, the energy needed to force the electron shells of reactants together, prior to the formation of products.

active site: the region of an enzyme molecule that binds substrates and performs the catalytic function of the enzyme.

active transport: the movement of materials across a membrane through the use of cellular energy, normally against a concentration gradient.

adaptation: a characteristic of an organism that helps it survive and reproduce in a particular environment; also, the process of acquiring such characteristics.

adaptive radiation: the rise of many new species in a relatively short time as a result of a single species that invades different habitats and evolves under different environmental pressures in those habitats.

adenosine diphosphate (a-den´-ō-sēn dī-fos´-fāt; ADP): a molecule composed of the sugar ribose, the base adenine, and two phosphate groups; a component of ATP.

adenosine triphosphate (a-den´-ō-sēn trī-fos´-fāt; ATP): a molecule composed of the sugar ribose, the base adenine, and three phosphate groups; the major energy carrier in cells. The last two phosphate groups are attached by "high-energy" bonds.

adipose tissue (a´-di-pōs): tissue composed of fat cells.

adrenal cortex: the outer part of the adrenal gland, which secretes steroid hormones that regulate metabolism and salt balance.

adrenal gland: a mammalian endocrine gland, adjacent to the kidney; secretes hormones that function in water regulation and in the stress response.

adrenal medulla: the inner part of the adrenal gland, which secretes epinephrine (adrenaline) and norepinephrine (noradrenaline).

adrenocorticotropic hormone (a-drēn-ō-kor-tik-ō-trō´-pik; ACTH): a hormone, secreted by the anterior pituitary, that stimulates the release of hormones by the adrenal glands, especially in response to stress.

aerobic: using oxygen.

age structure: the distribution of males and females in a population according to age groups.

agglutination (a-gloo-tin-ā´-shun): the clumping of foreign substances or microbes, caused by binding with antibodies.

aggression: antagonistic behavior, normally among members of the same species, often resulting from competition for resources.

aggressive mimicry (mim´ik-rē): the evolution of a predatory organism to resemble a harmless animal or part of the environment, thus gaining access to prey.

aldosterone: a hormone, secreted by the adrenal cortex, that helps regulate ion concentration in the blood by stimulating the reabsorption of sodium by the kidneys and sweat glands.

alga (al´-ga; pl., algae, al´-jē): simple aquatic plants that lack vascular tissue.

allantois (al-an-tō´-is): one of the embryonic membranes of reptiles, birds, and mammals; in reptiles and birds, serves as a waste-storage organ; in mammals, forms most of the umbilical cord.

allele (al-ēl´): one of several alternative forms of a particular gene.

allele frequency: for any given gene, the relative proportion of each allele of that gene in a population.

allergy: an inflammatory response produced by the body in response to invasion by foreign materials, such as pollen, that are themselves harmless.

allopatric speciation (al-ō-pat´-rik): speciation that occurs when two populations are separated by a physical barrier that prevents gene flow between them (geographical isolation).

alternation of generations: a life cycle, typical of plants, in which a diploid sporophyte (spore-producing) generation alternates with a haploid gametophyte (gamete-producing) generation.

altruism: a type of behavior that may decrease the reproductive success of the individual performing it but benefits that of other individuals.

alveolus (al-vē´-ō-lus; pl., alveoli): a tiny air sac within the lungs, surrounded by capillaries, where gas exchange with the blood occurs.

amino acid: the individual subunit of which proteins are made, composed of a central carbon atom bonded to an amino group ($-NH_2$), a carboxyl group ($-COOH$), a hydrogen atom, and a variable group of atoms denoted by the letter "R."

ammonia: NH_3; a highly toxic nitrogen-containing waste product of amino acid breakdown. In the mammalian liver, it is converted to urea.

amniocentesis (am-nē-ō-sen-tē´-sis): a procedure for sampling the amniotic fluid surrounding a fetus: A sterile needle is inserted through the abdominal wall, uterus, and amniotic sac of a pregnant woman; 10 to 20 milliliters of amniotic fluid is withdrawn. Various tests may be performed on the fluid and the fetal cells suspended in it to provide information on the developmental and genetic state of the fetus.

amnion (am´-nē-on): one of the embryonic membranes of reptiles, birds, and mammals; encloses a fluid-filled cavity that envelops the embryo.

G-1

amniote egg (am-nē-ōt´): the egg of reptiles and birds; contains an amnion that encloses the embryo in a watery environment, allowing the egg to be laid on dry land.

amoeba: a type of animal-like protist that uses a characteristic streaming mode of locomotion by extending a cellular projection called a pseudopod.

amoeboid cell: a protist or animal cell that moves by extending a cellular projection called a pseudopod.

amplexus (am-plek´-sus): in amphibians, a form of external fertilization in which the male holds the female during spawning and releases his sperm directly onto her eggs.

ampulla: a muscular bulb that is part of the water-vascular system of echinoderms; controls the movement of tube feet, which are used for locomotion.

amygdala (am-ig´-da-la): part of the forebrain of vertebrates that is involved in the production of appropriate behavioral responses to environmental stimuli.

amylase (am´-i-lās): an enzyme, found in saliva and pancreatic secretions, that catalyzes the breakdown of starch.

anaerobe: an organism whose respiration does not require oxygen.

anaerobic: not using oxygen.

analogous structures: structures that have similar functions and superficially similar appearance but very different anatomies, such as the wings of insects and birds. The similarities are due to similar environmental pressures rather than to common ancestry.

anaphase (an´-a-fāz): in mitosis, the stage in which the sister chromatids of each chromosome separate from one another and are moved to opposite poles of the cell; in meiosis I, the stage in which homologous chromosomes, consisting of two sister chromatids, are separated; in meiosis II, the stage in which the sister chromatids of each chromosome separate from one another and are moved to opposite poles of the cell.

androgen: a male sex hormone.

androgen insensitivity: a rare condition in which an individual with XY chromosomes is female in appearance because the body's cells don't respond to the male hormones that are present.

angina (an-jī´-nuh): chest pain associated with reduced blood flow to the heart muscle, caused by the obstruction of coronary arteries.

angiosperm (an´-jē-ō-sperm): a flowering vascular plant.

angiotensin (an-jē-ō-ten´-sun): a hormone that functions in water regulation in mammals by stimulating physiological changes that increase blood volume and blood pressure.

annual ring: a pattern of alternating light (early) and dark (late) xylem of woody stems and roots, formed as a result of the unequal availability of water in different seasons of the year, normally spring and summer.

antagonistic muscles: a pair of muscles, one of which contracts and in so doing extends the other; an arrangement that makes possible movement of the skeleton at joints.

anterior: the front, forward, or head end of an animal.

anterior pituitary: a lobe of the pituitary gland that produces prolactin and growth hormone as well as hormones that regulate hormone production in other glands.

anther (an´-ther): the uppermost part of the stamen, in which pollen develops.

antheridium (an-ther-id´-ē-um): a structure in which male sex cells are produced, found in the bryophytes and certain seedless vascular plants.

antibiotic resistance: the ability of a mutated pathogen to resist the effects of an antibiotic that normally kills it.

antibody: a protein, produced by cells of the immune system, that combines with a specific antigen and normally facilitates the destruction of the antigen.

anticodon: a sequence of three bases in transfer RNA that is complementary to the three bases of a codon of messenger RNA.

antidiuretic hormone (an-tē-dī-ūr-et´-ik; ADH): a hormone produced by the hypothalamus and released into the bloodstream by the posterior pituitary when blood volume is low; increases the permeability of the distal tubule and the collecting duct to water, allowing more water to be reabsorbed into the bloodstream.

antigen: a complex molecule, normally a protein or polysaccharide, that stimulates the production of a specific antibody.

aphotic zone: the region of the ocean below 200 m, where sunlight does not penetrate.

apical dominance: the phenomenon whereby a growing shoot tip inhibits the sprouting of lateral buds.

apical meristem (āp´-i-kul mer´-i-stem): the cluster of meristematic cells at the tip of a shoot or root (or one of their branches).

appendicular skeleton (ap-pen-dik´-ū-lur): the portion of the skeleton consisting of the bones of the extremities and their attachments to the axial skeleton; the pectoral and pelvic girdles, the arms, legs, hands, and feet.

aqueous humor (ā´-kwē-us): the clear, watery fluid between the cornea and lens of the eye.

Archaea: one of life's three domains; consists of prokaryotes that are only distantly related to members of the domain Bacteria.

archegonium (ar-ke-gō´-nē-um): a structure in which female sex cells are produced; found in the bryophytes and certain seedless vascular plants.

arteriole (ar-tēr´-ē-ōl): a small artery that empties into capillaries. Contraction of the arteriole regulates blood flow to various parts of the body.

artery (ar´-tuh-rē): a vessel with muscular, elastic walls that conducts blood away from the heart.

artificial selection: a selective breeding procedure in which only those individuals with particular traits are chosen as breeders; used mainly to enhance desirable traits in domestic plants and animals; may also be used in evolutionary biology experiments.

ascus (as´-kus): a saclike case in which sexual spores are formed by members of the fungal division Ascomycota.

asexual reproduction: reproduction that does not involve the fusion of haploid sex cells. The parent body may divide and new parts regenerate, or a new, smaller individual may form as an attachment to the parent, to drop off when complete.

association neuron: in a neural network, a nerve cell that is postsynaptic to a sensory neuron and presynaptic to a motor neuron. In actual circuits, there may be many association neurons between individual sensory and motor neurons.

atherosclerosis (ath´-er-ō-skler-ō´-sis): a disease characterized by the obstruction of arteries by cholesterol deposits and thickening of the arterial walls.

atom: the smallest particle of an element that retains the properties of the element.

atomic nucleus: the membrane-bound organelle of eukaryotic cells that contains the cell's genetic material.

atomic number: the number of protons in the nuclei of all atoms of a particular element.

atrial natriuretic peptide (ā´-trē-ul nā-trē-ū-ret´-ik; ANP): a hormone, secreted by cells in the mammalian heart, that reduces blood volume by inhibiting the release of ADH and aldosterone.

atrioventricular (AV) node (ā´-trē-ō-ven-trik´-ū-lar nōd): a specialized mass of muscle at the base of the right atrium through which the electrical activity initiated in the sinoatrial node is transmitted to the ventricles.

atrioventricular valve: a heart valve that separates each atrium from each ventricle, preventing the backflow of blood into the atria during ventricular contraction.

atrium (ā´-trē-um): a chamber of the heart that receives venous blood and passes it to a ventricle.

auditory canal (aw´-di-tor-ē): a canal within the outer ear that conducts sound from the external ear to the tympanic membrane.

auditory nerve: the nerve leading from the mammalian cochlea to the brain, carrying information about sound.

autoimmune disease: a disorder in which the immune system produces antibodies against the body's own cells.

autonomic nervous system: the part of the peripheral nervous system of vertebrates that synapses on glands, internal organs, and smooth muscle and produces largely involuntary responses.

autosome (aw´-tō-sōm): a chromosome that occurs in homologous pairs in both males and females and that does not bear the genes determining sex.

autotroph (aw´-tō-trof): "self-feeder"; normally, a photosynthetic organism; a producer.

auxin (awk´-sin): a plant hormone that influences many plant functions, including phototropism, apical dominance, and root branching; generally stimulates cell elongation and, in some cases, cell division and differentiation.

axial skeleton: the skeleton forming the body axis, including the skull, vertebral column, and rib cage.

axon: a long extension of a nerve cell, extending from the cell body to synaptic endings on other nerve cells or on muscles.

bacillus (buh-sil´-us; pl., **bacilli):** a rod-shaped bacterium.

Bacteria: one of life's three domains; consists of prokaryotes that are only distantly related to members of the domain Archaea.

bacterial conjugation: the exchange of genetic material between two bacteria.

bacteriophage (bak-tir´-ē-ō-fāj): a virus specialized to attack bacteria.

bacterium (bak-tir´-ē-um; pl., **bacteria):** an organism consisting of a single prokaryotic cell surrounded by a complex polysaccharide coat.

balanced polymorphism: the prolonged maintenance of two or more alleles in a population, normally because each allele is favored by a separate environmental pressure.

ball-and-socket joint: a joint in which the rounded end of one bone fits into a hollow depression in another, as in the hip; allows movement in several directions.

bark: the outer layer of a woody stem, consisting of phloem, cork cambium, and cork cells.

Barr body: an inactivated X chromosome in cells of female mammals, which have two X chromosomes; normally appears as a dark spot in the nucleus.

base: (1) a substance capable of combining with and neutralizing H^+ ions in a solution; a solution with a pH of more than 7; (2) in molecular genetics, one of the nitrogen-containing, single- or double-ringed structures that distinguish one nucleotide from another. In DNA, the bases are adenine, guanine, cytosine, and thymine.

basic: with an H^+ concentration less than that of OH^-; combining with H^+.

basidiospore (ba-sid´-ē-ō-spor): a sexual spore formed by members of the fungal division Basidiomycota.

basidium (bas-id´-ē-um): a diploid cell, typically club-shaped, formed by members of the fungal division Basidiomycota; produces basidiospores by meiosis.

basilar membrane (bas´-eh-lar): a membrane in the cochlea that bears hair cells that respond to the vibrations produced by sound.

basophil (bas´-ō-fil): a type of white blood cell that releases both substances that inhibit blood clotting and chemicals that participate in allergic reactions and in responses to tissue damage and microbial invasion.

B cell: a type of lymphocyte that participates in humoral immunity; gives rise to plasma cells, which secrete antibodies into the circulatory system, and to memory cells.

behavior: any observable response to external or internal stimuli.

behavioral isolation: the lack of mating between species of animals that differ substantially in courtship and mating rituals.

bicuspid valve: the atrioventricular valve between the left atrium and the left ventricle of the heart.

bilateral symmetry: a body plan in which only a single plane through the central axis will divide the body into mirror-image halves.

bile (bīl): a liquid secretion, produced by the liver, that is stored in the gallbladder and released into the small intestine during digestion; a complex mixture of bile salts, water, other salts, and cholesterol.

bile salt: a substance that is synthesized in the liver from cholesterol and amino acids and that assists in the breakdown of lipids by dispersing them into small particles on which enzymes can act.

binary fission: the process by which a single bacterium divides in half, producing two identical offspring.

binocular vision: the ability to see objects simultaneously through both eyes, providing greater depth perception and more-accurate judgment of the size and distance of an object from the eyes.

biodegradable: able to be broken down into harmless substances by decomposers.

biodiversity: the total number of species within an ecosystem and the resulting complexity of interactions among them.

biological clock: a metabolic timekeeping mechanism found in most organisms, whereby the organism measures the approximate length of a day (24 hours) even without external environmental cues such as light and darkness.

biological magnification: the increasing accumulation of a toxic substance in progressively higher trophic levels.

biomass: the dry weight of organic material in an ecosystem.

biome (bī´-ōm): a terrestrial ecosystem that occupies an extensive geographical area and is characterized by a specific type of plant community: for example, deserts.

biosphere (bī´-ō-sfēr): that part of Earth inhabited by living organisms; includes both living and nonliving components.

biotechnology: any industrial or commercial use or alteration of organisms, cells, or biological molecules to achieve specific practical goals.

biotic (bī-ah´-tik): living.

biotic potential: the maximum rate at which a population could increase, assuming ideal conditions that allow a maximum birth rate and minimum death rate.

birth control pill: a temporary contraceptive method that prevents ovulation by providing a continuing supply of estrogen and progesterone, which in turn suppresses LH release; must be taken daily, normally for 21 days of each menstrual cycle.

bladder: a hollow muscular storage organ for storing urine.

blade: the flat part of a leaf.

blastocyst (blas´-tō-sist): an early stage of human embryonic development, consisting of a hollow ball of cells, enclosing a mass of cells attached to its inner surface, which becomes the embryo.

blastopore: the site at which a blastula indents to form a gastrula.

blastula (blas´-tū-luh): in animals, the embryonic stage attained at the end of cleavage, in which the embryo normally consists of a hollow ball with a wall one or several cell layers thick.

blind spot: see *optic disc.*

blood: a fluid consisting of plasma in which blood cells are suspended; carried within the circulatory system.

blood clotting: a complex process by which platelets, the protein fibrin, and red blood cells block an irregular surface in or on the body, such as a damaged blood vessel, sealing the wound.

blood vessel: a channel that conducts blood throughout the body.

bone: a hard, mineralized connective tissue that is a major component of the vertebrate endoskeleton; provides support and sites for muscle attachment.

book lung: a structure composed of thin layers of tissue, resembling pages in a book, that are enclosed in a chamber and used as a respiratory organ by certain types of arachnids.

boom-and-bust cycle: a population cycle characterized by rapid exponential growth followed by a sudden massive die-off, seen in seasonal species and in some populations of small rodents, such as lemmings.

Bowman's capsule: the cup-shaped portion of the nephron in which blood filtrate is collected from the glomerulus.

bradykinin (brā´-dē-kī´-nin): a chemical, formed during tissue damage, that binds to receptor molecules on pain nerve endings, giving rise to the sensation of pain.

brain: the part of the central nervous system of vertebrates that is enclosed within the skull.

branch root: a root that arises as a branch of a preexisting root, through divisions of pericycle cells and subsequent differentiation of the daughter cells.

bronchiole (bron´-kē-ōl): a narrow tube, formed by repeated branching of the bronchi, that conducts air into the alveoli.

bronchus (bron´-kus): a tube that conducts air from the trachea to each lung.

bryophyte (brī´-ō-fīt): a simple nonvascular plant of the division Bryophyta, including mosses and liverworts.

bud: in animals, a small copy of an adult that develops on the body of the parent and eventually breaks off and becomes independent; in plants, an embryonic shoot, normally very short and consisting of an apical meristem with several leaf primordia.

budding: asexual reproduction by the growth of a miniature copy, or bud, of the adult animal on the body of the parent. The bud breaks off to begin independent existence.

buffer: a compound that minimizes changes in pH by reversibly taking up or releasing H^+ ions.

bulbourethral gland (bul-bō-ū-rē´-thrul): in male mammals, a gland that secretes a basic, mucus-containing fluid that forms part of the semen.

bulk flow: the movement of many molecules of a gas or fluid in unison from an area of higher pressure to an area of lower pressure.

bundle-sheath cell: one of a group of cells that surround the veins of plants; in C_4 (but not C_3) plants, bundle-sheath cells contain chloroplasts.

C_3 cycle: the cyclic series of reactions whereby carbon dioxide is fixed into carbohydrates during the light-independent reactions of photosynthesis; also called *Calvin-Benson cycle*.

C_4 pathway: the series of reactions in certain plants that fixes carbon dioxide into oxaloacetic acid, which is later broken down for use in the C_3 cycle of photosynthesis.

calcitonin (kal-si-tōn´-in): a hormone, secreted by the thyroid gland, that inhibits the release of calcium from bone.

calorie (kal´-ō-rē): the amount of energy required to raise the temperature of 1 gram of water by 1 degree Celsius.

Calorie: a unit of energy, in which the energy content of foods is measured; the amount of energy required to raise the temperature of 1 liter of water 1 degree Celsius; also called a kilocalorie, equal to 1000 calories.

Calvin-Benson cycle: see C_3 *cycle*.

cambium (kam´-bē-um; pl., cambia): a lateral meristem, parallel to the long axis of roots and stems, that causes secondary growth of woody plant stems and roots. See *cork cambium; vascular cambium*.

camouflage (cam´-a-flaj): coloration and/or shape that renders an organism inconspicuous in its environment.

cancer: a disease in which some of the body's cells escape from normal regulatory processes and divide without control.

capillary: the smallest type of blood vessel, connecting arterioles with venules. Capillary walls, through which the exchange of nutrients and wastes occurs, are only one cell thick.

capsule: a polysaccharide or protein coating that some disease-causing bacteria secrete outside their cell wall.

carbohydrate: a compound composed of carbon, hydrogen, and oxygen, with the approximate chemical formula $(CH_2O)_n$; includes sugars and starches.

carbon fixation: the initial steps in the C_3 cycle, in which carbon dioxide reacts with ribulose bisphosphate to form a stable organic molecule.

cardiac cycle (kar´-dē-ak): the alternation of contraction and relaxation of the heart chambers.

cardiac muscle (kar´-dē-ak): the specialized muscle of the heart, able to initiate its own contraction, independent of the nervous system.

carnivore (kar´-neh-vor): literally, "meat eater"; a predatory organism that feeds on herbivores or on other carnivores; a secondary (or higher) consumer.

carotenoid (ka-rot´-en-oid): a red, orange, or yellow pigment, found in chloroplasts, that serves as an accessory light-gathering molecule in thylakoid photosystems.

carpel (kar´pel): the female reproductive structure of a flower, composed of stigma, style, and ovary.

carrier: an individual who is heterozygous for a recessive condition; displays the dominant phenotype but can pass on the recessive allele to offspring.

carrier protein: a membrane protein that facilitates the diffusion of specific substances across the membrane. The molecule to be transported binds to the outer surface of the carrier protein; the protein then changes shape, allowing the molecule to move across the membrane through the protein.

carrying capacity: the maximum population size that an ecosystem can support indefinitely; determined primarily by the availability of space, nutrients, water, and light.

cartilage (kar´-teh-lij): a form of connective tissue that forms portions of the skeleton; consists of chondrocytes and their extracellular secretion of collagen; resembles flexible bone.

Casparian strip (kas-par´-ē-un): a waxy, waterproof band, located in the cell walls between endodermal cells in a root, that prevents the movement of water and minerals into and out of the vascular cylinder through the extracellular space.

catalyst (kat´-uh-list): a substance that speeds up a chemical reaction without itself being permanently changed in the process; lowers the activation energy of a reaction.

catastrophism: the hypothesis that Earth has experienced a series of geological catastrophes, probably imposed by a supernatural being that accounts for the multitude of species, both extinct and modern, and preserves creationism.

cell: the smallest unit of life, consisting, at a minimum, of an outer membrane that encloses a watery medium containing organic molecules, including genetic material composed of DNA.

cell body: the part of a nerve cell in which most of the common cellular organelles are located; typically a site of integration of inputs to the nerve cell.

cell cycle: the sequence of events in the life of a cell, from one division to the next.

cell division: in eukaryotes, the process of reproduction of single cells, usually into two identical daughter cells, by mitosis accompanied by cytokinesis.

cell-mediated immunity: an immune response in which foreign cells or substances are destroyed by contact with T cells.

cell plate: in plant cell division, a series of vesicles that fuse to form the new plasma membranes and cell wall separating the daughter cells.

cellular respiration: the oxygen-requiring reactions, occurring in mitochondria, that break down the end products of glycolysis into carbon dioxide and water while capturing large amounts of energy as ATP.

cellular slime mold: a funguslike protist consisting of individual amoeboid cells that can aggregate to form a sluglike mass, which in turn forms a fruiting body.

cellulase: an enzyme that catalyzes the breakdown of the carbohydrate cellulose into its component glucose molecules; almost entirely restricted to microorganisms.

cellulose: an insoluble carbohydrate composed of glucose subunits; forms the cell wall of plants.

cell wall: a layer of material, normally made up of cellulose or celluloselike materials, that is outside the plasma membrane of plants, fungi, bacteria, and some protists.

central nervous system: in vertebrates, the brain and spinal cord.

central vacuole: a large, fluid-filled vacuole occupying most of the volume of many plant cells; performs several functions, including maintaining turgor pressure.

centriole (sen´-trē-ōl): in animal cells, a short, barrel-shaped ring consisting of nine microtubule triplets; a microtubule-containing structure at the base of each cilium and flagellum; gives rise to the microtubules of cilia and flagella and is involved in spindle formation during cell division.

centromere (sen´-trō-mēr): the region of a replicated chromosome at which the sister chromatids are held together until they separate during cell division.

cephalization (sef-ul-ī-zā´-shun): the increasing concentration over evolutionary time of sensory structures and nerve ganglia at the anterior end of animals.

cerebellum (ser-uh-bel´-um): the part of the hindbrain of vertebrates that is concerned with coordinating movements of the body.

cerebral cortex (ser-ē´-brul kor´-tex): a thin layer of neurons on the surface of the vertebrate cerebrum, in which most neural processing and coordination of activity occurs.

cerebral hemisphere: one of two nearly symmetrical halves of the cerebrum, connected by a broad band of axons, the corpus callosum.

cerebrospinal fluid: a clear fluid, produced within the ventricles of the brain, that fills the ventricles and cushions the brain and spinal cord.

cerebrum (ser-ē´-brum): the part of the forebrain of vertebrates that is concerned with sensory processing, the direction of motor output, and the coordination of most bodily activities; consists of two nearly symmetrical halves (the hemispheres) connected by a broad band of axons, the corpus callosum.

cervical cap: a birth control device consisting of a rubber cap that fits over the cervix, preventing sperm form entering the uterus.

cervix (ser´-viks): a ring of connective tissue at the outer end of the uterus, leading into the vagina.

channel protein: a membrane protein that forms a channel or pore completely through the membrane and that is usually permeable to one or to a few water-soluble molecules, especially ions.

chaparral: a biome that is located in coastal regions but has very low annual rainfall.

chemical bond: the force of attraction between neighboring atoms that holds them together in a molecule.

chemical equilibrium: the condition in which the "forward" reaction of reactants to products proceeds at the same rate as the "backward" reaction from products to reactants, so that no net change in chemical composition occurs.

chemiosmosis (ke-mē-oz-mō´-sis): a process of ATP generation in chloroplasts and mitochondria. The movement of electrons down an electron transport system is used to pump hydrogen ions across a membrane, thereby building up a concentration gradient of hydrogen ions across the membrane; the hydrogen ions diffuse back across the membrane through the pores of ATP-synthesizing enzymes; the energy of their movement down their concentration gradient drives ATP synthesis.

chemoreceptor: a sensory receptor that responds to chemicals from the environment; used in the chemical senses of taste and smell.

chemosynthetic (kēm-ō-sin-the-tik): capable of oxidizing inorganic molecules to obtain energy.

chemotactic (kēm-ō-tak´-tik): moving toward chemicals given off by food or away from toxic chemicals.

chiasma (kī-as´-muh; pl., chiasmata): a point at which a chromatid of one chromosome crosses with a chromatid of the homologous chromosome during prophase I of meiosis; the site of exchange of chromosomal material between chromosomes.

chitin (kī´-tin): a compound found in the cell walls of fungi and the exoskeletons of insects and some other arthropods; composed of chains of nitrogen-containing, modified glucose molecules.

chlamydia (kla-mid´-ē-uh): a sexually transmitted disease, caused by a bacterium, that causes inflammation of the urethra in males and of the urethra and cervix in females.

chlorophyll (klor-ō-fil): a pigment found in chloroplasts that captures light energy during photosynthesis; absorbs violet, blue, and red light but reflects green light.

chloroplast (klor´-ō-plast): the organelle in plants and plantlike protists that is the site of photosynthesis; surrounded by a double membrane and containing an extensive internal membrane system that bears chlorophyll.

cholecystokinin (kō´-lē-sis-tō-ki´-nin): a digestive hormone, produced by the small intestine, that stimulates the release of pancreatic enzymes.

chondrocyte (kon´-drō-sīt): a living cell of cartilage. With their extracellular secretions of collagen, chondrocytes form cartilage.

chorion (kor´-ē-on): the outermost embryonic membrane in reptiles, birds, and mammals; in birds and reptiles, functions mostly in gas exchange; in mammals, forms most of the embryonic part of the placenta.

chorionic gonadotropin (CG): a hormone, secreted by the chorion (one of the fetal membranes), that maintains the integrity of the corpus luteum during early pregnancy.

chorionic villus (kor-ē-on-ik; pl., chorionic villi): in mammalian embryos, a fingerlike projection of the chorion that penetrates the uterine lining and forms the embryonic portion of the placenta.

chorionic villus sampling (CVS): a procedure for sampling cells from the chorionic villi produced by a fetus: A tube is inserted into the uterus of a pregnant woman, and a small sample of villi are suctioned off for genetic and biochemical analyses.

choroid (kor´-oid): a darkly pigmented layer of tissue, behind the retina, that contains blood vessels and pigment that absorbs stray light.

chromatid (krō´-ma-tid): one of the two identical strands of DNA and protein that forms a replicated chromosome. The two sister chromatids are joined at the centromere.

chromatin (krō´-ma-tin): the complex of DNA and proteins that makes up eukaryotic chromosomes.

chromosome (krō´-mō-sōm): in eukaryotes, a linear strand composed of DNA and protein, located in the nucleus of a cell, that contains the genes; in prokaryotes, a circular strand composed solely of DNA.

chronic bronchitis: a persistent lung infection characterized by coughing, swelling of the lining of the respiratory tract, an increase in mucus production, and a decrease in the number and activity of cilia.

chyme (kīm): an acidic, souplike mixture of partially digested food, water, and digestive secretions that is released from the stomach into the small intestine.

ciliate (sil-ē-et): a protozoan characterized by cilia and by a complex unicellular structure, including harpoonlike organelles called trichocysts. Members of the genus *Paramecium* are well-known ciliates.

cilium (sil´-ē-um; pl., cilia): a short, hairlike projection from the surface of certain eukaryotic cells that contains microtubules in a 9 + 2 arrangement. The movement of cilia may propel cells through a fluid medium or move fluids over a stationary surface layer of cells.

circadian rhythm (sir-kā´-dē-un): an event that recurs with a period of about 24 hours, even in the absence of environmental cues.

citric acid cycle: see *Krebs cycle*.

class: the taxonomic category composed of related genera. Closely related classes form a division or phylum.

classical conditioning: a training procedure in which an animal learns to perform a response (such as salivation) to a new stimulus that did not elicit that response originally (such as a sound); accomplished by pairing a stimulus that elicits the response automatically (in this case, food) with the new stimulus.

cleavage: the early cell divisions of embryos, in which little or no growth occurs between divisions; reduces the cell size and distributes gene-regulating substances to the newly formed cell.

climate: patterns of weather that prevail from year to year and even from century to century in a given region.

climax community: a diverse and relatively stable community that forms the endpoint of succession.

clitoris: an external structure of the female reproductive system; composed of erectile tissue; a sensitive point of stimulation during sexual response.

clonal selection: the mechanism by which the immune response gains specificity; an invading antigen elicits a response from only a few lymphocytes, which proliferate to form a clone of cells that attack only the specific antigen that stimulated their production.

clone: offspring that are produced by mitosis and are therefore genetically identical to each other.

cloning: the process of producing many identical copies of a gene; also the production of many genetically identical copies of an organism.

closed circulatory system: the type of circulatory system, found in certain worms and vertebrates, in which the blood is always confined within the heart and vessels.

club fungus: a fungus of the division Basidiomycota, whose members (which include mushrooms, puffballs, and shelf fungi) reproduce by means of basidiospores.

clumped distribution: the distribution characteristic of populations in which individuals are clustered into groups; may be social or based on the need for a localized resource.

cnidocyte (nīd´-ō-sīt): in members of the phylum Cnidaria, a specialized cell that houses a stinging apparatus.

coccus (kah´-kus; pl., cocci): a spherical bacterium.

cochlea (kahk´-lē-uh): a coiled, bony, fluid-filled tube found in the mammalian inner ear; contains receptors (hair cells) that respond to the vibration of sound.

codominance: the relation between two alleles of a gene, such that both alleles are phenotypically expressed in heterozygous individuals.

codon: a sequence of three bases of messenger RNA that specifies a particular amino acid to be incorporated into a protein; certain codons also signal the beginning or end of protein synthesis.

coelom (sē´-lōm): a space or cavity that separates the body wall from the inner organs.

coenzyme: an organic molecule that is bound to certain enzymes and is required for the enzymes' proper functioning; typically, a nucleotide bound to a water-soluble vitamin.

coevolution: the evolution of adaptations in two species due to their extensive interactions with one another, such that each species acts as a major force of natural selection on the other.

cohesion: the tendency of the molecules of a substance to stick together.

cohesion–tension theory: a model for the transport of water in xylem, by which water is pulled up the xylem tubes, powered by the force of evaporation of water from the leaves (producing tension) and held together by hydrogen bonds between nearby water molecules (cohesion).

coleoptile (kō-lē-op´-tīl): a protective sheath surrounding the shoot in monocot seeds, allowing the shoot to push aside soil particles as it grows.

collagen (kol´-uh-jen): a fibrous protein in connective tissue such as bone and cartilage.

collar cell: a specialized cell lining the inside channels of sponges. Flagella extend from a sievelike collar, creating a water current that draws microscopic organisms through the collar and into the body, where they become trapped.

collecting duct: a conducting tube, within the kidney, that collects urine from many nephrons and conducts it through the renal medulla into the renal pelvis. Urine may become concentrated in the collecting ducts if ADH is present.

collenchyma (kŏl-en´-ki-muh): an elongated, polygonal plant cell type with irregularly thickened primary cell walls that is alive at maturity and that supports the plant body.

colon: the longest part of the large intestine, exclusive of the rectum.

colostrum (kō-los´-trum): a yellowish fluid, high in protein and containing antibodies, that is produced by the mammary glands before milk secretion begins.

commensalism (kum-en´-sal-iz-um): a symbiotic relationship in which one species benefits while another species is neither harmed nor benefited.

communication: the act of producing a signal that causes another animal, normally of the same species, to change its behavior in a way that is beneficial to one or both participants.

community: all the interacting populations within an ecosystem.

compact bone: the hard and strong outer bone; composed of osteons.

companion cell: a cell adjacent to a sieve-tube element in phloem, involved in the control and nutrition of the sieve-tube element.

competition: interaction among individuals who attempt to utilize a resource (for example, food or space) that is limited relative to the demand for it.

competitive exclusion principle: the concept that no two species can simultaneously and continuously occupy the same ecological niche.

complement: a group of blood-borne proteins that participate in the destruction of foreign cells to which antibodies have bound.

complementary base pair: in nucleic acids, bases that pair by hydrogen bonding. In DNA, adenine is complementary to thymine and guanine is complementary to cytosine; in RNA, adenine is complementary to uracil, and guanine to cytosine.

complement reaction: an interaction among foreign cells, antibodies, and complement proteins that results in the destruction of the foreign cells.

complement system: a series of reactions in which complement proteins bind to antibody stems, attracting to the site phagocytic white blood cells that destroy the invading cell that triggers the reactions.

complete flower: a flower that has all four floral parts (sepals, petals, stamens, and carpels).

compound: a substance whose molecules are formed by different types of atoms; can be bro-ken into its constituent elements by chemical means.

compound eye: a type of eye, found in arthropods, that is composed of numerous independent subunits called ommatidia. Each ommatidium apparently contributes a single piece of a mosaiclike image perceived by the animal.

concentration: the number of particles of a dissolved substance in a given unit of volume.

concentration gradient: the difference in concentration of a substance between two parts of a fluid or across a barrier such as a membrane.

conclusion: the final operation in the scientific method; a decision made about the validity of a hypothesis on the basis of experimental evidence.

condom: a contraceptive sheath worn over the penis during intercourse to prevent sperm from being deposited in the vagina.

conducting portion: the portion of the respiratory system in lung-breathing vertebrates that carries air to the lungs.

cone: a cone-shaped photoreceptor cell in the vertebrate retina; not as sensitive to light as are the rods. The three types of cones are most sensitive to different colors of light and provide color vision; see also *rod*.

conifer (kon´-eh-fer): a member of a class of tracheophytes (Coniferophyta) that reproduces by means of seeds formed inside cones and that retains its leaves throughout the year.

connective tissue: a tissue type consisting of diverse tissues, including bone, fat, and blood, that generally contain large amounts of extracellular material.

constant region: the part of an antibody molecule that is similar in all antibodies.

consumer: an organism that eats other organisms; a heterotroph.

contest competition: a mechanism for resolving intraspecific competition by using social or chemical interactions.

contraception: the prevention of pregnancy.

contractile vacuole: a fluid-filled vacuole in certain protists that takes up water from the cytoplasm, contracts, and expels the water outside the cell through a pore in the plasma membrane.

control: that portion of an experiment in which all possible variables are held constant; in contrast to the "experimental" portion, in which a particular variable is altered.

convergence: a condition in which a large number of nerve cells provide input to a smaller number of cells.

convergent evolution: the independent evolution of similar structures among unrelated organisms as a result of similar environmental pressures; see *analogous structures*.

convolution: a folding of the cerebral cortex of the vertebrate brain.

copulation: reproductive behavior in which the penis of the male is inserted into the body of the female, where it releases sperm.

coral reef: a biome created by animals (reef-building corals) and plants in warm tropical waters.

cork cambium: a lateral meristem in woody roots and stems that gives rise to cork cells.

cork cell: a protective cell of the bark of woody stems and roots; at maturity, cork cells are dead, with thick, waterproofed cell walls.

cornea (kor´-nē-uh): the clear outer covering of the eye, in front of the pupil and iris.

corona radiata (kuh-rō´-nuh rā-dē-a´-tuh): the layer of cells surrounding an egg after ovulation.

corpus callosum (kor´pus kal-ō´-sum): the band of axons that connect the two cerebral hemispheres of vertebrates.

corpus luteum (kor´-pus loo´-tē-um): in the mammalian ovary, a structure that is derived from the follicle after ovulation and that secretes the hormones estrogen and progesterone.

cortex: the part of a primary root or stem located between the epidermis and the vascular cylinder.

cotyledon (kot-ul-ē´don): a leaflike structure within a seed that absorbs food molecules from the endosperm and transfers them to the growing embryo; also called *seed leaf*.

coupled reaction: a pair of reactions, one exergonic and one endergonic, that are linked together such that the energy produced by the exergonic reaction provides the energy needed to drive the endergonic reaction.

covalent bond (kō-vā´-lent): a chemical bond between atoms in which electrons are shared.

crab lice: an arthropod parasite that can infest humans; can be transmitted by sexual contact.

creationism: the hypothesis that all species on Earth were created in essentially their present form by a supernatural being and that significant modification of those species—specifically, their transformation into new species—cannot occur by natural processes.

crista (kris´-tuh; pl., cristae): a fold in the inner membrane of a mitochondrion.

crop: an organ, found in both earthworms and birds, in which ingested food is temporarily stored before being passed to the gizzard, where it is pulverized.

cross-bridge: in muscles, an extension of myosin that binds to and pulls on actin to produce muscle contraction.

cross-fertilization: the union of sperm and egg from two individuals of the same species.

crossing over: the exchange of corresponding segments of the chromatids of two homologous chromosomes during meiosis.

cultural evolution: changes in the behavior of a population of animals, especially humans, by learning behaviors acquired by members of previous generations.

cuticle (kū´-ti-kul): a waxy or fatty coating on the exposed surfaces of epidermal cells of many land plants, which aids in the retention of water.

cyanobacterium: a photosynthetic prokaryotic cell that utilizes chlorophyll and releases oxygen as a photosynthetic by-product; sometimes called *blue-green algae*.

cyclic AMP: a cyclic nucleotide, formed within many target cells as a result of the reception of amino acid derivatives or peptide hormones, that causes metabolic changes in the cell; often called a second messenger.

cyclic nucleotide (sik´-lik noo´-klē-ō-tīd): a nucleotide in which the phosphate group is bonded to the sugar at two points, forming a ring; serves as an intracellular messenger.

cyst (sist): an encapsulated resting stage in the life cycle of certain invertebrates, such as parasitic flatworms and roundworms.

cystic fibrosis: an inherited disorder characterized by the buildup of salt in the lungs and the production of thick, sticky mucus that clogs the airways, restricts air exchange, and promotes infection.

cytokinesis (sī-tō-ki-nē´-sis): the division of the cytoplasm and organelles into two daughter cells during cell division; normally occurs during telophase of mitosis.

cytokinin (sī-tō-kī´-nin): a plant hormone that promotes cell division, fruit growth, and the sprouting of lateral buds and prevents the aging of plant parts, especially leaves.

cytoplasm (sī´-tō-plaz-um): the material contained within the plasma membrane of a cell, exclusive of the nucleus.

cytoskeleton: a network of protein fibers in the cytoplasm that gives shape to a cell, holds and moves organelles, and is typically involved in cell movement.

cytotoxic T cell: a type of T cell that, upon contacting foreign cells, directly destroys them.

day-neutral plant: a plant in which flowering occurs as soon as the plant has grown and developed, regardless of daylength.

decomposer: an organism, normally a fungus or bacterium, that digests organic material by secreting digestive enzymes into the environment, in the process liberating nutrients into the environment.

deforestation: the excessive cutting of forests, primarily rain forests in the tropics, to clear space for agriculture.

dehydration synthesis: a chemical reaction in which two molecules are joined by a covalent bond with the simultaneous removal of a hydrogen from one molecule and a hydroxyl group from the other, forming water; the reverse of hydrolysis.

deletion mutation: a mutation in which one or more pairs of nucleotides are removed from a gene.

dendrite (den´-drīt): a branched tendril that extends outward from the cell body of a neuron; specialized to respond to signals from the external environment or from other neurons.

denitrifying bacterium (dē-nī´-treh-fī-ing): a bacterium that breaks down nitrates, releasing nitrogen gas to the atmosphere.

density-dependent: referring to any factor, such as predation, that limits population size more effectively as the population density increases.

density-independent: referring to any factor, such as freezing weather, that limits a population's size and growth regardless of its density.

deoxyribonucleic acid (dē-ox-ē-rī-bō-noo-klā´-ik; DNA): a molecule composed of deoxy-

ribose nucleotides; contains the genetic information of all living cells.

dermal tissue system: a plant tissue system that makes up the outer covering of the plant body.

dermis (dur´-mis): the layer of skin beneath the epidermis; composed of connective tissue and containing blood vessels, muscles, nerve endings, and glands.

desert: a biome in which less than 25 to 50 centimeters (10 to 20 inches) of rain falls each year.

desertification: the spread of deserts by human activities.

desmosome (dez´-mō-sōm): a strong cell-to-cell junction that attaches adjacent cells to one another.

detritus feeder (de-trī´-tus): one of a diverse group of organisms, ranging from worms to vultures, that live off the wastes and dead remains of other organisms.

deuterostome (doo´-ter-ō-stōm): an animal with a mode of embryonic development in which the coelom is derived from outpocketings of the gut; characteristic of echinoderms and chordates.

development: the process by which an organism proceeds from fertilized egg through adulthood to eventual death.

diabetes mellitus (dī-uh-bē´-tēs mel-ī´-tus): a disease characterized by defects in the production, release, or reception of insulin; characterized by high blood glucose levels that fluctuate with sugar intake.

dialysis (dī-āl´-i-sis): the passive diffusion of substances across an artificial semipermeable membrane.

diaphragm (dī´-uh-fram): in the respiratory system, a dome-shaped muscle forming the floor of the chest cavity that, when it contracts, pulls itself downward, enlarging the chest cavity and causing air to be drawn into the lungs; in a reproductive sense, a contraceptive rubber cap that fits snugly over the cervix, preventing the sperm from entering the uterus and thereby preventing pregnancy.

diatom (dī´-uh-tom): a protist that includes photosynthetic forms with two-part glassy outer coverings; important photosynthetic organisms in fresh water and salt water.

dicot (dī´-kaht): short for dicotyledon; a type of flowering plant characterized by embryos with two cotyledons, or seed leaves, modified for food storage.

differentially permeable: referring to the ability of some substances to pass through a membrane more readily than can other substances.

differential reproduction: differences in reproductive output among individuals of a population, normally as a result of genetic differences.

differentiated cell: a mature cell specialized for a specific function; in plants, differentiated cells normally do not divide.

differentiation: the process whereby relatively unspecialized cells, especially of embryos, become specialized into particular tissue types.

diffusion: the net movement of particles from a region of high concentration of that particle

to a region of low concentration, driven by the concentration gradient; may occur entirely within a fluid or across a barrier such as a membrane.

digestion: the process by which food is physically and chemically broken down into molecules that can be absorbed by cells.

dinoflagellate (dī-nō-fla´-jel-et): a protist that includes photosynthetic forms in which two flagella project through armorlike plates; abundant in oceans; can reproduce rapidly, causing "red tides."

dioecious (dī-ē´-shus): pertaining to organisms in which male and female gametes are produced by separate individuals rather than in the same individual.

diploid (dip´-loid): referring to a cell with pairs of homologous chromosomes.

direct development: a developmental pathway in which the offspring is born as a miniature version of the adult and does not radically change in body form as it grows and matures.

directional selection: a type of natural selection in which one extreme phenotype is favored over all others.

disaccharide (dī-sak´-uh-rīd): a carbohydrate formed by the covalent bonding of two monosaccharides.

disruptive selection: a type of natural selection in which both extreme phenotypes are favored over the average phenotype.

distal tubule: in the nephrons of the mammalian kidney, the last segment of the renal tubule through which the filtrate passes just before it empties into the collecting duct; a site of selective secretion and reabsorption as water and ions pass between the blood and the filtrate across the tubule membrane.

disulfide bridge: the covalent bond formed between the sulfur atoms of two cysteines in a protein; typically causes the protein to fold by bringing otherwise distant parts of the protein close together.

divergence: a condition in which a small number of nerve cells provide input to a larger number of cells.

divergent evolution: evolutionary change in which the differences between two lineages become more pronounced with the passage of time.

division: the taxonomic category contained within a kingdom and consisting of related classes of plants, fungi, bacteria, or plantlike protists.

DNA–DNA hybridization: a technique by which DNA from two species is separated into single strands and then allowed to re-form; hybrid double-stranded DNA from the two species can occur where the sequence of nucleotides is complementary. The greater the degree of hybridization, the closer the evolutionary relatedness of the two species.

DNA fingerprinting: the use of restriction enzymes to cut DNA segments into a unique set of restriction fragments from one individual that can be distinguished from the restriction fragments of other individuals by gel electrophoresis.

DNA library: a readily accessible, easily duplicable complete set of all the DNA of a particular organism, normally cloned into bacterial plasmids.

DNA polymerase: an enzyme that bonds DNA nucleotides together into a continuous strand, using a preexisting DNA strand as a template.

DNA probe: a sequence of nucleotides that is complementary to the nucleotide sequence in a gene under study; used to locate a given gene within a DNA library.

DNA sequencing: the process of determining the chemical composition of a DNA molecule (in particular, the order in which the molecule's constituent nucleic acids are arranged).

domain: the broadest category for classifying organisms; organisms are classified into three domains: Bacteria, Archaea, and Eukarya.

dominance hierarchy: a social arrangement in which a group of animals, usually through aggressive interactions, establishes a rank for some or all of the group members that determines access to resources.

dominant: an allele that can determine the phenotype of heterozygotes completely, such that they are indistinguishable from individuals homozygous for the allele; in the heterozygotes, the expression of the other (recessive) allele is completely masked.

dopamine (dōp´-uh-mēn): a transmitter in the brain whose actions are largely inhibitory. The loss of dopamine-containing neurons causes Parkinson's disease.

dormancy: a state in which an organism does not grow or develop; usually marked by lowered metabolic activity and resistance to adverse environmental conditions.

dorsal (dor´-sul): the top, back, or uppermost surface of an animal oriented with its head forward.

dorsal root ganglion: a ganglion, located on the dorsal (sensory) branch of each spinal nerve, that contains the cell bodies of sensory neurons.

double covalent bond: a covalent bond in which two atoms share two pairs of electrons.

double fertilization: in flowering plants, the fusion of two sperm nuclei with the nuclei of two cells of the female gametophyte. One sperm nucleus fuses with the egg to form the zygote; the second sperm nucleus fuses with the two haploid nuclei of the primary endosperm cell, forming a triploid endosperm cell.

double helix (hē´-liks): the shape of the two-stranded DNA molecule; like a ladder twisted lengthwise into a corkscrew shape.

douching: washing the vagina; after intercourse, an attempt to wash sperm out of the vagina before they enter the uterus; an ineffective contraceptive method.

Down syndrome: a genetic disorder caused by the presence of three copies of chromosome 21; common characteristics include mental retardation, distinctively shaped eyelids, a small mouth with protruding tongue, heart defects, and low resistance to infectious diseases; also called *trisomy 21*.

duct: a tube or opening through which exocrine secretions are released.

echolocation: the use of ultrasonic sounds, which bounce back from nearby objects, to produce an auditory "image" of nearby surroundings; used by bats and porpoises.

ecological isolation: the lack of mating between organisms belonging to different populations that occupy distinct habitats within the same general area.

ecological niche (nitch): the role of a particular species within an ecosystem, including all aspects of its interaction with the living and nonliving environments.

ecology (ē-kol´-uh-jē): the study of the interrelationships of organisms with each other and with their nonliving environment.

ecosystem (ē´kō-sis-tem): all the organisms and their nonliving environment within a defined area.

ectoderm (ek´-tō-derm): the outermost embryonic tissue layer, which gives rise to structures such as hair, the epidermis of the skin, and the nervous system.

effector (ē-fek´-tor): a part of the body (normally a muscle or gland) that carries out responses as directed by the nervous system.

egg: the haploid female gamete, normally large and nonmotile, containing food reserves for the developing embryo.

electrolocation: the production of high-frequency electrical signals from an electric organ in front of the tail of weak electrical fish; used to detect and locate nearly objects.

electron: a subatomic particle, found in an electron shell outside the nucleus of an atom, that bears a unit of negative charge and very little mass.

electron carrier: a molecule that can reversibly gain or lose electrons. Electron carriers generally accept high-energy electrons produced during an exergonic reaction and donate the electrons to acceptor molecules that use the energy to drive endergonic reactions.

electron shell: a region within which electrons orbit that corresponds to a fixed energy level at a given distance from the atomic nucleus of an atom.

electron transport system: a series of electron carrier molecules, found in the thylakoid membranes of chloroplasts and the inner membrane of mitochondria, that extract energy from electrons and generate ATP or other energetic molecules.

element: a substance that cannot be broken down, or converted, to a simpler substance by ordinary chemical means.

embryo: in animals, the stages of development that begin with the fertilization of the egg cell and end with hatching or birth; in mammals in particular, the early stages in which the developing animal does not yet resemble the adult of the species.

embryo sac: the haploid female gametophyte of flowering plants.

embryonic disc: in human embryonic development, the flat, two-layered group of cells that separates the amniotic cavity from the yolk sac.

emergent property: an intangible attribute that arises as the result of complex ordered interactions among individual parts.

emigration (em-uh-grā´shun): migration of individuals out of an area.

emphysema (em-fuh-sē´-muh): a condition in which the alveoli of the lungs become brittle and rupture, causing decreased area for gas exchange.

endergonic (en-der-gon´-ik): pertaining to a chemical reaction that requires an input of energy to proceed; an "uphill" reaction.

endocrine gland: a ductless, hormone-producing gland consisting of cells that release their secretions into the extracellular fluid from which the secretions diffuse into nearby capillaries.

endocrine system: an animal's organ system for cell-to-cell communication, composed of hormones and the cells that secrete them and receive them.

endocytosis (en-dō-sī-tō´-sis): the process in which the plasma membrane engulfs extracellular material, forming membrane-bound sacs that enter the cytoplasm and thereby move material into the cell.

endoderm (en´-dō-derm): the innermost embryonic tissue layer, which gives rise to structures such as the lining of the digestive and respiratory tracts.

endodermis (en-dō-der´-mis): the innermost layer of small, close-fitting cells of the cortex of a root that form a ring around the vascular cylinder.

endogenous pyrogen: a chemical, produced by the body, that stimulates the production of a fever.

endometrium (en-dō-mē´-trē-um): the nutritive inner lining of the uterus.

endoplasmic reticulum (ER) (en-dō-plaz´-mik re-tik´-ū-lum): a system of membranous tubes and channels within eukaryotic cells; the site of most protein and lipid syntheses.

endorphin (en-dor´-fin): one of a group of peptide neuromodulators in the vertebrate brain that, by reducing the sensation of pain, mimics some of the actions of opiates.

endoskeleton (en´-dō-skel´-uh-tun): a rigid internal skeleton with flexible joints to allow for movement.

endosperm: a triploid food storage tissue in the seeds of flowering plants that nourishes the developing plant embryo.

endospore: a protective resting structure of some rod-shaped bacteria that withstands unfavorable environmental conditions.

endosymbiont hypothesis: the hypothesis that certain organelles, especially chloroplasts and mitochondria, arose as mutually beneficial associations between the ancestors of eukaryotic cells and captured bacteria that lived within the cytoplasm of the pre-eukaryotic cell.

energy: the capacity to do work.

energy-carrier molecule: a molecule that stores energy in "high-energy" chemical bonds and releases the energy to drive coupled endothermic reactions. In cells, ATP is the most common energy-carrier molecule.

energy level: the specific amount of energy characteristic of a given electron shell in an atom.

energy pyramid: a graphical representation of the energy contained in succeeding trophic levels, with maximum energy at the base (primary producers) and steadily diminishing amounts at higher levels.

entropy (en´-trō-pē): a measure of the amount of randomness and disorder in a system.

environmental resistance: any factor that tends to counteract biotic potential, limiting population size.

enzyme (en´zīm): a protein catalyst that speeds up the rate of specific biological reactions.

eosinophil (ē-ō-sin´-ō-fil): a type of white blood cell that converges on parasitic invaders and releases substances to kill them.

epicotyl (ep´-ē-kot-ul): the part of the embryonic shoot located above the cotyledons but below the tip of the shoot.

epidermal tissue: dermal tissue in plants that forms the epidermis, the outermost cell layer that covers young plants.

epidermis (ep-uh-der´-mis): in animals, specialized epithelial tissue that forms the outer layer of skin; in plants, the outermost layer of cells of a leaf, young root, or young stem.

epididymis (e-pi-di´-di-mus): a series of tubes that connect with and receive sperm from the seminiferous tubules of the testis.

epiglottis (ep-eh-glah´-tis): a flap of cartilage in the lower pharynx that covers the opening to the larynx during swallowing; directs food down the esophagus.

epinephrine (ep-i-nef´-rin): a hormone, secreted by the adrenal medulla, that is released in response to stress and that stimulates a variety of responses, including the release of glucose from skeletal muscle and an increase in heart rate.

epithelial cell (eh-puh-thē´-lē-ul): a flattened cell that covers the outer body surfaces of a sponge.

epithelial tissue (eh-puh-thē´-lē-ul): a tissue type that forms membranes that cover the body surface and line body cavities, and that also gives rise to glands.

equilibrium population: a population in which allele frequencies and the distribution of genotypes do not change from generation to generation.

erythroblastosis fetalis (eh-rith´-rō-blas-tō´-sis fē-tal´-is): a condition in which the red blood cells of a newborn Rh-positive baby are attacked by antibodies produced by its Rh-negative mother, causing jaundice and anemia. Retardation and death are possible consequences if treatment is inadequate.

erythrocyte (eh-rith´-rō-sīt): a red blood cell, active in oxygen transport, that contains the red pigment hemoglobin.

erythropoietin (eh-rith´-rō-pō-ē´-tin): a hormone produced by the kidneys in response to oxygen deficiency that stimulates the production of red blood cells by the bone marrow.

esophagus (eh-sof´-eh-gus): a muscular passageway that conducts food from the pharynx to the stomach in humans and other mammals.

essential amino acid: an amino acid that is a required nutrient; the body is unable to manufacture essential amino acids, so they must be supplied in the diet.

essential fatty acid: a fatty acid that is a required nutrient; the body is unable to manufacture essential fatty acids, so they must be supplied in the diet.

estrogen: in vertebrates, a female sex hormone, produced by follicle cells of the ovary, that stimulates follicle development, oogenesis, the development of secondary sex characteristics, and growth of the uterine lining.

estuary: a wetland formed where a river meets the ocean; the salinity there is quite variable but lower than in sea water and higher than in fresh water.

ethology (ē-thol´-ō-jē): the study of animal behavior in natural or near-natural conditions.

ethylene: a plant hormone that promotes the ripening of fruits and the dropping of leaves and fruit.

euglenoid (ū´-gle-noid): a protist characterized by one or more whiplike flagella that are used for locomotion and by a photoreceptor that detects light. Euglenoids are photosynthetic, but if deprived of chlorophyll, some are capable of heterotrophic nutrition.

eukaryotic (ū-kar-ē-ot´-ik): referring to cells of organisms of the domain Eukarya (kingdoms Protista, Fungi, Plantae, and Animalia). Eukaryotic cells have genetic material enclosed within a membrane-bound nucleus and contain other membrane-bound organelles.

Eustachian tube (ū-stā´-shin): a tube connecting the middle ear with the pharynx; allows pressure between the middle ear and the atmosphere to equilibrate.

eutrophic lake: a lake that receives sufficiently large inputs of sediments, organic material, and inorganic nutrients from its surroundings to support dense communities; murky with poor light penetration.

evergreen: a plant that retains green leaves throughout the year.

evolution: the descent of modern organisms with modification from preexisting life-forms; strictly speaking, any change in the proportions of different genotypes in a population from one generation to the next.

excretion: the elimination of waste substances from the body; can occur from the digestive system, skin glands, urinary system, or lungs.

excretory pore: an opening in the body wall of certain invertebrates, such as the earthworm, through which urine is excreted.

exergonic (ex-er-gon´-ik): pertaining to a chemical reaction that liberates energy (either as heat or in the form of increased entropy); a "downhill" reaction.

exhalation: the act of releasing air from the lungs, which results from a relaxation of the respiratory muscles.

exocrine gland: a gland that releases its secretions into ducts that lead to the outside of the body or into the digestive tract.

exocytosis (ex-ō-sī-tō´-sis): the process in which intracellular material is enclosed within a membrane-bound sac that moves to the plasma membrane and fuses with it, releasing the material outside the cell.

exon: a segment of DNA in a eukaryotic gene that codes for amino acids in a protein (see also *intron*).

exoskeleton (ex´-ō-skel´-uh-tun): a rigid external skeleton that supports the body, protects the internal organs, and has flexible joints that allow for movement.

exotic/exotic species: a foreign species introduced into an ecosystem where it did not evolve; such species may flourish and outcompete native species.

experiment: the third operation in the scientific method; the testing of a hypothesis by further observations, leading to a conclusion.

exponential growth: a continuously accelerating increase in population size.

extensor: a muscle that straightens a joint.

external ear: the fleshy portion of the ear that extends outside the skull.

external fertilization: the union of sperm and egg outside the body of either parent.

extinction: the death of all members of a species.

extracellular digestion: the physical and chemical breakdown of food that occurs outside a cell, normally in a digestive cavity.

extraembryonic membrane: in the embryonic development of reptiles, birds, and mammals, either the chorion, amnion, allantois, or yolk sac; functions in gas exchange, provision of the watery environment needed for development, waste storage, and storage of the yolk, respectively.

eyespot: a simple, lensless eye found in various invertebrates, including flatworms and jellyfish. Eyespots can distinguish light from dark and sometimes the direction of light, but they cannot form an image.

facilitated diffusion: the diffusion of molecules across a membrane, assisted by protein pores or carriers embedded in the membrane.

fairy ring: a circular pattern of mushrooms formed when reproductive structures erupt from the underground hyphae of a club fungus that has been growing outward in all directions from its original location.

fallopian tube: see *oviduct*.

family: the taxonomic category contained within an order and consisting of related genera.

fat (molecular): a lipid composed of three saturated fatty acids covalently bonded to glycerol; solid at room temperature.

fat (tissue): adipose tissue; connective tissue that stores the lipid fat; composed of cells packed with triglycerides.

fatty acid: an organic molecule composed of a long chain of carbon atoms, with a carboxylic acid (COOH) group at one end; may be saturated (all single bonds between the carbon atoms) or unsaturated (one or more double bonds between the carbon atoms).

feedback inhibition: in enzyme-mediated chemical reactions, the condition in which the product of a reaction inhibits one or more of the enzymes involved in synthesizing the product.

fermentation: anaerobic reactions that convert the pyruvic acid produced by glycolysis into lactic acid or alcohol and CO_2.

fertilization: the fusion of male and female haploid gametes, forming a zygote.

fetal alcohol syndrome (FAS): a cluster of symptoms, including retardation and physical abnormalities, that occur in infants born to a mothers who consumed large amounts of alcoholic beverages during pregnancy.

fetus: the later stages of mammalian embryonic development (after the second month for humans), when the developing animal has come to resemble the adult of the species.

fever: an elevation in body temperature caused by chemicals (pyrogens) that are released by white blood cells in response to infection.

fibrillation: rapid, uncoordinated, and ineffective contractions of heart muscle cells.

fibrin (fī´-brin): a clotting protein formed in the blood in response to a wound; binds with other fibrin molecules and provides a matrix around which a blood clot forms.

fibrous root system: a root system, commonly found in monocots, characterized by many roots of approximately the same size arising from the base of the stem.

filament: in flowers, the stalk of a stamen, which bears an anther at its tip.

filtrate: the fluid produced by filtration; in the kidneys, the fluid produced by the filtration of blood through the glomerular capillaries.

filtration: within Bowman's capsule in each nephron of a kidney, the process by which blood is pumped under pressure through permeable capillaries of the glomerulus, forcing out water, dissolved wastes, and nutrients.

fimbria (fim´-brē-uh; pl., fimbriae): in female mammals, the ciliated, fingerlike projections of the oviduct that sweep the ovulated egg from the ovary into the oviduct.

first law of thermodynamics: the principle of physics that states that within any isolated system, energy can be neither created nor destroyed but can be converted from one form to another.

fission: asexual reproduction by dividing the body into two smaller, complete organisms.

fitness: the reproductive success of an organism, usually expressed in relation to the average reproductive success of all individuals in the same population.

fixed action pattern: a stereotyped, rather complex behavior that is genetically programmed (innate); commonly triggered by a stimulus called a releaser.

flagellum (fla-jel´-um; pl., flagella): a long, hairlike extension of the plasma membrane; in eukaryotic cells, it contains microtubules arranged in a 9 + 2 pattern. The movement of flagella propel some cells through fluids.

flame cell: in flatworms, a specialized cell, containing beating cilia, that conducts water and wastes through the branching tubes that serve as an excretory system.

flexor: a muscle that flexes (decreases the angle of) a joint.

florigen: one of a group of plant hormones that can both trigger and inhibit flowering; daylength is a stimulus.

flower: the reproductive structure of an angiosperm plant.

fluid: a liquid or gas.

fluid mosaic model: a model of membrane structure; according to this model, membranes are composed of a double layer of phospholipids in which various proteins are embedded. The phospholipid bilayer is a somewhat fluid matrix that allows the movement of proteins within it.

follicle: in the ovary of female mammals, the oocyte and its surrounding accessory cells.

follicle-stimulating hormone (FSH): a hormone, produced by the anterior pituitary, that stimulates spermatogenesis in males and the development of the follicle in females.

food chain: a linear feeding relationship in a community, using a single representative from each of the trophic levels.

food vacuole: a membranous sac, within a single cell, in which food is enclosed. Digestive enzymes are released into the vacuole, where intracellular digestion occurs.

food web: a representation of the complex feeding relationships (in terms of interacting food chains) within a community, including many organisms at various trophic levels, with many of the consumers occupying more than one level simultaneously.

foraminiferan (for-am-i-nif´-er-un): an aquatic (largely marine) protist characterized by a typically elaborate calcium carbonate shell.

forebrain: during development, the anterior portion of the brain. In mammals, the forebrain differentiates into the thalamus, the limbic system, and the cerebrum. In humans, the cerebrum contains about half of all the neurons in the brain.

fossil: the remains of a dead organism, normally preserved in rock; may be petrified bones or wood; shells; impressions of body forms, such as feathers, skin, or leaves; or markings made by organisms, such as footprints.

fossil fuel: a fuel such as coal, oil, and natural gas, derived from the remains of ancient organisms.

founder effect: a type of genetic drift in which an isolated population founded by a small number of individuals may develop allele frequencies that are very different from those of the parent population as a result of chance inclusion of disproportionate numbers of certain alleles in the founders.

fovea (fō´-vē-uh): in the vertebrate retina, the central region on which images are focused; contains closely packed cones.

free-living: not parasitic.

free nerve ending: on some receptor neurons, a finely branched ending that responds to touch and pressure, to heat and cold, or to pain; produces the sensations of itching and tickling.

fruit: in flowering plants, the ripened ovary (plus, in some cases, other parts of the flower), which contains the seeds.

fruiting body: a spore-forming reproductive structure of certain protists, bacteria, and fungi.

functional group: one of several groups of atoms commonly found in an organic molecule, including hydrogen, hydroxyl, amino, carboxyl, and phosphate groups, that determine the characteristics and chemical reactivity of the molecule.

gallbladder: a small sac, next to the liver, in which the bile secreted by the liver is stored and concentrated. Bile is released from the gallbladder to the small intestine through the bile duct.

gamete (gam´-ēt): a haploid sex cell formed in sexually reproducing organisms.

gametic incompatibility: the inability of sperm from one species to fertilize eggs of another species.

gametophyte (ga-mēt´-ō-fīt): the multicellular haploid stage in the life cycle of plants.

ganglion (gang´-lē-un): a cluster of neurons.

ganglion cell: a type of cell, comprising the innermost layer of the vertebrate retina, whose axons form the optic nerve.

gap junction: a type of cell-to-cell junction in animals in which channels connect the cytoplasm of adjacent cells.

gas-exchange portion: the portion of the respiratory system in lung-breathing vertebrates where gas is exchanged in the alveoli of the lungs.

gastric inhibitory peptide: a hormone, produced by the small intestine, that inhibits the activity of the stomach.

gastrin: a hormone, produced by the stomach, that stimulates acid secretion in response to the presence of food.

gastrovascular cavity: a saclike chamber with digestive functions, found in simple invertebrates; a single opening serves as both mouth and anus, and the chamber provides direct access of nutrients to the cells.

gastrula (gas´-troo-luh): in animal development, a three-layered embryo with ectoderm, mesoderm, and endoderm cell layers. The endoderm layer normally encloses the primitive gut.

gastrulation (gas-troo-la´-shun): the process whereby a blastula develops into a gastrula, including the formation of endoderm, ectoderm, and mesoderm.

gel electrophoresis: a technique in which molecules (such as DNA fragments) are placed on restricted tracks in a thin sheet of gelatinous material and exposed to an electric field; the molecules then migrate at a rate determined by certain characteristics, such as length.

gene: a unit of heredity that encodes the information needed to specify the amino acid sequence of proteins and hence particular traits; a functional segment of DNA located at a particular place on a chromosome.

gene flow: the movement of alleles from one population to another owing to the migration of individual organisms.

gene pool: the total of all alleles of all genes in a population; for a single gene, the total of all the alleles of that gene that occur in a population.

generative cell: in flowering plants, one of the haploid cells of a pollen grain; undergoes mitosis to form two sperm cells.

genetic code: the collection of codons of mRNA, each of which directs the incorporation of a particular amino acid into a protein during protein synthesis.

genetic drift: a change in the allele frequencies of a small population purely by chance.

genetic engineering: the modification of genetic material to achieve specific goals.

genetic equilibrium: a state in which the allele frequencies and the distribution of genotypes of a population do not change from generation to generation.

genetic recombination: the generation of new combinations of alleles on homologous chromosomes due to the exchange of DNA during crossing over.

genital herpes: a sexually transmitted disease, caused by a virus, that can cause painful blisters on the genitals and surrounding skin.

genome (jē´-nōm): the entire set of genes carried by a member of any given species.

genotype (jēn´-ō-tīp): the genetic composition of an organism; the actual alleles of each gene carried by the organism.

genus (jē-nus): the taxonomic category contained within a family and consisting of very closely related species.

geographical isolation: the separation of two populations by a physical barrier.

germ layer: a tissue layer formed during early embryonic development.

germination: the growth and development of a seed, spore, or pollen grain.

gibberellin (jib-er-el´-in): a plant hormone that stimulates seed germination, fruit development, and cell division and elongation.

gill: in aquatic animals, a branched tissue richly supplied with capillaries around which water is circulated for gas exchange.

gizzard: a muscular organ, found in earthworms and birds, in which food is mechanically broken down prior to chemical digestion.

gland: a cluster of cells that are specialized to secrete (release) substances such as sweat or hormones.

glial cell: a cell of the nervous system that provides support and insulation for neurons.

global warming: a gradual rise in global atmospheric temperature as a result of an amplification of the natural greenhouse effect due to human activities.

glomerulus (glō-mer´-ū-lus): a dense network of thin-walled capillaries, located within the Bowman's capsule of each nephron of the kidney, where blood pressure forces water and dissolved nutrients through capillary walls for filtration by the nephron.

glucagon (gloo´-ka-gon): a hormone, secreted by the pancreas, that increases blood sugar by stimulating the breakdown of glycogen (to glucose) in the liver.

glucocorticoid (gloo-kō-kor´-tik-oid): a class of hormones, released by the adrenal cortex in response to the presence of ACTH, that make additional energy available to the body by stimulating the synthesis of glucose.

glucose: the most common monosaccharide, with the molecular formula $C_6H_{12}O_6$; most polysaccharides, including cellulose, starch, and glycogen, are made of glucose subunits covalently bonded together.

glycerol (glis´-er-ol): a three-carbon alcohol to which fatty acids are covalently bonded to make fats and oils.

glycogen (glī´-kō-jen): a long, branched polymer of glucose that is stored by animals in the muscles and liver and metabolized as a source of energy.

glycolysis (glī-kol´-i-sis): reactions, carried out in the cytoplasm, that break down glucose into two molecules of pyruvic acid, producing two ATP molecules; does not require oxygen but can proceed when oxygen is present.

glycoprotein: a protein to which a carbohydrate is attached.

goiter: a swelling of the neck caused by iodine deficiency, which affects the functioning of the thyroid gland and its hormones.

Golgi complex (gōl´-jē): a stack of membranous sacs, found in most eukaryotic cells, that is the site of processing and separation of membrane components and secretory materials.

gonadotropin-releasing hormone (GnRH): a hormone produced by the neurosecretory cells of the hypothalamus, which stimulates cells in the anterior pituitary to release FSH and LH. GnRH is involved in the menstrual cycle and in spermatogenesis.

gonad: an organ where reproductive cells are formed; in males, the testes, and in females, the ovaries.

gonorrhea (gon-uh-rē´-uh): a sexually transmitted bacterial infection of the reproductive organs; if untreated, can result in sterility.

gradient: a difference in concentration, pressure, or electrical charge between two regions.

Gram stain: a stain that is selectively taken up by the cell walls of certain types of bacteria (gram-positive bacteria) and rejected by the cell walls of others (gram-negative bacteria); used to distinguish bacteria on the basis of their cell wall construction.

granum (gra´-num; pl., grana): a stack of thylakoids in chloroplasts.

grassland: a biome, located in the centers of continents, that supports grasses; also called *prairie*.

gravitropism: growth with respect to the direction of gravity.

gray crescent: in frog embryonic development, an area of intermediate pigmentation in the fertilized egg; contains gene-regulating substances required for the normal development of the tadpole.

gray matter: the outer portion of the brain and inner region of the spinal cord; composed largely of neuron cell bodies, which give this area a gray color.

greenhouse effect: the process in which certain gases such as carbon dioxide and methane trap sunlight energy in a planet's atmosphere as heat; the glass in a greenhouse does the same. The result, global warming, is being enhanced by the production of these gases by humans.

greenhouse gas: a gas, such as carbon dioxide or methane, that traps sunlight energy in a planet's atmosphere as heat; a gas that participates in the greenhouse effect.

ground tissue system: a plant tissue system consisting of parenchyma, collenchyma, and sclerenchyma cells that makes up the bulk of a leaf or young stem, excluding vascular or dermal tissues. Most ground tissue cells function in photosynthesis, support, or carbohydrate storage.

growth hormone: a hormone, released by the anterior pituitary, that stimulates growth, especially of the skeleton.

growth rate: a measure of the change in population size per individual per unit of time.

guard cell: one of a pair of specialized epidermal cells surrounding the central opening of a stoma of a leaf, which regulates the size of the opening.

gymnosperm (jim´-nō-sperm): a nonflowering seed plant, such as a conifer, cycad, or gingko.

gyre (jīr): a roughly circular pattern of ocean currents, formed because continents interrupt the currents' flow; clockwise-rotating in the Northern Hemisphere and counterclockwise-rotating in the Southern Hemisphere.

habituation (heh-bich-oo-ā´-shun): simple learning characterized by a decline in response to a harmless, repeated stimulus.

hair cell: the type of receptor cell in the inner ear; bears hairlike projections, the bending of which causes the receptor potential between two membranes.

hair follicle: a gland in the dermis of mammalian skin, formed from epithelial tissue, that produces a hair.

halophile (hā´-lō-fīl): literally, "salt-loving"; a type of archaen that thrives in concentrated salt solutions.

haploid (hap´-loid): referring to a cell that has only one member of each pair of homologous chromosomes.

Hardy-Weinberg principle: a mathematical model proposing that, under certain conditions, the allele frequencies and genotype frequencies in a sexually reproducing population will remain constant over generations.

Haversian system (ha-ver´-sē-un): see *osteon*.

head: the anterior-most segment of an animal with segmentation.

heart: a muscular organ responsible for pumping blood within the circulatory system throughout the body.

heart attack: a severe reduction or blockage of blood flow through a coronary artery, depriving some of the heart muscle of its blood supply.

heartwood: older xylem that contributes to the strength of a tree trunk.

heat of fusion: the energy that must be removed from a compound to transform it from a liquid into a solid at its freezing temperature.

heat of vaporization: the energy that must be supplied to a compound to transform it from a liquid into a gas at its boiling temperature.

heliozoan (hē-lē-ō-zō´-un): an aquatic (largely freshwater) animal-like protist; some have elaborate silica-based shells.

helix (hē´-liks): a coiled, springlike secondary structure of a protein.

helper T cell: a type of T cell that helps other immune cells recognize and act against antigens.

hemocoel (hē´-mō-sēl): a blood cavity within the bodies of certain invertebrates in which blood bathes tissues directly; part of an open circulatory system.

hemodialysis (hē-mō-dī-al´-luh-sis): a procedure that simulates kidney function in individuals with damaged or ineffective kidneys; blood is diverted from the body, artificially filtered, and returned to the body.

hemoglobin (hē´mō-glō-bin): the iron-containing protein that gives red blood cells their color; binds to oxygen in the lungs and releases it to the tissues.

hemophilia: a recessive, sex-linked disease in which the blood fails to clot normally.

herbivore (erb´-i-vor): literally, "plant-eater"; an organism that feeds directly and exclusively on producers; a primary consumer.

hermaphrodite (her-maf´-ruh-dīt´): an organism that possesses both male and female sexual organs.

hermaphroditic (her-maf´-ruh-dit´-ik): possessing both male and female sexual organs. Some hermaphroditic animals can fertilize themselves; others must exchange sex cells with a mate.

heterotroph (het´-er-ō-trōf´): literally, "other-feeder"; an organism that eats other organisms; a consumer.

heterozygous (het-er-ō-zī´-gus): carrying two different alleles of a given gene; also called *hybrid.*

hindbrain: the posterior portion of the brain, containing the medulla, pons, and cerebellum.

hinge joint: a joint at which one bone is moved by muscle and the other bone remains fixed, such as in the knee, elbow, or fingers; allows movement in only two dimensions.

hippocampus (hip-ō-kam´-pus): the part of the forebrain of vertebrates that is important in emotion and especially learning.

histamine: a substance released by certain cells in response to tissue damage and invasion of the body by foreign substances; promotes the dilation of arterioles and the leakiness of capillaries and triggers some of the events of the inflammatory response.

homeobox (hō´-mē-ō-boks): a sequence of DNA coding for special, 60-amino-acid proteins, which activate or inactivate genes that control development; these sequences specify embryonic cell differentiation.

homeostasis (hōm-ē-ō-stā´sis): the maintenance of a relatively constant environment required for the optimal functioning of cells, maintained by the coordinated activity of numerous regulatory mechanisms, including the respiratory, endocrine, circulatory, and excretory systems.

hominid: a human or a prehistoric relative of humans, beginning with the Australopithecines, whose fossils date back at least 4.4 million years.

homologous structures: structures that may differ in function but that have similar anatomy, presumably because the organisms that possess them have descended from common ancestors.

homologue (hō-´mō-log): a chromosome that is similar in appearance and genetic information to another chromosome with which it pairs during meiosis; also called *homologous chromosome.*

homozygous (hō-mō-zī´-gus): carrying two copies of the same allele of a given gene; also called *true-breeding.*

hormone: a chemical that is synthesized by one group of cells, secreted, and then carried in the bloodstream to other cells, whose activity is influenced by reception of the hormone.

host: the prey organism on or in which a parasite lives; is harmed by the relationship .

human immunodeficiency virus (HIV): a pathogenic retrovirus that causes acquired immune deficiency syndrome (AIDS) by attacking and destroying the immune system's T cells.

humoral immunity: an immune response in which foreign substances are inactivated or destroyed by antibodies that circulate in the blood.

Huntington's disease: an incurable genetic disorder, caused by a dominant allele, that produces progressive brain deterioration, resulting in the loss of motor coordination, flailing movements, personality disturbances, and eventual death.

hybrid: an organism that is the offspring of parents differing in at least one genetically determined characteristic; also used to refer to the offspring of parents of different species.

hybrid infertility: reduced fertility (typically, complete sterility) in the hybrid offspring of two species.

hybrid inviability: the failure of a hybrid offspring of two species to survive to maturity.

hybridoma: a cell produced by fusing an antibody-producing cell with a myeloma cell; used to produce monoclonal antibodies.

hydrogen bond: the weak attraction between a hydrogen atom that bears a partial positive charge (due to polar covalent bonding with another atom) and another atom, normally oxygen or nitrogen, that bears a partial negative charge; hydrogen bonds may form between atoms of a single molecule or of different molecules.

hydrologic cycle: the water cycle, driven by solar energy; a nutrient cycle in which the main reservoir of water is the ocean and most of the water remains in the form of water throughout the cycle (rather than being used in the synthesis of new molecules).

hydrolysis (hī-drol´-i-sis): the chemical reaction that breaks a covalent bond by means of the addition of hydrogen to the atom on one side of the original bond and a hydroxyl group to the atom on the other side; the reverse of dehydration synthesis.

hydrophilic (hī-drō-fil´-ik): pertaining to a substance that dissolves readily in water, or to parts of a large molecule that form hydrogen bonds with water.

hydrophobic (hī-drō-fō´-bik): pertaining to a substance that does not dissolve in water.

hydrophobic interaction: the tendency for hydrophobic molecules to cluster together when immersed in water.

hydrostatic skeleton (hī-drō-stat´-ik): a body type that uses fluid contained in body compartments to provide support and mass against which muscles can contract.

hydrothermal vent community: a community of unusual organisms, living in the deep ocean near hydrothermal vents, that depends on the chemosynthetic activities of sulfur bacteria.

hypertension: arterial blood pressure that is chronically elevated above the normal level.

hypertonic (hī-per-ton´-ik): referring to a solution that has a higher concentration of dissolved particles (and therefore a lower concentration of free water) than has the cytoplasm of a cell.

hypha (hī´-fuh; pl., hyphae): a threadlike structure that consists of elongated cells, typically with many haploid nuclei; many hyphae make up the fungal body.

hypocotyl (hī´-pō-kot-ul): the part of the embryonic shoot located below the cotyledons but above the root.

hypothalamus (hī-pō-thal´-a-mus): a region of the brain that controls the secretory activity of the pituitary gland; synthesizes, stores, and releases certain peptide hormones; directs autonomic nervous system responses.

hypothesis (hī-poth´-eh-sis): the second operation in the scientific method; a supposition based on previous observations that is offered as an explanation for the observed phenomenon and is used as the basis for further observations, or experiments.

hypotonic (hī-pō-ton´-ik): referring to a solution that has a lower concentration of dissolved particles (and therefore a higher concentration of free water) than has the cytoplasm of a cell.

immigration (im-uh-grā´-shun): migration of individuals into an area.

immune response: a specific response by the immune system to the invasion of the body by a particular foreign substance or microorganism, characterized by the recognition of the foreign substance by immune cells and its subsequent destruction by antibodies or by cellular attack.

imperfect fungus: a fungus of the division Deuteromycota; no species in this division has been observed to form sexual reproductive structures.

implantation: the process whereby the early embryo embeds itself within the lining of the uterus.

imprinting: the process by which an animal forms an association with another animal or object in the environment during a sensitive period of development.

inclusive fitness: the reproductive success of all organisms that bear a given allele, normally expressed in relation to the average reproductive success of all individuals in the same population; compare with *fitness*.

incomplete dominance: a pattern of inheritance in which the heterozygous phenotype is intermediate between the two homozygous phenotypes.

incomplete flower: a flower that is missing one of the four floral parts (sepals, petals, stamens, or carpels).

independent assortment: see *law of independent assortment*.

indirect development: a developmental pathway in which an offspring goes through radical changes in body form as it matures.

induction: the process by which a group of cells causes other cells to differentiate into a specific tissue type.

inflammatory response: a nonspecific, local response to injury to the body, characterized by the phagocytosis of foreign substances and tissue debris by white blood cells and by the walling off of the injury site by the clotting of fluids that escape from nearby blood vessels.

inhalation: the act of drawing air into the lungs by enlarging the chest cavity.

inheritance: the genetic transmission of characteristics from parent to offspring.

inheritance of acquired characteristics: the hypothesis that organisms' bodies change during their lifetimes by use and disuse and that these changes are inherited by their offspring.

inhibiting hormone: a hormone, secreted by the neurosecretory cells of the hypothalamus, that inhibits the release of specific hormones from the anterior pituitary.

innate (in-āt´): inborn; instinctive; determined by the genetic makeup of the individual.

inner cell mass: in human embryonic development, the cluster of cells, on one side of the blastocyst, that will develop into the embryo.

inner ear: the innermost part of the mammalian ear; composed of the bony, fluid-filled tubes of the cochlea and the vestibular apparatus.

inorganic: describing any molecule that does not contain both carbon and hydrogen.

insertion: the site of attachment of a muscle to the relatively movable bone on one side of a joint.

insertion mutation: a mutation in which one or more pairs of nucleotides are inserted into a gene.

insight learning: a complex form of learning that requires the manipulation of mental concepts to arrive at adaptive behavior.

instinctive: innate; inborn; determined by the genetic makeup of the individual.

insulin: a hormone, secreted by the pancreas, that lowers blood sugar by stimulating the conversion of glucose to glycogen in the liver.

integration: in nerve cells, the process of adding up electrical signals from sensory inputs or other nerve cells to determine the appropriate outputs.

integument (in-teg´-ū-ment): in plants, the outer layers of cells of the ovule that surrounds the embryo sac; develops into the seed coat.

intensity: the strength of stimulation or response.

interferon: a protein released by certain virus-infected cells that increases the resistance of other, uninfected, cells to viral attack.

intermediate filament: part of the cytoskeleton of eukaryotic cells that probably functions mainly for support and is composed of several types of proteins.

intermembrane compartment: the fluid-filled space between the inner and outer membranes of a mitochondrion.

internal fertilization: the union of sperm and egg inside the body of the female.

internode: the part of a stem between two nodes.

interphase: the stage of the cell cycle between cell divisions; the stage in which chromosomes are replicated and other cell functions occur, such as growth, movement, and acquisition of nutrients.

interspecific competition: competition among individuals of different species.

interstitial cell (in-ter-sti´-shul): in the vertebrate testis, a testosterone-producing cell located between the seminiferous tubules.

interstitial fluid (in-ter-sti´-shul): fluid, similar in composition to plasma (except lacking large proteins), that leaks from capillaries and acts as a medium of exchange between the body cells and the capillaries.

intertidal zone: an area of the ocean shore that is alternately covered and exposed by the tides.

intervertebral disc (in-ter-ver-tē´-brul): a pad of cartilage between two vertebrae that acts as a shock absorber.

intracellular digestion: the chemical breakdown of food within single cells.

intraspecific competition: competition among individuals of the same species.

intrauterine device (IUD): a small copper or plastic loop, squiggle, or shield that is inserted in the uterus; a contraceptive method that works by irritating the uterine lining so that it cannot receive the embryo.

intron: a segment of DNA in a eukaryotic gene that does not code for amino acids in a protein.

invertebrate (in-vert´-uh-bret): an animal that never possesses a vertebral column.

ion (ī´-on): a charged atom or molecule; an atom or molecule that has either an excess of electrons (and hence is negatively charged) or has lost electrons (and is positively charged).

ionic bond: a chemical bond formed by the electrical attraction between positively and negatively charged ions.

iris: the pigmented muscular tissue of the vertebrate eye that surrounds and controls the size of the pupil, through which light enters.

islet cell: a cluster of cells in the endocrine portion of the pancreas that produce insulin and glucagon.

isolating mechanism: a morphological, physiological, behavioral, or ecological difference that prevents members of two species from interbreeding.

isotonic (ī-sō-ton´-ik): referring to a solution that has the same concentration of dissolved particles (and therefore the same concentration of free water) as has the cytoplasm of a cell.

isotope: one of several forms of a single element, the nuclei of which contain the same number of protons but different numbers of neutrons.

J-curve: the J-shaped growth curve of an exponentially growing population in which increasing numbers of individuals join the population during each succeeding time period.

joint: a flexible region between two rigid units of an exoskeleton or endoskeleton, allowing for movement between the units.

keratin (ker´-uh-tin): a fibrous protein in hair, nails, and the epidermis of skin.

keystone species: a species whose influence on community structure is greater than its abundance would suggest.

kidney: one of a pair of organs of the excretory system that is located on either side of the spinal column and filters blood, removing wastes and regulating the composition and water content of the blood.

kinesis (kin-ē´-sis): an innate behavior in which an organism changes the speed of its random movement in response to an environmental stimulus.

kinetic energy: the energy of movement; includes light, heat, mechanical movement, and electricity.

kinetochore (ki-net´-ō-kor): a protein structure that forms at the centromere regions of chromosomes; attaches the chromosomes to the spindle.

kingdom: the second broadest taxonomic category, contained within a domain and consisting of related phyla or divisions. This textbook recognizes four kingdoms within the domain Eukarya: Protista, Fungi, Plantae, and Animalia.

kin selection: a type of natural selection that favors a certain allele because it increases the survival or reproductive success of relatives that bear the same allele.

Klinefelter syndrome: a set of characteristics typically found in individuals who have two X chromosomes and one Y chromosome; these individuals are phenotypically males but are sterile and have several femalelike traits, including broad hips and partial breast development.

Krebs cycle: a cyclic series of reactions, occurring in the matrix of mitochondria, in which the acetyl groups from the pyruvic acids produced by glycolysis are broken down to CO_2, accompanied by the formation of ATP and electron carriers; also called *citric acid cycle*.

kuru: a degenerative brain disease, first discovered in the cannibalistic Fore tribe of New Guinea, that is caused by a prion.

labium (pl., labia): one of a pair of folds of skin of the external structures of the mammalian female reproductive system.

labor: a series of contractions of the uterus that result in birth.

lactation: the secretion of milk from the mammary glands.

lacteal (lak-tēl´): a single lymph capillary that penetrates each villus of the small intestine.

lactose (lak´-tōs): a disaccharide composed of glucose and galactose; found in mammalian milk.

large intestine: the final section of the digestive tract; consists of the colon and the rectum, where feces are formed and stored.

larva (lar´-vuh): an immature form of an organism with indirect development prior to metamorphosis into its adult form; includes the caterpillars of moths and butterflies and the maggots of flies.

larynx (lar´-inks): that portion of the air passage between the pharynx and the trachea; contains the vocal cords.

lateral bud: a cluster of meristematic cells at the node of a stem; under appropriate conditions, it grows into a branch.

lateral meristem: a meristematic tissue that forms cylinders parallel to the long axis of roots and stems; normally located between the primary xylem and primary phloem (vascular cambium) and just outside the phloem (cork cambium); also called *cambium*.

law of independent assortment: the independent inheritance of two or more distinct traits; states that the alleles for one trait may be distributed to the gametes independently of the alleles for other traits.

law of segregation: Gregor Mendel's conclusion that each gamete receives only one of each parent's pair of genes for each trait.

laws of thermodynamics: the physical laws that define the basic properties and behavior of energy.

leaf: an outgrowth of a stem, normally flattened and photosynthetic.

leaf primordium (pri-mor´-dē-um; pl., primordia): a cluster of meristem cells, located at the node of a stem, that develops into a leaf.

learning: an adaptive change in behavior as a result of experience.

legume (leg´-ūm): a member of a family of plants characterized by root swellings in which nitrogen-fixing bacteria are housed; includes soybeans, lupines, alfalfa, and clover.

lens: a clear object that bends light rays; in eyes, a flexible or movable structure used to focus light on a layer of photoreceptor cells.

leukocyte (loo´-kō-sīt): any of the white blood cells circulating in the blood.

lichen (lī´-ken): a symbiotic association between an alga or cyanobacterium and a fungus, resulting in a composite organism.

life cycle: the events in the life of an organism from one generation to the next.

ligament: a tough connective tissue band connecting two bones.

light-dependent reactions: the first stage of photosynthesis, in which the energy of light is captured as ATP and NADPH; occurs in thylakoids of chloroplasts.

light-harvesting complex: in photosystems, the assembly of pigment molecules (chlorophyll and accessory pigments) that absorb light energy and transfer that energy to electrons.

light-independent reactions: the second stage of photosynthesis, in which the energy obtained by the light-dependent reactions is used to fix carbon dioxide into carbohydrates; occurs in the stroma of chloroplasts.

lignin: a hard material that is embedded in the cell walls of vascular plants; provides support in terrestrial species; an early and important adaptation to terrestrial life.

limbic system: a diverse group of brain structures, mostly in the lower forebrain, that includes the thalamus, hypothalamus, amygdala, hippocampus, and parts of the cerebrum and is involved in basic emotions, drives, behaviors, and learning.

limnetic zone: a lake zone in which enough light penetrates to support photosynthesis.

linkage: the inheritance of certain genes as a group because they are parts of the same chromosome. Linked genes do not show independent assortment.

lipase (lī´-pās): an enzyme that catalyzes the breakdown of lipids such as fats.

lipid (lī´-pid): one of a number of organic molecules containing large nonpolar regions composed solely of carbon and hydrogen, which make lipids hydrophobic and insoluble in water; includes oils, fats, waxes, phospholipids, and steroids.

littoral zone: a lake zone, near the shore, in which water is shallow and plants find abundant light, anchorage, and adequate nutrients.

liver: an organ with varied functions, including bile production, glycogen storage, and the detoxification of poisons.

locus: the physical location of a gene on a chromosome.

long-day plant: a plant that will flower only if the length of daylight is greater than some species-specific duration.

long-term memory: the second phase of learning; a more-or-less permanent memory formed by a structural change in the brain, brought on by repetition.

loop of Henle (hen´-lē): a specialized portion of the tubule of the nephron in birds and mammals that creates an osmotic concentration gradient in the fluid immediately surrounding it. This gradient in turn makes possible the production of urine more osmotically concentrated than blood plasma.

lung: a paired respiratory organ consisting of inflatable chambers within the chest cavity in which gas exchange occurs.

luteinizing hormone (LH): a hormone, produced by the anterior pituitary, that stimulates testosterone production in males and the development of the follicle, ovulation, and the production of the corpus luteum in females.

lymph (limf): a pale fluid, within the lymphatic system, that is composed primarily of interstitial fluid and lymphocytes.

lymphatic system: a system consisting of lymph vessels, lymph capillaries, lymph nodes, and the thymus and spleen; helps protect the body against infection, absorbs fats, and returns excess fluid and small proteins to the blood circulatory system.

lymph node: a small structure that filters lymph; contains lymphocytes and macrophages, which inactivate foreign particles such as bacteria.

lymphocyte (lim´-fō-sīt): a type of white blood cell important in the immune response.

lysosome (lī´-sō-sōm): a membrane-bound organelle containing intracellular digestive enzymes.

macronutrient: a nutrient needed in relatively large quantities (often defined as making up more than 0.1% of an organism's body).

macrophage (mak´-rō-fāj): a type of white blood cell that engulfs microbes and destroys them by phagocytosis; also presents microbial antigens to T cells, helping stimulate the immune response.

magnetotactic: able to detect and respond to Earth's magnetic field.

major histocompatibility complex (MHC): proteins, normally located on the surfaces of body cells, that identify the cell as "self"; also important in stimulating and regulating the immune response.

maltose (mal´-tōs): a disaccharide composed of two glucose molecules.

mammary gland (mam´-uh-rē): a milk-producing gland used by female mammals to nourish their young.

mantle (man´-tul): an extension of the body wall in certain invertebrates, such as mollusks; may secrete a shell, protect the gills, and, as in cephalopods, aid in locomotion.

marsupial (mar-soo´-pē-ul): a mammal whose young are born at an extremely immature stage and undergo further development in a pouch while they remain attached to a mammary gland; includes kangaroos, opossums, and koalas.

mass extinction: the extinction of an extraordinarily large number of species in a short period of geologic time. Mass extinctions have recurred periodically throughout the history of life.

mast cell: a cell of the immune system that synthesizes histamine and other molecules used in the body's response to trauma and that are a factor in allergic reactions.

matrix: the fluid contained within the inner membrane of a mitochondrion.

mechanical incompatibility: the inability of male and female organisms to exchange gametes, normally because their reproductive structures are incompatible.

mechanoreceptor: a receptor that responds to mechanical deformation, such as that caused by pressure, touch, or vibration.

medulla (med-ū´-luh): the part of the hindbrain of vertebrates that controls automatic activities such as breathing, swallowing, heart rate, and blood pressure.

medusa (meh-doo´-suh): a bell-shaped, typically free-swimming stage in the life cycle of many cnidarians; includes jellyfish.

megakaryocyte (meg-a-kar´-ē-ō-sīt): a large cell type that remains in the bone marrow, pinching off pieces of itself that then enter the circulation as platelets.

megaspore: a haploid cell formed by meiosis from a diploid megaspore mother cell; through mitosis and differentiation, develops into the female gametophyte.

megaspore mother cell: a diploid cell, within the ovule of a flowering plant, that undergoes meiosis to produce four haploid megaspores.

meiosis (mī-ō´-sis): a type of cell division, used by eukaryotic organisms, in which a diploid cell divides twice to produce four haploid cells.

meiotic cell division: meiosis followed by cytokinesis.

melanocyte-stimulating hormone (me-lan´-ō-sīt): a hormone, released by the anterior pituitary, that regulates the activity of skin pigments in some vertebrates.

melatonin (mel-uh-tōn´-in): a hormone, secreted by the pineal gland, that is involved in the regulation of circadian cycles.

membrane: in multicellular organism, a continuous sheet of epithelial cells that covers the body and lines body cavities; in a cell, a thin sheet of lipids and proteins that surrounds the cell or its organelles, separating them from their surroundings.

memory cell: a long-lived descendant of a B cell or T cell that has been activated by contact with an antigen; a reservoir of cells that rapidly respond to reexposure to the antigen.

meninges (men-in´-jēz): three layers of connective tissue that surround the brain and spinal cord.

menstrual cycle: in human females, a complex 28-day cycle during which hormonal interactions among the hypothalamus, pituitary gland, and ovary coordinate ovulation and the preparation of the uterus to receive and nourish the fertilized egg. If pregnancy does not occur, the uterine lining is shed during menstruation.

menstruation: in human females, the monthly discharge of uterine tissue and blood from the uterus.

meristem cell (mer´-i-stem): an undifferentiated cell that remains capable of cell division throughout the life of a plant.

mesoderm (mēz´-ō-derm): the middle embryonic tissue layer, lying between the endoderm and ectoderm, and normally the last to develop; gives rise to structures such as muscle and skeleton.

mesoglea (mez-ō-glē´-uh): a middle, jellylike layer within the body wall of cnidarians.

mesophyll (mez´-ō-fil): loosely packed parenchyma cells beneath the epidermis of a leaf.

messenger RNA (mRNA): a strand of RNA, complementary to the DNA of a gene, that conveys the genetic information in DNA to the ribosomes to be used during protein synthesis; sequences of three bases (codons) in mRNA specify particular amino acids to be incorporated into a protein.

metabolic pathway: a sequence of chemical reactions within a cell, in which the products of one reaction are the reactants for the next reaction.

metabolism: the sum of all chemical reactions that occur within a single cell or within all the cells of a multicellular organism.

metamorphosis (met-a-mor´-fō-sis): in animals with indirect development, a radical change in body form from larva to sexually mature adult, as seen in amphibians (tadpole to frog) and insects (caterpillar to butterfly).

metaphase (met´-a-fāz): the stage of mitosis in which the chromosomes, attached to spindle fibers at kinetochores, are lined up along the equator of the cell.

methanogen (me-than´-ō-jen): a type of anaerobic archaean capable of converting carbon dioxide to methane.

microevolution: change over successive generations in the composition of a population's gene pool.

microfilament: part of the cytoskeleton of eukaryotic cells that is composed of the proteins actin and (in some cases) myosin; functions in the movement of cell organelles and in locomotion by extension of the plasma membrane.

micronutrient: a nutrient needed only in small quantities (often defined as making up less than 0.01% of an organism's body).

microsphere: a small, hollow sphere formed from proteins or proteins complexed with other compounds.

microspore: a haploid cell formed by meiosis from a microspore mother cell; through mitosis and differentiation, develops into the male gametophyte.

microspore mother cell: a diploid cell contained within an anther of a flowering plant, which undergoes meiosis to produce four haploid microspores.

microtubule: a hollow, cylindrical strand, found in eukaryotic cells, that is composed of the protein tubulin; part of the cytoskeleton used in the movement of organelles, cell growth, and the construction of cilia and flagella.

microvillus (mī-krō-vi´-lus; pl., microvilli): a microscopic projection of the plasma membrane of each villus; increases the surface area of the villus.

midbrain: during development, the central portion of the brain; contains an important relay center, the reticular formation.

middle ear: the part of the mammalian ear composed of the tympanic membrane, the Eustachian tube, and three bones (hammer, anvil, and stirrup) that transmit vibrations from the auditory canal to the oval window.

middle lamella: a thin layer of sticky polysaccharides, such as pectin, and other carbohydrates that separates and holds together the primary cell walls of adjacent plant cells.

mimicry (mim´-ik-rē): the situation in which a species has evolved to resemble something else—typically another type of organism.

mineral: an inorganic substance, especially one in rocks or soil.

mitochondrion (mī-tō-kon´-drē-un): an organelle, bounded by two membranes, that is the site of the reactions of aerobic metabolism.

mitosis (mī-tō´-sis): a type of nuclear division, used by eukaryotic cells, in which one copy of each chromosome (already duplicated during interphase before mitosis) moves into each of two daughter nuclei; the daughter nuclei are therefore genetically identical to each other.

mitotic cell division: mitosis followed by cytokinesis.

molecule (mol´-e-kūl): a particle composed of one or more atoms held together by chemical bonds; the smallest particle of a compound that displays all the properties of that compound.

molt: to shed an external body covering, such as an exoskeleton, skin, feathers, or fur.

monoclonal antibody: an antibody produced in the laboratory by the cloning of hybridoma cells; each clone of cells produces a single antibody.

monocot: short for monocotyledon; a type of flowering plant characterized by embryos with one seed leaf, or cotyledon.

monocyte: a type of white blood cell that travels through capillaries to wounds where bacteria have gained entry, leaves the capillaries, and differentiates into a macrophage.

monoecious (mon-ē´-shus): pertaining to organisms in which male and female gametes are produced in the same individual.

monomer (mo´-nō-mer): a small organic molecule, several of which may be bonded together to form a chain called a polymer.

monosaccharide (mo-nō-sak´-uh-rīd): the basic molecular unit of all carbohydrates, normally composed of a chain of carbon atoms bonded to hydrogen and hydroxyl groups.

monotreme: a mammal that lays eggs; for example, the platypus.

morula (mor´-ū-luh): in animals, an embryonic stage during cleavage, when the embryo consists of a solid ball of cells.

motor neuron: a neuron that receives instructions from the association neurons and activates effector organs, such as muscles or glands.

motor unit: a single motor neuron and all the muscle fibers on which it forms synapses.

mouth: the opening of a tubular digestive system into which food is first introduced.

mucous membrane: the lining of the inside of the respiratory and digestive tracts.

multicellular: many-celled; most members of the kingdoms Fungi, Plantae, and Animalia are multicellular, with intimate cooperation among cells.

multiple alleles: as many as dozens of alleles produced for every gene as a result of different mutations.

muscle fiber: an individual muscle cell.

mutation: a change in the base sequence of DNA in a gene; normally refers to a genetic change significant enough to alter the appearance or function of the organism.

mutualism (mu´-choo-ul-iz-um): a symbiotic relationship in which both participating species benefit.

mycelium (mī-sēl´-ē-um): the body of a fungus, consisting of a mass of hyphae.

mycorrhiza (mī-kō-rī´zuh; pl., **mycorrhizae):** a symbiotic relationship between a fungus and the roots of a land plant that facilitates mineral extraction and absorption.

myelin (mī´-uh-lin): a wrapping of insulating membranes of specialized nonneural cells around the axon of a vertebrate nerve cell; increases the speed of conduction of action potentials.

myofibril (mī-ō-fī´-bril): a cylindrical subunit of a muscle cell, consisting of a series of sarcomeres; surrounded by sarcoplasmic reticulum.

myometrium (mī-ō-mē´-trē-um): the muscular outer layer of the uterus.

myosin (mī´-ō-sin): one of the major proteins of muscle, whose interaction with the protein actin produces muscle contraction; found in the thick filaments of the muscle fiber; see also *actin*.

natural causality: the scientific principle that natural events occur as a result of preceding natural causes.

natural killer cell: a type of white blood cell that destroys some virus-infected cells and cancerous cells on contact; part of the immune system's nonspecific internal defense against disease.

natural selection: the unequal survival and reproduction of organisms due to environmental forces, resulting in the preservation of favorable adaptations. Usually, natural selection refers specifically to differential survival and reproduction on the basis of genetic differences among individuals.

near-shore zone: the region of coastal water that is relatively shallow but constantly submerged; includes bays and coastal wetlands and can support large plants or seaweeds.

negative feedback: a situation in which a change initiates a series of events that tend to counteract the change and restore the original state. Negative feedback in physiological systems maintains homeostasis.

nephridium (nef-rid´-ē-um): an excretory organ found in earthworms, mollusks, and certain other invertebrates; somewhat resembles a single vertebrate nephron.

nephron (nef´-ron): the functional unit of the kidney; where blood is filtered and urine formed.

nephrostome (nef´-rō-stōm): the funnel-shaped opening of the nephridium of some invertebrates such as earthworms; coelomic fluid is drawn into the nephrostome for filtration.

nerve: a bundle of axons of nerve cells, bound together in a sheath.

nerve cord: a paired neural structure in most animals that conducts nervous signals to and from the ganglia; in chordates, a nervous structure lying along the dorsal side of the body; also called spinal cord.

nerve net: a simple form of nervous system, consisting of a network of neurons that extend throughout the tissues of an organism such as a cnidarian.

nerve tissue: the tissue that make up the brain, spinal cord, and nerves; consists of neurons and glial cells.

net primary productivity: the energy stored in the autotrophs of an ecosystem over a given time period.

neural tube: a structure, derived from ectoderm during early embryonic development, that later becomes the brain and spinal cord.

neuromuscular junction: the synapse formed between a motor neuron and a muscle fiber.

neuron (noor´-on): a single nerve cell.

neuropeptide: a small protein molecule with neurotransmitter-like actions.

neurosecretory cell: a specialized nerve cell that synthesizes and releases hormones.

neurotransmitter: a chemical that is released by a nerve cell close to a second nerve cell, a muscle, or a gland cell and that influences the activity of the second cell.

neutral mutation: a mutation that has little or no effect on the function of the encoded protein.

neutralization: the process of covering up or inactivating a toxic substance with antibody.

neutron: a subatomic particle that is found in the nuclei of atoms, bears no charge, and has a mass approximately equal to that of a proton.

neutrophil: a type of white blood cell that feeds on bacterial invaders or other foreign cells, including cancer cells.

nitrogen fixation: the process that combines atmospheric nitrogen with hydrogen to form ammonium (NH_4^+).

nitrogen-fixing bacterium: a bacterium that possess the ability to remove nitrogen (N_2) from the atmosphere and combine it with hydrogen to produce ammonium (NH_4^+).

node: in plants, a region of a stem at which leaves and lateral buds are located; in vertebrates, an interruption of the myelin on a myelinated axon, exposing naked membrane at which action potentials are generated.

nodule: a swelling on the root of a legume or other plant that consists of cortex cells inhabited by nitrogen-fixing bacteria.

nondisjunction: an error in meiosis in which chromosomes fail to segregate properly into the daughter cells.

nonpolar covalent bond: a covalent bond with equal sharing of electrons.

norepinephrine (nor-ep-i-nef-rin´): a neurotransmitter, released by neurons of the parasympathetic nervous system, that prepares the body to respond to stressful situations; also called noradrenaline.

northern coniferous forest: a biome with long, cold winters and only a few months of warm weather; populated almost entirely by evergreen coniferous trees; also called *taiga*.

notochord (nōt´-ō-kord): a stiff but somewhat flexible, supportive rod found in all members of the phylum Chordata at some stage of development.

nuclear envelope: the double-membrane system surrounding the nucleus of eukaryotic cells; the outer membrane is typically continuous with the endoplasmic reticulum.

nucleic acid (noo-klā´-ik): an organic molecule composed of nucleotide subunits; the two common types of nucleic acids are ribonucleic acid (RNA) and deoxyribonucleic acid (DNA).

nucleoid (noo-klē-oid): the location of the genetic material in prokaryotic cells; not membrane-enclosed.

nucleolus (noo-klē´-ō-lus): the region of the eukaryotic nucleus that is engaged in ribosome synthesis; consists of the genes encoding ribosomal RNA, newly synthesized ribosomal RNA, and ribosomal proteins.

nucleotide: a subunit of which nucleic acids are composed; a phosphate group bonded to a sugar (deoxyribose in DNA), which is in turn bonded to a nitrogen-containing base (adenine, guanine, cytosine, or thymine in DNA). Nucleotides are linked together, forming a strand of nucleic acid, as follows: Bonds between the phosphate of one nucleotide link to the sugar of the next nucleotide.

nucleus (atomic): the central region of an atom, consisting of protons and neutrons.

nucleus (cellular): the membrane-bound organelle of eukaryotic cells that contains the cell's genetic material.

nutrient: a substance acquired from the environment and needed for the survival, growth, and development of an organism.

nutrient cycle: a description of the pathways of a specific nutrient (such as carbon, nitrogen, phosphorus, or water) through the living and nonliving portions of an ecosystem.

nutrition: the process of acquiring nutrients from the environment and, if necessary, processing them into a form that can be used by the body.

observation: the first operation in the scientific method; the noting of a specific phenomenon, leading to the formulation of a hypothesis.

oil: a lipid composed of three fatty acids, some of which are unsaturated, covalently bonded to a molecule of glycerol; liquid at room temperature.

olfaction (ōl-fak´-shun): a chemical sense, the sense of smell; in terrestrial vertebrates, the result of the detection of airborne molecules.

oligotrophic lake: a lake that is very low in nutrients and hence clear with extensive light penetration.

ommatidium (ōm-ma-tid´-ē-um): an individual light-sensitive subunit of a compound eye; consists of a lens and several receptor cells.

oncogene: a gene that, when transcribed, causes a cell to become cancerous.

one-gene, one-protein hypothesis: the premise that each gene encodes the information for the synthesis of a single protein.

oogenesis: the process by which egg cells are formed.

oogonium (ō-ō-gō´-nē-um; pl., **oogonia):** in female animals, a diploid cell that gives rise to a primary oocyte.

open circulatory system: a type of circulatory system found in some invertebrates, such as arthropods and mollusks, that includes an open space (the hemocoel) in which blood directly bathes body tissues.

operant conditioning: a laboratory training procedure in which an animal learns to perform a response (such as pressing a lever) through reward or punishment.

operculum: an external flap, supported by bone, that covers and protects the gills of most fish.

opioid (ōp´-ē-oid): one of a group of peptide neuromodulators in the vertebrate brain that mimic some of the actions of opiates (such as opium) and also seem to influence many other processes, including emotion and appetite.

optic disc: the area of the retina at which the axons of the ganglion cell merge to form the optic nerve; the blind spot of the retina.

optic nerve: the nerve leading from the eye to the brain, carrying visual information.

order: the taxonomic category contained within a class and consisting of related families.

organ: a structure (such as the liver, kidney, or skin) composed of two or more distinct tissue types that function together.

organelle (or-guh-nel´): a structure, found in the cytoplasm of eukaryotic cells, that performs a specific function; sometimes refers specifically to membrane-bound structures, such as the nucleus or endoplasmic reticulum.

organic: describing a molecule that contains both carbon and hydrogen.

organism (or´-guh-niz-um): an individual living thing.

organogenesis (or-gan-ō-jen´-uh-sis): the process by which the layers of the gastrula (endoderm, ectoderm, mesoderm) rearrange into organs.

organ system: two or more organs that work together to perform a specific function; for example, the digestive system.

origin: the site of attachment of a muscle to the relatively stationary bone on one side of a joint.

osmosis (oz-mō´-sis): the diffusion of water across a differentially permeable membrane, normally down a concentration gradient of free water molecules. Water moves into the solution that has a lower concentration of free water from a solution with the higher concentration of free water.

osmotic pressure: the pressure required to counterbalance the tendency of water to move from a solution with a higher concentration of free water molecules into a solution with a lower concentration of free water molecules.

osteoblast (os´-tē-ō-blast): a cell type that produces bone.

osteoclast (os´-tē-ō-klast): a cell type that dissolves bone.

osteocyte (os´-tē-ō-sīt): a mature bone cell.

osteon: a unit of hard bone consisting of concentric layers of bone matrix, with embedded osteocytes, surrounding a small central canal that contains a capillary.

osteoporosis (os´-tē-ō-por-ō´-sis): a condition in which bones become porous, weak, and easily fractured; most common in elderly women.

outer ear: the outermost part of the mammalian ear, including the external ear and auditory canal leading to the tympanic membrane.

oval window: the membrane-covered entrance to the inner ear.

ovary: in animals, the gonad of females; in flowering plants, a structure at the base of the carpel that contains one or more ovules and develops into the fruit.

oviduct: in mammals, the tube leading from the ovary to the uterus.

ovulation: the release of a secondary oocyte, ready to be fertilized, from the ovary.

ovule: a structure within the ovary of a flower, inside which the female gametophyte develops; after fertilization, develops into the seed.

oxytocin (oks-ē-tō´-sin): a hormone, released by the posterior pituitary, that stimulates the contraction of uterine and mammary gland muscles.

ozone layer: the ozone-enriched layer of the upper atmosphere that filters out some of the sun's ultraviolet radiation.

pacemaker: a cluster of specialized muscle cells in the upper right atrium of the heart that produce spontaneous electrical signals at a regular rate; the sinoatrial node.

pain receptor: a receptor that has extensive areas of membranes studded with special receptor proteins that respond to light or to a chemical.

palisade cell: a columnar mesophyll cell, containing chloroplasts, just beneath the upper epidermis of a leaf.

pancreas (pān´-krē-us): a combined exocrine and endocrine gland located in the abdominal cavity next to the stomach. The endocrine portion secretes the hormones insulin and glucagon, which regulate glucose concentrations in the blood. The exocrine portion secretes enzymes for fat, carbohydrate, and protein digestion into the small intestine and neutralizes the acidic chyme.

pancreatic juice: a mixture of water, sodium bicarbonate, and enzymes released by the pancreas into the small intestine.

parasite (par´-uh-sīt): an organism that lives in or on a larger prey organism, called a host, weakening it.

parasitism: a symbiotic relationship in which one organism (commonly smaller and more numerous than its host) benefits by feeding on the other, which is normally harmed but not immediately killed.

parasympathetic division: the division of the autonomic nervous system that produces largely involuntary responses related to the maintenance of normal body functions, such as digestion.

parathormone: a hormone, secreted by the parathyroid gland, that stimulates the release of calcium from bones.

parathyroid gland: one of a set of four small endocrine glands, embedded in the surface of the thyroid gland, that produces parathormone, which (with calcitonin from the thyroid gland) regulates calcium ion concentration in the blood.

parenchyma (par-en´-ki-muh): a plant cell type that is alive at maturity, normally with thin primary cell walls, that carries out most of the metabolism of a plant. Most dividing meristem cells in a plant are parenchyma.

parthenogenesis (par-the-nō-jen´uh-sis): a specialization of sexual reproduction, in which a haploid egg undergoes development without fertilization.

passive transport: the movement of materials across a membrane down a gradient of concentration, pressure, or electrical charge without using cellular energy.

pathogenic (path´-ō-jen-ik): capable of producing disease; refers to an organism with such a capability (a pathogen).

pedigree: a diagram showing genetic relationships among a set of individuals, normally with respect to a specific genetic trait.

pelagic (puh-la´-jik): free-swimming or floating.

penis: an external structure of the male reproductive and urinary systems; serves to deposit sperm into the female reproductive system and delivers urine to the exterior.

peptide (pep´-tīd): a chain composed of two or more amino acids linked together by peptide bonds.

peptide bond: the covalent bond between the amino group's nitrogen of one amino acid and the carboxyl group's carbon of a second amino acid, joining the two amino acids together in a peptide or protein.

peptide hormone: a hormone consisting of a chain of amino acids; includes small proteins that function as hormones.

peptidoglycan (pep-tid-ō-glī´-kan): a component of prokaryotic cell walls that consists of chains of sugars cross-linked by short chains of amino acids called peptides.

pericycle (per´-i-sī-kul): the outermost layer of cells of the vascular cylinder of a root.

periderm: the outer cell layers of roots and a stem that have undergone secondary growth, consisting primarily of cork cambium and cork cells.

peripheral nerve: a nerve that links the brain and spinal cord to the rest of the body.

peripheral nervous system: in vertebrates, the part of the nervous system that connects the central nervous system to the rest of the body.

peristalsis: rhythmic coordinated contractions of the smooth muscles of the digestive tract that move substances through the digestive tract.

permafrost: a permanently frozen layer of soil in the arctic tundra that cannot support the growth of trees.

petal: part of a flower, typically brightly colored and fragrant, that attracts potential animal pollinators.

petiole (pet´-ē-ōl): the stalk that connects the blade of a leaf to the stem.

phagocytic cell (fa-gō-sit´-ik): a type of immune system cell that destroys invading microbes by using phagocytosis to engulf and digest the microbes.

phagocytosis (fa-gō-sī-tō´-sis): a type of endocytosis in which extensions of a plasma membrane engulf extracellular particles and transport them into the interior of the cell.

pharyngeal gill slit (far-in´-jē-ul): an opening, located just posterior to the mouth, that connects the digestive tube to the outside environment; present (as some stage of life) in all chordates.

pharynx (far´-inks): in vertebrates, a chamber that is located at the back of the mouth and is shared by the digestive and respiratory systems; in some invertebrates, the portion of the digestive tube just posterior to the mouth.

phenotype (fēn´-ō-tīp): the physical characteristics of an organism; can be defined as outward appearance (such as flower color), as behavior, or in molecular terms (such as glycoproteins on red blood cells).

pheromone (fer´-uh-mōn): a chemical produced by an organism that alters the behavior or physiological state of another member of the same species.

phloem (flō´-um): a conducting tissue of vascular plants that transports a concentrated sugar solution up and down the plant.

phospholipid (fos-fō-li´-pid): a lipid consisting of glycerol bonded to two fatty acids and one phosphate group, which bears another group of atoms, typically charged and containing nitrogen. A double layer of phospholipids is a component of all cellular membranes.

phospholipid bilayer: a double layer of phospholipids that forms the basis of all cellular membranes. The phospholipid heads, which are hydrophilic, face the water of extracellular fluid or the cytoplasm; the tails, which are hydrophobic, are buried in the middle of the bilayer.

photic zone: the region of the ocean where light is strong enough to support photosynthesis.

photon (fō´-ton): the smallest unit of light energy.

photopigment (fō´-tō-pig-ment): a chemical substance in photoreceptor cells that, when struck by light, changes in molecular conformation.

photoreceptor: a receptor cell that responds to light; in vertebrates, rods and cones.

photorespiration: a series of reactions in plants in which O_2 replaces CO_2 during the C_3 cycle, preventing carbon fixation; this wasteful process dominates when C_3 plants are forced to close their stomata to prevent water loss.

photosynthesis: the complete series of chemical reactions in which the energy of light is used to synthesize high-energy organic molecules, normally carbohydrates, from low-energy inorganic molecules, normally carbon dioxide and water.

photosystem: in thylakoid membranes, a light-harvesting complex and its associated electron transport system.

phototactic: capable of detecting and responding to light.

phototropism: growth with respect to the direction of light.

pH scale: a scale, with values from 0 to 14, used for measuring the relative acidity of a solution; at pH 7 a solution is neutral, pH 0 to 7 is acidic, and pH 7 to 14 is basic; each unit on the scale represents a tenfold change in H⁺ concentration.

phycocyanin (fī-kō-sī´-uh-nin): a blue or purple pigment that is located in the membranes of chloroplasts and is used as an accessory light-gathering molecule in thylakoid photosystems.

phylogeny (fī-lah´-jen-ē): the evolutionary history of a group of species.

phylum (fī-lum): the taxonomic category of animals and animal-like protists that is contained within a kingdom and consists of related classes.

phytochrome (fī´-tō-krōm): a light-sensitive plant pigment that mediates many plant responses to light, including flowering, stem elongation, and seed germination.

phytoplankton (fī´-tō-plank-ten): photosynthetic protists that are abundant in marine and freshwater environments.

pilus (pil´-us; pl., pili): a hairlike projection that is made of protein, located on the surface of certain bacteria, and is typically used to attach a bacterium to another cell.

pineal gland (pī-nē´-al): a small gland within the brain that secretes melatonin; controls the seasonal reproductive cycles of some mammals.

pinocytosis (pi-nō-sī-tō´-sis): the nonselective movement of extracellular fluid, enclosed within a vesicle formed from the plasma membrane, into a cell.

pioneer: an organism that is among the first to colonize an unoccupied habitat in the first stages of succession.

pit: an area in the cell walls between two plant cells in which secondary walls did not form, such that the two cells are separated only by a relatively thin and porous primary cell wall.

pith: cells forming the center of a root or stem.

pituitary gland: an endocrine gland, located at the base of the brain, that produces several hormones, many of which influence the activity of other glands.

placenta (pluh-sen´-tuh): in mammals, a structure formed by a complex interweaving of the uterine lining and the embryonic membranes, especially the chorion; functions in gas, nutrient, and waste exchange between embryonic and maternal circulatory systems and secretes hormones.

placental (pluh-sen´-tul): referring to a mammal, possessing a placenta (that is, species that are not marsupials or monotremes).

plankton: microscopic organisms that live in marine or freshwater environments; includes phytoplankton and zooplankton.

plant hormone: the plant-regulating chemicals auxin, gibberellins, cytokinins, ethylene, and abscisic acid; somewhat resemble animal hormones in that they are chemicals produced by cells in one location that influence the growth or metabolic activity of other cells, typically some distance away in the plant body.

plaque (plak): a deposit of cholesterol and other fatty substances within the wall of an artery.

plasma: the fluid, noncellular portion of the blood.

plasma cell: an antibody-secreting descendant of a B cell.

plasma membrane: the outer membrane of a cell, composed of a bilayer of phospholipids in which proteins are embedded.

plasmid (plaz´-mid): a small, circular piece of DNA located in the cytoplasm of many bacteria; normally does not carry genes required for the normal functioning of the bacterium but may carry genes that assist bacterial survival in certain environments, such as a gene for antibiotic resistance.

plasmodesma (plaz-mō-dez´-muh; pl., plasmodesmata): a cell-to-cell junction in plants that connects the cytoplasm of adjacent cells.

plasmodial slime mold: see *acellular slime mold*.

plasmodium (plaz-mō´-dē-um): a sluglike mass of cytoplasm containing thousands of nuclei that are not confined within individual cells.

plastid (plas´-tid): in plant cells, an organelle bounded by two membranes that may be involved in photosynthesis (chloroplasts), pigment storage, or food storage.

platelet (plāt´-let): a cell fragment that is formed from megakaryocytes in bone marrow and lacks a nucleus; circulates in the blood and plays a role in blood clotting.

pleiotropy (plē´-ō-trō-pē): a situation in which a single gene influences more than one phenotypic characteristic.

pleural membrane: a membrane that lines the chest cavity and surrounds the lungs.

point mutation: a mutation in which a single base pair in DNA has been changed.

polar body: in oogenesis, a small cell, containing a nucleus but virtually no cytoplasm, produced by the first meiotic division of the primary oocyte.

polar covalent bond: a covalent bond with unequal sharing of electrons, such that one atom is relatively negative and the other is relatively positive.

polar nucleus: in flowering plants, one of two nuclei in the primary endosperm cell of the female gametophyte; formed by the mitotic division of a megaspore.

pollen grain: the male gametophyte of a seed plant; also called *pollen*.

pollination: in flowering plants, when pollen grains land on the stigma of a flower of the same species; in conifers, when pollen grains land within the pollen chamber of a female cone of the same species.

polygenic inheritance: a pattern of inheritance in which the interactions of two or more functionally similar genes determine phenotype.

polymer (pah´-li-mer): a molecule composed of three or more (perhaps thousands) smaller subunits called monomers, which may be identical (for example, the glucose monomers of starch) or different (for example, the amino acids of a protein).

polymerase chain reaction (PCR): a method of producing virtually unlimited numbers of copies of a specific piece of DNA, starting with as little as one copy of the desired DNA.

polyp (pah´-lip): the sedentary, vase-shaped stage in the life cycle of many cnidarians; includes hydra and sea anemones.

polyploidy (pahl´-ē-ploid-ē): having more than two homologous chromosomes of each type.

polysaccharide (pahl-ē-sak´-uh-rīd): a large carbohydrate molecule composed of branched or unbranched chains of repeating monosaccharide subunits, normally glucose or modified glucose molecules; includes starches, cellulose, and glycogen.

pons: a portion of the hindbrain, just above the medulla, that contains neurons that influence sleep and the rate and pattern of breathing.

population: all the members of a particular species within an ecosystem, found in the same time and place and actually or potentially interbreeding.

population bottleneck: a form of genetic drift in which a population becomes extremely small; may lead to differences in allele frequencies as compared with other populations of the species and to a loss in genetic variability.

population cycle: out-of-phase cyclical patterns of predator and prey populations.

population genetics: the study of the frequency, distribution, and inheritance of alleles in a population.

positive feedback: a situation in which a change initiates events that tend to amplify the original change.

post-anal tail: a tail that extends beyond the anus; exhibited by all chordates at some stage of development.

posterior: the tail, hindmost, or rear end of an animal.

posterior pituitary: a lobe of the pituitary gland that is an outgrowth of the hypothalamus and that releases antidiuretic hormone and oxytocin.

postmating isolating mechanism: any structure, physiological function, or developmental abnormality that prevents organisms of two different populations, once mating has occurred, from producing vigorous, fertile offspring.

postsynaptic neuron: at a synapse, the nerve cell that changes its electrical potential in response to a chemical (the neurotransmitter) released by another (presynaptic) cell.

postsynaptic potential (PSP): an electrical signal produced in a postsynaptic cell by transmission across the synapse; it may be excitatory (EPSP), making the cell more likely to produce an action potential, or inhibitory (IPSP), tending to inhibit an action potential.

potential energy: "stored" energy, normally chemical energy or energy of position within a gravitational field.

prairie: a biome, located in the centers of continents, that supports grasses; also called *grassland*.

preadaptation: a feature evolved under one set of environmental conditions that, purely by chance, helps an organism adapt to new environmental conditions.

prebiotic evolution: evolution before life existed; especially, the abiotic synthesis of organic molecules.

precapillary sphincter (sfink´-ter): a ring of smooth muscle between an arteriole and a capillary that regulates the flow of blood into the capillary bed.

predation (pre-dā´-shun): the act of killing and eating another living organism.

predator: an organism that kills and eats other organisms.

premating isolating mechanism: any structure, physiological function, or behavior that prevents organisms of two different populations from exchanging gametes.

pressure-flow theory: a model for the transport of sugars in phloem, by which the movement of sugars into a phloem sieve tube causes water to enter the tube by osmosis, while the movement of sugars out of another part of the same sieve tube causes water to leave by osmosis; the resulting pressure gradient causes the bulk movement of water and dissolved sugars from the end of the tube into which sugar is transported toward the end of the tube from which sugar is removed.

presynaptic neuron: a nerve cell that releases a chemical (the neurotransmitter) at a synapse, causing changes in the electrical activity of another (postsynaptic) cell.

prey: organisms that are killed and eaten by another organism.

primary cell wall: cellulose and other carbohydrates secreted by a young plant cell between the middle lamella and the plasma membrane.

primary consumer: an organism that feeds on producers; an herbivore.

primary endosperm cell: the central cell of the female gametophyte of a flowering plant, containing the polar nuclei (normally two); after fertilization, undergoes repeated mitotic divisions to produce the endosperm of the seed.

primary growth: growth in length and development of the initial structures of plant roots and shoots, due to the cell division of apical meristems and differentiation of the daughter cells.

primary oocyte (ō´-ō-sīt): a diploid cell, derived from the oogonium by growth and differentiation that undergoes meiosis, producing the egg.

primary phloem: phloem in young stems produced from an apical meristem.

primary root: the first root that develops from a seed.

primary spermatocyte (sper-ma´-tō-sīt): a diploid cell, derived from the spermatogonium by growth and differentiation, that undergoes meiosis, producing four sperm.

primary structure: the amino acid sequence of a protein.

primary succession: succession that occurs in an environment, such as bare rock, in which no trace of a previous community was present.

primary xylem: xylem in young stems produced from an apical meristem.

primate: a mammal characterized by the presence of an opposable thumb, forward-facing eyes, and a well-developed cerebral cortex; includes lemurs, monkeys, apes, and humans.

primer pheromone: a chemical produced by an organism that alters the physiological state of another member of the same species.

primitive streak: in reptiles, birds, and mammals, the region of the ectoderm of the two-layered embryonic disc through which cells migrate, forming mesoderm.

prion (prē´-on): a protein that, in mutated form, acts as an infectious agent that causes certain neurodegenerative diseases, including kuru and scrapie.

producer: a photosynthetic organism; an autotroph.

product: an atom or molecule that is formed from reactants in a chemical reaction.

profundal zone: a lake zone in which light is insufficient to support photosynthesis.

progesterone (prō-ge´-ster-ōn): a hormone, produced by the corpus luteum, that promotes the development of the uterine lining in females.

prokaryotic (prō-kar-ē-ot´-ik): referring to cells of the domains Bacteria or Archaea. Prokaryotic cells have genetic material that is not enclosed in a membrane-bound nucleus; they lack other membrane-bound organelles.

prolactin: a hormone, released by the anterior pituitary, that stimulates milk production in human females.

promoter: a specific sequence of DNA to which RNA polymerase binds, initiating gene transcription.

prophase (prō´-fāz): the first stage of mitosis, in which the chromosomes first become visible in the light microscope as thickened, condensed threads and the spindle begins to form; as the spindle is completed, the nuclear envelope breaks apart, and the spindle fibers invade the nuclear region and attach to the kinetochores of the chromosomes. Also, the first stage of meiosis: In meiosis I, the homologous chromosomes pair up and exchange parts at chiasmata; in meiosis II, the spindle re-forms and chromosomes attach to the microtubules.

prostaglandin (pro-stuh-glan´-din): a family of modified fatty acid hormones manufactured by many cells of the body.

prostate gland (pros´-tāt): a gland that produces part of the fluid component of semen; prostatic fluid is basic and contains a chemical that activates sperm movement.

protease (prō´-tē-ās): an enzyme that digests proteins.

protein: an organic molecule composed of one or more chains of amino acids.

Protista (prō-tis´-tuh): a taxonomic kingdom including unicellular, eukaryotic organisms.

protocell: the hypothetical evolutionary precursor of living cells, consisting of a mixture of organic molecules within a membrane.

proton: a subatomic particle that is found in the nuclei of atoms, bears a unit of positive charge, and has a relatively large mass, roughly equal to the mass of the neutron.

protonephridium (prō-tō-nef-rid´-ē-um; pl., **protonephridia):** an excretory system consisting of tubules that have external opening but lack internal openings; for example, the flame-cell system of flatworms.

proto-oncogene (prō-tō-onk´-ō-jēn): a normal cellular gene that can cause cancer if a mutation transforms it into an oncogene.

protostome (prō´-tō-stōm): an animal with a mode of embryonic development in which the coelom is derived from splits in the mesoderm; characteristic of arthropods, annelids, and mollusks.

protozoan (prō-tuh-zō´-an; pl., **protozoa):** a nonphotosynthetic or animal-like protist.

proximal tubule: in nephrons of the mammalian kidney, the portion of the renal tubule just after the Bowman's capsule; receives filtrate from the capsule and is the site where selective secretion and reabsorption between the filtrate and the blood begins.

pseudocoelom (soo´-dō-sēl´-ōm): "false coelom"; a body cavity that has a different embryological origin than a coelom but serves a similar function; found in roundworms.

pseudoplasmodium (soo´-dō-plaz-mō´-dē-um): an aggregation of individual amoeboid cells that form a sluglike mass.

pseudopod (sood´-ō-pod): an extension of the plasma membrane by which certain cells, such as amoebae, locomote and engulf prey.

pulmonary circulation: the circulation of blood from the body, through the right atrium and right ventricle, and to the lungs.

punctuated equilibrium: a model of evolution, stating that morphological change and speciation are rapid, simultaneous events (the "punctuation") separated by long periods during which a species remains unchanged (the "equilibrium").

Punnett square method: an intuitive way to predict the genotypes and phenotypes of offspring in specific crosses.

pupa: a developmental stage in some insect species in which the organism stops moving and feeding and may be encased in a cocoon; occurs between the larval and the adult phases.

pupil: the adjustable opening in the center of the iris, through which light enters the eye.

purine: a type of nitrogen-containing base found in nucleic acids that consists of two fused rings; includes adenine and guanine in both DNA and RNA.

pyloric sphincter (pī-lor´-ik sfink´-ter): a circular muscle, located at the base of the stomach, that regulates the passage of chyme into the small intestine.

pyrimidine: a type of nitrogen-containing base found in nucleic acids that consists of a single ring; includes cytosine (in both DNA and RNA), thymine (in DNA only), and uracil (in RNA only).

pyruvate: a three-carbon moelcule that is formed by glycolysis and then used in fermentation or cellular respiration.

quaternary structure (kwat´-er-nuh-rē): the complex three-dimensional structure of a protein composed of more than one peptide chain.

queen substance: a chemical, produced by a queen bee, that can act as both a primer and a releaser pheromone.

radial symmetry: a body plan in which any plane along a central axis will divide the body into approximately mirror-image halves. Cnidarians and many adult echinoderms have radial symmetry.

radioactive: pertaining to an atom with an unstable nucleus that spontaneously disintegrates, with the emission of radiation.

radiolarian (rā-dē-ō-lar´-ē-un): an aquatic protist (largely marine) characterized by typically elaborate silica shells.

radula (ra´-dū-luh): a ribbon of tissue in the mouth of gastropod mollusks; bears numerous teeth on its outer surface and is used to scrape and drag food into the mouth.

rain shadow: a local dry area created by the modification of rainfall patterns by a mountain range.

random distribution: distribution characteristic of populations in which the probability of finding an individual is equal in all parts of an area.

reactant: an atom or molecule that is used up in a chemical reaction to form a product.

reaction center: in the light-harvesting complex of a photosystem, the chlorophyll molecule to which light energy is transferred by the antenna molecules (light-absorbing pigments); the captured energy ejects an electron from the reaction center chlorophyll, and the electron is transferred to the electron transport system.

receptor: a cell that responds to an environmental stimulus (chemicals, sound, light, pH, and so on) by changing its electrical potential; also, a protein molecule in a plasma membrane that binds to another molecule (hormone, neurotransmitter), triggering metabolic or electrical changes in a cell.

receptor-mediated endocytosis: the selective uptake of molecules from the extracellular fluid by binding to a receptor located at a coated pit on the plasma membrane and pinching off the coated pit into a vesicle that moves into the cytoplasm.

receptor potential: an electrical potential change in a receptor cell, produced in response to the reception of an environmental stimulus (chemicals, sound, light, heat, and so on). The size of the receptor potential is proportional to the intensity of the stimulus.

receptor protein: a protein, located on a membrane (or in the cytoplasm), that recognizes and binds to specific molecules. Binding by receptor proteins typically triggers a response by a cell, such as endocytosis, increased metabolic rate, or cell division.

recessive: an allele that is expressed only in homozygotes and is completely masked in heterozygotes.

recognition protein: a protein or glycoprotein protruding from the outside surface of a plasma membrane that identifies a cell as belonging to a particular species, to a specific individual of that species, and in many cases to one specific organ within the individual.

recombinant DNA: DNA that has been altered by the recombination of genes from a different organism, typically from a different species.

recombination: the formation of new combinations of the different alleles of each gene on a chromosome; the result of crossing over.

rectum: the terminal portion of the vertebrate digestive tube, where feces are stored until they can be eliminated.

reflex: a simple, stereotyped movement of part of the body that occurs automatically in response to a stimulus.

regeneration: the regrowth of a body part after loss or damage; also, asexual reproduction by means of the regrowth of an entire body from a fragment.

releaser: a stimulus that triggers a fixed action pattern.

releaser pheromone: a chemical produced by one organism that alters the behavior of another member of the same species.

releasing hormone: a hormone, secreted by the hypothalamus, that causes the release of specific hormones by the anterior pituitary.

renal artery: the artery carrying blood to each kidney.

renal cortex: the outer layer of the kidney; where nephrons are located.

renal medulla: the layer of the kidney just inside the renal cortex; where loops of Henle produce a highly concentrated interstitial fluid, important in the production of concentrated urine.

renal pelvis: the inner chamber of the kidney; where urine from the collecting ducts accumulates before it enters the ureters.

renal vein: the vein carrying cleansed blood away from each kidney.

renin: an enzyme that is released (in mammals) when blood pressure and/or sodium concentration in the blood drops below a set point; initiates a cascade of events that restores blood pressure and sodium concentration.

replacement-level fertility (RLF): the average birth rate at which a reproducing population exactly replaces itself during its lifetime.

replication: the copying of the double-stranded DNA molecule, producing two identical DNA double helices.

reproductive isolation: the failure of organisms of one population to breed successfully with members of another; may be due to premating or postmating isolating mechanisms.

reservoir: the major source and storage site of a nutrient in an ecosystem, normally in the abiotic portion.

resource partitioning: the coexistence of two species with similar requirements, each occupying a smaller niche than either would if it were by itself; a means of minimizing their competitive interactions.

respiratory center: a cluster of neurons, located in the medulla of the brain, that sends rhythmic bursts of nerve impulses to the respiratory muscles, resulting in breathing.

resting potential: a negative electrical potential in unstimulated nerve cells.

restriction enzyme: an enzyme, normally isolated from bacteria, that cuts double-stranded DNA at a specific nucleotide sequence; the nucleotide sequence that is cut differs for different restriction enzymes.

restriction fragment: a piece of DNA that has been isolated by cleaving a larger piece of DNA with restriction enzymes.

restriction fragment length polymorphism (RFLP): a difference in the length of restriction fragments, produced by cutting samples of DNA from different individuals of the same species with the same set of restriction enzymes; the result of differences in nucleotide sequences among individuals of the same species.

reticular formation (reh-tik´-ū-lar): a diffuse network of neurons extending from the hindbrain, through the midbrain, and into the lower reaches of the forebrain; involved in filtering sensory input and regulating what information is relayed to conscious brain centers for further attention.

retina (ret´-in-uh): a multilayered sheet of nerve tissue at the rear of camera-type eyes, composed of photoreceptor cells plus associated nerve cells that refine the photoreceptor information and transmit it to the optic nerve.

retrovirus: a virus that uses RNA as its genetic material. When it invades a eukaryotic cell, a retrovirus "reverse transcribes" its RNA into DNA, which then directs the synthesis of more viruses, using the transcription and translation machinery of the cell.

reverse transcriptase: an enzyme found in retroviruses that catalyzes the synthesis of DNA from an RNA template.

Rh factor: a protein on the red blood cells of some people (Rh-positive) but not others (Rh-negative); the exposure of Rh-negative individuals to Rh-positive blood triggers the production of antibodies to Rh-positive blood cells.

rhizoid (rī´-zoid): a rootlike structure found in bryophytes that anchors the plant and absorbs water and nutrients from the soil.

rhizome (rī´-zōm): an undergound stem, usually horizontal, that stores food.

rhythm method: a contraceptive method involving abstinence from intercourse during ovulation.

ribonucleic acid (rī-bō-noo-klā´-ik; RNA): a molecule composed of ribose nucleotides, each of which consists of a phosphate group, the sugar ribose, and one of the bases adenine, cytosine, guanine, or uracil; transfers hereditary instructions from the nucleus to the cytoplasm; also the genetic material of some viruses.

ribosomal RNA (rRNA): a type of RNA that combines with proteins to form ribosomes.

ribosome: an organelle consisting of two subunits, each composed of ribosomal RNA and protein; the site of protein synthesis, during which the sequence of bases of messenger RNA is translated into the sequence of amino acids in a protein.

ribozyme: an RNA molecule that can catalyze certain chemical reactions, especially those involved in the synthesis and processing of RNA itself.

RNA polymerase: in RNA synthesis, an enzyme that catalyzes the bonding of free RNA nucleotides into a continuous strand, using RNA nucleotides that are complementary to those of a strand of DNA.

rod: a rod-shaped photoreceptor cell in the vertebrate retina, sensitive to dim light but not involved in color vision; see also *cone.*

root: the part of the plant body, normally underground, that provides anchorage, absorbs water and dissolved nutrients and transports them to the stem, produces some hormones, and in some plants serves as a storage site for carbohydrates.

root cap: a cluster of cells at the tip of a growing root, derived from the apical meristem; protects the growing tip from damage as it burrows through the soil.

root hair: a fine projection from an epidermal cell of a young root that increases the absorptive surface area of the root.

root system: the part of a plant, normally below ground, that anchors the plant in the soil, absorbs water and minerals, stores food, transports water, minerals, sugars, and hormones, and produces certain hormones.

rough endoplasmic reticulum: endoplasmic reticulum lined on the outside with ribosomes.

runner: a horizontally growing stem that may develop new plants at nodes that touch the soil.

sac fungus: a fungus of the division Ascomycota, whose members form spores in a saclike case called an ascus.

sapwood: young xylem that transports water and minerals in a tree trunk.

saprobe (sap´-rōb): an organism that derives its nutrients from the bodies of dead organisms.

sarcodine (sar-kō´-dīn): a nonphotosynthetic protist (protozoan) characterized by the ability to form pseudopodia; some sarcodines, such as amoebae, are naked, whereas others have elaborate shells.

sarcomere (sark´-ō-mēr): the unit of contraction of a muscle fiber; a subunit of the myofibril, consisting of actin and myosin filaments and bounded by Z lines.

sarcoplasmic reticulum (sark´-ō-plas´-mik re-tik´-ū-lum): the specialized endoplasmic reticulum in muscle cells; forms interconnected hollow tubes. The sarcoplasmic reticulum stores calcium ions and releases them into the interior of the muscle cell, initiating contraction.

saturated: referring to a fatty acid with as many hydrogen atoms as possible bonded to the carbon backbone; a fatty acid with no double bonds in its carbon backbone.

savanna: a biome that is dominated by grasses and supports scattered trees and thorny scrub forests; typically has a rainy season in which all the year's precipitation falls.

scientific method: a rigorous procedure for making observations of specific phenomena and searching for the order underlying those phenomena; consists of four operations: observation, hypothesis, experiment, and conclusion.

scientific name: the name of an organism formed from the two smallest major taxonomic categories—the genus and the species.

scientific theory: a general explanation of natural phenomena developed through extensive and reproducible observations; more general and reliable than a hypothesis.

sclera: a tough, white connective tissue layer that covers the outside of the eyeball and forms the white of the eye.

sclerenchyma (skler-en´-ki-muh): a plant cell type with thick, hardened secondary cell walls, that normally dies as the last stage of differentiation and both supports and protects the plant body.

scramble competition: a free-for-all scramble for limited resources among individuals of the same species.

scrotum (skrō´-tum): the pouch of skin containing the testes of male mammals.

S-curve: the S-shaped growth curve that describes a population of long-lived organisms introduced into a new area; consists of an initial period of exponential growth, followed by decreasing growth rate, and, finally, relative stability around a growth rate of zero.

sebaceous gland (se-bā´-shus): a gland in the dermis of skin, formed from epithelial tissue, that produces the oily substance sebum, which lubricates the epidermis.

secondary cell wall: a thick layer of cellulose and other polysaccharides secreted by certain plant cells between the primary cell wall and the plasma membrane.

secondary consumer: an organism that feeds on primary consumers; a carnivore.

secondary growth: growth in the diameter of a stem or root due to cell division in lateral meristems and differentiation of their daughter cells.

secondary oocyte (ō´-ō-sīt): a large haploid cell derived from the first meiotic division of the diploid primary oocyte.

secondary phloem: phloem produced from the cells that arise toward the outside of the vascular cambium.

secondary spermatocyte (sper-ma´-tō-sīt): a large haploid cell derived by meiosis I from the diploid primary spermatocyte.

secondary structure: a repeated, regular structure assumed by protein chains held together by hydrogen bonds; for example, a helix.

secondary succession: succession that occurs after an existing community is disturbed—for example, after a forest fire; much more rapid than primary succession.

secondary xylem: xylem produced from cells that arise at the inside of the vascular cambium.

second law of thermodynamics: the principle of physics that states that any change in an isolated system causes the quantity of concentrated, useful energy to decrease and the amount of randomness and disorder (entropy) to increase.

second messenger: an intracellular chemical, such as cyclic AMP, that is synthesized or released within a cell in response to the binding of a hormone or neurotransmitter (the first messenger) to receptors on the cell surface; brings about specific changes in the metabolism of the cell.

secretin: a hormone, produced by the small intestine, that stimulates the production and release of digestive secretions by the pancreas and liver.

seed: the reproductive structure of a seed plant; protected by a seed coat; contains an embryonic plant and a supply of food for it.

seed coat: the thin, tough, and waterproof outermost covering of a seed, formed from the integuments of the ovule.

segmentation (seg-men-tā´-shun): an animal body plan in which the body is divided into repeated, typically similar units.

segmentation movement: a contraction of the small intestine that results in the mixing of partially digested food and digestive enzymes. Segmentation movements also bring nutrients into contact with the absorptive intestinal wall.

segregation: see *law of segregation.*

self-fertilization: the union of sperm and egg from the same individual.

selfish gene: the concept that genes promote their own survival in individuals through innate self-sacrificing behavior that enhances the survival of others that carry the same genes; helps explain the evolution of altruism.

semen: the sperm-containing fluid produced by the male reproductive tract.

semiconservative replication: the process of replication of the DNA double helix; the two DNA strands separate, and each is used as a template for the synthesis of a complementary DNA strand. Consequently, each daughter double helix consists of one parental strand and one new strand.

semilunar valve: a paired valve between the ventricles of the heart and the pulmonary artery and aorta; prevents the backflow of blood into the ventricles when they relax.

seminal vesicle: in male mammals, a gland that produces a basic, fructose-containing fluid that forms part of the semen.

seminiferous tubule (sem-i-ni´-fer-us): in the vertebrate testis, a series of tubes in which sperm are produced.

senescence: in plants, a specific aging process, typically including deterioration and the dropping of leaves and flowers.

sensitive period: the particular stage in an animal's life during which it imprints.

sensory neuron: a nerve cell that responds to a stimulus from the internal or external environment.

sensory receptor: a cell (typically, a neuron) specialized to respond to particular internal or external environmental stimuli by producing an electrical potential.

sepal (sē´-pul): the set of modified leaves that surround and protect a flower bud, typically opening into green, leaflike structures when the flower blooms.

septum (pl., septa): a partition that separates the fungal hypha into individual cells; pores in septa allow the transfer of materials between cells.

serotonin (ser-uh-tō´-nin): in the central nervous system, a neurotransmitter that is involved in mood, sleep, and the inhibition of pain.

Sertoli cell: in the seminiferous tubule, a large cell that regulates spermatogenesis and nourishes the developing sperm.

sessile (ses´-ul): not free to move about, usually permanently attached to a surface.

severe combined immune deficiency (SCID): a disorder in which no immune cells, or very few, are formed; the immune system is incapable of responding properly to invading disease organisms, and the individual is very vulnerable to common infections.

sex chromosome: one of the pair of chromosomes that differ between sexes and normally determine the sex of an individual; in humans, females have similar sex chromosomes (XX) whereas males have dissimilar ones (XY).

sex-linked: referring to a pattern of inheritance characteristic of genes located on one type of sex chromosome (for example, X) and not found on the other type (for example, Y); also called X-linked. In sex-linked inheritance, traits are controlled by genes carried on the X chromosome; females show the dominant trait unless they are homozygous recessive, whereas males express whichever allele is on their single X chromosome.

sexually transmitted disease (STD): a disease that is passed from person to person by sexual contact.

sexual recombination: during sexual reproduction, the formation of new combinations of alleles in offspring as a result of the inheritance of one homologous chromosome from each of two genetically distinct parents.

sexual reproduction: a form of reproduction in which genetic material from two parent organisms is combined in the offspring; normally, two haploid gametes fuse to form a diploid zygote.

sexual selection: a type of natural selection in which the choice of mates by one sex is the selective agent.

shoot system: all the parts of a vascular plant exclusive of the root; normally aboveground, consisting of stem, leaves, buds, and (in season) flowers and fruits; functions include photosynthesis, transport of materials, reproduction, and hormone synthesis.

short-day plant: a plant that will flower only if the length of daylight is shorter than some species-specific duration.

sickle-cell anemia: a recessive disease caused by a single amino acid substitution in the hemoglobin molecule. Sickle-cell hemoglobin molecules tend to cluster together, distorting the shape of red blood cell shape and causing them to break and clog capillaries.

sieve plate: in plants, a structure between two adjacent sieve-tube elements in phloem, where holes formed in the primary cell walls interconnect the cytoplasm of the elements; in echinoderms, the opening through which water enters the water-vascular system.

sieve tube: in phloem, a single strand of sieve-tube elements that transports sugar solutions.

sieve-tube element: one of the cells of a sieve tube, which form the phloem.

simple diffusion: the diffusion of water, dissolved gases, or lipid-soluble molecules

through the phospholipid bilayer of a cellular membrane.

single covalent bond: a covalent bond in which two atoms share one pair of electrons.

sink: in plants, any structure that uses up sugars or converts sugars to starch and toward which phloem fluids will flow.

sinoatrial (SA) node (sī´-nō-āt´-rē-ul): a small mass of specialized muscle in the wall of the right atrium; generates electrical signals rhythmically and spontaneously and serves as the heart's pacemaker.

skeletal muscle: the type of muscle that is attached to and moves the skeleton and is under the direct, normally voluntary, control of the nervous system; also called *striated muscle.*

skeleton: a supporting structure for the body, on which muscles act to change the body configuration; may be external or internal.

skin: the tissue that makes up the outer surface of an animal body.

slime layer: a sticky polysaccharide or protein coating that some disease-causing bacteria secrete outside their cell wall; helps the cells aggregate and stick to smooth surfaces.

small intestine: the portion of the digestive tract, located between the stomach and large intestine, in which most digestion and absorption of nutrients occur.

smooth endoplasmic reticulum: endoplasmic reticulum without ribosomes.

smooth muscle: the type of muscle that surrounds hollow organs, such as the digestive tract, bladder, and blood vessels; normally not under voluntary control.

sodium–potassium pump: in nerve cell plasma membranes, a set of active-transport molecules that use the energy of ATP to pump sodium ions out of the cell and potassium ions in, maintaining the concentration gradients of these ions across the membrane.

solvent: a liquid capable of dissolving (uniformly dispersing) other substances in itself.

somatic nervous system: that portion of the peripheral nervous system that controls voluntary movement by activating skeletal muscles.

source: in plants, any structure that actively synthesizes sugar and away from which phloem fluid will be transported.

spawning: a method of external fertilization in which male and female parents shed gametes into the water, and sperm must swim through the water to reach the eggs.

speciation: the process whereby two populations achieve reproductive isolation.

species (spē´-sēs): all of the organisms that are potentially capable of interbreeding under natural conditions or, if asexually reproducing, are more closely related to one another than to other organisms within a given genus; the smallest major taxonomic category.

specific heat: the amount of energy required to raise the temperature of 1 gram of a substance by 1°C.

sperm: the haploid male gamete, normally small, motile, and containing little cytoplasm.

spermatid: a haploid cell derived from the secondary spermatocyte by meiosis II; differentiates into the mature sperm.

spermatogenesis: the process by which sperm cells form.

spermatogonium (pl., spermatogonia): a diploid cell, lining the walls of the seminiferous tubules, that gives rise to a primary spermatocyte.

spermatophore: in a variation on internal fertilization in some animals, the males package their sperm in a container that can be inserted into the female reproductive tract.

spermicide: a sperm-killing chemical; used for contraceptive purposes.

spicule (spik´-ūl): a subunit of the endoskeleton of sponges that is made of protein, silica, or calcium carbonate.

spike initiation zone: on a neuron, the site where the action potential begins; where the axon leaves the cell body.

spinal cord: the part of the central nervous system of vertebrates that extends from the base of the brain to the hips and is protected by the bones of the vertebral column; contains the cell bodies of motor neurons that form synapses with skeletal muscles, the circuitry for some simple reflex behaviors, and axons that communicate with the brain.

spindle fiber: a football-shaped array of microtubules that moves the chromosomes to opposite poles of a cell during anaphase of meiosis and mitosis.

spiracle (spi´-ruh-kul): an opening in the abdominal segment of insects through which air enters the tracheae.

spirillum (spi´-ril-um; pl., spirilla): a spiral-shaped bacterium.

spleen: an organ of the lymphatic system in which lymphocytes are produced and blood is filtered past lymphocytes and macrophages, which remove foreign particles and aged red blood cells.

spongy bone: porous, lightweight bone tissue in the interior of bones; the location of bone marrow.

spongy cell: an irregularly shaped mesophyll cell, containing chloroplasts, located just above the lower epidermis of a leaf.

spontaneous generation: the proposal that living organisms can arise from nonliving matter.

sporangium (spor-an´-jē-um; pl., sporangia): a structure in which spores are produced.

spore: a haploid reproductive cell capable of developing into an adult without fusing with another cell; in the alternation-of-generation life cycle of plants, a haploid cell that is produced by meiosis and then undergoes repeated mitotic divisions and differentiation of daughter cells to produce the gametophyte, a multicellular, haploid organism.

sporophyte (spor´-ō-fīt): the diploid form of a plant that produces haploid, asexual spores through meiosis.

sporozoan (spor-ō-zō´-un): a parasitic protist with a complex life cycle, typically involving more than one host; named for their ability to form infectious spores. A well-known sporozoan (genus *Plasmodium*) causes malaria.

stabilizing selection: a type of natural selection in which those organisms that display extreme phenotypes are selected against.

stamen (stā´-men): the male reproductive structure of a flower, consisting of a filament and an anther, in which pollen grains develop.

starch: a polysaccharide that is composed of branched or unbranched chains or glucose molecules; used by plants as a carbohydrate-storage molecule.

start codon: a codon in messenger RNA that signals the beginning of protein synthesis on a ribosome.

startle coloration: a form of mimicry in which a color pattern (in many cases resembling large eyes) can be displayed suddenly by a prey organism when approached by a predator.

statolith: in plants, a starch-filled organelle in the stem and root cap that sinks to the downward side of cells and acts as a sensor for gravity.

stem: the portion of the plant body, normally located above ground, that bears leaves and reproductive structures such as flowers and fruit.

sterilization: a generally permanent method of contraception in which the pathways through which the sperm (vas deferens) or egg (oviducts) must travel are interrupted; the most common form of contraception.

steroid: see *steroid hormone*.

steroid hormone: a class of hormone whose chemical structure (four fused carbon rings with various functional groups) resembles cholesterol; steroids, which are lipids, are secreted by the ovaries and placenta, the testes, and the adrenal cortex.

stigma (stig´-muh): the pollen-capturing tip of a carpel.

stoma (stō´-muh; pl., stomata): an adjustable opening in the epidermis of a leaf, surrounded by a pair of guard cells, that regulates the diffusion of carbon dioxide and water into and out of the leaf.

stomach: the muscular sac between the esophagus and small intestine where food is stored and mechanically broken down and in which protein digestion begins.

stop codon: a codon in messenger RNA that stops protein synthesis and causes the completed protein chain to be released from the ribosome.

striated muscle: see *skeletal muscle*.

stroke: an interruption of blood flow to part of the brain caused by the rupture of an artery or the blocking of an artery by a blood clot. Loss of blood supply leads to rapid death of the area of the brain affected.

stroma (strō´-muh): the semi-fluid material inside chloroplasts in which the grana are embedded.

style: a stalk connecting the stigma of a carpel with the ovary at its base.

subatomic particle: the particles of which atoms are made: electrons, protons, and neutrons.

subclimax: a community in which succession is stopped before the climax community is reached and is maintained by regular disturbances—for example, tallgrass prairie maintained by periodic fires.

substrate: the atoms or molecules that are the reactants for an enzyme-catalyzed chemical reaction.

subunit: a small organic molecule, several of which may be bonded together to form a larger molecule. See also *monomer*.

succession (suk-seh´-shun): a structural change in a community and its nonliving environment over time. Community changes alter the ecosystem in ways that favor competitors, and species replace one another in a somewhat predictable manner until a stable, self-sustaining climax community is reached.

sucrose: a disaccharide composed of glucose and fructose.

sugar: a simple carbohydrate molecule, either a monosaccharide or a disaccharide.

suppressor T cell: a type of T cell that depresses the response of other immune cells to foreign antigens.

surface tension: the property of a liquid to resist penetration by objects at its interface with the air, due to cohesion between molecules of the liquid.

survivorship curve: a curve resulting when the number of individuals of each age in a population is graphed against their age, usually expressed as a percentage of their maximum life span.

symbiosis (sim´-bī-ō´sis): a close interaction between organisms of different species over an extended period. Either or both species may benefit from the association, or (in the case of parasitism) one of the participants is harmed. Symbiosis includes parasitism, mutualism, and commensalism.

symbiotic: referring to an ecological relationship based on symbiosis.

sympathetic division: the division of the autonomic nervous system that produces largely involuntary responses that prepare the body for stressful or highly energetic situations.

sympatric speciation (sim-pat´-rik): speciation that occurs in populations that are not physically divided; normally due to ecological isolation or chromosomal aberrations (such as polyploidy).

synapse (sin´-aps): the site of communication between nerve cells. At a synapse, one cell (presynaptic) normally releases a chemical (the neurotransmitter) that changes the electrical potential of the second (postsynaptic) cell.

synaptic cleft: a tiny space that separates the presynaptic and postsynaptic neurons.

synaptic terminal: a swelling at the branched ending of an axon; where the axon forms a synapse.

syphilis (si´-ful-is): a sexually transmitted bacterial infection of the reproductive organs; if untreated, can damage the nervous and circulatory systems.

systematics: the branch of biology concerned with reconstructing phylogenies and with naming and classifying species.

systemic circulation: the circulation of blood from the left ventricle, through the rest of the body (except the lungs), and back to the heart.

taiga (tī´-guh): a biome with long, cold winters and only a few months of warm weather; dominated by evergreen coniferous trees; also called *northern coniferous forest.*

taproot system: a root system, commonly found in dicots, that consists of a long, thick main root and many smaller lateral roots, all of which grow from the primary root.

target cell: a cell on which a particular hormone exerts its effect.

taste: a chemical sense for substances dissolved in water or saliva; in mammals, perceptions of sweet, sour, bitter, or salt produced by the stimulation of receptors on the tongue.

taste bud: a cluster of taste receptor cells and supporting cells that is located in a small pit beneath the surface of the tongue and that communicates with the mouth through a small pore. The human tongue has about 10,000 taste buds.

taxis (taks´-is; pl., **taxes):** an innate behavior that is a directed movement of an organism toward or away from a stimulus such as heat, light, or gravity.

taxonomy (tax-on-uh-mē): the science by which organisms are classified into hierarchically arranged categories that reflect their evolutionary relationships.

Tay-Sachs disease: a recessive disease caused by a deficiency in enzymes that regulate lipid breakdown in the brain.

T cell: a type of lymphocyte that recognizes and destroys specific foreign cells or substances or that regulates other cells of the immune system.

T-cell receptor: a protein receptor, located on the surface of a T cell, that binds a specific antigen and triggers the immune response of the T cell.

tectorial membrane (tek-tor´-ē-ul): one of the membranes of the cochlea in which the hairs of the hair cells are embedded. In sound reception, movement of the basilar membrane relative to the tectorial membrane bends the cilia.

telophase (tēl´-ō-fāz): in mitosis, the final stage, in which a nuclear envelope re-forms around each new daughter nucleus, the spindle fibers disappear, and the chromosomes relax from their condensed form; in meiosis I, the stage during which the spindle fibers disappear and the chromosomes normally relax from their condensed form; in meiosis II, the stage during which chromosomes relax into their extended state, nuclear envelopes reform, and cytokinesis occurs.

temperate deciduous forest: a biome in which winters are cold and summer rainfall is sufficient to allow enough moisture for trees to grow and shade out grasses.

temperate rain forest: a biome in which there is no shortage of liquid water year-round and that is dominated by conifers.

template strand: the strand of the DNA double helix from which RNA is transcribed.

temporal isolation: the inability of organisms to mate if they have significantly different breeding seasons.

tendon: a tough connective tissue band connecting a muscle to a bone.

tendril: a slender outgrowth of a stem that coils about external objects and supports the stem; normally a modified leaf or branch.

tentacle (ten´-te-kul): an elongate, extensible projection of the body of cnidarians and cephalopod mollusks that may be used for grasping, stinging, and immobilizing prey, and locomotion.

terminal bud: meristem tissue and surrounding leaf primordia that are located at the tip of the plant shoot.

territoriality: the defense of an area in which important resources are located.

tertiary consumer (ter´-shē-er-ē): a carnivore that feeds on other carnivores (secondary consumers).

tertiary structure (ter´-shē-er-ē): the complex three-dimensional structure of a single peptide chain; held in place by disulfide bonds between cysteines.

test cross: a breeding experiment in which an individual showing the dominant phenotype is mated with an individual that is homozygous recessive for the same gene. The ratio of offspring with dominant versus recessive phenotypes can be used to determine the genotype of the phenotypically dominant individual.

testis (pl., testes): the gonad of male mammals.

testosterone: in vertebrates, a hormone produced by the interstitial cells of the testis; stimulates spermatogenesis and the development of male secondary sex characteristics.

tetany: smooth, sustained maximal contraction of a muscle in response to rapid firing by its motor neuron.

thalamus: the part of the forebrain that relays sensory information to many parts of the brain.

theory: in science, an explanation for natural events that is based on a large number of observations and is in accord with scientific principles, especially causality.

thermoacidophile (ther-mō-a-sid´-eh-fīl): an archaean that thrives in hot, acidic environments.

thermoreceptor: a sensory receptor that responds to changes in temperature.

thick filament: in the sarcomere, a bundle of myosin that interacts with thin filaments, producing muscle contraction.

thin filament: in the sarcomere, a protein strand that interacts with thick filaments, producing muscle contraction; composed primarily of actin, with accessory proteins.

thorax: the segment between the head and abdomen in animals with segmentation; the segment to which structures used in locomotion are attached.

thorn: a hard, pointed outgrowth of a stem; normally a modified branch.

threshold: the electrical potential (less negative than the resting potential) at which an action potential is triggered.

thrombin: an enzyme produced in the blood as a result of injury to a blood vessel; catalyzes the production of fibrin, a protein that assists in blood clot formation.

thylakoid (thī´-luh-koid): a disk-shaped, membranous sac found in chloroplasts, the membranes of which contain the photosystems and ATP-synthesizing enzymes used in the light-dependent reactions of photosynthesis.

thymosin: a hormone, secreted by the thymus, that stimulates the maturation of cells of the immune system.

thymus (thī´-mus): an organ of the lymphatic system that is located in the upper chest in front of the heart and that secretes thymosin, which stimulates lymphocyte maturation; begins to degenerate at puberty and has little function in the adult.

thyroid gland: an endocrine gland, located in front of the larynx in the neck, that secretes the hormones thyroxine (affecting metabolic rate) and calcitonin (regulating calcium ion concentration in the blood).

thyroid-stimulating hormone (TSH): a hormone, released by the anterior pituitary, that stimulates the thyroid gland to release hormones.

thyroxine (thī-rox´-in): a hormone, secreted by the thyroid gland, that stimulates and regulates metabolism.

tight junction: a type of cell-to-cell junction in animals that prevents the movement of materials through the spaces between cells.

tissue: a group of (normally similar) cells that together carry out a specific function; for example, muscle; may include extracellular material produced by its cells.

tonsil: a patch of lymphatic tissue consisting of connective tissue that contains many lymphocytes; located in the pharynx and throat.

trachea (trā´-kē-uh): in birds and mammals, a rigid but flexible tube, supported by rings of cartilage, that conducts air between the larynx and the bronchi; in insects, an elaborately branching tube that carries air from openings called spiracles near each body cell.

tracheid (trā´-kē-id): an elongated xylem cell with tapering ends that contains pits in the cell wall; forms tubes that transport water.

tracheophyte (trā´-kē-ō-fīt): a plant that has conducting vessels; a vascular plant.

transcription: the synthesis of an RNA molecule from a DNA template.

transducer: a device that converts signals from one form to another. Sensory receptors are transducers that convert environmental stimuli, such as heat, light, or vibration, into electrical signals (such as action potentials) recognized by the nervous system.

transfer RNA (tRNA): a type of RNA that binds to a specific amino acid by means of a set of three bases (the anticodon) on the tRNA that are complementary to the mRNA codon for that amino acid; carries its amino acid to a ribosome during protein synthesis, recognizes a codon of mRNA, and positions its amino acid for incorporation into the growing protein chain.

transformation: a method of acquiring new genes, whereby DNA from one bacterium (normally released after the death of the bacterium) becomes incorporated into the DNA of another, living, bacterium.

transgenic: referring to an animal or a plant that expresses DNA derived from another species.

translation: the process whereby the sequence of bases of messenger RNA is converted into the sequence of amino acids of a protein.

transpiration (trans´-per-ā-shun): the evaporation of water through the stomata of a leaf.

transport protein: a protein that regulates the movement of water-soluble molecules through the plasma membrane.

transverse (T) tubule: a deep infolding of the muscle plasma membrane; conducts the action potential inside a cell.

trial-and-error learning: a process by which adaptive responses are learned through rewards or punishments provided by the environment.

trichocyst (trik´-eh-sist): a stinging organelle of some protists.

trichomoniasis (trik-ō-mō-nī´-uh-sis): a sexually transmitted disease, caused by the protist *Trichomonas*, that causes inflammation of the mucous membranes than line the urinary tract and genitals.

tricuspid valve: the valve between the right ventricle and the right atrium of the heart.

triglyceride (trī-glis´-er-īd): a lipid composed of three fatty-acid molecules bonded to a single glycerol molecule.

triple covalent bond: a covalent bond that occurs when two atoms share three pairs of electrons.

trisomy 21: see *Down syndrome*.

trisomy X: a condition of females who have three X chromosomes instead of the normal two; most such women are phenotypically normal and are fertile.

trophic level: literally, "feeding level"; the categories of organisms in a community, and the position of an organism in a food chain, defined by the organism's source of energy; includes producers, primary consumers, secondary consumers, and so on.

tropical deciduous forest: a biome with pronounced wet and dry seasons and plants that must shed their leaves during the dry season to minimize water loss.

tropical rain forest: a biome with evenly warm, evenly moist conditions; dominated by broadleaf evergreen trees; the most diverse biome.

true-breeding: pertaining to an individual all of whose offspring produced through self-fertilization are identical to the parental type. True-breeding individuals are homozygous for a given trait.

tubal ligation: a surgical procedure in which a woman's oviducts are cut so that the egg cannot reach the uterus, making her infertile.

tube cell: the outermost cell of a pollen grain; digests a tube through the tissues of the carpel, ultimately penetrating into the female gametophyte.

tube foot: a cylindrical extension of the water-vascular system of echinoderms; used for locomotion, grasping food, and respiration.

tubular reabsorption: the process by which cells of the tubule of the nephron remove water and nutrients from the filtrate within the tubule and return those substances to the blood.

tubular secretion: the process by which cells of the tubule of the nephron remove additional wastes from the blood, actively secreting those wastes into the tubule.

tubule (toob´-ūl): the tubular portion of the nephron; includes a proximal portion, the loop of Henle, and a distal portion. Urine is formed from the blood filtrate as it passes through the tubule.

tumor: a mass that forms in otherwise normal tissue; caused by the uncontrolled growth of cells.

tumor-suppressor gene: a gene that encodes information for a protein that inhibits cancer formation, probably by regulating cell division in some way.

tundra: a biome with severe weather conditions (extreme cold and wind and little rainfall) that cannot support trees.

turgor pressure: pressure developed within a cell (especially the central vacuole of plant cells) as a result of osmotic water entry.

Turner syndrome: a set of characteristics typical of a woman with only one X chromosome: sterile, with a tendency to be very short and to lack normal female secondary sexual characteristics.

tympanic membrane (tim-pan´-ik): the eardrum; a membrane, stretched across the opening of the ear, that transmits vibration of sound waves to bones of the middle ear.

unicellular: single-celled; most members of the domains Bacteria and Archaea and the kingdom Protista are unicellular.

uniform distribution: the distribution characteristic of a population with a relatively regular spacing of individuals, commonly as a result of territorial behavior.

uniformitarianism: the hypothesis that Earth developed gradually through natural processes, similar to those at work today, that occur over long periods of time.

unsaturated: referring to a fatty acid with fewer than the maximum number of hydrogen atoms bonded to its carbon backbone; a fatty acid with one or more double bonds in its carbon backbone.

upwelling: an upward flow that brings cold, nutrient-laden water from the ocean depths to the surface; occurs along western coastlines.

urea (ū-rē´-uh): a water-soluble, nitrogen-containing waste product of amino acid breakdown; one of the principal components of mammalian urine.

ureter (ū´-re-ter): a tube that conducts urine from each kidney to the bladder.

urethra (ū-rē´-thruh): the tube leading from the urinary bladder to the outside of the body; in males, the urethra also receives sperm from the vas deferens and conducts both sperm and urine (at different times) to the tip of the penis.

uric acid (ūr´-ik): a nitrogen-containing waste product of amino acid breakdown; a relatively insoluble white crystal excreted by birds, reptiles, and insects.

urine: the fluid produced and excreted by the urinary system of vertebrates; contains water and dissolved wastes, such as urea.

uterus: in female mammals, the part of the reproductive tract that houses the embryo during pregnancy.

vaccination: an injection into the body that contains antigens characteristic of a particular disease organism and that stimulates an immune response.

vacuole (vak´-ū-ōl): a vesicle that is typically large and consists of a single membrane enclosing a fluid-filled space.

vagina: the passageway leading from the outside of a female mammal's body to the cervix of the uterus.

variable: a condition, particularly in a scientific experiment, that is subject to change.

variable region: the part of an antibody molecule that differs among antibodies; the ends of the variable regions of the light and heavy chains form the specific binding site for antigens.

vascular: describing tissues that contain vessels for transporting liquids.

vascular bundle (vas´-kū-lar): a strand of xylem and phloem in leaves and stems; in leaves, commonly called a *vein*.

vascular cambium: a lateral meristem that is located between the xylem and phloem of a woody root or stem and that gives rise to secondary xylem and phloem.

vascular cylinder: the centrally located conducting tissue of a young root, consisting of primary xylem and phloem.

vascular tissue system: a plant tissue system consisting of xylem (which transports water and minerals from root to shoot) and phloem (which transports water and sugars throughout the plant).

vas deferens (vaz de´-fer-enz): the tube connecting the epididymis of the testis with the urethra.

vasectomy: a surgical procedure in which a man's vas deferens are cut, preventing sperm from reaching the penis during ejaculation, thereby making him infertile.

vector: a carrier that introduces foreign genes into cells.

vein: in vertebrates, a large-diameter, thin-walled vessel that carries blood from venules back to the heart; in vascular plants, a vascular bundle, or a strand of xylem and phloem in leaves.

ventral (ven´-trul): the lower side or underside of an animal whose head is oriented forward.

ventricle (ven´-tre-kul): the lower muscular chamber on each side of the heart, which pumps blood out through the arteries. The right ventricle sends blood to the lungs; the left ventricle pumps blood to the rest of the body.

venule (ven´-ūl): a narrow vessel with thin walls that carries blood from capillaries to veins.

vertebral column (ver-tē´-brul): a column of serially arranged skeletal units (the vertebrae) that enclose the nerve cord in vertebrates; the backbone.

vertebrate: an animal that possesses a vertebral column.

vesicle (ves´-i-kul): a small, membrane-bound sac within the cytoplasm.

vessel: a tube of xylem composed of vertically stacked vessel elements with heavily perforated or missing end walls, leaving a continuous, uninterrupted hollow cylinder.

vessel element: one of the cells of a xylem vessel; elongated, dead at maturity, with thick, lignified lateral cell walls for support but with end walls that are either heavily perforated or missing.

vestigial structure (ves-tij´-ē-ul): a structure that serves no apparent purpose but is homologous to functional structures in re-lated organisms and provides evidence of evolution.

villus (vi´-lus; pl., villi): a fingerlike projection of the wall of the small intestine that increases the absorptive surface area.

viroid (vī´-roid): a particle of RNA that is capable of infecting a cell and of directing the production of more viroids; responsible for certain plant diseases.

virus (vī´-rus): a noncellular parasitic particle that consists of a protein coat surrounding a strand of genetic material; multiplies only within a cell of a living organism (the host).

vitamin: one of a group of diverse chemicals that must be present in trace amounts in the diet to maintain health; used by the body in conjunction with enzymes in a variety of metabolic reactions.

vitreous humor (vit´-rē-us): a clear, jellylike substance that fills the large chamber of the eye between the lens and the retina.

vocal cord: one of a pair of bands of elastic tissue that extend across the opening of the larynx and produce sound when air is forced between them. Muscles alter the tension on the vocal cords and control the size and shape of the opening, which in turn determines whether sound is produced and what its pitch will be.

waggle dance: a symbolic form of communication used by honeybee foragers to communicate the location of a food source to their hivemates.

warning coloration: bright coloration that warns predators that the potential prey is distasteful or even poisonous.

water mold: a funguslike protist that includes some pathogens, such as the downy mildew, which attacks grapes.

water-vascular system: a system in echinoderms that consists of a series of canals through which seawater is conducted and is used to inflate tube feet for locomotion, grasping food, and respiration.

wax: a lipid composed of fatty acids covalently bonded to long-chain alcohols.

weather: short-term fluctuations in temperature, humidity, cloud cover, wind, and precipitation in a region over periods of hours to days.

Werner syndrome: a rare condition in which a defective gene causes premature aging; caused by a mutation in the gene that codes for DNA replication/repair enzymes.

white matter: the portion of the brain and spinal cord that consists largely of myelin-covered axons and that give these areas a white appearance.

withdrawal: the removal of the penis from the vagina just before ejaculation in an attempt to avoid pregnancy; an ineffective contraceptive method.

working memory: the first phase of learning; short-term memory that is electrical or biochemical in nature.

xylem (zī-lum): a conducting tissue of vascular plants that transports water and minerals from root to shoot.

yolk: protein-rich or lipid-rich substances contained in eggs that provide food for the developing embryo.

yolk sac: one of the embryonic membranes of reptilian, bird, and mammalian embryos; in birds and reptiles, a membrane surrounding the yolk in the egg; in mammals, forms part of the umbilical cord and the digestive tract but is empty.

Z line: a fibrous protein structure to which the thin filaments of skeletal muscle are attached; forms the boundary of a sarcomere.

zona pellucida (pel-oo´-si-duh): a clear, noncellular layer between the corona radiata and the egg.

zooflagellate (zō-ō-fla´-jel-et): a nonphotosynthetic protist that moves by using flagella.

zooplankton: nonphotosynthetic protists that are abundant in marine and freshwater environments.

zoospore (zō´-ō-spor): a nonsexual reproductive cell that swims by using flagella; formed by members of the protistan division Oomycota.

zygospore (zī´-gō-spor): a fungal spore, produced by the division Zygomycota, that is surrounded by a thick, resistant wall and forms from a diploid zygote.

zygote (zī´-gōt): in sexual reproduction, a diploid cell (the fertilized egg) formed by the fusion of two haploid gametes.

zygote fungus: a fungus of the division Zygomycota, which includes the species that cause fruit rot and bread mold.

Photo Credits

Back Cover
Philip van den Berg, HPH Photography

Frontmatter
ii Christoph Burki/Tony Stone Images v Audesirk

Chapter 1
Opener NASA Headquarters **1-1a** Andrew Syred/SPL/Photo Researchers, Inc. **1-1b** Craig Tuttle/The Stock Market **1-1c** Kim Taylor/Bruce Coleman, Inc. **1-3** E.H. Newcomb/Biological Photo Service **1-4** Anua & Mahoj Shah/Animals Animals/Earth Scenes **1-5** Johnny Johnson/DRK Photo **1-6** Lawrence Livermore National Laboratory/Photo Researchers, Inc. **1-7a** CNRI/SPL/Photo Researchers, Inc. **1-7b** Dr. M. Rohde, GBF/SPL/Photo Researchers, Inc. **1-7c** Eric V. Grave/Photo Researchers, Inc. **1-7d** Richard L. Carlton/Photo Researchers, Inc. **1-7e** Patti Murray/Animals Animals/Earth Scenes **1-7f** Jeff Rotman/Tony Stone Images **1-8** John Durham/SPL/Photo Researchers, Inc. **1-9** Doug Perrine/DRK Photo **1-10** François Gohier/Photo Researchers, Inc. **E1-2** Luiz C. Marigo/Peter Arnold, Inc. **E1-2** (inset) Gunter Ziesler/Peter Arnold, Inc. **1-12** Audesirk

Unit One
Opener Manfred Kage/Peter Arnold, Inc.

Chapter 2
Opener Stephen Dalton/Photo Researchers, Inc. **E2-1c** NIH/Science Source/Photo Researchers, Inc. **2-9a** David Dennis/Tom Stack & Associates **2-9b** Audesirk

Chapter 3
Opener Michael Skott/The Image Bank **3-2a** Jeremy Burgess, M.D./SPL/Photo Researchers, Inc. **3-3** (left) Larry Ulrich/DRK Photo **3-3** (middle) Jeremy Burgess, M.D./SPL/Photo Researchers, Inc. **3-3** (right) Biophoto Associates/Photo Researchers, Inc. **3-4** Richard Kolar/Animals Animals/Earth Scenes **3-5a** Jean-Michel Labat/Auscape International Pty. Ltd. **3-5b** Donald Specker/Animals Animals/Earth Scenes **E3-1** David Lee Waite/SportsChrome, Inc. **3-8a** (p. xxii) Robert Pearcy/Animals Animals/Earth Scenes **3-8b** Jeff Foott/DRK Photo **3-8c** Nuridsany et Perennou/Photo Researchers, Inc.

Chapter 4
Opener Sylvain Grandadam/Tony Stone Images **4-1** Photograph by Dr. Harold E. Edgerton. ©The Harold E. Edgerton 1992 Trust. Courtesy of Palm Press, Inc.

Chapter 5
Opener Jan Hinsch/SPL/Photo Researchers, Inc. **5-1** Farrell Grehan/Photo Researchers, Inc. **5-6** Dr. Joseph Kurantsin-Mills **5-9** M. M. Perry **5-10** Linda Hufnagel **5-13** Biophoto Associates/Science Source/Photo Researchers, Inc. **5-14** Stephen J. Krasemann/DRK Photo

Chapter 6
Opener Dennis Kunkel, University of Washington **6-3a** Secchi-Lecaque-Roussel-UCLAF/CNRI/SPL/Photo Researchers, Inc. **6-3b** Moredun Animal Health Ltd/SPL/Photo Researchers, Inc. **E6-1** Cecil Fox/Science Source/Photo Researchers, Inc. **E6-1** (inset) Biophoto Associates/Photo Researchers, Inc. **E6-2a** M.I. Walker/Photo Researchers, Inc. **E6-2b** David M. Phillips/Visuals Unlimited **E6-2c** Karl Aufderheide/Visuals Unlimited **E6-2d** Dr. K. Tanaka **6-6b** E. Guth, T. Hashimoto, and F. Conti **6-7** (p. xxiii) Phototake/Carolina Biological Supply Co. **6-8**

Omikron/Science Source/Photo Researchers, Inc. **6-9** Dr. Barry F. King/Biological Photo Service **6-10** D.W. Fawcett/Photo Researchers, Inc. **6-12** Thomas Eisner **6-13** Holt Studios/Nigel Cattlin/Photo Researchers, Inc. **6-14** Keith R. Porter **6-15** W.P. Wergin/Biological Photo Service **6-16** Biophoto Associates/Photo Researchers, Inc. **6-18** (upper) William L. Dentler/Biological Photo Service **6-18** (lower) Dr. de Harven/Photo Researchers, Inc. **6-19a** Ellen R. Dirksen/Visuals Unlimited **6-19b** Yorgos Nikas/Tony Stone Images

Chapter 7
Opener Alfred Pasieka/Peter Arnold, Inc. **7-1b** (p. xxiv) Colin Milkins/Oxford Scientific Films/Animals Animals/Earth Scenes **7-2a** Ken W. Davis/Tom Stack & Associates **7-2b,c** E.H. Newcome/Biological Photo Service

Chapter 8
Opener ©1985 Robert A. Tyrrell **8-3a** Focus on Sports Inc. **8-3b** Z. Leszczynski/Animals Animals/Earth Scenes

Unit Two
Opener Ron Kimball Photography

Chapter 9
Opener Kenneth Eward/BioGrafx-Science Source/Photo Researchers, Inc. **9-1a** Rosalind Franklin/Photo Researchers, Inc. **9-1b** CSHL Archives/Peter Arnold, Inc. **E9-1** A. Barrington Brown/Photo Researchers, Inc.

Chapter 10
Opener Frederic Jacana/Photo Researchers, Inc. **10-5a** Oscar L. Miller, Jr., University of Virginia **E10-2** Howard W. Jones, Jr., M.D., Eastern Virginia Medical School **E10-3** George M. Martin, M.D. **10-9** Science Source/Photo Researchers, Inc. **10-10a** Murray L. Barr **10-10b** Audesirk

Chapter 11
Opener Andrew J. Martinez **11-2** Dr. Gopal Murti/SPL/Photo Researchers, Inc. **11-3** CNRI/SPL/Photo Researchers, Inc. **11-4** Biophoto Associates/Photo Researchers, Inc. **11-5** CNRI/SPL/Photo Researchers, Inc. **11-7** Andrew S. Bajer **11-9a** T.E. Schroeder/Biological Photo Service **11-11a** (p. xxv) M. Abbey/Photo Researchers, Inc. **11-11b** John Durham/SPL/Photo Researchers, Inc. **11-11c** Carolina Biological Supply Co./Phototake NYC **11-12** Audesirk **E11-1** Photo courtesy of The Roslin Institute.

Chapter 12
Opener David Allan Brandt/Tony Stone Images **12-1** Archiv/Photo Researchers, Inc. **12-2** Audesirk **12-8** Biophoto Associates/Photo Researchers, Inc. **12-13** Jane Burton/Bruce Coleman, Inc. **12-15a** John Watney/Science Source/Photo Researchers, Inc. **12-15b** K.H. Switak/Photo Researchers, Inc. **12-15c** Gary Retherford/Photo Researchers, Inc. **12-16a** Dennis Kunkel/Phototake NYC **12-16b** Francis Leroy/Biocosmos/SPL/Photo Researchers, Inc. **12-17a** Hart-Davis/SPL/Photo Researchers, Inc. **12-18** Gunn & Stewart/Mary Evans/Photo Researchers, Inc. **E12-1** Nick Kelsh/Peter Arnold, Inc. **12-19a** CNRI/SPL/Photo Researchers, Inc. **12-19b** Lawrence Migdale/PIX

Chapter 13
Opener Abraham Menasche, Inc. **13-1a** Professor Stanley N. Cohen/SPL/Photo Researchers, Inc. **13-6** Hank Morgan/Photo Researchers, Inc. **E13-1** Cellmark Diagnostics, Germantown, MD. ©1989, ICI Americas Inc. **13-9a** Keith V. Wood/Science VU/Visuals Unlimited **13-10a** Norm

Thomas/Photo Researchers, Inc. **13-10b** Monsanto Company, Agricultural Group **13-11** Photo by Ned S. Gilmore/Steven Kurth. **13-12a** (p. xxvi) Courtesy of Bayer AG, Germany **13-12b** GenPharm International/Peter Arnold, Inc. **E13-2a** (p. xxvi) Gary Braasch Photography **E13-2b** Joseph R. Newhouse, Ph.D. **E13-3** Dr. Choy Hew, Hospital for Sick Children, Toronto, Ontario, and Dr. Garth Fletcher, Memorial University of Newfoundland, St. John's, Newfoundland **Table 13-3** Slim Films **13-13a** AP/Wide World Photos **13-14** Bruce Dale/National Geographic Society **13-15** David Stoecklein/The Stock Market **E13-4a** Mehau Kulyk/SPL/Photo Researchers, Inc.

Unit Three
Opener Martin Dohm/Photo Researchers, Inc.

Chapter 14
Opener Nicholas Parfitt/Tony Stone Images **Opener** (inset) Staffan Widstrand/The Wildlife Collection **14-1** Jim Steinberg/Photo Researchers, Inc. **14-3a** John Cancalosi/DRK Photo **14-3b** David M. Dennis/Tom Stack & Associates **14-3c** Chip Clark **E14-1** Corbis–Bettmann **E14-2a** Frans Lanting/Photo Researchers, Inc. **E14-2b** Christian Grzimek/OKAPIA/Photo Researchers, Inc. **14-7a,b** Stephen Dalton/Photo Researchers, Inc. **14-7c** Douglas T. Cheeseman, Jr./Peter Arnold, Inc. **14-7d** Frans Lanting/Minden Pictures **14-10a** (p. xxvii) Keith Gillett/Animals Animals/Earth Scenes **14-10b** Dan McCoy/Rainbow **14-10c** C. Eldeman/Petit Format/Science Source/Photo Researchers, Inc. **14-10d** Carolina Biological Supply Co./Phototake NYC **14-11a** Stephen J. Krasemann/DRK Photo **14-11b** Timothy O'Keefe/Tom Stack & Associates **14-12** M.W.F. Tweedie/Bruce Coleman, Inc.

Chapter 15
Opener Hans Pfletschinger/Peter Arnold, Inc. **15-3b** Luiz C. Marigo/Peter Arnold, Inc. **15-3c** Y.R. Tymstra/Valan Photos **15-4** Gregory Dimijian/Photo Researchers, Inc. **E15-1** Frans Lanting/Minden Pictures **15-5** Dr. E.R. Degginger **15-6** W. Perry Conway/Tom Stack & Associates **15-7** D. Cavagnaro/DRK Photo **15-10a** M.P.L. Fogden/Bruce Coleman, Inc. **15-10b** Tim Davis/Photo Researchers, Inc. **15-12** Robert & Linda Mitchell/Robert & Linda Mitchell Photography **15-13** Jeff Lepore/Photo Researchers, Inc. **15-14** Tom McHugh/Steinhart Aquarium/Photo Researchers, Inc. **15-15** Jim Kern Expeditions **15-18** NASA Headquarters **15-19** Thomas A. Wiewandt, Ph.D.

Chapter 16
Opener Courtesy Dr. Michael Hadfield, Department of Zoology, University of Hawaii/Photo by Frank LaBua **16-1a** Wayne Lankinen/Valan Photos **16-1b** Edgar T. Jones/Bruce Coleman, Inc. **16-3** Pat & Tom Leeson/Photo Researchers, Inc. **16-4** Guy L. Bush, Michigan State University **16-8** Tim Laman/NGS Image Collection **16-9** Joy Spurr/Bruce Coleman, Inc. **16-10** Dr. Paul V. Loiselle

Chapter 17
Opener David M. Sanders/David Sanders/DS Photography **17-3** Sidney Fox/Science VU/Visuals Unlimited **17-5** M. Abbey/Visuals Unlimited **17-7a** Milwaukee Public Museum, Photograph Collection **17-7b** James L. Amos/Photo Researchers, Inc. **17-7c** Carolina Biological Supply Co./Phototake NYC **17-7d** Douglas Faulkner/Photo Researchers, Inc. **17-8** Ludek Pesek/SPL/Photo Researchers, Inc. **17-10** Bob Gossington/Bruce Coleman, Inc. **17-11** Photo Researchers, Inc. **17-12a** Tom McHugh/Chicago Zoological Park/Photo Re-

searchers, Inc. **17-12b** Frans Lanting/Minden Pictures **17-12c** Nancy Adams/Tom Stack & Associates **17-13** Tim D. White/David L. Brill Photography **17-15** Neanderthal Museum, Mettmann, Germany **17-16** Jerome Chatin/Gamma-Liaison, Inc. **17-18** NASA/Johnson Space Center

Chapter 18
Opener Dr. E.R. Degginger **18-2a** C. Steven Murphree, Department of Entomology, Auburn University **18-2b** Dr. Greg Rouse, Department of Invertebrate Zoology, National Museum of Natural History, Smithsonian Institution **18-2c** Dr. Jeremy Burgess/SPL/Photo Researchers, Inc. **18-4a** Hans Gelderblom/Tony Stone Images **18-4b** Courtesy of Dr. W. Jack Jones, reprinted by permission of Springer-Verlag from W.J. Jones, J.A. Leigh, F. Mayer, C.R. Woese, and R.S. Wolfe. "*Methanococcus jannaschii* sp. nov., an extremely thermophilic methanogen from a submarine hydrothermal vent." *Archives of Microbiology* (1983) 136:254–261. ©1983 by Springer-Verlag. **18-7** Zig Koch/Kino Fotoarquivo/Discover Magazine

Chapter 19
Opener EM Unit, VLA/SPL/Photo Researchers, Inc. **19-2b** CDC/Photo Researchers, Inc. **19-4** Lee Simon/Stammers/SPL/Photo Researchers, Inc. **19-5** Stanley B. Pruisner **19-6a** David M. Phillips/Visuals Unlimited **19-6b** Dr. Tony Brain & David Parker/SPL/Photo Researchers, Inc. **19-6c** CNRI/SPL/Photo Researchers, Inc. **19-7** Photo Lennart Nilsson/Albert Bonniers Forlag **19-8a** Dr. Tony Brain/SPL/Photo Researchers, Inc. **19-9** A.B. Dowsett/SPL/Photo Researchers, Inc. **19-10** CNRI/SPL/Photo Researchers, Inc. **19-11** (p. xxvii) Dennis Kunkel/Phototake NYC **19-12** Yellowstone National Park **19-13a** Biophoto Associates/Photo Researchers, Inc. **19-13b** Biophoto Associates/Science Source/Photo Researchers, Inc. **19-14** C.P. Vance/Visuals Unlimited **19-15a** Carolina Biological Supply Co./Phototake NYC **19-15b** Eric Grave/Science Source/Photo Researchers, Inc. **19-16** William Merrill **19-17a** P.W. Grace/Science Source/Photo Researchers, Inc. **19-17b** Cabisco/Visuals Unlimited **19-18** Cabisco/Visuals Unlimited **19-19** David M. Phillips/Visuals Unlimited **19-20** Kevin Schafer/Peter Arnold, Inc. **19-21** Tony Stone Images **19-23** Eric V. Grave/Photo Researchers, Inc. **19-24** P.M. Motta & F.M. Magliocca/SPL/Photo Researchers, Inc. **19-25** M.I. Walker/Photo Researchers, Inc. **19-26** M.I. Walker/Science Source/Photo Researchers, Inc. **19-27a** Dr. E.R. Degginger **19-27b** Manfred Kage/Peter Arnold, Inc. **19-30** Biophoto Associates/Science Source/Photo Researchers, Inc.

Chapter 20
Opener Jeff Lepore/Photo Researchers, Inc. **20-1a** Robert & Linda Mitchell/Robert & Linda Mitchell Photography **20-1b** Elmer Koneman/Visuals Unlimited **20-2** Michael Fogden/DRK Photo **20-3b** Carolina Biological Supply Co./Phototake NYC **20-3c** Breck P. Kent **20-4a** W.K. Fletcher/Photo Researchers, Inc. **20-4b** David Dvorak, Jr. **20-5** David M. Phillips/Visuals Unlimited **20-6a** Scott Camazine/Photo Researchers, Inc. **20-6b** David M. Dennis/Tom Stack & Associates **20-6c** William J. Weber/Visuals Unlimited **20-6d** David M. Dennis/Tom Stack & Associates **20-8** Robert & Linda Mitchell Photography **20-9** Audesirk **20-11a** Jeff Foott Productions **20-11b** Robert & Linda Mitchell Photography **20-12** Stanley L. Flegler/Visuals Unlimited **20-13** M. Viard/Jacana/Photo Researchers, Inc. **20-14** G.L. Baron/Biological Photo Service **20-15** Cabisco/Visuals Unlimited

Chapter 21
Opener Tom & Pat Leeson/Photo Researchers, Inc. **21-1** Photo Researchers, Inc. **21-2** Jeffrey L. Rotman **21-3a** D.P. Wilson/Eric & David Hosking/Photo Researchers, Inc. **21-3b** Bob Evans/Peter Arnold, Inc. **21-3c** Lawrence E. Naylor/Photo Researchers, Inc. **21-4** Harry Rogers/Photo Researchers, Inc. **21-5a** John Gerlach/Tom Stack & Associates **21-5b** John Shaw/Tom Stack & Associates **21-6** Carolina Biological Supply Co./Phototake NYC **21-7a** Dwight Kuhn Photography **21-7b** Milton Rand/Tom Stack & Associates **21-7c** Larry Ulrich/DRK Photo **21-8** Milton Rand/Tom Stack & Associates **21-9c** Andy Roberts/Tony Stone Images **21-9d** John Kaprielian/Photo Researchers, Inc. **21-10a** Maurice Nimmo/A–Z Botanical Collection, Ltd. **21-10b** Audesirk **21-11** (upper) Dr. William M. Harlow/Photo Researchers, Inc. **21-11** (lower) Photo Researchers, Inc. **21-12a** Dwight Kuhn Photography **21-12b** David Dare Parker/Auscape International Pty. Ltd. **21-12b** (inset) Matt Jones/Auscape International Pty. Ltd. **21-12c** Dwight Kuhn Photography **21-12d** Audesirk **21-12e** Larry West/Photo Researchers, Inc. **21-14** T. Kitchin/Tom Stack & Associates

Chapter 22
Opener David Scharf/David Scharf Photography **22-4a** Larry Lipsky/DRK Photo **22-4b** Brian Parker/Tom Stack & Associates **22-4c** Charles Seaborn/Odyssey Productions **22-6a** Gregory Ochocki/Photo Researchers, Inc. **22-6b** Charles Seaborn/Odyssey Productions **22-06c** Audesirk **22-06d** John D. Cunningham/Visuals Unlimited **22-9a** Carolina Biological Supply Co./Phototake NYC **22-10** (upper) Martin Rotker/Phototake NYC **22-10** (lower)Stanley Flegler/Visuals Unlimited **22-11** Tom E. Adams/Peter Arnold, Inc. **22-12a** Carolina Biological Supply Co./Phototake NYC **22-12b** Howard Shiang, D.V.M., *J. Am. Vet. Med. Assoc.* 163:981, Oct. 1973 **22-14a** Kjell B. Sandved/Butterfly Alphabet, Inc. **22-14b** David Bull Photography **22-14c** J.H. Robinson/Photo Researchers, Inc. **22-15** Audesirk **22-16** Dwight Kuhn Photography **22-18** David Scharf/Peter Arnold, Inc. **22-19a** Carolina Biological Supply Co./Phototake NYC **22-19b** (p. xxix) Peter J. Bryant/Biological Photo Service **22-19c** Stephen Dalton/Photo Researchers, Inc. **22-19d** Werner H. Muller/Peter Arnold, Inc. **22-19e** Stanley Breeden/DRK Photo **22-20a** Dwight Kuhn Photography **22-20b** Tim Flach/Tony Stone Images **22-20c** Audesirk **22-21a** Tom Branch/Photo Researchers, Inc. **22-21b** Peter J. Bryant/Biological Photo Service **22-21c** Carolina Biological Supply Co./Phototake NYC **22-21d** Alex Kerstitch **22-23a** Ray Coleman/Photo Researchers, Inc. **22-23b** Alex Kerstitch **22-24a** Fred Bavendam/Peter Arnold, Inc. **22-24b** Ed Reschke/Peter Arnold, Inc. **22-25a** Fred Bavendam/Peter Arnold, Inc. **22-25b** Kjell B. Sandved/Photo Researchers, Inc. **22-25c** Alex Kerstitch **22-26a** Audesirk **22-26b** Jeff Foott Productions **22-26c** Chris Newbert/Bruce Coleman, Inc. **22-27b** Michael Male/Photo Researchers, Inc. **22-28** John Giannicchi/Science Source/Photo Researchers, Inc. **22-29** Tom McHugh/Photo Researchers, Inc. **22-30a** Tom McHugh/Steinhart Aquarium/Photo Researchers, Inc. **22-30b** Tom Stack & Associates **22-30b** (inset) Breck P. Kent **22-31a** Jeffrey L. Rotman **22-31b** David Hall/Photo Researchers, Inc. **22-32a** Peter David/Planet Earth Pictures **22-32b** Mike Neumann/Photo Researchers, Inc. **22-32c** Stephen Frink/Tony Stone Images **22-32d** Peter Scoones/Planet Earth Pictures **22-33** Breck P. Kent/Animals Animals/Earth Scenes **22-33b** Joe McDonald/Tom Stack & Associates **22-33c** Cosmos Blank/National Audubon Society/Photo Researchers, Inc. **E22-1** Stanley Breeden/DRK Photo **22-34a** David G. Barker/Tom Stack & Associates **22-34b** Roger K. Burnard/Biological Photo Service **22-34c** Frans Lanting/Minden Pictures **22-35** Carolina Biological Supply Co./Phototake NYC **22-36a** Walter E. Harvey/Photo Researchers, Inc. **22-36b** Carolina Biological Supply Co./Phototake NYC **22-36c** Ray Ellis/Photo Researchers, Inc. **22-37** Tom McHugh/Photo Researchers, Inc. **22-38a** Flip Nicklin/Minden Pictures **22-38b** Jonathan Watts/SPL/Photo Researchers, Inc. **22-38c** C. & M. Denis-Huot/Peter Arnold, Inc. **22-38d** S.R. Maglione/Photo Researchers, Inc. **22-39a** Yves Kerban/Jacana/Photo Researchers, Inc. **22-39b** Mark Newman/Auscape International Pty. Ltd. **22-39b** (inset) D. Parer & E. Parer-Cook/Auscape International Pty. Ltd.

Unit Four
Opener John Shaw/Tom Stack & Associates

Chapter 23
Opener Charles Krebs/The Stock Market **23-4a** Jeremy Burgess/SPL/Photo Researchers, Inc. **23-4b** John D. Cunningham/Visuals Unlimited **23-5** George Wilder/Visuals Unlimited **23-8a** Lynwood M. Chace/Photo Researchers, Inc. **23-8b** Dwight Kuhn Photography **23-9** (upper) Ed Reschke/Peter Arnold, Inc. **23-9** (lower) E.R. Degginger/Animals Animals/Earth Scenes **23-10** Audesirk **23-12** J. Robert Waaland, University of Washington/Biological Photo Service **23-13** Ed Reschke/Peter Arnold, Inc. **23-16a** D.C. Cunningham/Visuals Unlimited **23-16b** Stephen J. Krasemann/Photo Researchers, Inc. **23-17** John D. Cunningham/Visuals Unlimited **23-19** François Gohier/Photo Researchers, Inc. **23-20a** Kenneth W. Fink/Photo Researchers, Inc. **23-20b** Belinda Wright/DRK Photo **23-21a** Audesirk **23-21b** Carr Clifton/Minden Pictures **23-21c** Audesirk **23-21d** Ken W. Davis/Tom Stack & Associates **23-23a** Biophoto Associates/Science Source/Photo Researchers, Inc. **23-23b** R. Roncadori/Visuals Unlimited **23-24** Hugh Spencer/National Audubon Society/Photo Researchers, Inc. **E23-1** Michael Fogden/DRK Photo **23-26a** Ray Simon/Photo Researchers, Inc. **23-26b** Dr. Jeremy Burgess/SPL/Photo Researchers, Inc. **23-27a** E.R. Degginger/Animals Animals/Earth Scenes **23-27b** Dr. Martin H. Zimmermann/Harvard Forest

Chapter 24
Opener F. Stuart Westmorland/Photo Researchers, Inc. **24-3** Robert & Linda Mitchell Photography **24-4b** James L. Castner **24-5** Audesirk **24-6** Audesirk **24-7** Thomas Eisner **24-8** Nuridsany et Perennou/The National Audubon Society Collection/Photo Researchers, Inc. **24-9** Anthony Mercieca/Photo Researchers, Inc. **24-10a** (p. xxx) Joe McDonald/DRK Photo **24-10b** Merlin D. Tuttle/Bat Conservation International/Photo Researchers, Inc. **24-10c** Oxford Scientific Films/Animals Animals/Earth Scenes **24-11** Dr. J.A.L. Cooke/Oxford Scientific Films/Animals Animals/Earth Scenes **24-12a** Tom Bean/DRK Photo **24-12b** Michael Fogden/DRK Photo **24-13** Carolina Biological Supply Co./Phototake NYC **24-14** Dennis Kunkel, University of Washington **24-16** Carolina Biological Supply Co./Phototake NYC **E24-1a** George Holton/Photo Researchers, Inc. **E24-1b** Merlin D. Tuttle/Bat Conservation International/Photo Researchers, Inc. **24-20** Audesirk **24-21a** Audesirk **24-21b** Carolina Biological Supply Co./Phototake NYC **24-22** Audesirk **24-23** Scott Camazine/Photo Researchers, Inc. **24-24** Fogden Natural History Photographs

Chapter 25
Opener Merlin D. Tuttle/Bat Conservation International/Photo Researchers, Inc. **25-5a** Runk/Schoenberger/Grant Heilman Photography, Inc. **25-10** John Cancalosi/DRK Photo **25-11** Biophoto Associates/Photo Researchers, Inc. **25-13** Carolina Biological Supply Co./Phototake NYC

Unit Five
Opener Randy Wells/Tony Stone Images

Chapter 26
Opener Bryan & Cherry Alexander **26-3** Biophoto Associates/Photo Researchers, Inc. **26-4** Manfred Kage/Peter Arnold, Inc. **26-5** John Kennedy/Biological Photo Service/Tony Stone Images **26-6** Ray Simons/Photo Researchers, Inc. **26-7** L. Bassett/Videosurgery/Visuals Unlimited

Chapter 27
Opener Warren Rosenberg/Tony Stone Images **27-8** Dennis Kunkel **27-11** Lennart Nilsson/Bonnier Alba AB **27-12** Discover Text & Photo Syndication **27-13b** (p. xxxi) CNRI/SPL/Photo Researchers, Inc. **27-17** Lennart Nilsson **E27-1a** Lou Lainey/Time Life Syndication **E27-1c** Biophoto Associates/Photo Researchers, Inc. **27-21** John D. Cunningham/Visuals Unlimited

Chapter 28
Opener Bruce Watkins/Animals Animals/Earth Scenes **28-1a** Jeffrey L. Rotman **28-1b** Brian Parker/Tom Stack & Associates **28-1c** Dr. E.R. Degginger **28-5a** E.R. Degginger/Bruce Coleman, Inc. **28-5b** Breck P. Kent **28-5c** David M. Dennis/Tom Stack & Associates **28-6b** H.R. Duncker **E28-1** Bill Travis, M.D./National Cancer Institute **E28-2** O. Auerbach/Visuals Unlimited

Chapter 29
Opener John Gerlach/DRK Photo **29-1** Charles Krebs/Tony Stone Images **29-3** Tobi Zausner **E29-1** De Keerle-Parker/Gamma-Liaison, Inc. **29-5a** E.R. Degginger/Photo Researchers, Inc. **E29-2** St. Bartholomew's Hospital/SPL/Photo Researchers, Inc. **29-12b** Lennart Nilsson

Chapter 30
Opener Gerard Lacz/Animals Animals/Earth Scenes **30-4b** (p. xxxii) Photo Researchers, Inc. **30-9** Allstock/Tony Stone Images

Chapter 31
Opener Photo Lennart Nilsson/Albert Bonniers Forlag **31-1** CNRI/SPL/Photo Researchers, Inc. **31-2** Lennart Nilsson/Bonnier Alba AB **31-4** NIBSC/SPL/Photo Researchers, Inc. **31-6a** Ken Eward/Biografx/Science Source/Photo Researchers, Inc. **31-10a** Secchi-Lecaque/Roussel/CNRI/Science Source/Photo Researchers, Inc. **31-10b** CNRI/SPL/Photo Researchers, Inc. **31-11** Photo Lennart Nilsson/Albert Bonniers Forlag AB. ©Boehringer Ingelheim International GmbH. **31-15b** Lennart Nilsson/Bonnier Alba AB **E31-2** CNRI/SPL/Photo Researchers, Inc.

Chapter 32
Opener Richard Hutchings/Photo Researchers, Inc. **32-5** Corbis-Bettmann **32-7b** Biophoto Associates/Photo Researchers, Inc.

Chapter 33
Opener Dennis Kunkel **33-3a** Journal of Physiology **33-3b** Carolina Biological Supply Co./Phototake NYC **33-18** Hanna Damasio, M.D. **E33-2** Dr. Marcus Raichle/Washington University School of Medicine **E33-3a** Dr. William J. Powers **E33-3b** Monte S. Buchsbaum, M.D. **33-20a** Robert S. Preston/Prof. J.E. Hawkins **33-21a** (p. xxxiii) Dennis Kunkel/MicroVision **33-24** Susan Eichorst **33-25a** Frans Lanting/Minden Pictures **33-25b** L. West/Photo Researchers, Inc. **33-29a** Charles E. Mohr/Photo Researchers, Inc. **33-29b** Brian Parker/Tom Stack & Associates

Chapter 34
Opener Andrew D. Bernstein/NBA Photos **34-1f** Dr. Brian Eyden/SPL/Photo Researchers, Inc. **34-3b** Don Fawcett/Science Source/Photo Researchers, Inc. **34-4a** Tom McHugh/Steinhart Aquarium/Photo Researchers, Inc. **34-4b** Lawrence E. Naylor/Photo Researchers, Inc. **34-6** Photo Researchers, Inc. **34-8b** Manfred Kage/Peter Arnold, Inc. **E34-2a** Michael Klein/Peter Arnold, Inc. **E34-2b** Yoav Levy/Phototake NYC

Chapter 35
Opener Wayne Lynch **35-1** Andrew J. Martinez **35-3** P.J. Bryant/University of California at Irvine/Biological Photo Service **35-4-a** Tom McHugh/Photo Researchers, Inc. **35-4b** Audesirk **35-4c** Peter Harrison/Audesirk **35-5** Anne et Jacques Six **35-6** Michael Fogden/Oxford Scientific Films/Animals Animals/Earth Scenes **35-7a** Stephen J. Krasemann/DRK Photo **35-7b** Tom McHugh/Photo Researchers, Inc. **35-7c** Frans Lanting/Minden Pictures **35-10b** CNRI/SPL/Photo Researchers, Inc. **35-14** Gary Martin/Visuals Unlimited **35-17a** Lennart Nilsson **35-17b** Andrew Syred/Tony Stone Images **35-18a** Lennart Nilsson **E35-1** Juana Anderson/AP/Wide World Photos **E35-2** Hank Morgan/Photo Researchers, Inc.

Chapter 36
Opener Petit Format/Nestle/Photo Researchers, Inc. **36-3a** Image Quest 3-D/Natural History Photographic Agency **36-3a** Jim Bain/Natural History Photographic Agency **36-3b** Hans Pfletschinger/Peter Arnold, Inc. **36-3b** L. West/Photo Researchers, Inc. **36-4a** J.A.L. Cooke/Animals Animals/Earth Scenes **36-4b** K.H.Switak/Photo Researchers, Inc. **36-4c** S. Nielsen/DRK Photo **36-4d** Frans Lanting/Minden Pictures **E36-1** Oliver Meckes/Photo Researchers, Inc. **36-12a** Lennart Nilsson **36-12b** Petit Format/Nestle/Science Source/Photo Researchers, Inc. **36-12c** Lennart Nilsson **36-14** Audesirk **E36-2** Leonard McCombe/Life Magazine/Time Warner, Inc. **E36-3** Sterling K. Clarren, M.D., Children's Hospital and Medical Center, University of Washington School of Medicine

Chapter 37
Opener David J. Sams/Tony Stone Images **37-1** E. & D. Hosking/Frank Lane Picture Agency, Ltd. **37-3a** Frans Lanting/Minden Pictures **37-4a** Manfred Kage/Peter Arnold, Inc. **37-5** Boltin Picture Library **37-8** Thomas McAvoy/Life Magazine/Time Warner, Inc. **37-10** Renee Lynn/Photo Researchers, Inc. **37-11** Ken Cole/Animals Animals/Earth Scenes **37-12** Richard K. LaVal/Animals Animals/Earth Scenes **37-13** (p. xxxiv) Nuridsany et Perennou/Photo Researchers, Inc. **37-14** Robert & Linda Mitchell/Robert & Linda Mitchell Photography **37-15a** Stephen J. Krasemann/Photo Researchers, Inc. **37-15b** Hans Pfletschinger/Peter Arnold, Inc. **E37-1a** Barbara Webb, Department of Artificial Intelligence, University of Edinburgh, Scotland **E37-1b** Gilbert Grant/Photo Researchers, Inc. **37-16a** M.P. Kahl/DRK Photo **37-16b** Marc Chamberlain/Tony Stone Images **37-17** Ray Dove/Visuals Unlimited **37-18** William Ervin/Natural Imagery **37-19** Michael K. Nichols/National Geographic Image Collection **37-20** John D. Cunningham/Visuals Unlimited **37-22** Joe McDonald/Visuals Unlimited **37-23a** Norbert Wu/Peter Arnold, Inc. **37-23b** Frans Lanting/Minden Pictures **37-24** Anne et Jacques Six **37-26a** Dwight Kuhn Photography **37-26b** Audesirk **37-27** Jen & Des Bartlett/Bruce Coleman, Inc. **37-28** Fred Bruemmer/Peter Arnold, Inc. **37-31** Animals Animals/©Raymond A. Mendez **37-32** Photo Lennart Nilsson/Albert Bonniers Forlag **37-33** Dr. William Fifer, New York State Psychiatric Institute **37-34a** Frans Lanting/Minden Pictures **37-34b** Daniel J. Cox/Gamma-Liaison, Inc. **37-34c** Michio Hoshino/Minden Pictures

Unit Six
Opener "Paradise," Part One of "The Trilogy of the Earth" by Suzanne Duranceau/Illustratrice, Inc.

Chapter 38
Opener Ferrero-Labat/Auscape International Pty. Ltd. **38-1a** Jack W. Ekin/Jack Ekin Photography **38-1b** Gary Milburn/Tom Stack & Associates **38-1c** Rod Planck/Tom Stack & Associates **38-1d** Gary R. Zahm/DRK Photo **38-1e** Robert & Linda Mitchell/Robert & Linda Mitchell Photography **38-2a** Pr. S. Cinti/Ancona-U/CNRI/Phototake NYC **38-2b** Thomas Kitchin/Tom Stack & Associates **38-7** Bryan & Cherry Alexander **38-8** Tom McHugh/Photo Researchers, Inc. **38-9** Department of Natural Resources, Queensland, Australia **38-10** Hellio-Van Ingen/Auscape International Pty. Ltd. **38-11a** Robert & Linda Mitchell Photography **38-11b** Tom Bean/DRK Photo **38-11c** Gary Braasch Photography **38-13** NASA/Johnson Space Center **E38-4a** Frans Lanting/Minden Pictures **E38-4b** Andy Holbrooke/The Stock Market

Chapter 39
Opener Frans Lanting/Minden Pictures **39-3a** Henry Ausloos/Animals Animals/Earth Scenes **39-3b** Stephen Dalton/Oxford Scientific Films/Animals Animals/Earth Scenes **39-4a** Art Wolfe/Art Wolfe, Inc. **39-4b** Frans Lanting/Minden Pictures **39-5a** Marty Cordano/DRK Photo **39-5b** Paul A. Zahl/Photo Researchers, Inc. **39-5c** Ray Coleman/Photo Researchers, Inc. **39-6** Robert & Linda Mitchell Photography **39-7a** Ferrero-Labat/Auscape International Pty. Ltd. **39-7b** Zig Leszczynski/Animals Animals/Earth Scenes **39-8** James L. Castner **39-9a** Zig Leszczynski/Animals Animals/Earth Scenes **39-9b** Breck P. Kent **39-9c,d** Dr. E.R. Degginger **39-10** Zig Leszczynski/Animals Animals/Earth Scenes **39-11a,b** James L. Castner **39-11c** Jeff Lepore/Photo Researchers, Inc. **39-12a** Robert P. Carr/Bruce Coleman, Inc. **39-12b** Monica Mather **39-13a** Dan Aneshausley, Cornell University/Thomas Eisner **39-13b** Frans Lanting/Minden Pictures **E39-1** Gilbert Grant/Photo Researchers, Inc. **E39-1** (inset) Carol Hughes/Bruce Coleman, Inc. **39-14** (p. xxxv) W. Gregory Brown/Animals Animals/Earth Scenes **39-15a** Jim Zipp/Photo Researchers, Inc. **39-15b** Stephen J. Krasemann/Photo Researchers, Inc. **39-16a** Gary Braasch Photography **39-16b** (left) Dennis Oda/Tony Stone Images **39-16b** (right) Krafft-Explorer/Photo Researchers, Inc. **39-16c** (left) Wendy Shattil/Bob Rozinski/Tom Stack & Associates **39-16c** (right) Jeff Foott Productions **39-19** Robert & Linda Mitchell Photography **E39-2a** Ron Peplowski/The Detroit Edison Company **E39-2b** Chuck Pratt/Bruce Coleman, Inc. **E39-2c** Robert & Linda Mitchell Photography

Chapter 40
Opener Roberto Osti **40-8** Thomas A. Schneider **40-13** Bob Hallinen/Gamma-Liaison, Inc. **40-14** V. Leloup/Figaro/Gamma-Liaison, Inc. **40-15** William E. Ferguson **40-17** Will McIntyre/Photo Researchers, Inc.

Chapter 41
Opener Franklin J. Viola/Viola Photo Visions, Inc. **E41-1** NOAA/SPL/Photo Researchers, Inc. **41-2b** NASA/Johnson Space Center **41-6a** Audesirk **41-6b** Tom McHugh/Photo Researchers, Inc. **41-9a** Loren McIntyre/Photo Researchers, Inc. **41-9b** Schafer & Hill/Peter Arnold, Inc. **41-9c** Bruce Lyon/Valan Photos **41-9d** Michael Fogden/DRK Photo **41-9e** Frans Lanting/Minden Pictures **41-10** Jacques Jangoux/Peter Arnold, Inc. **41-11a** W. Perry Conway **41-11b** Aubry Lang/Valan Photos **41-11c** Stephen J. Krasemann/DRK Photo **41-11d** Wendy Bass **41-12** James Hancock/Photo Researchers, Inc. **41-13a** Don & Pat Valenti/DRK Photo **41-13b** Brian Parker/Tom Stack & Associates **41-13c** FPG International/©S. Maimone **41-13d** Tom McHugh/Photo Researchers, Inc. **41-14** William H. Mullins/Photo Researchers, Inc. **41-15** Thierry Rannou/Gamma-Liaison, Inc. **41-16** Brian Parker/Tom Stack & Associates **41-17** Harvey Payne **41-18a** Tom Bean/DRK Photo **41-18b** W. Perry Conway/Tom Stack & Associates **41-18c** Nicholas Devore III/Photographers/Aspen, Inc. **41-18d** Bob Gurr/Valan Photos **41-18e** Jim Brandenburg/Minden Pictures **41-19** Audesirk **41-20a** Gary Braasch Photography **41-20b** Thomas Kitchin/Tom Stack & Associates **41-20c** Stephen J. Krasemann/DRK Photo **41-20d** Joe McDonald/Animals Animals/Earth Scenes **41-21a** Susan G. Drinker/The Stock Market **41-21b** Audesirk **41-21c** Jeff Foott Productions **41-21d** Richard Thom/Tom Stack & Associates **41-22a** Audesirk **41-22b** Marty Stouffer Productions/Animals Animals/Earth Scenes **41-22c** Wm. Bacon III/National Audubon Society/Photo Researchers, Inc. **41-23** Gary Braasch Photography **41-24a** Stephen J. Krasemann/DRK Photo **41-24b** Jeff Foott/Tom Stack & Associates **41-24c** Michael Giannechini/Photo Researchers, Inc. **41-24d** James Simon/Photo Researchers, Inc. **41-27a** T.P. Dickinson/Comstock **41-27b** Audesirk **41-27b** (inset) Walter Dawn/National Audubon Society/Photo Researchers, Inc. **41-27c** Audesirk **41-27c** (inset) Thomas Kitchin/Tom Stack & Associates **41-27d** Robert Given **41-27d** (inset) Stan Wayman/Photo Researchers, Inc. **E41-2** Roy Morsch/The Stock Market **41-28a** Charles Seaborn/Odyssey Productions **41-28b** Alex Kerstitch **41-28c** Charles Seaborn/Odyssey Productions **41-28d** Alex Kerstitch **41-29a** Stephen J. Krasemann/DRK Photo **41-29b** Paul Humann/Jeffrey L. Rotman **41-29c** Fred McConnaughey/Photo Researchers, Inc. **41-29d** James D. Watt/Planet Earth Pictures **41-29e** Biophoto Associates/Science Source/Photo Researchers, Inc. **41-29f** Robert Arnold/Planet Earth Pictures **41-30** Dr. Frederick Grassle

Index